Outer Solar System

Viorel Badescu · Kris Zacny
Editors

Outer Solar System

Prospective Energy and Material Resources

Springer

Editors
Viorel Badescu
Candida Oancea Institute
Polytechnic University of Bucharest
Bucharest
Romania

Kris Zacny
Honeybee Robotics Spacecraft
Mechanisms Corporation
Pasadena, CA
USA

ISBN 978-3-030-08868-2 ISBN 978-3-319-73845-1 (eBook)
https://doi.org/10.1007/978-3-319-73845-1

Printed on acid-free paper

This Springer imprint is published by the registered company Springer International Publishing AG part of Springer Nature
The registered company address is: Gewerbestrasse 11, 6330 Cham, Switzerland

The high minded man must care more
for the truth than for what people think.

Aristotle (384–322 BCE)

Foreword

The frontier is hard and unknown, dangerous, and full of risk. The frontier is not a trail blazed with signposts that guide us to the summit; it is a horizon beyond which we cannot see. The discoveries that beckon beyond the frontier compel us to explore but exploration is never easy. These statements are as true today as they have been throughout the history of human civilization. And throughout history, technological innovations have propelled us forward, making exploration possible. From stone tools and fire, which enabled a migratory hominid population in Africa, to Fridtjof Nansen's rugged and reinforced hull on the Fram, our human capability to make new discoveries has always been tethered to our ability to innovate for new environments. Today, some of the greatest discoveries yet to be made may lie in the outer reaches of our solar system, a frontier filled with challenges that must be overcome.

The discovery that motivates my interest, and the interest of many of my colleagues, is the question of whether or not life exists beyond Earth. Has a second origin of life occurred in our solar system? Does biology work beyond Earth? The answer to these questions could lie beneath the icy shells of moons of the outer solar system—ocean worlds that harbor vast quantities of liquid water and in which life could exist today. These ocean worlds, distant from the warmth of the sun, maintain their liquid water oceans largely through the tidal energy dissipation that occurs as they orbit their massive parent planets (e.g., Jupiter or Saturn).

Europa and Enceladus stand out as premier targets in this search, as their liquid water oceans may be in contact with rocky seafloors that could help build and power life. Titan too is a compelling target, both for its liquid water ocean and for the prospect of "weird life" within the hydrocarbon lakes and seas that populate its icy surface. Along with these three moons, worlds like Pluto, Ganymede, Callisto, and Triton all provide intriguing clues to subsurface liquid environments—perhaps water mixed with ammonia, salts, and other antifreeze components. As further enticement for exploration of the outer solar system, the ice giant Uranus presents numerous puzzles for planet formation and hosts several moons for which we only have a few tantalizing images. Might these moons also harbor oceans? We just do not know…yet.

All of these worlds present severe challenges for spacecraft, robotics systems, and instrumentation. In some cases, the type of exploration—e.g., melting through the ice shells of ocean worlds—is something completely new in the realm of solar system exploration. When it comes to planetary exploration, our mantra has long been "flyby, orbit, land". In the decades to come, these distant oceans will push us to add "melt and swim" to this list. These are new frontiers and they will push us to advance exploration on Earth as we build the bridge to exploring worlds in the outer solar system.

Technological innovation, such as that presented throughout the chapters in this collected volume, pulls science forward and makes the great discoveries possible. The outer solar system requires us to think in new ways about spacecraft power and propulsion, as the missions must operate for long periods of time at far distances from the sun. The surface properties of many of the moons we will explore are largely unknown, requiring creative and robust solutions that are adaptable to a wide variety of possible terrains. And the exploration of oceans deep within these moons will require autonomy embedded within every aspect of the robotic system, as no engineer is going to be able to "joy stick" an underwater vehicle to safety from a billion kilometers away. Coupled with these challenges is the opportunity to let Mother Nature help us where she can—winds on Titan and within the atmospheres of ice giant planets could serve to power robotic vehicles; cracks in the ice shells of ocean worlds could serve as conduits to the oceans within; and magnetic fields and radiation belts could help drive power systems that help sustain spacecraft. The innovation and creativity need to push this grand frontier in our solar system is well-captured by the many ideas and concepts presented here. It is my hope—or rather prediction—that with the advances that have been made to date, and with many of the advances described in this book, humanity could, within the next 20 years, discover whether or not life exists beyond Earth.

Pasadena, CA, USA
October 2017

Dr. Kevin Peter Hand
NASA Jet Propulsion Laboratory
California Institute of Technology

Preface

This is the fifth volume of a Springer book series making an inventory of the material and energy resources of our solar system. The first four books, referring to resources existent on Mars, Moon, asteroids, and inner solar system, were published in 2009, 2012, 2013, and 2015, respectively.

This book presents a present-day perspective on the energy and material resources in the outer solar system for prospective human use. One investigates the advantages and limitations of various systems thought-out for future mankind utilization. The book collects together recent proposals and innovative options and solutions.

The book is structured along logical lines of progressive thought and may be conceptually divided into five parts.

The first part deals with *properties of planetary regolith* and contains two chapters. After the introductory Chap. 1, showing what we presently know about the surface composition of Pluto, Chap. 2 refers to the physical properties of icy materials.

The second part of the book deals with *resource, mining, and subsurface access* and contains five chapters. Chapter 3 enters the details of atmospheric mining in the outer solar system; Chap. 4 describes laser-powered cryobots and other methods for penetrating deep ice on ocean worlds under the framework of project Valkyrie. Chapter 5 describes the Europa Drum Sampler, Chap. 6 presents drilling mechanisms using piezoelectric actuators developed at Jet Propulsion Laboratories, while Chap. 7 focuses on ultrasonically assisted penetration of granular and cemented materials.

The third part of the book, dealing with the *missions and mission concepts,* consists of eight chapters. Chapter 8 deals with the flight in the outer solar system and interstellar travel, Chap. 9 describes the Triton Hopper, and Chap. 10 treats sub-ice autonomous underwater vehicle architectures for ocean world exploration and life search. The Titan submarine is presented in Chap. 11, while Chaps. 12 and 13 deal with two novel concepts: the WindBots and Enceladus Vent Explorer, respectively. Chapter 14 focuses on the exploration and exploitation of Kuiper Belt Object resources in the future, and Chap. 15 describes a sample return mission by an Unmanned Interplanetary Spaceship UNIS.

The fourth part of the book, referring to *enabling technologies,* consists of four chapters. Chapter 16 describes spacecraft power systems for the far reaches of the solar system, while Chap. 17 presents hybrid nuclear spacecraft for the outer planets. Chapter 18 refers to the exploration of the outer solar system, reviewing missions and their power systems, and Chap. 19 deals with multi-rendezvous solar electric propulsion mission opportunities to Jupiter Trojans.

The fifth part of the book, dealing with *business cases for resource utilization,* consists of two chapters. Chapter 20 deals with business in the outer solar system, while Chap. 21 focuses on several sources of energy.

Most of the chapters are interdisciplinary by nature, and some of them might be equally well included into more than one part. More details about the 21 chapters of the book are given below.

Chapter 1, by Cathy Olkin and Will Grundy, states that during the summer of 2015, the New Horizons spacecraft flew past Pluto at a distance of $\sim$12,500 km after a journey of more than 3 billion miles. This encounter transformed our understanding of Pluto's geology, surface composition, and atmosphere. Near-infrared spectroscopic imaging allows us to identify compositional units across Pluto's surface. There is a latitudinal pattern across Pluto with dark red complex organic macromolecules dominating the equatorial terrain and methane ice at Pluto's North Pole. The prominent bright area that is the western lobe of Pluto's "heart" is a basin of ices of nitrogen, methane, and carbon monoxide. These ices can coexist in both gas and solid forms at Pluto temperatures ($\sim$40 K). Water ice was also detected on Pluto's mountains and cliff faces. With the high-resolution infrared spectroscopic observations (better than 5 km in some locations), we can see the diversity of compositional units across Pluto.

Chapter 2, by Craig Pitcher and Yang Gao, shows that there is evidence that water ice exists on a number of bodies in the solar system. As ice deposits may contain biomarkers that indicate the presence of life, or can be used as a consumable resource for future missions, confirming these observations with in situ measurements is of great interest. Missions aiming to do this must consider how the presence of water ice in regolith affects both the regolith's properties and the performance of the instruments that interact with it. The properties of icy lunar and Martian regolith simulants in preparation for currently planned missions are examined in this chapter. These results can be used in future instrumentation testing and missions designed to explore other icy bodies in the solar system. The testing of icy lunar regolith simulants is summarized, before focusing on experiments demonstrating the change in properties of frozen NU-LHT-2M, a simulant of the highlands regolith found at the lunar poles, as water is added. Further tests showed a critical point of $5 \pm 1\%$ water mass content where the penetration resistance significantly increases. The addition of water to Martian regolith simulants was also examined, with the presence of salts resulting in the formation of cemented crusts under simulated Martian conditions. Additional tests with the ExoMars PSDDS demonstrated how increased internal cohesion caused by the water resulted in the failure of the instrument.

Chapter 3, by Bryan Palaszewski, states that atmospheric mining in the outer solar system has been investigated as a means of fuel production for high energy propulsion and power. Fusion fuels such as helium 3 (3He) and hydrogen can be wrested from the atmospheres of Uranus and Neptune and either returned to Earth or used in situ for energy production. Helium 3 and hydrogen (deuterium, etc.) were the primary gases of interest with hydrogen being the primary propellant for nuclear thermal solid core and gas core rocket-based atmospheric flight. A series of analyses were undertaken to investigate resource capturing aspects of atmospheric mining in the outer solar system. This included the gas capturing rate for hydrogen helium 4 and helium 3, storage options, and different methods of direct use of the captured gases. Additional supporting analyses were conducted to illuminate vehicle sizing and orbital transportation issues.

Chapter 4, by William Stone, Bart Hogan, Vickie Siegel, John Harman, Chris Flesher, Evan Clark, Scott Lelievre, Josh Moor, Bryce Rothhammer, Keith Huffstutler, Omkar Pradhan, Albin Gasiewski, Steve Howe, and Troy Howe, states that ocean worlds (Europa, Enceladus, Titan, and elsewhere) hold the greatest potential in humanity's search for extant life as well as a second, independent origin of life in the solar system beyond Earth. The most likely location for possible life is beneath the surface of these worlds—through ice caps kilometers in thickness—and into full-planet oceans warmed by gravitational tidal heating. Accessing these places and autonomously discovering life, there is one of the greatest engineering and robotics challenges of our age. This chapter discusses several novel options for penetrating kilometers of ice with a starting condition of vacuum and 100K temperature. Such devices will be the essential enablers for the delivery of behavior-based autonomous underwater vehicles capable of full subsurface ocean exploration and life search, which is the subject of Chap. 10. In this chapter, the authors briefly review work to date in the area of ice penetrators. The authors present designs for several types of cryobot vehicles and specifically focus on novel power source options both for analog testing of fully integrated systems here on Earth as well as a range of possible nuclear options for actual flight missions to ocean worlds. The authors describe the four primary impediments to a successful cryobot mission (including the "starting problem" of inserting the cryobot into an ice cap at 100 K temperatures and hard vacuum; the "cruise phase problem" of transferring heat to the vehicle sidewalls to prevent refreeze entrapment; the "debris problem"—that of penetrating impact debris fragments and brine layers embedded within an ice cap; and the "breakthrough problem" where a cryobot first enters a subterranean ocean) and offer possible solutions to these highly unusual problems.

Chapter 5, by Kris Zacny, shows that recently NASA approved the Flagship-class Clipper mission to explore Jupiter's Moon Europa. Europa is the most likely planetary body that could harbor life. The mission will also be augmented with a $\sim$350 kg lander to ground-truth orbital measurements and enhance concentration and detection limits. The lander will be launched as a standalone spacecraft rather than add on to the Clipper flyby. The lander requires a sampling system to capture surface and subsurface material for analysis by onboard instruments. The threshold mission includes chemical analyses of three samples from 10 cm depth or deeper.

Europa's surface topography is unknown on the scale of a lander and even less on the scale of the sampling system itself. This requires the sampling system to be compliant with variable surface features. The Europa surface could be composed of cryogenic water ice of different densities, salt, or frozen sulfuric acid. As such, the sampling system needs to be able to work with any of these materials. The strength of cryogenic water ice is equivalent to the strength of basalt. As such, sampling time, forces, and life of the sampling system could prove to be challenging. Presence of sulfuric acid limits the use of traditional aerospace materials. Since local gravity is 1.3 m/s^2, the practical limit of downforce available for the sampling system is in the 10s of Newtons range. This chapter presents a sampling system based on a continuous miner/roadheader design. The cutter drum of the Europa Drum Sampler (EDuS) is compliant with highly variable surface roughness. It is also extremely robust and has been shown to cut through weak and hard rocks. The system is currently being developed.

Chapter 6, by Yoseph Bar-Cohen, Stewart Sherrit, Mircea Badescu, Hyeong Jae Lee, Xiaoqi Bao, and Zensheu Chang, shows that drilling mechanisms are widely used as means of penetrating objects and formations in such fields as domestic, medical, industrial, military, geology, and extraterrestrial applications. Many types of drills have been developed over the years with the majority based on mechanical motion (rotary and/or percussive) of a cutting tool. Mechanical drills use a bit having a tip that interacts with the drilled material and applies forces over a small area to cause large shear and/or impact stresses for cutting or breaking the material. There is a wide variety of bit types that have been developed commercially, which can be readily purchased at local hardware stores. Increasingly, developers of drills for in situ exploration missions are seeking capabilities that address the complex challenges involved with extreme environments found at the planetary bodies where subsurface penetration is needed. The use of piezoelectric actuators offers effective drilling capabilities, particularly in extreme environments. In the last two decades, significant developments have been made in using piezoelectric actuation to perform percussive drilling. In these designs, the cutting surface is fractured by high- and low-frequency impacts that enhance the penetration. The performance is again significantly increased by rotating the bit to introduce shearing and a mechanism to allow the cuttings to be removed from the created borehole. This chapter is focused on the drilling mechanisms that are driven by piezoelectric actuators, which were developed by the authors at the Jet Propulsion Lab (JPL), Pasadena, CA.

Chapter 7, by David Firstbrook, Patrick Harkness, Xuan Li, Ryan Timoney, and Kevin Worrall, states that ultrasonic vibration can be applied to a number of penetration and drilling scenarios, either directly through a penetrator or ultrasonically assisted tool or through the ultrasonic–percussive technique and a physically decoupled drill bit. Considering these different approaches, this chapter describes experiments which indicate that, in granular materials, the addition of ultrasonics can sometimes reduce both the force and power required to reach a given depth. Ultrasonically assisted drills are shown to proceed more quickly than their unassisted counterparts, and ultrasonic–percussive systems are shown to extract samples of permafrost simulant and even Antarctic snowpack in a still-frozen state. In the latter case, the ultrasonic–percussive drill has been combined with a control system and sample-caching architecture to create a workable sample acquisition concept for cold space environments.

In Chap. 8, Alexander Bolonkin states that it is extremely difficult with current and near-future technology to achieve interstellar flight, even in our nearest star system, Alpha Centauri. Such an exploration and launch would require enormous energy, expensive equipment, and long travel time. Current conventional nuclear and thermonuclear onboard reactors cannot solve these problems. The author discusses and estimates the parameters needed to achieve such flight via a mini, automatic probe. Three new possible perspective propulsion systems are proposed: multi-reflex light system using new sell-multi-reflex mirrors and lasers, a cold plasma beam from Earth, and an onboard Micro Black Hole (MBH) nuclear photon rocket. In all approaches, technological innovations to current systems are offered to make it possible to implement within the current technological restraints. The requested launch system (i.e., laser multi-reflect propulsion, cold plasma beam propulsion, or MBH nuclear propulsion), onboard equipment, energy installation (generator and accelerator), environmental constraints, drag allowances, interstellar microparticles, and communication ability with Earth are all described, estimated, and validated. Analysis shows that the most realistic interstellar launch system is a laser beam using the cell reflective mirror or ultra-cold plasma beam.

Chapter 9, by Steven Oleson and Geoffrey Landis, states that Neptune's moon Triton is a fascinating object, a dynamic moon with an atmosphere, and geysers. Triton is unique in the outer solar system in that it is most likely a captured Kuiper belt object (KBO), a leftover building block of the solar system. When Voyager flew by it was the coldest body yet found in our solar system (33 K) and had volcanic activity, geysers, and a thin atmosphere. It is covered in ices made from nitrogen, water, and carbon dioxide, and shows surface deposits of tholins, organic compounds that may be precursor chemicals to the origin of life. Exploring Triton will be a challenge well beyond anything done in previous missions but the unique environment of Triton also allows some new possibilities for mobility. A conceptual design of a Triton Hopping probe was developed that both analyzes the surface and collects it for use to propel its hops. The Hopper would land near the South Pole in 2040 where geysers have been detected. Depending on the details of propulsion chosen, the Hopper should be able to jump over 300 km in 60 hops or less, exploring the surface and thin atmosphere on its way. This craft will autonomously carry out detailed scientific investigations on the surface, below the surface (drilling) and in the upper atmosphere to provide unprecedented knowledge of a KBO turned moon and expanding NASA's existing capabilities in deep space planetary exploration to include Hoppers using different ices for propellant.

Chapter 10 by William Stone, Kristof Richmond, Chris Flesher, Bart Hogan, and Vickie Siegel states that ice-covered oceans are found across the solar system. On Earth, such environments are known to harbor life. On some ocean worlds such as Europa, the unique combination of an actively recycled ice shell and rocky, possibly magmatic interior, may give rise to a geochemical system suitable to life. The entry into sub-ice oceans of ocean worlds enabled by new cryobot technologies calls for the development of autonomous underwater vehicles to explore these water bodies as the next phase of exploration. The most fruitful places to search for life will be at energy sources provided by physical and chemical gradients, which may not necessarily occur at the breakthrough location of a cryobot. This implies exploration using a mobile platform, and this in turn—due to extremely limited bandwidth and hours-long round-trip transmission delay—must be an autonomous platform. To achieve this, an intelligent underwater robotic explorer must be developed that can travel in an ice-covered ocean; identify signs of biological activity; home in on, acquire, and analyze samples; and return to the cryobot to upload data for subsequent transmission to Earth. This chapter investigates technologies required for such an under-ice rover. The authors describe in detail several approaches that have been developed and field-tested at Stone Aerospace during the past 15 years. The authors draw upon and extend technologies that have been developed for oceanography and oil prospecting, then address the unique problems of under-ice exploration where no prior knowledge exists. What is unique about the problems posed by ocean worlds is that it requires a high degree of system robustness, onboard operational and science decision-making, and navigation which has yet to be demonstrated as a complete system in terrestrial oceanographic or even research systems. The goal of a releasing a self-contained system into a completely unknown environment and expecting it to perform sensible exploration is still elusive, but definite progress is being made and the path forward is becoming more clear. The authors describe procedures and algorithms that have been tested in analog environments in Antarctica and elsewhere and suggest where research may be most effectively invested to advance operational capabilities for under-ice autonomous systems.

Chapter 11, by Steven R. Oleson, Jason Hartwig, Jeffrey Woytach, Michael Martini, Anthony Colozza, Robert Jones, Thomas Packard, Paul Schmitz, Amy Stalker, Ralph D. Lorenz, Michael V. Paul, and Justin Walsh, states that the conceptual design of a submarine for Saturn's moon Titan was a funded NASA's Innovative Advanced Concepts (NIAC). The effort investigated what science a submarine for Titan's liquid hydrocarbon ~93 K (180 °C) seas might accomplish and what that submarine might look like. Focusing on a flagship-class science system (~100 kg), it was found that a submersible platform can accomplish extensive and exciting science both above and below the surface of the Kraken Mare. The submerged science includes mapping using side-looking sonar, imaging, and spectroscopy of the sea at all depths, as well as sampling of the sea's bottom and shallow shoreline. While surfaced, the submarine will not only sense weather conditions (including the interaction between the liquid and atmosphere) but also image the shoreline, as much as 2 km inland. This imaging requirement pushed the landing date to Titan's next summer period (~2047) to allow for continuous lighted conditions, as well as Direct-to-Earth (DTE) communication, avoiding the need for a separate relay orbiter spacecraft. Submerged and surfaced investigations are key to understanding the hydrological cycle of Titan as well as gather hints to how life may have begun on Earth using liquid/sediment/chemical interactions. An estimated 25 Mb of data per day would be generated by the various science packages. Most of the science packages (electronics at least) can be safely kept inside the submarine pressure vessel and warmed by the isotope power system.

Chapter 12 by Adrian Stoica, Virgil Adumitroaie, Marco Quadrelli, Georgios Matheou, Marcin Witek, Marco Cipolato, Marco Dolci, James Roggeveen, Kyle Petersen, Kristina Andreyeva, Hunter Hall, Benjamin Donitz, and Leon Kim examines the concept of Wind Robots aiming to achieve long-term in situ science exploration of Gas Giants, Jupiter and Saturn. These planets are made almost entirely of hydrogen and helium, and have no hard surface to land on. Their low-temperature atmospheres are characterized by strong winds, at least in the observed upper atmosphere. The atmosphere in the areas of interest has low density, where it is very difficult to stay afloat, especially in conditions of higher gravity than on Earth. Combined with the desire to operate using only locally harvested energy, these aspects make the design of WindBots extremely challenging. As a general mission goal, one aims to operate slightly above and below the clouds region, which is between 0.3 bar and 10 bar on Jupiter, for a year-long-duration mission, in strong (potentially turbulent) winds. For example, WindBots would operate in the eyewall of the Great Red Spot, using the high wind and updrafts of the anticyclone, as well as horizontal gusts. The chapter summarizes the results of a study funded by a Phase I NASA Innovative Advanced Concepts (NIAC). The study looked at both naturally buoyant and winged solutions, as well as hybrids of the two. The chapter examines means of maintaining a long-duration mission and providing controls for the WindBots. It discusses several body configurations and deployment, as well as investigates energy recovery methods using active systems controlling the WindBots and passive systems to harvest naturally occurring wind, pressure, and thermal gradients in the atmosphere. It also reviews autonomy architecture and control schemes to be applied to the autonomous WindBots.

Chapter 13, by Masahiro Ono, Karl Mitchel, Aaron Parness, Kalind Carpenter, Saverio Iacoponi, Ellie Simonson, Aaron Curtis, Mitch Ingham, Charles Budney, Tara Estlin, Carolyn Parcheta, Renaud Detry, Jeremy Nash, Jean-Pierre de la Croix, Jessie Kawata, and Kevin Hand, explores a concept of sending small robots into the erupting vents on Enceladus. Descending into the vent is of great scientific interest not just because it could be a low-energy pathway to the subsurface ocean that may harbor life but also because the vent itself could be habitable. The two greatest challenges for the feasibility of the concept are the dynamic pressure due to the upward flow and the size of vent. The preliminary characterization of Enceladus vent suggests that these two parameters largely depend on the unknown eruption mechanism. With relatively calm "boiling" hypotheses, the dynamic pressure is up to $\sim 10^4$ Pa and the vent width could be in the order of 1 m, while more dynamic "cryovolcanic" hypotheses result in 10^3–10^7 Pa dynamic pressure and 1–30 cm vent width. Meanwhile, the system trade study concluded that the most robust design would be a highly automated limbed robot with ice screw end effector, powered by a separate surface module through a tether. The authors developed a prototype design of the robot, which is $\sim 5 \times 10 \times 30$ cm in size, 3 kg in mass, and can move against $\sim 10^5$ Pa of dynamic pressure at a speed of ~ 5.5 m/hr. Therefore, the concept is likely feasible assuming the "boiling" models while the feasibility is undetermined with the current best knowledge under the "cryovolcanic" models.

Chapter 14, by Volker Maiwald, states that the Kuiper Belt is a distant region of our solar system and by now, the discovery of several small worlds, like Pluto, Makemake, or Haumea, have shown that the trans-Neptunian region of the solar system is indeed more inhabited than first meets the eye. The large distance to the Sun makes investigating these worlds challenging but it is known that the Kuiper Belt contains methane, nitrogen, ethane, even water, and other volatiles in some quantity. While the Kuiper Belt is far away and a continuous presence there beyond our current technological capabilities, it is also clear that this large region of the solar system contains a number of useful resources. These could be of use in future endeavors of humankind, establishing a more extended presence (robotic and human) in the solar system. This chapter explains a scenario how the Kuiper Belt can be opened up for humanity and become a ground for exploration and resource exploitation in the future. The chapter discusses an approach involving self-replication of autonomous probes capable to explore Kuiper Belt Objects and harvest their resources, i.e., to replicate and also to transfer resources to other parts of the solar system. It elaborates on the relevant topics like power generation and trajectories and at the same time points out technological open issues that need to be closed for such a scenario to work.

Chapter 15, by Werner Grandl, Ákos Bazsó, and Andreas F. Felsenstein, states that the Outer Solar System (OSS) contains a large number of different objects, e.g., gas giant planets like Jupiter and Saturn, dwarf planets like Pluto, and countless comets and asteroids. In the past, space missions, like Pioneer, Voyager, and more recently New Horizons, have visited and explored some of these remote objects. Although the OSS hosts some of the currently most interesting potential mission targets—among them Trojan asteroids and small Kuiper belt objects—yet there is a lack of ambitious plans for a sample return mission to objects in the OSS. On Earth, we only have access to a very limited sample of rocks from extraterrestrial objects, mostly in the form of lunar rocks collected by the Apollo missions, or meteorites from the Moon or Mars. Meteorites also delivered material from the inner and middle parts of the asteroid belt but the OSS is very sparsely sampled (if at all). For a more complete picture of the formation and early evolution of the solar system, we would also need samples from the OSS including remote objects definitively originating from the Kuiper belt. If we want to utilize the natural resources of the OSS, we should bring back soil and rock samples of celestial bodies in the OSS for chemical analysis to estimate the resources of metals, water, and other useful materials out there. The authors have designed a preliminary concept for a sample return mission to two minor bodies in the OSS, the Jupiter Trojan asteroid (624) Hektor and the Centaur group asteroid (2060) Chiron. The spacecraft UNIS (Unmanned Interplanetary Spaceship) for this long-duration mission (probably lasting some decades) would be assembled in Low Earth Orbit. To reach the targets, it would need to perform several gravity assist maneuvers via planetary encounters in the Inner solar system. After arriving at the targets, a robotic lander (SPIDER) would be activated to descend to and sample the surface. With its six legs, the

SPIDER can walk across the asteroid's surface in any direction. Finally, the stored samples will be returned to Earth to analyze them in laboratories.

Chapter 16, by Robert Cataldo, states that missions to the outer planets present a unique challenge to power systems. Long trip times and the duration of science investigations place a lifetime demand on systems. In addition, extreme low temperatures can require excess power demands on spacecraft thermal management systems for maintaining proper temperatures for avionics and liquid propellants. Historically, missions beyond Mars have utilized nuclear power in the form of Radioisotope Thermoelectric Generators (RTG) to provide sufficient power in the cold and dark regions of space. From the 1970s with Pioneer and Voyager to 2006 on Pluto New Horizons, RTGs have provided both electric power and heat to a host of very successful missions. RTGs carry its own energy source, namely plutonium 238, that provides the heat source for ultimate conversion into electricity. Thus, the power available is independent of the sun. Over the past decade, however, solar cell technologies have made great strides in efficiency from under 20% to over 30% allowing missions like Juno, for use in Jupiter polar orbit. However, the Sun's light intensity at Jupiter is only ~5% than it is at Earth and ~0.1% at Pluto's orbit. The physical size of solar panels becomes very large and difficult to accommodate the further away from the Sun. Every mission is unique in spacecraft design, trip times, mission duration, science instrument requirements, etc. Therefore, the power system of choice is really the result of possible trades with other subsystem's requirements for mass, volume, and area. General guidelines for power system selection are discussed to provide insight into various technology considerations that should be evaluated in performing power system trade studies for the unique conditions of outer planet missions.

Chapter 17, by Mark A. Stull and Ricky Tang, states that chemical rockets are never going to allow us to exploit outer solar system resources, that is going to require massive spacecraft capable of transporting heavy loads of cargo. It is also going to require reducing trip durations to economically viable times, and this must be done at a cost sufficiently low that resources from beyond the asteroid belt can compete with those obtainable within it. Chemical fuels simply cannot produce the power density and specific impulse needed to meet these requirements. Nuclear fuels can provide orders of magnitude greater power density but conventional nuclear thermal propulsion still limits specific impulse to insufficient values. However, a hybrid nuclear reactor can employ a fusion reaction at less than 5% of breakeven in a gas dynamic mirror thruster to generate a flux of neutrons that can drive a subcritical fission reaction in a fissile fuel, or via the thorium cycle, in which fissile U_{233} is bred. The combined fusion and fission reactions can, in theory, generate specific impulse as high as several hundred thousand seconds, with continuous power levels in excess of 100 terawatts. If achievable in practice, this would enable a ship with economically viable cargo capacity to travel under continuous acceleration to all outer solar system bodies, including distant Eris, in economically acceptable times—days or weeks—without requiring prohibitively high amounts of propellant. But the bulk of the power would come from the fission reaction, and converting all or most of the potential fission energy to thrust presents a daunting challenge that present and foreseeable future technology may not be fully able to meet. Nevertheless, a more modest capability appears possible and could still enable high-capacity cargo ships to reach Jupiter and possibly Saturn in commercially viable times.

Chapter 18, by Simon Fraser, summarizes more than 45 years of outer solar system exploration to Saturn, Jupiter, Uranus, Neptune, Pluto, to their moons, and beyond. All past and present exploration missions to the outer solar system are briefly presented and discussed with respect to spacecraft, payload, and scientific objectives. In this, a special emphasis is put on describing the power systems applied with the different mission elements. Radioisotope Thermoelectric Generators (RTGs) are the power sources of choice with all but one mission launched to the outer solar system. RTGs are robust, do not have any rotating or moving parts, a long life, and predictable output power levels. This makes them particularly attractive for multiyear missions into regions of the solar system where solar insolation is very low. The first exception in this preference of RTGs is the most recent mission, where electrical energy was provided by large solar panels instead of a radioisotope-fuelled power source. The second exception consisted of two probes, which were piggybacked to their destination, and only had to be self-sufficient for a very limited period of time; batteries were chosen with both of these probes.

Chapter 19, by Volker Maiwald, states that Jupiter Trojans are a group of small bodies accumulated at the stable L_4 and L_5 libration points of the Jupiter–Sun system. They are numerous and are scientifically interesting regarding the formation and evolution of the solar system as a whole. Located at stable positions in Jupiter's solar distance, they are currently theorized to be dormant comets, containing volatiles, including water. A dedicated mission to this population of small bodies could verify hypotheses about their origin, thus shed light on the solar system history all-together and at the same time reveal details about their composition. This chapter details how such a mission could be accomplished, including a medium-sized spacecraft (launch mass < 1600 kg) and transfer trajectories, allowing a multi-rendezvous mission within a time frame of about 15 years. The spacecraft design, based on solar electric propulsion, contains details about the probe's subsystems and scientific payload, its mass and power budget, and also its configuration. The latter is dominated by the large deployable solar arrays, required to provide several kilowatts of power for the electrical thrusters. The spacecraft is able to reach three to four Trojans altogether, allowing a diverse review of Trojan properties and in general the gathering of a large amount of scientific data—directly facilitating any resource harvesting missions in future years.

Chapter 20, by Mike H. Ryan and Ida Kutschera, shows that the possibilities of future business opportunities that might exist within the outer solar system are explored. Fictional accounts of development within space are used to provide context between imagined paths of exploration and what is more likely within the foreseeable future. Additional observations as to the nature and extent of how business might develop are discussed. Although business creation within the realm of the outer solar system is likely to remain a distant possibility, it is not without terrestrial precedents. It is also probable that business opportunities would occur at a pace faster than expected if either the resources discovered among the planets and moons proved to be of greater value than anticipated or if technology made exploitation less difficult and more cost-effective. It is not if but when and how the resources of the outer solar system will be utilized in the future. In the near-term, business among the outer solar system will continue to be implausible until both opportunity and technology change significantly and catch up with the intriguing vision set forth by science fiction and business visionaries.

In Chap. 21, Alexander Bolonkin details two new mini thermonuclear reactors (cumulative and impulse) and a novel ultra-cold method of getting nuclear energy by cold thermonuclear synthesis. The new reactors are smaller and cheaper than the current and known future reactors by several orders of magnitude. They may be R&D in 2–3 years, may be used for transport vehicles (ships, trains, aircrafts, rockets, cars, and electric stations), and use cheap nuclear fuel.

***

The book allows the reader to acquire a clear understanding of the scientific fundamentals behind specific technologies to be used in the outer solar system region in the future. The principal audience consists of researchers (engineers and physicists) involved or interested in space exploration in general and outer solar system exploration in particular. Also, the book may be useful for industry developers interested in joining national or international space programs. Finally, it may be used for undergraduate, postgraduate, and doctoral teaching in faculties of engineering and natural sciences.

Bucharest, Romania — Viorel Badescu
Pasadena, USA — Kris Zacny

Acknowledgements

A critical part of writing any book is the review process, and the authors and editors are very much obliged to the following researchers who patiently helped them read through subsequent chapters and who made valuable suggestions: Phillip Abel (NASA Glenn Research Center, USA), Gary Bennett (NASA Headquarters, retired scientist, USA), Mirel I. Birlan (Observatoire de Paris, France), Isaiah Blankson (NASA Glenn Research Center, USA), Kevin Breisacher (NASA Glenn Research Center, USA), Bill Emrich (NASA Marshall Space Flight Center, USA), James Fincannon (NASA Glenn Research Center, USA), Dominic Fortes (Science and Technologies Facility Council, UK), Joseph Friedlander (Israel), Craig S. Galbraith (University of North Carolina, USA), Farid Gamgami (OHB System, Germany), Viktor Hacker (Graz University of Technology, Austria), Patrick Harkness (University of Glasgow, UK), Mark Krinker (City University of New York, USA), Bruce Montgomery (Magplane Technology Group of Companies, USA), Shmuel Neumann (Jerusalem University, Israel), Francis Nimmo (University of California Santa Cruz, USA), Marie Nowacki (Honeybee Robotics, USA), Ramesh Malla (University of Connecticut, USA), Chris McKay (NASA Ames Research Center USA), Marco Pavone (Stanford University, USA), Oleg Pensky (Perm State University, Russia), Marco Quadrelli (NASA Jet Propulsion Laboratory, USA), Max Schautz (European Space Agency), Wolfgang Seboldt (DLR Cologne, Germany), Scott Sherman (Texas A&M University-Corpus Christi, USA), Stewart Sherrit (NASA Jet Propulsion Laboratory, USA), Jean Souchay (Observatoire de Paris, France), John Stansberry (Space Telescope Science Institute, USA), Hans Thomas (Monterey Bay Aquarium Research Institute, USA), Harry Varvoglis (Aristotle University of Thessaloniki, Greece), and Dana Yoerger (Woods Hole Oceanographic Institution, USA).

Substantial help and guidance have been received from Thomas Ditzinger, from the Springer team, to whom our thanks are kindly addressed.

The editors, furthermore, owe a debt of gratitude to all authors. Collaborating with these stimulating colleagues has been a privilege and a very satisfying experience.

Viorel Badescu
Kris Zacny

Contents

Contributors

Virgil Adumitroaie Jet Propulsion Laboratory, California Institute of Technology, Pasadena, CA, USA

Kristina Andreyeva Jet Propulsion Laboratory, California Institute of Technology, Pasadena, CA, USA

Mircea Badescu Jet Propulsion Laboratory, California Institute of Technology, Pasadena, CA, USA

Xiaoqi Bao Jet Propulsion Laboratory, California Institute of Technology, Pasadena, CA, USA

Yoseph Bar-Cohen Jet Propulsion Laboratory, California Institute of Technology, Pasadena, CA, USA

Ákos Bazsó University of Vienna, Vienna, Austria

Alexander Bolonkin C&R, Brooklyn, NY, USA

Charles Budney Jet Propulsion Laboratory, California Institute of Technology, Pasadena, CA, USA

Kalind Carpenter Jet Propulsion Laboratory, California Institute of Technology, Pasadena, CA, USA

Robert Cataldo NASA Glenn Research Center, Cleveland, OH, USA

Zensheu Chang Jet Propulsion Laboratory, California Institute of Technology, Pasadena, CA, USA

Marco Cipolato Jet Propulsion Laboratory, California Institute of Technology, Pasadena, CA, USA

Evan Clark Stone Aerospace, Del Valle, TX, USA

Anthony Colozza NASA Glenn Research Center, Cleveland, OH, USA

Aaron Curtis Jet Propulsion Laboratory, California Institute of Technology, Pasadena, CA, USA

Jean-Pierre de la Croix Jet Propulsion Laboratory, California Institute of Technology, Pasadena, CA, USA

Renaud Detry Jet Propulsion Laboratory, California Institute of Technology, Pasadena, CA, USA

Marco Dolci Jet Propulsion Laboratory, California Institute of Technology, Pasadena, CA, USA

Benjamin Donitz Jet Propulsion Laboratory, California Institute of Technology, Pasadena, CA, USA

Tara Estlin Jet Propulsion Laboratory, California Institute of Technology, Pasadena, CA, USA

Andreas F. Felsenstein University of Vienna, Vienna, Austria

David Firstbrook School of Engineering, University of Glasgow, Glasgow, UK

Chris Flesher Stone Aerospace, Del Valle, TX, USA

Simon D. Fraser Energy Systems Engineering, Graz, Austria

Yang Gao Surrey Space Centre, University of Surrey, Guildford, UK

Albin Gasiewski University of Colorado Boulder, Boulder, CO, USA

Werner Grandl Consulting Architect, Tulln, Austria

Will Grundy Lowell Observatory, Flagstaff, USA

Hunter Hall Jet Propulsion Laboratory, California Institute of Technology, Pasadena, CA, USA

Kevin Hand Jet Propulsion Laboratory, California Institute of Technology, Pasadena, CA, USA

Patrick Harkness School of Engineering, University of Glasgow, Glasgow, UK

John Harman Stone Aerospace, Del Valle, TX, USA

Jason Hartwig NASA Glenn Research Center, Cleveland, OH, USA

Bart Hogan Stone Aerospace, Del Valle, TX, USA

Steve Howe Howe Industries, Tempe, AZ, USA

Troy Howe Howe Industries, Tempe, AZ, USA

Saverio Iacoponi Jet Propulsion Laboratory, California Institute of Technology, Pasadena, CA, USA

Mitch Ingham Jet Propulsion Laboratory, California Institute of Technology, Pasadena, CA, USA

Robert Jones NASA Glenn Research Center, Cleveland, OH, USA

Jessie Kawata Jet Propulsion Laboratory, California Institute of Technology, Pasadena, CA, USA

Leon Kim Jet Propulsion Laboratory, California Institute of Technology, Pasadena, CA, USA

Ida Kutschera Rubel School of Business, Bellarmine University, Louisville, KY, USA

Geoffrey Landis NASA Glenn Research Center, Cleveland, OH, USA

Hyeong Jae Lee Jet Propulsion Laboratory, California Institute of Technology, Pasadena, CA, USA

Xuan Li School of Engineering, University of Glasgow, Glasgow, UK

Ralph D. Lorenz Applied Physics Laboratory, Johns Hopkins University, Laurel, MA, USA

Volker Maiwald German Aerospace Center (DLR), Institute of Space Systems, Bremen, Germany

Michael Martini NASA Glenn Research Center, Cleveland, OH, USA

Georgios Matheou Jet Propulsion Laboratory, California Institute of Technology, Pasadena, CA, USA

Karl Mitchel Jet Propulsion Laboratory, California Institute of Technology, Pasadena, CA, USA

Jeremy Nash Jet Propulsion Laboratory, California Institute of Technology, Pasadena, CA, USA

Steven R. Oleson NASA Glenn Research Center, Cleveland, OH, USA

Catherine Olkin Southwest Research Institute, Boulder, USA

Masahiro Ono Jet Propulsion Laboratory, California Institute of Technology, Pasadena, CA, USA

Thomas Packard NASA Glenn Research Center, Cleveland, OH, USA

Bryan Palaszewski NASA John H. Glenn Research Center, Lewis Field, Cleveland, OH, USA

Carolyn Parcheta Jet Propulsion Laboratory, California Institute of Technology, Pasadena, CA, USA

Aaron Parness Jet Propulsion Laboratory, California Institute of Technology, Pasadena, CA, USA

Michael V. Paul Applied Physics Laboratory, Johns Hopkins University, Laurel, MA, USA

Kyle Petersen Jet Propulsion Laboratory, California Institute of Technology, Pasadena, CA, USA

Craig Pitcher Surrey Space Centre, University of Surrey, Guildford, UK

Omkar Pradhan University of Colorado Boulder, Boulder, CO, USA

Marco Quadrelli Jet Propulsion Laboratory, California Institute of Technology, Pasadena, CA, USA

Kristof Richmond Stone Aerospace, Del Valle, TX, USA

James Roggeveen Jet Propulsion Laboratory, California Institute of Technology, Pasadena, CA, USA

Mike H. Ryan Rubel School of Business, Bellarmine University, Louisville, KY, USA

Paul Schmitz NASA Glenn Research Center, Cleveland, OH, USA

Stewart Sherrit Jet Propulsion Laboratory, California Institute of Technology, Pasadena, CA, USA

Vickie Siegel Stone Aerospace, Del Valle, TX, USA

Ellie Simonson Jet Propulsion Laboratory, California Institute of Technology, Pasadena, CA, USA

Amy Stalker NASA Glenn Research Center, Cleveland, OH, USA

Adrian Stoica Jet Propulsion Laboratory, California Institute of Technology, Pasadena, CA, USA

William Stone Stone Aerospace, Del Valle, TX, USA

Mark A. Stull The Law Office of Mark A. Stull, Manchester, NH, USA

Ricky Tang Sandia National Laboratories, Albuquerque, NM, USA

Ryan Timoney School of Engineering, University of Glasgow, Glasgow, UK

Justin Walsh The Pennsylvania State University, State College, USA

Marcin Witek Jet Propulsion Laboratory, California Institute of Technology, Pasadena, CA, USA

Kevin Worrall School of Engineering, University of Glasgow, Glasgow, UK

Jeffrey Woytach NASA Glenn Research Center, Cleveland, OH, USA

Kris Zacny Honeybee Robotics Spacecraft Mechanisms Corporation, Pasadena, CA, USA

Part I
Properties of Planetary Regolith

Chapter 1
A Survey of Pluto's Surface Composition

Catherine Olkin and Will Grundy

Pluto was discovered less than 100 years ago (in 1930) by Clyde Tombaugh at Lowell Observatory in Flagstaff, AZ. The founder of the observatory, Percival Lowell, had initiated a systematic search of the sky to find the putative Planet X in 1905 (Tombaugh 1960). The existence of Planet X was predicted because of deviations of the motions of Neptune and Uranus from expected values. Other teams were also searching for Planet X including a team lead by William H. Pickering of Harvard Observatory. Lowell observatory continued the search after the founder's death in 1916, commissioning a new search telescope in 1927 that would allow astronomers to search a larger area of sky (12° × 15° using 14 × 17-in glass plates, Giclas 1980) to a fainter magnitude (16–17th magnitude)—the 13-inch astrograph. Each clear night near new moon, Clyde Tombaugh used the new telescope to image the sky near opposition. Observing the sky near opposition was essential because he was searching for Planet X by looking for its motion against the fixed stars and the motion is a maximum at opposition. An object out beyond Neptune would have a motion in the plane of the sky of ~1 arcmin/day as compared to asteroids which would move at ~15 arcmin/day. Clyde uses a machine, called a Zeiss blink comparator, to inspect the plates for moving objects by comparing plates taken a few nights apart. On the afternoon of February 18, 1930, he was comparing images of the sky near Delta Geminorium when he discovered Pluto—a 15th magnitude object that had moved 3.5 mm (about 8.5 arcmin) on two plates taken 6 days apart (Tombaugh 1960). He continued for 14 more years searching the sky for other objects in our solar system. While he

C. Olkin (✉)
Southwest Research Institute, Boulder, USA
e-mail: catherine.olkin@swri.org

W. Grundy
Lowell Observatory, Flagstaff, USA
e-mail: grundy@lowell.edu

V. Badescu and K. Zacny (eds.), *Outer Solar System*,
https://doi.org/10.1007/978-3-319-73845-1_1

discovered many asteroids, a comet and six galactic star clusters, he did not discover any more objects out beyond the orbit of Neptune.

Work by Kenneth E. Edgeworth published in the 1940s and by Gerard P. Kuiper in the 1950s on the origin of the solar system indicated that there might be a large number of objects beyond the orbit of Neptune in what is now called the Kuiper Belt (Davies et al. 2008). As technology improved, more distant small bodies were discovered. In 1977, Charles Kowal discovered Chiron (Kowal 1989), a minor planet whose orbit crosses the orbits of Saturn and Uranus. This object was another harbinger of the presence of the Kuiper Belt, as its orbit is unstable and it likely migrated to its current orbit from further out in the solar system.

It wasn't until 1992 when the next Kuiper Belt Object (KBO), after Pluto, was discovered (Jewitt and Luu 1993). In the 1990's, technology advanced to allow discovery of many more distant objects in the solar system. We had larger aperture telescopes, wide-field CCD cameras, and computer algorithms to search the digital images for moving objects. Dedicated searches for KBOs were undertaken, such as the Deep Ecliptic Survey from 1998 to 2003, which searched 550 deg^2 and discovered more than 300 KBOs (Elliot et al. 2005), the Canada–France Ecliptic Plane Survey (CFEPS) from 2003 to 2007, which searched 321 deg^2 and discovered 169 KBOs (Petit et al. 2011) and the on-going Outer Solar System Origins Survey (OSSOS) which reports initial results from the first 42 deg^2 of 85 KBOs (Bannister et al. 2016).

We now know that Pluto is one of more than 2500 known KBOs and it is the largest known object of this population with a radius of 1188.3 km (Nimmo et al. 2017). Pluto was one of the best-studied KBOs, even before the flyby of the New Horizons spacecraft. From color light curves and mutual occultations and eclipses of Pluto and its large moon Charon, it was known that Pluto had large albedo and color variegation. In 1976, using filter photometry in the 1–2 um band, Cruikshank et al. (1976) showed there was methane ice on Pluto's surface. In 1993, Owen et al. used near infrared spectroscopy to identify the signature of N_2 ice and CO ice on Pluto. Using Pluto's 6.4-day rotation period, Grundy et al. (2013) investigated longitudinal variations of Pluto's surfaces ices and found that CO ice on Pluto's surface was preferentially located on Pluto's anti-Charon hemisphere, but our knowledge of the distribution of volatiles (molecules on the surface that have an appreciable vapor pressure) on its surface was limited. Cruikshank et al. (2015) presents a review of the state of knowledge about Pluto's surface composition prior to the New Horizons encounter.

Even using the best telescopes available, Earth-based observations provide at most a few resolution elements across Pluto's disk. To significantly advance our understanding of Pluto, a spacecraft was needed. There were a number of Pluto mission studies initiated in the 1990's and NASA released an Announcement of Opportunity for a mission to Pluto and the Kuiper Belt in January 2001. By November 2001, the New Horizons mission was selected by NASA to explore Pluto and its large moon Charon.

The mission's three main objectives were to characterize the geology of Pluto and Charon, to map the surface composition of Pluto and Charon and to characterize the neutral atmosphere of Pluto and its escape rate. These and additional secondary and tertiary science objectives (see Young et al. 2008; Weaver et al. 2008) drove the selection of 7 science instruments on New Horizons. There is a high-resolution black-and-white imager that observed the most detailed images of Pluto's surface. Another instrument is a combined wide-field visible imager and shortwave mapping spectrometer (Reuter et al. 2008). The wide-field imager has both 4-color channels and black-and-white channels. The shortwave mapping spectrometer, LEISA, was used for mapping the surface composition of Pluto and Charon and is the primary instrument discussed here. Other instruments included 2 in situ plasma instruments, a dust detector, a UV imaging spectrometer and a radio science experiment.

When it came time to plan the details of the Pluto encounter, one of the first tasks was to determine which hemisphere we would image near closest approach. Because of the flyby nature of the mission design, the encounter (or closest approach) hemisphere would have much better resolution than the non-encounter hemisphere is seen at highest resolution from a range of 4.5 million km instead of 13,691 km, the closest approach distance from New Horizons to the center of Pluto (Stern et al. 2015). There were a number of scientific drivers in the choice of the encounter hemisphere including known color variegation and surface composition. As mentioned previously, absorption from CO had been seen on the anti-Charon hemisphere of Pluto, but not on the Charon-facing hemisphere. This is one reason that the New Horizons mission team chose to flyby the anti-Charon hemisphere of Pluto at closest approach. Now we know that the CO is preferentially located in the large depression (Moore et al. 2016), known as Sputnik Planitia [some features names are informal] (see Figs. 1.1 and 1.2).

Figure 1.1 shows the extreme albedo and color variegation across Pluto from the dark and red Cthulhu Macula (see Fig. 1.2) to the much brighter and gray Sputnik Planitia, to the bluish terrain in eastern Tombaugh Regio. The red coloring on Pluto has been attributed to the presence of organic molecules (Stern et al. 2015) which are the result of chemical reactions (initiated by charge particle or energetic photon bombardment) of the carbon and nitrogen bearing species to produce complex hydrocarbon molecules, commonly called tholins (Cruikshank et al. 2005).

The color-color plot in Fig. 1.3 is constructed from the three broadband colors and the narrow band methane channel of the color imager from the closest-approach color Pluto image (Fig. 1.1). Neutral terrain would be located at the (1,1) point on the graph, with bluer colors plotting above and to the right. Pluto has a generally reddish color as demonstrated in Fig. 1.3 by the Blue/Red ratio less than 1. In the Red to NIR, there are regions with a nearly flat slope between these two passbands (pixels with values of Red/NIR near 1.0) and there is also terrain with a red slope extending into the NIR (pixels with values of Red/NIR much less than 1.0).

Mapping from the color-color plot (Fig. 1.3) back to the globe demonstrates where similar color units on the surface are located, as seen in Fig. 1.4. The location of different composition units can be directly seen in these figures. The yellow terrain in the north is distinct from other terrains both visually (see Fig. 1.1) and in the distribution of points in the color-color plot. This source of the coloring agent is currently unknown. It could be an intermediate photolysis product of methane and

Fig. 1.1 Enhanced color image of Pluto taken ~40 min before closest approach at a distance of 35,000 km from Pluto's center. The color is a combination of the blue, red and NIR channels and has been enhanced to emphasize the different color terrains on the surface

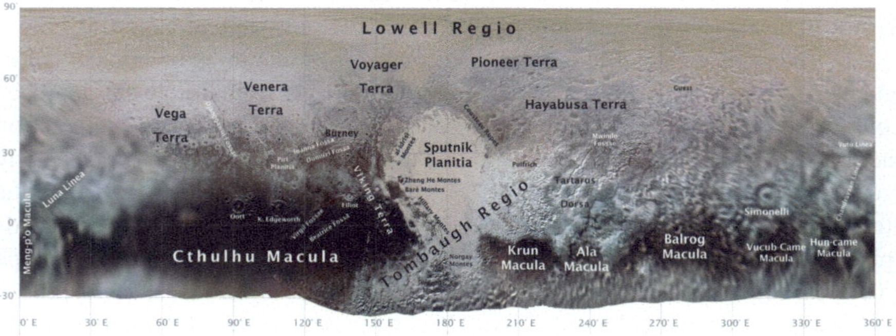

Fig. 1.2 Map of Pluto with informal and formal feature names. The large region in the center of the map labeled Sputnik Planitia is depressed by 3–4 km from the surrounding terrain (Moore et al. 2016). To the east of Sputnik Planitia is a bluish terrain called Eastern Tombaugh Regio. The equatorial region is dominated by dark red maculae

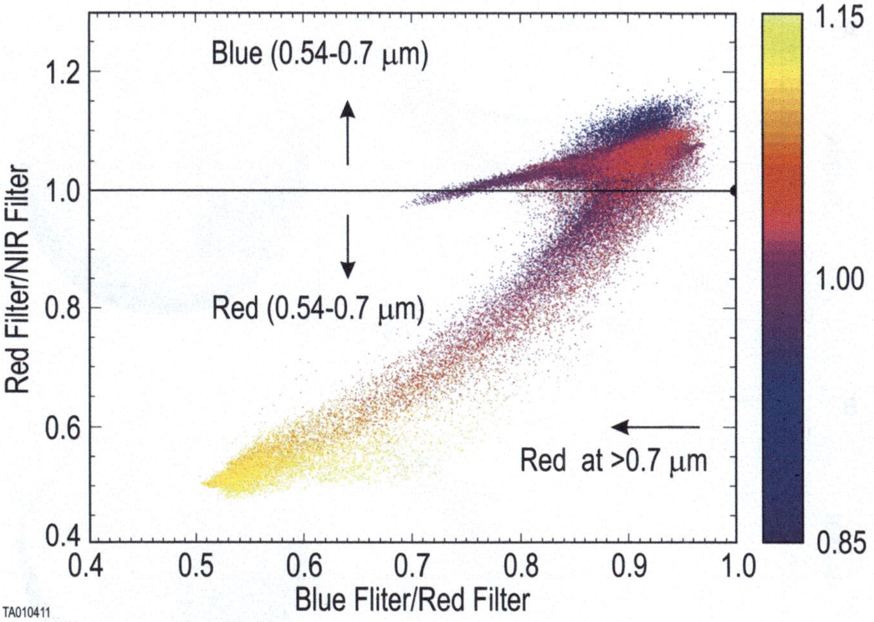

Fig. 1.3 A plot of color ratios in 3 filters (blue, red and near-IR) for pixels in the highest-resolution color observation of Pluto's encounter hemisphere. Red coloration increases to the lower left. The color of each point indicates the ratio of the reflectances in the methane filter to the near-IR filter. Red points indicate no methane; blue points indicate methane absorption at 890 nm

nitrogen but currently Pluto's atmosphere is thick enough that Lyman-alpha photons do not penetrate to the surface.

Panel B shows the least red terrain on Pluto as indicated by the black points at the top and right part of the color-color plot and it is methane rich. These pixels have a blue slope from the red to NIR wavelengths and a red slope from the blue to red wavelengths. This unique coloration could be the result of scattering from small particles. This region shows a sharp delineation between the east and west parts of Tombaugh Regio. Geologically, this delineation is coincident with the boundary of the glacial region on Pluto and in selected areas along this boundary there is evidence that ices are flowing from east Tombaugh Regio into Sputnik Planitia (Moore et al. 2016). This is also seen in the equivalent width maps for the 890 nm methane band constructed from the color images (Grundy et al. 2016).

The red end member of the color mixing line that dominates the color-color plot contains pixels in the center of Cthulhu Macula (see Panel C). This dark red color is consistent with the idea that this region is covered in tholins (complex organic macromolecules) as the result of long-term exposure to insolation due to the obliquity of Pluto's pole. Fine scale structure in Cthulhu Macula can be seen in the globe in panel C. Specifically there are linear features (dark lines in the white area

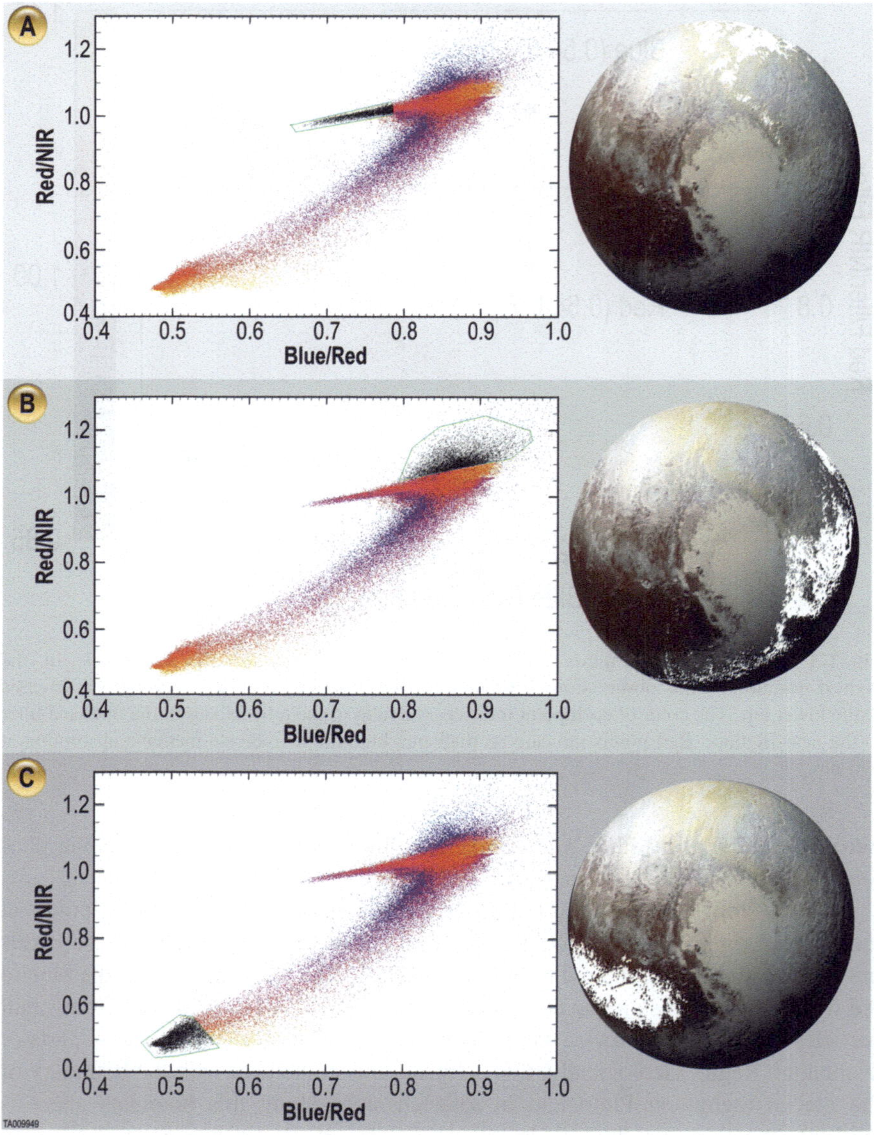

Fig. 1.4 For each of the three panels, the points in black enclosed in a green line on the color-color plot (left) correspond to the white pixels in the corresponding image (right). **A** Points that extend sharply to the left in the color-color plot correspond to the yellow terrain unit (see Fig. 1.1) in the north. **B** The points near the top of the color-color plot map to the bluish terrain predominantly located in eastern Tombaugh Regio and have significant methane absorption in the 890 nm band. **C** Points at the lower left of the color-color plot map mostly to dark red Cthulhu Macula, and exhibit less methane absorption at 890 nm

of the Panel C globe) corresponding to 2 networks of fossae (Virgil and Beatrice) and a linear mountain chain that appear as dark regions in Cthulhu Macula. Brighter methane ice is seen in all three of these features, at the high altitudes of the mountains in Cthulhu Macula and in the low elevations of the fossae.

The primary composition-mapping tool aboard New Horizons is the imaging spectrometer which records near-IR lights with wavelengths from 1.25 to 2.5 microns, dispersed across 256 spectral channels (Reuter et al. 2008). These wavelengths provide optimal remote sensing potential for the volatile ices known to be abundant on Pluto's surface (CH_4, N_2, and CO) as well as many other cosmically abundant compounds including H_2O, CO_2, C_2H_6, CH_3OH, and NH_3. These simple molecules are easy to detect and identify via the absorption of specific wavelengths of reflected near-infrared light, as shown in Fig. 1.5.

To avoid moving parts, wedge filters are attached directly to the imaging spectrometer's 256 × 256 detector array, such that each column of the array is sensitive to a specific wavelength of infrared light. The array is scanned across the scene to be studied by slewing the spacecraft and recording a series of images during the scan. Each part of the scene is eventually imaged at each of the wavelengths. The resulting sequence of 256 × 256 pixel images is then reconstructed into a spectral image cube where each pixel of the scene is a spectrum. Example spectra from different terrains across Pluto are shown in Fig. 1.6. From these spectra, it is possible to estimate the surface composition for each pixel.

Maps showing where ices are especially prominent on Pluto's encounter hemisphere reveal several intriguing patterns in the spatial distributions of the volatile ices N_2, CO, and CH_4 (Grundy et al. 2016). These patterns exhibit spatial scales from global to local, as shown in Fig. 1.7.

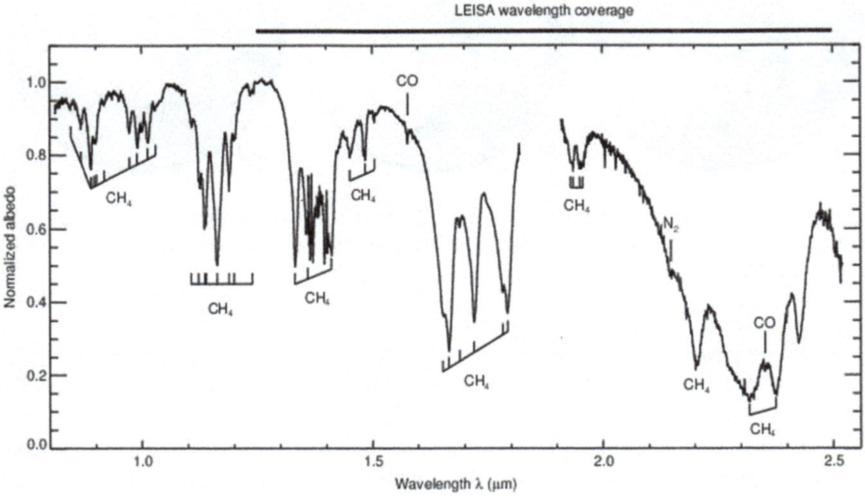

Fig. 1.5 Pluto's disk integrated spectrum obtained at NASA's Infrared Telescope Facility using the SpeX instrument. Absorption features of CH_4, CO and N_2 dominate

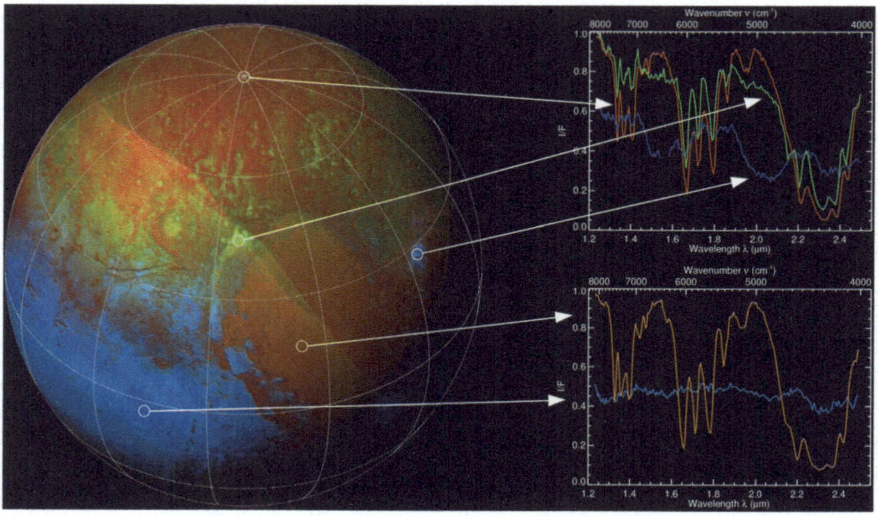

Fig. 1.6 The diversity of composition units across Pluto (left) and selected spectra (right). At Pluto's north pole, the spectrum (red in upper right panel) shows absorption features consistent with CH_4 ice. The spectrum in green (upper right panel) shows absorptions features consistent with CH_4 and N_2. The spectrum in blue (upper right panel) comes from a region near Pulfich crater on Pluto and is consistent with water ice absorption features. The spectrum from Cthulhu Macula (blue line in lower right panel) is relatively featureless as expected for long chain hydrocarbons

Fig. 1.7 Volatile ice maps for CH_4, N_2 and CO (from Grundy et al. 2016). The ices are predominantly located in Sputnik Planitia. The major species on Pluto's North pole and on East Tombaugh Regio is CH_4

It was expected from Earth-based observations and models of insolation that Pluto's volatile ices would migrate on seasonal time scales. In fact, at a global scale, the distribution of composition units across Pluto shows latitude-dependent patterns likely related to the gradual migration of its volatile ices sublimating and condensing in response to changing seasonal patterns of sunlight. The volatile ices are not seen in a dark, planet-girdling equatorial band, typified on the encounter hemisphere by Cthulhu Macula and Krun Macula. In Lowell Regio, the region surrounding the northern (current summer) pole, there is abundant CH_4 ice. CH_4 is the least volatile of the three volatile ices and thus is the last to sublimate away. The more volatile N_2 and CO ices are seen chiefly at mid-latitudes, and also in Tombaugh Regio, Pluto's bright heart-shaped feature.

Tombaugh Regio typifies a distinct regional scale pattern of ice distribution that interrupts the latitude pattern described above. Sputnik Planitia, Tombaugh's western half, is a deep basin partially filled with N_2, CO, and CH_4 ices. The depth is sufficient that these ices undergo solid-state convection, plastically deforming and creating a network of cells of upwelling material (McKinnon et al. 2016). The ice mixture turns over rapidly enough to refresh its surface and erase any sign of impact craters. The volatile ices, especially CH_4, are also present in the higher elevation terrain in the eastern half of Tombaugh, but not at similar altitudes just to the west of Sputnik Planitia. The difference could be another regional scale effect relating to prevailing wind patterns influenced by Sputnik.

At smaller, local scales, volatile ice distributions are evidently also controlled by topography. For instance, in the north of Cthulhu Macula, methane ice is seen on north-facing slopes inside craters. It is also evident along the crest of a mountain range. Further north, many craters host bright CH_4 ice on their rims, giving them halo-like appearances. This distribution of CH_4 contrasts with the distribution of the more volatile N_2 and CO ices, which are preferentially found on crater floors and in valleys.

Although Pluto's seasonally-mobile volatile ices form a veneer over much of the surface, New Horizons LEISA observations had sufficient spatial resolution to reveal for the first time localized outcrops of the H_2O ice bedrock, as shown in Fig. 1.8.

H_2O ice is seen along the western edge of Sputnik including in the mountains of Baré, Hillary, and near al-Idrisi. These mountains have a distinctive form, with similar sized blocks occurring near one another. They could have been undercut and rotated via the action of the convecting N_2 ice in Sputnik Planitia (Moore et al. 2016). The density of H_2O ice is slightly less than that of the dominant N_2 ice in Sputnik, so H_2O ice blocks could have been buoyantly floated and dislodged by the vast nitrogen reservoir surrounding them.

Several steep cliffs in Virgil and Beatrice fossae also exhibit prominent H_2O ice absorptions. The areas with the strongest H_2O absorptions show a distinctive orange color in Fig. 1.1. Further east, a region near Pulfrich crater in Tombaugh Regio also shows strong H_2O absorptions, but not associated with the orange coloration, implying a different, and possibly recent emplacement mechanism.

Photochemistry in Pluto's atmosphere converts CH_4 gas into heavier hydrocarbons such as C_2H_6, C_2H_4, and C_2H_2. These condense into haze particles and gradually settle to the surface (Gladstone et al. 2016). They are inert at Pluto surface temperatures in the 40–50 K range, so they are readily buried beneath the seasonally-mobile volatile ices. But in areas where the volatile ices are not present, we would expect to see these photochemical products. That they are not evident in the vicinity of the H_2O spot near Pulfrich is consistent with deposit being relatively young. In Cthulhu and the other equatorial dark regions, they appear to have

Fig. 1.8 H_2O ice outcrops on Pluto. The most significant spectral signature of water ice is seen near Pulfrich crater (upper right of figure). Water ice is also seen on steep cliffs (left) and the western edge of Sputnik Planitia

accumulated, and slowly been processed by cosmic rays into the dark, reddish, organic macromolecules mentioned earlier. H_2O ice bands are also seen in the equatorial dark regions like Cthulhu Macula and Krun Macula.

As we have seen on other planets, initial reconnaissance by spacecraft provides a huge leap forward in our understanding, and the New Horizons encounter with Pluto demonstrated this very significantly. Before the New Horizons flyby, we had a unresolved view of Pluto with almost no details on geology and composition mapping. The very capable New Horizons payload and encounter planning provided a wealth of new information about Pluto. We now know there is a large reservoir of volatile ices in a basin on Pluto. We see distinct color units on the surface of Pluto that correspond to different composition units, such as the dark tholin coated Cthulhu Macula and the methane-rich eastern Tombaugh region. Water ice was discovered on the surface of Pluto from the near infrared spectra of the Ralph/LEISA instrument.

While it will be a long time before we could harness material resources or energy from Pluto or other Kuiper Belt Objects, the New Horizons mission has informed our understanding of Pluto's surface composition and the distribution of molecular species across the planet.

References

Bannister, M. et al.: (2016) The outer solar system origins survey: I. Design and first-quarter discoveries. Astron. J. (submitted)

Cruikshank, D.P., Pilcher, C.B., Morrison, D.: Pluto: evidence for methane ice. Science **194**, 835–837 (1976)

Cruikshank, D.P., Imanaka, H., Dalle Ore, C.M.: Tholins as coloring agents on outer solar system bodies. Adv. Space Res. **36**, 178–183 (2005)

Cruikshank, D.P., et al.: The surface compositions of Pluto and Charon. Icarus **246**, 82–92 (2015). https://doi.org/10.1016/j.icarus.2014.05.023

Davies, G., et al.: The early development of ideas concerning the transneptunian region. In: Barucci, H., et al. (eds.) The Solar System Beyond Neptune, pp. 11–23. University of Arizona Press, Tucson (2008)

Elliot, J.L., et al.: The Deep Ecliptic Survey: a search for Kuiper belt objects and centaurs. II. Dynamical classification, the Kuiper belt plane, and the core population. Astron. J. **129**, 1117–1162 (2005)

Giclas, H.L.: History of the 13-inch photographic telescope and its use since the discovery of Pluto. Icarus **44**, 7–11 (1980). https://doi.org/10.1016/0019-1035(80)90047-0

Gladstone, G.R., et al.: The atmosphere of Pluto as observed by New Horizons. Science **351**, 1280 (2016). https://doi.org/10.1126/science.aad8866

Grundy, W.M., et al.: Near-infrared spectral monitoring of Pluto's ices: spatial distribution and secular evolution. Icarus **223**(2), 710–721 (2013). https://doi.org/10.1016/j.icarus.2013.01.019

Grundy, W.M., et al.: Surface compositions across Pluto and Charon. Science **351**(6279), 1283 (2016). https://doi.org/10.1126/science.aad9189

Jewitt, D., Luu, J.: Discovery of the candidate Kuiper belt object 1992 QB1. Nature **362**, 730–732 (1993)

Kowal, C.T.: A solar system survey. Icarus **77**, 118–123 (1989). https://doi.org/10.1016/0019-1035(89)90011-0

McKinnon, W.B., et al.: Convection in a volatile nitrogen-ice-rich layer drives Pluto's geological vigour. Nature **534**, 82–85 (2016). https://doi.org/10.1038/nature18289

Moore, J.M., et al.: The geology of Pluto and Charon through the eyes of New Horizons. Science **351**(6279), 1284–1293 (2016). https://doi.org/10.1126/science.aad7055

Nimmo, F., et al.: Mean radius and shape of Pluto and Charon from New Horizons images. Icarus **287**, 12–29 (2017). https://doi.org/10.1016/j.icarus.2016.06.027

Petit, J.-M., et al.: The Canada–France Ecliptic Plane Survey—full data release: the orbital structure of the Kuiper belt. Astron. J. **142**(131), 1–24 (2011)

Reuter, D.C., et al.: Ralph: a visible/infrared imager for the New Horizons Pluto/Kuiper belt mission. Space Sci. Rev. **140**(1–4), 129–154 (2008). https://doi.org/10.1007/s11214-008-9375-7

Stern, A.S., et al.: The Pluto system: initial results from its exploration by New Horizons. Science **350**, 292 (2015). https://doi.org/10.1126/science.aad1815

Tombaugh, C.W.: Reminiscences of the discovery of Pluto. Sky and Telescope **19**, 264 (1960)

Weaver, H.A., et al.: Overview of the New Horizons science payload. Space Sci. Rev. **140**(1–4), 75–91 (2008). https://doi.org/10.1007/s11214-008-9376-6

Young, L.A., et al.: New Horizons: anticipated scientific investigations at the Pluto system. Space Sci. Rev. **140**(1–4), 93–127 (2008). https://doi.org/10.1007/s11214-008-9462-9

Chapter 2
Physical Properties of Icy Materials

Craig Pitcher and Yang Gao

2.1 Introduction

The presence of water and ice on planetary bodies is of significant interest for robotic and human exploration. Ice deposits may contain organic molecular biomarkers that indicate the presence of life (Schulz-Makuch et al. 2012), and have great potential for use as an in situ resource in the form of propellants and consumables that can be used in future exploration activities (Sanders and Larson 2011; Carpenter et al. 2014).

To qualify a system for spaceflight, stringent tests must be performed in accurately simulated conditions to ensure that it is able to operate within the expected environment. Testing of the mechanisms that will interact with a planetary body's surface and subsurface will therefore require the use of a regolith simulant. This must have well-defined properties that mimic as best as possible the properties of the regolith that will likely be encountered. This can be achieved by developing simulants based upon the soil characteristics made from remote observations and/or previous sample acquisition missions. For missions that will explore areas where water ice is believed to reside, the properties of icy regolith, and its effect on the systems that will interact with it, must also be determined.

This chapter looks at how the presence of water ice affects the properties of lunar and Martian regolith. The focus is first given to the physical properties of the NU-LHT-2M lunar highland simulant when mixed with various water mass percentages, from dry to nearing saturation. The emphasis is then placed upon the

C. Pitcher (✉) · Y. Gao
Surrey Space Centre, University of Surrey, Stag Hill, Guildford GU2 7XH, UK
e-mail: craig.pitcher@open.ac.uk

Y. Gao
e-mail: yang.gao@surrey.ac.uk

V. Badescu and K. Zacny (eds.), *Outer Solar System*,
https://doi.org/10.1007/978-3-319-73845-1_2

phenomenon of cementation of Martian regolith, and how it can affect the operation of sampling instruments.

There is strong evidence of the presence of water in a number of bodies in the outer solar system. Of particular interest are the moons Europa and Enceladus. Plumes of water vapour have been observed erupting from both bodies (Hansen et al. 2011; Roth et al. 2014; Sparks et al. 2016), which has led to the suggestion that liquid water may exist within their outer shells (Čadek et al. 2016; Kalousová et al. 2016), and as a result may contain the conditions needed to support life (McKay et al. 2014). Any future missions that may attempt to perform in situ experiments of these bodies (Gowen et al. 2011) would have to contend with surface and subsurface material in which water ice is very likely to be present. As such, the experiments performed and the effects of the presence of water in regolith discussed here could be used as a reference for the testing of instruments designed to explore these outer solar system bodies.

2.2 Presence of Ice in Lunar Regolith

Significant evidence now exists to suggest that vast quantities of water ice are present within the lunar subsurface regolith, especially in the permanently shadowed craters at the polar regions. The LCROSS mission, which struck such a region in the south pole crater Cabeus and observed the resulting ejecta, detected concentrations of water ice in the regolith of $5.6 \pm 2.9\%$ by mass (Colaprete et al. 2010; Mitrofanov et al. 2010). Further evidence of surface water ice has been provided by measurements from the Lunar Orbiter Laser Altimeter and Diviner Lunar Radiometer Experiment (Fisher et al. 2017), and by previous observations made by the Visual and Infrared Mapping Spectrometer on board Cassini (Clark 2009).

To confirm the presence of water ice and other volatiles on the Moon, direct in situ investigations of the regolith are required. This has become a major goal of future lunar exploration missions (Fisackerly et al. 2014; Neal and Lawrence 2017), with the focus being placed upon the south polar region.

2.2.1 Testing of Icy Lunar Simulants

To date, more than 30 lunar regolith simulants have been developed, each designed to mimic the properties of the targeted lunar soil and to be used in hardware testing (Taylor 2010). The first of these was the MLS-1 analogue, which simulated the mineralogy, chemistry and texture of the lunar mare regolith (Seiferlin et al. 2008). The JSC-1 simulant, developed by NASA's Johnson Space Center, was subsequently designed to both complement the MLS-1 and approximate the lunar soil found at the Apollo 14 site (McKay et al. 1994). When the supply of JSC-1 was depleted, the JSC-1A series was developed and intended as a direct replacement.

Numerous experiments have been performed to determine the properties of icy regolith simulants. The effects of water ice mass concentrations ranging from zero to 11% on the excavatability and strength of JSC-1 have been explored (Gertsch et al. 2006). These experiments demonstrated the change in the regolith's behaviour, which ranged from weak coal when it contained 0–1.5% water mass, to moderate-strength limestones and strong sandstone with ~8.4 and ~10.6% water mass respectively.

Another study investigated penetration mechanics using a percussive cone penetrometer, in which JSC-1A was mixed with ice contents ranging incrementally from 0 to 8% by mass (Metzger et al. 2011). These tests were able to determine a correlation between increasing ice content and decreasing penetration rate. From this, it was proposed that the penetration mechanics for dry and icy regolith differ, with the restriction of grain movement in frozen soils resulting in fractured ice being trapped, thus preventing the penetrometer from descending further. Other examples of tests in icy regoliths can be found in Mantovani et al. (2014).

2.3 Lunar Highland Simulant Series NU-LHT

As the properties of a body's regolith are not constant, with composition, density and water content varying depending on its location, it is vital that the correct simulant for the targeted area is used when designing and testing instrumentation. As the composition of the regolith encountered at the south pole differs to the lunar mare analogues, their respective properties may differ. This could create significant errors during testing when using unrepresentative simulants, which could result in mechanisms that are unable to function correctly or fail entirely. As a result, it is critical for future lunar polar missions that the properties of the polar regolith when water ice is present are determined.

As a response to this growing interest in the exploration of the lunar polar regions, the Marshall Flight Space Center (MSFC) agreed to develop a polar regolith simulant series produced by the United States Geological Survey (USGS) (Stoeser et al. 2010). To the best of current knowledge, the soil covering the lunar polar regions is best approximated by the highland regolith. The only mission to have taken samples from an area representative of the lunar highlands is Apollo 16, whose landing site was situated over 200 km away from the nearest mare. Whilst the presence of mare, potassium and rare-earth elements prevent it from being a true representation of the highland material, the Apollo 16 samples are considered compositionally close. The 64001/64002 cores were subsequently chosen as the reference material for initial Figure of Merit calculations used for characterising the developed highland simulants (Schrader et al. 2009, 2010).

The average compositions of these samples were used to create the Lunar Highlands Type simulant, NU-LHT-2M, which was designed to represent the whole highlands regolith, as opposed to a single area (Stoeser et al. 2010). Although there have been numerous experiments investigating the properties of icy lunar simulants, as discussed in Sect. 2.1, there have yet to be any studies performed with regards to the properties of wet highlands regoliths. This section describes the first tests performed with icy NU-LHT-2M, and provides an initial characterisation of its physical properties.

2.3.1 Physical Properties

Although the LCROSS observations have provided an estimate of the water ice content in the lunar polar regolith, as discussed in Sect. 2.2, there exists no quantitative information concerning the exact extent and distribution of ice. As such, rovers and instruments could potentially encounter regolith with water contents ranging from dry to fully saturated.

The degree of saturation has a significant bearing on the physical properties of regolith. This is demonstrated in Fig. 2.1, in which six small samples of NU-LHT-2M were mixed with water, incrementally increasing by 6 ml (Pitcher et al. 2016). The resulting water mass percentage contents of the samples are given in Table 2.1. The samples were frozen overnight, after which they were qualitatively examined by pushing a pencil into them. While the two driest samples could easily be penetrated to the bottom, the difficulty to do so increased with water content. Sample three was difficult to penetrate to the base, while sample four could only be penetrated to a few millimetres, and the final two samples were impossible to penetrate. The solid nature of these final two samples also indicated that the saturation point occurs with a water mass content in the range of 13–17%, while there also appears to be a significant change in physical properties when the water content increases from 5 to 9%.

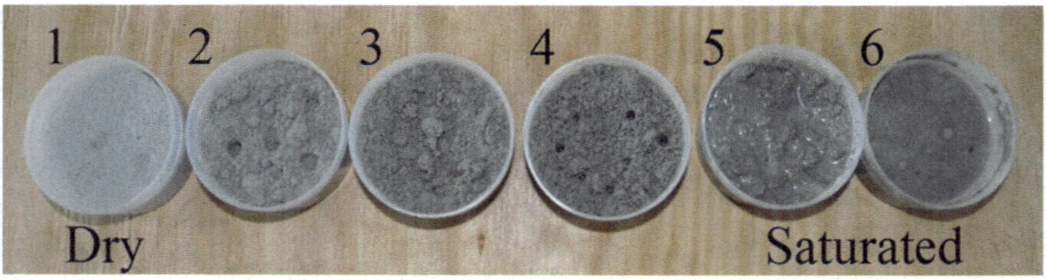

Fig. 2.1 Six frozen samples of NU-LHT-2M, with incrementally increasing volumes of water added

Table 2.1 Water content of the frozen samples

Sample	1	2	3	4	5	6
Volume of water (mL)	0	6	12	18	24	30
Water mass %	0.2	4.6	9.0	12.8	17.3	20.3

2.3.2 Drilling in Icy NU-LHT-2M

The effect of this change in properties is illustrated by a series of experiments, demonstrating how the presence of ice in regolith significantly increases the time taken to drill into a sample. The dual-reciprocating drilling (DRD) technique and test rig set-up (Pitcher and Gao 2015) was used to drill into a 33 cm deep sample of the NU-LHT-2M simulant. The drilling set-up and sample is shown in Fig. 2.2.

Drilling was first performed with the dry, room temperature NU-LHT-2M, after which it was re-poured, frozen and drilled into again. Water was then added incrementally into the regolith, giving it water mass contents of approximately 5, 7.5, 10 and 12.5% respectively (Pitcher 2017). These were also drilled into both at room temperature and after being frozen. For each experiment, the time taken for the drill to reach the sample base was recorded, and the results are presented in Fig. 2.3.

The results show the time taken to drill through room temperature samples with increasing water contents is fairly constant. The time increases as the ice content rises from 5 to 10%, reflecting the increased penetration difficulty seen in the observations made in Sect. 3.1. There is also a clear critical point between 10 and 12.5% ice content where the time taken rises substantially, which is likely due to the saturation point of the NU-LHT-2M being approached. This, along with the difficulty in penetrating Sample 4 in Fig. 2.1, suggests that 12.5% water content is the approximate point at which saturation begins to occur, and thus is the limit at which penetration can be achieved.

Fig. 2.2 The set-up for the NU-LHT-2M drilling experiments, showing the start of a dry run and end of a wet run. The drill head used has a diameter of 36 mm

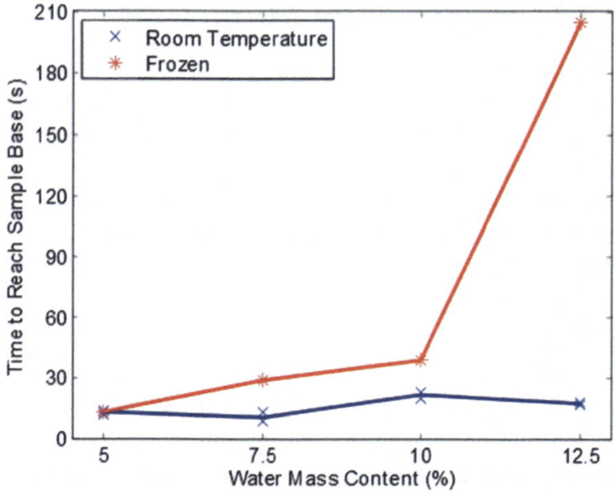

Fig. 2.3 Graph of the time required to reach the base of the NU-LHT-2M for different water contents and temperature conditions (Pitcher 2017)

2.3.3 Penetration Resistance

To study these changes in greater detail, frozen samples of NU-LHT-2M with controlled water contents were subjected to penetration resistance tests using a handheld penetrometer, with measurements made first through the centre of the sample, then at various points near the boundary. The penetrometer was slowly pushed into the simulant, and the resistance it experienced as a function of distance travelled was recorded (Pitcher et al. 2016). Figure 2.4 shows the penetration resistance profiles given by each test for all samples. The results support the observations made in Sect. 3.1, with a very clear decrease in penetration resistance as the water mass percentage decreases past a certain point. Each sample also showed a notable increase in resistance with depth. This is clearly seen with the 7–8% sample, with penetration becoming increasingly more difficult with depth. The same was also seen with the 4–5% sample, though the top 3 cm is very easy to penetrate, and only below this soft layer does the resistance increase to that experienced for the 7–8% sample. However the resistance in the 3–4% sample, despite increasing with depth, is much smaller than those of the other samples. This again suggests that the penetration resistance of icy NU-LHT-2M experiences a sharp change when the water mass content approaches 5%.

2.3.4 Compressive Strength

Uniaxial compression tests were also performed with frozen NU-LHT-2M cylindrical samples with water contents of 3, 6 and 12%. The samples were confined by the compression table and lid, shown in Fig. 2.5a, before being squeezed by the upwards motion of the table. The axial resistive force of the sample against this

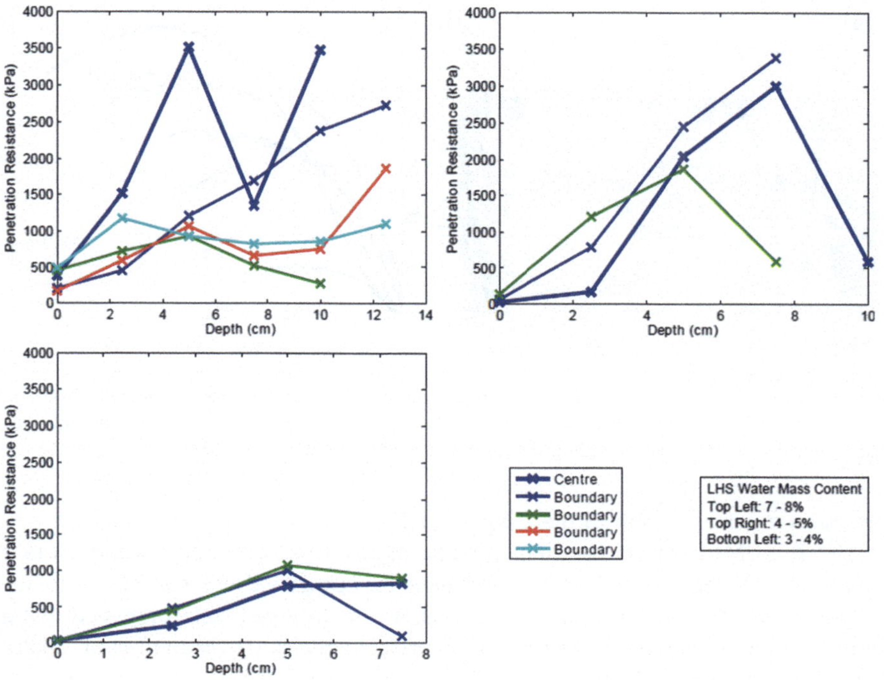

Fig. 2.4 Results of the penetration tests of the frozen NU-LHT-2M samples (Pitcher et al. 2016)

movement was measured against the deformation of the sample, defined as the percentage change in its axial height compared to its original height, until after the maximum axial stress has been achieved, when the sample begins to fail mechanically. The results of the tests performed, along with the average maximum compressive stresses achieved, are given in Fig. 2.5b. The graph shows that increasing the water mass content results in both a greater maximum compressive strength and larger deformation achieved before failure. These results highlight the increased cohesiveness and structural integrity created by the addition of ice in the regolith (Pitcher et al. 2016).

2.4 Evidence of Cementation on the Martian Surface

While the presence of water on Mars is well documented (Bandfield 2007; Heldmann et al. 2014), the chemistry and composition of the Martian regolith presents its own unique challenges in the form of soil induration, cementation and the formation of surface crusts. Evidence of this phenomenon has been seen in every successful lander mission to date, with the Viking landers the first to observe layers of lightly cemented fine-grained sediments 1–2 cm thick (Mutch et al. 1977), and a crusty to cloddy layer

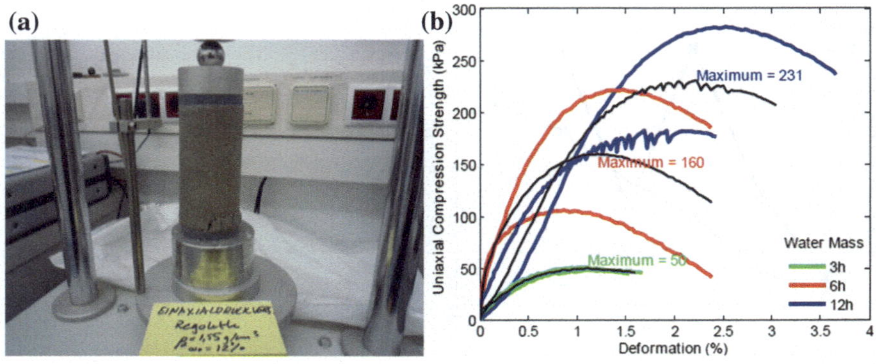

Fig. 2.5 **a** Experimental set-up for the uniaxial compression tests; the frozen sample has begun to fail, as evidenced by the cracks at the base. **b** Graph of the compressive strength of the three samples with deformation, including the averaged values (Pitcher et al. 2016)

was seen to cover ~86% of the Viking Lander 2 sample field (Moore and Jakosky 1989). Both MER rovers also found evidence of crusts, with Spirit seeing surface crusts on soil deposits and wheel track disturbances (Arvidson et al. 2004), while Opportunity's Microscopic Imager observations of disturbed soils suggested that a crust was formed through cementation of surface particles (Herkenhoff et al. 2004). Trenches excavated at the Phoenix landing site also revealed clods and crusting of the upper few centimetres of soil, with properties similar to those at the VL2 landing site (Smith et al. 2009). This evidence suggests that crusting is a near-ubiquitous occurrence on the Martian surface.

Understanding the cementation process, and the effect it has on the properties of the regolith, is necessary for the testing and design of instruments that will be handling Martian regolith samples for significant periods of time. This is of particular interest for the development of the *Powdered Sample Dosing and Distribution System* (PSDDS) (Redlich et al. 2016), a subsystem of the *Sample Preparation and Distribution System* (SPDS) of the ExoMars rover (Richter et al. 2015).

2.4.1 Composition and Properties of Duricrusts

Although the composition and chemistry of duricrusts has not been conclusively defined, evidence of clays in the soil has been inferred in a number of studies, while analysis of the Viking, Pathfinder and MER landing sites have demonstrated high sulphur concentrations of up to 10% (Chevrier and Mathe 2007), as well as the presence of chlorine (Sharp and Malin 1984).

The presence of these elements and their compounds, along with other cohesive materials, suggests that crusts are the result of a salt-cementation process caused by transient liquid water (Cooper and Mustard 2001). As the atmospheric pressure generally lies above the triple point of water, the daytime temperatures of over 0 °C allow frost deposits to warm and form a liquid phase. This water can then travel through the soil, transporting with it any dissolved salts present. When the liquid evaporates, these salts are deposited, cementing the dust and sand grains and forming the duricrust (Landis et al. 2004). Repetition of this process and transportation of water over time allows the cementation of Martian regolith to occur over vast distances. While the liquid phase does not need to occur regularly, the length of time it is present per year dictates the distribution and thickness of the crust. There is also evidence that only a limited amount of water is required, with the crusts on the Gusev plains believed to have been formed from low water activity and mineral deposition (Cabrol et al. 2006).

The properties of duricrusts have been examined in numerous experiments. Salt crusts created with magnesium sulphate, albeit with significantly more water than would be present on Mars, were used to examine diffusion barriers and vapour transport, and in particular the effect of salt content on porosity (Hudson and Aharonson 2008). Other experiments have examined the effects of cementation on the soil's spectral properties (Cooper and Mustard 2001) and the thermal inertia of the Martian surface (Piqueux and Christensen 2009), and investigated how the crust's physical properties are affected by its composition (Bishop et al. 2002). In this study, it was found that a high smectite content resulted in harder, thicker crusts, while palagonitic soil produced a thinner, more viable crust.

2.5 Cementation and Duricrust Formation in a Simulated Martian Environment

To continue the study into the properties of duricrusts, an experiment to chemically cement the surface of Mars regolith under conditions similar to that of the Martian environment was performed. The S7 analogue was used for this experiment. This is an unconsolidated clay/salt regolith simulant verified for use in the testing of the sample handling instruments of ExoMars, whose properties are given in Table 2.2.

The S7 sample was first wetted, using a water disperser that produced very fine water vapour, with the droplets directed into the sample via a vapour flow tube, to account for the presence of water in the soil. The sample was then placed on a cold plate and subjected to an atmospheric pressure of 6–8 mbar in a vacuum chamber regulated by a CO_2 inlet. The cold plate then underwent several cooling and heating cycles, bringing the sample's temperature below and above the freezing point of water (Kömle et al. 2017).

Table 2.2 Composition of the S7 regolith simulant (Durrant and Baglioni 2013)

Simulant component	Quantity (%)
Montmorillonite	67
Magnesium sulphate heptahydrate (Epsom salt)	30
Magnesium perchlorate hexahydrate	3

Fig. 2.6 The S7 analogue before (**a**) and after (**b**) the experiment, with part of the crust removed (**c**) to highlight the difference between the cemented crust and dry material. The container has a diameter of 72 mm and height of 33 mm (Kömle et al. 2017)

Figure 2.6a, b shows the S7 before and after the experiment. It can be seen that a consolidated crust of clearly cemented material approximately 5 mm thick has formed. Removing a section of this crust, as shown in Fig. 2.6c, reveals that the material below it is largely unchanged from the original powder-like texture. The crust could be removed without being destroyed mechanically, demonstrating the internal cohesiveness of the material. The cracks on the crust formed gradually during the cooling and heating cycles, through which gas trapped below the surface was able to escape.

2.5.1 Properties of the Component Materials

In order to understand the cementation process that results in this duricrust formation, the component parts of the analogue material, and their roles in the extent and speed of the soil cementation, were analysed. For this, the S8 analogue, consisting of 50% of the S7 mix and 50% of the fine-grained quartz Schwarzl UK4 sand, was used. Separate small samples of the Magnesium Sulphate, Montmorillonite and UK4 component parts and the S8 mixture were held within a steel container, closed by a lid with a central hole, into which the vapour flow tube was inserted. The samples were again wetted using the water vapour disperser for two hours, creating a water-saturated environment. The water mass contents of the samples were measured before and after the experiment, and the extent of the wetting was also approximately measured, with the results given in Table 2.3.

The extent of the wetting for each component material is dependent on their properties, hence the different levels measured. The low permeability of the Montmorillonite, caused by its particles swelling when taking on water, quickly blocks diffusion to the lower layers, resulting in only the top few millimetres being wetted (Kömle et al. 2017). Unless the sample is easily permeated, like the Magnesium Sulphate, the distribution of water throughout a sample will not be homogeneous when simply placed in a water-saturated environment. The presence of Montmorillonite is thus the main reason for the S8 mixture only experiencing wetting of the top 10 mm. This provides further evidence of the role of loamy

Table 2.3 Description and water content of the S8 and its component materials

Sample	Extent of wetting	Water content (%)	
		Before	After
Magnesium sulphate	Fully wetted to the sample base	36.049	37.153
Montmorillonite	Top 4–5 mm	9.121	N/A
UK4 sand	Wetted to ~4 mm above base	0.397	4.655
S8 mix	Top ~10 mm	9.855	18.339

materials and salts in the creation of crusts seen in the Martian surface, and explains the rapid transition from cemented to dry granular material.

2.5.2 *Effects of Cementation on Sampling Instruments*

The effect of soil cementation must also be taken into consideration with regards to sample handling instruments, such as the PSDDS described in Sect. 2.4. Samples taken by a drill can spend several days contained within a sampling subsystem before being delivered to an instrument. Frost deposits held within these subsurface samples could become exposed to atmospheric pressures and temperatures above the triple point of water. The resultant transient water will allow any dissolved salts to be deposited in and interact with the soil, and could thus result in the samples becoming cemented. This could result in operational difficulties for the mechanisms that will handle the sample.

To study this, dosing experiments with a qualification model of the PSDDS, shown in Fig. 2.7, were performed. The system first collects the sample in an entrance funnel (1). A portion of this is captured within a small opening, which is rotated 180° by a dosing mechanism (2), allowing the material to fall through an exit funnel (3). Transportation of the material can also be assisted by operating a built-in piezo vibrator (4).

These tests were performed with the S8 material, which was mixed with controlled volumes of water to produce samples with 5 and 7.5% water mass contents (Kömle et al. 2017). The PSDDS was placed in the vacuum chamber on the cold plate and brought to Mars atmospheric pressure, after which the temperature of the cold plate was lowered, bringing the sample temperature as low as −20 °C.

The majority of tests failed, with the material being unable to travel through the system. Removing the sample from the funnels manually revealed that small clumps of coagulated material of up to 5 mm width had formed. While the dry S8 would also tend to coagulate, even small water mass contents of 5% resulted in these clumps developing a significant internal strength under Mars conditions, which is maintained even after the sample is brought back to room temperature. The strength of these globules was enough to block the funnels and stop the dosing mechanism, effectively creating a single point of failure in the system.

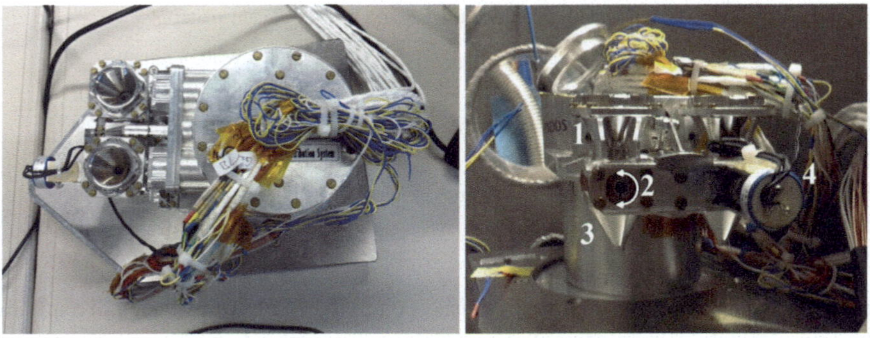

Fig. 2.7 The PSDDS qualification model. The funnels have a volume of approximately 1×10^{-5} m^3 (Kömle et al. 2017)

2.6 Conclusions

This chapter has looked at the physical properties of icy lunar and Martian regolith simulants. Focus was placed upon the lunar highland simulant NU-LHT-2M, as part of the preparation for future missions to the lunar polar regions, and the effect of cementation and crusts forming on the surface of Martian regolith on sample handling instruments, as part of the testing of the PSDDS to be used on ExoMars. The experiments previously performed on icy lunar mare simulants and the observations of duricrust from lander missions and investigations into its composition and properties were also summarised.

The NU-LHT-2M was shown to have a saturation point between 13 and 17% water content. Qualitative penetration tests demonstrated that the icy simulant experiences a sharp increase in penetration resistance in the range of 5–9% water mass. The effects of this were demonstrated in a series of drilling experiments. Cone penetration tests explored this further, demonstrating a critical point of ~5%, from which samples go from having a low resistance to being difficult to penetrate, and showing the relationship between penetration resistance and depth. Uniaxial compression tests also showed how increasing water content results in a greater strength and degree of deformation before failure.

It was possible to create a clearly defined crust in simulated Martian conditions, with a cohesive layer several millimetres thick covering the powder-like material underneath. The extent and depth of the wetting in the S8 analogue was found to be largely dependent on the Montmorillonite component material, due to the swelling of the particles as they take on water, consequently blocking further diffusion deeper into the sample. Tests with the PSDDS also showed that even small amounts of water in the sample give a significant internal cohesion to any cemented or clumpy material when frozen. The strength of this was enough to block the dosing funnels of the PSDDS, causing the mechanism to fail.

These experiments have further highlighted the impact of ice on the properties of regoliths, and the potential effect they may have on instruments they will interact with. The physical characteristics can change significantly, which must be taken into account in the design of instrumentation, to avoid critical failures as seen with the PSDDS tests. The results presented here can also be used as a step towards the characterisation of these icy regolith simulants, from which further tests can be performed to better define their properties for future testing.

References

Arvidson, R.E., Anderson, R.C., Bartlett, P., Bell III, J.F., Blaney, D., Christensen, P.R., Chu, P., Crumpler, L., Davis, K., Ehlmann, B.L., Fergason, R., Golombek, M.P., Gorevan, S., Grant, J. A., Greeley, R., Guinness, E.A., Haldermann, A.F.C., Kerkenhoff, K., Johnson, J., Landis, G., Li, R., Lindemann, R., McSween, H., Ming, D.W., Myrick, T., Richter, L., Seelos IV, F.P., Squyres, S.W., Sullivan, R.J., Wang, A., Wilson, J.: Localization and physical properties experiments conducted by spirt at Gusev Crater. J. Sci. **305**, 821–824 (2004)

Bandfield, J.L.: High-resolution subsurface water-ice distributions on Mars. J. Nat. **447**, 64–67 (2007)

Bishop, J.L., Murchie, S.L., Pieters, C.M., Zent, A.P.: A model for formation of dust, soil, and rock coatings on Mars: physical and chemical processes on the Martian surface. J. Geophys. Res. **107**, E11 (2002)

Cabrol, N.A., Farmer, J.D., Grin, E.A., Richter, L., Soderblom, L., Li, R., Herkenhoff, K., Landis, G.A., Arvidson, R.E.: Aqueous processes at Gusev crater inferred from physical properties of rocks and soils along the Spirit traverse. J. Geophys. Res. **111**, E02S20 (2006)

Čadek, O., Tobie, G., Van Hoolst, T., Massé, M., Choblet, G., Lefèvre, A., Mitri, G., Baland, R., Běhounková, M., Bourgeouis, O., Trinh, A.: Enceladus's internal ocean and ice shell constrained from Cassini gravity, shape and libration data. J. Geophys. Res. Lett. **43**, 5653–5660 (2016)

Carpenter, J.D., Barber, S., Cerroni, P., Fisackerly, R., Fumagalli, A., Houdou, B., Howe, C., Magnani, P.G., Morse, A., Monchieri, E., Reiss, P., Richter, L., Rizzi, F., Sheridan, S., Waugh, L, Wright, I.P.: Accessing and assessing lunar resources with PROSPECT. In: Annual Meeting of the Lunar Exploration Analysis Group, Maryland (2014)

Chevrier, V., Mathé, P.E.: Minearology and evolution of the surface of Mars: a review. J. Planetary Space Sci. **55**, 289–314 (2007)

Clark, R.N.: Detection of Adsorbed water and hydroxyl on the Moon. J. Sci. **326**, 562–564 (2009)

Colaprete, A., Schultz, P., Heldmann, J., Wooden, D., Shirley, M., Ennico, K., Hermalyn, B., Marshall, W., Ricco, A., Elphic, R., Goldstein, D., Summy, D., Bart, G., Asphaung, E., Korycansky, D., Landis, D., Sollitt, L.: Detection of water in the LCROSS ejecta plume. J. Sci. **330**, 463–467 (2010)

Cooper, C.D., Mustard, J.F.: Spectroscopy of loose and cemented sulfate-bearing soils: implications for duricrust on Mars. J. Icarus **158**, 42–55 (2001)

Durrant, S., Baglioni, P.: ExoMars rover module drill & SPDS sample materials for verification. In: EXM-RM-TNO-ESA-00003, ESA ESTEC, Noordwijk (2013)

Fisackerly, R., Carpenter, J., Houdou, B., Visentin, G., Savoia, M., Rizzi, F., Magnani, P., Barber, S., Reiss, P., Richter, L.: Accessing, drilling and operating at the Lunar South Pole: Status of European plans and activities. In: Earth and Space Science 2014: Engineering for Extreme Environments, St. Louis (2014)

Fisher, E.A., Lucey, P.G., Lemelin, M., Greenhagen, B.T., Siegler, M.A., Mazarico, E., Aharonson, O., Williams, J.P., Hayne, P.O., Neumann, G.A., Paige, D.A., Smith, D.E., Zuber, M.T.: Evidence for surface water ice in the lunar polar regions using reflectance measurements from the Lunar Orbiter Laser Altimeter and temperature measurements from the Diviner Lunar Radiometer Experiment. J. Icarus **292**, 74–85 (2017)

Gertsch, L., Gustafon, R., Gertsch, R.: Effect of water ice content on excavatability of lunar regolith. In: El-Genk, M. (ed.) Space Technology and Applications International Forum 813, Melville, pp. 1093–1100 (2006)

Gowen, R.A., Smith, A., Fortes, A.D., Barber, S., Brown, P., Church, P., Collinson, G., Coates, A. J., Collins, G., Crawford, I.A., Dehant, V., Chela-Flores, J., Griffiths, A.D., Grindrod, P.M., Gurvits, L.I., Hagermann, A., Hussmann, H., Jaumann, R., Jones, A.P., Joy, K.H., Karatekin, O., Miljkovic, K., Palomba, E., Pike, W.T., Prieto-Ballesteros, O., Raulin, F., Sephton, M.A., Sheridan, S., Sims, M., Storrie-Lombardi, M.C., Ambrosi, R., Fielding, J., Fraser, G., Gao, Y., Jones, G.H., Kargl, G., Karl, W.J., Macagnano, A., Mukherjee, A., Muller, J.P., Phipps, A., Pullan, D., Richter, L., Sohl, F., Snape, J., Sykes, J., Wells, N.: Penetrators for in situ subsurface investigations of Europa. J. Adv. Space Res. **48**, 725–742 (2011)

Hansen, C.J., Shemansky, D.E., Esposito, L.W., Stewart, A.I.F., Lewis, B.R., Colwell, J.E., Hendrix, A.R., West, R.A., Waite Jr., J.H., Teolis, B., Magee, B.A.: The composition and structure of the Enceladus plume. J. Geophys. Res. Lett. **38**, L11202 (2011)

Heldmann, J.L., Schurmeier, L., McKay, C., Davila, A., Stoker, C., Marinova, M., Wilhelm, M.B.: Midlatitude ice-rich ground on Mars as a target in the search for evidence of life and for in situ resource utilization on human missions. J Astrobiology **14**, 102–118 (2014)

Herkenhoff, K.E., Squyres, S.W., Arvidson, R., Bass, D.S., Bell II, J.F., Bertelsen, P., Ehlmann, B. L., Farrand, W., Gaddis, L., Greeley, R., Grotzinger, J., Hayes, A.G., Hviid, S.F., Johnson, J.R., Jolliff, B., Kinch, K.M., Knoll, A.H., Madsen, M.B., Maki, J.N., McLennan, S.M., McSween, H. Y., Ming, D.W., Rice Jr., J.W., Richter, L., Sims, M., Smith, P.H., Soderblom, L.A., Spanovich, N., Sullivan, R., Thompson, S., Wdowiak, T., Weitz, C., Whelley, P.: Evidence from opportunity's microscopic imager for water on Meridiani Planum. J. Sci. **306**, 1727–1730 (2004)

Hudson, T.L., Aharonson, O.: Diffusion barriers at Mars surface conditions: salt crusts, particle size mixtures, and dust. J J. Geophys. Res. **113**, E09008 (2008)

Kalousová, K., Souček, O., Tobie, G., Choblet, G., Čadek, O.: Water generation and transport below Europa's strike-slip faults. J. Geophys. Res. **121**, 2444–2462 (2016)

Kömle, N., Pitcher, C., Gao, Y., Richter, L.: Study of the formation of duricrusts on the Martian surface and their effect on sampling equipment. J. Icarus **281**, 220–227 (2017)

Landis, G.A., Blaney, D., Cabrol, N., Clark, B.C., Farmer, J., Grotzinger, J., Greeley, R., McLennan, S.M., Richter, L., Yen, A.: Transient liquid water as a mechanism for induration of soil crusts on Mars. In: Lunar and Planetary Science XXXV, Houston (2004)

Mantovani, J., Swanger, A., Townsend III, I., Sibille, L., Galloway, G.: Characterizing the physical and thermal properties of planetary regolith at low temperatures. In: Gertsch, L., Malla, R. (eds.) Earth and Space 2014, Missouri, pp. 43–51 (2014)

McKay, C.P., Anbar, A.D., Porco, C., Tsou, P.: Follow the plume: the habitability of Enceladus. J. Astrobiol. **14**, 352–355 (2014)

McKay, D.S., Carter, J.L., Boles, W.W., Allen, C.C., Allton, J.H.: JSC-1: a new lunar soil simulant. In: Galloway, R.G., Lokaj, S. (eds.) Engineering, Construction and Operations in Space IV, pp. 857–866. American Society of Civil Engineers, New York (1994)

Metzger, P., Galloway, G., Mantovani, J., Zacny, K., Craft, J. Low force icy regolith penetration technology. In: NASA Technical Memorandum 2011-216302, NASA Center for Aerospace Information, Hanover (2011)

Mitrofanov, I., Sanin, A., Boynton, W., Chine, G., Garvin, J., Golovin, D., Evans, L., Harshman, K., Kozyrev, A., Litvak, M., Malakhov, A., Mazarico, E., McClanahan, T., Milikh, G., Mokrousov, M., Nandikotkur, G., Neumann, G., Nuzhdin, I., Sagdeev, R., Shevchenko, V., Shvetsov, V., Smith, D., Starr, R., Tretyakov, V., Trombka, J., Usikov, D., Varenikov, A., Vostrukhin, A., Zuber, M.: Hydrogen mapping of the lunar south pole using the LRO neutron detector experiment LEND. J. Sci. **330**, 483–486 (2010)

Moore, H.J., Jakosky, B.M.: Viking landing sites, remote-sensing observations, and physical properties of Martian surface materials. J. Icarus **81**, 164–184 (1989)

Mutch, T.A., Arvidson, R.E., Binder, A.B., Guinness, E.A., Morris, E.C.: The geology of the Viking Lander 2 site. J. Geophys. Res. **82**, 4452–4467 (1977)

Neal, C.R., Lawrence, S.J.: A Multi-decadal sample return campaign will advance lunar and solar system science and exploration. In: European Lunar Symposium, Münster (2017)
Piqueux, S., Christensen, P.: A model of thermal conductivity for planetary soils: 2. Theory for cemented soils. J. Geophys. Res. **114**, E09006 (2009)
Pitcher, C.: Advancing the dual reciprocating drill design for efficient planetary subsurface exploration. Ph.D. thesis, University of Surrey (2017)
Pitcher, C., Gao, Y.: Analysis of drill head designs for dual-reciprocating drilling technique in planetary regoliths. J. Adv. Space Res. **56**, 1765–1776 (2015)
Pitcher, C., Kömle, N., Leibniz, O., Morales-Calderon, O., Gao, Y., Richter, L.: Investigation of the properties of icy lunar polar regolith simulants. J. Adv. Space Res. **57**, 1197–1208 (2016)
Redlich, D., Paul, R., Ott, S., Richter, L., Mühlbauer, Q., Thiel, M., Tattusch, T., Weisz, H., Musso, F., Durrant, S.: Development and testing of a "Backlash-Free" gas-tight high-precision sample dosing mechanism for the ExoMars 2018 Rover. In: 43rd Aerospace Mechanisms Symposium, NASA Ames Research Center, pp. 51–66 (2016)
Richter, L., Carianni, P., Durrant, S., Hofmann, P., Mühlbauer, Q., Musso, F., Paul, R., Redlich, D.: Progress report on development of the ExoMars 2018 sample processing and distribution subsystem (SPDS) and related OHB sample handling studies. In: 13th Symposium on Advanced Space Technologies in TRobotics and Automation (ASTRA), Noordwijk (2015)
Roth, L., Saur, J., Retherford, K.D., Strobel, D.F., Feldman, P.D., McGrath, M.A., Nimmo, F.: Transient water vapor at Europa's South Pole. J Science **343**, 171–174 (2014)
Sanders, G.B., Larson, W.E.: Integration of In-Situ resource utilization into Lunar/Mars exploration through field analogs. J. Adv. Space Res. **47**, 20–29 (2011)
Schrader, C.M., Rickman, D.L., McLemore, C.A., Fikes, J.C., Stoeser, D.B., Wentworth, S.J., McKay, D.S.: Lunar regolith characterization for simulant design and evaluation using figure of merit algorithms. In: 47th AIAA Aerospace Sciences Meeting, American Institute of Aeronautics and Astronautics, Reston (2009)
Schrader, C.M., Rickman, D.L., McLemore, C.A., Fikes, J.C.: Lunar regolith simulant user's guide. In: NASA Technical Memorandum 2010-216446, NASA Center for Aerospace Information, Hanover (2010)
Schultz-Makuch, D., Head, J.N., Houtkooper, J., Knoblauch, M., Furfaro, R., Fink, W., Fairén, A. G., Vail, H., Sears, S.K., Daly, M., Deamer, D., Schmidt, H., Hawkins, A., Sun, H.J., Lim, D. S.S., Dohm, J., Irwin, L.N., Davila, A.F., Mendez, A., Andersen, D.: The biological oxidant and life detection (BOLD) mission: a proposal for a mission to Mars. J. Planetary Space Sci. **67**, 57–69 (2012)
Seiferlin, K., Ehrenfreund, P., Garry, J., Gunderson, K., Hütter, E., Kargl, G., Maturilli, A., Merrison, J.P.: Simulating Martian regolith in the laboratory. J. Planetary Space Sci. **56**, 2009–2025 (2008)
Sharp, R.P., Malin, M.C.: Surface geology from viking landers on Mars: a second look. J. Geol. Soc. Am. Bull. **95**, 1398–1412 (1984)
Smith, P.H., Tamppari, L.K., Arvidson, R.E., Bass, D., Blaney, D., Boynton, W.V., Carswell, A., Catling, D.C., Clark, B.C., Duck, T., DeJong, E., Fisher, D., Goetz, W., Gunnlaugsson, H.P., Hecht, M.H., Hipkin, V., Hoffman, J., Hviid, S.F., Keller, H.U., Kounaves, S.P., Lange, C.F., Lemmon, M.T., Madsen, M.B., Markiewicz, W.J., Marshall, J., McKay, C.P., Mellon, M.T., Ming, D.W., Morris, R.V., Pike, W.T., Renno, N., Staufer, U., Stoker, C., Taylor, P., Whiteway, J.A., Zent, A.P. H_2O at the phoenix landing site. J. Sci **325**, 58–61 (2009)
Sparks, W.B., Hand, K.P., McGrath, M.A., Bergeron, E., Cracraft, M., Deustua, S.E.: Probing for evidence of plumes on Europa with HST/STIS. Astrophys. J. **829**, 121 (2016)
Stoeser, D., Wilson, S., Rickman, D.: Design and specifications for the highland regolith prototype simulants NU-LHT-1M and -2M. In: NASA Technical Memorandum 2010-216438 NASA Centre for Aerospace Information, Hanover (2010)
Taylor, L.: Status of lunar regolith simulants and demand for Apollo samples. In: Lunar Exploration Analysis Group (LEAG) Analysis Reports. NASA. http://www.lpi.usra.edu/leag/reports/SIM_SATReport2010.pdf (2010)

Part II
Resource, Mining, and Subsurface Access

Chapter 3
Atmospheric Mining in the Outer Solar System: Resource Capturing, Storage, and Utilization

Bryan Palaszewski

3.1 Atmospheric Mining in the Outer Solar System

Atmospheric mining of the outer solar system is one of the options for creating nuclear fuels, such as helium 3 (3He), for future fusion powered exploration vehicles or powering reactors for Earth's planetary energy needs Palaszewski (2005, 2006, 2007, 2008, 2009, 2010, 2011; Frisbee 2003). Uranus' and Neptune's atmospheres would be the primary mining sites, and robotic vehicles would wrest these gases from the hydrogen-helium gases of those planets. While preliminary estimates of the masses of the mining vehicles have been created (Palaszewski 2005, 2006, 2007, 2008, 2009, 2010, 2011), additional supporting vehicles may enhance the mining scenarios.

There are vast reserves of potential fuels and propellants in the outer planets (Palaszewski 2005, 2006, 2007, 2008, 2009, 2010, 2011). While the idea of mining outer planet atmospheres is indeed enticing, the challenges to designing mining vehicles may be somewhat daunting. While past studies related to the Daedalus Project Palaszewski (2005) have assumed the use of fusion propulsion for the aerostat and aerospacecraft that mine the atmosphere and carry the fuel to Jupiter's orbit, nuclear thermal rockets may also allow a more near term propulsion option. While the mass of the NTP options will, in most cases, be higher than the fusion powered options, the more near term NTP vehicle may still be attractive (Frisbee 2003; Dunn 1987; Noca et al. 2002; Hunt et al. 1997), although closed cycle gas core nuclear rockets may provide high specific impulse and high thrust without invoking fusion rockets (Starr et al. 2006; Bussard 1971; Bussard and Jameson 2006; Borowski et al. 1999; Kendall and Stoeffler 1972; Latham and Rodgers 1972; Latham 1971; Rodgers and Latham 1972; Troutman et al. 2003; Adams et al. 2003).

B. Palaszewski (✉)
NASA John H. Glenn Research Center, Lewis Field, Mail Stop 5-10,
21000 Brookpark Road, Cleveland, OH 44135, USA
e-mail: bryan.a.palaszewski@nasa.gov

V. Badescu and K. Zacny (eds.), *Outer Solar System*,
https://doi.org/10.1007/978-3-319-73845-1_3

During the helium 3 capturing, large amounts of hydrogen and helium 4 are produced. Analyses were conducted to quantify the mass production rates of these other potential fuels. Also, capturing the hydrogen and helium 4 to fuel additional exploration and exploitation vehicles was addressed. New options for fleets of small and large aerospacecraft for exploration and exploitation missions are discussed.

3.2 Resource Capturing Studies

Studies of the gas capture rate and its influence on mining time in the atmosphere were conducted. Aerospacecraft cruisers have been identified as a "best" solution for atmospheric mining (Palaszewski 2005, 2006, 2007, 2008, 2009, 2010, 2011). To power these vehicles, atmospheric hydrogen gas would be liquefied and used a rocket propellant for the ascent to orbit. Gaseous or liquid hydrogen would be use to power the engines during atmospheric mining operations. Figure 3.1 shows an overall schematic of a closed cycle gas core rocket propulsion option. Helium 3 (3He) would be separated from the atmospheric hydrogen and helium (4He) captured, liquefied and stored as a payload that would be returned to orbit. Table 3.1 provides the fraction of 3He in the outer planet atmospheres.

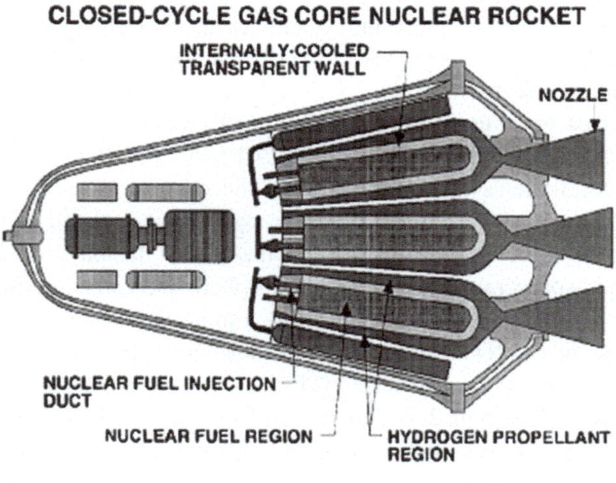

Fig. 3.1 Gas core rocket propulsion for the mining cruiser Frisbee (2003)

Table 3.1 Fraction of helium 3 and helium 4 in outer planet atmospheres

	Uranus	Neptune
Amount of 3He in 4He	1.00×10^{-4}	1.00×10^{-4}
Amount of 4He in atmosphere	0.152	0.19
Amount of 3He in atmosphere	1.52×10^{-5}	1.90×10^{-5}

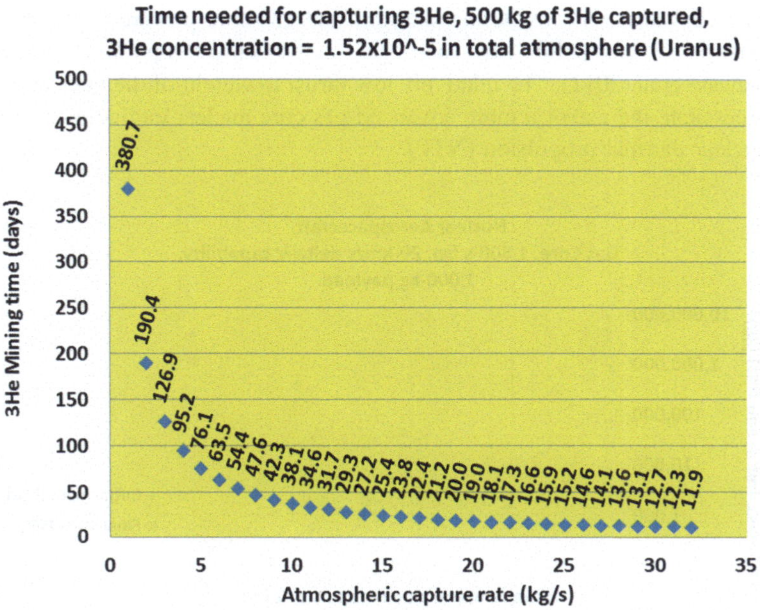

Fig. 3.2 Mining time versus the capture rate for Uranus

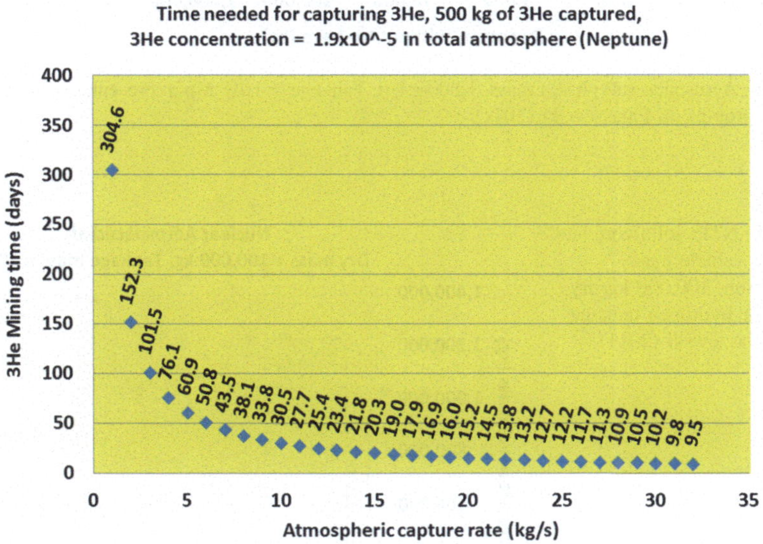

Fig. 3.3 Mining time versus the capture rate for Neptune

Figures 3.2 and 3.3 show the helium 3 mining time versus the atmospheric capture rate for Uranus and Neptune, respectively. A 500-kg payload of 3He is captured during the mining time.

Figures 3.4 and 3.5 provide the sizing of the gas core powered vehicles and a comparison of the solid core and gas core vehicle options, respectively (Palaszewski et al. 2011). The relatively low thrust to weight of the nuclear engines may necessitate the use of a more advanced gas core nuclear engine over the solid core nuclear thermal propulsion (NTP).

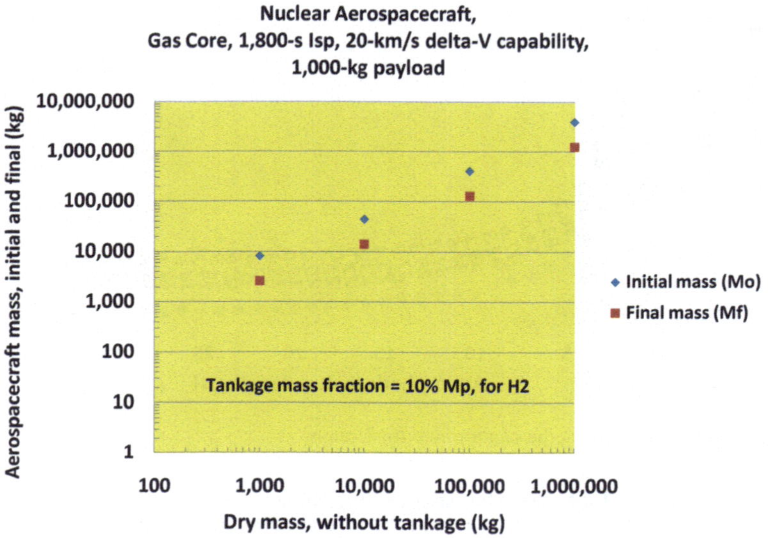

Fig. 3.4 Aerospacecraft (ASC) mass, 1,800-s Isp, Tankage = 10% Mp, representative of gas core nuclear propulsion Palaszewski (2011)

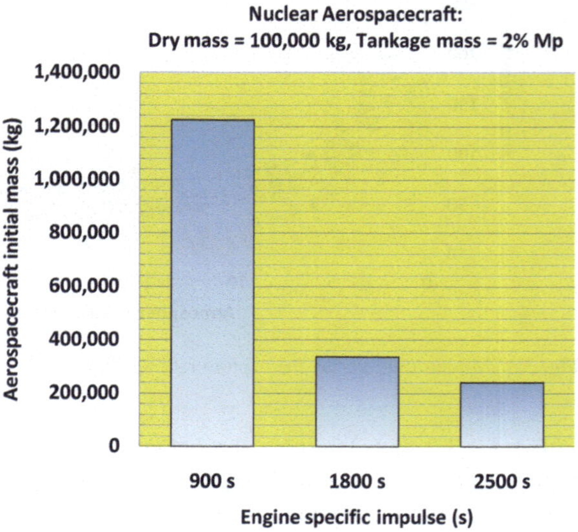

Fig. 3.5 NTP: solid core and gas core vehicle mass comparison, 100,000 kg dry mass, 2% hydrogen tankage mass Palaszewski (2011)

3.3 Fueling and Refueling Options

After completing the analyses of the time for propellant capture it became clear that a large amount of liquid hydrogen was produced each day of 3He production. Figures 3.6 and 3.7 depict the relatively large mass fractions of hydrogen and helium 4 that are processed to extract the desired helium 3. It is clear that such large masses will be useful for not only refueling the mining cruiser aerospacecraft, but may be important for other related applications.

Figure 3.8 shows the 3He capture time (for 500 kg), the mass of hydrogen produced per day, and the hydrogen needed to fuel gas core rocket powered aerospacecraft (ASC) at a specific impulse of 1800 and 2500 s, all as a function of atmospheric gas capture rate. In this case, the 3He in the atmosphere is = $1.52e^{-5}$ (a case for Uranus), and the ASC dry mass = 100,000 kg. As an example, of the atmospheric capture rate were 4 kg/s, there required amount of 500 kg of 3He would be captured in 95.2 days. During that time, 293,000 kg of hydrogen would be produced per day. To fully fuel an 1800-s Isp gas core ASC is 270,000 kg. A hydrogen propellant load of 148,000 kg is needed for the 2500-s Isp gas core powered ASC. Similarly, if the atmospheric capture rate were 10 kg/s, the time for capturing the 500 kg of 3He would be 38.1 days. During those 38.1 days, 732,600 kg of hydrogen would be produced per day. Thus, more than two (2) 1800-s gas core ASC vehicles could be refueled per day. While the mining vehicle (ponderously and politely) continues its 3He capturing, additional vehicles could flit about far from the mining ASC and gather needed information on potential storms or other disturbances that the mining ASC must avoid.

For a 1,000,000 kg dry mass, the mining case also show significant hydrogen benefits. In the case for Neptune (3He = $1.9e^{-5}$), at an atmospheric capture rate of 22 kg/s, there is enough hydrogen produced to refuel a 2500-s ASC every day. At that capture rate, it takes 13.8 days to mine the required 500 kg of 3He. So 13 orbital missions could be conducted or numerous sorties in the atmosphere by UAVs requiring smaller hydrogen propellant loads could be completed (Fig. 3.9).

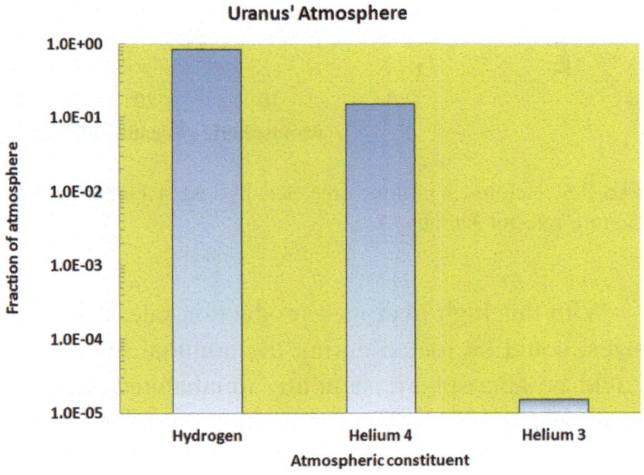

Fig. 3.6 Fractions of atmospheric gases for Uranus

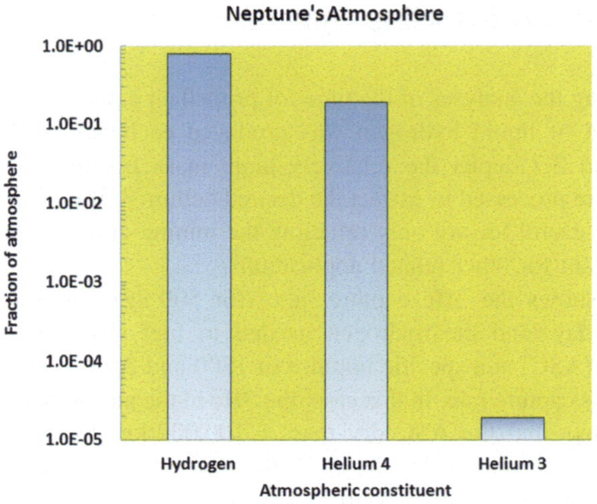

Fig. 3.7 Fractions of atmospheric gases for Neptune

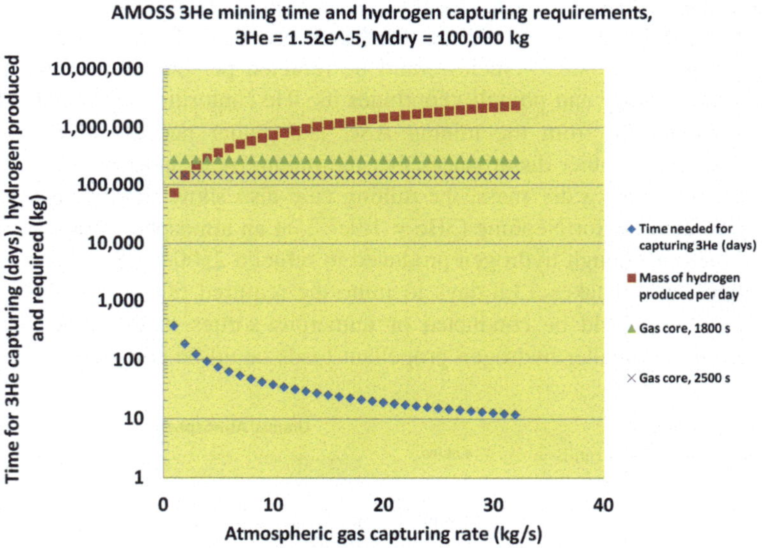

Fig. 3.8 Helium 3 mining time and hydrogen capture (mass per day) versus atmospheric gas capture rate for Uranus

With this high hydrogen production rate, fleets of aerospacecraft, of a variety of sizes, could be fueled during the nominal time of capturing the 3He. Such a fleet could be atmospheric sampling uninhabited aerial vehicles (UAVs), small orbital missions, or UAVs for in situ planetary meteorological studies.

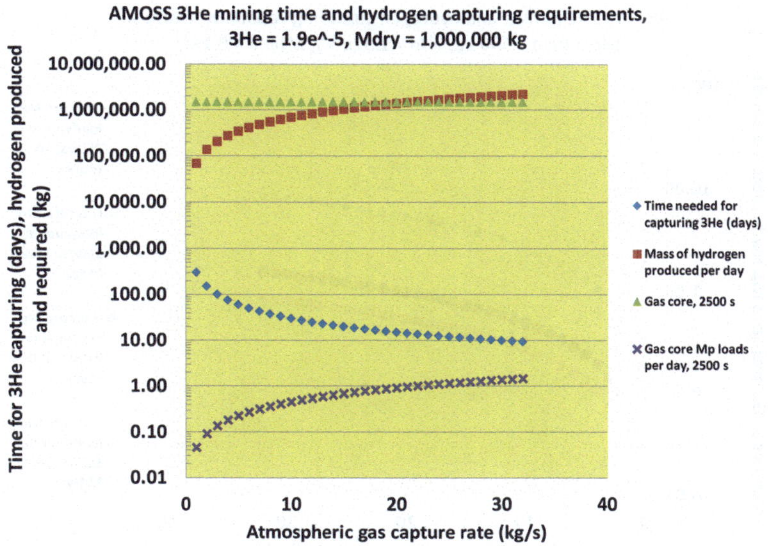

Fig. 3.9 Helium 3 mining time and hydrogen capture (mass per day) versus atmospheric gas capture rate for Neptune

Refueling of cryogenic ASC vehicles will no doubt be a challenge (in robotic aerial refueling, etc.), and there will be additional cryogenic transfer losses and propellant tank chilldown requirements, however, the mass of hydrogen produced is quite impressive and is a ripe area for investigating hydrogen usage options.

Figure 3.10 compares all of the hydrogen capturing cases for Uranus. In the chart, the number of gas core rocket hydrogen propellant loads captured is as high at 15.8 for the 2500-s GCR cases (with a 100,000 kg dry mass, 32 kg/s capture rate). At a 10 kg/s atmospheric capture rate, the maximum number of hydrogen loads is 4.95 (or just less than 5). The lowest value is 0.27 hydrogen loads per day.

Similar analyses are shown for the other vehicle designs for Neptune in Fig. 3.11: 1800 and 2500 s Isp nuclear gas core rocket (GCR) aerospacecraft (ASC) with 100 and 1000 MT dry masses. With the Neptune analysis, the rates of hydrogen capture are slightly lower, and the capture rates are very similar to the Uranus cases.

While capturing helium 3 and hydrogen, there is also a very significant amount of helium 4 than can be captured. Figures 3.12 and 3.13 provide the helium 4 capture capability per day. The capture capability of the helium 4 is expressed in the equivalent masses of hydrogen to fuel the gas core rockets. This equivalent figure of merit of GCR propellant loads makes for a more direct comparison of the masses of hydrogen and helium 4. The helium 4 capture masses are approximately 15–19% of the hydrogen capture masses.

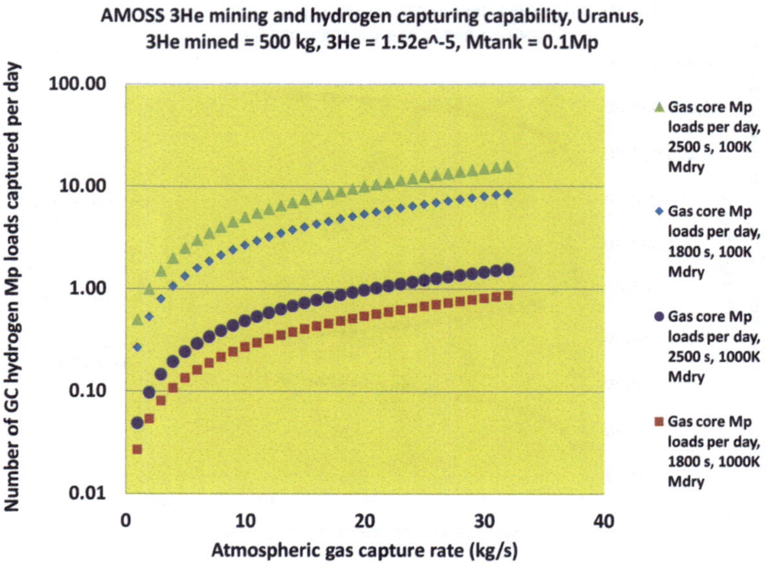

Fig. 3.10 Number of gas core rocket hydrogen propellant loads captured per day versus atmospheric gas capture rate—Uranus

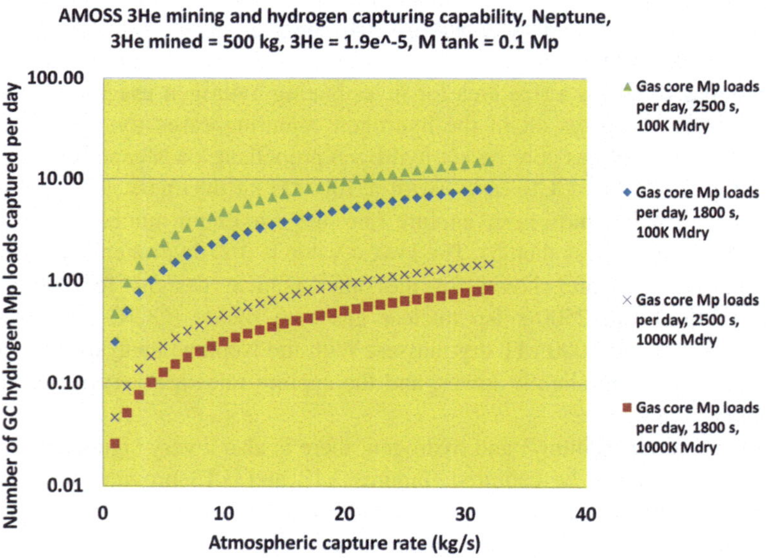

Fig. 3.11 Number of gas core rocket hydrogen propellant loads captured per day versus atmospheric gas capture rate—Neptune

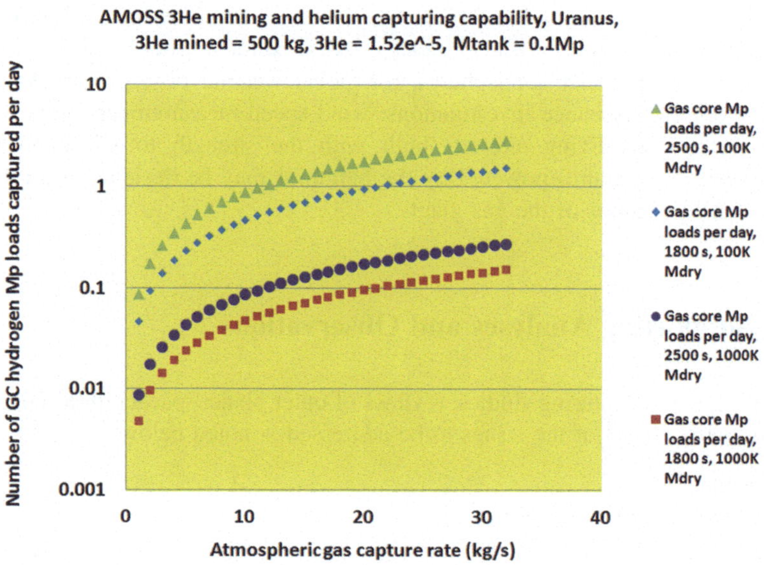

Fig. 3.12 Number of gas core rocket (mass equivalent hydrogen) propellant loads of helium 4 captured per day versus atmospheric gas capture rate—Uranus

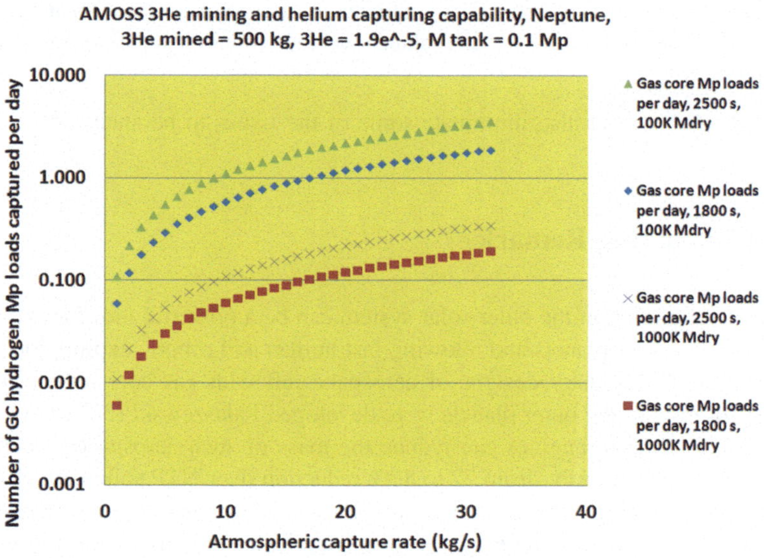

Fig. 3.13 Number of gas core rocket (mass equivalent hydrogen) propellant loads of helium 4 captured per day versus atmospheric gas capture rate—Neptune

With this added helium 4 resource, many vehicles could be fueled. Entire fleets of aerospacecraft or other aerial vehicles (UAVs, balloons, rockets, etc.) could fly through the outer planet atmospheres, for global weather observations, localized storm or other disturbance investigations, wind speed measurements, polar observations, etc. Deep-diving aircraft (built with the strength to withstand many atmospheres of pressure) powered by the helium 4 may be designed to probe the higher density regions of the gas giants.

3.4 Supporting Analyses and Observations

In addition to the capturing studies, reviews of outer planet spacecraft design issues were initiated. A list of the issues to be addressed is noted below:

- Mission planning.
- Cryogenic fuel storage issues.
- Cryogenic dust (outer planet moons, ice migration). Mass concentrations (mascons) on the moons, etc.
- Drilling into ice, walkers on ice-dust surfaces.
- Possible power generation using electro dynamic tethers (EDT), cutting across the outer planet magnetic field lines.
- Global Positioning System (GPS) vehicles in outer planet orbits for navigation.
- Observational satellite for outer planet weather monitoring, diverting cruisers from harm.

Figures 3.A1–3.A4 also illuminate some of the issues to be analyzed.

3.5 Concluding Remarks

Atmospheric mining in the outer solar system can be a powerful tool for extracting fuels from the outer planets and allowing fast human and robotic exploration of the solar system. Preliminary designs of aerospacecraft with gas core rocket nuclear engines for mining the outer planets were developed Palaszewski (2011). Analyses showed that gas core engines can reduce the mass of such aerospacecraft mining vehicles very significantly: from 72 to 80% reduction over NTP solid core powered aerospacecraft mining vehicles. While this mass reduction is important in reducing the mass of the overall mining system, the complexity of a fissioning plasma gas core rocket is much higher than the more traditional solid core NTP engines. Additional analyses were conducted to calculate the capture rates of hydrogen and helium 4 during the mining process. Very large masses of hydrogen and helium 4 are produced every day during the often lengthy process of helium 3 capture and gas separation. Typically, these very large additional fuel masses can dwarf the requirements needed for hydrogen capture for ascent to orbit. Thus, the potential for fueling small and large fleets of additional exploration and exploitation vehicles exists. Additional aerospacecraft or other aerial vehicles (UAVs, balloons, rockets, etc.) could fly through the outer planet atmospheres, for global weather observations, localized storm or other disturbance investigations, wind speed measurements, polar observations, etc. Deep-diving aircraft (built with the strength to withstand many atmospheres of pressure) powered by the excess hydrogen or helium 4 may be designed to probe the higher density regions of the gas giants.

Based on these analyses, there will likely be several possible future avenues for effective use of the gases of the outer planets for exciting exploration missions. When focusing on Uranus and Neptune, these planets offer vast reservoirs of fuels that are more readily accessible than those from Jupiter and Saturn and, with the advent of nuclear fusion propulsion, may offer us the best option for the first practical interstellar flight.

Appendix: Issues for Cryogenic Operations

See Figs. 3.A1, 3.A2, 3.A3, 3.A4.

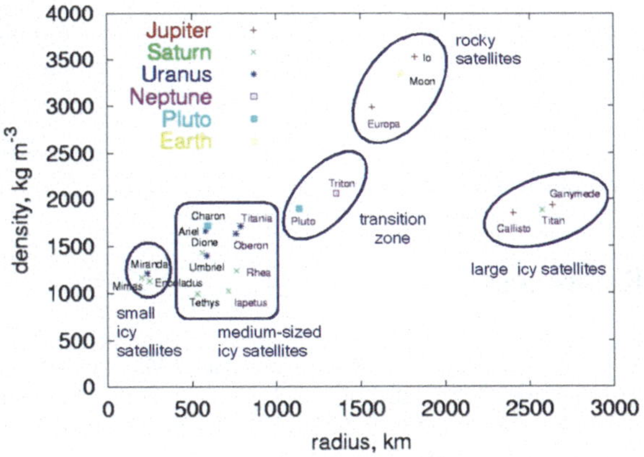

Fig. 3.A1 Outer planet moon densities Hussmann et al. (2006)

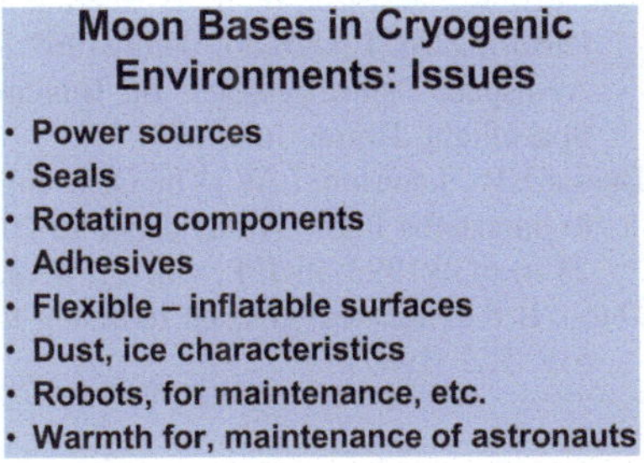

Fig. 3.A2 Issues for cryogenic outer planet moon surface operations [RASC, HOPE study (Troutman et al. 2003; Adams et al. 2003)]

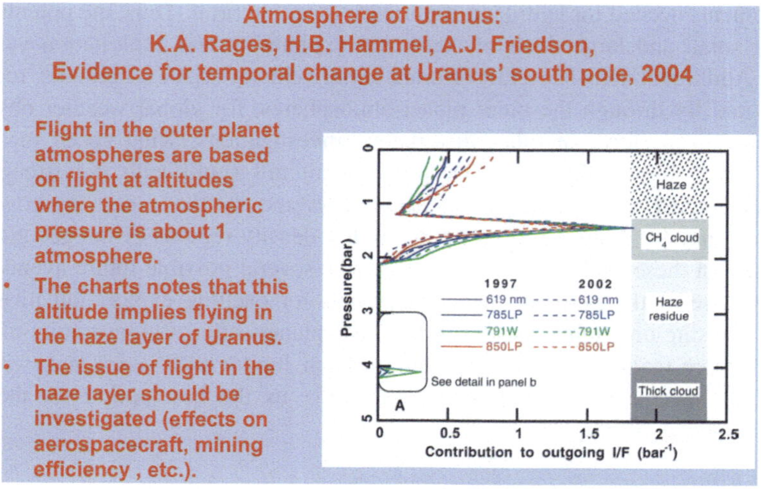

Fig. 3.A3 Uranus atmospheric structure, haze phenomena (Rages et al. 2004)

Fig. 3.A4 Atmospheric mining issues

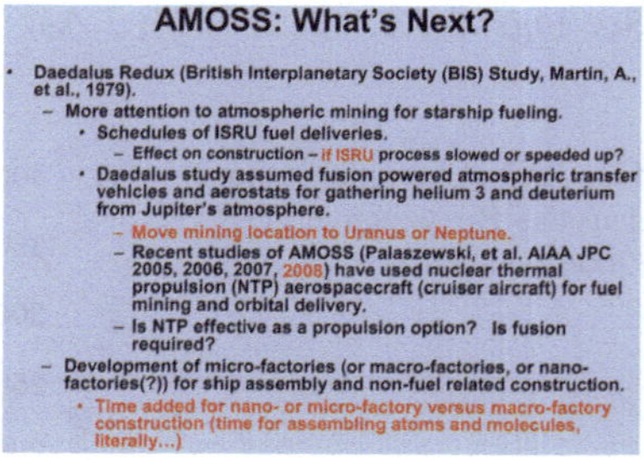

References

Adams, R.B., Alexander, R.A., Chapman, J.M., Fincher, S.S., Hopkins, R.C., Philips, A.D., Polsgrove, T.T., Litchford, R.J., Patton, B.W., Statham G., White, P.S., Thio, Y.C.F.: Conceptual Design of In-Space Vehicles for Human Exploration of the Outer Planets. NASA/TP—2003-212691, Nov 2003

Borowski, S.K., Dudzinski, L.A., McGuire, M.L.: Artificial gravity vehicle design option for NASA's human mars mission using bimodal NTR propulsion. AIAA Paper 99–2545 (1999)

Bussard, R.: ASPEN II: Two Staging and Radiation Shielding Effects on ASPEN Vehicle Performance, LA-26-80, 09/06/1967 and in ASPEN: Nuclear Propulsion for Earth to Orbit Aerospace Plane Vehicles. In: Bussard RW (eds.) Proceedings International Conference on Spaceflight, Rome, June 1971

Bussard, R., Jameson, L.W.: The QED engine spectrum: fusion electric propulsion for airbreathing to interstellar flight. AIAA paper 93-2006, 29th Joint Propulsion Conference, Monterey, CA 6/28 to 6/30/1993; in JPP, vol. 11, no.2, pp. 365–372

Dunn, B.P.: High-energy orbit refueling for orbital transfer vehicles. J. Spacecraft Rockets **24**(6), 518–522 (1987)

Frisbee, R.: Advanced space propulsion for the 21st century. J. Propul. Power **19**(6) Nov–Dec 2003

Hunt, J.L., Laruelle, G., Wagner, A.: Systems challenges for hypersonic vehicles; AGARD interpanel symposium on future aerospace technology in service to the alliance. NASA-TM-112908, AGARD-Paper-C37 (1997)

Hussmann, H., Sohl, F., Spohn, T.: Subsurface oceans and deep interiors of medium-sized outer planet satellites and large trans-neptunian objects. Icarus **185**(1), 258–273 (2006)

Kendall, J.S., Stoeffler, R.C.: Conceptual design studies and experiments related to cavity exhaust systems for nuclear light bulb configurations. Report Number: L-910900-15; NASA-CR-129298 (1972)

Latham, T.S.: Summary of the performance characteristics of the nuclear light bulb engine, Report Number: AIAA PAPER 71-642 (1971)

Latham, T.S., Rodgers, R.J.: Small nuclear light bulb engines with cold beryllium reflectors. Report Number: AIAA PAPER 72-1093 (1972)

Noca, M., Polk, J.E.: Ion thrusters and LFAs for outer planet exploration. In: AAAF 6th International Symposium, Versailles, France, May 2002

Palaszewski, B.: Atmospheric mining in the outer solar system. AIAA 2005-4319, July 2005

Palaszewski, B.: Atmospheric mining in the outer solar system: issues and challenges for mining vehicle propulsion AIAA 2011-6041, Aug 2011

Palaszewski, B.: Atmospheric mining in the outer solar system: mining design issues and considerations. AIAA 2009-4961, Aug 2009

Palaszewski, B.: Atmospheric mining in the outer solar system: mission scenarios and options for in-situ resource utilization. AIAA 2007-5598, July 2007

Palaszewski, B.: Atmospheric mining in the outer solar system: orbital transfer vehicles and outer planet moon base options. AIAA 2008-4861, July 2008

Palaszewski, B.: Atmospheric mining in the outer solar system: university studies of mining vehicles and propulsion. AIAA 2010-6573, Aug 2010

Palaszewski, B.: Atmospheric mining in the outer solar system: vehicle sizing issues. AIAA 2006-5222, July 2006

Rages, K.A., Hammel, H.B., Friedsond, A.J.: Evidence for temporal change at Uranus' south pole. Icarus **172**, 548–554 (2004)

Rodgers, R.J., Latham, T.S.: Analytical design and performance studies of the nuclear light bulb engine, Report Number: L-910900-16; NASA-CR-129295 (1972)

Starr, B.R., Westhelle, C.H., Masciarelli, J.P.: Aerocapture performance analysis for a neptune-triton exploration mission. NASA/TM-2006-214300, Apr 2006

Troutman, P.A., Bethke, K., Stillwagen, F., Caldwell, D.L. Jr., Manvi, R., Strickland, C., Krizan, S.A.: Revolutionary Concepts for Human Outer Planet Exploration (HOPE), Space Technology & Applications International Forum (STAIF—2003), Albuquerque, New Mexico, 2–6 Feb 2003

Chapter 4
Project VALKYRIE: Laser-Powered Cryobots and Other Methods for Penetrating Deep Ice on Ocean Worlds

William Stone, Bart Hogan, Vickie Siegel, John Harman, Chris Flesher, Evan Clark, Omkar Pradhan, Albin Gasiewski, Steve Howe and Troy Howe

4.1 Introduction

The existence of water beneath the ice of outer Solar System moons has opened up new targets for the search for extant life (Chyba and Phillips 2001; McKay et al. 2014), making sub-ice oceans of these bodies among the most likely places to be successful in this search.

W. Stone (✉) · B. Hogan · V. Siegel · J. Harman · C. Flesher · E. Clark
Stone Aerospace, 3511 Caldwell Lane, Del Valle, TX, USA
e-mail: billstone@stoneaerospace.com

B. Hogan
e-mail: bart.hogan@stoneaerospace.com

V. Siegel
e-mail: vickie.siegel@stoneaerospace.com

J. Harman
e-mail: john.harman@stoneaerospace.com

C. Flesher
e-mail: chris.flesher@stoneaerospace.com

E. Clark
e-mail: evan.clark@jpl.nasa.gov

O. Pradhan · A. Gasiewski
University of Colorado Boulder, 425 UCB, Boulder, CO, USA
e-mail: omkarpradhan@gmail.com

A. Gasiewski
e-mail: Al.Gasiewski@colorado.edu

S. Howe · T. Howe
Howe Industries, 140 E Rio Salado Pkwy, #902 Tempe, Tempe, AZ, USA
e-mail: steve@howeindustries.net

T. Howe
e-mail: troy@howeindustries.net

V. Badescu and K. Zacny (eds.), *Outer Solar System*,
https://doi.org/10.1007/978-3-319-73845-1_4

The discovery of life off Earth would be a watershed event. The proof that we are not alone will be a more powerful revelation than the fact of a manned moon landing in 1969. Inherent in this accomplishment will be the realization of extraordinary advances in aerospace, robotics, microbiology, cryo-engineering, compact power systems, and many other fields.

Recent excitement in this area has been fueled by observations of plumes issuing from both Enceladus and Europa, and sampling these via fly-by missions has been proposed as a "cheap" way to find extraterrestrial life. Meanwhile a precursor fly-by mission (the Europa Clipper) will carry ice-penetrating radar to Europa in 2022, and a lightweight lander has been mandated by the U.S. Congress to be launched two years later. In reality, none of these missions stands much chance of actually discovering extant life—microbial or otherwise. The odds of detecting life at the outermost surface of Europa are vanishingly small due to the extant radiation environment (Teodoro et al. 2016; Paranicas et al. 2007, 2009). Complete irradiation of surface ice occurs over short timescales. Even in an optimistic scenario wherein recent upwelling in chaos terrain transports microbes from the ocean up to the top several meters of ice, all biomarker evidence will likely be destroyed by radiation.

The reality is that finding living microbes, multicellular organisms, or macro organisms means going to where the liquid water is and searching for energy sources that would support life. By far the most likely place to find life will be at the floor of the 100-kilometer-deep Europan ocean at hydrothermal vents powered by Jovian gravitational tidal dynamics. To a lesser extent there is the possibility of extremophile microbes existing deep within the ice column, well below the levels of destructive radiation at the surface.

There have been many depictions of what the ultimate sub-surface Ocean World exploration mission might entail. We envision that Ocean World exploration would

Fig. 4.1 A 2-stage cryobot-AUV mission lands on Europa

Fig. 4.2 The cryobot, powered by a 100 kW_{th} fission power source and using CCHWD, bores through 30 km of 100 K ice

follow a progression of steps, each supporting the next: (1) remote observations by orbiters or fly-by missions (particularly to ascertain ice thickness); (2) characterization of surface ice with landers; (3) the novel concept of sub-surface exploration by a robotic ice-penetrating vehicle—known as a cryobot; and finally, (4) sub-surface exploration by a cryobot containing a marsupial autonomous underwater vehicle (AUV) rover that explores the broader ocean and contains sophisticated sensors for the detection and sampling of life (for onboard

Fig. 4.3 Obstacles are to be expected. The cryobot IceSAR system probes the ice out to one kilometer range while resolving objects large enough to cause problems. The vehicle deviates its trajectory to bypass the obstacle by using lateral cutting jets from the CCHWD hot water jets

characterization prior to data uplink). See Figs. 4.1, 4.2, 4.3, 4.4, 4.5 and 4.6 for our visualization of these stages.

It can be argued that the 4th of these mission classes can only be conducted following an initial cryobot mission that acquires information on ocean currents, water chemistry, temperature, and density. There is a real possibility that unpredicted features and variations of the ocean environment may, once known, drive AUV design in an unexpected direction. Thus, any cryobot mission to an Ocean World must also consider the ice penetration as only the first half of the mission: the

Fig. 4.4 Breakthrough: the cryobot penetrates the Europan sub-surface ocean, releasing a radioisotope-powered long range AUV at the floor of the ocean

Fig. 4.5 The long range AUV may optionally carry marsupial microbots that investigate high energy, dangerous, locations. The microbots sample likely sources of microbial life

Fig. 4.6 Hydrothermal vents, produced by Jovian gravitational tidal flexing of Europa's silicate core, are the most likely features that would provide energy to support microbial life. Microbots sample the vents before bringing their payload back to the AUV mothership

equally important second half will be to deploy the cryobot (or a sub-section of the vehicle acting as an oceanographic sonde) in a controlled vertical descent through the ocean to acquire this critical design information for the subsequent AUV mission, analogous to the requirement for the flyby missions' ice penetrating radar providing information on ice characteristics and thickness prior to a cryobot mission.

The ice caps of the outer planets have been speculated to vary from as thin as one kilometer to as much as 30 km with commonly quoted numbers of 10 or 20 km. Similarly, estimates for the ocean depth on Europa vary widely from 30 to 100 km (Chyba and Phillips 2001; McKay et al. 2014).

The starting conditions at the surface are hard vacuum and temperatures around 100 K. An intelligent probe designed to work in these environments must therefore deal with high radiation, hard vacuum, extreme cold, and very high pressures (even at 0.134 g gravity, the bottom of the Europan ocean could see pressures as high as 1700 bar, greater than that at the bottom of the Challenger Deep on Earth).The remainder of this chapter deals with the issue of how to get through the ice. We will describe in detail several approaches to do so that were developed at Stone Aerospace under NASA funding during the past 14 years. Specifically we will be discussing the design of cryobots—intelligent, autonomous systems designed to be self-contained devices that can navigate vertically through ice much the same way as autonomous underwater vehicles are able to maneuver independently in 3D through an ocean. Intelligence, self-contained power, and autonomy are what separate true cryobots from traditional ice penetrators (sometimes referred to as "melt probes"). When one seriously attempts to go down this design path the use of nuclear energy as the power source for an Ocean World cryobot is inescapable. We will discuss this in detail in Sect. 4.4. Ideally, the way to rapidly advance discussion on how to best carry out an Ocean World sub-surface mission would be to develop and field test designs for a 2-stage nuclear-powered cryobot that delivered a marsupial AUV payload (also nuclear-powered) into deep, unexplored Antarctic lakes (the existence of which is now known thanks to tests of the Europa flyby mission ice-penetrating radar). This immediately brings up a problem: While the Antarctic Treaty presently only forbids the presence of nuclear weapons in Antarctica, the reality is that use of a fission reactor would have to meet with approval of an international environmental review board that, while perhaps in the end permitting the mission, would certainly set back a nuclear-powered Antarctic demo mission by many years. It thus makes more sense to use non-nuclear surrogate power sources for cryobot Antarctic mission field testing, but to also design the vehicle to be ready for a drop-in nuclear power source for a flight mission to Europa that is being developed in parallel to the surrogate Antarctic tests. Since this reality has affected our research from the start we will discuss in detail (see Sects. 4.5, 4.6 and 4.7) these clean, surface-supplied power source alternatives for conducting full-scale cryobot tests on Earth, in Antarctica.

While we will cite illustrative other research in this area (and will provide references thereto) this chapter is not intended to be a universal state-of-the-art review of all work in the area of cryobot design. Rather, we wish to give the serious systems designer some insight into where we have gone in the area of cryobot design; what we have learned, errors that were made; concepts that worked and did not work; and data from fielded cryobot systems that illuminate possibilities for future flight vehicles. There have been several epiphany moments during the course of this research, perhaps the most profound of which is that a successful cryobot mission to the Ocean World moons may in fact require the use of hybrid vehicles to deal with the different regimes of environment. We will elaborate on this topic in subsequent sections.

4.2 Getting Through Ice

We first briefly review the available options for ice penetration before presenting and contrasting the designs developed at Stone Aerospace during the course of the VALKYRIE and SPINDLE projects [VALKYRIE (Very-deep Autonomous Laser-powered Kilowatt-class Yo-yoing Robotic Ice Explorer) Phase 1 NASA Grant NNX10AE29G (2010); VALKYRIE Phase 2 NASA Grant NNX11AJ89G (2011–2016); SPINDLE (Sub-glacial Polar Ice Navigation, Descent, and Lake Exploration) Phase 1A NASA Grant NNX15AT32G (2015–2017)]. Ultimately, for a system to be viable for either an analog or flight system it has to be capable of multi-kilometer descents through ice. Practical design criteria for such systems would be those that are capable of accessing Antarctic sub-glacial lakes as analog test sites for Ocean World moons. Actual flight vehicles must be able to do the same mission within the environmental conditions previously described for Ocean Worlds. Ultimately, meeting those flight criteria will require the construction of a Europa Simulator facility of significant size that maintains a large column of ice at 100 K temperatures in hard vacuum. This is the essence of the “Starting Problem” (which we will discuss below) and there is no way to simulate that on Earth short of a specialized “Cryo-Vac” facility. Many of the systems discussed in this section do not meet all those criteria but we will mention them because they are relevant reference points from which to advance designs towards flight vehicles.

4.2.1 Mechanical Drilling

Traditional rock drilling technology (Vasiliev et al. 2011; Goodge and Severinghaus 2014) converted for deep ice access is effective, but comes at a price. The infrastructure is enormous—mainly because it is necessary to carry along kilometers of drill stem in addition to power and fuel. Furthermore, the hole has to be back filled with anti-freeze to prevent the drill stem from freezing in and to prevent the hole from closing due to hydrostatic pressure from the surrounding ice. In Antarctica, the hole re-freeze/collapse rate is about 1 cm of radius per hour if no steps are taken to maintain the hole. The antifreeze that is required to prevent that for traditional rock drilling equipment adds dramatically to the deployment mass and would not pass planetary protection requirements. The Russians, working at Lake Vostok and after several decades of effort (Vasiliev et al. 2011), have succeeded in accessing that lake at a depth of 3800 m using traditional rock drilling equipment (see Fig. 4.7). A good example of a modern mobile deep ice Antarctic rock drilling system put to the purpose of making deep holes in ice is the RAID drill (Goodge and Severinghaus 2014).

RAID (Fig. 4.8) is capable of reaching depths up to 3 km to bedrock (it is not permitted to be used for sub-glacial lake access due to forward contamination issues arising from the use of drilling fluids). The deployment "footprint" for RAID for a three kilometer penetration is on the order of 500 metric tonnes. By stark comparison, it is unlikely for the near future that landed payloads on Europa will exceed several hundred kilograms. [The science payload for the lightweight Europa Lander is being limited to 35 kg and this is making use of the predicted performance of the

Fig. 4.7 The Russian mechanical drilling outpost at Lake Vostok, Antarctica (from Vasiliev et al. 2011)

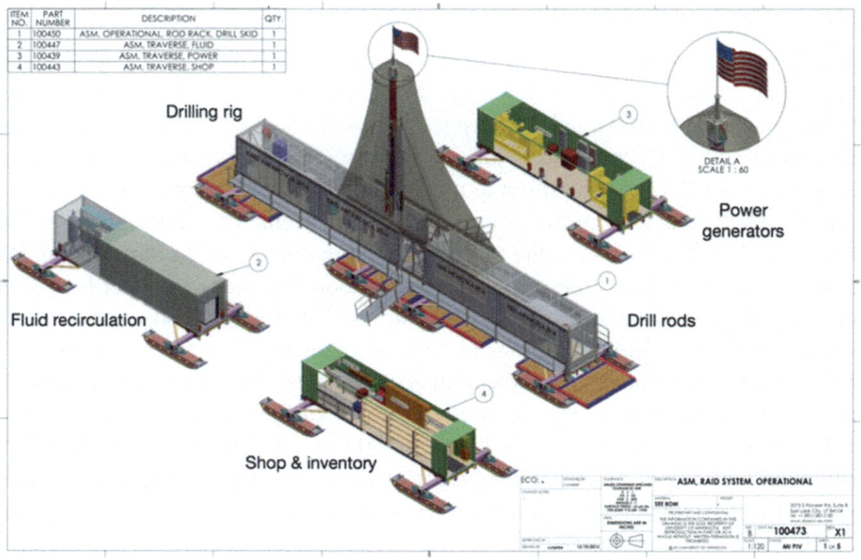

Fig. 4.8 The RAID drill as planned for deployment in the 2016–2017 Antarctic season (from Goodge and Severinghaus 2014)

Block 1 SLS launch system which does not yet exist. The reality of serious Ocean World missions in our lifetime will depend on a paradigm shift away from single fully integrated launch vehicles to larger systems assembled in LEO from several component and fuel supply launches from Earth.]. There are, of course, other issues affecting such a flight drill even if it could be magically transported to the surface of Europa: at 100 K temperatures there is no known compound, short of some binary gasses (e.g. LN_2) that would remain liquid and keep the hole open for drilling.

Specialized rock coring drills that have been developed for Mars sample extraction (Vago et al. 2006; Zacny et al. 2013; Chu et al. 2014) might be modified and considered for a Europa mission. However, the mechanical complexity, system mass, and very limited achievable depth (on the order of centimeters) precludes this approach as being viable for reaching the sub-surface oceans.

There has been at least one investigation of a self-contained hybrid mechanical/thermal drill (Weiss et al. 2008; Hsu 2010) and some have suggested that a mechanical drill could feet chippings to a melter. These concepts purport to address issues of getting through debris layers but omit details on what would be done with that material. However, those tests that have been attempted along these lines have all failed when they have encountered actual debris layers in ice. Chippings are by nature less dense than the base material and there is no volume allocated nor available in a self-contained drill to place this displaced, less dense material (surface-based drills solve this by using drilling fluids that extract the tailings, but then that requires all the infrastructure, previously mentioned, including drill stems and drilling fluids). Further, melting ice in addition to mechanical drilling represents a significant duplication of effort (and therefore waste of precious power). The increased complexity of a mechanical moving system that is dynamically interacting with the environment (and thus subject to shock loading) creates a higher probability of system failure. Replacement of a drill bit on Earth is a costly, time consuming endeavor, even for production oil drilling units; replacement on Europa would be impossible. Thus, while the idea of using self-contained mechanical drilling elements—even if only for penetration of debris layers—seems like a logical idea at first there are no measures for where to put that displaced, expanded material and the decreased mean-time-between-failure for a mechanical drilling system means a much higher likelihood of a full mission failure during the course of the months a likely ice cap penetration mission will take on Europa.

4.2.2 Hot Water Drills

By far the most successful ice penetrating systems presently in use on Earth employ hot-water drilling (HWD). Recent examples include the IceCube (Benson et al. 2014) and WISSARD (Rack 2014) drilling platforms (see also Figs. 4.9 and 4.10) used at the south pole and the Whillans Ice Stream, respectively, in the past few

Fig. 4.9 The vast IceCube hot water drill complex at Amundsen-Scott South Pole Station (from Benson et al. 2014)

Fig. 4.10 The mobile WISSARD hot water drill on the Whillan's Ice Stream, Antarctica in 2014 (courtesy Frank Rack)

years. The principle behind this approach is deceptively simple: heat water to 90 °C and pressures in excess of 100 bar and jet the hot water out a pendulum-stabilized weighted nozzle that is being lowered into the ice. Actual practice is more complex, of course, but this is the basic principle. Enormous power levels have been used for the above projects. IceCube used a 5 MW heat source (diesel burners and pumps). Importantly, the IceCube project used two separate types of drills: a passive "Firn Drill" for getting through the top 80 m of the ice cap and the Enhanced Hot Water Drill for the remainder of the 2800 m deep hole. A total of 86 holes were made (and seeded with sensors to construct the IceCube neutrino array). From this data, including melt rates, an important comparison metric was made possible: the passive Firn Drill, which utilized a conical shaped melt head through which hot glycol was recirculated and which operated at a power level of 150 kW (Fig. 4.11) achieved a melt rate of 380 kWH/m^3. Meanwhile, the Enhanced Hot Water Drill, powered at a level of 5 MW (Fig. 4.12), achieved a melt rate metric of 198 kWH/m^3, nearly double that of the passive system. This is an obscure metric that has significant implications for Ocean World ice penetration: it implies that for a given descent rate a hot water drill system would require half the power, all other things being equal, allowing for a smaller vehicle and thus reduced, possibly enabling, launch mass. However, there is an even more compelling advantage of hot water drilling over the "hot penny" approach. We have alluded above that there are stages of descent in a Europa mission. Specifically there are four stages:

Fig. 4.11 The IceCube Firn drill at south pole station. The Firn drill used copper tubing through which heated glycol was passed. This was used to penetrate the unconsolidated first 80 m of the ice cap prior to use of the hot water drill (from Benson et al. 2014)

- *Stage 1*: The Starting Problem: how to get the cryobot into the ice when starting in hard vacuum and 100 K environmental conditions. As will be discussed below this is not a simple problem
- *Stage 2*: Cruise Phase: descending through clean ice at temperatures starting at 100 K and warming up to 273 K at the ocean interface. This phase assumes that the cryobot has descended sufficiently for sublimate to refreeze inside the descent hole and eventually close it off to the point where vapor pressure can build and eventually support the existence of liquid phase water. The vast majority of the descent will be under these conditions.
- *Stage 3*: The Obstacle Problem: it would be folly to assume that the ice on any of the Ocean Worlds is flawlessly clean. In fact, it is highly probable that impact debris may be randomly scattered throughout the ice shell. It is equally probable that brine layers from up-welling events have formed within the ice, as they do on Antarctic ice shelves.
- *Stage 4*: Ocean Penetration: without some form of positive vertical control of the cryobot, breakthrough into the sub-surface ocean could initiate an uncontrolled free fall and loss of the mission. For this reason, as will be discussed in Sect. 4.6, we are developing a system (SPINDLE) that is capable of bi-directional vertical travel using servo-controlled spoolers for communications and a separate strength element that allows for precision velocity control of the vehicle, regardless of whether it is in ice or water.

Regarding Stage 3, this is the second instance where the "hot point" thermal probe melter will fail (see the discussion in Sect. 4.2.3 for why it will also fail the

Fig. 4.12 The IceCube hot water drill at south pole station. It used a 5 MW diesel heater and pump system to create 2.5 km deep holes, 60 cm in diameter, in 48 h

Starting Problem). As a thermal probe penetrates a debris-loaded layer the debris will continue to fall out of suspension and collect below the heating element. Eventually this serves as a heat shield and the penetrator stalls. This has been commonly observed in the U.S. Antarctic program with coil-type glycol heater probes used to melt holes into Dry Valley permanently ice covered lakes. A hot water drill, in pronounced contrast, will power through such layers even including volcanic tephra layers (e.g. Fig. 4.13) approaching a meter in thickness (e.g. see Thorsteinsson et al. 2008). As will be discussed in Sect. 4.5 it is possible to use a hot water drill to cut side pockets, into which turbulent flow will sweep debris, thus clearing the pathway ahead.

Fig. 4.13 Icelandic tephra interspersed in layers of the glacier on the slopes of the Grimsvotn caldera in Iceland is one example of how debris and sediment can become stratified within an ice sheet

For use on a flight mission a nuclear powered closed-cycle hot water drill (CCHWD) appears at this time to be the lead candidate for enhanced probability of success. There is one weakness of the CCHWD: it can only operate where ice can be liquefied, and hence will not work in vacuum environment at Europa's surface. The Starting Problem must be solved in a different way, which we will discuss in detail in Sect. 4.7.

4.2.3 Thermal Probes

The first melt probe is generally attributed to Philberth (1962; see also Fig. 4.14). Its principles—a high voltage power supply; a self-contained probe that paid out its power tether from a spooler; and a resistive heater block ("hot point") at the nose of the cryobot—have propagated to the present day with only minor changes in the approach.

Examples of modern high-voltage melt probes include: a NASA Jet Propulsion Laboratory (JPL) cryobot designed to emulate a 1 kW Radiothermal Power Source (RPS) (Zimmerman 2000; Fig. 4.15); Ice Diver, a long, narrow, javelin-shaped probe (Winebrenner et al. 2013; Fig. 4.16); Ice Mole, a square cross section cryobot

Fig. 4.14 A profile section of the original Philberth probe, the archetype for high voltage thermal probes (from Philberth 1962)

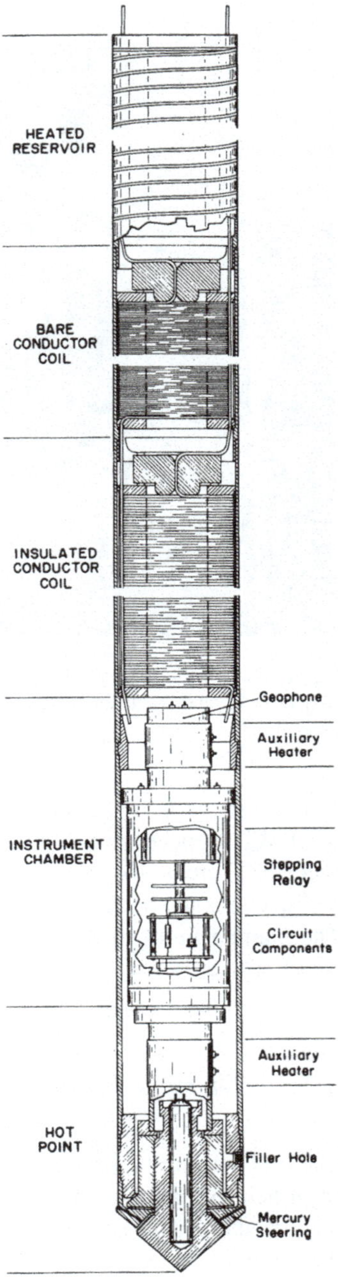

that uses an ice screw for forward motion (Dachwald et al. 2014; Fig. 4.17), and IceShuttle Teredo (Wirtz and Hildebrandt 2016; Fig. 4.18) which is a passive melt probe that is being designed to transport an AUV through shallow sea ice. All share the common mode of hot point passive heating at the nose. As such, all share a

Fig. 4.15 A small diameter, electrically heated, cryobot developed at JPL in the late 1990s. It had an onboard pump and a nose jet and represented the first CCHWD (from Zimmerman 2000)

Fig. 4.16 IceDiver, a direct descendent of the Philberth probe developed by Washington APL (from Winebrenner et al. 2013)

common failure mode in an Ocean World surface scenario. Hot point melt probes tested under vacuum have shown extreme difficulty initiating penetration (Kaufmann et al. 2009) as there is virtually no conductive thermal contact between

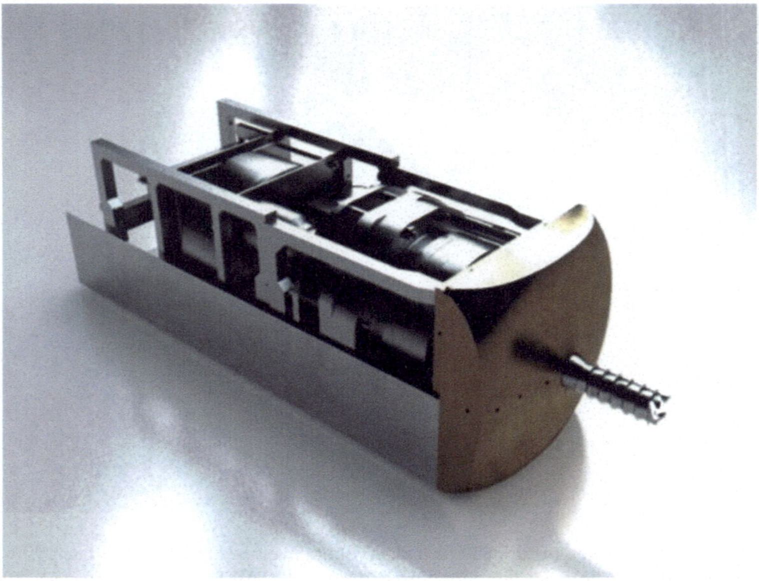

Fig. 4.17 IceMole, an electrically heated thermal probe which uses a rotating ice screw at the melt head to pull the vehicle forward. It uses a square cross section to provide resistance to rotational torque while the ice screw turns (Dachwald et al. 2014)

Fig. 4.18 IceShuttle Teredo, a concept vehicle designed to transport a torpedo-class AUV through sea ice (from Wirtz and Hildebrandt 2016)

the probe nose and rough ice surface due to a lack of liquid water (Fig. 4.19). As it is heated, the ice simply sublimates and any transferred heat is quickly dissipated due to the low power density and extreme cold. The current status of the above technologies was summarized by Ulamec et al. (2007) and Biele et al. (2011). At

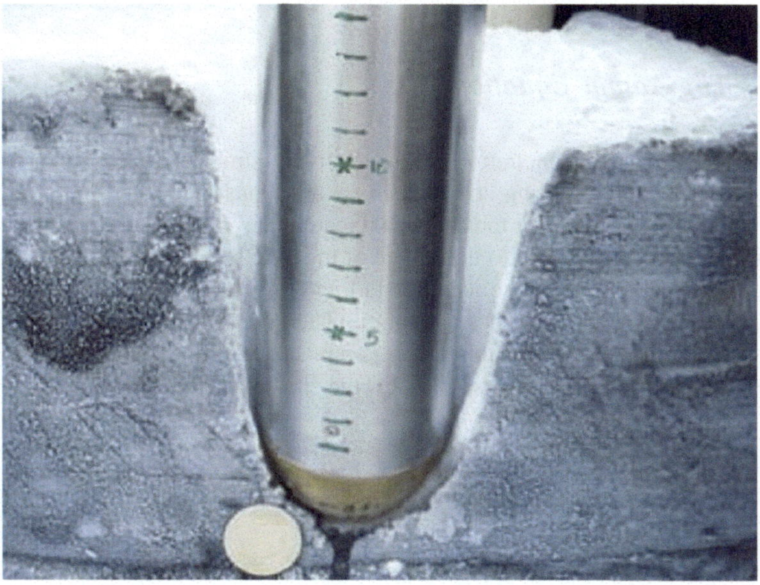

Fig. 4.19 Shows a unique vacuum chamber experiment involving a Philberth-class electrically heated thermal probe in 243 K (−15 °C) ice with a 6.35 cm diameter probe at 80 W input power after 48 h. Importantly the ice is well above Europan temperatures and yet it is clear that the thermal probe has stalled out (from Kaufmann et al. 2009)

that time, no one had considered a hybrid approach in which a self-contained cryobot utilized any other approach than passive heating as the mechanism of descent.

Two other references in the thermal probe literature are worth mentioning. Lorenz (2012) performed an unusual calculation dealing with a crash landing of a Pu^{238} radiothermal heat source on Europa. The motivating concern was whether the heat source would melt its way through the ice and contaminate the Europan ocean. The calculations showed that in 100 K ice the nominally spherical 1 kW probe would not be able to penetrate the ice. An unusual thermal probe design suggested for glaciology use in Antarctica by Talalay et al. (2014) investigated the possibility of recovering a high-voltage-powered probe at the conclusion of a mission by heating the aft elements of the vehicle and reeling the vehicle back using an onboard winch. All prior thermal probe designs were expected to be one-use devices in which the probe was left in the glacier at the conclusion of a mission. A review of the Talalay device by the authors indicated that there would be problems with potential high voltage shorting and with mechanical design of the retrieval mechanism given the very small diameter, and very high aspect ratio (length/diameter). These issues were specifically addressed in the SPINDLE project and are reported in Sect. 4.6.

4.3 The Physics of Melting Ice

4.3.1 The Aamot Equation

The thermophysics of penetrating ice is relatively straightforward for a thermal point source. The ice, starting at 100 K in vacuum must be heated to the point of sublimation. The input power to achieve a fixed descent rate (e.g. in m/h) during the Starting Stage is proportional to the square of the radius of the probe, the density of the ice, the specific heat of the ice, the starting temperature, the sublimation temperature, and the latent heat of sublimation. The equation is the same for Cruise Stage except that the heat of sublimation is replaced with the latent heat of fusion.

The above situation is summarized by the first component of Aamot thermal probe equation (Aamot 1967):

$$P = P_0(R_{probe}, v) + P_{cond}(R_{probe}, L_{probe}, v) \tag{4.1}$$

The first part of Eq. (4.1) can be expanded as:

$$\mathrm{P_0} = (\pi \mathrm{R}^2_{\mathrm{probe}}) \nu\, \rho_{\mathrm{ice}}(\mathrm{c_{ice}}(\mathrm{T_m} - \mathrm{T_0}) + \mathrm{H_m}) \tag{4.2}$$

where: P_0 = the power that must be applied, in Watts; R_{probe} = the probe radius, in cm; ν = the vehicle speed, in m/s; ρ_{ice} = the density of the ice, in kg/m^3; C_{ice} = 2093 J kg^{-1} K^{-1} is the specific heat capacity of water ice; T_m = the melt (or sublimation) temperature of the ice, K; T_0 = the starting temperature of the ice, K; H_m (or H_s) = the latent heat of fusion (or sublimation, respectively) = 3.35×10^5 J kg^{-1} for H_m and 2.8×10^6 J kg^{-1} for H_s.

It was this portion of the equation that was used by (Lorenz 2012) to determine the penetration status of a spherical radiothermal heat source. It is also this portion of the equation that accurately determines the speed of descent of a cryobot under a fixed power input in temperate ice (ice that does not refreeze behind the vehicle).

It is readily apparent that downward speed under fixed power input is inversely proportional to the square of the vehicle radius (or diameter). This fact of cryo-physics is the dominant driving factor in the design of an ice penetrator and it forces one towards the smallest possible diameter cryobot because power has traditionally been considered to be a limited commodity in the field. Unfortunately, all physically realizable cryobots require internal volume for such things as guidance, navigation, and control systems; environmental and other more sophisticated *science payload* sensors; spoolers for power delivery cables; nose and wall heaters; jet pumps; heat exchangers; science sampling systems; obstacle avoidance systems; and a host of other sub-systems, not to mention the possibility of carrying a second stage vehicle (e.g. an AUV) within. All of these add length to a vehicle. The smaller the diameter selected by a designer, the longer the vehicle gets and for many devices there are fixed geometries that serve to set a limit on the degree to which the diameter can be reduced. We will discuss several of these practical limitations below.

The cryo-physics implications of a long vehicle are significant, and this is clearly illustrated by the second term in the Aamot equation, which deals with the power needed to prevent the sidewalls of the vehicle from freezing to the wall of the ice shaft before the vehicle can achieve a steady state descent rate. We are speaking now of the general case, in non-temperate ice, where water behind the vehicle will eventually freeze and close off the hole preventing a direct retrieval of the vehicle. See Humphrey and Echelmeyer (1990) for estimates of closure rates.

The power that must be directed to sidewall heating is given (Aamot 1967; Ulamec et al. 2007; Kaufmann et al. 2009) as Eq. (4.3):

$$P_{cond} = \frac{4\lambda_{ice}(T_m - T_0)}{R_{probe}\pi^2}(2\pi R_{probe}) \int_0^{L_{probe}} \int_0^{\infty} \frac{e^{-\kappa u^2 s/v}}{u\left[J_0^2(R_{probe}u) + Y_0^2(R_{probe}u)\right]} du ds \tag{4.3}$$

where: λ_{ice} = thermal conductivity of ice; L_{probe} = the melt proble length, cm; k = heat diffusion coefficient = $\lambda_{ice}\,(v\rho_{ice}\, c_{ice})^{-1}$; u = the integration argument for the Bessel functions J_0 and Y_0; s = the spatial coordinate along the length of the probe.

The direct numerical solution of the above equations is difficult. An alternative that we have taken, that is a more powerful, generalized approach that allows for complex vehicle geometry and discrete heating element modeling, is to use a finite element-based computational transient (time based) thermal conduction simulation.

Our procedure is iterative and it models conductive losses of a moving vehicle, even though there is no "motion" of the probe or ice in the finite element model: The first step is to perform a transient (time based) thermal heat conduction study of a horizontal "slice" of ice that the cryobot will pass through. The cryobot is assumed to be a flat nosed cryobot as shown in Aamot's estimates and the shell is assumed to be cylindrical. As the cryobot passes through this "slice" of ice and the flat nose removes the ice directly in front of it, and the shell encounters ice that has not been warmed yet (even though some conduction from nose melting will help here, this is neglected at this stage) we fix the surface of the ice temperature at 0 °C (melting temperature). If the initial conditions for the leading edge of the shell are for the bulk ice to be at its ambient far field temperature and the surface of the ice fixed at 0 °C, then the finite element solution yields the heat flux required to keep the ice from freezing. The time based transient analysis calculates the time based flux required to keep the surface at 0 °C. The time stamp for each step corresponds to a point along the length of the shell that must be provided this same flux to keep the walls from freezing. This can be then integrated numerically (using a spreadsheet e.g.) and it thus generates the conductive losses required to keep the probe from freezing in place, just as Aamot was attempting.

The surprising result of these analyses is that for Europan ice at 100 Kelvins the thermal power that must be directed to preventing freezing of the vehicle sidewall to the ice (and thus entrapment of the vehicle) represents a large majority (typically > 80%) of the total thermal power budget to maintain a fixed downward velocity through the ice. The longer the vehicle the more pronounced this factor will become, so while there is a penalty for increasing the vehicle diameter, so too is there a penalty for too long a vehicle.

4.3.2 Results of Simulations

As part of the VALKYRIE project, but particularly during the SPINDLE project, we conducted a series of parametric analyses to determine power requirements for a Europan cryobot operating primarily during Cruise Phase (in the ice with vapor pressure present to allow water formation and no obstacles). We examined realistic vehicles ranging from 12.5 cm in diameter up to 75 cm in diameter with a fixed aspect ratio of 15:1 (length over diameter) penetrating 100 K ice. Results from these calculations are shown in Figs. 4.20, 4.21, 4.22 and 4.23. Figure 4.20 is a plot of penetration velocity (in units of kilometers per month) as a function of the input power (in kW) for the previously mentioned vehicle diameters.

A different variation of this type of plot is shown in Fig. 4.21 in which the total energy (in Megawatt-Days) is shown as a function of the total ice thickness (in kilometers) to reach a sub-surface ocean. For anyone who has been tasked with a serious investigation of the design of a functional Ocean Worlds cryobot these numbers are like being hit with a bucket of cold water. The power levels needed to run these missions are very, very large. A cryobot large enough to carry a long

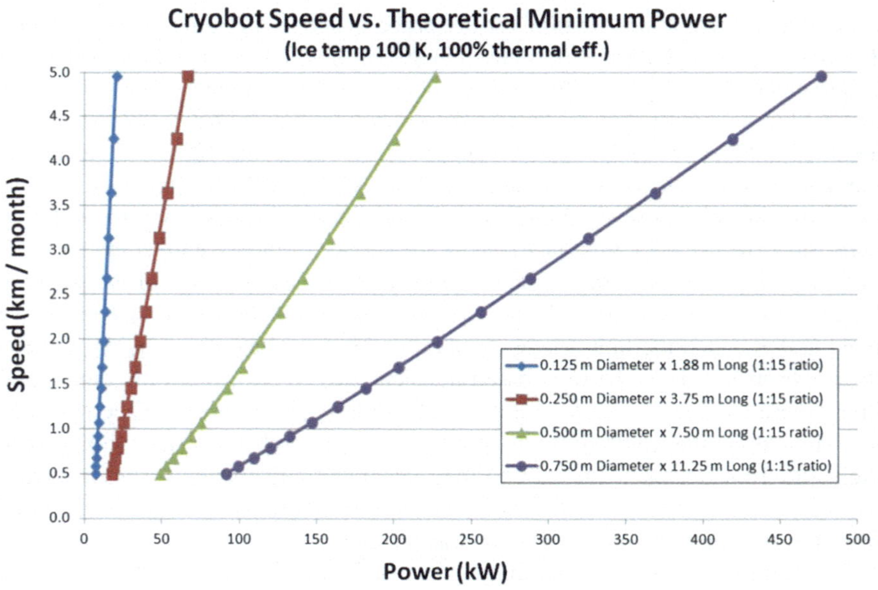

Fig. 4.20 Cryobot speed of descent (in kilometers per month) is shown as a function of input power in kilowatts for various diameter vehicles at Europan surface ice temperature

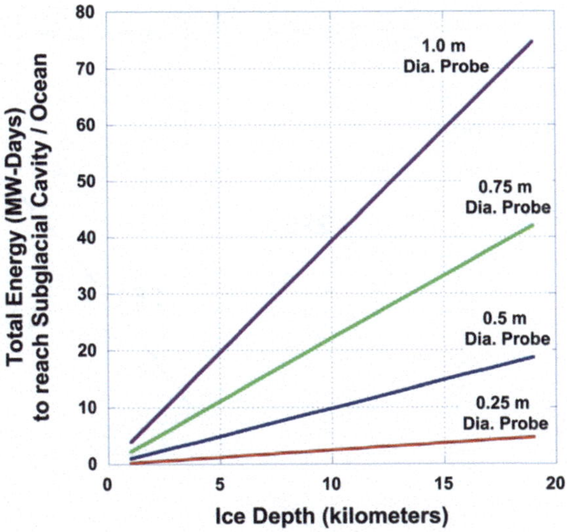

Fig. 4.21 Cryobot performance (at 100 K ice temperature) depicted in terms of total energy expended (in megawatt-days) to penetrate a specified ice thickness for various diameter vehicles

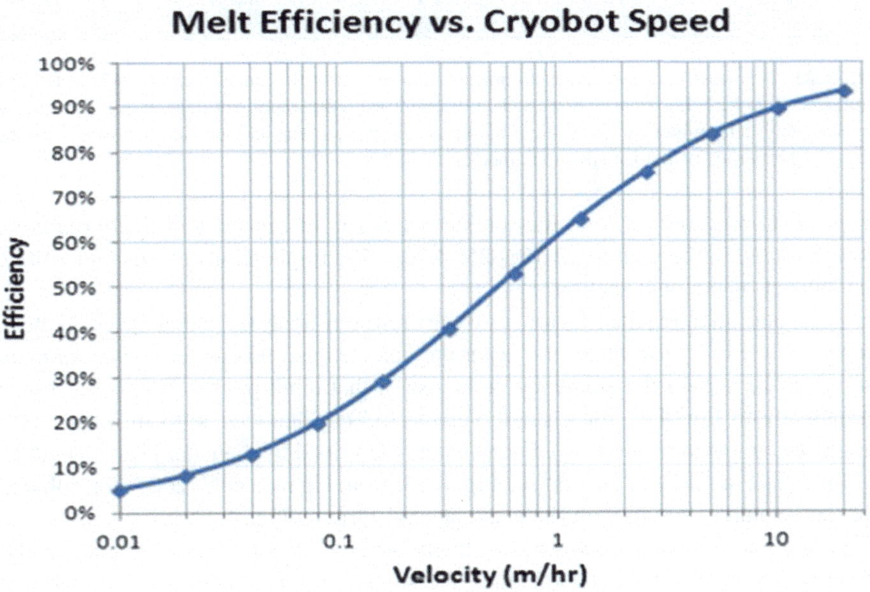

Fig. 4.22 Cryobot melting efficiency (the percentage of total energy expended that goes directly into melting) as a function of vehicle speed. Calculations assumed a 53 cm vehicle diameter with a length of 15 m and an ice temperature of 213 K. At slower speeds most of the energy goes into heating of the sidewalls

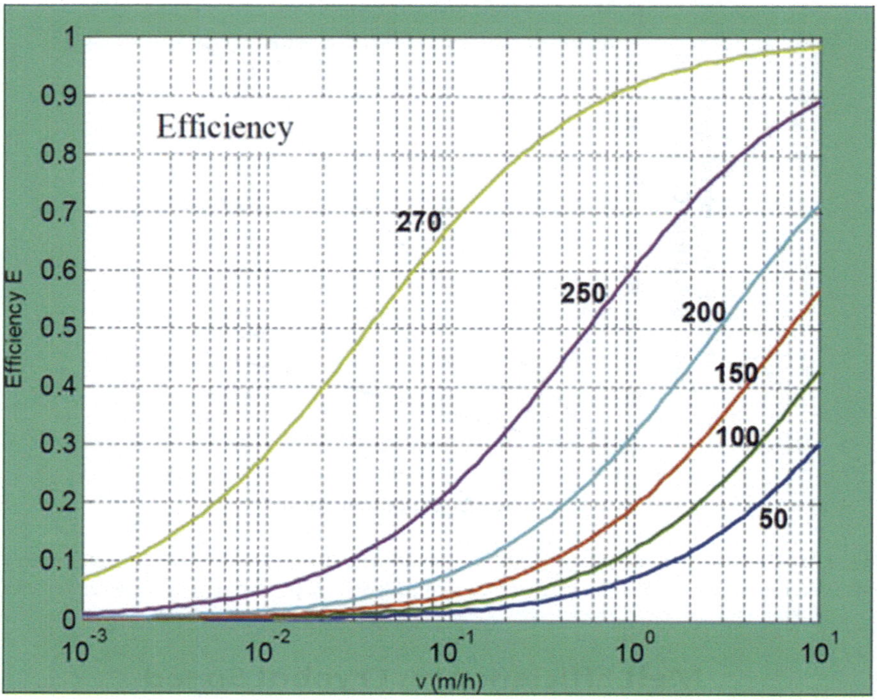

Fig. 4.23 It is possible to generate an array of curves using the same analysis techniques that produced Fig. 4.22. Each curve represents a different ice temperature (in degrees K). The curves are uniquely determined by the vehicle diameter, length, ice temperature, input power, and the physical properties of the ice (from Ulamec et al. 2007)

range RTG-powered AUV to the sub-surface ocean of Europa will likely require a power source in the vicinity of 100–250 kW_{th}. The implications of this fact will be the subject of the next section.

A second unanticipated factor resulting from our analysis is that speed matters. For a vehicle in Cruise Phase the faster the vehicle descends under a given amount of power the greater the efficiency in the use of that input power. This is illustrated graphically in Fig. 4.22 for a single vehicle (53 cm diameter x 15 m length) at a single ice temperature (−60 °C). Each vehicle (described by diameter D and length L) will generate an array of curves (see Fig. 4.23; from Ulamec et al. 2007) for different ice temperatures, but all with the same general form as Fig. 4.22.

A minimum viable downward velocity is around 1 m/hour (or, conversely, roughly a year to pass a 10 km thick ice shell). Faster velocities are possible at higher power levels, smaller vehicle diameters and shorter vehicle lengths.

4.4 Power Sources

4.4.1 Background

Although it may be possible to send a very small diameter probe deep into the Europan ice cap, this will not have a high probability of the detection of life due to the very limited 3D volume that can be assessed by a cryobot due to its constraint to 2D travel. To significantly expand the exploration volume (and improve the probability of successful search for life) a long range AUV needs to be carried as a second stage system, deployed once the ocean is reached, and preferably deployed on the ocean floor to enhance the search for hydrothermal energy sources that might support life. There are limiting physical constraints to the diameter of an AUV and related deployment mechanisms. During the SPINDLE project (Sect. 4.6) we converged on a design with a 40 cm diameter cryobot and a 100 kW_{th} power source as a viable test article for a proof-of-concept 2-stage (cryobot and AUV) deployment to Antarctica. On Europa, then, where do we get that 100 kW_{th}?

Solar-electric power is not an option for a number of reasons (Fig. 4.24). The incident solar radiation at the surface of Europa is only 50 W/m^2. If we assumed, generously at this time, a 20% photovoltaic conversion efficiency that would imply a surface solar array 100 m on a side—the deployment of which would not be feasible for a self-contained lander. Furthermore, the minimum launch mass of this system would be on the order of 1000 kg, not counting power conditioning; high voltage conversion and transmission systems; tether; and an onboard vehicle spooler to pay out the umbilical to the surface-based power source. A further factor

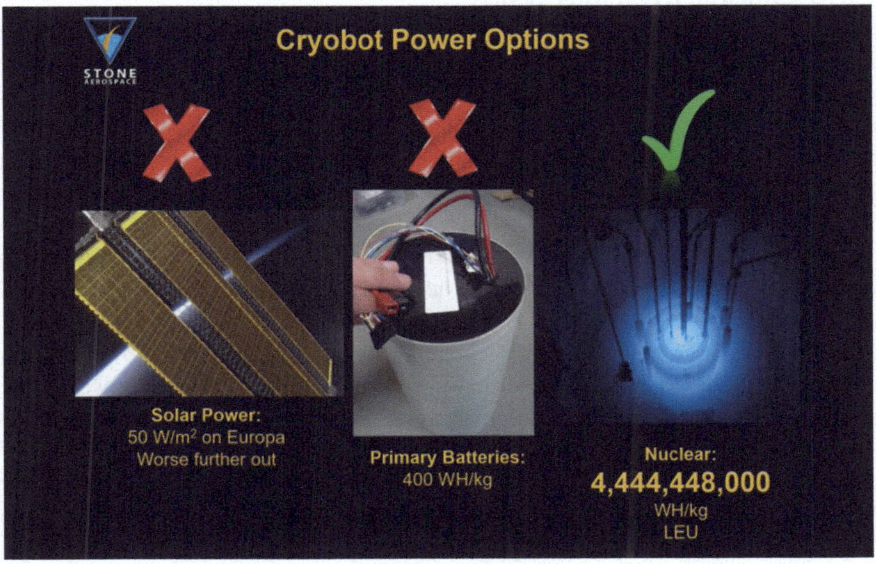

Fig. 4.24 A comparison of three possible power sources for cryobot missions to the Ocean Worlds

against solar power is that the radiation environment of Europa consists largely of highly energetic protons and electrons that would rapidly destroy the polymeric-based solar array substrate.

The present lightweight Europa lander tentatively scheduled for 2024 launch is planned as a battery-operated system. The present best power density available using batteries is about 400 WH/kg. The present impossibility of using batteries for a cryobot mission is readily demonstrated by simple mathematics: a half-meter diameter vehicle penetrating a 10 km ice cap at 100 K ice temperature will require 10 MW-days of power. Using the above-listed power density for batteries (the best Li-ion technology available today) this amounts to a battery mass alone of 600 metric tonnes, which is five times the predicted launch capacity of the SLS booster, just for the batteries.

In contrast, nuclear material (e.g. the fissioning of uranium) yields 4.4×10^9 WH/kg of energy storage. A small diameter (less than 25 cm) cryobot could easily be powered by a large radiothermal power source (RPS) and a cryobot capable of delivering an AUV to the Europan ocean could be powered by an extremely small low enriched uranium (LEU) fission thermal source. There is thus no escaping the use of nuclear power for a sub-surface Ocean World mission. Over the past two years our lab has done preliminary design of two power systems for candidate cryobot missions: a 25 kW_{th} Sr^{90} radiothermal generator (RTG) and an ultra-compact 100 kW_{th} fission reactor that takes advantage of ice-reflected neutrons to reduce mass. Both designs have back-power conversion to generate electricity to run the onboard guidance, navigation and control, hot water drilling, and other systems.

4.4.1.1 Nuclear RPS Power Source

The general shortage of Pu^{238} in the United States RPS inventory is presently severely restricting planning for any robotic planetary missions beyond the Mars 2020 rover. Taking this into account during the SPINDLE Phase A1 study we began investigations of alternative RPS power sources that may be of utility to a small-scale cryobot for initial ice cap penetration and ocean characterization. We focused on a generic new design that could be scaled up to 25 kW_{th}.

The proposed device is a deep space power source for use in icy moon missions such as Europa. It is designed to provide heat and electrical power to systems within the SPINDLE probe dimensions. This particular design uses Strontium-90 (Sr90) as a heat source, uses thermoelectric generators to produce electrical power, and is designed to produce 25 kW of thermal power and 1 kW of electric power at the beginning of the mission. Thermal power from this device is used primarily to melt the ice of Europa in order to allow the SPINDLE vehicle to traverse the icy crust of the planet and penetrate into the ocean below. The electrical subsystems provide power to pumps, computers, and other components that may require electricity to function. The device (Fig. 4.25) has a diameter of 16 cm, a length of 76 cm, and a mass of 280 kg (with a shadow shield, see below). Comparatively, the Kilopower space reactor (Gibson et al. 2015) has a mass of roughly 1500 kg.

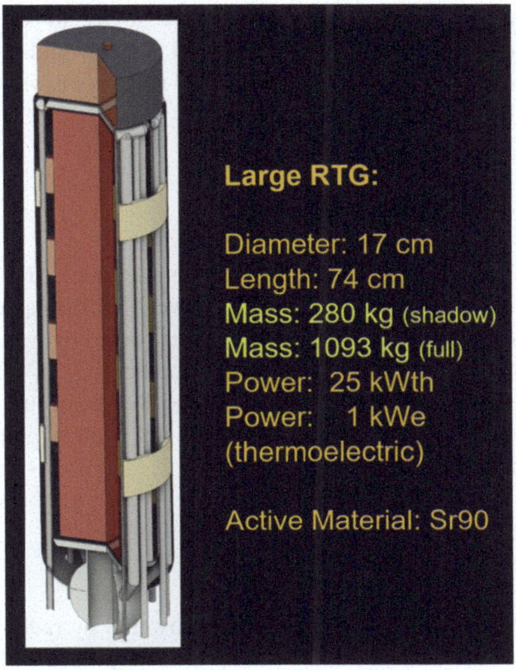

Fig. 4.25 A 3D computer model of a large radiothermal generator based on the use of plentiful Sr90 from reactor waste. A 25 kW_{th} design was achieved with a total power system mass of 280 kg if a shadow type shield can be used (this is possible with proper spacecraft design). Because RPS (radiothermal power systems) are always on there would be a need for thermal radiators on the spacecraft to deal with the continuous 25 kW_{th} output. However, this represents a potential alternative path to bypass the problems of limited supply of Pu^{238}

4.4.1.2 Sr^{90} Fuel

Strontium-90 is a radioisotope with a specific power of ~500 W_{th} per kilogram in the form of strontium titanate ($SrTiO_3$). This form is used to increase the operating temperature as well as reduce the water solubility of the isotope. It has a half-life of 28.8 years, and decays through alpha decay to yttrium-90. Yttrium-90 then decays through beta emission, which often creates gamma rays through bremsstrahlung radiation. Due to this decay chain, Sr^{90} often requires more shielding than Pu^{238}, the other favored radioisotope for space exploration. However, the cost of Pu^{238} limits its usefulness as a large scale power producer. The SPINDLE Sr^{90}-based RPS device uses a *cermet* core to encapsulate the radioisotope fuel. Cermet materials take their name from the concatenation of the terms ceramic and metal, which in this case are $SrTiO_3$ and tungsten, respectively. Cermet materials have a number of benefits, including high operating temperature, very high strength, and customizable geometry.

The fuel form also provides a certain level of internal shielding. However, due to the amount of gamma rays produced by Sr^{90}, extra shielding will be required. The tungsten shield must have a thickness of at least 7 cm to reduce the radiation field to acceptable levels, but a "shadow shield" can be used to protect only the sensitive components of the SPINDLE craft. This is achieved by placing the shield on just the end of the power source in between the unit and any electronic systems, thus creating a "shadow" in the incident radiation. Alternately, a full, 7 cm thick shield can be placed around the entire device to keep radiation from escaping in any direction. In this case, the total mass would increase to 1093 kg and is likely a mission de-enabler. The shadow-shield-only version appears attractive for a feasible lander mission.

4.4.1.3 Power Conversion

The SPINDLE RPS device uses thermoelectric generators (TEGs) to produce the necessary 1 kW of electric power. The hot side of the TEGs are attached to the cermet fuel, while the cold side is cooled by water produced from melting the ice of the environment. The required efficiency of 4% is an achievable goal for TEG systems, as current radioisotope thermoelectric generators operate at 5.5%. The fluid flow is provided by hot water drilling pumps used by SPINDLE to descend through the ice of the moon.

4.4.1.4 Potential Issues

The SPINDLE RPS device is powered by radioisotope material, and thus it has no on/off switch. When in transit, it may need to be actively cooled to ensure overheating does not damage the unit. Previous systems have used a cover gas or active cooling during loading and transit as necessary.

Having a shadow shield will require special care in transit as well, as any components aboard the spacecraft will be exposed to high levels of radiation. To combat this, the device can be surrounded with shielding during transit and removed from its housing after landing, or placed away from sensitive components. It may be possible to keep the device on an extended boom for the duration of the trip, thus reducing radiation levels.

A thermal system of this size may require special effort to reduce temperature peaking within the fuel. Although the cermet can withstand very high temperatures, it may be beneficial to introduce internal flow channels to keep the center of the core from becoming too hot. This can be incorporated into the manufacturing of the cermet with little difficulty, and can be determined by thermal modeling.

Cermet fuel forms have been investigated as a potential fuel form for nuclear thermal rockets, but are not currently used in missions that have been cleared for flight. This will require a period of testing and approval to demonstrate that the fuel is safe. There is currently no alternative fuel to the NASA General Purpose Heat Source (GPHS) which specifically uses plutonium fuel in a carbon aeroshell. The GPHS unit is incompatible with large scale radioisotope power, and any new fuel form, or geometry will also require qualification.

4.4.2 *Nuclear Fission Power Source:* Europa Cryo Reactor

Space based fission reactors have two major design requirements which significantly increase the mass of the system: shielding and waste heat rejection. The former presents issues due to very high power levels creating a large amount of neutron and gamma radiation, which can cause damage to onboard systems. Thus, large masses of shielding or large separation distances are required. Waste heat rejection requirements are an issue due to the large power levels involved, requiring large heat rejection systems. In situations without adequate environmental methods for cooling—such as space flight and/or planets with no atmosphere—this results in very large, very heavy radiators.

In order to make fission power viable, we developed a design for a new reactor —The *Europa Cryo Reactor*—with a greatly reduced mass which operates at lower temperatures and utilizes lighter materials. This is achieved by the following novel design elements:

- The surrounding ice on Europa serves as an effective neutron reflector, reducing the critical nuclear fuel mass to operate the reactor, thus reducing overall reactor mass.
- We changed the moderator material from the more commonly used graphite to lithium hydride (LiH). LiH is a very good neutron moderator, and allows for the core mass to be greatly reduced. However, the operational temperature of LiH is also much lower than that of graphite, and so the operational temperature of the reactor must also be reduced.
- We use the environment for reactor cooling. Contrary to space-faring reactors that must dissipate waste heat through massive radiators, the surrounding 100 K ice makes for a fantastic heat sink.
- High efficiency electrical power generation cycles (using heat engines) can be used thanks to the presence of gravity and a suitable waste heat rejection system (to the same 100 K ice sink). The proposed system would use a standard Rankine cycle for electrical power conversion.

The end result is a very lightweight power system capable of delivering 100 kW of thermal power and 5 kW of electric power, with a total mass of under 500 kg, including a shadow shield. Importantly, the reactor can be maintained in an un-fired state during launch, transit trajectory, and landing. The baseline design for the Europa Cryo Reactor (ECR) is shown in Fig. 4.26.

The ECR is designed to use low enriched uranium (LEU) which is much safer to use and easier to handle than highly enriched uranium (HEU). Because of this, the regulations on the fuel are also greatly reduced. Private entities can own LEU, as seen by its use in commercial power reactors, and the shipping, storage, and handling can be taken care of much more easily. In addition, use of LEU means that an expensive (of order $50 M) Category 1 security facility is not required at the launch site.

Fig. 4.26 A 3D computer model of a low enriched uranium (LEU) fission reactor that takes advantage of ice reflection neutronics to sustain a critical reaction in a 100 kW_{th} power reactor. The total mass, including shielding is 515 kg, substantially lower than any previously designed power reactor for space missions. The mass includes a 5 kW Rankine power cycle electric generator which also takes advantage of the ice as the cold side for the power cycle. The reactor can be launched cold and activated upon reaching the surface and thus requires no thermal control radiators

4.4.3 Integration

Because the reactor can be turned on and off, there does not need to be any significantly difficult handling requirements for launch vehicle integration. The reactor subsystems can be installed at any point in time during the spacecraft construction without posing a danger to the facility or personnel. In contrast, radioisotope sources (RPS, as described in Sect. 4.2) continuously emit some levels of radiation, and are always producing thermal power. This makes integration of an RPS into the launch system complex, as it may need to be installed at the end of the construction process.

4.4.4 Radiation Effects

Reactor systems can be constructed, installed, and shipped without producing any power or radiation field. Therefore, the only danger of radiation occurs when the system is deployed and turned on. Although once operational, the system will generate a considerable radiation field, there are ways of mitigating these effects.

The radiation field intensity changes proportionally to the inverse of the square of the distance from the source which is why in many depictions of nuclear thermal rockets for human spaceflight the crew pod is located on a long boom as far forward of the reactor as possible. Large separation distances for a cryobot are not possible so shielding of some form is required to protect onboard electronics, machinery, and instruments.

The least massive approach to radiation shielding to use a selective "shadow shield", which blocks the field only in the solid angle in which the payload is exposed. The shield—which can be as simple as a few centimeter thick disk of tungsten—effectively casts a "shadow" on the payload by blocking radiation. This results in a greatly reduced mass compared to a shield that encompasses the entire core. Since the reactor will be dormant during integration, launch, cruise, and landing, and since there will be no humans present when the reactor fires up, the use of a shadow shield seems to strike the best engineering compromise.

It will be possible to dynamically adjust the output power of the reactor (from 0 to 100 kW) thus providing options for safe deployment and recovery, e.g., of a second stage AUV. Residual radiation levels for the ECR are expected to be under 2% of full power within 6 min of shut down, and dropping (Khan and Ahmad 1997).

4.4.5 Power System/Heat Dissipation

Space power generation is somewhat more complex than terrestrial generation because there is often a lack of environmental contribution in the zero-gravity vacuum of space. Waste heat rejection, for example, is a major problem because during space flight all waste heat must be rejected through thermal radiation. This mechanism works best at high temperatures, whereas the power cycle depends on the heat rejection side to be at low temperatures. Additionally, having no gravity often results in difficulties with flow, especially two phase flow, because the working fluids in a dynamic cycle do not separate based on density without the aid of a gravitational field.

Many space-based power systems use thermoelectric conversion due to the lack of moving parts and long lifespan of these generators. Yet the conversion efficiency is often quite low, with many systems reaching only 6–8%. Due to this, large thermal power levels must be achieved to produce relatively modest electrical power, and the waste heat associated with thermoelectric systems is staggering.

Yet the prospect of an icy moon destination allows for major heat rejection capabilities, because the ice around the power plant can be melted and used as a massive heat sink. An icy moon mission will also include a reactor with the capability for large scale thermal power generation, so thermoelectric conversion remains a possibility because the high power levels and good heat rejection capabilities are present.

However, the aspects involved with an icy moon mission also lend themselves well to a more traditional Rankine cycle. In the past, Rankine cycles have been downplayed for space exploration due to the large amounts of waste heat and difficulties with two phase flow in space. However, now that the waste heat rejection system is potentially very effective, and icy moon destinations will provide a gravitational field for flow separation, this option becomes more favorable. Considerations for starting/stopping and long term operation will still require the cycle to be modified for long term use, but it has the potential to greatly reduce the system mass and provide very high levels of electrical power.

Specific aspects to address in regards to the power system include: minimizing working fluid leakage, minimizing damage to the turbine to ensure long life, ensuring the waste heat rejection water/ice bath does not evaporate, dealing with cold temperatures and temperature changes during transit, and starting the cycle once arrived at the destination. Although many of these aspects are not usually major hazards for Rankine cycles, the technology is mature enough that many of these concerns have already been addressed, and the design of a very long lived, autonomous power cycle is a viable possibility.

Preliminary calculations by Howe Industries predict the shielding necessary for a 100 kW-th LiH moderated LEU core at a 3 m standoff distance and a 30 cm^2 shadow area would be on the order of 288 kg. The core, including fuel, moderator, and reflector, would total 67 kg. The power cycle required to produce ~5 kW-e would have a mass of roughly 100 kg. This places the total mass for the reactor and power system at ~455 kg.

4.4.6 End of Life and Estimated Cost

Once the mission is complete, the reactor may shut down and greatly reduce the power levels. The reactor will continue to produce heat at a small fraction, roughly 2% initially after shutdown, of the original power due to decay products, but the time for these levels to reach non-threatening levels is on the order of weeks. Once a suitable amount of time has passed, the reactor can be deposited in the ice or left on the surface without danger of contaminating the environment.

The cost of LEU is incredibly low compared to HEU or radioisotopes. The cost for conversion can be calculated using current prices for Separative Work Units (SWU): each SWU is currently $60–160 (World Nuclear Association 2016). This price was verified at the (U.S. Energy Information Administration 2016) as well. To enrich to 20% requires 45 SWU/kg, and so the price per kg becomes $7200/kg maximum. For the ~33.42 kg of LEU required for the LiH core, the price is ~$240,000. This is consistent with existing TRIGA reactors, which cost ~$7M/MW, or $280,000 for 40 kW-th (including housing, buildings, etc.). Development is expected to cost on the order of $2M in order to design a functioning core out of the new materials. Additionally, licensing and approvals are expected to add between $1–5M. Assuming an average of $3M for such costs, the total cost to get the first unit built and approved will be near $7M. Reliability is a major factor and will require thorough testing and updated designs.

4.4.7 Timeline

Metallic uranium fuel is currently used by the US Navy, lithium hydride is readily accessible, and the control system and reactor design are extensions of proven, related designs. Still, the core will need to be designed from the ground up, and testing of the ECR to verify calculations will need to be done at a reactor facility. It is expected that the ECR reactor final design and testing effort will take roughly two years. Following that, construction of the reactor will take another 2–3, for a total of 4–5 years. Simultaneously, the design and testing of an ultra-robust Rankine cycle power conversion system can be undertaken in parallel. Because the goal of this effort is to increase longevity and not efficiency, no new scientific or engineering breakthroughs are necessary. We expect this effort to take no more than four years, and can be performed at the same time as the reactor effort. Administrative approval times can vary based on the interest of the governing party, but it is likely that tests will need to be done on a completed prototype unit to achieve the approval to launch. This can also run in parallel with the design effort. However, approval and testing will also likely require analysis of the final unit. This will likely add on a year to the end of the project. The total time required for development, construction, and approval of the lightweight ECR reactor power system could be on the order of 5–6 years. Thus, while many government space agencies have predicted Ocean World cryobot missions not being possible until the 2050–2060 time frame, analysis of the critical power systems indicate that this time line could be shortened by 35 years were there a political will to undertake the mission.

4.5 Project VALKYRIE [Very-Deep Autonomous Laser-Powered Kilowatt-Class Yo-Yoing Robotic Ice Explorer]

4.5.1 Origin of Concept

During the field work with the DEPTHX autonomous underwater vehicle (Stone et al. 2014) at Cenote Zacaton (the world's deepest known hydrothermal spring) in northern Mexico in the spring of 2007 we maintained a 1 Gbps data link to the vehicle by means of a thin (2 mm diameter) armored fiber optic thread, which was paid out from the surface as the vehicle descended. The fiber was mainly there to allow Mission Control personnel to observe vehicle behaviors and intervene if the pre-established mission goals and behavior-based code was not performing as expected. The data were transmitted using an Ethernet-to-fiber converter, which had at its core a laser diode, which sent photons through a 25 μm fused silica fiber (the remainder of the 2 mm diameter data thread was padding and armor for the silica core). While thinking about this one day one of us (Stone) posed the question: "how many photons at once could you send down such a fiber (i.e. what power could be optically transmitted in that fashion)?".

Co-author Hogan spent a portion of the summer of 2007 investigating this question and the answer surprised all of us: the theoretical limit of power that could be transmitted through a 25 micron fused silica fiber was controlled by the self-focusing limit (Kelley 1965), which occurred at a staggering power level of 4.6 MW (megawatts). Our first effort was to recheck our numbers but they came out the same. There were, of course, many details to be investigated, but the thought of sending power on the order of megawatts to a remote vehicle via a miniature thread —much, much smaller than a high voltage conductor—suddenly opened options for a practical means for testing Ocean World cryobots on the Antarctic plateau, as it would represent an ultra-clean power source.

From 2010-2016 NASA funded the development of the VALKYRIE cryobot (Stone et al. 2014; Stone 2016), which used a high energy laser as the primary power source coupled to the vehicle by a fiber optic tether (the original concept is shown in Fig. 4.27). In this approach, 5 kW of laser light is transmitted from a surface-based laser (Fig. 4.28) and injected into an optical waveguide (Fig. 4.29) that is spooled out from the descending cryobot (Fig. 4.30). Light exits the end of the fiber, travels through diverging optics, and strikes a beam dump at the head of the vehicle (Fig. 4.31), which channels thermal power to hot water jets that melt the descent hole. Effectively, VALKYRIE is a self-contained hot water drill wherein the energy used for heating is provided optically. The vehicle eventually included an aft Science Payload extension (Fig. 4.32) that included autonomous life detection and sampling sub-systems.

In June 2014, the vehicle penetrated 30.5 m into Alaska's Matanuska Glacier, proving the concept of an optically powered cryobot (Fig. 4.33). The descent was stopped only due to time limitations. The vehicle, measuring 25.4 cm in diameter by 2.5 m length, achieved average descent rates of 1 m/h when driven by a 5 kW laser, matching the theoretically predicted descent rate in temperate ice.

VALKYRIE used an integrated protein fluorescence spectrometer to detect microorganisms and dust deposits during vehicle descent. Positive detection activated on-board samplers (top section of vehicle shown in Figs. 4.32 and 4.33). This was the first practical demonstration of an autonomous decision-to-collect life detection system used on a cryobot. The VALKYRIE work proved the ability to send significant amounts of optical power through a very thin silica glass fiber. Significant amounts of power can be transferred 40 km or more, either within ice or under sea, using VALKYRIE's laser-over-fiber technology. The physics principles utilized in the design make the vehicle scalable and enables operation of the vehicle to extraordinary distances. For further in-depth discussion of the novel elements of

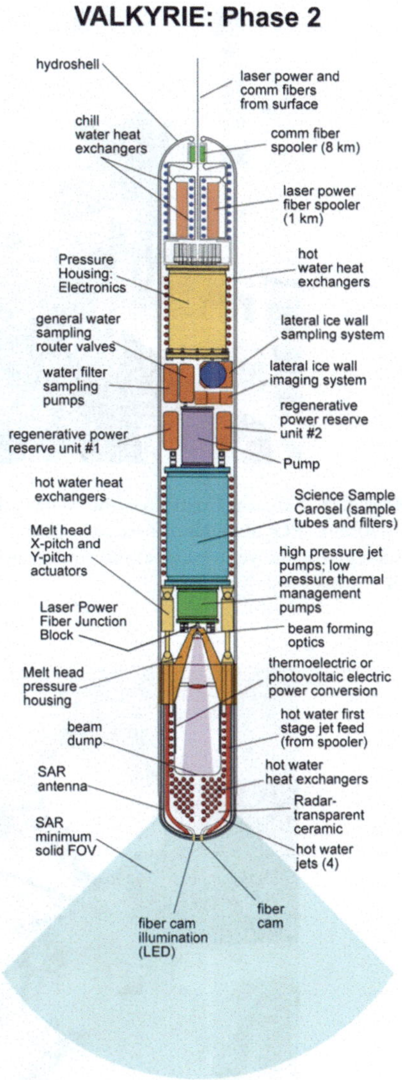

Fig. 4.27 An early depiction of the VALKYRIE vehicle. The novel elements included the remote laser power source; an on-vehicle bare fiber spooler; and a forward beam dump where photons were converted to heat and electricity. The vehicle also included one forward and four lateral hot water drill jets. The drive pumps for the jets were onboard. An onboard computer system controlled the vehicle and provided guidance and navigation. It includes conformally-mapped ice penetrating end-fire radar surrounding the beam dump

VALKYRIE technology the reader is referred to the following U.S. patents (Stone and Hogan 2015, Stone 2016).

The initial phase of project VALKYRIE had the principal objective of conducting a laboratory testing program to assess fiber optic power transmission loss rates and limiting state characteristics with respect to high energy optical fiber power transmission over long ranges. Experimental variables include laser wavelength (1070 and 1555 nm); bare fiber diameter; and fiber bend radius. Secondary variables included variations in glass doping agents and fiber exterior coatings. The objective of these experiments was to prove or disprove the key VALKYRIE tenet

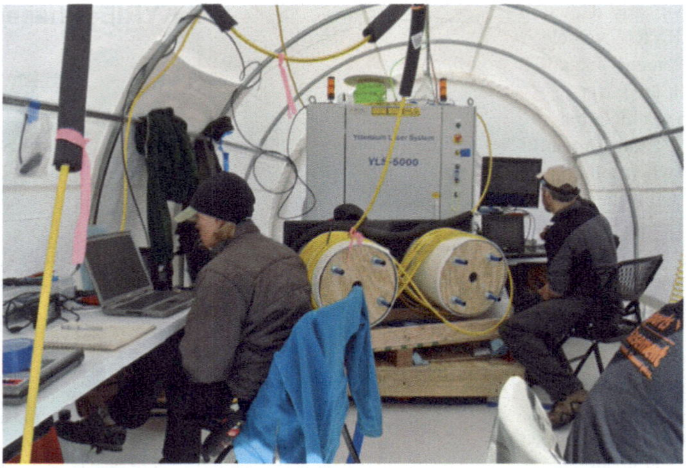

Fig. 4.28 Mission Control on the Matanuska Glacier, Alaska in June of 2014 shows the 5 kW 1070 nm IPG fiber laser used to power VALKYRIE. We chose to use an armored optical waveguide (the yellow cable) for sending power to the vehicle as a risk reduction measure for the initial field tests

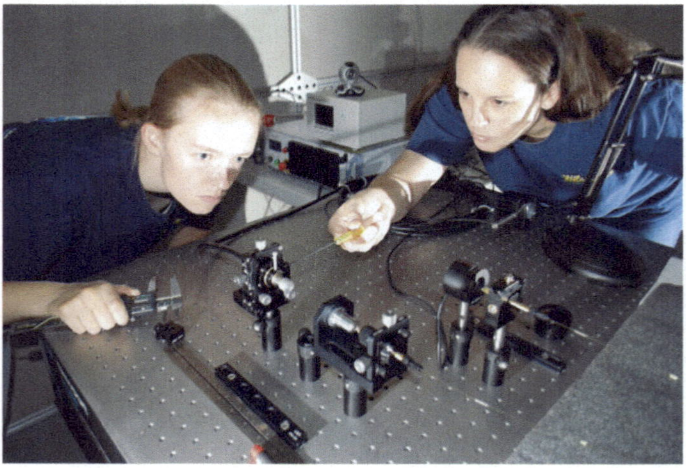

Fig. 4.29 Initial experiments at Stone Aerospace with empirical determination of bare fiber bending power loss as a function of on-vehicle fiber spooler radius involved injecting the beam into the test fiber using an XY stage for aligning the beam with the fiber. Output from the fiber spooler was sent to the hockey-puck-size optical power meter at center right. In this fashion we were able to test a large number of fibers and spools without having to resort to fusion splicing of the fibers

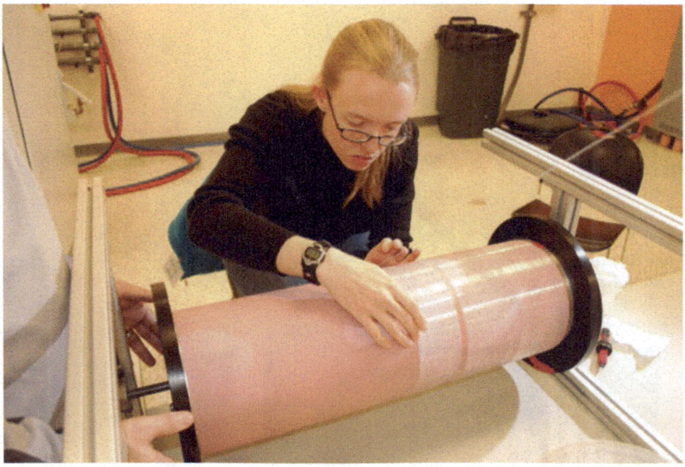

Fig. 4.30 Bare fiber transmission tests on a vehicle-compatible 200 mm diameter spooler were conducted in January of 2012 using a 5 kW laser. The tests showed that losses were low enough that the spooler could be run in air with drum temperatures not exceeding 30 °C

Fig. 4.31 The 2014 field test vehicle in Mission Control on the Matanuska glacier in Alaska. One of the four lateral cutting jets can be seen on the black anodized ring immediately behind the nose cone. X-shaped slots on the nose allowed for the vehicle pumps to draw melt water up through the black heat exchanger to the hot water jet pumps

Fig. 4.32 VALKYRIE Stage II in May of 2015 prior to deployment to the Matanuska. The vehicle has doubled in length with the addition of an autonomous life detection and sampling science payload. The vehicle diameter remained at 25.4 cm and thus downward velocity at the full 5 kW power level saw similar descent velocities to the shorter 2014 vehicle since in the temperate ice of the glacier the hole did not refreeze behind the melt head

Fig. 4.33 Team VALKYRIE on the Matanuska glacier in July of 2015 with the final test vehicle with all access panels attached

of being able to power a cryobot remotely via fiber-optically transmitted energy over a mission range sufficient to reach any Antarctic sub-glacial lake.

4.5.2 High Power Laser Transmission Test Over Long-Distance Fiber Optics

It is important for the reader to understand that in 2010 when the VALKYRIE project was initially funded by NASA that the prevailing opinion was that sending a high power laser beam down a long fiber that was wrapped on a spool carried by a vehicle was folly. We were told to expect that the instant the laser was turned on there would be a "fiber fuse" initiated at the first sharp bend in the fiber where all the energy would focus and thus melt the fiber backwards towards the laser, eventually destroying the laser. Confirming or dispelling this concern was our first objective.

High power fiber transmission tests were conducted at IPG Photonics facilities in Oxford, Massachusetts under experiment design and direction of Stone Aerospace (SAS) engineers during the first week of July, 2010 (Fig. 4.34). The test light guide was a custom 1.05 km-long × 400 μm core diameter silica fiber that was spooled onto a 1 meter-diameter polyethylene tray. The spooling diameter was designed to be above the theoretical threshold for significant power loss due to fiber bending effects (which were to be investigated separately in a series of low power tests). The 1-m-diameter fiber coil was water cooled during testing.

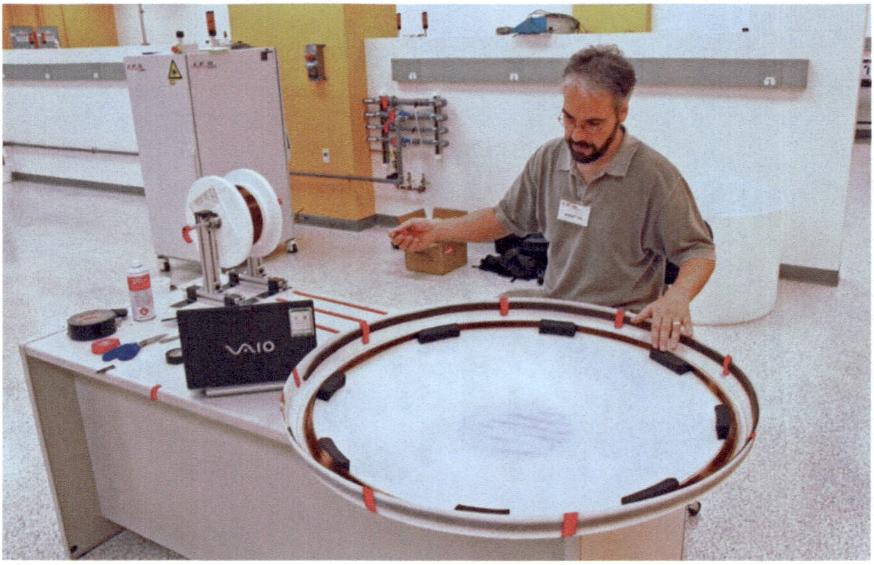

Fig. 4.34 The first high power-over-fiber spool test was conducted in the summer of 2010. 1050 m of bare 300 μ transmission fiber was wrapped in a 1 m diameter coil which was immersed in water prior to injecting 11 kW of laser light into the fiber

The custom-made power transmission fiber was specified by SAS and manufactured by Polymicro. The fiber consisted of a multi-mode step index pure silica core with fluorine doped cladding (Fig. 4.35). It had the following layer structure: 400 μm core diameter, 440 μm OD cladding, and 480 μm OD polyimide coating (for prevention of water intrusion). The composite fiber had a numerical aperture (NA) of 0.22. IPG provided a 20 kW multi-mode fiber laser, 1070–1080 nm, Model # YLS-20000. The laser was calibrated using a collimator and a Primes calorimeter to measure CW (continuous) power. It was calibrated from 40 W to 20 kW.

Figure 4.36 shows the experiment architecture. Connection of modular high power laser elements requires the use of special connectors in which the fiber is terminated into a fused quartz block which is contained within a water cooled housing (Fig. 4.37).

Two connectors can be joined within a beam coupler (Fig. 4.38), which contains internal optics that expand and collimate the beam, as well as align it with the adjacent connector optics. The coupler housing is also water cooled and contains several sensor interlocks to provide automated shut down of the laser in the event of over-heating of the coupler. The interlocks are required since the power levels are sufficient to initiate a fiber "fuse" in the event of a thermal runaway in a connector or a spot in the fiber that becomes lossy and overheats. In such a case all of the laser energy (11 kW peak in the case of the tests conducted by SAS) is dumped into the

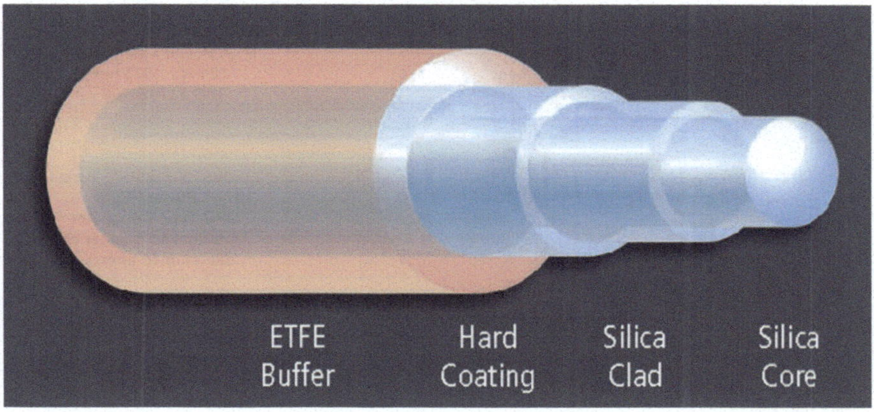

Fig. 4.35 Cross section of the multi-clad fused silica fiber that was used in the first high power transmission test. The core was 300 μm diameter. The OD was approximately 480 μm

failure point, which then fuses the silica fiber. This fuse then propagates rapidly back toward the laser.

We used a fusion fiber splicer to join the test fiber coil to the lab connectors. Ultimately it was determined that the use of a 300 μm core diameter connector fiber (smaller than the high power SAS transmission fiber) led to better coupling of power between the laser and the test coil. The details of the splice, and the methods to reduce the power that is lost in the splice, remains a black art that requires much additional lab research. The power limit reached on the July 2010 tests was defined by the stray light that made its way into the multi-layer cladding, where heat would be absorbed and ultimately lead to fuse formation. We ameliorated this by stripping the cladding in the vicinity of the splice and immersing the splice on a heat sink plate covered with thermal compound that had very low infrared (IR) light absorption characteristics (see Fig. 4.39). This allowed most of the stray light in the vicinity of the splice to escape without heating the splice. Nonetheless, this crude approach eventually led to a burn through at the splice at a power level of 11 kW after several minutes. As this chapter is being written tests are planned for the SPINDLE project to qualify a similar fiber and improved splice system up to 65 kW CW power at 1070 nm wavelength.

Importantly, the formal testing, planned for cycling from 0 to 10 kW through the SAS test fiber, was completed successfully. Once those were completed we attempted to "push" the system to see where it would break. As previously mentioned there were significant issues with the splice overheating both due to cladding losses (poorly aligned switch/coupler, poorly matched fiber) and some unavoidable splice loss. One of the interesting results is that the polyimide coating on the custom SAS fiber had a visibly dark pigment which seemed to absorb more IR radiation than the IPG fiber, which had a visibly clear acrylate coating. This may be one clue for further investigation for reduction of power loss.

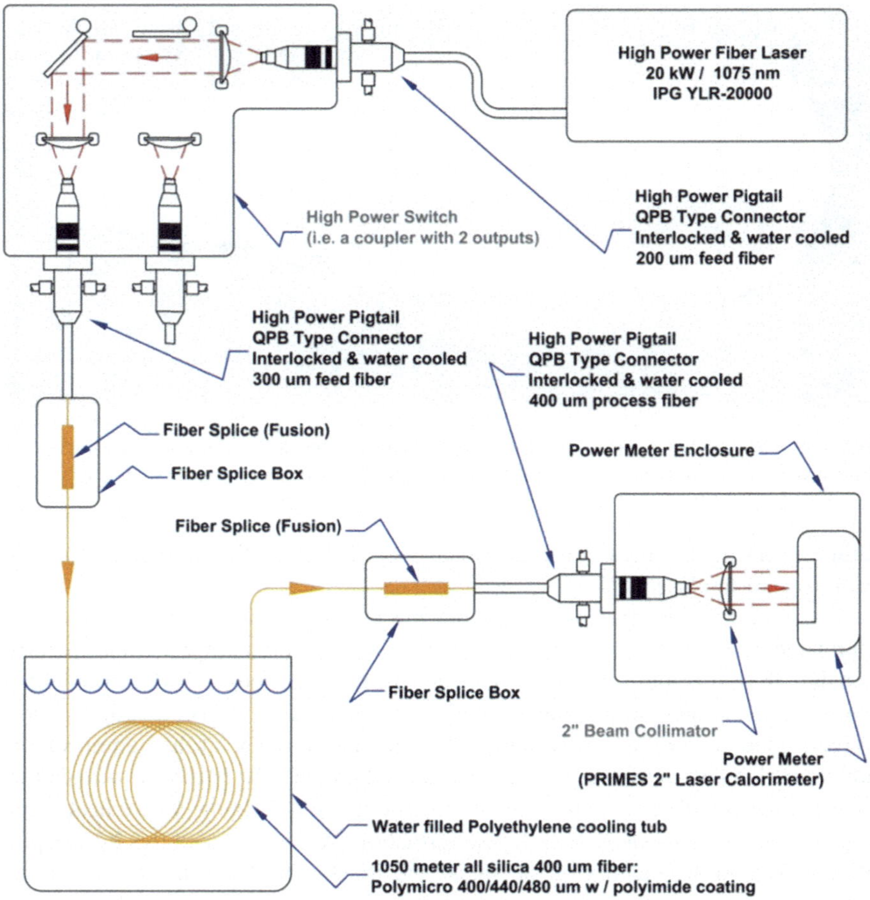

Fig. 4.36 Schematic of the first high power fiber transmission test. Power was lost at each connector and at each fusion splice. Cumulatively around 1% of transmitted power was lost in this fashion

The loss in the fiber was calculated at various laser output power points from 40 W to 10 kW. The fiber transmitted 86.6% (0.625 db/km) of the injected power, at the 10 kW laser setting, through the 1050 m-long fiber coil. The 10 kW laser setting produced 9.955 kW from the laser, but this was reduced to an estimated 9.589 kW after the switch/coupler and including splice losses (see Fig. 4.38). The switch/coupler loss was calibrated with a 400 μm standard high power delivery cable, but we needed to switch to a 300 μm pigtail for our test, as the 400 μm pigtail coupled too much power to the cladding layer. We had no good way to calibrate the loss with the smaller 300 μm pigtail we were using, or to calibrate the actual splice loss. If we had additional time, we could have severed our fiber at each end, re-spliced (with low loss because it is the same fiber) and re-calibrated. There was not time for this, so the loss was estimated based on various pieces of

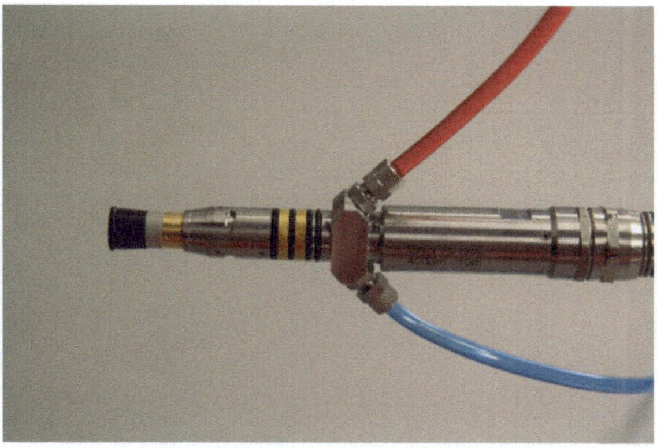

Fig. 4.37 Close-up of a commercial QBH high power fiber laser connector. The black cap at far left covers a fused silica cylinder with optically planar ends. The fiber is fused to the right side of the block. Red and blue hoses carry coolant water

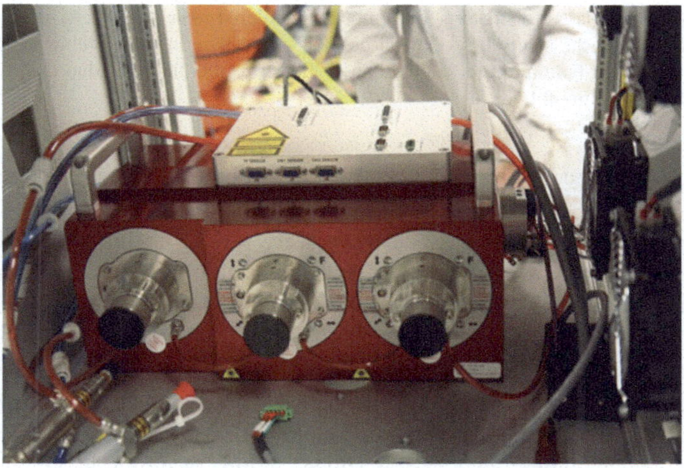

Fig. 4.38 Close-up of a fiber laser junction switch. This block can divert the input laser beam to one of two outputs. Internal workings include beam expanders followed by beam collimators and switching mirrors. Electronics surrounding the block are connected to stray-light sensors located inside the block which detect if any of the internal optics are failing, which would lead to catastrophic destruction of the block in a short period of time. This block handles up to 20 kW of power

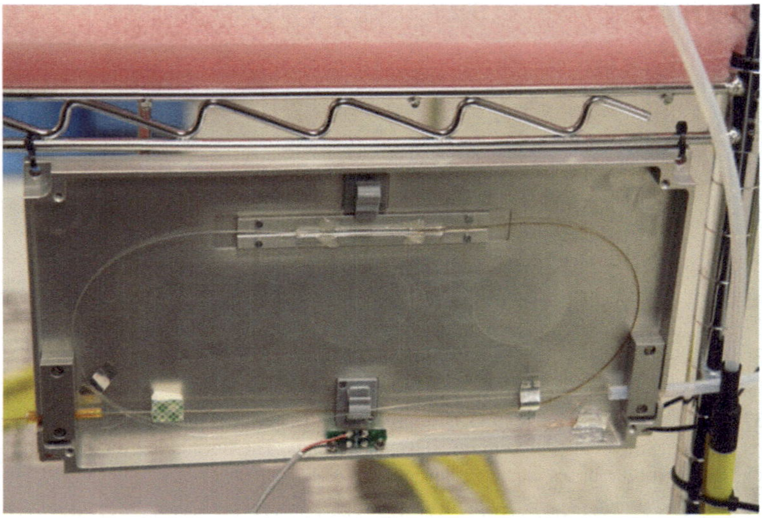

Fig. 4.39 Close-up of a high power fiber fusion splice. The laser output fiber (clear fiber on the right) is fused to the larger diameter spooler fiber (tan colored fiber on the left) at the aluminum channel at the top of the cavity box. Optical gel holds the splice in place within the box. Green printed circuit card at bottom center contains stray-light detectors that will shut down the laser if the splice fails, initiating a fiber fuse

information combined with comments by IPG personnel. This composite estimate was 1.5% loss; this can be seen in the tabulated results (plotted in Fig. 4.40). The overall loss was very consistent, changing very little with injected power level, although at very low power levels the signal-to-noise ratio created more uncertainty (the calorimeter functions better at high power). There was a slight trend showing a small increase in transmission as the power increased. This may not just be a result of a better measurement due to a greater signal to noise ratio at higher powers. It is possible that the modal distribution of the laser output changed with increasing power which resulted in a change to the coupling of light to the cladding occurring inside the switch/coupler and at the spliced joints resulting in an increase in transmission efficiency. Either way, this is a very small effect of only +0.7 and −0.5% from the average loss, if the first two very low power points are removed (38 & 95 W). The loss for the fiber was 13.4% per kilometer (0.625 db/km), which compares well to the theoretical limit of 12.0% per kilometer (0.557 db/km) based on data for OFS fiber and transmission for 1070 nm light through pure silica. One area of concern during the testing was the possibility of encountering Stimulated Raman Scattering (SRS), which is a non-linear effect that can limit power transmission. Calculations were made to anticipate the onset level, which was

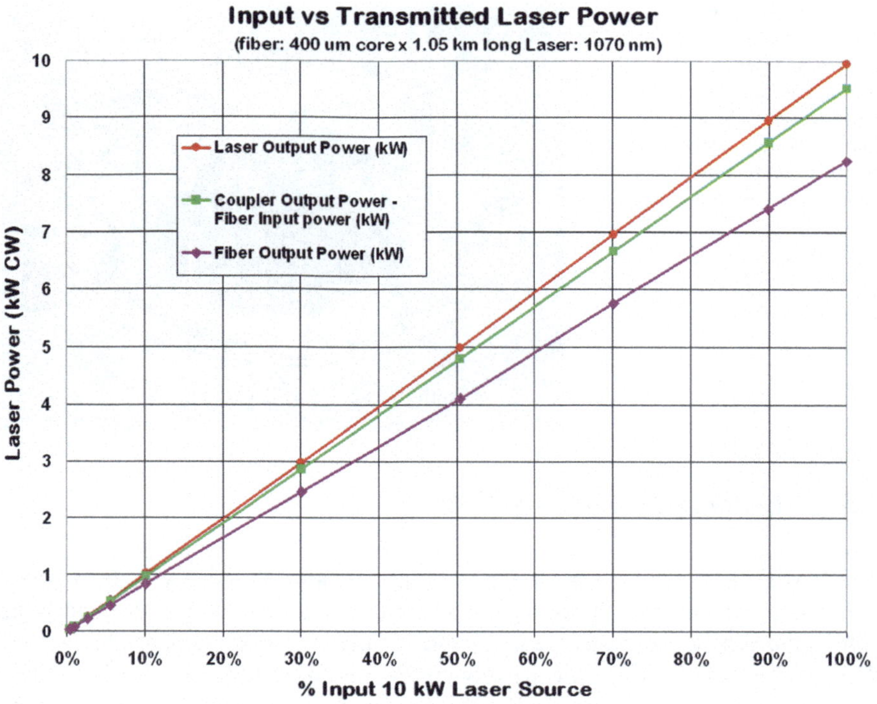

Fig. 4.40 Empirical data from the high power transmission test showing power levels measured after various connections and ultimately after the fiber spool (lowest purple line) indicating the relative losses that occur at couplers and fusion splices

conservatively predicted to at occur at 20 kW. No SRS was seen during the testing through 11 kW input power, which would have been apparent from a roll off in power transmission efficiency.

The results of the testing were successful; we were able to transmit 10 kW CW (1070 nm) laser power through 1.05 km of 400 μm core all silica 0.22 NA step index optical fiber, with a loss of 13.4% at 1 km equivalent length (0.625 db/km). Losses compared well to the theoretical limit for transmission of 1070 nm light through pure silica (12.0%/km, 0.557 db/km). Heating of the fiber coil never exceeded 30 °C at maximum transmitted power (Fig. 4.41). We also gained insight into equipment and techniques required for fusion splicing in high power applications.

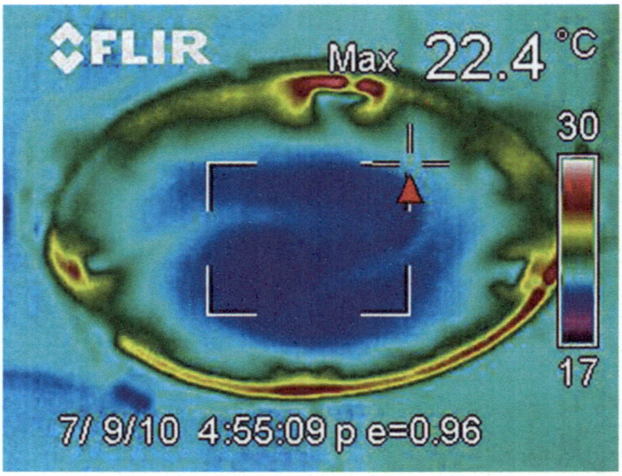

Fig. 4.41 FLIR infrared image of the fiber spool during the high power transmission test at 11 kW input power, indicating maximum fiber temperature of 30 °C, well within continuous operating range

4.5.3 Bending Power Loss Tests

The VALKYRIE vehicle was designed so that it could self-deploy its own power tether (deployment from the surface of an ice cap is not possible as the melt wake behind the vehicle will re-freeze in a matter of minutes or hours depending on the thermal state of the ice). An actively-cooled on-board fiber spooler is used to achieve this. However, there is a limit to how small (in diameter) the spool can be made due to the decreasing ability to maintain total internal reflection of the laser light within the fiber core with decreasing spooler radius.

We developed an optical bench-based experiment to test fiber optic transmission under various conditions, with a particular focus on bend losses associated with wrapping transmission fiber onto a deployment spooler. Bend loss has both linear and non-linear components, both of which were investigated. We employed an IPG YLR-40 (40 W CW) Ytterbium fiber laser, fiber optic injection, extraction and measurement equipment as well as an optical table and ancillary equipment. While significant research has been conducted by industry into bending loss, particularly for single mode communications fiber, limited research has been conducted into bend loss for kilowatt-class multimode lasers. Available research and modeling predicted losses in the range of 10–30%, for various fiber types and applicable spooler size range. Various parameters contribute to bending losses. Here are some key facts:

- Bend loss depends on bend radius, increasing rapidly (non-linearly) as the bend radius is decreased.
- Bend loss depends on fiber radius, increasing rapidly (non-linearly) as the fiber radius increases, for a given bending radius.
- Fiber losses decease (non-linearly) as the fiber numerical aperture (NA) increases.

- Bending losses are highly dependent on launch/source conditions, particularly the source NA, and source modal distribution.
- Under-filling the fiber will significantly reduce the bend losses. Under-filling occurs when the launch condition of the source have a lower NA than the NA of the fiber itself.
- Higher-order propagating modes have higher losses, so the mode distribution of the source will effect the bending losses.

We are able to control all these parameters to some extent; the most difficult is the modal distribution of the source. This is done by using a mode filter or a mode scrambler. The key testing variables were the fiber radius, bend radius and fiber NA.

An optical fiber is typically constructed with a fiber core, fiber cladding, fiber coating and frequently a fiber jacket. The fibers we tested did not have a jacket as it would occupy an unacceptable amount of volume, as the jacket is often more volume than the fiber itself. The following is a list of fibers that we tested (9 different fibers, 3 each type):

(1) Ceramoptec: NA 0.22, 0.28 & 0.37, 400 μm/440 μm/480 μm (core/cladding/coating): multimode, step index, pure silica core, fluoride doped silica cladding, polyimide coating.
(2) Ceramoptec: NA 0.22, 0.28 & 0.37, 300 μm/330 μm/360 μm (core/cladding/coating): multimode, step index, pure silica core, fluoride doped silica cladding, polyimide coating.
(3) Ceramoptec: NA 0.22, 0.28 & 0.37, 200 μm/220 μm/240 μm (core/cladding/coating): multimode, step index, pure silica core, fluoride doped silica cladding, polyimide coating.

We have completed the testing of a number of fibers and the results at high power have been consistent with the available low power (communications) data on the subject, although there is no published bend loss for our particular set of parameters. Additionally, we have a single data point for bending loss from the fiber used for the high power testing, which is consistent with and corroborates our in-house testing. This single data point came from Polymicro (the manufacturer), and was taken from the 400/440/480 μm fiber × 1050 m long while wrapped on an 105 mm diameter shipping spool. This measurement will have both macro bending (from the mandrel) and micro bending from the relatively low quality spooling of the fiber, causing the fiber to overlay itself and create micro-bends. The loss is somewhat higher, which may be indicative of the degree of micro-bending effect (which would not be included in our bend loss testing because the fiber was constrained to one layer on the test spools). Precision winding and sparse winding will eliminate micro-bending in field practice.

Figure 4.42 shows bend loss vs. bend radius for different fibers. The fibers have a pure silica core and fluoride-doped silica cladding with polyimide coating. The only difference in the fibers is the numerical aperture, one with NA of 0.22 (common) and one with NA of 0.28 (less common). This corresponds to an acceptance angle of 25.4° and 32.5° respectively (for full angle). The experimental results were obtained from 50 m long fibers fully wrapped around testing mandrels of different sizes and the loss was extrapolated out to 1 km. The fibers were pre-tested at approximately a 1 m diameter which was assumed to generate very low (undetectable) bending loss and then wrapped around various test mandrels to determine the bend loss for each diameter tested. As noted previously, the launch conditions can have dramatic effect on the bending loss, and "steady state" bending loss is only achieved after beam equilibrium conditions have occurred. "Launch" in the sense of fiber optics means to inject the laser beam from the laser into the bare fiber. This can be done using discrete couplers as previously mentioned or by using a precision XY optical stage to align the laser output fiber with a planar-sheared transmission fiber with a very small air gap. There is some loss associated with this method but it proved faster than fusion splicing for the purposes of establishing minimum bend radii. The bend loss was tested at various lengths during the spooling process to determine if we had reached steady state loss or if it was still in transition. This can be seen more clearly for certain sets of parameters and less clearly for others. There was an attempt to determine the steady state loss, as this is what can be extrapolated out to multi-kilometer lengths. Obviously, from Fig. 4.42, the higher NA fibers exhibit a dramatically lower bend loss. This same lower loss is also seen for micro bending, which often occurs when the fiber is wrapped over very small imperfections.

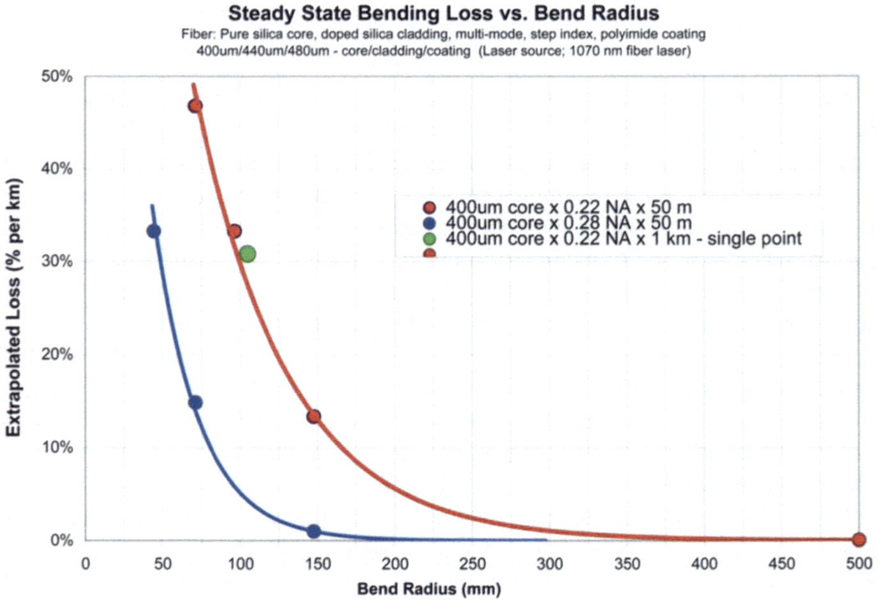

Fig. 4.42 Optical power loss per kilometer as a function of bend radius for 400 micron fibers with varying numerical apertures

On the basis of these experiments we conservatively selected the VALKYRIE spooler radius to be 200 mm and used higher NA fibers. Depending on the fiber core diameter, NA, and laser wavelength it may be possible to considerably reduce the threshold radius, allowing for the possibility of small diameter, low power probes designed for only short penetration distances (see, specifically the discussion in Sect. 4.7 on this subject).

For larger diameter two-stage cryobots capable of delivering an AUV there will be minimal issues with bending losses. This is a particularly useful finding for the design of laser-powered Antarctic penetrator systems. The discoveries presented in Sect. 4.7 are scalable and present a novel alternative sub-glacial access method.

4.5.4 Fiber Strength

One of the more significant challenges for the VALKYRIE project was to insure the fiber did not fail. This can occur due to thermal issues (overheating during power transmission) or mechanical issues (breaking). The ultimate tensile strength of fused silica fiber is theoretically around 14 GPa, but due to flaws in the fiber, the actual failure strength (stress) is typically below 5 GPa. Polymicro fiber (used for our test at IPG) is proof tested to 700 MPa. The fiber tensile strength varies significantly and has a distribution due to unknown crack formation during fabrication. Typically, it has a 5% chance of breaking at 4.3 GPa and 90% failure at 4.7 GPa (OFS 2016). This is for new, unstressed fiber (unbent & untwisted) at room temperature. Fiber loses strength over time (often referred to as "static fatigue," "stress corrosion," "aging," etc.), flaws enlarge and creep/propagate due to penetration of water or water vapor and weakening of the silica bonds occur. A hermetically sealed fiber will resist fatigue and water intrusion. A common approach is to use a very thin layer of carbon which is very effective, but this is only available through limited manufacturers and it needs to be explored as to the absorption characteristics at 1070 nm, as a coating with high absorption can cause increased probability of fiber failure under bending.

Optical power running through a fiber can also reduce strength significantly, depending on fiber type and conditions. OFS, a leading fiber manufacturing company, found that the polymer coatings could initiate a failure mode that could reduce strength dramatically under tight bending. It appears this mode of failure is initiated by the coating absorbing the light lost from bending and causing a focal point or hot spot on the cladding, which in turn caused mechanical failure.

In a cryobot spooler the fiber will see combined stress from torsion, tension and bending as well as potential strength reduction due to high power/high flux operation. In the process of wrapping the bare fiber on the cryobot spooler a pre-twist must be applied so that the unspooled fiber will have no hockling and will be

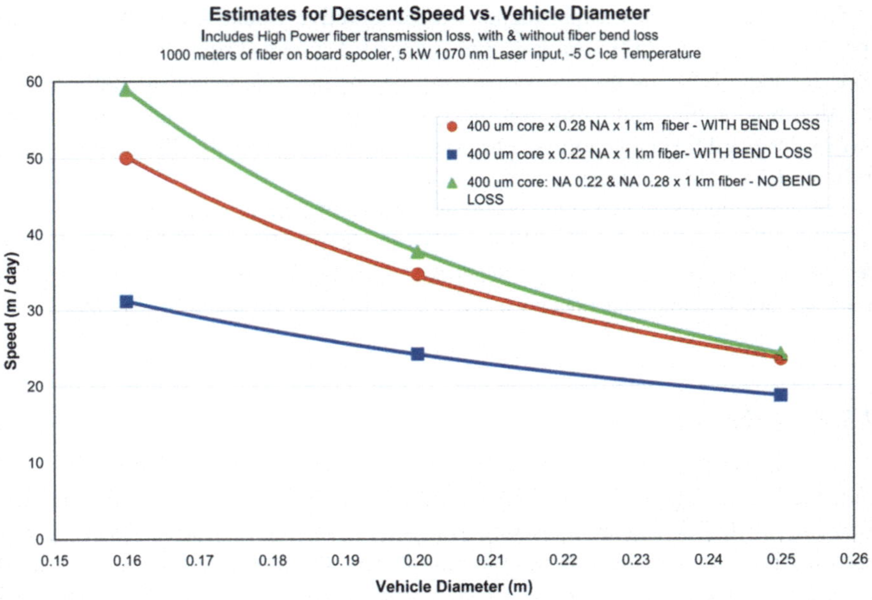

Fig. 4.43 Various cryobot performance curves relating to the type of optical fiber used. In general, the smaller the fiber numerical aperture (NA) the more power that is delivered to melting (and thus enhanced melt rate). This figure shows the dependence of vehicle speed to vehicle diameter for different fibers. VALKYRIE performance is predicted at the extreme right of the figure (25.4 cm diameter). Actual field tests in temperate ice showed average descent speeds at 5 kW power of approximately 1 m per hour using an NA of 0.28

torsion free. The preliminary test results shown in Fig. 4.42, however, indicate that deep Antarctic missions are feasible without any special changes to the design. More efficient spoolers, however, will lead to the possibility of faster, smaller, and lighter weight cryobots. Further research into the area of specialty fibers and custom wavelengths designed to handle tight bend radii and high power levels will be beneficial to advancing photon-powered cryobots.

4.5.5 Descent Rate Estimates with Fiber Bend Loss for VALKYRIE

Figure 4.43 shows estimates for vehicle descent rates based on our current experimental data and analysis. This plot includes fiber losses due to bending and high power transmission loss based on the previously described lab tests and is based on a 5 kW 1070 nm laser source. The plot compares two 400 μm core, all silica, step index, multimode fibers with different fiber NA values, one with 0.28 and one with 0.22. Note that the bending losses dramatically reduce the descent rate for small diameter vehicles; bending losses only modestly affect the descent speed for larger

Table 4.1 Wavelength, cost, and power for various lasers

Fiber laser type	λ (nm)	Laser size kW	$ per watt	Total cost of Laser ($k)	Laser efficiency (%)	Theoretical loss db (km)	% Transmitted (km)	Power delivered @ 5 km (kW)
Ytterbium Diode pumped -SM	1070	10	50	$500	~30	0.557	0.88	5.3
Ytterbium Diode pumped-MM	1070	10	50	$500	~30	0.557	0.88	5.3
Raman fiber laser	1310	7	175	$1225	~20	0.244	0.945	5.3
Raman fiber laser	1480	6	200	$1200	~15	0.153	0.965	5
Erbium Diode pumped (MM)	1550	3	200	$600	~15	0.144	0.967	2.5

diameter vehicles—this is of substantial importance for an actual Europa cryobot or for an Antarctic laser-powered penetrator. Bending loss can be substantially ameliorated by developing a fiber with a significantly higher numerical aperture. While bending losses increase rapidly as the diameter is reduced, the descent rates continue to increase (for all cases) due to a greater reduction in required power necessary for the smaller frontal area/melt head diameter. Fibers with smaller core diameters and larger fiber NA values are currently being investigated at Stone Aerospace.

4.5.6 *Wavelength, Cost, Power, and Performance Issues for VALKYRIE*

Table 4.1 and Fig. 4.44 summarize, respectively, the current theoretical and actual available laser source hardware and their associated cost metrics and theoretical power attenuation as a function of wavelength. From Fig. 4.44 it is clear that a 1550 nm wavelength laser has distinct advantages from the viewpoint of optical power transmission. However, such high power lasers do not yet exist. Using the existing concepts for Erbium 1550 nm diode-pumped, CW multi-mode fiber lasers, it is possible at this time (but has never been done) to build a one-off, custom 3 kW device.

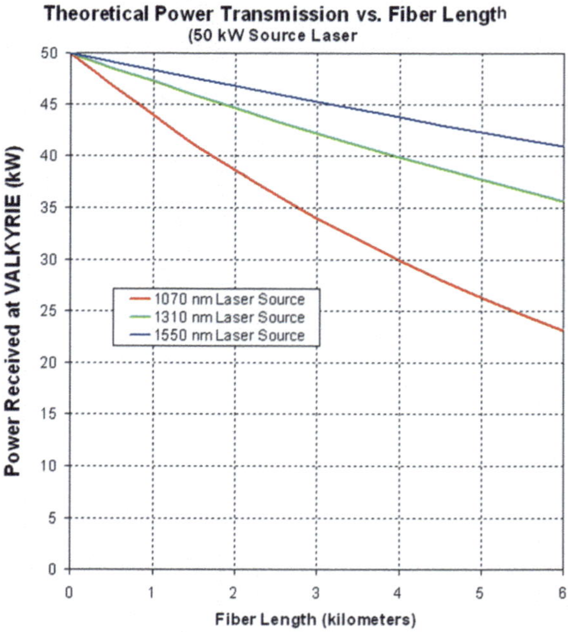

Fig. 4.44 Theoretical power transmission as a function of fiber length (in kilometers) for various wavelength lasers. While 1070 nm wavelength is readily available today in high power fiber lasers, the more optimal 1550 nm wavelength is well behind in terms of both maximum available power and in cost per watt of optical power

The highest power 1550 nm fiber laser that has been built and operated to date is 300 W. The cost per watt is nearly 4X that for the Ytterbium 1070 nm diode pumped fiber laser. IPG has built up to 100 kW CW multimode and 20 kW single mode with up to 100,000 h diode lifetimes and a delivered cost of under $25/W in 2017. Raman lasers fall in between these two metrics with likely current maximum power levels (again, for a custom, one-off device using current technology) of 7 kW for a 1310 nm system and 6 kW for a 1480 nm system. Currently, only the 1070 nm systems have sufficient power to enable a full depth (4000 m) Antarctic ice cap mission, but it would be prudent from the standpoint of operating expense (particularly considering diesel-fueled generator operating costs) to continue investigating the feasibility of 1550 nm lasers. Currently the best efficiency (from wall plug to photons) is running slightly over 50% for 1070 nm fiber lasers.

4.5.7 Lab Testing of a Full-Scale Fiber Spooler

In Sect. 4.5.3 we discussed low power tests conducted at 40 W power to determine bend loss effects in fiber. A design was subsequently developed for a 220 mm diameter bare fiber spooler for the VALKYRIE field test vehicle. High power transmission and bend loss testing using a full scale spooler was conducted under the direction of Stone Aerospace at IPG Photonics high power laser laboratory in January of 2012. This included a full scale spooler mock-up to determine realistic fiber transmission loss rates and a comparison to predicted results was performed, as well as assessing the thermal conditions/performance of the spooler. High power fiber splices were also tested in conjunction with the spooler with excellent success. The experiment setup was identical to that shown in Fig. 4.36 with the exception that the full-scale spooler mockup was substituted for the original 1-meter-diameter coil.

We wound 250 m of fiber on the spooler—a length sufficient to enable all field campaign work envisioned in Phase 2 of the VALKYRIE project. The fiber was fabricated by OFS and had a 365 um pure silica core, with a 400 um doped silica cladding (NA = 0.22) and HCS/ETFE outer coatings (see Fig. 4.35). The spooler (Fig. 4.30) had a diameter of 0.22 m and a length of 0.5 m with an aluminum mandrel and Delrin flanges. A layer of Bergquist thermal foam was used to line the spooler surface to enhance heat transfer and reduce fiber micro-bending. For this test series the bare fiber was hand wound onto the spooler drum to ensure that there was only one fiber layer and that successive wraps did not overlie prior wraps. Micro-bending due to such fiber overlap has been reported to lead to premature failure of telcom fibers and in a high power transmission scenario fiber over-crossing is to be explicitly avoided. The winding method was "drum-to-drum" so that no fiber twisting occurred during the spooler winding process. However, for vehicle spool-out a pre-twist to each fiber wrap is required so that it will deploy axially off one end of the spooler without initiating hockling.

During our previous high power testing, the fiber splice joints ran quite hot (enough to terminate the test through fiber fusing). We improved the splice technique and only moderate temperature rise (to about 34 °C steady state) was noted during this set of tests (see Fig. 4.45).

The beam enters the fiber on the spooler (Fig. 4.46) with a numerical aperture (NA) of ~0.10 (11.5° full angle of divergence), and leaves the fiber with an NA of ~0.16 (18.4° full angle).

This divergence is principally a result of mode conversion to higher order modes due to the bending of the fiber around the spooler. In the actual cryobot implementation the beam must be extracted from the fiber and rapidly diverged to allow for the shortest length beam dump. Additionally, the beam profile must also be shaped to optimize the electrical power generation system (Sect. 4.5.10). We began the spooler test with a compact collimator, having a diameter of 25 mm and focal length of 60 mm. However, the above described beam divergence caused a portion of the beam to impinge on support elements inside the collimator. It thus heated rapidly beyond 60 °C and the test was aborted. We then changed to a larger collimator 50 mm diameter × 100 mm focal length) with internal water cooling. We were able to measure the output from the fiber for both power and beam profile. The power was measured with a PRIMES Power Monitor, and showed very low loss, much lower bend loss than expected. It appears there is a critical length to fully develop the beam and from various measurements it appears this may occur at a fiber length of 500 m or so with the current fiber design we are using, at which point the bend loss will start to slowly increase, but it will still be well controlled for our chosen spool diameter of 0.22 m.

Approximately 1% of the power loss comes from the present fiber coupler. Approximately 3.6% of the loss comes from the basic fiber loss for 250 m, then about ~0.1% loss for the collimator and ~0.1% loss for the splice, which results in an estimated average value of ~0.4% bend loss. As noted, the bend loss will not really be seen until the beam has fully developed, which is estimated to occur at

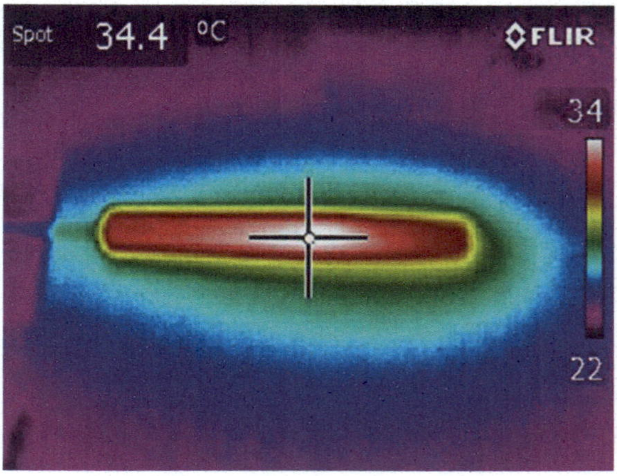

Fig. 4.45 A FLIR thermal image of one of the two power fiber fusion splices during the January 2012 full-scale fiber spooler test for VALKYRIE indicating acceptable heat buildup at 5 kW continuous operating power

Fig. 4.46 Infrared imager photo of the VALKYRIE full scale fiber spooler test at 5 kW power level, in air. This graphically illustrates how power is radiated through the fiber walls (through Raman scattering) to a greater degree in the spooled fiber than in the feed fiber (leading to the upper left of the photo). Below a critical spooler radius, which is a function of both the laser wavelength and the fiber characteristics, energy loss through the fiber walls will build up exponentially. The use of appropriately designed heat exchangers surrounding the spooler will allow most of this energy loss to be captured and used strategically where it is needed, at least during the initial stages of descent. Once the fiber is paid out there is still attenuation loss that will be radiated uselessly into the ice, but that will be at a lower rate per meter than when on the spool

~500 m. Even so, this is still a manageable situation for much longer distances using a 1070 nm source laser.

It should be noted that during the January 2012 spooler tests the initial testing was done with the spool submerged in water (Fig. 4.46). The spooler temperature (and water bath temperature) never rose much above room temperature. We then conducted the same tests with the spooler in air. The steady state result is shown in Fig. 4.47 which indicated a maximum temperature of 36 °C in the fiber—an encouraging result that provides a margin of safety during initial vehicle power up for field campaigns.

We had originally been expecting to have to create a melt water shaft in which to submerge the vehicle before powering up the laser.

On the basis of these tests it was decided to design the VALKYRIE vehicle around a 5 kW fiber laser source (1070 nm) and set the vehicle diameter at 25 cm. The length for the core test vehicle was 1.5 m for the 2014 field tests and it grew (with the addition of the science payload) to 2.5 m in 2015.

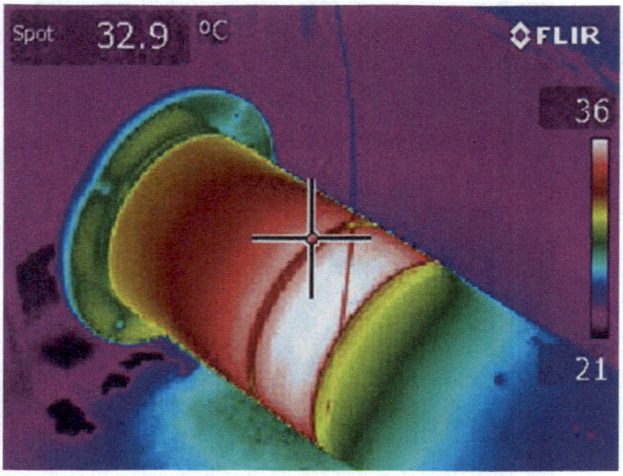

Fig. 4.47 FLIR thermal image of the VALKYRIE full scale spooler test showing fiber temperature is within acceptable limits (at 36 °C maximum) with the spooler operating in air at 5 kW input power

4.5.8 Beam Dump and Heat Exchanger

The core of the VALKYRIE vehicle consisted of a beam dump, where optical power from the laser was converted to heat, and a means for converting some of that power to electricity. Figures 4.27 and 4.48 show our initial concepts for a laser-powered cryobot. As we progressed down the path towards actually building a working prototype it became clear that more effective means than thermoelectric conversion would be needed to generate the roughly 160 W of onboard electrical power needed to run the computers and pumps. Section 4.5.9 discusses the solution that was implemented. Figure 4.49 shows a cutaway of the as-built vehicle illustrating how some of the theoretical designs of Fig. 4.48 were actually implemented. An early decision in the design process involved the separation of the testing of two critical beam dump elements: the heat exchanger (and its associated beam dispersion optics and hot water jet systems that extracted heat from the heat exchanger) and the power generation system (consisting of a tuned-wavelength heat-tolerant photovoltaic array that would see incident laser flux on the order of 10 W/cm^2. The latter topic is discussed in Sect. 4.5.9.

The beam dump heat exchanger (Fig. 4.50) is arguably the most complicated single component in the vehicle. It has to perform four complex tasks all in one very compacted volume: to efficiently collect heat from the laser beam; to transfer that heat to the water being sent to the ice melting jets at the nose of the vehicle; to handle switching piping lines that allow for the vehicle to select which jets will be in use at any given moment (for direct descent or for turning); to pass electrical sensing lines between the nose of the vehicle and the onboard computer; and to serve as the interface barrel for the synthetic aperture radar (SAR) obstacle avoidance antennae (Sect. 4.5.10). The interior (Fig. 4.50 (left)) contains scores of forward-facing beveled circular grooves that serve to absorb radiant energy and prevent it from being reflected back towards the beam dispersion optics. A non-off-gassing "optical black" ceramic coating was anodized onto the metal.

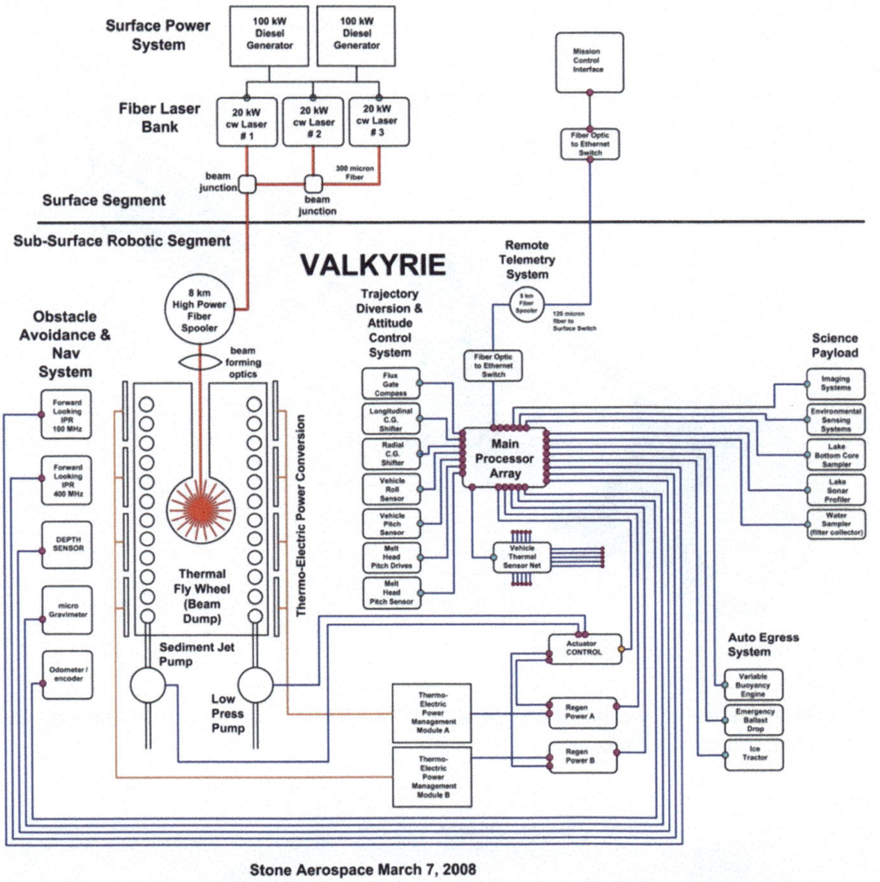

Fig. 4.48 An early block diagram of the VALKYRIE vehicle internal systems. Note that in this instance separate fibers are used for power and communications. It is possible to multiplex optical data onto the power fiber. Depending on the mission depth it would be possible to use redundant power spoolers with multiplexed data to increase survivability. The use of ultra-thin walled "K tubes" for protection of fiber while still being able to spool out the tether from the vehicle allows the fibers to withstand hydrostatic and shear forces generated by the ice

Heat transfer is maximized through the use of a microchannel heat exchanger consisting of hundreds of wire-EDM-milled vertical channels (Fig. 4.50 (right)) that transfer the returning melt water through the hot sidewalls.

Diaphragm pumps (discussed later) then pump that heated water through much larger channels (shown with pressed-in tubular oring joining seals in the outer circumference) to reach the nose-mounted melting and turning jet nozzles.

The forward end of the beam dump radiates heat into the sidewalls through the use of an infrared reflector (Fig. 4.51). The back side of the reflector cap (Fig. 4.52) contains a series of planar heat exchanger channels that serve to cool the reflector.

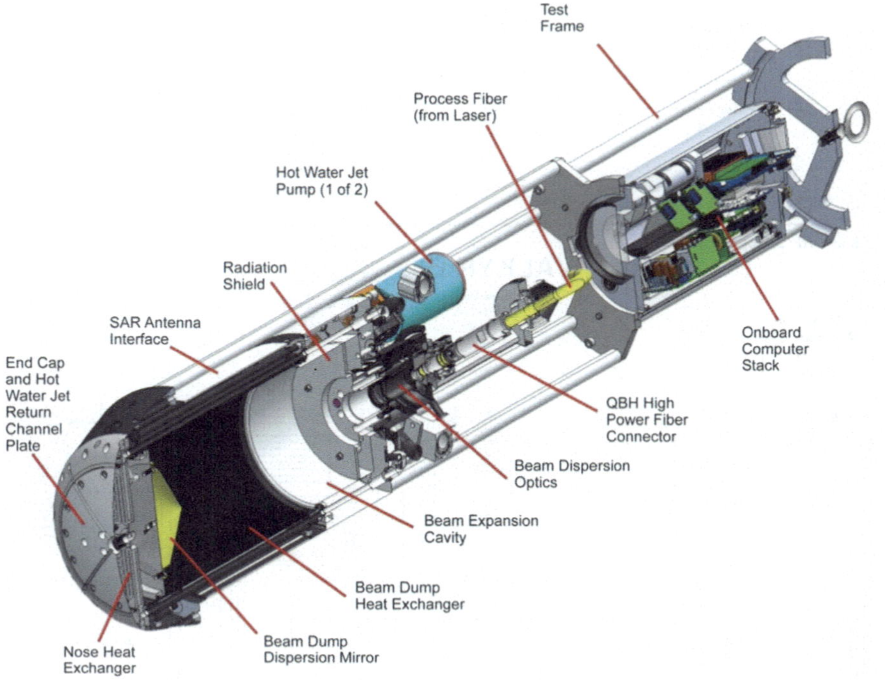

Fig. 4.49 Solid model section of the Stage 1 VALKYRIE vehicle in 2014 showing key components in the beam dump area

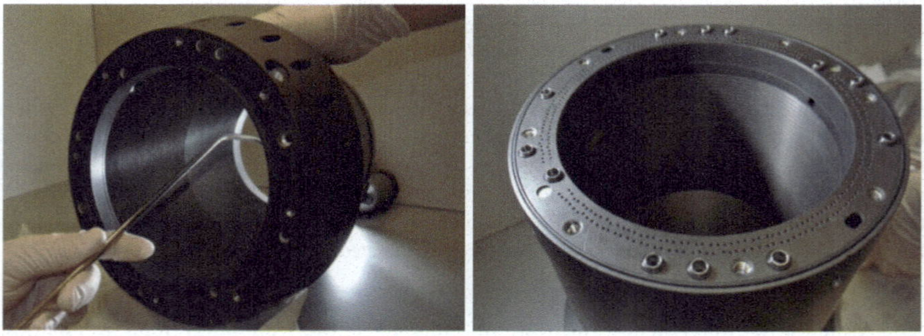

Fig. 4.50 (left) Front view of the interior of the VALKYRIE beam dump heat exchanger. Radial grooves faced forward to prevent reflection of radiation backwards towards the beam expansion optics. The walls are anodized using an "optical black" process to enhance absorption of infrared energy. (right) Back side view of the VALKYRIE heat exchanger. Small holes are wire-EDM milled micro-channels for absorbing heat from the beam dump as melt water is drawn from the front of the vehicle and back to the hot water drill jet pumps. Large tube connections represent five independent output jet lines (from each jet pump to any of five output nozzles) and two independent melt water intake lines

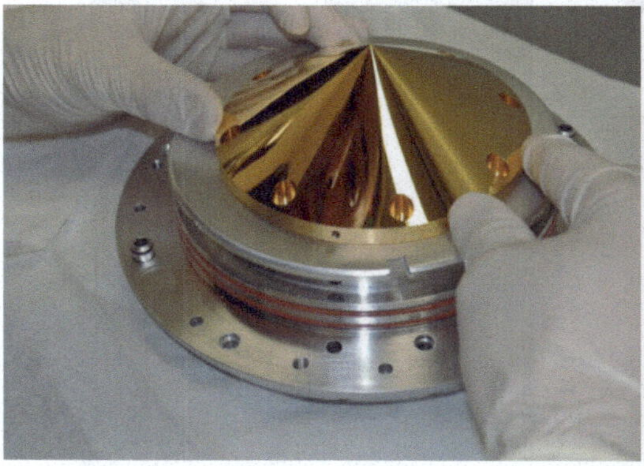

Fig. 4.51 (left)Infrared reflector at the bottom of the beam dump sends nearly uniform radiation to the internal sidewalls of the heat exchanger in the beam dump cavity. (right) Clean room assembly of the beam dump. After assembly the cavity is purged with ultra-pure dry nitrogen

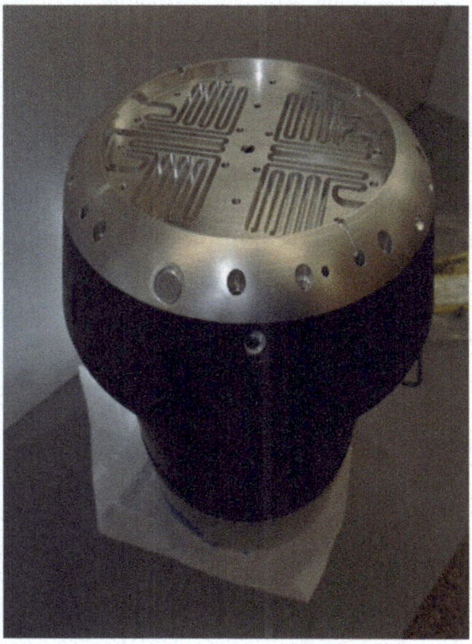

Fig. 4.52 The back side of the beam dump infrared mirror is cooled by running melt water through a planar heat exchanger prior to entry into the primary beam dump heat exchanger

A final nose cone cap plate (shown on the front of the vehicle in Fig. 4.53) contains the forward melt jet and cruciform-shaped melt water return channels that draw water back through the heat exchanger. In the black-colored housing of the heat exchanger, behind the bare aluminum front caps in Fig. 4.53 (left), two of the

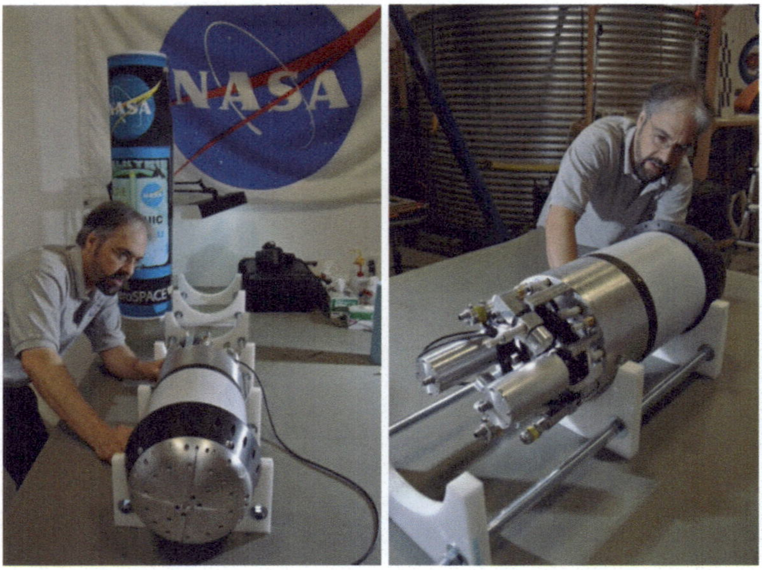

Fig. 4.53 (left) First assembly of the Stage 1 VALKYRIE vehicle showing the bow section. The central (downward) jet is located at the apex and two lateral cutting jets can be seen at the transition to the black anodized ring. The milled cross slots in the nose channel melt water to intake ducts leading to the heat exchangers. (right) Rear view of the Stage 1 VALKYRIE closed cycle hot water drill. The two large cylinders are diaphragm jet pumps

lateral turning water jets can be seen, mounted in the black anodized cone. The rear of the beam dump (Fig. 4.53 (right)) is populated with a number of fluid control systems including redundant diaphragm pumps (for driving the hot water jets), pre-pump intake filters and over-pressure regulators. The final design included servo-controlled pump output routing valves to steer flow to lateral cutting jets vice the forward central nozzle (see Sect. 4.5.12).

There are three primary thermal states that could arise during the operation of the VALKYRIE cryobot on an actual in-ice mission:

- *CCHWD Mode*: full laser power (5 kW) and the hot water jets operating;
- *Passive Mode*: full laser power and shut down (intentional or otherwise) of both water pumps; and
- *Recirculation Mode*: full laser power with the jets disabled and water being recirculated through the beam dump heat exchanger with the pumps functioning.

We analytically examined each of these states in detail using FEA thermal modeling procedures (SolidWorks Flow Professional). Our initial thinking centered on the idea of using beryllium oxide (BeO) as the beam dump material, but toxicity issues and the complexity of the necessary piping precluded this option in the end on the grounds of fabrication problems. We investigated copper as a material but it is not available in large billet sizes and similarly cannot be cast with the very complex internal channels that are required. This finally led to using a machined block of T6061 T6 aircraft alloy aluminum and a heat-resistant, photon-absorbing, non-off-gassing coating. With that basis we began investigating the concept of using wire EDM milled micro-channels as the basis for the heat exchanger. This had the highest heat flow rate, and EDM machining could bore hundreds of straight, small diameter holes the length of the planned 30 cm long cavity.

The analyses, then, focused on the spacing, number, and diameter of the microchannels. The final microchannel count was 328 × 1.6 mm diameter vertically machined holes (Fig. 4.50 (right)). All of the simulations were performed using a 0.000322 (K m^2)/W resistive contact between all mating surfaces. A Heat Flux of 85,000 W/m^2 was applied to the end face of the beam dump. The internal diameter of the beam dump has a heat flux input of 35,000 W/m^2. The outside surface of the nose cone was set to a static temperature of 0 °C (i.e. meltwater).

The peak steady state wall temperature for *CCHWD Mode* was less than +10 °C (peak temperature anywhere in the beam dump was 39 °C at the end face, but this is not effective in melting—it is the heat picked up in the sidewall that is of importance provided that the end face internal temperature does not exceed the temperature limit for the photovoltaic power conversion array. If the pumps fail at full laser power (*Passive or "Hot Penny" Mode*) the peak wall temperature reaches 29 °C; for the *Recirculation Mode*, which is the mode used while taking water samples from the front of the vehicle for the science payload in 2015, the peak side wall temperature is less than 25 °C. All of these are modeled based on the use of a 5 kW laser and are scale-able with power input. These analyses indicate that the heat transfer through the walls of the beam dump is efficient. None of these temperatures are close to service limits of internal materials nor electrical components, and thus portend nominal, long service life for the beam dump.

4.5.9 Vehicle High Power Optics

Perhaps the single most sensitive design element for the vehicle is the point where the fiber that carries coherent light from the laser has to enter the beam dump. For robustness our design has focused around modularity so that failed components can be replaced in the field. That meant having a "quick connect" style high power laser coupling at the beam dump rear bulkhead. We used QBH class connectors that fuse the core of an armored fiber to a solid quartz optical window (Fig. 4.37). For an actual field deployment the bare fiber spooler uses QBH connectors spliced at the two ends of the spool. The problem is that these connectors are not waterproof.

We designed a pressure proof housing (Fig. 4.54) as follows: an optical junction block has a QBH female bulkhead coupler mounted on one end that allows field coupling of a QBH connector. The quartz block end optic is thus mounted inside the optics junction block such that its beam will travel down the hollow core of the junction block to where it encounters a lens array that expands the beam dispersion angle to achieve a cone-like beam that then impinges on the end of the beam

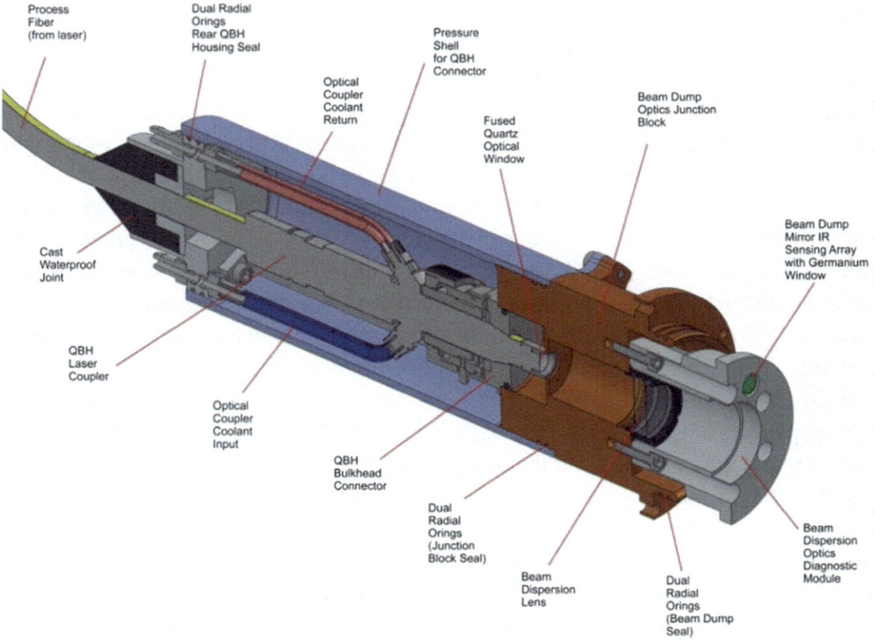

Fig. 4.54 The beam dump optics system had to allow for field connection of the laser fiber, allow for temperature and radiation backscatter diagnostic sensors, and be waterproof to 200 m underwater depth. Presented here is a 3D section view from the laser interface cable to the dispersion optics elements

dump. However, the back side of the QBH coupler is not waterproof and, furthermore, the coupling requires liquid cooling.

As shown in Fig. 4.55 the incoming fiber that feeds the coupler is bonded to an end fitting by means of a waterproof casting (3 M Type 4). Both the cable and the casting housing were cleaned and primed prior to casting to prevent debonding. The fitting contains a dual o-ring radial seal. Similarly the optics junction block contains a dual o-ring radial seal. A cylindrical pressure shell (Fig. 4.55) then slides down over the two o-ring sealing sets and bolts to the optics junction block. Liquid cooling lines from the QBH connector pass through the rear sealing block via machined fittings that are also waterproof and are isolated from the internal volume of the connector. The cavity is then purged with dry nitrogen during assembly to prevent condensation.

When working with multi-kilowatt lasers it is important to understand that beam energy flux intensities can be extraordinary—sufficient that if a speck of dust or a condensate spot is present anywhere in the optics chain those locations will absorb large amounts of energy, sufficient to lead to rapid destruction of the optics. Because of this it is essential to monitor temperatures and levels of scattered radiation and to tie those readings to an automated shut-down system. We used three infrared sensors in the VALKYRIE optics stack. One tracks the temperature

Fig. 4.55 The integrated beam dump optics stack. The waterproof pressure shell (200 m water service depth) is not shown; it consists of a cylindrical tube with precision-turned o-ring sealing surfaces on each end and a milled cross-shaped connection bracket on the bottom that matches the projection pattern on the optics junction block. The dual o-ring seals (top and bottom) have different diameters to allow the tube to be raised above the upper seal set for attachment of the QBH connector. The cavity then created after sealing is purged with dry nitrogen

of the lens mount; a second discrete sensor tracks the dispersion lens temperature. Both of these are narrow beam (12°) MLX90614ESF-DCH sensors from Melexis Technologies. The third sensor, a MLX90620ESF-BAB also from Melexis Technologies, is a 16 × 4 thermopile array (3.75° by 3.75° per pixel) which measures most of the interior of the beam dump. Backscatter radiation in the 1070 nm wavelength regime is intense so strong filtering is required to enable meaningful stand-off temperature measurement. We successfully employed a Germanium window (Fig. 4.56)—a 10 mm diameter × 1.5 mm thick disk with antireflective coating from Edmund Optics (part 62–643). To further enhance signal clarity the laser is turned off for a brief instant (1 ms) during temperature readings.

High power transmission tests were conducted as shown in Fig. 4.57. We used a PRIMES power meter (at left in Fig. 4.57) to determine the actual transmitted power levels. Of critical interest during these tests was the temperature of the lens system. As added protection during initial testing we used a FLIR camera that was set up inside the radiation chamber. The FLIR unit was aimed at the transmission

Fig. 4.56 Germanium window used to filter backscatter radiation and allow for direct temperature sensing of beam dump optics

lens in the optics block. Peak temperatures at 5 kW transmission power never exceeded steady state values of 42.9 °C in air, thus qualifying the design for continuous use. During actual missions this value was much lower owing to the optics block being surrounded by melt water.

We designed the optics block to handle two different lenses: one for photovoltaic (PV) array testing and one for high power mirror testing of the beam dump heat exchanger. For the PV array tests we selected a fused silica, laser-grade lens with anti-reflective (AR) coatings on both sides for 1064 nm. The lens is plano-concave with the convex side having a 13.1 mm radius of curvature, and a lens OD of 25.4 mm. The beam divergence is 8° at the exit point from the QBH quartz block. The lens further increases the divergence angle to 35°. The final divergence is affected by the axial location of the lens relative to the QBH connector. The hard part is getting all of the beam through the lens while also creating a large divergence angle (the F number of the lens). We selected the lowest F number available in laser grade fused silica. For the high power mirror tests we used a fused silica laser-grade lens with 1064 nm anti-reflective coating. It is a 25.4 mm diameter × 6 mm thick flat lens. Beam divergence is not effected and remains the same as the exit beam from the QBH quartz block: 8°.

Fig. 4.57 High power testing of the optics block. The diverging beam from the optics block lens (assembly at right) impinged on a PRIMES power meter (left) which recorded the actual transmitted output power. The temperature of the beam dispersion lens was monitored using a FLIR infrared camera digitally linked to a remote display. The entire test setup was mounted on an optical bench housed in an absorbent black steel case to prevent stray radiation

4.5.10 Onboard Power Generation

Fiber optic power transmission allows laser-powered cryobots to simultaneously operate at high power while utilizing a fiber tether that can be much smaller than an electrical tether operating at the same power. Inside the cryobot optical energy is converted to thermal energy in an optically black cavity referred to as a “beam dump.” The beam dump is effectively the hot side of a heat exchanger, with the cold side being the ice in front of the vehicle. To operate at maximum efficiency meltwater is pulled into the cryobot and circulated through a series of small fluid pathways where it absorbs the heat from the beam dump. The water is then ejected at high pressure from a port on the nose cone of the vehicle. Electric pumps are required to circulate the meltwater. These pumps and the attendant solenoid routing valves represent the bulk of electrical power consumed by the vehicle. This power is on the order of 2–4% of the total power of the vehicle. In the case of VALKYRIE the total electrical power consumption was 120 W-electric with a total input power of 5000 W-optical. It is possible to design an integrated optical/electrical tether to provide both the optical melt power and the electrical pumping power from the surface. However, this solution results in a tether that sacrifices many of the advantages of an optical system including tether reliability, bend radius, diameter and weight.

To solve this problem and maintain a monolithic fiber tether a back-power conversion scheme must be implemented wherein a small part of the optical energy is intercepted by a high-efficiency photovoltaic array. The VALKYRIE cryobot design utilized an annular ring of space-grade, wavelength-tuned Indium Gallium Arsenide (InGaAs) photovoltaic cells as depicted in Fig. 4.58. The cells are grown on a Germanium substrate to minimize lattice mismatch defects which are a major source of inefficiency in photovoltaic cells. These cells achieve efficiencies between 40 and 50%. The cells are arranged in an annular ring at the far end of the beam dump where they intercept about 4% of the laser beam. Electrically, the cells produce an open circuit voltage of roughly 0.7–0.75 V depending on irradiance. At full 5000 W optical power the cells produce a short circuit current of approximately 8 A. This results in a maximum power of 4.1 W/cell at 5 kW optical irradiance. The maximum power point of the cells occurs at approximately 0.5 V at 8 A. The VALKYRIE design utilized 36 cells in series. This configuration provided a maximum power of approximately 145 W at 18–20 V. Each cell has a bypass diode that becomes active in the event the cell fails open. In this scenario the output voltage and maximum output power of the system are reduced, but the system continues to provide power. Interconnects between the cells utilize PTFE (Teflon) insulated conductors. PTFE is a high temperature fluropolymer with the added benefit that it has a high optical transmission coefficient at 1070 nm.

Figures 4.59 and 4.60 show the test setup for determining the actual output power of a single InGaAs photovoltaic cell under laser illumination. The range from the optics block and illumination flux density are set to duplicate the conditions in the actual beam dump. Figure 4.61 shows the expected performance for the final 36 chip PV array (Fig. 4.58) based on the measured single-cell performance. The implementation shown in Fig. 4.58 is predicted to generate 160 W continuous electrical power. Successful laboratory tests of the 160 W array took place in 2017 as part of the SPINDLE project (Sect. 4.6).

Raw power generation is only one facet of providing reliable backpower for the cryobot. Attendant management circuity (Fig. 4.62) must be included to produce a steady output voltage and to maintain and monitor the health of the PV array. Figure 4.62 is a block diagram depiction of the power management system. The PV array intercepts between 300 and 350 optical watts, of which 40–50% is converted to electrical energy when the array is operating at peak power. However, when the array is not operating at peak power the excess optical power must be dissipated as heat by the array. Simply moving the operating point of the cells from peak power to open circuit increases the thermal load on the cells from approximately 150 W to the full 300–350 W. Therefore, to prevent thermal damage to the cells it is important to constantly operate the cells at maximum power output. Depending on the operational mode the cryobot electrical load can vary from 10 to 110% of the rated PV output. To account for this an adaptive, variable load must be included to keep the cells operating at peak power. This scheme is similar to maximum power point tracking (MPPT) technology used in solar generation to maximize the power

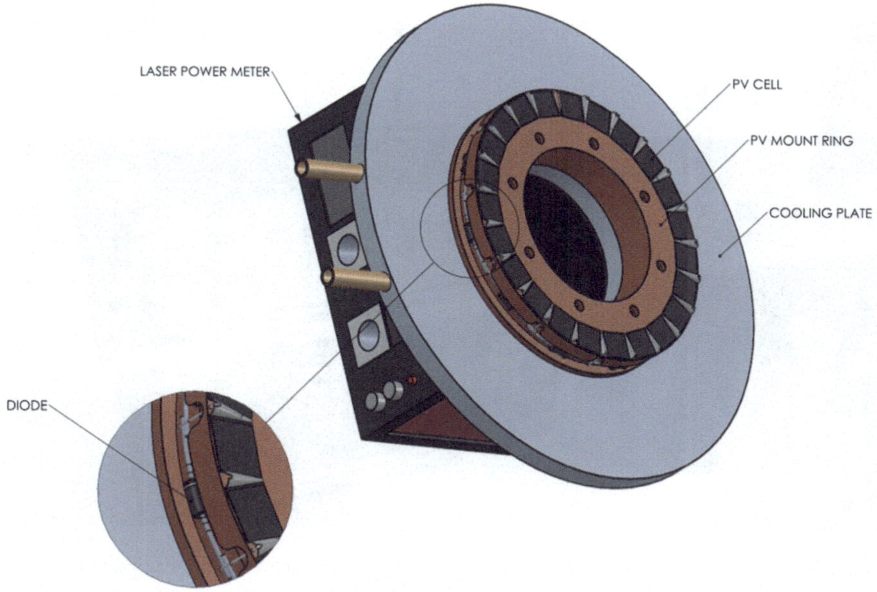

Fig. 4.58 3D model of the 36-chip, 160 W tuned-wavelength, high temperature ceramic photovoltaic back power generation array test that will be conducted as part of project SPINDLE in the spring of 2017

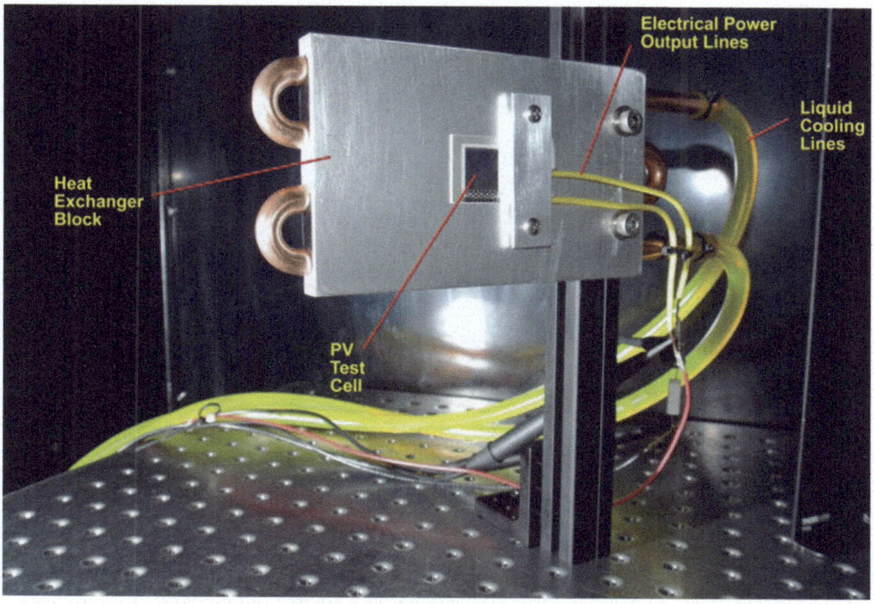

Fig. 4.59 Prototype photovoltaic (PV) power conversion test cell mounted to its heat exchanger backing, conducted during the VALKYRIE project

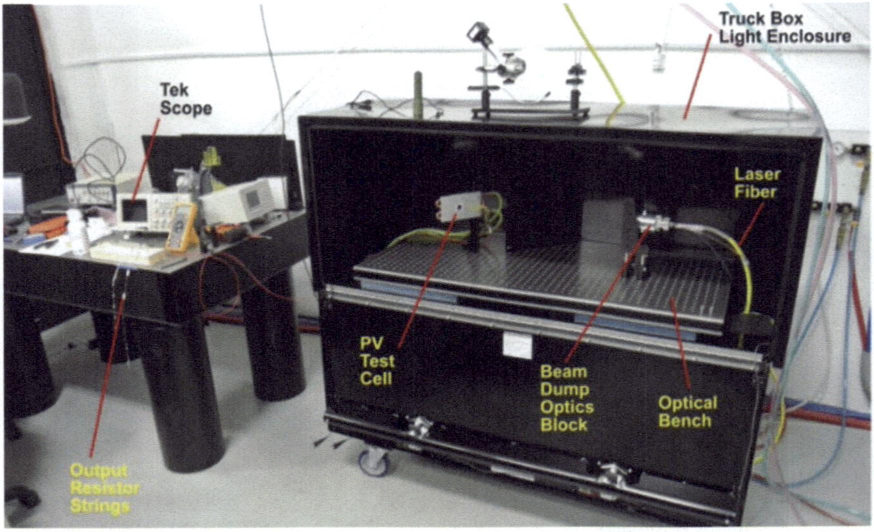

Fig. 4.60 Overall view of the single PV chip test setup. Laser power was channeled to the beam dump optics block via a high power process fiber. The diverging beam illuminated the PV cell from a distance identical to that of the actual beam dump. The PV cell is inclined at a 45° angle to the beam, as it will be when mounted in the beam dump. The entire setup, mounted on a portable optical bench, is enclosed in a light tight metal case for safety during actual beam firing. Precision output resistors (left) were used to measure PV output current through a Tek scope

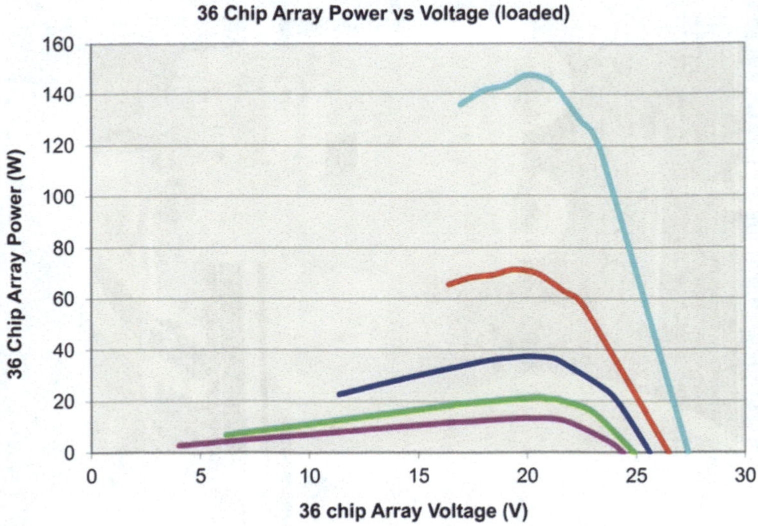

Fig. 4.61 Extrapolated power output for the 36 chip array shown in Fig. 4.58 based on the results of the single chip tests. The planned bus voltage for VALKYRIE and SPINDLE class cryobots is 24 VDC, indicating that approximately 140 W of usable continuous power will be available from the annular array

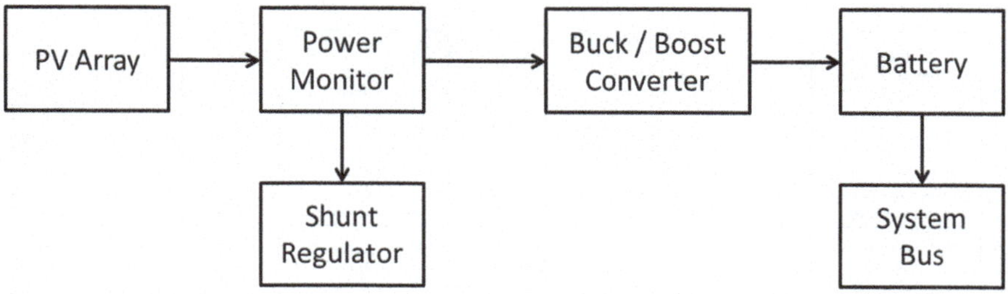

Fig. 4.62 Back-power generation control system for the SPINDLE 160 W array test

output from a PV cell regardless of irradiance with one critical difference; the VALKYRIE scheme seeks to modulate the total load to keep the PV array operating at maximum power while commercial MPPT schemes create an impedance match between a solar array and a mismatched load. The adaptive load designed for VALKYRIE uses multiple emitter balanced Darlington transistors that are configured as a shunt regulator. The shunt regulator can modulate the output voltage of the PV array from open circuit to short circuit. A power monitoring circuit allows the instantaneous power as a function of array voltage to be logged. This allows a full current-voltage (I-V) sweep to be performed on the array. The health status of the array can be assessed once the peak power is known. This I-V sweep is performed on regular intervals and the health status of the array is reported back to the main vehicle computer. Should the peak power change for a constant irradiance a health status flag is set and reported back to mission control for human intervention. The output of the shunt regulator serves as the input to a high power buck/boost switching power supply that provides a constant 24 V system bus output. The bus receives parallel power from a buffer battery that can seamlessly provide surge power and can also fully power the cryobot during periods when the PV output is disconnected for health status checks or when the laser has been turned off.

The VALKYIRE power system represents a step forward in electrical back-power generation from a fiber laser source. The technology has application outside of cryobots, including long-range autonomous underwater vehicles (AUVs) and "clean" rovers where the radioisotope thermoelectric generator must be divorced from the rover to prevent environmental contamination.

4.5.11 Onboard Electronics and Control Software

Figure 4.63 shows a block diagram of the internal hardware and electronics systems for the as-deployed 2015 VALKYRIE vehicle (Fig. 4.33). The main components (dashed boxes) contain the following systems:

Main Electronics Housing: The main electronics housing (Fig. 4.64) contains the processors used to interface with all sensors and actuators on the vehicle.

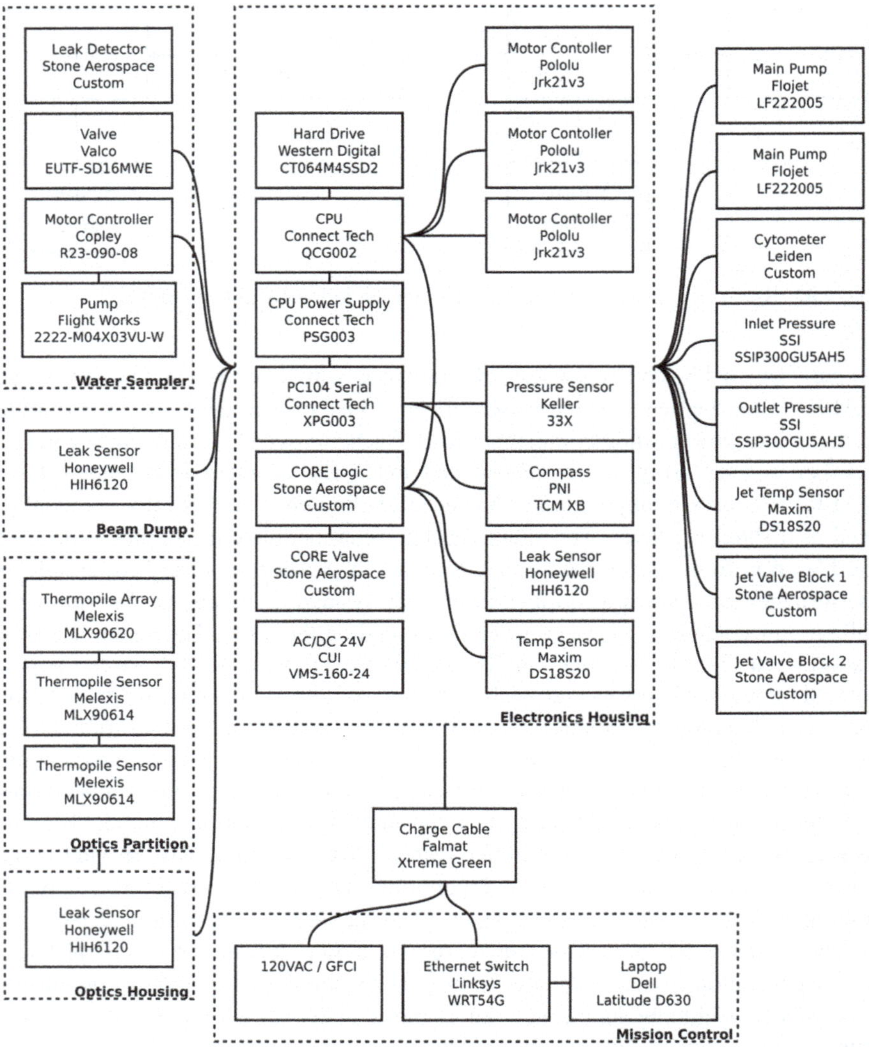

Fig. 4.63 Electronics block diagram for VALKYRIE. Block diagram for the Main Housing lists the PC104 form factor CPU hardware stack elements

Generally speaking it was a multi-board stack of standard PC104 class industrial controller cards. In particular one PC104-form factor component—The CORE Logic board—is the heart of the robot. It is a custom designed printed circuitry board (PCB) designed by Stone Aerospace to interface with most of the sensors on the vehicle and quickly respond to critical errors. The microcontroller (Atmel ATmega328) interfaced with external sensors (see Table 4.2) primarily using an 8-port I2C hub in addition to other digital I/O pins. It monitored the temperature of the lens, lens holder, and optics block as well as the water in front of the nose and

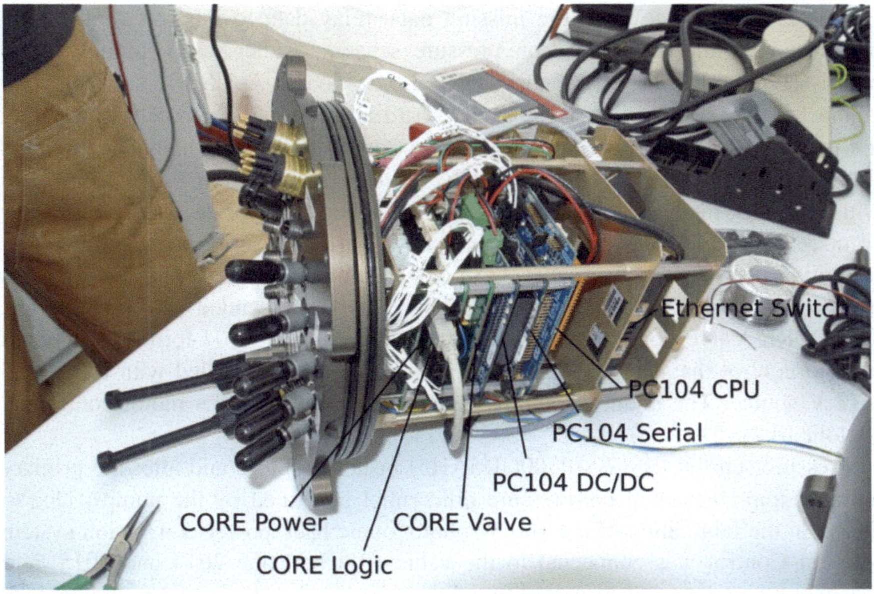

Fig. 4.64 Oblique view of the VALKYRIE Stage 2 CPU stack showing back side of the pressure bulkhead end plate. A total of 25 port connections for controlling various vehicle systems interface at this bulkhead, including Science Payload elements (flow cytometer, water sampler), and the IceSAR obstacle avoidance system

Table 4.2 VALKYRIE microcontroller external sensors

Humidity sensor	Honeywell	HIH6120
Thermopile sensor	Melexis	MLX90614
Thermopile array	Melexis	MLX90620
Pressure sensor	SSI	SSIP300GU5AH5
Temperature sensor	Maxim	DS18S20
Valve control	Microchip	MCP23017

MOSFETs on the motor controllers (Pololu Jrk21v3) for each pump. In addition it was designed to break the connection on the laser interlock in case of an over-temperature fault, and it controlled the CCHWD jet diversion valves. This board was also used to interface with humidity sensors installed in each housing to detect leaks.

Most sensors communicated with the CORE Logic microcontroller using the I2C bus. The transmission failure rate was significant over long cables for some of the sensors. As a workaround we installed I2C buffer boards throughout the vehicle to boost I2C transmission voltage from 3.3 to 12 V. This allowed us to get past the issue, however there may be better ways of dealing with this problem. We suspect that an improved Schmidt trigger may allow low voltage I2C communications over longer distances without voltage level shifting. The CPU board (Connect Tech

QCG002) was also used to log mission data, relay data to mission control, and interface with the high precision pressure sensors (Keller 33X), pump control boards (Pololu Jrk21v3), and the CORE Logic board. A USB-to-serial adapter connected the CPU to the CORE Logic board allowed us to remotely re-program the microcontroller without opening the electronics housing. A magnetic compass (PNI TCM XB) was used to accurately determine orientation of the vehicle to within 0.1 deg pitch and 0.5 deg heading. At the beginning of each field season the compass was calibrated on-site to correct for hard and soft iron errors. A high resolution pressure sensor (Keller 33X) (0.01% full scale precision at 10 bar) was used to determine vehicle depth and speed as it descended through the ice. The CORE Valve board was another custom breakout PC104-form factor board used to control the steering jet valves. Each valve was controlled with a 200 mA ±12 V source. The main jet pumps were driven by two COTS motor controllers (Pololu Jrk21v3).

Pressure sensors (SSI SSIP300GU5AH5) installed before and after the primary jetting pumps measured the pressure differential generated by the pumps. Due to delays in the fabrication of the 160 W photovoltaic back-power-conversion system Mission Control was connected to the vehicle (in both the 2014 and 2015 field seasons) by a tether that was used to deliver 120 VAC and ethernet to the vehicle. An AC/DC converter transformed 120 VAC to 24 VDC to power various systems. The ethernet cable we used was from a previous project and not designed specifically for the mission. We had to slow the ethernet speed to 10BaseT because of significant electrical noise coupling between the 120 VAC and ethernet lines.

Optics Housing/Beam Dump: To prevent condensation and leaks from occurring in the optics housing we purged the housings with dry nitrogen and then monitored relative humidity using a Honeywell HIH6120 relative humidity sensor. We looked at several sensors and had difficulty finding one that performed well in the low humidity, low temperature environment we expected to operate in. The HIH6120 had the best accuracy (±1.0%) at 10% relative humidity and 0 °C of the sensors we looked at. For future projects we recommend installing a 1 bar pressure sensor and pulling a slight vacuum in order to detect very slow leaks. The beam dump mechanics were quite complicated with scores of orings and plugs required to seal it and on several occasions prior to field deployment we discovered leaks in the beam dump that were time consuming to diagnose. A vacuum sensor would have reduced the amount of time required to pinpoint leaks.

4.5.11.1 Vehicle Software Architecture

Figure 4.65 shows a block diagram for the guidance, navigation, and control (GN&C) onboard software architecture for the as-fielded VALKYRIE vehicle in 2015.

Mission control consisted of a laptop connected to an external monitor. An operator at the laptop controlled actuators on the vehicle using a graphical user interface (control panel). A real-time visualization (IR array viewer) displayed

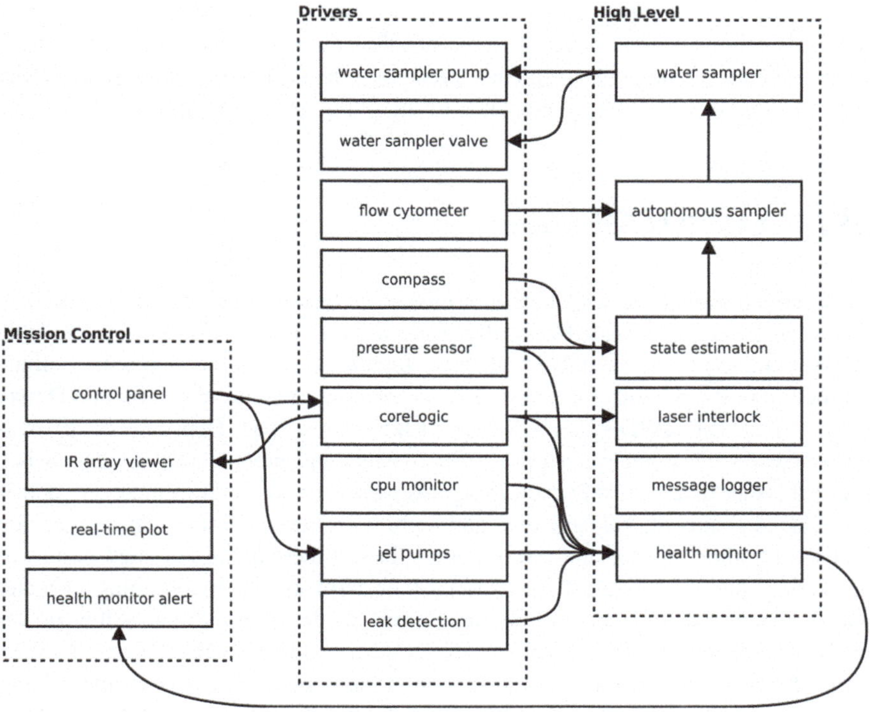

Fig. 4.65 VALKYRIE Stage 2 software block diagram

temperatures reported by the thermopile array pointed toward the beam dump. Sensor data including depth, velocity, orientation of the vehicle, scientific data, etc. were displayed using real-time plots of each sensor. To notify the operator in case of a fault a text-to-speech program was used (health monitor alert). Several software drivers were written to interface the vehicle computer to hardware. Most devices onboard the vehicle connected with a microcontroller on the vehicle (CoreLogic) including the pump pressure sensors, thermopile sensors, valves, and leak sensors.

Data from navigation sensors (compass, pressure sensor) were fed to a pose estimation algorithm (state estimation). The vehicle estimated depth by multiplying tared pressure sensor measurements with an empirically tuned constant used to account for the volume difference of ice transitioning to water and other un-modeled effects. Dead-reckoned position was obtained by integrating depth and orientation measurements.

To prevent damage to the vehicle a health monitor program monitored the status of various sensors including humidity, pressure, temperature, and reported faults. A laser interlock program automatically switched off power to the laser when the lens temperature became too high.

The autonomous sampler (Sect. 4.8) decided when to take water samples based on measurements from a protein fluorescence flow cytometer and estimated position of the vehicle. Commands from this process were sent to a custom-built water sampler module which sent commands to the pump and valve drivers.

4.5.12 CCHWD Subsystems

As described earlier VALKYRIE is a closed-cycle hot water drill (CCHWD) in which the power is provided optically using a laser. Implementation of a reliable CCHWD means having reliable high flow, high pressure pumps that will continue to function in the presence of debris. This is far more of an issue in a CCHWD than it is for traditional HWD systems, where a Rodwell reservoir (Benson et al. 2014) is the source for the input water prior to the heaters and pumps and thus debris is a minimal issue. In a CCHWD everything takes place onboard the vehicle and debris (e.g. glacial sediment, volcanic dust and tephra) are serious issues. These preclude the use of traditional gear and piston pumps. In fact there are no off-shelf pumps of any variety that are designed for extreme long term use in the pumping of water laden with glacial fines and debris and further, to do so at extremely low power levels. The pumps are the primary power consumers on VALKYRIE and so their design for low current, long life under continuous operation has been the driving engineering design issue. Figure 4.66 shows 3D details of the present design of the VALKYRIE jet pumps. We custom-developed several generations of diaphragm pumps for VALKYRIE and tested each for hundreds of hours of continuous use in the lab and similarly during the field missions in 2014 and 2015. The diaphragm valve approach was selected over all others due to the reduced sensitivity of passing glacial fines through the pump over a long period of time. The pumps were driven by a Maxon 310009 DC motor. The completed pumps measured 4.7 cm diameter $\times$ 20 cm length and were capable of pumping 6 L/min at 345 kPa over ambient pressure. Without oil filling the pumps were pressure rated to 100 m underwater; if oil filled they could be used to unlimited depth. Each pump had to be capable of routing to five jets on the front of the vehicle—direct down, and four lateral turning jets—plus entering into passive melting and recirculation modes; all of these required valves, filters, pressure sensors, and over-pressure valves. Thus small size, high power, high efficiency, and high reliability were essential. We tried and abandoned many alternatives before settling on the Maxon 310009 motor. The traditional underwater pump shaft seal design was abandoned for a magnetic coupling arrangement. This consists of two discs loaded with magnets of opposite polarities (Fig. 4.66). We were then able to transmit the rotation of the motor through a solid metal bulkhead eliminating any chance for water leaks into the motor housing. With one of the discs located in water a ceramic bearing needed to be used to support the spinning magnetic disc. Standard metal bearings fail when submerged in water.

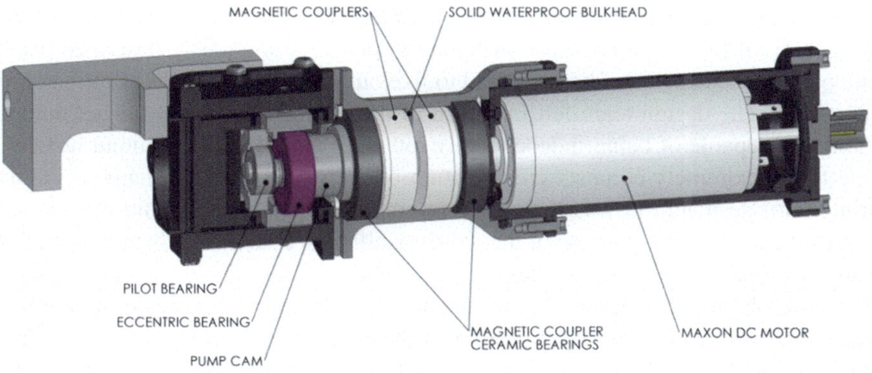

Fig. 4.66 3D solid model cut away of the VALKYRIE Stage 2 hot water jet pump. Key improvements are the addition of magnetic couplers to isolate the pump cavity from the motor cavity; the use of ceramic (silicon nitride) bearings for the wet and dry sides of the magnetic couplers, and the addition of a pilot bearing to the diaphragm pump

Fig. 4.67 Beginning of system integration of the two valve stacks used in VALKYRIE to divert hot water from the jet pumps to any of the five output jets. There is one bank of valves for each of the redundant hot water jet pumps. The tan-colored base of the valve array is a 3D-printed manifold that would not have been physically possible using traditional machining technology

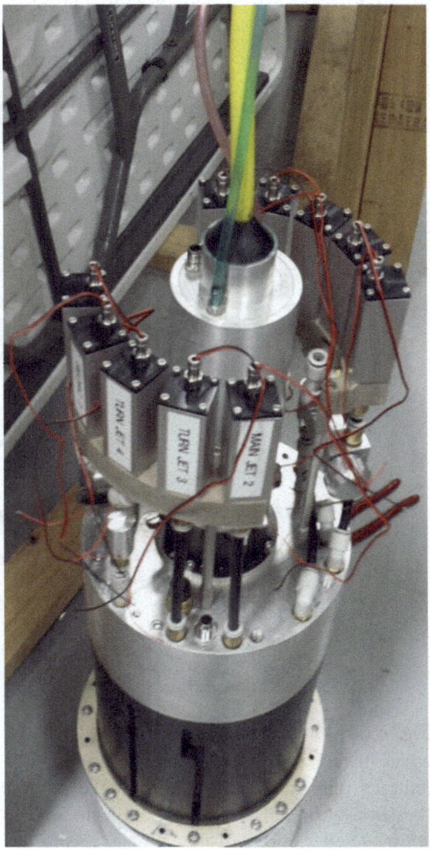

An equally difficult problem with VALKYRIE was the design, fabrication, and integration of the jet pump valve switching system. As previously described, each pump had to be capable of being routed to five output hot water jet nozzles (e.g. see Fig. 4.53). The design team felt that each of the independent hot water jet pumps should be capable of being independently routed to any of the five output nozzles. Due to the extremely compact cross section of the vehicle (25 cm diameter) a 3D printed nozzle manifold was developed to compress functionality into the limited volume available. The purpose of the valve manifold is to control where intake melt water and heated water are directed.

It controls the main jet, jetting to turn the vehicle and the ability to recirculate water internally to prevent over heating of the Beam Dump and possible contamination of water samples during science sampling. Eventually these two valve banks can also be used to divert water to shell heating for better turning of the vehicle and prevention of freeze-in. The valves consist of a 10 mm travel linear actuator in a ported manifold block. The actuator moves a piston to close off and open ports in the manifold to control water flow in the vehicle. There are two banks of five valves each. They were packaged this way to conserve space and keep the valves near the ports they interacted with on the beam dump. The manifold was 3D printed using an SLS (Selective Laser Sintering) process out of Nylon. 3D printing allowed us to

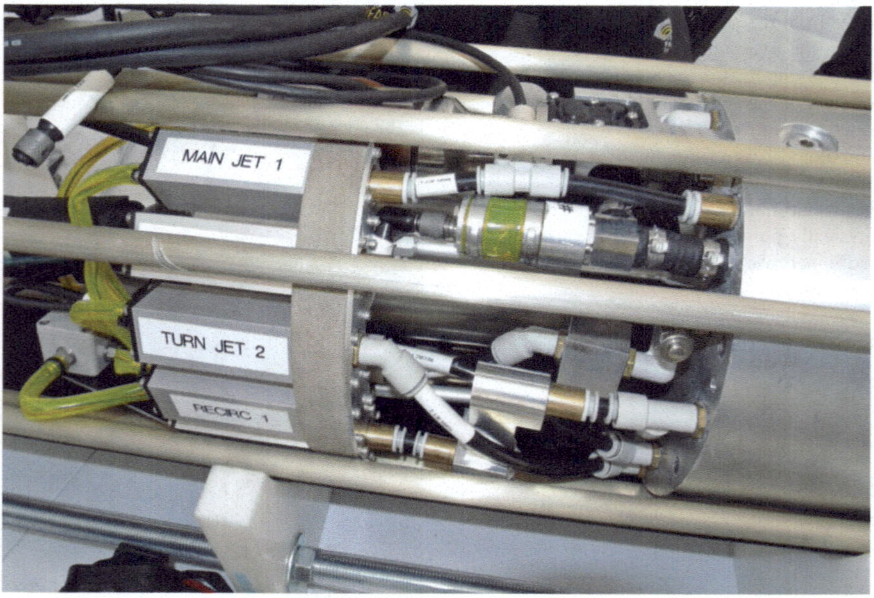

Fig. 4.68 Photo of the completed VALKYRIE Stage 1 field test vehicle showing the core section adjacent the beam dump (the solid aluminum body to the right of the photo). Line filters (for removal of glacial till from the intake ports at the front of the vehicle), pressure sensors, over pressure blow off valves, and electrical cabling and couplers make this one of the most complicated sections of the vehicle

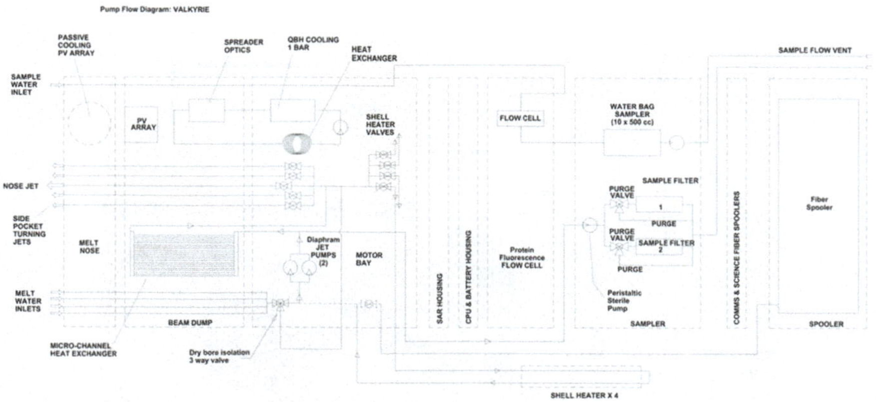

Fig. 4.69 VALKYRIE fluid systems schematic

make internal passages that could not be made with any other fabrication process. The specialized passages cut the amount of external plumbing required substantially. We were also able to package the valves in a much smaller overall package. See Figs. 4.67 and 4.68 for details. This design was tested at the Matanuska glacier during the 2014 and 2015 field campaigns and was proven to work reliably. The overall pump and fluid control schematic for the vehicle is shown in Fig. 4.69.

4.5.13 Field Deployment Results

Team VALKYRIE spent a total of 3 months on the Matanuska glacier, Alaska in 2014 and 2015. The field site was located 15 km up-glacier from the Alaska Highway (Fig. 4.70). The Matanuska was selected for two reasons: first it was possible to drive there from our lab in Texas, which simplified the transport of a substantial amount of equipment and second it is a temperate glacier, which meant that the melt hole would not refreeze behind the vehicle. The second attribute was useful in that it allowed for testing the basic functionality of the vehicle while not having to yet address the issue of vehicle recovery from an Antarctic site where the hole would freeze closed behind the vehicle. The field configuration for VALKYRIE is shown in Fig. 4.71 wherein a 30 kW, 480 VAC, 3-phase diesel generator was used as the field power plant. This generator drove a 5 kW IPG laser and its associated thermal control system. The total weight of this system was approximately 1800 kg with most components being acquired in 2012. Dramatic advances in laser technology during the intervening years indicate that this deployment mass for the same effect could be achieved in one third the mass or less in 2017.

Our 2014 deployment was intended as an engineering shakedown exercise. Our primary objectives were to determine the descent characteristics of the vehicle

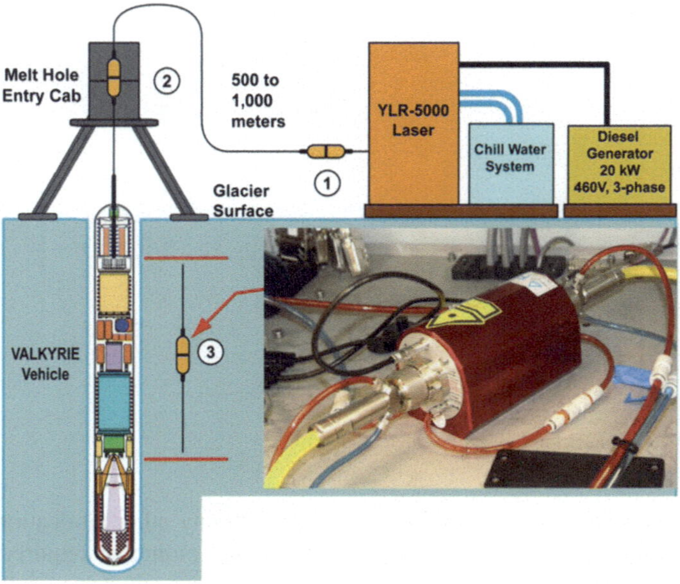

Fig. 4.70 Schematic of the field setup for VALKYRIE showing key infrastructure elements

under full power (5 kW) and the ability to deviate from vertical path (in order to avoid obstacles that may be detected ahead of the vehicle by its synthetic aperture radar). Figure 4.72 shows a typical vertical velocity plot as a function of mission clock time. The ripples on the mean velocity are real. This appears to be a feature of the current hot water jet configuration on VALKYRIE and is a function of jet water temperature, jet pressure and flow rate. A close up of the last several hours of this test show that the average period of the velocity cycling is approximately 26 min. This repeatable pattern begins with the forward HWD jet (3 mm in diameter and carrying water at 14 °C at 345 kPa over ambient pressure with a jet velocity of 13 m/s) water in a downward, tight stream into hard ice. The stream cuts a narrow diameter cylinder in the ice below the vehicle beam dump to a depth of approximately 1.5 vehicle diameters. The jet then plumes and begins to expand the lowest portions of the hole as a spherical cavity. Eventually the cavity is breached by passive melting from the beam dump and the vehicle then commences to slowly drop into the new chamber as it melts the remaining rim at the top of the chamber. This type of behavior has been reported in other HWD field use (Thorsteinsson et al. 2008) and can be made more uniform by increasing both water temperature and pressure.

Figure 4.73 shows a 3D trajectory for a test run involving the use of the lateral turning jets. We were able to run up to two lateral jets (at 90 degrees from one another) simultaneously. The vehicle was able to achieve a deviation angle from vertical of up to 7° and then be brought back to vertical, thus proving the ability of the vehicle to divert in the presence of obstacles.

Fig. 4.71 Photo of the actual field implementation showing 30 kW 480 V generator at left; laser thermal control module in the middle; and Mission Control, with the 5 kW laser, to the right. The cryobot gantry was previously shown in Fig. 4.33

Fig. 4.72 Close-up plot of the data for VEHICLE Velocity plot during the last five hours of the mission trajectory plotted in Fig. 4.73, showing a pronounced periodicity

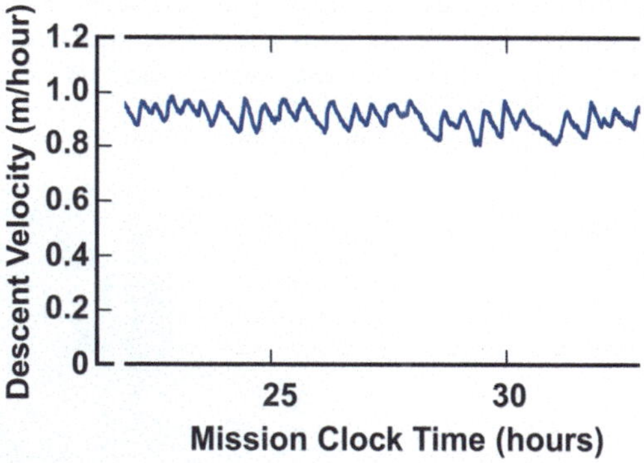

Figures 4.74, 4.75 and 4.76 show, respectively, a normal vertical jet descent; a descent with one lateral jet firing along with the primary vertical jet; and two lateral jets firing but the forward jet disabled. The vehicle was able to turn away from vertical in the case shown in Fig. 4.75 but not in the case of Fig. 4.76 where the lateral jets cut fairly deep slots as the vehicle passively descended at a slower rate, but the slots were not large enough for the vehicle to slide into (and thence be able to deviate laterally). It was concluded that the lateral jets should have a wider dispersion angle (more like a shower head) to enhance vehicle turning. Essentially,

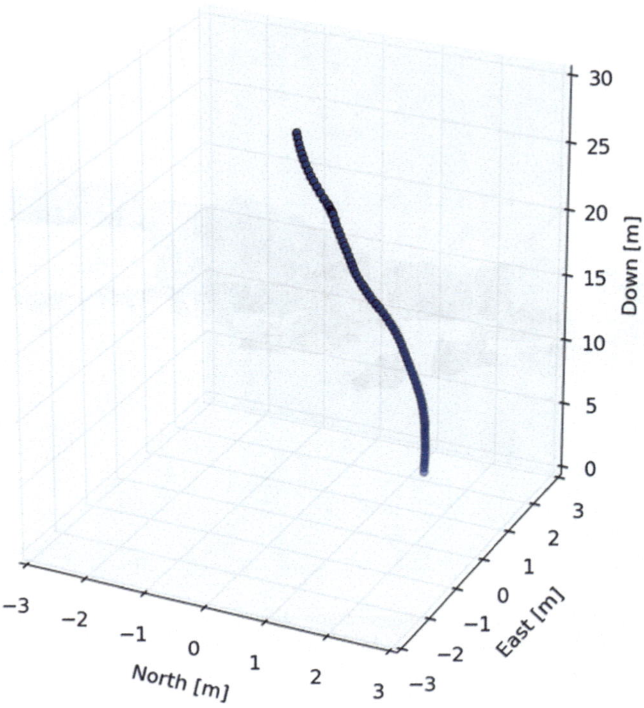

Fig. 4.73 3D Descent Trajectory for the June 22–23, 2014 VALKYRIE mission. The vehicle reached a depth of 30.52 m. The vehicle was functioning nominally, maintaining an average 0.9 m/h descent rate. The test was terminated only due to lack of a longer test umbilical

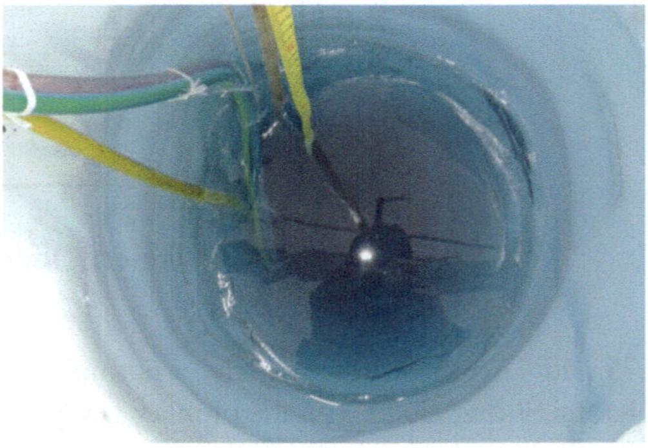

Fig. 4.74 Typical down-hole view following the direct jetting passage of VALKYRIE during the 2014 field campaign. The various umbilical components are: pink tubing (inlet temperature control fluid for laser coupler); blue tubing (not visible behind the green); green cable (ethernet and CPU power); yellow (laser fiber); 9 mm nylon rope (safety and recovery); and a metric survey tape for absolute vertical position measurement. The hole is nominally 25 cm diameter but due to heated water rising over the course of a mission the hole enlarges significantly towards the surface

Fig. 4.75 Looking down hole following a test of the firing of a single lateral hot water jet in conjunction with the primary downward jet. The lateral cavity at 7 o'clock uniformly continues down the hole underwater—the hole was partially pumped out to allow this photo of the lateral cavity

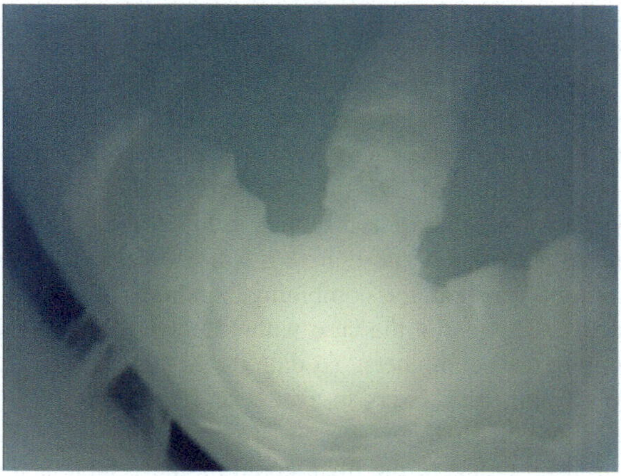

Fig. 4.76 The result of a protracted VALKYRIE mission in which only two lateral hot water jets were fired with no forward jet. The result was the formation of two lobe-shaped cavities that are parallel to but offset from the main vehicle descent axis. The ice bridge clearly visible at top center of this image prevented the vehicle from slipping into the lateral cavities (which would have led to strong turning of the vehicle off the vertical axis). Further, it can be seen that the lateral holes are deeper than the bottom of the main shaft hole. Descent was thus caused by passive melting in the downward direction while the parallel slots were being cut—this resulted in a significant reduction in downward velocity of the vehicle. Future designs will use a wider fan turning jet so that individual lateral cavities will be capable of displacing the vehicle after one vehicle-length of descent

the vehicle must cut a parallel slot that is at least one vehicle-length deep and at least the diameter of the vehicle for the cryobot to slip into the lateral slot and initiate lateral turning.

4.5.14 Lessons Learned from VALKYRIE

VALKYRIE was a fundamentally risky, envelope-pushing concept: to use a laser to power a robot by means of a long fiber optic waveguide that would permit operation of the robot at distances of many kilometers from the power source. Between 2010 (the beginning of Phase 1 lab tests) and completion of the Phase 2 field campaign in 2015 we learned a lot. Here are some of the key findings:

- The fundamental novel concept of sending large amounts of power through a long fused silica fiber (optical waveguide) works. The fiber does not "explode" or "fuse" the second the laser is fired, as many skeptics originally opined.
- The most effective means (compact and efficient in terms of wall-plug-to-coherent-photons conversion) of generating optical power suitable for remote vehicle power operation is via the use of fiber lasers. These are currently at 52% wall plug efficiency and climbing each year and the volume and mass of equivalent units are shrinking each year, as is the cost metric ($/watt). At this time (2017) a board level fiber laser is running around $30/W for kilowatt + category lasers and dropping.
- The most compact means for deploying an optically-powered vehicle is for the vehicle to pay out a bare fiber behind it as it explores. This could be accomplished through the use of a precision-wound axial spooler or from a traditional drum spooler and a fiber optic rotary joint. For a 1070 nm wavelength laser, the attenuation is of the order of 12% per kilometer of fiber. This can be improved dramatically if a wavelength of 1550 nm is used. However, at present 1070 nm lasers dominate the market because there is a direct means for fiber-pumping the laser and a simple method for combining hundreds of source lasers is now standard practice. This is not the case with 1550 nm lasers, which still require complex pumping and combiners. To date the highest power 1550 nm laser is about 300 W. Commercial lasers in excess of 100 kW at 1070 nm are now a 6-month lead time, off shelf.
- The minimum bend radius of the fiber stored on the vehicle is a function of the fiber material and layup, the wavelength of the laser, and the diameter of the fiber. Tests conducted at Stone Aerospace suggest that power loss (through Raman scattering) begins to accelerate at bend radii less than 100 mm. This sets a lower limit on cryobot diameter for missions to great depth (but smaller diameter cryobots are still feasible at lower power levels and shallower mission depths).
- The theoretical limit to power that can be sent down an optical waveguide is staggeringly large: on the order of 4.6 MW (megawatts), determined by the

"self-focusing" limit for fiber lasers. The maximum power tested during the VALKYRIE project was 11 kW during laboratory tests. For practical reasons, including the size and weight of the laser and support equipment, 5 kW was the power chosen to run the initial field tests. From a limiting implementation standpoint the engineering crux is the transition from the laser power source to the vehicle fiber. This must take place in a fiber fusion splice wherein the (smaller) laser feed fiber is fused to the (larger diameter) vehicle spooler fiber. Further tests are needed to develop an efficient, reliable, and economical splicing method for ultra-high-power lasers (or an alternative method using beam expanding optical coupling technology) in order to enable practical high power levels (in the 100 kW range) to be transmitted to a remote robotic system. These would be essential to master if a laser-powered cryobot of significant diameter (0.3–0.4 m) were to be designed for Ocean World ice cap penetration or for general sub-glacial lake access in Antarctica. This is not an intractable problem, and the engineering approach for solving it is well in hand, but the work remains to be done to bring it to a flight technology readiness level (TRL). For less ambitious objectives (e.g. sending 100–300 W of power to a clean rover near the Martian polar ice cap from an RTG-powered lander) the methods developed and proven during the VALKYRIE project are sufficient to justify flight mission systems development.

- The Science Payload (see Sect. 4.9) and the autonomous decision-to-collect algorithm operated on a basis comparable to that of a trained microbiologist. This method of using a flow-through fluorescence cytometer and microbe detection algorithm thus presents a viable down-select system for autonomously assessing in situ samples as a cryobot descends through an ice column. This system is presently at TRL6 and would form a useful flight tool for discriminating feed stock prior to delivery of a melt sample to, e.g., a 3D optical microscope or to a compact DNA sequencer.

4.6 Spindle

In 2015 NASA approved a Phase A systems definition study of SPINDLE (Sub-glacial Polar Ice Navigation, Descent, and Lake Exploration)—an autonomous 2-stage cryobot consisting of a robotic ice-penetrating carrier vehicle (cryobot) and a marsupial, hovering autonomous underwater vehicle (HAUV). The cryobot was to descend through an ice body into a sub-ice aqueous environment and deploy the HAUV to conduct reconnaissance, life search, and sample collection. The HAUV was to return to, and auto-dock with, the cryobot at the conclusion of the mission for subsequent data uplink and sample return to the surface. The cryobot was ultimately to be designed for a 4 km penetration through a terrestrial ice sheet and the HAUV was to be designed for persistent exploration and science presence in Antarctic sub-glacial lakes. In addition, as previously reported in

Sect. 4.4, explicit steps were taken to develop initial designs for the nuclear power systems that would enable such a 2-stage flight vehicle.

Taken to completion in a three-phase program, SPINDLE was to deliver an integrated and field-tested system that would be directly transferable into a Flagship-class mission to either the hypothesized shallow lakes of Europa (Schmidt et al. 2011), the sub-surface ocean of Ganymede (McGrath et al. 2013), or the geyser/plume sources on both Europa and Enceladus (Roth et al. 2014; Hsu et al. 2015). During the Phase A study a number of significant events took place. The team used the ARTEMIS HAUV vehicle to test new behavior algorithms and hardware for autonomous exploration, rendezvous and docking under the Ross Ice Shelf when deployed vertically through a 1.2 m diameter drilled shaft (Kimball et al. 2016), proving the ability of an autonomous rover to perform significant exploration missions (the longest ARTEMIS mission was more than 10 km under ice with deployment and recovery through a 10 m deep access shaft as the only route of entry and egress and with open ocean more than 50 km distant) followed by fully automated retrieval and data upload. Two other important breakthroughs relating to cryobot technology were made during SPINDLE Phase A, both dealing with methods of transferring surface power to a test vehicle in Antarctica. The first of these will be discussed in this section while the second (a spin-off of the VALKYRIE project) will be discussed in Sect. 4.7.

4.6.1 Design of a 2-Stage Cryobot

Following a year of simulations, thermal analyses, and laboratory proof-of-concept tests, a design emerged (Fig. 4.77) that appeared feasible. Among the key features were:

- A cryobot diameter of 40 cm and length of 7 m
- A bi-directional power spooler onboard the vehicle
- A bi-directional independent data fiber spooler onboard the vehicle
- A switching CCHWD that could bore downward, laterally, and upwards
- Sizing of the vehicle to allow inclusion of a viable small 6 degree-of-freedom (DOF) HAUV that could be deployed and retrieved by the cryobot
- Sizing of the vehicle around a viable 100 kW_{th} nuclear fission power source
- Accommodation of both high voltage electrical and optical energy transfer from the surface

Designing a non-nuclear SPINDLE prototype for analog testing required us to develop a means of transmitting energy from a surface power plant to the vehicle. Importantly, the SPINDLE design had to pay out its own power tether as it descended, since the hole would re-freeze behind the vehicle during Antarctic operations, and then reel itself back in while melting its way upward at the conclusion of a mission. The SPINDLE design trade competition pitted the VALKYRIE high

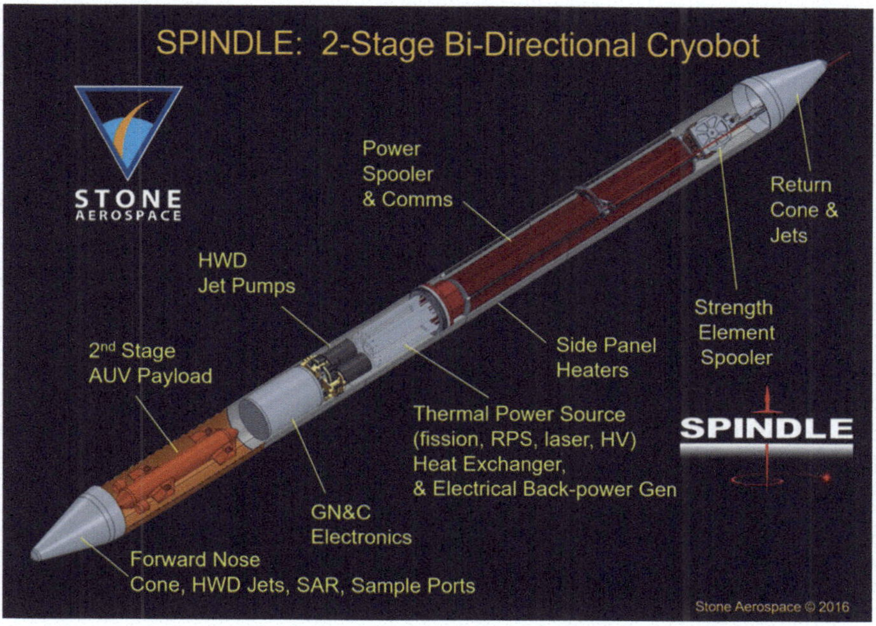

Fig. 4.77 The SPINDLE 2-stage, bi-directional cryobot was designed for a round trip to the bottom of Antarctic subglacial lakes. Principal module elements (from the bow) included: forward CCHWD jets; AUV cargo bay; guidance, navigation and control electronics pod; redundant CCHWD jet pumps; the thermal power source (either nuclear, laser, or high voltage) and back-power conversion systems; power and communications tether spooler; elevation control servo spooler; and aft CCHWD jets. The system was designed for a 100 kW_{th} power source. The vehicle measures 40 cm in diameter and 8 m long

power laser approach (using a 100 kW 1070 nm fiber laser) against an effort to push the limits of high-voltage, high-power transmission via small diameter conductors. The results of this study for a field-deployable, small-footprint bi-directional cryobot currently favors use of a surface-based high voltage generator over the laser approach.

However, that situation is ephemeral: each year the wall plug-to-photons conversion efficiency is rising and while the VALKYRIE approach of channeling laser power into a beam dump is less efficient today than an improved high voltage system, the novel alternative way of using laser power (described in Sect. 4.7) will shortly be competitive as a general deep ice access method.

4.6.2 *Breakthroughs in High Voltage Transmission*

The primary technical problem with using high voltage power transmission is the risk of arc-over. This was the limiting problem with the original Philberth probe: as the voltage is increased the diameter of the conductor can be decreased to deliver the same power, but the size of the cable increases due to the larger thickness of insulation required. Eventually the size of the cable dominates the vehicle architecture and becomes unwieldy. SPINDLE was being designed as a bi-directional cryobot for 4 km of ice thickness and even with a 40 cm diameter hull the length of the vehicle was growing to over 20 m with traditional, proven, high voltage ROV cables used in offshore oil work. The key to moving forward was to reduce the diameter of the cable. Lab tests were conducted for a wide variety of novel wire insulators because as voltage increases, insulation eventually controls the amount of power that can be transmitted before arc over and for each insulator there is a characteristic curve (e.g. Fig. 4.78).

The left cross-section in Fig. 4.79 shows a traditional ROV power cable designed for 30 kW; the right graphic shows the lab tested diameter reduction achieved with SPINDLE for a 50 kW cable by custom fabricating conductors with Kapton insulation. This tether diameter reduction enables the design a vehicle of reasonable deployment size that is capable of reaching any of Earth's deep subglacial lakes.

4.6.3 Power, Communications, and Strength Spoolers

The SPINDLE design uses two servo-controlled tether spoolers: a dedicated strength spooler for descent and ascent and a dedicated power/comms spooler (Fig. 4.80). This approach reduces overall vehicle diameter; reduces internal resistive heat buildup in the power spooler; and reduces differential thermal stress. While the use of two tethers in a single hole may appear to pose a challenge, the approach is not unprecedented. The IceCube project (Benson et al. 2014; and

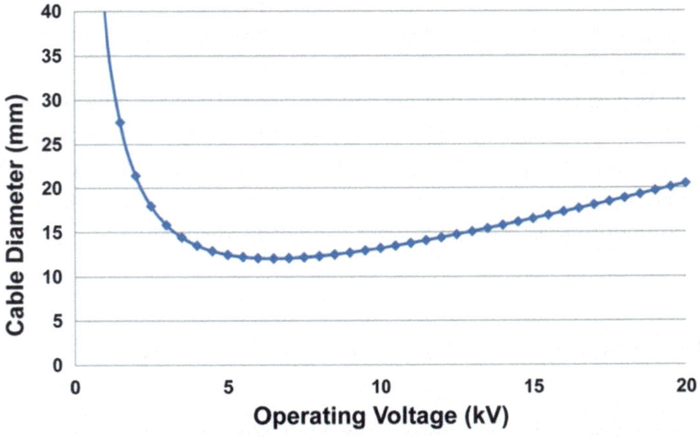

Fig. 4.78 Characteristic curve of cable diameter vs drive voltage (in kV)

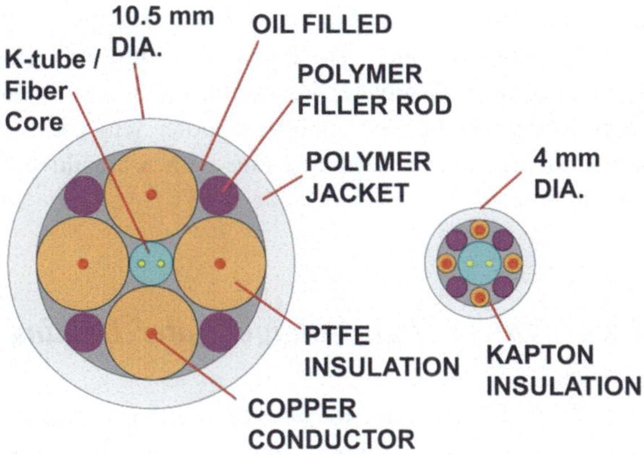

Fig. 4.79 Traditional 30 kW ROV high voltage cable (left) and a 50 kW-rated SPINDLE cable (right), achieved through the use of bare Kapton insulation

Fig. 4.80 Servo-controlled Power/communication and Strength tether spoolers for SPINDLE. These geometries were sized for the physically-realizable cables needed for a 2500 m-deep Antarctic sub-glacial initial test of SPINDLE within a 40 cm diameter cryobot

personal communications with Dennis Duling, lead field engineer) had multiple down-hole lines and did not have twisting problems in the course of drilling 86 two-kilometer-deep holes. To mitigate any twisting SPINDLE contained yaw (axial torque) thrusters that maintain vehicle yaw angle at all times.

Based on the high voltage lab work conducted during Phase A the SPINDLE tether consists of four Kapton-insulated conductors arranged in a square cross-section. The dramatic size reduction is shown in Fig. 4.80 for a 100 kW tether (a 50 kW tether measuring 3 mm in diameter was successfully laboratory tested at Stone Aerospace in Phase A). The center of the cross section is occupied by the armored K-tube fiber optic link. All interstitial spaces are filled with a low viscosity, high dielectric strength oil to reduce friction during flexure and provide additional voltage standoff. A polyethylene jacket provides mechanical integrity and abrasion resistance. The resulting design will resist hydrostatic stresses and the stainless core will prevent ice compression and shear damage to the critical communications fibers. The strength spooler will use a compression and shear insensitive Spectra 4 mm recovery line with a tensile safety factor of 25 over vehicle weight in air.

4.6.4 Direct High Voltage Water Heating

SPINDLE will be able to achieve high penetration rates by using high-pressure water jets to rapidly transfer heat from a novel direct high voltage heater system (Figs. 4.81, 4.82 and 4.83) to the ice. High pressure jets are essential to enabling penetration through debris layers (Thorsteinsson et al. 2008). This novel architecture (Fig. 4.83) has not been implemented before because of the inherent challenges in making large volumes of hot water from a high voltage power source. Standard electro-resistive heating elements operate at relatively low voltage and high current,

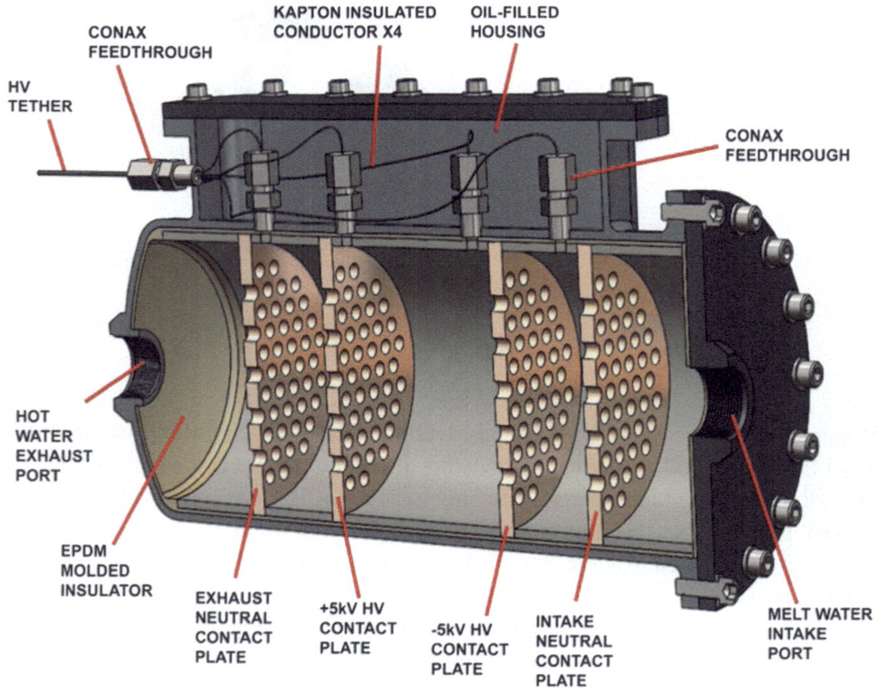

Fig. 4.81 Close-up section of direct HV water heater. A conductive fluid passes through high voltage AC current electrodes designed to inject power and leave no residual charge in the fluid. Because the voltage is AC there is no electrolysis of the working fluid

taking advantage of I^2R heating in the element. An electro-resistive element capable of operating at high voltage with sufficient surface area to allow rapid heat transfer forces the heater design into an impossible geometry. The SPINDLE project found a novel solution by passing high-voltage, low current, AC power through a moving conducting fluid. This generates resistive heating in the fluid with 100% efficiency without inducing electrolysis. The resistivity of the process fluid can be tuned over a wide range by controlling the concentration of polar molecules in the fluid. This tunable resistivity allows unprecedented power densities to be achieved. Laboratory testing conducted during SPINDLE Phase A achieved power densities of over 600 kW/l. SPINDLE thus uses a closed-loop heater system that is in thermal communication with an open-loop hot water drill via a heat exchanger (Fig. 4.82). The primary heater loop uses a process fluid with a depressed freezing point so the vehicle can restart even after being frozen in the ice. Primary loop circulation is accomplished by a high-volume, low-pressure centrifugal pump. Process fluid transits through the high-voltage heater core and into the primary side of the heat exchanger. Meltwater enters inlet ports aft of the nose cone and is pumped through the secondary side of the heat exchanger by a series of high pressure, high volume

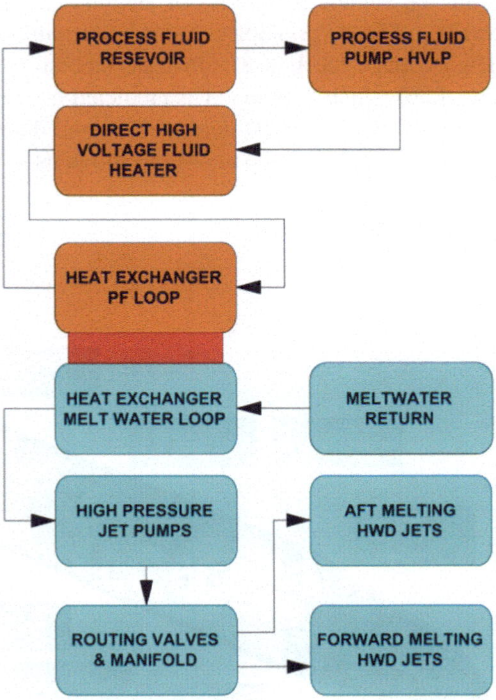

Fig. 4.82 Major components of the CCHWD using direct high voltage water heating. It should be noted that the heating system is fully closed cycle while the melt water portion of the system is open loop

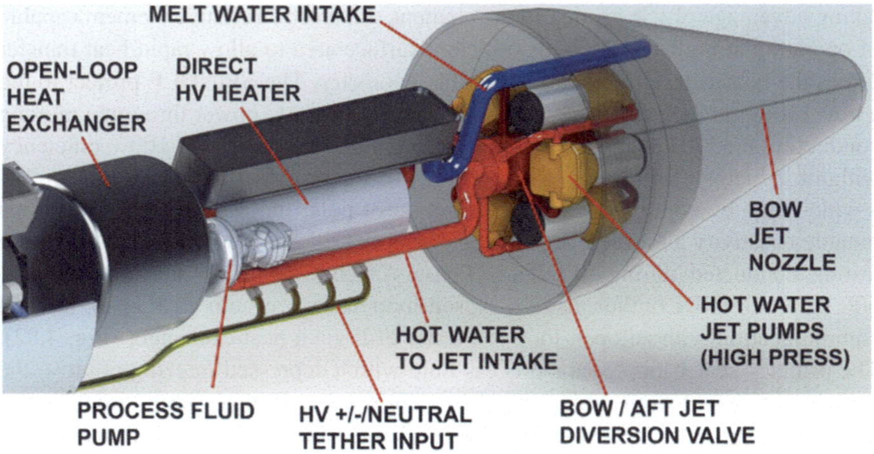

Fig. 4.83 3D Mockup of nose section with high pressure CCHWD elements

diaphragm pumps. After the water travels through the heat exchanger it is ejected from the vehicle in a series of jets that can be controlled via a series of solenoid valves. These principles were successfully tested in the PHIL (Prototype High-voltage Ice Lance) sub-scale laboratory prototype cryobot (Figs. 4.84, 4.85 and 4.86) which functioned as predicted from previous component tests and theory. The combination of extraordinarily small diameter power transmission line with a means to reliably dump enormous amounts of power in a small location open new

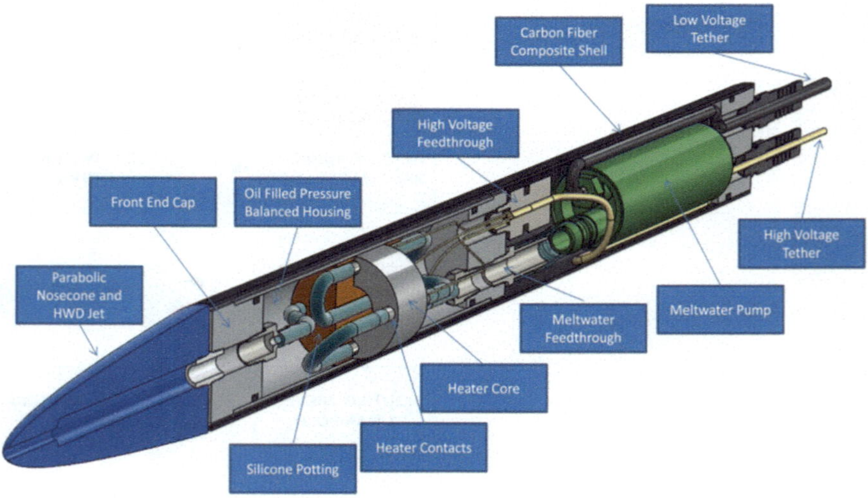

Fig. 4.84 A 3D model of PHIL (Prototype High-voltage Ice Lance) showing key components. This device was built and successfully tested in the summer of 2016

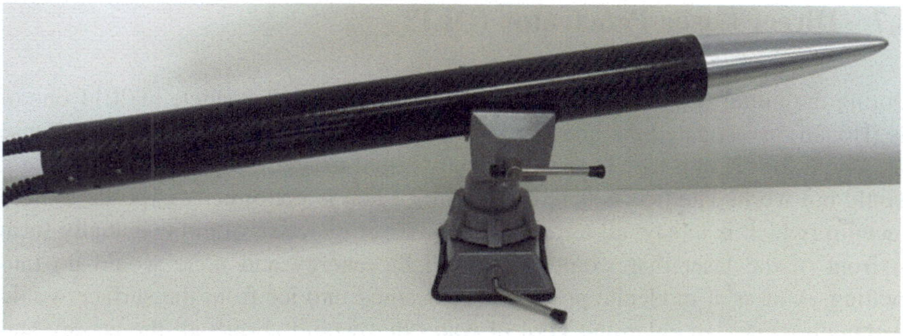

Fig. 4.85 The physical test prototype for PHIL. The top cable at left in the image is the 10 kV AC power input line (4 wires); the lower cable at left is lab power to operate the onboard jet pump (for speed and simplicity of testing the core invention we left the onboard back-power generation elements to a later prototype)

Fig. 4.86 PHIL in ice during initial testing at Stone Aerospace in August of 2016

possibilities for terrestrial ice penetrators and provide clean access options for planetary sub-ice exploration where the nuclear power source remains with the lander. In general this would be a scenario in which small diameter probes were powered by lander-based RTG systems and penetrations would be on the order of tens to hundreds of meters. The scenario of a 2-stage cryobot-HAUV system remains possible for an Ocean World mission only with the use of a 100–250 kW_{th} fission reactor onboard the cryobot.

4.7 Direct Laser Penetrator (DLP)

During the initial vehicle design phase for the VALKYRIE vehicle in 2011 one of us (Hogan) investigated the question of whether it would be feasible to directly use laser light for drilling ice. There were many initial reasons why this seemed like it would not work. The first is that, at least in ice that is not directly exposed to hard vacuum (which is to say, all terrestrial ice) a layer of water would eventually form in front of the laser that would absorb all the energy and none would go into melting. A laser at moderate power being beamed into ice from the surface would eventually create an enlarging pool of water that would absorb all the energy and thus prevent focused energy from reaching the bottom of the hole. If the laser end optics were lowered down the hole on a fiber (as with the VALKYRIE vehicle) and the power levels were high then the water in front of the optics head was likely to flash, potentially creating a pressure wave or explosion that would damage the cryobot. This supposition was not unfounded as at least one team had previously investigated this possibility using a CO_2 laser (Sajurai et al. 2015) and confirmed that in a vertical melting orientation the water did indeed flash and the only way for them to continue testing was to ensure that the test ice block was tilted down and the beam was impinging on the side of the block so the water generated could run out leaving only exposed ice. The desired behavior is obviously to have all the energy be absorbed by the ice and little if any by the water.

As part of the SPINDLE project we investigated in detail a number of related questions:

- What is the transmission efficiency of laser light through water and through ice?
- What optimal laser wavelength provides low loss in water and gradual absorption in ice?
- What amount of reflection occurs at optics-water boundaries and ice-water boundaries?

4.7.1 Initial Laboratory DLP Tests

Detailed research by Stone Aerospace into the absorption of electromagnetic radiation into both ice and water (e.g. Kedenburg et al. 2012; Warren 1984 and others) was undertaken in 2016. Figures 4.87 and 4.88 indicate that below 700 nm wavelength both material phases are nearly 100% transparent, and above about 1200 nm both are 100% opaque. However, as illustrated in Fig. 4.87 (for liquid water) and Fig. 4.88 (for water ice), there is an extraordinary anomaly at a wavelength of 1070 nm. Figure 4.89 shows a plot of energy (at 1070 nm) absorbed in liquid water (blue line) and water ice (red line) as a function of penetration distance into each respective media. The optimal frequency for melting ice thus appears to

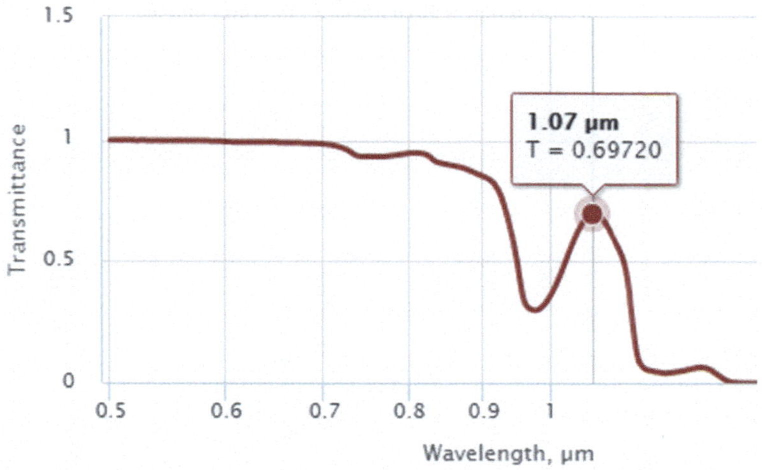

Fig. 4.87 Transmittance spectra for liquid water (from Kedenburg et al. 2012)

be close to 1070 nm, a wavelength at which electrical-to-optical conversion efficiency in existing lasers is high (>50% and rapidly increasing).

In May 2016, a series of lab tests were conducted (Figs. 4.90, 4.91, 4.92, 4.93, 4.94 and 4.95) using a 3.2 cm diameter direct-laser probe (DLP) operating at varying power levels from a 1070 nm Ytterbium fiber laser.

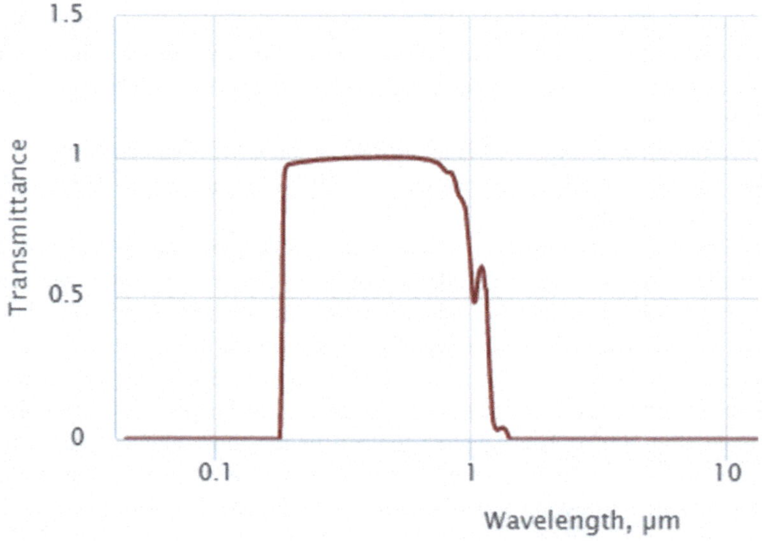

Fig. 4.88 Transmittance spectra for water ice (from Warren 1984)

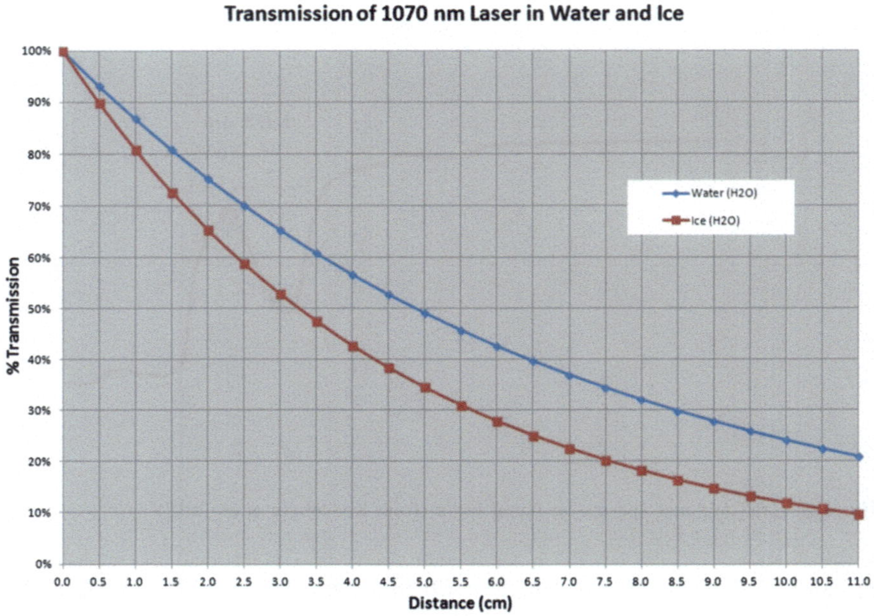

Fig. 4.89 Transmission of 1070 nm laser light in water and ice as a function of distance from the point of optical entry into the medium

Figure 4.90 shows the internal configuration of the optics chain. For expediency we used a long tube to serve as the cryobot body and placed the collimating and focusing optics (Fig. 4.91) at the end. A laser power coupler, beam collimator, and optical alignment stage (XY) were included at the top of the probe (Fig. 4.92). The system was fired at successively increasing power levels (Fig. 4.93) and the rates of descent into an ice block at −26 °C were recorded.

The purple glow in Fig. 4.93 is an artifact of the CMOS camera we were using to record the test. The lens did not have an NIR filter and so 1070 nm radiation appears as bright purple.

The initiation of descent can be seen in the visible spectrum in Fig. 4.94 in one of the earlier tests in which a parallel set of beam optics were employed. The ice crazing that can be seen ahead of the cryobot optics is where most of the energy is being deposited. We later used a set of focusing optics that created a 6° internal angle beam which deposited the majority of the optical energy approximately 15–20 cm ahead of the cryobot, leading to a narrow cone-shaped melt front ahead of the vehicle (Fig. 4.95). This appears to lead to the most efficient deposition of energy and the fastest descent speed although further work in this area is merited.

Effectively linear descent speed response was obtained in 247 K (−26 °C) ice over a range power levels from 50 W to 2.5 kW. The descent speed at 2.5 kW was

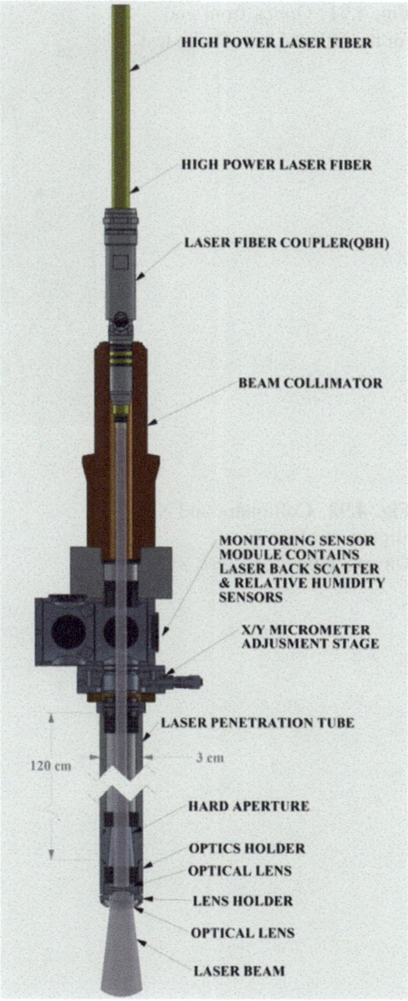

Fig. 4.90 Cross section of Direct-Laser-Probe test conducted at Stone Aerospace May 26, 2016

greater than 12 m/hour and accelerating, but ice block size limited the test to 1 m of penetration. The temperature was monitored at the nose cone and peaked at only 30 °C at the highest power level tested (2.5 kW). Later, in August of 2016, tests at 5 kW laser power reached penetration rates of 22 m/h in identical ice blocks and temperatures as the earlier tests. The present largest commercial 1070 nm fiber laser delivered is 100 kW. Using such a power source and the above described probe, with no further optimizations, we can extrapolate that DLP could access the deepest ice on Earth in under eight hours.

Fig. 4.91 Optics front end for the Direct-Laser-Probe test

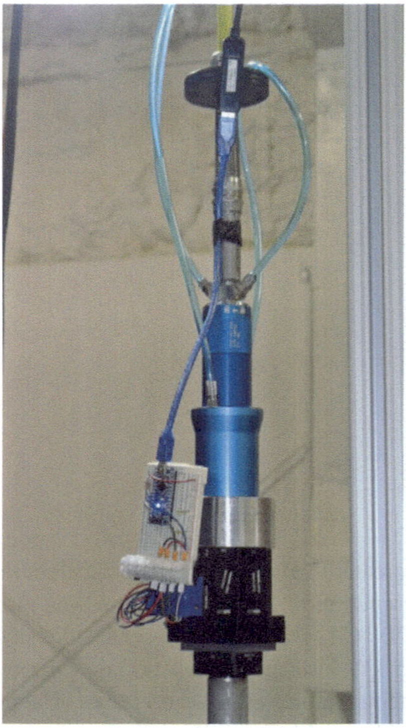

Fig. 4.92 Collimator and XY alignment stage for Direct-Laser-Probe test

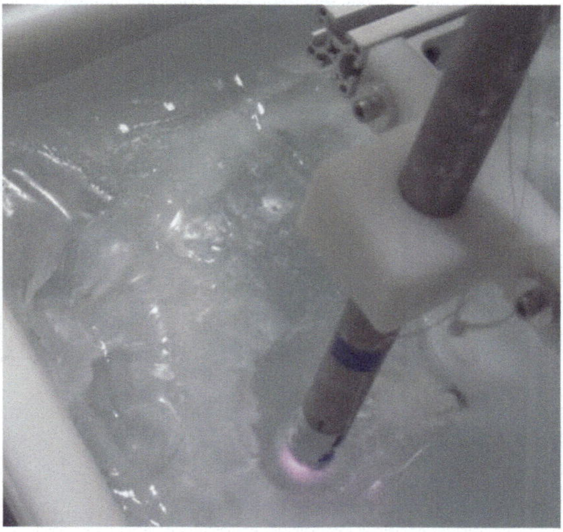

Fig. 4.93 DLP entering ice at a power level of 50 W. The violet glow is from the 1070 nm laser beam (Near-IR, not visible to human eye) and interpreted by the camera's CMOS sensor as violet, due to lack of Near-IR filtering

Fig. 4.94 Initiation of laser firing at 2.5 kW power for the May 26, 2016 DLP test. The interior of the cone ahead of the lens is already liquid. The crazing of the ice ahead of the point of the cone was observed to become more pronounced at higher power levels

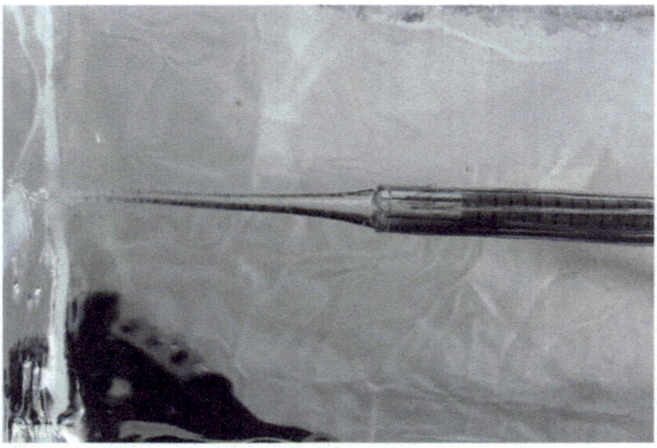

Fig. 4.95 High power test in August 2016 at 5 kW with different front end optics. In this test the beam was focused to a point some 20 cm ahead of the probe, creating this lance-like melt zone ahead of the probe. The speed at this power level was 22 m/h

4.7.2 Implications for Ocean World Missions

Results of the initial DLP tests, performed at 1070 nm, showed efficient absorption in ice with only modest losses through water and at reflective surfaces. Specifically we observed:

- Modest transmission loss in water allowing penetration of water pockets forward of vehicle
- Absorption of laser energy in ice was near optimal, allowing for deposit of energy into a controlled volume of ice ahead of the vehicle
- Reflections at the output optic-water boundary are very low due to the index of refraction for fused silica being well matched to that of water with no antireflective coating (Fresnel loss of 0.21%)
- Reflections at the ice-water boundary are extremely low due to the index of refraction of ice being well matched to that of water (fresnel loss of 0.01%)

In vacuum at 100 K, DLP will deposit 100% of its energy directly into the ice, sublimating the ice in the process and venting to the vacuum environment. As the probe descends, sublimate refreeze will eventually close the hole behind the probe. Following hole closure, continued melting will produce a thin layer of liquid water just ahead of the probe. 1070 nm laser light will pass through this water layer with low loss (0.1% loss at 0.1 mm, 1% loss at 1 mm, 13% loss at 10 mm), and the bulk of the energy will be directly deposited into the volume of ice ahead of the probe. With proper focus and collimation the melt area will rapidly conform to the probe nose cone. This is significant because with a traditional melt probe, significant amounts of energy are lost in the process of coupling the energy to the ice (manifesting as a larger-than-needed borehole diameter). The optical absorption anomaly at 1070 nm is extraordinary because it also coincides with the wavelength of readily-available fiber lasers, which have recently seen dramatic improvements in power, size and mass reduction and electrical-to-photon conversion efficiency.

In December of 2016 NASA funded the ARCHIMEDES project, which will advance DLP to the level of a TRL6 pre-flight level instrument for penetration of the shallow Europan ice cap (Fig. 4.96). Plans are currently underway to test the probe's fused silica fiber optic tether and output lens at cryogenic temperatures ($\sim$100 K) in a Cryo-Vac ice chamber and measure the response of these glass components under actual mission conditions. Perhaps surprisingly, optical fibers remain flexible at cryogenic temperatures. It has been shown that operation at cryogenic temperatures has little effect on the fiber: it retains flexibility, shows a negligible increase in focal ratio degradation, and the transmission remains constant. Furthermore, demonstrated breaking strain under bending improves by 280% for silica fibers at cryogenic temperatures. Bending has been demonstrated down to 6 mm diameter before fracture in an off-the-shelf laser transmission fiber at room temperature, and transmitting 100 W of power. A fiber can handle more power as the bend diameter is increased—hundreds of watts at bend diameters of 10–15 mm and potentially over 1 kW above 25 mm diameter. All of this implies that a small diameter probe that will function at cryogenic temperatures can be achieved. In the special case of Europa techniques for radiation-hardening of the fiber and other optical components will need to be investigated.

The ARCHIMEDES direct-laser penetrator technology has critical benefits over conventional hot point melt probes and mechanical drills. It is particularly advantageous for the extreme cold and vacuum environment on the surface of ocean

Fig. 4.96 Development of ARCHIMEDES, a 4 cm diameter DLP probe targeted for the Europa lander mission, was approved by NASA in December of 2016. It also carries fiber spectroscopy sensors for detection of biomarkers in the ice

worlds and is an immediate solution to the "Starting Problem" described earlier. Currently our direct laser probe is targeted for very efficient penetration of 1–10 m depth and scales well for penetration of 10s to 100s of meters. The estimated power consumption on Europa for various diameter DLP probes for a 5 m deep penetration in hard vacuum and 100 K ice temperature is shown in Fig. 4.97.

With direct laser heating no thermal contact is needed and 100% of the laser power is deposited directly into the ice volume ahead of the probe with no thermal handling losses. 1070 nm radiation has greater absorption in ice than in water, limiting attenuation through a water pocket that will eventually form in front of the probe once the hole seals shut behind the vehicle (sublimate will eventually freeze to the hole and seal it). Another benefit of depositing energy into a volume of ice, rather than upon a surface, is a lower temperature for sampling compared to hot point probes. Hot point (or "hot penny") probes must generate large temperature gradients to force heat through layers of ice and water. In initial DLP testing, we observed only moderate temperature rises at the ice-water interface, implying that samples will not be excessively heated.

The ARCHIMEDES probe's fiber optic tether, along with a dedicated sensor fiber, enables optical access to the subsurface environment. This access can be utilized by a lander's onboard instruments or instruments onboard the penetrator. These sensors can search for biomarkers and characterize the radiation/light environment for subsurface habitability, including photosynthetic potential and

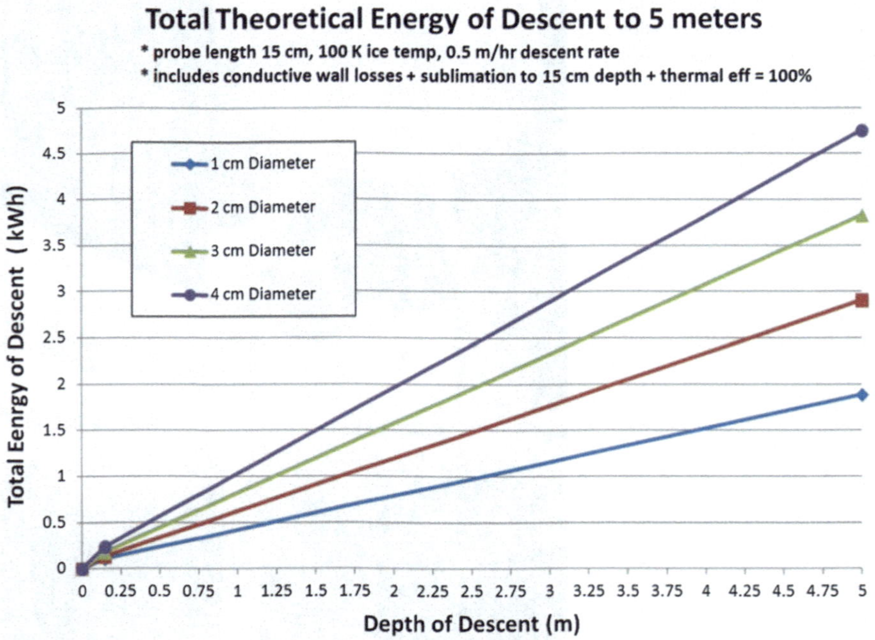

Fig. 4.97 Theoretical total energy requirements at 100 K in vacuum for various DLP probe diameters as a function of penetration distance on Europa

radiation environment as a source for energy and damage. This combination—laser penetrator and integrated fiber instruments—could be a powerful new lightweight tool for an ocean world lander that obviates both the mass and complexity of a "grab and transfer" materials-handling system.

4.8 Obstacle Avoidance

Thus far we have confined discussion to ice penetration physics and methods and designs for ice penetrating vehicles. A closely related issue, particularly in areas of active volcanism and glaciation in mountainous regions on Earth, and specifically for planetary ice caps like those of the Ocean World moons, is that of obstacle avoidance. Anyone considering an ice penetration mission to Europa and similar targets would be risking the success of the mission if they assumed that there will be clear, clean ice from the surface to the sub-surface ocean. We have already demonstrated the ability of a cryobot to deviate from vertical with the VALKYRIE vehicle and directed CCHWD jetting. The SPINDLE vehicle is designed to be bi-directional and can thus "retreat" if faced with an impenetrable obstacle. The missing piece, then, is the ability to look sufficiently far ahead through the ice to determine in advance whether the vehicle needs to deviate from a nadir descent path as dictated by gravity. Part of the VALKYRIE project was directed towards development and field testing of an initial design for an ice-penetrating synthetic aperture radar (SAR) compatible with the geometry and thermal environment of VALKYRIE. The objective of this work was to construct a bench prototype and test the field strength and achievable resolution of such a look-ahead instrument and to use the prototype on an actual glacier to frame the achievable performance limits (range, resolution, power consumption) for a SAR system that is compatible with a cryobot mission.

In this section we describe a forward looking (end fire) synthetic aperture radar (SAR) system, called IceSAR. The design of IceSAR involved analysis of novel conformal log periodic antennas for the radar and the definition of the forward looking SAR ambiguity function. Fabrication and laboratory characterization of the antennas system design and in situ testing of the SAR system are also discussed in detail.

The need for radar sounding of Europan sub-surface is motivated by the imaging results from the Galileo mission (Moore 2000). Radar sounding for Europa in the upcoming Europa fly-by mission has been discussed in Blankenship et al. (2009). However, the Europa Flyby Mission ice penetrating radar will not have sufficient resolution to identify objects that would stop a cryobot. The IceSAR system described in this chapter will provide critical forward-looking (end fire) obstacle avoidance capabilities for an Ocean Worlds cryobot. Testing and characterization of the antenna and radar system was carried out at Stone Aerospace in Austin, Texas and on the Matanuska Glacier in Alaska.

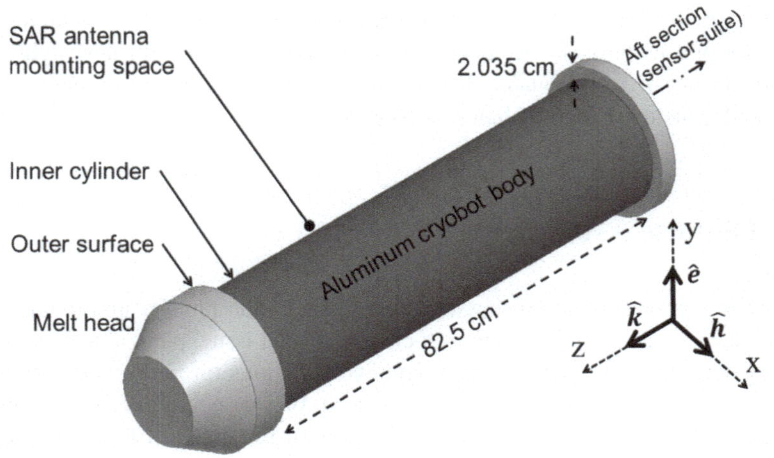

Fig. 4.98 A to-scale model of the section of the cryobot cylinder indicating the volume available for the SAR antenna is shown. The electric and magnetic field unit vectors $\hat{e}$ and $\hat{h}$ and wave propagation unit vector $\hat{k}$ are shown with bold lines on the cryobot coordinate axis

The primary purpose of IceSAR is to provide forward looking obstacle avoidance and ice feature mapping capability for the cryobot. The development of the IceSAR system consists of two parts. First is the design of a conformal log-periodic folded slot array (LPFSA) antenna that forms an agile radiating element for the SAR. The choice of operating frequency and antenna type is influenced by electromagnetic propagation inside ice as well as the mounting space available on the cryobot for the antenna. The antenna is to be flush mounted between the melt head and the sensor suit section (Fig. 4.98) and must fit on the outer surface of a cylinder. For the VALKYRIE cryobot this meant a working diameter of 25.4 cm and within a maximum axial length of 82.5 cm. The maximum radial depth available was 2.035 cm. The active melting of the cryobot creates a thin layer of fresh ice melt water covering the flush mounted antennas. The antenna design must therefore take into account the effects of this dielectric layer. The second part involves the analysis of forward SAR resolution characteristics and design of the radar analog and digital system that include RF signal generation, transmit-receive isolation and digital SAR data processing using a combination of ARM processor and field programmable gate arrays (FPGA).

4.8.1 SAR Antenna Design

The LPFSA (Nurnberger et al. 1994) is designed by extensive full wave simulations that model the flush mounted antenna and the cryobot along with the melt water layer formed around cylinder. Each conformal folded slot antenna (FSA) element

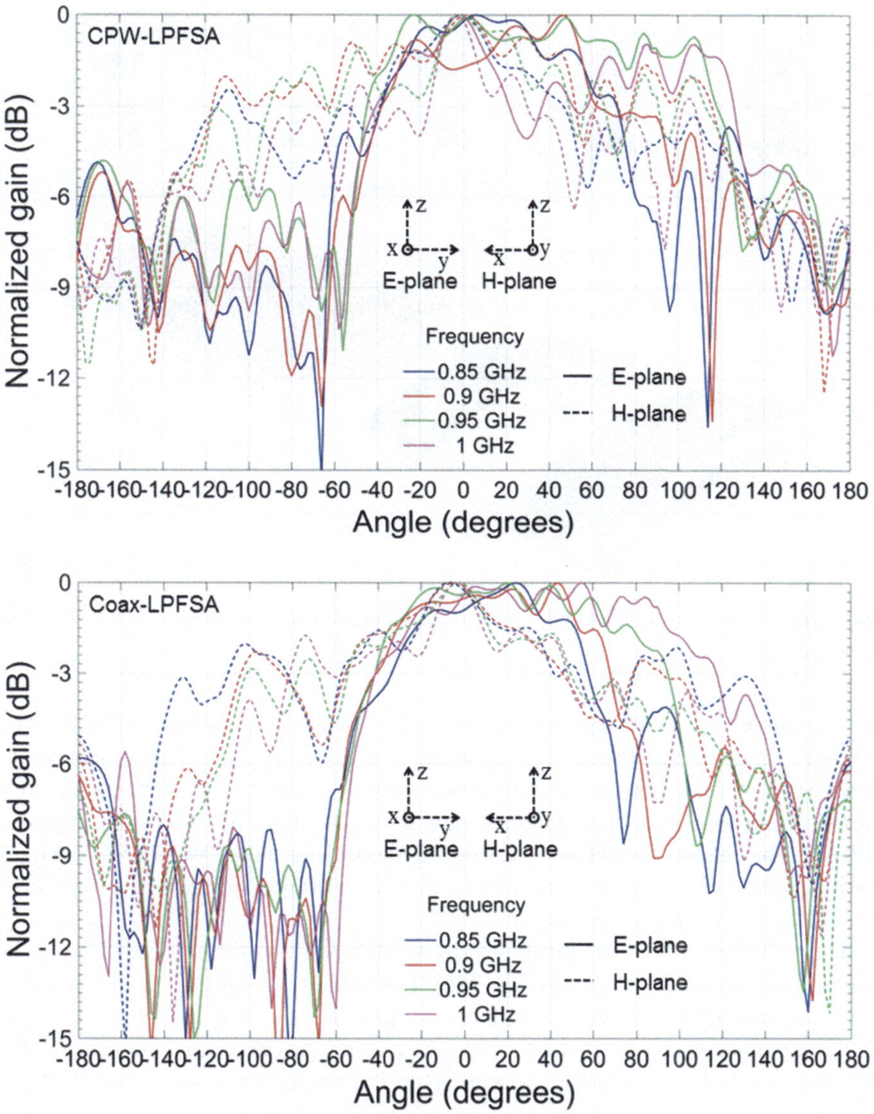

Fig. 4.99 In (top) and (bottom) the E and H plane gain patterns are shown for the chosen design constants τ and σ

making up the LPFSA is backed by a metallic cavity with small cuts in the corners so as to allow the surrounding fresh melt water to flood the volume inside.

The cavity depth is designed to be equal to a quarter wavelength of the dominant TE_{10} mode inside a fresh water filled waveguide. Two types of feeds are used to interconnect the FSA antennas, a co-planar waveguide (CPW) and a semi-rigid coaxial cable. The two versions are named CPW-LPFSA and Coax-LPFSA. The

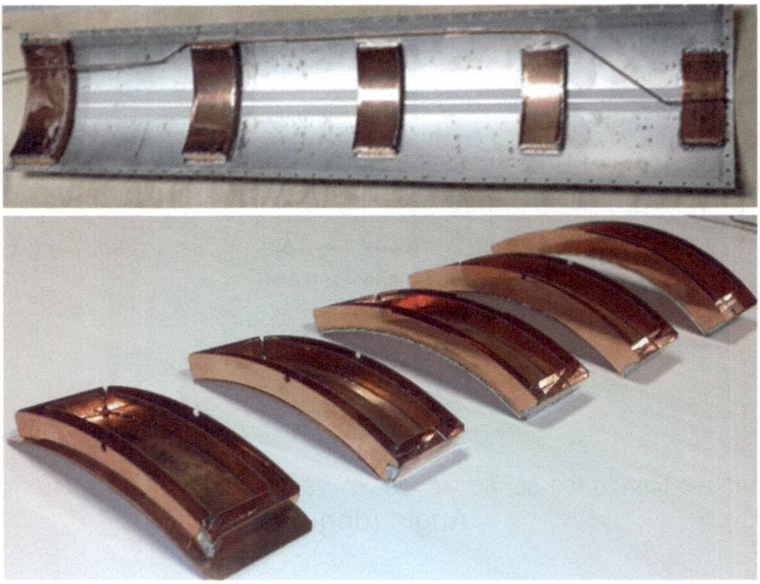

Fig. 4.100 A fabricated CPW-LPFSA panel is shown in (top) along with the metallic cavities in (bottom)

dimensions of the CPW and coaxial feeds are chosen so as to have a 50 Ω characteristic impedance. The design constant τ is used to control the logarithmic periodicity of the resonant antenna elements and is selected to be 0.86 for both types of antennas so as to allow a bandwidth of 0.54–1 GHz for a five element LPFSA. The spacing constant σ is selected to be 0.75 for the CPW-LPFSA and 0.55 for the Coax-CPW so as to cause constructive interference and therefore maximize directivity in the end fire direction.

The resulting LPFSA design shows strong main lobes in the end fire direction in both E and H planes as shown in Fig. 4.99. The field of view (FOV) can be noted to cover an included pan angle (H-plane) of 120 degress and included tilt angle (E-plane) of between 120° and 200°. The FOV is in terms of an angle range because the antenna patterns are not constant over the frequency range. Two LPFSA antennas of each type are fabricated by soldering metallic cavities to the antenna shape printed onto a Rogers RO4003C substrate (Fig. 4.100). The four antennas can be flush mounted onto the cryobot cylinder with each occupying a quarter section of the surface.

4.8.2 IceSAR Back End Design

A functional block diagram of the radar system is shown in Fig. 4.101.

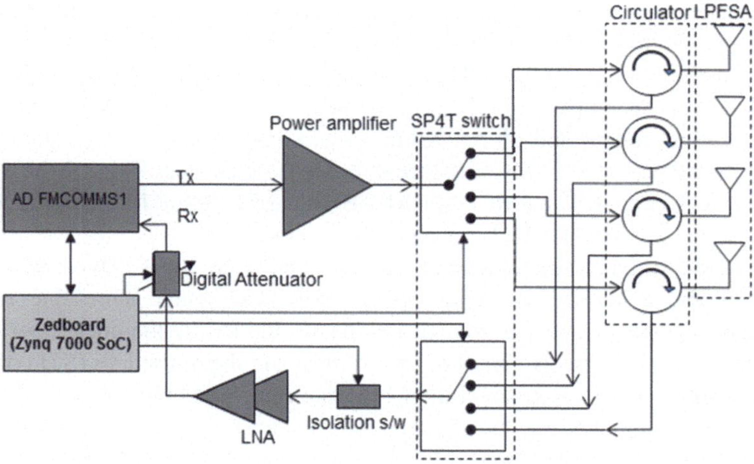

Fig. 4.101 Functional block diagram of the SAR transmit/receive system

The SAR back end (Fig. 4.102, top) is designed with commercial off the shelf (COTS) components in a compact assembly to achieve signal generation,transmit/ receive and data storage (Zuagg 2006). In the transmit path the power amplifier allows up to 40 dB of gain. The two Single Pole 4 Throw (SP4T) switches enable switching between the antenna elements thus allowing for 16 possible transmit-receive antenna pairs. The digital processing system leverages the

Fig. 4.102 (top) Microwave components assembled on the top tray of SAR module' and (bottom) the Zedboard and FMCOMMS1 assembly attached to the bottom tray

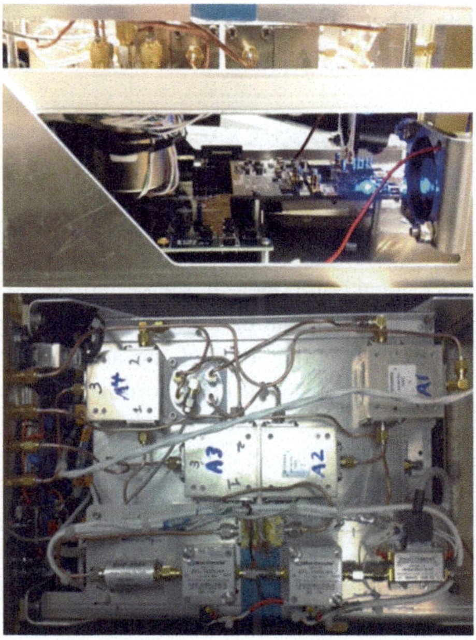

sophisticated yet easy to use FPGA development platform called Zedboard (Fig. 4.102, bottom) which is based on the Xilinx Zynq 7000 all programmable system on chip (SoC) processor. The Zynq SoC uses an ARM + FPGA architecture core processor. Specifically the Zynq 7000 features a dual-core ARM® CortexTM – A9 processing system and 28 nm Xilinx programmable logic in a single device. The Zedboard provides a FMC interface to the Analog Devices FMCOMMS1-EBZ radio transceiver card. The FMCOMMS1-EBZ has a 512 MS/s 16 bit digital-analog converter and a 250 MS/s 14 bit analog-digital converter.

The IceSAR back end hardware implemented with the COTS components described above allow a range of up to 1 km. The range achievable for the Matanuska campaign was 125 m and was mainly limited by the noise floor of the FMCOMMS1 RF transceiver card of 65–70 dBm and the return loss (S11) of 8–10 dB at the antenna feed terminals (Fig. 4.113) that can exceed the maximum average RF power at the input of the digital attenuator. However a modified transmit/receive scheme using the fast switching isolation switch (Fig. 4.101) and signal averaging can improve this range to 2 km for the same noise floor. A maximum aperture length of 9 m was achieved during the Matanuska campaign, however longer aperture lengths also allow for a linearly proportional improvement in signal to noise ratio thus allowing for additional range of detection. A maximum allowable SD memory of 16 GB imposes additional limits on maximum possible aperture length and therefore the maximum range as well.

The IceSAR back end hardware electronics are integrated into a development board and as such the individual components on the development boards are rated to work as low as −50 °C (323 K). In order to allow nominal operation of the back end hardware, selected components can be heated using heat recycled from the cryobot operation itself. The electronics will also be space qualified at the CU-Laboratory for Atmospheric and Space Physics (LASP).

4.8.3 Endfire SAR Resolution

The resolution of a radar can be studied by analyzing its ambiguity function (Cook and Bernfeld 1968). A radar ambiguity function is a complex valued function, and is a critical component in the systematic search for radar waveforms and processing techniques that can be used to optimize the resolving capability. The forward SAR geometry and the form of the ambiguity function is shown in Fig. 4.103 and is described in a cylindrical coordinate system. The radial resolution for the end fire SAR system is obtained by calculating the normalized magnitude of the ambiguity function for different axial range values of a point scatterer and then computing the full width at half power of its main lobe. Since in the present analysis a monostatic configuration is considered, the SAR cannot resolve in the azimuthal plane (x-y plane). The target is thus assumed to lie directly in front of the radar and therefore is in the so-called end fire direction of the radar antenna.

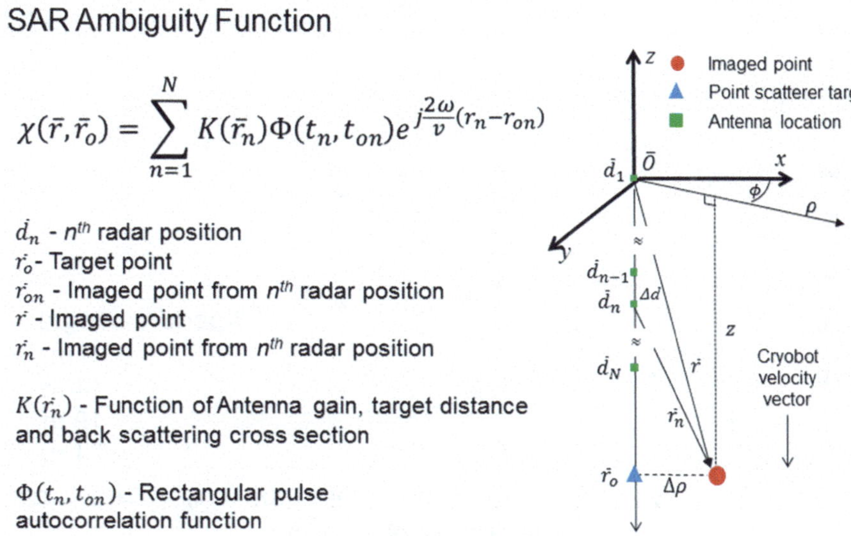

Fig. 4.103 The forward SAR imaging geometry

From the plots in Fig. 4.104, firstly, a linear degradation of the radial resolution is observed for a fixed synthetic aperture length as the axial range (along the negative z direction in Fig. 4.103) of the point scatterer is increased from 25 to 525 m. Secondly, increasing the synthetic aperture length improves the radial resolution for a fixed axial range of the point scatterer. For example for a 9 m aperture length (as was achieved during the Matanuska campaign) a theoretical radial resolution of 4 and 15 m can be achieved at an axial range of 25 and 100 m respectively. For an aperture length of 24 m a radial resolution of 10 m can be achieved with the IceSAR hardware remaining the same. The initial laboratory test rig is shown in Fig. 4.105.

The range (axial) resolution remains constant because it only depends upon the width of the transmit pulse for a monotonic transmitter. In this case a pulse width of 1.5 ms is assumed resulting in an axial resolution of ~25 m. Range resolution can be improved by implementing a chirped (linear frequency modulated) waveform. The direct digital synthesis procedure (implemented in the FPGA) used to generate the I/Q transmit signal can be appropriately modified to produce a chirped pulse. The IceSAR antennas allow for a range resolution of as low as 0.42 m for an antenna impedance bandwidth of 200 MHz (Figs. 4.104, 4.107 and 4.108).

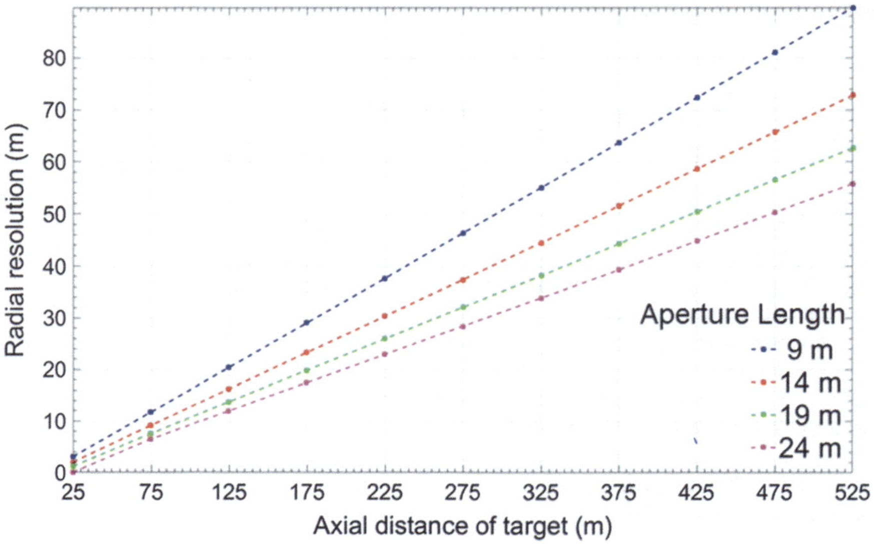

Fig. 4.104 The radial resolution obtained from the half power full width of the main lobe of the SAR ambiguity function is plotted in (b) for successively increasing point target distances along the cryobot axis from the origin O^{-}

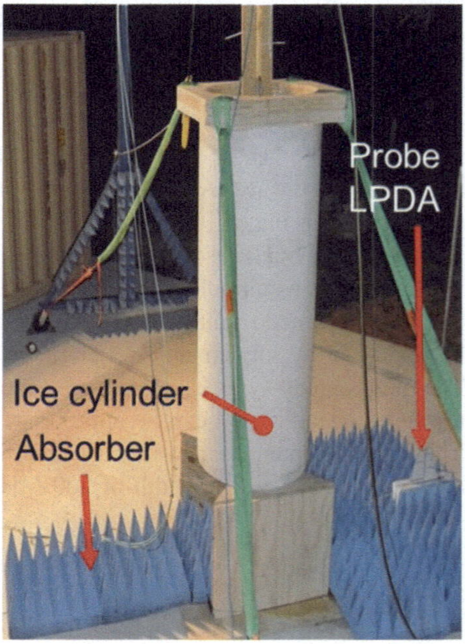

Fig. 4.105 Antenna measurement setup at Stone Aerospace Inc

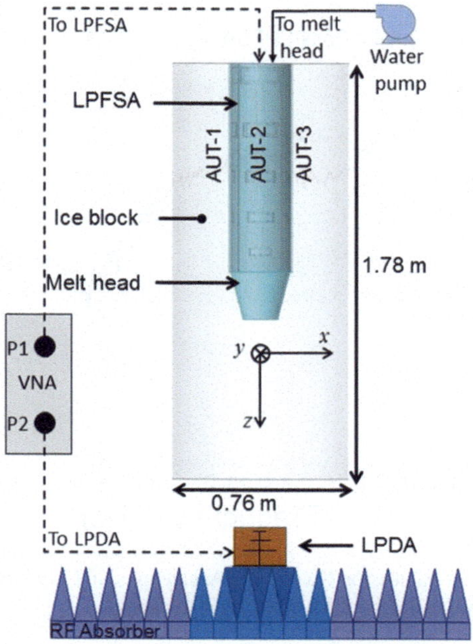

Fig. 4.106 Simulation model of the test setup used in full wave simulations

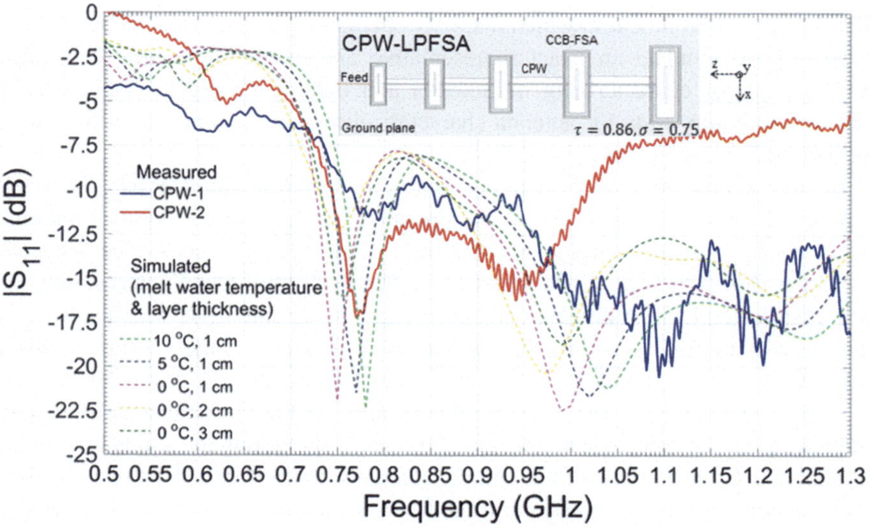

Fig. 4.107 Measured S_{11} values for CPW-LPFSA antennas and are compared with simulated values for various melt water temperatures and layer thickness

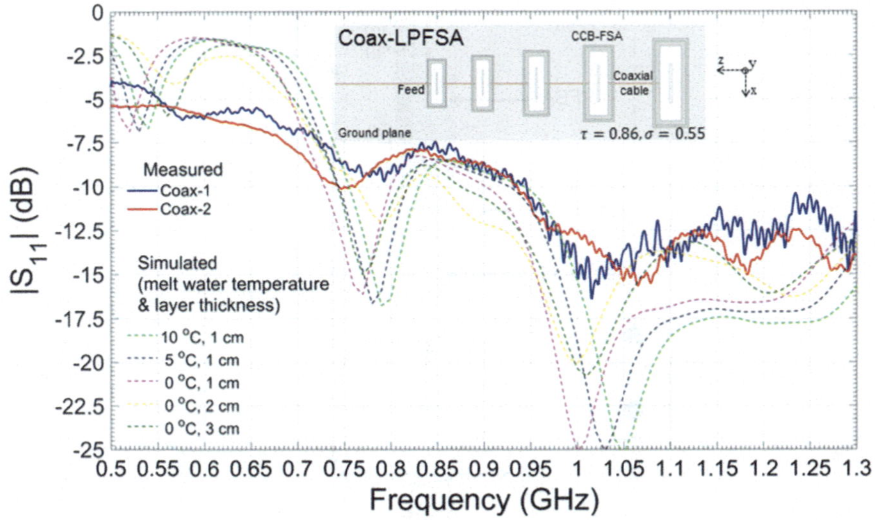

Fig. 4.108 Measured S11 values for Coax-LPFSA antennas and are compared with simulated values for various melt water temperatures and layer thickness

4.8.4 SAR Antenna Measurements and SAR Image Reconstruction

The atypical environment in which the SAR operates makes it is very difficult if not impossible to emulate an exact representative and at the same time controlled environment for characterizing its antenna and radar performance. Nevertheless outdoor laboratory tests for antenna characterization were carried out at the Stone Aerospace Inc. facility in Austin, Texas.

In order to emulate glacial operating conditions a 0.76 m diameter by 1.78 m tall cylinder of ice (weighing $\sim$ 1 ton) was prepared and the antenna mounted onto the cryobot was gradually lowered into the ice cylinder along its axis. The test configuration shown in Figs. 4.106 and 4.111 facilitated return loss S_{11} measurements of the fabricated LPFSA antennas shown in Figs. 4.107 and 4.108 which were then compared with full wave simulations carried out over a variety of melt water thickness values (1, 2 and 3 cm) and melt water temperature 0, 5 and 10 °C.

A well characterized log periodic dipole array (LPDA) antenna is used as a reference probe and is translated underneath the cryobot along the x axis in order to measure S_{21} between the LPFSA and the reference probe (Figs. 4.109 and 4.110). The S_{21} thus obtained is a measure of the E-plane radiation in the end fire direction for the LPFSA under test. In situ testing of the LPFSA antennas and SAR radar system was carried out on the Matanuska Glacier, Alaska near 61° 42′ 9.3″ N latitude and 147 °C 37′ 23.2″ W longitude.

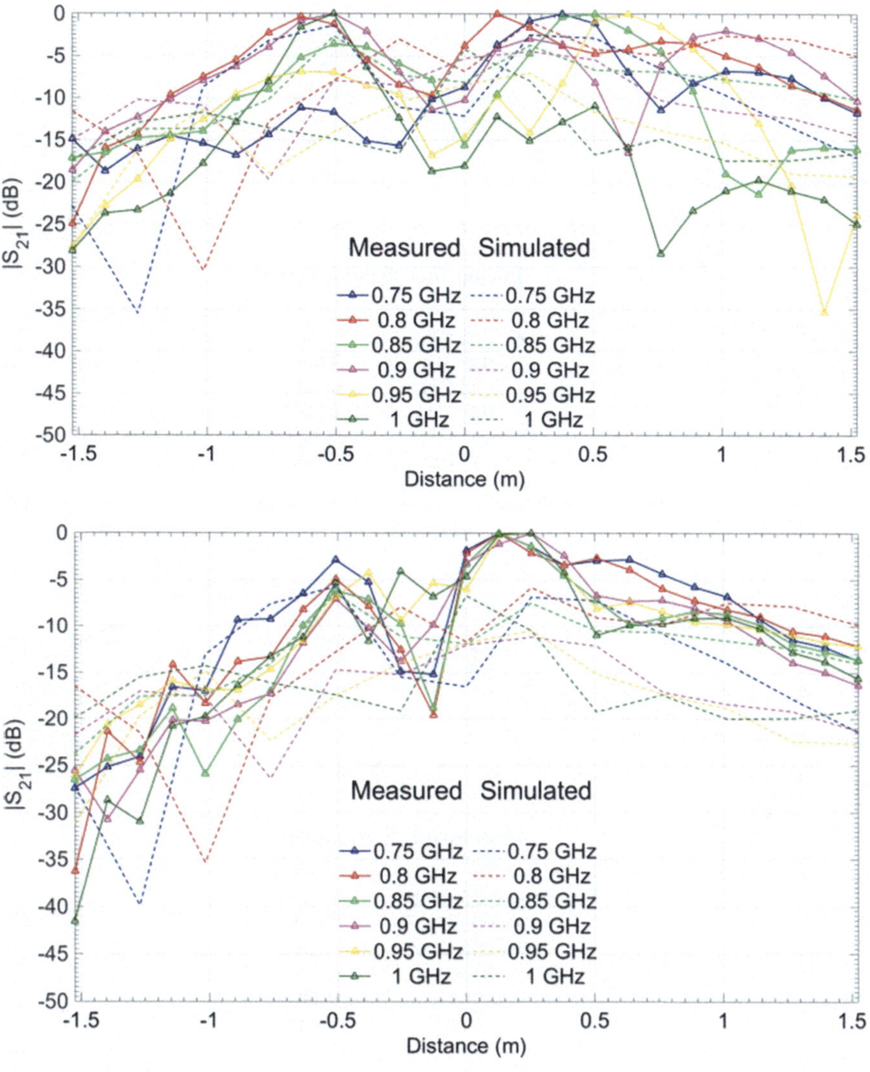

Fig. 4.109 The LPDA reference probe is used to measured S_{21} values, plotted along with the simulated values for the pair of CPW-LPFSA antennas

Ice boring was achieved using a hot water ("Hotsy") heater/pump along with the same melt head attached to the antenna as was used at Stone Aerospace. The antenna structure was suspended from its aft end by a gantry with a pulley arrangement for deliberate lowering down or raising up of the antenna (Fig. 4.111). A tape measure attached to a fixed point on the gantry was used to accurately measure the depth of the antenna. An older melt hole which was bored in a previous expedition and subsequently filled up with soil is at a distance of ~7.5 m from

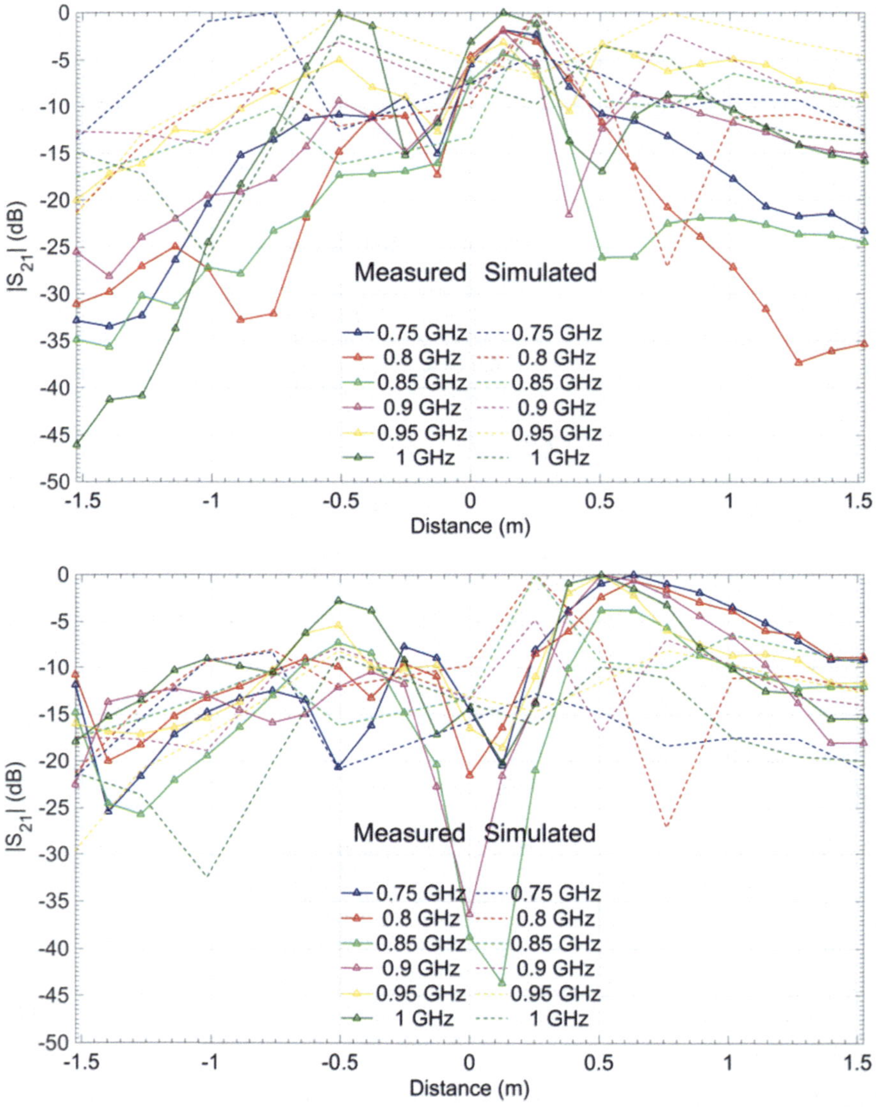

Fig. 4.110 The LPDA reference probe is used to measured S_{21} values, plotted along with the simulated values for the pair of COAX-LPFSA antennas

borehole in which the radar was operated. Apart from this other discontinuities in ice also act as scatterers of the transmitted radiation. The measured and simulated antenna S_{11} values are shown in Fig. 4.112.

A reconstructed SAR image in terms of the normalized power at the receiver is shown in Fig. 4.113. As noted before the cryobot moves downward along the negative z axis and the x axis is the radial distance from the path of the cryobot. The

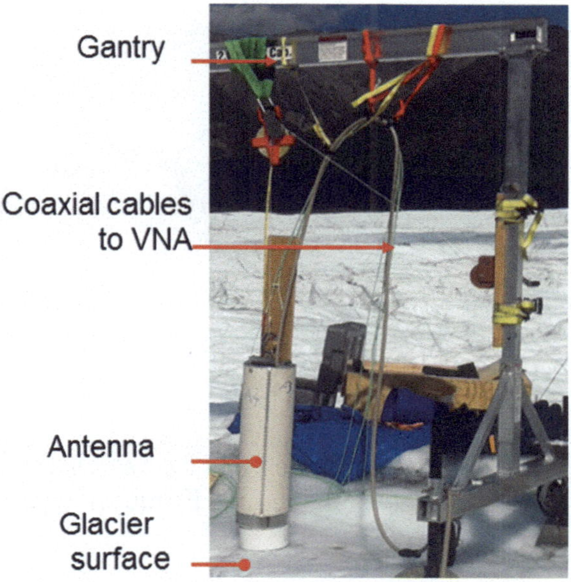

Fig. 4.111 Antenna measurement setup on the Matanuska glacier, AK, showing the VALKYRIE cryobot antenna section suspended from the gantry

image is formed by coherent aggregation of the match-filtered received signals over a 5 m long synthetic aperture. The four LPFSA antennas arranged around the cryobot can be used to differentiate between the azimuth coordinates of the radar echos by implementing phase sensitive interferometric processing applied to the received signal at each antenna (Rosen et al. 2000).

4.9 Closing Thoughts

There have been a number of significant technological innovations since the initial thermal ice penetrator probe invented by Philberth in the 1960s. Some of the crucial new findings in just the past few years—all a result of NASA cryobot research funding—include:

- **CCHWD**: The development and employment of closed-cycle hot water drilling (CCHWD) technology. This approach adopts the impressive capability of polar hot water drilling systems developed for Antarctic deep drilling projects to exhibit both a greater melt rate metric (in m^3/kWH) over passive melting systems as well as to offer a powerful means for punching through brine and debris layers that would otherwise stop a pure thermal ("hot penny") probe. Because a CCHWD can use diversion valves and lateral nozzles, jet pumps can use hot water to both steer a cryobot off vertical as well as to elegantly and simply create pockets where non-soluble brine and debris could be pushed aside while allowing the vehicle to continue downward. This is a powerful capability not

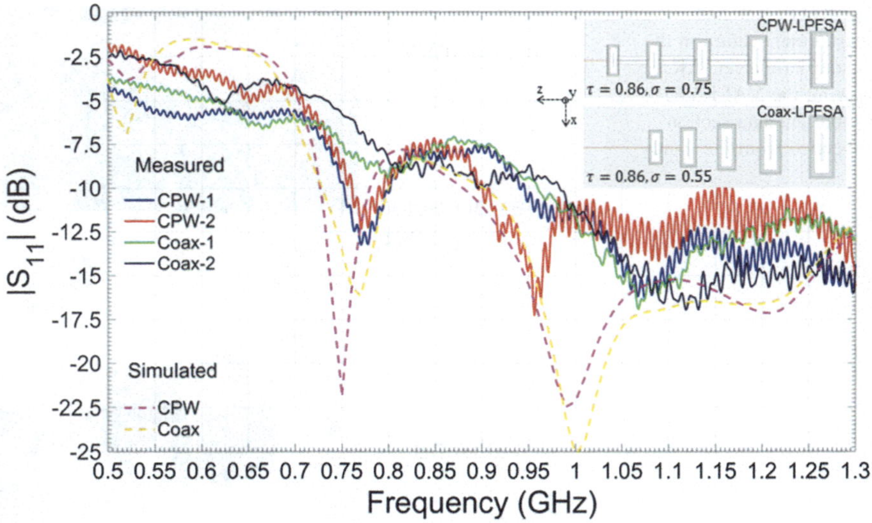

Fig. 4.112 S_{11} measurements

Fig. 4.113 Reconstructed radar image

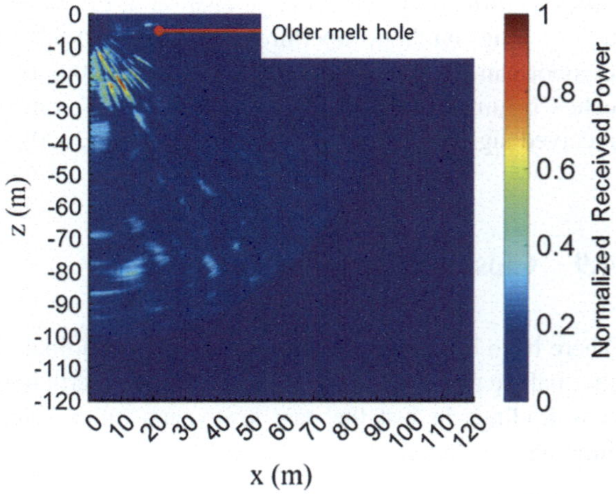

present in any of the other approaches that have been investigated for cryobot systems.

- **Direct Laser Probe** (DLP) technology presents a game changing option. For clean ice it offers the situation where 100% of transmitted power goes into the melting of ice in a controllable geometry ahead of the vehicle. This leads to overall efficiency gains because the vehicle can move faster and less waste heat goes into the heating of sidewall ice (as would be the case, e.g. with a thermal probe). This offers a powerful alternative to HWD and mechanical drilling systems as well for Antarctic sub-glacial access on Earth, where the power generation systems can be conveniently de-coupled from the cryobot and based on the surface, with the cryobot reeling out a thin power fiber as it descends. DLP is the one technology that appears best suited to solve the "Starting Problem" for Ocean World ice penetration, where a penetrator will begin at the ice surface in 100 K temperatures and in hard vacuum. Despite these conditions the optical energy is still completely deposited into the ice and into the sublimate, thus allowing a DLP penetrator to efficiently drill in hard vacuum at any ice temperature with no moving parts. It thus bypasses the complexities of a mechanical drill, which might be otherwise considered for addressing the starting problem. The primary disadvantage to DLP at this time is energy conversion efficiency from the primary power source and the need for relatively clean ice. On Earth, operating in an Antarctic setting, diesel fuel will need to be converted to electricity which will in turn be converted to photons. The efficiency of the conversion from electrons to photons is currently over 50% and increasing each year, making DLP laser drills more attractive for deep Antarctic drilling. For Ocean World settings DLP is ideal for solving the Starting Problem as well as short to medium range (several hundreds of meters) penetrations where the power source remained on the lander. With the rapid rate of size reduction and rising energy conversion efficiency it is entirely possible that DLP could also be used in a fully-integrated system on a nuclear-powered cryobot.

- **Power-over-Fiber (POF)**: VALKYRIE demonstrated that enormous amounts of optical power can be sent to a remote mobile agent over a very thin (25–300 μm diameter) fused silica fiber. There is attenuation of power over distance: about 12% per kilometer for fused silica and a laser wavelength of 1070 nm. However, at 1550 nm the attenuation reduces to approximately 1.3% per kilometer. Project VALKYRIE successfully sent 11 kW of 1070 nm laser power through a 1.05 km fiber, in line with predictive theory. Using the power-over-fiber approach, power levels as high as 100 kW at 1070 nm can now be injected into a fiber. It follows that industrial amounts of power can be made available as far away as 40 km distant (using 1550 nm wavelengths) using power-over-fiber. Back-power systems will be needed at the point of application with the most efficient method presently being tuned-wavelength high temperature ceramic photovoltaic systems. POF represents an interesting general capability whose application is not limited to cryobot power delivery. Its most effective use in planetary missions will be as a power delivery mechanism for probes (e.g. a Starting Problem DLP system) that will only be penetrating short distances (several hundreds of meters of ice) or to powering remote rovers that must have ultra clean power sources due to planetary protection considerations (e.g., the ice cap of Mars).
- **High Voltage and Super Insulation**: As part of the SPINDLE project an extensive effort was made to determine if high voltage power cables—such as those used in commercial oil patch ROV systems—might be made dramatically smaller. The effort focused on insulators, rather than conductors, as these dominated the design at higher voltages. Eventually an aerospace plastic, not

normally thought of as an insulator—Kapton—was investigated. Normally this material comes as aluminized polyimide sheets for spacecraft thermal insulation but the pure material is an extraordinary insulator. Custom-fabricated high voltage power cables tested at Stone Aerospace during project SPINDLE showed a 93% reduction in tether cross sectional area for a cable capable of transporting 50 kW of power at 10 kV. This opens the possibility for, at least, alternative terrestrial sub-glacial access power transfer.

- **Direct High-Voltage Water Heating**: One of the weak points in a traditional (i.e. Philberth-class) thermal probe is the use of electro-resistive heating elements that operate at relatively low voltage and high current, taking advantage of I^2R heating in the element. An electro-resistive element capable of operating at high voltage with sufficient surface area to allow rapid heat transfer forces the heater design into an impossible geometry. The SPINDLE project found a novel solution by passing high-voltage, low current, AC power through a moving conducting fluid. This generates resistive heating in the fluid with 100% efficiency without inducing electrolysis at power densities of over 600 kW/l. This immediately opens a door to effective CCHWD power direction (from amorphous heat to directed hot water jetting). The combination of an extraordinarily small diameter power transmission line (as described above) with a means to reliably dump enormous amounts of power in a small location open new possibilities for terrestrial ice penetrators and provide clean access options for planetary sub-ice exploration where the nuclear power source remains with the lander.
- **Ultra-compact Fission Power**: Nuclear energy is the inescapable power source for sub-ice Ocean World missions. SPINDLE broke new ground in reactor design by looking at the ability of 100 K ice to reflect neutrons. Extensive neutronic modeling and reactor design studies conducted during SPINDLE showed that a 100 kW_{th} reactor could be built in one of the smallest configurations ever achieved (16 cm diameter × 76 cm length), five times smaller than present planetary power systems in design and testing, by making use of ice-reflected neutrons to sustain criticality. This new class of reactor—the Europa Cryo Reactor—is capable of being operated at variable power from shut down to 100 kW_{th}. It can be transported in an un-fired state, thus eliminating the need for extensive space/vacuum thermal radiators (which would be required, e.g., for a large scale radioisotope-based power source).

4.9.1 What We Can and Cannot Do on Ocean Worlds

Within the laws of physics there are solutions for reaching the silicate cores of Ocean Worlds; now, not in the next century. That places us squarely in one of the most exciting times in history: the possibility of the creation of a sophisticated, full-autonomous behavior-based collection of robotic systems that can reach extraordinarily remote places and to seek out life, without human guidance or intervention. The discovery of microbial life in the Europan Ocean would not only be one of the greatest scientific achievements in history, it would also have societal implications exceeding those of the first manned lunar landing. Many place the occurrence of this event beyond 2050 or 2060. We at Stone Aerospace, and many of our collaborators, see things differently. The engineering, science, and programming tools are in place where such a mission could be launched in ten years. It will take two things to make that happen on such a time schedule: first, a resolute commitment on the part of government(s) to provide the financial resources and red-tape-cutting to enable it; alternatively, a single individual in today's era of stratified wealth could also enable the mission; and second, a change in the way we go about sending things to other worlds. The present fixation with building bigger rockets to launch larger payloads—a retro nod to things that previously worked in different circumstances in the past—is misguided. The better approach is to focus on development of a new capability—orbital staging in LEO—that takes advantage of economies of scale in the block purchase of larger numbers of smaller launch vehicles and permits the assembly of arbitrarily large payloads for delivery to Ocean World targets. Ultimately orbital staging will be further leveraged by the availability of lunar sourced propellant, but the concept is achievable now using existing ELV systems. It will require a mind-set change in the existing national space agencies to do so, however. Alternatively, a privately-funded mission could bypass all the bureaucratic resistance and do it independently. Launch mass of the system leaving for Europa will be the controlling issue. A full-scale, 2-stage nuclear fission-powered cryobot/AUV mission will likely represent a 2 metric tonne vehicle apart from the launch bus and lander but such landed mass could be achieved with orbital staging.

4.9.2 *Hybrid Cryobots*

All discussion to date on possible cryobot missions has assumed a one-technology approach: investigators have looked at thermal probes; hot water drills; mechanical drills; and now, lasers. Each group has their favored design. Yet, none of these independently address all of the three discrete phases that are likely to occur on any given Ocean World mission: the Starting Problem (hard vacuum; 100 K ice); the cruise stage (clean but cold ice); and the Debris Problem (brine or impact debris). To this we should also add the Ocean Stage—because a cryobot that is not designed to handle the problem of what happens upon breakthrough to a sub-surface ocean is doomed to failure. Each of these regimes is sufficiently unique that the design space for a successful cryobot needs to—indeed must—consider ultimately that no single technology will work for all, and that a hybrid machine that uses two or more of the above technologies must be combined into a more sophisticated system. A further consideration that has heretofore not been attempted is that a large scale Ocean World simulation chamber is needed in which to test, at full-scale, integrated systems under hard vacuum at 100 K temperature and with obstacle layers present. No amount of systems tests in polar regions will obviate the need for this type of testing. Polar regions testing of fully-integrated, full-scale concepts is the absolutely essential first step to showing what might work. Only by follow on cryo-vac testing will we know whether it will work, e.g., on Europa.

4.9.3 The Ice Elevator

Almost all cryobot and thermal probe research and field work has dealt with gravity-driven, unidirectional missions, which is to say: downward with no recovery. While this may be acceptable for an initial short distance penetration in ice to determine environmental conditions, it is insufficient when we consider the need to deal with the ocean interface. A free falling cryobot (nadir pendulum stabilized) will drop uncontrolled to the floor of the ocean, yielding, at best, a one-transit set of measurements of the ocean. Oceans, however, are dynamic and anyone thinking to design a responsive, survivable AUV system for subsequent long range search for hydrothermal vents, e.g., needs to know the temporal response of the ocean: how do the currents and tides vary with the orbital period; how does density, temperature, salinity vary with depth over time? Are there periodic up-wellings or radical shifts in these parameters. We need this information to (a) determine the buoyancy of the AUV; (b) determine how much automated ballast compensation will be needed; (c) determine how much power/thrust will be needed to deal with ocean currents; (d) the types of materials that will be compatible with potentially corrosive seas; and (e) the controlling maximum hydrostatic pressure. One way to deal with this, and, as well, to address issues of obstacle avoidance response options open to a cryobot, would be to develop a general purpose "Ice Elevator", as opposed to a uni-directional cryobot. Such a device would be capable of traveling both downward through the ice and water under servo control and also ascending back up through the ocean and ice shell on demand (this requires bi-directional melting capability, but that is precisely what the SPINDLE project was about). Optimized algorithms—dynamically adapted to predicted ocean dynamics as more and more data are acquired—would help remove unknowns about the ocean (and the ice column) that would then dramatically increase the chances for successful deployment of a long range life search AUV. The AUV stage is ultimately necessary because the odds of a single, vertical penetration line intersecting a hydrothermal vent or other discrete energy source would be exceedingly small. Design work on a bi-directional 40 cm diameter cryobot during the SPINDLE project indicated that such a design is inherently feasible at larger cryobot diameters. In such a design separate communications and strength servo spoolers would need to be developed to potentially handle up to 30 km within ice and up to 100 km of subsequent ocean depth. This is feasible using extensions to existing technology (e.g. UHMW polyethylene-based high strength polymers and armored data fiber) and is not unprecedented: the ARTEMIS vehicle (Kimball et al. 2016) carried a 15 km data fiber deployment spooler for its explorations in 2015 under the McMurdo Ice Shelf.

Acknowledgements The majority of the work reported herein was conducted under NASA grants NNX10AE29G (VALKYRIE Phase 1); NNX11AJ89G (VALKYRIE Phase 2); and NNX15AT32G (SPINDLE).

We would like to thank NASA program managers Dr. John Rummel; Dr. Mary Voytek; and Dr. James Gaier, Dr. Curt Niebur, and Joan Salute, respectively, for their help and guidance during the conduct of these projects.

Stone Aerospace founder Dr. Bill Stone was PI for all of these projects. The following individuals were integral to the development and deployment of these vehicles and the interpretation of the resulting field data:

VALKYRIE: Nathan Bramall (Co-I, Leiden Measurement Systems); Dr. Peter Doran (Co-I, LSU); Dr. Brent Christner (Co-I, UF); Dr. Slawek Tulaczyk (UCSC); Dr. Jason Gulley (Co-I, USF); and Stone Aerospace team members Corey Hackley; Scott Lelievre; Josh Moor; Bryce Rothhammer; Keith Huffstutler; and Pamela Minor.

SPINDLE: Nathan Bramall (Co-I, Leiden Measurement Systems); Dr. Peter Doran (Co-I, LSU); Dr. Brent Christner (Co-I, UF); Dr. Slawek Tulaczyk (UCSC); Dr. Britney Schmidt (Co-I, Georgia Tech); Dr. Chris McKay (NASA ARC); Dr. Kevin Hand (NASA, JPL); Dr. Jason Gulley (Co-I, USF); Dr. Stacy Kim (Co-I, SJSU), and Stone Aerospace team members Scott Lelievre; Josh Moor; Bryce Rothhammer; Keith Huffstutler; and Pamela Minor.

References

Aamot, H.W.C.: Heat transfer and performance analysis of a thermal probe for glaciers. In: CRREL-TR-194. Cold Regions Research and Engineering Lab, Hanover (1967)

Benson, T., Cherwinka, J., et al.: IceCube enhanced hot water drill functional description. Ann. Glaciology **55**(68) (2014). https://doi.org/10.3189/2014aog68a032

Biele, J., Ulamec, S., Hilchenbach, M., Komle, N.I.: In situ analysis of Europa ices by short-range melting probes. Adv. Space Res. **48**, 755–763 (2011). https://doi.org/10.1016/J.ASR.2010.02.029

Blankenship, D., Young, D.A., Moore, W., Moore, J.C.: Radar Sounding of Europa's Subsurface Properties and Processes: The view from Earth. University of Arizona Press, Tucson (2009)

Chu, P., Spring, J., Zacny, K.: ROPEC—rotary percussive coring drill for Mars Sample Return. In: Proceedings of the 42nd Aerospace Mechanisms Symposium, pp. 195–210. NASA Goddard Space Flight Center, 14–16 May 2014

Chyba, C.F., Phillips, C.B.: Possible ecosystems and the search for life on Europa. Proc. Natl. Acad. Sci. USA **98**, 801–804 (2001)

Cook, C., Bernfeld, M.: Radar Signals An Introduction to Theory and Application, 2nd edn. Academic Press Inc., New York (1968)

Dachwald, B., et al.: IceMole: a maneuverable probe for clean in situ analysis and sampling of subsurface ice and subglacial aquatic ecosystems. Ann. Glaciology **55**(65), 14–22 (2014)

Gibson, M., Mason, L., Bownam, C., Poston, D., Creasy, J., Robinson, C.: Development of NASA's small fission power system for science and human exploration. In: NASA/ TM—2015-218460, June 2015

Goodge, J.W., Severinghaus, J.P.: Rapid access ice drill: a new tool for exploration of the deep antarctic ice sheets and subglacial geology. In: American Geophysical Union Fall Meeting 2014, Abstract C53A-0279 (2014)

Hsu, J.: Dual drill designed for Europa's ice. Astrobiology Magazine, April 15, 2010. http://www.astrobio.net/news-exclusive/dual-drill-designed-for-europas-ice/

Hsu, H.-W., Postberg, F., Sekine, Y., Shibuya, T., Kempf, S., et al.: Ongoing hydrothermal activities within Enceladus. Nature **519**, 207–210 (2015)

Humphrey, N., Echelmeyer, K.: Hot-water drilling and bore-hole closure in cold ice. J. Glaciol. **36** (124), 287–298 (1990)

Kaufmann, E., et al.: Melting and sublimation of planetary ices under low pressure conditions: laboratory experiments with a melting probe prototype. Earth Moon Planet **105**, 11–29 (2009). https://doi.org/10.1007/s11038-009-9296-9

Kedenburg, S., Vieweg, M., Gissibl, T., Giessen, H.: Linear refractive index and absorption measurements of nonlinear optical liquids in the visible and near-infrared spectral region. Opt. Mat. Express **2**, 1588–1611 (2012)

Kelley, P.L.: Self-focusing of optical beams. Phys. Rev. Lett. **15**, 10051008 (1965)

Khan, L.A, Ahmad, N.: A comparative study of decay heat generation rates in low enriched uranium fueled research reactors. Nucl. Sci. J. **34**(6), 396–403 (1997). ISSN 0029-5647

Kimball, P.W., et al.: Antarctic deployment of the ARTEMIS under-ice AUV docking system. J. Field Robot. (Fall 2016)

Lorenz, R.D.: Thermal drilling in planetary ices: an analytic solution with application to planetary protection problems of radioisotope power sources. Astrobiology **12**, 799–802. https://doi.org/10.1089/ast.2012.0816

McGrath, M.A., Jia, X., Retherford, K., Feldman, P.D., Strobel, D.F., et al.: Aurora on Ganymede. J. Geophys. Res. [Space Phys.] **118**, 2043–2054 (2013)

McKay, C.P., Anbar, A.D., Porco, C., Tsou, P.: Follow the plume: the habitability of Enceladus. Astrobiology **14**(4), 352–355 (2014)

Moore, J.: Models of radar absorption in Europan Ice. Icarus **147**(1), 292–300 (2000)

Nurnberger, M.W., Volakis, J.L., Mosko, J.A., Ozdemir, T.: Analysis of the log-periodic folded slot array. In: Proceedings of IEEE Antennas and Propagation Society International Symposium and URSI National Radio Science Meeting, vol. 2, pp. 1282–1285 (1994)

OFS/Furukawa Company, F18941, 550-22 HCXtreme High Power Delivery Optical Fiber, 3 pp. www.ofsoptics.com (2016)

Paranicas, C., Mauk, B.H., Khurana, K., Jun, I., Garrett, H., Krupp, N., Roussos, E.: Europa's near-surface radiation environment. Geophys. Res. Lett. **34**, L15103 (2007). https://doi.org/10.1029/2007gl030834

Paranicas, C., Cooper, J.F., Garrett, H.B., Johnson, R.E., Sturner, S.J.: Europa's Radiation Environment and Its Effects on the Surface, p. 529. University of Arizona Press, Tucson (2009). ISBN: 9780816528448

Philberth, K.: Une Methode Pour Mesurer Les Temperatures a L'interieur d'un Inlandis. Comptes Rendus des Seances de l'Academie des Sci. **254**(22) (1962)

Rack, F.R.: Enabling clean access into Subglacial Lake Whillans: development and use of the WISSARD hot water drill system. Philos. Trans. **A374**, 20140305 (2014). https://doi.org/10.1098/rsta.2014.0305

Rosen, P.A., Hensley, S., Joughin, I.R., Li, F.K., Madsen, S.N., Rodriguez, E., Goldstein, R.M.: Synthetic aperture radar interferometry. Proc. IEEE **88**(3), 333–382 (2000)

Roth, L., Saur, J., Retherford, K.D., Strobel, D.F., Feldman, P.D., et al.: Transient water vapor at Europa's south pole. Science **343**, 171–174 (2014)

Sajurai, T., Chosrowjan, H., Somekawa, T., Fujita, M., Motoyama, H., Watanabe, O., Izawa, Y.: Studies of melting ice using CO_2 laser for drilling. In: Cold Regions Science and Technology COLTEC 2174 (2015). https://doi.org/10.1016/j.coldregions.2015.09.014

Schmidt, B.E., Blankenship, D.D., Patterson, G.W., Schenk, P.M.: Active formation of "chaos terrain" over shallow subsurface water on Europa. Nature **479**, 502–505 (2011)

Stone, W.C.: Accessing oceans beneath ice. In: Ocean Worlds 2 Conference, Center for Marine Robotics, Woods Hole Oceanographic Institute, 25–26 Aug 2016

Stone, W.C., Hogan, B., Siegel, V., Lelivre, S., Flesher, C.: Progress towards an optically powered cryobot. Ann. Glaciol. **55**(66) (2014). https://doi.org/10.3189/2014aog65a200

Stone, W.C., Hogan, B.: Optical energy transfer and conversion system. United States Patent 9,090,315 (2015)

Stone, W.C., Hogan, B.: Optical power transfer system for powering a remote mobility system for multiple missions. United States Patent 9,383,520 (2016)

Talalay, P.G., et al.: Recoverable autonomous sonde for environmental exploration of antarctic subglacial lakes: general concept. Ann. Glaciol. **55**(66) (2014). https://doi.org/10.3189/2014aog65a200

Teodoro, L.F.A., Davila, A.F., McKay, C.P., Dartnell, L.R., Elphic, R.C.: Ionizing radiation on the surface of Europa: implications for the search for evidence of life. In: 47th Lunar and Planetary Science Conference (2016)

Thorsteinsson, T., et al.: A hot water drill with built-in sterilization: design, testing, and performance. JÖKULL **57**, 71–82 (2008)

Ulamec, S., et al.: Access to glacial and subglacial environments in the solar system by melting probe technology. Rev. Environ. Sci. Biotechnol. **6**, 71–94 (2007). https://doi.org/10.1007/s11157-006-9108-x

U.S. Energy Information Administration.: Uranium marketing annual report, May 2016. http://www.eia.gov/uranium/marketing/

Vago, J., et al.: ExoMars: searching for life on the red planet,. ESA Bull. **126** (2006)

Vasiliev, N.I., Talalay, P.G., et al.: Twenty years of drilling the deepest hole in ice. Progress Report (2011). https://doi.org/10.2204/iodp.sd.11.05.2011

Warren, S.G.: Optical constants of ice from the ultraviolet to the microwave. Appl. Opt. **23**, 1206–1225 (1984)

Weiss, P., et al.: Study of a thermal drill head for the exploration of subsurface planetary ice layers. Planet. Space Sci. **56**(9), 1280–1292 (2008)

Winebrenner, D. P., et al.: A thermal ice-melt probe for exploration of Earth-analogs to Mars, Europa and Enceladus. In: Lunar and Planetary Science Conference, vol. 44 (2013)

Wirtz, M., Hildebrandt, M.: IceShuttle Teredo: an ice-penetrating robotic system to transport an exploration AUV into the ocean of Jupiter's Moon Europa. In: 67th International Astronautical Congress (IAC), Guadalajara, Mexico, 26–30 Sept 2016. (IAC-16-A3.5.2)

World Nuclear Association.: Uranium enrichment. http://www.world-nuclear.org/information-library/nuclear-fuel-cycle/conversion-enrichment-and-fabrication/uranium-enrichment.aspx. 11 Mar 2016

Zacny, K., et al.: Wireline deep drill for exploration of Mars, Europa, and Enceladus. In: Aerospace Conference. IEEE. (2013)

Zimmerman, W.: Extreme electronics for in situ robotic/sensing systems. In: Aerospace Conference Proceedings, vol. 7. IEEE (2000)

Zuagg, E.C., Hudson, D.L., Long, D.G.: The BYU SAR: a small student-built SAR for UAV operation. In: 2006 International Geoscience and Remote Sensing Syposium, pp. 411–414 (2006)

Chapter 5
Europa Drum Sampler (EDuS)

Kris Zacny

5.1 Introduction

Historically, the majority of planetary exploration has been limited to 2D – surface or near surface. However, a vast wealth of information related to past climate and geologic history as well as past or present life can be revealed by exploring in 3D—beneath the surface. Of particular interest to Astrobiology are Ocean Worlds; that is planetary bodies with oceans. *Oceans could exist in diverse forms on moons and dwarf planets, offering clues in the quest to discover life beyond our home planet.*

Recently, NASA approved the Flagship-class Clipper mission to explore Jupiter's Moon Europa. Europa is the most likely planetary body that could harbor life. That is also the reason why a Europa mission has been the top priority of the past two National Research Council's Decadal Surveys. The mission will also be augmented with a ~350 kg lander to ground-truth orbital measurements and enhance concentration and detection limits. The lander will be launched as a standalone spacecraft rather than add on to the Clipper flyby. The lander requires a sampling system to capture surface and subsurface material for analysis by onboard instruments. The threshold mission includes chemical analyses of 3 samples from 10 cm depth or deeper (Hand et al. 2017).

Because of the environment that includes extremely low temperatures (surface temperature is 110 K at the equator and 50 K at the poles) as well as great distance from the Sun—which limits effectiveness of solar-derived energy—it is very likely the surface mission will be short lived (e.g. days rather than weeks or months), akin to Rosetta's Philae life duration. This is unlike most other missions, such as the Mars Phoenix which survived over 3 months, or the Mars InSight mission that is scheduled to survive over two years on the Martian surface.

K. Zacny (✉)
Honeybee Robotics Spacecraft Mechanisms Corporation,
398 W Washington Blvd., Pasadena, CA, USA
e-mail: zacny@honeybeerobotics.com

V. Badescu and K. Zacny (eds.), *Outer Solar System*,
https://doi.org/10.1007/978-3-319-73845-1_5

Europa's surface topography is unknown on the scale of a lander and even less on the scale of the sampling system itself. One of the best imagers of the Europa surface was taken by the camera onboard NASA's Galileo spacecraft. It is a very high resolution view of the Conamara Chaos region and shows an area where icy plates have been broken apart and moved around (Fig. 5.1). The top of this image shows corrugated area ending in cliffs over a hundred meters high. The debris that are piled at the base of the cliffs are the size of a house.

The unknown terrain requires the sampling system to be compliant to variable surface features. The Europa surface could be composed of cryogenic water ice of different densities (very dense to very porous), salt, or frozen sulfuric acid. As such, the sampling system needs to be able to work with any of these materials. Figure 5.2 shows potential terrestrial analogs: Devil Gold Course in Death Valley, CA. It can be seen that both the rough texture of the large halite salt crystal formations as well as very steep crevices in the glacier would be very challenging for both the lander and sampling system.

Europa surface material can vary from water ice (of various densities and in turn strength) to salt, frozen sulphuric acid and combination of the three. In addition, dust or even rocks (e.g. meteorites) could be present in various concentrations. As such, ice could range from pure ice to dirty ice to ice cemented ground, for example. There is limited data on strength of these materials at cryogenic temperatures. Compressive strength is the primary parameter driving sampling system energy and force requirements.

Figures 5.3 and 5.4 list properties of icy-soil and ice as a function of water saturation (for soil) and temperature (for ice).

It can be seen that the strength for icy-soil and ice at cryogenic temperature is similar to the strength of basalt, which has 3 times the strength of concrete. The strength of rock salt (sodium chloride) is approximately 40 MPa at room temperature, and lowers from there with drops of temperature (Sriapai et al. 2012). Increase in strength with drop in temperature has also been observed for very dense rocks such as basalts (Zacny and Cooper 2007). It is also possible for the surface of Europa

Fig. 5.1 The image covers an area 1.7 km by 4 km. The resolution is 9 m per pixel

Fig. 5.2 Devil's Golf course

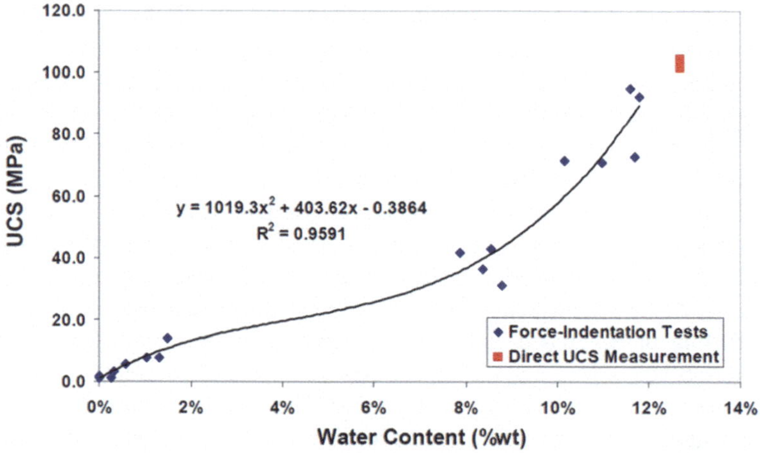

Fig. 5.3 Uniaxial compressive strength versus water content of compacted JSC lunar simulant at ~77 K (Gertsch et al. 2008)

to be covered by crusty (high porosity) ice and snow formed when plumes of salty water freezes and drops down to the surface. Hence, the sampling system would need to be able to handle the entire spectrum of material porosities and strengths. Presence of sulphuric acid also limits the use of traditional aerospace materials.

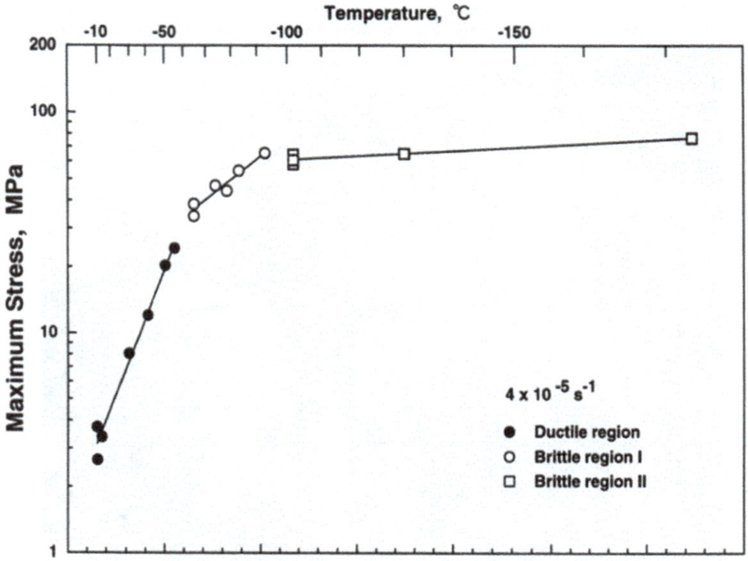

Fig. 5.4 Strength of ice as a function of temperature (Arakawa and Maeno 1997)

Additional challenges on material selection for the sampler are issues such as contamination control and planetary protection. For example, sterilization through Dry Heat Microbial Reduction requires exposure to 125 °C for 35 h. This high temperature limits the selection of otherwise promising materials and also reduces the strength of many alloys, which anneal at these temperatures. Since local gravity is 1.3 m/s^2, the maximum force the 350 kg lander could provide to a drilling system is therefore approximately 450 N. Given the sampler will be deployed from a robotic arm, the practical limit will be in the 10 s of Newtons range.

Handling icy chips can be very difficult for two reasons: icy chips could potentially stick and ice could potentially sublime away. The issue of sticking was first observed during acquisition of icy soil on Mars Phoenix mission. This stickiness was attributed to the scoop being exposed to direct Sun and warming up too much. Testing in Antarctica and freezers, however, have shown that as long as temperature is below freezing, drilled cuttings from ice and icy-soil will behave just like dry sand on the beach (Zacny et al. 2013a). The issue of sublimation is actually very challenging. When sampling occurs in hard vacuum and temperatures increase above ~150 K, ice sublimes extremely fast, on a timescale of seconds. In fact, sublimation rate of ice in vacuum is purely function of temperature. Below approximately 150 K, it's slow (on the order of hours). However, as soon as the temperature starts increasing, sublimation rate follows exponential curve (Andreas 2006). Figure 5.5 shows fraction of mass loss for 1 micron particles as a function of temperature. At 120 K, ice particles take 10^8 s to lose 20% of their mass, while at 170 K, they completely sublime away in seconds.

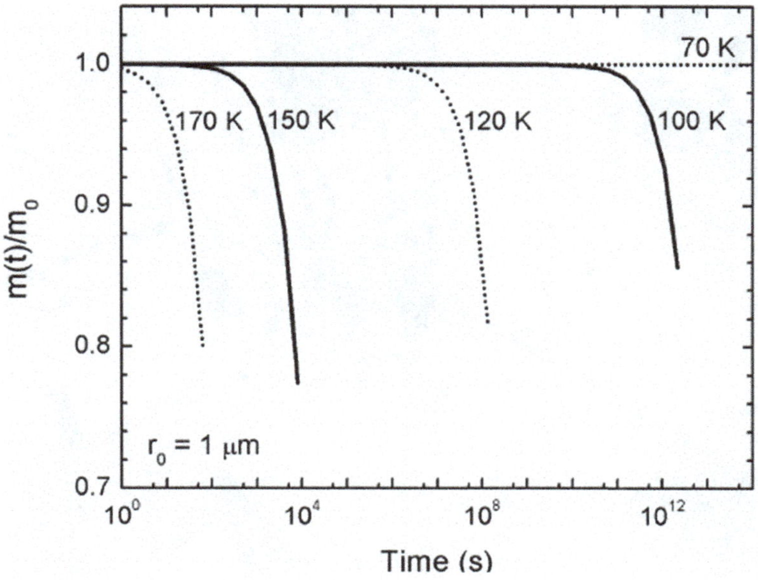

Fig. 5.5 The fraction of initial mass (*m*0) remaining after time *t* when spherical samples of pure water ice with initial radius of 1 μm are exposed at the indicated temperature

5.2 Example of Planetary Sample Acquisition Systems

Over the past few decades, a number of sampling technologies have been developed for various planetary bodies (Bar Cohen and Zacny 2009; Ball et al. 2007; Zacny et al. 2013b). The vast majority of these developments focused on the exploration of Mars and the Moon, while the status of sampling technology for Icy Bodies remained relatively low. The following sections describe some of the technologies for acquisition of cryogenic samples. It should be noted that some of the sampling approaches for comets could be adopted for Europa surface sampling, as well.

5.2.1 Percussive Scoop and Mars Phoenix Icy Soil Acquisition Device

It is known that hammering or percussion drastically reduces excavation forces. For this reason, a number of digging tests were conducted in frozen (–20 °C) water saturated granular material. The digging system employed a Surveyor replica scoop with a 45 mm wide blade. At water concentration of only 5 wt%, the scoop managed to penetrate ~1 cm when a 3500 N force was applied to it. With an addition of a 150 W percussive system and application of 3500 N force the same scoop managed to penetrate barely 5 cm depth (Fig. 5.6). These experiments have

Fig. 5.6 Tests with a Surveyor replica scoop in JSC-1a with various water wt% and at −20 °C

effectively demonstrated that excavating ice-bound regolith, even with an impact-driven scoop, will be nearly impossible.

Mars Phoenix dealt with icy soil by incorporating a small drill bit, called rasp, into its scoop assembly (Fig. 5.7). The rasp, when using only a few Watts of power, successfully penetrated millimeters into the Martian subsurface and ballistically transferred icy-chips into the scoop (Bonitz et al. 2008). However, for rasp to be effective, the surface area had to be cleared of overburden regolith and leveled. For this reason, the scoop included blades for scraping the surface flat. Since the knowledge of Europa surface is poor, this approach should not be used on the very first landed mission.

5.2.2 Trenchers

Another option for sample acquisition is trenching (Fig. 5.8). A saw or a chainsaw can relatively easily cut through hard materials and does not need lubrication. However, sample transfer is more difficult. Hence, a trenching system would require some kind of a sample capture system or a separate scoop.

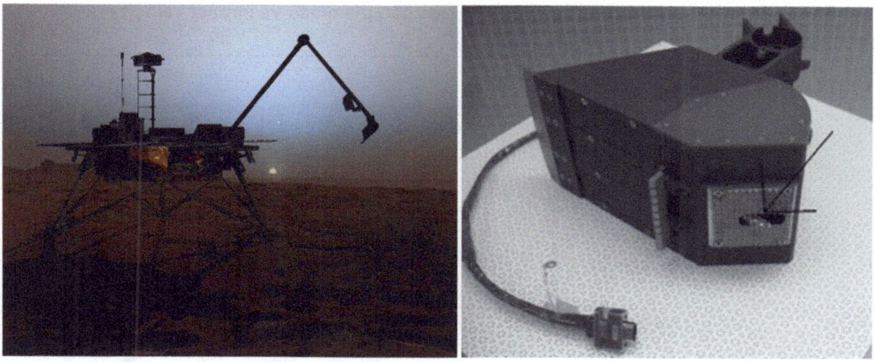

Fig. 5.7 Mars phoenix icy soil acquisition device

Fig. 5.8 Chainsaws are frequently used to cut and sculpture ice

5.2.3 Harpoon Samplers

Harpoon samplers can be fired into the surface, capture a sample during the course of penetration into the subsurface, and reeled back into the spacecraft using a tether. Hence, they could potentially be deployed from landers. Honeybee developed a number of harpoon concepts for variety of formations (Fig. 5.9). The final breadboard was tested in ice at ~ -150 °C and captured samples when impacting at up to 45° off vertical (Bar Cohen and Zacny 2009).

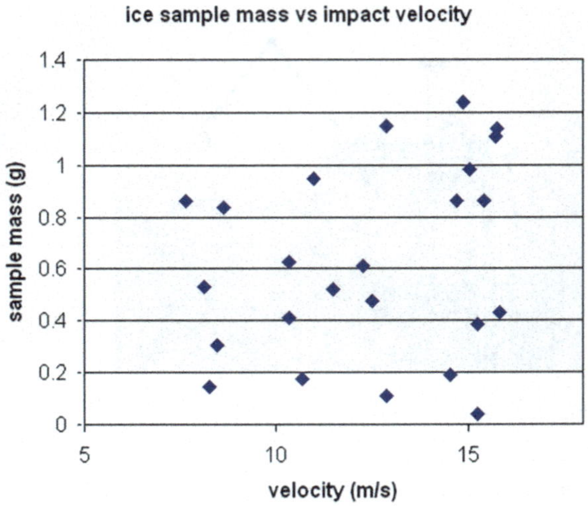

Fig. 5.9 Data from Honeybee Robotics harpoon samplers in cryogenic ice (Bar Cohen and Zacny 2009)

5.2.4 *Rotary-Percussive Drills*

There have been several rotary-percussive core and powder acquisition drills developed over the past decade. These include Resource Prospector drill design for capturing volatile rich samples from the Moon (Paulsen et al. 2017; Zacny et al. 2016) and Rotary-Percussive Corer (RoPeC) designed to capture cores and chips (Fig. 5.10).

These systems have been designed for near surface (<10 cm) and deep (1 m) applications and ranged in mass from 1 kg for near surface to 15 kg for deeper version. These efforts have shown that it is possible to capture cores and cuttings out of weak and very strong materials alike.

5.2.5 *Melt Probes*

Meltprobes use heat to melt, and sometimes sublime, water-ice in order to penetrate water-ice. Melt probes are very robust since they just need heaters to operate. As such, they are believed to be the optimum approach for penetrating deep into Europa's ices. Meltprobes have been originally developed by the Cold Regions Research and Engineer Labs (CRREL), but they never found application in Antarctica due to slow penetration rates and higher energy needed to penetrate. Melt probes also require >10x more energy than mechanical systems. The majority of heat (90%) is in fact lost to a highly conductive ice (ice at 100 K is 3x more conductive than ice at 270 K), rather than used to melt or sublime ice (Ulamec et al. 2007). Treffer (2006) conducted ice melting tests in vacuum chamber in ice at

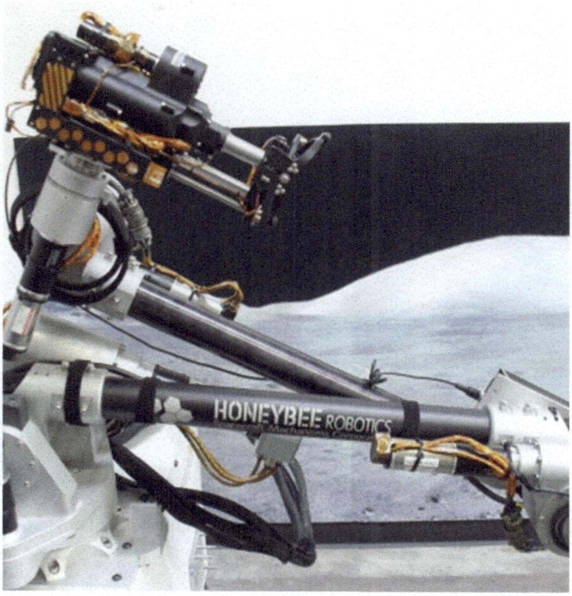

Fig. 5.10 Rotary-percussive drills were developed for capture of cores and cuttings (Chu et al. 2014)

220 K. A 2 cm diameter sphere needed 60 Watts to penetrate at 3.4 cm/h in ice at 220 K. In 100 K ice, this penetration rate would drop to 0.25 mm/h.

5.3 Europa Drum Sampler (EDuS)

It will be difficult to adapt existing samplers for the Europa mission. Scoops, and even percussive scoops, will not penetrate cryogenic ice. Trenchers might have problems with sample capture and delivery, drills require much better position control and will not work easily in very friable materials (drills will easily drill through it but might have problems capturing samples), and melt probes are too energy intensive. Other options such as piercing blades, harpoons etc., will not penetrate cryogenic ice.

Europa Drum Sampler (EDuS) is an alternative approach designed to capture samples of ice, salt, or any other material from $\sim$10 cm depth (Fig. 5.11). The sampling system is based on a continuous miner/roadheader design, used in road construction and mining. The advantage of the roadheader cutter drum design is that it's compliant to highly variable surface roughness. It's also extremely robust and have been shown to cut through weak and hard rocks (continuous miners have been field tested in iron ore mines, cutting >200 MPa iron ore formation).

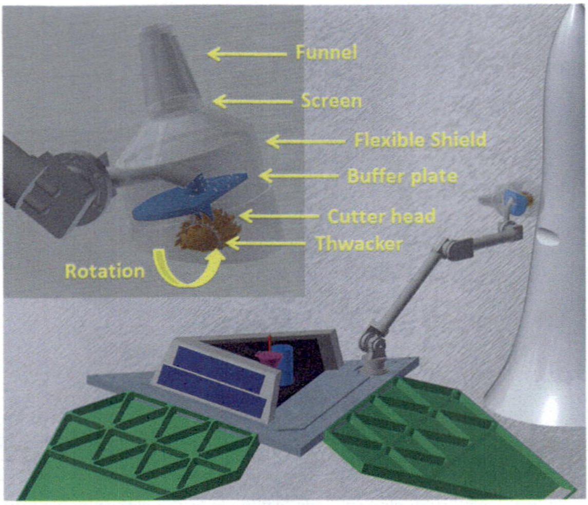

Fig. 5.11 A concept of a Europa Drum Sampler (EDuS) is based on terrestrial roadheader

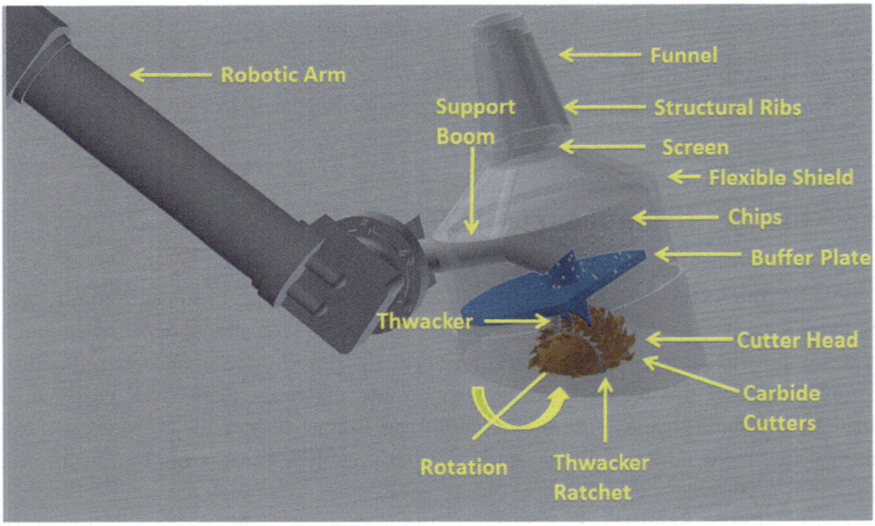

Fig. 5.12 Details of the notional Europa Drum Sample (EDuS) design

Using roadheader cutter drum as a starting point, the Europa sampler requires several features to fit the mission profile and requirements. The Europa Drum Sampler (EDuS) shown in Fig. 5.12 consists of a Cutter Head with carbide tipped teeth enclosed in a Flexible Shield. The Cutter Head engages the surface while spinning in a counter clockwise direction. This has a tendency to pull the Robotic

Arm down towards the surface, reducing required excavation forces. The Cutter Head has a passive Thwacker (a ratchet) that works whenever the drum spins. The Thwacker-induced vibrations provide micro-hammering to the cutters and in turn lower excavation forces. Excavated chips ballistically fly out and up—some of them will hit the Flexible Shield and fall behind the Buffer Plate. The Shield is flexible to allow for the EDuS to penetrate deeper into material; while the Cutter Head moves forward, the Shield deflects and wraps around the surface features. The Shield could be made of transparent Mylar to allow visual assessment of sample volume. The Buffer Plate is used to prevent the chips from falling out during the excavation process and Robotic Arm motions. The Flexible Shield ends with a Screen and a Funnel. The Screen size will depend on maximum particle sizes that instruments require. As such, for now this is a placeholder. The reason for having a Screen is that it will be very difficult to control particle sizes during excavation—these can range from microns to cm-size chunks, depending on material properties. To aid in the sieving process, the Cutter Head will rotate to engage the Thwacker. The shape and size of the Funnel can be determined based on the location of the instrument inlet port, and kinematic position of the Robotic Arm.

The EDuS structural components include Support Boom, Buffer Plate, and Cutter Head. The Support Boom is hollow and its position and length can be adjusted depending on the required excavation depth and volume on the lander. The Support Boom could be telescopic (spring deployed) if there is no space on the lander, but the system is required to dig deeper. As such, during launch, the Boom can be compressed and spring-extend upon landing. A spring loaded Boom has an added advantage of mitigating Thwacker vibrations to the Robotic Arm. The Buffer Plate is, again, a structural member whose main purpose is to prevent chips from falling out. The gap between the top of the Plate and the Shield can be adjusted after running the tests and determining the optimum gap size.

The Cutter Head is the most important part of the system. Design of the Cutter Head and its performance will affect the rest of the system the most. The Cutter Head has been designed in a shape of a typical cylindrical pressure vessel. The teeth are placed on all rotating surfaces, including the convex sides. This shape can deal with a range of surface topographies from flat to very jagged. The width of the Cutter Head can be adjusted depending on the position accuracy of the robotic arm (if robotic arm has very low positioning accuracy, the drum should be wider). The teeth are very sharp to reduce cutting forces and are made of carbide to increase the life. The exact shape and length of the teeth will be driven by maximum torque the Head can generate—shorter teeth require less torque. The Cutter Head has a ring of teeth (Thwacker Ratchet) as part of the Thwacker system. The amplitude and vibration of Thwacker can be adjusted: higher frequency is increased via adding more teeth and higher amplitude (energy) is applied by increasing spring stiffness of the Thwacker hammer that pushes against the teeth. Thwacking will reduce cutting forces and aid in sample delivery.

The critical design trade will include actuation system for the Cutter Head. There are notionally two options. In the first one, an actuator can be housed inside the Head itself. The actuator would probably be vibration isolated from the structure and would need to be heated (either through self-heating or film heaters). This has advantage of just passing wires through the Support Boom. However, a disadvantage is that actuator size, and in turn power, will be limited somewhat to the available volume. Plus, the drum will be somewhat 'warm' from the 'warm' actuator. The second option is to place the actuator in the Support Boom itself and run a drive shaft to bevel gears inside the Cutter Head. Other options include belt drives, chain drives and son on. The advantage of this approach is that the actuator size is independent of the Cutter Head size and the source of heat (actuator) is placed away from the Cutter Head. A disadvantage, however, is that now additional subsystems (draft shaft, belt drive etc.) are required, and this leads to more mass and frictional losses.

The Shield, Screen, and Funnel are all flexible, probably made of Mylar (although this can be changed). The reason for using Mylar is that Mars Phoenix biobarrier used Mylar and successfully passed DHMR tests. If Mylar is made transparent, we can determine volume of material inside the Shield prior to delivery. The reason for flexible Shield is that it will allow cutter head to dig deeper, while flexible Shield will simply deform and take the shape of the surface features.

The Screen size can be adjusted based on the required particle size. Since it is virtually impossible to determine particle sizes that will be generated during excavation (this depends not only on the cutter type but also on material), the screen is a necessary subsystem. To aid in sieving process, the Cutter Head will need to be rotated to provide Thwacking vibrations.

The concept for operation with subsystem details is shown in Fig. 5.13. Upon landing, the Robotic Arm will move the EDuS to a 'ready' position. Once digging location has been selected, EDuS will start spinning and will be placed against the surface (it's important to start spinning before contacting the surface). The head is spinning in a counter clockwise direction (i.e. undercuts material) to put downward forces on the arm and the lander. EDuS can cut at any orientation and into variable topography. The diameter of the Cutter Head will determine the maximum depth for a single pass. The initial depth should be no more than half the drum diameter. However, the optimum operation should be in a sweeping motion, moving slowly towards the material to dig deeper. In this way, the maximum depth is limited by the robotic arm extension only. The sweeping motion has already been used on MSL's Dust Removal Tool.

Once a sample has been captured, the arm positions EDuS above an instrument inlet cup. At that point, the Cutter Head starts rotating which generates vibration (Thwacking). Hence, volume of sample dispensed will be a function of Thwacking time.

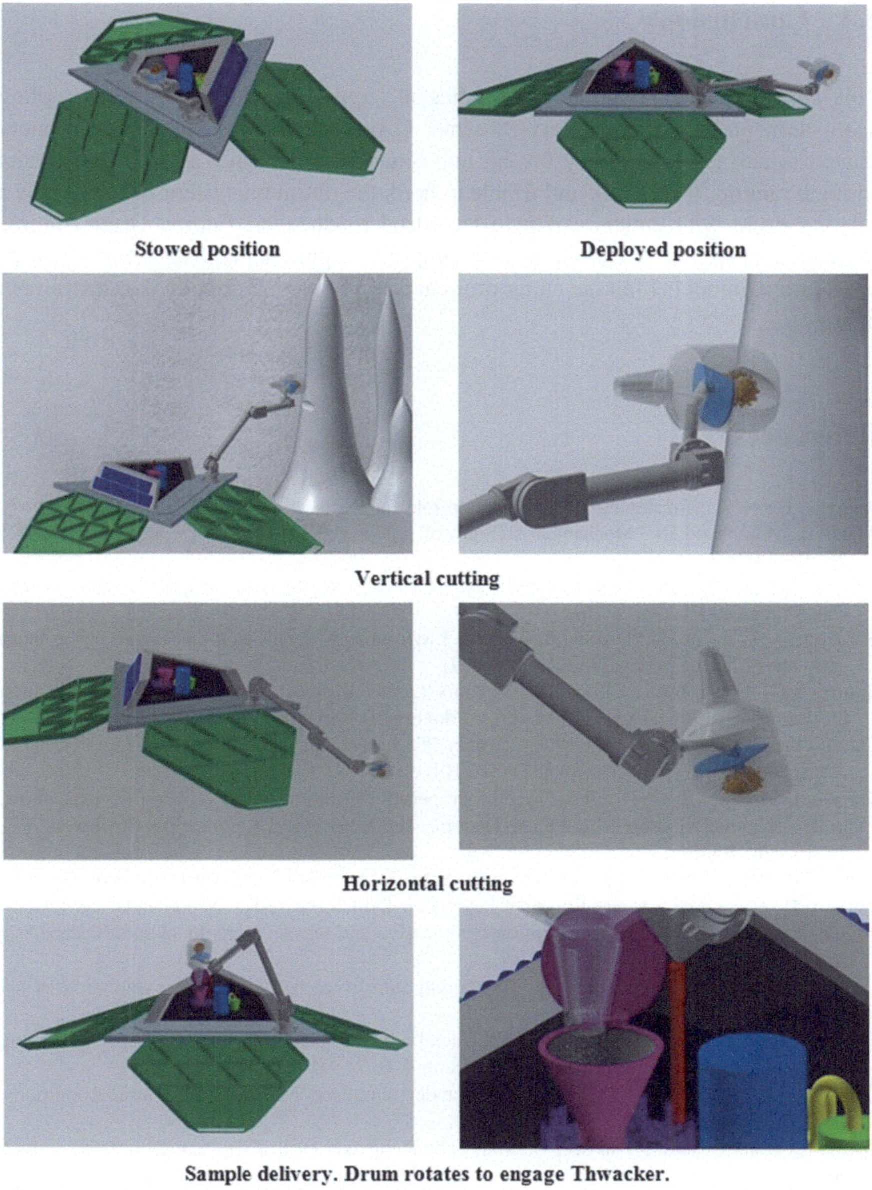

Fig. 5.13 Europa Drum Sample (EDuS) sequence of operations

5.4 Conclusions

This chapter presented numerous studies in support of Europa lander sampling acquisition and sample delivery systems. Given the Europa surface uncertainty related to surface topography on the micro and macros scale, as well as material strength ranging from weak and friable to hard, the optimum system will resemble a rotating drum with cutters—akin to terrestrial roadheader. Europa Drum Sample (EDuS) sampler meets Europa lander sampling requirements. The sample is captured into a funnel behind the cutter drum and can be sieved directly into instrument inlet ports.

References

Andreas, E.: New estimates for the sublimation rate for ice on the Moon, Icarus (2006)

Arakawa, M., Maeno, N.: Mechanical strength of polycrystalline ice under compression. CRST **26**, 1997 (1997)

Ball, A., et al.: Planetary Landers and Entry Probes, 1st edn. Cambridge University Press, Cambridge (2007)

Bar-Cohen, Y., Zacny, K.: Drilling in Extreme Environments Penetration and Sampling on Earth and Other Planets. Wiley, New York (2009)

Bonitz, R.G., et al.: NASA Mars 2007 Phoenix Lander Robotic Arm and Icy Soil Acquisition Device. JGR **113**, E00A01. (2008). https://doi.org/10.1029/2007je003030

Chu, P., Spring, J., Zacny, K.: ROPEC—ROtary PErcussive Coring Drill for Mars Sample Return. AMS, 14–16 May 2014, NASA GSFC (2014)

Gertsch, L.S., et al.: Review of lunar regolith properties for design of low power lunar excavators. In: International Conference on Case Histories in Geotechnical Engineering, Arlington, VA, 11–16 Aug 2008

Hand, K.P., et al.: Report of the Europa Lander Science Definition Team. Posted in February 2017

Paulsen, P., et al.: TRL6 Lunar Resource Prospector Drill, LPSC 2017. Woodlands, TX (2017)

Sriapai, T., et al.: Effect of temperature on compressive and tensile strengths of salt. ScienceAsia **38**(2012), 166–174 (2012)

Treffer, M., et al.: Preliminary studies concerning subsurface probes for the exploration of icy planetary bodies, PSS (2006)

Ulamec, S., et al.: Access to glacial and subglacial environments in the Solar System by melting probe technology. Rev. Environ. Sci. Biotechnol. **6**, 71–94 (2007)

Zacny, K., Cooper, G.: Coring Basalt rock under simulated Martian atmospheric conditions. Mars J. **3**, 1–11 (2007)

Zacny, K., et al.: Reaching 1 m deep on Mars: The Icebreaker Drill. Astrobiology (2013a). https://doi.org/10.1089/ast.2013.1038

Zacny, K., et al.: Asteroids: Anchoring and sample acquisition approaches in support of science, exploration, and in situ resource utilization. In: Badescu, V. (ed.) Asteroids: Prospective Energy and Material Resources, pp. 287–343. Springer, Heidelberg (2013b)

Zacny, K., Paulsen, G., Yaggi, B., Mellerowicz, B., Wang, A., Hyman, C., Hedlund, M., Kleinhenz, J., Smith, J., Quinn, J.: Development of the Lunar Drill for the Resource Prospector Mission, IEEE Aerospace Conference, 5–12 March 2016, Big Sky, MT (2016)

Chapter 6
Drilling Mechanisms Using Piezoelectric Actuators Developed at Jet Propulsion Laboratory

Yoseph Bar-Cohen, Stewart Sherrit, Mircea Badescu, Hyeong Jae Lee, Xiaoqi Bao and Zensheu Chang

6.1 Introduction

Drilling mechanisms are widely used in many diverse fields including domestic, medical, industrial, military, geology and extraterrestrial applications (Bar-Cohen and Zacny 2009). Generally, scientists and engineers have developed many types of drills with the majority of designs based on mechanical motion (rotary and/or percussive) of a cutting tool. These drills have been the result of the effort to deal with the challenges presented by the large variety of materials that need to be penetrated. Mechanical drills use a bit having a tip that interacts with the drilled material and applies forces over a small area to cause large shear and/or impact stresses for cutting or breaking the material. There is a wide variety of bit types that have been developed commercially, which can be readily purchased at local hardware stores.

Y. Bar-Cohen (✉) · S. Sherrit · M. Badescu · H. J. Lee · X. Bao · Z. Chang
Jet Propulsion Laboratory, California Institute of Technology, MS 67-119,
4800 Oak Grove Dr, 91109-8099 Pasadena, California, USA
e-mail: yoseph.bar-cohen@jpl.nasa.gov

S. Sherrit
e-mail: stewart.sherrit@jpl.nasa.gov

M. Badescu
e-mail: mircea.badescu@jpl.nasa.gov

H. J. Lee
e-mail: Hyeong.Jae.Lee@jpl.nasa.gov

X. Bao
e-mail: xiaoqi.bao@jpl.nasa.gov

Z. Chang
e-mail: zensheu.chang@jpl.nasa.gov

V. Badescu and K. Zacny (eds.), *Outer Solar System*,
https://doi.org/10.1007/978-3-319-73845-1_6

For space science, one of the most important questions is "whether life has ever arisen anywhere else in the universe" (Gershman and Wallace 1999). In order to answer this question sampling tools based on a variety of mechanism have been developed to acquire samples for instrument interrogation. Since water is a critical resource for life-as-we-know-it, NASA prioritized the targets of future exploration to the Solar System planetary bodies, which are currently known to have or have had flowing liquid water. In the most recent Planetary Decadal Survey [Vision and Voyages for Planetary Science in the Decade 2013–2022 (NRC 2013)], Mars, Europa, and Enceladus were specifically called out for future exploration because of the accessibility to their aqueous regions. These bodies have the highest likelihood of having extant or extinct life and they require sample acquisition from below the surface. However, each of these bodies presents its own drilling challenges. On Mars, the drilled materials would consist of dry regolith with physical properties (i.e. tensile strength, hardness, etc.) whose characteristics may potentially vary several orders of magnitude as a function of depth. On Enceladus and Europa, drilling would be through ice at temperatures below 100 K and in a low gravitational field. In addition, on Enceladus and Europa high surface radiation can also pose a problem for electronics and instrumentation. Among the strategies of drilling into these bodies, there are common requirements including penetration of the subsurface, capture of samples, and delivery of unaltered samples to instruments for analysis.

Increasingly, developers of drills for in situ exploration missions are seeking capabilities that address the complex challenges involved with extreme environments found at the planetary bodies where subsurface penetration is needed. The use of piezoelectric actuators offers effective drilling capabilities, particularly in extreme environments. In the last two decades, significant developments have been made in using piezoelectric actuation to perform percussive drilling. In these designs, the cutting surface is fractured by high and low frequency impacts that enhance the penetration. The performance is again significantly increased by rotating the bit to introduce shearing and a mechanism to allow the cuttings to be removed from the created borehole.

This chapter is focused on the drilling mechanisms that are driven by piezoelectric actuators, which were developed by the authors at the Jet Propulsion Lab (JPL), Pasadena, CA. The developed mechanisms are operated at ultrasonic frequencies as well as some that involve the use of a free-mass that transforms the vibration frequencies ultrasonic to the sonic range. The mechanisms that involved the use of free-mass include the Ultrasonic/Sonic Driller/Corer (USDC) and the many novel designs that were derived from its concept. This Section of the chapter presents an introduction, while Sect. 6.2 covers the USDC and its various components and mechanisms. The piezoelectric actuated drills that do not involve the use of free-mass are covered in Sect. 6.3. The drills that were developed for operation at high temperatures are covered in Sect. 6.4 and the designs for low temperature are covered in Sect. 6.5. The USDC-based deep drills are described in Sect. 6.6. Finally, Sect. 6.7 presents the information on Lab-on-a-drill.

6.2 The Ultrasonic/Sonic Driller/Corer (USDC)

Since 1998, the authors have been developing piezoelectric actuated drills that address the need for drilling mechanisms that can be operated in low gravity using low preload (Sherrit et al. 1999; Bao et al. 2003; Bar-Cohen and Zacny 2009). Operating with low axial load allows for drilling from lightweight robotic platforms such as rovers. The first piezoelectric actuated drill that was developed is the ultrasonic/sonic driller/corer (USDC). It uses an intermediate free-mass between the ultrasonic horn (actuator) and the bit converting high frequency vibrations to low frequency hammering (Fig. 6.1). Unlike typical ultrasonic drills where the bit is acoustically coupled to the horn, in the USDC the actuator drives a free-mass that converts the ultrasonic vibrations to hammering at sonic frequencies. This mechanism enables imparting higher energy blows onto the bit. In 2000, the USDC received the R&D Magazine award as one of the 100 most innovative instruments of that year.

The USDC consists of three key components: actuator, free-mass, and bit (Bar-Cohen et al. 1999). The actuator generates ultrasonic vibrations that transfer energy to the free-mass and in turn impacts the bit that imparts stress impulses onto the drilled medium. These impulses are fracturing rocks when their ultimate strength is exceeded. The actuator consists of a piezoelectric stack, backing, and horn, where a stress bolt between the backing and the horn keeps the piezoelectric stack in compression. The backing and the stress bolt are free to expand and they enhance the delivered power to the horn, while the horn amplifies the generated displacements. The USDC, in its nominal configuration, is designed with the piezoelectric actuator driven at resonance in a frequency of about 20 kHz. To maintain the operation at resonance, software or hardware control is used to drive

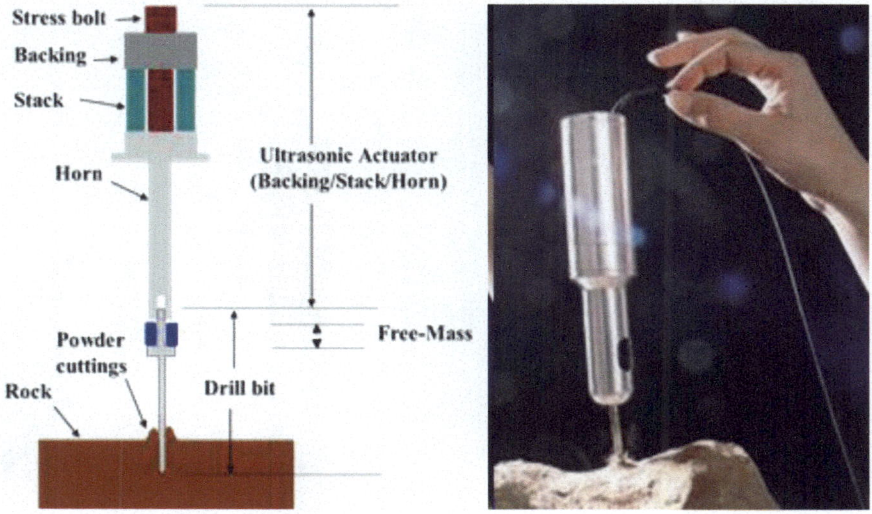

Fig. 6.1 Schematic cross-section view (left) of the USDC and a photo showing its ability to drill with minimum axial force (right)

the actuator and ensure maximum electric current input. This control adjusts the resonance frequency to compensate for the effect of the drilled medium to reduce the Q (the Quality Factor, which is the ratio of the mechanical energy stored divided by the energy loss in each cycle) of the resonator and shifts the resonance frequency as the actuator is slightly heated during operation. In addition, attention is given to the impacts that cause time variations in the current signal. The resonance frequency shifting effect is minimized using various control algorithms such as hill climbing, extremum seeking, and others (Aldrich et al. 2006a, b, 2008a, b).

The USDC has been implemented in various configurations and sizes. The first large diameter USDC based drill was developed in 2001 which had a cylinder shaped drill with a diameter of about 2.5 cm and length of about 30 cm. It was called the Ultrasonic/Sonic (U/S) Gopher (Fig. 6.2). The drill is a wireline mechanism that is tethered for power and movement up and down the borehole. The drill was demonstrated to penetrate limestone to a depth of about 11 cm. It was then followed by development of a series of USDC based Gopher technology for deep drilling. The first progeny of the U/S Gopher was an ice drill, called Ultrasonic/Sonic (U/S) Ice-Gopher, which was demonstrated in Antarctica to drill as deep as about 2-m. The lessons learned from the development of the Ice-Gopher has been implemented in the development of the Auto-Gopher drill, where the first generation was developed to operate as a semi-autonomous corer. Currently, the Auto-Gopher-2 is being developed as a fully autonomous wireline system. It is developed with (1) Formation fracture, (2) Sample capture, (3) Sample ejection, (4) Embedded electronics, and (5) Autonomous drilling with fault detection. Once this system is fully developed,

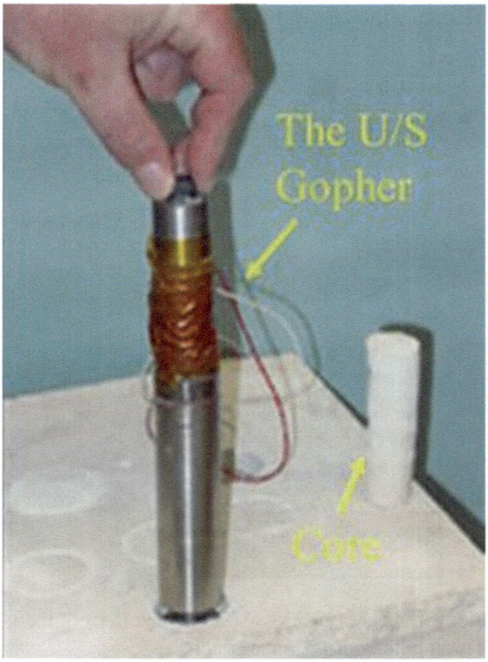

Fig. 6.2 Photographs of the Ultrasonic/Sonic Gopher prototype drill and a limestone core drilled by the device

the prototype will be tested to drill rocks, permafrost, and ice in an environmental chamber that has a simulated Martian ambient environment.

One of the major limitations of drilling on planetary bodies is the available power generated by the lander or rover. The proven power sources for *in situ* exploration missions with a lander are solar panels and the Radioisotope Thermoelectric Generators (RTG). In order for the Auto-Gopher to drill ice effectively, the test data suggests that two to three hundred Watts of power is necessary. In contrast, an ice-melting probe would likely require kW levels of thermal power Smith et al. (2006).

Since Gopher is a wireline type drill, one of the major concerns is borehole collapse that can take place above the drill. In addition, as the depth increases the hole can get smaller due to the increasing pressure, which can cause deformation and creep and in turn causes pinching of the borehole. However, the latter may not be an issue for ice drilling on extremely cold, low gravity bodies, since the ice deformation and creep are a function of the ice temperature and gravity (Patterson 1977). The gravity and ice temperature on Mars, Europa and Enceladus are quite low, and therefore minimal creep may take place as the penetration depth scales with the inverse of the gravity. Moreover, the Auto-Gopher is periodically retracted to extract the cuttings and a mechanism of counter rotating the drill can be used as a reamer and maintain the borehole diameter.

Following the development of the USDC, numerous novel drill designs were conceived and have been disclosed in NASA New Technology Reports (NTR) as well as registered patents. The references to some of the NTRs and Patents include (e.g., Badescu et al. 2006a, b, 2009a, b, c, 2010; Bao et al. 2004, 2015; Bar-Cohen et al. 2001, 2002, 2003a, b, 2004, 2005a, b, c, 2007a, b, 2008, 2010a, 2012, 2014a, b; Dolgin et al. 2001; Sherrit et al. 2001a, b, 2002, 2003, 2004a, b, 2005b, 2006, 2008, 2009, 2010, 2012, 2016). These include the Ultrasonic/sonic Rock Abrasion Tool (URAT) (Dolgin et al. 2001), Ultrasonic/Sonic Gopher and the Auto-Gopher for deep drilling (Badescu et al. 2011a, b; Bar-Cohen et al. 2012), the Lab-on-a-drill (Chang et al. 2003), and many other piezoelectric drilling mechanisms (Bar-Cohen and Zacny 2009). Further, the USDC was demonstrated to drill ice and various rocks including granite, diorite, basalt, and limestone. In the configuration where the bit vibrated longitudinally and the drill was not augmented with bit rotation, sensors (such as thermocouple and fiber optics) were integrated into the bit and were used to analyze the formed borehole wall while drilling. Another benefit of using a non-rotating bit has been the ability to produce non-round shape cores such as rectangles and triangles. To predict and optimize the performance of the drill analytical models and analysis tools were developed.

6.2.1 Principle of Operation of the USDC

The heart of the USDC-based drills is a vibratory mechanism that is driven at ultrasonic frequency. Its main attribute includes the low axial preload required to drill various formations. High impact hammering blows are realized by the principle of the piezoelectric effect, where the input AC electrical signal on a piezoelectric transducer leads to mechanical vibrating resonant system, generating large mechanical forces on the target (Bao et al. 2003). The level of output mechanical forces of the ultrasonic hammer system is closely related to its vibration velocity that is proportional to the excitation frequency and displacement amplitude. Therefore, components that allow a high vibration velocity are a key enabling technology for large hammer actions. To meet the demands of high performance hammer actions, the piezoelectric stack configuration is implemented with a combination of the optimum design of ultrasonic horn (Chang et al. 2004; Sherrit et al. 2004a, 2010). Various mechanisms give rise to the high vibration velocity. The presence of a stepped horn amplifies the displacement induced by the piezoelectric stack by a ratio of the area of the horn base to the horn tip when it is driven at its fundamental half-wavelength axial mode resonance frequency. In addition to this geometric amplification, there is also a resonance amplification due to the Q of the resonance. Finally, for a given voltage the piezoelectric stack gives the increased piezoelectric strain that is proportional to the number of the piezoelectric layers.

The energy generated by the ultrasonic horn is then transferred to a free mass, converting a high-frequency, low-amplitude excitation into a low-frequency high impact energy on the bit, realizing a relatively low voltage, and high efficiency ultrasonic hammer system. This allows for the effective conversion of high frequency, high force vibration of the horn tip into low frequency, high displacement hammering blows. These impacts create a large enough stress in the target to enable efficient drilling. A simple collision model can be applied to explain the basic mechanism of the horn/free mass interaction. If we assume that the energy loss and time duration of the impact is negligible and perfectly elastic collision, the post collision velocities can be calculated using following equations based on the conservation of momentum and energy:

$$m_1 v_{\mathrm{i}1} + m_2 v_{\mathrm{i}2} = m_1 v_{\mathrm{f}1} + m_2 v_{\mathrm{f}2} \tag{6.1}$$

$$\frac{1}{2} m_1 v_{\mathrm{i}1}^2 + \frac{1}{2} m_2 v_{\mathrm{i}2}^2 = \frac{m_1 v_{f1}^2}{2} + \frac{m_2 v_{f2}^2}{2} \tag{6.2}$$

where v_{f1} and v_{f2} are post collision velocities of horn and free-mass, and v_{i1} and v_{i1} are pre-collision velocities of horn and free mass, respectively. The effective mass of the horn, m_1, and the mass of free-mass, m_2. The equations leads to the post collision velocities by solving for v_{f1} and v_{f2}:

$$v_{f1} = \frac{(m_1 - m_2) v_{i1} + 2 m_2 v_{i2}}{m_1 + m_2} \tag{6.3}$$

$$v_{f2} = \frac{m_1 (2 v_{\mathrm{i}1} - v_{\mathrm{i}2}) + m_2 v_{\mathrm{i}2}}{m_1 + m_2} \tag{6.4}$$

Assuming that the effective mass of the horn (m_1) is much higher than the mass of free mass (m_2), i.e., $m_1 \gg m_2$, the post collision velocities of free mass can be simplified as $v_{f2} \approx -v_{i2} + 2v_{i1}$.

These equations imply that when the masses of free mass are much lower than those of piezoelectric actuator or probe bit and target, the free mass bounces back and forth between the horn and the probe, and its post collision velocities increase after interaction with the vibration tip. During the process, the momentum after the impact is transferred to the probe bit at each impact, delivering high impact forces on the target (Bao et al. 2003; Sherrit et al. 2000). Other analytical studies of elasto-plastic contact force due to impacts include Vila and Malla (2016a;, b,c).

Since the horn tip is not phased to the free mass we need to consider that the impacts may happen at different phases of the tip vibration i.e. the tip velocity may be positive or negative. A computer program was developed to simulate the collision process (Bao et al. 2003). The speeds of the free mass after impacts as a function of tip vibration phase for different incoming speeds and tip vibration amplitudes are shown in Fig. 6.3. The results show the speeds of the free mass has a larger chance of being increased than being decreased. As shown in Fig. 6.4, the average increase of the free mass kinetic energy is a function of the ratio of incoming speed over the horn tip vibration amplitude. This result has also been confirmed experimentally (Harkness et al. 2011). The vibrating horn tip could effectively deliver energy to the free mass when the free mass speed is less than ~2 times the velocity amplitude of the tip vibration.

The drive software that controls the operation of the piezoelectric transducer of the percussive mechanism is a critical part of the performance optimization. This requires ability to monitor the drive frequency and adjust for its drift to maintain operation at resonance. It is well known that the electromechanical efficiency of the piezoelectric transducer is maximum when driven at its resonant frequency, where the same amplitude of displacement can be achieved at small drive field (applied voltage) at this frequency due to the resonance build up (Sherrit et al. 1999).

Extremum seeking control is a well developed control method which tracks a varying maximum or minimum of a performance function. The main advantage of this method is that the piezoelectric system can keep a performance at its extremum value by seeking operating set-points that maximize the vibration velocity, which occurs at its resonant frequency. The algorithm that was implemented is the "hill-climbing method", where the software determines a favorable direction in each iteration based on the comparison between the previous and current admittance values. The control inputs continue to update in this manner until the performance output begins "climbing" up one of the resonant peaks of the device. This iterative process can be illustrated using the following formula:

$$x_{k+1} = x_k + \alpha . p \quad (6.5)$$

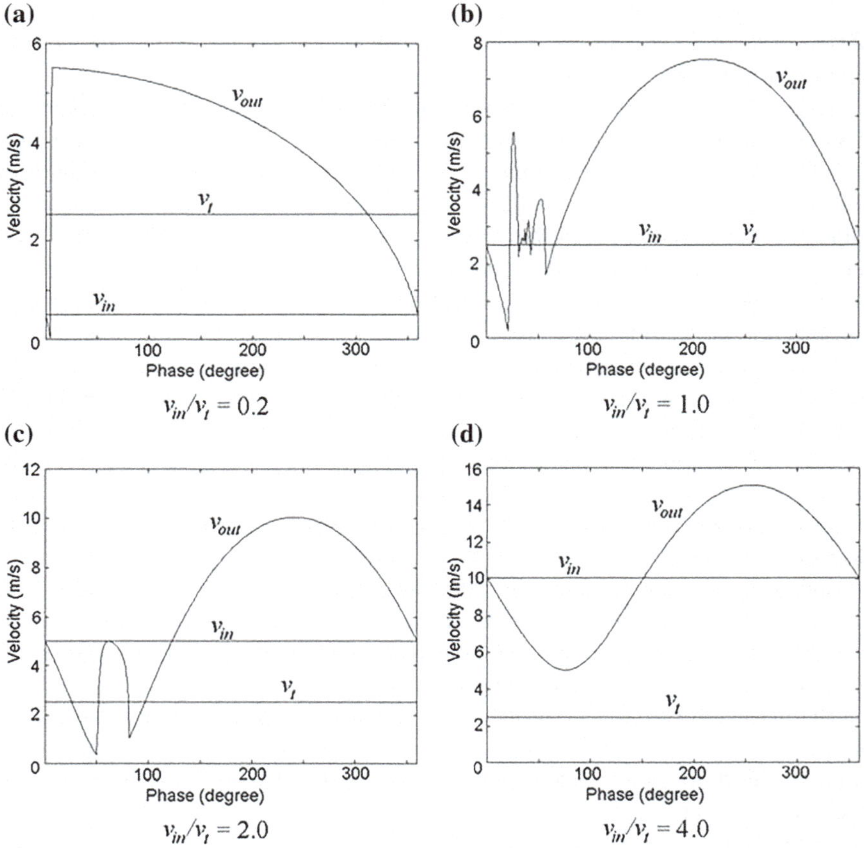

Fig. 6.3 The speeds of the free-mass after impacts, v_{out}, versus relative tip vibration phase for different incoming free mass speeds, v_{in}. The solid horizontal lines indicate the levels of the incoming speed, v_{in}, and the amplitude of tip velocity, v_t

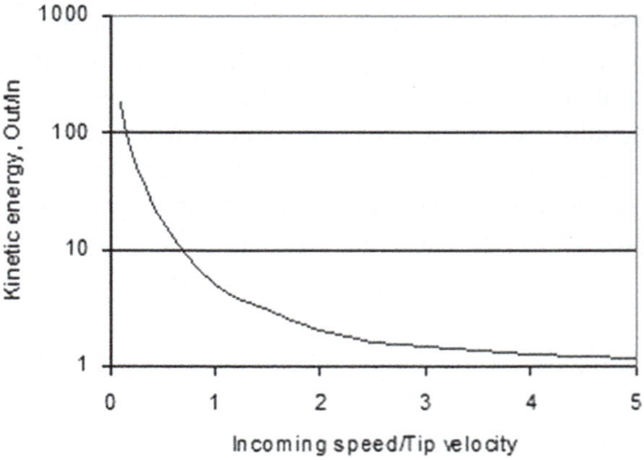

Fig. 6.4 The average increase of the kinetic energy of free mass after a collision with tip of the horn

with

$$p = \begin{cases} 1 & if \quad dy_k/dx_k > 0 \\ -1 & if \quad dy_k/dx_k < 0 \end{cases} \tag{6.6}$$

where $dy_k = y_k - y_{k-1}$ is the change in the parameter being tracked and $dx_k = x_k - x_{k-1}$, is the change in the parameter being adjusted. α is the user-defined step size and p is a search direction. The variables x and y correspond to frequency and admittance, respectively. Note that a good step-size control is critical in this algorithm; although a large step size increases the convergence rate to the peak, it leads to the high oscillations around the peak, resulting in jump back and forth on both sides about the peak. In contrast, a small step-size decreases the oscillations around the peak, but it is possible to get trapped in a spurious local maximum if the step size is too small (Aldrich et al. 2006a).

6.2.2 The Key Components of the USDC

6.2.2.1 Various Horn Configurations

The horn is an important part of the USDC actuator amplifying the produced vibration amplitude (Chang et al. 2004; Sherrit et al. 1999, 2001a, 2004a, 2010, 2016). In the majority of horn designs, that the authors have developed, a stepped horn was used. This horn is made as a step from the diameter of the piezoelectric stack elements to the tip diameter with the amplification increasing as the ratio of the square of the diameters (areas ratio). In many cases, specific requirements of the application have produced modifications of the stepped horn configuration. Over the years, the authors developed many novel horn designs and examples include folded and small volume as in the Ultrasonic Rock Abrasion Tool or versions that generate large impact forces, as is the case of the various Gopher dog-bone horn designs (see Sect. 6.6).

Folded Horn

The length of the horn in the USDC actuator can be a concern when there are specifications with volume constraints and the reduction of its size can be critical to the integration of the unit as a sampler. A compact shape horn was developed with hollow configuration that amplifies vibrations of high power actuation mechanisms (Sherrit et al. 2001a). A graphic illustration of a folded horn configuration is shown in Fig. 6.5 and a photo that shows a folded horn versus a straight one is shown in Fig. 6.6. Folded horns can be configured in axis-symmetric and planar shapes to provide manufacturing options. According to the Finite Element Analysis that was conducted, the tip displacement of a folded horn increases due to the bending effect

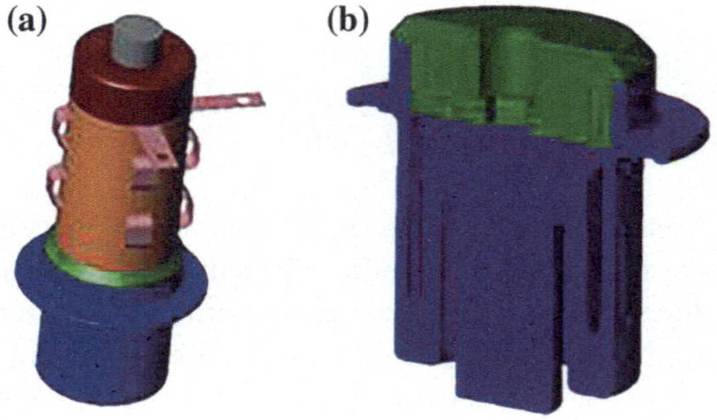

Fig. 6.5 Solid model assembly drawing of: **a** folded horn with piezoelectric stack actuator, and backing. **b** Cross section of the folded horn. **c** Monolithic folded horn design

Fig. 6.6 Photograph of a folded horn (16 kHz) is shown on the left and, for comparison, a straight horn with approximately the same frequency (20 kHz) is shown on the right

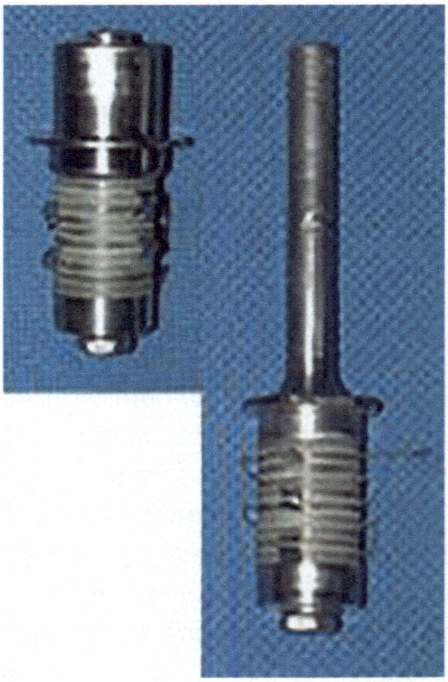

of the folded horn (Sherrit et al. 2001b). However, tests of the assembled folded horn showed the mechanical Q and the tip displacement were much lower compared to the straight horn. It is the authors' assessment that there has not been effective energy transfer across the screw threads to the base that holds the horn. Another design based on a single monolithic horn that is producing larger tip displacements is shown in Fig. 6.5c. This horn was used in order to develop the powder sampler that is shown in Fig. 6.20 (Chang et al. 2005).

Dog-Bone Shape Horn

To improve the drilling performance by increasing the impact momentum, a dog-bone design was conceived and demonstrated. This horn design also allowed the benefit of mounting the bit to the actuator and operating with dual free-mass to permit both forward and reverse hammering, where the reverse hammering could prevent sticking of the drill bit in the borehole. For this purpose, various horn designs were examined analytically and compared to conventional and solid shape horns and it was shown that the dog-bone design has a superior performance (Chang et al. 2004; Sherrit et al. 2004c). To demonstrate the capability of the dog bone horn, a finite element model was used to determine the control parameters and showed the excitation of superior tip displacement and velocity. A piezoelectric actuator design with a dog bone horn based on this analysis is shown in Fig. 6.7.

6.2.2.2 USDC with Various Free-Masses

The free-mass is a critical element in the activation of the ultrasonic/sonic percussive mechanism. It performs the frequency reduction from the ultrasonic (or high frequencies) to the sonic (or low frequencies). The reduction ratio reaches orders of magnitude where, in the case of the original USDC design, it reduces the frequency from about 20 kHz to about 60–1000 Hz. The original configuration of the USDC is shown in Fig. 6.8 on the left—it consisted of a steel disk with a hole that was used with a rod insert to provide guiding and retaining. The rod, which was part of the top of the drill bit, was partially inserted into the horn and acted as a shaft for the bit, free mass, and horn alignment. The free-mass in this case had a diameter of 0.350 in. and it was cut in thickness to meet the mass requirements of the specific designs. Since the free-mass tended to jam (terminating the operation making it difficult to remove the drill bit), it was necessary to modify this design. In a subsequent design, the rod was eliminated and instead a cup shape was used with a solid cylinder shape free-mass. This design significantly improved the durability of the bit. A cross-section illustration of both, the support with a shaft and the one with the cup, are shown in Fig. 6.8.

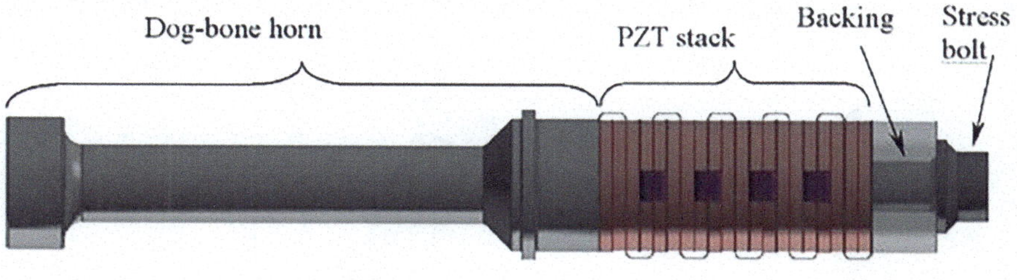

Fig. 6.7 Illustration of a dog-bone horn configuration

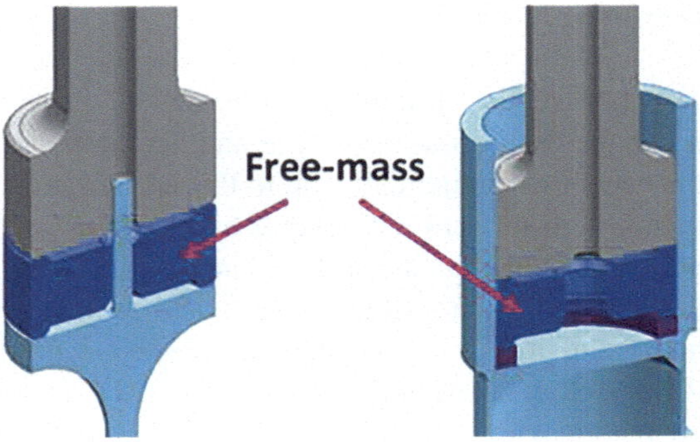

Fig. 6.8 Left—rod configuration for the retention of the free-mass; right—cup configuration for the free-mass retention

In an effort to determine guiding parameters in the design of effective free-mass configurations, tests were made with free-masses having weights from 1 to 8 g, in increments of 0.5 g up to 4 g, and in 1 g up to 8 g. To avoid sharp edges on the free-mass configuration, which may dig into the inside diameter of the drill bit as the mass travels up and down, 0.050-in chamfer was made around the edges (see Fig. 6.9). As the material for the free-masses, Tool Steel having RC-60 hardness was selected that matched the bit material.

A series of drilling tests into dolomite were made using different free-mass materials and thicknesses. The test results show dependence of the drilling rate on the characteristics of the free-mass (Fig. 6.10). The data suggests that the effectiveness of the free-mass is less dependent on the material type and length to first order and is primarily controlled by the mass. In this test, for the horn and the bit used in these experiments, the optimum mass was found to be 2.2 g.

The free-mass was cylindrical and its geometry led to jamming inside the bit and, in order to eliminate this jamming, a donut shape free-mass was produced as shown in Fig. 6.11. This led to systematic designs of free-masses where the impact force is applied over minimal area. In normal operation, the free-mass bounces between the horn and the bit, i.e. encounter impacts with both the horn and the bit.

The impact stress profile, maximum stress, contact time duration, and, in sequence, the required yielding stress for the materials of the free-mass, bit and horn are affected by the contact area. A small contact area will have higher contact pressures, and is therefore more likely to go into local yield, negatively affecting restitution (maybe invalidating the conservation of momentum assumptions) and

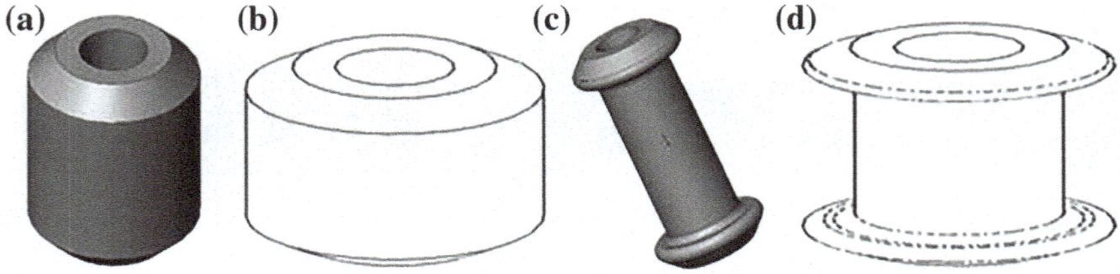

Fig. 6.9 Four different shapes of free-mass with chamfers

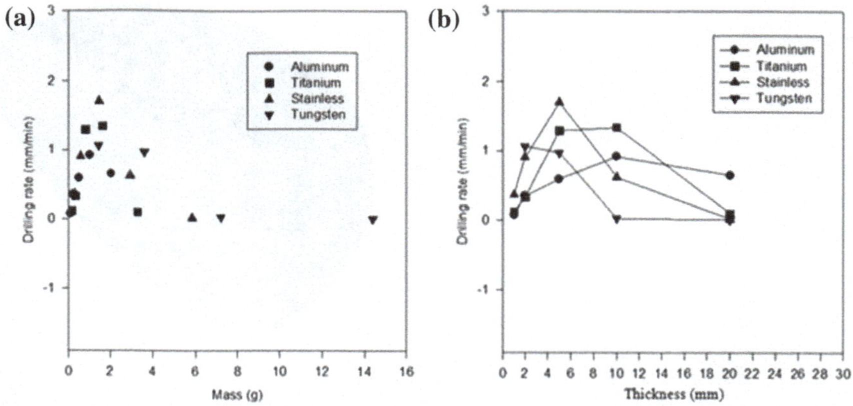

Fig. 6.10 The drilling rate in dolomite as a function of the **a** mass and **b** thickness of the free-mass

Fig. 6.11 Top and side view of three different donut shape free-masses that were made for the high temperature drill

reducing free-mass lifetime. On the other hand, the larger contact area will result in lower stress in the contact region and avoid yielding the free-mass materials and its interfaced areas. However, before the excitation voltage is applied to the transducer the horn, free-mass and the bit are pressed together. Larger contact area results in a stronger coupling of the bit to the horn transducer. A strong coupling will greatly change the vibration characteristics of the transducer and make the USDC difficult to start. To obtain optimum performance, trade-offs need to be done in designing the free-mass. For this purpose, the following novel designs were conceived (Bao et al. 2015):

- Point contacts—For this purpose, spherical, elliptical or pill shapes are designed to assure that the impact locations are points that are in the top and bottom of the free-mass (Fig. 6.12). The surface curvature at the top and bottom can be used to control actual contact area in impacts.
- Circular contact—Using spherical, elliptical or pill shapes with shaped or chamfered parallel surfaces on the top and bottom leaves circular contact surfaces with the horn and the bit.

Fig. 6.12 Free-masses with point contacts

- Point/Circular contacts—The design of the free-mass can be made with one side as a point, while the other surface is circular.
- Line contacts—The donut shape (the right mass in Fig. 6.11) represents one of the designs under this category. The curvature at contact line in cross section can be used to control actual contact area in impacts.
- Ring contacts—A free-mass with ring interface areas on the horn and the bit sides (Fig. 6.13).
- Line/Ring contacts—The design of the free-mass can be made with a line on one side and ring on the other as shown in Fig. 6.14.
- Dashed line or ring contacts—All the line and ring contacts mentioned above can be dashed along the length or circle.
- Depending on the ratio of the diameter to the height, and the free-mass retention method used (the cup or rod), the free-mass can be configured with one or more sliding surfaces on the outside or inside diameter surface or both.
- Matching horn tips and free mass as shown in Fig. 6.15 may also offer some utility in maximizing the stress pulse.
- Matching free-mass to bit's cup contact area can be done similar to the matching of the free-mass to the horn tip. The illustration shown in Fig. 6.15 is applicable to this design where the free-mass is placed on top of the interface.

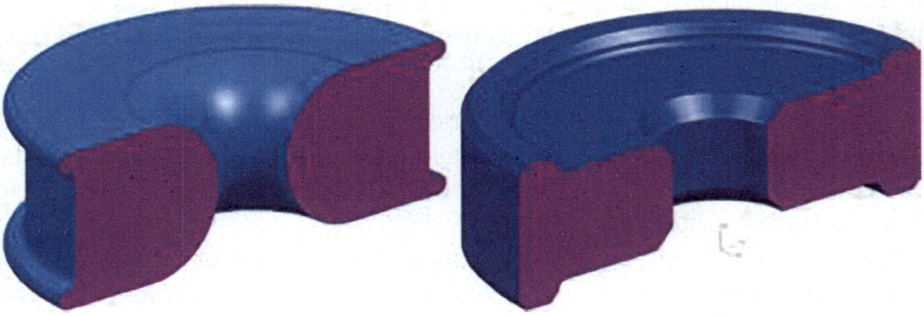

Fig. 6.13 Free-masses with ring contact areas

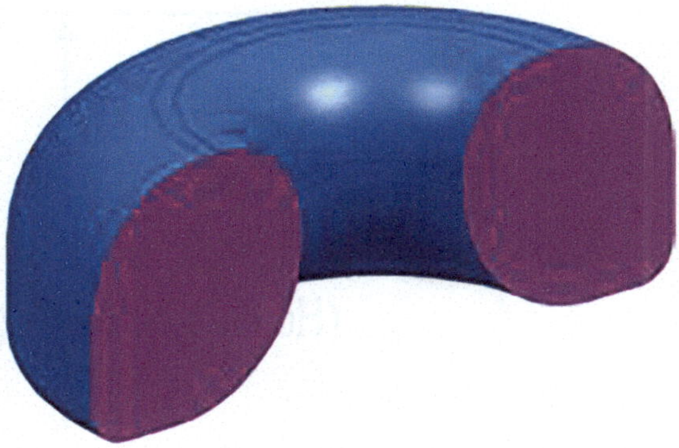

Fig. 6.14 Free-masses with line/ring contact areas

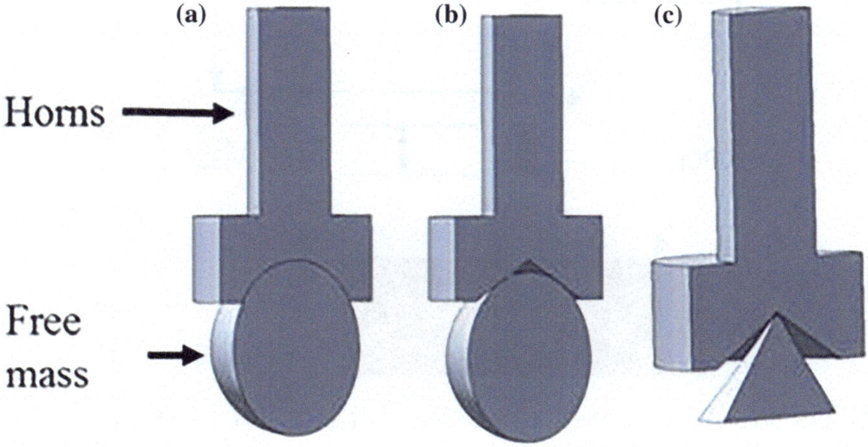

Fig. 6.15 Examples of horn tip and free-mass integrated design: **a** increase in contact area; **b** ring contact; and **c** point contact

6.2.2.3 USDC with Various Bits

Abrasion Bits

Abrasion of the surface of a rock using low axial force and limited average power is needed to remove weathered layers and expose pristine sections of rock samples. Conventional rotating mechanisms require high axial loads and may pose contamination risk since they may "shed" lubricants and filings from their motor gearbox. The USDC offers the advantages of requiring low axial force, low average power, low number of components, and the capability to produce a mechanism of powered cuttings removal. In its original configuration, the USDC was designed to drill or core and it was not designed to remove layers of weathered material from

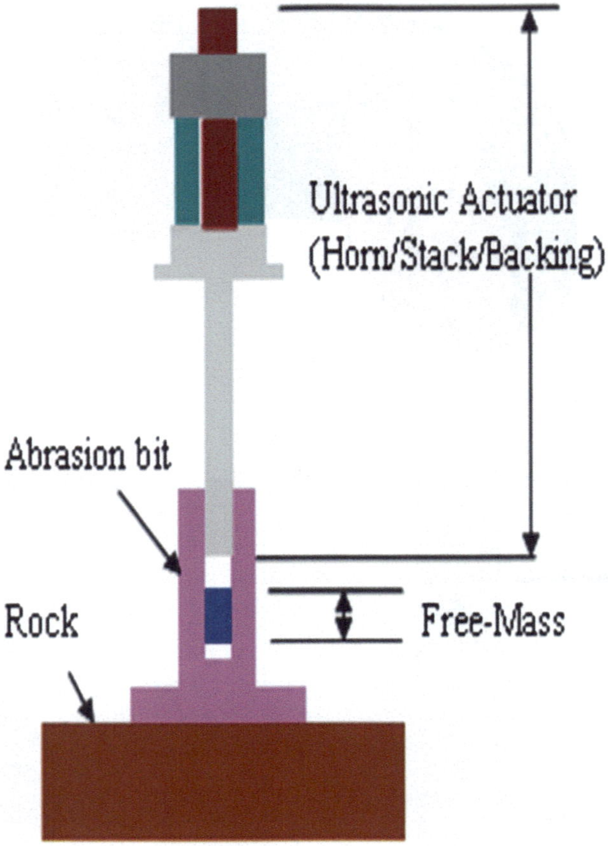

Fig. 6.16 Schematics of the Ultrasonic Rock Abrasion Tool (URAT). (*Note* The details of the Ultrasonic Actuator are shown in Fig. 6.1)

rocks. To address the need for abrasion tool, an Ultrasonic Rock Abrasion Tool (URAT) was developed (Dolgin et al. 2001) and its abrasion bit was designed similar to the hammering surface of a meat tenderizer. Teeth were machined onto the bottom of the disk that is part of the abrasion bit. These teeth amplify the loading force from the drill and enhance the action of the URAT.

A schematic cross-section of the URAT is shown in Fig. 6.16 illustrating its components and the compact structure. A photo of an abraded basalt is shown in Fig. 6.17. The developed abrasion tool consisted of 40 mm (1.6 in.) diameter disk that was attached to a shank that fitted around the horn. The free-mass is placed inside the shank between the horn and the bottom of hole along the inner part of the bit. On the bottom of the abrading disk, teeth were machined in the form of pins that stick out of the disk.

Fig. 6.17 Photographic view of basalt that was abraded by the URAT. The footprint of the abrasion bit can be seen on the surface

Interchangeable Bit

One of the most promising benefits of the USDC is the simple interface it provides to the bit, and the simplicity of the shape of the bit itself. The use of multiple bits can be essential in future missions as it is simply not practical to fly a multitude of samplers to support various instruments. Using a single actuator that utilizes multiple bits for drilling, coring, surface preparation, and sampling, offers a practical approach where the bits can be exchanged as needed (Bar-Cohen et al. 2002). The bit does not require sharpening, but if a bit is damaged for whatever reason, it could be replaced. Using the ability to exchange bits of the USDC, the JPL team worked jointly with Alliance Spacesystems, LLC to employ their robotic arm with a USDC unit. A bit interchange mechanism and a tool caddy of bits were developed. The system was called Mars Integrated Drilling and Sampling (MIDAS) system. A photograph of the system that includes the robotic arm, the USDC, and a tool caddy of bits is shown in Fig. 6.18. The robotic arm was designed to interchange bits from the caddy by snapping them in and out as needed.

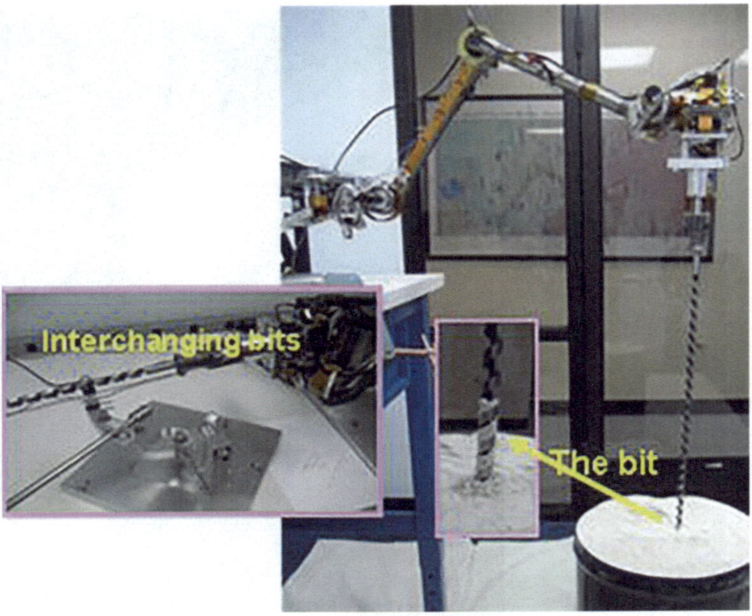

Fig. 6.18 The Mars integrated drilling and sampling (MIDAS) system with an interchangeable bit mechanism for the USDC. (Photographed at Alliance Spacesystems, LLC by Lori Shiraishi, JPL/Caltech/NASA and published with permission.)

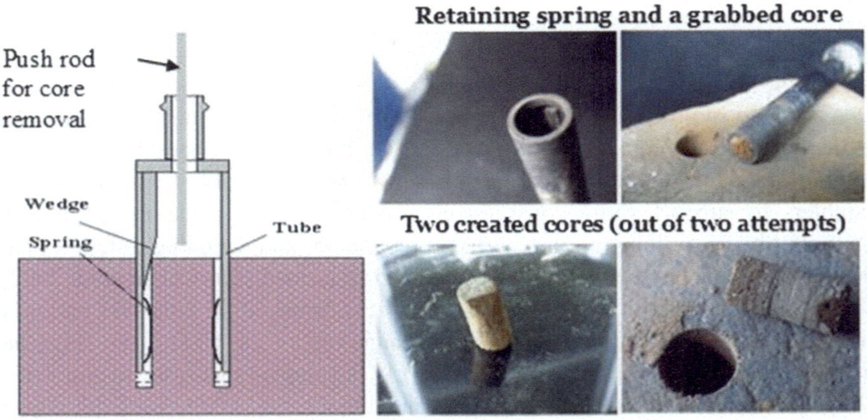

Fig. 6.19 A cross-section illustration of the all-in-one bit is shown on the left. The bit and some formed core are shown on the right. (Courtesy NASA/JPL.)

All-in-One Bit

The use of multiple bits requires a manipulation system to exchange bits as needed. If a manipulation system is not available, it is highly desirable to accomplish as many functions as possible using a single bit. For this purpose, an all-in-one bit was developed and produced as well as demonstrated (Bar-Cohen et al. 2002). The bit consists of a tube with a wedge at the top of the inner surface, a set of springs near the tip and a push rod that is inserted thru a center hole in the bit (see Fig. 6.19).

Once a core is produced at a length of the inner section of the bit, the wedge introduces transverse forces at the top of the core to cause maximum stress near the root and shear fracture. Side springs hold the sheared core for removal from the borehole. The core is extracted from the bit using the push rod from the top of the bit. Figure 6.19 shows photographs of an all-in-one coring bit and the produced cores.

Bit for Acquiring Powdered Cuttings

In the sampling process, USDC generates powder in the form of fine cuttings (Sherrit et al. 2001a, 2003). Tests of the produced powder particles have shown that their sizes are ideal for high quality X-ray diffraction (XRD) spectra (Blake et al. 2003; Chipera et al. 2003). The XRD patterns are essentially indistinguishable from powders that were made using laboratory Retsch mills. Further, the distribution of the particle sizes is quite comparable to the one obtained from a standard laboratory Retsch mill.

In order to produce cuttings sampler that is compact and lightweight, a bit was configured that consists of cavities which trap the powder. The cavities acquire the ejected particles that enter a hollow inner cavity at the bottom of the bit (see Fig. 6.20 on the left) and retained until they need to be delivered to an instrument or disposed of, in which case the actuation is turned on and the powder is shaken out of the collector volume (Sherrit et al. 2006). Using Finite Element and solid models, a compact horn configuration was designed that allowed for the development of a 265-g device (Chang et al. 2004). The bit was made hollow with an end-effector section that is brazed on the bit and has teeth to enhance the cutting performance.

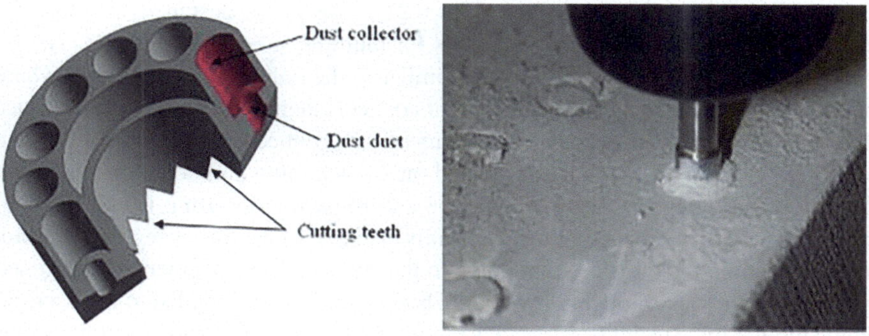

Fig. 6.20 A cross section of the bit of the powdered-cuttings sampler (left) and a photo of the sampler in action drilling a limestone and accumulating cuttings inside the bit (right). Courtesy NASA/JPL

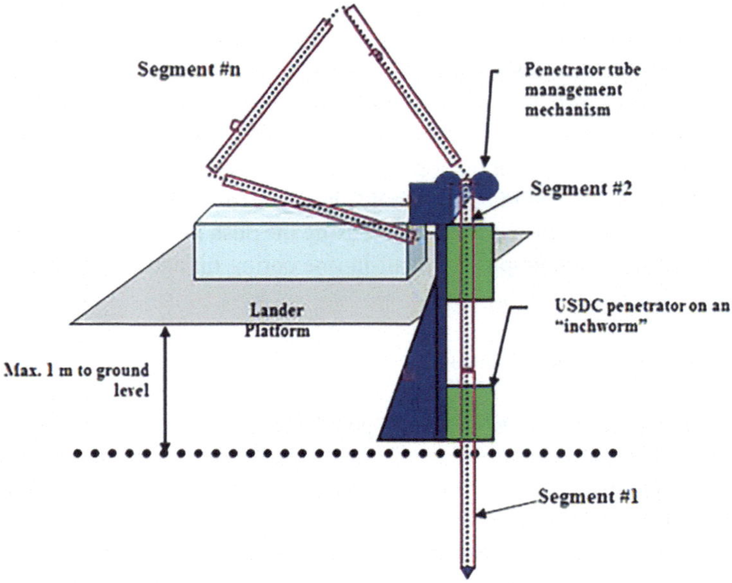

Fig. 6.21 The foldable bit mechanism where multiple segments are unfolded while penetrating the ground

Foldable Bits

Drilling several tens of centimeters or more of planetary subsurface using a single bit requires a long bit and its launch and deployment can be a challenge. To simplify the package and stowage of the required bit, a foldable configuration can be implemented allowing for packaging by stowing the segments when launched and unfolding upon deployment. The bit can be integrated with sensors to perform various measurements including detection and profiling the presence of liquid water/ice in regolith of the subsurface (Bar-Cohen et al. 2007b). An unfolding mechanism transforms the bit into a long rigid instrumented device that can percussively penetrate the subsurface (Fig. 6.21). The bit is designed to allow for extraction and insertion at other locations for multiple sites measurements. The bit contains a center string and is unfolded similar to the poles of camping tents where hollow segments have an elastic cable that connects them to the bit two ends. Once the cable is pulled from one end, the segments are connected and are turned into a long rigid rod. The penetrator is mounted on a positioning and deployment system that allows bringing the bit to the surface and impacting/inserting it into the subsurface. The USDC impacting mechanism is clamped to the top of segment #1 and, upon activation, the bit is brought down to the surface. Then, segment #2 is aligned with segment #1 and brought down while both segments are coupled to form a rigid single tubular bit. The bit is released using an "inchworm" mechanism and it pushes the assembled segments while connecting them until the last bit segment penetrates

Fig. 6.22 Bit with bi-stable coiled configuration that is extended by unrolling the tape shape

the subsurface or the required penetration depth has been reached. The process can be reversed to fold the bit for stowage and for insertion in other locations.

Another foldable configuration of the bit can be the use of a bi-stable structure, which can be designed to be stable in both the extended and coiled forms. Figure 6.22 shows the folded or coiled shape where it is partially being rolled out to start forming the extended configuration of a rigid tube. Once the unrolling is completed, the bit takes the unfolded extended configuration having the shape of a tube with the full length. The rolling and extending of the bit can be done as desired providing a compact configuration that can be inserted into the ground without collapsing. A suit of sensors can be imbedded into the material matrix.

6.2.3 Various USDC Configurations

6.2.3.1 Ultrasonic Rock Abrasion Tool (URAT)

One of the first applications of the USDC Technology has been the development of the Ultrasonic Rock Abrasion tool (URAT). The URAT (Fig. 6.23) was designed as a device that uses the hammering action to abrade a 4.5 cm circular pattern to a depth of 5 mm and thus expose rock surfaces (basalt, andesite, etc.). The abraded

Fig. 6.23 Photograph of the URAT

surface is required for instrument access to pristine rock material below the weathering layer. The device is based on the same technology that drives the Ultrasonic/Sonic Driller/Corer (USDC). The stress generated by the piezoelectric stack is amplified by a stepped ultrasonic horn and the horn tip oscillates at ≈20 kHz with a velocity that ranges from 1 to 10 m/s depending on the mechanical loading and amplitude of the drive voltage. The abrading surface is composed of pyramidal teeth on a circular base. The combination of the horn amplification, piezoelectric stack and drive at resonance produce impact forces that crush the rock under the pyramidal teeth.

The horn tip contacts a mass that is free to move along the long axis of the horn and is in proximity to the horn tip. When the mass touches the tip it captures momentum from the horn tip and is ejected from the tip towards the base of the bit where it impacts to create a stress wave that travels along the bit to the drill tip. At the drill bit tip, the stress wave breaks the underlying rock when the stress exceeds the ultimate compressive (soft rocks) or tensile (hard rocks) stress. The devices can be designed to be lightweight (<0.5 kg) and low power (<15 W) making them ideal for mounting on the end of a rover arm. The URAT was developed as a backup to the Rock Abrasion Tool that was flown on the Mars Exploration Rovers and it is shown in Fig. 6.24. Figure 6.24 shows the device in operation abrading a basalt and brick sample. As part of the development of the URAT, power and control electronics were produced and the mechanical design was breadboarded and tested (Dolgin et al. 2001) along with designs to shorten the horn length (Sherrit et al. 2004a).

6.2.3.2 Rotary Augmentation of the USDC-Percussive Action

Percussion and rotation have long been the most widely used methods of penetrating materials. Percussion is very effective in fracturing hard, brittle materials like stone and ceramics, while rotation is more effective on soft and/or ductile materials such as wood, plastics, and metals. One advantage of rotation is the removal of cuttings from the hole along the flutes on the bit. Percussion fractures the material but, unless the cuttings are removed, it continues to hammer at the loose material in the hole creating

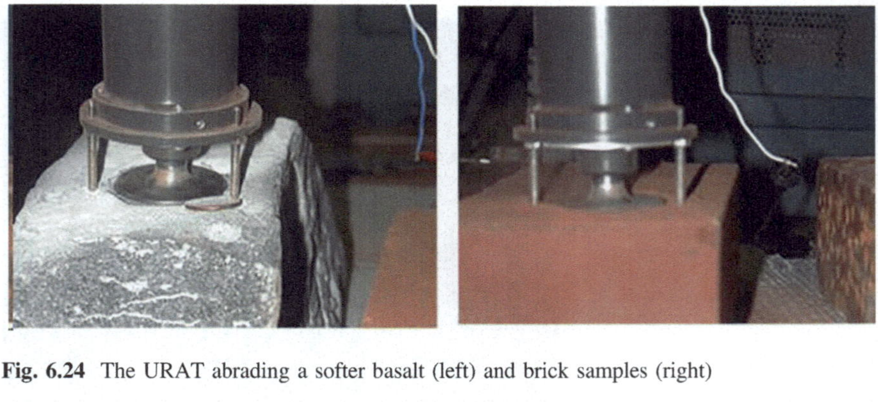

Fig. 6.24 The URAT abrading a softer basalt (left) and brick samples (right)

finer and finer particles until a limit is reached. This wastes energy that could be used to penetrate the drilled medium. Combining rotation and percussion produces a highly effective penetration mechanism. Existing hammer-drills produce their hammering pneumatically or mechanically via coupling to a rotary motor. In the reported research, it was found that one could produce novel augmentation of rotary drills by combining the USDC and rotary action. This approach allowed for separate rotation and percussion power supplies and the ability to adjust the power in different power combinations to test the benefits of the augmenter breadboards over rotation only.

At shallow depths of less than 2–2.5 cm, the powdered cuttings that are generated by the USDC travel along the bit partially because of the air pressure from the vibrating bit and partially due to the bit vibration itself. There are particular situations where the borehole orientation with respect to gravity vector can lead to larger penetration depth before powder removal saturation. One may drill into a porous material where the created cuttings are not enough to plug all pores and so the penetration is not stopped even after more than 7 cm. It is also possible that, in low gravity low-pressure environment, the cuttings removal would continue to take place when reaching larger depths if the cuttings float out of the hole.

To enable greater depth of penetration, pressurized gas was introduced from the center of the bit. Using this approach, basalt core that is about 10-cm long was produced. To further increase the rate of penetration and avoid the use of pressurized gas, which is not expected to be practical in planetary exploration missions, the rotary-hammer capability was developed. This has been the result of the lessons learned from a field test of the Ultrasonic/Sonic gopher drill in Antarctica, which have shown that powdered cuttings removal is a significantly limiting factor in rapid drilling to depths beyond few centimeters (Badescu et al. 2006a, b). In order to improve the cuttings removal and the drilling speed, rotation was introduced creating a rotary-hammer mechanism (Badescu et al. 2007). The hammering and rotation actuations were generated by different mechanisms and their independent drilling operation provides redundancy in case of failure of one of these two mechanisms.

The bit was designed with flutes allowing the cuttings to be travel upward along the outer surface of the bit. A photograph of the bit, the drilled borehole, and the powdered cuttings are shown in Fig. 6.25. For the demonstration of the capability of the developed rotary-hammer drill a 14 mm diameter bit was used to penetrate a limestone. The USDC was operated with ~100 W continuous power and ~29 N

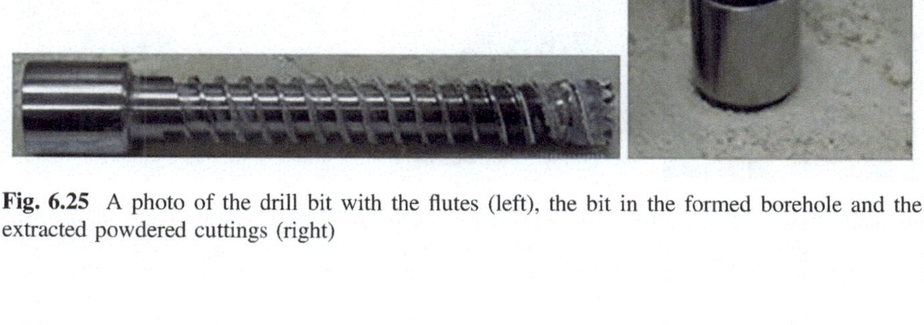

Fig. 6.25 A photo of the drill bit with the flutes (left), the bit in the formed borehole and the extracted powdered cuttings (right)

(6.5 lb) weight-on-bit and it reached a depth of approximately 8.5-cm in a total continuous drilling time of 5 min.

Actuating the augmenter by piezoelectric stack to produce the hammering action provided many benefits (Badescu et al. 2007; Bar-Cohen and Zacny 2009). These characteristics of operation under low power with small tip displacement and low axial preload are beneficial to sample acquisition tasks in planetary exploratory missions with very tight power budget on planets with low gravity. For the construction and remodeling industry, the drill is easier to use as well as quieter compared to other hammer drills.

6.2.3.3 USDC for Reaching 0.5 m Deep

Taking advantage of the combination of the USDC and rotating action, Honeybee Robotics and JPL jointly developed a sampler that was designed to reach in regolith a depth of 0.5 m. The focus of this task was to produce a sample with an all-in-one type bit that is capable of coring, retaining, and transferring samples. A photograph of the drill is shown in Fig. 6.26. The drill was able to produce samples in the form of cored (9 mm diameter and 51 mm long) and powdered cuttings. This regolith sampling drill creates a borehole that is 14.5 mm in diameter and has a sample chamber allowing the capture of 0.25 cc of material. The system was integrated with a load cell to sense preload, which provided closed-loop control of the feed-rate and Weight on Bit (WOB) of 10–12 N.

The drill was designed to have an additional core break-off tube, inside the primary drill tube, that can separate the core from the base material via a rotary shearing action. The shearing force is created by the relative rotation of two non-concentric components and the action is taking place at the lower end of the

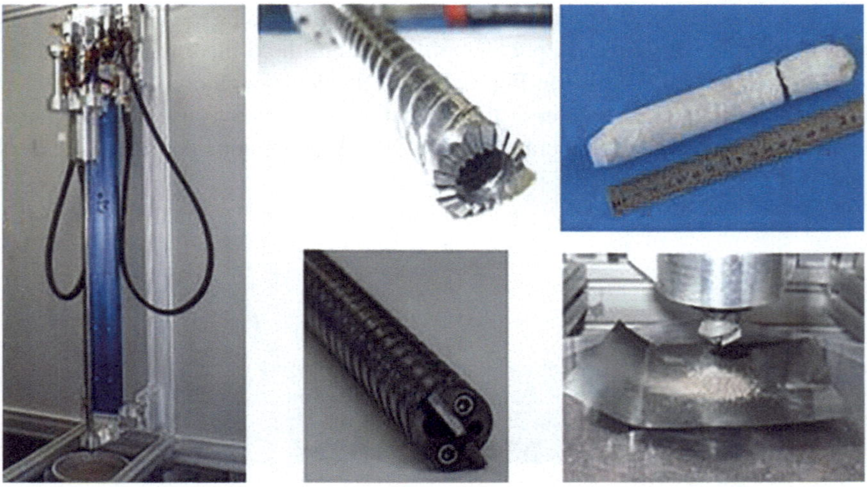

Fig. 6.26 Photos of Subsurface Sampler and parts. Clockwise from: full assembly; coring drill string; core sample; regolith sample; regolith sampling drill string. (Courtesy Honeybee Robotics.)

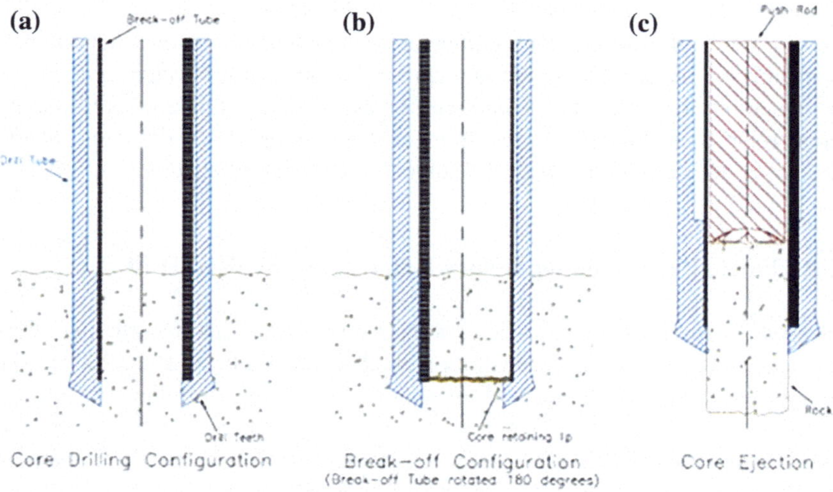

Fig. 6.27 Subsurface sampler core break-off and ejection. (Courtesy Honeybee Robotics.)

core near the cutting tip of the drill bit. These components also act to retain a consolidated core in the break-off configuration (Fig. 6.27, middle). During core drilling and core ejection operations, the center bore axes of these two components are aligned (Fig. 6.27, left and right).

The regolith sampling drill can penetrate both consolidated and unconsolidated material and it has internal cavity for capturing either regolith or powder cuttings. The delivered sample is in the form of sand or fine powder, depending on the material being sampled. Once the sample has been acquired and brought to surface, the mechanism has the capability of pushing out the fine cuttings that are collected inside the sample cavities. Twin push rods translate up and down inside their respective sample cavities when handing over the sample to an external cache.

Testing the drill operation included capturing both consolidated cores and powder samples in laboratory ambient and Mars-relevant environments. Coring in ambient conditions showed that a 5.08 cm (2-in) limestone core could be created and captured in 3.7 h using a 20% duty cycle on the piezoelectric stack. Drilling into Mojave Mars Simulant (MMS) sample in ambient conditions showed a capture of 696 mg of MMS in 3 h at a depth up to 4.5 in. using a 20% duty cycle. Drilling into stacked layers of colored plaster verified the systems capability to selectively sample from a given depth and retain stratigraphy. Environmental testing was done in a vacuum chamber at 3-Torr CO_2 at −20 °C and showed the capability to core to a depth of 0.465 in. in 4 h. No core was captured since the breadboard could not autonomously perform the capture operation at the time the testing was performed. It should be noted that decreased, but satisfactory, performance was noticed compared to ambient performance. It was concluded that this was due to a shift in the natural frequency of the USDC piezoelectric stack due to the cold temperatures. This frequency shift has been shown significant effect on the drilling performance.

This suggested the need for frequency tracking software or hardware as a method of improving the performance. This particular implementation of bit rotation, percussion and core break-off was later implemented in other drill configuration. In this implementation, the drill rotation and core break-off tube actuation were done by different rotary actuators. The powertrain of the actuator is not discussed herein since it does not constitute part of the subject of the present chapter.

6.2.3.4 Single Piezo-Actuator Rotary-Hammering (SPaRH) Drill

It has been shown that drilling via a combination of rotation and hammering offers an efficient and effective rapid penetration mechanism (Badescu et al. 2007). The rotation provides an intrinsic method for removal of cuttings from the borehole via flutes while the impact and shear forces aid in the fracturing of the penetrated medium. Conventional drills that use a single actuator are based on a complex mechanism with many parts and their use in future mission involves greater risk of failure and/or may require lubrication that can introduce contamination. The SPaRH drill (Sherrit et al. 2009, 2012) is a compact drill that uses a single piezoelectric actuator to produce hammering and rotation of the bit. A horn with asymmetric grooves was designed to impart a longitudinal (hammering) and transverse force (rotation) to a keyed free mass. The drill requires low axial pre-load since the hammering-impacts fracture the rock under the bit kerf and rotate the bit to remove the powdered cuttings while augmenting the rock fracture via shear forces. The vibrations 'fluidize' the powdered cuttings inside the flutes reducing the friction with the auger surface. This action reduces the consumed power and heating of the drilled medium helping to preserve the pristine content of the acquired samples. The drill consists of an actuator that simultaneously impacts and rotates the bit by applying force and torque via a single piezoelectric stack actuator without the need for a gearbox or lever mechanism.

Figure 6.28 shows a schematic view of two actions of the horn on the free mass that can produce rotation. The first is an extension that is off the horn axis that

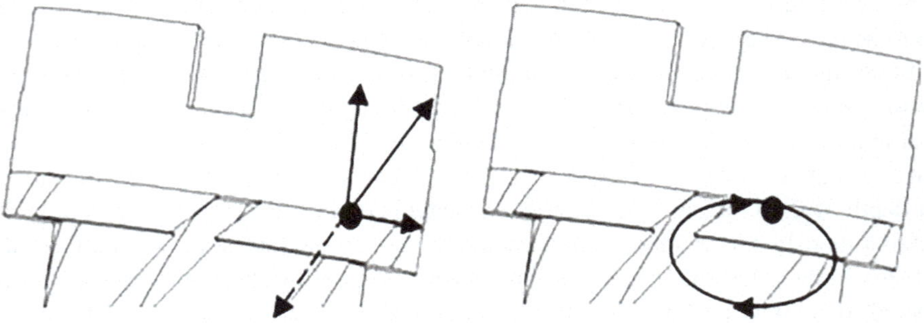

Fig. 6.28 View of the two potential horn motions (helical extension or bending) that can develop an impact and torque at a point (black dot) on the free mass. Black arrow and ellipse show contact

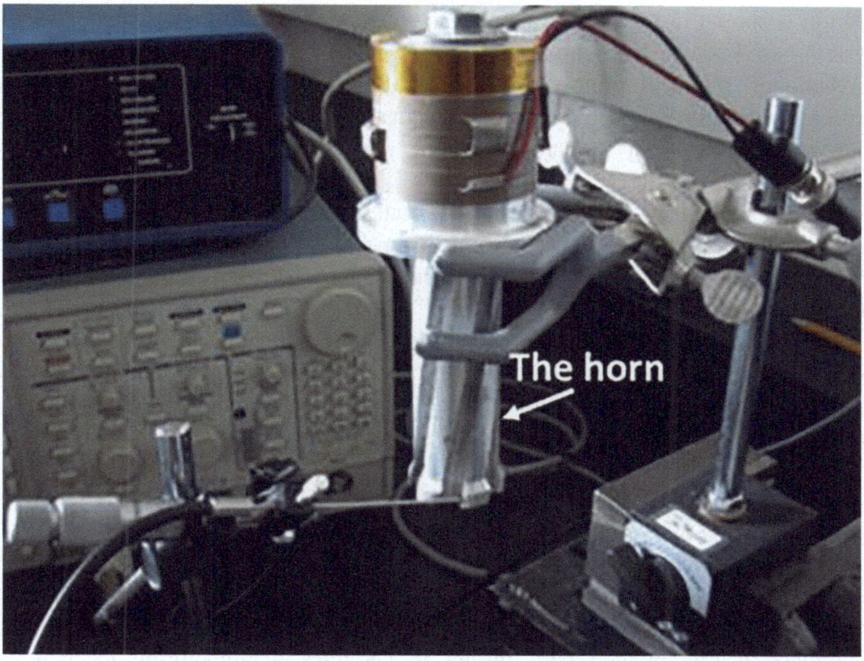

Fig. 6.29 A horn design that produced the ultrasonic motions to create hammering and rotation

pushes and twists the free mass. The second is an elliptical motion of the contact point that pushes the free mass and drags it to the right.

Contact pushes the free-mass up and drags it to right in each case. The normal force is much lower on retraction or on the bottom portion of the ellipse. The net force on the left produces a prolonged rotation of the bit as well as an impact. The bending ellipse shown on the right produces a rotation. A mode that rotates in the opposite direction around the ellipse would induce reverse rotation. Various horn design were studied and the configuration that was found to produce the required

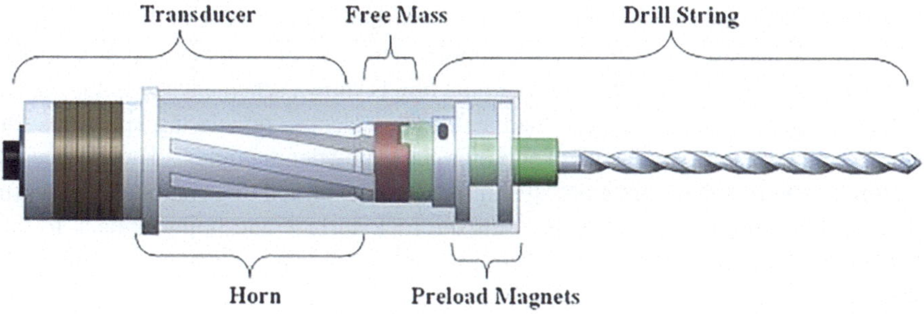

Fig. 6.30 The CAD model of the final transducer assembly of the SPaRH Drill

motion is shown in Fig. 6.29. Figure 6.30 shows a CAD design on a SPaRH drill system based on the horn shown in Fig. 6.29.

Although the drill prototype shown in Fig. 6.30 was un-optimized, higher drilling rates can be expected in future designs. The drill rate was increased from 2 to 8 mm/min in Santa Barbara Limestone simply by decreasing the free mass weight by 25%.

Other test results worth noting is that rotation in the opposite direction (clockwise) could be produced when the transducer was driven at its second resonance frequency at about 22.4 kHz. At this frequency minimal extension and hammering action was present and therefore the "drill" could possibly be designed to act as a pure rotary motor. A transducer, which can produce linear actuation and rotary actuation in two directions, could prove to be a promising tool on future NASA missions. In circumstances where both a linear and rotary actuator would typically be required, they could be replaced by a single solid-state actuator.

6.2.3.5 Ultrasonic/Sonic Crusher for Crushing, Milling, and Powdering

Various situations exist where one wants to reduce a large sample into powder. Examples of this include ball milling in ceramics manufacturing to increase homogeneity and powdering samples for powder X-ray diffraction or fluorescence to determine crystal structure, composition etc. Abrasion of the surface of a rock using low axial force and limited average power is considered a critical technology for future NASA planetary exploration missions. Current techniques use ball millers, grinders, in the manual case in the lab a mortar, and pestle to grind small samples into powders. Conventional rotating (grinding) or shaking (ball milling) mechanisms require high axial loads, significant amount of power and they may require contamination sources such as lubricants and filings from grinding wheels. The use of the ultrasonic/sonic driller/corer (USDC) offers significant advantages in requiring low axial force, low average power, low number of components, and the capability to produce a mechanism for the production of powered samples from cores or pebble sized rocks. The device crushes and powders rock, ceramic and other brittle materials for various applications including analysis.

A lightweight low power ultrasonic/sonic crushing device was built and demonstrated to require lower power to crush samples into powder. A schematic of a device for powdering and collecting the powder on a film is shown schematically in Fig. 6.31.

The crusher requires low axial load and it is an electric powdering apparatus for turning stone/pebble-sized rocks or cores into powder samples. The device is essentially a USDC with the horn tip and free mass confined to a chamber. The horn tip excites the free mass, which then hits the rock surface and bounces back to the horn where it acquires more energy from the tip and it is again driven into the rock sample.

Examples of powder generated using the crusher are shown in Fig. 6.32. Powder distribution sizes ranged from millimeter size particles to micron particles. Powder

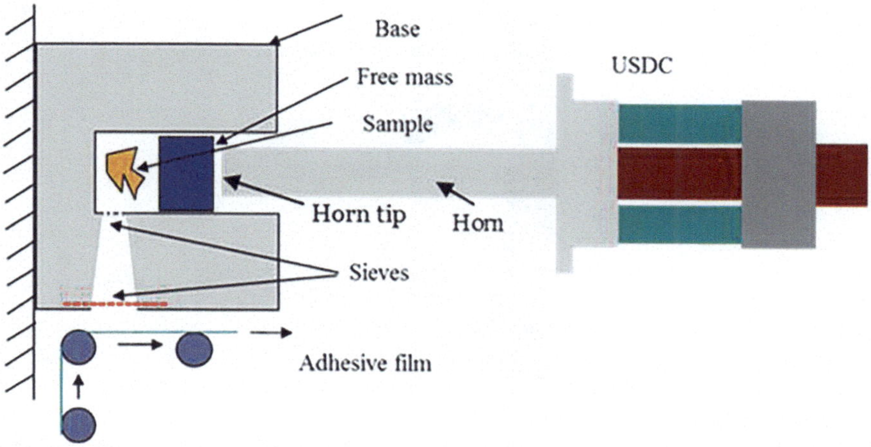

Fig. 6.31 A schematic diagram of a rock crusher implemented as a system to collect rock powder for in-situ measurements or sample return

Fig. 6.32 Examples of rock types that were crushed using the USDC crusher. Top row of photos shows the generated powder and the bottom row shows the original rock material that was drilled

distribution may be adjusted by the use of screens, modifying the free mass surface or adjusting the time the sample is in the powdering chamber.

6.2.3.6 Ultrasonic/Sonic Anchor

To support requirements for anchoring of legged and wheeled rovers, inflatable structures, and landers, the USDC offers significant advantages due to its ability to operate via low axial load.

Such a capability using a low mass device and relatively low power is needed to support platforms that would operate in low gravity environments and accessing steep slopes in rugged extraterrestrial terrains. Using a modified USDC mechanism, an Ultrasonic/Sonic Anchor was designed and fabricated. The anchor was used to drill a limestone in an angle and enabled a legged robotic mechanism to climb walls

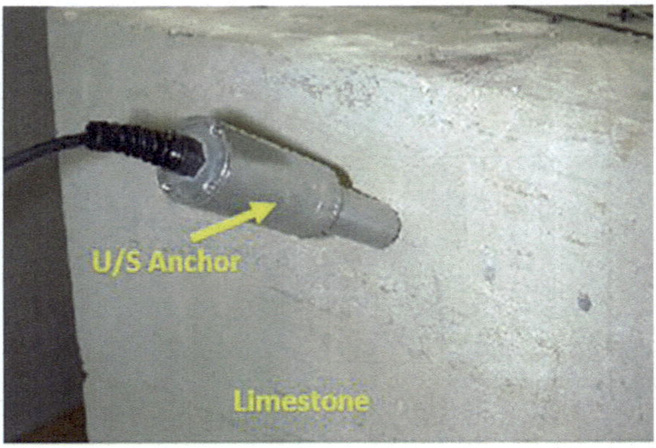

Fig. 6.33 The Ultrasonic/Sonic anchor mounted in an angle on a limestone block

(Fig. 6.33) (Bar-Cohen and Sherrit 2007). Operating the hammering action of the USDC in reverse allows its extraction from the medium onto which it was anchored and thus avoiding possible jamming.

6.2.3.7 One-Meter USDC-Based Packed Soil Penetrator

The ability to penetrate packed soil using low axial load was demonstrated by developing a USDC based penetrator with a 1-meter-long 3.18 mm diameter bit. Inserting the bit into the soil by a pushing force alone required several hundred pounds, which could easily result in bit buckling. Using the Ultrasonic/Sonic impacting mechanism, the developed penetrator was demonstrated to reduce greatly the required push force. Experimentally, it was demonstrated that the required push force to penetrate highly packed soil down to about 1-m was reduced from 90.7 kg (200 lb) to 3.2 kg (7 lb) (Bao et al. 2004). This effort involved harmonic modal analyses of the actuator and system impact allowing for the effective design of the penetrator. The modal analysis was used in the actuator design and the harmonic analysis to predict the actuator performance. The actuator design parameters were determined such that the location of the mounting structure coincides with the actuator neutral plane. The impact analysis was used to derive an optimal weight of the free-mass by determining the interaction between the free-mass, the ultrasonic horn, and the bit.

6.2.3.8 Mars Integrated Drilling and Sampling (MIDAS) System

The Mars Integrated Drilling and Sampling (MIDAS) System for Low Mass, Mobile, Robotic Platforms is a lightweight system developed by Alliance Spacesystems, LLC (Alliance) (Zacny et al. 2009). It enables automatic retrieval of multiple samples from regolith up to a depth of 0.5 m. MIDAS combines a 5 degree-of-freedom (DOF) robotic arm (derived from the Mars Exploration Rovers

Fig. 6.34 The MIDAS USDC and bit interchange mechanism assembly. (Courtesy NASA/JPL and Alliance Spacesystems—Now Space Division of MDA Information Systems, Inc.)

(MER)) and an Ultrasonic/Sonic Driller/Corer (USDC) (Bar-Cohen et al. 2001) that was developed by the authors. MIDAS was developed with an interchangeable bit mechanism and bits tool caddy capable of retrieving samples from 0.5 m and depositing them into an instrument for further analysis or a storage container for return to Earth. The complete system that was developed has approximately 1 m length and it weighs less than 6 kg. The MIDAS USDC and Bit Interchange Mechanism assembly are shown in Fig. 6.34.

The MIDAS system (Fig. 6.35) was designed to acquire, handle, and deliver samples for storage or analysis. Its 5 DOF robotic arm is based on the Instrument Deployment Device (IDD) that was developed by Alliance Spacesystems for the Mars Exploration Rovers launched in June 2003, which is still functioning on Mars Opportunity rover as of June 2016.

This low-mass robotic arm provides very precise positioning and feedback that allows MIDAS to:

- Control the rate of the drilling progress and weight-on-bit.
- Sense the depth of drilling.
- Automate the acquisition of multiple samples.

The major benefits of utilizing the USDC are:

- Power-effective and mass-efficient drilling.
- Low force requirements that allow operation from flexible, low-mass platforms.
- Relatively simple bits that are reliable, do not need sharpening, and reduce contamination.
- Simple mechanical bit interface making possible the use of multiple specific-function bits to accomplish a wide variety of tasks.

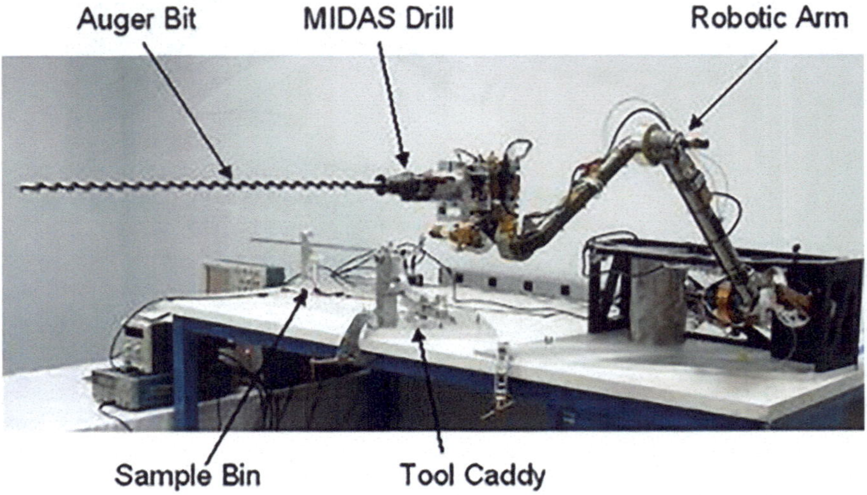

Fig. 6.35 The MIDAS system with the tool caddy. (Courtesy NASA/JPL and Alliance Spacesystems—Now Space Division of MDA Information Systems, Inc.)

The interchangeable bit mechanism and a tool caddy enable the system to:

- Repeatedly return to the worksite to change bits or to collect multiple samples.
- Perform change-out of multiple bits enabling to perform variety of tasks.
- Replace damaged or worn-out bits.
- Place different tools and instruments on the target.

The use of the USDC mechanism combined with the dexterity of the robotic arm make multiple, interchangeable tool bits a practical means of accomplishing a wide variety of tasks and provide a very powerful tool for exploring the Martian subsurface. Utilizing different tool bits is an efficient way to perform many tasks. However, this approach has traditionally required an overly complicated mechanism. MIDAS solves this problem by providing a simple interface for transferring the mechanical energy of the USDC to the bit. Rotation of the bit was introduced to provide a fresh work surface to the bit and aid in debris transfer, minimizing the need to transfer high torque or have complicated couplings. The interface is essentially a spring-loaded detent that allows the bit to be attached and detached with ease, then locked in place to provide sufficient holding force for all operations.

One of the most promising benefits of the USDC is the simple bit interface (see Fig. 6.36). The use of multiple bits is essential for multi-functional sampling as it is simply not practical to accomplish the multitude of tasks envisioned for future missions with a reasonable probability of success with a single tool. These tasks can, however, be accomplished by a single actuator that utilizes multiple bits for drilling, coring, surface preparation, and sampling. The USDC and bit interchange mechanism makes multiple bits and the exchange of those bits very practical. Many of the bits are as simple as a tube of a single material joined to the interface of the

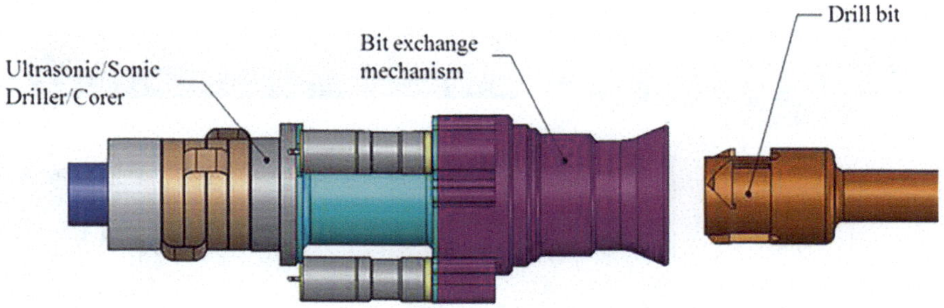

Fig. 6.36 The interfacing of the bit and the drill via snap-on design. (Courtesy NASA/JPL and Alliance Spacesystems, which was the Space Division of MDA Information Systems, Inc.)

standard bit interchange mechanism. The bits typically do not require sharpening but, if bit is damaged or jammed, it can be readily disengaged and replaced. The latching mechanism that allows the attachment and release of bits to the USDC's horn is shown in Fig. 6.36. The interface is not required to transmit high torque, nor is it required to have power or communications connections. The interface must simply keep the bit straight, free in the axial direction, and allow a very light axial force to be applied to the bit by the USDC. Extended development testing was performed on the first MIDAS prototype system and it has successfully acquired samples from 0.5 m deep in a simulated Martian regolith. With a more powerful actuator and further development, the system could potentially perform the additional tasks of rock drilling, coring, and abrading.

6.3 Piezoelectric Actuated Drills Driven at Ultrasonic Frequencies

6.3.1 Percussive Augmenter of Rotary Drills (PARoD)

The Percussive Augmenter of Rotary Drills (PARoD) was designed to turn rotary drills into effective rotary-hammering drills (Aldrich et al. 2008b). The breadboard was shaped as a percussive bit that can be mounted onto commercially available rotary drills and augment their operation. The augmenter was designed in two steps where in Step I, the actuator was sized to drive a 0.64 cm (0.25-in) diameter bit. Later, the augmenter was scaled to drive 5.08 cm (2.0-in) diameter bits. A slip ring is used to provide electric power to the piezoelectric actuator with minimal electric noise while being rotated by the commercial electric motor. In designing the augmenter, two options of bits were considered:

- A bit with a free-mass: This bit design allows rotating the tip and transmitting the impacts while the complete bit turns as a single unit.

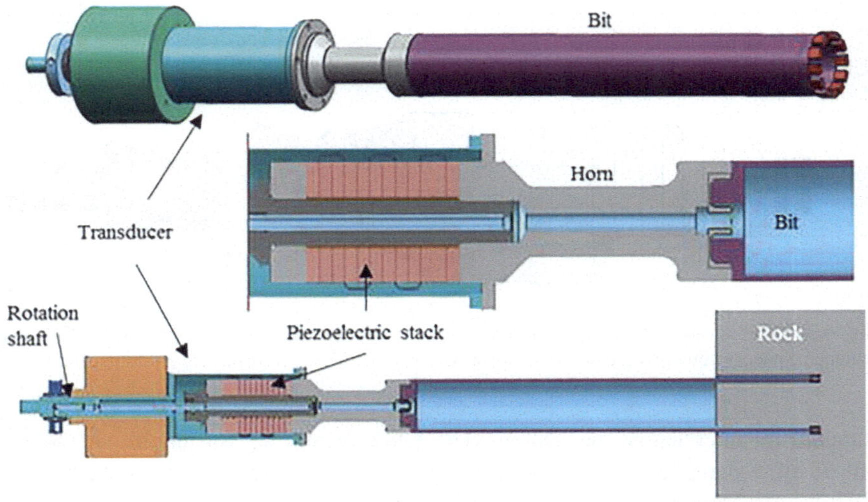

Fig. 6.37 Graphic illustration of the augmenter without free-mass. The bit cross-section is shown in close-up and full views

- A bit without a free-mass: The ultrasonic hammering action is generated by a piezoelectric stack and the rotation from the drill is applied directly to the concrete wall (Fig. 6.37).

After preliminary analysis and ergonomic considerations, it was determined that it is more effective to develop the augmenter without the use of a free-mass in its configuration. To optimize the operation of the augmenter, a control algorithm was incorporated to make logical decisions based on past and present sensor inputs regarding which particular control mode the system should be operated. For this purpose, an Extremum-Seeking Control (ESC) algorithm (Aldrich et al. 2006a) was used to maximize the power output of the PARoD actuator by tracking its resonant frequency. The resonant frequency of the actuator is temperature- and load-sensitive, and it was tracked in real-time. The ESC algorithm included a parameter estimator that determines the operation efficiency of drilling based on a subset of the most recent input and output measurements of the drill. This information is used to adjust the drive frequency and the duty cycle of the breadboard so that it operates at the most efficient drilling rate subject to the constraint that the piezoelectric actuator is not driven beyond its capability.

6.3.1.1 Designing the PARoD

The design of the augmenter bit required the ability to predict its performance. For this purpose, a combination of numerical and finite element models (developed with ANSYS^{TM}) was used to determine the vibration modes of the actuator (Badescu et al. 2012). The actuator was designed such that it would induce minimal vibrations at the nodal plane that serves as the mounting section on the actuator structure. This choice allows a connection from the actuator to the housing without transmitting vibrations from the actuator. The driving frequency was selected to operate at the resonance frequency of the piezoelectric stack/horn actuator. The actuator was fabricated using a stack of piezoelectric rings with a backing layer for forward power delivery and a horn that amplifies the induced displacement.

Since two of the key areas for potential improvements of the performance of the actuator are the horn design and the type of piezoelectric stack, various alternative designs were investigated. MultiPhysics Finite Element Package was used to account for various effects including electromechanical coupling and thermal shifts in the resonance frequency (Aldrich et al. 2008b). In developing the analytical model, we took into account the operation of the bit in rotation as a modulation of the vibrations of the bit.

The ultrasonic actuator was used for driving the Percussive Augmenter of Rotary Drills (PARoD) with a commercially available 6.4-mm diameter drill bit (Fig. 6.38).

The bit was connected through an adapter, which was custom-designed and fabricated to maximize the performance of the drill. A connection shaft at the base of the ultrasonic actuator was used to mount the PARoD to the commercial rotary drill and a slip ring was used to provide electric power to the actuator. The details of the assembled drill are shown in Fig. 6.38.

6.3.1.2 Analytical Modeling of the Piezoelectric Actuator with 6.4-mm Diameter Bits

The design of the PARoD requires its nodal plane coincident with its mounting plane as is implemented in the model that predict its performance. Thus, the

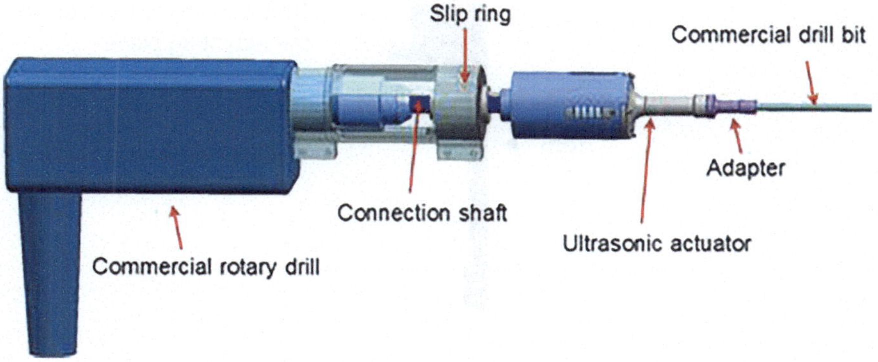

Fig. 6.38 Graphic illustration of the assembled PARoD, which is an ultrasonic percussive-rotary drill

vibration transmitted to the rotary drill mounting fixture is minimized. The ultrasonic actuator is axisymmetric but only axial resonance modes were of interest. The ANSYS axisymmetric elements were used to model half of the cross-section of the transducer. Compared to a 3D model, this 2D model using the axisymmetric elements greatly reduced the size of the FEM model and calculation time. Thus, the model allowed performing quicker analysis and more design iterations to optimize the design. The resonance frequencies, mode shapes, and nodal planes of the actuator were predicted. These were accomplished by performing modal analysis. As a part of this analysis, the length and diameter of the ultrasonic horn were adjusted to make the nodal plane of the fundamental mode coincident with the designed mounting plane (Fig. 6.39).

The resonance frequency (f_r) was determined to be 19,240 Hz, and the anti-resonance frequency (f_a) as 19,514 Hz. The electro-mechanical coupling factor (κ) was determined as

$$\kappa = \sqrt{\frac{f_a^2 - f_r^2}{f_a^2}} = 0.167 \tag{6.7}$$

After the modal analysis was completed and the dimension had been decided, harmonic analysis was performed to predict the horn tip displacement and electric

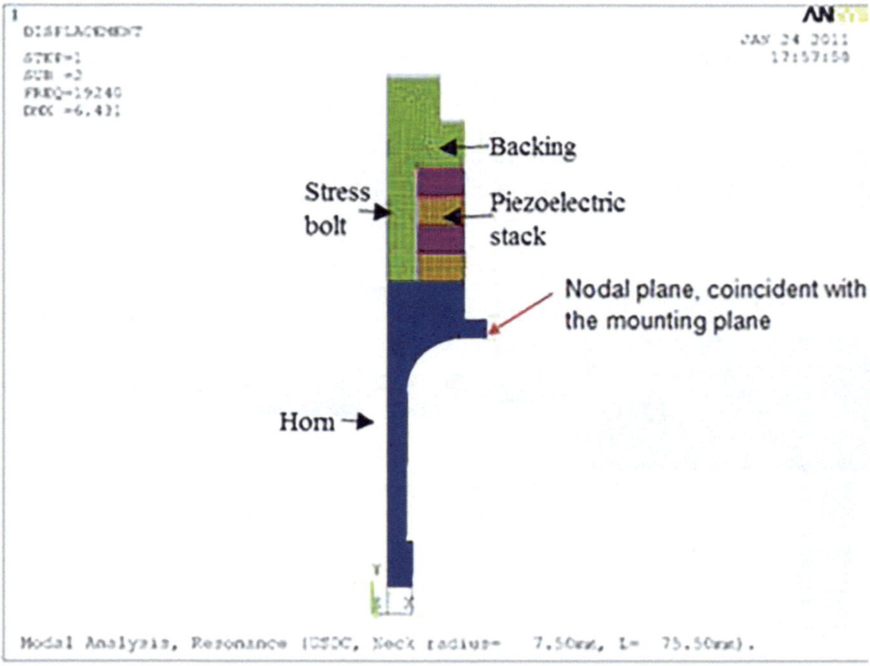

Fig. 6.39 Modal analysis of the piezoelectric actuator

power consumption of the actuator. Then the adapter and the bit were added to the FEM model to create the full PARoD model. The length and the diameter of the adapter were adjusted to make the nodal plane coincident with the mounting plane. The use of iterative analysis showed coincidence of the nodal and mounting planes are achieved with a length of 80 mm and diameter of 10 mm. Further analysis suggested that the third axial mode, at 19,734 Hz, was an effective mode for the actuation of the PARoD. The anti-resonance frequency was 19,816 Hz. The electro-mechanical coupling factor (κ) was determined by Eq. (6.1) to be 0.09. The harmonic analysis predicted the drill bit tip displacement and electric power consumption of the PARoD.

The displacement at the tip of the bit is relative small (3.66 μm @ 100 V). To increase the tip displacement, we changed the length of the adapter and reanalyzed the PARoD. It was found that shorter adapters generate larger tip displacements with higher coupling factors and higher power consumption. The adapter with a length of 10 mm generates a tip displacement of 6.65 μm for 100-volt input, which is about 100% greater than the displacement for the 80 mm adapter. However, the modal analysis of the shorter adapter shows a very large distortion around the mounting plane. This means stronger bolts and nuts for the mounting plane may be needed. Several configurations were considered in the design of the augmenter with the 6.4-mm diameter bit including one-stack, three-stack, and inverted horn designs. Larger 5.08-cm diameter PZT rings were also investigated instead of the 3.8-cm diameter rings.

6.3.1.3 Fabrication of Augmenter with 6.4-mm Diameter Bit

The augmenter is designed to provide hammering to any commercial rotary drill, and is mounted between the drill and the bit. The bit is inserted into an adapter and the casing is connected through the slip ring to fit a three-point chuck of the drill (Fig. 6.40). This arrangement allows the augmenter to be used with a variety of rotary drills. As mentioned previously, the augmenter is driven by a piezoelectric stack and it consists of a backing material, an amplifying horn, an adapter piece, and a slip ring. A bolt from the backing to the horn maintains the piezoelectric stack in compression, which is needed since piezoelectric ceramics have lower tensile strength then compressive strength. The stack consists of 6 piezoelectric rings connected electrically with alternating positive and negative electrodes to the supply power. The horn diameter is smaller than the piezoelectric rings diameter and this way the displacement at the tip is amplified (Sherrit 2005a). The length of the adapter was determined through modal analysis to minimize deflection at the mounting plane. The slip ring is used to transfer power from stationary to rotating components.

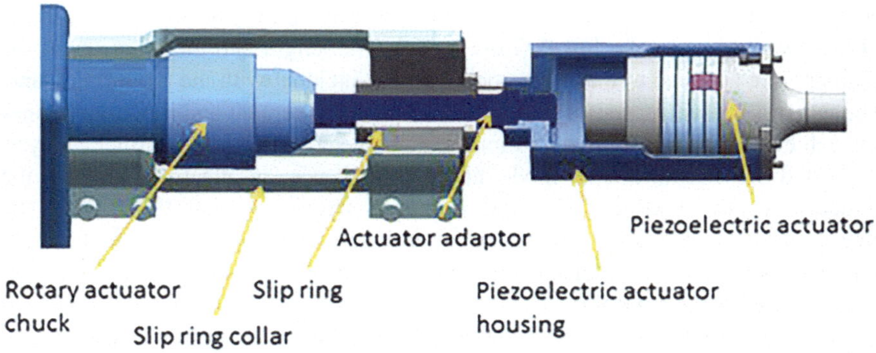

Fig. 6.40 The augmenter components with the slip ring section on the left and the piezoelectric actuator on the right with rotation shaft connecting them

6.3.1.4 Testing the Augmenter with 6.4-mm Diameter Bit

Using the results of the analysis, the augmenter design with 80-mm long adapter was chosen. The PARoD was assembled and its impedance spectra was measured. The PARoD was then attached to a Bosch rotary drill (Hammerdrill 1199VSR) (Badescu et al. 2012). The fixed weight-on-bit (2.72 kg), the ultrasound frequency (22.51 Hz), and the drilling time (30 s) were recorded for each drill test. It was noted that the augmenter was most effective when it was driven at its piezoelectric actuator resonant frequency. This produced the largest tip displacement and resulted in more energy being imparted to the drilled medium.

A control program was needed because the resonance frequency is not completely stable due to shifts when the temperature changes or the applied force or contact with the side of the hole being drilled, and imperfect electrical transmission through the slip ring. Creating a list of constant drive frequencies to apply at every possible case would be inefficient and not effective. To maintain the resonance frequency, a control program, was written to read the current and voltage and compare the magnitudes to the previous step by either increasing or decreasing the drive frequency. Initially, it had only obtained the peak value to decide the next frequency and stepped by 4 Hz. The problem is that current and voltage become out of phase when the drill is not in resonance, so the peak values are not occurring at the same time. The small frequency step was another problem; due to the instability in the signal, the small step size led to shifts based on noise.

The program was adjusted to calculate power used by the augmenter. It logged simultaneous arrays of voltage and current, multiplied them element by element, and averaged by the length of the array. The length of the array is determined by the drive frequency to contain a set number of whole periods. Power is a better value to optimize and is important test data. The step size was increased to 20 Hz. The larger step size quickly approaches resonance, but oscillates above and below it. The program monitors past frequencies to recognize this pattern and reduce the step size when needed. Power values were also monitored and the step size was increased if there was a significant drop in power. The ability to log test data in a text file was also added.

To perform these tests, a preliminary test setup was fabricated to control and monitor the power sent to the Bosch rotary drill as well as the power and frequency of vibration sent to the PARoD prototype. These drill tests were performed in order to assess quickly the performance of the PARoD prototype. To have a better control of the preload and the alignment of the drill with the drilled media as well as to have a more precise measurement of the drilling rate, a testbed was constructed.

Test results showed the percussive augmenter greatly increased the drill rate over rotation alone. All of the test data were based on the maximum power drawn by the augmenter during the tests. The average power used was actually much lower than the maximum.

The power versus drilling rate data (Fig. 6.41) shows increase of almost two times when using the percussive augmenter compared to rotary drill only. The power in these tests was kept constant at 160-W and the weight-on-bit is 4.2 kgf. Using a fixed power of 50-W to the rotary drill and increasing the power to the augmenter showed an increase in drilling rate about over 5 times (Fig. 6.42). In Fig. 6.42 the drilling rates are shown for limestone vs. total power used, where the rotary power is ~50 W and the weight-on-bit is 4.2 kgf.

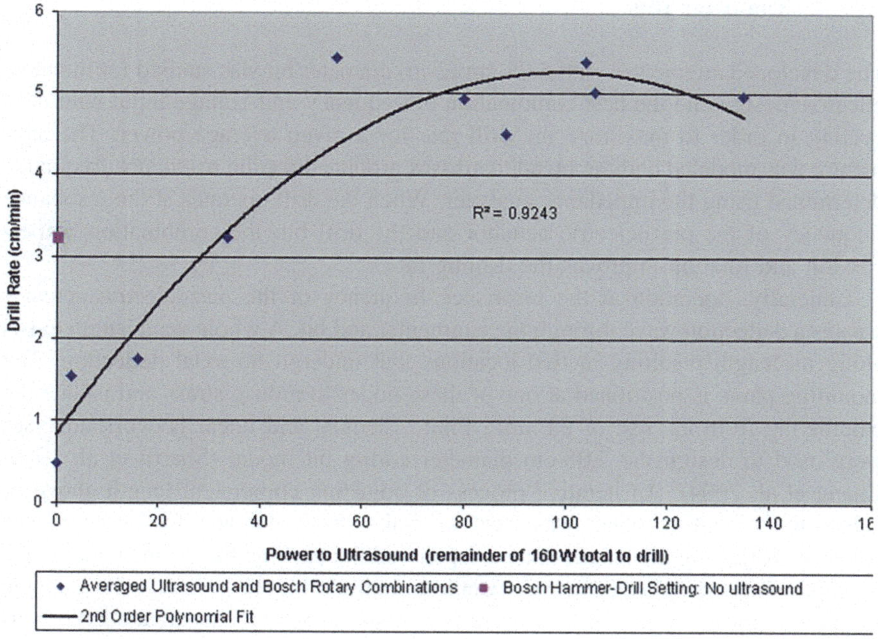

Fig. 6.41 Tests of the PARoD with 6.4-mm diameter bit with the combined rotation and hammering

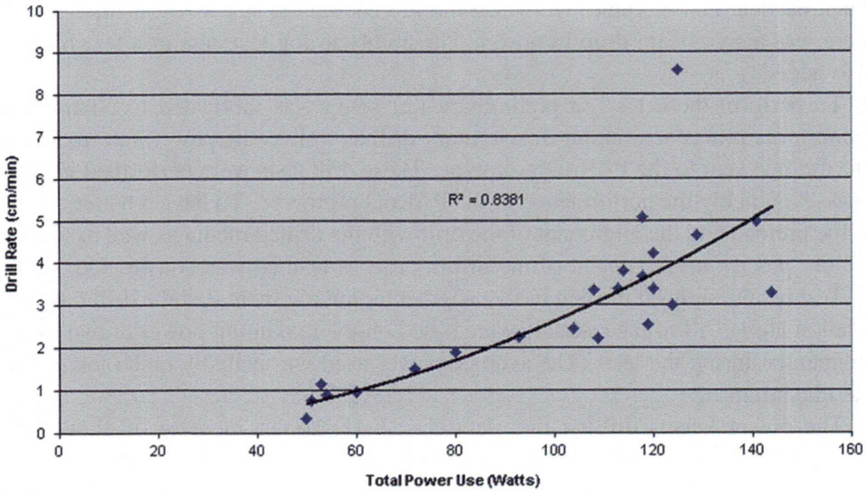

Fig. 6.42 Tests of the PARoD with 6.4-mm diameter bit using fixed power to rotation while increasing the power to hammering (the solid line shows the best fit graph using 2nd order polynomial fit, while the diamond dotes show the actual measurements)

6.3.1.5 Modeling of Augmenter Configurations with 5.08-cm Diameter Bits

The developed augmenter with 5.08-cm (2-in) diameter bit was studied for the most effective design and the best combination of frequency and voltage input with duty cycling in order to maximize the drill rate for a given average power. The augmenter was modeled and the breadboard was produced for the resonance frequency determined using the impedance analyzer. When the drill operates at the resonance frequency of the piezoelectric actuator and the drill bit, the combination of percussion and rotation improves the drilling rate.

Generally, operation at the resonance frequency of the piezoelectric actuator creates a deflection wave through the augmenter and bit. A whole wavelength exists along its length resulting in two locations that undergo no axial deflection. The mounting plane is positioned at one of these nodes to reduce stress and isolate the hammering from the rest of the drill. Finite Element and linear network analyses were used to design the 5.08-cm diameter coring bit model (Sherrit et al. 1999; Chang et al. 2004). An iterative process of adjusting component length above or below the mounting plane was used to make the nodal and mounting planes coincide. Adjusting the length of the adapter and the backing moves the nodal plane by shortening the deflected wave. To allow multiple bits to be used with a single augmenter, the backing and horn were held constant while a separate adapter was tuned to each bit. The goal is to identify a series of tuned adapters and bits that can all be used with the same horn since having multiple bit choices is important to construction applications.

6.3.1.6 Testing the 5.08-cm Bit Augmenter Breadboard

A testbed was designed and assembled to support the 5.08-cm (2-in) bit augmenter, the rotary drill and the related fixtures. A commercially available Hilti drill (Model TE50) was reconfigured and fit with an adapter to contain a piezoelectric transducer in line with the drill bit. The transducer was made of a stack of 5.08-cm diameter disks of PZT ceramic material preloaded in compression to prevent fracture during the voltage cycling. The tests were performed at resonance for maximum displacement rates and were conducted at different power levels to the transducer and the drill motor over a range of frequencies for best volume removal rates. The testbed is shown in Fig. 6.43 and it consisted as follows: A pulley system (A) changes the weight-on-bit. The power to the Hilti drill (B) is controlled by a Variac Transformer (C) that causes rotation between 32 and 60% input. Any lower and rotation ceases; any higher and the test stand becomes less stable and/or the rotation speed maxes out. The piezoelectric transducer is located at the end of the drill (D), and the drill bit (E) attaches at the end of the transducer.

In order to determine the capabilities of the ultrasonic transducer, tests with rotation-only were performed to have a baseline comparison with rotation and percussion. The rotation speed was kept the same in all tests for comparison. All pertinent data was recorded including the weight-on-bit, and the power to the ultrasonic transducer and the rotary drill. The test results are shown in Fig. 6.44 where the volume removed data for W = 0 W represent the data for rotation only while the rest of the data are showing the performance under augmented levels of ultrasonic drilling. The results have shown improvement in the removed volume at the level of up to 4 times when using 300-W total power and 2 times when using 600-W total power.

6.3.2 The Portable Rapid and Quiet Drill (PRAQD)

There are applications where the drilling needs to be done quietly. The use of a piezoelectric actuated drill that is operated at frequencies that are higher than 20 kHz offers this capability. For this purpose, an ultrasonic drill that is driven by piezoelectric stack at 25 kHz was developed (Sherrit et al. 2005a). The actuator housing was designed and fabricated using rapid prototyping to produce an easy to handle tool as shown in Fig. 6.45. In this figure, the housing is shown with a cut out to reveal the inside components of the drill. The ultrasonic drill has as its main components a transducer, housing, and drill bit. The transducer includes a stack of piezoelectric discs and electrodes, backing layer, a stress bolt and a horn. The stress bolt, the horn, and the drill bit have been designed with a central hole that is intended for blowing pressurized gas to eliminate the dust particles created in the drilling process. The use of pressurized gas is needed more in the case of drilling vertically where the dust may stay inside the formed hole.

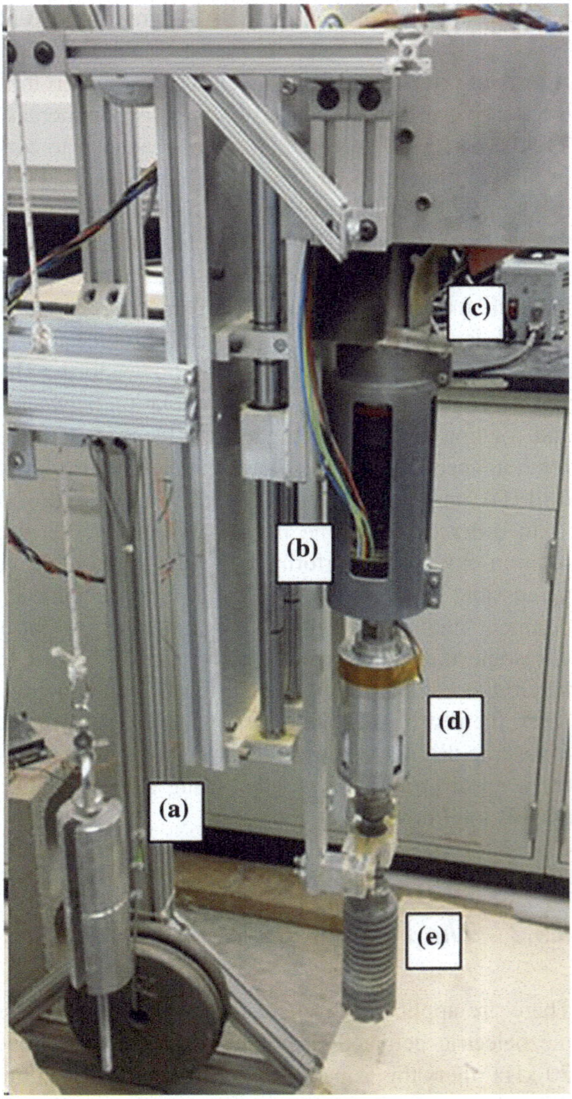

Fig. 6.43 The augmenter and the testbed using a Hilti as a rotary drill on the test stand

The bit was modified to include a bigger diameter threaded end for mounting onto the horn and a hardened carbide crown at the cutting end. The use of thread allows replacement of bits in the field if they fail providing a quick and inexpensive method of dealing with their possible fracture. Low velocity compressed air was blown through a center hole in the actuator and drill bit to help eliminate accumulation of powder. A view of the developed drill prototype is shown in Fig. 6.46 (a) without and (b) with the housing. A close-up view of the bit while in action drilling a concrete block is shown in Fig. 6.47.

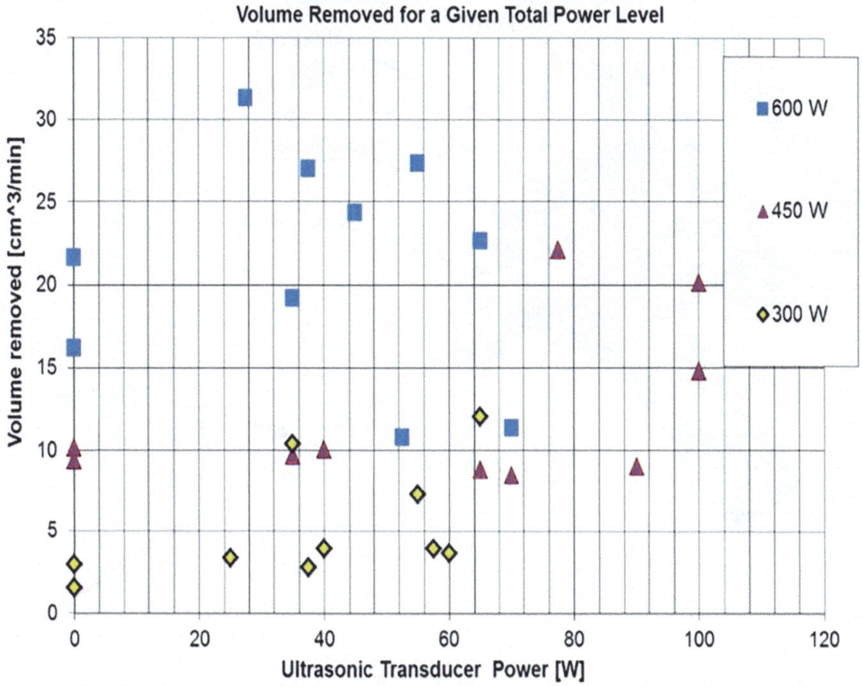

Fig. 6.44 Test results comparing the rotation only (W = 0 W) versus rotation and augmentation by the PARoD with 5.08-cm diameter bit

The current noise level is at the level of a scratching sound as the tip of the bit grinds down the concrete surface. The sources of sound that are causing emission of audible noise will be investigated including vibrations of the drill components in the audible spectrum and the wall material fracturing in order to understand how to further mitigate sound generation. Another possible approach is to increase the actuator resonance frequency above 45 kHz, which is above dogs hearing range.

6.4 Piezoelectric Actuated High Temperature Drills

NASA strategic goals and science objectives are increasingly requiring future missions to be involved with robotic exploration of planets where the environment poses greater challenges to existing technologies. These extreme environments include very high temperatures such as that found on Venus (460 °C) and very low temperatures as on Europa and Titan (−180 °C). In addition, potential mission may require sampling in very low gravity as is found on asteroids, and comets as well as at high pressure environments such as on Venus (90 bars). These challenges to the

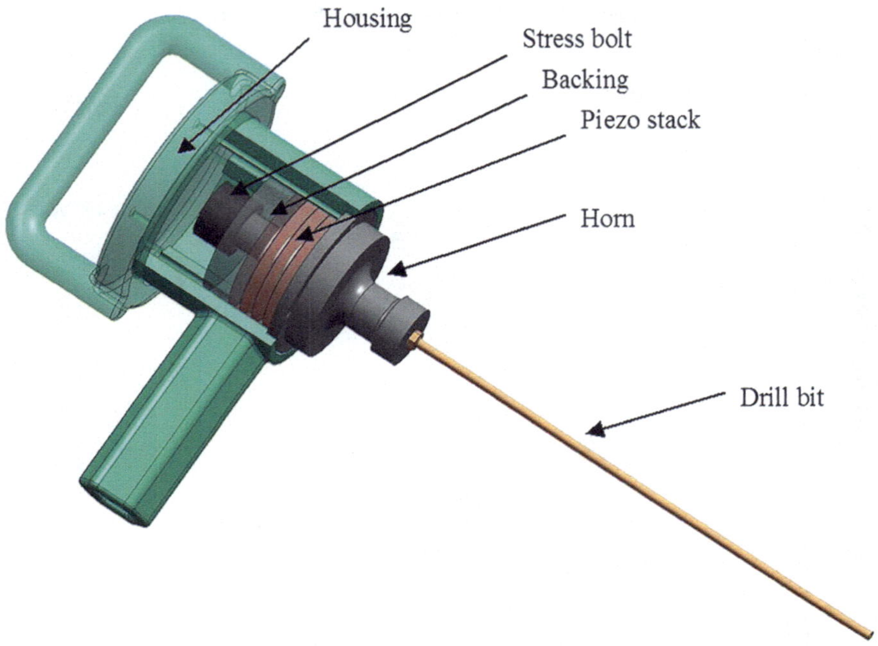

Fig. 6.45 The Portable Rapid and Quiet Drill (PRAQD) design and its components

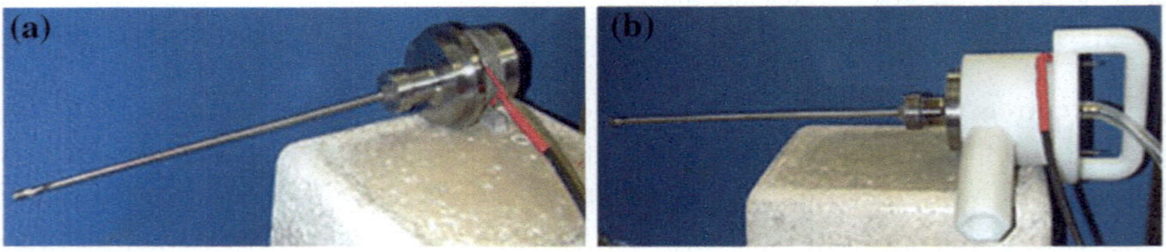

Fig. 6.46 The prototype Portable Rapid and Quiet Drill (PRAQD) without (**a**) and with (**b**) housing

required technologies limit the in situ sampling possibilities and they are requiring new approaches using capabilities that are not available. While there is significant development of capabilities for low temperature applications, the technologies for high temperature applications are still limited. For the low temperature range, the USDC was demonstrated to drill at temperatures as cold as −180 °C. To address the need for a drill that can operate on Venus, a USDC-based high temperature sampler was developed. For this purpose, piezoelectric materials with a Curie temperature that is higher than 500 °C have been used. These included such materials as $LiNbO_3$ and modifications of Bismuth Titanate and $Bi(MgTi)O_3$-$PBTiO_3$ and BST.

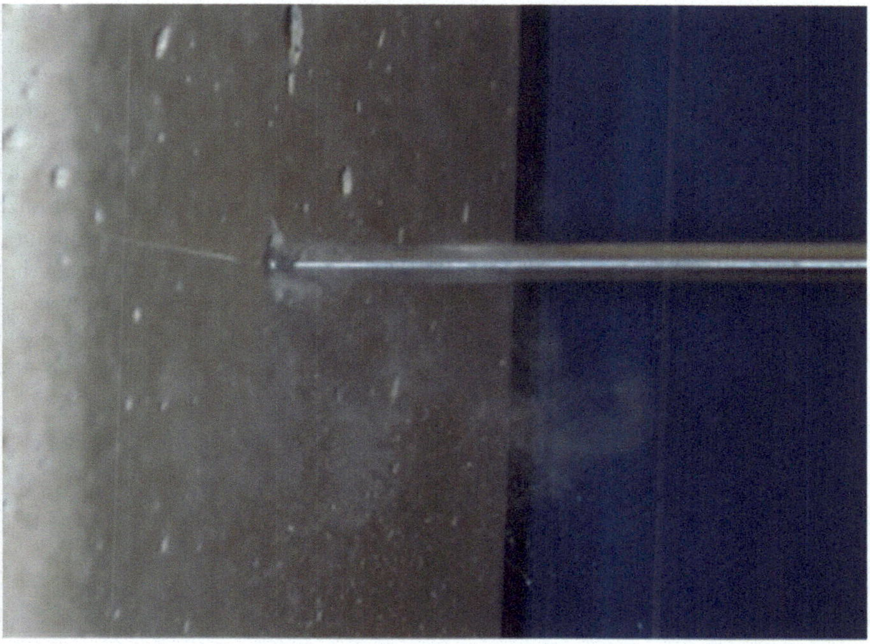

Fig. 6.47 Close up view of the drill while drilling

6.4.1 High Temperature Piezo-Ceramic Actuators

Generally, there are many piezoelectric materials with high Curie temperature, which dictates the highest temperature to which such materials can be heated and still maintain functionality as actuators. The single crystal $LiNbO_3$ has the highest known Curie temperature and it is more than 1100 °C. A variety of new piezoelectric materials at a lower temperature levels, with a high Curie temperature are being developed (Sherrit et al. 2014). To determine the aging characteristics of $LiNbO_3$ at 500 °C, an isothermal test was done for 1000 h (Bar-Cohen et al. 2011, 2012, 2014c; Bao et al. 2010; Sherrit et al. 2014) and, to the level of the measurement error, no appreciable change in properties was observed demonstrating its high stability. Since the electrodes on the piezoelectric actuator need to sustain the high temperature environment as well, three different types of electrode materials were tested up to 500 °C, including sputtered platinum film, gold-palladium thick film, and sputtered gold film. The results of the electrodes exposure to 500 °C showed that the sputtered gold electrodes showed severe degradation while the Pd-Au and Pt electrodes operated quite effectively after high temperature exposure. Based on these results and the fabrication ease, the electrodes made of sputtered platinum films were chosen for the actuators.

In order to fabricate piezoelectric stacks for driving high temperature percussive drills, a series of bismuth titanate discs with various composition levels of titanate and dopants (W, Fe, Ca, Sr, and Mn) were made at Penn State University (PSU) (Bar-Cohen et al. 2010b). The discs were tested to measure the electrical characteristics at room and high temperatures of up to 500 °C. The test results have shown significantly better electromechanical conversion capability compared to the previously known materials. Following these tests, efforts were made to optimize the piezoelectric responses using a tungsten dopant into the powder mixture. Effective mixtures were identified and HT piezoelectric rings were made using hot isostatic pressing to ensure the production of robust low porosity ceramic compositions. Specifically, bismuth titanate rings were made with diameters of 25.4 and 38.1 mm (1.0 and 1.5 in.) and, in parallel, $LiNbO_3$ rings with 25.4 and 50.8 mm (1.0 and 2.0 in.) diameter were used.

6.4.2 Materials for Fabricating the Drill

Many materials can be used to produce devices for operation at high temperatures. Making a percussive drill also requires that the non-piezoelectric materials can sustain impacts and this is a challenging issue that can affect the durability of the drill (Fergus and Hoffmann 2014; Smialek and Jacobson 2014). Several materials were used to produce the bits that were found effective at room temperatures, including CPM-3V a tool steel with hardness (HRC 59). This CPM-3V alloy is made by the Crucible Particle Metallurgy process, designed to provide maximum resistance to breakage and chipping in a highly wear-resistant steel. However, the bits that were made of Crucible hardened CPM-3V and exposed to 500 °C have demonstrated degradation during drilling. Tungsten Carbide was also tested and this material is considered one of the hardest commercially available materials having hardness of HRA 92. To ensure that the actuator can be operated at 500 °C, the horn was fabricated of Titanium while the stress-bolt and backing were made of stainless steel. In order to reduce thermal instabilities, Inconel Belleville washers were added between the stress bolt and backing. This allowed for increasing the thermal stability and thermal expansion matching, while maintaining the pre-stress of the piezoelectric stack as the actuator was heated to 500 °C.

The testing of the performance of the high temperature drills was performed in three materials: limestone, basalt, and bricks. Based on a study by Schultz (1993), basalt rock with an unconsolidated compressive strength of 262 MPa at room temperature was found to have a compressive strength of 210 MPa at 450 °C. Generally, large variability was found between the different rocks that were drilled as well as along the depth while drilling. Clay bricks were used in the end to drill samples with more uniform properties at high temperatures in order to compare drilling performance from test to test.

6.4.3 Modeling and Analysis of Actuation Transducers

The piezoelectric transducer that drives the USDC is the key to its operation and its maximal performance. Therefore, it is essential to optimize its design. For this purpose, finite element (FE) and equivalent circuit models along with impedance spectrum measurements were used to predict, analyze, and characterize the performances of various transducer configurations. The ANSYS finite element package, which is capable of dealing with piezoelectric materials, was used to evaluate and optimize various transducer designs. The FE modeling was first performed using the available room temperature materials data and then for verification the corresponding equivalent circuit was calculated with lumped components (Bao et al. 2003). The verification was done by impedance spectrum measurement at room temperature. Then, the transducer impedance data were measured at high temperature to obtain the ratio of changes of the circuit components. For various transducer designs of similar configurations, the corresponding equivalent circuits calculated by FE were modified using the changing ratios determined using data from the first fabricated transducer to estimate the performance at high temperature. Details of the analysis and the experimental corroboration are given in (Bar-Cohen et al. 2014c).

6.4.4 Testbed Setup

In order to test the developed samplers at high temperatures, a testbed chamber was used. The chamber provided a controlled elevated temperature environment and enabled placement of the drill and samples inside the chamber. The tests were performed while controlling the preload and tracking the drilling rate as a function of various test parameters. The testbed consisted of a commercial horizontal tube shape furnace (made by Carbolite, UK) with customized insulation side caps having a center hole (Fig. 6.48).

This hole was used in the initial studies for inserting the bit through the side cap while the drill was kept outside the chamber. For the initial testing, rock samples were placed inside the chamber and were drilled at various temperatures while the USDC was placed outside and the preload was controlled pneumatically. Later, the drill was inserted completely inside the chamber while the cable to its drive electronics was inserted through the hole. To control the load on bit and the feed rate, a pushrod was used (Fig. 6.49) that was controlled by a pneumatic cylinder. This fixture allows sliding the drill while driving it and securing the drilled rocks along the bit path. The drill was pushed from outside the chamber using a controlled preload system.

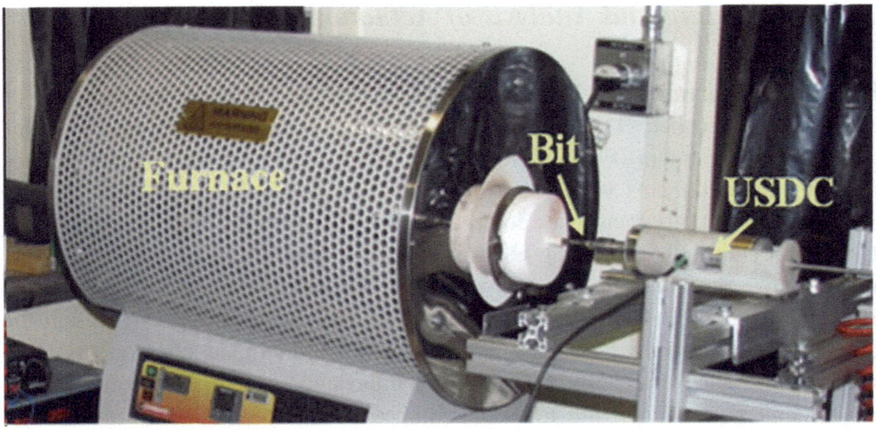

Fig. 6.48 Photographic view of the HT testbed

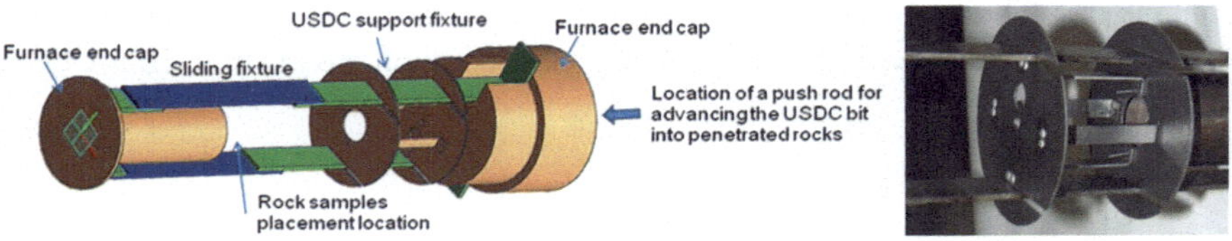

Fig. 6.49 The support fixture for testing the USDC inside the HT chamber (left) and the close-up photograph of the USDC mount (right)

6.4.5 High Temperature Piezoelectric Actuated Drills

6.4.5.1 Basic Designs of the HT Piezo-Actuated Drills

Components of a USDC that can operate at high temperatures were produced to support the testing for the developed piezoelectric stacks. The key components of the USDC that were considered for high temperature operation are the actuator, free-mass and bit. The use of titanium allows for producing a low mass device that has high mechanical strength and good thermal stability. The piezoelectric materials that were tested included; bismuth titanate (BiT) polycrystalline ceramics made by Ferroperm (PZ46), tungsten (W) doped BiT ceramics (PSU), and $LiNbO_3$, single crystals. Preliminary data showed that the samplers that were driven by both BiT ceramics from Ferroperm and PSU worked at 500 °C, however the BiT made by PSU were found to be relatively robust and showed better performance in terms of drilling rates. Based on the test results it was decided to make larger diameter samplers using PSU BiT and $LiNbO_3$. $LiNbO_3$ single crystals and BiT ceramic discs (PSU) were mounted as actuators and held in compression by the stress bolt to prevent fracture of the discs during operation. The analytical results were used to guide the design of transducers with 2.5 cm (1 in.), 38.1 mm (1.5 in.) and a 48.3 mm (2 in.) diameter.

6.4.5.2 2.54 cm (1 in.) Diameter Samplers

Samplers with 2.54 cm (1 in.) actuators were tested drilling brick at both room temperature and 500 °C. A photograph of the $LiNbO_3$ actuator after being exposed to the 500 °C environment is shown in Fig. 6.50.

Several cracks were found on the crystal after test and it is believed to be due to overstressing the crystals while tightening the stress bolt. Note that the single crystals are relatively fragile compared to polycrystalline ceramics and, therefore, care must be taken during fabrication and operation. However, the test results were encouraging since the resonance characteristics of the $LiNbO_3$ transducer indicated good response in spite of the fracture of the individual discs. To overcome the fracture issue, the discs were manufactured with beveled edges to minimize stresses on the corners and softer electrodes were used to avoid stress concentrators. In addition, a more stringent assembly procedure was initiated. Another issue that was identified was the electric shorting to the stress bolt from the electrodes through the mica film electrical insulation. To prevent applying transverse stresses on the mica insulation and possibly damaging the film and causing a short, great efforts were made to assure the alignment of the electrodes. In addition, a thicker layer of mica was used to increase its insulation capability.

6.4.5.3 3.81 cm (1.5 in.) Diameter Bismuth Titanate Samplers

A 3.81 cm (1.5 in.) diameter bismuth titanate driven USDC-based sampler was assembled and tested on a 26 mm thick brick at 460 °C. The weight-on-bit was

Fig. 6.50 Photograph of the first HT USDC actuator breadboard

20 N and the preload for the free-mass was 17 N. The sampler was driven in a duty cycle of 50% (i.e., half the time on and half off). The bit was manually rotated slowly at a speed of about 4–5 RPM. It was noted that it was much easier to rotate when the sampler was activated. The torque needed for the rotation was estimated to be less than 0.02 Nm (0.2 in.-lb). A starting hole of about 1 mm deep was drilled initially at room temperature and the drilling was restarted at high temperature. The driving frequency was 20.9–21.0 kHz, and the averaged power was 26 W. The power factor was low (approximately ~0.094) and the sampler drilled through the sample in 21 active minutes and reached the other end of the brick sample. The average drilling rate was found to be 1.2 mm/min.

6.4.5.4 5.08 cm (2 in.) Diameter $LiNbO_3$ Samplers

In the larger drill design A 5.08 cm (2 in.) diameter $LiNbO_3$ actuator was assembled and the stack was compressed circumferentially by wrapping high temperature fiberglass around it. In order to provide protection of the $LiNbO_3$ stack from cracking during assembly two de-poled PZT rings were used on the either ends of the $LiNbO_3$ stack. In order to create rotation in the oven a self-rotating bit assembly was included. Tests that were conducted at room temperature with a rate of 7–8 mm/minute in a brick sample were recorded. The drill was then tested at high temperature (460 °C). The result from this test is given in Fig. 6.51. The driving frequency, voltage and weight-on-bit were adjusted during the test in order to search for the best operating parameters. In the last 5 min of the test, the driving frequency was 19.9 kHz, the average power was 78 W, the weight-on-bit was 18.7 N, the preload for free-mass was 40 N, and the average drilling rate was found to be 0.43 mm/min. After the test, cracks were seen on the stack accompanied with a large degraded drilling performance at room temperature. This result indicated

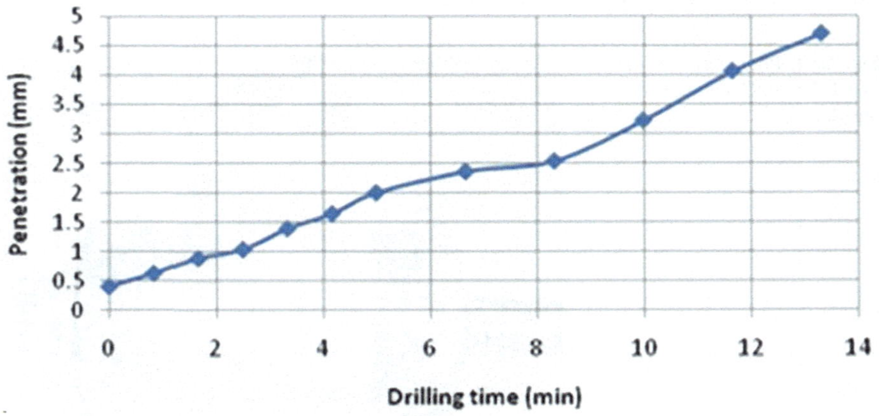

Fig. 6.51 The penetration depth as a function of drilling time at 460 °C for the drill with 5.08 cm (2 in.) diameter $LiNbO_3$ actuator

that other methods to inhibit cracking in the $LiNbO_3$ single crystals at high temperature needed to be implemented in order to solve this problem. One approach to avoid cracking is to use high temperature 1–3 $LiNbO_3$/epoxy composites. This configuration allows for improved mechanical flexibility and more robustness in thermal cycling, while maintaining piezoelectric properties. This approach will be considered for future actuator designs.

6.5 Piezoelectric Actuated Low Temperature Drills

The Ultrasonic/Sonic (U/S) Ice-Gopher was developed in 2005 under a NASA Astrobiology for Science and Technology for Exploring Planets (ASTEP) task (Doran et al. 2003; Bar-Cohen et al. 2004; Badescu et al. 2006a; Bar-Cohen and Zacny 2009). The drill was tested in field demonstrations at Mt. Hood, Oregon, as well as in Lake Vida, McMurdo Dry valleys, Antarctica. The goal of the task has been to establish the capability to investigate the presence of ice and fluids near the surface of Mars.

The major components of this drill were the piezoelectric actuator, drill bit, and a free mass as in other USDC designs and a pump with housing, preload weights and pneumatic and electrical tethers (Fig. 6.52). For modeling, the weights of the transducer housing, the pump and its housing, and the preload were combined and treated as a total weight of the transducer. The analytical model of the USDC has been described in details at previous publications (Sherrit et al. 2000; Bao et al. 2003). In producing an optimal U/S Ice-Gopher, the three key components of the USDC mechanism were carefully analyzed in order to optimize the displacement at resonance. In order to have a high velocity of the bit surface we need an actuator with a high mechanical Q, and operating at or near its first longitudinal resonance frequency.

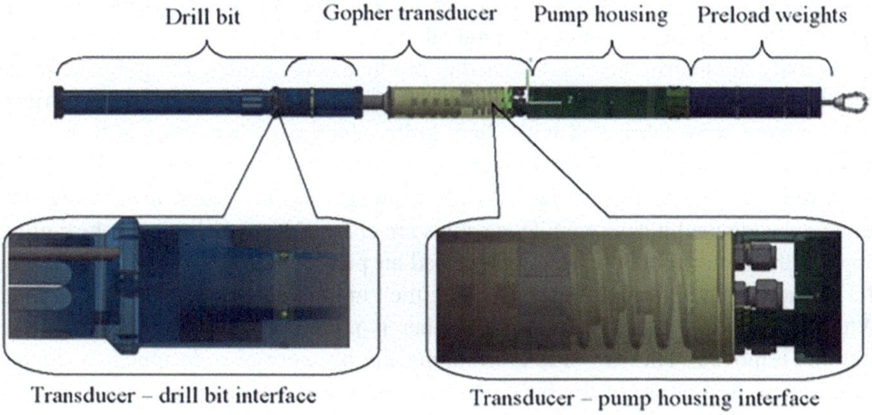

Fig. 6.52 A general schematic view of the U/S Gopher system design

The parameters that were investigated include the number and the size of the piezoelectric rings, the size and characteristics of the pre-stress bolt, geometry and dimensions of the horn, location of the neutral plane to coincide with the mounting plane, weight of the free-mass, geometry and dimensions of the coring bit. A finite element analysis was used to analyze the transducer configuration and the stack and horn shape. The design parameters were modified until the mounting plane on the horn coincides with the neutral plane of the transducer. The free-mass is excited by the actuator to bounce back and forth between the horn and the coring bit at frequencies varying from tens of Hz to 1 kHz. This low frequency vibration results in the free-mass gaining energy from the actuator and transferring it to the coring bit in sharp stress pulses. These interactions were analyzed to determine the impact frequency, the momentum after the impact and the energy transferred to the bit at each impact.

6.5.1 Design and Fabrication of the U/S Ice-Gopher

A series of designs with various free mass sizes and horn shapes were considered, and analyzed. The free mass varied from 125 to 500 g and the horns that were analyzed included standard straight horn and "dog-bone" shaped horn (Sherrit et al. 2004a). A straight horn has the same diameter throughout its length, while the "dog-bone" horn has a base, a neck, and a head. The neck portion is longer and has a smaller diameter. It was introduced to increase the vibrations amplitude of the horn head. It was determined that a design having a neck with small diameter and using a free mass of 250 g transfers the most power to the drill bit under the same input power. The horn provides a mounting flange for the transducer housing and it is done on the nodal plane of the transducer minimizing vibration transfer to the housing. In order to ensure the mounting on the nodal plane, the flange was designed to coincide with the nodal plane of the transducer. A through-hole that was drilled through the actuator allowed for the transmission of compressed air down to the drill bit for ice chips removal.

A cavity in the drill bit was designed to produce cores while cutting the ice at the bottom by bit two concentric rings that were connected by radial teeth. Both rings and the teeth were fabricated with sharp knife-edges at their cutting surfaces. The bit was also designed to allow through two side air tubes for compressed air to be delivered onto the cutting surface. While drilling, the bit creates an ice core that remains inside the bit cavity while ice chips are channeled away between the cutting rings of the bottom section. The compressed air pushes ice chips up the hole, along the outside of the bit, and they fall inside the cuttings cavity through an opening. When the cuttings cavity is full, the gopher is pulled up from the holes and the cuttings removed. The two side tubes along the bit were also designed to sample liquid from the bottom of the hole when reaching wet ice or brine.

The gopher was tested in the lab and at two field sites: on a glacier at Mt. Hood, Oregon and in Antarctica. Based on the lessons learned from the first field test, design changes and additions were made to include temperature sensor that was embedded into the housing cap for measuring the ambient temperature in the borehole during the drilling operation. Moreover, a sediment-sampling bit was retrofitted with an adaptor to allow mounting onto the actuator as a replaceable fixture. The sediment sampler allows sampling and retaining sediments in the form of unconsolidated sample. The adaptor allows attaching the sediment sampler to the dog-bone horn of the actuator and it was made as an add-on fixture for use to acquire samples from the bottom of the lake. To assure backup in case of failures in the field, a series of six complete actuators, three drill bits, two pumps, two pump housings and a set of preload weights were fabricated, assembled, tested in the lab and prepared for use in the field in Antarctica. The drive electronics used consisted of a function generator and a power amplifier. The frequency control could be performed manually or automatically by a LabVIEW control program. In the automatic mode a digital oscilloscope, a computer and an in-house LabVIEW software program were used to control and monitor the actuator performance. The oscilloscope reads and averages the peak voltage and current across the actuator. The computer is used to run a control interactive software program that uses a hill-climbing algorithm (Aldrich et al. 2006b). The computer program uses an extremum seeking algorithm and maximizes the current through the actuator by controlling the frequency of the signal generated by the function generator. A support frame with an electric winch and a steel cable were designed and adapted to lower the U/S Gopher down the borehole and lift it after a drilling session.

6.5.2 Test Results and Lessons Learned

For deep drilling, the USDC mechanism was used to develop a "Gopher" that can acquire core samples using a bit diameter as large as 6.4 cm, which is larger than the USDC actuator housing. The device is shown schematically in Fig. 6.53, where a core is formed up to the length of the internal size of the bit and it is brought to the surface and removed from the borehole to create room for a new one.

This process is repeated until the desired depth is reached. To demonstrate the capability of the Gopher it was tested on a glacier at Mt. Hood, OR and the lessons learned were implemented into the design. The ambient temperature was above freezing however, the glacier provided a good testbed for the first demonstration of the U/S Gopher capability. This test pointed out design flaws and inefficiencies in the design that were corrected for the Antarctica field test. Since the top layer was fluffy snow, it was necessary to remove about two and half meters of the snow before there was sufficiently hard packed snow to conduct the needed tests. During this test session, a total of 1.25 m drilling was accomplished in a total drilling of 5 h with an average drilling rate of 0.25 m/h. Heating of the transducer was observed and there were difficulties removing the ice chips due to limited pressure of the compressed air. There were also failures of the actuator transducers mainly due to improper location of electrodes soldering and heating due to inefficient frequency

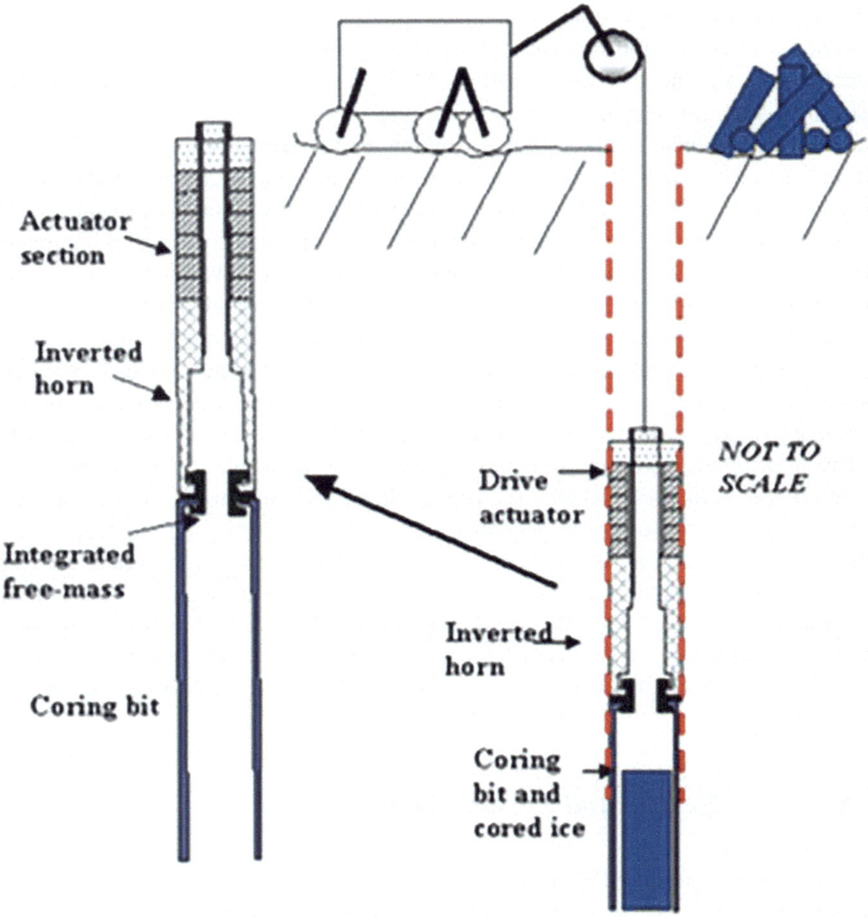

Fig. 6.53 Schematic view of the ultrasonic-gopher operating inside the borehole. The cores are identified in blue

control. These problems were addressed prior to the departure to Antarctica with a final redesign of the U/S Gopher hardware and software control.

The field test in Antarctica was conducted at Lake Vida, which is located in the Dry Valleys, Antarctica, which is a lake with ice layer thickness ranging from 15 to 20 m and the temperature was from −5 °C at the surface to −12 °C at 16 m depth. It provided an important opportunity to demonstrate the feasibility of this technology. In addition, this field-testing allowed for determining the associated challenges and requirements to enhance the drill capability for future drilling objectives. The U/S Ice-Gopher unit successfully reached 1.76-m deep, which was significantly deeper than the 1.6-m length of the entire Gopher assembly.

As part of the tests in the drill was used to penetrate also sea ice McMurdo Sound, where the ambient temperature ranged from −15 to −5 °C depending on the

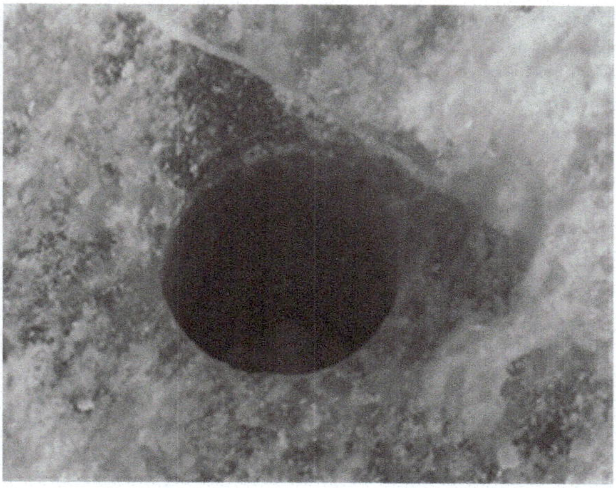

Fig. 6.54 Sea ice drilled borehole entry

day, however the sea ice temperature was not measured. The tracking frequency control was initially done using the "hill-climbing" software program and later it was done manually. Bits with and without core traps were tested and a total depth of 35 cm was drilled. One of the drill bits was modified by removing four of the cutting teeth and sharpening the remaining ones. The drilling reached maximum depth of about 35 cm and a photo of the drilled hole is shown in Fig. 6.54.

Various duty cycles and drilling times were tested with both manual and automatic drive frequency control. In addition, attempts were made to determine more accurately the drilling rate. The data provided by the driving frequency control program was saved and analyzed later in the lab at JPL. Airflow for removing the ice chips was used at all times during drilling where the air pressure in front of the filter was maintained at 70 psi.

As can be seen from Fig. 6.55, the U/S Gopher was completely below the ice surface in the drilled borehole.

This has an important milestone since once the drill reaches this level of depth it effectively can drill as deep as logistically possible without other physical limitations (e.g. change of drilling material, tether length etc.). The "hill-climbing" program control software that we developed was mainly used with fixed frequency to control the duty cycle. During the manual control of the frequency, it was found that the voltage-current phase matching approach provided a better drilling performance than current maximization algorithm. A duty cycle of 30% was used to limit the increase in temperature of the transducer.

Other observations included the fact that the chip removal mechanism worked up to a depth of approximately 2 or 3 cm of ice penetration and needed the removal of the core pucks and chips in order to penetrate further. One of the reasons this happened is that the ice melts inside the borehole during drilling. This melting caused the ice chips to get wet and stick to the bit surface above the cutting ring. In addition, if the ice pucks were big enough they packed inside the drill bit. This required frequent extraction of U/S Gopher from the borehole for chips removal.

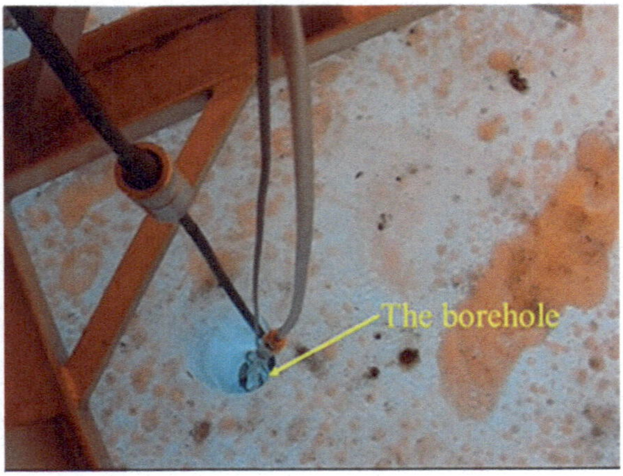

Fig. 6.55 The gopher in the drilled hole

After reaching the 1.76 m mark, efforts were made to optimize the duty cycle, however, the delay in making the setting caused the U/S Gopher was frozen inside the borehole and could not be removed using the winch and "hammer out" mechanism. Eventually, the Gopher was removed by digging around it using chainsaws.

Other areas where the robustness of the Gopher design could be improved include the following: (a) the screws of the retaining ring need to be increased, (b) the voltage feedthroughs need to be redesign for higher power, and (c) controlling the phase matching seems to provide better drilling performance than total current maximization. It was difficult running the actuator for a long period without transducer heating. Ice melting is to be avoided to prevent the U/S Gopher from refreezing inside the borehole during the off time of the duty cycle. Chips removal using compressed air works well for above 100-psi pressure of compressed air and having the ice chips maintained frozen inside the hole. Any melting or pressure and flow rate drop can lead to reduced efficiency of the chips removal and packing above and inside the cutting ring. In order to maintained pressure above 100 psi, either a more powerful air compressor is used or a flow control valve needs to be added to the air circuit to turn off the airflow during idle time and allow the pressure build-up in the compressor tank.

6.6 The USDC-Based Gopher as a Deep Drill

Planetary drilling has different challenges than found when drilling on Earth. These include limited system mass, power, and energy as well as low pressure (or vacuum) and low temperature environments. From the mass standpoint, unless drill strings are made of low-density material, adding drill sections to reach greater depths makes the entire system quickly very heavy. In addition, the system needs a robotic drill string feeding mechanism such as a carousel and mating connections between each drill string. This not only increases the mass but also the system complexity and, therefore, increases the risk of the system failure. If the drill system requires sensors such as thermocouple on the bit for monitoring the environment around the drill to avoid freezing (that would be catastrophic to the mission), the drill strings would also need to have an electrical feedthrough mating system. A robotic system with autonomous drill string management and downhole power/data capability has previously been built for planetary applications. However, it was quite complex (Zacny et al. 2008).

Since using water or mud is difficult or impossible in planetary exploration due to low pressure and low temperature conditions, drilled cuttings need to be conveyed to the surface using an auger (e.g. screw) or some other means (e.g. pneumatic, mechanical bailers, pocket sampler etc.). The parasitic drag of the rotating auger strings against the borehole coupled with even larger frictional drag produced by the cuttings as they are moved up the hole require prohibitively large torques and power. This is the most important factor limiting the penetration depth of the continuous drill string when it approaches about 10 m deep. A possible solution is to employ a single long bit (Bar-Cohen et al. 2012; Zacny et al. 2013). In this approach, the drill is periodically retracted to clear the formed chips. However, auger drag due to cuttings removal and the parasitic losses due to auger rubbings can limit the bit length to about 1 m. The main drawback to the strings approach is its management system that has to be very robust to cope with countless drill string connections and disconnections during the course of drilling a deep hole.

The use of wireline approach solves most of the problems associated with continuous drill string. In this case, the drill is suspended on a tether and all the motors and mechanisms are built into a tube that ends with the drill bit. The tether provides the power, data communication and the mechanical connection to a planetary platform on a surface. The first generation of this type of USDC-based deep drill was implemented in the Ultrasonic/Sonic Ice-Gopher. This drilling system is shown in Fig. 6.56.

Generally, wireline systems involve the mechanical complexity of packaging motors and actuators into a slim tube. In addition, as opposed to a continuous drill string system, where the Weight on Bit (WOB) also known as a preload, is provided by a lander or a rover, the WOB in a wireline system is provided by anchoring the

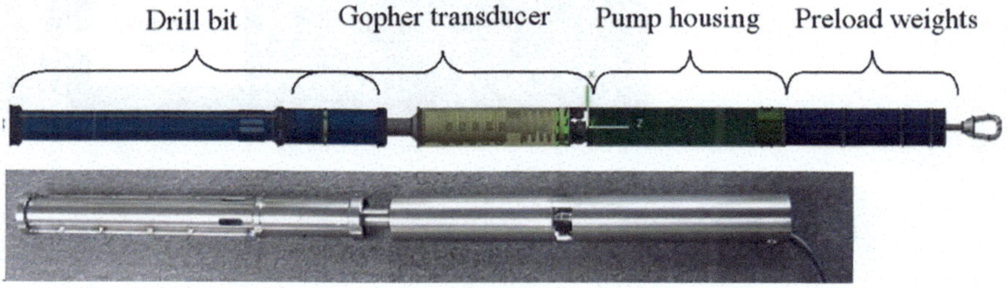

Fig. 6.56 The Ultrasonic/Sonic Ice-Gopher illustration (top) and photo (bottom)

drill to the borehole wall by locking the upper section of the drill. This approach requires the use of a linear feed using an internal screw to push on the drilling mechanism and the drill bit itself during drilling.

6.6.1 The Auto-Gopher-1

The Auto-Gopher-1 drill was developed as a joint effort of the JPL's NDEAA laboratory and Honeybee Robotics Corp. The drill incorporates a rotary motion onto a percussive USDC-based mechanism with a fluted bit (Badescu et al. 2006a; Bar-Cohen and Zacny 2009; Bar-Cohen et al. 2014c; Zacny et al. 2013). The Auto-Gopher is a stand-alone drilling system (Fig. 6.57) requiring no additional actuation from the surface to perform drilling except for the tether management.

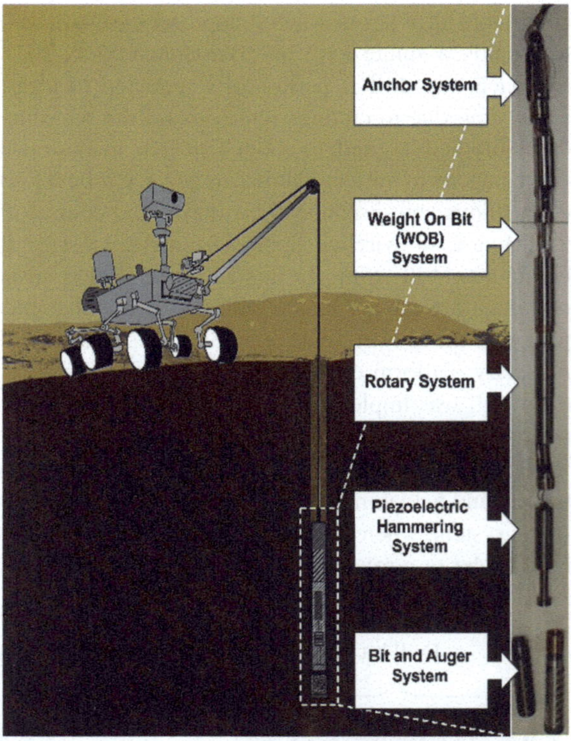

Fig. 6.57 Illustration of the Auto-Gopher, concept as a wireline deep drill

The drill consists (from top to bottom, Fig. 6.57) of an Anchor, Weight on Bit (Preload) Drive, Rotary System (Auger Drive), Hammer/Percussive System, and a Bit with an Auger. The Anchor uses a set of three compliant shoes to push against a borehole and anchor itself to the drilled hole with a force of 1600 N. This force is sufficient to provide a resistance to rotary torque from cutting bit as well as vertical force from the Weight on Bit. The WOB Drive is provided by internally actuated ball-screw and is designed for WOB of 1000 N. An integrated load cell provides a force feedback for WOB control. The Rotary system uses a cluster of 3 actuators with a combined electrical power up to 360 W for rotating a coring bit and an auger. Accounting for electrical and mechanical (e.g. gearbox) losses, the system can generate a torque of 15.5 N-m at 100 rpm. The hammer system employs a piezoelectric actuated percussive mechanism for providing impacts via free mass. The impact energy imparted to the bit is stochastic with a range of frequencies: Lower energy blows of the order of 0.1 J have frequencies in the range of hundreds of Hz while the higher energy blows of the order of 0.4 J have a frequency range of e 10 Hz. The hammering action is independent of the rotary motion and hence can be engaged when the formation becomes too hard for rotary action alone to cut through or when the tungsten carbide teeth get dull. In addition, a Percussive system allows the cuttings within the bailer bucket above the core to compact more and in turn occupy less volume.

In the Auto-Gopher-1 drilling mechanism, a coring bit was used to allow for acquisition of 60 mm diameter 100 mm long cores. The outside diameter of the coring bit is 71 mm. Drilled cuttings are moved up the auger flutes and fall into the cuttings chamber above the core chamber. The integrated bailer above the core chamber can accommodate all the cuttings produced during the 100 mm drilling; accounting for a cuttings volume expansion factor of 3 (i.e. as rock is being drilled, the resultant cuttings will occupy up to 3x the volume of the base rock/ice). Upon drilling the 10 cm long core, the drill is retracted and the cuttings chamber is emptied.

Figure 6.58 shows the field deployed Auto-Gopher in a gypsum quarry of the US Gypsum Company outside Borrego Springs, California and the drilled 3.07 m deep hole. The purpose of the field test was to demonstrate the drilling to a depth more than the length of the drill bit, perform core recovery, and obtain drilling telemetry to extrapolate the drill time and energy required to drill at greater depth.

The rate of penetration (ROP) of Auto-Gopher as a function of percussive power by varying the level of duty cycle is shown in Fig. 6.59. Although an increase in the duty cycle resulted in an increase in the rate of penetration, this might cause the device overheating and damage the piezoelectric elements for long-term use. For both safe and efficient operation of the device, 50% of duty cycle would be optimal as any generated heat is dissipated during off time reducing the temperature rise during operation. From Fig. 6.59, it can be observed that when the duty cycle was reduced to 50% (5 s on and 5 s off), the rate of penetration was down to 80 cm/h. However, the rate of penetration was found to increase with the decrease in the duration of the ON-OFF cycle (1 s ON and 1 s OFF), offering the same level of penetration rate as 100% duty cycle operation.

Fig. 6.58 The drilled deployed in the field with the extracted cores (left) and a close-up of the drilled hole (right)

6.6.2 *The Auto-Gopher-2*

The Auto-Gopher-2 represents the latest generation in electro-mechanical deep drilling based on the USDC technology (Bar-Cohen and Zacny 2009). It builds on the previous successful technology demonstrations that were enabled by the Auto-Gopher-1. The new drill diameter has been reduced since the breakoff forces are proportional to core diameter and smaller breakoff forces are desired/required. Instead of a coring bit, an auger is being designed to reduce the potential risk that is associated with the failure of coring process. The drill deployment system includes the tether management (drum) which enables the goal to develop a fully autonomous system that lowers the drill to the surface (initially) and then to the bottom of the hole, and upon sample capture, returns the drill to the surface. The tether, in addition to the electrical conductors also includes a structural cable with a breaking strength of 10x the drill's weight to address the possibility of jamming and borehole

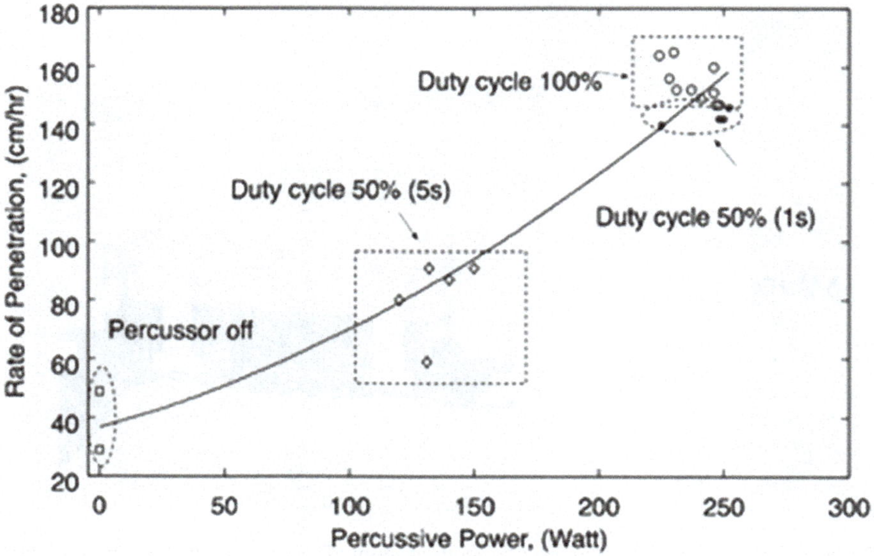

Fig. 6.59 Penetration rate as a function the percussive power and duty cycle

friction. To maintain the drilling direction, the possibility of integrating two anchoring systems along the Auto-Gopher is being considered where each with three individually activated shoes and guide the drill gravitationally.

Creating a core is only the first step in performing deep penetration and acquisition of samples. The second step is to break the core at its base and capture it within the coring bit. The core break-off system needs to be reusable and allows for easy core ejection by the core handling system on the surface. There are several ways that a core can be detached from a parent rock. These include: (1) pulling (breaking the rock in tension), (2) twisting (breaking the rock in slow shear), (3) impact shear (breaking the rock by twisting it at high speed—sort of an impact twist), (4) shearing (breaking the rock by "cutting" as in pinching), or (5) bending (breaking the rock in bending at the base by applying a side force on top). The preferred method for core breaking is not necessarily chosen based on the lowest required force but rather on the complexity of implementation into the design. The Auto-Gopher-1 did not include robotic core breakoff and catching capabilities since the focus of that development has been on demonstrating wireline technology using anchoring.

To resolve the issues that were mentioned above, various solutions were considered for the Auto-Gopher-2. An example of a solution is to employ a floating wedge that engages the core only after a certain core length has been reached. Figure 6.60 shows the bit with three chambers: core, cuttings and free mass. The core chamber includes at its upper end a mechanism for core break-off. This mechanism breaks the core in bending, as it is one of the lowest force mechanisms required for core break-off. In testing this solution, efforts were made to avoid

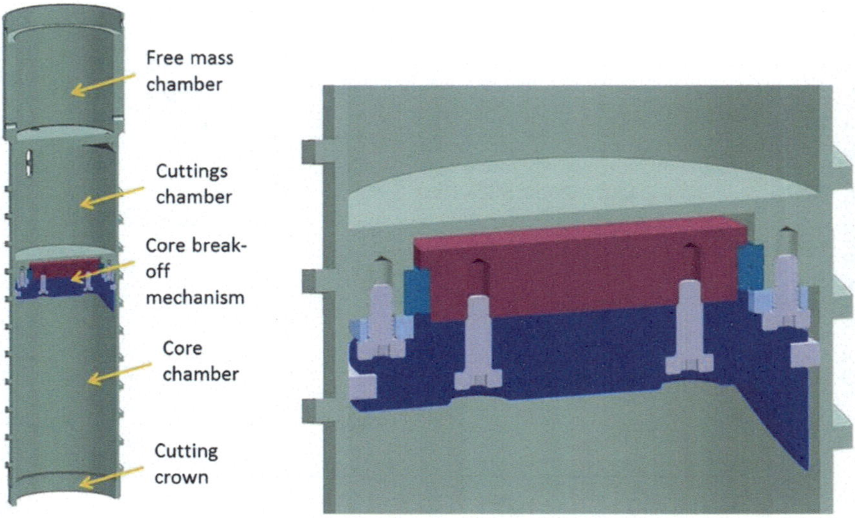

Fig. 6.60 Bit with integrated core break-off mechanism. The bit is shown on the left and a close-up of the Floating Wedge is shown on the right

engaging the core and forcing it to rotate. During the field tests, it was observed that a broken core tends to jam the bit if it tries to rotate (Fig. 6.61). This Floating Wedge solution tries to decouple the rotation of the core break-off wedge from the bit rotation and it is a passive solution, as it does not require any additional actuators or control.

The core break-off system needs to be reusable and allow easy core ejection by the core handling system on the surface. The testing of the mechanism was done using a standalone, manually deployable system that includes a tube with internal wedge (Fig. 6.62).

The wedge engages the core only after a certain core length has been reached and tests have shown successful breakoff and retention of the generated cores (Fig. 6.63). It should be noted that not all cores sheared at the bottom and some broke further up in the core chamber.

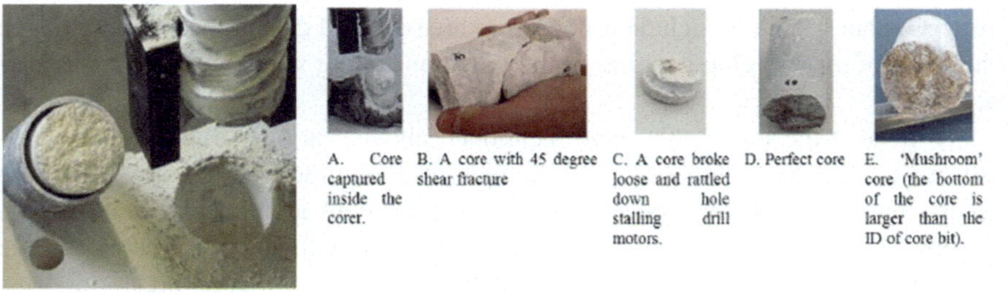

Fig. 6.61 Left: a separate core break off system was used to break-off and capture the core, right: core breakoff system needs to be able to work with various cores

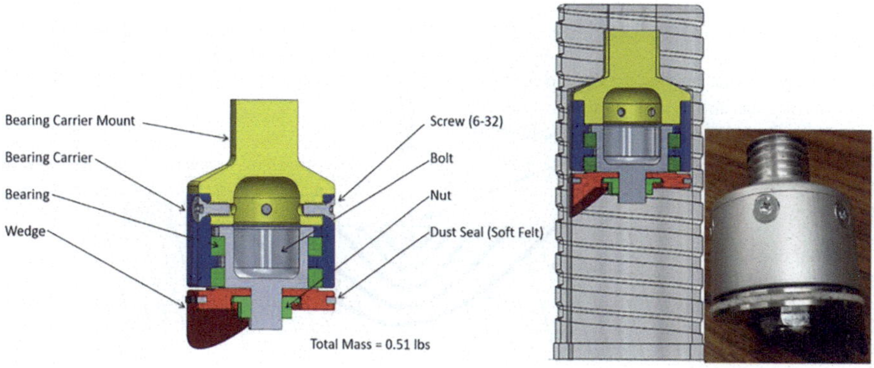

Fig. 6.62 The break-off and retrieval breadboard. In the middle graphics, the core breaker is illustrated inside the bit and its photo is shown on the right

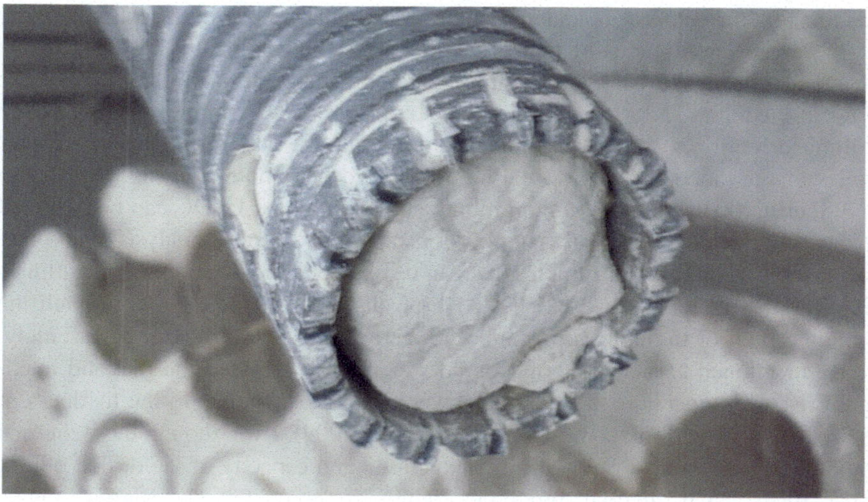

Fig. 6.63 Successful core sample break-off and removal

6.7 Lab-on-a-Drill

The ultrasonic/sonic driller/corer (USDC) is an effective tool that allows both sampling and probing the sampled medium while it carries sensors for in situ analysis. Integrating these system capabilities has been the basis for establishing a USDC device as a Lab-on-a-drill (Fig. 6.64). Vibrations, which are generated by the device upon impact of the bit with an object or the ground, impart elastic waves and provide a sounding mechanism for noninvasive probing of the impacted medium. This sounding capability combined with detectors can provide information about the ground subsurface-layered structure, mechanical properties, and presence

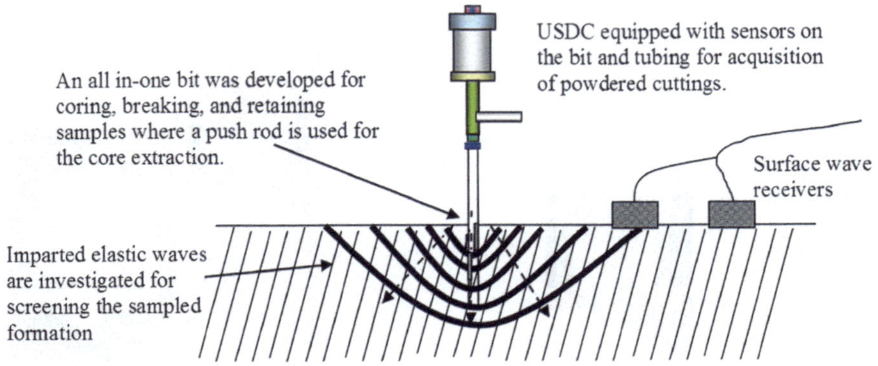

Fig. 6.64 The USDC as a lab-on-a-drill

of geological cavities. Further, pressing the device onto various materials or objects modifies the mechanical boundary condition and changes the electrical impedance of the piezoelectric actuator that drives the device. Gauging the changes in the electrical impedance can provide information for estimating the stiffness of the pressed medium. These two probing capabilities allow screening objects prior to drilling or coring them. In addition, equipping the bit with such sensors as fiber optics or thermocouple allows real-time on-board or remote analysis (Bar-Cohen and Sherrit 2003b). In addition to probing and sensing, the USDC can be designed to acquire various forms of samples for on-board or post analysis and other applications. The types of samples that can be extracted include cores, and cuttings. The sampled material can be collected from the front of the device, through the hollow center, or through various tubes that are connected to the device assembly.

As mentioned earlier, a thermocouple was used to measure the rate and maximum rise of temperature. These values were found to correlate to the hardness of the rock that were drilled (Bar-Cohen and Zacny 2009). Even though these thermal variables are dependent on the heat conductivity and capacity of the drilled object, most rocks have thermal properties within a comparatively narrow range. Compiling temperature rise rate and maxima as a function of time for a variety of drilled materials has demonstrated the feasibility of using a thermocouple-on-the-bit as a means of assessing the drilled medium hardness (Fig. 6.65) (Bar-Cohen and Sherrit 2003b). However, there may be an estimate error if the rock contains water or ice that could be evaporated and causes cooling or has a high iron content like Magnetite.

Using an optical fiber that is integrated into the bit provided a sensing capability. A fiber with approximately 160 μm diameter probe head was imbedded into a 10-mm diameter coring-bit with a 1-mm wall thickness (Fig. 6.66 left). Reflection data in the wavelength range of 400–1200 nm were used to examine the weight percent of a fluorescent powder in a mixture. The test result is shown in Fig. 6.66 on the right. This result has great significance in the search for life since the use of

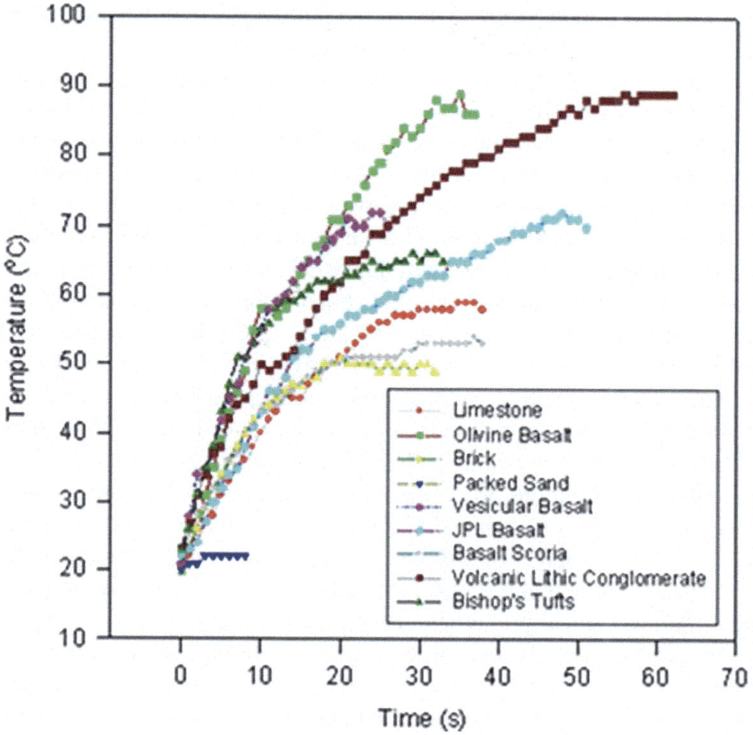

Fig. 6.65 Temperature-versus-time data were obtained by use of a thermocouple embedded near a USDC tool bit drilling through materials of various hardness levels

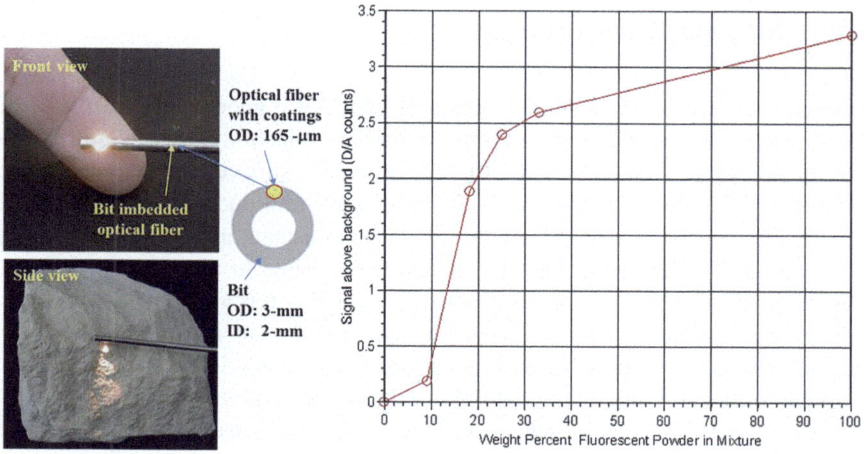

Fig. 6.66 The imbedded fiber-optic in the bit (left) and the response in the range of 545 and 700 nm (right)

fiberoptics with UV light in the range of 200-nm wavelength has the potential for identifying biological markers using biofluorescence.

The USDC induced hammering action provides a sounding mechanism for noninvasive probing of the ground geology to provide information about its subsurface structure, and mechanical properties. To take advantage of the generated elastic waves, accelerometers can be used to sense the imparted waves and their interactions to support analysis of the received wave characteristics. As a diagnostic tool, this probing method can provide information about the soil mechanical properties, geological anisotropy, and layered characteristics as well as detect, locate and characterize geological cavities. Such a capability can be useful in such areas as construction and geological excavation. The principle behind such geophysical probing is that the induced elastic waves travel through rocks and subsurface, encountering various interfaces that produce signatures can be analyzed. The presence of layers causes dispersive wave characteristics, small discontinuities cause scattering and attenuation, and large discontinuities (rocks, cavities, etc.) produce reflections. Methods are available to extract geophysical information from the imparted waves (Bar-Cohen and Nazarian 1997), offering effective probing and imaging for in situ geophysical probing providing valuable information about the stiffness constants of soil and rocks. Generally, the use of noninvasive methods to determine the mechanical properties of the ground is an outgrowth of the field of seismology. These methods transmit elastic waves through a medium and analyze the received waves after interacting with the various geophysical features, layer characteristics, material/ground physical properties and flaws in the host media. The reported studies have shown the ability to get repeatable quantitative surface wave velocity data for a variety of rocks.

6.7.1 Surface Wave Velocity

As mentioned earlier, the hamming action of USDC is a good source of ultrasonic elastic wave energy. The first test of the probing capability was to utilize this acoustic energy generation feature of USDC to induce surface wave onto a sample and measure the received signal with an elastic wave transducer. The surface wave velocity of a media is determined by its elastic constants and mass density. By measuring the surface wave velocity, it may be possible to identify the type and characteristics of the sample or aid to determine whether it is a hard or soft material.

The surface wave velocity is modeled by applying a point source to a half space. This half space assumption is a good approximation and it holds well as long as the sample is sufficiently thick to avoid the reflection from bottom and no interference with the signals of interest will occur. Figure 6.67 schematically illustrates the various types of waves that are generated by a point source loading onto a half-space and the waves include P-wave, S-wave, and surface wave. The P-wave carries about 7% of the total energy, the S-wave about 26%, and the surface wave

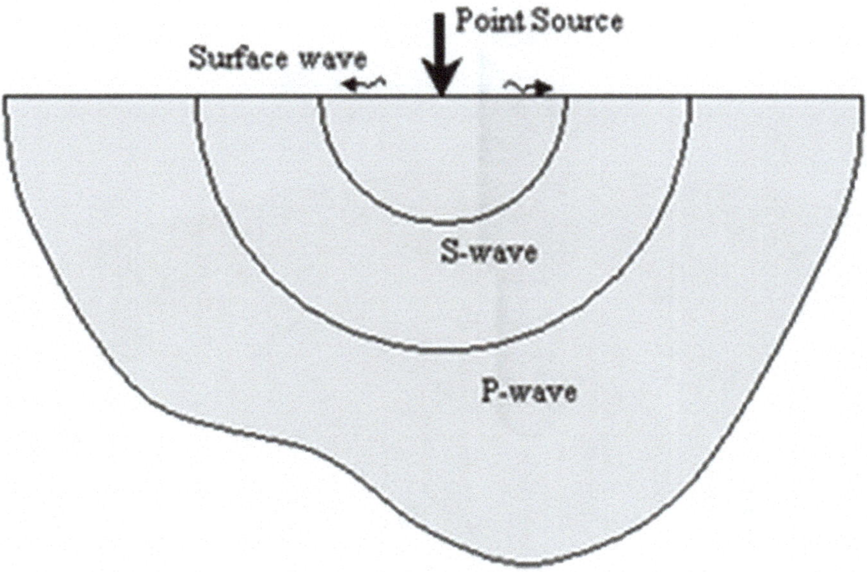

Fig. 6.67 Waves generated by a point source on a half-space

the rest, about 67%. That means that the majority of energy of the signals is dominated by the surface wave (Graff 1991).

A schematic diagram of the experimental setup is shown in Fig. 6.68, which depicts the use of USDC as the source of surface waves. When USDC is operated at its normal working condition, the drill bit impacts the sample at a random frequency, which ranges from 60 to 1000 Hz. If we use this normal operating mode as the wave source, the signals generated will be too random to do a consistent analysis. Instead, a function generator is used to provide a controlled signal to the USDC. A sinusoidal tone burst is sent to the USDC at its resonance frequency. The number of cycles that were induced onto the actuator should be small enough so as not to induce any ultrasonic/sonic transformation in the USDC to cause a drilling action. The key is to keep the drill bit, the free mass, and the horn in contact and vibrating at the nominal 20 kHz ultrasonic frequency. The receiving transducers used for the experiments are broadband transducers B1080 made by Digital Wave Corporation.

Typical signals generated by USDC on a limestone block using controlled signals from a function generator are shown in Fig. 6.69. The two transducers are placed in line with the USDC and separated by a distance of 2 in. To calculate the velocity of surface wave, we need to know the time-of-flight for the signal to travel from the first transducer to the second. For complicated signals such as those shown in Fig. 6.69, the best way to determine the time-of-flight is to do cross-correlation analysis C_{12} of the signals as given by:

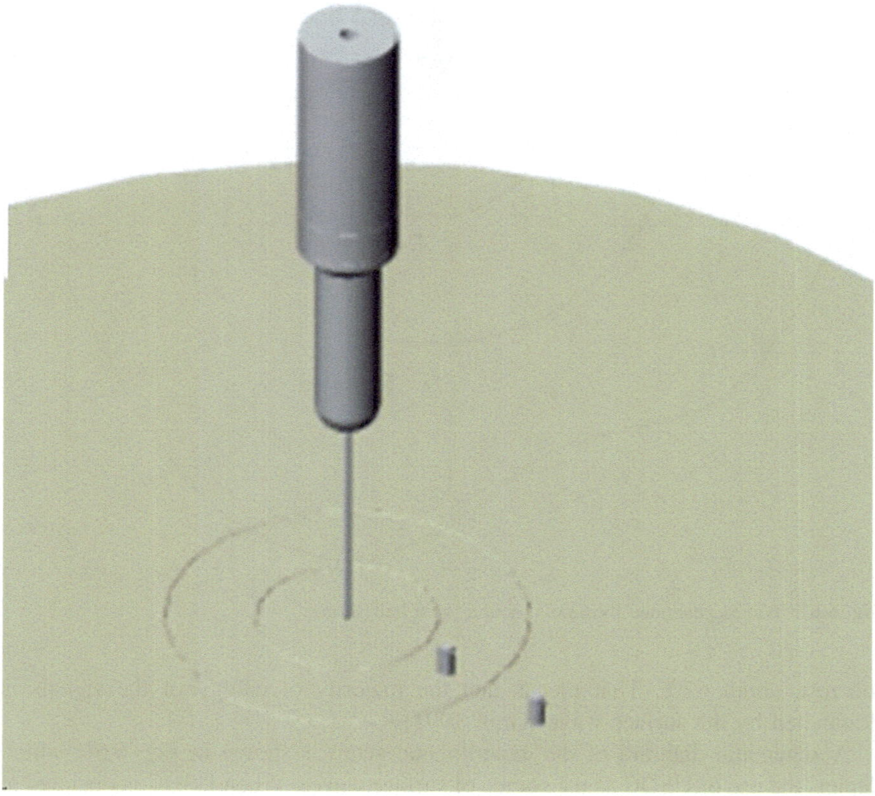

Fig. 6.68 Using USDC as the source of surface waves

$$c_{12}(\tau) = \frac{1}{t}\int_{0}^{t} S_1(t)S_2(t+\tau)\mathrm{d}t \tag{6.8}$$

where S_1 and S_2 are the first and the second signal, τ is the shift of time.

Equation (6.8) shows the formula for calculation of the cross-correlation. Since the functional form for the signal is not available, a discrete form of Eq. (6.8) was used for numerical analysis:

$$c_{12}(n\Delta\tau) = \frac{1}{t}\sum_{t} S_1(t)S_2(t+n\Delta\tau)\Delta t \tag{6.9}$$

The result of a cross-correlation analysis is shown in Fig. 6.70. The highest peak in the figure provides a measure of the time-of-flight between the two transducers.

A surface wave velocity of 1814 m/s was calculated using this time of flight, which is in excellent agreement with the results obtained from the pencil-break

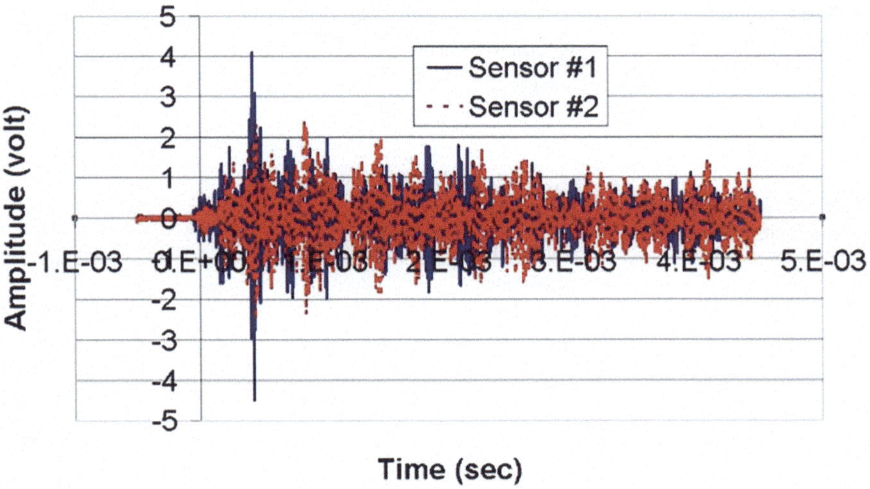

Fig. 6.69 Typical signals generated by USDC on a limestone block

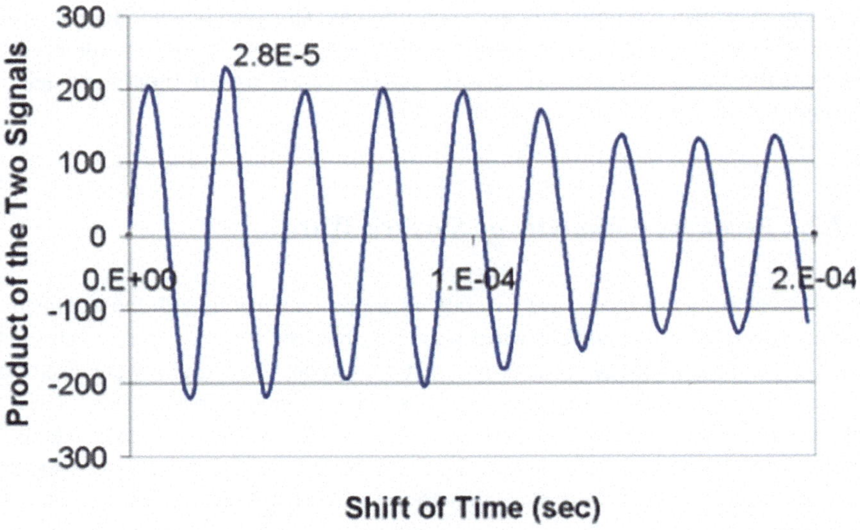

Fig. 6.70 Results of cross-correlation analysis using signals shown in Fig. 6.69 (limestone)

experiments that were performed earlier as a baseline for verifying the results of USDC experiments. Table 6.1 below shows the comparison of results from pencil-break and USDC tests for both limestone and sandstone indicating significant degree of agreement. This result is encouraging in regards with the use of the USDC as a sounding source for the measurement of the surface wave velocity.

Table 6.1 Comparison of results from pencil-break and USDC experiments

	Pencil-break (m/s)	USDC (m/s)	Difference (%)
Limestone	1814	1814	0
Sandstone	2230	2120	4.9

Table 6.2 Measured surface wave velocities for various media

Rock type	Surface wave velocity (km/sec)
Limestone	1.81
Sandstone	2.23
Concrete	2.54
Brick	1.49
Talc	2.07
Basalt	2.06

Using the approach outlined above, a series of rock materials were tested and the surface wave velocities are presented in Table 6.2. It is interesting to note out that the experimental results are repeatable and the velocities for most of the materials tested are distinct. However, the velocities for basalt and talc are very close, although their material properties are very different. This phenomenon indicates the fact that there is a need for additional effective probing algorithms, and the need to combine the results with the surface wave velocity test results, in order to be able to distinguish between different types of rocks.

6.7.2 Frequency Analysis of Surface Wave

In addition to the surface wave velocity measurement mentioned above, the frequency spectrum of the surface wave was also analyzed for rock type recognition. For the purpose of proof of concept, materials other than rocks were used for preliminary experiments. Figure 6.71 shows the typical frequency spectra for 5 different materials, including an aluminum plate, brick, tile, and two keystones used in construction. It is easily seen that the 5 spectra are very distinct. However, the task of constructing a robust algorithm for rock recognition remains a challenge. The idea was to establish a spectrum database for various types of rocks. The spectrum derived from surface wave tests on an unknown sample is then compared to all the spectra in the database through cross-correlation analysis. Theoretically, the best match provides the answer to what type of rock the sample is made of. During the preliminary experiments, two events of surface wave were sent to the 5 samples mentioned earlier. The first event from each sample was compared to the second event from the same sample and the first events from other samples. The results of the cross-correlation analysis are shown in Table 6.3.

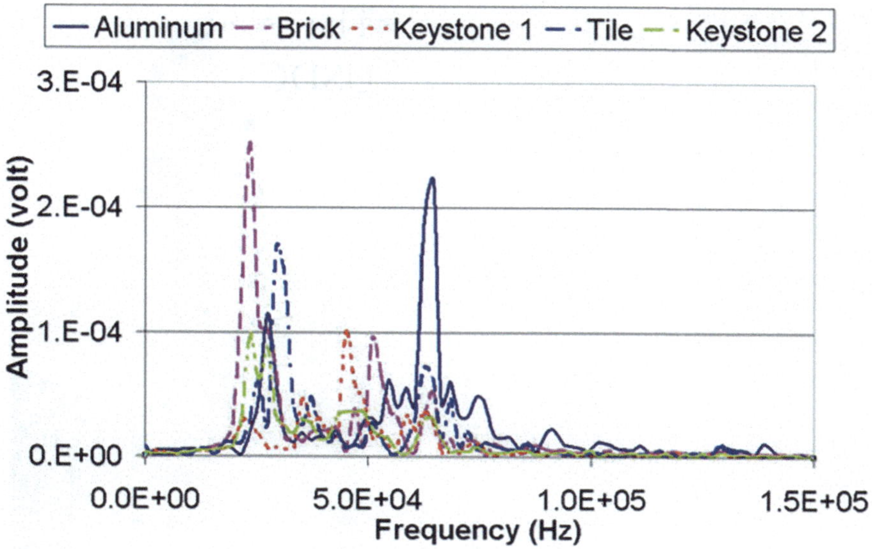

Fig. 6.71 Frequency spectra of surface waves

Table 6.3 Results of cross-correlation analysis on surface wave spectra

	Aluminum	Brick	Keystone 1	Tile	Keystone 2
Aluminum	0.881	0.438	0.458	0.599	0.571
Brick	0.438	0.903	0.451	0.473	0.862
Keystone 1	0.458	0.451	0.903	0.391	0.663
Tile	0.599	0.473	0.391	0.900	0.585
Keystone 2	0.571	0.862	0.663	0.585	0.830

The numbers on the diagonal axis are the comparison between two events from the same sample (Table 6.3). In general, the results conform to what we would expect except for the brick/keystone#2 element in the second row of the last column. The largest value should always be the diagonal element. The reason why there is an exception is likely due to un-optimized experimental conditions. In order to reduce these exceptions and increase the confidence in the results, other transducers are being tested and the experiment is being redesigned f or various ranges of frequency and different sizes of rocks.

6.7.3 Resonance and Impedance Analysis

The source of acoustic energy of the USDC actuator is the stack of piezoelectric ceramics (PZT) that are pre-stressed between the horn and the backing. In addition to

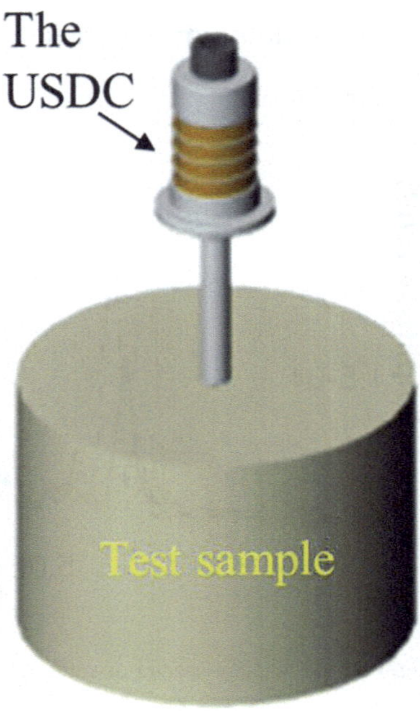

Fig. 6.72 Resonance and impedance test using USDC actuator

the converse piezoelectric effect, piezoelectric materials also have a direct piezoelectric effect, which allows the material to function as sensors. When a USDC actuator touches different types of rocks, the resonance frequency and the electric impedance of the actuator changes due to the change of the mechanical boundary condition on the horn tip. As shown in Fig. 6.72, the horn/PZT assembly was used as a resonator with the tip of the horn attached to a rock sample. The resonance frequency and the impedance of the assembly were monitored by an impedance analyzer.

The impedance analysis was carried out using the HP 4194A Impedance Analyzer, where the real and imaginary parts of the admittance of the USDC were recorded (Fig. 6.73). The reference value of the resonance frequency is 22.77 kHz with a Q of 4484 and it is measured when the USDC is not loaded. Once the USDC

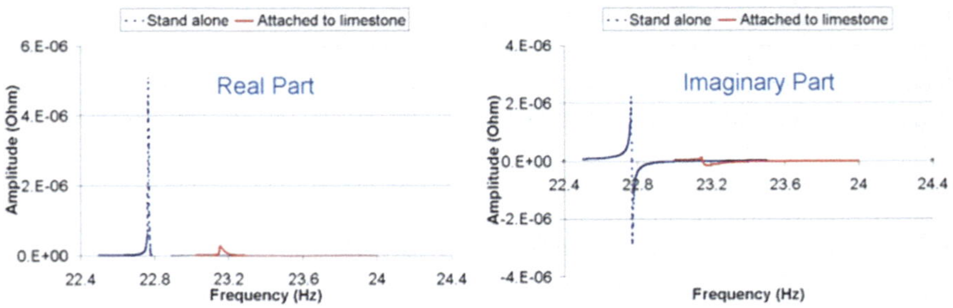

Fig. 6.73 Resonance and impedance analysis for USDC actuator when standing alone and attached to a limestone

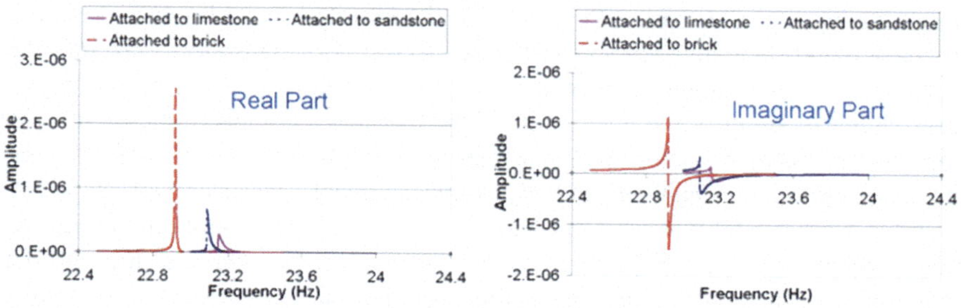

Fig. 6.74 Comparison of resonance and impedance analysis between limestone, sandstone, and brick

was attached to a limestone block, the resonance frequency was shifted to 23.15 kHz and Q value s shifted to 368. Both the resonance frequency and the impedance of the USDC actuator changed due to the loading by the limestone. Using this approach, the resonance and impedance tests were also conducted for sandstone and brick and the results are shown in Fig. 6.74. It is observed that the resonance frequencies are 23.15, 22.92, and 23.09 kHz while attached to limestone, brick, and sandstone respectively. The Q values are 368, 2103, and 589, respectively.

These results indicate that the resonance frequency and the impedance of the USDC actuator are sensitive to the material of the object to which it is attached. However, the measured data was found to be highly sensitive to the coupling condition between the USDC actuator, and the sample. The characteristics of rocks that include irregular shapes, surface roughness, etc. make it extremely difficult to keep the coupling conditions the same for each test. In order to produce a predictable and controllable load the interface between the rock and the horn tip needs to be standardized. In the initial tests, we found large fluctuations of the results depending upon the quality of the rock/horn coupling.

6.8 Summary and Conclusions

Piezoelectric actuated mechanisms have been designed and tested for drilling, coring, abrading, penetrating, sample handling, crushing and powdering. The USDC drilling mechanism, which uses a free-mass, has been applied in many mechanisms where the small high frequency generated displacement is converted to lower frequencies with larger stress at the bit/rock interface. Through extensive research and innovation, the USDC mechanism has been modified to allow many designs and configurations that enable cutting into geological formations. Various such tools have been used for sample acquisition or sample preparation.

To allow effective design and construction of the various modifications of the USDC an analytical model was developed to predict its behavior towards the goal of optimizing its performance. Physical models were developed for each section of the device and their interactions. The piezoelectric horn transducer was analyzed using finite element models and with some simplifications converted to an equivalent circuit to simplify the interaction of the free-mass and electronic driver. The horn tip free-mass interaction was analyzed by a simple collision theory to explore the basic drive mechanism and by finite element approach for accuracy. A spring-mass model was developed to obtain time efficient solutions. Finite element models were also applied to the free-mass/drill bit and the drill bit/rock interactions. The program simulating the operation of the device was integrated from the models of the main parts and the interactions. The developed models allowed for the investigation of the various interactions of the USDC. It was shown that, by using the free-mass, the continuous high frequency vibration of the horn could be converted to low frequency high mechanical power impacts. These impacts created large enough strain in the rock to enable efficient drilling. Using an analytical/computation based simulation of the operation of the USDC; the characteristics of the USDC performance were investigated. The USDC simulation has been used to guide the design of a variety of prototype devices. The drilling rates calculated from the model were found to be in general agreement with drilling rate data measured on a variety of rock samples. Initial studies on using the acoustic properties of the USDC to measure rock properties were also discussed.

The in-depth understanding of the actuation mechanism earned from the studies described in this article has resulted in the design of a large suite of novel devices based on the USDC technology that can drill, core, abrade, penetrate, powder or crush rock surfaces using low pre-load, power and mass.

Acknowledgements Research reported in this Chapter was conducted at the Jet Propulsion Laboratory (JPL), California Institute of Technology, Pasadena, CA, under a contract with National Aeronautics Space Administration (NASA). Some of the research is funded by the NASA's Maturation of Instruments for Solar System Exploration (MatISSE) program as well as many other tasks. Reference made in this Chapter to any specific commercial product, process, or service by trade name, trademark, manufacturer, or otherwise, does not constitute or imply its endorsement by the United States Government or the Jet Propulsion laboratory, California Institute of Technology.

The authors would like to thank Kris Zacny, Honeybee Robotics, for his contributions to the development of the Auto-Gopher system. For the work on PARoD, the authors would like to thank Placidus Technologies, LLC (Contact: Joshua Leavitt), for funding its development. For the work on the high temperature drilling, the author would like to thank Tom Shrout, Penn State University, for his contributions. Also, the authors would like to express their thanks to the current and former interns of the group who helped in producing and testing the various developed piezoelectric actuated mechanisms that are reported in this Chapter. The interns include David Freeman, William Cervantes, Jennifer Hasenoehrl, Patrick Ostlund, Melissa Troutt, Nicholas Morris, Louisa Taylor Avellar, Scott Widholm, Aaron Ressa, Chris Donnelly, Lukas Domm, Robert Rovai, Beck Chen, Daniel Geiyer, James Scott, Phillip Allen, Kate Boudreau, Andy Hedge, Steve Kassab, Randy Stark, James Wincentsen, Lauren Wessel, and Tao Stettler.

In addition, the authors would like to express their appreciation of Ramesh B. Malla, University of Connecticut, Storrs, CT, and Patrick Harkness, University of Glasgow, Scotland, UK, for their thorough review of the chapter and the great comments and suggestions.

References

Aldrich, J., Sherrit, S., Badescu, M., Bar-Cohen, Y., Bao, X., Chang, Z.: Controller using extremum-seeking to drive heavily perturbed electroactive actuators at resonance. In: New Technology Report (NTR), Docket No. 43519, JPL, Pasadena, CA (2006a)

Aldrich, J.B., Sherrit, S., Bar-Cohen, Y., Bao, X., Badescu, M., Chang. Z.: Extremum-seeking control of Ultra-sonic/Sonic Driller/Corer (USDC) driven at high-power. In: Proceedings of the SPIE Modeling, Signal Processing and Control Conference, vol. 6166, SPIE, Bellingham, WA (2006b)

Aldrich, J.B., Badescu, M., Jones, C., Bar-Cohen, Y.: Resonance seeking control in an event-triggered discrete-time domain. In: Proceedings of the 2008 American Control Conference (ACC), held in Seattle, Washington (2008a)

Aldrich, J., Bar-Cohen, Y., Sherrit, S., Badescu, M., Bao, X., Scott, J.: Percussive augmenter of rotary drills (PARoD) for operating as a Rotary-Hammer Drill. In: New Technology Report (NTR), Docket No. 46550, JPL document, Pasadena, CA (2008b)

Badescu, M., Sherrit, S., Olorunsola, A., Aldrich, J., Bao, X., Bar-Cohen, Y., Chang, Z., Doran, P.T., Fritsen, C.H., Kenig, F., McKay, C.P., Murray, A., Du, S., Peterson, T., Song, T.: Ultrasonic/sonic Gopher for subsurface ice and brine sampling: analysis and fabrication challenges, and testing results. In: Proceedings of the SPIE Smart Structures and Materials Symposium, Paper #6171–07, SPIE, Bellingham WA (2006a)

Badescu, M., Sherrit, S., Bar-Cohen, Y., Bao, X., Kassab, S.: Ultrasonic/Sonic Rotary-Hammer Drill (USRoHD). In: U.S. Patent No. 7,740,088, 22 June 2010. NASA New Technology Report (NTR) No. 44765, JPL document, Pasadena, CA (2006b)

Badescu, M., Kassab, S., Sherrit, S., Aldrich, J., Bao, X., Bar-Cohen, Y., Chang, Z.: Ultrasonic/sonic driller/corer as a hammer-rotary drill. In: Proceedings of the 14th International Symposium on: Smart Structures and Materials & Nondestructive Evaluation and Health Monitoring, International Society for Optics and Photonics, Paper No. 65290S, SPIE, Bellingham WA (2007)

Badescu, M., Bickler, D.B., Sherrit, S., Bar-Cohen, Y., Bao, X., Hudson, N.H.: Rolling tooth core break-off and retention mechanism. In: NTR Docket No. 47354. NASA Tech Briefs, vol. 35, no. 6, 2011, pp. 50–51. U.S. Patent No. 8,657,027, 25 Feb 2014, 19 Oct 2009a

Badescu, M., Sherrit, S., Bar-Cohen, Y., Bao, X., Backes, P.G.: Scoring dawg—core break-off and retention mechanism. In: NTR Docket No. 47355, JPL document, Pasadena, CA. Also, NASA Tech Briefs, vol. 35, no. 6, 2011, pp. 48–49, 19 Oct 2009b

Badescu, M., Sherrit, S., Bar-Cohen, Y., Bao, X., Lindemann, R.A.: Praying mantis—bending core break-off and retention mechanism. In: NTR Docket No. 47356, JPL document, Pasadena, CA. Also, vol. 35, No. 6, 2011, NASA Tech Briefs, Tech Briefs Media Group, New York, pp. 49–50, 19 Oct 2009c

Badescu, M., et al.: Ultrasonic/sonic Gopher for subsurface ice and brine sampling: analysis and fabrication challenges, and testing results. In: Mechanical and Aerospace Systems 2011 Sensors and Smart Structures Technologies for Civil, SPIE, Bellingham WA (2011a)

Badescu, M., Sherrit, S., Bao, X., Bar-Cohen, Y., Chen, B.: Auto-gopher—a wire-line rotary-hammer ultrasonic drill. In: Proceedings of SPIE—the International Society for Optical Engineering, Paper No. 7981, SPIE, Bellingham WA (2011b)

Badescu, M., Bar-Cohen, Y., Sherrit, S., Bao, X., Chang, Z., Donnelly, C., Aldrich, J.: Percussive augmenter of rotary drills (PARoD). In: Proceedings of SPIE—the International Society for Optical Engineering. Paper No. 8345–121, SPIE, Bellingham WA (2012)

Bao, X., Bar-Cohen, Y., Chang, Z., Dolgin, B.P., Sherrit, S., Pal, D.S., Du, S., Peterson, T.: Modeling and computer simulation of ultrasonic/sonic driller/corer (USDC). IEEE Trans Ultrason Ferroelectr Freq Control (UFFC) **50**(9), 1147–1160 (2003)

Bao, X., Bar-Cohen, Y., Chang, Z., Sherrit, S., Stark, R.: Ultrasonic/sonic impacting penetrator (USIP). In: NASA NTR No. 41666, JPL document, Pasadena, CA (2004)

Bao, X., Scott, J., Bar-Cohen, Y., Sherrit, S., Widholm, S., Badescu, M., Shrout, T., Jones, B.: Ultrasonic/sonic drill for high temperature application. In: SPIE Sensors and Smart Structures Technologies for Civil, Mechanical, and Aerospace Systems, Smart Structures and Materials Symposium, Paper No. 7647–116, SPIE, Bellingham WA (2010)

Bao, X., Sherrit, S., Badescu, M., Bar-Cohen, Y., Askins, S., Ostlund, P.: Free-mass and interface configurations of hammering mechanisms. US Patent No. 8,960,325 (2015)

Bar-Cohen, Y., Sherrit, S., Dolgin, B., Peterson, T., Pal, D., Kroh, J.: Smart-ultrasonic/sonic driller/corer. U.S. Patent No. 6,863,136, March 8, 2005, NASA NTR No. 20856 (1999)

Bar-Cohen Y., Nazarian, S.: Nonintrusive geophysical probing end effector (NGPE) for Mars exploration. Mars Drilling Workshop, held at JPL, Pasadena, CA, January 28 (1997)

Bar-Cohen, Y., Sherrit, S., Dolgin, B.P., Bridges, N., Bao, X., Chang, Z., Yen, A., Saunders, R.S., Pal, D., Kroh, J., Peterson, T.: Ultrasonic/sonic driller/corer (USDC) as a sampler for planetary exploration. In: Proceedings of the IEEE Aerospace Conference on the Topic of "Missions, Systems, and Instruments for in Situ Sensing," (Session 2.05) (Mar 2001)

Bar-Cohen, Y., Randolph, J., Ritz, C., Cook, G., Bao, X., Sherrit, S.: Sample preparation, acquisition, handling and delivery (SPAHD) system using the ultrasonic/sonic driller/corer (USDC) with interchangeable bits. In: NTR Docket No. 30640, 1 May 2002

Bar-Cohen, Y., Sherrit, S.: Self-mountable and extractable ultrasonic/sonic anchor (U/S-Anchor). In: NASA NTR No. 40827 (2003a)

Bar-Cohen, Y., Sherrit, S.: Thermocouple-on-the-bit a real time sensor of the hardness of drilled objects. In: NASA NTR No. 40132, JPL document, Pasadena, CA (2003b)

Bar-Cohen, Y., Sherrit, S., Chang, Z., Wessel, L., Bao, X., Doran, P.T., Fritsen, C.H., Kenig, F., McKay, C.P., Murray, A., Peterson, T.: Subsurface ice and brine sampling using an ultrasonic/sonic gopher for life detection and characterization in the mcmurdo dry valleys. In: Industrial and Commercial Applications of Smart Structures Technologies Conference, SPIE Smart Structures and Materials Symposium, Paper #5388-32, JPL document, Pasadena, CA (15–18 Mar 2004)

Bar-Cohen, Y., Sherrit, S., Dolgin, B., Peterson, T., Pal, D., Kroh, J.: Smart-ultrasonic/sonic driller/corer. U.S. Patent No. 6,863,136 (2005a)

Bar-Cohen, Y., Sherrit, S., Dolgin, B., Bao, X., Askins, S.: Ultrasonic/sonic mechanism of deep drilling (USMOD). U.S. Patent No. 6,968,910 (2005b)

Bar-Cohen, Y., Sherrit, S., Dolgin, B., Bao, X., Askins, S.: Ultrasonic/sonic mechanism of deep drilling (USMOD). U.S. Patent No. 6,968,910, 29 Nov 2005c

Bar-Cohen, Y., Sherrit, S.: Self-mountable and extractable ultrasonic/sonic anchor. U.S. Patent No. 7,156,189, 2 Jan 2007

Bar-Cohen, Y., Sherrit, S., Bao, X., Badescu, M., Aldrich, J., Chang, Z.: Subsurface sampler and sensors platform using the ultrasonic/sonic driller/corer (USDC). In: Paper #6529-18, SPIE Smart Structures and Materials Symposium, San Diego, CA, JPL document, Pasadena, CA, 19–22 Mar 2007a

Bar-Cohen, Y., Badescu, M., Iskenderian, T., Sherrit, S., Bao, X., Lindemann, R.: Penetrator with foldable bit for packing in a compact configuration and unfolding for deep drilling. In: NTR Docket No. 45289, JPL document, Pasadena, CA. Also, NASA Tech Briefs, vol. 34, no. 5, 2010, pp. 50–51, 16 July 2007b

Bar-Cohen, Y., Badescu, M., Sherrit, S.: Rapid Rotary-Percussive Auto-Gopher for deep subsurface penetration and sampling. In: NASA NTR No. 45949, JPL document, Pasadena, CA (2008)

Bar-Cohen, Y., Zacny, K. (eds.): Drilling in extreme environments—penetration and sampling on earth and other planets. Wiley—VCH, Hoboken, NJ, ISBN-10: 3527408525, ISBN-13: 9783527408528, 827 p (2009)

Bar-Cohen, Y., Badescu, M., Sherrit, S.: Acquisition and retaining granular samples via rotating coring bit. In: NASA NTR No. 47606, JPL document, Pasadena, CA (2010a)

Bar-Cohen, Y., Bao, X., Scott, J., Sherrit, S., Widholm, S., Badescu, M., Shrout, T., Jones, B.: Drilling at high temperatures using ultrasonic/sonic actuated mechanism. In: ASCE's Earth and Space 2010 Conference, Honolulu, HI, 14–17 Mar 2010b

Bar-Cohen, Y., Badescu, M., Bao, X., Sherrit, S., Shrout, T., Zhang S., Jones, B.: High temperature piezoelectric actuated sampler for operation on Venus. In: Final Report for the PIDDP Task, NASA WBS No. 811073.02.06.55.13, Period of Performance from January 2007–January 2011, Report No. D-68419 (Jan 27, 2011)

Bar-Cohen, Y., Badescu, M., Sherrit, S., Zacny, K., Paulsen, G., Beegle, L., Bao, X.: Deep drilling and sampling via the wireline auto-gopher driven by piezoelectric percussive actuator and EM rotary motor. In: Proceedings SPIE Smart Structures and Materials/NDE Symposium, SPIE, Bellingham WA (2012)

Bar-Cohen, Y., Badescu, M., Lee, H.J., Sherrit, S., Zacny, K., Paulsen, G.L, Beegle, L., Bao, X.: Auto-Gopher—a wireline deep sampler driven by piezoelectric percussive actuator and EM rotary motor. In: Proceedings of the ASCE Earth and Space 2014 Conference, Symposium 2: Exploration and Utilization of Extraterrestrial Bodies, ASCE, Reston, VA (2014a)

Bar-Cohen, Y., Sherrit, S., Herz, J.L.: Ultrasonic/sonic jackhammer (USJ). US Patent No. 8,910,727, 16 Dec 2014b

Bar-Cohen, Y., Bao, X., Badescu, M., Sherrit, S., Zacny, K., Kumar, N., Shrout, T.R., Zhang, S.: Chapter 14: High-temperature drilling mechanisms. In: Bar-Cohen Y. (ed.), High Temperature Materials and Mechanisms, ISBN 13: 9781466566453, ISBN 10: 1466566450, CRC Press, Taylor & Francis Group, Boca Raton, Florida (2014c) pp. 427–466

Blake, D.F., Sarrazin, P., Chipera, S.J., Bish, D.L., Vaniman, D.T., Bar-Cohen, Y., Sherrit, S., Collins, S., Boyer, B., Bryson, C., King, J.: Definitive Mineralogical Analysis of Martian Rocks and Soil Using the CHEMIN XRD/XRF Instrument and the USDC Sampler. In: Proceedings of the Sixth International Conference on Mars (2003)

Chang, Z., Sherrit, S., Bao, X., Bar-Cohen, Y.: In-situ rock probing using the ultrasonic/sonic driller/corer (USDC). In: Proceedings of the SPIE Smart Structures and Materials Symposium, Paper 5056–73, SPIE, Bellingham WA, 3–6 Mar 2003

Chang, Z., Sherrit, S., Bao, X., Bar-Cohen, Y.: Design and analysis of ultrasonic horn for USDC (Ultrasonic/Sonic Driller/Corer). In: Proceedings of the SPI Smart Structures Conference, vol. 5388–34, SPIE, Bellingham WA, Mar 2004

Chang, Z., Sherrit, S., Badescu, M., Bao, X., Bar-Cohen, Y.: Design and analysis of ultrasonic actuator in consideration of length reduction for a USDC (ultrasonic/sonic driller/corer). Smart Structures and Materials, pp. 72–79. International Society for Optics and Photonics, SPIE, Bellingham WA (2005)

Chipera, S.J., Bish, D.L., Vaniman, D.T., Sherrit, S., Bar-Cohen, Y., Blake, D.F.: Use of an ultrasonic/sonic driller/corer to obtain sample powder for CHEMIN—a combined XRD/XRF instrument. In: Proceedings of the 34th Lunar and Planetary Science Conference, League City, TX, 17–21 Mar 2003

Dolgin, B., Sherrit, S., Bar-Cohen, Y., Rainen, R., Askins, S., Lewis, D., Carson, J., Dawson, S., Bao, X., Chang, Z., Peterson, T., Bickler, D.: Ultrasonic rock Abrasion tool (URAT). NTR Docket No. 30403, JPL document, Pasadena, CA. Also, NASA Tech Briefs, New York, vol. 30, no. 7, 2006, pp. 58, 12 Oct 2001

Doran, P.T., Bar-Cohen, Y., Sherrit, S., Fritsen, C., Kenig, F., McKay, C.P., Murray, A.: Life detection and characterization of subsurface ice and brine in the Mcmurdo dry valleys using an Ultrasonic Gopher: a NASA ASTEP Project. In: Third International Conference on Mars Polar Science and Exploration, Alberta, Canada, October 13–17, 2003

Fergus, J.W., Hoffmann, W.P.: Chapter 3: Refractory metals, ceramics, and composites for high-temperature structural and functional applications. In: Bar-Cohen Y. (ed.), High temperature materials and mechanisms, ISBN 13: 9781466566453, ISBN 10: 1466566450, CRC Press, Taylor & Francis Group, Boca Raton, Florida (2014) pp. 39–68

Gershman, R., Wallace, R.A.: Technology needs of future planetary missions. Acta Astronautica **45**(4–9), 329–335 (1999)

Graff, K.F.: Chapter 6: Waves in semi-infinite media. In: Wave Motion in Elastic Solids. Dover Publications, New York (1991)

Harkness, P., Lucas, M., Cardoni, A.: Maximization of the effective impulse delivered by a high-frequency/low-frequency planetary drill tool. IEEE Trans. Ultrason. Ferroelectr. Freq. Control **58**(11), 2387–2396 (2011). https://doi.org/10.1109/tuffc.2011.2096

NRC: Vision and Voyages for Planetary Science in the Decade 2013–2022. Committee on the Planetary Science Decadal Survey, Space Studies Board Division on Engineering and Physical Sciences, National Research Council (NRC), The National Academies Press, Washington, D.C. (2013) http://solarsystem.nasa.gov/docs/Vision_and_Voyages-FINAL.pdf, (retrieved on 9 Jan 2017)

Paterson, W.S.B.: Secondary and tertiary creep of glacier ice as measured by borehole closure rates. Rev. Geophys. **15**(1), 47–55 (1977)

Schultz, R.A.: Brittle strength of basaltic rock masses with applications to Venus. J. Geophys. Res. **98**(E6), 10,883 (1993). http://doi.org/10.1029/93JE00691

Sherrit, S., Dolgin, B.P., Bar-Cohen, Y., Pal, D., Kroh, J., Peterson, T.: Modeling of horns for sonic/ultrasonic applications. In: Proceedings of the IEEE International Ultrasonics Symposium, pp. 647–651 (Oct 1999)

Sherrit, S., Bao, X., Chang, Z., Dolgin, B., Bar-Cohen, Y., Pal, D., Kroh, J., Peterson, T.: Modeling of the ultrasonic/sonic driller/corer: USDC. 2000 IEEE Int. Ultrason. Symp. Proc. **1**, 691–694 (2000)

Sherrit, S., Askins, S.A., Gradziel, M., Dolgin, B.P., Bar-Cohen, Y., Bao, X., Cheng, Z.: Novel ultrasonic horns for power ultrasonics. In: NASA NTR No. 30489, JPL document, Pasadena, CA. Also, NASA Tech Briefs, vol. 27, no. 4, 2003, pp. 54–55 (2001a)

Sherrit, S., Askins, S.A., Gradziel, M., Dolgin, B.P., Bar-Cohen, Y., Bao, X., Cheng, Z.: Novel ultrasonic horns for power ultrasonics. NASA NTR No. 30489, submitted on December 6, 2001, JPL document, Pasadena, CA. Also, NASA Tech Briefs, vol. 27, no. 4, 2003, pp. 54–55 (2001b)

Sherrit, S., Bar-Cohen, Y., Dolgin, B., Bao, X., Chang, Z.: Ultrasonic crusher for crushing, milling, and powdering. In: NASA NTR No. 30682, JPL document, Pasadena, CA (2002)

Sherrit, S., Bar-Cohen, Y., Bao, X., Chang, Z., Blake, D., Bryson, C.: Ultrasonic/sonic rock powdering sampler and delivery tool. In: NASA NTR No. 40564, JPL document, Pasadena, CA (2003)

Sherrit, S., Badescu, M., Bao, X., Bar-Cohen, Y., Chang, Z.: Novel horn designs for power ultrasonics. In: Proceedings of the IEEE International Ultrasonics Symposium, Montreal, Canada, 24–27 Aug 2004a

Sherrit, S., Bao, X., Bar-Cohen, Y., Chang, Z.: Resonance analysis of high temperature piezoelectric materials for actuation and sensing. In: SPIE Smart Structures and Materials Symposium, Paper #5388-34, SPIE, Bellingham WA, 15–18 Mar 2004b

Sherrit, S., Bar-Cohen, Y., Chang, Z., Bao, X.: Dog bone piezoelectric actuation horns. In: NTR Docket No. 40949, JPL document, Pasadena, CA. Also, NASA Tech Briefs, vol. 31, no. 12, 2007, pp. 48, 50, 25 Feb 2004c

Sherrit, S.: Smart material/actuator needs in extreme environments in space. In: Proceedings of the SPIE Smart Structures Conference, SPIE vol. 5761–48, SPIE, Bellingham WA (Mar 2005a)

Sherrit, S., Badescu, M., Bar-Cohen, Y., Chang, Z., Bao, X.: Portable rapid and quiet drill (PRAQD). U.S. Patent No. 7,824,247, 4 Nov 2010. NASA NTR No. 42131, JPL document, Pasadena, CA (2005b)

Sherrit, S., Bar-Cohen, Y., Badescu, M., Bao, X., Chang, Z., Jones, C., Aldrich, J.: Compact non-pneumatic powder sampler (NPPS). In: NASA NTR No. 43614, JPL document, Pasadena, CA (2006)

Sherrit, S., Badescu, M., Bar-Cohen, Y.: Miniature low-mass drill actuated by Flextensional Piezo-Stack (DAFPiS). In: NASA NTR No. 45857, JPL document, Pasadena, CA (2008)

Sherrit, S., Bao, X., Badescu, M., Bar-Cohen, Y.: Single piezo-actuator rotary-hammering (SPaRH) drill. In: NASA NTR No. 47216, JPL document, Pasadena, CA (2009)

Sherrit, S., Bao, X., Badescu, M., Bar-Cohen, Y., Ostlund, P., Allen, P., Geiyer, D.: Planar rotary piezoelectric motor using ultrasonic horns. In: A Provisional Patent was filed on July 7, 2011, NASA NTR No. 47813, JPL document, Pasadena, CA (2010)

Sherrit, S., Domm, L., Bao, X., Bar-Cohen, Y., Chang, Z., Badescu, M.: Single piezo-actuator rotary-hammering (SPaRH) drill. SPIE Smart Structures and Materials + Nondestructive Evaluation and Health Monitoring, pp. 83452B–83452B. International Society for Optics and Photonics, SPIE, Bellingham WA (2012)

Sherrit, S., Lee, H.J., Zhang, S., Shrout, T.R.: Chapter 11: High temperature electromechanical actuators. In: High Temperature Materials and Mechanisms, ISBN 13: 9781466566453, ISBN 10: 1466566450. CRC Press, Taylor & Francis Group, Boca Raton, Florida (2014)

Sherrit, S., Bao, X., Badescu, M., Bar-Cohen, Y., Allen, P.: Monolithic flexure pre-stressed ultrasonic horns. US Patent No. 9,352,358 B2 (2016)

Smialek, J.L., Jacobson, N.S.: Chapter 5: Oxidation of high-temperature aerospace materials. In: Bar-Cohen Y. (ed.), High Temperature Materials and Mechanisms, ISBN 13: 9781466566453, ISBN 10: 1466566450, CRC Press, Taylor & Francis Group, Boca Raton, Florida (2014) pp. 95–162

Smith, M., Cardell, G., Kowalczyk, R., Hecht, M.H.: The Chronos thermal drill and sample handling technology. In: The 4th International Conference on Mars Polar Science and Exploration, Davos, Switzerland, Abstract 8095 (2006)

Vila, L., Malla, R.B.: Analytical model of the contact interaction between the components of a special percussive mechanism for planetary exploration. Acta Astronaut. J. **118**, 158–167. (Jan–Feb 2016a). (http://dx.doi.org/10.1016/j.actaastro.2015.09.016; published Online: 11 Nov. 2015)

Vila, L., Malla, R.B.: Dynamic impact force on a special drilling mechanism for planetary exploration. ASCE J. Aerosp. Eng. **29**(4), 10 pages. (2016b). (http://dx.doi.org/10.1061/(ASCE)AS.1943-5525.0000601; published online: 24 Feb. 2016)

Vila, L., Malla, R.B.: Elasto-plastic contact force due to repeated longitudinal impacts on a novel percussive drill. J. Eng. Mech. **142**(10), 10 pages. (2016c). (http://dx.doi.org/10.1061/(ASCE)EM.1943-7889.0001132; Published online: 08 July 2016)

Zacny, K., Bar-Cohen, Y., Brennan, M., Briggs, G., Cooper, G., Davis, K., Dolgin, B., Glaser, D., Glass, B., Gorevan, S., Guerrero, J., McKay, C., Paulsen, G., Stanley, S., Stoker, C.: Drilling Systems for Extraterrestrial Subsurface Exploration. Astrobiol. J. (2008). https://doi.org/10.1089/ast.2007.0179

Zacny K., Bar-Cohen, Y., Davis, K., Coste, P., Paulsen, G., Sherrit, S., George, J., Derkowski, B., Gorevan, S., Boucher, D., Guerrero, J., Kubota, T., Thomson, B.J., Stanley, S., Thomas, P., Lan, N., McKay, C., Onstot, T.C., Stoker, C., Glass, B., Wakabayashi, S., Whyte, L., Visentin, G., Re, E., Richter, L., Badescu, M., Bao, X., Fincher, R., Hoshino, T., Magnani, P., Menon, C.: Extraterrestrial drilling and excavation. In: Bar-Cohen, Y., Zacny, K. (eds.) Drilling in Extreme Environments—Penetration and Sampling on Earth and Other Planets, ISBN-10: 3527408525, ISBN-13: 9783527408528. Wiley—VCH, Hoboken, NJ (2009)

Zacny K., Paulsen, G., Mellerowicz, B., Bar-Cohen, Y., Beegle, L., Sherrit, S., Badescu, M., Corsetti, F., Craft, J., Ibarra, Y., Bao, X.: Wireline deep drill for exploration of Mars, Europa, and Enceladus. In: Proceedings of the 2013 IEEE Aerospace Conference, Big Sky, Montana (2013)

Chapter 7
Ultrasonically-Assisted Penetration of Granular and Cemented Materials

David Firstbrook, Patrick Harkness, Xuan Li, Ryan Timoney and Kevin Worrall

7.1 Introduction

Granular material can often be penetrated by the application of high-frequency vibrations. This effect may be seen in loosely packed granular material, in permafrost where the discrete grains exist in an icy matrix, and even where those grains have been compacted and cemented to form a sedimentary rock. For space applications, the vibrations may be reasonably generated by a Langevin transducer and their energy delivered to the target material by a number of different mechanisms, depending on the nature of the target and the depth or bore diameter of the desired drill campaign. The application of such vibrations is generally associated with reductions in weight-on-bit and power requirements when compared to more traditional techniques.

Loosely-packed granular materials, such as sands and soils, can be tackled by direct penetration. This relatively simple technique uses ultrasonic excitation of the penetrator to fluidise the material around the tip of the probe and reduce the resistance accordingly. This can be so effective that it has even been suggested that total power consumption may be reduced as a consequence.

D. Firstbrook (✉) · P. Harkness · X. Li · R. Timoney · K. Worrall
School of Engineering, University of Glasgow, Glasgow, UK
e-mail: David.Firstbrook@glasgow.ac.uk

P. Harkness
e-mail: Patrick.Harkness@glasgow.ac.uk

X. Li
e-mail: lixuangarnett@gmail.com

R. Timoney
e-mail: r.timoney.1@research.gla.ac.uk

K. Worrall
e-mail: kevin.worrall@glasgow.ac.uk

V. Badescu and K. Zacny (eds.), *Outer Solar System*,
https://doi.org/10.1007/978-3-319-73845-1_7

More compacted materials, including rocks, may be tackled by using ultrasonic vibration to excite the longitudinal mode of a more traditional rotating drillbit, in a technique called ultrasonically-assisted drilling (UAD). This approach can increase the rate of progress by an order of magnitude under certain conditions.

However, both these direct approaches require that the cutting tool and the sonotrode be part of the same tool, which ultimately limits their flexibility. An alternative technique is the ultrasonic-percussive approach (as described under the name 'USDC' elsewhere in this book), whereby the ultrasonic assembly transmits impulse to the drilling bit through an intermediate free-mass. This technique has been applied to both soils and rocks, and has the advantage that the isolated drillstring can be more generalised in form and can therefore be assembled from a number of discrete elements using relatively simple mechatronics and interfaces. This makes it possible to assemble a drilling and sampling tool that can reach depths which exceed the practical length of a resonant penetrator without the need to send the transducer itself downhole.

This chapter shall consider these three fundamental approaches ultrasonic drilling: direct penetration, ultrasonically-assisted drilling, and the ultrasonic-percussive technique.

7.2 Direct Penetration

Direct penetration of a granular material is achieved by simply forcing a longitudinally-resonant probe, without rotation, to the desired depth.

7.2.1 The Technique

The use of ultrasonics to facilitate direct penetration requires a probe specially designed to operate as a resonant structure. It has been shown that reductions in both penetration force and total power requirements can thus be achieved, with a trend towards stronger effects in lower gravity (Firstbrook et al. 2017).

The device used to demonstrate this is a titanium sonotrode, designed to resonate at 20 kHz. This is, in turn, connected to an ultrasonic transducer (L500, from Sonic Systems of Ilminster, UK) and housed in an experimental rig to be accommodated in the gondolas of the Large Diameter Centrifuge (ESTEC, Noordwijk, Netherlands). Using this rig, experiments can be conducted at a number of apparent gravitational accelerations (see Fig. 7.1) and thus possible performance in space may be examined.

The rig is instrumented such that ultrasonic amplitude and penetrator position may be externally commanded and logged, while a shaker can be engaged to reset the packing of the targeted granular material to a known and homogeneous ground state between experimental runs. When the transducer is energized at the resonant

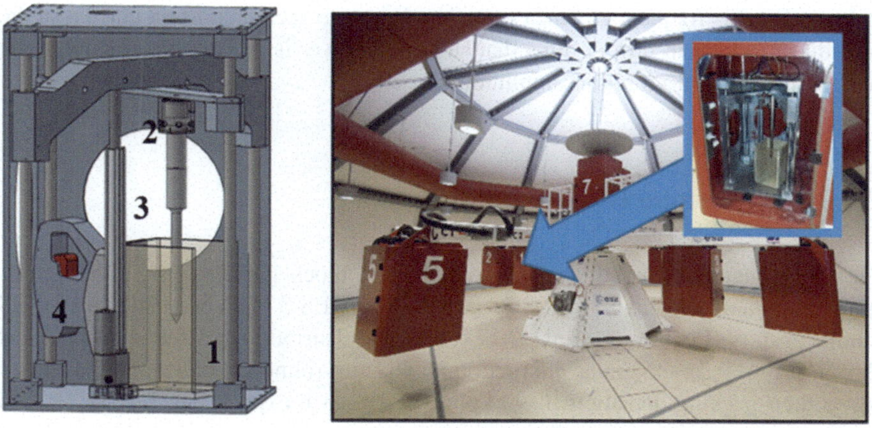

Fig. 7.1 Left: the resonant penetrator and the rig in which it is housed. Note the receptacle for the granular material (1), the transducer (2), the linear actuator which provides the penetration force (3), and the shaker used to reset the granular material before each run (4). Right: the rig is then placed in a gondola and spun in a centrifuge. The gondolas are hinged to ensure that the apparent gravity vector remains aligned with the penetrator axis as the system rotates

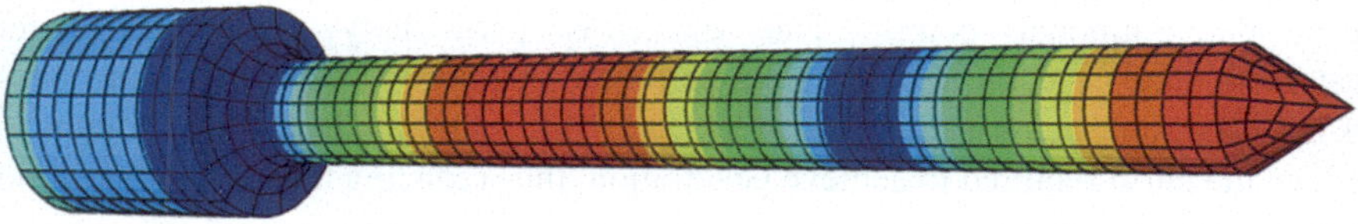

Fig. 7.2 The operating mode of the penetrator, which is 261 mm long overall. Warm colors indicate higher vibration amplitudes

frequency of the probe, it will dissipate electrical power (in W) to provide a displacement (in μm) to the base of the penetrator.

The standing wave thus generated provides a proportionally higher displacement amplitude at the tip (and also at one other antinodal region) as indicated by warm colors of Fig. 7.2, which represents a finite-element based modal analysis of the 261-mm long penetrator used in these experiments.

Due to the shape of the sonotrode, the gain has a value of 3.5 from base to tip. Given that typical excitation amplitudes are on the order of 1 μm (as set out in the results which follow) the typical amplitude at the tip is significantly below the grain size of the target material. The ultrasonic fluidization of the granular material, which is thought to be crucial to the performance of this technique, therefore takes place through relatively small motions that can be combined to facilitate global flows of material away from the probe.

During operations in the centrifuge, the resonant frequency is tracked by a control loop during each run, with small changes to the nominal 20 kHz value being made in real time to compensate for mechanical perturbations. Voltage delivered at this frequency is then modulated so as to seek a constant current in the transducer, which in turn ensures that sufficient power is delivered to maintain the desired mechanical vibration amplitude.

7.2.2 *Performance*

The experiments are conducted in lower-hazard block paving sand, although regolith simulants such as SSC-1, SSC-2, SSC-3 and ES-3 have been shown to yield similar results (Firstbrook et al. 2017). The size distribution of the sand peaks at 250 μm (Fig. 7.3), the grains are predominantly sub-rounded quartz, and the shaker resets their packing density to 1790 kg/m^3 before each experiment. This is verified by checking the fill level on the container.

7.2.2.1 General Observations Before Penetration

When the penetrator is placed on the surface of block paving sand and energized with ultrasonic vibration, the boundary layer is quasi-fluidised as grains flow away from the antinodal points. This effect is particularly pronounced with water-saturated sand, as clumped grain particles have been noted to climb the sides of the penetrator against the force of gravity. This quasi-fluidisation is thought to reduce the effort required to achieve penetration, thus reducing the force, power and energy needed to insert the probe.

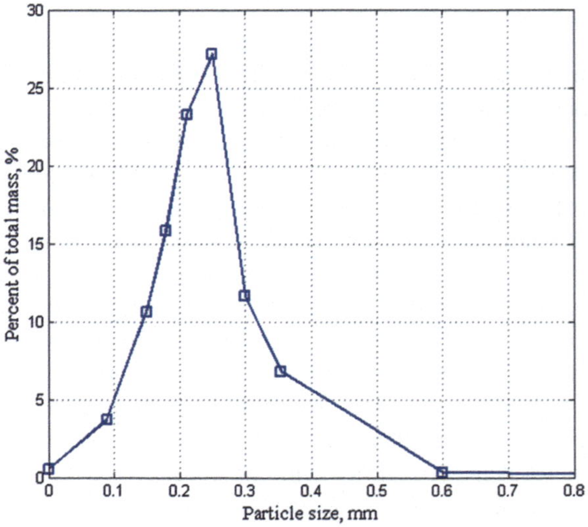

Fig. 7.3 The particle size distribution of block paving sand

Fig. 7.4 Block paving sand exhibiting an active boundary layer around the tip of the penetrator. This appears to be associated with reduced penetration forces for the penetrator as a whole

As an illustration of this, Fig. 7.4 shows the agitation of block paving sand near the tip of the activated probe. The surface of the granular material has become sunken near the tip while individual grains may be observed in flight nearby. Further along from the tip, the grains are seen to travel along the sides of the penetrator towards the nodal region.

However, the authors note that this test, and the further tests conducted in the centrifuge, are carried out under atmospheric conditions with air trapped between the grain boundaries. The effect of this air is unknown, because to our knowledge ultrasonic fluidization of granular materials has not yet been attempted in a vacuum. This (and other considerations such as inter-granular electrostatic attraction) may influence the performance of the technique in space.

7.2.2.2 Effects on Force Requirements

To quantify the effects on the penetration force requirements, a series of experiments are conducted where the force required to maintain a rate of 9 mm/s through the block paving sand is recorded. In each case, the sand is reset by the shaker and the centrifuge spun up to the desired g-loading. The penetrator is then energized to the desired ultrasonic level and driven into the target by constant-rate retraction of the linear actuator in Fig. 7.1, while the force transducer measures the force required to achieve this progress. Finally, when the apparent weight of the

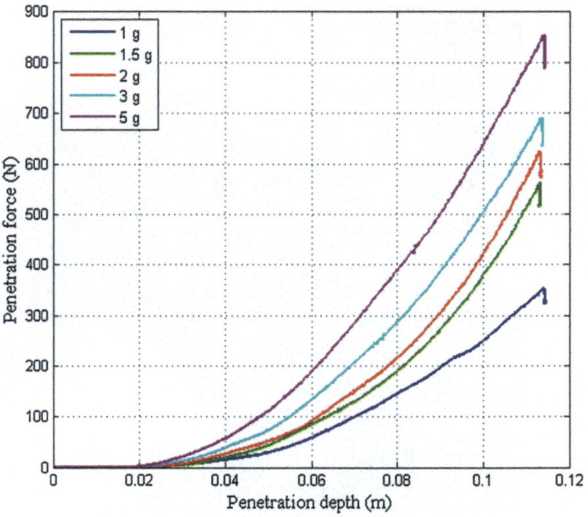

Fig. 7.5 The penetration forces encountered by the unenergized penetrator at a range of apparent gravities. No runs were conducted above 5 g with unenergized penetrators, due to the high forces anticipated

transducer and penetrator assembly is mathematically subtracted (based on their combined mass of 1.7 kg), the net penetration force can be found.

The next step is to conduct a control experiment in which the ultrasonic excitation is not applied. In these studies, the force requirements are found to increase with penetration depth and also with apparent gravity, as shown in Fig. 7.5, but two data points (at 7 and 10 g) are omitted due to the large (and indeed damaging) internal forces anticipated.

In analyzing and comparing these results, it is perhaps the peak force encountered at maximum depth that represents the most useful indicator of the overall profile. Therefore, only the peak forces encountered at approximately 11 cm depth are presented in Fig. 7.6 such that the effects of ultrasonic vibration across a wider parameter space can become clear. It is immediately apparent that peak penetration forces are significantly reduced when even low ultrasonic vibration is applied, and the greatest percentage force reductions are observed in the lower gravity experiments.

7.2.2.3 Effects on Power and Energy Requirements

Reducing the force requirements for penetrators facing granular materials is obviously attractive in low-gravity space applications, but we must also consider the power and energy required to achieve these reductions.

There are two separate energy inputs to consider: the ultrasonic power and the linear actuation power. Naturally, when we seek a greater ultrasonic amplitude to reduce the peak resistance of the target material, the transducer requires more power. On the other hand, when we calculate the peak power required by the linear actuator (by multiplying the peak resistance force by the penetration velocity) it is

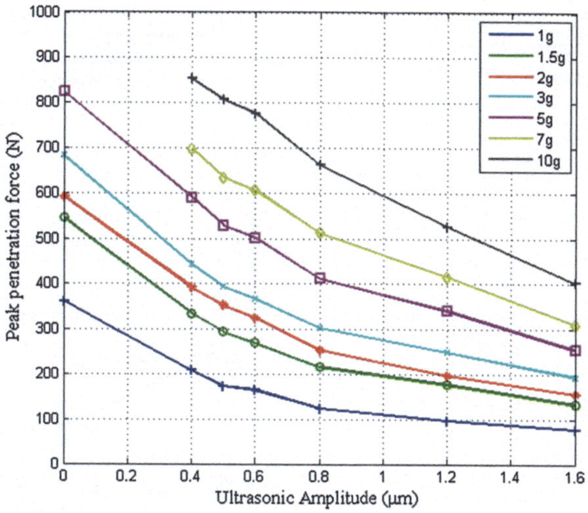

Fig. 7.6 The maximum penetration forces encountered (invariably at 11 cm depth) at a range of ultrasonic amplitudes and apparent gravities. The omitted runs are apparent on the 7 and 10 g traces

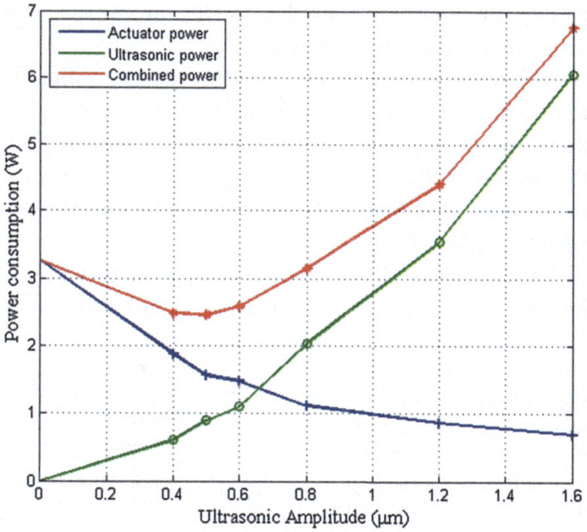

Fig. 7.7 The peak power required to achieve full-depth penetration in 1 g, at a penetration rate of 9 mm/s

immediately apparent that ultrasonically-assisted reduction of that resistance force will pay its own dividends.

Taking the 1 g experiment as an example (Fig. 7.7), we can observe that exploitation of this tradeoff can actually yield a modest reduction (around 20%) in the peak power required to achieve penetration.

The peak power consumed, however, is only part of the total energy requirement. By integrating the power curves along the entire penetration run, instead of considering the maximum force point alone, the total energy expended may be found. This result, shown in Fig. 7.8 for the 1 g experiment, indicates that—yet

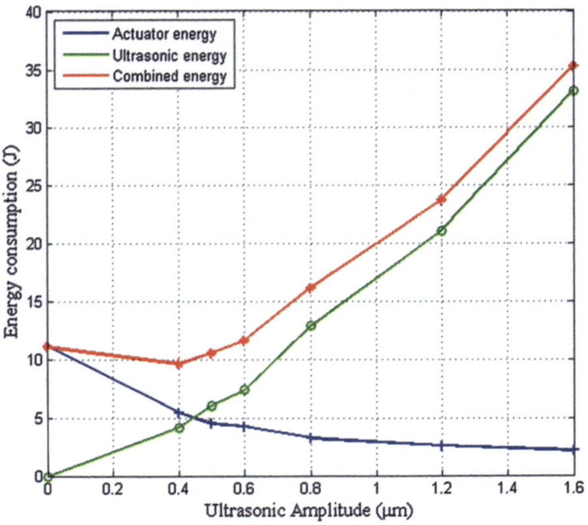

Fig. 7.8 The total energy required to achieve full-depth penetration in 1 g

again—there can be a modest saving in the total energy required at the optimum ultrasonic vibration amplitude.

When these experiments are repeated at different apparent gravities, the magnitude of the power savings and the ultrasonic vibration amplitude which yields them remains approximately constant. This can be observed by normalizing both the peak power and total energy curves with respect to a non ultrasonically-assisted run, namely the data obtained at 0 µm excitation, and comparing from one gravitational setting to the next.

When this is done, power savings of around 20% are consistently observed between 0.4 and 0.6 µm vibration amplitude, as set out in Fig. 7.9. However,

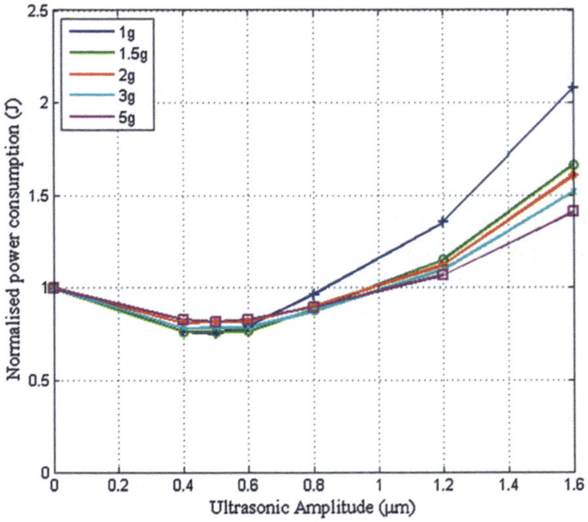

Fig. 7.9 Peak power requirements (the sum of ultrasonic and actuator demand) across a range of apparent gravities, normalized with respect to zero ultrasonic excitation at 1 g

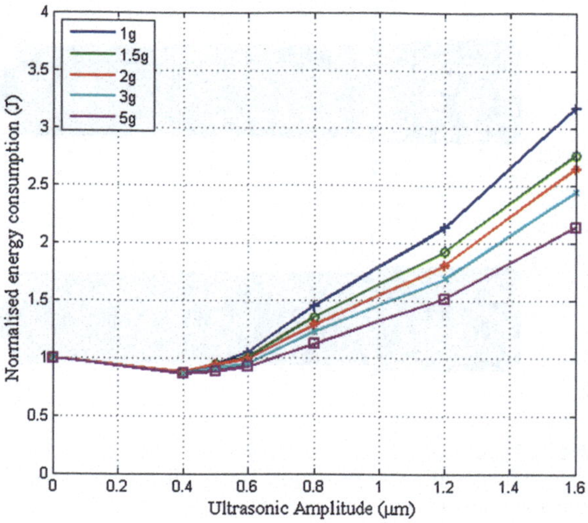

Fig. 7.10 Total energy requirements (the sum of ultrasonic and actuator demand) across a range of apparent gravities, normalized with respect to zero ultrasonic excitation at 1 g

Fig. 7.10 illustrates that the total energy savings are slightly more modest, and appear to peak at a slightly lower vibration amplitude.

7.2.3 *Summary*

Taken together, these results indicate that the ultrasonically-assisted direct penetration technique can reduce the forces needed to penetrate granular materials, particularly at lower gravity levels. In fact, ultrasonic assistance can reduce penetration forces so significantly that peak power and total energy demands may, under optimum conditions, also be reduced.

This finding has significant applicability for penetrators, but also for walking robots that may need extra purchase (or, alternatively, reduced extrication forces) and regolith handling systems.

7.3 Ultrasonically Assisted Drilling

Some drill scenarios are best addressed with a traditional rotating drillbit. However, the rate of progress can often be greatly increased if ultrasonic vibrations are superimposed upon the rotations of the drill, such that a high-frequency hammering action is applied to the very small frontal area of the drillbit itself. This technique, which is already widely-used in milling and manufacturing applications, is called ultrasonically-assisted drilling.

Fig. 7.11 Operating mode displacement maps for transducer assemblies holding two drillbits (3 mm (top) and 5 mm (bottom)). The drillbits are screwed into the sonotrode and the operating frequency is 20 kHz

7.3.1 The Technique

To test this technique, the 20 kHz ultrasonic transducer previously described is mounted so that it can rotate whilst power is supplied through a slip ring. The sonotrode, meanwhile, is shaped to accept either a 3 or 5 mm diameter drillbit, while the whole assembly is mounted such that it can be advanced towards target materials. The mode shape is shown in Fig. 7.11, where warm colours represent the greatest displacement, and the rig is shown in Fig. 7.12.

7.3.2 Performance

The tests are conducted in Locharbriggs sandstone, which has a compressive strength of around 50 MPa, (BRE 2000) and in Blaxter sandstone, which has a compressive strength of around 40 MPa (BRE 1997).

7.3.2.1 Observations

The application of ultrasonics to the drills results in an immediate increase in the rate of progress, with debris violently ejected from the drill site, as shown in Fig. 7.13.

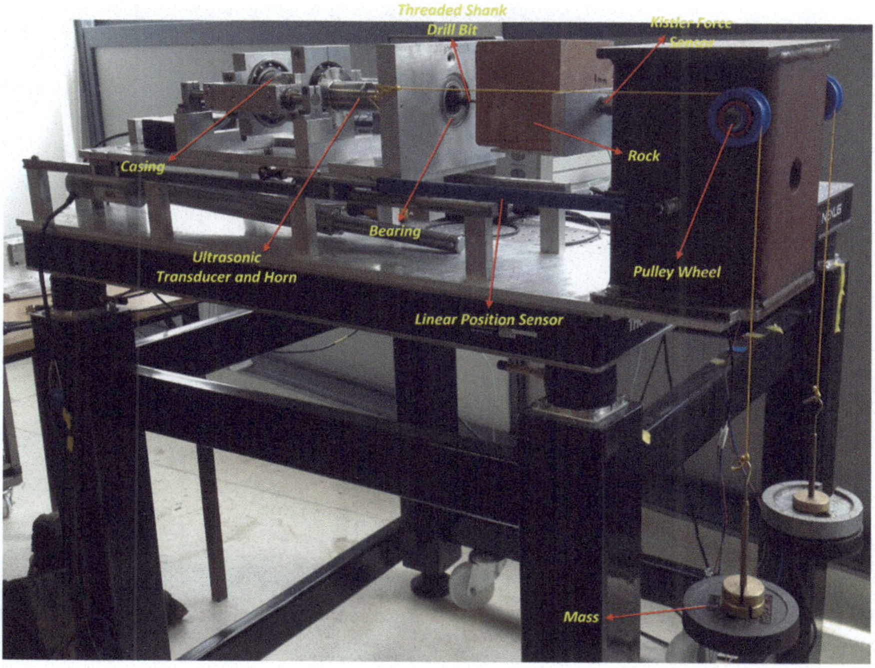

Fig. 7.12 The ultrasonically assisted drill facing a block of Locharbriggs sandstone. Note the force transducer behind the rock and the linear position sensor for measuring progress, which is driven by two weights on the yellow strings (out of shot). There is also a (disconnected) linear actuator for constant-rate experiments

7.3.2.2 Effects on Rate-of-Progress

To quantify the effects on rate-of-progress, the transducer is rotated at 15 rpm whilst being held against the rock sample by a weight-on-bit of 15 N. Trials are conducted firstly with the ultrasonic vibration turned off, as a control, with 3 and 5 mm drillbits. The power consumed by the UAD system, as a sum of the demand of the rotation motor and the ultrasonic system, is logged throughout. These results are presented in Fig. 7.14.

The experiments are then repeated with 5 μm of vibration amplitude, with the second set of results presented in Fig. 7.15.

It is apparent from the results that the application of ultrasonic vibration increases the rate of progress many times over, although there is also a net increase in the power drawn. Furthermore, the ultrasonically assisted systems appear to experience a decline in rate-of-progress at a much greater depth, if indeed it happens at all.

Fig. 7.13 The 3 mm ultrasonically assisted drill in Locharbriggs sandstone. Note the fine spoil being ejected from the drill site

7.3.2.3 Effects on Power Requirements

The application of ultrasonic vibration results in a net increase in the power requirements because the ultrasonic signal must be added to the baseline power draw of the rotational motor (12 W).

However, an increase from the 12 W baseline to ~20 W overall must be set against progress multipliers of up to approximately 20x (3 mm bit, Locharbriggs sandstone). Although some experimental scenarios only give a few-fold increase in performance, the specific energy expended is nonetheless significantly reduced in every case.

7.3.3 Summary

The ultrasonically-assisted drilling technique offers increased rate-of-progress, and although it is likely that the total power requirement will increase compared to traditional drilling systems—at least within the parameter set here considered—there are significant efficiency savings to be made. These savings are likely to be particularly accessible when smaller drilling bits are used and rapid access is required.

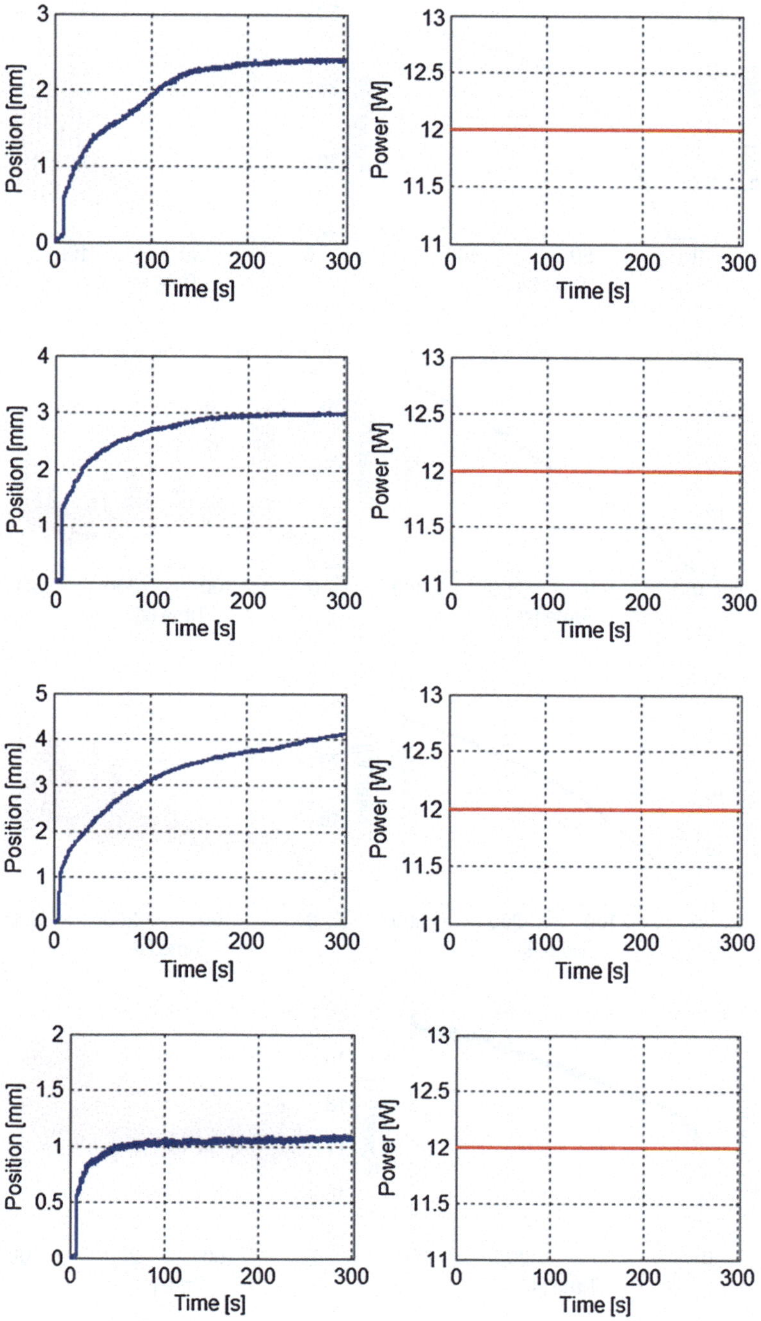

Fig. 7.14 Non-ultrasonic drilling, progress and power curves for (top row) 3 mm drillbits into Locharbriggs sandstone, (second row) 3 mm into Blaxter sandstone, (third row) 5 mm into Locharbriggs sandstone, and (bottom row) 5 mm into Blaxter sandstone

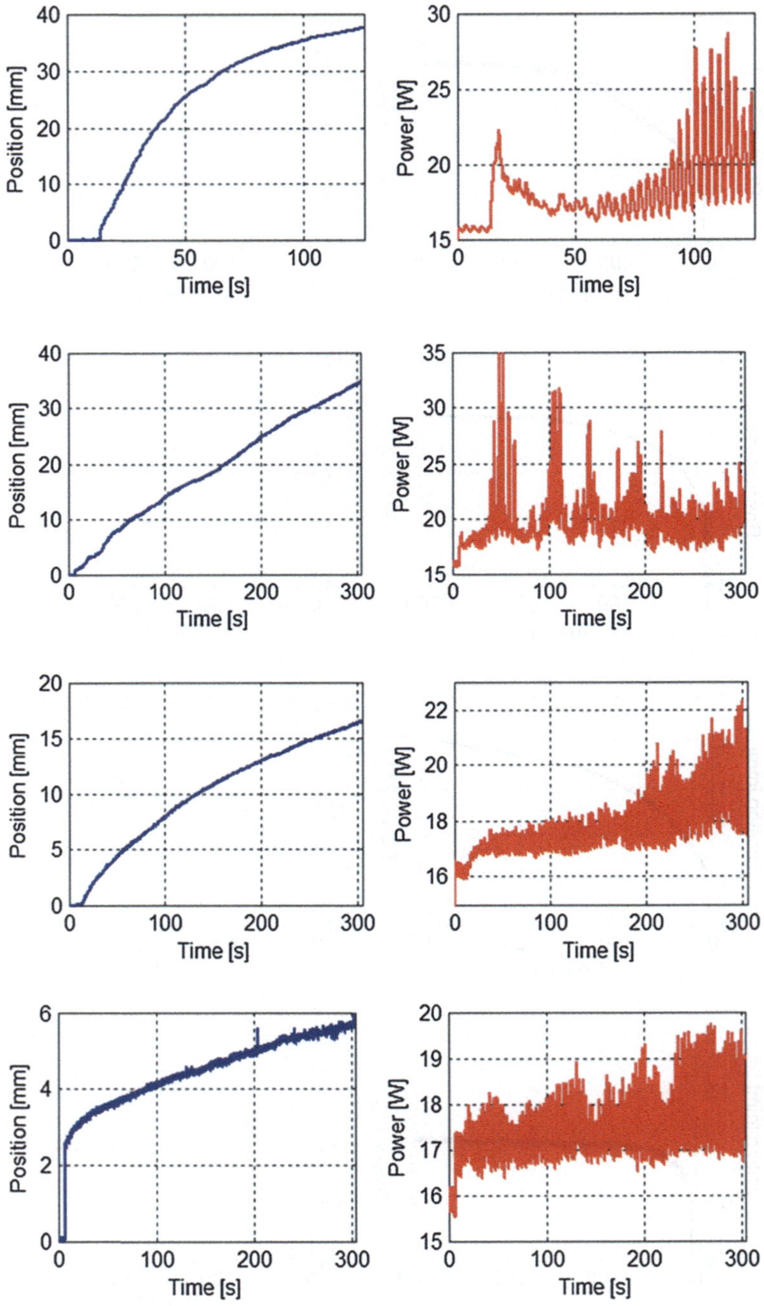

Fig. 7.15 Ultrasonic drilling, progress and power curves for (top row) 3 mm drillbits into Locharbriggs sandstone, (second row) 3 mm into Blaxter sandstone, (third row) 5 mm into Locharbriggs sandstone, and (bottom row) 5 mm into Blaxter sandstone

7.4 The Ultrasonic-Percussive Technique

Direct application of ultrasonic vibration is effective, but it requires that the transducer be placed relatively close to the cutting face, and to go deeper the transducer must cut a bore at least equal to its own cross-section. This may be challenging in some circumstances, and therefore a different approach is sometimes needed to provide the option of decoupling the transducer and the cutting bit while still transmitting cutting forces to either granular (Bao et al. 2006) or more cohesive (Bar-Cohen et al. 2000) materials.

7.4.1 The Technique

One such approach is to use an oscillating free-mass to transform high-frequency ultrasonic vibrations into lower-frequency impulses which can then be transmitted through a detached cutting bit, thus delivering a series of percussive impacts. This means that the drilling bit no longer acts as a sonotrode, but simply as a hammering element. The rock-cutting performance of these blows may be expressed by the rate at which they deliver 'effective impulse', or impulse at a force level exceeding the compressive strength of the target. This technique was developed under the name Ultrasonic/Sonic Drilling/Coring (USDC) at JPL during the late 1990s (Bar-Cohen et al. 2000).

One implementation of the ultrasonic-percussive technique is the Ultrasonic Planetary Core Drill (UPCD) in Fig. 7.16, which uses a half-wavelength transducer to excite a half-wavelength sonotrode (although alternatives use an overall half-wavelength system). Front and rear springs provide internal preload which ensures that the ultrasonic gear and the cutting bit come back together again after each cycle of the free-mass, while an eccentric motor drives the spoil extraction auger via a spline assembly.

The change of momentum applied to the free-mass at each impact must be maximized (within a practical parameter set) by a high degree of mechanical gain in

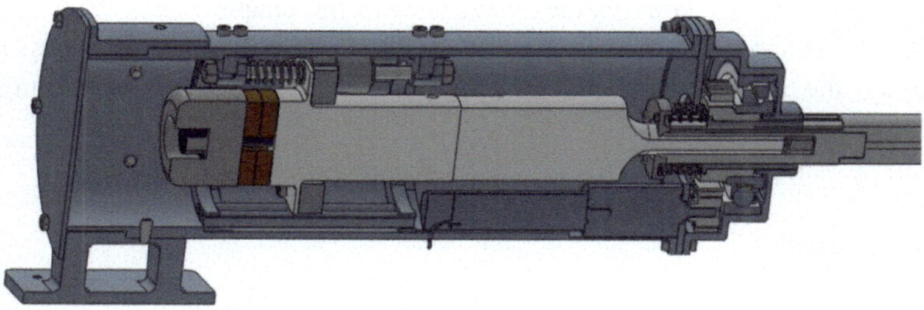

Fig. 7.16 The internal layout of a typical ultrasonic-percussive drill: the Ultrasonic Planetary Core Drill

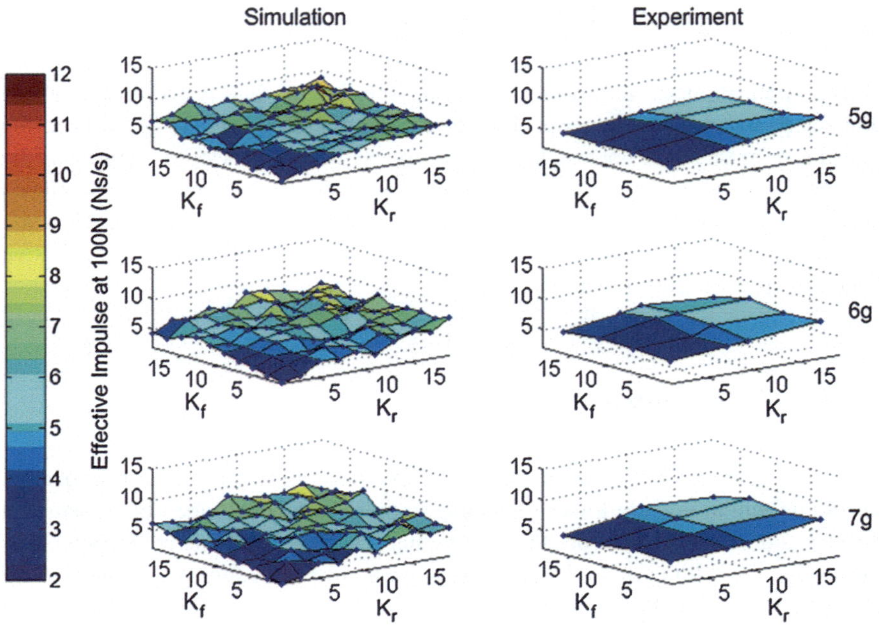

Fig. 7.17 Simulated and experimental data showing the effect of varying front and rear spring stiffnesses (k_f and k_r) on the delivery rate of effective impulse (100 N+) of a UPCD setup as the free-mass is varied from 5 to 7 g. The drillbit mass is 80 g and the transducer mass is 1.925 kg. The preload is 10 N and the base amplitude is 5 μm

the ultrasonic horn, although bulbous shapes have also been considered by other researchers (Harkness et al. 2011).

The effect of the masses of the component elements on the device dynamics must also be considered, where the drillstring is particularly significant because its mass and terrain damping will vary with depth, and the stiffness and internal preload between the front and back springs is also important.

Increasingly stiff springs will increase the impulse delivery rate (Figs. 7.17, 7.18 and 7.19), at the price of increasing the force needed to displace the transducer assembly inside the casing. This displacement is needed to facilitate some of the drillsting assembly maneouvres described later in the chapter.

On the other hand, excessively high internal preload will consume excess power and reduce the rate at which effective impulse is delivered to the cutting face (Fig. 7.20) (Li et al. 2017).

7.4.1.1 Control

The weight-on-bit applied to an ultrasonic-percussive device also affects its internal dynamics, and miscalculation can terminate the oscillation of the free-mass.

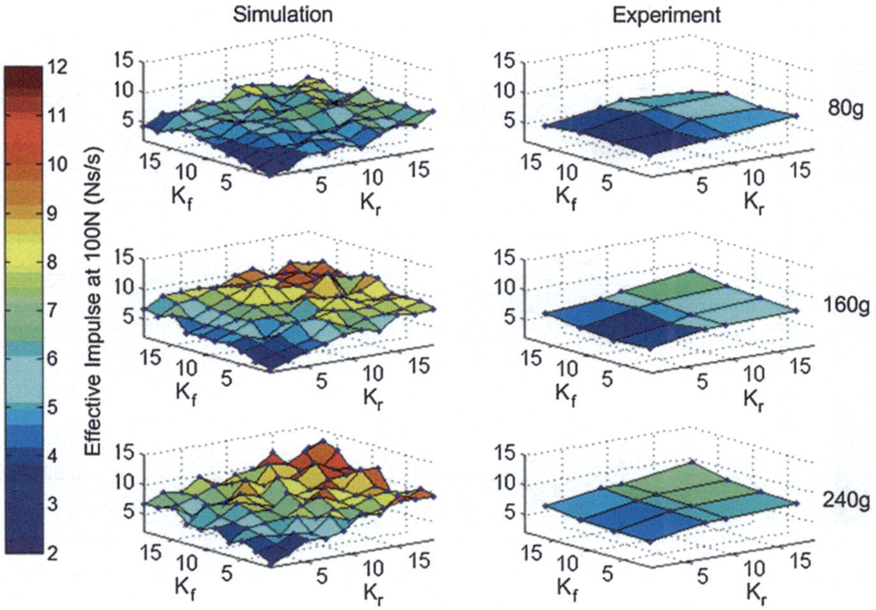

Fig. 7.18 Simulated and experimental data showing the effect of varying front and rear spring stiffnesses (k_f and k_r) on the delivery rate of effective impulse (100 N+) of a UPCD setup as the drillbit mass is varied from 80 to 240 g. The free-mass is 6 g and the transducer mass is 1.925 kg. The preload is 10 N and the base amplitude is 5 μm

The dynamics are relatively forgiving, such that self-weight wirelining with a single coring bit is an effective option, but for long drillstring operations the weight-on-bit must be managed in real time.

One technique which can achieve this is transducer amplitude feedback (Li et al. 2015). A resonance tracking system, such as the one previously described in relation to direct penetration, can be augmented by a further loop which seeks to advance the drill tool when the power required to maintain a preset amplitude falls below a certain threshold. This indicates that the oscillations of the free-mass have become less intense due to the drillbit having pulverized the material providing resistance, and so the transducer may be advanced to compensate. Some internal preload is maintained at all times due to the springs, however, which ensures that percussion does not cease.

A typical drill progression under such a loop is set out in Fig. 7.21. The instantaneous power demanded by the transducer to maintain a preset amplitude is shown in (a), which is low-pass filtered in (b). This is compared to an upper or lower threshold (dotted lines), and when the lower threshold is crossed the drill is advanced (c). Weight-on-bit is increased, and the power demand increases until the cycle repeats.

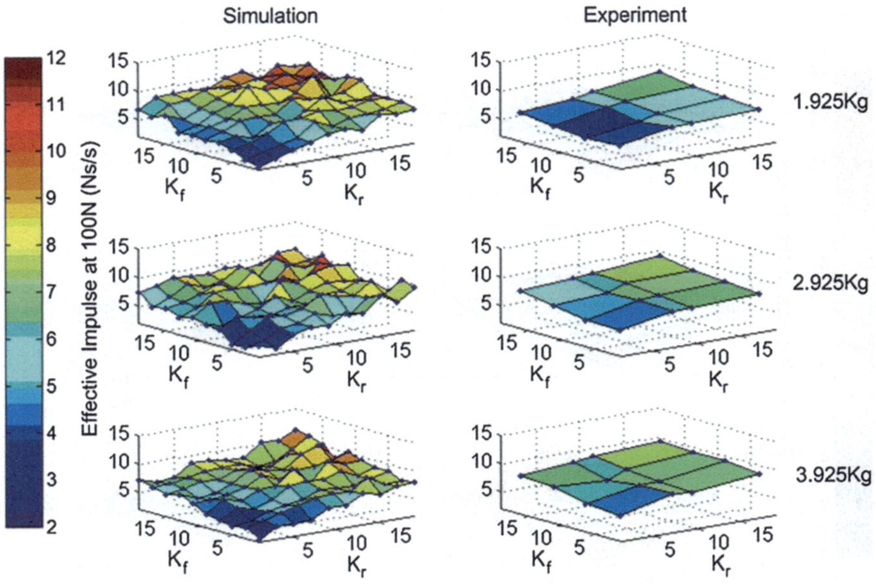

Fig. 7.19 Simulated and experimental data showing the effect of varying front and rear spring stiffness (k_f and k_r) on the delivery rate of effective impulse (100 N+) of a UPCD setup as the transducer mass is varied from 1.925 to 3.925 kg. The free-mass is 6 g and the drillbit mass is 80 g. The preload is 10 N and the base amplitude is 5 μm

7.4.1.2 Compatible Drillstring and Sampling Architectures

The drillstring itself must be made up from multiple elements if a relatively small surface package is envisioned. Screw threads are a common interface between drillstring elements, and even though ultrasonic percussion can act to break the torque-tightening of screw threads, the torque transmitted through the normal rotation of an operational drillstring can prevent the interfaces from becoming undone. However, this means that reversing the rotation of the drillstring for unjamming or disassembly purposes can cause unwanted separation of the interfaces. The use of bayonet-style interfaces (Fig. 7.22), which can lock in either direction, has been shown to address this problem and still transmit percussion effectively.

As bayonet-style drillstring elements with L-type female parts can be built up by a drill tool with a T-shaped interface, it is possible to assemble a drillstring that can be rotated in either direction without separation, so long as positive weight on bit is maintained. Furthermore, if such a drillstring is clamped at ground level, additional drillstring elements can be added or removed by the drill tool itself, using a rotating carousel system to store unused bits or cache sample-bearing ones. This is the operational concept of the Ultrasonic Planetary Core Drill (Fig. 7.23) (Timoney et al. 2015).

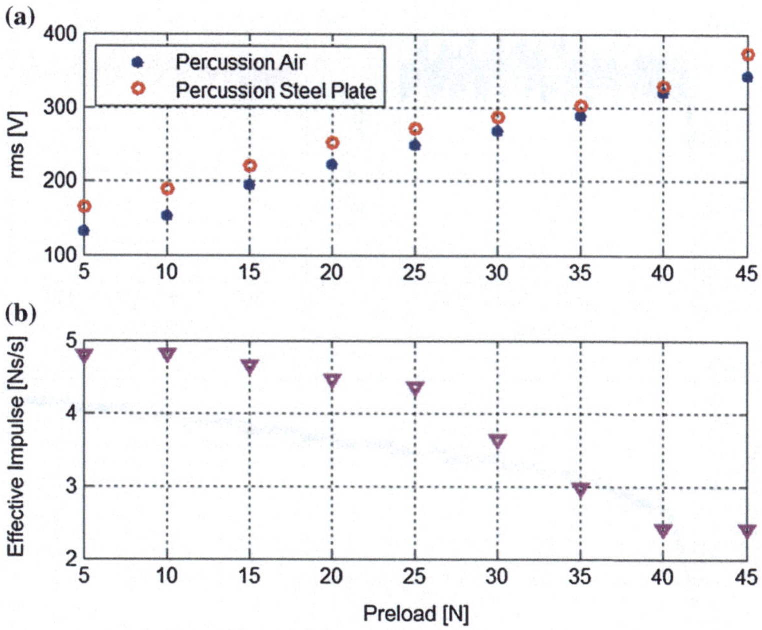

Fig. 7.20 The experimental voltage required to maintain a 5 μm excitation under percussion increases with internal preload (**a**); while the effective impulse falls over the same parameter sweep (**b**)

7.4.2 Performance

The performance of this technique has been tested using the UPCD device in two different materials, namely Locharbriggs sandstone and a permafrost simulant made up from block paving sand and water at typically 10–20% by mass, frozen to −30 °C. The Locharbriggs sandstone has a compressive strength of around 50 MPa, but permafrost hardness is highly variable. Reasonable estimates are in the 3–8 MPa range (Li and Zhu 2003).

7.4.2.1 Observations

Drilling is conducted using a drilling bit with two cutting teeth and a two-scroll auger, as shown in Fig. 7.24.

The Locharbriggs sandstone is easily drilled and extracted, but the permafrost samples are prone to drillbit freeze-in when cold material is drilled at or near room temperature. Figure 7.25 shows a bore after a freeze-in event, where a resolidified layer of material is apparent at the sides of the shaft. The bond this layer forms can prove almost impossible to break with the percussion and torque available.

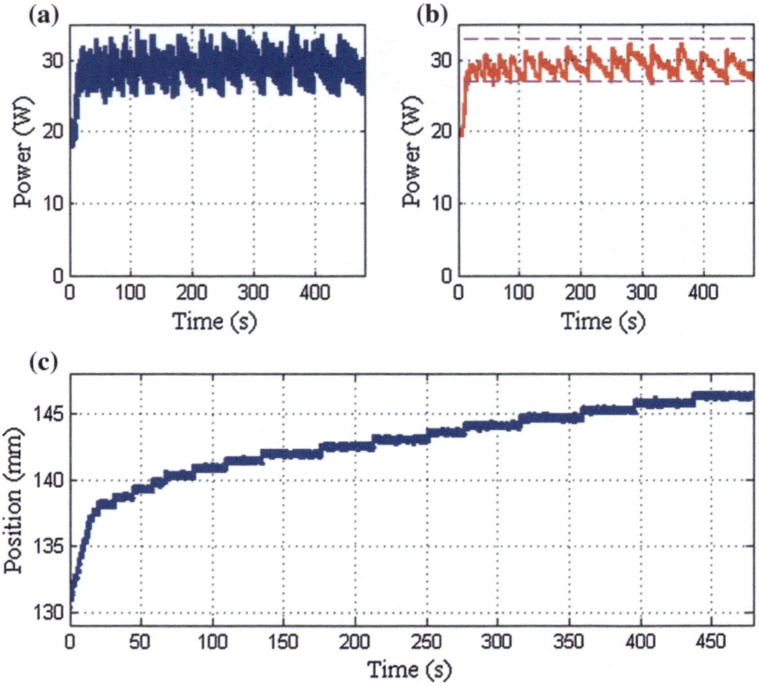

Fig. 7.21 The instantaneous power demanded by a transducer operating an ultrasonic-percussive drillbit (**a**), filtered and compared to two thresholds (**b**). When the lower threshold is crossed, the drill in question advances through, in this case, a Sherwood sandstone target (**c**). The bit diameter is 21 mm

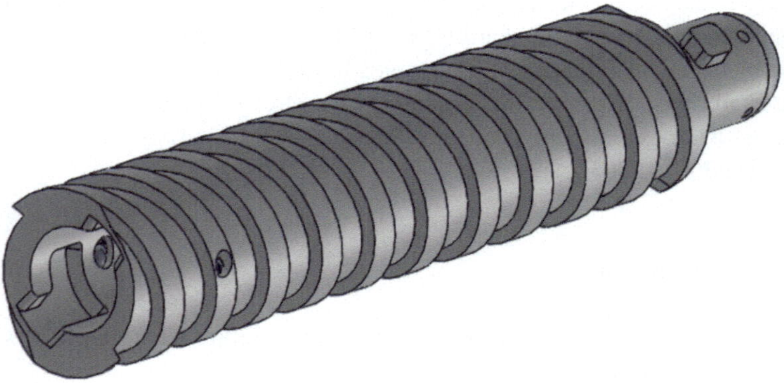

Fig. 7.22 A drillstring element for ultrasonic-percussive drilling. Note the L-shaped female interface (1), which (when mated to a male part) will transmit torque in either direction but release when the rotation is driven backwards under tension. Two detents are provided at the female end (2), but the male part has four detent sockets (3). This allows it to lock, in either direction, onto alternative T-shaped female interfaces which will not release when rotated in *either* direction under tension, but only when the rotation is *reversed* under tension

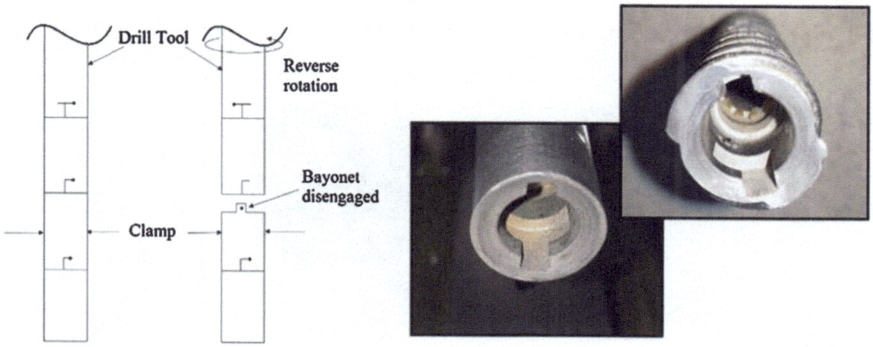

Fig. 7.23 Left: the concept for drillstring disassembly: clamping one bit allows a drill tool equipped with a T-style interface to add or remove new bits above that point. Right: a drill tool equipped with a T-style interface, and a drill bit with an L-style interface. This means that a top section can be added or removed from the drillstring at will, for assembly and caching purposes

Fig. 7.24 The cutting bit tip. Two angled tungsten carbide teeth are used, which are silver-soldered into place. Left: as new, and right: with the teeth badly worn and a coating of sandstone dust

However, even under room temperature drilling conditions the inside of the drillbit remains relatively cool, and still-frozen samples of permafrost material can easily be extracted (Fig. 7.26).

7.4.2.2 Drilling and Augering Performance

With the delivered ultrasonic power threshold set to 10 W, and additional draws (auger rotation, vertical actuation, and signal generation) amounting to around 35 W, progress into permafrost runs at approximately 8.3 mm/min while progress

Fig. 7.25 A 2 cm diameter, 10 cm deep hole drilled in permafrost simulant. Note the lighter layer of resolidified material. This can trap a drillbit almost instantly, sometimes with only a few seconds' warning in the form of increased ultrasonic power or torque demands

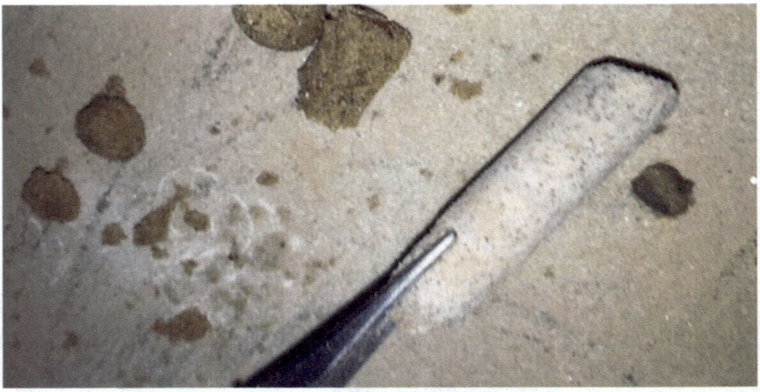

Fig. 7.26 An 8 mm diameter core, still frozen, extracted after ultrasonic-percussive drilling at room temperature. Previous cores nearby have since melted

into Locharbriggs sandstone runs at an average of 2.2 mm/min. Figure 7.27 illustrates progress through two 20 cm blocks of sandstone, although we note that a slightly softer second block likely explains the apparent acceleration at 100 min when the drill broke through from the upper bock and into the lower one (see also Fig. 7.29).

Of course, in drilling both materials there are three separate challenges: cutting or fracturing the material, feeding the auger, and then lifting the material away while maintaining the stability of the shaft.

In the case of the architecture used in this chapter, cutting the material through the ultrasonic-percussive technique proves to be straightforward. Nonetheless, feeding the auger is challenging in sandstone, and if it is not properly achieved the buildup of cuttings in the hole causes progress to halt at around 4 cm. This limit is

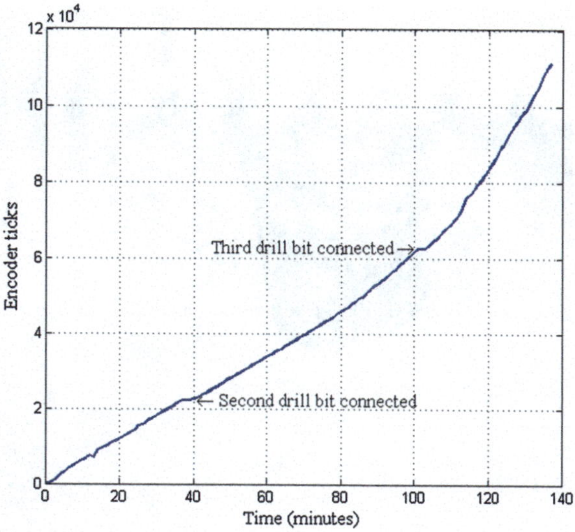

Fig. 7.27 Progress of the ultrasonic-percussive drill in Locharbriggs sandstone. There are approximately 400 encoder ticks per mm

reached when the shaft is too deep for spoil to be ejected by the sonic action alone, and compares well to the progress stalls seen elsewhere in the literature (Badescu et al. 2007).

To illustrate this stall phenomenon Fig. 7.28 shows two identical augers, operating at the same rotational speed. The image on the left is not feeding and an annular spoil heap has built up about 1 cm from the hole, where the majority of the ejected debris lands. This drill attempt will come to a halt at about 4 cm, and it is already visually apparent that the spoil is very fine and light in colour, indicating that the drill is expending energy to regrind pre-existing cuttings rather than make progress.

On the other hand the image on the right is feeding well, as is apparent from the relatively coarse material clearly being brought up by the auger flute. This drill attempt will proceed to more than 30 cm, with almost constant rate of progress throughout.

Although it is very difficult to observe the flow path of particles in the vicinity of the cutting teeth, their behavior between the teeth and the auger entrances, through channels called junk slots, is vital if feeding is to occur. In the geometry here tested, it is found that new teeth (such as those shown in Fig. 7.28 right) will result in a self-feeding auger. Similarly, worn teeth (Fig. 7.28 left) will result in a self-feeding auger so long as they are operating well below the critical depth of 4 cm, but crucially, worn teeth cannot penetrate the 4 cm barrier from above. In these circumstances, the annulus between the bit and the sides of the borehole becomes ‘dry’ of cuttings, progress stalls, and regrinding begins.

Fig. 7.28 Left: failure to feed, and right: successful auger feeding

If feeding is achieved, however, the rpm required to lift the spoil in ultrasonic-percussive drillstrings does seem to conform well to expectations for traditional, rotary-only systems (Zacny and Cooper 2007).

In permafrost, there are seldom problems with failure to feed, and rates of progress are usually excellent. Rather, the primary cause of drill failure in permafrost is freeze-in. This can take place when the local environment is close to the melting point of the permafrost, and appears to be foreshadowed by two signatures: a change in the nature of the spoil heap, and increasing power requirements.

Figure 7.29 illustrates this change in spoil behavior, from cloddy radial strips to a more granular heap. However, this is highly subjective and a better warning sign is increasing downhole torque. When this sign is acted upon, and the drillstring immediately extracted, freeze-in can usually be avoided.

7.4.2.3 Integrated System Testing

The Ultrasonic Planetary Core Drill was tested at Coal Nunatak, on Alexander Island, with the support of the British Antarctic Survey. The terrain was observed to show features reminiscent of the slope linea and patterned ground that have been observed on Mars, as shown in Fig. 7.30.

The regolith at the test site was found to be around 0 °C, which resulted in very wet spoil that did not auger easily. Progress to 6 cm took almost two hours at an

Fig. 7.29 Drilling in permafrost simulant. The nature of the spoil heap has suddenly changed, from horizontally-growing radial strips to a vertically-climbing tube, which can (empirically) indicate that conditions are changing such that freeze-in is imminent

Fig. 7.30 NASA Mars images (left), and Coal Nunatak (right)

average power draw of 30 W, as shown in Fig. 7.31. No greater depth was achieved in the rocky terrain with this device.

The drillstring assembly, sample acquisition, sample-caching, and drillstring disassembly operations of the UPCD all worked as planned during a separate test run in the snowpack, demonstrating the effectiveness of the bayonet interface system. Figure 7.32 illustrates the general layout of the system: the idealized green drill tool can be raised or lowered on the vertical rail, and the idealized yellow

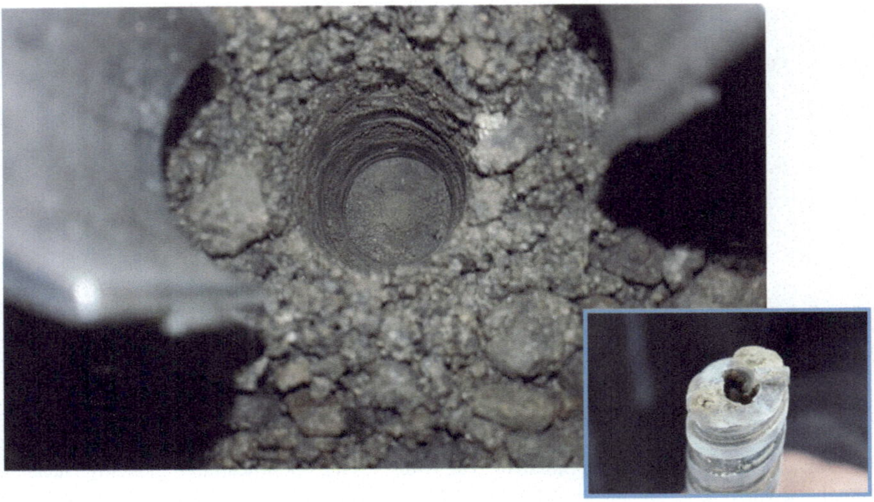

Fig. 7.31 Drilling in the regolith at Coal Nunatak, December 2016. Inset—the drill bit, showing muddy spoil in the auger openings

Fig. 7.32 The UPCD rig drilling to 30 cm (three drillstring sections) in the Antarctic snowpack under human-in-the-loop control, but purely mechanical operation. Inset: an ice sample after drilling, extraction, caching, and manual opening of the silo

Fig. 7.33 Left: lids on top of the carousel, showing their top bayonets for manipulation by the drill tool and screw thread for torque-tightening into the silos, and right: a lid being lowered onto a drill bit

carousel can offer up new drillbits in silos. These can be selected one at a time and assembled downhole, before being capped with the idealized purple lids and replaced in their silos with core samples intact.

All of these maneuvers can be completed using bayonet connections, with the exception of closing the silos with the lids to retain the sample. The lids instead have a screw thread, to allow the drill tool to torque-tighten them down against a Viton seal to ensure that volatiles are retained (Fig. 7.33).

7.5 Conclusion

Experimental campaigns have demonstrated that the application of ultrasonic vibration to probes and drillbits can reduce the drill forces, increase penetration speeds, and sometimes even save power and energy. Ultrasonic vibration can also be used to excite ultrasonic-percussive hammering, which can be controlled by semi-autonomous feedback loops and used to reach greater depths when combined with traditional augering systems. There are also feasible robotic systems that can be used to build drillstrings and sampling architectures compatible with the use of the ultrasonic-percussive technique.

These findings can be considered alongside terrestrial applications of ultrasonics in dust-handling and transportation, as well as cutting and drilling, to suggest applications in ultrasonically-assisted penetration, drilling, sample-handling and locomotion, particularly in low-gravity environments.

References

Badescu, M., Kassab, S., Sherrtt, S., Aldrich, J., Bao, X., Bar-Cohen, Y., Chang, Z.: Ultrasonic/sonic driller/corer as a hammer drill. In: Proceedings of SPIE, vol. 6529 (2007)

Bao, X., Bar-Cohen, Y., Chang, Z., Sherrit, S., Badescu, M., Du, S., Song, T., Peterson, T.: USDC-based rapid penetrator of packed soil. In: Proceedings of SPIE, vol. 6171, pp. 57–62 (2006)

Bar-Cohen, Y., Sherrit, S., Dolgin, B., Pal, D., Peterson, T., Kroh, J., Krahe, R.: Ultrasonic/sonic drilling/coring (USDC) for in-situ planetary applications. In: Proceedings SPIE, vol. 3992, pp. 661–668 (2000)

BRE: British Stone List Blaxter Sandstone Technical Data Sheet. (1997) http://projects.bre.co.uk/stonelist/dunhblax.html

BRE: British Stone List Locharbriggs Sandstone Technical Data Sheet. (2000) http://projects.bre.co.uk/ConDiv/stonelist/locharbriggs.html

Firstbrook, D., Worrall, K., Timoney, R., Harkness, P., Suñol, F., Gao, Y.: An experimental study of ultrasonic vibration and the penetration of granular material. In: Proceedings of the Royal Society A, vol. 473, Issue 2198 (2017)

Harkness, P., Lucas, M., Cardoni, A.: Maximization of the effective impulse delivered by a high-frequency/low-frequency planetary drill tool. IEEE Trans. Ultrason. Ferroelectr. Freq. Control **58**, 2387–2396 (2011)

Li, X., Worrall, K., Timoney, R., Harkness, P., Bolhovitins, A., Lucas, M.: A motion control system design for the Ultrasonic Planetary Core Drill (UPCD) Unit. In: AIAA Space 2015 Conference, Pasadena, USA, 2015

Li, H.P., Zhu, Y.L.: Uniaxial Compressive Strength of Saturated Frozen Silt, Permafrost, pp. 679–684. A.A. Balkema, Leiden, Netherlands (2003)

Li, X., Harkness, P., Worrall, K., Timoney, R., Lucas, M.: A parametric study for the design of an optimised ultrasonic-percussive planetary drill tool. IEEE Trans. Ultrason. Ferroelectr. Freq. Control **64**, 577–589 (2017)

Timoney, R., Harkness, P., Worrall, K., Li, X., Bolhovitins, A., Lucas, M.: The development of the European Ultrasonic Planetary Core Drill (UPCD), In: AIAA Space 2015 Conference, Pasadena, USA, 2015.

Zacny, K., Cooper, G.A.: Methods for cuttings removal for holes drilled on Mars. Mars **3**, 42–56 (2007)

Part III
Missions and Missison Concepts

Chapter 8
Flight in the Outer Solar System and Interstellar Travel

Alexander Bolonkin

8.1 Introduction: Review of Main Problems with Interstellar Flight

Interstellar travel is the term used for hypothetical piloted or unpiloted travel between stars. Interstellar travel will be much more difficult than interplanetary spaceflight; the distances between the planets in the Solar System are less than 30 astronomical units (AU), whereas the distances between stars are typically hundreds of thousands of AU. We typically express these distances in light-years. Because of the vastness of those distances, interstellar travel would require a high percentage of the speed of light, or huge travel time, lasting from decades to millennia or longer. There have been multiple interstellar projects in the past, dating back to 1958, such as:

- Project Orion, manned interstellar ship (1958–1968)
- Project Daedalus, unmanned interstellar probe (1973–1978)
- Starwisp, unmanned interstellar probe (1985)
- Project Longshot, unmanned interstellar probe (1987–1988)
- Starseed/launcher, fleet of unmanned interstellar probes (1996)
- Project Valkyrie, manned interstellar ship (2009)
- Project Icarus, unmanned interstellar probe (2009–2014)
- Sun-diver, unmanned interstellar probe

The speeds required for a project that would allow interstellar travel in a human lifetime far exceed what current methods of spacecraft propulsion can provide. Even with a hypothetically perfectly efficient propulsion system, the kinetic energy

A. Bolonkin (✉)
C&R, 1310 Avenue R, #6-F, Brooklyn, NY 11229-2838, USA
e-mail: abolonkin@gmail.com

V. Badescu and K. Zacny (eds.), *Outer Solar System*,
https://doi.org/10.1007/978-3-319-73845-1_8

corresponding to those speeds is enormous by today's standards of energy production. Moreover, collisions with cosmic dust and gases could produce very dangerous effects to both passengers and the spacecraft itself.

A number of strategies have been proposed to deal with these problems—ranging from giant arks that would carry entire societies and ecosystems, to microscopic space probes. Many different spacecraft propulsion systems have been proposed to give spacecraft the required speeds: nuclear propulsion, beam-powered propulsion, and methods based on speculative physics.

In April 2016, scientists announced Breakthrough Starshot, a Breakthrough Initiatives program, to develop a proof-of-concept fleet of small centimeter-sized, gram-scale, light sail spacecraft. These would be named *StarChip*, capable of making the journey to Alpha Centauri, our nearest extrasolar star system. The ultra-light StarChip robotic nanocrafts are planned to travel at speeds of 20 and 15% of the speed of light, taking between 20 and 30 years to reach the star system, respectively, and about 4 years to notify Earth of a successful arrival. There is also discussion of a Future Micro Interstellar Project to take place between 2030 and 2040. These missions will have to carefully consider the many issues that will be faced (explained in more detail in the subsequent sections), and plan accordingly.

8.1.1 Interstellar Distances

Because of the vast travel time required, distances between stars are usually expressed in light-years—defined as the distance that a ray of light travels in a year. Light in a vacuum travels around 300,000 km (186,000 miles) per second, thus translating to some 9.46 trillion kilometers (5.87 trillion miles) or 63,241 AU in a year. Proxima Centauri is 4.243 light-years away.

Another way of understanding the vastness of interstellar distances is by scaling: one of the closest stars to the Sun, Alpha Centauri A, is a Sun-like star.

8.1.2 Required Energy

The velocity for a manned, round trip flight spanning a few decades is several thousand times greater than those of present space vehicles. This means that due to the v^2 term in the kinetic energy formula, millions of times as much energy is required. As a barometer, the world energy consumption in 2008 was 143,851 terawatt-hours. Accelerating one ton to one-tenth of the speed of light requires at least 450 PJ or 4.5×10^{17} J, or 125 terawatt-hours, without factoring in efficiency of the propulsion mechanism. This energy must also be generated either from on-board from stored fuel, fuel harvested from the interstellar medium, or energy projected over immense distances.

8.1.3 Interstellar Medium

A thorough knowledge of the properties of the interstellar dust and gas through which the vehicle must pass will be essential for the design of any interstellar space mission. A major issue with traveling at extremely high speeds is that interstellar dust may cause considerable damage to the craft, due to the high relative speeds and large kinetic energies involved.

8.1.4 Travel Time

An interstellar ship would face manifold hazards found in interplanetary travel, including vacuum, radiation, weightlessness, and micrometeoroids; faced for long periods of time. Even the minimum multi-year travel times to the nearest stars are beyond current manned space mission design experience.

8.1.5 Communications

Round-trip delay time is the minimum time between an observation by the probe and the moment the probe can receive instructions from Earth, reacting to the observation. Information can travel no faster than the speed of light. For the Voyager 1 round-trip delay was about 36 h, but near Proxima Centauri, it would be pproximately 8 years. In an unmanned vehicle, faster reactions would have to be pre-programmed to be carried out automatically. Of course, in the case of a manned flight, the crew can respond immediately to their observations. However, the round-trip delay time would still makes any crew extremely isolated from Earth (analogous to how past long distance explorers were similarly isolated before the invention of the electric telegraph). Thus, interstellar communication is still problematic—even if a probe could reach the nearest star, its ability to communicate efficiently with Earth would be difficult given the extreme distance.

8.1.6 Prime Targets for Interstellar Travel

There are 59 known stellar systems within 20 light years of the Sun, containing 81 visible stars. There are a couple which could be considered prime targets for interstellar missions. The closest star system to our Solar System is Alpha Centauri, at a distance of 4.3 light year (ly). This system has three stars (G2, K1, M5).

Component A is similar to the Sun (a G2 star). Alpha Centauri B was originally thought to have one confirmed planet, but this was a false positive. The second closest star is Barnard's Star, at a distance of 6 light years. One is small, low-luminosity M5 red dwarf.

8.1.7 Rocket Concepts and Propulsion Systems

All rocket concepts are limited by the rocket equation. This sets the characteristic velocity available as a function of exhaust velocity and mass ratio—the ratio being initial (M_0, including fuel) to final (M_1, fuel depleted) mass.

Very high specific power, the ratio of thrust to total vehicle mass, is required to reach interstellar targets within sub-century time-frames. Some heat transfer is inevitable, and a tremendous heating load must be adequately handled. Thus, for interstellar rocket concepts of all technologies, a key engineering problem (seldom explicitly discussed) is limiting the heat transfer from the exhaust stream back into the vehicle.

8.1.8 Light Beamed Propulsion

The power per thrust required for a perfectly collimated output beam is 300 MW/N (half this if it can be reflected off the craft). Power sources with high energy-density would be required to provide reasonable thrust without unreasonable weight. The specific impulse of a photonic rocket is hard to define, since the output has no (rest) mass and is not expended fuel; if we take the momentum per inertia of the photons, the specific impulse is just c, which is impressive. However, when considering the mass of the source of the photons (e.g., atoms undergoing nuclear fission), the specific impulse is brought down to significantly, to 300 km/s (c/1000) or less. Adding the infrastructure for a reactor into the equation (some of which also scales with the amount of fuel) just reduces the value further. Finally, any energy lost via conduction away by engine supports (versus loss through radiation which is redirected precisely to aft), radiation in some other direction, or neutrinos, will further degrade the efficiency.

A light sail or magnetic sail powered by a massive laser or particle accelerator in the home star system could potentially reach even greater speeds than rocket propulsion or pulse propulsion methods. A sail method would not need to carry its own reaction mass, and therefore, would only need to accelerate the craft's payload, thus allowing for higher speeds.

8.1.9 Breakthrough Starshot Initiative

Each *StarChip* nanocraft is expected to carry miniaturized cameras, navigation gear, communication equipment, photon thrusters, and a power supply. Additionally, each nanocraft would be fitted with a meter-scale light sail, made of lightweight materials, on a gram-scale mass.

Four sub-gram scale digital cameras—each with a minimum 2-megapixels resolution—are envisioned for the nanocrafts. Four sub-gram scale processors should also be included. There is also planning for the inclusion of four sub-gram scale photon thrusters, each minimally capable of performing at a 1 W diode laser level. A 150 mg atomic battery, powered by plutonium-238 or americium-241, should adequately act as a fuel source. A coating, possibly made of beryllium copper, is planned to protect the nanocraft from dust collisions and atomic particle erosion. The lightsail is projected to be no larger than 4 by 4 m (13 by 13 ft), possibly of composite graphene-based material. The material would have to be very thin and somehow be able to reflect the laser beam without absorbing any of its thermal energy. Absorbing said thermal energy would lead to the vaporization of the sail.

8.2 Research of Space Flight in the Outer Solar System

Humanity has always sought to learn about the world and the space surrounding it. An important part of this knowledge is knowledge of the universe, the search for other intelligent beings, information sharing, and the extension of the Mind existence. This can be found that in the billions of solar systems which make up our universe. It is reasonable to assume that among these planets there will be other intelligent beings with which contact could be established for an exchange of acquired knowledge.

8.2.1 Nearest Stars

There are 5 known stellar systems within 12 light years of the Sun, containing 7 visible stars. Table 8.1 shows what could be considered as prime targets for future interstellar missions.

The speed at which light travels is $c = 3 \times 10^8 \mathrm{m/s}$ with the runs in one year being $t_1 \approx 31.45 \times 10^6$ s. One light year is thus a distance of approximately 10^{13} km ($D = ct \approx 10^{13}$ km). The time of getting information can be found using Eq. (8.1) below:

Table 8.1 Closest candidates for interstellar travel

Stellar system	Distance (light years)	Brief Information of stellar system
Alpha Centauri (three stars)	4.3	Closest system. Three stars (G2, K1, M5). Component A is similar to the Sun (a G2 star). Alpha Centauri B was thought to have one confirmed planet, but this was a false positive
Barnard's Star (2 stars)	6	Small, low-luminosity M5 red dwarf. Second closest to Solar System
Sirius (2 stars)	8.7	Large, very bright A1 star with a white dwarf companion
Epsilon Eridani (colder Sun)	10.8	Single K2 star slightly smaller and colder than the Sun. It has two asteroid belts, might have a giant and one much smaller planet, and may possess a solar-system-type planetary system
Tau Ceti (similar Sun)	11.8	Single G8 star similar to the Sun. High probability of possessing a solar-system-type planetary system: current evidence shows 5 planets with potentially two in the habitable zone

$$T = \left(\frac{c}{v} + 1\right) d \tag{8.1}$$

where T is flight time in year; c = $3 \times 10^8 \mathrm{m/s}$; v is probe speed (m/s); d is distance to star (light years). If we want to be able to receive any information in reasonable time period (i.e. 40 years), the relative probe speed must reach 15–25% of the speed of light. This equates to an impressive speed (v) of between 45,000 and 75,000 km/s. this goal is inconceivable at this time. For relative speed of 15% light speed (v/c = 0.15), the flight time and information passing between Earth and, our closest candidate, Alpha-Centauri equates to approximately 33 years.

8.2.2 The Closest Star System

Alpha Centauri consists of three stars: the pair Alpha Centauri A and Alpha Centauri B, and a small and faint red dwarf (Alpha Centauri C), better known as Proxima Centauri. Alpha Centauri A (α Cen A) has 110% of the mass and 151.9% the luminosity of the Sun. Alpha Centauri B (α Cen B) is smaller and cooler, at 90.7% of the Sun's mass and 44.5% of its visual luminosity. During the pair's 79.91-year orbit about a common center, the distance between them varies from about that between Pluto and the Sun to that between Saturn and the Sun. Proxima is at the slightly smaller distance of 1.29 parsecs, or 4.24 light years, from the Sun, making it the closest star to the Sun, even though it is not visible to the naked eye. The separation of Proxima from Alpha Centauri AB is about 0.06 parsecs (i.e. 0.2 light years or 15,000 astronomical units (AU)), equivalent to 500 times the size of Neptune's orbit.

Until the 1990s, technologies did not exist that could detect planets outside our Solar System. Since then, exoplanet-detection capabilities have steadily improved to the point where Earth-mass planets can be detected. Alpha Centauri is envisioned as a likely first target for manned or unmanned interstellar exploration. Crossing the huge distance between the Sun and Alpha Centauri using current spacecraft technologies would take several millennia, though the possibility of nuclear pulse propulsion or laser light sail technology, as considered in the Breakthrough Starshot program, could reduce the journey time to a matter of decades.

8.2.3 *Efficiency from Innovations and Explorations*

In any given case, is very difficult to estimate the efficiency of this profit for humanity. The efficiency from Innovations and Explorations may be approximately estimated by Eq. (8.2) below:

$$\mathrm{E} = \mathrm{P}/\mathrm{C} \tag{8.2}$$

where E = coefficient efficiency; P = estimation of the future profit; C = estimation of the R&D. We could only estimate the massive expenses for R&D of this exploration; these estimations come in at hundreds of billions the USA dollars. The main problems with the efficacy of this mission lie in the following questions:

1. How do we launch the nanocraft so that they can reach the very high speeds required?
2. What useful information could the Micro-probe get about Alfa Centauri in a fly by?
3. How would we try to pass information collected back to Earth?

These questions must be researched and answered as we move into an interstellar mission. The following sections will look in close detail at some of these important issues.

8.2.4 *Request Energy for Interstellar Launch*

Consider the simplest case of the constant acceleration:

$$\begin{aligned} &S = \frac{at^2}{2}, \quad V = at, \quad S = \frac{V^2}{2a}, \quad a = \frac{V^2}{2S}, \quad F = am, \\ &E = FS = \frac{mV^2}{2}, \quad P = \frac{E}{t}, \quad N_1 = cF. \end{aligned} \tag{8.3}$$

Table 8.2 Result of computation Eq. (8.3) for the probe mass m = 0.1 kg and the final speed V/c = 0.15, V = 0.45 × 10^8 m/s, g = 10 m/s^2 via distance of acceleration

S, m	10^5	10^6	10^7	10^8	10^9	10^{10}
a = V^2/2S, m/s^2	10^{10}	10^9	10^8	10^7	10^6	10^5
a, g	10^9	10^8	10^7	10^6	10^5	10^4
t = V/a, s	4.5 × 10^{-3}	4.5 × 10^{-2}	4.5 × 10^{-1}	4.5 × 10^0	4.5 × 10^1	4.5 × 10^2
F = ma, N	10^8	10^7	10^6	10^5	10^4	10^3
P = E/t, W	2.2 × 10^{15}	2.2 × 10^{14}	2.2 × 10^{13}	2.2 × 10^{12}	2.2 × 10^{11}	2.2 × 10^{10}
N_1 = cF, W	3 × 10^{16}	3 × 10^{15}	3 × 10^{14}	3 × 10^{13}	3 × 10^{12}	3 × 10^{11}
N_1, MkW	3 × 10^7	3 × 10^6	3 × 10^5	3 × 10^4	3 × 10^3	3 × 10^2

where S is distance of acceleration (m); a is acceleration (m/s^2); t is time of acceleration (sec); V is final speed (m/s); F is force (N); m is mass of probe (kg); E is requested energy for acceleration (J); P is need power for acceleration (W); N_1 is need power of laser (electric station) for single (one) reflection (conventional mirror) without a mirror loss (W), for laser efficiency: 0.1; $c = 3 \times 10^8 \mathrm{m/s}$ is the light speed (m/s).

If we take the probe mass m = 0.01 kg, and the final speed V/c = 0.15, V = 0.45 × 10^8 m/s, the request minimum energy is about $E \approx 10^{13}$ J = 10^7 MJ. The result of the computation of Eq. (8.3) for the probe mass m = 0.01 kg, and the final speed V/c = 0.15, V = 0.45 × 10^8 m/s, are presented in Table 8.2.

Now, the power of the powerful electric station is about 10 MkW. That means if we accelerate our probe 0.01 kg at distance 10 mln.km with acceleration 10^4 g by laser and conventional mirror, we need in power 30 strong electric stations in during 450 s (i.e. 7.5 min). The acceleration, 10^4 g, has projectile ability comparable to a big gun. The most current lasers have efficiency about 0.02–0.06. If in the future, a good laser has efficiency 0.1, we would need it in 300 powerful electric stations.

8.2.5 *Possible Launch Nuclear Propulsion*

Many people think that nuclear propulsion can solve the space travel. That is a correct assumption for travel into Solar system, but is not correct for interstellar flight. The reasoning behind this can be shown through the following equation. Take the kinetic energy of mass, and the speed equation of a rocket in the rocket system coordinate,

$$\text{from } E = \frac{mW^2}{2}, \text{ we have } W = \left(\frac{2E}{m}\right)^{0.5} = E_s^{0.5}, \quad \Delta V = -W \ln \frac{M_f}{M_0}, \tag{8.4}$$

where E is energy of fuel (J); m is mass of fuel (kg); E_s specific energy of fuel (J/kg); W—exhaust (ejection) velocity of fuel (m/s); ΔV is rocket speed (m/s); M_f is final mass of rocket; M_0 is initial mass of rocket.

For chemical fuel E_s = (4–16) MJ/kg and W = 2–4 km/s. For typical $M_f/M_0 = 0.1$, ln 0.1 = −2.3, the rocket speed is 4–9 km/s (We need speed of 45,000 km/s).

The estimate of speed that could be reached by the rocket having thermonuclear reactor can be seen in the most perspective reaction equation below:

$$D + T = {}^4H_e\ (3.5\ \text{MeV}) + n\ (14.1\ \text{MeV}),$$
$$1\ \text{eV} = 1.6 \times 10^{-19}\text{J}, E = 17.6\ \text{MeV} = 28.2 \times 10^{-13}\text{J} \tag{8.5}$$

The energy of neutron, neutrino, and gamma rays is very difficult to use because they request a large thickness (mass) of materials to absorb the neutrons or gamma rays.

Assume we use it anyway. The fuel mass for Eq. (8.5) is $m = \mu m_n = 5.1.67\ 10^{-27} = 8.35 \times 10^{-27}$ kg. Here, μ is number of nucleons take part in reaction, and m_n is mass of one nucleon. From Eq. (8.4) we get a fuel exhaust speed $W = 26 \times 10^3$ km/s, and a rocket speed (for multi-stagy rocket) $V = 60 \times 10^3$ km/s. We need only $V = 45 \times 10^3$ km/s (see above), but we cannot get the thermonuclear energy now. The installation for it (ITER) is very complex and expensive (>$15 B), has a mass of many thousands of tons, and will may take until 2040 until there is an industrial application.

There are perspective proposals of cheap, small thermonuclear cumulative/ impulse, and ultra-cold compression reactors of mass about 100–300 kg, but they still need R&D (Milonni and Eberly 1988; Driggers 2001). There is an interesting perspective nuclear reactor called a Micro Black Hole (MBH) which can convert any matter to energy with 100% efficiency (Bolonkin 2004c). Currently, we only hope to get MBH via Large Hadron Collider.

Fission nuclear reactors are far along in development, and there are a lot of space projects which use them. Most all of these projects/reactors have a large mass; more sum tons plus their nuclear energy in 2–4 times less that a fusion reaction. At this point, they not acceptable for a macro space probe (0.01 kg).

There is also the idea of transferring the energy in space over a long distance by plasma or electron beam (Kantrowitz 1972; Collection "Beamed Energy Propulsion" 2003; Metzger and Landis 2001). Similarly, this idea still needs R&D.

There are good innovations with isotopes in developed energy and propulsion systems (Kantrowitz 1972). Their total summary energy may be more than that of a fission reaction in the long travel times required. Their main flaw is the small power output and uncontrollability. They may have an application for correction trajectory and getting energy in long time space flight, though.

8.2.6 Acceleration Space Probe by Laser Beam

Many scientists believe that the option of the laser beam can solve the problem of acceleration in space, being that it would not require the launch fuel and energy in the probe. A thin, lightweight sail would reflect the laser beam, and the light pressure could then accelerate the small probe at the needed speed. There is a lot of research about how to use the solar light for flight in our Solar System. However, the previous section (Table 8.2) shows the possibility for acceleration of the probe to impressive interstellar speeds of 45 thousand km/s. The estimated laser beam required is expanding, and thus requires a large sail and laser diameter. The beam has a maximal distance of acceleration of about 10 million kilometers, but this distance requests a special Continuous Wave (CW) large laser with a power of more than N_1 = 3000 MW, for 100% efficiency (see last column in Table 8.2). For 10% efficiency, the requested power is ten times higher. Currently, the conventional continuous wave operation laser produces 3 kW, while the most impressive impulse power laser installation in the World (NIF - National Ignition Facility) has in impulse energy of 120 kJ. NIF costs $3.5 billion. Using that information, we can estimate how much it would cost for the conventional launch beam laser system.

Some examples of pulsed systems with high peak power:

- 700 TW (700×10^{12} W)—National Ignition Facility, a 192-beam, 1.8-megajoule laser system adjoining a 10-meter-diameter target chamber.
- 1.3 PW (1.3×10^{15} W)—world's most powerful laser as of 1998, located at the Lawrence Livermore Laboratory.

8.3 Multi-reflex Light Launch Propulsion Systems

It is well-known that solar light is pressing on any and every surface it reaches. In 1900, the Russian scientist P. Lebedev measured light pressure. It was very small, at 4×10^{-6} Pa. In 1982 Bolonkin offered and researched the idea of reflecting a laser beam off of a special, highly reflective cell mirror having for different lightwaves (Bolonkin 2004b). This would theoretically allow an increase of efficiency of the mirror in both time and increase in light pressure. He also offered the laser engine and accelerator as viable options. Later in 2004, Bolonkin also researched the application of this idea for space launch and energy transfer over long distances (Collection "Beamed Energy Propulsion" 2003; Bolonkin 1985).

The purpose of this work is to develop and to draw attention to the revolutionary idea of a light multi-reflection by cell mirror. This idea allows the design of new engines, space and air propulsion systems, and energy transmission over millions of kilometers, allowing for creation of new weapons, etc. This method and the main innovations were first offered in 1982 while he resided in the former USSR (Bolonkin 2004b). Now, the author Bolonkin shows the immense possibilities of this idea in many fields of engineering—astronautics, aviation, energy, optics, direct conversion of light (laser beam) energy to mechanical energy (light engine), to name a few. The following sections consider the multi-reflex propulsion systems for space and energy transmission over long distances in interstellar travel.

8.3.1 History

The relatively conventional way to send a spacecraft on an interstellar journey is to use the solar sail or a laser sail (Garwin 1958; Forward 1962). This method is not effective because the light intensity is very low, with only one reflection. There has been a lot of research in this area, and into solar sails in general.

Kantrowitz offered the conventional method for using a laser beam for space propulsion (Kantrowitz 1972). He transferred energy using a laser beam to a space vehicle, converted light energy into heat, and evaporated a material to then obtain thrust from the gas pressure of said evaporated material. There has been extensive research on this method (Collection "Beamed Energy Propulsion" 2003). However, it is complex, has low efficiency, has limited range (divergence of the laser beam), requires special material located on board the space ship, and requires a very powerful laser.

In 1983, Bolonkin offered another method: that of using light beam energy, then the direct conversion of light energy into mechanical pressure (for an engine) or thrust (for launchers and propulsion systems) by multiple reflections (Bolonkin 1985). There was only one other work related to this topic, published in 2001 (Metzger and Landis 2001). The suggested system has several innovations which make the proposed method possible improve its parameters millions of times, different from the other related work. The difference between our suggested system and the previous system is analyzed and discussed below.

The reflection of light is the most efficient method to use for powering a propulsion system; it gives the maximum possible specific impulse (light speed is 3×10^8 m/s). The system also does not expend mass. However, the light intensity in full reflection is very small, only about 0.6×10^{-6} kg/kW. In 1983, the idea of increasing the light intensity by a multi-reflex method (multiple reflection of the light beam by special cell mirror) was suggested, and Bolonkin offered some innovations to dramatically decrease the losses in mirror reflection (including a cell mirror and reflection by a super–conducting material) (Bolonkin 2004b). This allows the system to make millions of reflections to gain some Newtons of thrust per kW of beam power. This allows for the design of many important devices (in particular, beam engines) which convert light directly into mechanical energy and solve many problems in aviation, space, energy and energy transmission (Bolonkin 2004b).

In the last few years, achievements in optic materials and lasers have decreased the losses from reflection, allowing Bolonkin to return to this topic, and make it his primary area of research. He solved the main problems: the design of a highly efficient reflector (special cell mirror), a light lock, focusing prismatic lightweight mirrors and lenses, a laser ring, and a beam transfer over very long distances (millions of km) with only very small beam divergence, light storage, a beam amplifier, a modulator of light frequency, balloon suspension of mirrors, etc. (Bolonkin 2004b).

8.3.2 Light and Light Devices

A short description of electromagnetic radiation can be found in the publication Inexpensive Mini Thermonuclear Reactor (Driggers 2001). A conventional mirror can reflect a maximum of 98–99% of the incident light energy of some bands of light waves. This gives a maximum of 200–300 reflections which is not enough for propulsion systems and engines. Because the light pressure is so low (about 0.6×10^{-6} kg/kW), we need at least a million reflections.

There is a well-known method for increasing mirror reflection. The layers of a quarter-wave optical thickness of high and low refractive-index materials increase the reflectance. After more than 12 layers, the reflective efficiency of a dielectric mirror approaches 100%, with virtually no absorption or scattering. Maximum reflectance occurs only in a region around the design wavelength. The size of the region depends on the design of the stack of multiple dielectric coatings. Outside this region, the reflectance is reduced. For example, at one-half the design wavelength it falls to the level of the uncoated substrate. The dielectric mirror is also designed for use at a specific angle of incident radiation. At other angles, the performance is reduced, and the wavelength of maximum reflectance is shifted.

Unfortunately, this dielectric mirror method is not suitable for mirrors moving relative to each other, as the reflected frequency is shifted slightly, and this frequency shift accumulates over multiple reflections. Also, conventional mirrors tend to reflect the beam in an unwanted direction if the mirrors are not kept in perfect alignment to the beam. The proposed cell mirror reflects the beam in the same direction, which is very important for decreasing the beam divergence. The small cells provide high reflectance and offer small absorption.

A narrow laser beam is the most suitable for a light engine and use in light propulsion. There are many different types of lasers with different powers (peak power up to 10^{12} W), wavelengths (0.2–700 μm), efficiencies (1% up to about 95%), and pulse rates (up to some thousands of impulses per second) or continuous operations. In publications in the References of this report, the reader will find a brief description of the laser in more detail (Milonni and Eberly 1988; Driggers 2001).

At the present time, we are seeing significant advances in high-power weapons-class lasers (Milonni and Eberly 1988). The laser power reaches 1 million watts. For our computation, the beam divergence is very important. The laser beam divergence (see 8.6) is shown by the following equation:

$$\theta = \frac{2}{\sqrt{\pi}} \frac{\lambda}{D} = 1.13 \frac{\lambda}{D} \tag{8.6}$$

where θ is the angle of divergence (rad), λ is the wavelength (m), D is an aperture diameter (m).

In particular, the diameter of the laser beam may be increased by an optical lens for reducing the beam divergence. The aperture diameter may be also increased by a laser ring (Fig. 8.1). The reflex capacity may be improved by using a super conductive material (this idea needs additional research). More detailed information can be found in various publications in the references (Milonni and Eberly 1988; Driggers 2001; Bolonkin 2004b).

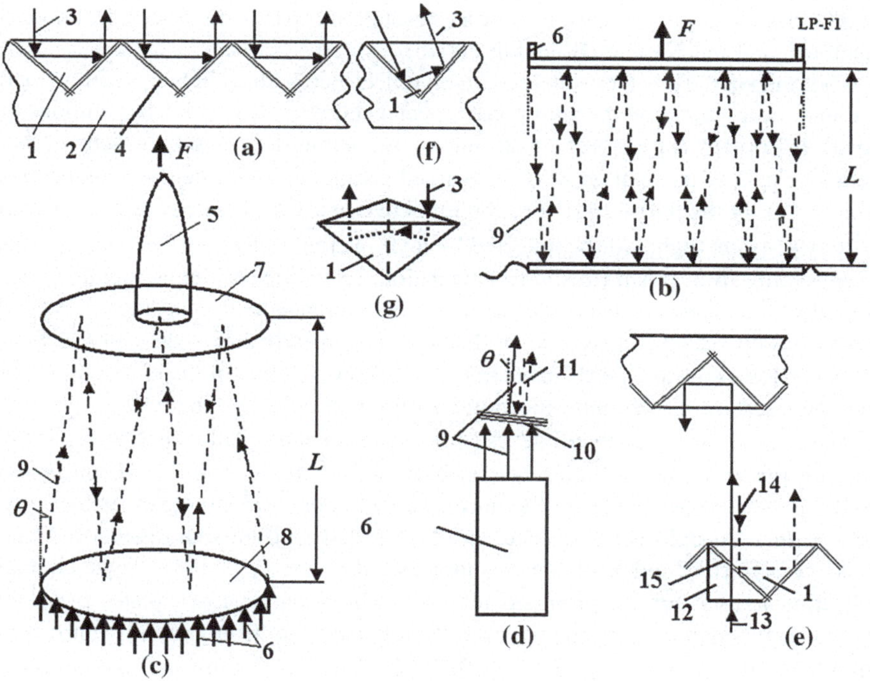

Fig. 8.1 Space launcher. Notations: 1—prism, 2—mirror base, 3—laser beam, 4—mirror after chink (optional), 5—space vehicle, 6—lasers (ring set of lasers), 7—vehicle (ship) mirror, 8—planet mirror, 9—laser beam, 10—multi-layer dielectric mirror, 11—laser beam after multi-reflection (wavelength $\lambda_{11} > \lambda_9$), 12—additional prism, 13—entry beam, 14—return beam, 15—variable chink between main and additional prisms. **a** Prism (cell, corner cube) reflector. **b** Beam multi-reflection. **c** Launching by multi-reflection. **d** The first design of the light lock. **e** The second design of the light lock. **f** Reflection in the same direction when the beam is not perpendicular to mirror surface. **g** Mirror cell (retroreflector cell or cube corner cell)

8.3.3 Description of Multi-reflex Launch Installation

In a multiple reflection propulsion system, a set of well-defined tasks appear: how to increase a mirror's reflectivity, how to decrease the light dispersion (from mirror imperfections and non-parallel surfaces), how to decrease the beam divergence, how to inject the beam between the mirrors (while keeping the light between the mirrors for as long as possible), how to decrease the attenuation (a mirror, prism material, etc.), how to increase the beam range, and how much force the system can have.

To address these problems, the author proposes a special "cell mirror" which is very reflective and reflects light in the same direction from which it came, a "laser ring" which decreases the beam divergence, "light locks" which allows the light beam to enter but keep it from exiting, a "beam transfer", a "focusing prismatic thin lens", prisms, a set of lenses, mirrors located in space, on asteroids, moons, satellites, and much more (Bolonkin 2004b).

Cell mirrors. To achieve the maximum reflectance, reduce light absorption, and preserve beam direction the author uses special cell mirrors which have millions of small 45° prisms (in Fig. 8.1). Cell mirrors are retroreflector cells or cube corner cells. A light ray incident on a cell is returned parallel to itself after three reflections (Fig. 8.1g). In the mirror, provided the refractive index of the prism is greater than $\sqrt{2}(\cong 1.414)$, the light will be reflected by total internal reflection. The small losses may be only from prism (medium) attenuation, scattering, or due to small surface imperfections and Fresnel reflections at the entrance and exit faces. Fresnel reflections do not result losses when the beam is perpendicular to the entry surface. No entry losses occur where the beam is polarized in parallel of the entry surface or the entry surface has an anti-reflection coating with reflective index $n_1 = \sqrt{n_0 n_2}$.

Here, n_0, n_2 are reflective indexes of the vacuum and prism respectively. These cell mirrors turn a beam (light) exactly back at 180° if the beam deviation is less 5–10° from a perpendicular to the mirror surface. For incident angles greater than $\sin^{-1}(n_1/n_2)$, no light is transmitted, an effect called total internal reflection. Here, n is the refractive index of the medium and the lens ($n \approx 1$–4). Total internal reflection is used for the proposed reflector, which contains two plates (mirrors) with a set of small corner cube prisms reflecting the beam from one side (mirror) to the other side (mirror) (Fig. 8.1b, c, f). Each plate can contain millions of small (30–100 μm) prisms from highly efficient optic material used in optical cables (Bolonkin 2003b). For this purpose a superconductivity mirror may also be used (Bolonkin 1985).

Laser ring. The small lasers are located in a round ring (Fig. 8.1c). A round set of lasers allows us to increase the aperture, resulting in a smaller divergence angle θ. The entering round beam (9 in Fig. 8.1a) has slip θ (or θ/2) to the vertical. The beam is reflected millions of times, as is shown in Fig. 8.1b, c, and creates a repulsive force F. This force may be very high; tens of N/kW (see the computation below) for motionless plates. In a vacuum, it is limited only by the absorption (dB) of the prism material (see below) and beam divergence. For the mobile mirror (as for a launch vehicle), the wavelength increases and beam energy decreases as the mirrors move apart.

This system can be applied to a space vehicle launch on a planet that has no atmosphere and small gravity (for example, the Moon; high gravity requires high beam power) (Bolonkin 1992a).

Light lock. The first design of a light lock allows the laser beam to enter, but closes the exit of a returned ray. The beam (9 in Fig. 8.1d) of a continuous laser, passes through a multi-layer dielectric mirror (10 in Fig. 8.1d). The entering beam runs the full length between mirrors (Fig. 8.1b, c), reflects a million times, and enters from the other side (11 in Fig. 8.1d). For moving (separating) mirrors, the wavelength is changed because the beam gives up energy to the moving mirrors (see computations below). As a result, the wavelength increases ($\lambda_{11} > \lambda_9$) when the distance increases, and the wavelength decreases ($\lambda_{11} < \lambda_9$) when the distance decreases. The mirror (10 in Fig. 8.1d) is designed to pass the laser beam (9 in Fig. 8.1d) and to reflect back the "used" ray (11 in Fig. 8.1d). If the beam is not reflected by the mirror (10 in Fig. 8.1d), it enters into the laser and will be reflected back by the laser's internal mirror.

The second design of the light lock is shown in Fig. 8.1e. This contains an additional prism 12 and an impulse laser. When laser beam 13 enters the system, the additional prism 12 is pushed into the main prism 1. While the beam runs between the mirrors, the additional prism is disconnected from the main prism and the return beam 14 cannot go back in. It travels inside the reflected mirrors with a lot of reflections if the mirrors have the right focuses. The chink (15) between the additional and main prisms may be very small, about a light wavelength (1 μm). A piezoelectric plate can be used to move the additional prism.

A continuous or pulse laser may be used for the first light lock, and a pulse laser may be used for the second lock. We can easily compute average laser power. The details of attenuation of light propagating through an optical material are considered in physics textbooks. To increase the number of reflections, we use a set of very small prisms and a highly efficient optical material (dB = 0.1–0.5).

Space beam transfer. Space beam transfer is shown in Fig. 8.2a. The first lens has a large aperture for the laser beam, and then focuses the beam, which decreases the divergence angle θ. The other Fresnel's lens then continues to further focus the beam (Fig. 8.2a).

Non-focused beam loses intensity through diffracted rays, but beam transfer has a special focusing lens. If the focus is located at a distance $S_1 = D/2\theta$, the beam does not have losses through up to a diffracted rays in this distance S, but after the distance S the divergence angle becomes 2θ (Fig. 8.2b). If we need to transmit energy at a distance L less than S (for example, in launching), this method is fine, since the distance between the mirrors $L \ll S$ and the beam is reflected many times without loss. If we want to transfer the energy over very long distance, the method shown in Fig. 8.2c may be better. In this method the beam is focused on point at a distance $S_2 = D/\theta$. The beam has small amounts of diffraction everywhere, but the losses are smaller after a distance $1.5S_1$ than in the case of Fig. 8.2b. If an

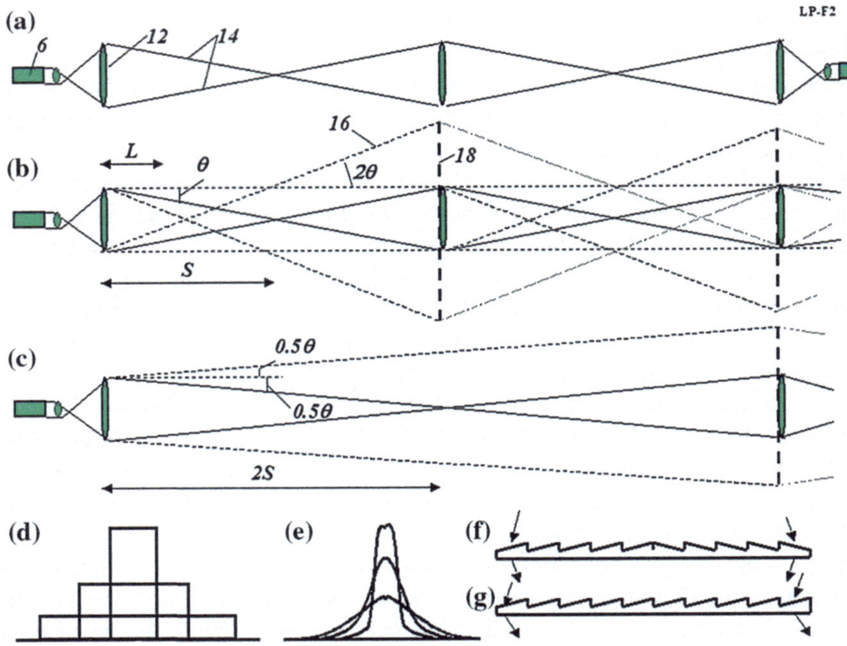

Fig. 8.2 Laser beam long-distance transfer. Notations: 12—lens, 14—bounds of laser rays, 15—light receiver, 16—divergence ray. **a** focused beam, **b** focused beam with angle θ which has part S without divergence, **c** focused beam with angle 0.5 θ which has minimum divergence at a long distance, **d** beam with a plate wave front, **e** Gaussian beam with normal distribution of beam front, **f** Fresnel's (prism) lens, **g** lens for changing the beam direction

intermediate lens with a much larger diameter than the initial lens (Fig. 8.2b, c) is added midway, it is possible to decreases the beam diffraction energy losses to a very small value.

The distribution of energy in a gross section area of the beam is also important for divergence and diffraction losses. The plate front of the wave (Fig. 8.2a), and plate distribution of energy and divergence (Fig. 8.2d) are worst, and give the maximum of energy losses. A normal distribution of beam energy and a Gaussian beam is better because the losses of beam energy through diffraction are reduced at the edges (Fig. 8.2e).

Energy transfer is done in the following ways. First the Fresnel's lenses (collimators) (Fig. 8.2f), Fresnel's prisms (Fig. 8.2g), and mirrors are (permanently) located in space (Fig. 8.3a). Their trajectories and the receiving space vehicle's trajectory in space are known. Through commands from Earth, a space ship, or the vehicle's computer, the mirrors and lenses are turned to the required operational angles (angular position). A small pilot ray may be used for aiming and focusing. The required angular changes are small (for focusing and small corrections in direction) and may be made by piezoelectric controlled plates. After the pilot ray reaches the space vehicle as required, the full power beam is transmitted to the

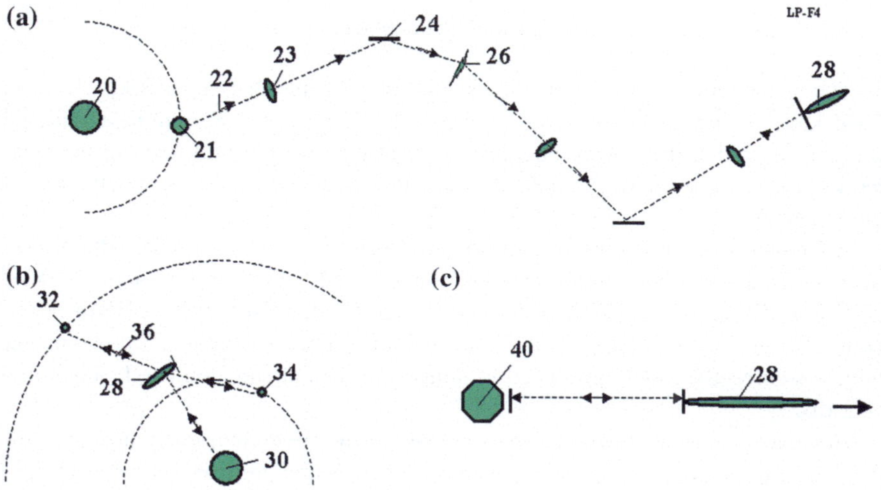

Fig. 8.3 Space energy transfer over long distance. **a** Transferring thrust from Earth to space ship by laser beam. **b** Using of satellites (or moons) to change the vehicle's trajectory **c** Using of asteroid for launching of ship. Notations: 20—Sun, 21—Earth, 22—laser beam, 23—Fresnel's lens, 24—mirror, 26—Fresnel's prism, 28—space vehicle, 30—planet, 32, 34—planet satellite, 36—multi-reflection, 40—asteroid

space vehicle. This beam may be used to launch vehicles from an asteroid or small mass planetary satellites (Fig. 8.1c), to change the vehicle's trajectory (Fig. 8.3b), or to increase the acceleration of the space vehicle near an asteroid (Fig. 8.3c) using the multi-reflex method (Fig. 8.3a–c). This beam energy may be also used by the space vehicle for its rocket engine and internal power requirements. The distance between lenses may reach tens of millions of kilometers (see computation below). The average distances of the nearest planets from the Sun are: Venus 108×10^6 km, Earth 150×10^6 km, Mars 228×10^6 km. Transfer efficiency of system may be about 0.7–0.9 (see computation below).

8.3.4 Theory (Estimation) of Multi-reflex Launching and Light Beam Transfer

Special theory and methods for this case are developed below. Such will include the attenuation of beam, beam power changes, focusing, and efficiency. This is analyzed via computation.

Attenuation of beam. The attenuation of light passing and propagating through an optical material is caused either by absorption or scattering. In both cases, the power is lost over a distance, *z*, from the power *N*(*z*), propagating at that point. Thus, we expect an exponential decay, as modeled below:

$$N(z) = N(0)\exp(-yz) \tag{8.7}$$

The attenuation coefficient, y, is normally expressed in dB km^{-1}, with 1 dB km^{-1} being the equivalent of 2.3×10^{-4} m^{-1}. Absorption is a material property in which the optical energy is normally converted into heat. In the scattering processes, some of the optical power in the guided modes is radiated out of the material.

Attenuation in some current and potential very-low-loss materials, which have been created for fiber communication, have a dB value of up to $a = 0.0001$ (Fig. 8.4; Bolonkin 1992b). We use conventional values of 0.1–0.4 dB/km in our computation. Clean air has $\xi = 0.333 \times 10^{-6}$ m^{-1}. The conventional optical matter which is currently widely produced in industry has an attenuation coefficient equal to 2 dB.

However, some of these materials are highly reactive chemically and are thus mechanically unsuitable for drawing into a fiber. Some are used as infrared light guides; none are presently used for optical communication, but may be useful for our purposes. Our mechanical property and wavelength requirements are less stringent than those for optical communications. We use $a = 0.1$–0.4 dB km^{-1} in our computation. The conventional optical material widely produced by industry for optical cables has an attenuation coefficient of 2 dB km^{-1}.

Change in beam power. The beam power will be reduced if one (or both) reflector is moved, because the wavelength changes. The total relative loss of the beam energy in one double cycle (when the light ray is moved to the reflector and back) is shown in the equation below:

$$q = 1 - (1-2\gamma)(1-2\xi)(1 \pm 2v)\varsigma \tag{8.8}$$

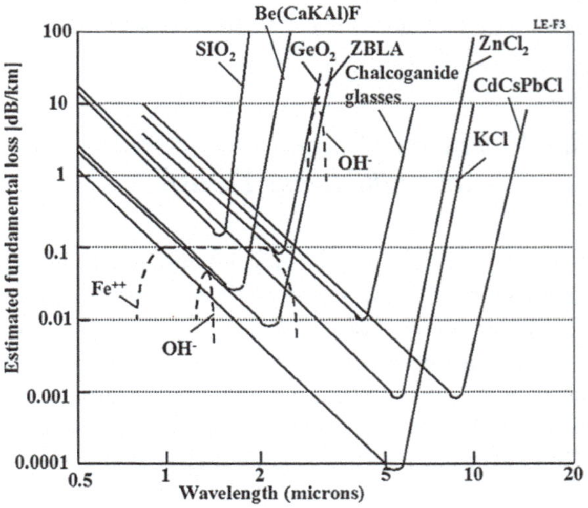

Fig. 8.4 The estimation of basic attenuation of some very low loss materials

Here, $v = V/c$, V is the relative speed of the mirrors [m/s], $c = 3 \times 10^8$ m/s is the speed of light. We take the "+" when the distance is reduces (braking) and take "–" when the distance is increased (as in launching, a useful work for light), γ is the light loss through prism attenuation, ξ is the loss (attenuation) in the medium (air) (in clean air $\xi = 0.333 \times 10^{-6}\ \text{m}^{-1}$), v is the loss (useful work) through relative mirror (lens) movement, ς is the loss through divergence and diffraction.

Multi-reflex light pressure. The light pressure, T, of two opposed high reflectors after a series of reflections, n, to one another is demonstrated in the equation below:

$$T_0 = \frac{2N_0}{c}, \quad T_1 = \frac{2N_0}{c} q, \quad T_2 = \frac{2N_0}{c} q^2, \quad T_3 = \frac{2N_0}{c} q^3, \ldots, T_{n-1} = \frac{2N_0}{c} q^n. \tag{8.9}$$

When q = constant, this is a geometric series. The sum of n members of the geometric series is

$$T = \frac{2N_0}{c}\frac{1 - q^n}{1 - q}. \quad \textit{If} \quad n = \infty, \quad \textit{then} \quad T_\infty = \frac{2N_0}{c}\frac{1}{1 - q}, \quad q < 1. \tag{8.10}$$

The efficiency coefficient, η, may be computed using the equation below:

$$\eta = TV/N_0 \tag{8.11}$$

Focusing the beam. If the lens used in focused at a range S_1, the distance, S, without ray divergence is (Fig. 8.2a):

$$S = \frac{D}{2\theta}, \quad \theta = \frac{2}{\sqrt{\pi}}\frac{\lambda}{D}, \quad S = \frac{\sqrt{\pi}}{4}\frac{D^2}{\lambda} = 0.443\frac{D^2}{\lambda}. \tag{8.12–8.14}$$

Here, D is the diameter of the lens or mirror [m]. This distance is equal to the lens focus distance for the case in Fig. 8.2b ($S_1 = S$). In the case Fig. 8.2c (transfer over very long distance), the optimal focus distance is $S_2 = 2S_1$.

Computations and Representations. The computation of Eq. (8.14) is presented by graphical representation in Figs. 8.5 and 8.6. As you will see, the necessary focus distance may be high.

The values in Eq. (8.3) can be computed as:

$$\gamma = yz = 0.00023al, \quad l = m\lambda, \quad m \geq 1, \quad \xi = 0.333 \times 10^{-6} L, \tag{8.15}$$

a is the attenuation coefficient in dB [km^{-1}] (Fig. 8.4), m is initial value of the wavelength which can be located in cell size l [m] (Bolonkin 2004b).

The loss through divergence, ς, for the case, shown in Fig. 8.2b, d is:

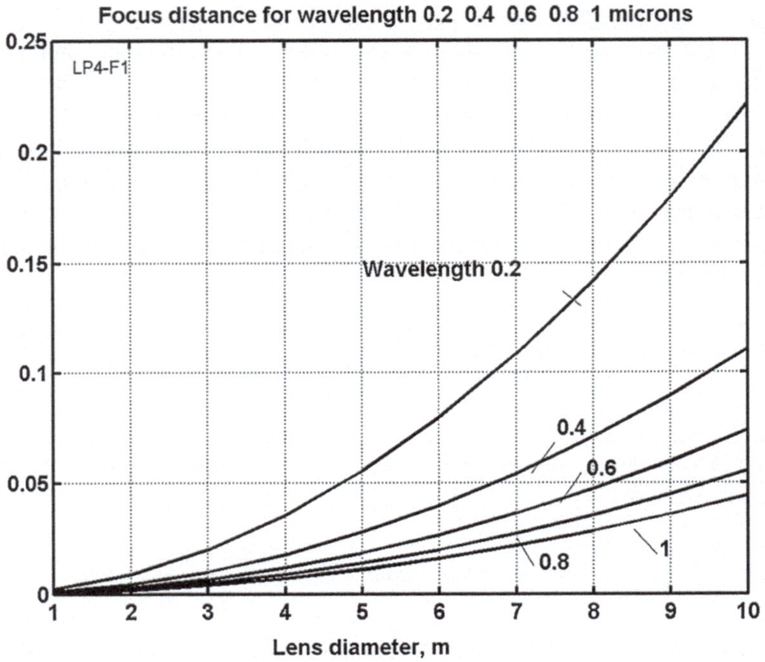

Fig. 8.5 Focus distances [10^6 km, million kilometers] versus lens diameters 1–10 m and wavelength λ = 0.2–1 μm

$$\varsigma = \frac{\pi (D/2)^2}{\pi (D/2 + 2\gamma(L-S))^2}, \quad 2\gamma(L-S) = \frac{4k}{\sqrt{\pi}}\frac{\lambda(L-S)}{D},$$
$$\varsigma = 1/\left(1 + \frac{8k\lambda(L-S)}{\sqrt{\pi}D^2}\right)^2 \textit{for } L > S. \tag{8.16}$$

In this case, L is the distance between the mirrors (lenses) [m], and, k is the focus coefficient. In the case in Fig. 8.2b (where the focus distance is $D/2\theta$) $k = 0$ when $L < S$ (for transfer) or $n < S/L$ (for reflection) and $k = 2$ when $L > S$, or $n > S/L$; in the case in Fig. 8.2c (S is absent, $S = 0$) $k = 0.5$ if the focus distance is D/θ; $k = 1$ if focus distance is infinity (i.e. no focusing).

The relative beam power along its trajectory for plate power distribution as in Fig. 8.2d is:

$$\overline{N} = N/N_0 = 1 \quad when \quad L \le S_1 \quad and \quad \overline{N} = \varsigma \quad when \quad L > S_1 = D/2\theta \tag{8.17}$$

The force coefficient, A, shows how many times the initial light pressure is increased. For $L < S_1$ it is:

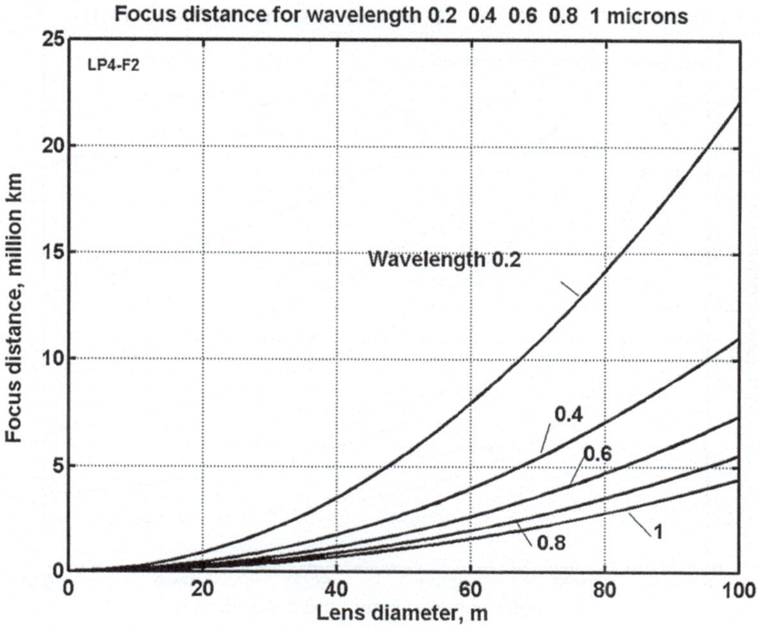

Fig. 8.6 Focus distances versus lens diameters D = 1–100 m and wavelength λ = 0.2–1 μm

$$A = \frac{1 - q^n}{1 - q} \tag{8.18}$$

The multi-reflex launch of a space vehicle from a small planet with low gravity, or without an atmosphere (the Moon or an asteroid), may be computed using the following equations (for focusing—Fig. 8.2b; and beam distributions—Fig. 8.2d):

$$T = \frac{2N_0}{c}\frac{q^{n_1} - 1}{q - 1} + \frac{2N_0}{c}q^{n_1}\frac{q_1^{n_2} - 1}{q_1 - 1}, \quad n_1 = \frac{S_1}{L}, \quad n_2 = n_1 - n_3, \quad n_3 = \frac{\ln m}{2v},$$
$$q = 1 - (1 - 2\gamma)(1 - 2v), q_1 = q\varsigma, \quad \Delta V = \left(\frac{T}{M} - g\right)\Delta t, \quad V_{i+1} = V_{i_i} + \Delta V,$$
$$\Delta L = V_i \Delta t. \quad L_{i+1} = L_i + \Delta L, \quad t_{i+1} = t_i + \Delta t. \tag{8.19}$$

Here, the first element in T is the thrust when the beam runs the distance S_1 without divergence. The second element in T is the thrust when the beam runs the distance with divergence. M is space vehicle mass [kg], while g is the planet's gravity [m/s^2]. When $n_3 < n_1$, we take $n = n_3$ and compute T using Eq. (8.10). If $n_3 > n_1$, we compute T using Eq. (8.19).

Computation of the efficiency co-efficient, η, in Eq. (8.13), are presented graphically in Fig. 8.7.

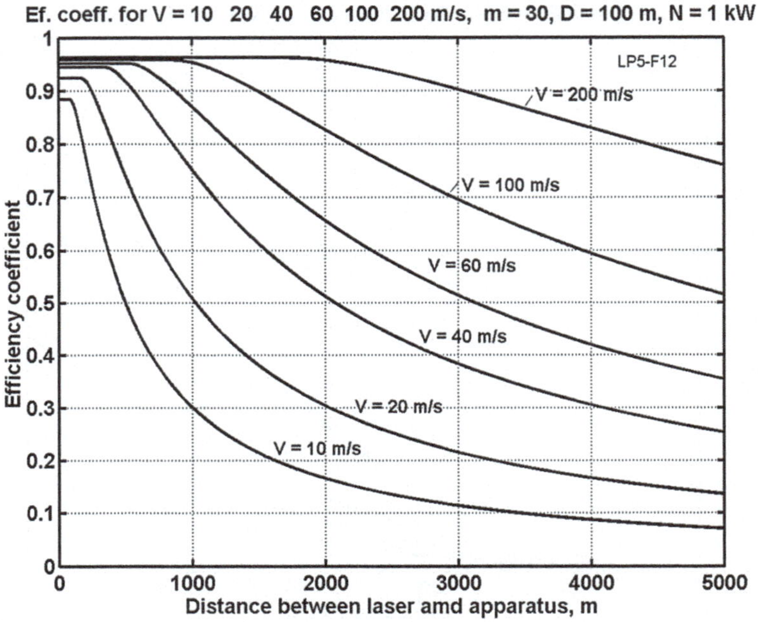

Fig. 8.7 Efficiency coefficient versus distance [m] for vehicle speed V = 10–200 m/s, attenuation coefficient a = 0.5 dB, cell size m = 30, mirror diameter D = 100 m, beam power N = 1 kW

The thrust and the efficiency coefficient decrease when the distance is above some critical value, then a portion of the energy beam leaves the space between the mirrors through diffraction.

The mirror diameter is large because small mirror diameters decrease the attainable speed. Starting the process from an asteroid or a planet's moon with low gravity improves the attainable speed of the probe. Unfortunately, the multi-reflex launch from planets with an atmosphere does not work well because the multi-reflected rays travel long distances in a gas medium and lose a lot of energy.

Below is the equation for computing the beam power from the divergence and distance when the Gaussian beam has normal distribution (Fig. 8.2f):

- For case 1 (the focus is into point $2S_1$, Fig. 8.2c)

$$\overline{N}_1 = 2\psi\left[s\left(\frac{D}{D+\theta L}\right)^2\right], \quad \theta = \frac{2}{\sqrt{\pi}}\frac{\lambda}{D}, \quad S = 0. \tag{8.20}$$

Above, ψ is the probability function of normal distribution.

- For case 2 (the focus is located at point S, Fig. 8.2b)

$$\textit{When} \quad L \leq S_1, \quad \overline{N}_2 = 1. \quad \textit{When} \quad L > S_1 \quad \overline{N}_2 = 2\psi\left[s\left(\frac{D}{D + 4\theta(L - S_1)}\right)^2\right] \tag{8.21}$$

Here, s is a relative distribution value. The results of computations for space (vacuum) are presented in Fig. 8.8. It is shown that the focused beam travels without major losses if the distance between the mirrors (for mirror diameter D = 100–200 m) is 10–18 million kilometers, and may travel up to 100 million kilometers with an efficiency of about 0.2. This means that the focused beam can permanently transfer (without losses) energy from Earth to the Moon or back (a distance of 0.4×10^6 km); and for 2–3 months (with efficiency 0.2) every two years, transfer to Mars at a distance of $60–150 \times 10^6$ km.

For computation of the relative beam power in air at altitude H, we may use Eqs. (8.20) and (8.21) corrected for air attenuation. This can be seen below:

$$\overline{N}_{a1} = \overline{N}_1(1 - b), \quad \overline{N}_{a2} = \overline{N}_2(1 - b), \quad \textit{where} \quad b = 0.334 \times 10^{-6}\frac{\rho_H}{\rho_0}L \tag{8.22}$$

With this, ρ_H, *and* ρ_o, are the air density at altitudes H, and H = 0, respectively.

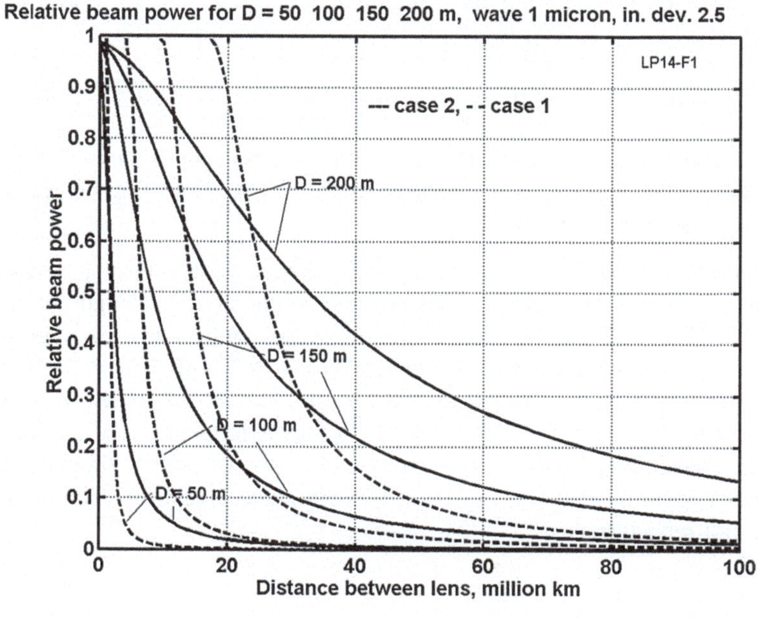

Fig. 8.8 Relative beam power of the normal (Gaussian) distribution (s = σ = 2.5) in a vacuum versus distance in million kilometers between lenses for focusing at D/2θ (- -, case 1) and D/θ (-, case 2)

The computed parameters are not optimal. Our purpose is to demonstrate the method of computation. Computations (Figs. 8.7 and 8.8) are made for a beam power $N_0 = 1$ kW. For beam power $N_0 = 10, 100, 1000$ kW we must multiply the force in Figs. 8.7 and 8.8 by 10, 100, and 1000 respectively.

8.3.5 Estimations for High Speed and Long Distance

1. **Maximum decrease of request energy from multi-reflexing the light-beam**

As we have seen in Table 8.2 the requested energy (power) for acceleration relativistic probe is very high. Multi-reflexing allows significantly decrease it. Let us separately estimate loss and benefits (increasing the thrust from multi-reflection) from cell mirror in atmosphere, space and from material of the mirror cell.

(a) *Energy Loss in Earth's Atmosphere.* It is known in clean atmosphere on Earth surface, the light beam losses the part of energy $\xi = 0.333 \times 10^{-6}$ 1/m. The Earth atmosphere has the pressure $p = 10^4$ kg/m^2 and density $\rho = 1.225$ kg/m^3 (on Earth surface). If Earth atmosphere has constant density its thickness is $H = p/\rho = 10^4/1.225 = 8163$ m ≈ 8.2 km. The laser installation may locate on area/altitude 2 km (mountain up 4 km, or the light beam will passed in vacuum tube of artificial tower/mast up 10–60 km).

If we take the altitude 5 km, the rest altitude will be about 8–5 = 3 km. The loss will be

$$\xi = 0.333 \times 10^{-6} \cdot 3 \times 10^3 = 10^{-3} \cdot n \approx 1/2\xi = 1/2 \times 10^{-3} = 500. \quad (8.23)$$

This means [refer to Eq. (8.15)] expecting full reflections through the Earth atmosphere, which increases the light pressure by 500 times (i.e. decreasing the request energy in 500 times).

(b) *Energy Loss in the Cell Mirror.* Let us to estimate the loss energy (reflectivity) the cell mirror [Eq. (8.15)]. Assume the minimum wavelength of light is $\lambda_o = 0.2$ μm $= 0.2 \times 10^{-6}$ m, maximum wavelength is $\lambda = 10^{-5}$, $m = 10^{-5}/0.2 \times 10^{-6} = 50$, absorption coefficient is $a = 10^{-2}$ [dB/km] $= 10^{-5}$ [dB/m], length of cell is $l = m\lambda_o = 50 \times 0.2 \times 10^{-6} = 10^{-5}$ m, reflectivity of cell mirror is $\gamma_1 = al = 10^{-5} \times 10^{-5} = 10^{-10}$. That is in hundreds of million better than conventional mirror ($\gamma_1 = 10^{-2}$) and many thousands time more than multi-layer mirror ($\gamma_1 = 5 \times 10^{-4}$). We can neglect this loss. The number of the full reflection is:

$$n \approx 1/2\gamma_1 = 1/2 \times 10^{-10} = 5 \times 10^9. \quad (8.24)$$

(c) *Loss from a Moving Probe.* Assume we want to accelerate probe up $V = 8$ km/s. The average speed is $V_a = 8/2 = 4$ km/s. The relative speed is

$v = V_d/c = 4/3 \times 10^5 = 1.33 \times 10^{-5}$. That means the number of full reflection is $n = 1/2v = 1/2.1.33 \times 10^{-5} = 3.76 \times 10^4$. If speed of probe is high, the efficiency of cell mirror significantly decreases. In our case of interstellar probe maximal speed is $V = 0.15c = 45 \times 10^3$ km/s, average relative speed is $v = 0.15/2 = 0.075$. Number of full reflection is $n = 1/2v = 1/2.0.075 = 6.67$.

Conclusion: The offered cell mirror is very efficiency for intersolar launch and travel, but less efficiency for interstellar launch.

2. ***Heating of the Reflect Mirror***

Let us estimate the heating (temperature) of the mirror. The temperature of the mirror is

$$T = 100\left(\frac{P_s}{C_s}\right)^{1/4}, \quad \text{where} \quad P_s = \frac{P\gamma_1}{2s} \tag{8.25}$$

Here, T is the temperature of the mirror, K; P_s is absorbed power, W/m^2; $C_s = 5.67$ W/m^{2}K^4 is absorbed coefficient; P is power delivered by laser to mirror, W (see Table 8.1); γ_1 is loss coefficient for one reflection; s is area of mirror, m^2.

Let us take the $s = 5$ m^2, the power light beam $P = 2.2 \times 10^{11}$ W (see last column in Table 8.1). For the cell mirror, $\gamma_1 = 10^{-10}$ and $T = 80$ K. For a conventional mirror, $\gamma_1 = 10^{-2}$ and T = 8000 K. For multi-layer mirrors, the best is $\gamma_1 = 5 \times 10^{-4}$ (one wavelength) and $T = 2100$ K. As you can see only cell mirror is acceptable for interstellar probe.

3. ***Loss probe speed from gravity field of Earth and Sun.***

The probe losses speed for start from Earth surface and arriving to Earth orbit around Sun: 11.2 km/s. The probe losses speed to arriving from Earth orbit around Sun to space out the solar system: 42.1 km/s. If probe will use the Earth orbit speed (that limits the start time up 2–3 month every year), we can save 30 km/s. In the last case, we have a loss on gravitation only 11.2 + 42.1 − 30 = 23.2 km/s. Radial velocity at Alpha-Centauri Star A is −21.4 km/s, while Star B is −18.6 km/s. All of these velocities are small in comparison to the requested Interstellar velocity of 45,000 km/s.

4. **Interstellar flight drag of environment.**

(a) *Shortly Information about interstellar medium.* In astronomy, the interstellar medium (ISM) is the matter that exists in the space between the star systems in a galaxy. This matter includes gas in ionic, atomic, and molecular forms, as well as dust and cosmic rays. It fills interstellar space and blends smoothly into the surrounding intergalactic space. The energy that occupies the same volume, in the form of electromagnetic radiation, is the interstellar radiation field.

In all phases, the interstellar medium is extremely tenuous by terrestrial standards. In cool, dense regions of the ISM, matter is primarily in molecular form, and reaches number densities of 10^6 molecules per cm^3 (1 million molecules per cm^3). In hot, diffuse regions of the ISM, matter is primarily ionized, and the density may be as low as 10^{-4} ions per cm^3. By mass, 99% of the ISM is gas in any form, and 1% is dust (Driggers 2001). Of the gas in the ISM, by number, 91% of atoms are hydrogen and 9% are helium, with 0.1% being atoms of elements heavier than hydrogen or helium.

Stars form within the densest regions of the ISM and molecular clouds, and replenish the ISM with matter and energy through planetary nebulae, stellar winds, and supernovae. The Warm Ionized Medium (WIM) holds approximately 205 to 50% of the interstellar volume, has a scale of 1000 pc, a temperature of about 8000 K, and a density ranging from about 0.2–0.5 ionized atoms in cm^3.

The Sun is currently traveling through the Local Interstellar Cloud (LIC), a denser region in the low-density Local Bubble.

(b) Let us take our estimation the interstellar density where $\gamma = 1$ H/cm^3 = 10^6 H/m^3 (where here, H is hydrogen atom). One Light Year (ly) has time $t = 31.54$ 10^6 s. Light speed is $c = 3 \times 10^8$m/s. Light runs (in 1 ly) the distance $L = ct \approx 10^{16}$ m/ly = 10^{13} km/ly. For probe speed $v = 0.15c = 45 \times 10^6$ m/s, the number of atoms getting the 1 m^2 of reflector is $N = \gamma L = 10^6 \times 10^{16} = 10^{22}$. The mass of atoms is $m = m_p N = 1.67 \times 10^{-27}$ $10^{22} = 1.67 \times 10^{-7}$ kg/m^{2}ly. Energy is $E = mv^2/2 = 1.67 \times 10^8$ J/m^{2}ly. If the atoms which will be stopped by the mirror have mass $m_{\text{m}} = 0.01$ kg/m^2, then the loss of speed by probe will be:

$$\Delta V \approx \frac{mv}{m_m} = \frac{1.67 \times 10^{-7} \cdot 45 \times 10^6}{10^{-2}} = 750 \ [\text{m/s} \cdot \text{m}^2 \cdot ly] \tag{8.26}$$

Full probe speed loss for mirror area $s = 5$ m^2 and 4.3 light years of flight is $\Delta V = 0.75 \cdot 5 \cdot 4.3 = 16$ km/s. It is a permissible part for speed 45,000 m/s. For breakdown of the mirror having the surface mass density 10 g/m^2 is enough energy 0.5 meV (Bolonkin 2004c, p. 935). Atom of medium for speed $V = 45{,}000$ km/s has energy about 10 meV. That means the most of atom will fly through the mirror and loss only 5% its energy.

The loss probe speed decreases by 20 times. The density of atoms in our Solar system at Earth orbit is about 20 H/cm^3. Estimation gives the loss speed of the probe in the Solar system to approximately 20 m/s. We can neglect it. There is no problem with the interstellar atom drag.

(c) *Problem the interstellar dust.* Cosmic dust can be further distinguished by its astronomical location: intergalactic dust, interstellar dust, interplanetary dust. By one estimate, as much as 40,000 tons of cosmic dust reaches the Earth's surface every year.

The interstellar dust has particles $d = 0.01$–0.2 μm. The mass of dust is about 1% of the gas mass. Particles composed from this consist of graphite, silicon carbide. Their density is about 3 g/cm^3. The drag from dust can be neglected.

Let us to estimate the holes created from dust. Take the average size of particles: 0.1 μm = 10^{-7} m, volume 10^{-21} m^3, mass one particle is $m_1 = 3 \times 10^{-18}$ kg. Total mass of the particles is $M = 1.67 \times 10^{-9}$ kg/m^{2}ly (1% of gas mass). The total number of particles is $N = M/m_1 = 5.33 \times 10^8$ 1/m^{2}ly. If one particle made the hole area $s_1 = d^2 = 10^{-14}$ m^2, the area total hole area will be approximately $s = s_1N = 5.33 \times 10^{-6}$ 1/m^{2}ly. During flight time $t = 4.3$ ly, the damage will be $S = 2.3 \times 10^{-5}$1/m^2. In reality, damage may be in an order of 1–2 more. We can still neglect it. If interstellar drag is large, the mirror can be folded.

8.3.6 Discussion

Here we compare the "Multi-Bounce Laser-Based Sail" system to the proposed method—the "Multi-Reflex Propulsion System".

1. The "Multi-Bounce Sail" uses the well-known multi-layer mirror which has high reflectance only in a region around the design wavelength. Outside this region, the reflectance is reduced. For example, at one-half the design wavelength, it falls to that of the uncoated substrate. As shown in this work, the wavelength changes by a small amount at each reflection in the mobile mirror. This means that after enough reflections the multi-layer mirror has lost its high reflectivity. It is impossible to use the multi-layer mirror for a multi-bounce space sail that is moving. The author has proposed the innovative new cell-mirror for which the reflectivity does not depend on wavelength for wavelengths that are less than a cell length.
2. The multi-layer mirror is extremely large (1 km^2), extremely thin (1600 nm), and with a density of 10 gm/m^2 and weight of 7850 kg (Metzger and Landis 2001). A very small angle of deviation at the multi-layer mirror surface (one thousandth of a degree) under beam pressure, leads to complete defocusing at a distance of some millions of kilometers. This means the mirror will make only one reflection (Bolonkin 1992c). The average mirror angle will also be changed permanently for a moving space ship.

It is impossible to exactly control (turn) the orientation of this gigantic and very thin sail. The new cell-mirror reflects the laser beam back in exactly the same direction if the surface and sail deviation are less than 5–10°. This means the mirror directorial control is not necessary on the space craft. Also, there may be imperfections in the surface film, thus the mirror control is not necessary.

3. The maximum reflection at multi-layer mirror is 99.95 (Metzger and Landis 2001). The reflection of the cell-mirror is $(1 - 0.4 \times 10^{-9})$ or 10^8 times better than the multi-layer mirror. The maximum reflection value of the multi-layer

mirror is only 1000 (Metzger and Landis 2001). The value for reflections off the cell-mirror is in the millions.
4. The diameter of the multi-layer mirror is 1 km, the size of our cell mirror is 100 m for large and heavy, manned ships, and 2 m for a micro probe.
5. The gigantic multi-layer solar mirror gives an acceleration of only 0.33 m/s^2. This is not enough to even launch itself from Earth (Earth's gravity is 9.8 m/s^2), Mars (3.72 m/s^2), or the Moon (1.62 m/s^2). The author's solar cell-mirror gives an acceleration of 20 m/s^2 (laser up 10^5g), and its size is 100 times smaller. If we were to make a solar cell-mirror 1 km in diameter, the capability of a space ship would be fantastic.

The author shows here, only some of the advantages of one innovation (changing from the well-known multi-layer mirror to the new *cell-mirror*). There are many deficiencies of the previous system which make its application virtually impossible (Metzger and Landis 2001). For example, with the multi-layer mirror the laser is located on the Earth's surface and its beam moves (from the laser to the ship and back to the laser) through the Earth's atmosphere a lot of times. The computation shows that the beam's energy will quickly be lost due to absorption and scattering by the Earth's (or Mars) atmosphere when it travels a long distance though it. In our system, the beam moves through the atmosphere only once time and reflects between the Moon mirror and the space ship for all other times. This is insured by the innovation of the light lock.

Another deficiency of the laser-based sail system is that when the space ship is close to Earth, the sail will reflect the beam back to the laser. If the efficiency of the propulsion system were sufficient, the laser might be damaged or destroyed. This problem is absent in the proposed system because it uses a light lock, which closes the return path of laser beam.

The suggested laser ring (a set of small lasers located in a circle), beam transfer, and self-focused mirror and Fresnel's lens decrease the beam divergence, and increase the beam transfer distance. It is possible to install the cell-mirror on the Moon or on Mars and transfer a laser beam to them, and then to make a space ship decelerate. The other system requires a nuclear electric power station (of several Giga Watts Power) to be built and delivered, as well as a super powerful laser on Mars.

General discussion. The offered multi-reflex light launcher, space, and air focused energy transfer system is very simple (needing only special mirrors, lenses, and prisms). It also has a high efficiency. One can directly transfer the light beam into space acceleration and mechanical energy. A distant propulsion system can obtain its energy from the Earth; however, we need very powerful lasers. Sooner or later the industry will create these powerful lasers (and cell mirrors), and the ideas presented here will become possible. The research on these problems should be started now.

Multi-reflex engines may also be used in aviation, as the energy can be transferred from the power stations on the ground to the aircraft using laser beams (Bolonkin 2004b). The aircraft would no longer carry fuel and the engine would be lighter in weight, so its load capability would essentially double. The industry currently produces a one Megawatt (1000 kW) laser. This is the right size for mid-weight aircraft (10–12 tons).

The linear light engine does not have a limit to its speed and may be used to launch space equipment and space ships in non-rockets method (Milonni and Eberly 1988; Driggers 2001; Bolonkin 1983, 1985, 1987, 1990, 1991, 2004c, b; Kantrowitz 1972; Collection "Beamed Energy Propulsion" 2003; Metzger and Landis 2001; Garwin 1958; Forward 1962). This method is certain to have many military applications.

8.4 Plasma Beam Space Propulsion for Interstellar Flight

This section presents and examines another revolutionary method—non-rocket transfer of energy and thrust into Space, with distances of millions kilometers. Theory and computations have been developed and completed. The method is more efficient than transmission of energy by high-frequency waves. This method may be used for space launch and for acceleration the spaceship and probes for very high speeds, up to relativistic speed by current technology. Research also contains prospective projects which illustrate the possibilities of using the suggested method.

8.4.1 Introduction

Transportation of energy, matter, or impulse is very important for long period space trips especially for lengthy distance voyages. The spaceship crew or astronauts on planets can need additional energy or ship thrust. Most people think that is impossible to transfer energy a long distance in outer space except electromagnetic waves. Unfortunately, electromagnetic waves have large divergence and cannot be used for a long distance (millions of kilometers) transfer. However, the space vacuum is very good medium for offered method and special transfer of energy and momentum.

Brief History. About 40 years ago, scientists received plasma flow having speeds up to 1000 km/s, power of 10 kW, mass consumption of 0.1 g/s, and electric current up to a million amperes. However, the application of plasma beams in space needs a series of inventions, innovations, and research. In particular, they include methods of decreasing the plasma divergence, discharging, dispersion of velocity, collection of the plasma beam in space at long distance from source, conversion of the beam energy into electricity and other types of energy, conversion of plasma impulse (momentum) in space apparatus thrust, conversion of plasma into matter, control, etc.

The solutions of the main noted problems and innovations are suggested by Bolonkin in early (1982–1983) patent applications and given articles. In particularly, the main innovations are:

1. Using neutral plasma (not charged beam);
2. Using ultra-cool plasma or particle beam in conventional temperature;
3. Control electrostatic collector which separates and collects the ions at spaceship;
4. Control electrostatic generator which convert the ion kinetic energy into electricity;
5. Control electrostatic ramjet propulsion;
6. Special control electrostatic mirror-reflector;
7. Recombination photon engine;
8. Recombination thermo-reactor;
9. Research made for both conventional and relativistic particle speeds.

About 20 years ago, scientists received ultra-cold plasma having the ion temperature lower than 1×10^{-3} K. Velocity dispersion was $10^{-4} \times 10^{-6}$, and beam divergence for conventional temperatures was 10^{-3} rad. If a plasma accelerator is specially designed for receiving the ultra-cold plasma, its temperature may be appreciably decreased. There is not a huge problem in getting cold ions from solid electrodes, or cold electrons from solid points, when molecular speed is small.

8.4.2 *Description of Innovation*

Innovative installation for transfer energy and impulse (Fig. 8.9) includes: the ultra-cold plasma injector, electrostatic collector, electrostatic electro-generator-thruster-reflector, and space apparatus. The plasma injector creates and accelerates the ultra-cold low density plasma.

The installation works the following way: the injector-accelerator forms and injects the cold neutral plasma beam with high speed in spaceship direction. When the beam reaches the ship, the electrostatic collector of spaceship collects and separates the beam ions from large area and passes them through the engine-electric generator or reflects them by electrostatic mirror. If we want to receive the thrust in the near beam direction (~90°) and electric energy, the engine works as thruster

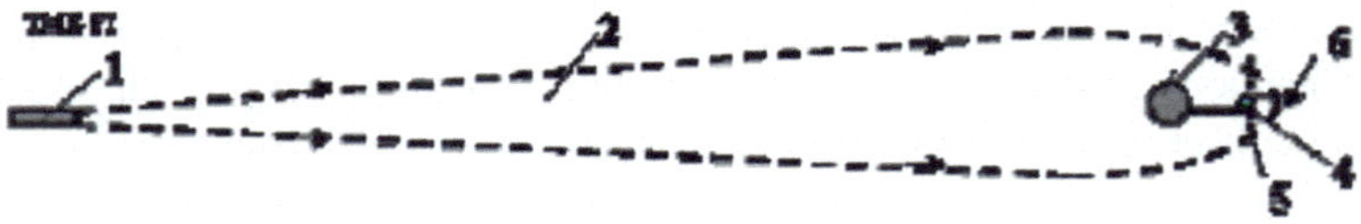

Fig. 8.9 Long distance space transfer of electric energy, matter, and momentum (thrust). Notation: 1—injector-accelerator of neutral ultra-cold plasma (ions and electrons), 2—plasma beam, 3—space ship or planetary team, 4—electrostatic ions collector (or magnetic collector), 5—braking electric nets (electrostatic electro-generator-thruster-reflector), 6—thrust

(accelerator of spaceship and breaker of beam) in the beam direction, and electric generator. If we want to get thrust in the opposite of beam direction, the space engine must accelerate the beam ions and expend energy. If we want to have maximum thrust in beam direction, the engine works as full electrostatic mirror and produces double thrust in the beam direction (full reflection of beam back to injector). The engine does not expend energy for full reflection.

The thrust is controlled by the electric voltage between engine nets; the thrust direction is controlled by the engine nets angle to beam direction (Bolonkin 2003b). Note, the thrust can slow the ship (decrease the tangential ship speed) and a far ship (located out of Earth orbit) can return to the Earth via Sun gravity.

Also note, the Earth's atmosphere absorbs and scatters the plasma beam, and the beam injector must be located on an Earth space mast or tower (up 40–60 km), or the Moon (Bolonkin 2003a, c). Only high energy beams can break through the atmosphere with little divergence. The advantage: the injector has a reflector and when the ship locates not far from the injector the beam will be reflected a lot of times and thrust increases in thousand times at start (Fig. 8.10) similar example in Bolonkin (2004a).

The proposed engine may be also used as AB-ramjet engine, utilizing the Solar wind or interstellar particles (Bolonkin 2003b).

The electrostatic collector and electrostatic generator-thruster-reflector proposed and described in Bolonkin (2003b). The main parts are presented below.

A Primary Ramjet propulsion engine is shown in Bolonkin (2006a; Fig. 8.1) and (Bolonkin 2006c; Chap. 2). Such an engine can work in a charged environment. For example, the surrounding region of space medium contains positive charge particles (protons, ions). The engine has two plates (1, 2), and a source of electric voltage and energy storage (3). The plates are made from a thin dielectric film covered by a conducting layer. The plates may be a net. The source can create an electric voltage U and electric field (electric intensity E) between the plates. One also can collect the electric energy from plate as an accumulator.

The engine works in the following way: apparatus are moving (in left direction) with velocity V (or particles 4 are moving in right direction). If voltage U is applied

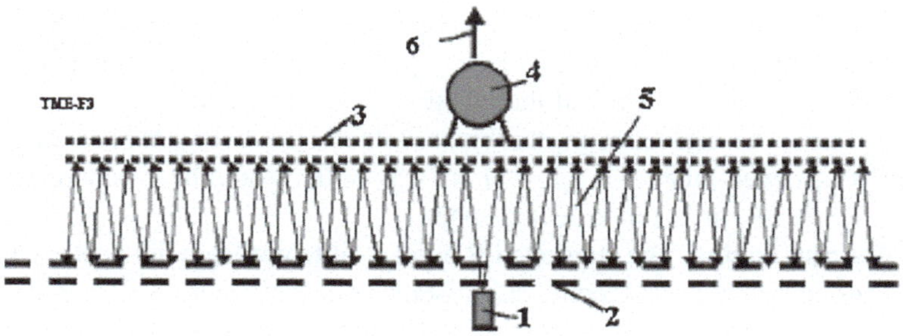

Fig. 8.10 Multi-reflection start of the spaceship having proposed engine. Notation: 1—injector-accelerator of cold ions or plasma, 2, 3—electrostatic reflectors, 4—space ship, 5—plasma beam, 6—the thrust

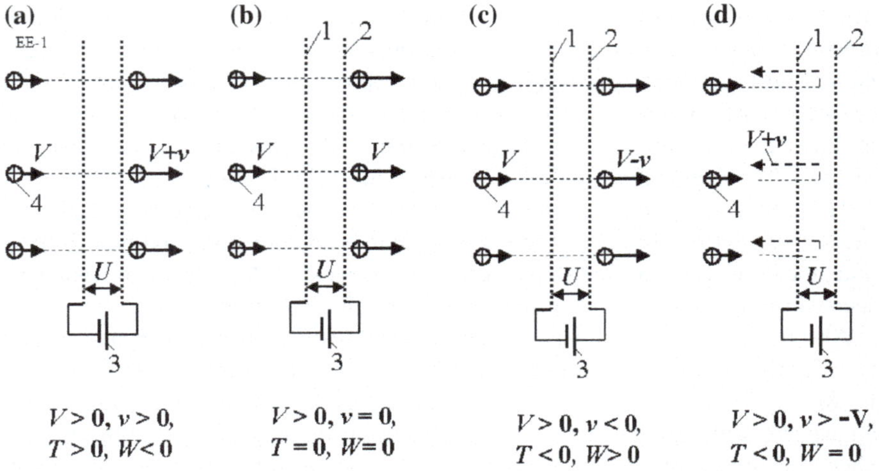

Fig. 8.11 Explanation of primary Space Ramjet propulsion (engine) and electric generator (in braking). Work in regime thrust. **b** Idle. **c** Work in regime brake. **d** Work in regime strong brake (full reflection). Notation: 1, 2—plate (film, thin net) of engine; 3—source of electric energy (voltage U); 4—charged particles (protons, ions); V—speed of apparatus or particles before engine (solar wind); v—additional speed of particles into engine plates; T—thrust of engine; W—energy (if W < 0 we spend energy)

to the plates, it is well-known that main electric field is only between plates. If the particles are charged positive (protons, positive ions) and the first and second plate are charged positive and negative, respectively, then the particles are accelerated between the plates and achieve the additional velocity $v > 0$. The total velocity will be $V + v$ behind the engine (Fig. 8.11a). This means that the apparatus will have thrust $T > 0$ and spend electric energy $W < 0$ (bias, displacement current). If the voltage $U = 0$, then $v = 0$, $T = 0$, and $W = 0$ (Fig. 8.11b). If the first and second plates are charged negative and positive, respectively, the voltage changes sign.

Assume the velocity v is satisfying $-V < v < 0$. Thus, the particles will brake, and the engine (apparatus) will have drag and will also brake. The engine transfers braked vehicle energy into electric (bias, displacement) current. That energy can be collected and used. Note that velocity v cannot equal $-V$. If v were equal to $-V$, that would mean that the apparatus collected positive particles, accumulated a big positive charge and then repelled the positive charged particles.

If the voltage is high enough, the brake is the highest (Fig. 8.11d). Maximum braking is achieved when $v = -2\ V$ ($T < 0$, $W = 0$). Note, the v cannot be more then $-2\ V$, because it is full reflected speed.

AB-Ramjet engine. The suggested Ramjet is different from the primary ramjet. The suggested ramjet has a specific electrostatic collector 5 (Fig. 8.12 a, c–g). Other authors have outlined the idea of space matter collection, but they did not describe or research the principal design of a collector. Really, for charging of a collector, we must move away from apparatus the charges. The charged collector attracts the

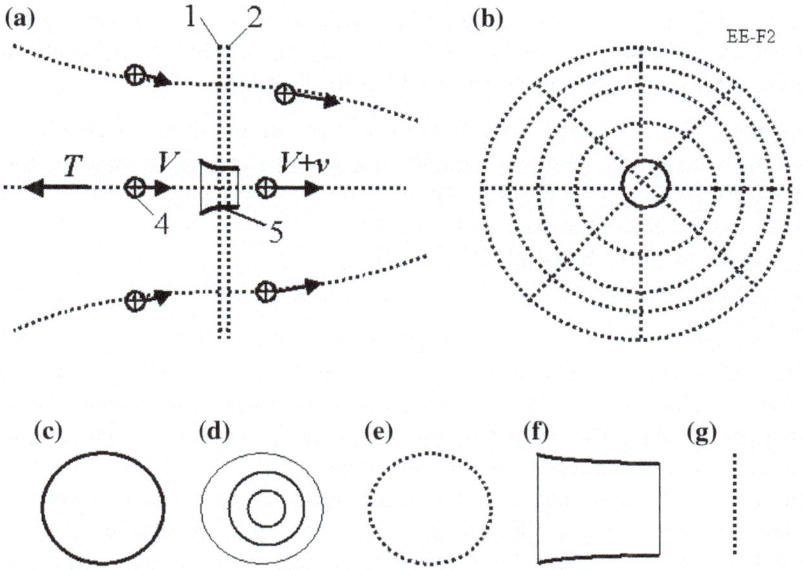

Fig. 8.12 Space AB-Ramjet engine with electrostatic collector (core). **a** Side view. **b** Front view. **c** Spherical electrostatic collector (ball). **d** Concentric collector. **e** Cellular (net) collector. **f** Cylindrical collector without cover butt-ends. **g** Plate collector (film or net)

same amount of the charged particles (charged protons, ions, electrons) from the space medium. When discharged from the collector, work will be idle. That cannot be useful.

The electrostatic collector cannot adsorb matter (as offered by some inventors), because it can adsorb *only* opposed charges particles, which will be discharged the initial charge of collector. Physic law of conservation of charges does not allow us to change charges of particles.

The suggested collector and ramjet engine have a special design (thin film, net, special form of charge collector, particle accelerator). The collector and engine passes the charged particles *across* (through) the installation and changes their energy (speed), deflecting and focusing them. That is why we refer to this engine as the AB-Ramjet engine. It can create thrust or drag, extract energy from the kinetic energy of particles or convert the apparatus' kinetic energy into electric energy, and deflect and focus the particle beam. The collector creates a local environment in space because it deletes (repeals) the same charged particles (electrons) from apparatus and allows the Ramjet to work when the apparatus speed is close to zero. Bolonkin developed the theory of the electrostatic collector and published it (Bolonkin 2005). The conventional electric engine cannot work in the usual plasma without the main part of the AB-engine—the special pervious electrostatic collector.

The plates of the suggested engine are different from the primary engine. They have concentric partitions which create additional radial electric fields (electric intensity) (Fig. 8.12b). They straighten, deflect and focus the particle beams and improve the efficiency coefficient of the engine. The central charge can have a different form (core) and design (Fig. 8.12 c–h). It may be:

1. a sphere (Fig. 8.12c) having a thin cover of plastic film and a very thin (some nanometers) conducting layer (aluminum), with the concentric spheres inserted one into the other (34, Fig. 8.12d);
2. a net formed from thin wires (Fig. 8.12e);
3. a cylinder (without butt-end) (Fig. 8.12f); or
4. a plate (Fig. 8.12g).

The design is chosen to produce minimum energy loss (maximum particle transparency—see "Theory" section). The safety (from discharging, emission of electrons) electric intensity in a vacuum is 10^8 V/m for an outer conducting layer and negative charge. The electric intensity is more for an inside conducting layer, and thousands of times more for positive charge.

The engine plates are attracted one to the other (see theoretical section). They can have various designs (Fig. 8.13a–d). In the rotating film or net design (Fig. 8.13a), the centrifugal force prevents contact between the plates. In the inflatable design; the low pressure gas prevents plate contact. A third design has (inflatable) rods supporting the film or net. The fourth design is an inflatable toroid which supports the distance between plates or nets (Fig. 8.13d).

As a note, the AB-ramjet engine can work using the neutral plasma. The ions will be accelerated or braked; the electrons will be conversely, braked or accelerated. The mass of the electrons is less than the mass of ions, by thousands of times, and the AB-engine will produce the same thrust or drag.

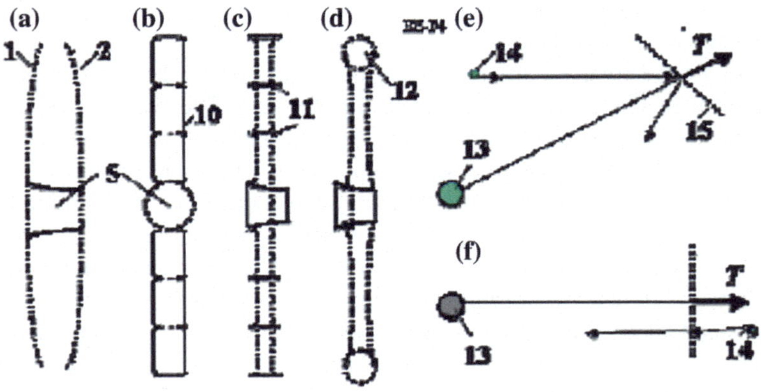

Fig. 8.13 Possible design of the main part of ramjet engine. **a** Rotating engine. **b** Inflatable engine (filled by gas). **c** Rod engine. **d** Toroidal shell engine. **e** AB-Ramjet engine in brake regime. **f** AB-Ramjet engine in thrust regime. Notation: 10—film shells (fibers) for support thin film and creating a radial electric field; 11—rods for a support the film or net; 12—inflatable toroid for support engine plates; 13—space apparatus; 14—particles; 15—AB-Ramjet

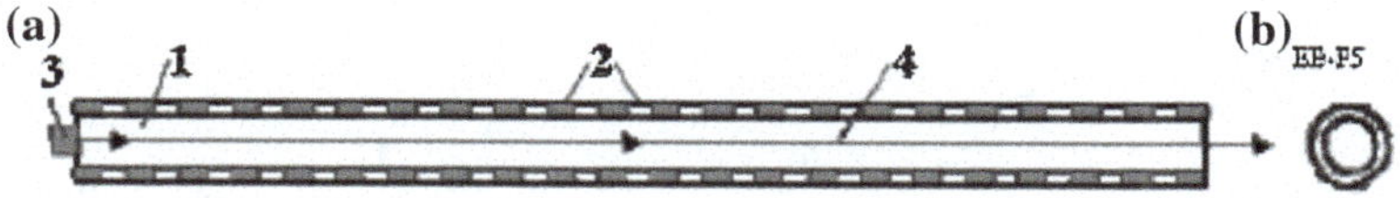

Fig. 8.14 Electric gun for charging AB-Ramjet engine and transfer charges (energy) in long distance. **a** Side view. **b** Front view. Notations: 1—gun tube, 2—opposed charged electrodes, 3—source of charged particles (ions, electrons), 4—particles beam

Plasma accelerator. The simplest linear plasma accelerator (principle scheme of linear particle accelerator) for the plasma beam is presented in Fig. 8.14. The design is a long tube (up to 10 m) which creates a strong electric field along the tube axis (100 MV/m or more). The accelerator consists of the tube with electrical isolated cylindrical electrodes, ion source, and voltage multiplier. The accelerator increases the speed of ions, but in the end of the tube, electrons are injected into ion beam. This plasma accelerator can accelerate charged particles up to 1000 meV. Electrostatic lens and special conditions allow the creation of a focusing and self-focusing beam which can transfer the charge and energy long distances into space. The engine can be charged from a satellite, a spaceship, the Moon, or a top atmosphere station (space tower) (Bolonkin 1991, 2006d). The beam may also be used as a particle beam weapon.

Approximately ten years ago, the conventional linear pipe accelerated protons up to 40 meV, with a beam divergence of 10^{-3} rad. However, acceleration of the multi-charged heavy ions may result in significantly more energy.

At present, the energy gradients as steep as 200 GeV/m have been achieved over millimeter-scale distances using laser pulses. Gradients approaching 1 GeV/m are being produced on the multi-centimeter-scale with electron-beam systems, in contrast to a limit of about 0.1 GeV/m for radio-frequency acceleration alone. Existing electron accelerators such as SLAC <http://en.wikipedia.org/wiki/SLAC> could use electron-beam afterburners to increase the intensity of their particle beams. Electron systems, in general, can provide tightly collimated, reliable beams while laser systems may offer more power and compactness.

The cool plasma beam carries three types of energy: kinetic energy of particles, ionization, and dissociation energy of ions and molecules. These also carry particle mass and momentum. The AB-Ramjet engine can utilize only kinetic energy of plasma particles and momentum. The particles brake and produce an electric current and thrust or reflected and produce only thrust in the beam direction. If we want to collect a plasma matter and to utilize also the ionization energy of plasma (or space environment) ions and dissociation energy of plasma molecules we must use the modified AB-Ramjet engine described below (Fig. 8.15).

The modified AB-engine has magnetic collector (option), three nets (two last nets may be films), and issue voltage (that also may be an electric load). The voltage, U, must be enough for full braking of charged particles. The first two nets brake the electrons and precipitate (collect) the electrons on the film 2 (Fig. 8.15).

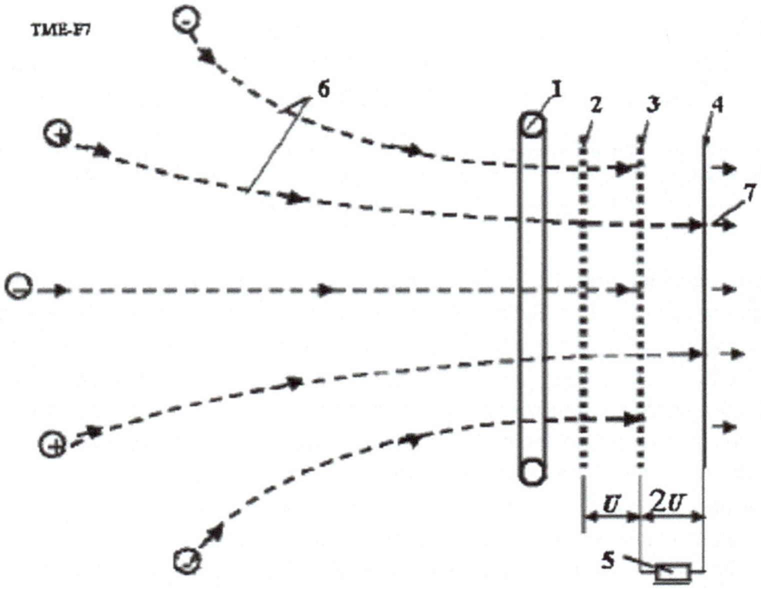

Fig. 8.15 AB-engine with collected matter of plasma beam, kinetic energy of particles, energy ionization and dissociation; Notations: 1—magnetic collector; 2—4—plates (films, nets) of engine; 5—e; electric load; 6—particles of plasma; 7—radiation; U- voltage between plates (nets)

The last couple of films (2, 3 in Fig. 8.11) brake and collects the ions. The first couple of nets accelerate the ions; that is why the voltage between them must be double.

The collected ions and electrons have the ionized and dissociation energy. This energy is significantly more powerful (up 20–150 times) than the chemical energy of rocket fuel (see Table 8.1), but significantly less than the kinetic energy of particles (ions) equal U (in eV—U may be millions volts). They may be used by ship. The ionization energy is conventionally picked out in photons (light, radiation), which easy are converted in a heat (in closed vessel); the dissociation energy is conventionally picked out in heat.

The light energy may be used in the photon engine as thrust (Fig. 8.16a), or in a new power laser (Fig. 8.16b). The heat energy may be utilized in a conventional way (Fig. 8.16c). The offered new power laser (Fig. 8.16b) has an ultra-cool rare plasma with short lifetime, located in a cylinder. If we press it (decrease density of plasma), the electrons and ions will connect and produce photons of very closed energy (laser beam). If we compress very quickly by explosion, the power of the beam will be high. The power is only limited by the amount of plasma energy.

After recombination if ions and electrons, we receive the conventional matter. This matter may be used as nuclear fuel (in thermonuclear reactor), medicine, food, drink, oxidizing agent for breathing, etc.

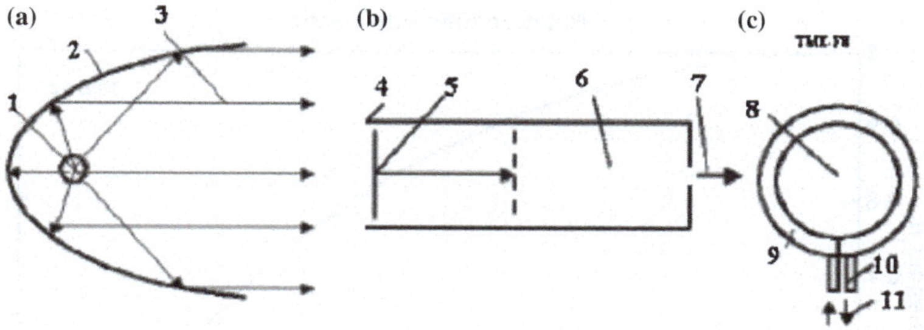

Fig. 8.16 Conversion of ionization energy into radiation and heat; a—photon engine; b—power laser (light beamer); c—heater; Notations: 1—recombination reactor; 2—mirror; 3—radiation (light) beam; 5—piston; 6—volume filled by cold rare plasma; 7—beam; 8—plasma; 9—heat exchanger; 10—enter and exit of hear carrier; 11—heat carrier

8.4.3 Transfer Theory of the High Speed Neutral Ultra-Cold Plasma and Particles

Below are the main equations and computations of neutral, ultra-cold plasma beams having a velocity up to relativistic speed. These equations were created from conventional mechanics and relativistic theory. Note the ratios:

$$\beta = \frac{V}{c}, \quad \beta_S = \frac{V_S}{c} \tag{8.27}$$

where V is plasma beam speed, m/s; $c = 3 \times 10^8 \mathrm{m/s}$ is light speed, m/s; V_s is projection of a ship speed in beam direction.

1. **Relative relativistic time**, $\bar{t}$, for observer moving together with beam is

$$\bar{t}' = \frac{t'}{t} = \sqrt{1 - \beta^2} \tag{8.28}$$

where t' is time for observer moving together with beam (system coordinate connected with beam)[s], t is time for Earth's observer[s]. Computation of Eq. (8.28) is presented in Fig. 8.17. The beam time decreases for relativistic speed. That means the beam divergence is also decreased and beam energy may be passed for long distance.

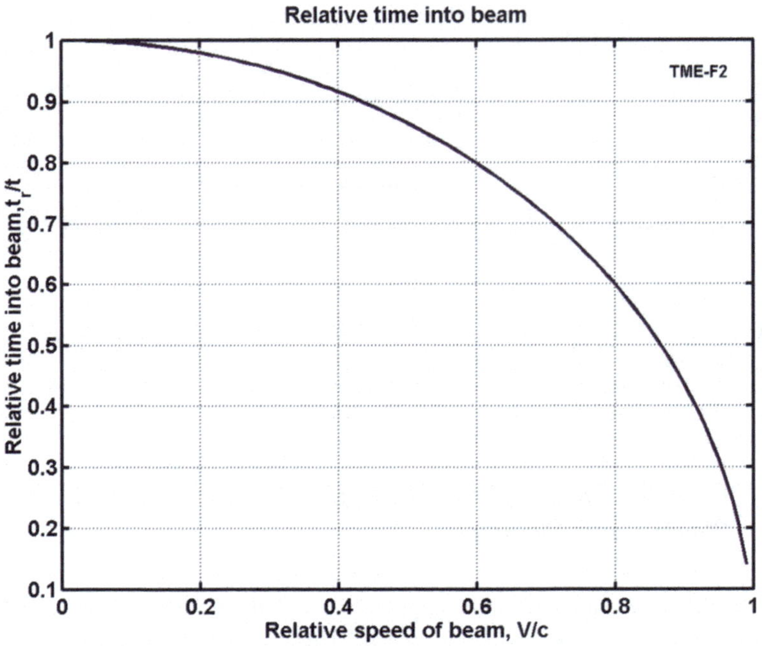

Fig. 8.17 Beam relative time versus beam relative speed for high relativistic beam speed

2. **The power spent for acceleration plasma beam** in Earth for efficiency = 1 (kinetic power of particle beam) is

$$P_B = \frac{M_0 c^2}{2} \frac{\beta^2}{\sqrt{1-\beta^2}}, \quad \text{or for} \quad \beta \ll 1 \quad P_B = \frac{M_0 V^2}{2} \ [\mathrm{W}] \tag{8.29}$$

where M_0 is mass flow of beam, kg/s in Earth system of coordinate. The computations of Eq. (8.29) for the intervals (0–0.1)c and (0–0.95)c are presented in Figs. 8.18 and 8.19. The relativistic speed needs very high power in any method because the relativistic beam requires this energy.

3. **The power P_i of dissociation and single ionization** of one nucleon is:

$$P_i = 1.6 \times 10^{-19} \frac{M_0}{m_p n} e_i \quad [\mathrm{J/s}] \quad \text{or} \quad P_i = \frac{M_0}{m_p n} e_i \quad [\mathrm{eV/s}] \tag{8.30}$$

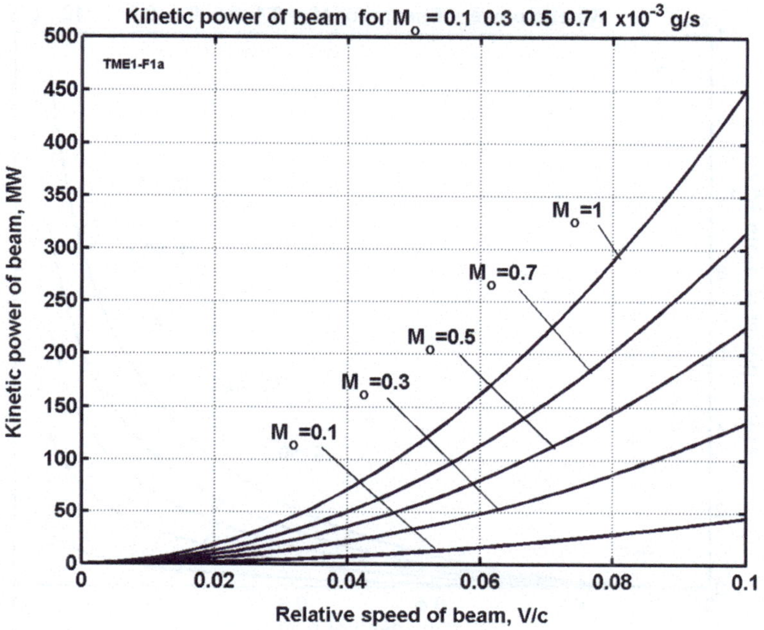

Fig. 8.18 Power for the beam acceleration via beam flow mass and relative beam speed for interval (0–0.1)c

where $m_p = 1.67 \times 10^{-27}$ kg is mass of proton, n is number of nucleon in nucleus, e_i is energy of dissociation, ionization, or molecular breakup respectively. The energy of the first ionization (ion lost one electron) approximately equals from 2 to 14 eV. Magnitudes of this energy for some molecules and ions are in Table 8.3.

If speed is relativistic, this energy is small in comparison with the kinetic energy of the beam. For interplanetary speed (V_S = 8–15 km/s), the energy of ionization reaches 15–50% from kinetic energy of the beam. That decreases the coefficient of efficiency for the launch installation. If we used the heavy ions or a charged matter, the ionization energy decreases but voltage increases. For interplanetary vehicles, it is not important because required voltage for low speeds is small ($U \approx$ 5–20 V). Figure 8.20 shows the required energy for different cases.

4. **The maximal thrust (drag)** from the full reflected one charged plasma beam, for Earth's observer and relativistic speed and non-relativistic speed may be estimated by following equations:

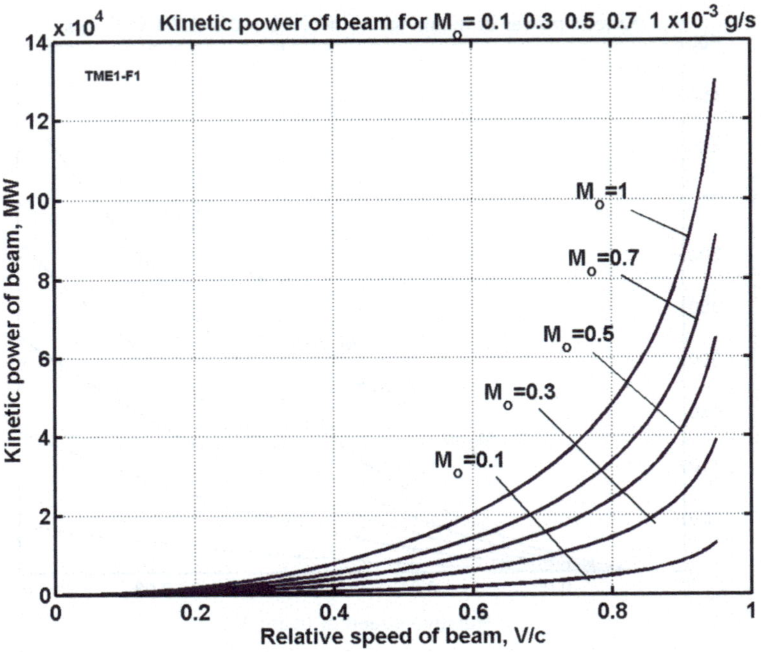

Fig. 8.19 Power for the beam acceleration via beam flow mass and relative beam speed for interval (0–0.95)c

Table 8.3 Energy ionization, dissociation, and molecular breakup of some molecules and ions in eV

Molecular breakup	$H_2O \rightarrow H_2 + O$	2 eV	$CO_2 \rightarrow C + O_2$	0.093 eV
Dissociation	$H_2 \rightarrow H + H$	4.48 eV	$O_2 \rightarrow O + O$	5.1 eV
Ionization	$H \rightarrow H^+$	13.6 eV	$H_2 \rightarrow H_2^+$	2.65 eV
Ionization	$O_2 \rightarrow O_2^+$	6.7 eV		

$$T_{\max} = 2M(V \mp V_S) \approx \frac{2M_0 c(\beta \mp \beta_S)}{\sqrt{1-\beta^2}}, \quad \text{or}$$

$$\text{for} \quad \beta_S \ll 1 \quad \text{the thrust is} \quad T_{\max} = \frac{2M_0 c\beta}{\sqrt{1-\beta^2}}, \tag{8.31}$$

$$\text{for} \quad \beta \ll 1, \quad \beta_S \ll 1, \quad \text{the thrust is} \quad T_{\max} = 2M_0(V - V_S)$$

Here, M is calculated mass of a moving relativistic particle flow, kg/s; M_0 is mass of the particle flow measured by Earth's observer, kg/s. [Note: If the space ship move along the beam in same direction, the thrust is decreased (sign is "−"); if that moves in opposed direction, the drag is increased (sing is "+"). This drag

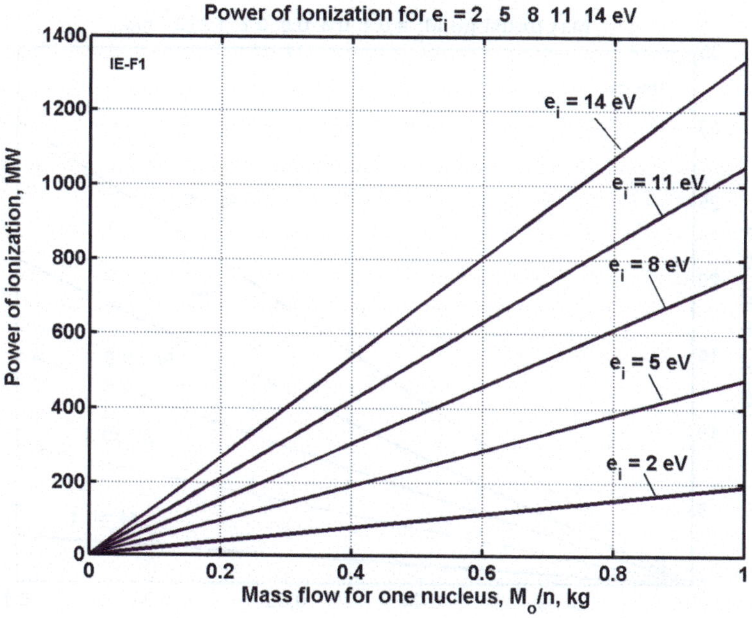

Fig. 8.20 Power of ionization versus mass flow and ionization potential (Eq. (8.30))

(thrust) is not requested the ship propulsion energy]. The result of computation for intervals (0/0.1)c, (0/0.95)c, $V_S = 0$ are presented in Figs. 8.21 and 8.22.

5. **The divergence of beam** is a very important quantity. If divergence is small, we can pass our energy in long distance S:

$$D = \frac{ut'}{Vt} = \frac{u}{c}\frac{\sqrt{1-\beta^2}}{\beta}S, \quad \overline{D} = \frac{D}{S}, \quad S = c\beta t \tag{8.32}$$

where u is maximal radial speed, m/s; D is maximal radial distance (radius of plasma beam), m; $\overline{D}$ is relative divergence (angle of divergence, $\theta = 2\overline{D}$ rad); t is time of beam moving, s.

The computation of Eq. (8.32) is shown in Fig. 8.23. We need the u (ultra-cold plasma) for decreasing of divergence to be as small as possible (u = 0.01–1 m/s). In this case, we can transfer energy in the large distance and accelerate a ship for relativistic speed. The plasma is mixture of ions and electrons. If it is low-density, it can exist a long time. The cold plasma can be emitted from solid electrodes. [Note: Eqs. (8.28) and (8.32) shows when V = c, then t′ = 0 and deviation D = 0. That means the deviation can be small as we want but we need a big power for it.] The corresponding temperature is:

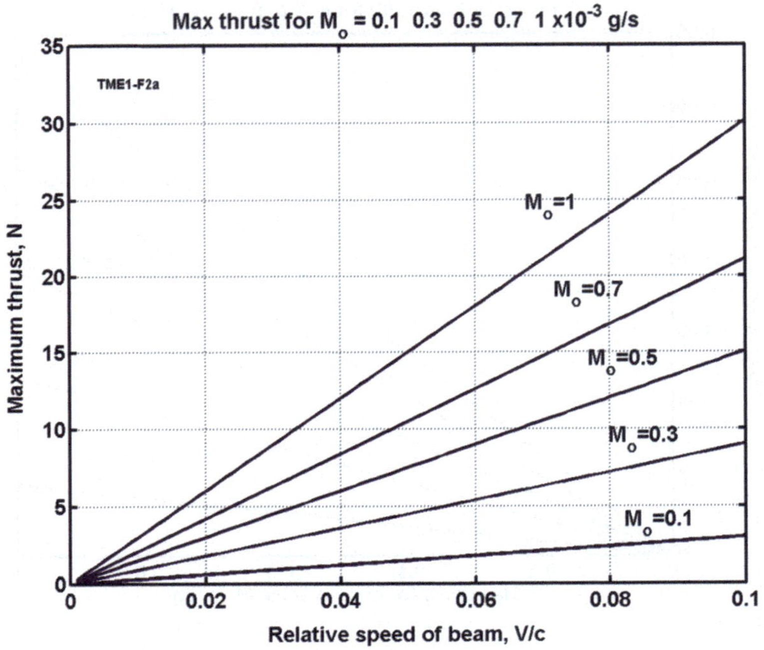

Fig. 8.21 Maximum thrust (drag) is produced by beam in space ship for $V_S = 0$ and the interval (0/0.1)c

$$T_c = \frac{mu^2}{ik}, \tag{8.33}$$

where m is mass of molecule (ion) [kg]; $m = m_p n$, here $m_p = 1.67 \times 10^{-27}$ is mass of proton, n is number of nucleons into nucleus; $i = 3$ for single ion (for example O^+), $i = 5$ for double molecule (for example O_2^+), $i = 6$ for multi-molecular ions, $k = 1.39 \times 10^{-23}$ is Boltzmann constant. For $u = 0.1$–1 m/s the temperature is about 10^{-3} K, the relative divergence is 10^{-9}.

6. **Accelerate voltage is**

$$U = \frac{mV^2}{2q} = \left(\frac{m_p}{q}\right)\frac{nc^2}{2}\frac{\beta^2}{\sqrt{1-\beta^2}} \tag{8.34}$$

where $q = 1.6 \times 10^{-19}$ C is the electron (ion) charge. The computations are presented in Figs. 8.24 and 8.25. The needed voltage may be reduced by Z times if the ion has Z charges (delete Z electrons from ion).

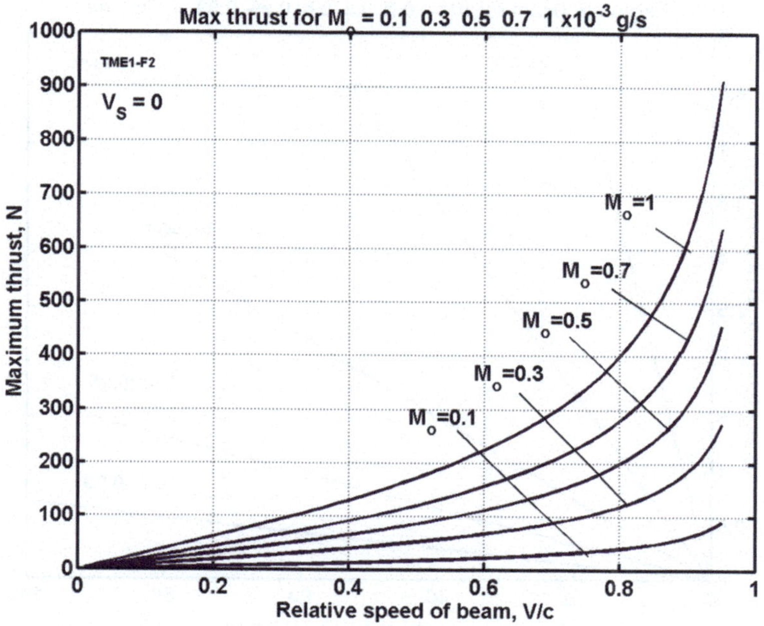

Fig. 8.22 Maximum thrust (drag) is produced by beam in space ship for $V_S = 0$ and interval (0/0.95)c

7. **The speed V_s and distance of space ship, S**, can be computed by a conventional method (Earth's observer):

$$V_S = at, \quad S = \frac{at^2}{2}, \quad S = \frac{V_S^2}{2a}, \quad a = \frac{T}{M_S} \tag{8.35}$$

where a is ship acceleration, m/s^2. M_s is ship mass, kg, V_S is ship speed measured by Earth's observer, m/s.

8. **Relative beam speed** for a ship observer is:

$$\beta_{BS} = \frac{\beta \pm \beta_S}{1 + \beta\beta_S} \tag{8.36}$$

where β, β_s is relative speed of beam and space ship respectively measured by Earth's observer. The sign "−" is used for same direction of speeds.

9. **Loss energy of the beam** in the Earth atmosphere may be estimated by the following way:

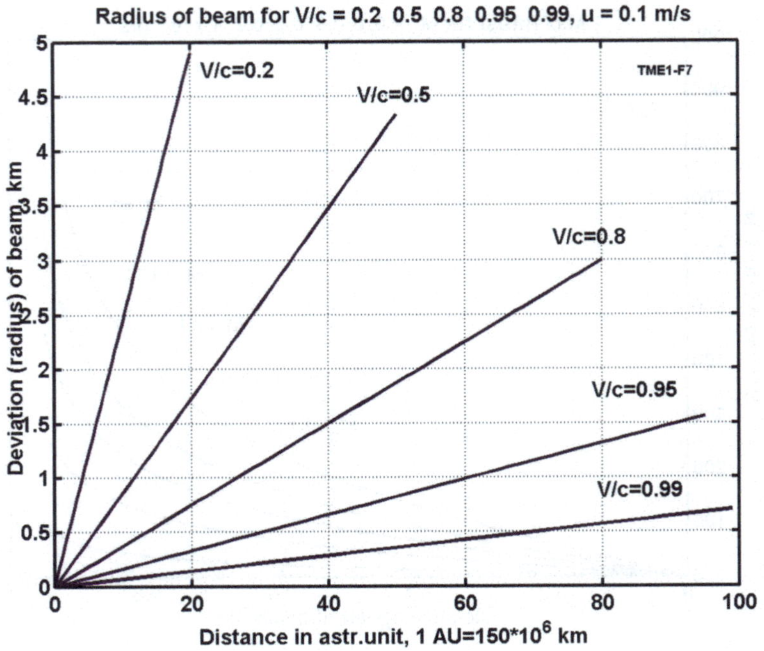

Fig. 8.23 Radius of beam divergence via distance and ratio V/c

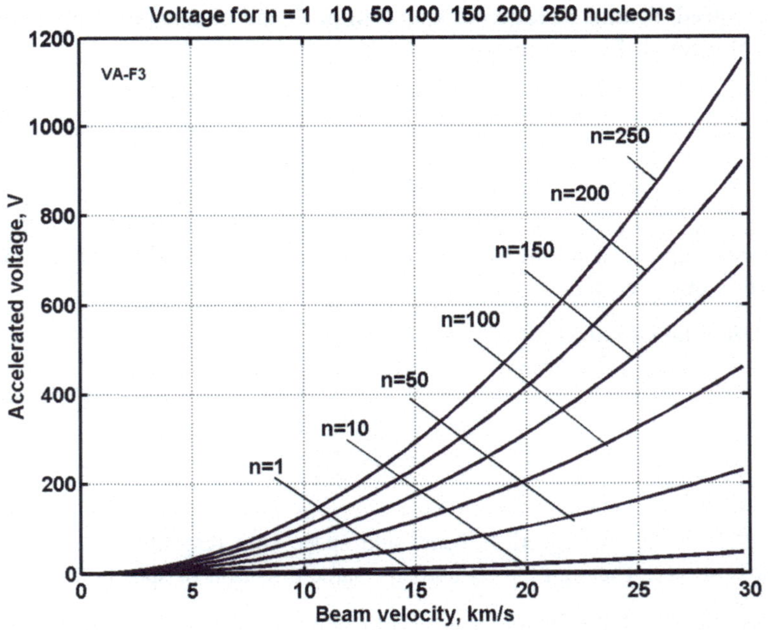

Fig. 8.24 Accelerated voltage versus the conventional beam speed and number of nucleons

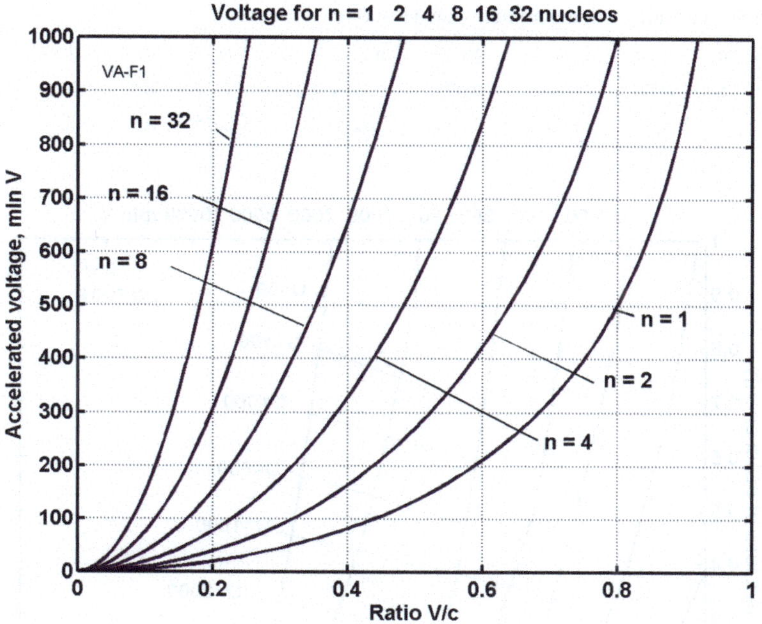

Fig. 8.25 Accelerated voltage versus a relativistic speed ratio V/c and number of nucleons

$$\tau = \frac{100H_0\rho_0\bar{\rho}(h)\bar{p}(h)}{R_t(U)}, \quad R_t = \frac{m}{m_p}R_t\left(\frac{m_p}{m}U\right) \tag{8.37}$$

where $H_0 = P_a/\rho = 10^4/1.225 = 8163$ m is thickness (height) of Earth atmosphere having constant density $\rho = 1.225$ kg/m^3, $P_a = 10^4$ kg/m^2 is the atmospheric pressure; $\bar{\rho}(h)$ is relative atmosphere density; $\bar{p}(h)$ is relative atmosphere pressure; R_t is particle track in atmosphere [cm]; m is mass of particle, kg; h is altitude, m; U is beam energy, MeV; $\rho_o = 0{,}001225$ g/cm^3 is atmosphere density; 100 is transfer coefficient meter into cm. Quantities R_t, $\bar{\rho}(h)$, $\bar{p}(h)$ for proton are given below in Tables 8.4 and 8.5.

Results of computation of Eq. (8.37) are presented in Figs. 8.26, 8.27 and 8.28.

Table 8.4 Value R_t [g/cm^2] versus energy of proton in MeV

U [MeV]	0.1	1	10	50	100	200	300	400	500
R_tg/cm^2	1×10^{-4}	1.09×10^{-2}	0.99×10^{-1}	2.56	8.835	29.64	58.08	93.73	133.3
U	600	700	800	1000	2000	3000	5000	7000	10,000
R_t	176	222	270	370	910	1363	2543	3583	5081

Table 8.5 Standard Earth atmosphere (Landis 2004)

h km	0	5	10	20	40	60	100
$\bar{\rho}(h)$	1	0.661	0.338	0.072	3.27×10^{-3}	2.71×10^{-4}	4.41×10^{-7}
$\bar{\rho}(h)$	1	0.533	0.261	0.054	2.92×10^{-3}	8.35×10^{-4}	3.20×10^{-7}

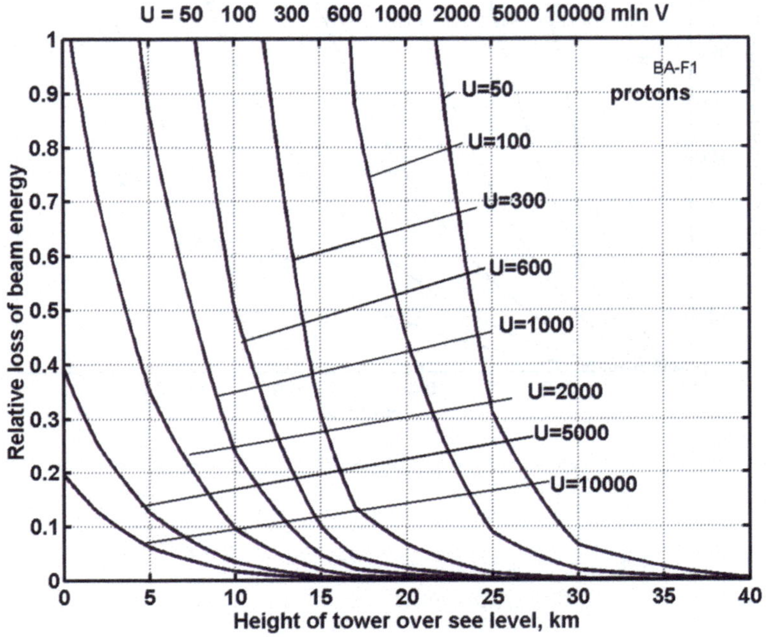

Fig. 8.26 Relative energy loss of the proton particle beam via a tower altitude in Earth atmosphere. Accelerate voltage U is in millions volts

Evidently, only high energy particle beams break up the Earth atmosphere. There is no problem if the particle beam starts from a space tower of 40–80 km altitude or from the Moon (Bolonkin 2003a, 2004c).

The last formula in Eq. (8.37) allows recalculation by the particle track for any atom. For example, we want to calculate the particle track for oxidizer particle having $m = 16m_p$ and energy 8000 meV. We take the R_t from Table 8.4 for $U = 8000/16 = 500$ meV and multiple by 16. The result is $R_t = 133 \times 16 = 2128$. The particle track $T_r = R_t/\rho_o = 2128/0.001225 = 1737142$ cm = 17.4 km in the air having density 1.225 kg/m^3. That is enough to break the Earth's atmosphere of the constant density 8.163 km, but the loss of energy will be 8.163/17.4 = 0.47 (47%). The divergence may also be increased by atmosphere. Loss and divergence may be improved if the beam station is located on a mountain or special tower having the height about 40–60 km.

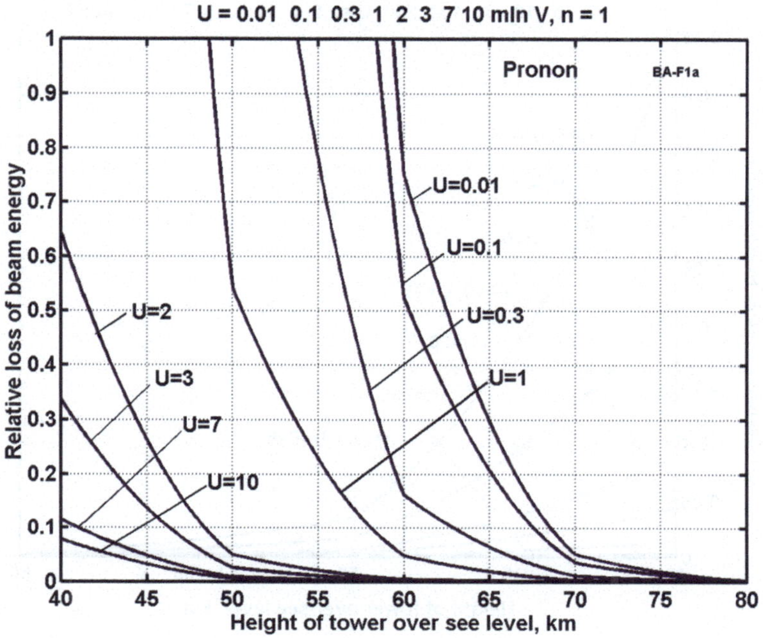

Fig. 8.27 Relative energy loss of the proton particle beam via a tower altitude in Earth atmosphere. Accelerate voltage U are in millions volts. Angles in curve are result of the linearization data of Tables 8.4 and 8.5

10. **Multi-reflex launch and landing** (Fig. 8.17): In a starting or braking period, the thrust (braking) can be increased if we use the multi-reflect method (Bolonkin 2005). Multi-reflect in launching does not increased the installation power (thrust is increased by increasing of efficiency); multi-reflex in braking converts the apparatus kinetic energy into the electric energy which can be utilized by apparatus or operated station. The theory of multi-reflection is described below see (Bolonkin 2005).
11. **Change in beam power**: The beam power will be reduced if one (or both) reflector is moved, because the beam speed changes. The total relative loss, q, of the beam energy in one double cycle (when the beam is moved to the reflector and back) is:

$$q = (1-2\gamma)(1-2\xi)(1 \pm 2v)\varsigma,\ q > 0, \tag{8.38}$$

where v is the loss (useful work) through relative mirror (lens) movement, $v = V_S/V$, V_S is the relative speed of the electrostatic mirrors (space apparatus) [m/s], $V =$ is the speed of the beam (in system of coordinates connected with an power operating

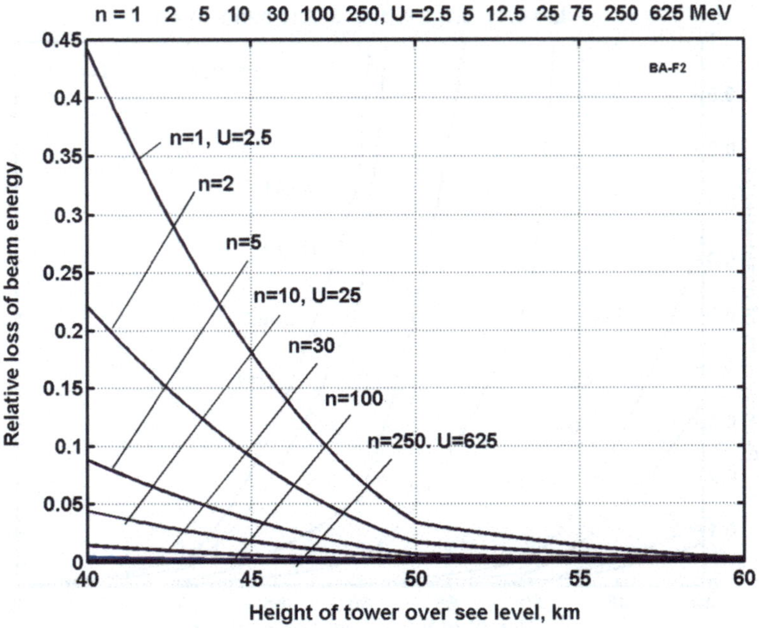

Fig. 8.28 Relative energy loss of the particle beam via the tower altitude in Earth atmosphere and number n of nucleons in nucleus. Accelerate voltage U in millions volts

station). We take the "+" when the distance is reduces (braking) and take "−" when the distance is increased (as in launching, a useful work for beam). γ is the coefficient reflectivity of an electrostatic mirror (the loss of beam energy through the electrostatic reflector); ξ is the loss (attenuation) in the medium (air) (see point 8). If no atmosphere exists, $\xi = 0$; ς is the loss through beam divergence ($\varsigma = 1$ if $D < D_r$, where D_r is diameter of the electrostatic mirror). For a wire net electrostatic reflector $\varsigma = 2d_w/l$ where d_w is diameter of wire, l is size of mesh. For example, for the net having a mesh 0.1×0.1 m, $l = 0.1$ m and a wire $d_w = 0.0001$ m, then $\varsigma = 0.002$.

The beam pressure, T, of two opposed high reflectors after a series of reflections, N, to one another is:

$$T(V_S) = T_0 + \frac{2P_B}{V}\sum_{j=1}^{N} q^j(V_S), \quad N(V_S) = \sqrt{\frac{kD_rV}{uS}}. \quad P_B = \frac{M_0V^2}{2}. \qquad (8.39)$$
$$T_0 = 2M_0(V - V_S) \geq 0$$

where S is distance between electrostatic reflectors of the station and ship [m], $k = 1$ to 1.5 is the correction coefficient for the case when $D > D_r$. For primary estimation, $k = 1$. V_S is small and V is high, the multi-reflex T may be large. For example, if $V_S = 10$ m/s, $V = 30$ km/s, S is small, the number of reflection may reach

n = 30,000/10/2 = 1500 times more than regular thrust. That is well for ship trip starting and braking.

12. **Limitation of reflection number**: If the reflector is moved away, the maximum number of reflections, N, is limited by $q > 0$, $V_S < 0.5\ V$ (see Eq. (8.38)). At ship launch or braking, the maximum thrust is limited by a safety acceleration or deceleration. The propulsion efficiency coefficient, η, (without loss for ionization) may be computed using the equation

$$\eta = TV_S/P_B \tag{8.40}$$

For full reflection, Eq. (8.40) has the form:

$$\eta = \frac{4(V - V_S)V_S}{V^2}, \quad \eta_{\max} = 1 \quad \text{for} \quad V = 2V_S \tag{8.41}$$

Computation of launch and landing trajectories computed by the usual method of integration is:

$$a_i = T/M_S, \quad t_{i+1} = t_i + \Delta t, \quad V_{S,i+1} = V_{S,i} + a_i \Delta t, \quad S_{i+1} = S_i + V_{S,i}\Delta t \tag{8.42}$$

8.4.4 Project: Interstellar Probe Having Speed 30.000 Km/S, Weight 0.1 Kg

Let us assume we want to estimate an interstellar probe which can reach the nearest solar systems. As known, they are located about 4–60 light years from our Sun. That means the apparatus having speed V_S = 0.1c (V_S = 30,000 km/s) can reach them in 40–60 years. Reminder, "Voyager-1" was flown for 30+ year, sending information up to present time. But it has speed only 20 km/s and was reached only the boundary of Solar system (about 2 billion km).

Assume, the weight of interstellar probe is 0.1 kg. If the distance of acceleration is $S = 1.5 \times 10^{11}$ m (1 AU), the acceleration and acceleration time must be:

$$a = \frac{V_S^2}{2S} = \frac{9 \times 10^{14}}{2 \cdot 1.5 \times 10^{11}} = 3 \times 10^3 \text{ m/s}^2,$$

$$t = \frac{V_S}{a} = \frac{3 \times 10^7}{3 \times 10^3} = 10^4 \text{ s} = 167 \text{ min} = 2.78 \text{ h}$$

The thrust, requested acceleration energy and power are

$$T = aM_S = 3 \times 10^3 \cdot 0.1 = 300\,N, \quad W = \frac{M_S V_S^2}{2} = \frac{0.1 \cdot 9 \times 10^{14}}{2} = 4.5 \times 10^{13}\ \text{J},$$
$$P = \frac{W}{t} = \frac{4.5 \times 10^{13}}{10^4} = 4.5\ \text{GW}$$

The mass of the beam flow, and energy spent by beam station are (Eqs. (8.29) and (8.31)):

$$M_0 = \frac{T\sqrt{1-\beta^2}}{2c(\beta-\beta_S)} = \frac{300\sqrt{1-0.01}}{2 \cdot 3 \times 10^8(0.1-0.05)} \approx 10^{-5}\ \text{kg/s},$$
$$P_B = \frac{M_0 c^2}{2}\frac{\beta^2}{\sqrt{1=\beta^2}} = \frac{10^{-5} \cdot 9 \times 10^{16}\, 0.01}{2\sqrt{1-0.01}} = 4.5\ \text{GW}$$

Here, β_S = 0–0.1. We take the average value $\beta_S = 0.05$. Notice that $P_B = P$; this means our installation transfer the station energy to ship with efficiency = 1. Unfortunately, this energy is very high. Tens of electric power stations must accelerate this probe in 23 days. We cannot decrease this amount by any methods because that is a minimum energy required by space probe. Divergence D for $u = 0.01$ m/s, voltage U ($n = 1$), jet speed $\beta = 0.1$ and plasma temperature T_p are

$$D = \frac{u}{c}\frac{\sqrt{1-\beta^2}}{\beta}S \approx \frac{0.01}{3 \times 10^8}\frac{1.5 \times 10^{11}}{0.1} = 500\ \text{m},$$
$$T_p = \frac{mu^2}{ik} = \frac{1.67 \times 10^{-27}10^{-4}}{3 \cdot 1.38 \times 10^{-23}} = 0.4 \times 10^{-8}\text{K},$$
$$U = \left(\frac{m_p}{q}\right)\frac{nc^2}{2}\frac{\beta^2}{\sqrt{1-\beta^2}} \approx \left(\frac{1.67 \times 10^{-27}}{1.6 \times 10^{-19}}\right)\frac{1 \cdot 9 \times 10^{16}}{2}\frac{0.01}{1} = 4.7 \times 10^6\quad \text{V}.$$

For jet speed $\beta = 0.9$ divergence $D = 25$ m, for jet speed $\beta = 0.99$ divergence $D = 7$ m. The power of dissociations ($H_2 \to H + H$, 2.2 eV) and ionization (HH^+, 13.6 eV) are equal to:

$$E_i = 1.6 \times 10^{-19}(e_d + 2e_i)\frac{M_0}{m_p n} = 1.6 \times 10^{-19}(2.2 + 2 \cdot 13.6)\frac{10^{-5}}{1.67 \times 10^{-27} \cdot 2}$$
$$= 14\ \text{kW}$$

In given case (comparison with P_B above) this value is small and we can negligee it. But into planetary flight (V_s = 8–30 km/s and large M_0) this energy is essential.

8.4.5 Discussion

G.A. Landis writes about using particle beams for interstellar flight (Landis 2004). The beam is braked by a magnetic sail. His work also contains only common speculations; no theory, no mathematical models, no computations. For more than ten years, authors investigated the magnetic sail, but not its theory, so no formulas allowing for correct calculation or estimation the magnetic sail drag were formed. Moreover, most MagSail works contain a common mistake. Landis offered the beam temperature of 45 K. The theory in this article shows that this temperature gives the beam divergence, which does not allow for interstellar flights. The estimations show that for high speed particles, especially relativistic particles, the affective diameter equals meters, and magnetic field must be powerful. In addition, the magnetic sail is impossible to create at the present time: electric ring needs in cryogenic temperature and spaceship must have power cryogenic equipment because the Sun will warm the ring for any heat insulation; for starting the ring needs a power electric station; a special equipment is necessary for displacing the ring of 100 km diameter into space; if the ring temperature exceeds a critical cryogenic temperature in any ring place, the ring explodes. The ring weight is also high (22 tons for diameter 100 km), and the produced magnetic field is very weak (10^{-6} T). The magnetic sail thus does not have active control. This means that the ship will move in one (non-controlled) direction, and a ship mission will be useless. These obvious defects make the application of the magnetic sail impossible, with little or no progress in solutions to these problems since 1988.

The offered idea to use the AB-Ramjet engine suggested by Bolonkin in 1982 (Milonni and Eberly 1988; Driggers 2001; Kantrowitz 1972; Collection "Beamed Energy Propulsion" 2003; Metzger and Landis 2001; Bolonkin 1987, 1991, 1992a, c). He developed detail in Bolonkin (2006d). The installation contains an electrostatic particle collector. The propulsion-reflected system is made of a light net from thin wire, which can have a large area (tens km), and thus allows controlled thrust and controlled thrust direction without the turning of net (Fig. 8.9). This new method uses the ultra-cold full neutral relativistic plasma, and having small divergence. The method may be used for the acceleration of space apparatus (up relativistic speed) for launch and landing on small planets (asteroids, satellites) without atmosphere. For Earth, the offered method will be efficient if a tower (mast) of about 40–80 km in height is built (Bolonkin 2003b). At the present time, it is the most realistic method for relativistic probe.

The method suggested here does not require a magnetic sail. The electrostatic sail and AB-Ramjet engine offered as solutions by Bolonkin (2005). This sail is light (100–300 kg), cheap, and has tens kilometers (hundreds km for low beam density) of the effective radius. For example, for solar wind the magnetic effective radius decreases proportional $1/R^2$ (where R is distance of the sail from the Sun), electrostatic effective radius decreases approximately $1/R$ see (Bolonkin 2005). That is very important advantage.

8.5 Converting of Matter to Nuclear Energy, and Photon Rockets for Space Flight

Bolonkin offers a new nuclear generator which allows to convert any matter to nuclear energy in accordance with the Einstein equation $E = mc^2$. The method is based upon tapping the energy potential of a Micro Black Hole (MBH) and the Hawking radiation created by this MBH. As is well-known, the vacuum continuously produces virtual pairs of particles and antiparticles, in particular, the photons and anti-photons. The MBH event horizon allows separating them. Anti-photons can be moved to the MBH and be annihilated; decreasing the mass of the MBH, the resulting photons leave the MBH neighborhood as Hawking radiation. The offered nuclear generator (named by author as AB-Generator) utilizes the Hawking radiation and injects the matter into MBH and keeps MBH in a stable state with near-constant mass.

The AB-Generator can produce gigantic energy outputs and should be very small and cheaper than a conventional electric station by a factor of hundreds of times. One also may be used in aerospace as a photon rocket or as a power source for many vehicles.

Many scientists expect the Large Hadron Collider at CERN will produce one MBH every second. A technology to capture them may follow; than they may be used for the AB-Generator.

8.5.1 Black Hole

In general relativity, a black hole is a region of space in which the gravitational field is so powerful that nothing, including light, can escape its pull. The black hole has a one-way surface, called the event horizon, into which objects can fall, but out of which nothing can come out. It is called "black" because it absorbs all the light that hits it, reflecting nothing; this is like a perfect blackbody in thermodynamics.

Despite its invisible interior, a black hole can reveal its presence through interaction with other matter. A black hole can be inferred by tracking the movement of a group of stars that orbit a region in space which looks empty. Alternatively, one can see gas falling into a relatively small black hole, from a companion star. This gas spirals inward, heating up to very high temperature and emitting large amounts of radiation that can be detected from earthbound and earth-orbiting telescopes. Such observations have resulted in the general scientific consensus that, barring a breakdown in our understanding of nature, black holes do exist in our universe.

It is impossible to directly observe a black hole. However, it is possible to infer its presence by its gravitational action on the surrounding environment; particularly with microquasars and active galactic nuclei, where material falling into a nearby black hole emits a large amount of X-ray radiation due to heating. This observation method allows astronomers to detect their existence. The only objects in science that meet these observations, and are consistent within the framework of general relativity, are black holes.

A black hole has only three independent physical properties: mass, charge and angular momentum. In astronomy black holes are classed as:

- Supermassive—containing hundreds of thousands to billions of solar masses and are thought to exist in the center of most galaxies, including the Milky Way.
- Intermediate—containing thousands of solar masses.
- Micro (also *mini black holes*)—having masses much less than that of a star. At these sizes, quantum mechanics is expected to take effect. There is no known mechanism for them to form via normal processes of stellar evolution, but certain inflationary scenarios predict their production during the early stages of the evolution of the universe.

According to some theories of quantum gravity, black holes may also be produced in the highly energetic reaction produced by cosmic rays hitting the atmosphere, or even in particle accelerators such as the Large Hadron Collider. The theory of Hawking Radiation predicts that such black holes will evaporate in bright flashes of gamma radiation. NASA's Fermi Gamma-ray Space Telescope satellite (formerly GLAST), launched in 2008, is searching for such flashes.

The defining feature of a black hole is the appearance of an *event horizon*; a boundary in spacetime beyond which events cannot affect an outside observer. Since the event horizon is not a material surface but rather merely a mathematically defined demarcation boundary, nothing prevents matter or radiation from entering a black hole, only from exiting one.

For a non-rotating (static) black hole, the *Schwarzschild radius* delimits a spherical event horizon. The Schwarzschild radius of an object is proportional to the mass. Rotating black holes have distorted, non-spherical event horizons. The description of black holes given by general relativity is known to be an approximation, and it is expected that quantum gravity effects become significant near the vicinity of the event horizon. This allows observations of matter in the vicinity of a black hole's event horizon to be used to indirectly study general relativity and proposed extensions to it.

Though black holes themselves may not radiate energy, electromagnetic radiation and matter particles may be radiated from just outside the event horizon via Hawking radiation. At the center of a black hole lies the singularity, where matter is crushed to infinite density. There, the pull of gravity is infinitely strong, and space-time has infinite curvature. This means that a black hole's mass becomes entirely compressed into a region with zero volume. This zero-volume, infinitely dense region at the center of a black hole is called a gravitational singularity. The singularity of a non-rotating black hole has zero length, width, and height; a rotating black hole's is smeared out to form a ring shape lying in the plane of rotation. The ring still has no thickness and hence no volume.

The photon sphere is a spherical boundary of zero thickness, such that photons moving along tangents to the sphere will be trapped in a circular orbit. For non-rotating black holes, the photon sphere has a radius 1.5 times the Schwarzschild radius. The orbits are dynamically unstable, hence any small perturbation (such as a particle of in falling matter) will grow over time, either setting it on an outward trajectory escaping the black hole or on an inward spiral eventually crossing the event horizon.

Rotating black holes are surrounded by a region of space-time in which it is impossible to stand still, called the ergosphere. Objects and radiation (including light) can stay in orbit within the ergosphere without falling to the center. Once a black hole has formed, it can continue to grow by absorbing additional matter. Any black hole will continually absorb interstellar dust from its direct surroundings and omnipresent cosmic background radiation. Much larger contributions can be obtained when a black hole merges with other stars or compact objects.

8.5.2 *Hawking Radiation*

In 1974, Stephen Hawking showed that black holes are not entirely black, but emit small amounts of thermal radiation (Hawking 1974). He came to this result by applying quantum field theory in a static black hole background. The result of his calculations shows that a black hole should emit particles in a perfect black body spectrum. This effect has become known as Hawking radiation. Since Hawking's result, many other scientists have verified the effect through various methods. If his theory of black hole radiation is correct, then black holes are expected to emit a thermal spectrum of radiation, and thereby lose mass, because according to the theory of relativity mass is just highly condensed energy ($E = mc^2$). Black holes will shrink and evaporate over time. The temperature of this spectrum (Hawking temperature) is proportional to the surface gravity of the black hole, which in turn, is inversely proportional to the mass. Large black holes, therefore, emit less radiation than small black holes.

On the other hand if a black hole is very small, the radiation effects are expected to become very strong. Even a black hole that is heavy compared to a human would evaporate in an instant. A black hole the weight of a car ($\sim 10^{-24}$ m) would only take a nanosecond to evaporate, during which time, it would briefly have luminosity more than 200 times that of the sun. Lighter black holes are expected to evaporate even faster; for example, a black hole of mass 1 TeV/c^2 would take less than 10^{-88} s to evaporate completely. Of course, for such a small black hole, quantum gravitation effects are expected to play an important role and—although current developments in quantum gravity do not indicate so—could hypothetically make such a small black hole stable.

8.5.3 Micro Black Holes

Gravitational collapse is not the only process that could create black holes. In principle, black holes could also be created in high energy collisions that create sufficient density. Since, classically, black holes can take any mass, one would expect micro black holes to be created in any such process, no matter how low the energy. However, to date, no such events have ever been detected, either directly or indirectly, as a deficiency of the mass balance in particle accelerator experiments. This suggests that there must be a lower limit for the mass of black holes.

Theoretically this boundary is expected to lie around the Planck mass ($\sim 10^{19}$ GeV/c^2, $m_p = 2.1764 \times 10^{-8}$ kg), where quantum effects are expected to make the theory of general relativity break down completely. This would put the creation of black holes firmly out of reach of any high energy process occurring on or near the Earth. Certain developments in quantum gravity however suggest that this bound could be much lower. Some brane-world scenarios for example put the Planck mass much lower, maybe even as low as 1 TeV. This would make it possible for micro black holes to be created in the high energy collisions occurring when cosmic rays hit the Earth's atmosphere, or possibly in the new Large Hadron Collider at CERN. These theories are however very speculative, and the creation of black holes in these processes is deemed unlikely by many specialists.

8.5.4 The Smallest Possible Black Hole

To make a black hole, one must concentrate mass or energy sufficiently so that the escape velocity from the region in which it is concentrated exceeds the speed of light. This condition gives the Schwarzschild radius, $r_o = 2GM/c^2$, where G is Newton's constant and c is the speed of light, as the size of a black hole of mass M. On the other hand, the Compton wavelength, $\lambda = h/Mc$, where h is Planck's constant, represents a limit on the minimum size of the region in which a mass M, at rest, can be localized. For sufficiently small M, the Compton wavelength exceeds the Schwarzschild radius, and no black hole description exists. Thus, the smallest mass for a black hole is approximately the Planck mass, which is about 2×10^{-8} kg or 1.2×10^{19} GeV/c^2.

Any primordial black holes of sufficiently low mass will Hawking evaporate to near the Planck mass within the lifetime of the universe. In this process, these small black holes radiate away matter. A rough description is that pairs of virtual particles emerge from the vacuum near the event horizon, with one member of a pair being captured, and the other escaping the vicinity of the black hole. The net result is the black hole loses mass (due to conservation of energy). According to the formulae of black hole thermodynamics, the more the black hole loses mass, the hotter it becomes and the faster it evaporates, until it approaches the Planck mass. At this stage, a black hole would have a Hawking temperature of $T_P/8\pi$ (5.6×10^{32} K), which means an emitted Hawking particle would have an energy comparable to the mass of the black hole. Thus, a thermodynamic description breaks down. Such a mini-black hole would also have an entropy of only 4πnats, approximately the minimum possible value.

At this point, the object can no longer be described as a classical black hole, and Hawking's calculations also break down. Conjectures for the final fate of the black hole include total evaporation and production of a Planck mass-sized black hole remnant. If intuitions about quantum black holes are correct, then close to the Planck mass the number of possible quantum states of the black hole is expected to become so few and so quantized that its interactions are likely to be quenched out. It is possible that such Planck-mass black holes, no longer able either to absorb energy gravitationally like a classical black hole because of the quantized gaps between their allowed energy levels, nor to emit Hawking particles for the same reason, may in effect be stable objects. They would in effect be WIMPs, weakly interacting massive particles; this could explain dark matter.

8.5.5 Creation of Micro Black Holes (MBH)

Production of a black hole requires concentration of mass or energy within the corresponding Schwarzschild radius. In familiar three-dimensional gravity, the minimum such energy is 10^{19}GeV, which would have to be condensed into a region of approximate size 10^{-33} cm. This is far beyond the limits of any current technology; the large hadron collider (LHC) has a design energy of 14 TeV. This is also beyond the range of known collisions of cosmic rays with Earth's atmosphere, which reach center of mass energies in the range of hundreds of TeV. It is estimated that for to collide two particles to within a distance of a Planck length with currently achievable magnetic field strengths would require a ring accelerator about 1000 light years in diameter to keep the particles on track.

Some extensions of present physics posit the existence of extra dimensions of space. In higher-dimensional space-time, the strength of gravity increases more rapidly with decreasing distance than in three dimensions. With certain special configurations of the extra dimensions, this effect can lower the Planck scale to the TeV range. Examples of such extensions include large extra dimensions, special cases of the Randall-Sundrum model, and String theory configurations. In such scenarios, black hole production could possibly be an important and observable effect at the LHC.

8.5.6 Virtual Particles

In physics, a virtual particle is a particle that exists for a limited time and space, introducing uncertainty in their energy and momentum due to the Heisenberg Uncertainty Principle. Vacuum energy can also be thought of in terms of virtual particles (also known as vacuum fluctuations) which are created and destroyed out of the vacuum. These particles are always created out of the vacuum in particle-antiparticle pairs, which shortly annihilate each other and disappear. However, these particles and antiparticles may interact with others before disappearing.

The net energy of the Universe remains zero so long as the particle pairs annihilate each other within Planck time. Virtual particles are also excitations of the underlying fields, but are detectable only as forces. The creation of these virtual particles near the event horizon of a black hole has been hypothesized by physicist Stephen Hawking to be a mechanism for the eventual "evaporation" of black holes. Since these particles do not have a permanent existence, they are called virtual particles, or vacuum fluctuations of vacuum energy.

An important example of the "presence" of virtual particles in a vacuum is the Casimir effect. Here, the explanation of the effect requires that the total energy of all of the virtual particles in a vacuum can be added together. Although the virtual particles themselves are not directly observable in the laboratory, they do leave an observable effect: their zero-point energy results in forces acting on suitably arranged metal plates or dielectrics. Thus, virtual particles are often popularly described as coming in pairs, a particle and antiparticle, which can be of any kind.

The evaporation of a black hole is a process dominated by photons, which are their own antiparticles and are uncharged. The uncertainty principle in the form $\Delta E \Delta t \geq \hbar$ implies that in the vacuum one or more particles with energy ΔE above the vacuum may be created for a short time Δt. These virtual particles are included in the definition of the vacuum.

Vacuum energy is an underlying background energy that exists in space even when devoid of matter (known as free space). The vacuum energy is deduced from the concept of virtual particles, which are themselves derived from the energy-time uncertainty principle. Its effects can be observed in various phenomena (such as spontaneous emission, the Casimir effect, the van der Waals bonds, or the Lamb shift), and it is thought to have consequences for the behavior of the Universe on cosmological scales.

8.5.7 AB-Generator of Nuclear Energy

A simplified explanation of MBH radiation and work of an AB-Generator can be seen in Fig. 8.29. As known, the vacuum continuously produces virtual pairs of particles and antiparticles, in particular, photons and anti-photons. In conventional space, they exist only for a very short time, then annihilate and return back to nothingness. The MBH event horizon, having very strong super-gravity, allows separation of the particles and anti-particles, in particular, photons and anti-photons. Part of the anti-photons move into the MBH and annihilate with photons decreasing the mass of the MBH and return back a borrow energy to vacuum. The free photons

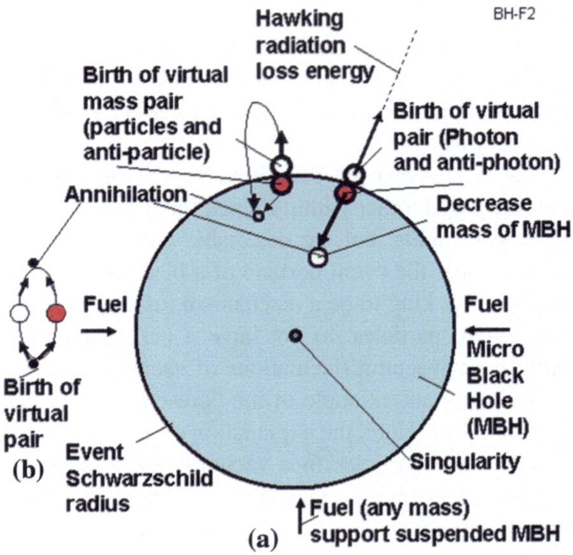

Fig. 8.29 Hawking radiation. **a** Virtual particles at even horizon. **b** Virtual particles out even horizon (in conventional space)

leave from the MBH neighborhood as Hawking radiation. That way the MBH converts any conventional matter to Hawking radiation which may be converted to heat or electric energy by the AB-Generator. This AB-Generator utilizes the produced Hawking radiation and injects the matter into the MBH, while maintaining the MBH in stable suspended state. [Note: The photon does NOT have rest mass. Therefore a photon can leave the MBH's neighborhood (if it is located beyond the event horizon). All other particles having a rest mass and speed less than light speed cannot leave the Black Hole. They cannot achieve light speed because their mass at light speed equals infinity and requests infinite energy for its escape—an impossibility.]

8.5.8 Description of the AB-Generator

The offered nuclear energy AB-Generator is shown in Fig. 8.30. That includes the Micro Black Hole (MBH) suspended within a spherical radiation reflector and heater. The MBH is supported (and controlled) at the center of sphere by a fuel (plasma, proton, electron, matter) gun. This AB-Generator also contains the heat engine (for example, gas, vapor turbine), electric generator, coolant (heat transfer agent), an outer electric line, internal electric generator (antenna) with customer.

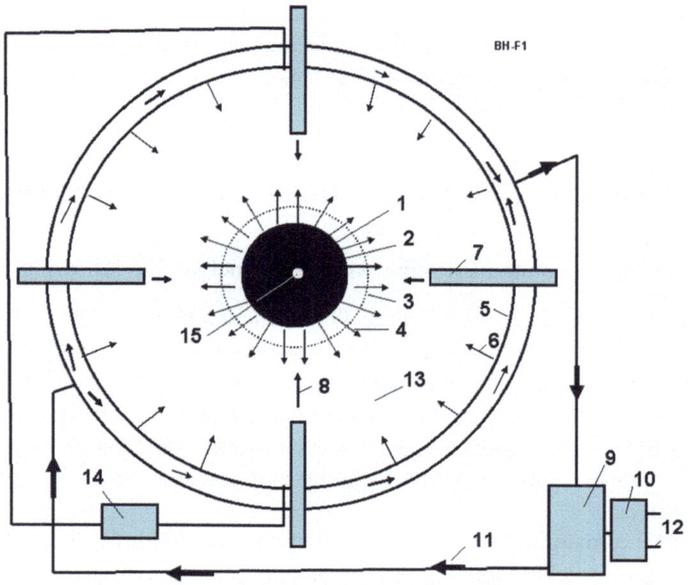

Fig. 8.30 Offered nuclear-vacuum energy AB-Generator. Notations: 1—Micro Black Hole (MBH), 2—event horizon (Schwarzschild radius), 3—photon sphere, 4—black hole radiation, 5—radiation reflector, antenna and heater (cover sphere), 6—back (reflected) radiation from radiation reflector 5, 7—fuel (plasma, protons, electrons, ions, matter) gun (focusing accelerator), 8—matter injected to MBH (fuel for Micro Black hole), 9—heat engine (for example, gas, vapor turbine), 10—electric generator connected to heat engine 9, 11—coolant (heat transfer agent to the heat machine 9), 12—electric line, 13—internal vacuum, 14—customer of electricity from antenna 5, 15—singularity

8.5.9 Working Capabilities

The generator works the following way: MBH, by selective directional input of matter, is levitated in captivity and produces radiation energy. That radiation heats the spherical reflector-heater. The coolant (heat transfer agent) delivers the heat to a heat machine (for example, gas, vapor turbine). The heat machine rotates an electric generator that produces the electricity to the outer electric line. Part of MBH radiation may accept by sphere (as antenna) in form of electricity.

The control fuel guns inject the matter into MBH, and do not allow bursting of the MBH. This action also supports the MBH in isolation, suspended from dangerous contact with conventional matter. They also control the MBH size and the energy output.

Any matter may be used as the fuel, for example, accelerated plasma, ions, protons, electrons, micro particles, etc. The MBH may be charged and rotated. In this case the MBH may has an additional suspension by control charges located at the ends of fuel guns or (in case of the rotating charged MBH) may have an additional suspension by the control electric magnets located on the ends of fuel guns or at points along the reflector-heater sphere.

8.5.10 Innovations, Features, Advantages, and Research

Some problems and solutions dealing with MBH include the following:

(1) A practical method (the MBH being obtained and levitated, details of which are beyond the scope of this chapter) and installation for converting any conventional matter to energy in accordance with Einstein's equation $E = mc^2$.
(2) MBHs may produce gigantic amounts of energy, and this energy is in the form of dangerous gamma radiation. This dangerous gamma radiation Doppler shifts when it moves against the MBH gravity and converts to safely tapped short radio waves.
(3) The MBH of marginal mass has a tendency to explode (through quantum evaporation, very quickly radiating its mass in energy). The AB-Generator automatically injects metered amounts of matter into the MBH and keeps the MGH in a stable state or grows the MBH to a needed size, or decreases that size, or temporarily turns off the AB-Generator (decreases the MBH to a Planck Black Hole).
(4) The radiation flux exposure of AB-Generator (as result of MBH exposure) is not dangerous because the generator cover sphere has a vacuum, and the MBH gravity gradient decreases the radiation energy.
(5) The MBH may be supported in a levitated (non-contact) state by generator fuel injectors.

8.5.11 Theory of AB-Generator

Below there are main equations for computation the conventional black hole (BH) and AB-Generator.

General theory of Black Hole

1. *Power produced by BH* is:

$$P = \frac{\hbar c^6}{15360\pi G^2} \frac{1}{M^2} \approx 3.56 \times 10^{32} \frac{1}{M^2} \text{ [W]} \tag{8.43}$$

where $\hbar = h/2\pi = 1.0546 \times 10^{-34}$ J/s is reduced Planck constant, $c = 3 \times 10^8$ m/s—light speed, $G = 6.6743 \times 10^{-11}$ m^3/kg s^2 is gravitation constant, M—mass of BH, kg.

2. *Temperature of black body* corresponding to this radiation is:

$$T = \frac{\hbar c^3}{8\pi G k_b}\frac{1}{M} \approx 1.23 \times 10^{23}\frac{1}{M} \tag{8.44}$$

where $k_b = 1.38 \times 10^{-23}$ J/K is Boltzmann constant.

3. *Energy Ep [J] and frequency ν_0 of photon at event horizon* are:

$$E_p = \frac{hc^3}{16\pi G}\frac{1}{M}, \quad \nu_0 = \frac{E_p}{h} = \frac{c^3}{16\pi G}\frac{1}{M} = 8.037 \times 10^{33}\frac{1}{M}, \quad \lambda_0 = \frac{c}{\nu_0} = 3.73 \times 10^{-26} M \tag{8.45}$$

where $c = 3 \times 10^8$ m/s is light speed, λ_o is wavelength of photon at even radius, m. h is Planck constant.

4. *Radius of BH event horizon* (Schwarzschild radius) is:

$$r_0 = \frac{2G}{c^2}M = 1.48 \times 10^{-27} M\,[\mathrm{m}] \tag{8.46}$$

5. *Relative density* (ratio of mass M to volume V of BH) is:

$$\rho = \frac{M}{V} = \frac{3c^2}{32\pi G^3}\frac{1}{M^2} \approx 7.33 \times 10^{79}\frac{1}{M^2}\,\left[\mathrm{kg/m^3}\right] \tag{8.47}$$

6. *Maximal charge of BH* is:

$$Q_{\max} = 5 \times 10^9 eM \approx 8 \times 10^{-10} M\,[\mathrm{C}] \tag{8.48}$$

where $e = -1.6 \times 10^{-19}$ is charge of electron, C.

7. *Life time of BH* is:

$$\tau = \frac{5120\pi G^2}{\hbar c^4}M^3 = 2.527 \times 10^{-8} M^3, \mathrm{s}. \tag{8.49}$$

8. *Gravitation around BH* (r is distance from center) and on event horizon is:

$$g = \frac{GM}{r^2}, \quad g_0 = \frac{c^4}{4G}\frac{1}{M} = 3 \times 10^{42}\frac{1}{M}, \quad ms^{-2} \tag{8.50}$$

Developed Theory of AB-Generator

Below are research and the theory developed by author for estimation and computation of facets of the AB-Generator.

9. *Loss of energy of Hawking photon in BH gravitational field.* It is known the theory of a redshift allows estimating the frequency of photon in central gravitational field when it moves TO the gravity center. In this case the photon increases its frequency because photon is accelerated the gravitational field (wavelength decreases). But in our case the photon moves FROM the gravitational center, the gravitational field brakes it and the photon loses its energy. That means its frequency decreases and the wavelength increases. Our photon gets double energy because the black hole annihilates two photons (photon and anti-photon). That way, the equation for photon frequency at distance $r > r_o$ from center we can write in form:

$$\frac{\nu}{\nu_0} \approx 1 + \frac{2\varphi}{c^2} \tag{8.51}$$

where $\Delta\varphi = \varphi - \varphi_o$ is difference of the gravity potential. The gravity potential is:

$$\Delta\varphi = \varphi - \varphi_0, \quad \varphi = \frac{GM}{r}, \quad \varphi_0 = \frac{GM}{r_0}, \quad r_0 = \frac{2GM}{c^2} \tag{8.52}$$

Let us substitute (8.52) in (8.51), we get:

$$\frac{\nu}{\nu_0} \approx 1 + \frac{r_0}{r} - \frac{r_0}{r_0}, \quad \text{or} \quad \frac{\nu}{\nu_0} = \frac{\lambda_0}{\lambda} \approx \frac{r_0}{r} \tag{8.53}$$

It is known, the energy and mass of photon is:

$$E_f = \hbar\gamma, \quad E_f = m_f c^2, \quad m_f = E_f/c^2 \tag{8.54}$$

The energy of photon linear depends from its frequency. [Reminder: The photon does not have a rest mass].

The relative loss of the photon radiation energy ξ at distance r from BH and the power P_r of Hawking radiation at radius r from the BH center is:

$$\xi = \frac{r_0}{r}, \quad v = \xi v_0, \quad P_r = \xi P \tag{8.55}$$

The r_o is very small and ξ is also very small and $v \ll v_o$.

The result of an energy loss by Hawking photon in the BH gravitational field is very important for AB-Generator. The energy of Hawking radiation is very big; we need to decrease it by many orders of magnitude. The initial Hawking photon is gamma radiation that is dangerous for people and matter. In r distance, the gamma radiation may be converted into conventional light or radio radiation, which are not dangerous and may be reflected, focused or a straightforward way converted into electricity by antenna.

10. *Reflection of Hawking radiation back to MBH.* For further decreasing the MBH produced energy the part of this energy may be reflected to back in MBH. A conventional mirror may reflect up 0.9–0.99 of radiation (ξ_r = 0.01–0.1, ξ_r is a loss of energy in reflecting), the multi layers mirror can reflect up 0.9999 of the monochromatic light radiation ($\xi_r = 10^{-3}$–10^{-5}), and AB-mirror from cubic corner cells allows reflection of non-monochromatic light radiation with efficiency up $\xi_r = 10^{-13}$ strong back to source. In the last case, the loss of reflected energy is (Bolonkin 2006b):

$$\xi_r = 0.00023al, \quad l = m\lambda, \quad m \geq 1, \tag{8.56}$$

where l is size of cube corner cell, m; m is number of radiation waves in one sell; λ is wavelength, m; a is characteristic of sell material see (Bolonkin 2006b, Fig. A8.3). Minimal value $a = 10^{-2}$ for glass and $a = 10^{-4}$ for KCl crystal.

The reflection of radiation to back in MBH is may be important for MBH stabilization, MBH storage and MBH ‘switch off’.

11. *Useful energy of AB-Generator.* The useful energy P_u[J] is taken from AB-Generator is:

$$P_u = \xi \xi_r P \tag{8.57}$$

12. *Fuel consumption* is:

$$\dot{M} = P_u / c^2 \tag{8.58}$$

The fuel consumption is very small. AB-Generator is the single known method in the World now which allows full converting reasonably practical conversion of (any!) matter into energy according the Einsteinian equation $E = mc^2$.

13. *Specific pressure on AB-Generator cover sphere, p* [N/m^2] and on the surface of MBH p_o is:

$$p = \frac{kP_r}{Sc} = \frac{kP_r}{4\pi r^2 c} = 2.65 \times 10^{-10} \frac{kP_r}{r^2},$$
$$p_0 = \frac{P}{S_0 c} = \frac{\hbar c^9}{15360 \cdot 16\pi^2 G^4} \frac{1}{M^4} = 8.57 \times 10^{76} \frac{1}{M^4} \qquad (8.59)$$

where k = 1 if the cover sphere absorbs the radiation and $k \approx 2$ if the cover sphere high reflects the radiation, S is the internal area of cover sphere, m^2; S_0 is surface of event horizon sphere, m^2; p_o is specific pressure of Hawking radiation on the event horizon surface. [Note: the pressure p on the cover sphere is small (see Project), but pressure p_o on event horizon surface is very high.]

14. *Mass particles produced on event surface.* On the event horizon surface, particles may be also produced with speed $V < c$. Let us take the best case (for leaving the BH) when their speed is radially vertical; they cannot leave the BH because their speed V is less than light speed c. The maximal radius of lifting r_m [m] is:

$$dV = -gdt, \quad dV = -\frac{g}{V}dr = -\frac{GM}{V}\frac{dr}{r^2}, \quad r_m = \frac{2GM}{c^2 - V_0^2} = \frac{r_0}{1-(V/c)^2} \qquad (8.60)$$

where g is gravitational acceleration of BH, m/s^2; t is time, sec.; r_o is BH radius, m; V_0 is particle speed on event surface, m/s^2. If the r_m is less than radius of the cover sphere, the mass particles return to BH and do not influence the heat flow from BH to cover sphere. That is in the majority of cases.

15. *Explosion of MBH.* The MBH explosion produces the radiation energy:

$$E_e = Mc^2 \qquad (8.61)$$

MBH has a small mass. The explosion of MBH having $M = 10^{-5}$ kg produces 9×10^{11} J. That is energy of about 10 tons of good conventional explosive (10^7 J/kg). However, there is a vacuum into the cover sphere and this energy is presented in radiation form. In reality, only very small part of explosion energy reaches the cover sphere, because the very strong MBH gravitation field brakes the photons, and any mass particles. Find the energy which reaches the cover sphere via:

$$\mathrm{d}E = \xi c^2 \mathrm{d}M, \quad \xi = \frac{r_0}{r}, \quad r_0 = \frac{2G}{c^2}M, \quad \mathrm{d}E = \frac{2G}{r}M\mathrm{d}M,$$
$$E = \frac{G}{r}M^2 = 6674 \times 10^{-11}\frac{M^2}{r} \tag{8.62}$$

As you see in Eq. (8.62) the powerful MBH gravity field is the strong brake of the products of MBH explosion. For example, there was an accident and our MBH reactor (see section "Project …" below). Reactor was destroyed in orbit or on Earth. The mass of our MBH is $M = 10^{-5}$ kg. Let us take the radius one molecule ($r = 10^{-9}$ m). In this area the MBH explosive energy is:

$$E = \frac{G}{r}M^2 = 6674 \times 10^{-11}\frac{(10^{-5})^2}{10^{-9}} = 6674 \times 10^{-12}\ \mathrm{J} \tag{8.63}$$

This very small energy can damage only the nearest molecules of environment. That means: no dangers for ships and Earth in explosion of our MBH. The time and explosion energy is very small. We lose only MBH. If we can produce MBH or save them, we can shut of our reactor at any time.

The specific exposure radiation pressure of MBH pressure p_e [N/m^2] on the cover sphere of radius $r < r_o$ may be computed by the way:

$$p_e = \frac{E}{V} = \frac{3G}{4\pi}\frac{M^2}{r^3} = 1.6 \times 10^{-11}\frac{M^2}{r^3}, \quad r > r_0 \tag{8.64}$$

where $V = 3/4\ \pi r^3$ is volume of the cover sphere.

That way the exposure radiation pressure on sphere has very small value and presses very short time. Conventional gas balloon keeps pressure up 10^7 N/m^2 (100 atm). However, the heat impact may be high and AB-Generator design may have the reflectivity cover and automatically open windows for radiation.

Your attention is requested toward the next important result following from Eqs. (8.62) to (8.64). Many astronomers try to find (detect) the MBH by a MBH exposure radiation; but this radiation is small, may be detected but for a short distance, does not have a specific frequency, and has a variably long wavelength. This may be why during more than 30 years nobody has successfully observed MBH events in Earth environment though the theoretical estimation predicts about 100 of MBH events annually. Observers take note!

16. *Supporting the MBH in suspended (levitated) state.* The fuel injector can support the MBH in suspended state (no contact the MBH with any material surface). The maximal suspended force equals:

$$F = qV_f, \quad q = \frac{P_u}{c^2}, \quad F = \frac{P_u V_f}{c^2} \tag{8.65}$$

where q is fuel consumption, kg; V_f is a fuel speed, m/s. The fuel (plasma) speed $0.01c$ is conventionally enough for supporting the MBH in suspended state.

17. *AB-Generator as an electric generator.* When the Hawking radiation reaches the cover as radio microwaves they may be straightforwardly converted to electricity because they create a different voltage between different isolated parts of the cover sphere as in an antenna. Maximal voltage which can produces the radiation wave is:

$$w = \frac{\varepsilon\varepsilon_0 E^2}{2} + \frac{\mu\mu_0 H^2}{2}, \quad w = \frac{P_r}{c}, \tag{8.66}$$

where w is density of radiation energy, J/m^3; E is electric intensity, V/m; H is magnetic intensity, T; $\varepsilon_o = 8.85 \times 10^{-12}$ F/m is the coefficient of the electric permeability; $\mu_o = 4\pi \times 10^{-7}$ N/A^2 is the coefficient of the magnetic permeability; $\varepsilon = \mu = 1$ for vacuum. Let us take moment when $H = 0$, then:

$$E = \sqrt{\frac{2w}{\varepsilon_0}} = \sqrt{\frac{2P_r}{\varepsilon_0 c}} = 2.73\sqrt{P_r} \quad U \approx b\pi DE, \quad b = \frac{\pi D}{0.5\lambda} \leq 1,$$
$$P_e \approx bP_r, \quad \lambda = \lambda_0 \frac{r}{r_0} = 16r, \quad b = \frac{4\pi r}{16r} = \frac{\pi}{4}, \tag{8.67}$$

where E is electric intensity, V/m; U is voltage of the AB-generator, V; b is relative size of antenna, D is diameter of the cover sphere if the cover sphere is used as a full antenna, m; P_e is power of the electric station, W.

As you see, about $\pi/4$ of total energy produced by an AB-Generator can be received in the form of electricity, and $(1 - \pi/4)$ reflects back to MBH; we may tap heat energy which convert to any form of energy by conventional (heat engine) methods. If we reflect the most part of the heat energy back into the MBH, we can have only electricity and do not have heat flux. If we will use the super strong and super high temperature material AB-material offered in (Bolonkin 2009a) the conversion coefficient of heat machine may be very high.

18. *Critical mass of MBH located in matter environment.* Many people are afraid of MBH experiments because BH can absorb the Earth. Let us find the critical mass of MBH which can begin uncontrollably to grow into the Earth environment. That will happen when BH begins to have more mass than mass of Hawking radiation. Below is the equation for the critical mass of initial BH. The educated reader will understand the equations below without detailed explanations:

$$
\begin{aligned}
&\mathrm{d}V = g\mathrm{d}t, \quad g = \frac{GM}{r^2}, \quad \mathrm{d}t = \frac{\mathrm{d}r}{V}, \quad V\mathrm{d}V = g\mathrm{d}r, \quad \int_V^c V\mathrm{d}V = \int_r^{r_o} \frac{GM}{r^2}\mathrm{d}r,\\
&r_0 = \frac{2G}{c^2}M, V^2 = c^2\frac{r_0}{r}, V = c\sqrt{\frac{r_0}{r}}, \quad \mathrm{d}t = \frac{\sqrt{r}\mathrm{d}r}{c\sqrt{r_0}}, \quad \int_t^0 \mathrm{d}t = \frac{1}{c\sqrt{r_0}}\int_r^{r_o}\sqrt{r}\mathrm{d}r,\\
&t = \frac{2}{3c\sqrt{r_0}}\left(r^{3/2} - r_0^{3/2}\right) \approx \frac{2r^{3/2}}{3cr_0^{1/2}}, \quad r = \left(\frac{3c\sqrt{r_0}}{2}t\right)^{3/2}, r = 1.65G^{1/2}M^{1/3}t^{2/3},\\
&\dot{M} = \frac{P}{c^2} = \frac{\hbar c^4}{15360\pi G^2}\frac{1}{M^2} = 4\times 10^{15}\frac{1}{M^2}, \text{for} \quad t = 1\ s,\\
&\dot{M}_c = \frac{4}{3}\pi r^3\gamma = 6\pi\gamma G^{3/2}M \approx 10^{-4}\gamma M, \quad M = M_c e^{6\pi\gamma G^{3/2}t} \approx M_c e^{10^{-4}\gamma t},\\
&t = \frac{1}{6\pi\gamma G^{3/2}}\ln\frac{M}{M_c} \approx \frac{10^4}{\gamma}\ln\frac{M}{M_c},
\end{aligned}
\tag{8.68}
$$

where V is speed of environment matter absorbed by MBH, m/s; g is gravity acceleration of MBH, m/s; r is distance environment matter to MBH center, m; t is time, sec; $\dot{M}$ is mass loss by MBH, kg; $\dot{M}_c$ is mass taken from Earth environment by MBH, kg; γ is density of Earth environment, kg/m^3; M_c is critical mass of MBH when one begin uncontrollable grows, kg; t is time, sec.

Let us to equate the mass $\dot{M}$ radiated by MBH to mass $\dot{M}_c$ absorbed by MBH from Earth environment, we obtain the critical mass M_c of MBH for any environment:

$$
M_c^3 = \frac{\hbar c^4}{92160\pi^2 G^3}\frac{1}{\gamma} = 3.17\times 10^{24}\frac{1}{\gamma}, \quad \text{or} \quad \gamma = 3.17\times 10^{24}\frac{1}{M_c^3} \tag{8.69}
$$

If MBH having mass $M = 10^7$ kg (10 thousand tons) is put in water (γ = 1000 kg/m^3), this MBH can begin uncontrollable runaway growth and in short time (~74 s) can consume the Earth into a black hole having diameter ~9 mm. If this MBH is located in the sea level atmosphere (γ = 1.29 kg/m^3), the initial MBH must has critical mass $M = 10^8$ kg (100 thousand tons). The critical radius of MBH is very small. In the first case ($M = 10^7$ kg) $r_o = 1.48 \times 10–^{20}$ m, in the second case ($M = 10^8$ kg) $r_o = 1.48 \times 10^{-19}$ m. Our MBH into AB-Generator is not dangerous for Earth because it is located in vacuum and has mass thousands to millions times less than the critical mass.

However, in a moment of extreme speculation, if far future artificial intelligence (or super-small reasoning) beings will be created from nuclear matter, they can convert the Earth into a black hole to attempt to access quick travel to other stars (Solar systems), past and future Universes, and even possibly past and future times (Bolonkin 2009a). [Note: We got our equations in assumption $\lambda/\lambda_o = r/r_o$. If $\lambda/\lambda_o = (r/r_o)^{0.5}$ or other relation, the all above equations may be easy modified.]

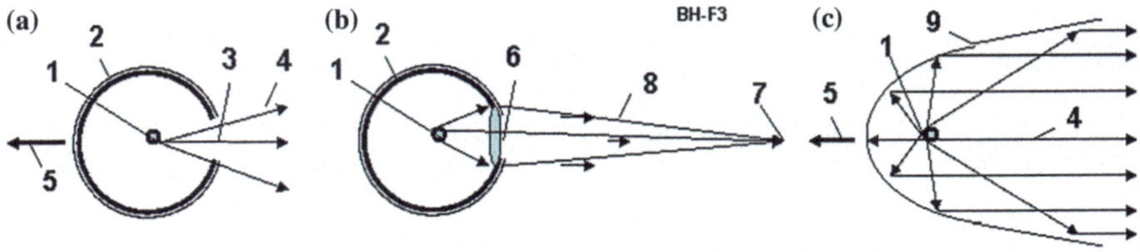

Fig. 8.31 AB-Generator as photon rocket and radiation (photon) weapon. **a** AB-Generator as a simplest photon rocket. **b** AB-Generator as focused radiation (photon, light or laser) weapon. **c** Photon rocket with Micro-Black Hole of AB-Generator; Notations: 1—control MBH; 2—spherical cover of AB-Generator; 3—window in spherical cover; 4—radiation of BH; 5—thrust; 6—lens in window of cover; 7—aim; 8—focused radiation; 9—parabolic reflector

8.5.12 AB-Generator as Photon Rocket

The offered AB-Generator may be used as the most efficient photon propulsion system (photon rocket). The photon rocket is the dream of all astronauts and space engineers, a unique vehicle, which would make practical interstellar travel. But a functioning photon rocket would require gigantic energy. The AB-Generator can convert any matter in energy (radiation) and gives the maximum theoretical efficiency.

The some possible photon propulsion system used the AB-Generator is shown in Fig. 8.31. In simplest version (***a***) the cover of AB generator has window 3, the radiation goes out through window and produces the thrust. More complex version (***c***) has the parabolic reflector, which sends all radiation in one direction and increases the efficiency. If an insert in the AB-Generator covers the lens 6 which will focuses the radiation in a given direction, at the given point the temperature will be a billions degree (see Eq. (8.44)) and AB-Generator may be used as a photon weapon.

The maximal thrust T of the photon engine having AB-Generator may be computed (estimated) by equation:

$$T = \dot{M}c, \ [\mathrm{N}]$$

For example, the AB-generator, which spends only 1 g of matter per second, will produce a thrust 3×10^5 N or 30 tons.

8.5.13 Photon Rockets and Relativistic Flight

A photon rocket is a hypothetical rocket that uses thrust from emitted photons (radiation pressure by emission) for its propulsion (Bolonkin 2006b). Photons could be generated by onboard generators, as in the nuclear photonic rocket.

The speed an ideal photon rocket will reach, in the absence of external forces, depends on the ratio of its initial and final mass:

$$v = c\frac{(m_i/m_f)^2 - 1}{(m_i/m_f)^2 + 1} \tag{8.70}$$

where m_i is the initial mass and m_f is the final mass: $c = 3 \times 10^8$ m/s is light speed. The gamma factor corresponding to this speed has the simple expression:

$$\gamma = 0.5(m_i/m_f + m_f/m_i) \tag{8.71}$$

The Lorentz factor or Lorentz term is the factor by which time, length, and relativistic mass change for an object while that object is moving. The expression appears in several equations in special relativity, and it arises in derivations of the Lorentz transformations. The Lorentz factor is defined as Bolonkin (2006b):

$$\gamma = \frac{1}{\sqrt{1 - v^2/c^2}} = \frac{1}{\sqrt{1 - \beta^2}} = \frac{dt}{d\tau} \tag{8.72}$$

where v is the relative velocity between inertial reference frames, β is the ratio of v to the speed of light c, τ is the proper time for an observer (measuring time intervals in the observer's own frame), t is coordinate time, $c = 3 \times 10^8$ m/s is the speed of light in a vacuum.

The simplest case of a Lorentz transformation is a boost in the x-direction, which describes how space-time coordinates change from one inertial frame using coordinates (x, y, z, t) to another (x', y', z', t') with relative velocity v:

$$t' = \gamma\left(t - \frac{vx}{c^2}\right), \quad x' = \gamma(x - vt) \tag{8.73}$$

Corollaries of the above transformations are the results:

Time dilation: The time ($\Delta t'$) between two ticks as measured in the frame in which the clock is moving, is longer than the time (Δt) between these ticks as measured in the rest frame of the clock:

$$\Delta t' = \gamma \Delta t \tag{8.74}$$

Length contraction: The length ($\Delta x'$) of an object as measured in the frame in which it is moving, is shorter than its length (Δx) in its own rest frame:

$$\Delta x\prime = \Delta x/\gamma \tag{8.75}$$

Applying conservation of momentum and energy leads to these resussslts:

Relativistic mass: The mass of an object m in motion is dependent on Υ and the rest mass m_0:

$$m = \gamma m_0 \tag{8.76}$$

Relativistic momentum: The relativistic momentum relation takes the same form as for classical momentum, but using the above relativistic mass:

$$\vec{p} = m\vec{v} = \gamma m_0 \vec{v} \tag{8.77}$$

Relativistic kinetic energy: The relativistic kinetic energy relation takes the slightly modified form:

$$E_k = E - E_0 = (\gamma - 1)m_0 c^2 \tag{8.78}$$

The Lorentz factor has the Maclaurin series:

$$\gamma = \frac{1}{\sqrt{1-\beta^2}} \approx 1 + \frac{1}{2}\beta^2 + \cdots \tag{8.79}$$

which is a special case of a binomial series. The approximation $\gamma \approx 1 + {}^1/_2\, \beta^2$ may be used to calculate relativistic effects at low speeds. It holds to within 1% error for $v < 0.4$ c ($v <$ 120,000 km/s), and to within 0.1% error for $v < 0.22$ c ($v <$ 66,000 km/s).

The truncated versions of this series also allow physicists to prove that special relativity reduces to Newtonian mechanics at low speeds. For example, in special relativity, the following two equations hold:

$$\vec{p} = \gamma m\vec{v}, \quad E = \gamma mc^2 \tag{8.80}$$

For $\gamma \approx 1$ and $\gamma \approx 1 + {}^1/_2\, \beta^2$, respectively, these reduce to their Newtonian equivalents:

$$\vec{p} = m\vec{v}, \quad E = mc^2 + \frac{1}{2}mv^2 \tag{8.81}$$

The Lorentz factor equation can also be inverted to yield:

$$\beta = \sqrt{1 - \frac{1}{\gamma^2}} \tag{8.82}$$

This has an asymptotic form of:

$$\beta = 1 - \frac{1}{2}\gamma^{-2} - \frac{1}{8}\gamma^{-4} - \frac{1}{16}\gamma^{-6} - \frac{5}{128}\gamma^{-8} + \cdots \tag{8.83}$$

The first two terms are occasionally used to quickly calculate velocities from large γ values. The approximation $\beta \approx 1 - {}^1/_2\gamma^{-2}$ holds to within 1% tolerance for $\gamma > 2$, and to within 0.1% tolerance for $\gamma > 3.5$.

Project of AB-Generator for Photon Rocket

Let us to estimate the possible energy production of an AB-Generator. This is not optimal, but rather an example of computation and possible parameters. Let us take the MBH mass $M = 10^{-5}$ kg and radius of the cover sphere $r = 5$ m, with no reflection. Using the Eqs. (8.43)–(8.67) we receive:

$$\begin{aligned}
&P = 3.56 \times 10^{32}/M^2 = 3,56 \times 10^{42} \quad \text{W}, \\
&r_0 = 1.48 \times 10^{-27}M = 1.48 \times 10^{-32} \quad \text{m}, \\
&\xi = r_0/r = 2.96 \times 10^{-33}, \\
&P_r = \xi P = 1.05 \times 10^{10}, \quad P_u = \xi\xi_r P = P_r, \quad \text{W}, \quad \xi_r = 1. \\
&\lambda_0 = 3.73 \times 10^{-26}M = 3.73 \times 10^{-31} \quad m. \\
&\lambda = 16 \cdot r = 80 \quad \text{m}. \\
&p = \frac{P_r}{4\pi c r^2} = 0.111 \quad \frac{N}{m^2}, \quad c = 3 \times 10^8 \quad \text{m/s}, \\
&\dot{M} = P_u/c^2 = 1.17 \times 10^{-7} \quad \text{kg/s}, \\
&p_e = 1.6 \times 10^{-11}\frac{M^2}{r^3} = 1.28 \times 10^{-23} \quad \text{N/m}^2
\end{aligned} \tag{8.84}$$

The main notations in Eq. (8.84): $P_r = P_u = 1.05 \times 10^{10}$ W is the useful energy (π/4 of this energy may be taken as electric energy by cover antenna, the rest is taken as heat); $\lambda = 80$ m is wavelength of radiation at cover sphere (that is not dangerous for people); $\dot{M} = 1.17 \times 10^{-7}$kg/s is fuel consumption; $r_o = 1.48$ 10^{-32} m is radius of MBH; $p_e = 1.28 \times 10^{-23}$ N/m^2 is explosion pressure of MBH.

The explosion pressure is very small. That is less than radiation pressure on the cover surface $p = 0.111$ N/m^2. It is no wonder because BH takes back the energy spent for acceleration of the matter by then eating the matter. There are no dangers from explosion of MBH.

Heat transfer and internal electric power are:

$$\begin{aligned}
&q = \frac{P_u}{S} = \frac{P_u}{4\pi r^2} = 3.34 \times 10^7 \quad \frac{\text{W}}{\text{m}^2}, \\
&\quad \text{For} \quad \delta = 2 \times 10^{-3} \text{ m}, \quad \lambda_h = 100, \quad \Delta T \approx q\delta/\lambda_h = 668\text{K}, \\
&E = 2.73\sqrt{P_r} = 2.8 \times 10^5 \quad \text{V/m}, \quad U = E \cdot 2r = 2.8 \times 10^6 \quad \text{V}, \\
&P_e = P_r/8 = 1.31 \times 10^9 \quad \text{W},
\end{aligned} \tag{8.85}$$

where q is specific heat transfer through the cover sphere, S is internal surface of the cover sphere, m^2; δ is thickness of the cover sphere wall, m; λ_h is heat transfer coefficient for steel; ΔT is difference temperature between internal and external walls of the cover sphere; E is electric intensity from radiation on cover sphere surface, V/m; U is maximal electric voltage, V; P_e is electric power, W.

We get the power, heat, and electric output of a AB-Generator similarly to how we would a very large complex of present day Earth's electric power stations ($P_r = 10^{10}$ W, ten billion of watts). The AB-Generator is cheaper by a hundred times than a conventional electric station, especially since, we may reflect heat energy back to the MBH and not built a heat engine with all the problems of conventional power conversion equipment (using only electricity from spherical cover as antenna). Hopefully the Large Hadron Collider at CERN can get the initial MBH needed for AB-Generator. The other way to obtain one is to find the Planck MBH (remaining from the time of the Big Bang and former MBH) and grow them to target MBH size.

8.6 Results

1. The method and installation for converting any conventional matter to energy according the Einstein's equation $E = mc^2$, where m is mass of matter, kg; $c = 3 \times 10^8$ is light speed, m/s.
2. The Micro Black Hole (MBH) is offered for this conversion.
3. Also is offered the control fuel guns and radiation reflector for explosion prevention of MBH.
4. Also is offered the control fuel guns and radiation reflector for the MBH control.
5. Also is offered the control fuel guns and radiation reflector for non-contact suspension (levitation) of the MBH.
6. For non-contact levitation of MBH the author also offers:
   (a) Controlled charging of MBH and of ends of the fuel guns.
   (b) Controlled charging of rotating MBH and control of electric magnets located on the ends of the fuel guns or out of the reflector-heater sphere.
7. The author researches show the very important fact: A strong gamma radiation produced by Hawking radiation loses energy after passing through the very strong gravitational MBH field. The MBH radiation can reach the reflector-heater as the light or short-wave radio radiation. That is very important for safety of the operating crew of the AB-Generator.
8. The matter particles produced by the MBH cannot escape from MBH and cannot influence the Hawking radiation.
9. Research shows another very important fact: the MBH explosion (hundreds and thousands of TNT tons) in radiation form produces a small pressure on the reflector-heater (cover sphere) and does not destroy the AB-generator (in a correct design of AB-Generator). That is very important for safety of the operating crew of the AB-generator.

10. The author researches show another very important fact: the MBH cannot capture by oneself the surrounding matter and cannot automatically grow to consume the planet.
11. As the initial MBH can be used the Planck's (quantum) MBH which may be everywhere. The offered fuel gun may to grow them (or decrease them) to needed size or the initial MBH may be used the MBH produce Large Hadron Collider (LHC) at CERN. Some scientists assume LHC will produce one MBH every second (86,400 MBH in day). The cosmic radiation also produces about 100 MBH every year.
12. The spherical dome of MBH may convert part of the radiation energy to electricity.
13. A correct design of MBH generator does not produce the radioactive waste of environment.
14. The attempts of many astronomers find (detect) the MBH by MBH exposure radiation will not be successful without knowing the following: the MBH radiation is small, may be detected only over a short distance, does not have specific frequency, and has a variable long wavelength.

8.6.1 Discussion

We got our equations in assumption $\lambda/\lambda_o = r/r_o$. If $\lambda/\lambda_o = (r/r_o)^{0.5}$ or other relation, the all above equations may be easy modified. The Hawking article was published 34 years ago (1974) (Hawking 1974). After this time the hundreds of scientific works based in Hawking work appears. No facts are known which creates doubts in the possibility of Hawking radiation but it is not proven either. The Hawking radiation may not exist. The Large Hadron Collider has the main purpose to create the MBHs and detect the Hawking radiation.

Some people think the Black Hole (BH) is dangerous for Earth. It is try for very big BH more 10,000 tons. We work with very small MICRO Black Hole (MBH) having mass 0.01 g. Our MBH into AB-Generator is not dangerous for Earth because it is located in vacuum steel sphere and has mass thousands to millions times less than the critical mass. Our MBH is controlled by fuel. When we stop to inject the fuel, our MBH is vaporised (explode). Author shown this explosion is small and not dangerous for the offered reactor.

The AB-Generator could create a revolution in many industries (electricity, car, ship, transportation, etc.). That allows designing photon rockets and flight to other star systems. The maximum possible efficiency is obtained and a full solution possible for the energy problem of humanity. These overwhelming prospects urge us to research and develop this achievement of science (Hawking 1974; Bolonkin 2006b, 2009a, b).

8.7 Overview and Conclusion

Interstellar flight is impossible at current time. It is in some orders more difficult than trip to Mars. The sending of small probe to the nearest star systems "Alfa-Centauri" requests gigantic energy (about 100 powerful electric station), expensive equipment (hundreds of billions dollars) and large trip time (30–40 years). The conventional nuclear and thermonuclear on-board reactors cannot also solve this problem. We are hoping to find an advanced civilization that will move our technology leap. But we do not know: there is planets at nearest star systems having conditions closed to our Earth? Is there a life on these planets? What is level their development? If they are above us, why do not they came to us, or at least do not send signals to us?

There are many skeptic questions. But we do not must wait, we must develop our science and technology. Early or later we reach a level when human/robot civilization colonize not only our galactic but all Universe and will create the new Universe and new intelligent civilization, higher mind (Hawking 1974; Bolonkin 2006b).

References

Bolonkin, A.A.: Method of stretching of thin film. Russian patent application #3646689/10 138085, 28 Sept 1983, Russian PTO (1983) (in Russian)

Bolonkin, A.A.: Light Pressure Engine. Patent (Author Certificate) #1183421, 1985, USSR, (priority on 5 January, 1983) (1985)

Bolonkin, A.A.: New way of thrust and generation of electrical energy in space. Report ESTI, 1987, (Soviet Classified Projects) (1987)

Bolonkin, A.A.: Aviation, Motor and Space Designs, Collection Emerging Technology in the Soviet Union, Delphic Ass., Inc., pp. 32–80 (English) (1990)

Bolonkin, A.A.: The development of Soviet rocket engines, Delphic Ass. Inc., Washington, 122 p. (in English) (1991)

Bolonkin, A.A.: A space motor using solar wind energy (magnetic particle sail). In: The World Space Congress, Washington, DC, USA, 28 Aug–5 Sept 1992, IAF-0615 (1992a)

Bolonkin, A.A.: Simple space nuclear reactor motors and electric generators running on radioactive substances. In: The World Space Congress, Washington, DC, USA, 28 Aug–5 Sept 1992. IAF-92-0573 (1992b)

Bolonkin, A.A.: Space electric generator, run by solar wing. In: The World Space Congress, Washington, DC, USA, 28 Aug–5 Sept 1992, IAF-92-0604 (1992c)

Bolonkin, A.A.: Asteroids as propulsion system of space ships. JBIS 56(3/4), 98–107 (2003a)

Bolonkin, A.A.: Optimal inflatable space towers with 3–100 km height. JBIS 56(3/4), 87–97 (2003b)

Bolonkin, A.A.: Non-rocket transport system for space travel. JBIS 56(7/8), 231–249 (2003c)

Bolonkin, A.A.: Hypersonic gas-rocket launcher of high capacity. JBIS **57**(5/6), 162–172 (2004a)

Bolonkin, A.A.: Light Multreflex engine. JBIS **57**(9/10), 353–359 (2004b)

Bolonkin, A.A.: Multi-reflex propulsion systems for space and air vehicles and energy transfer for long distance. JBIS **57**(11/12), 379–390, (2004c)

Bolonkin, A.A.: Electrostatic solar wind propulsion system, AIAA-2005-3653. In: 41-st Propulsion Conference, 10–12 July 2005, Tucson, Arizona, USA (2005)

Bolonkin A.A.: Electrostatic AB-Ramjet Space Propulsion, In: AIAA-2006-6173, AIAA/AAS Astrodynamics Specialist Conference, 21–24 Aug 2006, AEAT, USA, vol. 79, pp. 3–16 (2006a)

Bolonkin, A.A.: Non-Rocket Space Launch and Flight, Elsevier, 488 p. (2006b)

Bolonkin A.A.: New Concepts, Ideas and Innovations in Aerospace, Technology and Human Sciences, NOVA. Chap. 3. ISBN-13: 978-1-60021-787-6 (2006c)

Bolonkin A.A.: Optimal space tower (mast), Presented as paper AIAA-2007-0367 to 45th AIAA aerospace science meeting, Reno, Nevada, USA, 8–11 Jan 2007 (2006d)

Bolonkin, A.A.: Converting of matter to nuclear energy by AB-Generator. Am. J. Eng. Appl. Sci. **2**(2), 683–693 (2009a)

Bolonkin, A.A.: Femtotechnology. Nuclear AB-matter with fantastic properties. Am. J. Eng. Appl. Sci. 2 (2), 501–514 (2009b)

Collection "Beamed Energy Propulsion": First International Symposium on Beamed Energy Propulsion, American Institute of Physics, Melville, New York (2003)

Driggers, R.G. (ed.): Encyclopedia of Optical Engineering, Marcel Dekker, Inc., New York (2001)

Forward, R.L.: Pluto—the gateway to the stars. Missiles Rockets **10**, 70–75 (1962)

Garwin, R.L.: Solar sailing—a practical method of propulsion within the solar system. Jet Propuls. **28**, 188–190 (1958)

Hawking, S.W.: Black hole explosions? Nature **248**, 30–31 (1974)

Kantrowitz, A.: Propulsion to orbit by ground-based lasers. Astronaut. Aeronaut. **10**, 74–76 (1972)

Landis, G.A.: Interstellar flight by particle beam. Acta Astronaut. **55**, 931–934 (2004)

Metzger, R.A., Landis, G.: Multi-bounce laser-based sails. In: Proceedings of the Shape Technology and Application, Forum-2001, Albuquerque, New Mexico, 2001, AIP Conference Proceedings, vol. 552, p. 397 [File: STAIF 2001 B] (2001)

Milonni, P.W., Eberly, J.H.: Lasers. Wiley, New York (1988)

Chapter 9
Triton Hopper: Exploring Neptune's Captured Kuiper Belt Object

Steven R. Oleson and Geoffrey Landis

9.1 Summary

Neptune's moon Triton is a fascinating object (Figs. 9.1 and 9.2), a dynamic moon with an atmosphere, and geysers (Cruikshank 1995). Triton is unique in the outer solar system in that it is most likely a captured Kuiper belt object (KBO)—a leftover building block of the solar system (Cruikshank 1995). When Voyager flew by, it was the coldest body yet found in our solar system (33 K), yet had volcanic activity, geysers, and a thin atmosphere (Fig. 9.3). It is covered in ices made from nitrogen, water, and carbon dioxide, and shows surface deposits of tholins, organic compounds that may be precursor chemicals to the origin of life. At a distance of over 30 AU, it would be by far the most distant object ever landed on by a spacecraft.

The Triton Hopper effort set out to design a mission to not merely land, but repeatedly fly across the surface of Triton, utilizing the volatile surface ices (primarily nitrogen) as propellant to launch across the surface and explore all the moon's varied terrain. It was determined that such a Hopper can be developed using simple methods of collection and propulsion. Gathering 100 kg of surface nitrogen ice with either a robotic scoop, or atmospheric gas using a cryopump provides sufficient propellant to allow this ~300 kg, highly instrumented lander to hop 5 km across the surface once a month, using a low-temperature radioisotope-thermal engine. More sophisticated propulsion could increase the flight range. Two years of hopping would allow this lander to hop 150 km and visit 30 sites! The Triton Hopper is a Phase 1 mission concept study funded by the NASA Innovative

S. R. Oleson (✉) · G. Landis
NASA Glenn Research Center, 21000 Brookpark Rd, Cleveland, OH 44135, USA
e-mail: steven.r.oleson@nasa.gov

G. Landis
e-mail: geoffrey.landis@nasa.gov

V. Badescu and K. Zacny (eds.), *Outer Solar System*,
https://doi.org/10.1007/978-3-319-73845-1_9

Fig. 9.1 Triton Hopper Mission destination, the largest moon of Neptune

Fig. 9.2 Artist's concept of Triton Hopper

Advanced Concepts (NIAC) program. The final Hopper design is shown in Fig. 9.4.

The concept of operations for the Hopper is shown in Fig. 9.10 and detailed in Sect. 9.3.1.

The main focus of this NIAC effort was to determine feasible methods of gathering, processing and using in situ propellant. According to Voyager the surface is predominantly nitrogen, in the form of both ice and snow on the surface (Fig. 9.5). Triton has a very thin atmosphere (~ 1 Pa), again mostly of nitrogen. Both surface and atmosphere sources were analyzed for propellant collection.

Four main approaches to gathering propellant were considered, each having various options (e.g., shovels or drills or scrapers). The simplest in-situ propellant

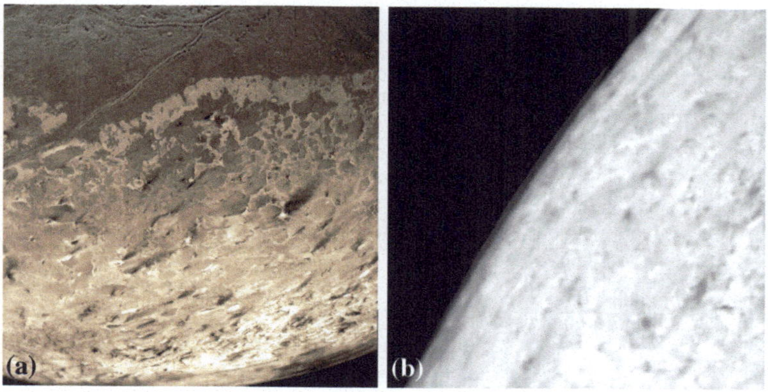

Fig. 9.3 **a** Voyager image of South Pole of Triton showing geysers. **b** Voyager image showing thin atmosphere

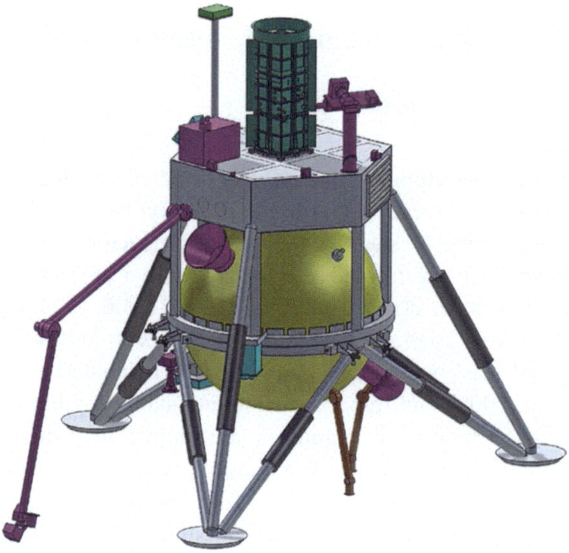

Fig. 9.4 Triton Hopper concept

found, both to process and to utilize, was cold gas nitrogen gathered from either the thin atmosphere or frozen on the surface.

The huge impact of such a Hopper is clear when considering that it replaces thirty single-site landers on Triton's surface. Even a Triton rover would be hard pressed to cover 150 km in two years. For comparison, the MER rover Opportunity covered 42 km in eleven years.

This chapter details the trades and a point design created for the Triton Hopper.

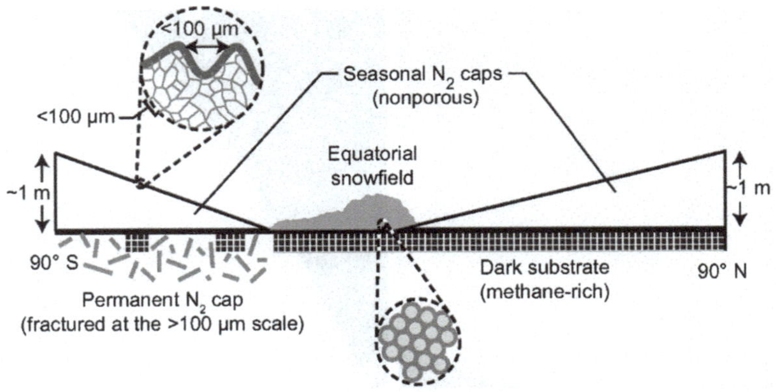

Fig. 9.5 Triton surface conditions. (Adapted from Eluszciewicz, 1991, in Neptune and Triton edited by D. P. Cruikshank. © 1995 The Arizona Board of Regents.) Reprinted by permission of the University of Arizona Press (Cruikshank 1995)

9.2 Study Background, Assumptions and Approach

9.2.1 Previous Studies

Before the Hopper concept was selected for detailed analysis, a large variety of alternate concepts for Triton exploration were evaluated, including remote sensing from orbiters; stationary landers; impactors; wheeled, walking, hopping, and hovercraft rovers; and several alternative approaches to rocket-powered Hoppers (Jerred et al. 2012; Shafirovich et al. 2006; Nothnagel et al. 2010; Davoodi and Acikmese 2014; Sercel et al. 1987; Meetin and Seifert 1974).

At least one past Neptune mission evaluated a Triton lander design (Bienstock et al. 2008), but remote sensing and stationary landers cannot achieve the baseline science, which will require in situ sample analysis and the ability to move from an initial landing site to a site featuring accessible surface organic compounds, which may not be available at all surface locations.

Because of the variety and global distribution of scientific targets, long-distance mobility is critical to achieving the scientific goals. The probe needs to be able to investigate multiple features: traversing to geyser regions and investigating plumes which will give access to samples from the interior; moving to tholin deposit regions to investigate primordial organic chemistry; investigating unique geological features such as the cantaloupe terrain. While Mars rovers have demonstrated record traverse distances of tens of kilometers on another world, Triton exploration demands orders of magnitude better traverse capability: we need pole to pole mobility to fully explore all the features of this fascinating moon. The rocket-propelled Hopper approach was the only one with the combination of long-distance mobility with the ability to take shorter hops to achieve the full range of desired science. As discussed later in Sect. 9.3.2, the ability to achieve multiple hops requires refueling. These considerations drive the selection of the in situ refueled rocket Hopper for exploration.

9.2.2 Triton Overview

Unique among the large moons of the solar system, Triton is in an inclined retrograde orbit, rather than the direct orbit of all but the smallest of solar system moons. This, and other clues, indicates that Triton is very likely a captured KBO. Thus, a mission to explore Triton will be the first-ever exploration of a KBO, giving us in situ insight into one of the least understood and most interesting kinds of body in the solar system. The only detailed information we have about the surface of Triton is from a single fly-by of Voyager 2 in 1989, which only imaged one hemisphere of the moon. Nevertheless, what we know shows that the object is complicated and interesting.

As a KBO, Triton is of interest because it contains material from the outer solar system, and thus allows study of composition of areas of the primordial solar system that are relatively unchanged by thermal alteration processes that occur nearer to the Sun. Of particular interest are surface deposits of tholins, complex organic materials that are dark red in color, and which are not present on the surfaces of inner solar system bodies, but ubiquitous on the larger KBOs. These organic deposits may hold keys to the biochemical origin of life.

In addition to its interest as a KBO, Triton is a dynamic object. It is one of the few bodies in the solar system that is seen to have active eruptions. Near the South Pole, cryo-volcanism is in evidence, with geysers shooting up to 8 km high; such geysers may be a link to a hypothesized subsurface ocean. The plumes of these geysers give access to the interior of the moon. Other regions of the surface show geological evidence of low-temperature volcanism. It is one of the few moons with a noticeable (although thin) atmosphere, with clouds believed to be composed of frozen nitrogen crystals at altitudes of 1–3 km from the surface, and a tropospheric haze, hypothesized to be composed of hydrocarbons and nitriles created by photochemical alteration of methane, ammonia, and nitrogen. Triton also includes surface features of unknown origin such as "cantaloupe terrain."

9.2.3 Triton Operational Environment

Triton, as shown in Fig. 9.6, is the largest moon of Neptune. It has some unique characteristics. Its surface is a mix of smooth planes near heavily cratered areas. The reason for this significant difference in terrain is not fully understood. Also, Triton is the only moon in the solar system that orbits its planet in a direction opposite to its rotation. It was also recently discovered, from Earth ground

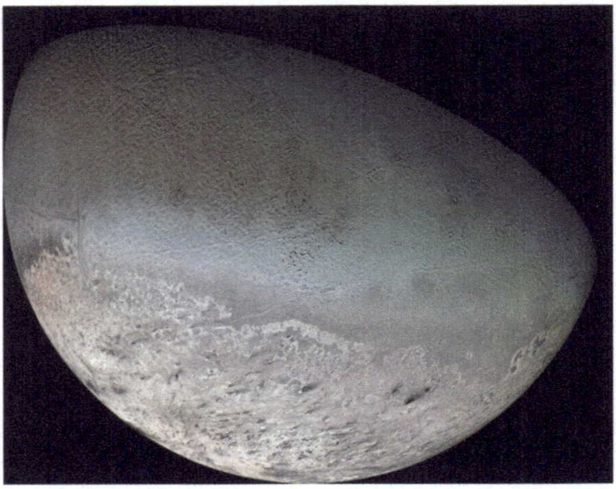

Fig. 9.6 Image of Triton (NASA 2004)

observations, that Triton experiences seasons causing its thin atmosphere to vary in thickness (Space.com 2010). Each season lasts approximately 40 Earth years due to the long solar orbital period of Neptune. The surface of Triton has a number of different materials, normally gaseous in Earth's environment, that are frozen and cover the surface. These include mainly nitrogen with some water, carbon monoxide and carbon dioxide. Triton's physical properties are summarized in Table 9.1.

Triton's rarified atmosphere is mostly nitrogen. However, methane and carbon monoxide have also been detected. It is believed that a layer of frozen nitrogen on the surface with higher concentrations of carbon monoxide, up to ten times greater than seen below the surface, vaporizes from the surface during the summer time causing the increase in seasonal atmospheric density that was detected.

Data from the Voyager 2 spacecraft indicated that cryo-volcanic activity is likely occurring on the surface of Triton in the form of gaseous nitrogen geysers. It is also believed that an ammonia-rich water ocean exists beneath the icy surface.

The operating environment of the Hopper will be on the surface of Triton where it will collect propellant and then use a ballistic hop to another location. The components on the Hopper have to withstand the deep space transit to Triton and operation within the cold surface environment. While on the surface the Hopper will collect material either from the atmosphere or by directly scooping it from the surface into a tank where it will be stored and used as propellant. The condition of the surface material is not completely known, and may range from a fluffy snow like consistency to solid ice. The thermal conductivities of the main surface components in solid form are given in Table 9.1.

The cold environment of Triton provides a number of challenges in the operation of equipment and materials. Operating within this environment requires all of the electronics and science equipment to be insulated and isolated as much as possible from the environment, or operate cold.

Table 9.1 Triton physical properties

Property	Value
Diameter	2707 km (about ¾ the diameter of Earth's Moon)
Solar flux at Triton	1.5 W/m^2
Average surface temperature	38 K
Solar orbital period	165 Earth years
Orbital period around Neptune	141.6 h (retrograde motion opposite planets rotation, same as rotation period (one Triton day), tidally locked with Neptune)
Atmosphere composition	Nitrogen, carbon monoxide, methane, carbon dioxide
Atmosphere pressure	14–40 microbar (depending on the season), (7.5–30 mTorr)
Gravitational acceleration	0.779 m/s^2
Surface composition	55% frozen nitrogen; 15–35% water ice; 10–20% frozen carbon dioxide
Surface material thermal conductivity[a]	Nitrogen: 0.25 W/mK at 38 K; Water ice: 23.9 W/mK at 38 K; Carbon dioxide: 5 W/mK at 38 K
Orbital inclination	To Neptune's equator: 156.9°; To the ecliptic: 129.8°
Orbital eccentricity	0.0
Atmosphere gas constant (R_a)	296.8 J/kg K
Atmosphere ratio of specific heats (γ_a)	1.4
Atmosphere specific heat (c_{pa})	1039 J/kg K

[a]Stachowiak et al. (1994); Rabin (2000); Sumarokov et al. (2003)

The gravitational acceleration on Triton (0.779 m/s^2) is less than half that of Earth's Moon. This low gravity presents challenges in using force to break into the icy surface and extract material for propellant. Due to the low gravity the spacecraft weight is low, limiting in turn the amount of downward force that can be applied. To overcome this, approaches such as scraping or using opposing scoops to dig into the surface could enable enough force to be generated to break into the surface ice.

9.2.4 Hopper Science

The science instrument package was chosen to address the major scientific enigmas of Triton. The selected package made use of the ability of the Hopper to sample widely varied terrain, including both horizontal sampling as well as vertical sampling through the plumes of geysers and through layers of the thin nitrogen atmosphere.

Experiences from the Mars Exploration Rovers and other missions have shown how critical visual photography is, not merely for operations planning, but as a capable scientific tool for understanding geomorphology and surface processes. Rover operations have also shown the value of stereo imagery. Thus, the first instrument is a highly capable camera. In addition to scientific use, the camera will also produce high-definition images for use in the public outreach, to bring the public into "virtual presence" on the surface of the icy moon. The camera will not only characterize the geology of the surface locations, it will also be used for in-flight imagery during the hops, allowing detailed aerial surveying of the geological context of each location and three-dimensional reconstruction of terrain from stereo imagery.

Likewise, visible and near-infrared spectrometry have been shown to be of great value for remote mineralogy. These tools will be the main remote-sensing instruments.

At a landed location, detailed measurements will be done using the in situ instruments. The vehicle is uniquely suited for surface analysis, since as a part of the refueling operation it will take surface samples into the vehicle, allowing complete characterization of surface material. A significant characterization will be the use of mass spectrometry. In addition, classic in situ instruments will be applied, including microscopic imaging, X-ray crystallography, and an alpha-particle X-ray spectrometer (APXS), all of which have proven to be of great value during rover missions.

Measurements when landed will be supplemented by measurements during flight, when the instruments will characterize the vertical profile of the atmosphere and cloud layers. The vehicle will also be able to fly through the plumes of geysers, taking samples of the geyser activity as a method of probing the deep interior of the moon.

Finally, three additional tools will give information on what is below the surface. A small ground-penetrating radar will allow a vertical profiling of the local surface. This will determine the amount of layering, which gives information about climate. Depending on the landing site, the frequency chosen, and the signal to noise ratio, it may also be possible to characterize the subsurface liquid reservoir which feeds the geyser activity.

Seismometry can also give valuable information. Geyser eruptions are typically accompanied with localized crustal seismic activity, and measuring this activity will yield information about the liquid reservoir feeding the geyser and about the mechanisms of geyser action. If Triton is subject to global seismic activity, measuring seismic motions may give information about the deep interior as well. The lander will be equipped with microelectromechanical (MEMS) accelerometers on the footpads, measuring three components of acceleration; accelerometers on each foot give redundancy to the system, as well as some ability to filter spacecraft induced disturbances out from seismic ground motion.

Finally, precision navigational measurement will give a measure of the wobble of Triton in its orbit, which is a powerful tool for determining size of the liquid core.

The major science instruments of the baseline science package are described in Sect. 9.4.1.

9.2.5 Design Trades

While getting the Hopper to Triton was considered outside the scope of the Phase I effort some work was performed to determine how difficult it is to get a lander onto Triton. All past concepts have had to rely on advanced propulsion techniques to reach Neptune in a timely fashion with existing launchers. Figure 9.7 summarizes the delivery trades. While use of nuclear electric propulsion would allow getting into low Triton orbit (and thereby minimizing the triton descent propulsion system size) it was felt that use of a Solar Electric Propulsion (SEP) stage and aerocapture system is far more synergetic with other current NASA technology efforts (especially the piloted Mars mission's use of SEP and aerocapture).

Using the SEP and aerocapture system produced a launch and delivery concept similar to that explored in "Mission Trades for Aerocapture at Neptune" (Bailey and Noca 2004). As such, the concept developed in the reference was assumed to be the delivery concept to Neptune. New trajectories to Neptune using SEP and aerocapture for a 2029 launch date were developed as was a notional mission and combined solid/bipropellant landing stage to get the Hopper to the surface. The baseline CONOPS is shown in Fig. 9.9.

The main focus of this effort was to determine feasible methods of gathering, processing and using in situ propellant. First the propellant must be gathered. According to Voyager the surface is predominantly nitrogen, whether in ice or snow form on the surface (see Fig. 9.5). As mentioned earlier Triton has a very thin atmosphere (~ 1 Pa), again mostly of nitrogen.

Since nitrogen was so plentiful and easily accessible (both frozen on the surface and in the atmosphere) it was considered the easiest propellant to utilize and process. But how to acquire it? Four main methods were considered, each having

Triton Hopper Delivery Trades	Chemical Capture	SEP/Chemical Capture	SEP/ Aerocapture	Nuclear Electric Propulsion

Trip time to Triton	~ 18 yr	~ 15 yr	~ 12 yr	~ 17 yr
Triton ΔV	~ 4 km/s	~ 3 km/s	~ 3 km/s	~ 1.2 km/s
Selection pros	Customer for SLS?	Only requires new SEP stage	Trip time excellent, technologies being developed	Smallest Hopper landing stage
Selection cons	Trip time	Landing stage ~ 3X mass of Hopper	Landing stage ~ 3X mass of Hopper	NEP not currently in development

Fig. 9.7 Triton delivery trades

Approach using similar mass/power to gather ~100 kg N_2	Pumping from ~ 1 Pa atmosphere	Cryopumping from ~ 1 Pa atmosphere	Subliming from surface using waste heat and pumping	Mechanically gathering from surface
Simplicity/heritage/ global accessibility	High	High	Low	Medium
Current knowledge of N_2 conditions at surface	Homogeneous atmosphere	Homogeneous atmosphere	Variable: Ice, snow N_2 ice conducts as well as stainless steel and is as hard as water ice (0 °C)	
Purity of N_2	High	High	Medium	Low (solids, other gases)
Speed	~ 3000 l/s	~ 3000 l/s		7 days
Mass/power	400 kg/2 kW	~10 kg/50 W[a]	TBD	10 kg/30 W average
Production rate	~ 15 kg N_2/day	~ 15 kg N_2/day	TBD	~17 kg N_2/day

[a]Assuming cryopump waste heat conductively transferred to surface

Fig. 9.8 Propellant collection trades

various options (e.g., shovels or drills or scrapers). These methods are shown in Fig. 9.8.

The obvious choice is to gather the propellant in its frozen form on the surface. Since the science system already had a scoop to gather samples for analyzing in an oven it is easy enough to reuse it to gather frozen nitrogen. Nitrogen should only be as hard as water ice (0 °C) and even easier to gather in the form of snow. However, the low gravity of Triton (~1/2 Earth's Moon) and icy terrain might make it difficult to gather nitrogen at all locations. Adding to this the issue of other contaminants caused the search for a better primary system. Use of a turbopump to gather the atmosphere and then allow it to freeze on the inside of the tank (at slightly higher pressure) will work, but to gather roughly 100 kg of nitrogen in a week's time would require too much power and mass. Assuming a 50 W, 10 kg limit a smaller turbopump could gather the 100 kg but would require over 6 months to do so. Cryopumping from the thin atmosphere will work regardless of where the Hopper is. In order to reduce power requirements for the cryopump it was determined that the cryopump should operate at very low temperatures and reject its waste heat directly to the surface (~90% of the 50 W of power) through a deployable arm. Good conduction through the arm and a large inlet opening (~10 cm diameter) are key to making the cryopumping option work, so mechanical options would be used as a backup to cryopumping if good surface thermal contact between the cryopump and the surface cannot be made. The large opening would double as the funnel for frozen nitrogen from the shovel. Either system can gather the amount of nitrogen in roughly a week with 50 W or less of power. Once filled, a unique interior door system will seal the tank and the pressure will keep it sealed.

Once the tank is filled the options of how to use the propellant were explored. Nitrogen gas has been used as a very effective if low performing (~60 s I_{sp})

propellant for both launch vehicles and spacecraft and was even used for the Skylab space station Reaction Control System (RCS). Besides the low I_{sp} the high pressure required for the system requires heavy tanks, but such high pressure simplifies the propellant feed system. The ~100 kg of nitrogen can be vaporized and heated to 300 °K and 2000 psi in about eleven days using 60 W via heaters. Use of waste heat piped from the radioactive power source was deemed complex and detrimental to power production. Use of radioisotope heater units would provide four times the waste heat, but would also be more challenging to integrate to the low temperature platform during propellant collection. Heating the tank will also heat the propulsion system so it will not need to operate at extremely low temperatures.

As an alternative the nitrogen can be heated through a resistojet to improve its specific impulse (>100 s I_{sp}), which requires large amounts of electrical power. Heating a thermal sink slowly over a week and then passing the nitrogen over it to improve I_{sp} could also provide such performance. While selection of the thermal sink material was conducted (and lithium or beryllium were chosen as candidates) and initial performance estimated (over 100 s I_{sp}), the simpler high-pressure system was chosen for the Phase I effort. Such improved I_{sp} performance has been shown to quadruple hop distance for only a doubling of I_{sp} so this approach will be further investigated in a Phase II.

9.3 Baseline Design

9.3.1 Concept of Operations (CONOPS)

As mentioned previously the main focus of the Triton Hopper design was in the operations on the surface and the hops. However, a notional launch and delivery scheme was adapted from a previous reference and the mission recalculated for the appropriate Triton Hopper launch window around 2030 when Jupiter will be available for gravity assist flybys. Figure 9.9 illustrates the launch and delivery conops. Using the past reference a heavy lift launcher (Delta IV Heavy or equivalent) launches the vehicle to escape where a SEP stage performs inner-solar system burns to enable an Earth flyby. Not long after, the SEP stage is jettisoned, since the distance to Neptune makes solar arrays ineffective. A Jupiter gravity assist is next which greatly accelerates the Triton orbiter/Hopper to a 2041 Neptune encounter. Another key technology is the use of an aerocapture system in lieu of a large chemical system to capture the orbiter/Hopper into an elliptical orbit around Neptune. This orbit is reduced/changed to have an apoapsis just below Triton by Neptune using aerobraking. At this point the Hopper is separated and uses a combination of solid and bipropellant stages to capture into a 200 km Triton orbit. From there the bipropellant 'sky platform' descends to 250 m where the Hopper can perform its first hop to the surface. The sky platform then crashes on the surface, though given more mass for legs and landing equipment, the sky platform

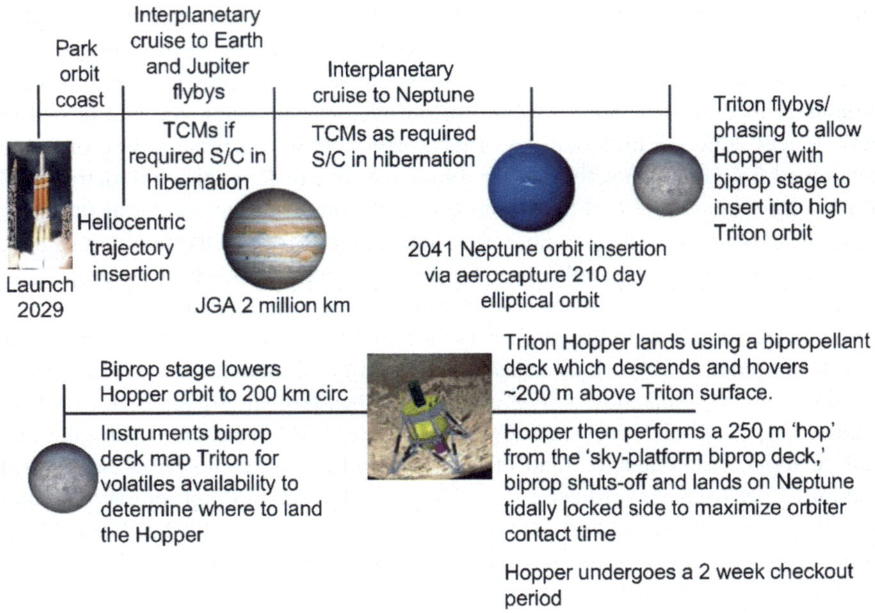

Fig. 9.9 Hopper launch and delivery CONOPs

could make a single landing from which the Hopper could take off. Once on the surface the Hopper begins its science, propellant collection/processing, and hopping cycle supported by the Neptune orbiter.

The cycle of Triton science, propellant gathering/processing and hopping is illustrated in Fig. 9.10. With limited power ($\sim$50 W once housekeeping is allotted for) it is estimated that propellant gathering (either by the cryopump or arm/scoop) will require eight days, as will the propellant heating and vaporization to pressure. Since the orbiter will make a close approach every twelve days, four days of every twelve were allotted for science with eight for propellant. So after landing the Hopper will gather propellant and perform science with an orbiter flyby to relay science and assess how much propellant is gathered. Key components to propellant collection are two thermal conducting legs, one from the tank and one from the cryocooler to keep the tank/cryocooler near surface temperatures ($\sim$35 °K). The orbiter's close pass (<200,000 km) should last 20 hours so that controllers on Earth can receive the signals, evaluate them and send back next step instructions in that time period, including next hop instructions.

During the next pass the Hopper will inform ground controllers of its propellant state (pressure and temperature). Controllers will then signal a go ahead and the Hopper will hop its $\sim$2 min, 5 km horizontal, 1 km high trajectory gathering atmospheric data and relaying it and housekeeping data to the orbiter during the entire hop. Final approach will be programmed from the Earth, based on high-resolution orbiter images, but final touchdown and collision avoidance of

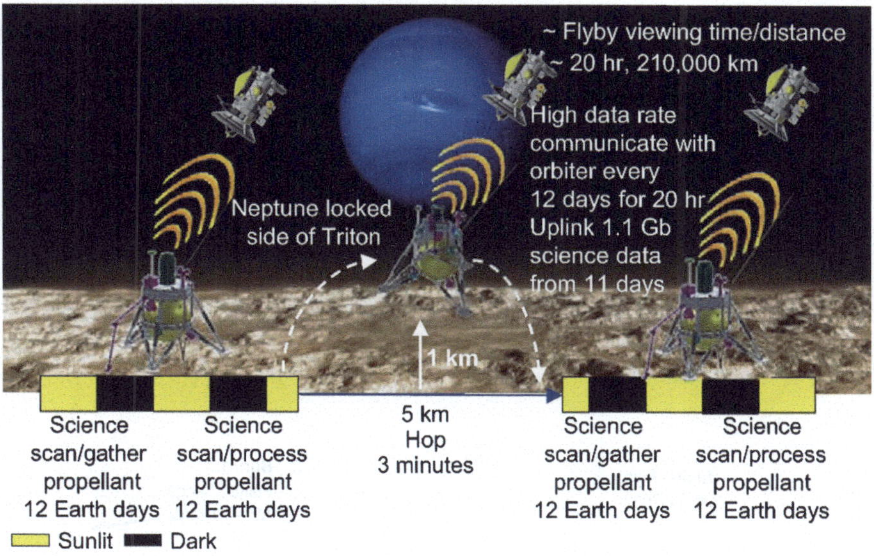

Fig. 9.10 Triton Hopper surface CONOPS

greater than 10° slopes and greater than 10 cm debris will be calculated real-time by the Hopper lidar and main computer. Once down the Hopper will deploy its cooling legs for the tank and cryocooler and begin the process again.

At this rate, over a two year mission, the Hopper should be able to hop roughly thirty times for a distance of 150 km. Thirty locations could thus be sampled and evaluated by a single lander. Further hops could be made by improving the I_{sp} of the propulsion system or reducing the Hopper mass. Depending upon Hopper health it should be able to continue hopping given an extended mission duration.

9.3.2 Mission

Launch Vehicle Ascent, Park Orbit and TTI

The trans-Triton injection (TTI) burn would be performed with the launch vehicle, a Delta IV Heavy. On October 26, 2029, the Delta-V Heavy could deliver 8685.5 kg to a C_3 of 4.3 km^2/s^2 to begin the interplanetary cruise phase to Triton.

Earth to Triton Cruise

The craft will be placed in hibernation for the majority of the cruise phase except during periods of thrusting, planetary encounters for gravity assists, planned health assessments, or when any trajectory correction maneuvers must be performed. The Earth to Triton trajectory is shown in Fig. 9.11.

Following TTI, the spacecraft will use the SEP thrusters to maneuver in heliocentric space. Prior to capturing in orbit around Neptune, the spacecraft performs

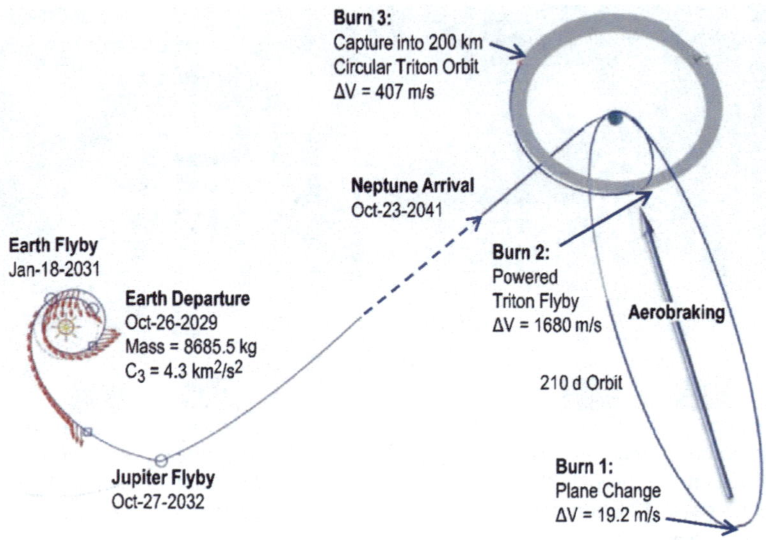

Fig. 9.11 Interplanetary and delivery trajectory

Earth and Jupiter gravity assist maneuvers to further increase its velocity. The spacecraft arrives at Neptune on October 23, 2041. The total interplanetary cruise phase of the mission requires twelve years and 2610 kg of Xe propellant to complete.

After arriving at Neptune, the spacecraft will aerocapture into a 210-day orbit around Neptune. The net delivered mass to Neptune *prior* to performing the Neptune aerocapture is 4475 kg, in addition to the now dry 1600 kg SEP vehicle (see Table 9.2). Following the aerocapture, a plane-change burn will be performed at apogee to adjust the spacecraft inclination. The spacecraft will then perform aerobraking to reduce the apogee altitude from the initial 210-day orbit to an elliptical orbit with apogee at Triton altitude. A powered flyby of Triton followed by a short coast will reduce the hyperbolic arrival velocity and adjust the arrival velocity vector to minimize the ΔV required to capture into a circular 200 km Triton orbit.

Triton Entry, Descent, Landing and Operation

All mission analyses for the Triton Hopper's surface operational scenarios were performed using the Optimal Trajectories by Implicit Simulation (OTIS) program,

Table 9.2 Interplanetary delivery summary

Parameter	Value
EOL power to EP system (kW)	30
I_{sp} (s)	~2000
Arrival mass at Neptune (kg)	6075
SEP vehicle dry mass (kg)	1600
Pre-NOI net delivered mass (kg)	4475

version 4. The exception is for the ΔV estimates of the Earth departure using an expendable launch vehicle (ELV); those were taken from analyses with the Copernicus trajectory program.

Hopper Delivery—From 200 km Circular Triton Orbit: This initial delivery scenario assumes the Hopper has been delivered as described in Sect. 9.3.2 to a 200 km circular orbit at Triton. The delivered vehicle stack consists of the Hopper vehicle and a bipropellant stage. This Hopper vehicle represents a more mature configuration and is consistent with the baseline operational vehicle described in the next section. The bipropellant stage has a thrust level of 14,095 N and an I_{sp} of 320 s. While in the 200 km orbit, the bipropellant stage performs a de-orbit burn followed by a coast along a suborbital arc (i.e., fall towards surface). Next, the bipropellant stage performs a slowdown burn to reduce velocity. Finally, the Hopper performs a controlled landing hop on Triton's surface. The delivered Hopper wet mass is approximately 356 kg which accounts for a dry Hopper of 325 kg and a partial propellant load, enough to perform the initial landing hop to the surface 250 m downrange. The Hopper's propulsion system is described (top level) in the next section.

Operation of Triton Hopper

As described above, once on Triton's surface, the Hopper collects resources and produces propellant that is used to perform its hops. The Hopper vehicle has a wet (fully loaded) mass of 450 kg. This includes a dry Hopper weighing 325 kg and a full propellant load of 125 kg with a 2% unusable propellant margin since nitrogen liquefies below a propellant remaining fraction of 0.02. The propulsion system consists of cold gas rockets; these rockets have been used in the mission analysis with an initial tank pressure of 2000 psia, temperature = 300 K. The propulsion system is described in detail in Sect. 9.4. The thrust level of the Hopper, derived from the baseline case, and optimized for maximum range of that case, is 1203.8 N (270.6 lbf).

Initial analyses have been performed to ascertain the required thrust level for the Hopper propulsion system, various hopping distances and propellant loads. The baseline case is a result of using a cold gas nitrogen rocket and maximizing the hopping distance using a 125 kg propellant limit (with 2% unusable). The sequence of events for each operational hop is as follows:

- Vertical takeoff followed by a pitch-over during which flight path angle rate is not to exceed −0.175 rad/s (approximately −10°/s). Constant maximum thrust.
- Ballistic coast with a non-impulsive flow at 5% of the nominal maximum burn flow rate to account for the attitude control system (ACS) maneuver's propellant usage
- Landing profile: Propulsive landing during which the thrust is allowed to throttle down to 10% of maximum thrust, if needed. Final "touchdown" throttle setting, which may range from 0 to 0.3 (1/s), is derived from requiring the vehicle to remain in a vertical orientation with a final altitude ≤ 0.1 m and the final velocity ≤ 0.01 m/s
- Vehicle ends in a vertical position

Figures 9.12 and 9.13 show the altitude versus time and versus range, respectively, for the initial delivery from a 200 km Triton orbit (only the final landing is performed by Hopper) in blue and the baseline hop operational hop while on the Triton surface in red. The resulting thrust-to-weight ratio is over three.

A summary of the mission ΔV's can be seen in Table 9.3. The propellant numbers in Table 9.3 represent usable propellant. Additional propellant, 5% of the usable, is carried as margin and held as inert mass in both the descent stage and lander. This margin is assumed available for terminal descent collision avoidance with ground debris. No margin on propellant was carried in the solid stage, rather the inert mass of that stage was assumed to be 8% of the total mass of the stage. The propellant required for the Hopper to hop off of the descent stage and perform the initial landing on Triton, 29.9 kg of main propellant and 0.7 kg of ACS propellant, is pre-loaded on the Hopper, while the propellant required for each hop, 114.9 kg of main propellant and 2.5 kg of ACS propellant, is processed and loaded onto the Hopper while on the surface of Triton. The totals for the Lander in the last row of Table 9.3 represent the amount of propellant the Hopper requires to perform the initial landing and one 5 km hop.

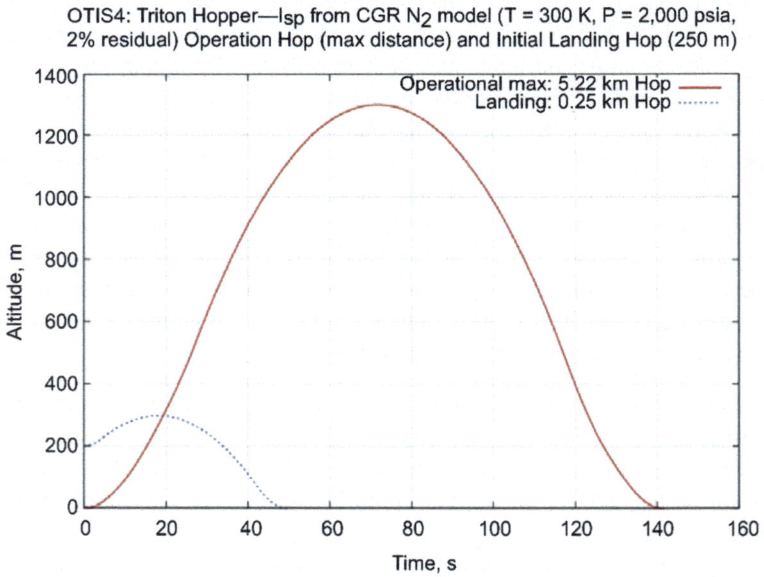

Fig. 9.12 Altitude versus time for delivery landing from 200 km Triton orbit (blue) and baseline operational hop

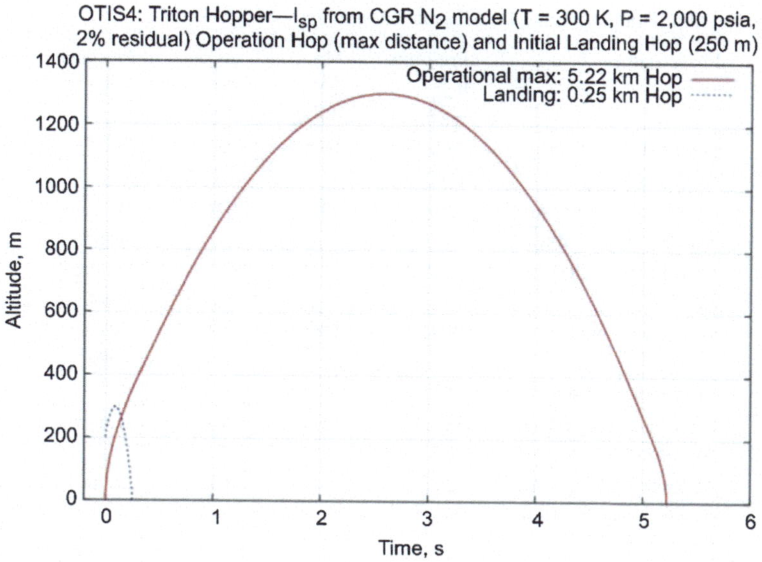

Fig. 9.13 Altitude versus range for delivery landing from 200 km Triton orbit (blue) and baseline operational hop

9.3.3 System Level Summary

The Triton Hopper system uses a radioisotope power system that powers two In-Situ Resource Utilization (ISRU) systems to generate the nitrogen propellant necessary to propel the Hopper across Triton. Figure 9.14 shows the Triton Hopper block diagram and illustrates how all of the major vehicle subsystems are integrated.

The Triton Hopper Master Equipment List (MEL) contains the mass roll up; including a subsystem aggregate mass growth allowance (MGA), for each vehicle subsystem. The Triton Hopper MEL is in Table 9.4.

The MEL shown in Table 9.5 captures the bottoms-up current best estimate (CBE) and growth percentage of the Triton Hopper that the subsystem designers calculated for each line subsystem. In order to meet the total required mass growth of 30% an allocation is necessary for growth on basic dry mass at the system level, in addition to the growth calculated on each individual subsystem. This additional system-level mass is counted as part of the inert mass to be flown along the required trajectory. Therefore, the additional system-level growth mass impacts the total propellant required for the mission design.

Table 9.3 Triton Hopper—Case 1—Mission ΔV summary

Phase no.	Phase name	Pre-burn mass (kg)	Main ΔV (m/s)	ACS ΔV (m/s)	Propulsive element	I_{sp} (s)
1	Powered Triton flyby	2442.5	1680	0.0	Solid stage	294
2	Drop solid stage	1362.5	–	–	–	–
3	Capture into 200 km circular Triton orbit	1268.6	407	0.0	Descent stage	323
4	De-orbit from 200 km circular	1115.7	37	0.0	Descent stage	323
5	Slow down burn	1102.8	1066	0.0	Descent stage	323
6	Drop descent stage	787.8	–	–	–	–
7	Landing burn, 250 m hop off of descent stage	389.8	41	0.0	Lander	52
8	ACS for landing burn	359.9	0	0.1	Lander	52
9	Process propellant for 5 km hop	359.2	–	–	–	–
10	~5 km hop	476.6	141	0.0	Lander	52
11	ACS for 5 km hop	361.7	0	3.5	Lander	52
Total, solid stage			1680	0.0		
Total, descent stage			1509	0.0		
Total, lander			182	4.5		

Concluded

Phase no.	Phase name	Main prop (kg)	ASC prop (kg)	Post burn mass (kg)	Change in mass (kg)
1	Powered Triton flyby	1080	0.0	1362.5	−1080.0
2	Drop solid stage	–	–	1268.6	−93.9
3	Capture into 200 km circular Triton orbit	153	0.0	1115.7	−153.0
4	De-orbit from 200 km circular	13	0.0	1102.8	−12.8
5	Slow down burn	315	0.0	787.8	−315.0
6	Drop descent stage	–	–	389.8	−398.0
7	Landing burn, 250 m hop off of descent stage	30	0.0	359.9	−29.9
8	ACS for landing burn	0	0.7	359.2	−0.7
9	Process propellant for 5 km hop	−117	0.0	476.6	117.4
10	~5 km hop	115	0.0	361.7	−114.9
11	ACS for 5 km hop	0	2.5	359.2	−2.5
Total, solid stage		1080	0.0		
Total, descent stage		481	0.0		
Total, lander		145	3.2		

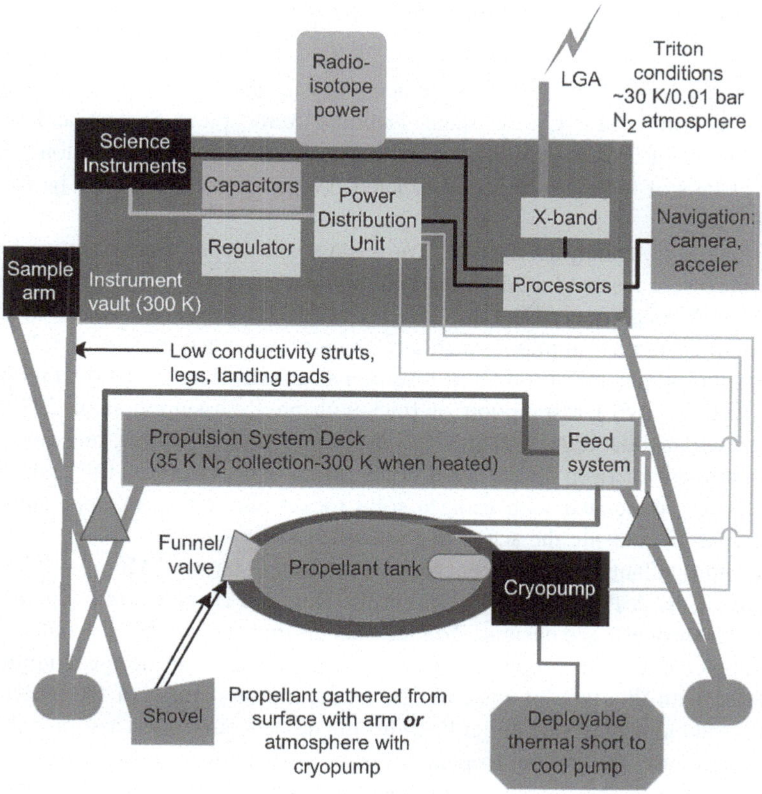

Fig. 9.14 Triton Hopper block diagram

Table 9.4 Triton Hopper MEL

Case 1 CD-2015-127	Basic mass (kg)	Total mass, with growth (kg)
Triton Hopper	1206.8	1256.1
Hopper	402.8	445.3
Science payload	23.0	23.9
Attitude Determination and Control Subsystem	14.6	15.8
Command and Data Handling Subsystem	14.7	19.0
Communications and tracking	4.2	4.6
Electrical power subsystem	39.5	44.0
Thermal control (non-propellant)	33.0	38.0
Propulsion (chemical hardware)	87.2	102.7
Propellant (chemical)	127.1	127.1
Structures and mechanisms	59.5	70.2

9.3.4 Triton Entry System/Cruise Stage

Four primary architecture elements are necessary for the Triton Hopper to reach the surface of Triton: SEP Cruise Stage, Neptune Aerocapture System and Orbiter, Solid Propellant Triton Orbit Insertion Stage and a Bipropellant Triton Descent Stage. Details of the mission and associated ΔV for each element can be found in Sect. 9.3.2.

During mission analysis, the SEP Cruise stage and Neptune Aerocapture System and orbiter were resized based on the assumptions presented in the paper "Mission Trades for Aerocapture at Neptune" (Bailey and Noca 2004). Additional details on the system design assumptions, with any changes are noted in Sect. 9.3.2.

The solid propellant Triton Orbit Insertion stage was sized based on a propellant mass fraction (PMF) assumption of 0.93 with an I_{sp} assumption of 293 s. The commercially available ATK Star 37GV is close to what is needed, but would need to be modified for deep space use. Further work is necessary to fully address all design issues associated with using a solid rocket motor for this application, but those issues are outside the scope of the current study.

The Bipropellant Triton Descent Stage was sized to have a PMF of ~0.55 using an I_{sp} of 303 s. A PMF of 0.55 is within the expected range for this type of stage based on historical stage designs. To meet this desired stage PMF, a detailed design was completed for the propulsion system to understand the design complexities associated with this type of system. The remaining inert mass of the system was then adjusted to achieve the target PMF for the descent stage. It is assumed that this stage would use the Triton Hopper avionics and landing sensor systems. It is therefore assumed the descent stage would contain only the propulsion, structure

Table 9.5 Triton hopper system summary

Main subsystems	Basic mass (kg)	Total mass with growth (kg)
Hopper	402.8	445.3
Science payload	23.0	23.9
Attitude Determination and Control Subsystem	14.6	15.8
Command and Data Handling Subsystem	14.7	19.0
Communications and tracking	4.2	4.6
Electrical power subsystem	39.5	44.0
Thermal control (non-propellant)	33.0	37.9
Propulsion (chemical hardware)	87.2	102.8
Propellant (chemical)	127.1	127.1
Propulsion (EP hardware)	0.0	0.0
Propellant (EP)	0.0	0.0
Structures and mechanisms	59.5	70.2
Dry mass desired system level growth	275.7	358.4
Additional growth (carried at system level)	–	–
Total wet mass with growth	402.8	485.6

and minimal control avionics necessary to perform its mission. A dedicated follow-on design study would be necessary to fully work out the design details of this stage, as the detailed descent stage design was also outside the scope of this study.

Based on the mission assumptions used to deliver the Hopper to the Triton surface, Table 9.6 shows how the Triton Hopper mass compares to the available delivery mass at Neptune/Triton. Additional details on the various components used to deliver the Triton Hopper to the surface of Triton are listed in Sect. 9.3.4.

9.3.5 Triton Concept Drawings and Descriptions

This section focuses solely on the layout of the Triton Hopper in its landed configuration. Design of the launch vehicle packaging and all other elements of the mission architecture were beyond the scope of this study. The configuration of the Triton Hopper while on the surface can be seen in Fig. 9.15.

The Triton Hopper can be separated into three major elements: the bus, the ISRU system with propellant tank, and the landing structure with thrusters. Figure 9.16 shows the breakout of these three major elements. All of the Triton Hopper components not dedicated to the ISRU, propulsion, and landing structures are contained on the bus. The cryocooler and tank collection funnel for the ISRU system are mounted directly to the propellant tank itself. Finally, the landing element is comprised of the landing legs, thrusters, and the tank mounting structure.

These three major elements of the Triton Hopper are thermally isolated from one another through the structural layout and material properties of the structures. This isolation is required to prevent heat transfer between the "hot" and "cold" elements that comprise the Triton Hopper.

Table 9.6 Triton Hopper delivered mass summary

	Mass (kg)	Assumptions (if applicable)
Net mass delivered to Neptune orbit	4475.0	COMPASS team trajectory analysis
Aerocapture system	1119.0	From Bailey and Noca (2004)
Payload left in Neptune orbit	792.0	From Bailey and Noca (2004)
Available mass for Triton Hopper and descent stages	2564.0	
Triton orbit insertion stage, solid	1189.2	Solid stage I_{sp} = 293 s, PMF = 0.93
Triton descent stage, bipropellant	885.7	Bipropellant I_{sp} = 323 s, PMF = 0.55
Triton Hopper, mass at launch	391.4	Loaded propellant at launch enough for short landing hop, ~33 kg
Available launch margin	97.8	
Available launch margin (%)	3.8%	

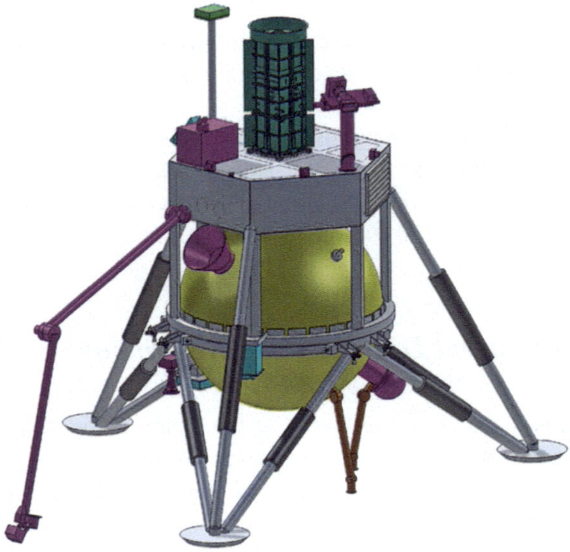

Fig. 9.15 Isometric view of the Triton Hopper

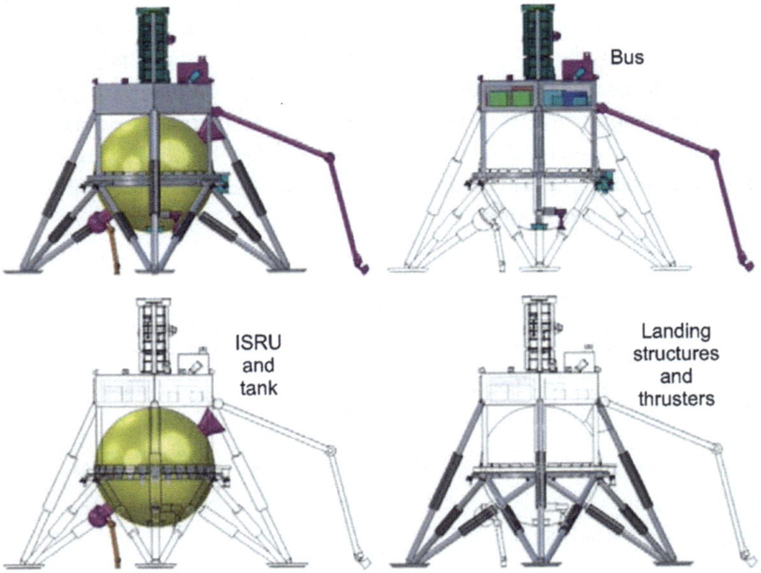

Fig. 9.16 Major elements of the Triton Hopper

All of the electronics are contained on the bus element and due to the waste heat generated by these electronics; the bus is considered the "hot" element of the Hopper. In order to prevent the heat from the bus entering the propellant tank while on Triton's surface, the only conductive path from the bus into the tank is down to the landing pads through the three longer landing legs struts, then up through the six shorter landing leg struts and into a square tubular ring that provides the interface to the mounting tabs contained on the tank. While this is a conductive path for heat transfer from the bus to the tank, the long path through low conductive structures keeps the transfer to a minimum. The structure of the bus is comprised of a hexagonal space frame with closeout panels that provide the interfaces for a majority of the components contained on the bus. The vertical members located at the vertices of this hexagonal space frame extend down to a tubular ring that goes around the circumference of the tank. This ring does not come in direct contact with the tank and "floats" above the second ring that interfaces directly to the mounting tabs contained on the tank while the Hopper is on the Triton surface. During the thrusting and landing phases of the mission, both of which do not require thermal isolation between the bus and tank, these two rings come together and allow the loads to be carried through the vertical members of the bus structure. Once landed on the surface, the three longer members of the landing legs will "push" the bus structure upwards slightly to provide the separation between the two rings, and thus provide the thermal isolation between the bus and tank. Additional structure extends outward from this ring to mount the LIDAR, and downward from this ring to mount the landing radar and the heritage Mars Descent Imager (MARDI) used for surface science. Details of the components contained on the bus are discussed later in this section while a more detailed description on the structural design of the Triton Hopper can be found in Sect. 9.4.8, Structures and Mechanisms.

The propellant tank, cryocooler, and tank collection funnel comprise the second major element of the Triton Hopper. The tank is mounted to the tubular ring to which the six shorter landing legs are integrated. Again, this provides a thermal barrier between the propellant tanks and the "hot" bus structure, with the only conductive path between the two being down the three longer landing legs, into the landing pads, and up the six shorter landing legs into the ring to which the tank is mounted. This conductive path minimizes the heat transfer between the bus and tank by utilizing low conductive materials for the structures. Mounted directly to the tank is the "cold" end of the cryocooler. Waste heat from the "hot" end of the cryocooler is transferred through a mechanical arm that extends down to the cold Triton surface. A tank collection funnel, located on the tank nearly opposite from the cryocooler, is used to allow frozen surface material from Triton to be dumped into the tank by a robotic arm with a scoop, in order to turn the frozen surface into propellant required for hopping to another location. The propellant tank is heated, when needed, by a series of heaters located on the tank (not shown in the CAD design), and is cooled, when needed, by a mechanical arm attached to the tank that transfers heat from the tank down to the cold Triton surface. Those components included on the propellant tank and ISRU element of the Triton Hopper can be seen in Fig. 9.17. A more detailed discussion on the thermal control for the tank can be found in Sect. 9.4.6, Thermal Control System.

The landing structure is the third major element that comprises the Triton Hopper. It is comprised of the three landing pads, landing legs, and square tubular

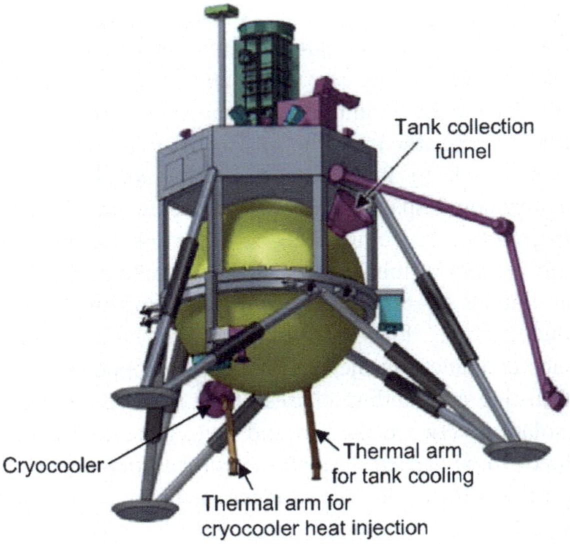

Fig. 9.17 Components contained on the propellant tank and ISRU element of the Triton Hopper

ring that contain the thrusters and to which the tank is mounted. The landing legs and pads provide a 3-m diameter footprint. Each of the three landing legs is comprised of one long member extending from the pad to the hexagonal bus structure, two shorter members that extend between the pads and the ring used to mount the tank, and a single 16-in. diameter landing pad. The six main thrusters and six RCS thrusters are mounted to structures contained on the ring to which the tank is mounted. The main thrusters are grouped into three pairs, each (pair) 120° from one another around the ring, and provide an upward thrust vector. The RCS thrusters are also grouped into three pairs, each (pair) 120° from one another around the ring, and provide a tangential thrust vector in opposing directions for each pair. Thruster locations and orientation can be seen in Fig. 9.18. A more detailed discussion on the thrusters can be found in Sect. 9.4.7, Propulsion System, while more details on the landing legs can be found in Sect. 9.4.8, Structures and Mechanisms.

Figure 9.19 shows two images of the lander with a color-coded breakdown of all the components by subsystem. Note that the components related to the ISRU system (cryocooler and tank collection funnel) are included in Sect. 9.4.1.

Components contained on the bus but located external to the bus structure include the Advanced Stirling Radioisotope Generator (ASRG), communications antenna, two radiator panels, two star trackers, six cameras that comprise the 360-camera, meteorology package, Mastcam, IR spectrometer, chemical analysis package, robotic arm (containing the APXS sensor head, scoop, and a camera), the MARDI camera, landing radar antenna, and the LIDAR. All of these components are called out in Fig. 9.20.

The ASRG is mounted to the center of the top closeout panel of the bus structure. This orientation, while adding height to the lander, allows the existing mounting interface to be utilized and mounted directly to the closeout panel, and

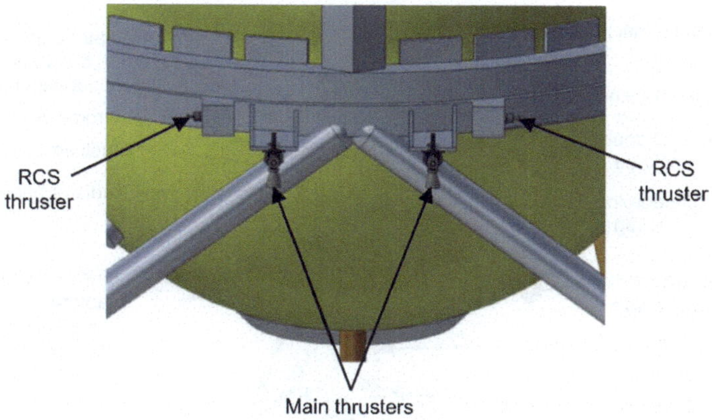

Fig. 9.18 Thruster location and orientation

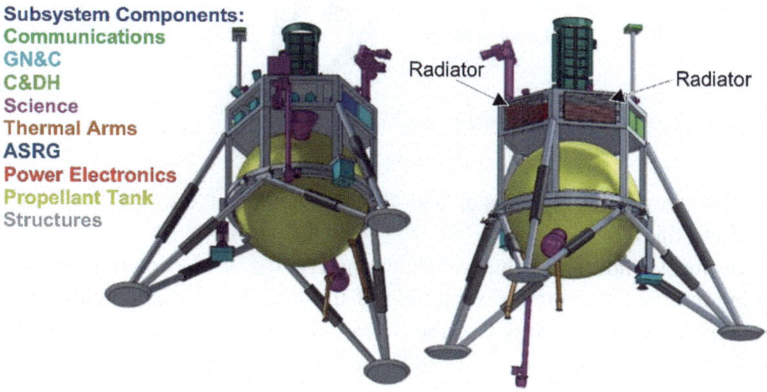

Fig. 9.19 Subsystem breakdown of the Triton Hopper

allows all of the radiator fins to be used for heat rejection, something that could not be done if the ASRG were "laid down" on the panel to minimize the lander height. A fixed boom is extended from the top closeout panel of the bus structure to which the X-band omni antenna is mounted to the top. The boom is required to keep the antenna above the ASRG with an unobstructed view of the sky.

The two radiator panels that provide the heat rejection for the internal electronics are mounted to adjacent closeout panels on the sides of the bus structure. While this location provides a view of the Triton surface, the cold temperature of the surface does not reduce the heat rejection capability of the radiators. Both star trackers are mounted out at the edge of the top closeout panel of the bus structure opposite from one of the radiator panels in order to avoid any potential thermal distortion of their view. Each star tracker is angled upward at a 45° angle and both are angled away from one another by 90°. This location and orientation ensures that each has an

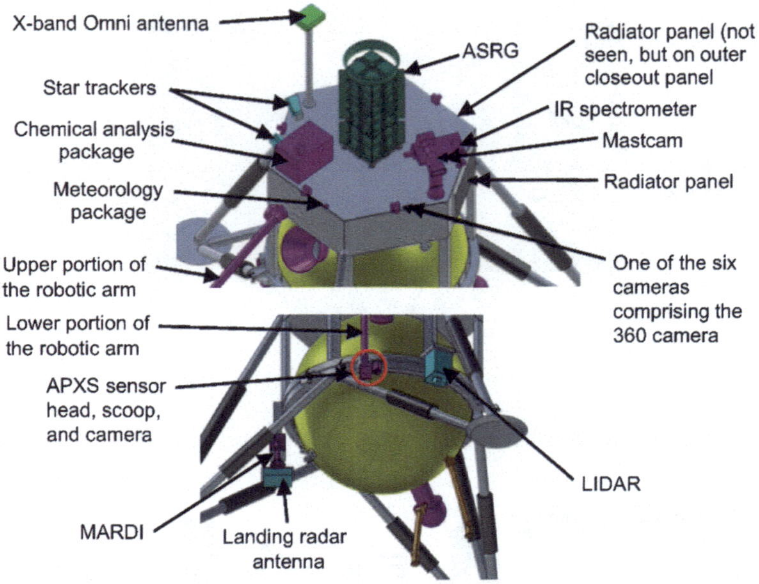

Fig. 9.20 Components located external to the Triton Hopper bus

unobstructed view of deep space, and that in the event one has the Sun in its field-of-view, the other will not.

Each of the six individual cameras that comprise the 360°-camera is mounted out at the edge and in the center of each side of the hexagonal shaped closeout panel. Each camera is pointed parallel to the Triton surface. These locations and orientation of each camera provide a full 360° view while on the surface of Triton. The meteorology package is also mounted to the top closeout panel out near the edge, in order to keep it away from the ASRG. The Mastcam and IR Spectrometer are both mounted to a platform located on the top of a mast extending up from the top closeout panel of the bus. The mast has the capability to rotate 360° providing a panning capability, while the platform on top of the mast provides a tilt capability. While the view of the Mastcam will be blocked when panned back towards the ASRG, the mast was located directly opposite the antenna boom so that this boom will not provide an additional blockage other than that from the ASRG. The chemical analysis package is also mounted directly to the top closeout panel and is located out near the edge of that panel. The side closeout panel relative to the chemical analysis package provides the interface for the base of the robotic arm. By locating these two components on the same side of the bus, the scoop contained on the end of the arm can be used to place surface samples into the chemical analysis package, in addition to the tank collection funnel, also located on this side of the bus, though mounted to the tank. A camera and APXS sensor head are also located on the end of the robotic arm.

The MARDI camera from the science subsystem, along with the landing radar antenna, is mounted to a tubular structure that extends down from the ring contained on the bus structure. This allows those components to have an unobstructed view of the surface while also maintaining a thermal barrier between them and the propellant tank. Finally, the LIDAR used by the landing subsystem is mounted off of the ring portion of the bus structure. Again, this allows for an unobstructed view of the surface and provides a thermal barrier between itself and the propellant tank.

The remaining subsystem components are all contained inside the hexagonal bus structure and are mounted to the inside face of the side closeout panels. These components are called out in Fig. 9.21. For the Power Subsystem, the ultracapacitor and DC/DC converter, Stirling controller, and the power management and distribution (PMAD) shunt are located on the faces directly behind the two radiator panels, as these components provide a majority of the waste heat. The enclosure of the command and data handling (C&DH) subsystem, and the landing radar and velocimeter for the landing subsystem both share a face. Both of the inertial measurement units (IMUs) are mounted to a single closeout panel, while the star tracker electronics are mounted to the closeout panel directly underneath the location of the star tracker optical heads. Finally, both X-band transceivers are mounted to a single closeout panel near the location of the boom containing the antenna. More detailed discussions on all of the subsystems can be found in their corresponding sections of this chapter.

The overall dimensions of the landed Triton Hopper are shown in Fig. 9.22.

9.4 Subsystem Breakdown

9.4.1 Science Payload and ISRU System

The science instrument package was chosen to address the major scientific enigmas of Triton. It will make use of the ability of the Hopper to sample widely varied terrain, including both horizontal sampling as well as vertical sampling through the plumes of geysers and through layers of the thin nitrogen atmosphere.

The major science investigations will include atmospheric science, remote measurements of the surface (including geomorphology and spectroscopy), and detailed in situ analysis of surface materials.

The electronics packages for the instruments will be in a warmed electronics enclosure heated during operation. Motors for pointing of instruments will incorporate electric heaters to allow the pointing motors to be preheated before use.

Science Requirements

The goal for the science payload on the Triton Hopper was to hop thirty times in two years a distance of 150 km total. Investigations on the surface and subsurface will be pursued as well as sensing and imaging during the hops. See Sect. 9.2.4 for a description of the science mission goals.

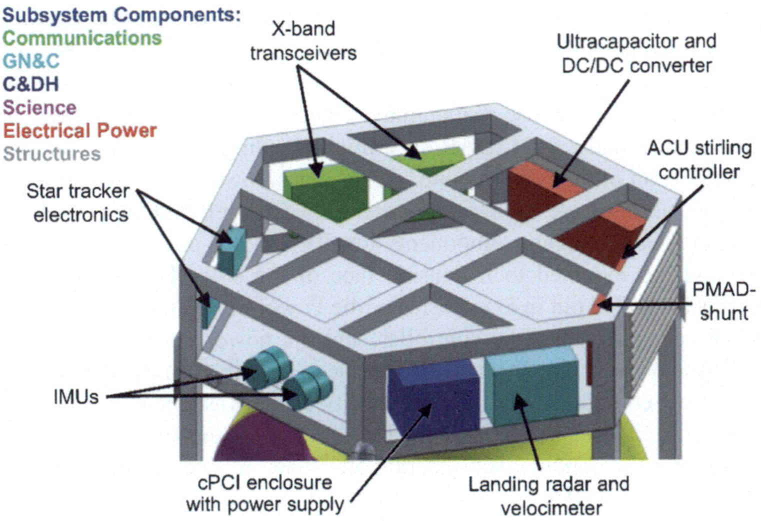

Fig. 9.21 Components located inside the Triton Hopper bus

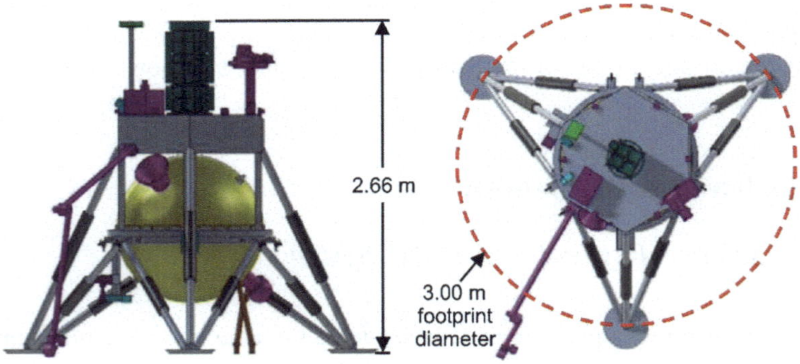

Fig. 9.22 Overall dimensions of the landed Triton Hopper

Science Instruments

Instead of a high-resolution color stereo camera *and* the high-resolution high-speed camera, a copy of the Mastcam from the Mars Science Laboratory was chosen along with a set of six small cameras to allow full visibility around the vehicle. The camera weighs about 1 kg and draws 13 W when imaging (Malin et al. 2005).

The landed science suite consists of

- **Descent imager** based on the Mars Descent Imager (MARDI). This will be used to provide Aerial photography during landing and hops. In addition, this downward-looking camera will be incorporated to take detailed images during landing, and also during each hop.

- The **landing-site chemical analysis package** will include a Quadrupole Mass Spectrometer (QMS), and Gas Chromatograph (GC), based on an evolved version of the instruments flown on the Surface Analysis at Mars (SAM) instrument suite on the Mars Curiosity rover. The QMS and the GC can work together for definitive identification of organic compounds.
- **Robotic arm (RA)** to acquire samples. The robotic arm will be based on the Phoenix robotic arm, but redesigned slightly for the lower gravity of Triton. The RA is 2.35 m in length with an elbow joint in the middle allowing the arm to trench about 0.5 m (1.6 ft) below the Triton surface, also allowing for collection of ice and snow to be processed for propellant. This instrument was designed to be able to penetrate through the ice/soil mixtures at Mars polar landing sites, but will need to be modified for Triton conditions similar to those expected at the landing site.
- **Mastcam (mast camera)** from the Mars Science Laboratory (MSL). The Mastcam will take color images and color video of the surface. The Mastcam will be mounted on a pointable mast, so that images can be stitched together to form a panoramic view of the camera's surroundings.
- **360° cameras**: In addition to the high-resolution camera, six small cameras will provide a 360° panoramic view in all directions, to show the lander's surroundings.
- **Spectroscopy**: V/UV/NIR point spectrometer based on an instrument designed at NASA Glenn, with optics added. Spectroscopy is the study of optical spectra, and will be used to examine the mineralogy of the surface materials, as well as the optical spectra of the Triton geyser ejecta. The V/UV/NIR spectrometer will be mounted on the same mast as, and co-pointed with, the Mastcam.
- **Meteorology package** to measure pressure, temperature and wind speeds. Pressure measurement devices: mass estimate one gram (plus electronics).
- **Alpha-particle X-ray spectrometer (APXS) instrument**. APXS has proved in Mars missions to be a remarkably effective, reliable, and lightweight instrument to measure elemental composition of both surfaces and atmosphere. The APXS will be mounted on the robotic arm, and determines the elemental chemistry of rocks and soils using alpha particles and X-rays. (Based on MER instrument, Rieder et al. 2003).

ISRU: Propellant Gathering Systems

Cryocooler for Triton Atmosphere Collection: One means of collecting the gas required for Hopper propulsion is to freeze out the nitrogen atmosphere inside the propulsion tank. To accomplish this a cold surface, both below the freezing point of the atmosphere in temperature and capable of removing the heat of sublimation from the solidifying nitrogen, is required. Designs similar to this known as "cryopumps" are used to trap residual nitrogen in vacuum chambers. Their main components are a mechanical cryocooler to cool the surface and remove heat, and an extended surface heat exchanger to trap and hold the frozen nitrogen. However, since the Triton atmosphere is much colder, a new design is required.

First Approach: The first design started with the "cryopump" concept, but replaced the conventional single stage cryocooler with a two stage cryocooler capable of reaching 30 K. A temperature of 30 K was picked as sufficiently below the atmospheric freezing point temperature to promote rapid freezing. It was calculated that a heat removal of 5 W was required to freeze sufficient nitrogen for propulsion (roughly 100 kg) within 8 days (the desired fill time). There was assumed an additional 5 W of heat load into the tank, but this was removed via a thermal link to the surface. Note the thermal link strategy cannot be used directly to freeze atmosphere, as the atmosphere and surface are quite close in temperature. Using the GRC Cryogenic Analysis Tool (CAT) (Guernsey et al. 2005; Plachta and Kittel 2003) to model a conventional room temperature based cryocooler resulted in the characteristics shown below:

- Cryocooler input power: 457 W
- Cryocooler mass: 18 kg
- Radiator area: 3.05 m^2 (281 K)
- Radiator mass: 16.8 kg
- Cryocooler control electronics: 5.3 kg

Figure 9.23 shows an image of a similar design (Plachta 2004).

Because the planned propellant tank is a composite overwrap design, and hence fairly low thermal conductivity, it was decided to include an extended surface to act as a tank internal heat exchanger. This was designed as a 6061-T6 aluminum (Al) vane structure. The overall structure was sized to hold the entire volume of solid nitrogen required for the propellant load with no more than 5 cm of nitrogen between vanes. The resultant design was 14 semicircular (11 cm radius) 6061-T6 Al vanes 0.5 mm thick mounted around a support tube. The cold head of the cryocooler was then thermally linked to the support tube. Mass of this element was 0.3 kg (estimated from the structure volume and an Al density of 2.7 g/cm^3). The element design has not been checked for launch and landing loads. Accommodating those loads may require mass growth.

Final Approach: In order to reduce the power requirements of the atmospheric collection system, a second design was undertaken. The second design retained the basic concepts of the first design, but replaced the radiator with a thermal link to the surface. This allowed for design of a single stage cryocooler operating from a cold platform at 60 K (due to limitations of CAT 60 K was the coldest temperature still analyzable) to the 30 K required for freezing nitrogen. This resulted in a cryocooler with the much more modest characteristics:

- Cryocooler input power: 53 W
- Cryocooler mass: 2.4 kg
- Cryocooler control electronics: 0.5 kg
- Thermal link: 2.1 kg (see details in Sect. 9.4.6)

A quick estimate of size based on similar cryocooler parts yielded the following physical dimensions:

Fig. 9.23 Two stage conventional 20 K flight cryocooler used in GRC ground testing

- 13-cm diameter by 15-cm long cylinder cold end heat exchanger
- 20 cm diameter by 15-cm high compressor dome
- 1.3-cm thick by 18-cm diameter interface ring acts as hot side heat exchanger

The cold end heat exchanger was estimated from the size of the second stage of a room temperature cryocooler with similar capacity. The compressor size was estimated from the warm compressor driving the second stage, although this estimate is less certain since a cold compressor might have different volume requirements (the higher density of cold gas will tend to make them smaller). Although many different materials are used within the cryocooler, the exterior of the cryocooler will be 300 series stainless steel. Any material mismatch within the cryocooler will have been accounted for in the cryocooler design.

Although the mass and size of the first design seemed feasible, providing an additional 457 W from a 110 W power system was a completely unacceptable solution. The second design proposed is the recommended design. Key risks of this design are as follows. Although working conditions are the same as the second stage of existing cryocoolers and cold compressors are part of existing designs, operation without a warm compressor will require a custom design. Integrating the cryocooler and tank internal heat exchanger into composite tank will be challenging, and may require embedding metal bosses into the tank weave.

ISRU Ice and Snow Collection System: Two primary methods are being considered to recover surface material for ISRU: scoop and drilling. Using a scoop has the potential to gather a greater quantity of material over time when considering loose material. However, it faces a limitation of material hardness. Harder, more consolidated materials may require an additional tool to break up the deposit so that is accessible to the scoop. The best example of this is the scoop on the Phoenix rover, which landed in the higher latitude region of Mars where permafrost exists near the surface. Retrieving the ice-cemented regolith required the use of a rasp tool in addition to the percussive scoop blade. This method was successful, but reduced the amount of material that could be recovered in each scoop. In terms of flight hardware, the phoenix scoop is the best baseline for this type of material recovery (most other examples involved loose regolith material on the Moon and Mars).

Drilling is more rugged and effective in harder materials. It also allows greater access to subsurface resources, which can be especially advantageous when overburden is significant. Several concepts are being developed for drilling resource recovery, but are still unproven in flight. This includes core drilling, where material is captured within the hollow center of the drill bit, and augering where drill cuttings are collected either on the auger flights or otherwise conveyed. Narrower drill bits have a higher rate of penetration, but result in less material recovered. For core drilling, the consistency of the core will reflect that of the deposit itself. For consolidated icy material, the sample will be solid or in large chunks, unlike the more granular material that would result from auguring. This would impact the selection of the resource processing method used.

For the Triton Hopper, the resource is available at or near the surface so scooping is feasible. The phoenix scoop was used for the initial trade study since it has been demonstrated in permafrost with strength in the range of what may exist at the Triton surface. However, resource recovery may prove slow, requiring 400 scoops to meet the system nitrogen needs.

Science Payload and ISRU System MEL

After going through several design iterations in order to reduce mass and power as far as possible and still maintain a high level of science experiments performed on the Hopper, Table 9.7 captures the final science payloads and ISRU System.

9.4.2 Communications and Tracking

Communications Requirements

The communications requirement for this scenario is the design of a science data uplink from the Triton Hopper Surface platform to an orbiting platform.

Communications Assumptions

The communications subsystem design (Fig. 9.24) will consist of a single fault tolerant space qualified small deep space transceiver (SDST) configured for X-band operations (8.0–8.4 GHz) mounted internal to the Hopper hexagonal platform structure. An omnidirectional antenna is used and mounted on top of the platform. The communications uplink is closed with a 4 m dish antenna mounted Triton nadir facing on the orbiter approximately 210,000 km distance (Fig. 9.25). Further assumptions are a standard 3 dB margin, included in the communications link for the link budget analysis, which is typical for space design applications due to the uncertainty of the components performance and available real effective isotropic radiated power. The Hopper platform will communicate with the orbiter every twelve days for 20 hours to uplink approximately 1.1 Gb of science data accumulated over the past eleven days. This Hopper subsystem link design does not include a space to Earth relay. All communications are assumed direct Hopper platform to orbiter. All Hopper platform hardware is assumed to be minimum TRL 5 or greater.

Table 9.7 Science payload masses

Description	Quantity	Unit mass (kg)	Basic mass (kg)	Total mass with growth (kg)
Science payload			23.0	23.7
Instruments			20.1	20.1
Descent imager	1	1.0	1.0	1.0
Robotic arm	1	8.0	8.0	8.0
Landing site chem. analysis pkg	1	9.7	9.7	9.7
Mastcam	1	1.0	1.0	1.0
360° camera	6	0.0	0.2	0.2
Spectroscopy	1	0.1	0.1	0.1
Meteorology package	1	0.1	0.1	0.1
APXS instrument	1	0.1	0.1	0.1
ISRU			2.9	3.8
Cryopump	1	2.4	2.4	3.1
Cryopump electronics	1	0.5	0.5	0.7

Communications Design and MEL

X-band is always an attractive option for higher capacity data downlinks because the data throughput matches or exceeds other comparable use frequency bands for the same size antenna, with minimal impact from atmospheric influences. Also, X-band has heritage use and less frequency crowding as compared to some other frequency bands for higher capacity downlink operations. In this case, the communications subsystem design is based around the use of a space qualified SDST configured for X-band operations. The SDST was chosen as the transmitter because it is highly configurable and can be modified to accommodate the necessary parameters, including proper data coding and modulation, to ensure mission success. It also has a built in beacon mode for ranging and emergency communications. The SDST is vibration and radiation tolerant, respectively to 25 g_{rms} and

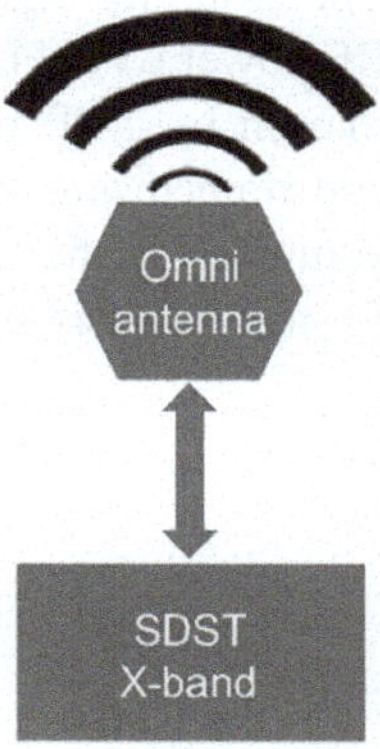

Fig. 9.24 Communications system representation

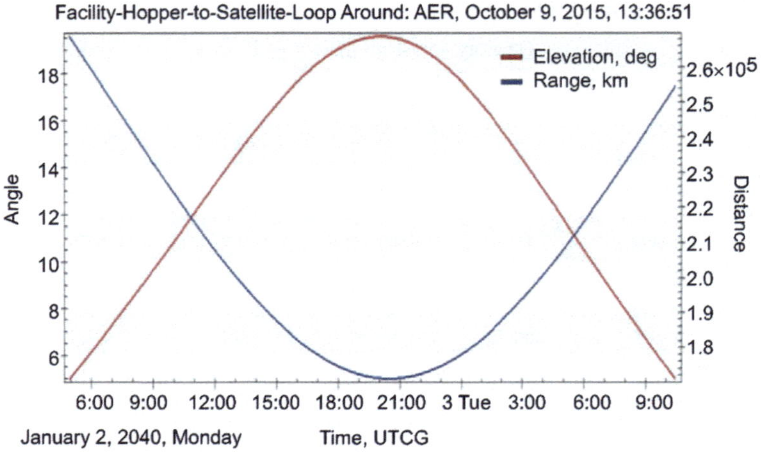

Fig. 9.25 Typical orbiter—Hopper data pass (every 12 Days)

100 krad and can be modular configured up to a maximum of 15 Mbps. An omnidirectional antenna with approximately 3 dBi of gain is mounted zenith facing on top of the Hopper platform. A 4 m antenna with approximately 47 dBi of gain is mounted Triton nadir facing on the orbiter platform. A slightly higher gain in both the Hopper platform and orbiter antenna is anticipated, although the minimum worst case is used. For analysis, we use the amount of data to be uplinked from the Hopper to the orbiter, and the orbiter period of viewing (usually minimum 5° elevation on each horizon) such that the minimum speed of the link is calculated as follows:

$$\begin{aligned}[1.1\ \text{Gb}/20\ \text{h}] \times [1\ \text{h}/3600\ \text{s}] &= [1.1\ \text{Gb}]/[20 \times 3600]\ \text{s} \\ &= 15277.78\ \text{b/s or} \sim 16\ \text{kbps}\end{aligned}$$

Link budget analysis shows that the X-band subsystem with 10 W (10 dBW) of radio frequency (RF) power can accommodate up to a 9 kbps data transfer rate. Therefore, for higher data rates to close the link and meet the required 16 kbps, the RF power would have to increase to a minimum of 17 W (12 dBW), reduce the link margin by 2 dB or use a 2:1 data compression ratio. Since the Hopper is power constrained and increased power is not an option, it is recommended that data compression be used rather than link margin reduction. The Communication System MEL is shown in Table 9.8.

9.4.3 Command and Data Handling (C&DH) System

The C&DH subsystem is used to provide computer control and data storage for the Triton Hopper.

C&DH Requirements

The C&DH System provides the processing capability and data storage to operate the equipment on the Triton Hopper. The system concept has a single fault tolerant main processor rated for 100-krad total dose radiation environment. The system has a science data storage capability large enough to accommodate 1 Gb every 12 days.

C&DH Assumptions

The C&DH enclosure will be maintained in a temperature environment of 300 K.

C&DH Design and MEL and Risks

The concept design solution includes a flight controller enclosure populated with electronic boards selected based on their capability to meet mission goals and to survive the environmental conditions. The main components are:

- Two processor cards utilizing Power PC 750 radiation hardened cards, or equivalent.
- Watchdog switcher.
- Solid-state memory card.
- cPCI enclosure with power supply.
- Atomic Clock module/Ultra oscillator module.
- Valve/Motor drivers.
- Data interface cards (RS422/485) for communications, science and navigation instruments.
- Wiring harness and connectors.

The system concept is illustrated in Fig. 9.26, and the C&DH MEL is shown in Table 9.9.

Table 9.8 Communication system MEL

Description	Quantity	Unit mass (kg)	Basic mass (kg)	Total mass with growth (kg)
Communications and tracking	–	–	4.2	4.6
X-Band system	–	–	4.2	4.6
X-band transceiver	2	2.0	4.0	4.4
X-band omni antenna	1	0.2	0.2	0.2

9.4.4 Guidance, Navigation, and Control (GN&C)

GN&C Requirements

The GN&C requirements for the Triton Hopper were as follows:

- Land on a slope no greater than 13°
- Vertical velocity at landing no greater than 1 m/s
- Horizontal velocity at landing no greater than 1 m/s
- Land within 50 m of the targeted area on the initial landing and subsequent hops
- Actively avoid landing on debris >10 cm

While the angle at which the Hopper would actually tip over is estimated to be 49°, the requirement of landing on a slope no greater than 13° was taken from the Apollo Lunar Lander and was related to being able to takeoff safely from the surface. The requirement for the vertical velocity at landing to be no more than 1 m/s is a derived requirement based off of a 3 g limit at landing assumed by the structures subsystem and a 20 cm displacement distance of the landing legs upon landing. The requirement for the horizontal velocity to be no more than 1 m/s is also a derived requirement based off of the assumption of landing on a 13° slope and applying a factor of safety to ensure that the Hopper does not tip over when landing.

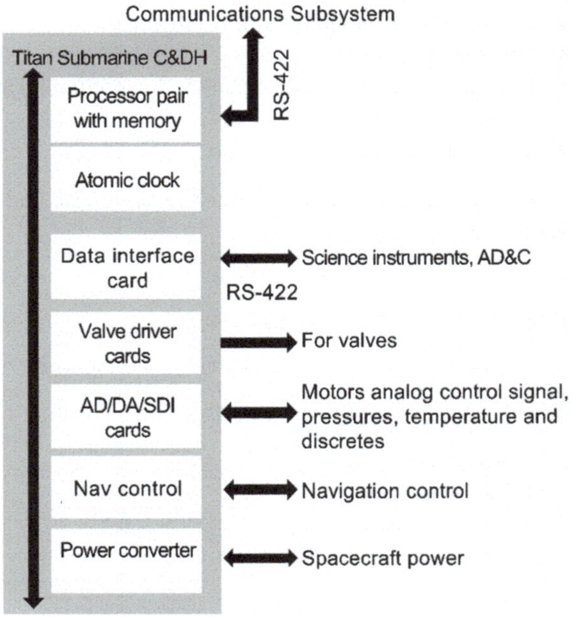

Fig. 9.26 Triton Hopper command and data handling system

Table 9.9 Command and data handling system MEL

Description	Quantity	Unit mass (kg)	Basic mass (kg)	Total mass with Growth (kg)
C&DH	–	–	14.7	19.0
C&DH hardware	–	–	12.7	17.0
Processor board	2	1.0	2.0	2.6
Data interface unit	0	0.3	0.0	0.0
Time generation unit	1	0.5	0.5	0.7
Command and control harness (data)	1	3.0	3.0	4.7
cPCI enclosure with power supply	1	3.0	3.0	3.6
Valve drivers	2	1.4	2.7	3.5
Separation drivers	1	1.5	1.5	2.0
Instrumentation and wiring	–	–	2.0	2.0
AD/DA/SDI card	2	0.5	1.0	1.0
Pressure and temperature sensors	20	0.1	1.0	1.0

GN&C Assumptions

The following assumptions were made in the design of the Triton Hopper:

- Orbital imagery would be performed with a 10 cm resolution to aid personnel on Earth in choosing safe landing locations
- Desired landing and subsequent hop locations would be uploaded from Earth, with the Hopper performing hazard avoidance to find a safe area to land within 50 m of the desired landing location
- Simple geometric shapes in estimating vehicle moments of inertia

GN&C Design Summary

The GN&C design of the Hopper is based off of heritage from the Mars Phoenix Lander and the Origins, Spectral Interpretation, Resource Identification, Security, Regolith Explorer (OSIRIS-REx) sample return spacecraft. The navigation hardware consists of:

- Two Micro-Advanced Stellar Compass Star trackers with one being redundant
- Two Northrop Grumman LN-200s Inertial Measurement Units (IMU) with one being redundant
- One landing radar and velocimeter with heritage from the Phoenix Lander
- One GoldenEye 3D Flash LIDAR Space Camera with heritage from OSIRIS-REx

The star tracker provides estimates of the vehicle's attitude in inertial space. The IMU contains three solid-state silicon accelerometers that are used to estimate propulsive ΔV's and three solid-state fiber optic gyros to estimate the vehicle's angular velocity and aid in attitude estimates. The landing radar and velocimeter provides estimates of the vehicle's altitude and ground relative velocity during landing. The Flash LIDAR is used for hazard avoidance, aiding in determining a safe landing site by generating elevation maps during landing.

Control of the Hopper is provided by 6 Moog 58-126, 222 N (50.0 lbf) thrusters for main propulsion which can be pulsed to provide pitch and yaw control, and 6 Moog 58-113, 4.5 N (1.0 lbf) thrusters that provide roll control. The GN&C MEL can be seen in Table 9.10, with the control thrusters being held in the propulsion subsystem MEL.

While the navigation hardware is located on the Hopper, the IMU(s) and Star Tracker(s) are also used during the following phases of the mission, providing estimates of the vehicle's position, velocity, attitude and body rates:

- Powered Triton Flyby—propulsively performed by the solid stage with control also provided by the solid stage
- Capture into 200 km circular Triton orbit—propulsively performed by the descent stage with control also provided by the descent stage
- Deorbit and descent—also performed by the descent stage with control also provided by the descent stage

After the descent stage performs the deorbit maneuver from the 200 km circular orbit about Triton, the descent stage/Hopper begin to descend to the surface. The altitude, vertical velocity and horizontal velocity are measured using the landing radar and Doppler velocimeter. At an altitude of approximately 200 m, the descent stage performs a burn to null both the horizontal and vertical velocity of the vehicle at which point the Hopper "hops" off of the descent stage using preloaded propellant. The descent stage falls to the surface as the Hopper begins a controlled descent, nominally up to 250 m away from the point that it detached from the descent stage. During the Hopper's descent, the flash LIDAR is used for hazard avoidance by generating elevation maps, aiding in determining a safe landing site. The Hopper lands safely on the surface assuming a slope of no greater than 13°. Its vertical velocity should be no greater than 1 m/s with a horizontal velocity no greater than 1 m/s. Once on the surface, all GN&C subsystem components are

Table 9.10 Guidance, navigation, and control MEL

Description	Quantity	Unit mass (kg)	Basic mass (kg)	Total mass with growth (kg)
GN&C	–		14.6	15.8
IMU	2	0.8	1.5	1.6
Flash LIDAR	1	6.5	6.5	6.7
Landing radar and velocimeter	1	5.0	5.0	5.9
Star tracker optical head	2	0.3	0.5	0.5
Star tracker electronics box	2	0.6	1.1	1.1

turned off to save power, however knowledge of the Hopper's latitude, longitude and heading in relation to a Triton body fixed coordinate frame is stored in the computer. Before a hop takes place, the navigation hardware is powered back on. The star tracker is powered on to estimate the Hopper's heading in inertial space by imaging the background stars, which is then translated to a Triton body fixed coordinate frame and used to verify the stored heading in the computer. The lander then performs up to a 5 km hop, with the vehicle's position, attitude and body rates estimated using the IMU, altitude and ground relative velocity estimated using the landing radar and Doppler velocimeter, and the flash LIDAR aiding in determining a safe landing site. Figure 9.27 shows the locations of the GN&C hardware on the Hopper.

The estimated maximum horizontal velocity that the Hopper can land with on a 0° slope without tipping over is 2.5 m/s. If a 13° slope is assumed, the maximum horizontal velocity is reduced to 1.9 m/s. By assuming a safety factor of 2, 0.95 m/s was assumed to be the maximum horizontal landing velocity. It is assumed that this is within the capability of the Doppler velocimeter to null the horizontal velocity to less than this limit since the Phoenix lander that landed on Mars had a maximum designed horizontal velocity of 1.4 m/s, with the actual value estimated to be ~0.06 m/s.

9.4.5 *Electrical Power*

The Triton Hopper requires a power system that can provide electricity to the Hopper during the twelve-year transit from Earth to Triton, through Triton descent

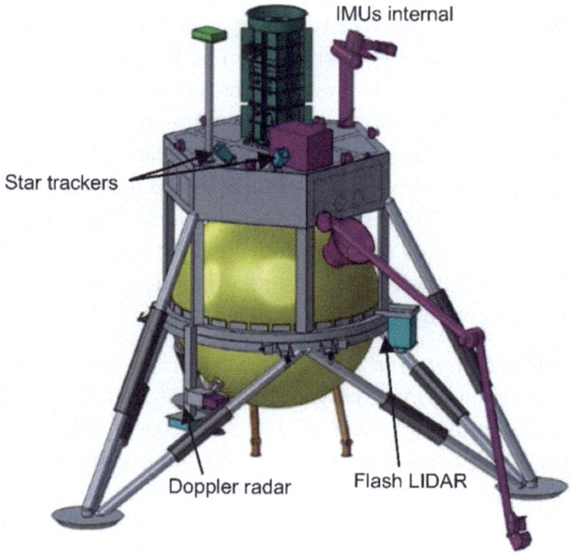

Fig. 9.27 Location of the GN&C hardware on the Hopper

and landing, and during the two years of surface operations. The Advanced Stirling Radioisotope Generator (ASRG) was selected as the electrical power source for spacecraft and lander. Though no longer an active NASA program it represents a reasonable estimate of how a future high performance Radioisotope Power System (RPS) should perform. In addition to the ASRG an ultra-capacitor based energy storage system was added to meet short duration high power output portions of the mission.

Power Requirements
Power varies from 25 W at launch to a peak of 599 W during the Landing Hop and Burn mission phase (3), see Table 9.11(a). Although the power requirement for the Landing Hop-Burn is high relative to nominal power requirements its duration is only 50 s. Nominal power during science is 108 W with communications and propellant collections requiring 67 and 116 W, respectively. The main Hop/Burn phase requires 594 W with duration of 60 s. The Main Hop-Coast and science phase requires 201 W for 120 s. Nominal power output from the ASRG is 117 W. This leads to three relatively short duration phases of the mission (Landing Hop-Burn, Main Hop-Burn and Main Hop-Coast + Science) requiring more power than the ASRG can provide. The relatively short duration of high required power along with the very long duration of dormant activity for the energy storage system required alternative battery technology (other than lithium-ion) to be used. For this power system Ultracapacitors were used to provide this power peaking capability.

Power Assumptions
It is assumed that the ASRG (Fig. 9.28) location on the vehicle allows it to have an effective sink temperature <100 K (Fig. 9.29). It was also assumed that the ASRG has degraded over its fourteen years of operation at a rate of 1.2% per year (Fig. 9.30). Estimated power output for the ASRG at end-of-mission is 115 W. The ASRG mass will be 32 kg with a volume envelope of 34- by 34- by 77-cm with a Stirling controller that will be located above the main body of the spacecraft. The energy storage system selected is based upon currently available Maxwell Ultracapacitors and that their "unlimited storage when discharged" duration specification is adequate for the twelve years of storage/transit, with availability and performance as advertised. It is assumed that the Triton Hopper will operate at a bus voltage of 28 V (±6) and excess power will be sent to a shunt. This is in addition to the ASRG shunt that is included, which is needed when the ASRG is removed from the spacecraft electrical system. The bus is assumed to lose 2% of its power in line losses and the Ultracapacitors have a 95% charge/discharge efficiency.

Power Design and MEL
The Triton Hopper power system uses a single ASRG connected to the spacecraft bus to provide power (Fig. 9.31). Sixteen Maxwell Ultracapacitor Model BCA P0650 are used with eight in series and two parallel strings with each string capable

Table 9.11 Power equipment list

(a) Power modes 1–5 (W)

Description	Power mode 1	Power mode 2	Power mode 3	Power mode 4	Power mode 5
Power mode name	Launch	Cruise	Landing hop, burn	Science	Communications
Power mode duration	60 min	10 year	50 s	4 day	20 h
Triton Hopper	19.4	26.4	460.7	83.4	51.4
Hopper	19.4	26.4	460.7	83.4	51.4
Science payload	0.0	0.0	17.0	57.0	6.0
Instruments	0.0	0.0	17.0	57.0	6.0
ISRU	0.0	0.0	0.0	0.0	0.0
AD&C	0.0	0.0	94.3	0.0	0.0
C&DH	18.5	18.5	22.5	18.5	18.5
Communications and tracking	0.0	5.0	24.0	5.0	24.0
Thermal control (non-propellant)	0.88	0.9	0.9	0.9	0.9
Propulsion (chemical hardware)	0.0	2.0	302.0	2.0	2.0
Bus power, system total	19.4	26.4	460.7	83.4	51.4
30% growth	5.8	7.9	138.2	25.0	15.4
Total bus power requirement	25.2	34.3	598.9	108.4	66.8

(b) Power modes 6–9 (W)

Description	Power mode 6	Power mode 7	Power mode 8	Power mode 9
Power mode name	Propellant collection	Propellant processing	Main hop, burn	Main hop,
Power mode duration	8 day	11 day	60 s	120 s
Triton Hopper	89.4	88.4	456.7	154.7
Hopper	89.4	88.4	456.7	154.7
Science payload	56.0	2.0	15.0	15.0
Instruments	2.0	2.0	15.0	15.0
ISRU	54.0	0.0	0.0	0.0
AD&C	0.0	0.0	94.3	94.3
C&DH	14.5	18.5	20.5	18.5
Communications and tracking	5.0	5.0	24.0	24.0
Thermal control (non-propellant)	11.9	60.9	0.9	0.9
Propulsion (chemical hardware)	2.0	2.0	302.0	2.0
Bus power, system total	89.4	88.4	456.7	154.7
30% growth	26.8	26.5	137.0	46.4
Total bus power requirement	116.2	114.9	593.7	201.1

of providing full power peaking. Each capacitor has a peak operating voltage of 2.7 V, maximum current draw of 25 A, and mass of 160 g. The energy storage envelope is 44- by 7- by 14-cm and has a total mass of 2.56 kg with an operational temperature range from −40 to 65 °C. Voltage range during peak operation during the Main Hop-Burn phase is from 21.6 to 16 V. Total energy stored in each string is 19 W h with approximately 10 W h used during the 60 s Burn/Hop. Total leakage current from each of the capacitors is 1.5 mA. Because the capacitors do not operate over the same voltage range as the spacecraft bus, a DC/DC convertor is required. This DC/DC convertor is 90% efficient and converts the nominal 28-V spacecraft power to the 21-V (max) capacitor voltage.

A mass summary for the power system is shown in Table 9.12. This shows the basic mass along with a column including the margin for each of the components.

9.4.6 Thermal Control System

The thermal control system for the Triton Hopper was basically separated into two segments: the thermal control of the electronics enclosure and the thermal control of the propellant tank. Because of the significantly different operating requirements both segments of the vehicle had to be thermally isolated from each other as much as possible. The thermal system design included devising an approach to maintain the electronics and science instruments within their desired temperature operating range and being able to cool the propellant tank to ambient conditions and then bring it up to the desired 300 K propellant temperature prior to a hop within the power budget and operational timeframe available.

The thermal environment on Triton has little variation over the course of the mission (see Sect. 9.2.3). Therefore the thermal system was designed for the remotely sensed steady-state environmental conditions on the surface of Triton. The main focus of the thermal design is in controlling the heat flow into and out of

Fig. 9.28 ASRG

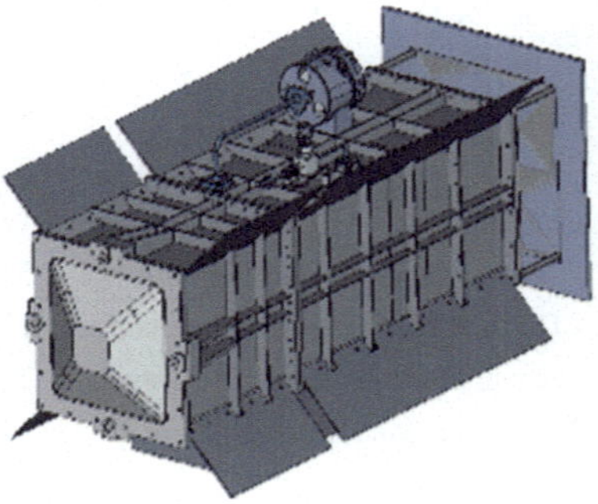

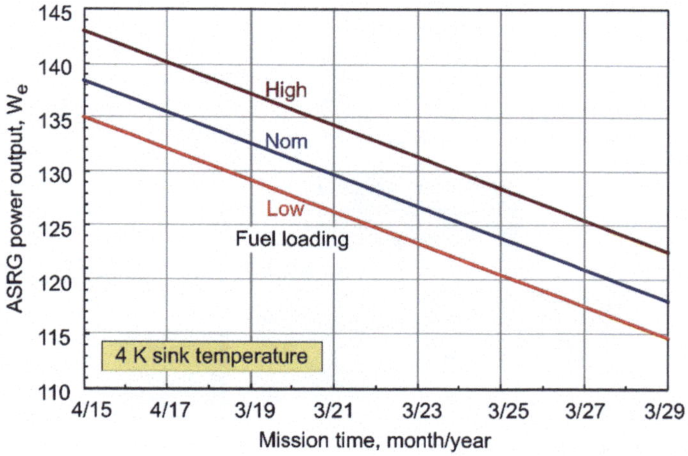

Fig. 9.29 ASRG power output versus mission year

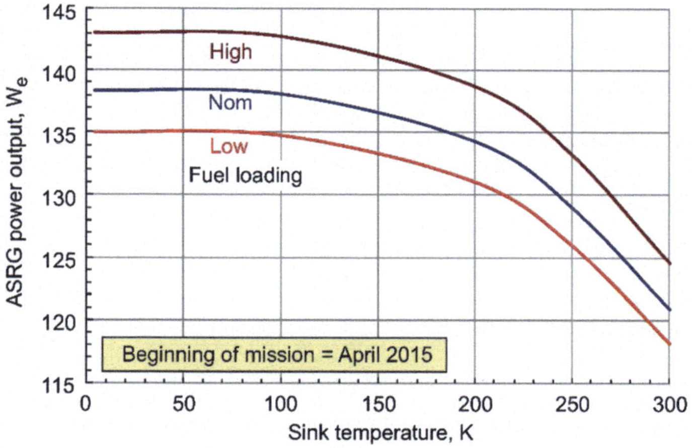

Fig. 9.30 ASRG power output versus sink temperature

the propellant storage tank in order to maintain the correct temperature for collecting, storing and heating the nitrogen propellant. Operating on Triton's surface is challenging due to the low temperature of the surroundings ($\sim$38 K). The surface is covered with solid nitrogen ice and snow. To control the temperature of the propellant storage tank during filling the tank will need to be cooled to near surface temperatures. Then during pressurization it will need to be warmed to near 300 K.

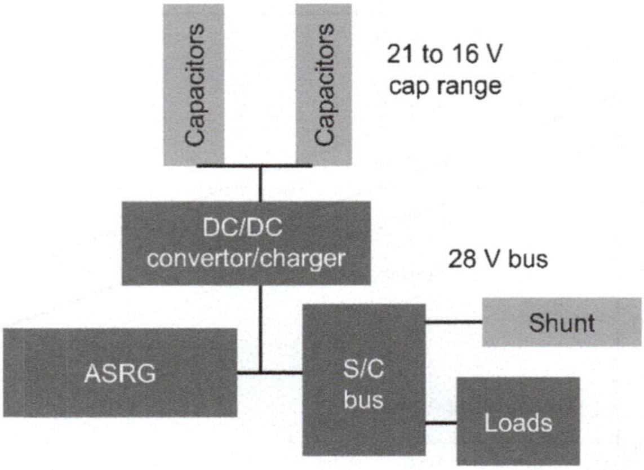

Fig. 9.31 Power schematic

Table 9.12 Power system MEL

Description	Quantity	Unit mass (kg)	Basic mass (kg)	Total mass (kg)
Electrical power subsystem	–		39.5	44.0
Power generation	–		32.0	35.2
ASRG	1	25.0	25.0	27.5
ACU-stirling controller	1	7.0	7.0	7.7
Power management and distribution	–		7.5	8.8
Harness	1	1.0	1.0	1.2
PMAD-shunt	1	3.0	3.0	3.6
Ultracapacitor and DC/DC convertor	1	3.5	3.5	4.0

This wide operating temperature range requires a heat management system that enables control of the tank temperature with a minimal amount of input power. To accomplish this control a movable conductive path is used in the form of a highly conductive arm which allows the tank to connect thermally with the surface during cooling and which is then removed during the heating process.

The main thermal system components are listed in Table 9.13 and illustrated in Fig. 9.32.

- Insulation: electronics and propellant tank
- Radiator for electronics
- Avionics, thermal management, (cooling plates, heat pipes)
- Temperature sensors, controllers, switches, data acquisition
- Conductive arms for tank thermal control
- Electric heaters for electronics thermal control.

Triton Hopper Thermal System Surface Operation

For the Triton Hopper to operate on the cold Triton surface their temperature has to be maintained within their desired operating range. The vehicle is separated into two distinct sections that have significantly different thermal requirements, the electronics plus science instrumentation enclosure and the propellant tank. The required temperature range, given in Table 9.14, is much higher than the temperature of the surrounding environment. To maintain the desired operating temperature electric heaters along with the waste heat from the isotope power system is utilized to warm the electronics enclosure interior as well as heat the propellant tank prior to a hop. Any excess heat generated during operation is rejected through a small radiator located on the side of the electronics enclosure. The radiator incorporates louvers so that it can be effectively turned off when not needed. The Hopper must be insulated to provide sufficient thermal control during operation to maintain the desired operating temperatures.

A thermal analysis was performed to determine the internal operating temperature and to size the required insulation layers and radiator needed to maintain the desired internal operating temperature. The heat losses from the interior were broken down into:

Table 9.13 Thermal control system MEL

Description	Quantity	Unit mass (kg)	Basic mass (kg)	Total mass (kg)
Thermal control (non-propellant)			33.0	38.0
Active thermal control			6.9	7.9
Heaters	25	0.2	5.0	5.8
Thermal control/ heaters circuit	4	0.2	0.8	0.9
Data acquisition	1	1.0	1.0	1.2
Thermocouples	10	0.0	0.1	0.1
Passive thermal control			24.6	28.3
Heat sinks	2	0.1	0.3	0.3
Heat pipes	2	0.5	1.0	1.1
Radiators	1	1.8	1.8	2.1
Multi-layer insulation (MLI)	1	7.4	7.4	8.5
Engine MLI	1	1.8	1.8	2.1
Propellant tank MLI	1	8.0	8.0	9.2
Tank conductive arm	1	2.3	2.3	2.6
Cryocooler conduction arm	1	2.1	2.1	2.4
Semi-Passive thermal control			1.5	1.7
Louvers	1	1.1	1.1	1.3
Thermal switches	2	0.2	0.4	0.5

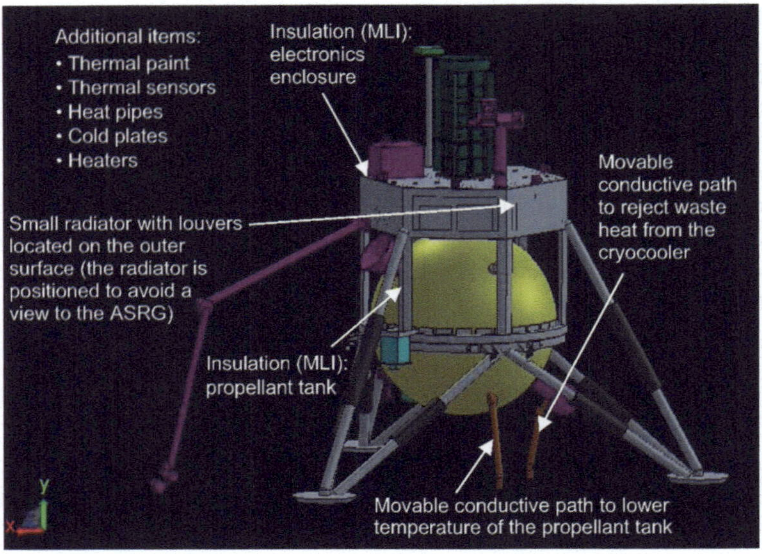

Fig. 9.32 Triton Hopper thermal system components

- Pass-through wires for operating external instruments
- Heat loss directly through the insulation
- Heat loss through the structure that passes through the insulation.

The radiator sizing was based on an energy balance analysis of the area needed to reject the identified heat load to space. From the area a series of scaling equations were used to determine the mass of the radiator. The radiator is located on the side deck of the Triton Hopper. This provides a good view to deep space with about a 0.5 view to the surface and no view to the radioisotope system radiators. There is insulation between the radiator and electronics enclosure body providing a single surface for radiating. The radiator is connected to the cold plates with heat pipes to move heat from the interior to the radiator. The radiator was sized to remove the waste heat from the electronics enclosure during operation at full power. The radiator will need to reject heat at power levels above 31 W.

Louvers were used on the radiator to help minimize heat loss during times when the internal power consumption and waste heat are low. If electronics thermal output decreases the louvers and heaters will be used to maintain the internal temperature of the electronics. The louvers are a passive device that opens up as the temperature of the radiator increases and then close as it cools. This maintains the desired heat flow from the radiator to keep the internal temperature within its operating limits. The addition of louvers increases the radiator area by approximately 30% over a radiator without louvers. The louver specific mass is 4.5 kg/m^2.

The radiator specifications are summarized in Table 9.15.

Cold plates and heat pipes are utilized to move heat from the electronics packages to the radiator. These components consist of conductive plates with

Table 9.14 Thermal system specifications

Specifications/ component	Value/approach
General dimensions:	
Propellant tank Electronics enclosure	1-m-diameter 1- by 1- by 0.3-m
Electronics system waste heat	100 W
Operating temperature	Electronics enclosure: 230–310 K (–43 to 37° C) Propellant tank: 38–300 K (–235 to 27 °C)
Insulation (MLI)	MLI is used for insulating the electronics enclosure and propellant tank
Radiator system	A radiator with louvers is used to reject the waste heat from the electronics as needed. Al (0.1- by 0.1- by 0.005-m), Water based heat pipes to move heat from the electronics packages
Passthroughs and conductive paths: Structure and science	Wires and ports for data collection and control are the assumed pass-through for the electronics enclosure. These passthroughs account for 5% of the insulated surface area Conductive path for rejecting heat from the cryocooler and propellant tank: Pyrolytic graphite (1600 W/mK)
Heating	Electric heaters and waste heat from the electronics and radioisotope power system are utilized to heat and maintain temperature within the spacecraft

integral heat pipes connected to them. The cold plates also incorporated heaters in order to maintain the desired electronics temperature throughout the mission. Due to the low temperatures the Hopper can experience during operation ammonia was selected as the working fluid. The specifications of the heat pipes and cold plates are given in Table 9.16.

Waste heat from the internal components as well as electric heaters is used to provide heat to the electronics when operating. Strip heaters are used to provide heat to the components within the electronics enclosure as well as the propellant tank. Flat plate heaters are used on the propellant tank and cold plates to provide heat to the electronics if necessary with heaters located on each cold plate. Heaters are also utilized within the propellant tank as a heat source for melting and gasifying the collected nitrogen and bringing the propellant up to 300 K. Thermal control is accomplished through the use of a network of thermocouples whose output is used to control the power to the various heaters. A data acquisition and control computer is used to operate the thermal system. Maximum heater power is approximately 100 W.

Due to the low atmospheric pressure radiation is the main mechanism for heat transfer within the Hopper and to the surroundings. Therefore MLI was selected as the means for insulating the Hopper components.

MLI was used to cover the exterior exposed portion of the electronics enclosure and propellant tank. The insulation was analyzed to determine the required number

Table 9.15 Radiator specifications

Variable/characteristic	Value
Radiator solar absorptivity	0.14
Radiator emissivity	0.84
Max radiator Sun angle	0°
View factor to the hopper body and RPS radiator	0
View factor to the surface	0.5
Radiator operating temperature range	223–310 K
Power dissipation	68 W (internal electronics)
Radiator area	0.23 m^2 (body mounted)

Table 9.16 Heat pipe and cold plate specifications

Variable/characteristic	Value
Cooling plate and heat pipe material and density	Al, 2770 kg/m^3
Number of cooling plates	Four (for electronics)
Cooling plate dimensions	0.1- by 0.1- by 0.005-m
Heat pipe working fluid	Ammonia

of layers and the corresponding mass and heat loss needed to maintain the average 300 K interior temperature during normal operation. A tradeoff was performed between the insulation mass and the required heater power cold time operation. It was determined that 25 layers would provide the best insulating option for this mission. The insulation model was based on radiation heat transfer analysis of the heat transfer from the spacecraft through the insulation to the surroundings. Note that there is a rarified nitrogen atmosphere on Triton. Although the gas pressure is very low this can affect the operation of the MLI by providing a conductive path between the layers. For atmosphere pressures between 7.5 and 30 mTorr, there will be degradation in the performance of the MLI (ASTM International 2013). To minimize this effect, it is proposed to design convection barriers between the MLI layers to minimize this convection. A 5% pass-through area was assumed for the insulation heat loss for items passing through the insulation.

The pass-throughs in the insulation around the electronics enclosure allow heat to leak out of the enclosure and therefore need to be minimized. One of the main heat leaks through the insulation are the structural legs that support the electronics enclosure, as illustrated in Fig. 9.33. Because of the very cold surrounding environment, conduction down the legs to the surface can be significant. To minimize this, the structural legs that go from the electronics enclosure to the surface are constructed from a titanium (Ti) alloy (Ti-6Al-4 V) with a low thermal conductivity of 6.7 W/mK. The heat loss, from the three legs to the surface, is 12.9 W.

The structure of the Triton Hopper is also used to isolate different segments of the spacecraft in order to help maintain thermal control. This is necessary since different sections have greatly different temperature requirements. The long support structure legs connected to the electronics enclosure also provide a long conductive path from the propellant tank to the enclosure. The structure holding the propellant tank is separated from the structure holding the electronics enclosure while on the surface.

While on the surface and the propellant is being collected, the tank is held to near ambient temperature conditions. A conductive rod that is lowered to the surface accomplishes this. This conductive rod is constructed from a high conductivity material, pyrolytic graphite, with a thermal conductivity of 1600 W/mK and 3 cm in diameter. The conductive rod, illustrated in Fig. 9.34, conducts any excess heat from the propellant tank during filling to the surface where it is rejected to maintain the tank temperature near ambient conditions. The path is sized to move approximately 3 W of heat from the tank to the surface with an approximate 0.7 K temperature difference between the tank and surroundings. The tank itself is also designed with highly conductive strips along its interior to move heat from over its surface to the conductive rod. The conductive rod details are listed in Table 9.17.

A conductive rod is also used to reject heat from the cryocooler, as shown in Fig. 9.34 that is used to collect and solidify nitrogen from the atmosphere. The cryocooler conductive rod works in a similar manner as that for the tank except that it operates at a much larger temperature difference and therefore can conduct significantly more heat. The details on the cryocooler conductive rod are also given in Table 9.17. The cryocooler is described in Sect. 9.4.1.

While on the surface during the propellant collection process, the tank is at near ambient conditions and needs to be isolated from the upper electronics enclosure.

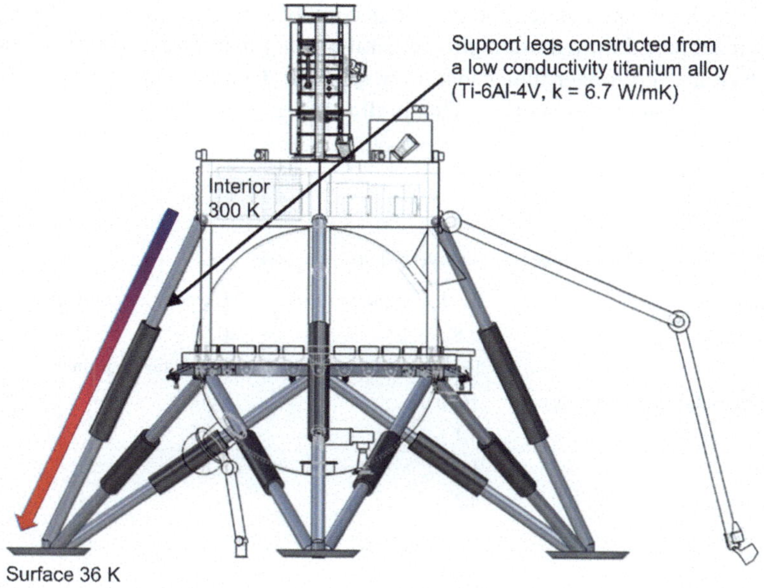

Fig. 9.33 Heat leak down electronics enclosure support legs

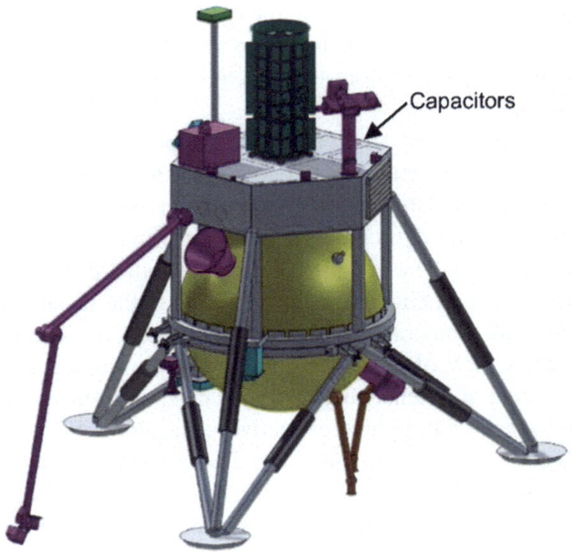

Fig. 9.34 Illustration of the conductive rods for rejecting excess heat from the cryocooler and propellant tank

To accomplish this the structural ring connected to the support structure of the electronics enclosure is separated from the structural ring connected to the support structure of the tank. This gap between these rings minimizes any direct conduction from the electronics enclosure to the tank. The only conductive path between them is through the support legs, which is a long path through a very low conductivity material.

During takeoff and landing the two rings compress against each other to provide structural support during these high structural loading maneuvers. But then once on the surface and no longer under load they separate. Compression springs between the two rings enable this motion. This is illustrated in Fig. 9.35.

Table 9.17 Tank and cryocooler conductive rod specifications

Characteristic	Tank conductive rod	Cryocooler conductive rod
Length (m)	0.5	0.5
Material	Pyrolytic graphite	Pyrolytic graphite
Thermal conductivity (W/mK)	1600	1600
Heat rejected (W)	3.1	56.6
Mass (kg)	1.53	0.89
Diameter (cm)	4.2	3.2
Temperatures		
Hot (K)	38.7	60
Cold (K)	38	38

9.4.7 Propulsion System

The propulsion system for this design consists of three major elements. There is a solid rocket motor based braking stage, a bipropellant descent staged that is used to lower the Hopper close to the Triton surface, and a nitrogen gas based monopropellant system used for terminal descent to Triton, initial landing, and subsequent surface hop maneuvers via replenishing the propellant with in situ resources. Due to funding constraints, complete integrated conceptual designs of the solid rocket motor braking stage and bipropellant descent stage were not conducted, but a conceptual level design of their propulsion systems was conducted in order to ascertain the impact of the required ΔV maneuvers on propellant load, and thus on the entire spacecraft that must be fitted to the launch vehicle.

Propulsion System Requirements

The solid rocket braking stage and bipropellant descent stage are required to provide both the braking ΔV required and adequate control required to place Hopper close enough to Triton's surface as to allow it to safely land using an initial quantity of nitrogen gas propellant that was brought from Earth. Once on the surface, the Hopper propulsion system is required to provide adequate performance for additional hop maneuvers via cold gas thrusters utilizing warmed nitrogen gas as propellant. The nitrogen used for these maneuvers is required to be obtained as an in situ resource.

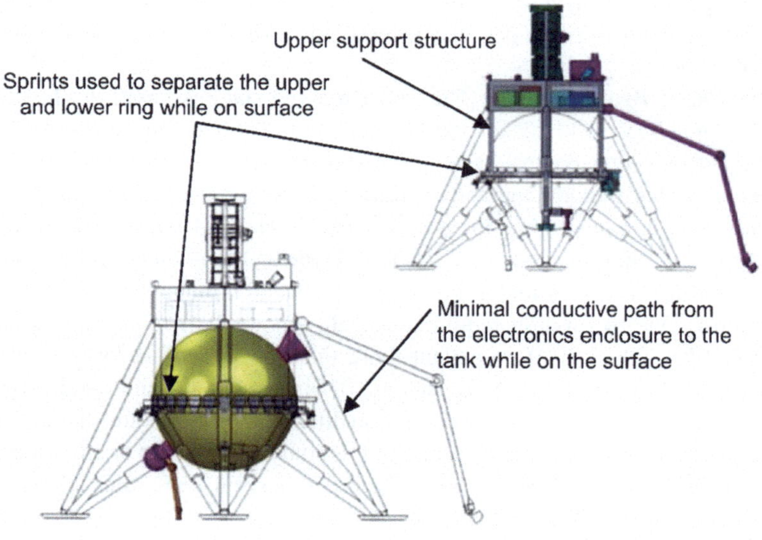

Fig. 9.35 Conductive paths between the hopper upper and lower structure

Propulsion System Assumptions

In order to reduce integration and thermal flow path complexity, it is assumed that the propulsion system will utilize a single large central propellant tank. Due to the long trip to Triton, it is assumed that pyrotechnic valves are used for propellant isolation during both the launch and cruise phases of the mission. The system is assumed to be single fault tolerant and utilize as many commercial off-the-shelf (COTS) or near COTS components as possible. Multiple thrusters will be located around the Hopper, thus thrust vector control is assumed to be accomplished via thruster pulse modulation. Although heat will be supplied to the Hopper tank to warm the nitrogen gas prior to any hop maneuver, it is assumed that the actual hop itself can be modeled as an adiabatic blow-down. Because nitrogen in the tank will cool and the pressure will decrease as propellant is expelled, it is assumed for analysis purposes that all useable propellant is consumed when either the critical temperature or critical pressure of nitrogen is reached, thus avoiding any potential condensing of the remaining propellant in the tank. The remaining nitrogen in the tank is thus treated as trapped residual propellant.

Propulsion Systems Designs

The propulsion system for this design consists of three major elements. First, there is a Star-37GV solid rocket motor based braking stage that is used for Triton orbital insertion, a monomethyl hydrazine (MMH) and nitrogen tetroxide (NTO) based bipropellant descent stage that is used to lower the spacecraft close to the Triton surface, and a nitrogen gas based monopropellant system used exclusively by the Hopper. This final system is used for both the initial Triton surface landing, and will be refueled via in situ resources such as the nitrogen gas present in the atmosphere or the suspected nitrogen snow on the surface. Once collected, the nitrogen is heated to form pressurized gas that is used for propulsive hop maneuvers with powered landings.

Solid Based Braking Stage: The initial braking stage is based on the existing ATK STAR-37GV solid rocket motor. This motor utilizes an electromechanical flex seal thrust vector control system that can provide $\pm 4°$ of thrust vectoring in both planes. This motor's inert mass fraction is 8%, the casing is comprised of graphite-epoxy composite, and the propellant is TP-H-3340, which is 18% Al, 71% ammonium perchlorate, and 11% HTPB (Hydroxyl-terminated polybutadiene). This motor delivers a nominal I_{sp} of 293.5 s at a nominal thrust of 56.9 kN (12.80 klb_f). Due to mission requirements, this motor will require an additional 8.7% (93.7 kg (206.5 lb_m)) propellant more than the current 1076 kg (2371 lb_m) propellant load listed in the ATK catalog. This quantity of additional propellant may require a minor redesign or even a requalification of the motor, the impacts of which, however, were beyond the scope of this study. One option to circumvent this issue, is to add additional propellant to the liquid descent stage, thus allowing it to provide the required additional ΔV.

Bipropellant Descent Stage: The descent stage used to lower the spacecraft to Tritons surface is a bipropellant stage utilizing MMH and NTO for propellants. This system has four main thrusters, sixteen RCS thrusters, two fuel tanks, two oxidizer tanks, a single fault tolerant feed system, and is pressurized with helium stored in a single COTS tank. The four main thrusters are Aerojet R-42s that deliver a nominal thrust of 890 N (200 lb_f) at a nominal I_{sp} of 303 s. The 16 RCS thrusters are Aerojet model R-6D, which produces a nominal thrust of 22 N (5.0 lb_f) at a nominal I_{sp} of 294 s, and are assumed to be mounted in four pods containing four thrusters each. The feed and pressurization systems consist of COTS components, a nominal instrumentation suite, and tank and line heaters.

The nominal oxidizer/fuel ratio for both the main and RCS thrusters is $\sim$1.65, which for MMH and NTO results in nearly identical fuel and oxidizer volumes. This allows identical tanks to be used for both propellants, which can be mounted in an opposing pair configuration to minimize center-of-gravity shift during operation. For this design, the selected ATK model 80309-1 tank is constructed of Ti-6Al-4 V, is polar boss mounted, and has a maximum operating pressure (MOP) of 2.19 MPa (318 psia). With a diameter of 56.2 cm (22.1 in.) and a length of 81.3 cm (32.0 in.), it has an integral vane type propellant management device, and operates at 78% rated capacity with the current propellant load. The helium pressurant gas for this system is stored in one COTS ATK model 80459-1 pressurant tank. This tank is a composite overwrapped pressure vessel (COPV) design with a MOP of 31.0 MPa (4500 psia), is 42.2 cm (16.36 in.) in diameter, and is 66.3 cm (26.1 in.) long. Figure 9.36 shows a preliminary piping and instrumentation diagram (P&ID) of the descent stage propulsion system.

Hopper Propulsion System: The propulsion system on the Hopper is a nitrogen-based cold gas system that is based around a single 116.7 cm (45.9 in.) diameter spherical custom COPV tank with a Ti alloy liner. The system has six large thrusters for axial propulsion that are located in three groups spaced equally around the circumference of the tank. Roll control is accomplished via six smaller thrusters that fire tangentially to the main thrusters and are located in three opposing pairs alongside the main thrusters.

The system is designed to be refueled via in situ propellant gathering. The primary method is to use a cryocooler to cool a heat exchanger surface inside the tank. Once cold enough, the ambient atmosphere will collect and freeze inside the tank. A backup system is to use the robotic arm to scoop nitrogen snow from the surface and feed it directly into the tank. To help facilitate propellant collection, a 15.25 cm (6.0 in.) diameter opening is located near the top of the tank. This opening has a collection funnel mounted to the outside of the tank that is designed to help funnel any snow scooped up by the robotic arm to fall into the tank. There is a door mounted inside the tank that closes before the propellant is heated and the tank pressurized. The placement of the collection funnel, thrusters, and the other major propulsion elements on the Hopper is shown in Fig. 9.37.

Once adequate nitrogen has been collected inside the tank, the door is closed, and heat is applied to the tank in order to vaporize the nitrogen and heat it to 300 K. For a 116.7 cm (45.9 in.) diameter tank, that equates to a pressure of 13.79 MPa (2000 psia) for a propellant load of 127.0 kg (279.9 lb_m). This warmed nitrogen gas is then fed to the thrusters via a nominal single fault tolerant feed system with line heaters, a nominal instrumentation suite, normally closed pyrotechnic valves

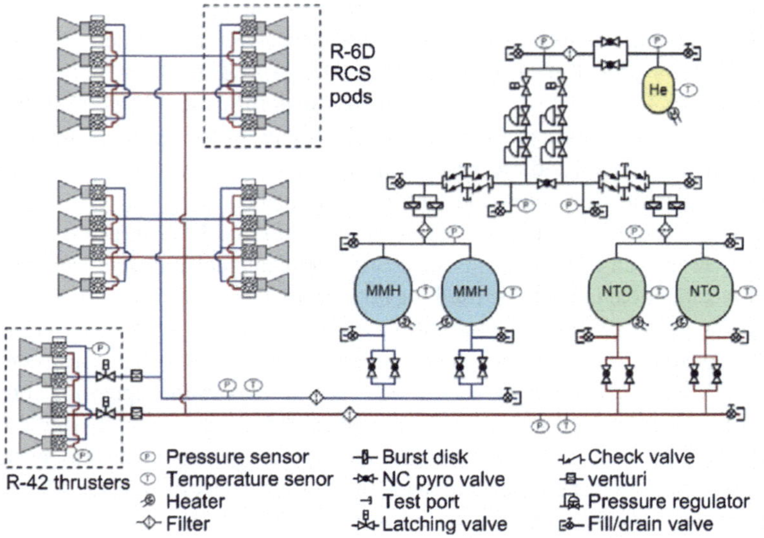

Fig. 9.36 Preliminary descent stage P&ID

for propellant isolation during launch and the coast to Triton, and a set of latching valves that can allow Triton atmosphere to flow into the tank in the event the tank door fails closed. This will greatly increase propellant gathering time, due to the small flow area, but should allow the hopper to continue its mission. The preliminary P&ID for the Hopper propulsion system is shown in Fig. 9.38.

There are a total of twelve thrusters on the Hopper, six main thrusters and six smaller roll control thrusters. The six main thrusters are designed to provide a nominal thrust of 223 N (50 lbf) and are based on Moog COTS thruster model 58-126. The six roll control thrusters provide a nominal thrust of 4.5 N (1.0 lbf) and are based on Moog model 58-118. Though all the thrusters are hard mounted to the spacecraft body, it is planned that the cold gas thrusters can be pulse modulated to

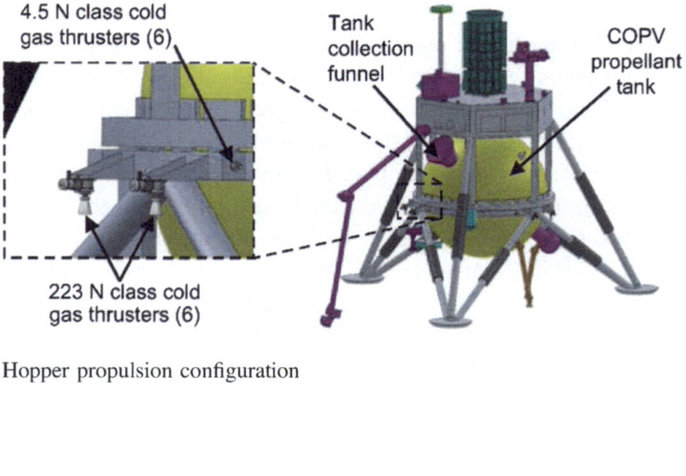

Fig. 9.37 Hopper propulsion configuration

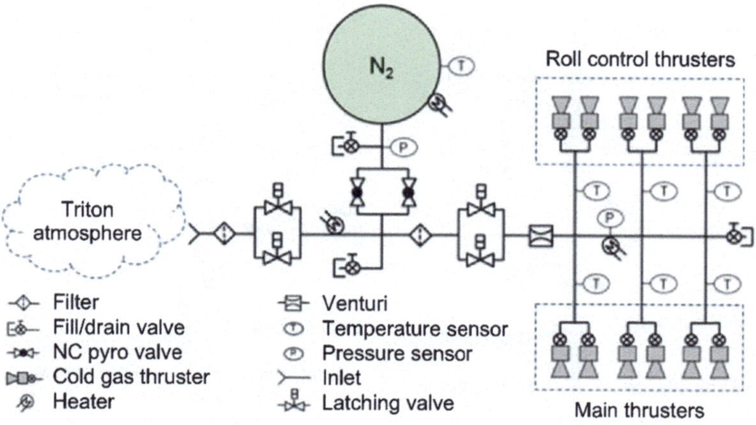

Fig. 9.38 Hopper preliminary P&ID

adjust the resultant thrust vector to that required by the spacecraft during hop maneuvers.

Propellant Tank Door: The internally mounted propellant tank door is designed to seal the 15.25 cm (6.0 in.) diameter opening into the tank and allow for propellant pressurization. The door actuation system is located inside the tank and has internal redundancy to meet the single fault tolerant requirement. The door has an ellipsoidal shape to support the internal tank pressure, is envisioned to retract into the tank when open, and has a seal surface around the rim that seals against a flange on the tank wall when closed. This design utilizes the internal tank pressure to help provide sealing force during propulsion system operation. There is a cylindrical funnel extension that extends into the tank to help guide nitrogen snow into the tank while protecting the seals from accidental contact and potential damage. There are two axisymmetric seals present to provide single fault tolerance. A schematic showing a cross sectional view of the door mechanism is shown in Fig. 9.39, and illustrates how the system would operate in both the open and closed positions.

Tank Door Seals: One key component to the proper operation of the propulsion system is the door seals. If they fail, the vehicle is stranded and the propulsion system rendered inoperative. For this reason, dual concentric seals will be used to provide a level of redundancy. Although the selection and detail design of the seal system is beyond the scope of this study, a survey of potential seal options from various manufactures was conducted to determine what options are commercially available. According to literature available from Parker, several types of metal seals can handle temperatures form cryogenic to 1144 K (1600 °F) and pressures up to 524.0 MPa (76,000 psi). A typical selection matrix for various seal types is shown in Fig. 9.40.

From the seal options evaluated during this study, the spring energized O-ring and wire ring designs seem to be the most applicable at this stage in the design, as they are typically used in applications requiring mild corrosion resistance and

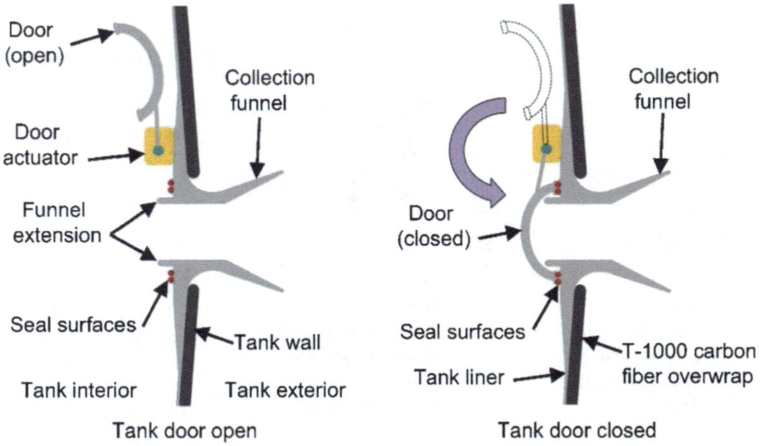

Fig. 9.39 Propellant tank door schematic

exposure to cryogenic to moderate temperatures. Both are readily available in 304/304L, 316/316L, 321, and 347 stainless steel alloys, with 304 and 321 being the preferred alloys for the metal O-ring and wire ring designs. These types of seals are typically plated depending on the operational environment, and the base material can be tempered various ways including annealing, work hardening, and solution heat-treating.

Hopper Propulsion System Performance
In preparation for a hop maneuver, the propellant tank is heated to 300 K and pressurizes to a design operating pressure of 13.79 MPa (2000 psia). This is the initial condition of the propellant in the tank for a hop maneuver, and as propellant is expelled through the thrusters, the temperature and pressure in the tank decrease. This complicates trajectory calculations since thruster I_{sp} is a function of

Seal type	Sealing requirements					
	High springback	Low load	High load	Low leak rate	Pressure capability	Low cost
Metal C-ring	◯	◯	◯	◉	●	◉
Metal E-ring	●	●	⊘	○	◯	○
Metal O-ring	○	⊘	◉	◉	●	◉
Metal U-ring	◉	●	⊘	○	◉	◯
Metal wire ring	⊘	⊘	●	◯	●	●
Spring energized C-ring	◯	⊘	●	●	●	○

Fig. 9.40 Typical seal selection matrix

temperature and changes as the propellant is consumed. For the purposes of this design, propellant consumption is modeled as an adiabatic blow-down process, which is limited to a critical point if either critical temperature or pressure is met. This critical point limit is in place to help reduce the risk of condensation in the tank or feed system, which could result in a sudden drop in propulsion system performance as the nitrogen propellant enters a region where phase change to a liquid can occur. The resulting nitrogen left in the tank at this point is treated as trapped residual propellant. Calculated tank conditions and estimated thruster I_{sp} are shown in Figs. 9.41 and 9.42 as functions of propellant mass fraction for the Hopper system.

Hopper Propulsion System MEL

A listing of the various components in the Triton Hopper propulsion system and their corresponding masses is shown in Table 9.18.

9.4.8 Structures and Mechanisms

Structures and Mechanisms Requirements

The Triton Hopper structures must contain the necessary hardware for research instrumentation, communications, and power while minimizing overall size. The structural components must be able to withstand applied loads from the launch vehicle, operational maneuvers, and landing. In addition, the structures must provide minimum deflections, sufficient stiffness, and vibration damping. The maximum axial load of 3 g is anticipated upon landing on the planet surface. Other parts

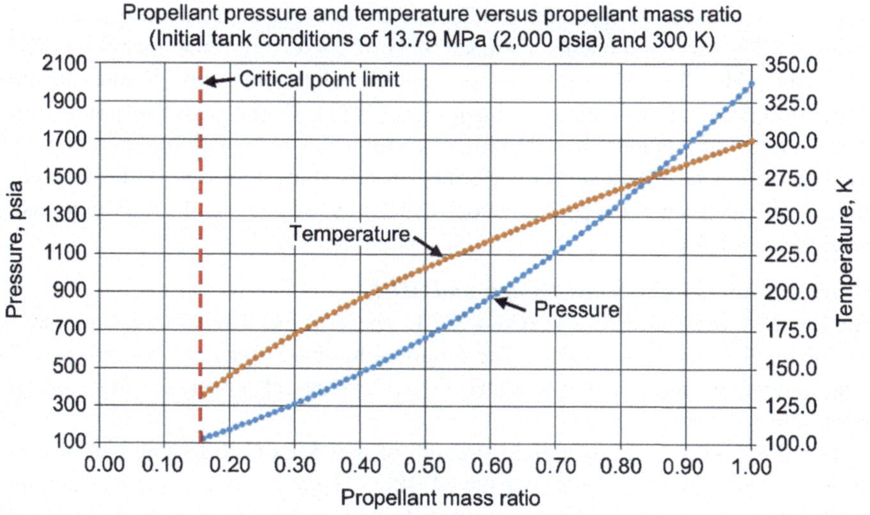

Fig. 9.41 Hopper tank conditions during propulsive operations

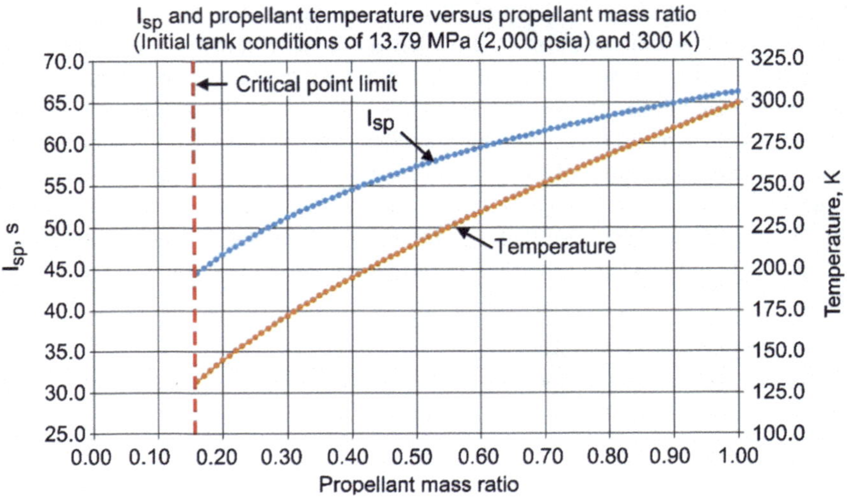

Fig. 9.42 Thruster I_{sp} and propellant temperature during propulsive operations

of the flight may impose a higher axial load with lateral loads. The goal of the design is to minimize mass of the components that comprise the structure of the spacecraft bus, and fit within the physical confines of the launch vehicle. In addition, the structures must provide low thermal conductivity in an effort to thermally isolate components within the lander and the surrounding environment. The Hopper is to operate in a low cryogenic temperature environment.

Structures and Mechanisms Assumptions

The main bus consists of a space frame with tubular and plate members that are assumed to provide the optimum architecture for housing the necessary operational hardware. The bus components are made of Ti-6Al-4 V, extra low interstitial (ELI), from the Metallic Materials Properties Development and Standardization (MMPDS), Federal Aviation Administration (2012). Figure 9.16 illustrates views of the Lander and its main bus. Joining of components is by welding, bonding, and threaded fasteners. The analysis performed in this study looks at the landing conditions with an assumed approach velocity of 1 m/s (39 in./s) and a maximum axial acceleration of 3 g.

Structures and Mechanisms Design and MEL

The Lander bus consists of a Ti-6Al-4 V ELI space frame. The various components are mounted to the space frame. The mounted hardware includes components for communications and tracking; C&DH; GN&C; electrical power; thermal management, and science.

The Ti-6Al-4 V ELI has a yield strength of 869 MPa (126 ksi) and an ultimate strength of 924 MPa (134 ksi) per the MMPDS (Federal Aviation Administration 2012). Safety factors are 1.25 on the yield strength and 1.4 on the ultimate strength per NASA–STD–5001 (NASA 1996), for a protoflight design. The resulting

Table 9.18 Propulsion system master equipment list

Description	Quantity	Unit mass (kg)	Basic mass (kg)	Total mass with growth (kg)
Propulsion (chemical hardware)			87.2	102.8
Primary chem. system hardware			87.2	102.8
Main engine hardware			86.6	102.2
Large cold gas thruster	6	1.5	9.0	11.3
Feed system	1	9.2	9.1	11.4
COPV tank	1	56.7	56. 7	65.2
Mounting hardware	1	2.0	2.0	2.1
Sample door assembly	1	6.0	6.0	7.5
Heat transfer vanes	1	0.3	0.3	0.4
Cryocooler mounting boss	1	3.5	3.5	4.4
RCS hardware			0.6	0.6
Cold gas thruster	6	0.1	0.6	0.6
Propellant (chemical)			127.1	127.1
Propellant			124.3	124.3
Fuel			124.3	124.3
Fuel useable	1	114.9	114.9	114.9
Fuel margin	1	5.8	5.8	5.8
Fuel residuals (unused)	1	3.6	3.6	3.6
Propellant			2.8	2.8
Fuel			2.8	2.8
Fuel usable	1	2.5	2.5	2.5
Fuel margin	1	0.3	0.3	0.2
Fuel residuals (unused)	1	0.1	0.1	0.1

allowable stress is 660 MPa (95.7 ksi) limited by the ultimate stress. The Ti alloy was selected due to its high strength and desired low thermal conductivity from ambient to near zero absolute temperatures. The thermal conductivity is 7.3 W/(m K) (51 Btu-in/(h ft^2 °F)). The material and bus architecture provides a technology readiness level (TRL) of six as per Mankins (1995).

The main bus consists of square tubular members with 51 mm (2.0 in.) sides and a wall thickness of 0.8 mm (0.030 in.). The legs use 64 mm (2.5 in.) OD tubing with a wall thickness of 0.8 mm (0.030 in.).

The landing gear radius relative to the center of gravity (CG) height is 1.15 times the CG height. This is comparable to the Apollo Lunar Module and a proposed Lunar Lander as reported by Epps (2006).

The MEL for the structures and mechanisms of the Triton Lander is shown in Table 9.19.

Structures and Mechanisms Analytical Methods

Preliminary structural analysis and modeling was performed using given launch and landing loads and the dimensions of the proposed spacecraft bus structure. Analytical methods utilizing a spreadsheet tool were employed to analyze the bus. A maximum axial load of approximately 3 g is anticipated on the spacecraft upon landing on Triton's surface.

A simple analysis was performed on the vertical members of the Lander bus. It was assumed that the members each support 100 kg and 3 g acceleration is applied due to the landing. The resulting axial stress is 3 MPa (470 psi). With an allowable stress of 660 MPa (95.7 ksi) the resulting margin is 219. The load is assumed to be shared equally among the vertical members.

The landing legs do use a spring/damper system to absorb the impact energy during landing. Limiting the peak acceleration of the bus to 3 g requires a displacement of 1.7 cm (0.7 in.) with an approach velocity of 1 m/s (39 in./s) and a constant deceleration. With an approximate bus mass of 500 kg (1100 lb) and the assumption that one leg may take the brunt of the impact the resulting stress in the main tubular member would be 420 MPa (61 ksi). The resulting margin is 0.57.

Table 9.19 Triton lander structures and mechanisms MEL

Description	Quantity	Unit mass (kg)	Basic mass (kg)	Total mass with Growth (kg)
Structures and mechanisms			59.5	70.2
Structures			53.0	62.6
Primary structures			51.7	61.0
Tank support ring	1	11.2	11.2	13.2
Upper bus	1	28.1	28.1	33.2
Long legs	3	1.2	3.6	4.3
Short legs	6	1.0	6.1	7.2
Landing pad	3	0.9	2.6	3.1
Secondary structures			1.3	1.6
LIDAR mount	1	0.6	0.6	0.8
Antenna mount	1	0.4	0.4	0.4
Camera mount	1	0.3	0.3	0.4
Mechanisms			6.5	7.6
Installations			6.5	7.6
AD&C	1	0.6	0.6	0.7
CD&H	1	0.6	0.6	0.7
Comm. and tracking	1	0.2	0.2	0.2
Electrical power	1	1.6	1.6	1.9
Thermal	1	0.9	0.9	1.0
Propulsion chem	1	2.7	2.7	3.2

An additional installation mass was added for each subsystem in the mechanisms section of the structures subsystem. These installations were modeled using 4% of the calculated best estimate dry mass of each of the subsystems. The 4% magnitude for an initial estimate compares well with values reported by Heineman (1994) for various manned systems. This is to account for attachments, bolts, screws, and other mechanisms necessary to attach the subsystem elements to the bus structure, and not book kept in the individual subsystems. An 18% growth margin was applied to the resulting installation mass. These margins are placed onto the subsystem elements prior to the additional margin that was added in order to reach the 30% MGA required on the dry mass elements.

9.5 Conclusions

The Triton Hopper conceptual design demonstrates that it is possible to design an innovative Hopper vehicle propelled by in situ resources that could accomplish a roving mission on Triton, a captured Kuiper belt object of high scientific interest.

The approach developed for the Triton Hopper concept can be adapted and utilized on other bodies, whether icy or not. By the use of a cryopump grounded to the cold surface, even thin atmospheres can be pumped to create propellants. Ices are even easier to collect and process given the right tools.

Follow-on efforts may involve exploring higher performing high temperature thrusters, lightening the lander, and/or speeding up propellant collection/processing in order to extend hop distance and frequency. A complete launch and delivery analysis to Triton will also be performed. Finally, use of a Triton-type Hopper on Pluto will be analyzed.

Acknowledgements The authors wished to thank the following team members both for their written contributions to this work: Julie Kleinhenz, David Chato, Ian Dux, Jeffrey Woytach, Waldy Sjauw, Laura Burke, Mike Martini, James Fittje, John Gyekenyesi, Tony Colozza, Paul Schmitz, Hector Dominguez, Robert Jones, Tom Packard, Elizabeth Turnbull, Tom Parkey. Other team members who contributed to the concept include: Les Balkanyi, Stan Grisnik, Vikram Shyam, Ian Dux, Mike Bur, Elizabeth Turnbull, Tom Parkey, and Anita Tenteris.

References

ASTM International: Standard Guide for Evacuated Reflective Insulation in Cryogenic Service. C740/C740M-13 (2013)

Bailey, R., Noca, M.: Mission Trades for Aerocapture at Neptune. AIAA–2004–3843 (2004)

Bienstock, B., Atkinson, D., Atreya, S., Baines, K., Wright, M., Masciarelli, J.: Neptune orbiter, probe, and Triton Lander mission. In: NASA Progress in Astronautics and Aeronautics Series, vol. 224. AIAA, Reston, VA (2008)

Cruikshank, D.P. (ed.): Neptune and Triton. The University of Arizona Press, Tucson, AZ (1995)

Davoodi, F., Acikmese, B.: REARM: Re-entry Hopper Spacecraft System on Mars. AIAA 2014–4377 (2014)

Epps, B.P.: Design of Spacecraft for Exploration of the Moon and Mars. Massachusetts Institute of Technology (2006)

Federal Aviation Administration: Metallic Materials Properties Development and Standardization (MMPDS) (2012)

Guernsey, C.S., Baker, R.S., Plachta, D., Kittel, P., Christie, R.J., Jurns, J.: Cryogenic Propulsion with Zero Boil-Off Storage Applied to Outer Planetary Exploration. JPL D-31783 (2005)

Heineman Jr, W.: Design Mass Properties II: Mass Estimating and Forecasting for Aerospace Vehicles Based on Historical Data. NASA JSC 26098 (1994)

Jerred, N.D., Colley, S., O'Brien, R.C., Howe, S.D., O'Brien, J.E.: The Mars Hopper: Development, Simulation and Experimental Validation of a Radioisotope Exploration Probe for the Martian Surface. AIAA 2012–5052 (2012)

Malin, M.C., et al.: The mast cameras and Mars descent imager (MARDI) for the 2009 Mars Science Laboratory. Lunar Planet Sci XXXVI (2005)

Mankins, J.C.: Technology Readiness Levels: A White Paper. NASA Office of Space Access and Technology, Advanced Concepts Office (1995)

Meetin, R.J., Seifert, H.S.: Propulsion dynamics of lunar hoppers. J Spacecraft **11** (1974)

NASA: Structural Design and Test Factors of Safety for Spaceflight Hardware. NASA–STD–5001 (1996)

NASA: Cassini-Huygens Saturn Arrival Press Kit (2004)

Nothnagel, S.L., Bailey, Z.J., Cunio, P.M., Hoffman, J.A., Cohanim, B.E., Streetman, B.J.: Development of a Cold Gas Spacecraft Emulator System for the TALARIS Hopper. AIAA 2010–8778 (2010)

Plachta, D.: Results of an Advanced Development Zero Boil-Off Cryogenic Propellant Storage Test. NASA/TM—2004-213390 (2004)

Plachta, D., Kittel, P.: An Updated Zero Boil-Off Cryogenic Propellant Storage Analysis Applied to Upper Stages or Depots in an LEO Environment. NASA/TM—2003-211691 (2003)

Rabin, Y.: The effect of temperature-dependent thermal conductivity in heat transfer simulations of frozen biomaterials. CryoLetters **21**, 163–170 (2000)

Rieder, R., Brueckner, J., Gellert, R.: The new Athena alpha particle X-ray spectrometer (APXS) for the Mars exploration rovers. J. Geophys. Res. **108**(E12), 8066 (2003). https://doi.org/10.1029/2003JE002150

Sercel, J.C., Blandino, J.J., Wood, K.L.: The Ballistic Mars Hopper: An Alternative Mars Mobility Concept. AIAA–87–1901 (1987)

Shafirovich, E., Salomon, M., Gokalp, I.: Mars hopper versus Mars rover. Acta Astronaut. **59**, 710–716 (2006)

Space.com Staff: Seasons Discovered on Neptune's Moon Triton (2010)

Stachowiak, P., Sumarokov, V.V., Jucha, J., Jezowski, A.: Thermal conductivity of solid nitrogen. Phys. Rev. B **50** (1994)

Sumarokov, V.V., Stachowiak, P., Jezowski, A.: Low-temperature thermal conductivity of solid carbon dioxide. Fizika Nizkikh Temp. **29**, 603–605 (2003)

Chapter 10
Sub-ice Autonomous Underwater Vehicle Architectures for Ocean World Exploration and Life Search

William Stone, Kristof Richmond, Chris Flesher, Bart Hogan and Vickie Siegel

10.1 Introduction

Ice-covered oceans are found across the Solar System. On Earth, such environments are known to harbor life. On some Ocean Worlds such as Europa, the unique combination of an actively recycled ice shell and rocky, possibly magmatic interior may give rise to a geochemical system suitable to life (e.g. Hand and Chyba 2009) and not so terribly different from the terrestrial cryosphere, where the ice may act as a suitable interface along which melt and freeze provide chemical gradients of which life can take advantage. The entry into sub-ice oceans of Ocean Worlds enabled by the cryobot technologies described in Chap. 4 calls for the development of autonomous underwater vehicle (AUV) rovers to explore these water bodies as the next phase of exploration. The most fruitful places to search for life will be at energy sources provided by physical and chemical gradients, which may not necessarily occur at the break-through location of a cryobot. This implies exploration using a mobile platform, and this in turn–due to extremely limited bandwidth and hours-long round-trip transmission delay–must be an autonomous platform. This forms the final step (element four in the vision for Ocean Worlds Exploration

W. Stone (✉) · K. Richmond · C. Flesher · B. Hogan · V. Siegel
Stone Aerospace, 3511 Caldwell Lane, Del Valle, TX 78617, USA
e-mail: billstone@stoneaerospace.com

K. Richmond
e-mail: kristof.richmond@stoneaerospace.com

C. Flesher
e-mail: chris.flesher@stoneaerospace.com

B. Hogan
e-mail: bart.hogan@stoneaerospace.com

V. Siegel
e-mail: vickie.siegel@stoneaerospace.com

V. Badescu and K. Zacny (eds.), *Outer Solar System*,
https://doi.org/10.1007/978-3-319-73845-1_10

presented in Chap. 4) of an Ocean Worlds life search program. An intelligent underwater robotic explorer which can travel in an ice-covered ocean; identify signs of biological activity; home in on, acquire, and analyze samples; and return to a docking station (the cryobot "mothership") to upload data and recharge is a powerful tool in the search for life off Earth.

This chapter investigates technologies required for such an under-ice swimming rover. We describe in detail several approaches that have been developed and field-tested at Stone Aerospace, mostly under NASA funding, during the past 15 years. Many areas of activity in the oceans and waters on Earth–from oceanographic monitoring to search and recovery to petroleum exploration–have employed AUVs to extend our reach into areas difficult or impossible for humans to access (Broad 1997). However, such vehicles are overwhelmingly restricted to open water, far from the bottom or other obstacles, and with a quick escape route to the water surface to terminate a mission. While the technologies and capabilities required for open water AUVs provide a solid foundation for a swimming rover on Ocean Worlds, there are particular capabilities required to perform an autonomous search for life in such remote waters, only accessible through an ice cover. Specifically, this chapter discusses the design of AUVs which are able to:

- be deployed through a restricted access hole in an ice ceiling;
- navigate independently in 3D through a water body with 3D surfaces and obstacles;
- autonomously find unmapped, unexplored areas and efficiently execute an exploration and 3D map creation strategy;
- search for signs of life;
- precisely maneuver to discovered targets and bring a scientific sensor probe in contact and/or take samples; and
- navigate their way back to the deployment location to dock for data upload, recharge, or recovery.

All of the above attributes require levels of onboard autonomy and behavior-based interaction that far exceed the "mow-the-lawn" scripted-mission characteristics of extant commercial, military and scientific AUVs. Table 10.1 summarizes how our AUV development program has addressed each of these critical technology areas.

As in Chap. 4, we will cite illustrative examples of other research on this topic. However, this chapter is again not intended to be a universal state-of-the-art review of all work in the area of under-ice underwater vehicle design. Our intention is to give the serious systems designer insight into our program of under-ice AUV design and present what we have learned, concepts that worked and did not work, and data from fielded AUV systems that illuminate possibilities for future flight vehicles. For completeness, we give a general overview of the state of the art in autonomous underwater vehicles in Sect. 10.2. Because Stone Aerospace has developed a number of vehicles over the past 15 years and fielded them on diverse missions, we first present a brief chronology of our development path followed by a detailed

Table 10.1 Stone Aerospace AUVs developing technologies for Ocean World exploration

AUV name	Deployment through ice ceiling	Autonomous navigation	Autonomous exploration	Life search	Probing and sample collection	Autonomous rendezvous and docking
DEPTHX		X	X	X	X	
ENDURANCE	X	X			X	X
ARTEMIS	X	X		X	X	X
SUNFISH®	X	X	X			X

engineering description of each vehicle in Sect. 10.3. In Sect. 10.4, we address key problems facing all Ocean World AUV designers and describe our approaches to solving these problems along with their effectiveness (or lack thereof) in field deployment. We summarize the most salient lessons we have learned along the way in Sect. 10.5. Finally, Sect. 10.6 presents the critical path going forward, and actions the Ocean Worlds community can take to bring the vision of discovery of life in the sub-ice oceans to reality in the next 20 years.

10.2 Sub-ice AUVs

Autonomous underwater vehicles (AUVs) are generally defined as submersible robotic vehicles which carry on board all power, sensing, and processing power necessary to complete their mission. AUVs were initially developed under military funding in the 1950s. The first AUV is generally recognized to be the SPURV vehicle developed at the University of Washington's Applied Physics Lab (UW-APL 1973). SPURV, and almost all present-day AUVs, are torpedo-class vehicles consisting of a cylindrical body, hemispherical nose cone, and tapered tail cones equipped with propellers (with and without Kort nozzles). Generally they are steered either by articulated servo-controlled tail fins or articulated thruster nozzles, are stable when in motion, but are not well suited for hovering or proximity operations.

While AUVs are able to perform unattended operations, the state-of-the-art AUV deployed for oceanographic research and commercial bathymetric survey is generally limited by its sensing and processing capabilities to a low level of environmental awareness. Thus, modern AUVs generally must operate in "safe" regions away from obstacles (e.g. the sea floor or ice ceilings), and are primarily employed to conduct oceanographic (ocean water) or bathymetric (bottom-topography) survey operations–without the possibility for close inspection or interaction with objects in their environment. In addition, vehicles are generally deployed in the open ocean, where recovery simply means coming to the surface to be recovered by a surface ship, and a failsafe abort is always available in the form of releasing a drop-weight. As highlighted in Sect. 10.1, these options are not available to a swimming rover deployed on an ice-covered Ocean World. To open up exploration of these new environments, a whole new class of capabilities must be developed to allow for the types of precise exploration, sampling, docking, and failure recovery required for extraterrestrial life search.

Despite their current limitations, the value of AUVs to explore under an ice roof was recognized early in their development, and initial work allowing under-ice operations has been performed by a variety of groups. The Dorado vehicle (Bellingham 1995) was deployed with standard open-ocean methods at the edge of ice pack in Arctic waters as part of the ALTEX project (Sibenac 2004). The AUV Autosub II (Millard 1998) was also deployed from open water and was one of the first vehicles to perform extended operations under an ice shelf (Nicholls 2006). These operations were focused on survey and long-range mapping, with vehicles deployed from surface vessels at the ice margin.

Several underwater vehicles have been deployed and recovered from the ice surface to perform work in ice-covered waters. The majority of these deployments were performed through ice holes large enough to allow the vehicles to be deployed in their nominal passively-stable orientations. Examples include tethered vehicle work by Doble (2009), Kunz et al. (2009), Bowen et al. (2014), as well as untethered vehicle work by Bellingham (1995), Kukulya (2010), and Kaminski (2010). Following the development of DEPTHX, ENDURANCE (see Sect. 10.3 of this chapter), and VALKYRIE (see Chap. 4), a number of groups also began to investigate deploying torpedo-shaped ROV vehicles vertically through ice bore holes. Examples include work by Cazenave (2011) and Spears (2016). Concurrently with VALKYRIE and ARTEMIS (both fully integrated systems fielded under ice), the IceShuttle Teredo project (Wirtz and Hildebrand 2016) developed and lab-tested several subsystems of an AUV deployment and docking system designed to deploy through a narrow vertical ice shaft, with the specific goal of developing technology for Europa.

10.3 AUV Development at Stone Aerospace

In the spring of 1995, one of us (Stone) was asked to consider how it might be possible to create automated, high-resolution 3D maps of subterranean tunnels formed from channelized flow conduits in karst aquifers, specifically multi-level labyrinthine spring systems in north Florida. The solution to this problem was complex and cut to the heart of the matter for any autonomous underwater system working in an "overhead" environment–one in which direct egress to the surface was not possible and in which external real-time navigation aiding would not be possible. Over the next three years funding for this project was raised from private patrons as well as through a grant from the National Geographic Society (National Geographic Society 1999). The fundamental problems posed were twofold: first, there was no extant system for creating geo-registered true 3D maps; and second, the tunnels to be investigated in this project–at Wakulla Springs, Florida—were unexplored and underwater, with no extant map or navigation data to guide exploration. This remained a chicken-or-egg paradox for several years. Given the limited funding available for the project an early decision was made to focus on design and fabrication of the segment of the device—the Digital Wall Mapper, or DWM—that would create the 3D maps. This was the automated portion of the project. The device used a high-grade ring laser gyro IMU, high resolution depth sensor, and a helical array of 32 narrow beam 300 kHz sonar transducers for creation and registration of the 3D map. True mobility autonomy was deferred to the future and cave divers were used to pilot the device, which was mounted to the front of a long-range propulsion vehicle (see Fig. 10.1).

Fig. 10.1 A piloted mapping mission into Wakulla Springs, Florida in 1999 using the Digital Wall Mapper

Decisions on which tunnel to explore and map, and guidance to return home were relegated to a human pilot. The map was ultimately geo-registered (and corrections made for onboard IMU drift) through the use of through-rock magnetic oscillator localization (Stone et al. 2000). The data produced by the DWM were unique in that they were true 3D (see Fig. 10.2 which shows geo-registered data from several overlaid mapping missions inside Wakulla Springs) wherein the geometry of the environment was changing radically in all directions as one progressed into unexplored space. The DWM was used to create the first fully 3D digital cave map, which included precise details of more than six kilometers of Wakulla Springs.

At the conclusion of the National Geographic expedition in February of 1999 there was an engineering meeting that ended with an unexpected and astonishing outburst. After listening patiently to a review of the diver's suggestions for improvement concerning the operation of the DWM, Nigel Jones, the designer of the onboard computer system, said, "if we had had another year and another million dollars we would not have needed the bloody divers." The implication was that the only thing the divers did was to steer the vehicle and make a decision as to which tunnel to go down. The DWM stored onboard a large amount of information from which 3D maps were eventually constructed. However, it would have been straightforward to have constructed those maps internally while the vehicle moved forward through the cave. Using those 3D maps for real-time navigation, vehicle control, and decision-making in a labyrinthine environment was the next logical step to which Jones had alluded and it had profound implications for intelligent robotic exploration in sub-ice environments.

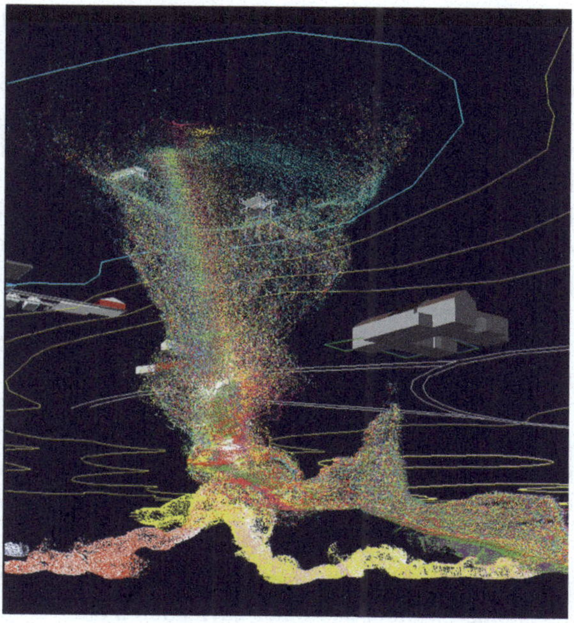

Fig. 10.2 A portion of the first 3D cave map, showing the 100 m diameter entrance funnel of Wakulla Springs, Florida, generated using data from the Digital Wall Mapper

In the remainder of this section, we briefly describe the goals, physical configuration, and deployment scenarios for the AUV systems developed by Stone Aerospace to fulfill this initial vision. These brief descriptions are intended to provide the context for the Ocean-World-relevant technology areas we identified in Table 10.1. The critical technologies, and how they have been developed on our AUV platforms, are then discussed in detail in Sect. 10.4.

10.3.1 DEPTHX

Three years after the initial inspiration in the Wakulla project discussion, NASA ASTEP funded the DEPTHX project. DEPTHX (for Deep Phreatic THermal Explorer) was designed to test autonomous navigation concepts and life detection strategies for a future sub-ice mission to Europa. The field test was conducted at Cenote Zacatón, the world's deepest hydrothermal spring. In May of 2007, pioneering a then-new navigation concept known as 3D simultaneous localization and mapping (SLAM), DEPTHX explored Zacatón to a depth of 335 m where the bottom, a sloping breakdown pile more than 140 m in diameter, was finally found. DEPTHX also implemented a hierarchical science autonomy behavior in which the vehicle would populate 3D voxel space with environmental data and follow chemoclines or thermoclines to wall surfaces where it would then commence machine vision proximity operations, looking for changes in background coloration that might suggest the presence of microbial communities of interest. It would then autonomously extend a science probe and collect both liquid and solid core samples. An auxiliary onboard microscope and microfluidics system allowed for real-time machine vision tracking of microbe trajectories. On May 26, 2007, DEPTHX explored into Cenote Zacatón where it encountered a threshold hydrogen sulphide (H_2S) concentration at a depth of 114 m. It autonomously made a decision to follow the chemocline to a nearby wall, collected a core sample, and returned home. The rock samples autonomously collected by DEPTHX at Zacatón (to depths of 280 m in follow on missions) subsequently resulted in the discovery of 4 new phyla of bacteria at a time when less than 100 were known on Earth.

The Zacatón sinkhole group is formed within a complex karst groundwater system that is fascinating both geologically and biologically. Figure 10.3 shows the primary and deepest of these (the Zacatón cenote).

This cenote harbors a diversity of life, varying with depth, and, importantly, at the time of the DEPTHX project, it represented previously unknown territory. This environment provided the context for developing and testing three of the most critical capabilities needed by any sub-ice Ocean World AUV:

- The capacity to explore completely unknown territory, create maps of that territory in three dimensions, and utilize those just-created maps to return itself "home" without any other navigation aids.
- The ability to seek out locations possessing a high likelihood of microbial life and make a decision to collect samples.
- The means to manipulate a science package to an autonomously selected target and collect samples, unambiguously demonstrating that science autonomy is feasible beneath the ice on an ocean world.

These three capabilities form fundamental underpinnings of the success of sub-ice Ocean World missions, and thus became the focus of the DEPTHX project.

To address Capability 1, a Simultaneous Localization And Mapping (SLAM) algorithm was developed to allow DEPTHX to both map an unknown area in 3D, and immediately navigate relative to the map without requiring external navigation aids (which would be limited to beacons on the penetration cryobot with limited

Fig. 10.3 View of the Cenote Zacatón sink in Tamaulipas, Mexico

line-of-sight range in an Ocean World scenario). This capability is described in depth in Sect. 10.4.2. To address Capability 2, a hierarchical life search algorithm was developed to look for targets with a high probability of containing the highest concentrations of biomass. This is further described in Sect. 10.4.4. To address Capability 3, DEPTHX contained a sampling arm to drive a corer into the wall and was designed to be hover-capable (passively stable in pitch and roll, but able to hold station, yaw to any heading, and translate horizontally and vertically) to be able to place the arm precisely at the target location found by the life search system. This capability is described in more detail in Sect. 10.4.5.

10.3.1.1 DEPTHX Vehicle Design

The vision for a sub-ice ocean exploration system that was presented in the first few figures of Chap. 4 addresses the problem of original exploration and life search in two stages: a long range, nuclear powered "prime mover" that performs initial ocean bottom mapping but stands clear of high energy zones that it detects (e.g. hydrothermal vents) and in such locations deploys a smaller, hovering craft to conduct the dangerous work of close approach in what may be very complex environments. Such a hovering vehicle would ideally be attitude independent; contain redundant propulsion systems; be capable of easily extracting itself from complex geometry even in the event of failure of portions of its propulsion system; be able to image its world in full 3D in real-time; and be able to construct emergency 3D egress routes to the carrier vehicle. Thus, the carrier vehicle would have much in common with long range present-day AUVs on earth (e.g. Bellingham et al. 2010) while the close approach vehicle would in theory be along the lines of a hovering AUV that would be rotationally symmetrical. It was this latter concept that drove the design of DEPTHX.

To our knowledge, DEPTHX represents the first AUV designed from the ground up to implement 3D SLAM. The vehicle design morphed dramatically during the course of the project and it will be worthwhile here to summarize how we arrived at the final design. Initially, the thinking had been to cannibalize a commercial ROV chassis and add science and navigation instrumentation to that platform (Fig. 10.4), mainly to cut costs and speed schedule.

This initial design (Stone et al. 2010) involved the use of clustered banks of sonar transducers with three clusters of 8 serving as forward, port, and starboard obstacle avoidance arrays and a single lateral-looking barrel sensor that was to be used for map building. A commercial ROV is not a very hydrodynamic device and an exterior "hydroshell" was envisioned to cover the active components to both reduce drag and to minimize the likelihood of a snagged vehicle during actual exploration in a labyrinthine overhead environment of the type that was anticipated at the Cenote Zacatón test site, as well as in the ice cracks in the surface crust of an Ocean World such as Europa or Enceladus. Figure 10.5 shows the "potato", a second-generation concept where the sonar clusters are un-grouped and the discrete transducers are mounted on the surface of the hydroshell, with shielded coaxial

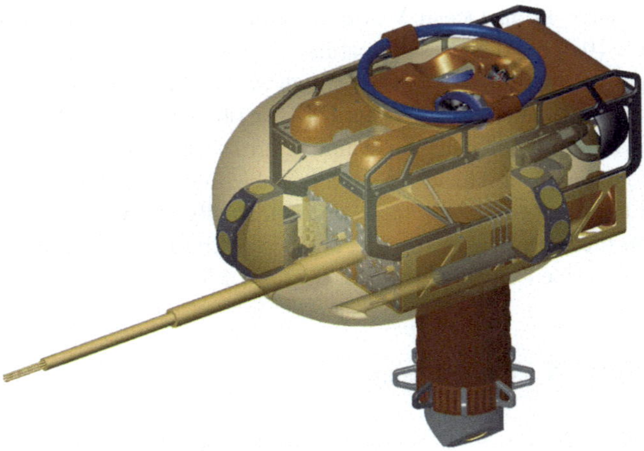

Fig. 10.4 Initial concept vehicle for DEPTHX

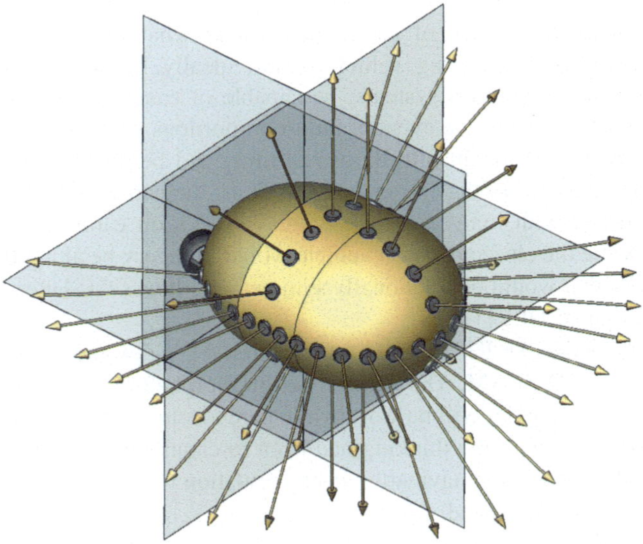

Fig. 10.5 Second-generation concept for DEPTHX showing three orthogonal planes of discrete sensors for SLAM map building

cables carrying the signal to a common digital signal processor array. This design began to approach the idea of a flexibly-deployed low-profile sensor system in which the AUV could approach near 4π steradian viewing (see Sect. 10.3.4.2 for the importance of this design criterion). However, in the potato design there still had to be a propulsion system and this occupied the stern of the craft, as in traditional ROV design.

Fig. 10.6 An early test of portions of the DEPTHX sensor array prior to being lowered into Cenote Zacatón, May 2005

This remained the design through the spring of 2005 as hardware and software development for the May 2005 field campaign to Zacatón dominated work on the project. Portions of the DEPTHX sensor array (Fig. 10.6) were tested in a drop sonde configuration to begin generation of SLAM test data from the actual site. While at Zacatón, there were a number of discussions to define the concept of operations for an actual exploration mission. The drop sonde developed for the May 2005 expedition (Stone et al. 2010) utilized the original barrel scanner concept for map building. The data coming in from Zacatón, however, began to evidence weaknesses in the field of view of a purely planar array in attempting to provide real-time 3D maps of the vehicle environment. A number of potential array concepts—spring boarding off the breakthrough potato skin idea—were sketched out at Zacatón. All of the sketches had the common concept of providing *true* 4π steradian viewing (for the moment we left out the issue of propulsion).

Later simulation in the context of the SLAM algorithm (see Sect. 10.4.2) in representative 3D labyrinthine environments favored one design that came to be called the "three Great Circles" concept. On paper there were three orthogonal, intersecting planes and one could envision the intersection of those three planes lying at the center of a sphere. The intersections of the planes with the sphere produced the "Great Circles" and it was on those lines that the discrete sonar elements would be placed. In this design, the concept of "mapping" and "obstacle avoidance" sonar became blurred and all sensors could be piped to both bodies of code through a publish-subscribe data system. During June of 2005, the initial design concepts gelled for a vehicle that incorporated this sensing approach in a compact geometry that had effectively no "stern" or "bow" and could therefore maneuver in any direction should it enter a blind corner in a sub-surface setting. Figure 10.7 shows a later version of this in July of 2005. Considerations for the placement of horizontal thrusters resulted in an equatorial band of thrusters with one in each quadrant, allowing any two to fail and yet preserve translational and rotational control of the vehicle. In Fig. 10.7, the purple-colored sonar units are 200 m range, 2° width beam elements for long range obstacle avoidance, while the blue transducers are narrow-beam 100 m range sensors. The placement of the thrusters on the equator forced the X-Y plane imaging sonars to be shifted off the centroidal plane.

The mass estimates for the overall vehicle came in at near 1 metric ton. This required more "syntactic" (glass sphere epoxy foam) floatation than was feasible within a true geometric ellipsoid. Further, the viewing angle for the return sonar pulse required a 15° clear cone in front of each transducer to reduce multipath reflections. These constraints led to the adoption of four expanded ellipsoidal quadrants on the top side of the vehicle, as shown in Fig. 10.8.

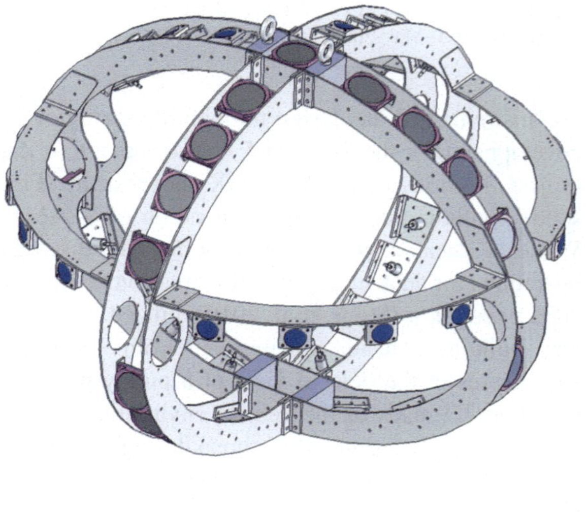

Fig. 10.7 The DEPTHX internal structural frame showing discrete SLAM mapping sensors. The horizontal circle of blue transducers was eventually placed near the bottom of the vehicle

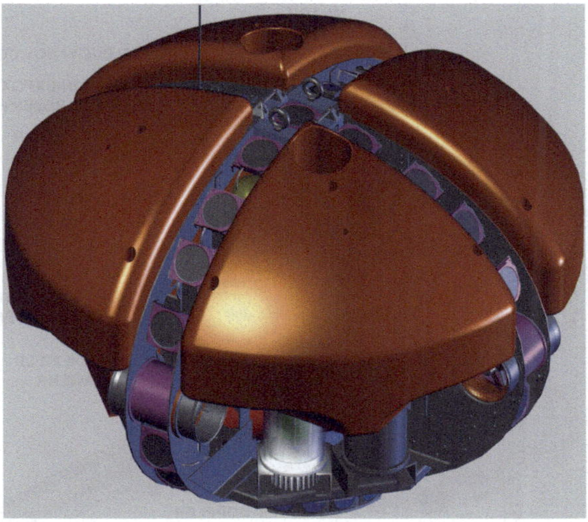

Fig. 10.8 DEPTHX final design, showing ellipsoidal geometry and syntactic floatation panels

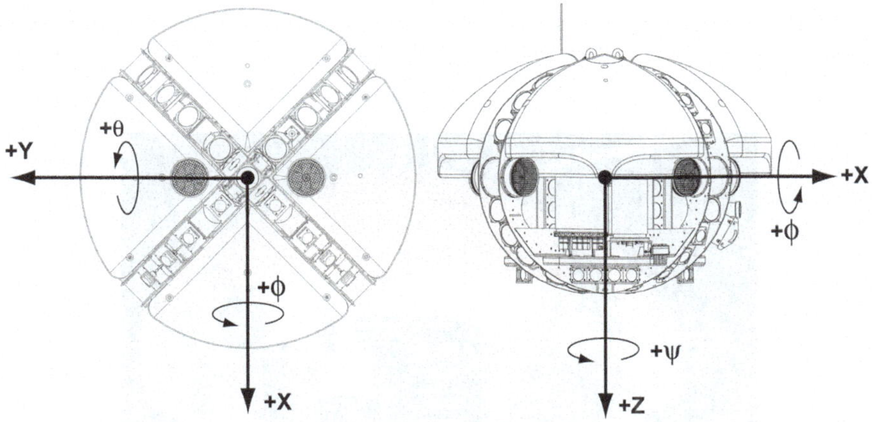

Fig. 10.9 DEPTHX vehicle coordinate system. Roll and Pitch were effectively pendulum damped and so this was a four degree-of-freedom vehicle

The vehicle was battery-powered with the heavy battery housings located at the very bottom of the ellipsoid. This led to a pendulum-stabilized vehicle with 4 degrees of freedom (DOF) of motion (Fig. 10.9).

The vehicle was able to yaw almost effortlessly, which enabled an unusual spin-mapping behavior, the results of which are shown later in Fig. 10.15.

From a systems debug viewpoint, we wanted a design that permitted all of the vehicular electronics and control systems to be accessible while still connected to the rest of the vehicle. This is frequently accomplished in ROV and AUV design

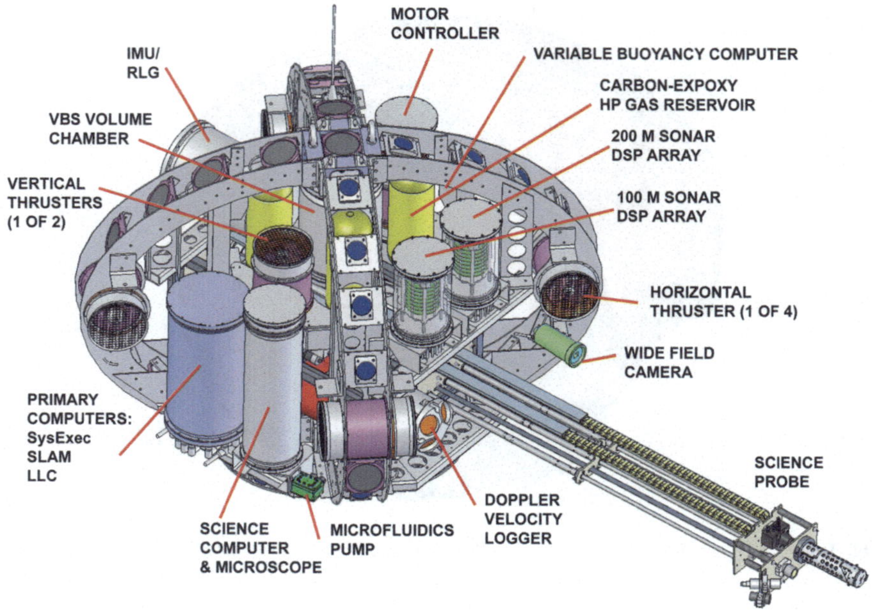

Fig. 10.10 An open-shell solid model view of the internals of the final DEPTHX design, showing vertical access electronics modules and the extensible sampling probe

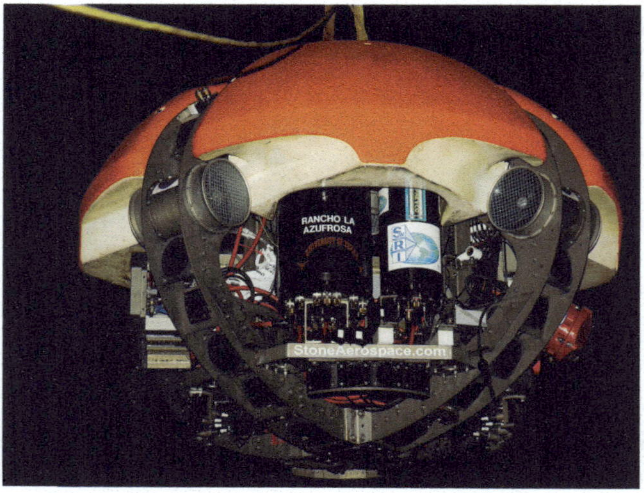

Fig. 10.11 The as-built DEPTHX, following a 10 h mission in Cenote Zacatón in May 2007

with swing-out housings that permit a bell-shaped pressure housing tube and top cap to be removed, leaving the bottom end plate, with its electronics interface cables still attached, connected to the vehicle. We spent considerable time working this problem before arriving at the solution shown in Figs. 10.10 and 10.11. This permitted overhead hoist removal of individual domes and allowed for precision re-fit when electronics needed maintenance or repair. The syntactic buoyancy quadrants were likewise designed for direct vertical removal. DEPTHX had 11 separate pressure housings. The service depth rating for the entire vehicle was 1000 m.

The overall vehicle diameter (major ellipsoid axis) was 2.13 m and the height (ellipsoid minor axis) was 1.52 m. Structural analyses of housings that saw full hydrostatic pressure, as well as all of the structural framing elements, were conducted using the Cosmos non-linear buckling FEA codes (a precursor of the codes presently integrated into SolidWorks Advanced Simulation). Individual housing electrical penetrations and connecting cables (of which there were 96 external power and communications lines comprising 250 m of cable and more than 400 electrical conductors) were largely achieved using dry-mate connectors from SeaCon. Figure 10.12 shows the system architecture for the overall vehicle.

The science mission for DEPTHX required the vehicle to be able to acquire both wall core samples and liquid samples from the water column. In order to do this

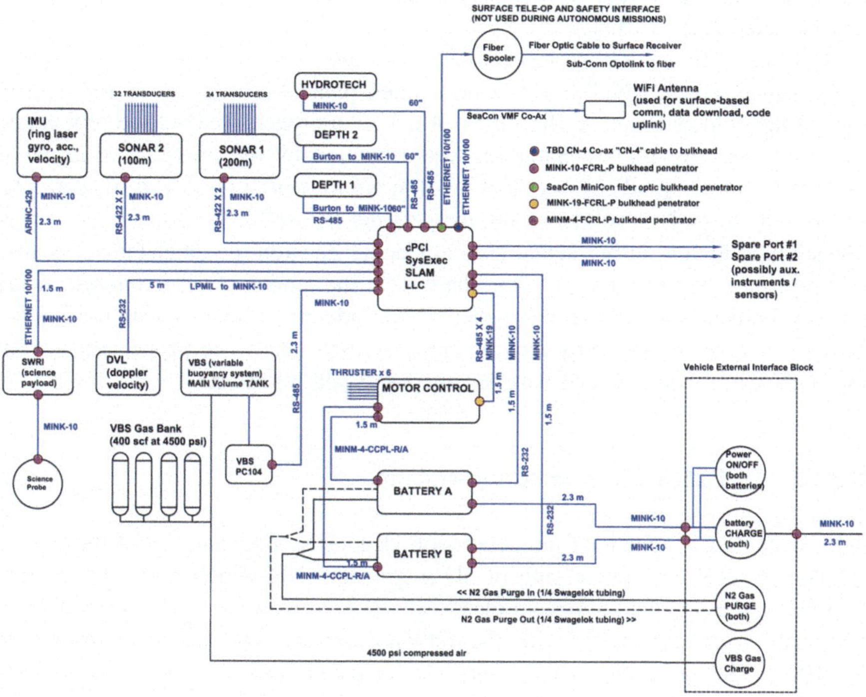

Fig. 10.12 System architecture for the DEPTHX vehicle

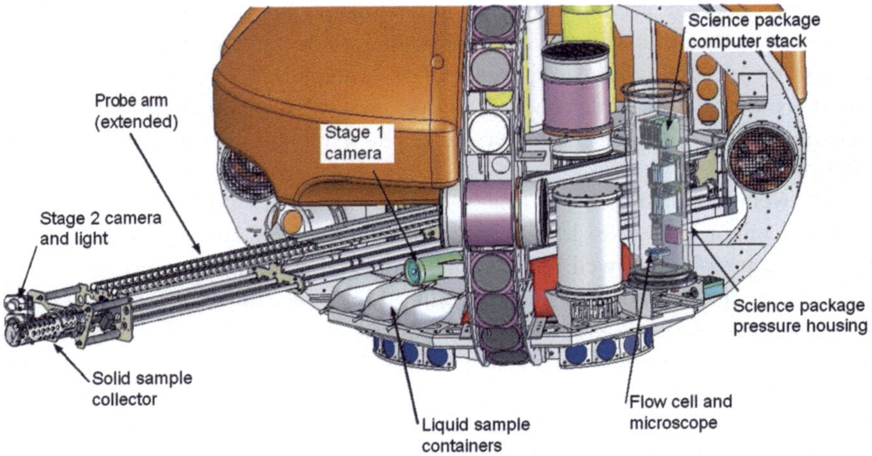

Fig. 10.13 Computer model of the extensible sampling probe for DEPTHX

without risking entrapment of the vehicle when approaching an overhung or jagged wall, an extendable probe was developed that was able to reach 1.5 m beyond the ellipsoid shell (Fig. 10.13).

Thus, the vehicle could come up to a 1.5 m safe standoff distance (as measured by the sonar arrays, which constructed a fitted plane from the geometric data in front of the vehicle to assess the safety of a wall approach), extend the probe, and obtain the needed samples. Sampling was triggered by the presence of an environmental variable gradient (see Sect. 10.4.5: we measured dissolved oxygen, pH, conductivity, dissolved solids, redox, temperature and sulphide) or changes in wall coloration, although during actual tests at Cenote Zacatón the latter failed because there were no significant wall color changes in the anoxic zone of interest to the project scientists. Environmental gradients (sulphide, in particular) were used as the basis for a science autonomy experiment described later. The probe arm system (Fig. 10.72, and Sect. 10.4.5) was designed and built by SwRI.

10.3.1.2 DEPTHX Electronics Sub-systems

There were 36 computational processors on DEPTHX. The majority of these were dedicated to real-time processing of 3D map sonar data which was subsequently supplied to the navigation system (see Sect. 10.4.2). However, a large number of processors were associated with the onboard sensors and actuators shown in Fig. 10.12. The vast majority of all inter-device data transfer was accomplished via serial communications using a number of protocols which were determined by the availability of appropriate sensors and actuators. Most data transfer was accomplished using RS-485 protocol but there were also numerous USB, ARINC-429, and 10/100 Ethernet devices. The primary vehicle computer was an industrial cPCI controller rack with three CPU boards, a serial I/O card, Ethernet I/O card, and inline power supply running off telcom voltage (48–60 Vdc). Most of the peripheral processing was handled either on PC104+ card stacks or custom-designed embedded PCBs and sensors. By far the largest impediment to implementing the vehicle architecture shown in Fig. 10.12 was device interfacing. Each device that did not have a communication protocol that matched the core cPCI computer system serial I/O or Ethernet card ports required a converter—a physical device that took up space and power. Each sensor and actuator (and often the data communications converters) required different types of power (different voltage requirements and different current levels). The main vehicle power bus ran nominally at 60 Vdc when fully charged and discharged to as low as 48 Vdc before power converters would begin shutting down. This meant that many different power supplies (usually solid state DC-DC converters) were needed. All of these power sources and data converters took up space and, in many cases, were sources of electrical noise that had to be meticulously tracked down and filtered. Figures 10.40 and 10.41 in Sect. 10.3.3 below give a good example of the problems that will face any group aiming to develop a flight vehicle AUV system when it comes to interfacing and controlling a wide array of sensors and actuators for work at ambient pressures up to 1600 bar (the floor of the European ocean, e.g.). DEPTHX, like the other vehicles described in later sections, used off-the-shelf components wherever possible to shorten development time and save cost in order to focus on the main objectives listed above. But this came at the expense of volume and weight—two resources that will be in very short supply on any sub-ice mission to Ocean Worlds. Development of flight versions will demand radical miniaturization of all sensors. It will also require system-wide standardization of power and data communications protocols in addition to board-level customization (ASIC and FPGA-based logic) in order to reduce the size and power draw of the onboard computational and control electronics.

10.3.1.3 DEPTHX Field Deployments

Figure 10.14 shows DEPTHX being lowered (by crane) into the Zacatón cenote. Initial tests of the SLAM algorithm took place in cenote La Pilita (Fig. 10.15), a 105 m deep hypogenic hourglass-shaped chamber. The vehicle was spun about the Z-axis (yaw) at a rate of approximately 1 rpm during descent to achieve the uniform 3D scan.

Fig. 10.14 Lowering DEPTHX into Cenote Zacatón in May of 2007. The floating mission control structure can be seen against the far wall of the sink

Zacatón itself (Fig. 10.16) maintained a giant silo cross section (90 m $\times$ 140 m) to a depth of over 300 m where a sloping talus pile leads to greater depths on the northwest wall. At the start of the DEPTHX missions in 2007 the entire Rancho Azufrosa (Zacatón) hydrothermal spring complex remained unexplored. DEPTHX produced the composite 3D map shown in Fig. 10.17.

The SLAM algorithm (see Sect. 10.4.2) utilized an egress algorithm that sought to maximize radial standoff distance to the nearest wall, thus effectively centering

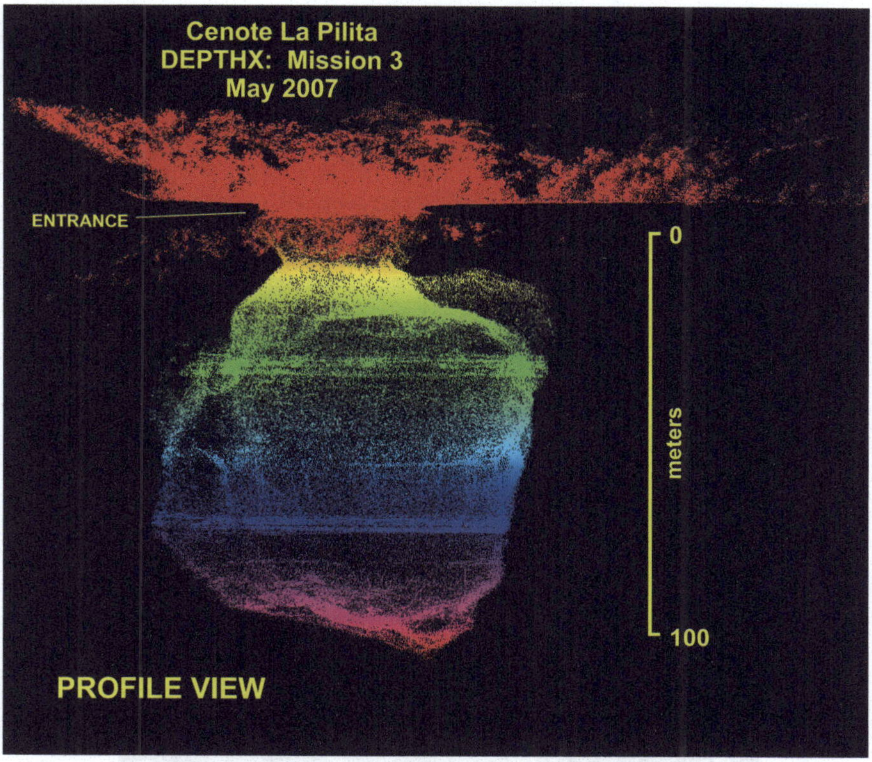

Fig. 10.15 Spin-generated map of Cenote La Pilita created by DEPTHX in May of 2007

the vehicle in the middle of the subterranean features as it rose to the surface. No significant horizontal tunnels were discovered that would have served to test the ability of the system to handle complex changes in environment topography. Experiments in autonomous exploration, life search and sampling are discussed in Sects. 10.4.4 and 10.4.5.

10.3.2 ENDURANCE

The ENDURANCE (Environmentally Non-Disturbing Under-Ice Robotic Antarctic Explorer) project was also funded by the NASA ASTEP program in order to test fundamental concepts relating to autonomous under-ice science and exploration that will be relevant to future missions to outer-planet icy moons such as Europa and Enceladus. The project was also supported by the National Science Foundation

Fig. 10.16 Profile view of the underwater portion of Cenote Zacatón map created by DEPTHX in May of 2007

(NSF) Office of Polar Programs (OPP), which arranged for deployment of the robot, its development team, and the science team to Antarctica.

A completely re-designed vehicle using the DEPTHX frame, ENDURANCE was designed to test under-ice navigation concepts and the ability to construct large-scale 3D chemistry models of sub-glacial lakes. In the austral summers of 2008/9 and 2009/10, the vehicle was deployed to its field location at the permanently ice-covered West Lake Bonney (WLB) in Taylor Valley, one of the Antarctic Dry Valleys (Doran 2002; Stone 2009; Murarka et al. 2009; Richmond

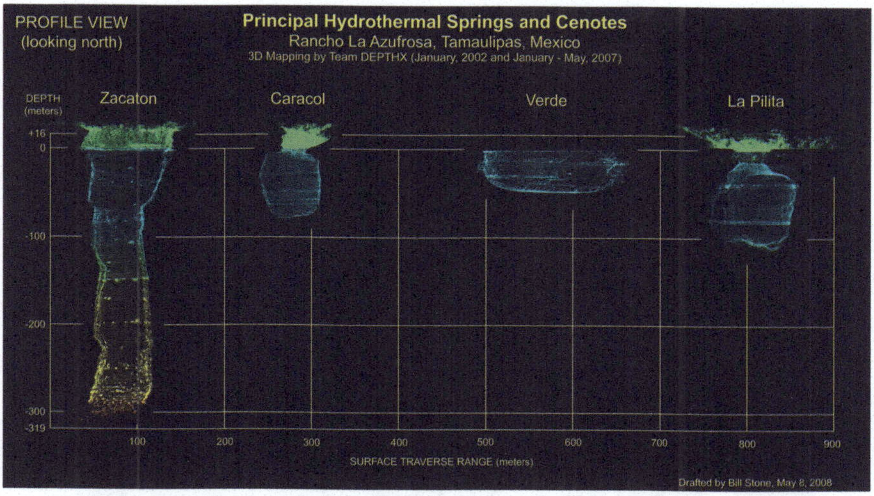

Fig. 10.17 Complete profile section of the known cenotes at Rancho la Azufrosa, Tamaulipas, Mexico, explored and mapped by DEPTHX. Light green-colored points at the top were acquired by lidar scans above the water surfaces

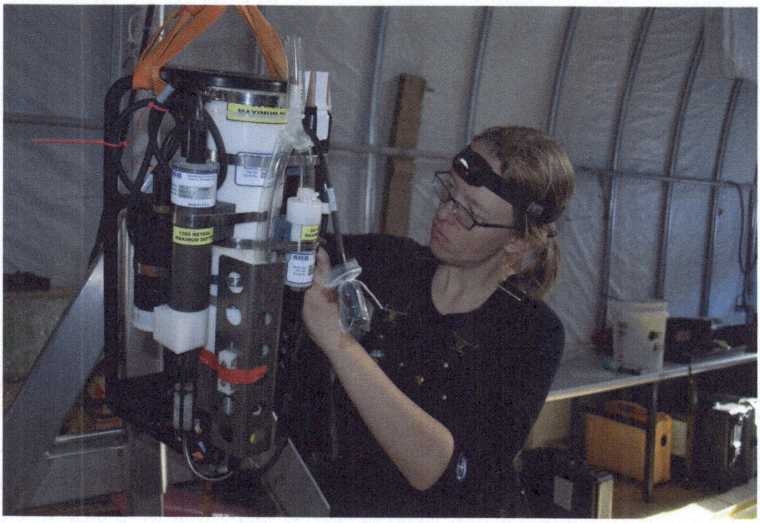

Fig. 10.18 The custom-designed deployable sonde developed for ENDURANCE to acquire vertical chemistry profiles in Lake Bonney, Antarctica

et al. 2009; Hogan et al. 2009). The new vehicle carried a science payload that consisted of a servo-controlled drop-sonde (Figs. 10.18, 10.19 and 10.20) which carried an array of sensors and imagers.

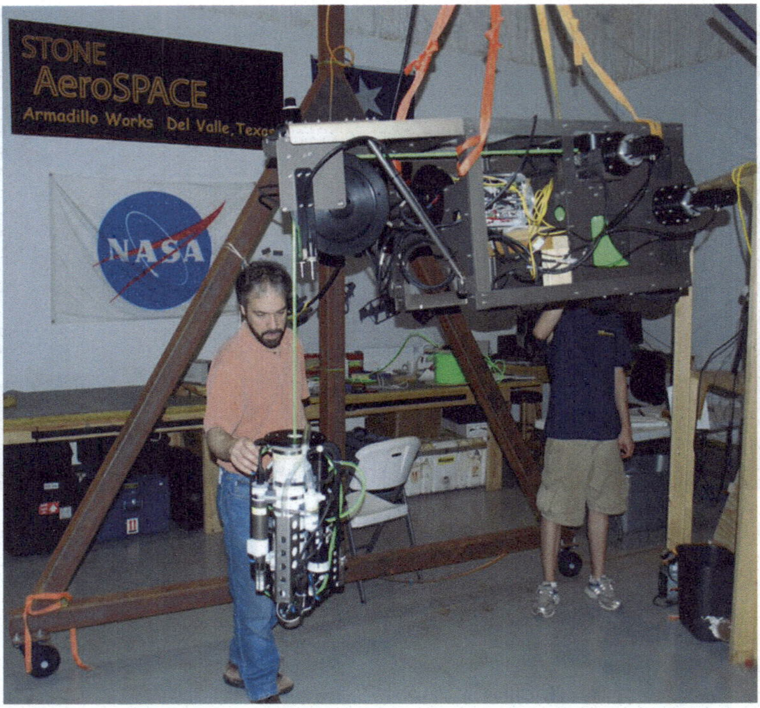

Fig. 10.19 The science payload for ENDURANCE replaced the sampling probe from DEPTHX with a servo-controlled instrument sonde that could be dropped to a depth of 60 m below the vehicle

Ultimately, ENDURANCE mapped all of West Lake Bonney and created 3D chemistry models which revealed the presence of a new hypersaline source at −5 °C, independent of the well-known Blood Falls site. Missions of more than 4 km round-trip were launched from a single melt hole in the 4 m thick ice cap. The vehicle used a 3-stage autonomous navigation and docking system to return to and rise up through the melt hole. Of historical note, ENDURANCE autonomously explored Scott Pass in the Bonney Riegel narrows between East and West Lake Bonney, where Capt. Robert Scott's party took measurements on foot in 1903. One hundred seven years later, ENDURANCE hovered where Scott once stood, and created a 3D map which revealed a water level rise of 16 m during that time period (see Fig. 10.30).

10.3.2.1 West Lake Bonney

The McMurdo Dry Valleys are the largest ice-free region in Antarctica with a total area of about 4800 km^2. These valleys shelter a polar desert environment with mean annual temperatures at the valley floor between −30 and −15 °C and precipitation

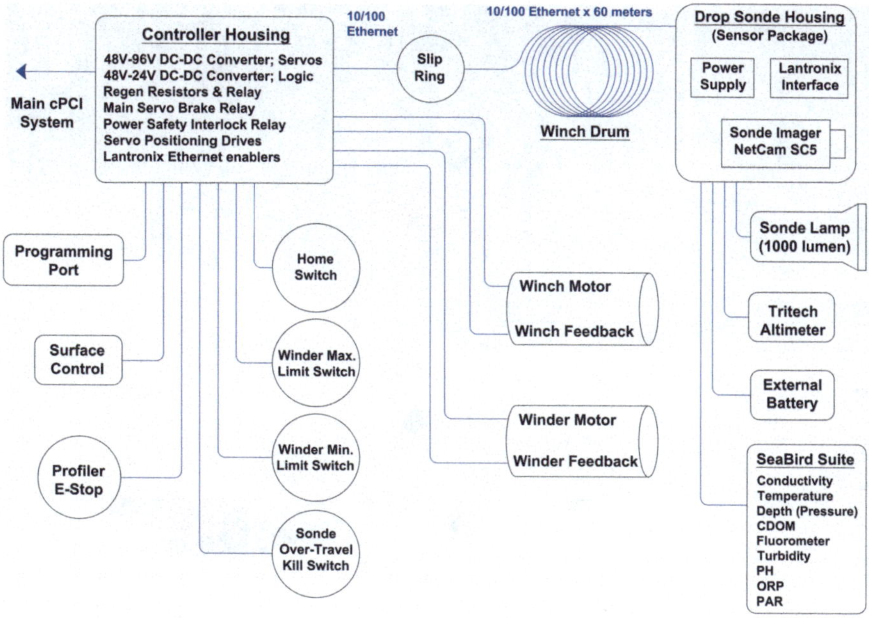

Fig. 10.20 Schematic for the ENDURANCE instrument sonde

of less than 10 cm per year (Doran 2002). There are about 20 lakes in the McMurdo Dry Valleys, almost all of which maintain a perennial ice cover (2.8–6.0 m) over liquid water. These lakes are remnants of glacial lakes believed to be as old as 4.6 million years (Doran et al. 1994).

West Lake Bonney (Figs. 10.21 and 10.22) presents a unique and challenging environment for exploration and study. The lake's permanent ice cover's thickness ranges from a completely open moat at the lake margins in summer to a seasonally-varying 3–4 m in the lake interior.

This ice cover prevents extensive exploration without the use of an underwater vehicle. The lake lobe is approximately triangular, with the "base" of the triangle formed by the face of the Taylor Glacier and the deltas of two glacial streams. Its "apex" is formed by the mouth of the Bonney Riegel Narrows separating the east and west lobes. The maximum lake width is ~ 800 m, while the distance from the glacier to the Narrows is ~ 2000 m. The maximum lake depth is ~ 42 m, the maximum depth at the glacier face is ~ 25 m and in the Narrows ~ 16 m. An additional primary physical feature of Lake Bonney is its unique, highly stratified chemistry. The west lobe comprises three distinct layers. At the bottom, extending up to a depth of ~ 15 m, sits a hypersaline body of water which has had minimal external interaction since the formation of the perennial ice cover. The salinity of this layer ranges from around twice that of seawater at the top to 4–5 times seawater at the greatest depths. On this dense saline layer floats a freshwater lens of around 10 m thickness composed of annual meltwater inflows. The ice cover forms the

Fig. 10.21 DEPTHX arrives by helicopter to the access hole on West Lake Bonney, November 2008

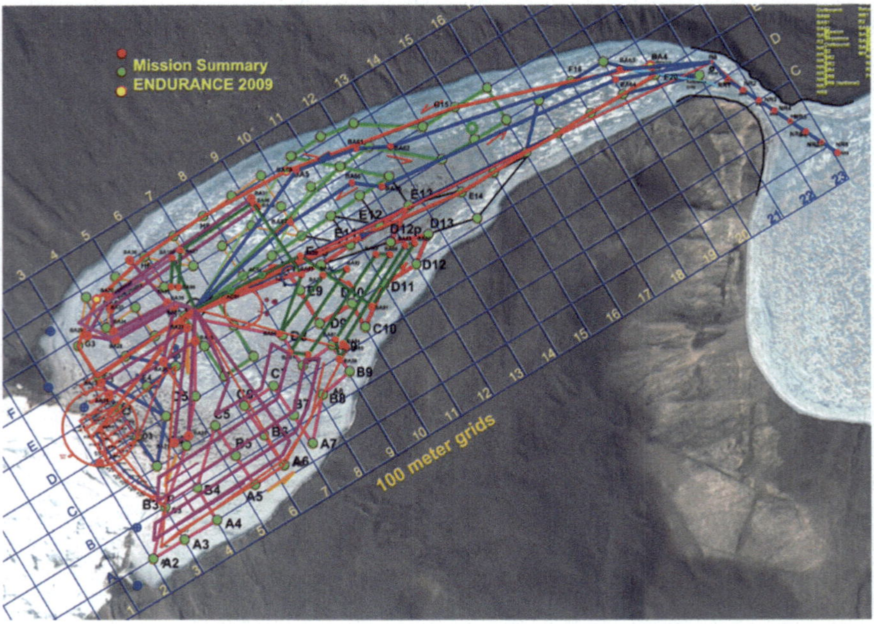

Fig. 10.22 Satellite image of West Lake Bonney with the DEPTHX mission trajectories superimposed

top-most layer. Deployment was effected by melting a vertical shaft into the ice cover and lowering the vehicle through the shaft. Recovery took place through the same hole.

To protect the unique Lake Bonney environment and to obtain reliable water chemistry data, disturbance of the lower saltwater layers was to be kept at a minimum during operations. Hence, ENDURANCE was restricted to the upper freshwater layer. However, the first campaign in 2008 revealed a complex 3D morphology at the glacier, including a large recessed area hidden under an ice shelf near its base (see Figs. 10.26 and 10.27). Exploration of this area—including the scientifically interesting glacier grounding line—was not immediately possible, as it required the vehicle to travel into the hypersaline lower lake layer. As the changes in vehicle buoyancy in the denser water were greater than the available ballasting could accommodate, and pending a review of the environmental effects and potential dangers to the vehicle of such a sub-glacial mission, exploration of this area was postponed until 2009 (see Sect. 10.3.2.3).

10.3.2.2 Project Goals

The scientific goals of both field campaigns were to:

- characterize the biogeochemistry of the lake in full 3D via automated sub-ice sonde casts and characterize year-to-year changes;
- create a high-resolution 3D map of the underwater interface between Taylor Glacier and WLB and, if possible, to visually image the glacier face; and
- create a high-resolution bathymetric map of the lake floor.

The primary engineering goal was to assess and improve the vehicle's ability to deploy autonomously down a melt hole 4 m deep with a diameter only slightly larger than itself, to ranges as far as 2.5 km from the melt hole, and to return home safely at the end of the mission. During the course of the two field seasons the team learned a great deal about how to operate this unique sub-ice science platform in the target environment.

10.3.2.3 ENDURANCE Vehicle

The ENDURANCE vehicle design began with the DEPTHX vehicle frame, to which extensive modifications were made. Exploration of Lake Bonney presented special challenges requiring changes to the hardware, navigation, and control systems from the DEPTHX configuration. In particular, the SLAM system employing the 4π-steradian sonar array in the confined environments of the cenotes was unsuitable for the Lake Bonney environment. The lack of physical relief throughout much of the lake, the significant multipath and beam bending issues due to the ice roof and water density profile, and the need for very precise navigation to ensure return to the 2 m diameter deployment melt hole all spoke against using the DEPTHX sonar-based SLAM for navigation. The primary hardware changes included:

- addition of an inverted Ultra-Short Baseline (iUSBL) system (Milne 1983) to allow the vehicle to determine the relative location of a transponder hung under the melt hole (see Sect. 10.4.2);
- replacement of the science sampling arm with a servo-operated spooler deploying a sonde on a tether to the bottom of the lake to enable vertical water profiling while keeping the vehicle in the freshwater lens;
- addition of "ice feet" to allow the vehicle to rest against the ice ceiling during profiling maneuvers;
- the addition of an up-look camera to enable visual docking with the melt hole using a blinking light shining down from the surface (see Sect. 10.4.6); and
- a battery retrofit in the second year (2009) to enable the vehicle to explore the entire extent of WLB in addition to the narrows leading to East Lake Bonney (ELB).

The wrap-around sonar array previously used primarily for SLAM navigation was employed for local obstacle avoidance, and a high-resolution multibeam sonar was used for mapping.

10.3.2.4 Field Operations with ENDURANCE

The primary scientific objective of the ENDURANCE project was to create 3D chemistry maps of West Lake Bonney over the course of two separate Antarctic seasons to determine if there were chemical and thermal changes taking place spatially throughout the highly stratified lake and, in addition, to detect the presence of anomalies in these variables—analogous to the existence of the well-known hypersaline source that exits Taylor Glacier at Blood Falls. The method by which this was achieved is shown in Figs. 10.18, 10.19, 10.20, 10.21, 10.22, 10.23 and 10.24.

A technique previously developed for computed tomography—the Marching Cubes algorithm (Lorensen and Cline 1987)—was to be used to extract iso-surfaces for chemical species. The algorithm requires a rectilinear array for efficient computation. To achieve this, West Lake Bonney was divided into a grid (see Fig. 10.22). Beneath the 4-m-thick ice cap was a lens of glacial melt water through which the vehicle was permitted to transit. When the vehicle reached one of the pre-designated grid points it would initiate an energy-conserving "ice picking" maneuver where it would float up (being positively ballasted by approximately 100 g) and contact the ice. Once stable it would release the sonde, which would descend under servo control at 0.1 m/s to within 1 m of the floor (detected by an altimeter as well as verified by the vehicle's down-look multi-beam sonar). There the sonde would take video images of the floor, and then rewind at 0.1 m/s until

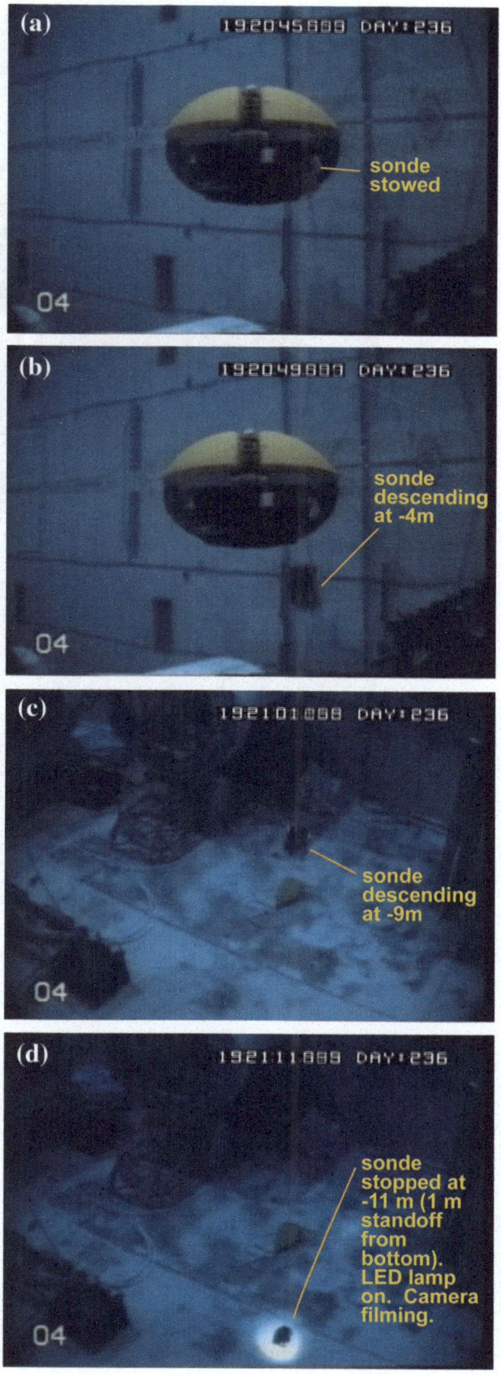

Fig. 10.23 Pre-deployment test of the ENDURANCE vertical water chemistry profiling sonde system at Johnson Space Center. The fully automated sonde cast was designed to stop at an altitude of 1 m above the lake floor and to take images of the bottom before being retrieved to the vehicle

Fig. 10.24 Pre-launch calibration of the sonde instruments (being lowered into the water of Lake Bonney) for an ENDURANCE mission in November 2009

docked with the vehicle (see Fig. 10.23 demonstrating this sequence at the Johnson Space Center Sonny Carter neutral buoyancy facility). Due to the limitations on range imposed by the onboard battery system, it was not possible to survey all of the grid points in one mission. In 2009, it took a total of 16 runs to complete all the sonde casts and lake-floor and under-ice bathymetry mapping (see Fig. 10.22).

The approach described above was able to be rezoned. Based on anomalies in temperature detected in the sonde missions of Fig. 10.22, a refined search area for a hypersaline source was defined at the head of Taylor glacier (Fig. 10.25). A 3D local map of the glacier front is shown in Fig. 10.26. A cross section at the location of a previously unknown −5 °C water source is shown in Fig. 10.27 and a plot of the temperature variation measured by the vehicle at the Taylor glacier grounding line (Spigel et al. 2017) is shown in Fig. 10.28.

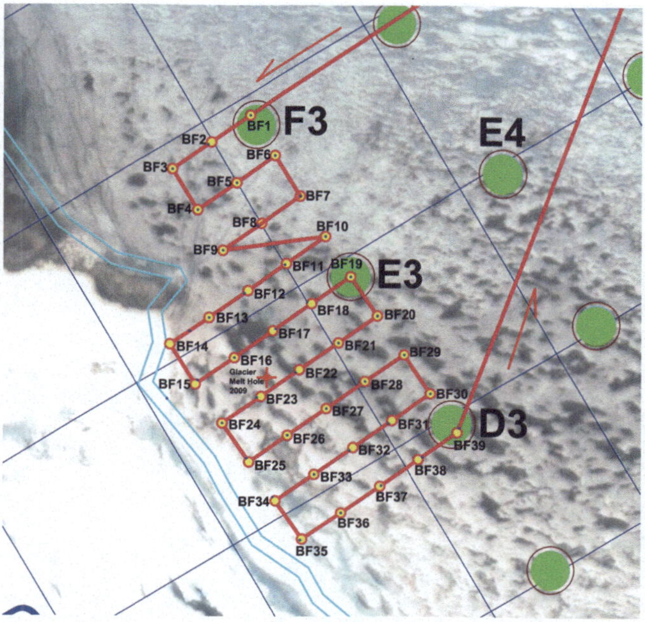

Fig. 10.25 A scripted fine-grid mission for ENDURANCE to discover hypersaline, cold water sources emanating from Taylor glacier (lower left). Green circles were previous sonde casts that suggested the need for a refined search (red lines)

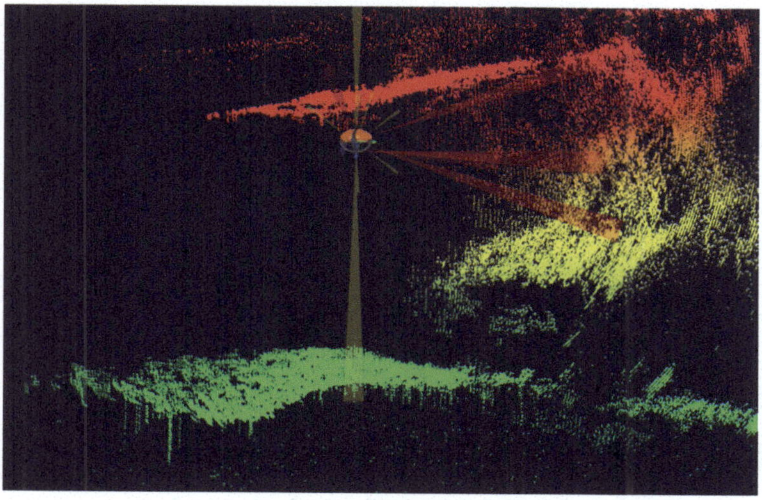

Fig. 10.26 Screen capture from the 3D situational awareness visualizer developed for ENDURANCE. This was a live image, relayed by a fiber optic data thread to Mission Control allowing for a supervised autonomous mission

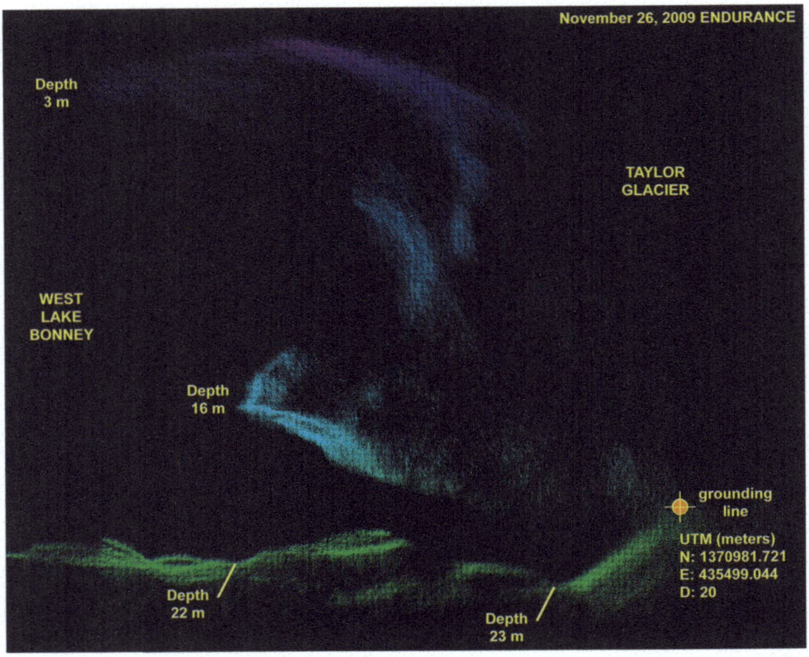

Fig. 10.27 Profile section of the Taylor glacier grounding line, mapped by ENDURANCE. The halocline begins at a depth of 16 m

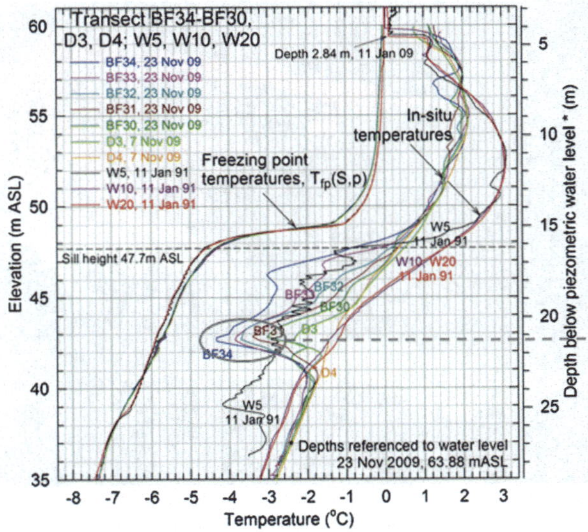

Fig. 10.28 Water temperature plots versus depth showing the detection of a cold hypersaline source at the grounding line of Taylor glacier (gray ellipse) discovered by ENDURANCE (from Spigel et al. 2017)

Fig. 10.29 Reviewing ENDURANCE data in the CAVE-2 facility at the University of Illinois Chicago (UIC). The view is looking northeast into the Bonney Riegel Narrows. Vertical white lines represent sonde casts. All data are interactive—clicking on a white line displays the depth profiles for all sonde instruments at that location

ENDURANCE carried 9 instruments on the sonde: depth; temperature, electrical conductivity, photosynthetically active radiation (PAR), turbidity, chlorophyll-a, colored Dissolved Organic Matter (cDOM), pH and redox (see Fig. 10.20 for a schematic of the sonde architecture). Several missions were conducted through the Bonney Riegel Narrows (upper right in Fig. 10.22). The results of those missions (Fig. 10.29) were used to extract a profile section through the site where Scott's men took measurements in 1903, thus evidencing the 16 m rise in water level (Fig. 10.30).

10.3.3 ARTEMIS

Two years following completion of the ENDURANCE missions, NASA funded Stone Aerospace to develop the ARTEMIS vehicle. ARTEMIS is a 6 degree-of-freedom, hovering autonomous underwater vehicle (AUV) designed and built as part of the NASA ASTEP SIMPLE project, which aimed to characterize the underside of the McMurdo Ice Shelf (see Figs. 10.31, 10.32 and 10.33).

The ARTEMIS vehicle (Fig. 10.34) is a heavily modified, long-range, low drag variant of the DEPTHX/ENDURANCE platform.

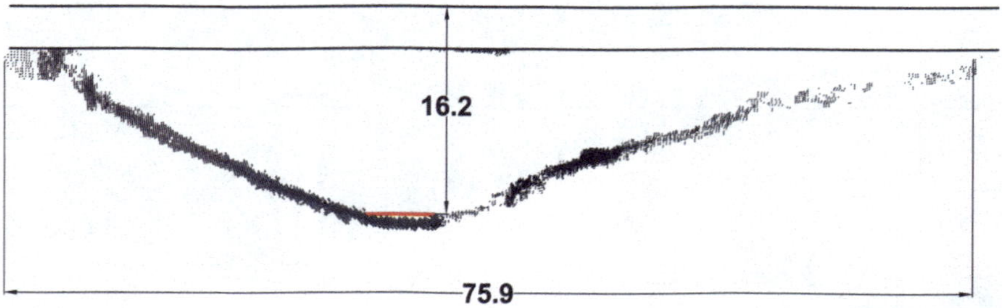

Fig. 10.30 Sonar cross section in the Bonney Riegel Narrows where Scott's 1903 exploration party recorded the width of the ice cap between east and west Lake Bonney (red line). Since then the water level has risen 16 m

Fig. 10.31 Site of mission control for the ARTEMIS/SIMPLE project. Basecamp was located on sea ice approximately 100 m from the edge of the McMurdo Ice Shelf. In the background is Mount Erebus

ARTEMIS was designed to be deployed from a sea-ice drill hole adjacent the McMurdo Ice Shelf (MIS) (Fig. 10.31 and Sect. 10.4.1) and to perform three classes of missions:

- exploration and high definition mapping of both the ice-water interface and the benthic interface to aid interpretation and calibration of data from airborne ice-penetrating radar (IPR) over-flights, with the goal of furthering the IPR being developed to send on the Europa Clipper mission;

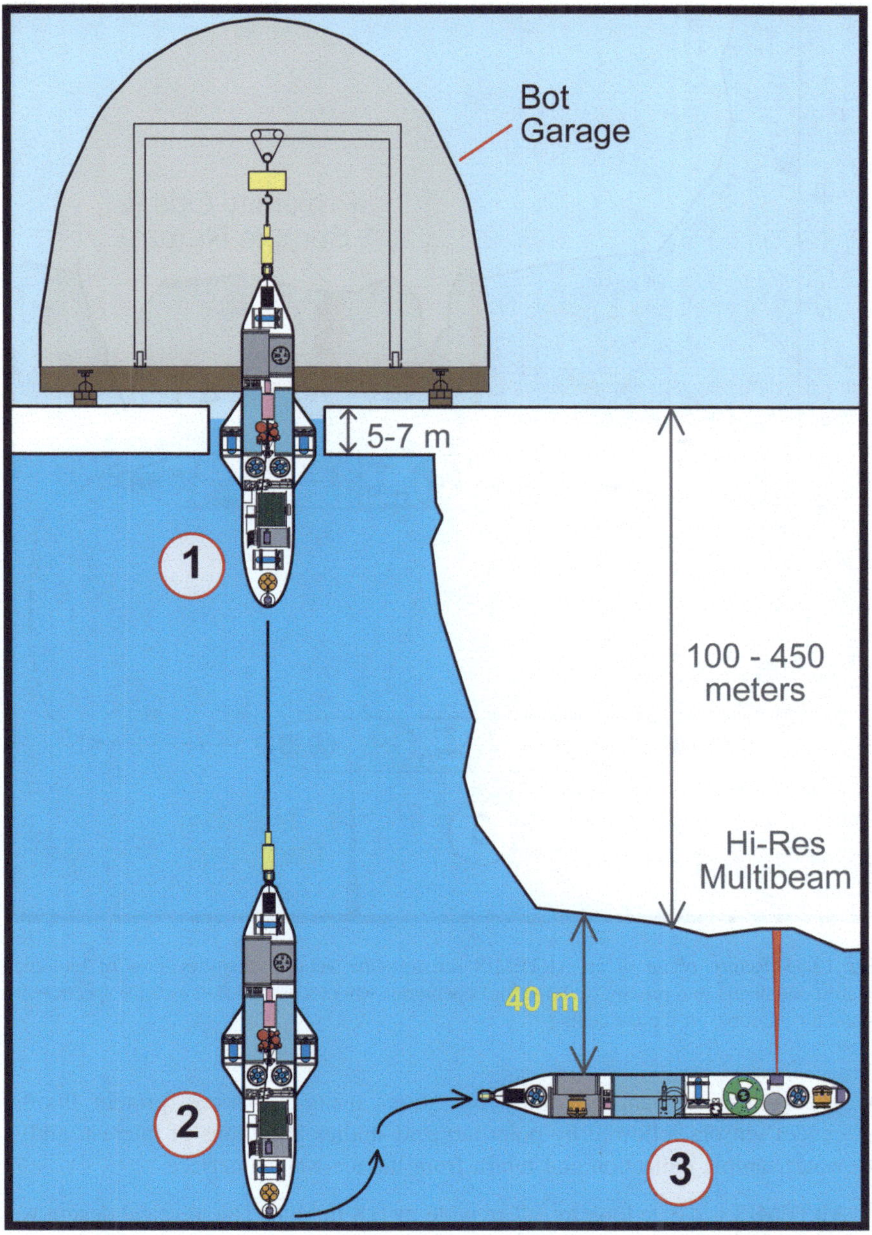

Fig. 10.32 Concept of operations for an ARTEMIS mission: the vehicle is dropped vertically through a drilled access shaft in sea ice; it then descends to a depth below the roof of the ice shelf and begins up and down-look bathymetric mapping on the inbound leg of the mission

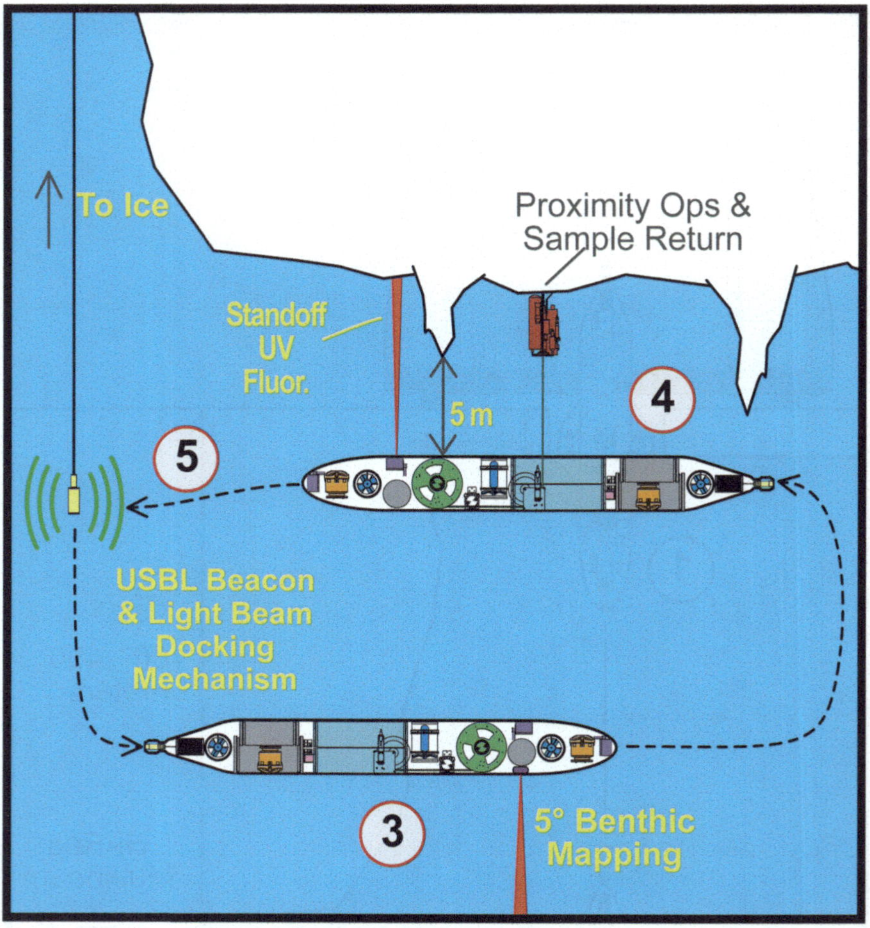

Fig. 10.33 Return phase of an ARTEMIS mission: the vehicle identifies sites of biological interest and enters into proximity (hovering) operations where a protein fluorescence spectrometer looks for microbes in the ice ceiling

- astrobiology-relevant proximity operations using bio-assay (protein fluorescence) sensors followed by point-targeted studies at regions of interest; and
- water sample collection and return from the ice-water interface.

ARTEMIS is 4.3 m long by 1.2 m wide by 0.8 m high. The physical design was constrained by the need to deploy and retrieve the vehicle through a 1.3 m diameter hole drilled through the sea ice adjacent to the ice shelf. ARTEMIS was outfitted with forward-, downward-, and upward-looking sonar instruments for obstacle avoidance, sea floor bathymetry, and ice ceiling mapping, respectively. Since egress was restricted to a single hole in the sea ice and the nearest open water was more than 50 km distant, a precise return-home procedure and docking system (Sect. 10.4.6) were needed to ensure recovery.

ARTEMIS incorporates an extensive suite of water chemistry instruments, mostly contained in a sonde package that could be extended above (and retracted into) the vehicle for proximity operations near the ice ceiling and to obtain undisturbed water sonde casts (Sect. 10.4.5). A protein fluorescence spectrometer designed to look for protein signatures in the ice ceiling, and the inlet tube for a water sampler able to collect 36 100 mL samples per mission are also located in the sonde. In the McMurdo Sound, a 15 km long, 1.5 mm diameter data fiber was deployed by the vehicle, linking it to Mission Control on the surface and enabling 1 Gbps data feedback on mission progress and environmental conditions (see Sect. 10.5 for details on this type of supervised autonomy). The vehicle carries a 10 W magnetic field generator that enabled surface tracking of the vehicle through both sea ice and the ice shelf. ARTEMIS conducted 20 under-ice missions under the McMurdo Ice Shelf, many with complex trajectories aimed at mapping the sea-ice/ice-shelf transition (see Figs. 10.35 and 10.36).

Maximum mission length was 10.1 km round trip and top speed was 1.05 m/s. Hovering proximity operation maneuvers were successfully developed to approach and investigate the ice ceiling. Sonde casts and water sampling were routinely

Fig. 10.34 Stern view of ARTEMIS, showing its twin prime mover thrusters. The thrusters were custom-designed and included dual independent windings, dual independent motor controllers for each winding, and power isolation leading to two separate vehicle batteries. Multiple system failures could occur within this architecture and still allow the vehicle to return home

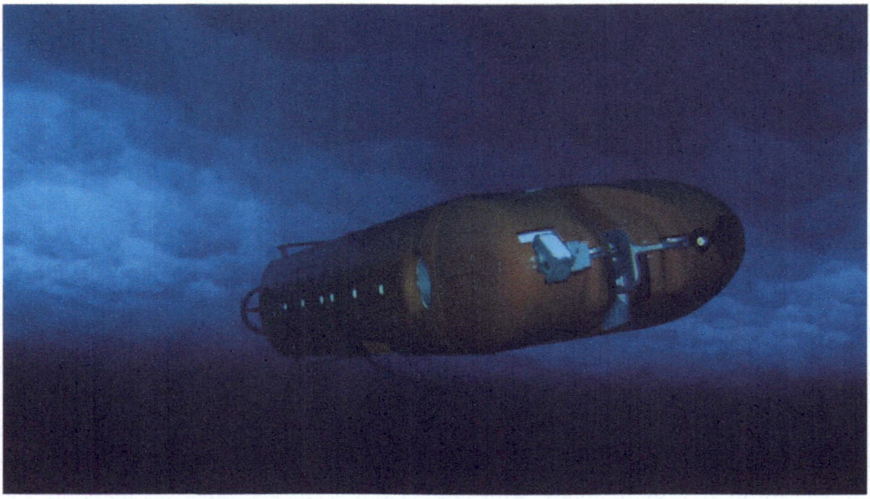

Fig. 10.35 ARTEMIS undergoing trapped air release roll and pitch maneuvers prior to initiation of a mission under the McMurdo Ice Shelf in November 2015

Fig. 10.36 A screen shot from the 3D vehicle visualizer from an ARTEMIS mission in which multiple transits were made between the sea ice and the McMurdo Ice Shelf. The transition from sea ice to ice shelf was much less abrupt than had been expected, possibly due to a thick (5 m+) layer of platelet ice under both the sea ice and ice shelf at the time of these missions (November–December 2015)

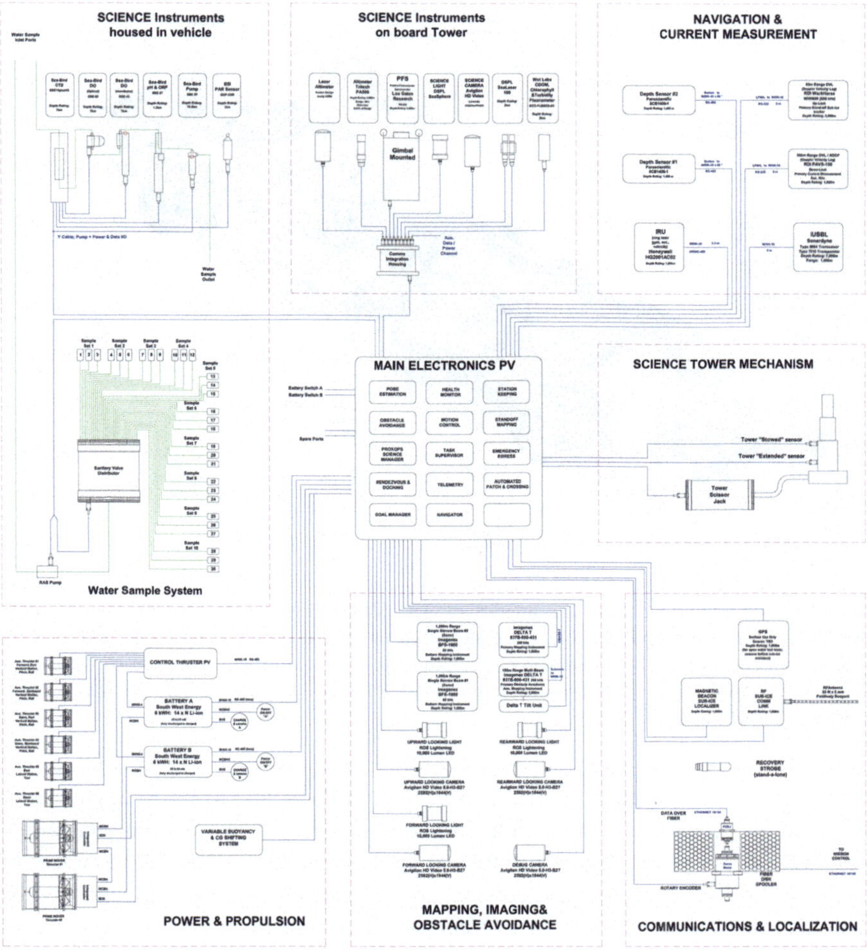

Fig. 10.37 System architecture for the ARTEMIS vehicle

conducted. The AUV successfully completed fully-automated return, rendezvous, and docking procedures, details of which will be discussed below. Figure 10.37 shows a schematic of the ARTEMIS vehicle. An annotated, 3D model of the vehicle is shown in Fig. 10.38 and a late stage assembly room image is presented in Fig. 10.39.

Due to the extensive science payload (Fig. 10.37) system integration of this vehicle was complex. As an example, Figs. 10.40 and 10.41 show the internal details of just one of the scores of pressure vessels mounted within the syntactic and ABS hydroshell. Computation was split up among several separate processors, and

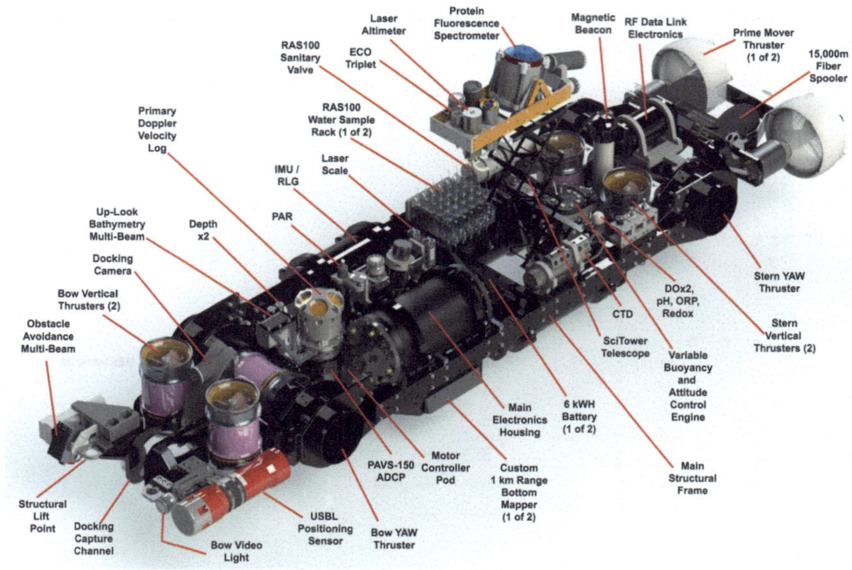

Fig. 10.38 Computer model of the interior of ARTEMIS. The vehicle used two prime mover thrusters for forward propulsion and six attitude thrusters for hovering, rendezvous, and docking. A servo-actuated scissor-lift allowed for proximity operations with the underside of the ice cap using an extensible sonde that could be retracted into the body of the vehicle for transit cruise between sites of interest

in some cases moved off to a local cluster (e.g. in the sonde tower to aggregate its instruments into one communications link) but the wide variety of communications protocols, sensor power requirements, and the large number of devices to either be read or controlled, leads to a compressed, complicated environment. Inventing ways to miniaturize this capability, to standardize all communications, to ruggedize all electrical connections, and to isolate noise (both acoustic and electrical) will again be the front and center challenge facing anyone designing a flight AUV destined for an Ocean World.

ARTEMIS was designed to be capable of both long-range cruise for mapping and survey, and precision positioning and interaction with the solid ice ceiling in a previously unexplored environment. The control system allowed for full 6-DOF hover maneuvering, enabling precision interaction with the solid ice ceiling, as well as aiding in rendezvous and docking at the conclusion of a mission.

Fig. 10.39 Assembly floor view of ARTEMIS as it was approaching completion and prior to attachment of syntactic panels and hydroshell skins

The ENDURANCE method of down-hole deployment and auto-docking return was extended to a vertically-deployed, horizontally-recovered concept that is depth independent and highly relevant to sub-ice deployment on an Ocean World (see Sect. 10.4.6).

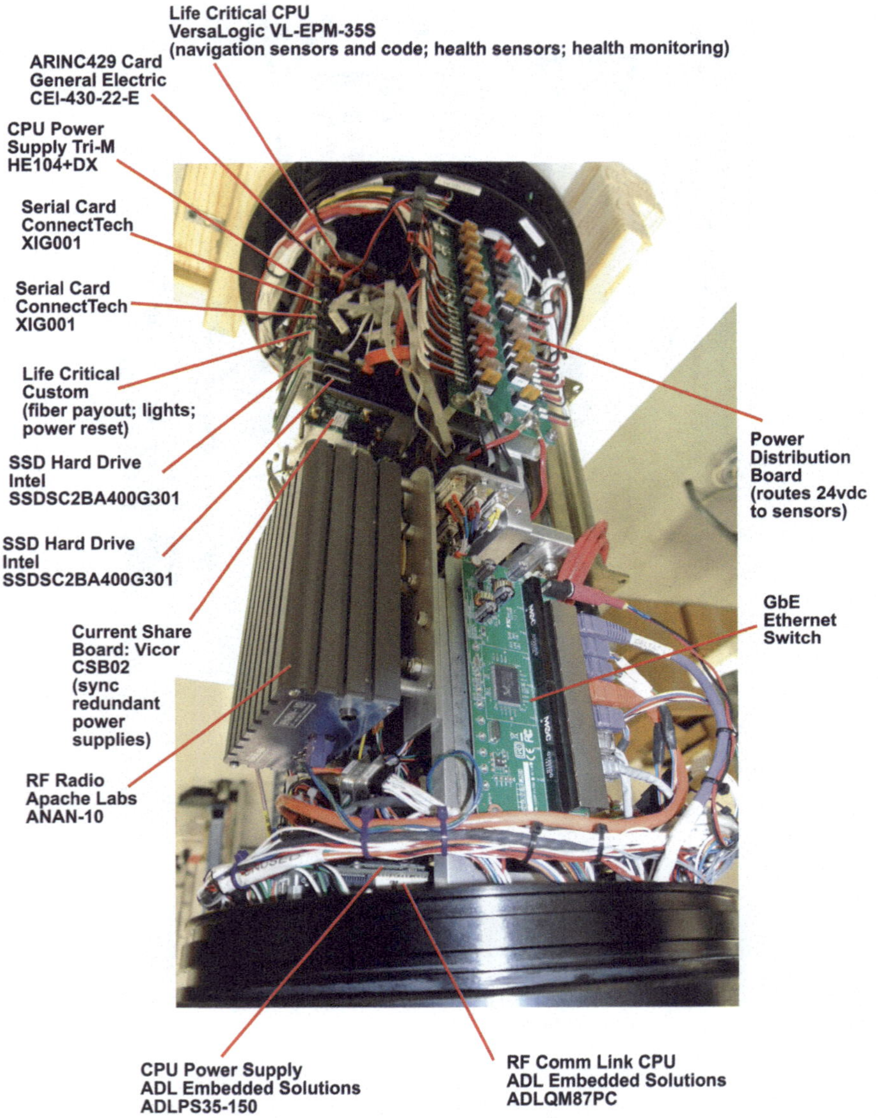

Fig. 10.40 Front view of the primary control module for ARTEMIS

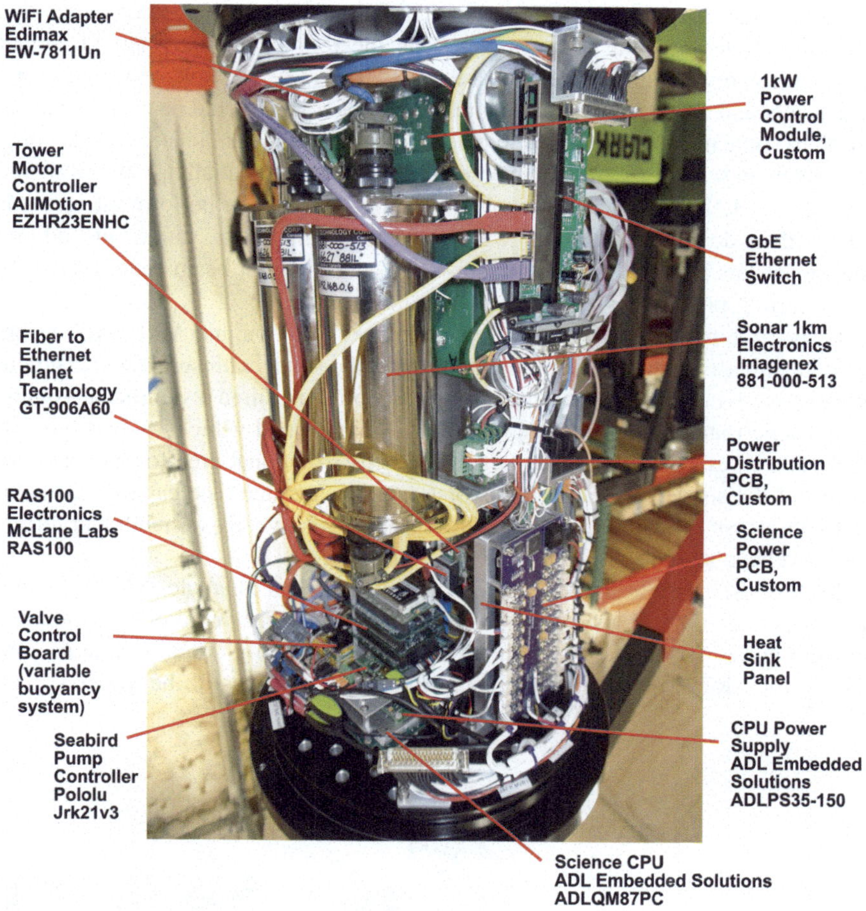

Fig. 10.41 Back view of the primary control module for ARTEMIS

10.3.3.1 ARTEMIS Vehicle Design

The primary design constraints on ARTEMIS included:

- deployment through a 1.3 m-diameter drilled shaft, with limitations on height imposed by an available Polar Haven field lab;
- a cruise range that allowed for relevant comparison with ice penetrating radar overflights;
- the ability to have the science instruments see undisturbed water during vertical profiling operations; and
- the ability to operate in close hovering proximity to the ice ceiling for protein fluorescence measurements.

At the time of the planning for the ARTEMIS mission there was conjecture that a freshwater lens would be encountered on the roof of the ice shelf (a similar situation can result with platelet ice—see Sect. 10.4.11). A vehicle cruising in −3 °C seawater and subsequently entering this fresh water layer could be expected to ice up. Thus some means was needed to place instruments against the ice while keeping the vehicle in seawater. The nominal vehicle configuration initially proposed was to "invert" the ENDURANCE concept of a vertical spooler to perform an upwards (rather than downwards) profile with an aqueous chemistry sonde. In addition several sensors were to be added: 1-km-range acoustic tight beam echosounders, a water current profiler, and a water sampling system.

Early on, however, two primary difficulties with the nominal ARTEMIS design were identified: (1) the floatation required to lift the sonde with sufficient tension to prevent overwrap or other tangling during respooling required a significant volume of syntactic foam, making packaging into the constrained vehicle dimensions difficult as well as presenting (2) a large sonde cross-sectional area for which simulations of the sonde performance in worst-case 0.3 m/s interface currents expected under the Ross Ice Shelf (Robinson 2004) indicated severely destabilizing effects on the vehicle's ability to station keep.

In response to these findings, we investigated several alternate vehicle geometries which could still meet the deployment and science requirements outlined above. Figures 10.42 and 10.43 show the two dramatically different options considered for ARTEMIS to meet the constraints and objectives of the mission.

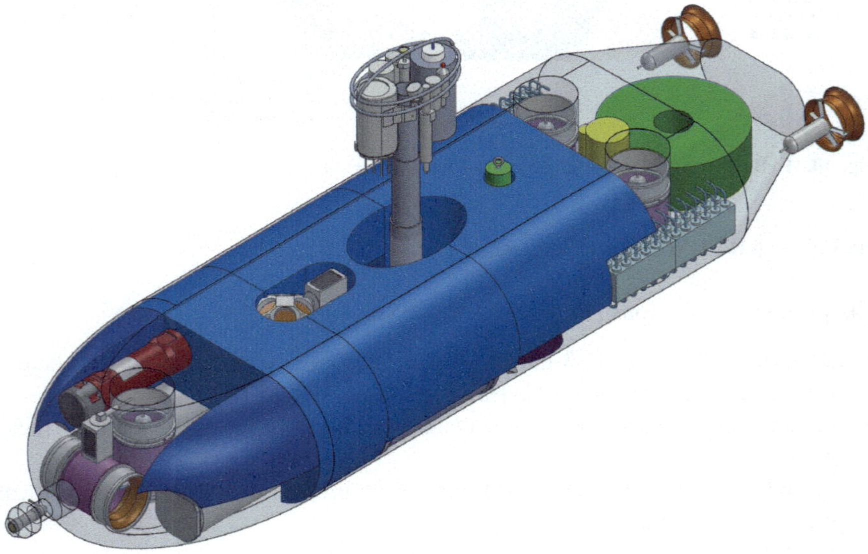

Fig. 10.42 Initial concept design for ARTEMIS. Green annular cylinder at the rear was a 15 km fiber spooler. This design used a mechanically more complex concept of an extensible/retractable sonde tower for under ice proximity operations

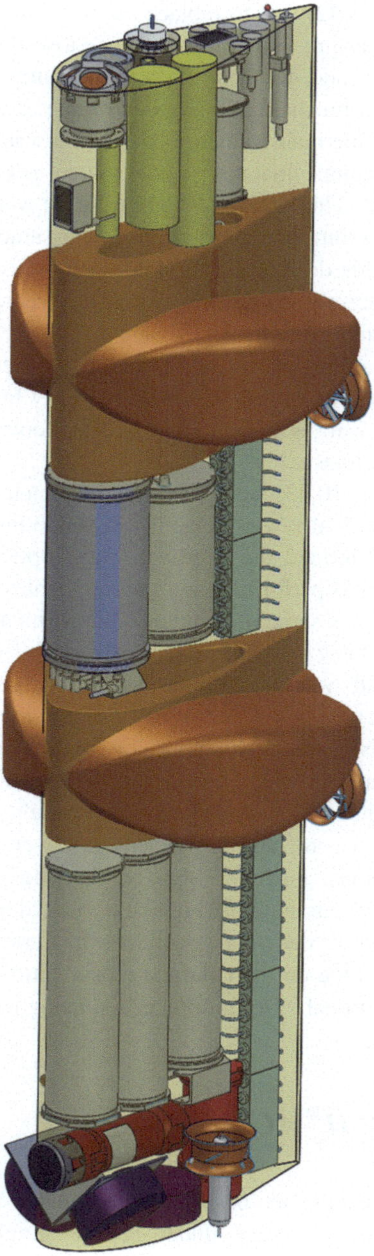

Fig. 10.43 Competing down-hole design for ARTEMIS known as the flying wing. Proximity operations instruments were mounted on top of the wing and propulsion units further down so as not to ice up in fresh water or platelet ice lenses under the ice cap

The first option (Fig. 10.42) was to replace the profiling spooler and cable with an extensible arm. This design, termed the "science tower" option, did require the entire vehicle to move up and down in the water column to perform the desired profiles, but allowed for a much smaller sonde volume as all flow-through instruments could be placed in the main vehicle body with a water inlet piped to the top of the tower, and only enough floatation was required to keep the minimal sonde roughly neutrally buoyant. This option also allowed the water sampling inlet to be piped adjacent to the flow-through sensors inlet, with samples being stored in the main vehicle body. For this design, the rigid connection to the vehicle meant that the sonde could be precisely controlled using the vehicle thrusters, allowing operations in any expected current. The extensible tower was required to keep thruster wash from contaminating the water being profiled, to keep the vehicle (super-cooled during transit in the −3 °C saltwater) away from any freshwater melt layer next to the ice, as well as to present a smaller footprint to contact the ice, which had unknown roughness.

The second option (Fig. 10.43) would essentially stretch and thin the DEPTHX/ENDURANCE design vertically while placing all sonde instruments on the top of the vehicle. This design, termed the "vertical wing" option, would have a large righting moment, and would perform all operations (deployment, transit, profiling, docking, and recovery) in the same orientation. This eliminated significant amounts of complexity in the vehicle: vehicle control would be significantly simplified (being essentially a 3-DOF vehicle); the intricate reorienting and re-ballasting for docking and recovery maneuvers required for the torpedo-type vehicles would be replaced with ENDURANCE-like vertical motion up a water-filled hole; and all the additional mechanisms and moving parts to support extension and retraction of a tower would be eliminated. This design cut the weight and component count estimates by nearly a third compared to the science tower option.

Analysis of both options, however, showed that the volume of components required for the ARTEMIS mission could not be packaged into a hydrodynamically efficient vertical wing. For all the weight, volume, drag coefficient, and complexity reduction, the total drag force on the enlarged cross-sectional area of the wing cut the vehicle range at least in half. Design proceeded using the science tower option.

10.3.4 The SUNFISH® AUV

Between 2010 and 2017 a compact, person-portable 6DOF AUV named SUNFISH was developed at Stone Aerospace under internal funding. It is a compact, light-weight 6-DOF controllable vehicle containing a basic sensor suite: inertial measurement unit (IMU), Doppler Velocity Logger (DVL), multi-beam acoustic mapper, conductivity-temperature-depth (CTD), camera and lights, and USBL acoustic beacon. It can perform a variety of missions: cruise survey operations, exploration and mapping of complex 3-D geometries, and precision inspection.

Fig. 10.44 Early prototype of SUNFISH being prepared for shipment to Antarctica to serve as an initial environmental test vehicle and later possible recovery vehicle for ARTEMIS

SUNFISH (Fig. 10.44) was our first attempt at miniaturizing concepts tested in DEPTHX and ARTEMIS. It was also deployed alongside ARTEMIS during development testing and under the McMurdo Ice Shelf, both to simplify development and validation of software behaviors on a logistically much simpler platform, and to provide a backup rescue capability should ARTEMIS require recovery. This prototype has also been used to develop autonomous exploration capabilities in unknown, unstructured environments such as underwater caves (see Fig. 10.45).

A primary purpose of SUNFISH is to extend exploration behavior and mission decision making beyond that of scripted missions to the point where the vehicle accepts only high level goal states and makes its own local decisions on how to achieve those objectives. One could argue that the 3D SLAM navigation test at Zacatón using the DEPTHX vehicle was a simple scenario—the cenote is basically a very deep vertical silo. But on 3 October 2016, SUNFISH became the first

Fig. 10.45 SUNFISH inside Peacock Springs, Florida, preparing for a fully autonomous exploration mission, October 2016

autonomous robot to truly cave dive, navigating its way through Peacock Springs, Florida—a 3D labyrinthine environment—picking its choice of passages with no human intervention, and producing a 3D map of what it explored. Such behavior-based high level autonomy will be the sine qua non of any sub-ice mobility system used on Ocean Worlds.

10.4 Enabling AUV Technologies for Ocean World Exploration

The above vehicles and their field missions gave us an opportunity to address many of the issues that will need to be confronted in the design of a new generation of swimming rovers destined for sub-ice missions to Ocean Worlds. Below, we discuss the most important of these, and show the steps that have been taken to advance this technology towards flight readiness.

10.4.1 Deployment Through an Ice Ceiling

An obvious requirement for any underwater vehicle traveling to an Ocean World in the outer Solar System is that it must be capable of deployment through the solid ice cover that protects the liquid water on any of these bodies. Most underwater vehicle work under thick ice has operated from open water at the ice edge—a luxury not available to Ocean World explorers. Section 10.2 of this chapter reviewed the few vehicles that have deployed through openings deliberately cut in the ice.

Due to the desire to re-use the existing vehicle architecture from DEPTHX, ENDURANCE required a 2.3 m diameter ice melt hole in the Lake Bonney ice cap —large enough to allow it to be deployed in its passively stable horizontal orientation while fitting its ellipsoidal geometry (see Fig. 10.46). For the relatively small thickness of ice (4 m), this was accomplished with standard portable "Hotsy" closed-cycle glycol melt system employed by the US Antarctic Program.

In general, on Earth, the deeper the access shaft the narrower the maximum diameter of the carrier cryobot and hence that of an internally-carried AUV. While a

Fig. 10.46 Down-hole deployment of ENDURANCE in West Lake Bonney, November 2008. The 2.3 m diameter hole was melted using diesel-heated glycol run through spiral heating tubes. In 2009 it took 10 days to melt the 4 m deep access hole

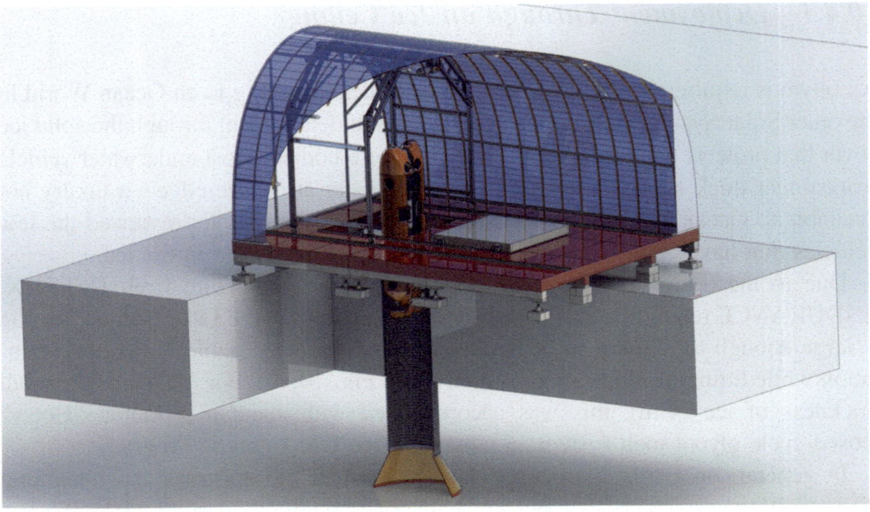

Fig. 10.47 ARTEMIS vertical launch and recovery concept

2.3 m diameter hole could be accommodated in Lake Bonney, the 1.2 m diameter limit on the ARTEMIS entrance diameter was set by available mechanical drilling equipment that could be deployed onto the McMurdo Sound sea ice (up to 8 m thick). Most hot water drills used to penetrate deeper sections of the Ross Ice Shelf are only capable of creating 0.3 m diameter shafts. Studies conducted during the SPINDLE project (see Chap. 4) indicate that an Ocean World cryobot will likely have a maximum diameter in the 0.4–0.5 m range with cryobot-only missions (no AUV deployment) likely targeting closer to 0.3–0.35 m diameter, owing to launch mass restrictions. So these are the numbers that have to be considered in the design of an AUV capable of exploring significant volumes of a sub-surface ocean.

The sub-ice deployment of ARTEMIS (see Figs. 10.47, 10.48, 10.49 and 10.50) required the vehicle to start in a vertical position in the hole and transition to horizontal for normal operations and cruise. Considering that the delivery cryobot is forced by physics to be a long, slender vertical cylinder the initial orientation of an AUV at sub-ice deployment will of necessity be vertical. However—as shown by the drag calculations for the ARTEMIS vertical wing concept discussed in Sect. 10.3.3—a vertical AUV is not suited for long range operations. Thus, one of the first engineering details for such a deployment will be to address pitch shifting the vehicle to a horizontal attitude following delivery to the ocean.

While a buoyancy-shifting mechanism was considered for ARTEMIS it was finally decided to use a weight hung from the rear of the vehicle during launch. ARTEMIS was designed as a 6DOF vehicle (meaning that it was pose-neutral and non-rotating in seawater but could assume any pose and maneuver in that pose) and it was possible to pitch up 90° at the conclusion of a mission once the vehicle was docked (see Sect. 10.4.6), and then pull the vehicle up the vertical shaft. However,

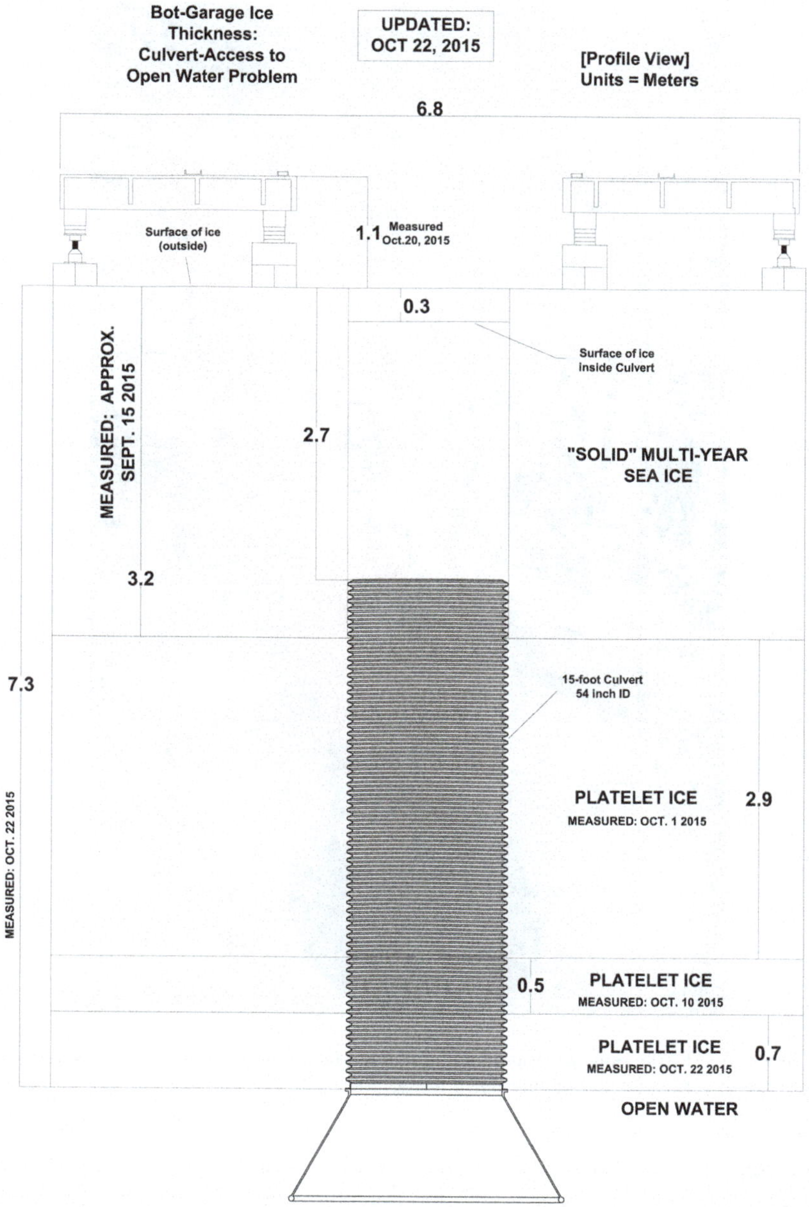

Fig. 10.48 Down-hole geometry for the ARTEMIS missions. Approximately 4 m of solid sea ice were underlain by 5+ m of semi-fluid platelet ice that would have constantly filled the access shaft. A corrugated tube was lowered through the access shaft to prevent platelet instrusion. A 4-part docking funnel, assembled underwater by divers, was used to give the vehicle an easier entry into the metal tube

Fig. 10.49 ARTEMIS launch consisted of lifting the vehicle from a horizontal stowed position, pitching up to vertical then dropping down the access shaft

on descent this was inherently unstable. Hence, a drop weight was used to avoid contact of the vehicle thrusters with the borehole wall. This weight was released from the vehicle (and retrieved to the surface) at the start of each mission. This guaranteed that the vehicle stayed vertical during descent through the borehole without requiring active control.

Fig. 10.50 Once in the water ARTEMIS was lowered down the hold on a loop of rope. A temporary ballast weight, attached to one of the rear stabilizer fins, was then removed by hoisting the rope it was attached to. The vehicle would then pitch down 90° to its horizontal cruise pose and the safety rope would be pulled up leaving the vehicle free to commence its mission

On a one-use remote deployment it may be possible to pendulum-stabilize the AUV in a horizontal pose such that once ejected from the cryobot, it naturally assumes that pose. Alternatively, a similar function could be accomplished through the use of a mechanically or chemically-triggered drop weight once the vehicle is ejected in a vertical pose.

As far as AUV recovery is concerned, it is assumed as we write this that deployment of an AUV in such a situation would be considered a one-time event and, while the vehicle will need to return to (and likely dock with) the cryobot for data transfer, there has not yet been any discussion of the necessity of retrieving the AUV from the ocean at the end of its useful life. Inclusion of full recovery and retrieval of the cryobot/AUV system to the surface at the conclusion of an Ocean Worlds mission would add significant mass and complexity to the mission. Such discussions will certainly take place to answer questions of planetary protection since both the cryobot and AUV will be nuclear powered. On Earth, where cryobots may prove to be superior alternatives for long-term sub-glacial access to traditional drilling (hot water or mechanical) techniques, there is no question of the desirability of recovery of the AUV system, both for cost effectiveness as well as return of samples autonomously collected by both the AUV and the cryobot (while in the ice column).

10.4.2 Navigation and Mapping Under Ice

Navigation for underwater vehicles poses several challenges beyond those encountered by land, sea, or space vehicles. Navigating under ice poses further significant complications, not the least of which is that, unlike oceanographic systems, one does not have the failure option of floating to the surface and being recovered in the event of a navigation error. An outer planet sub-ice ocean represents a remote, unstructured environment that will be the ultimate testing ground for unaided navigation. There will be no external navigation aids available—no equivalent of LORAN or GPS. Water quickly attenuates the electromagnetic radiation which forms the basis for most navigation systems above water on Earth. It is possible to set up acoustic beacon arrays (e.g. Austin 1984; Whitcomb 1999). However, their added mass, limited range, and the logistical complexity of deploying and calibrating them make this difficult and costly on Earth, and a complete non-starter for Ocean Worlds.

As a result of these environmental constraints, most underwater navigation is based on "dead reckoning"—the same compass and log-line technique as employed by mariners before the advent of modern navigation, except employing modern sensors. All of our AUVs have this type of navigation system as a core component. These navigation systems use as primary input sensor data from an inertial measurement unit (IMU) or higher grade inertial navigation system (INS, Fig. 10.51), a Doppler velocity log (DVL, Figs. 10.52 and 10.53), and pressure depth sensors (Figs. 10.53 and 10.54).

In a dead reckoning system (and any underwater navigation system in general), the vehicle orientation is obtained from the IMU or INS. This orientation comes from integrating angular rates from gyroscopes, and thus necessarily drifts as small measurement errors and biases in the rate measurements are added up over time. For a relatively low-cost fiber-optic gyroscope (FOG) such as employed on the SUNFISH® vehicle, this drift is $\sim 0.05°$ per hour in isothermal conditions. For a high-grade INS using stabilized ring-laser gyros (RLGs) as employed on DEPTHX, ENDURANCE, and ARTEMIS, the drift is $\sim 0.01°$ per hour. The IMU or INS also provides a measure of vehicle linear acceleration which can be integrated to estimate velocity, or double-integrated to determine position. Small tilt errors cause the velocity estimate to rapidly diverge from truth; the doppler velocity log (DVL) provides a way to correct these velocity estimates. The IMU or INS is used for position input only for brief periods (~ 10 s) in the absence of data from the DVL. A model of the vehicle dynamics taking as input the thruster RPMs can also be used to allow the vehicle to go for longer periods of time without DVL. Loss of DVL can occur if the vehicle is too far from any solid surface to get good acoustic reflection (from 60 m for the RDI Explorer DVL on the SUNFISH® vehicle to 500 m for the RDI PAVS-150 DVL on ARTEMIS). ARTEMIS employed two DVL instruments (Figs. 10.53 and 10.57) with the shorter-range instrument looking upward at the bottom of the ice cap and the long-range instrument looking downward to the floor of the Ross Sea. A DVL can also drop out (lose lock) if it is

Fig. 10.51 Ring laser gyro IMU used for DEPTHX, ENDURANCE, and ARTEMIS

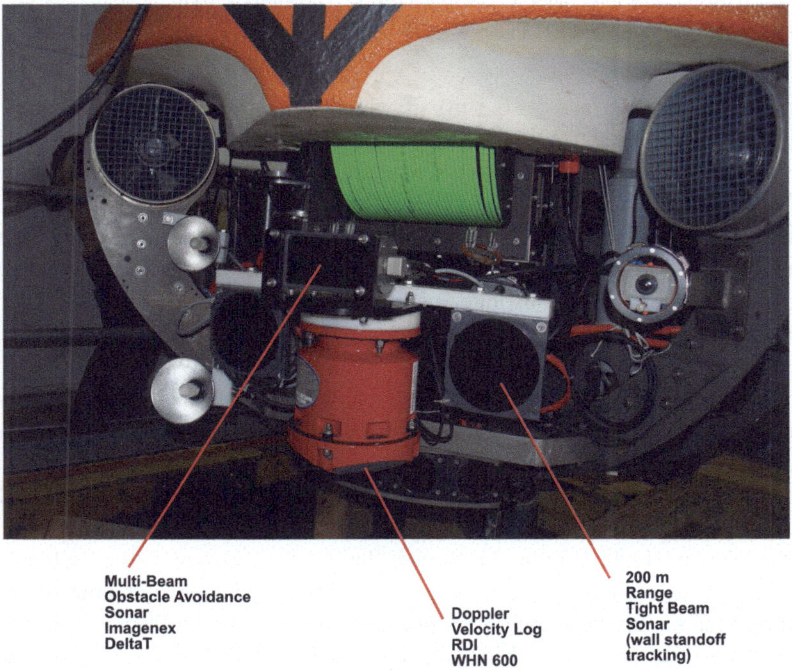

Fig. 10.52 ENDURANCE navigation, obstacle avoidance, and wall proxops sensors

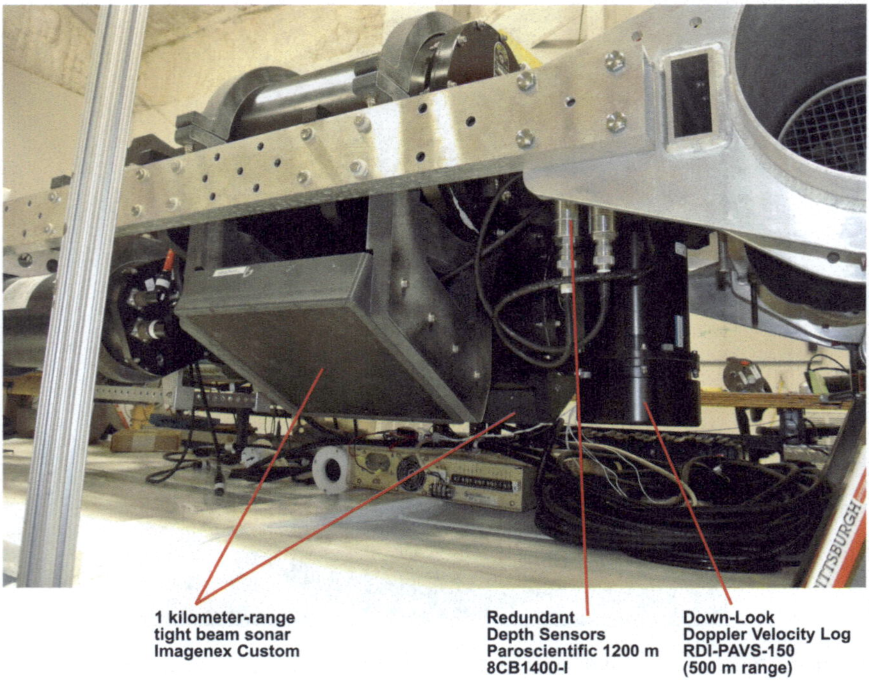

Fig. 10.53 ARTEMIS long range sensors included two down-look 1-km range tight beam sonar transducers and a 500 m range doppler velocity log

too close a target surface (~0.5 m) in which case the instrument transmit and receive channels are interfering. The Z-coordinate (depth) is best provided by absolute measurement from pressure sensors measuring hydrostatic pressure. All of our NASA vehicles have employed redundant high-accuracy depth sensors.

While such dead reckoning systems provide a reliable, continuous measurement of vehicle position and orientation, over the long term they suffer from a critical disadvantage: there is no absolute measure of horizontal position in the world incorporated into the navigation solution. This leads to a steady unavoidable build-up of position drift error as the vehicle moves about. With an INS and DVL rigidly mounted and referenced to the vehicle coordinate frame and an appropriate Kalman filter in use, error buildup can be bounded between 0.04 and 0.2% of distance traveled with smooth paths that return to the starting point tending towards the lower end of that spectrum, but it can never be absolutely bounded without dependence on vehicle trajectory length.

Under-ice navigation imposes another level of constraints on a vehicle and navigation system. Under a true remote ice cap, the only access to the outside world is through the deployment hole. Periodic surfacing to obtain GPS position updates to eliminate dead-reckoning drift as employed by many open-ocean underwater vehicles is out of the question. Deployment and calibration of acoustic navigation

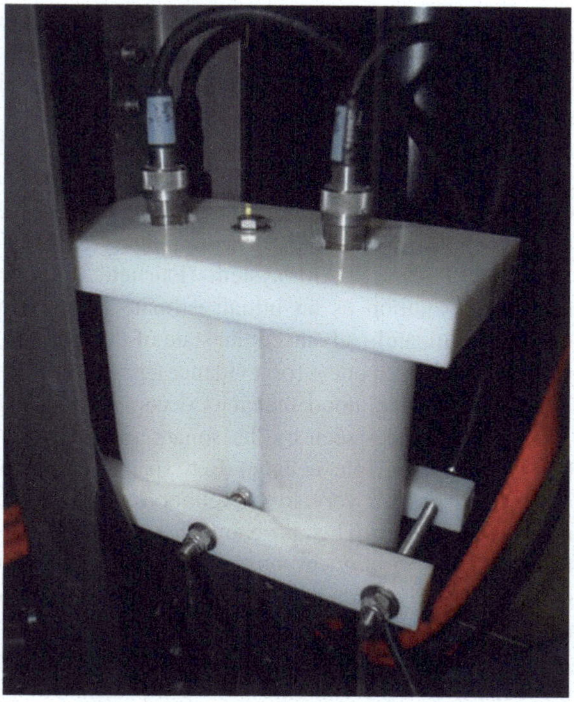

Fig. 10.54 Precision 60 m range depth sensors used on ENDURANCE

arrays around a theater of operations would consume an inordinate amount of vehicle power and payload. The vehicle must of necessity rely on its own capabilities—only what it can carry with it. External fiducial references will be limited to a single point at ice break-through. Below we will discuss two approaches that have been successfully implemented on our AUVs to extend and to some extent correct errors resulting from dead reckoning drift.

10.4.2.1 Simultaneous Localization and Mapping (SLAM)

DEPTHX was designed to develop capabilities to use the local 3D environment geometry as a position reference, developing significantly new capabilities in Simultaneous Localization and Mapping (SLAM). Here, the vehicle navigation matches the most recent data from mapping sensors to an internally-consistent map that is built up in real time as the vehicle moves. This allows it to localize itself within the map, and build the map as new areas are explored. For the acoustic ranging employed on DEPTHX, this form of navigation is particularly suited to environments with significant, distinguishable 3D structure. Obtaining range for measurements in all directions (true 4π steradian measurement) ensures that, while some ranges come from new, unexplored areas, there is always overlap with previously-seen terrain that can be used to place new information in the correct spatial relation. This forces a different solution than is typically employed on ocean-going AUVs where the predominant vehicle type is torpedo-shaped and the typical complement of sonars onboard consists of a downward or upward looking sidescan or multibeam unit and perhaps a forward looking wide beam obstacle avoidance transducer. The "great circle" acoustic sonar array design described in Sect. 10.3.1.1 provided the required ranging for the DEPTHX SLAM system. Caverns, cracks, canyons, fissures, spires or other high-relief terrain are ideal feature targets for this type of acoustic ranging SLAM.

The DEPTHX SLAM system (Fairfield 2009) employed the dead-reckoned navigation solution as an initial estimate for subsequent feature matching. This system significantly advanced the state of the art in SLAM. The system mapped the world as an occupancy (or evidence) grid—a 3D array of cells with each cell indicating the likelihood that it is occupied (Fairfield et al. 2005, 2006, 2007b). Thus solid surfaces seen by the sonars are labeled with a very high likelihood of being occupied while cells in free space (water) traversed by the sonar beam unhindered have a very low likelihood of occupancy. Unexplored areas have a neutral likelihood. This probabilistic representation allows for the noisy sonar measurements to be incorporated safely without becoming prematurely overconfident: a cell only takes on a definite occupied or unoccupied value once sufficient measurements have been made in it.

To determine the location of the vehicle *while* it is moving around in the map, the SLAM system combines the dead-reckoned navigation pose estimate with periodic comparisons of the latest sonar range readings with a hypothesized map. In fact, the combination of all possible vehicle trajectories and possible maps form one large hypothesis space, and it is the job of the SLAM system to determine which of these hypotheses best satisfy the (noisy) measurements taken by its dead-reckoning system and mapping sonars. The DEPTHX SLAM system actively tracked numerous (~ 100 s) hypotheses in what is termed a particle filter (Fairfield 2009), selecting the one having the greatest likelihood of being correct given all measurements of vehicle motion and sonar ranges, and knowledge of the error profiles of those sensors. The occupancy grid representation developed for DEPTHX was particularly suited to efficiently storing multiple, partially overlapping map hypotheses, and quickly comparing them to current sonar measurements.

This system allowed DEPTHX to autonomously navigate into and map previously unexplored hydrothermal springs (very deep cenotes) in northern Mexico. DEPTHX was able to move safely through this complex 3D environment with full knowledge of its surroundings. This is a fundamental capability that other more advanced behaviors (such as those described in Sect. 10.4.3) depend on. Figure 10.63 shows a single frame replay of the DEPTHX SLAM engine operating within Cenote Zacatón. Light purple conical rays represent discrete sonar pings that have resulted in confirmed range measurements during this particular real-time test cycle; yellow colored areas are voxels representing the highest probability for the presence of a true geometric boundary. The scale of this 3D map is 140 m left to right and 90 m top to bottom. The DEPTHX vehicle pose estimate (one of up to 400 estimates in this computational cycle) is shown at the convergence of the sonar beams, to scale, as an orange ellipsoid.

Importantly, 3D SLAM works best in "feature rich" environments. If there are no features—such as an endless almost-flat lake bottom—SLAM degrades and the navigation solution will be dominated by the dead reckoning solution. Furthermore, while the algorithm is elegant and powerful, it is also computation- and memory-hungry. A particle filter may utilize hundreds of 3D maps (each consisting of potentially tens of thousands of evidence grid voxels) representing possible world geometry at a particular location and orientation relative to the vehicle. The computing system must then determine which vehicle pose relative to those maps presents the least estimation error, and do this at a real-time update rate. The slower the update rate the faster the rate of true vehicle position error propagation. These computational round offs are analogous to that of vehicle position estimation error propagation from dead reckoning, such that an un-closed trajectory computed from SLAM will eventually drift from reality as well. But loop closures in the map will vastly improve accuracy and it is this feature that differs from a purely proprioceptive solution based on dead reckoning, which will continue to propagate unless some external correction is applied (which will not be possible on Ocean Worlds).

This immediately suggests a possible strategy for Ocean World exploration and navigation: start exploration where unique geometry exists. This will occur at two locations: the underside of the ice cap and the bottom of the lunar oceans. Given that it is the discovery of life that is driving the exploration of the outer planet moons the best strategy will thus be to utilize a servo-spooling cryobot deployment from the ocean breakthrough point (bottom of the ice cap) and proceeding directly to the bottom of the ocean floor (see Chap. 4). The AUV should be released there, where it can then radially build a SLAM map of the world, increasing the radius of the map with each foray out from the cryobot. A unique localization within the refined map—corrections to which can now be performed using standard 3D least squares survey loop closure algorithms (McKenzie 2016)—represents an absolute position correction (the equivalent of an INS position solution receiving a periodic GPS update for commercial aviation on Earth). The map can then be expanded radially indefinitely (for nuclear powered AUVs) until, in theory, the entire floor of the lunar ocean is mapped, and within which the likelihood of the discovery of a hydrothermal vent would be high. The same operational scenario could be used to explore the underside of the ice cap, but since there are no known energy sources there the probability of finding life there will be significantly less than in the vicinity of a hydrothermal vent on the ocean floor.

The SUNFISH® AUV brings new capabilities for SLAM using a standard multibeam sonar (see Figs. 10.65 and 10.66) in a much smaller, more maneuverable package than DEPTHX. A multibeam sonar provides a "fan" of sonar ranges over a defined angle. The Imagenex DeltaT 837B multibeam installed on the vehicle provides a fan of 120°, with beam spacing of 1°, 0.5°, or 0.25° (i.e. 120, 240, or 480 beams). The maximum range is 120 m, with a range resolution of 0.2% of maximum. The maximum range can be set to a smaller value (down to 5 m) to get higher range resolution. Though it provides high beam density within the swath covered by the fan, this sensor does not inherently provide the 4π steradian coverage required to allow for effective SLAM. Instead, the SUNFISH vehicle uses its maneuverability and 6-DOF control to physically scan the sensor, providing the forward-backward coverage required. The precise nature of this maneuvering can take various forms, primarily depending on the nature of the 3D geometry being mapped. For open-water operations, the DeltaT is placed in the "vertical fan" configuration, and the vehicle can periodically stop and yaw 360° to scan an area while maintaining DVL lock on the bottom to ensure a good base dead-reckoning solution.

The 4π steradian SLAM implementation on DEPTHX was a state of the art algorithm at the time it was developed but had several issues that prevented it from being used to explore large scale spaces in a robust way (e.g. CPU and memory requirements, limits on map size, particle depletion). The SLAM algorithm for the SUNFISH® AUV was redesigned to address these issues. The new algorithm registers scans using a featureless correlation-based method and pose graph optimization with a robust backend to stitch together the individual registration results with dead reckoned position estimates. This type of algorithm is more robust to gross errors in dead reckoning (it is not susceptible to particle depletion) and is able to generate global maps of arbitrary size (Cadena 2016; Hidalgo 2015; Ribas 2008; Mallios 2014).

Sonar data is significantly noisier than terrestrial sensors (camera, laser) and must be filtered before it can be used for obstacle avoidance or registration. The SUNFISH® AUV sonar filtering algorithm works as follows: Range estimates from the sonar are projected into a common frame (centered at the scan origin with axes aligned to north, east, down) to obtain a point cloud and a list of beam vectors that did not hit a target (we call these far range measurements). As the vehicle scans the area, measurements are stored in bins that are evenly distributed on a sphere, and each bin is median filtered to produce a spherical range map. The advantage of this representation is that it avoids any map projection error, facilitating movement in any direction without bias.

The SUNFISH® AUV uses a correlation based registration algorithm to match scan pairs as a basis for SLAM. In underwater environments featureless registration methods are attractive since sonar is not as precise or information rich as sensors used to explore environments located in air or vacuum. The algorithm works as follows:

- Project the spherical map into a voxel grid. While there are various ways to do this (e.g. occupancy, log likelihood, TDSF), we found that it is attractive to use a zero-sum voxel grid to speed up computation for our approach.
- Each voxel grid is converted into the spectral domain using a 3D discrete Fourier transform (DFT) and one is aligned to the other along the depth axis using sub-pixel image shifting (vehicle depth is well known using pressure sensor data).
- Each shifted voxel grid has the inverse Fourier transform applied to its depth axis to result in a stack of 2D DFT images. The Fourier-Mellin transform is used

to calculate cross correlation of each slice of the stack. The cross correlation for each stack is summed together to obtain cross correlation of the voxel grid pairs.

- The correlation maximum is determined for yaw angular offset. The maximum is refined to sub-pixel accuracy by calculating a high resolution cross correlation in a 1.5 pixel region around the maximum and determining the up-sampled image maximum.
- One voxel grid image is aligned to the other by rotating it (and scaling if the voxel grid resolutions are different), the stack of 2D DFT images are re-calculated for the transformed voxel grid.
- The correlation maxima is determined for north and east translational offset. The maximum is refined to sub-pixel accuracy by calculating a high-resolution cross correlation in a 1.5 $\times$ 1.5 pixel region around the maximum and determining the up-sampled image maximum.

For our implementation, each registration requires on the order of 0.5 s to execute on modern hardware (4th generation Intel i7 processor), allowing real-time operation. Registrations are run on parallel cores for additional speedup. Each scan is registered against some of its nearest neighbors and vehicle path is optimized using pose graph optimization with a robust backend (switchable constraints). Registration for the datasets we have looked at so far result in a 60–70% registration success rate, however this method produces a significant number of outlier measurements. Pairing this registration method with a robust graph optimization backend prevents outliers from corrupting the map, resulting in a robust SLAM algorithm that is able to run in real-time.

For operations in more confined geometries such as caves and tunnels, maintaining DVL lock is generally not an issue, and the SUNFISH vehicle is more free to scan. In such applications, the DeltaT is placed in the "horizontal fan" configuration, and scans by continuously rolling (Fig. 10.65) while transiting. By controlling the roll rate and forward velocity, this behavior can ensure an overlapping helical sonar swath providing forward-backward coverage in tunnel-like environments. In less tunnel-like environments, the vehicle can periodically stop and perform a 4π steradian scan similar to the open-water behavior. This scan-matching SLAM provides a corrected point cloud as a map, which allows for a wide variety of higher-level behaviors such as inspection and identification, wall-following, obstacle avoidance, path planning, and autonomous exploration (see Sect. 10.4.3) to work with the map and perform the custom processing they require.

10.4.2.2 Single-Beacon (iUSBL) Navigation

For AUVs deployed through a hole bored through the ice—e.g. by a cryobot—the presence of engineered hardware at the cryobot/dock location allows for the employment of a navigational aid consisting of a pair of acoustic transceivers on the dock and on the vehicle forming part of an Ultra-Short BaseLine (USBL) system. A USBL consists of one phased-array transceiver which pings an omnidirectional

Fig. 10.55 Omnidirectional USBL transponder used for both ENDURANCE and ARTEMIS missions. This was suspended at the centerline of the melt hole approximately 5 m below the roof of the ice cap for ENDURANCE. For ARTEMIS, the beacon was attached just above the lighted docking rod, also on centerline of the drilled access shaft

transponder beacon and determines its range and bearing by looking at the time of flight and signal direction of the response. In addition, both pings may be modulated to also carry (relatively low bandwidth) data as an acoustic modem. In a standard USBL configuration, the (smaller) transponder beacon (see Fig. 10.55) is placed on the vehicle and its location is tracked from the stationary transceiver.

In an "inverted" (iUSBL) configuration, the vehicle uses the transceiver to track a stationary transponder beacon (e.g. located at a dock or recovery point). This sort of absolute position reference is useful in areas that have unvariegated physical topography (such as the essentially flat ice roofs faced by ENDURANCE and

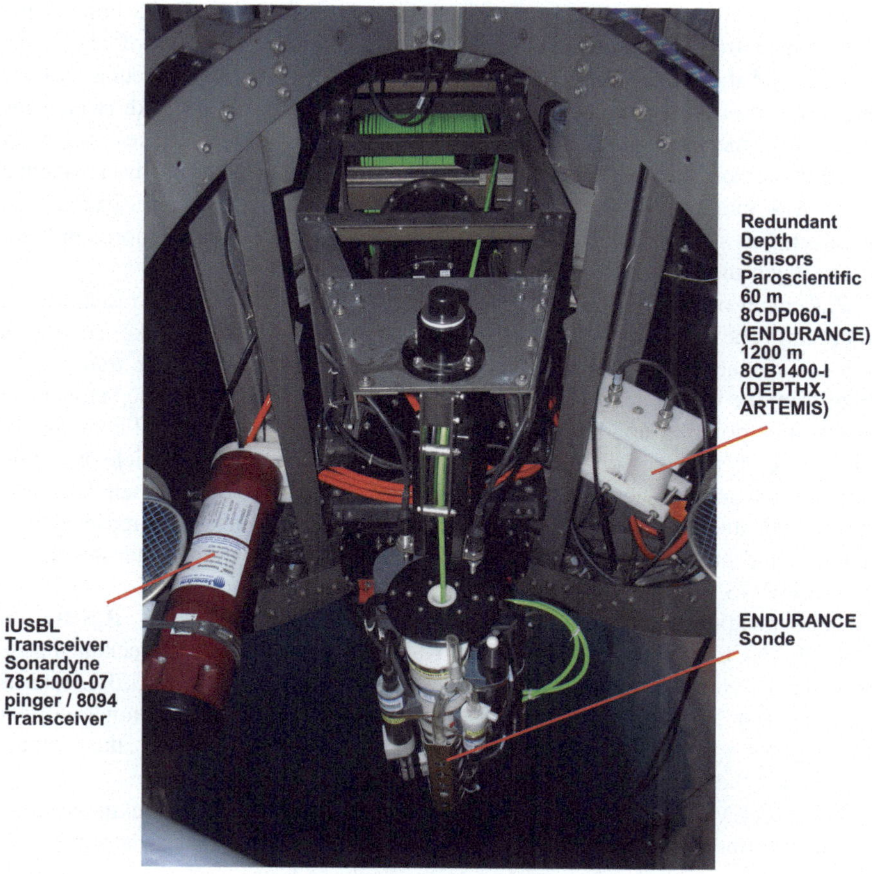

Fig. 10.56 Inverted iUSBL transceiver mounted inside the sonde bay on ENDURANCE. The black transducer element on the front end of the maroon-colored transceiver must be approximately pointed in the direction of the transponder in order to receive signal. Both ENDURANCE and ARTEMIS used a routine yawing search maneuver to locate the vector of maximum signal during return for rendezvous and docking

ARTEMIS) or complex sonar environments (such as the beam-bending Lake Bonney halocline faced by ENDURANCE) which make SLAM unreliable. They are also particularly useful in docking (McEwen 2008; Bellingham 2016), as the position error is proportional to range from the beacon, and thus navigation becomes more accurate as the vehicle approaches the dock. Both ENDURANCE and ARTEMIS employed such a system. ENDURANCE used a Sonardyne 8094 iUSBL system on the vehicle to track the transponder (Figs. 10.55 and 10.56), which was hung from a line at one side of the access shaft. The iUSBL returned measurements of range and bearing to the transponder beacon at a relatively low update rate (every 2–5 s).

The Lake Bonney environment was particularly challenging. Reflections of the sonar pings from a strong halocline below the vehicle, the ice roof above the vehicle, and the nearby shallows and glacier face all affected the accuracy of the measured position. Fortunately, for operation in the center of the lake (where the deployment melt hole was located), both the halocline and ice surfaces are nearly flat and horizontal, so that deflection of the USBL beam paths in the horizontal plane was minimal. The vertical component of USBL position measurements, however, was strongly affected, but this component could be safely ignored in favor of the much more accurate pressure-depth measurements.

Placement of the transponder beacon in this environment was also found to be important. The transponder field of view is a full hemisphere. In Lake Bonney, it was found that the optimal iUSBL configuration is one in which the transponder hung from the melt hole just above the halocline at 7 m depth, with the hemisphere central axis pointing up. This configuration gave full 360° horizontal coverage of the vehicle operational volume and maintained a good view of the vehicle operating near the ice roof. When the vehicle came to within 5 m of the melt hole, the transponder and line would be retrieved to prevent snagging. This retrieval procedure could be avoided if employing the type of docking system later developed for ARTEMIS (see Sect. 10.4.6).

This position (range and bearing) measurement from the iUSBL on ENDURANCE was combined in a linear Kalman filter with the dead-reckoned pose estimate to reduce accumulated dead-reckoning drift. Range-only USBL returns were ignored. The error in the USBL measurement is range-proportional, so that this system guaranteed progressively better position updates as the vehicle approached the melt hole during egress.

For ARTEMIS, the iUSBL formed a critical component of the docking system, which is further described in Sect. 10.4.6. The iUSBL provided an absolute reference of the vehicle position relative to the dock, and was used to eliminate drift error accumulated by dead-reckoned navigation over the course of a mission. The ARTEMIS dead-reckoning system was based on the same sensors used for DEPTHX (see Sect. 10.4.2.1), and the same iUSBL system as employed by ENDURANCE (see Figs. 10.57 and 10.58). Here, the USBL measurement was first filtered to eliminate multipath and other outliers before being used to update the navigation Kalman filter (Kimball et al. 2018).

Figure 10.58 shows a schematic of the overall ENDURANCE multi-stage navigation system. The final stage, described in Sect. 10.4.6, used a collimated down-hole oscillating light source as the target for a machine vision system

In the field under the McMurdo Ice Shelf, the ARTEMIS iUSBL reliably returned position updates from up to 500 m range from the dock, and up to 800 m under some conditions. Incorporating these position hits into the navigation

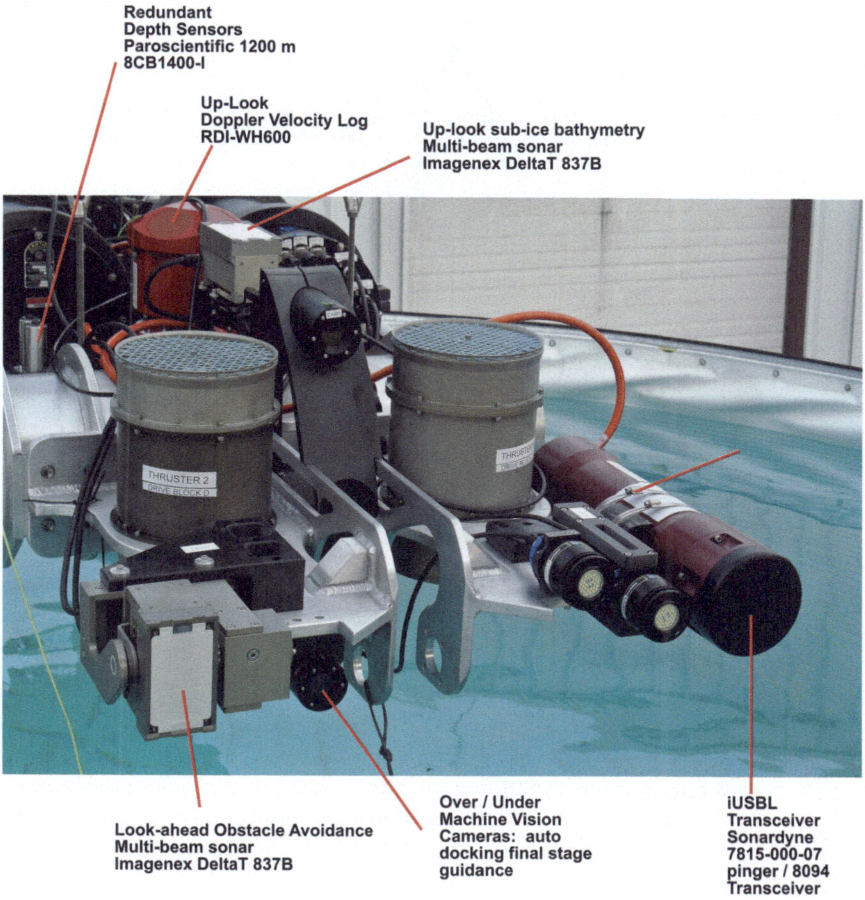

Fig. 10.57 View of the bow of ARTEMIS showing iUSBL transceiver, redundant machine vision docking cameras, obstacle avoidance multi-beam sonar and other navigation instruments

solution caused the error in the vehicle position estimate to decrease rapidly. In fact, the system fielded on ARTEMIS was shown to be able to reliably bring the vehicle back to the dock location even in the presence of large errors introduced into the dead-reckoned position estimate, into the position estimate confidence, and even into the physical alignment of the iUSBL transceiver head on the vehicle. The ability of the iUSBL-aided navigation system to achieve docking does not depend on perfect tuning and calibration. The final approach control for auto-docking with ARTEMIS was controlled by redundant machine vision cameras (Fig. 10.57) that will be discussed in Sect. 10.4.6.

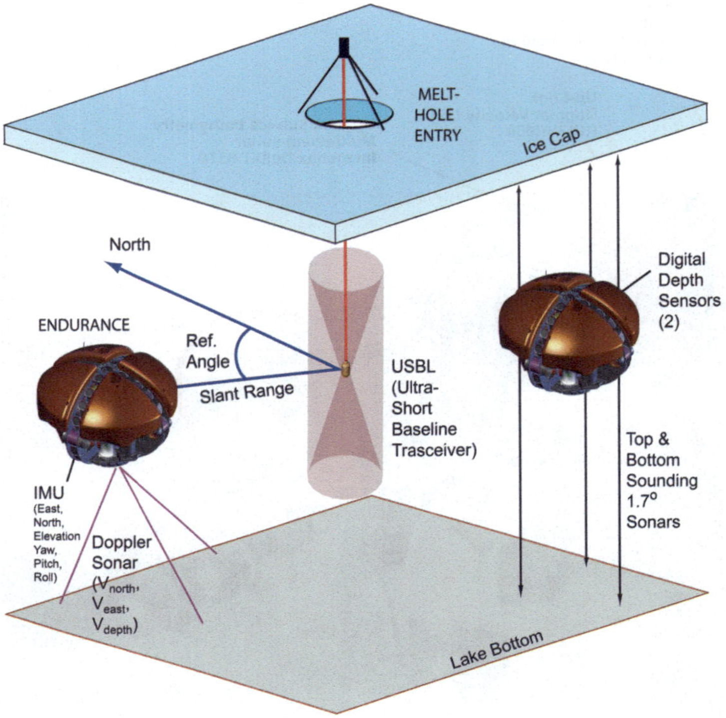

Fig. 10.58 Concept of operations for ENDURANCE, showing 3-stage egress navigation elements: dead reckoning sensors; iUSBL system; and collimated, oscillating light source shown down the access hole for an up-look machine vision camera on the apex of the vehicle to acquire for final docking

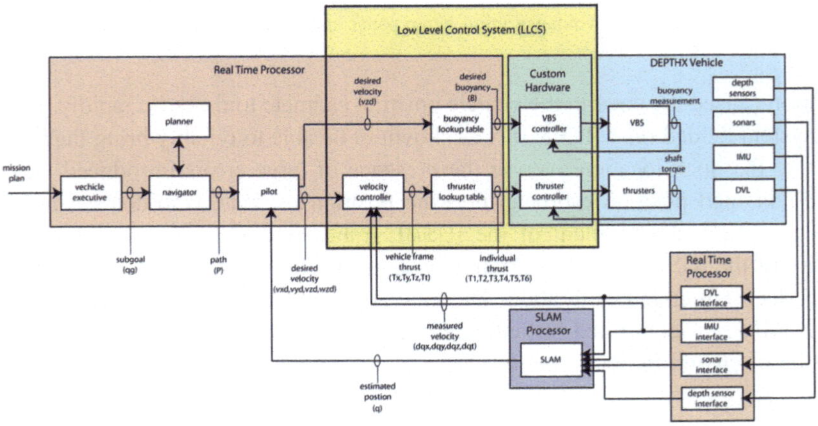

Fig. 10.59 Control system diagram for DEPTHX, illustrating SLAM integration into the navigation Kalman filter

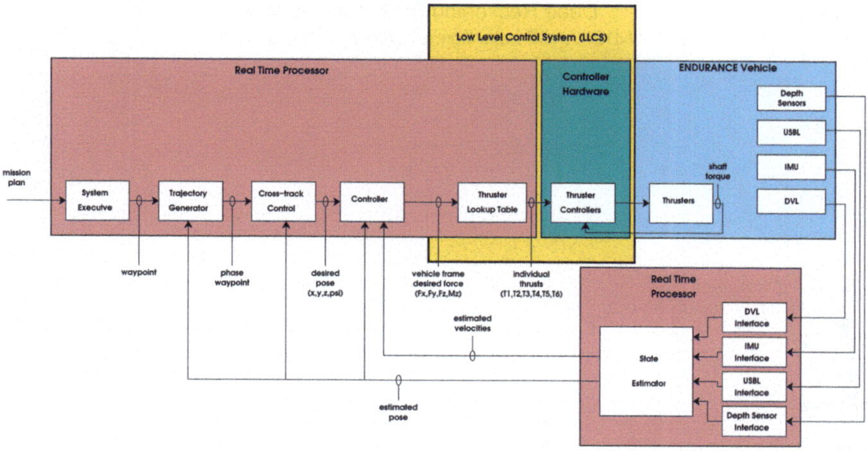

Fig. 10.60 Control system diagram for ENDURANCE, illustrating iUSBL integration into the navigation Kalman filter

The overall control system architectures for DEPTHX and ENDURANCE are shown in Figs. 10.59 and 10.60, respectively. General long term accuracy for the systems described above is shown in Fig. 10.61. The dead reckoning system error buildup is a function of range traveled and the solution is dominated by the quality of the DVL instrument. Absolute error in true position from an iUSBL system increases with range. For an automated sub-ice rendezvous and docking system to function successfully, the dead reckoning solution must be at least good enough to get the vehicle with range of the USBL transponder (up to an 800 m radius). At this point the iUSBL solution can bring the vehicle close to the capture point. A third stage of control is normally required at this point for fine approach and docking.

10.4.3 Autonomous Exploration

The communications delays and bandwidth limitations to Earth for any vehicle operating in ice-covered waters on Ocean Worlds implies a vastly more advanced autonomous capability from that practiced by state-of-the-art planetary rovers, such as those currently on Mars which rely on frequent high-bandwidth communications with Earth to plan out and verify their every move. In particular, such vehicles will need to be able to explore unknown, unstructured environments with only minimal goal-level external direction.

While the sub-ice navigation technologies such as 3D SLAM and USBL-aiding form a critical prerequisite for exploration, they only tell the vehicle where it is and where objects in its environment are. They do not tell a vehicle where to go: which

(a)

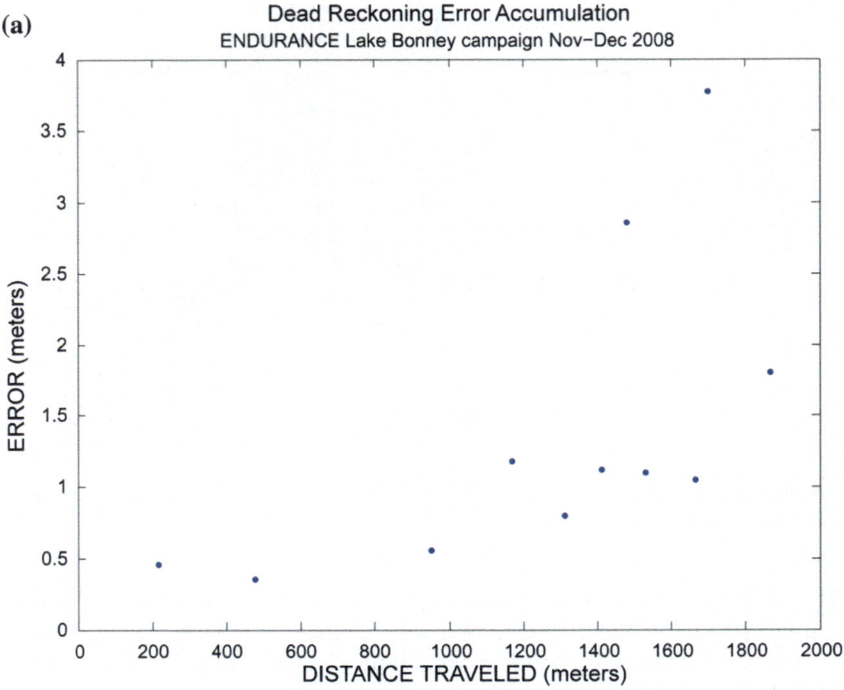

(b)

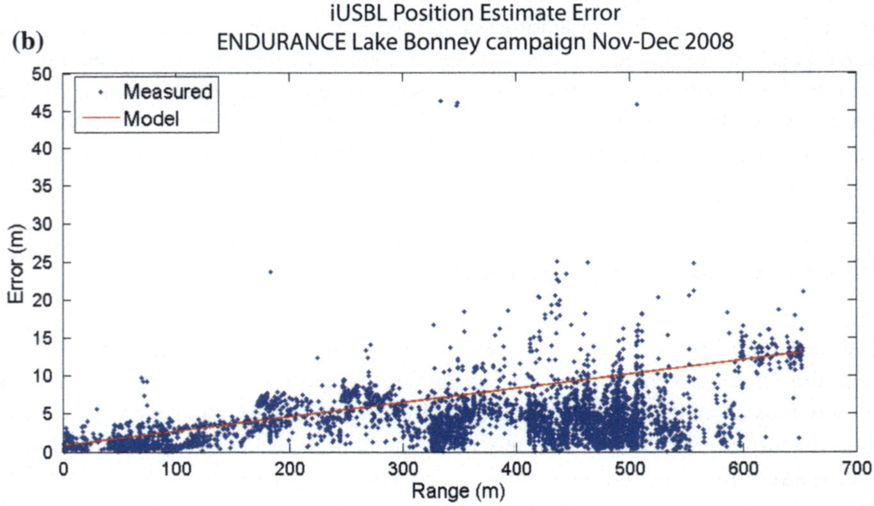

Fig. 10.61 Typical navigation accuracy achieved on ENDURANCE missions for a pure dead reckoning solution (top chart) and for a pure iUSBL solution (bottom chart)

areas may be dangerous and which are most fruitful for gaining new information about the world. These latter tasks are the domain of autonomous exploration.

Autonomous exploration has been investigated in more terrestrial contexts. Here, wheeled platforms with scanning laser ranging systems are able to take advantage of a very stable, high-precision sensing environment. As such, several autonomous exploration systems have been developed based on the concept of moving to the "frontier" between explored and unexplored areas using laser scanners in university campus environments (e.g. Gonzalez-Baños and Latombe 2002; Joho et al. 2007).

10.4.3.1 DEPTHX

While much of DEPTHX development was devoted to developing and demonstrating its mapping and navigation capabilities, it also demonstrated completely autonomous exploration. This was effected by instructing DEPTHX to avoid walls and obstacles seen by the sonars while completing a notional trajectory broadly described by high level objectives (e.g. "explore; attempt to stay in the middle of the world; do not approach obstacles closer than 10 m distant unless entering proxops; do not descend deeper than 500 m" etc.). Violation of any of those simple

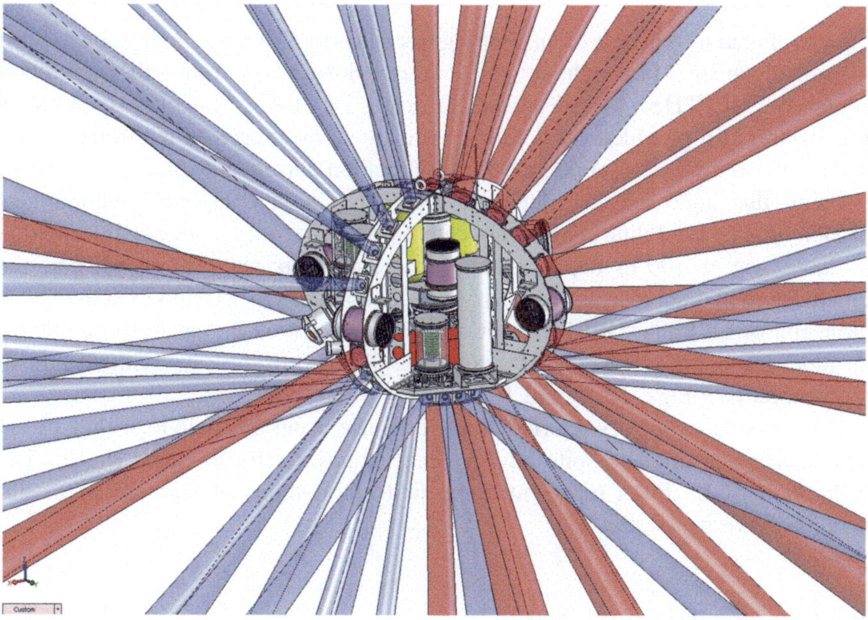

Fig. 10.62 Computer model of DEPTHX showing simulated sonar firings from its 56 discrete tight-beam acoustic transducers. Red beams are from 200 m range sensors; blue beams are from 100 m range sensors. In addition to providing primary geometric input to the SLAM engine these data were also used as input to behaviors dealing with obstacle avoidance, approach to surfaces, and for standoff control during proximity operations (life search and sampling)

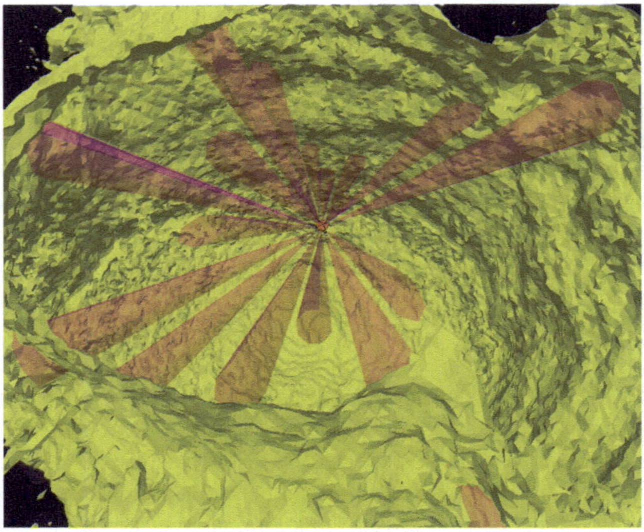

Fig. 10.63 Evidence grid 3D map of Cenote La Pilita constructed by DEPTHX. Valid sonar returns from the current scan cycle are shown as red cones emanating from the vehicle. Not all transducers received usable return signals

imperatives led to issuance of an abort egress behavior. This was a step above what is being done on Mars right now, but it is a long ways from full exploration autonomy. DEPTHX was designed to test 3D SLAM, but its sonar array (Fig. 10.62) also provided extraordinary situational awareness in real-time, particularly if the vehicle was set into a slow yaw rotation during an exploration mission.

Using this approach, we deployed DEPTHX on a completely autonomous mission in the La Pilita cenote (no data fiber). The only *a priori* information available was a 100 m depth reading that had been obtained by divers; the remainder of the submerged geometry was unknown. On its autonomous exploration run of La Pilita, DEPTHX not only simply verified the depth to be 100 m, but more importantly constructed a complete 3D map of the cave without human intervention (see Figs. 10.15 and 10.63).

The robot subsequently returned to the entrance using this map to within 15 cm of its starting position, despite having been underwater for as much as 4 h. The exploration of La Pilita represented the first autonomous exploration of a submerged cavern by a robot.

10.4.3.2 The SUNFISH® AUV

On SUNFISH, the scans employed by the SLAM system described in Sect. 10.4.2 also provide the basis for autonomous exploration behaviors. Obstacles are enlarged in the spherical range map by a safety radius to obtain a safe obstacle avoidance map. The enlargement operation can be performed in real-time by pre-calculating angular distance from a bin to each of its neighbors and storing it in a lookup table. Each range in the obstacle avoidance map indicates how far the vehicle can move in a particular direction before it will collide with an obstacle. Simple exploration behaviors can be developed using this map, for example moving in a random direction (to increase the size of knowledge of the world) or moving in a desired direction as far as possible, or choosing a path that represents the least cluttered option or, e.g. the largest tunnel within a 3D labyrinth (e.g. Fig. 10.64). Without a global map representation exploration behaviors like this will either take too long to explore an area or get stuck in a local minima. Fully autonomous exploration requires the use of the globally registered map paired with a global exploration algorithm.

The SUNFISH® AUV has demonstrated capabilities to autonomously explore complex 3-D environments. In the examples shown in Figs. 10.64, 10.65 and 10.66, SUNFISH was tasked with exploring into a branching underwater labyrinth. The goal state was to move as far as possible in a generalized direction (towards a distant UTM global coordinate) while choosing the least cluttered path (i.e. the largest choice if a junction were to be encountered—as it has in Fig. 10.64) while not violating any obstacle avoidance allowed proximity limits. This is quite different from simple obstacle avoidance. A map has to be constructed and a decision has to be made to advance the vehicle towards a geographic goal within a complex environment—and there is no assurance that a tunnel that is chosen continues. In that case the vehicle would have to return to a decision point and make a 3D

Fig. 10.64 SUNFISH commences a roll-scan maneuver at a tunnel junction in Peacock Springs, Florida in October 2016

trajectory decision as to how to advance the exploration in the most desirable manner. In the image shown SUNFISH is configured with a single 480-point, 120° fan multi-beam mapping sensor. Since the vehicle is able to maneuver in six degrees of freedom, however, it first completes a 360° roll maneuver (Fig. 10.65) to produce a laterally looking map of the immediate world (Fig. 10.66). This is insufficient, though, for decision making as what lays ahead remains unknown. The vehicle then performs a 90° yaw maneuver followed by a 360° roll, then a yaw back to its original heading. With this new data added to the map (Fig. 10.66) the vehicle is now able to select the larger (right hand) tunnel to proceed onward. Such behaviors are attitude independent—the vehicle might have detected a large dome in the ceiling and gone up to investigate it had that been the largest way onward and generally advancing toward the global desired end coordinates. Alternate maneuvers and sensor orientations can be used to similarly create a zone of knowledge about the vehicle that then forms the basis for making the next move.

10.4.3.3 Future Work

The above examples are exceedingly limited in scope, but they do give a glimpse of the paths that must be followed if we are to insert autonomous mobile systems into completely unknown 3D "overhead" environments. There is clearly a great need for expanding the capabilities of autonomous rovers to perform autonomous

Fig. 10.65 Following a roll maneuver in the scene illustrated in Fig. 10.64 a segment of the tunnel surrounding the SUNFISH vehicle is populated with map data. But this still leaves insufficient information for a forward exploration decision

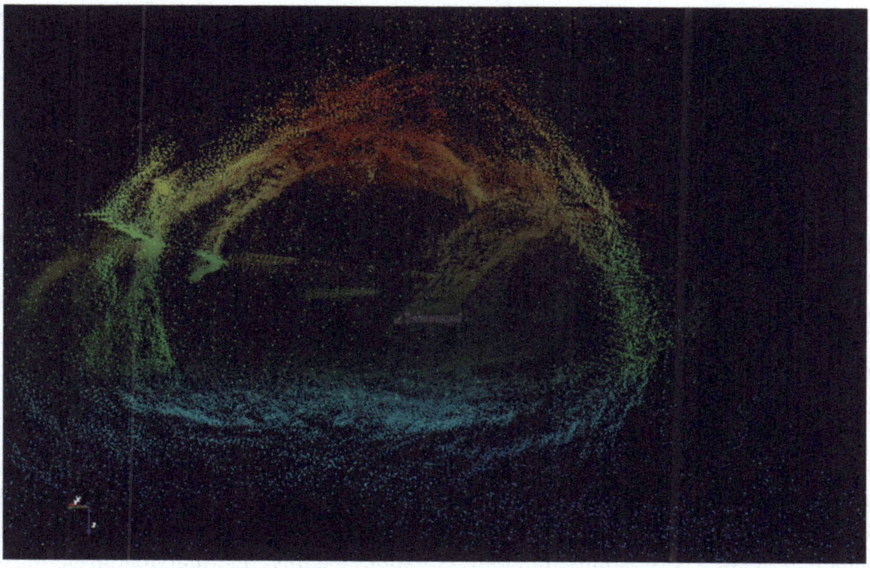

Fig. 10.66 The vehicle then yaws 90° and completes another 360° roll maneuver, creating a sufficiently

exploration in order to enable a successful Ocean Worlds swimming rover mission. Certainly porting some of the exploration techniques developed for terrestrial platforms referred to above to the less stable, noisier underwater environment should be investigated. It is likely that the constraints of this environment will bring about breakthrough developments in exploration autonomy. Communications abilities in such scenarios are simply far too limited and the environment is too unknown and possibly dynamic for the types of either highly supervised, slow single-stepped motions that are state of the art for Martian planetary rovers, or the fully pre-computed, pre-scripted flybys of planetary probes such as Cassini, Dawn, or the planned Europa Clipper. Current terrestrial underwater work in overhead environments is also far removed from the rigors of the complete autonomy in an unknown environment required for Ocean Worlds exploration, but they remain an excellent testbed for development of exploration strategies, sensors, and maneuvers. The steps we have laid out and taken to date form a groundwork upon with much further research and development will have to take place.

10.4.4 Autonomous Life Search

Beyond the autonomous exploration and mapping tasks described in Sect. 10.4.3, an Ocean Worlds explorer seeking life must also act as an autonomous robotic scientist. This will require independently sniffing out signs in the physical environment which are the most indicative of nearby life and to follow them to the most promising locations (e.g. energy sources) to sample and investigate in detail. Because the most likely form of life, if it exists on Ocean Worlds, will be in the form of single or multi-celled microbes, an autonomous system must be able to process and identify what is being seen at that scale. Moreover, because there is a finite risk that the microbes being identified may have come from Earth, riding on the spacecraft (despite significant pre-launch measures to reduce this possibility), the success bar is elevated: the AUV must be able to detect unique life that is not from Earth (e.g. by proving either visually, chemically, or via unique DNA signature, that what is being seen does not exactly fit with any taxa known to exist on Earth). To some microbiologists a single microscope image may be sufficient to make such a determination of uniqueness; to others, a video clip from a microscope; and to yet others, only the gold standard of a DNA sequence would be conclusive.

Figure 10.67 shows one possible sequential archetype for an autonomous life detection strategy, to be implemented on a sub-surface AUV via discrete sensor selection, real-time data stream analysis, vehicle mobility characteristics, and onboard behavioral code. All of these must be implemented within the context of launch dynamics (g-loading, random vibration loading, thermal loading); local surface environment (hard vacuum, 100 K temperatures); and final detection environment characteristics (hydrostatic pressures of 1600 bar, temperatures varying from −5 °C to potentially much greater than 100 °C near a hydrothermal vent). From an engineering sense these are not insurmountable obstacles in 2017, but they will require careful consideration and robust simulation and environmental testing to improve the chances for success, i.e. the first discovery of life off Earth.

As related in Sect. 10.4.3 exploration and mapping behaviors will be closely linked as they directly affect the quality of the navigation solution where the environment topography is pronounced and unique. As mentioned earlier, the best strategy for achieving this is to drop to the ocean floor and radially expand the navigational map while conducting a search for life. The primary purpose of the navigation engine will always be to ensure that the exploration AUV is able to relocate the cryobot for data uplink to the lander and from there to the Deep Space Network (secondarily, it will be to enable a revisit of a potentially interesting site that is discovered on an earlier mapping mission).

To be very clear, no one expects that Europa, or any of the Ocean World moons, will harbor anything close to the bio-density of Earth's oceans (where a single cubic centimeter of seawater can hold 10 million viruses, one million bacteria and about 1000 protists). Indeed, when planning for the 2015 ARTEMIS mission to investigate the underside of the McMurdo Ice Shelf for life, it was determined that the sensitive protein fluorescence spectrometer—designed for detection of life on Europa—would be overwhelmed, saturated, by exposure to seawater and its dense payload of life and that instead the vehicle would bring the sensor up against the bottom of the ice shelf and the sensor would be focused on an area several centimeters inside the ice, where the number of cells per cubic centimeter was sufficiently low as to serve as a useful test of the instrument. The strategy that will be most effective on Ocean Worlds is represented by steps 3 and 4 in Fig. 10.67. These

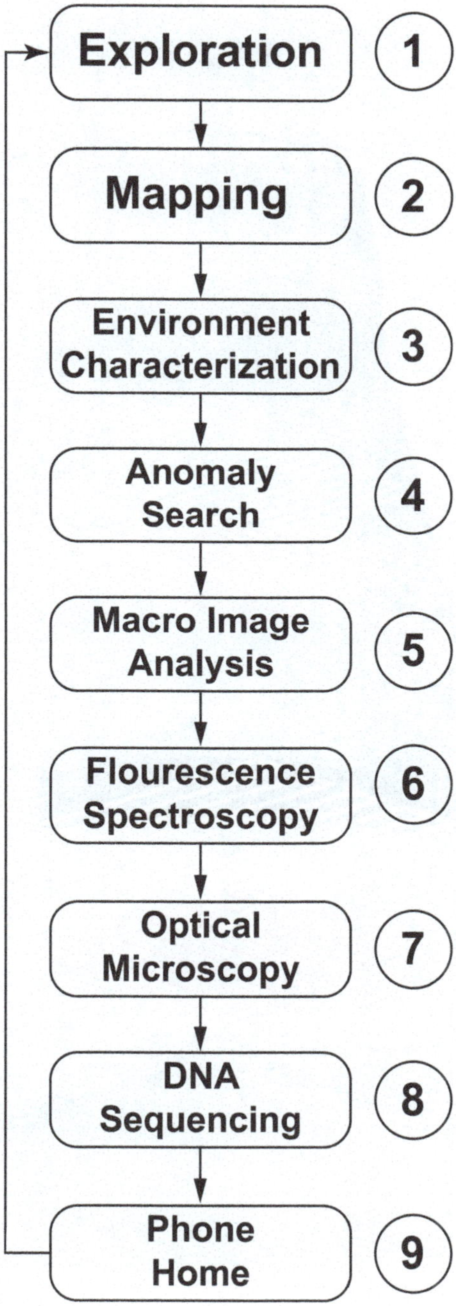

Fig. 10.67 One possible candidate for a hierarchy for autonomous detection of life on Ocean Worlds

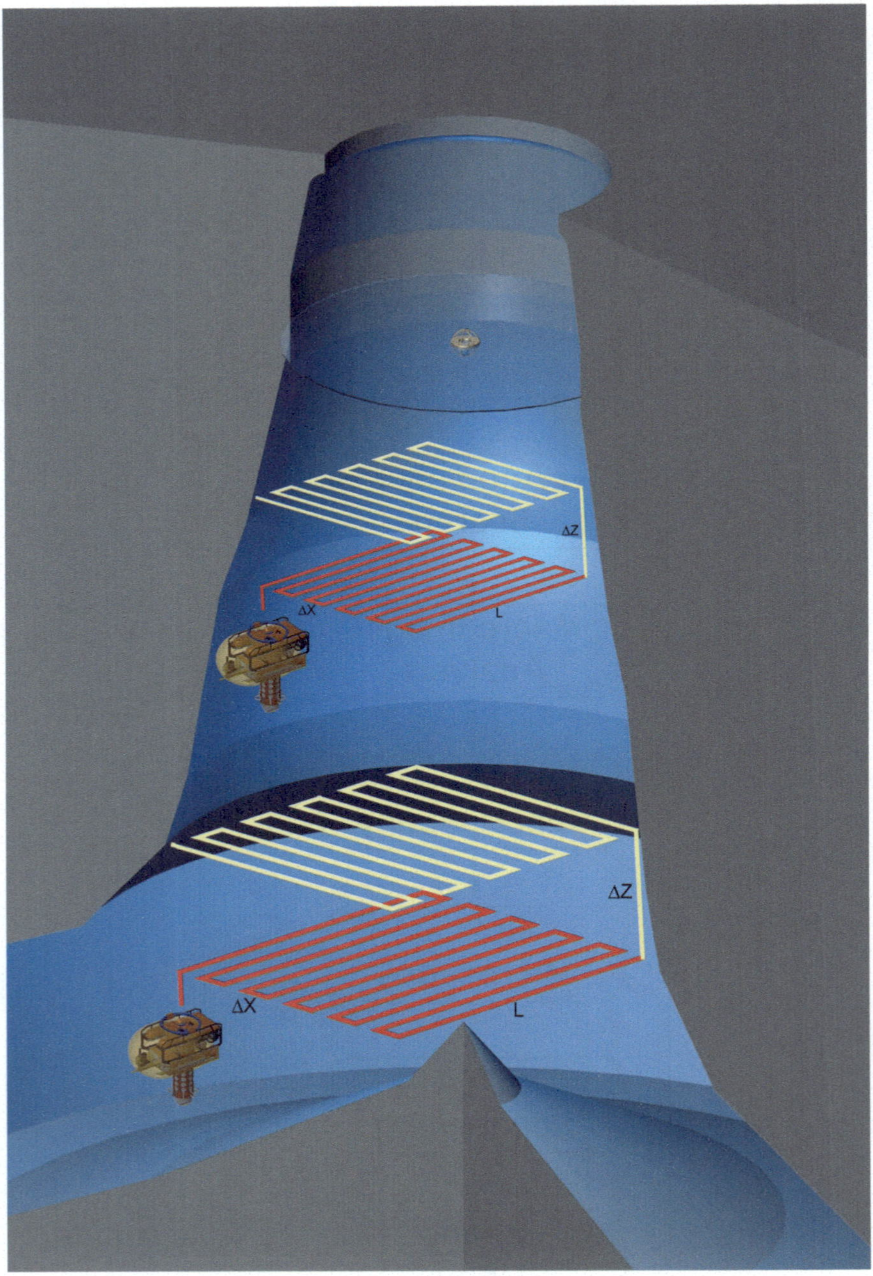
ΔZ
ΔX
L
ΔZ
ΔX
L

◂**Fig. 10.68** The third step in the hierarchy involves building 3D aqueous chemistry maps of the regions explored by the AUV. One simple depiction of how that might take place (using a model of Cenote Zacatón) would be to "mow the lawn" in 3D layers, building up information to populate a 3D voxel model of the world with water chemistry data that can be operated upon to sense gradients in species of interest, and to subsequently follow those to the source. Another way might be the approach taken by ENDURANCE in which the vehicle moves throughout a defined surface (planar or spherical) and periodically reels out an instrument sonde—potentially all the way to the bottom of the ocean)

are shown visually, respectively, in Figs. 10.68 and 10.69 in the context of the exploration actually conducted at Cenote Zacatón using the DEPTHX vehicle in 2007.

The approach for these two steps is not unique: a great deal of effort has gone into the development of ocean going (torpedo class) AUVs to search for ocean rift hydrothermal vents (see e.g. Yoerger et al. 2007; German et al. 2008a, b; Tao 2012; Petillo and Schmidt 2012; Tian 2014). Such algorithms seek to minimize the required transit range of a limited power AUV. If one could populate all 3D voxel space of a discretized ocean (e.g. by "mowing the lawn" in 3D, that is, driving the vehicle through each virtual voxel) with environmental data one could, as in the case of ENDURANCE, use an iso-surface extraction tool such as a Marching Cubes (or more advanced) algorithm and then compute gradients in a particular chemical species of interest (e.g. SO_2). A minimum-length trajectory could then be extracted leading to the source (greatest intensity/concentration) of the anomaly and programmed into the vehicle control system.

Once one reaches an anomaly (say, a hydrothermal vent or rift on the ocean floor), the most logical first step is image analysis of macroscopic color images. Aside from identification of motile elements in the field of vision one can also search for changes in background coloration that would suggest the presence of biomass or other biological-sourced material, for example using the approach first proposed by Malhalanobis (1936) and modern variants. Microbial communities do vary in color in optical wavelengths—viz ring colorations in hot springs in Yellowstone national park in the United States—where each different taxon inhabits contours of different temperature as the water cools when reaching the surface.

Fluorescence spectroscopy (step 6 in Fig. 10.67) uses an excitation light source at a specific wavelength, chosen because certain biomarkers will fluoresce if illuminated at certain wavelengths, and observes the magnitude of the optical fluorescence of the target. Chlorophyll-A, amino acids, and polyaromatic hydrocarbons (PAH) are common examples of biomarker targets. It is possible to construct spectrometers that operate at multiple wavelengths (e.g. Bramall 2007, 2009; Boqomolny 2013), thus enhancing the discrimination filter when assessing whether a potential sample contains bio-markers or fluorescing mineral particles. Raman spectroscopy (Ianoul et al. 2002; Marshall 2010; Beegle et al. 2015) can be similarly employed to produce similar results: that something biological, dead or alive, is being illuminated. Such instruments for underwater vehicles can be built in two formats: as a cytometer (in which fluid is drawn in from the environment and passed

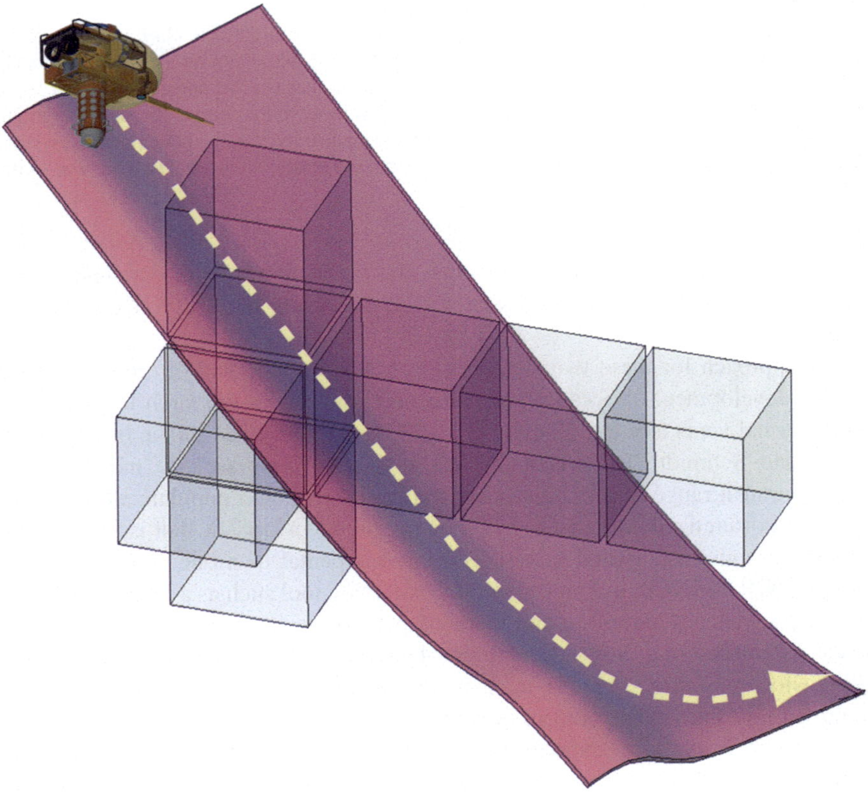

Fig. 10.69 A graphical depiction of an AUV following a 3D plume trajectory towards an energy source in the sub-surface ocean

through a sapphire transparent pressure cell through which as laser is shown and a receiving single-photon multiplier tube located on the opposing window) or as a direct illumination system where an optical window is presented against an object of interest (e.g. an ice surface, or a solid surface) and a similar stimulated fluorescence measurement is taken through the common window. In either case these are not definitive tests, and serve primarily as a gate-keeper to the next two instruments.

Having an optical microscope onboard brings with it many advantages: a trained microbiologist viewing a well-focused image will almost certainly be able to immediately detect the presence and class of microorganisms within the image field of view. The hard part is producing a sharp image of an object of interest. There are two parts to this challenge: first, getting a sample of a promising area into the microscope. The external pressure will be in the vicinity of 1600 bar and rapid decompression of sample captures to a 1-bar instrument cavity will almost certainly destroy the object of interest. The alternative is to use a sapphire flow cell, not unlike the concept behind the protein fluorescence cytometer mentioned above. Second, once a sample is introduced into a flow cell there will be a finite depth owing to structural mechanics of materials that can resist 1600 bar pressure. Microbes tend to move around and if something promising exists it typically will not remain in focus. A human scientist operating such an instrument on Earth can interact and refocus continuously on something of interest. Reducing such skill to an automated system will be a significant development effort, largely because it asks the question: what is important? This will require real-time machine vision assessment of possible microbe archetypes present in the instantaneous field of view and a prioritized decision matrix as to which object should be tracked and imaged.

DNA sequencing instruments are becoming more compact every year. As of the writing of this chapter the MinION sequencer from Oxford Scientific (Olsen 2016) represents the smallest, lightest weight device for field sequencing (a palm-sized device with a mass of 100 g). In this approach a protein nanopore is set in an electrically resistant polymer membrane. An ionic current is passed through the nanopore by setting a voltage across the membrane. If an analyte (e.g. a strand of DNA) passes through the pore, this event creates a characteristic disruption in current. Measurement of that current makes it possible to identify the molecule in question. In the case of DNA, the current is changed as the bases G, A, T and C pass through the pore in different combinations. We can anticipate that such technology will improve in resolution and reduce in mass and size during the planning phases for Ocean World sub-surface missions. This type of instrument will face similar problems to that of an optical microscope with regard to acquisition of samples. In this case there will be no avoiding a rapid decompression of the sample but that may be less of a problem for a DNA sequencer than an instrument designed to detect the geometry, color, texture, and other characteristics of an integral specimen.

10.4.4.1 DEPTHX Life Detection Experiments at Cenote Zacatón

We tested a significant portion of the algorithm pictorially described in the block diagram of Fig. 10.67 with the DEPTHX vehicle in the spring of 2007; only elements 6 and 8 were not attempted (largely because suitable field instruments did not exist at that time).

As described above, the most straightforward basis for determining suitable habitats where life will concentrate is to look for sharp gradients and interfaces (microbial life likes to have a surface to congregate upon). In the case of DEPTHX and the Zacatón cenote complex, these were expected to take the form of sulfur chemoclines or plumes in the water, the rock-water interface, and visual color gradients or edges on the rock wall (Franke et al. 2007). The gradients represent areas of available energy that life can exploit on the microbial scale, and the solid-liquid interfaces provide secure habitats for microbes to grow and survive.

In order to test the science autonomy capabilities of DEPTHX, water and wall sample collections were made in both La Pilita and Zacatón to characterize organisms at a particular depth so as to minimize cross contamination. The robotic sampling arm (see Sect. 10.4.5) and triggering system worked autonomously and this basic pattern was applied at various depths and locations to characterize the varying environmental conditions and biologic activity. Unfortunately, DEPTHX discovered that the environment was nearly homogeneous. Most sensor data showed insignificant variation in more than 300 m of depth in the majority of the aqueous chemistry sensors it carried onboard. The sole exception was the sulphide level, which climbed very gradually with increasing depth. Coloration of the walls of the hydrothermal spring was discovered to be a monochromatic orange-brown from just below the photic zone to the maximum depths reached by the vehicle (see Fig. 10.76). These two factors limited the degree to which we were able to exercise the complete life search algorithm.

There was a slowly increasing trend in the concentration of sulphide below the level of the photic zone. An arbitrary concentration of 0.042 ppm was selected as a trigger threshold for initiation of autonomous sample collection. On 26 May 2007, DEPTHX descended into Cenote Zacatón, mapping along the way and maintaining a centroidal XY position for safety. At a depth of 114 m, it detected the sulphide threshold and began maneuvering towards the wall, entered into proximity operations (servoing on a fixed wall stand off range), extended the probe, and impacted the wall using an automated core sampler (Sect. 10.4.5). It then initiated its nominal egress procedure and returned to the shaft centroid for ascent. The critical portion of the mission is shown in Fig. 10.70.

The thin yellow trace in Fig. 10.70 is the vehicle trajectory, which clearly shows the sudden detection of the sulphide threshold and the decision to traverse laterally to the wall at that depth. The orange dots are individual sonar wall hits. Figure 10.71 shows sulphide data from a previous mission that identified the increasing sulphide versus depth trend, and shows the trigger threshold for initiation of sampling behavior.

We repeated this maneuver twice more with differing trigger thresholds down to a depth of almost 300 m. The resulting core samples were sequenced over the following year and led to the discovery of four new phyla of bacteria (Sahl 2010, 2011). While not a complete test of the architecture described in Fig. 10.67, DEPTHX proved that the concepts are inherently capable of being developed and implemented into an autonomous hovering underwater vehicle.

10.4.5 Precision Control and Sampling

It is evident in the hierarchical life detection approach described in the preceding section that steps 6–8 will require a sample to be brought onboard for more refined sensors to operate upon. Collection of random water samples from the water column on an Ocean World will in general be ineffective, because in low energy

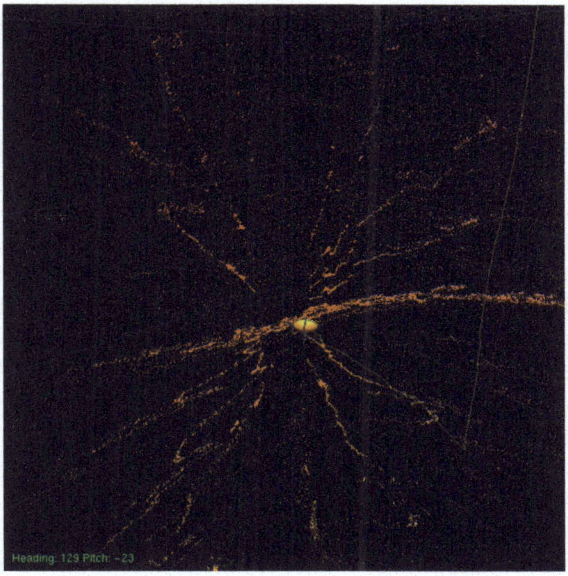

Fig. 10.70 A view from the 3D visualizer for DEPTHX showing the vehicle trajectory, descending, in yellow, then abruptly heading for the wall where a core sample was taken autonomously

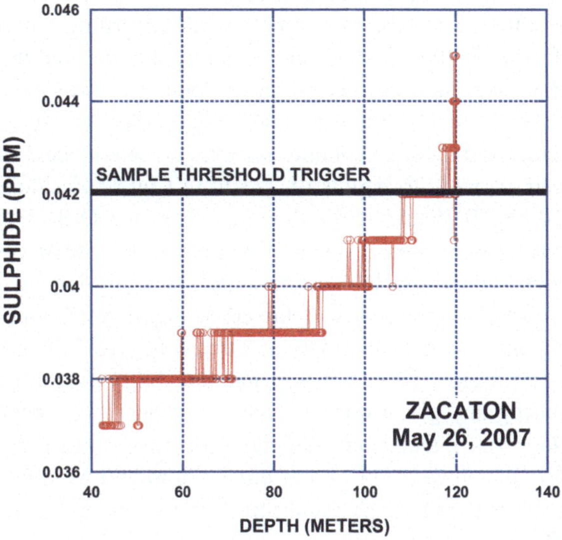

Fig. 10.71 Real-time data on sulphide concentration in Cenote Zacatón as a function of depth. A sharp buildup at 120 m depth triggered the proximity operations behaviour to fetch a wall core sample

environments life, if it exists, will be much more likely to be found harbored on a surface adjacent a local energy source (e.g. a hydrothermal vent). Thus, after exploring, mapping, searching for gradients in environmental variables, and looking for color anomalies on surfaces the vehicle must explicitly perform proximity operations (proxops): fine precision hover maneuvering, potentially in collaboration with the use of dextrous or extensible appendages, and sample acquisition systems (both for liquids and solids).

DEPTHX (as well as ENDURANCE, and later ARTEMIS) achieved controllable proxops using two specific behaviors: wall (or ceiling) approach, and wall stand-off tracking. Both behaviors use velocity control feedback (see Figs. 10.59 and 10.60). In the first behavior, an approach to a target wall is made based on, for example, a plume tracking trajectory while maintaining heading and preventing lateral drift. To do this, a plane is fitted to the forward looking sonar data and the vehicle servos on that plane until achieving the target stand off distance. Once so initialized, the vehicle can then move up/down or left/right along the wall at the designated standoff distance, either in differential displacement control or in constant velocity mode, while maintaining an angular lock to the wall plane. Proximity operations control for DEPTHX is discussed in greater detail in Fairfield et al. (2007a).

Some of the most complicated DEPTHX missions took place in Cenote La Pilita (Figs. 10.15 and 10.17). In repeated dives, DEPTHX ran untethered down to 40 m depth then maneuvered 30 m horizontally under the ceiling of La Pilita to ascend into a dome at 30 m depth (this can be seen in the profile map shown at bottom right in Fig. 10.15 as a rise in the roof leading to the right). There it automatically located the wall, moved into 3 m stand off, and extended its sampling arm. It flew toward the wall until the coring mechanism was in contact and while the robot held position, it took a core sample while recording video and drawing a water sample. The rover then backed off, retracting the arm and then dove down, out from under the dome and returned to the surface. This was executed autonomously without a tether, so the results were known only after the rover returned to the surface and the data examined. Eventually samples were collected throughout the La Pilita cenote and, as well, to depths of 273 m in Cenote Zacatón. Figures 10.72, 10.73, 10.74, 10.75, 10.76 and 10.77 show details of the DEPTHX sampling system including a custom-built spring loaded titanium tube core sampler which could acquire wall rock samples up to 6 cm in length $\times$ 1 cm diameter (Figs. 10.74 and 10.77).

The samples, once obtained at depth, were automatically capped underwater. Samples were manually transferred (Fig. 10.77) to specimen tubes at the conclusion of a mission. DEPTHX also had the ability to store up to five 1 L samples onboard during any given mission (but could only take one core sample). Standard procedure was to take water samples at the time of the core shot since any microbes in the initial centimeter or two of crust would be mixed into a cloud of debris around the core site (Fig. 10.76) and represented a possible auxiliary collection opportunity. The water sample intake is seen as the white tube (with white mesh cover) just below the core sampler in Fig. 10.72.

ENDURANCE also utilized wall-following (see Fig. 10.26) for detection of the grounding line for Taylor glacier in 2009. ARTEMIS also utilized similar stand off tracking maneuvers under the McMurdo Ice Shelf in 2015. The extendable science tower (Figs. 10.78 and 10.79) was designed to allow instruments (particularly the protein fluorescence spectrometer—PFS) to be in contact with the underside of the ice cap while the vehicle remained at a safe standoff. ARTEMIS used an off-shelf

Fig. 10.72 A close-up image of the extensible wall coring and water sampling probe arm developed for DEPTHX

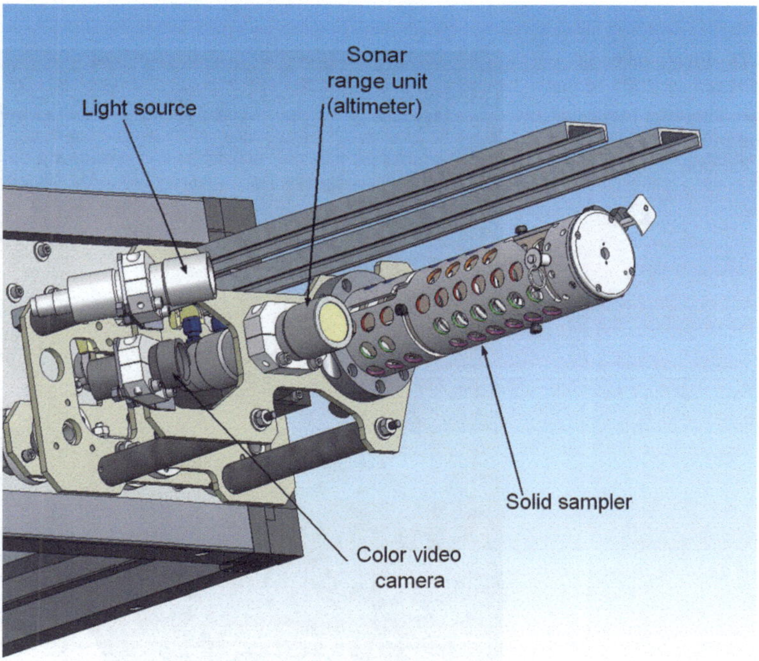

Fig. 10.73 A computer-generated model of the DEPTHX sampling probe showing key components

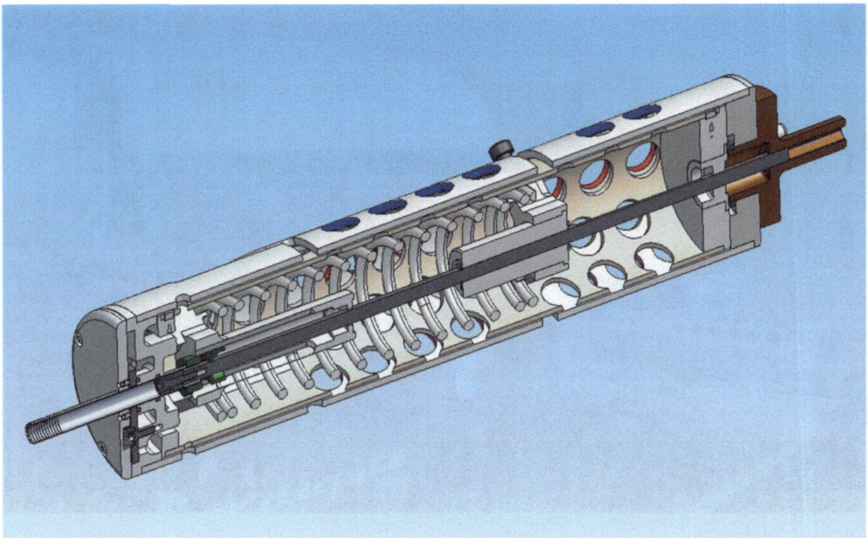

Fig. 10.74 A section view of a model of the DEPTHX coring tool. This was a spring-loaded device that could obtain one 6 cm long by 1 cm diameter wall rock core per mission. The core was automatically capped following extraction from the wall

Fig. 10.75 Image taken at night in February 2007 of the DEPTHX sampling probe operating in shallow depth at Cenote Zacatón

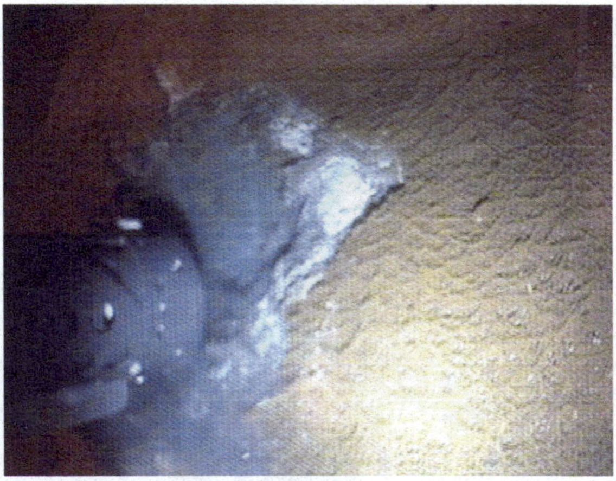

Fig. 10.76 Close up underwater image of a core sample being acquired at a depth of 283 m in Cenote Zacatón. The impact of the core tool has broken away part of the accumulated biomat material surrounding the tool

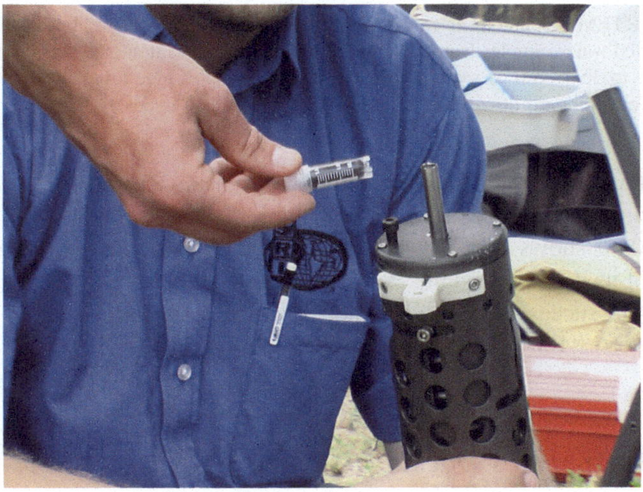

Fig. 10.77 Wall core sample from 273 m deep in Cenote Zacatón is removed from the coring tool following a DEPTHX mission in May 2007

Fig. 10.78 ARTEMIS sonde in its extended configuration. Light blue spiral tubing leading up to the sonde carried water samples from the ice ceiling surface to an automated collection system that could store up to 36 × 100 cc samples per mission; the smaller blue tube carried pumped water from the sonde to inboard flow-through aqueous water chemistry sensors

water sampling system (RAS100), modified to fit inside the vehicle. It was capable of taking up to 36 × 100 cc samples per mission (Fig. 10.81).

Figure 10.80 shows an early underwater test of the extension mechanism at about 30% of its deployed height. Water samples (eventually to be triggered by PFS readings) were successfully and routinely acquired using a heavily-re-engineered version of a RAS-100 oceanographic water sampler (Fig. 10.81).

10.4.6 Autonomous Docking

While it is possible to transfer limited amounts of data underwater using acoustic, optical, RF, and oscillating magnetic fields in a stand off, non-contacting scenario, the reliable bandwidth can be increased by orders of magnitude if a direct physical contact can be achieved. The data of greatest value collected during a sub-ice Ocean Worlds mission will be macroscopic and microscopic high resolution video as well as DNA sequence reads, all of which represent large data transfers. Automated rendezvous and hard docking will be required both to achieve the high bandwidth needed as well as to reduce power requirements over station keeping in a strong ocean current. Systems have been developed for terrestrial torpedo-class AUVs

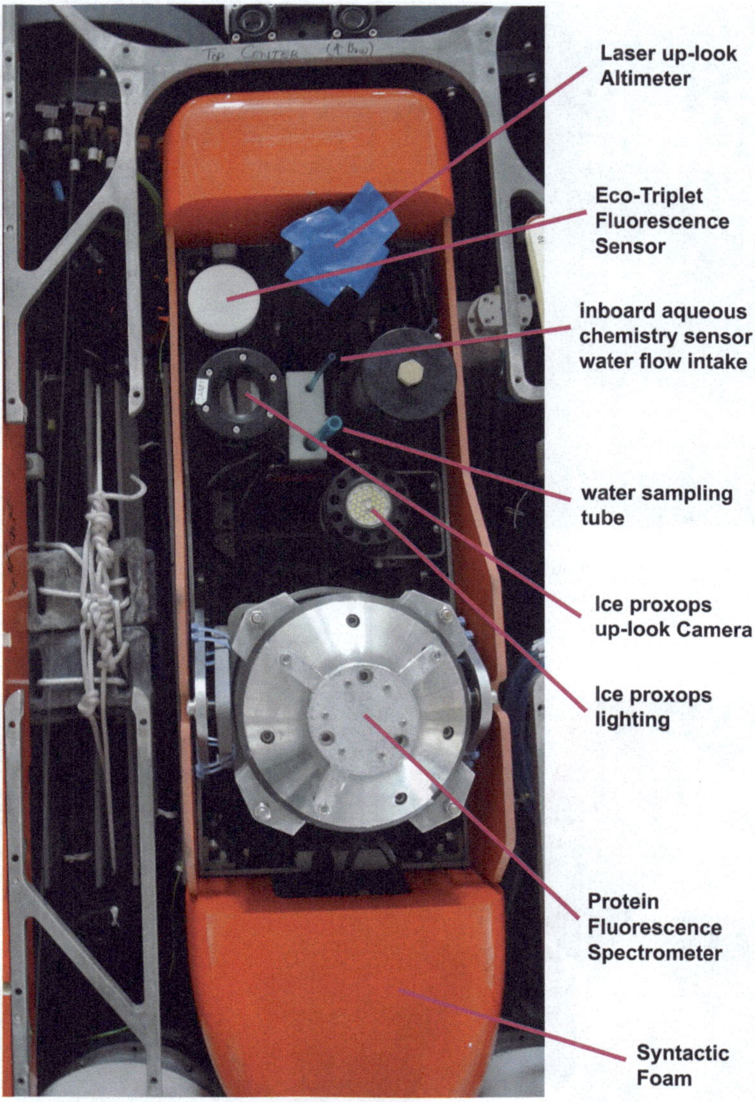

Fig. 10.79 Top view of the ARTEMIS sonde showing key components

(e.g. Bellingham 2016) that utilize large funnel targets (either fixed on the ocean floor or suspended in the water column) for capture of the AUV. This approach will not be feasible in an Ocean World scenario owing to both launch mass limitations and extreme volumetric limits imposed on the cryobot. A compact rendezvous and docking approach that can be reduced in mass demands that the AUV be hover capable. As mentioned in Sect. 10.4.2 any successful sub-ice underwater navigation system functioning on Ocean Worlds will likely have multiple inputs into a Kalman

Fig. 10.80 Video frame showing the ARTEMIS sonde tower deploying in the Ross Sea at a depth of 12 m during a test of the extension and retraction mechanisms

Fig. 10.81 Two custom-designed banks of 18 × 100 cc water samples are removed from ARTEMIS following

filter with weighting factors that depend upon range from a deployment site. A final stage navigation aid will be required that allows for precision approach—none of the methods described in Sect. 10.4.2 provide this. Below we describe experiments we have successfully conducted in sub-ice vehicle recovery that use oscillating optical sources for final stage rendezvous and docking. Variations on this theme will likely be of utility for planning of a cryobot-deployed AUV mission.

10.4.6.1 ENDURANCE

The ENDURANCE mission scenario presented an opportunity to develop autonomous docking capabilities to ensure vehicle return. The accuracy of the dead-reckoned navigation described in Sect. 10.4.2 was sufficient to bring the vehicle back to within 8 m of the melt hole, but bringing the vehicle into the melt hole required precision on the order of 0.1 m to allow the vehicle to safely center and ascend the hole. Thus, a novel vision-based docking algorithm was developed to enable the vehicle to locate and ascend the melt hole after reaching the nominal home position. For this, a downward-facing oscillating, tightly focused light was suspended centered above the melt hole (see Fig. 10.82).

The autonomous docking then proceeded as follows. When the ENDURANCE dead reckoning indicated that it had nominally reached the melt hole, a search behavior (Fig. 10.83) was initiated in which the vehicle moved in an outward spiral pattern to look for the light. After the light was detected, an ascent behavior was initiated.

The system used a zenith-pointing VGA video camera (Fig. 10.84) to search for the presence of the oscillating, collimated light source. The algorithm segmented the video frame into candidate intensity sources (green boundaries in Fig. 10.82).

The light detection algorithm needed to be able to distinguish the docking light target from other environmental light sources, such as light from the sun filtering through the ice as well as any reflections from the ice. To distinguish the target light source from these persistent sources, the light source was made to oscillate at a fixed frequency, and the algorithm tracked the oscillation signature of the target, without requiring synchronization between the camera and light. The algorithm made only the assumption that the vehicle did not move too quickly frame to frame. If it found the source oscillation signature, it locked onto that source (blue perimeter in Fig. 10.82) and held it horizontally in the camera center (using its station-keeping control, see Fig. 10.85).

For vertical control, an upward velocity was commanded only when the light was sufficiently centered in the image. The ascent controller stopped when the vehicle reached the water surface. The details of this algorithm are discussed in Murarka (2009) and Gulati (2010).

The vision-based docking algorithm was successfully employed in both the 2008/9 and 2009/10 ENDURANCE campaigns at Lake Bonney. The image sequence in Fig. 10.82 is taken from Mission 5 (ENDURANCE 2008/9) live video and shows (A) the vehicle crossing the melt hole edge and the first appearance (but

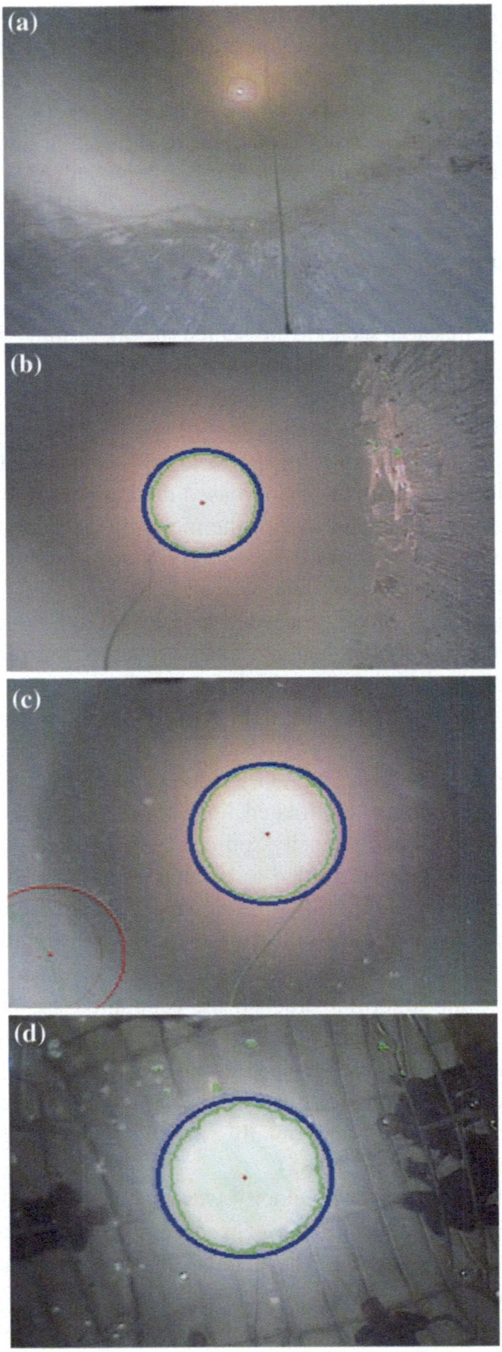

Fig. 10.82 Four frame captures from the machine vision auto-docking engine on ENDURANCE showing target verification of an oscillating light and the subsequent centering of the vehicle on the light beam and then rising up the center of the hole and surfacing

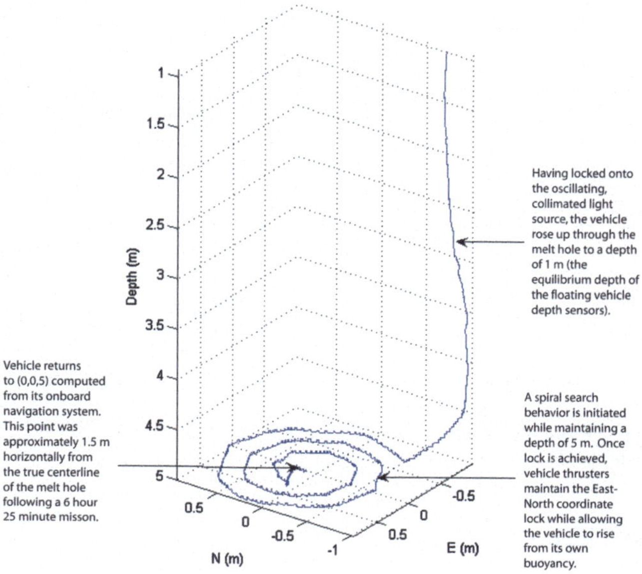

Fig. 10.83 Spiral search algorithm for locating the down-hole oscillating light following best return approach using dead reckoning and USBL guidance systems

Fig. 10.84 The ENDURANCE up-look machine vision camera for final rendezvous and docking

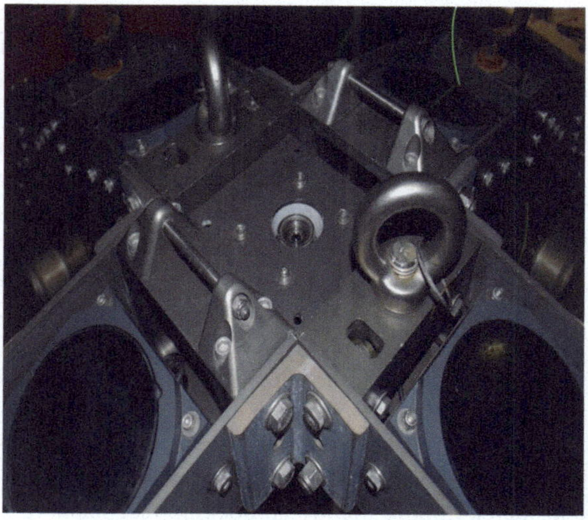

not detection) of the light source; (B) positive lock being acquired on the light source; (C) ENDURANCE rising up the hole; and (D) first breaking surface as the entire crew looks on. Operation of the system included completely automated transition from end-of-mission to docking and recovery.

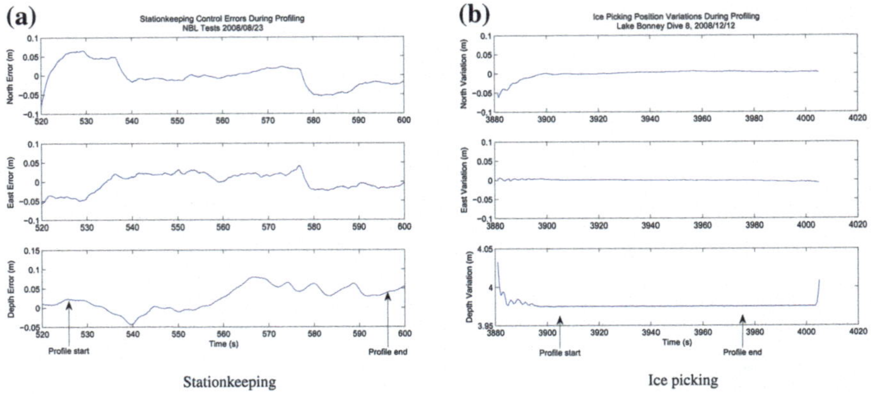

Fig. 10.85 Typical station keeping (hovering) performance for ENDURANCE during water chemistry (sonde) profiling

10.4.6.2 ARTEMIS

The requirements for docking were even more stringent for ARTEMIS: the vehicle range (and hence dead-reckoned error accumulation) was much larger; the McMurdo Sound deployment area had significant water currents; there was no open water within 50 km for an abort recovery; and the vehicle needed to reorient from horizontal to vertical to complete the end-of-mission recovery maneuver.

Building on the success with visual docking proven by ENDURANCE as well as the iUSBL system it tested, a hierarchical return-home strategy was developed for ARTEMIS. General navigation used the same dead reckoning system employed by DEPTHX/ENDURANCE, with 0.1% of distance traveled average drift error. For the maximum round-trip vehicle range of 20 km, this guaranteed return to within 20 m of the hole with very high confidence, as long as DVL lock was maintained. The next stage of the return-home strategy employed the iUSBL with a 1 km nominal range from the beacon placed at the melt hole. This sensor allowed the vehicle to correct dead-reckoning drift and make its way to within 2 m of the dock location. At this point, a visual (machine vision) homing routine took over and guided the vehicle to the actual docking apparatus with centimeter-level precision. Figures 10.86 and 10.87 describe the state machines devised to coordinate these multiple levels of navigation and control.

The actual docking hardware was also a significant area of development. We first considered a more traditional docking receptacle (e.g. probe-cone systems as used for spacecraft) with lighted fiducials, thrusters, multiple servo-actuated latches, and a large electromagnet to grab onto a probe on the vehicle nose. It was quickly decided that although this solution would provide a high degree of functionality for docking, a solution with so many subsystems and moving parts would not be suitable for remote deployment where robustness was of critical importance. Also,

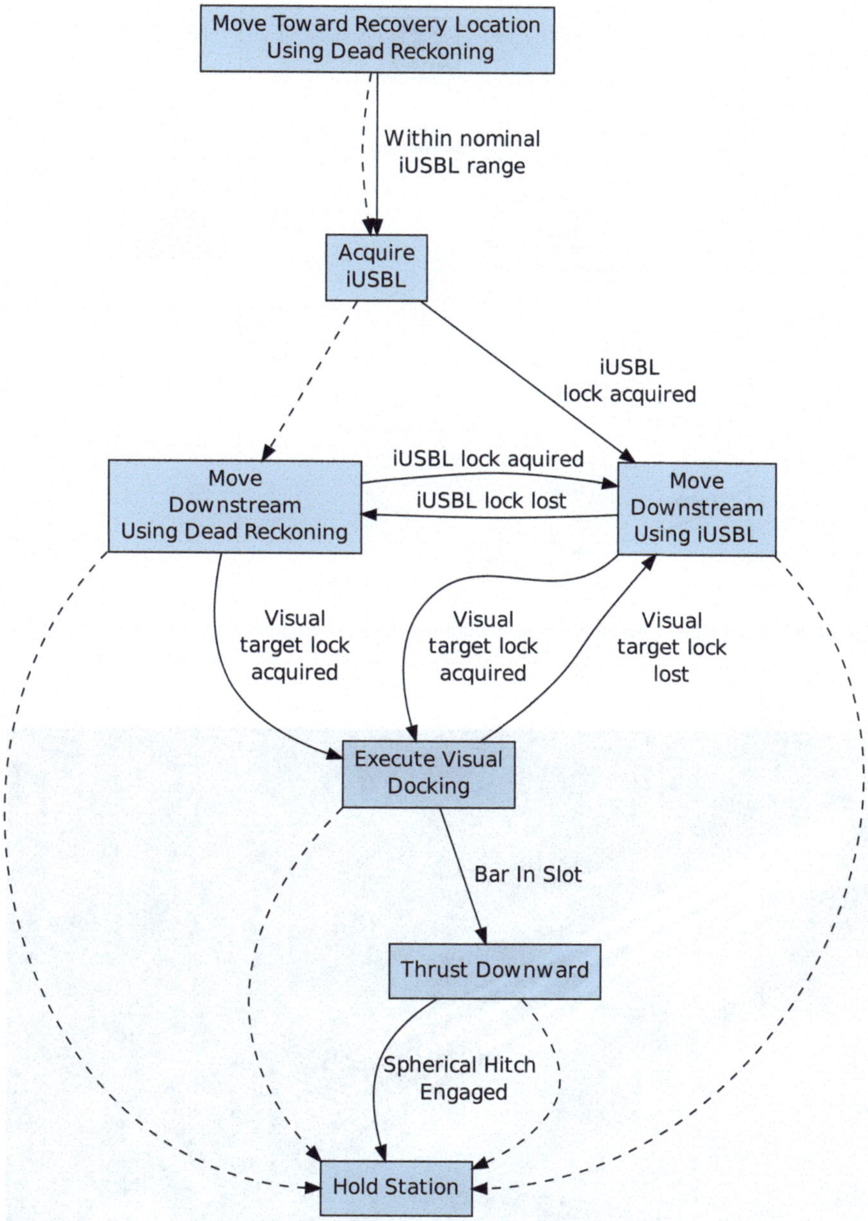

Fig. 10.86 Long range rendezvous and docking state machine for ARTEMIS

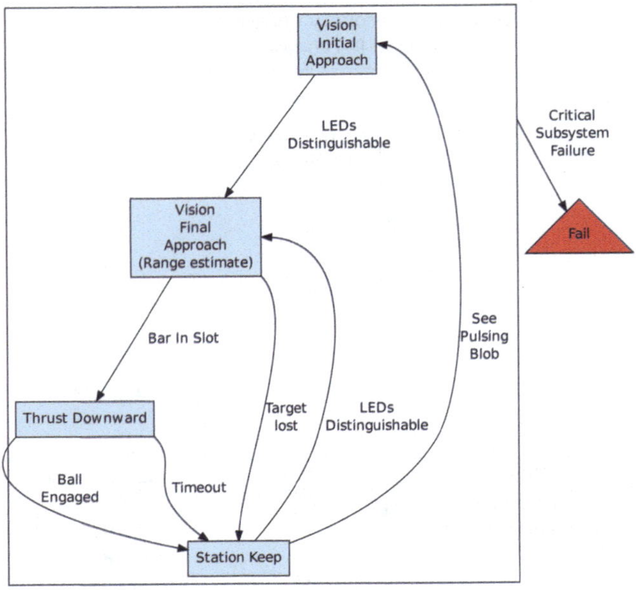

Fig. 10.87 Final stage (machine vision) rendezvous and docking state machine for ARTEMIS

Fig. 10.88 Close up view of the ARTEMIS docking bar with eight columns of 1-cm-spaced blue LED lights, without its polycarbonate tube cover and oil filling

there was little resemblance to a system that might be usable on an Ocean World—a critical aspect of the technology development program.

Instead, we developed a new design focused on electromechanical and operational simplicity. The final design for the ARTEMIS dock consisted of an illuminated vertical docking bar hanging below the borehole with an iUSBL beacon attached to the top (Figs. 10.88 and 10.89).

The bar is a transparent cylindrical oil-filled tube measuring 2 m long and 5 cm in diameter. Eight LED light strips, each containing 200 discrete LEDs spaced 1 cm apart, were placed inside the cylindrical housing to approximate a uniform distribution of light. The light intensity was switched between dim and bright at a frequency of 2 Hz. The docking bar fitted into a slot in the nose of ARTEMIS where it is captured by a spring-loaded mechanical latch (Fig. 10.90), although parallel designs had been prepared to achieve latching with a servo-operated pin.

After engaging the bar, ARTEMIS slid down its length until a spherical hitch at the bottom of the docking bar engaged with a spherical cavity at the back of the nose slot. The top of the bar has a universal joint which, combined with the weight of the ball hitch, keeps the bar vertical regardless of current towing the recovery cable. Finally, using thrusters, ARTEMIS pitches stern-down 90° on the spherical hitch so that it may be pulled vertically upward through the bore hole. This process is illustrated in Fig. 10.91 and the approach from 50 m out to capture is shown in Figs. 10.92, 10.93, 10.94 and 10.95.

Fig. 10.89 The ARTEMIS docking rod being prepared for a down hole test beneath the sea ice in McMurdo sound, October 2015

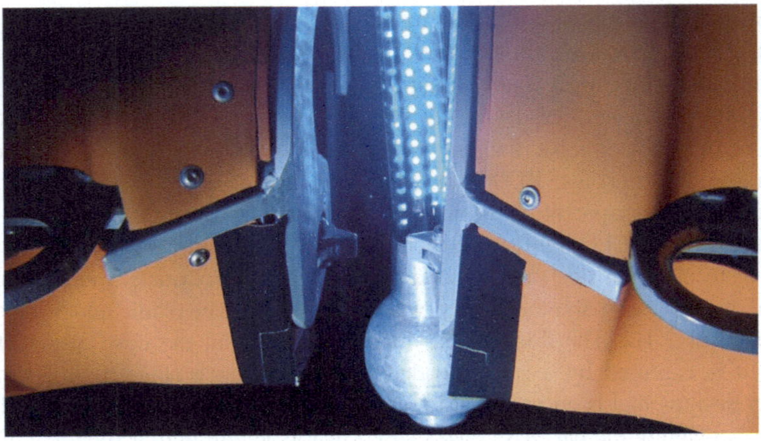

Fig. 10.90 Close up image of the bow of ARTEMIS showing the mechanical capture system. Once latched the vehicle can no longer leave the docking rod. The vehicle can then be pitched up and extracted vertically through the drilled access hole

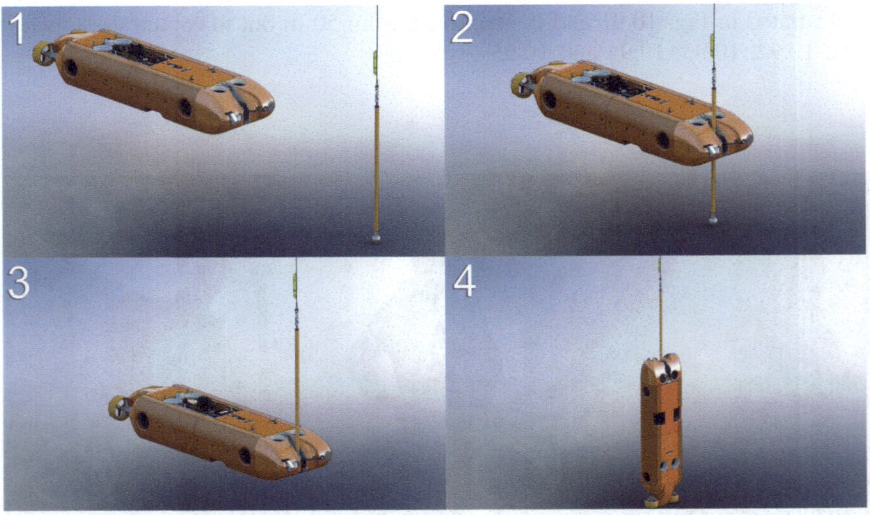

Fig. 10.91 Computer-generated sequence of automated rendezvous and docking with ARTEMIS. In Frame 1 the vehicle has acquired the docking rod within the field of view of the machine vision system; In Frame 2 the vehicle approaches the center of the docking rod until the rod is captured by the vehicle; In Frame 3 the vehicle performs a–Z translation (descends) until the docking rod sphere is engaged in a receptacle in the vehicle frame; and in Frame 4 the vehicle pitches 90° down until the vehicle is vertical and the rod and vehicle are extracted through the access shaft

Fig. 10.92 Vehicle-view of the docking rod, access shaft, and a diver observer from a range of approximately 50 ms during an auto-docking sequence

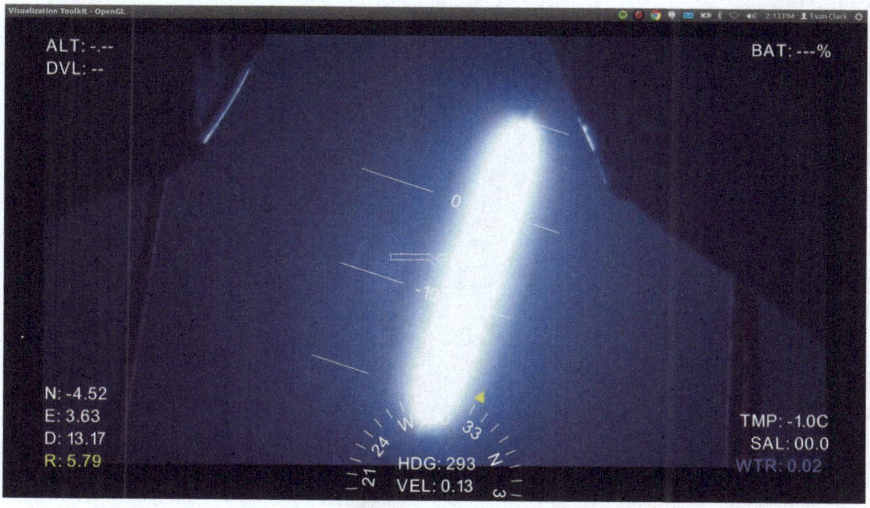

Fig. 10.93 Vehicle-view of the docking rod from a distance of 5 m

The view of the same scenes as interpreted by the machine vision systems with results of the state machine calculations superimposed, is shown in Figs. 10.96, 10.97 and 10.98.

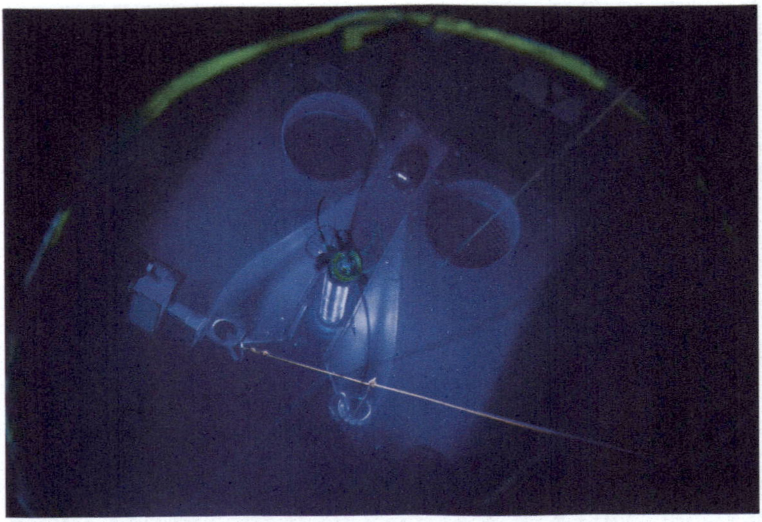

Fig. 10.94 Telephoto view looking down the drilled access shaft following a successful docking capture showing the vehicle and the docking rod

Fig. 10.95 The captured vehicle is successfully retrieved up the access shaft at the conclusion of a mission

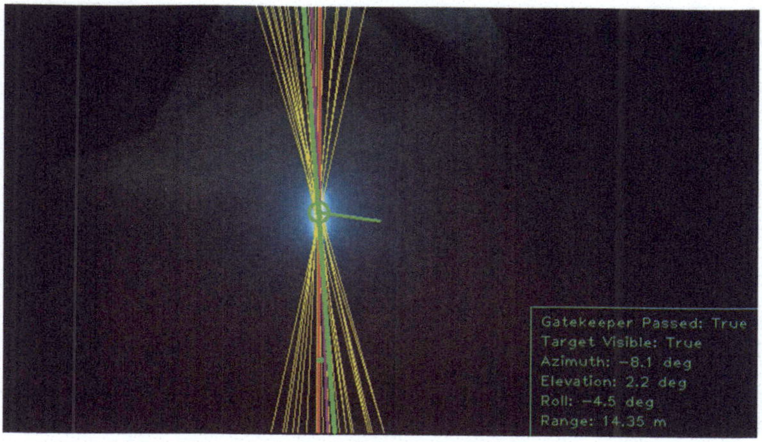

Fig. 10.96 Machine vision view of the docking rod from 50 m range with alignment metrics superimposed over the image

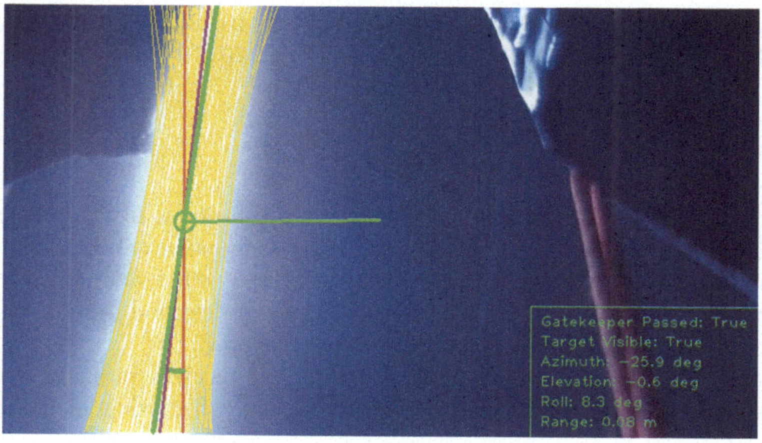

Fig. 10.97 Machine vision view of the docking rod from 2 m range with alignment metrics superimposed over the image

The ARTEMIS docking design has several advantages. It allows the AUV to approach from any direction and from a wide range of depths without requiring swivel mechanisms or active control in the dock. It is compact, and it would be straightforward to build this type of docking system into a cryobot-deployed vehicle. The lighting system provides three levels of guidance: as a "blob" of light —visible as far as 300 m distant in clear water in the Ross Sea—it served as a bearing to the dock; at closer range, the linear profile of the source (Fig. 10.97) provides a roll reference; and finally, at closer range (<5 m) the spacing of the

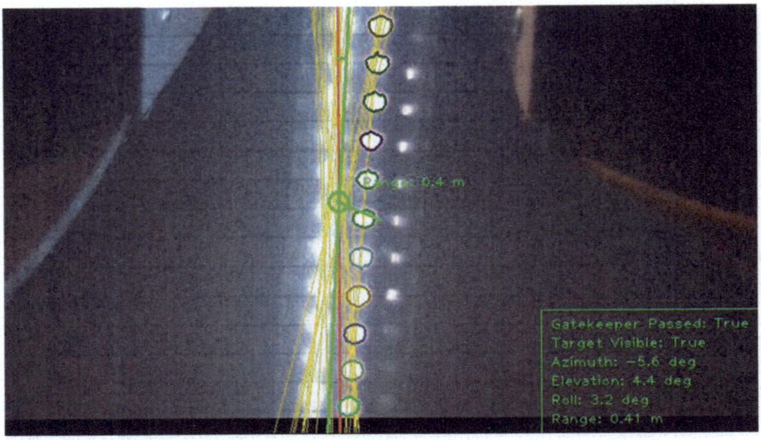

Fig. 10.98 Machine vision view of the docking rod from 30 cm range with alignment metrics superimposed over the image. Note the acquisition of individual LEDs and the parallax calculation of closing range

individual LEDs can be used to determine range (Fig. 10.98). For further details on the ARTEMIS docking algorithm (see Kimball et al. 2018).

10.4.6.3 Visual Docking with the SUNFISH® AUV

In 2016 and 2017 improvements were made to the methods first developed for ARTEMIS for auto docking. The new docking system employed the same target docking bar, although the light switching was replaced with a sinusoidal intensity sweep from full off to full on to improve the frequency detection. A mechanism to capture the dock with SUNFISH remains to be developed.

The improved SUNFISH dock detection algorithm receives an image from the machine vision camera and then detects the docking bar using the following algorithm:

- Determine camera pitch in the global North/East/Depth coordinate frame; if pitch magnitude is greater than 65° ignore the image, otherwise proceed.
- Undistort the image to compensate for radial and tangential lens distortion. Rotate the image so it is aligned with gravity. This allows us to compare sequential images in a gravity aligned reference frame.
- For each LED column visible in the rotated image, calculate the average intensity. This becomes the input vector for further processing. Record the input vector with timestamp. The algorithm buffer stores the most recent 20 input vectors and timestamps.
- Cross correlate historical input vectors and spatially align them with the most recent one. Apply the continuous-time Fourier transform at the target frequency to each column of the aligned input vectors and take the absolute value. We call

this the output vector and it is a representative measure of system lock and alignment with the oscillating linear dock.

- Find the column of the output vector that has the largest magnitude. If this magnitude is larger than a specified threshold we consider the target to be detected. The target bearing can then be directly obtained from the column index.

Note that it is possible to align sequential input vectors using motion compensation (by aligning previous images to the yaw angle of the most recent). This approach works when the vehicle is far away from the target, however it fails to align the input vectors at close range due to perspective distortions.

The visual docking algorithm searches until the target is within the field of view of the camera. Once found the algorithm commands the vehicle to orient the camera toward the target and slowly move toward it. Once the vehicle contacts the target, a sharp deceleration will register on the IMU, leading to successful termination of the mission.

The use of an oscillating light and real-time Fourier transform of the machine vision images has proven to be robust means of underwater dock detection (including both pose and range) which can then be used in conjunction with a hovering AUV to achieve a high level of capture success. We believe this type of approach will be a useful tertiary stage navigation aid for any Ocean World AUV rendezvous and docking scenario for transferring data collected by the AUV to the suspended cryobot (and subsequent uplink to the lander). The linear lighting array can be distributed on the body of the cryobot carrier vehicle so that no discrete docking bar, as described above, is needed. Instead, a small, discrete capture bar will take its place allowing for a much smaller, lower mass dock, but based on the principles described above.

10.5 Lessons Learned

We review here several of the most important lessons from our AUV development program and, most importantly, field testing, which are applicable to under-ice roving vehicles as would be employed on Ocean Worlds.

Lesson 1: Expect the Unexpected

On its very first powered descent into Lake Bonney, ENDURANCE discovered "microbubbles". The water around the vehicle instantly turned a dense milky white and tens of thousands of sub-millimeter bubbles adhered to all vehicle surfaces. Notably, their presence obscured optical ports and rendered inoperative all acoustic sensors. The bubbles were caused by reduced pressure being created by the vehicle thrusters in the presence of supersaturated gasses in the water. These gases are the result of ice accretion on the bottom of the Lake Bonney ice cap. As the water freezes, dissolved gasses are rejected, leading to increased concentration in the water. Even though we were aware that supersaturated gases were present in Lake Bonney (via LTER data) prior to the project, nobody anticipated the impact on vehicle performance.

From an engineering standpoint, the discovery of the "microbubble" issue is of extreme importance and will have a dramatic effect on design and control of any AUV going to Europa and other sub-glacial bodies of water on other planets. Microbubble release is caused by cavitation, which can be triggered either by flow over propulsion system elements (e.g. prop blades) or over the vehicle itself (protuberances from the shell of an otherwise laminar flow body can trigger cavitation beyond a certain velocity that is vehicle geometry and media dependent). The design of a laminar flow, non-cavitating vehicle, as well as strategies to deal with bubble or other debris accumulation will be a high priority for any AUV designed to map large-scale sub-glacial lakes and planetary oceans.

A more general observation is that the vehicle must either be extremely capable and flexible—able to learn and reconfigure itself on the spot, and in particular adjusting for significant changes in buoyancy—or possess detailed prior data on the target environment. Thus, the most accessible exploration path likely needs to take a stepped approach where multiple missions expand knowledge of the environment and vehicles' performance within it (this is directly recommended in Chap. 4 as one of the primary roles of a precursor cryobot mission that extends all the way to the ocean floor). With proper design, it is possible that some degree of buoyancy shift can be accommodated in the field. If the vehicle is flexible enough to accommodate the unknown conditions, and there is sufficient communication capability (including docking for an underwater vehicle) software updates can expand a vehicle's capabilities to a certain extent. This underscores the need for a robust data link with the vehicle and an attempt in the design phase to anticipate as many behavioral unknowns as possible, while equipping the vehicle with enough sensors to self-diagnose when a novel problem arises and to "phone home" with that data.

Lesson 2: The Utility of "Supervised Autonomy"

The vehicles developed by Stone Aerospace generally run in two modes: fully autonomous and "supervised autonomy." In the latter we connect a high bandwidth data fiber to the vehicle, purely for high-bandwidth communications (e.g. Fig. 10.99).

This allows operators, engineers, and users (especially scientists) to observe what the bot is seeing in real-time—generally including a live 3D map and live sensor data displays—during an otherwise autonomous mission and yet have the ability to intervene if merited. We have used this capability to advantage on multiple missions:

Fig. 10.99 Stern view of ARTEMIS with skins and syntactic removed showing the 15 km data fiber spooler

- To detect defective sensor readings that are difficult or impossible to characterize ahead of time and thus cannot be automatically detected, and then abort a mission if merited.
- To observe anomalies in the data that suggest the presence of something unusual, and cause the vehicle to go into a fine grid measurement mode to more precisely locate the anomaly, doing so during the conduct of the mission.
- To adjust settings on certain sensors to improve the quality of the data when working in a varying topography and material environment. This was particularly true in the case of bathymetric data (whose return signal is highly dependent on the material characteristics of the floor sediment).
- To develop advanced behaviors and test them in the field while "looking over the robot's shoulder" to allow for a drastically more aggressive development cycle.

The power of humans in the loop should not be underestimated. Generally in our field deployments, we could obtain all data without humans in the loop, but either data quality would suffer or deployment duration would become unreasonably long. Similarly, the development of new vehicle behaviors in a new environment is most efficient when being able to implement the develop-test-iterate cycle in real time. While it is possible to work out a good deal of development in simulation and in the lab, in our experience, humans in the loop in the field allow us to be far more aggressive with our development as well as science and get much higher return.

In summary, there is a full spectrum of autonomy, from servo-level joystick-controlled drones to completely autonomous agents that do exploration and research by themselves. An optimal development schedule and science return is obtained somewhere in the middle, where well-tested, established automation takes over low-level tedious tasks, and human reasoning is freed to concentrate on the big picture and responding to unexpected events, developing new capabilities on the fly.

We do not, however, envision the high-bandwidth communications we implement for Supervised Autonomy to be a requirement (or even possible) for a deep space flagship mission working beneath an ice cap. The greatest returns are for developing and proving the technology during field test campaigns on Earth. However, technologies to enable data communications through deep ice and great distances underwater which have sufficient bandwidth to provide some level of supervision and, naturally, data transfer, definitely remain to be developed and must receive high priority in the coming years if such a mission is to be a success. It is also worth considering that while long range sub-ice AUV missions on Ocean Worlds will certainly require completely autonomous operation, it may be worthwhile to equip the AUV with a limited range fiber optic (bare fiber) spooler for initial forays away from the cryobot, so as to allow high fidelity early feedback to mission controllers prior to stepping up to fully autonomous long range missions.

Lesson 3: Use Redundant Sensors for Critical Control Tasks

For critical automated systems that depend on discrete sensed data, it is prudent to employ an array of alternative sensors to assure a conservative decision datum. A case in point: the ENDURANCE Profiler/Sonde was equipped with a bottom altimeter that was designed to inform the spooler servos to stop descent when the sonde was 1 m above the lake bottom. While this solution worked during testing, we found that the returns from the extremely soft sediments at the bottom of Lake Bonney were too weak to make for a reliable stop. We developed a solution on site that involved construction of a data fusion filter that used the sonde altimeter, rotary encoder readings from the Profiler drum giving sonde pay-out, multi-beam down-look sonar, DVL bottom ranges, and discrete beam down-looking sonars. During periods of loss of sonde altimeter data, the remaining sensors were used to calculate an estimated distance from the sonde to the bottom to determine when the cable payout should stop to suspend the sonde at the required 1 m standoff range. This worked reliably on all remaining missions.

Similarly, we generally rely on multiple sensors and sensing modalities to arrive at a navigation position (and dynamic state) estimate. On DEPTHX, the SLAM solution complemented the pure dead-reckoned estimate using DVL and RLG INS. On ENDURANCE and ARTEMIS, where the available 3D geometry in the environment would not sufficiently constrain the vehicle horizontal position, the DVL/INS dead reckoning was complemented by the iUSBL, the visual homing, and magnetic beacon tracking systems (see Figs. 10.100, 10.101 and 10.102), each having complementary strengths and weaknesses.

The SUNFISH® AUV employs DVL/IMU dead reckoning complemented by an iUSBL, a scan-matching SLAM solution, visual homing, and GPS (when at the surface). Such combinations lead to a high probability of successful mission completion (as marked, e.g., by the vehicles' ability to return to dock). For a flagship mission AUV, additional navigational layers will be merited (including, e.g. 3D SLAM, bathymetric (terrain-relative) localization, image mosaic matching, etc.).

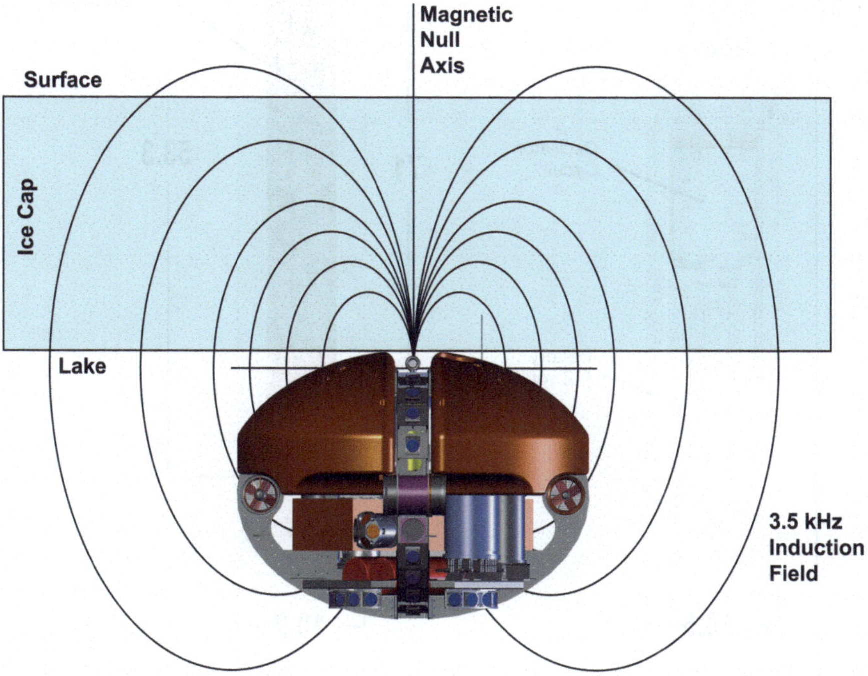

Fig. 10.100 Notional vision of the ENDURANCE through-ice tracking system. The magnetic oscillator carried by the vehicle generated a an AC magnetic field whose null (vertical) axis could be tracked through ice

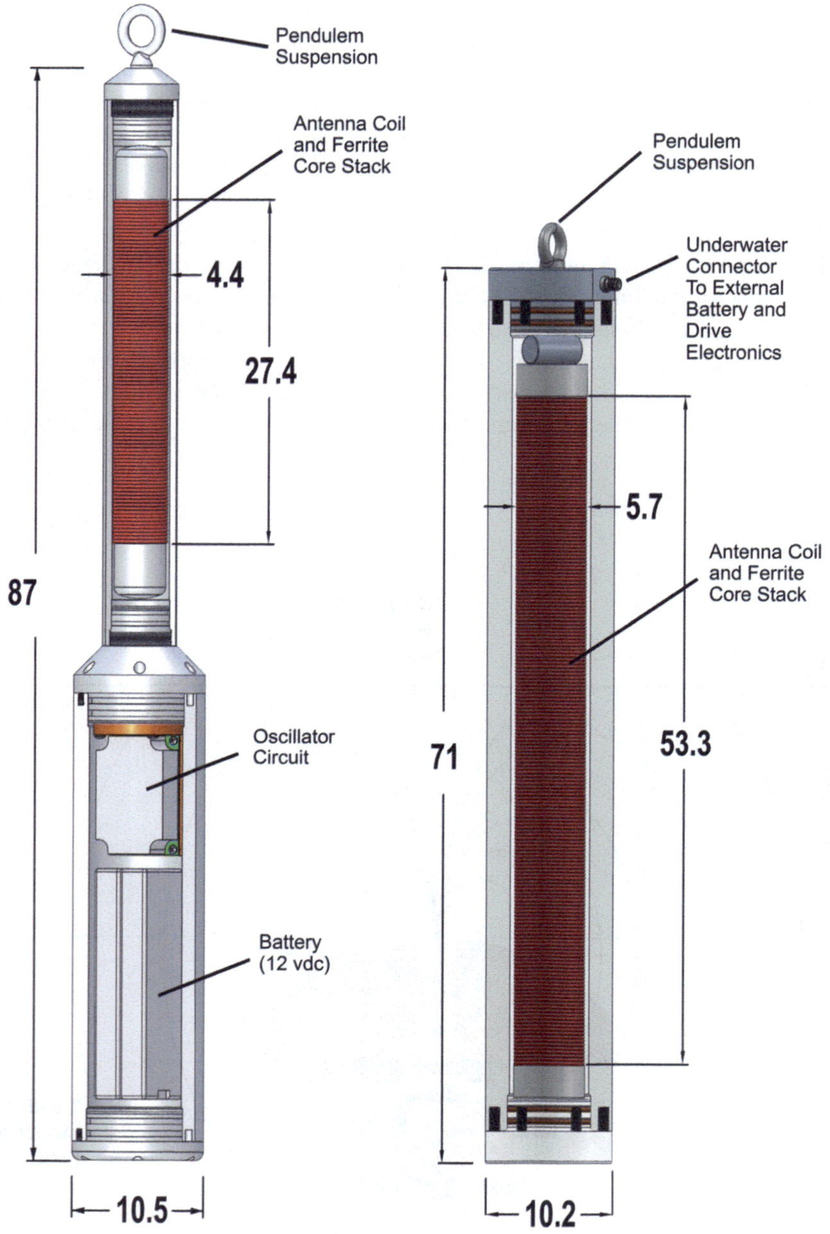

Fig. 10.101 Two versions of magnetic field generators used for ENDURANCE (left) and ARTEMIS (right). The ENDURANCE beacon operated at 1 W power level and, because Lake Bonney had a lense of fresh water that the vehicle was travelling within, this was easily detected as far as 600 m distant from the vehicle; The ARTEMIS beacon could be driven to 25 W and was able to be detected at a depth of 30 m of seawater in the Ross Sea beneath 4 m of sea ice and 5 m of planet ice (total 9 m)

Fig. 10.102 ARTEMIS tracking team consisting of the magnetic beacon localizer (center); GPS for absolute vehicle position lock (left); and radio operator for relaying vehicle positions to Mission Control. This method of tracking the vehicle can be used to penetrate hundreds of meters of ice and is a useful tool for validation of sub-ice autonomous systems performance

Lesson 4: Acoustic Sensors Work Under an Ice Cap

Several groups have demonstrated navigation with a DVL looking up at ice (Sibenac 2004) and up-look multibeam mapping of sea ice and icebergs (Kimball and Rock 2008). ARTEMIS was another demonstration of this kind of capability.

On the other hand, to our knowledge, before ENDURANCE, an iUSBLsystem had never been used beneath an ice cap. For that mission, the complications of the environment were further compounded by the presence of a strong chemocline (with associated large density gradient) in the lake. Both the underside of the ice cap (4 m thick), and the surface of the chemocline (at 16 m depth) served as acoustic reflectors. The vehicle was generally constrained to work within this 12 m thick sandwich, at distances of up to 2 km from the melt hole. The degree to which an iUSBL would be functional in such an environment was unknown. ENDURANCE was able to make effective use of the USBL navigation solution out to about 300 m from the acoustic beacon suspended at the melt hole, and in more open waters, ARTEMIS was able to clearly find and use the iUSBL from 800 m range. More powerful, compact beacons will need to be developed for a Europa homing solution. Ideally, such a beacon (suspended below the cryobot) would be able to be received from tens of kilometers so that a vehicle mapping the silicate core surface at the bottom of the European ocean could return to the vicinity of the dock location using dead reckoning and SLAM and successfully restore the data link with the lander.

Lesson 5: Redundancy is Essential

The DEPTHX/ENDURANCE chassis was designed with parallel power sources and parallel thrust systems, so that loss of any single power supply and loss of up to three thrusters would not prevent the vehicle from achieving its mission. Both of these capabilities were invoked during work at Lake Bonney: asymmetric power consumption led to automatic shut down of a power stack on two occasions. On a variation of this principle, ARTEMIS had separate power and control busses for maneuvering and attitude control thrusters, and independently-powered and -controlled double armatures for each prime mover, allowing the vehicle to return home (albeit at reduced power) if any single armature, power bus, or comms bus/controller failed.

10.6 Closing Thoughts

The depth of Ocean World sub-surface oceans, and the thickness of the ice caps that cover them remain widely debated unknowns. Estimates of the ice cap thickness vary from 1 to 30 km in the case of Europa with corresponding estimates of ocean depth varying from 30 to 100 km below that point. However, local gravity must be taken into account. Europa's 0.134g local surface gravity means that a 100 km deep ocean will have a hydrostatic pressure head of a little over 1300 bar (not counting the ice cap). This is not radically dissimilar to the 1100 bar pressures at the bottom of the Challenger Deep on Earth, an environment for which several vehicles (both manned and robotic) have explored and survived. There should be no illusions, however: a 2-stage, cryobot-deployed AUV search for life on any of the Ocean Worlds will comprise the most sophisticated devices yet conceived by mankind and represent one of the greatest intellectual and physical exploration challenges of our era. Yet the mission is within our grasp, within 20 years, if there were financial means to enable the mission. Below are a few elements required to prepare for such a mission.

10.6.1 Aggressively Pursue Terrestrial Analog Campaigns

Earth's polar regions, but in particular Antarctica, with its numerous unexplored sub-glacial lakes, present appropriate analogs for testing fully integrated cryobot-AUV systems at relevant scales. Efforts need to be undertaken to establish long term field laboratories that allow for persistent cryobot access to (and recovery from) these sub-glacial lakes—in effect to provide on-demand ice elevators for the deployment and test of persistent sub-surface AUV missions as well as to assess the long-term effects of sub-ice missions lasting years on the performance of both cryobot and AUV systems. It is only through extensive field testing of fully-integrated Ocean World systems in these unexplored sub-glacial lakes—Lake Vostok being the most well known, but there are scores of others that are equally, and in many cases more interesting than Vostok—that true reliability and validation of appropriate life search, detection, and characterization algorithms will be achieved. The 4 km of ice thickness in Antarctica represent the last 4 km of ice before ocean breakthrough on Europa. Presently, there is no dedicated nor well-supported program to enable routine fielding of fully-integrated prototype cryobot-AUV systems to Antarctica, and it remains an important funding gap in efforts to enable more advanced life search missions to Ocean Worlds.

10.6.2 Conduct Realistic Environmental Testing

While Antarctic sub-glacial lakes provide a full-scale laboratory for testing cryobot deployment of AUVs into sub-surface oceans, there is no terrestrial analog for the starting conditions on the surface of any of the Ocean Worlds. In order to prove out both a dedicated cryobot-only mission and integrated (2-stage cryobot-deployed-AUV) full scale systems a large scale test facility is required that will be capable of chilling a nominally 5 m diameter $\times$ 20 m high cylinder of water to 100 K temperature and pulling a hard vacuum. The exact size of the facility will depend on the general expectations for the size of the carrier cryobot for the most ambitious sub-surface AUV deployment missions. Both diameter and depth of such a facility are tied to two vehicle thermal characteristics: the degree to which power must be diverted to the sidewalls of the cryobot in order to prevent capture of the vehicle via refreeze before it can pass the zone just melted; and the speed of descent (which relates to the total power being dumped at the melt face and the diameter of the vehicle). The diameter of such a facility needs to be large enough that thermal diffusion from the exterior is insufficient to affect the environment the vehicle sees and, conversely, that the vehicle does not heat the ice boundary to an extent that would significantly deviate from that of an infinite heat sink; any failure to do so will compromise the validity of the test. Similarly, the height of the facility will be determined by the time to reach steady descent rate that can be instrumented. The faster the vehicle, the taller the test facility must be. Such a facility is not unprecedented—NASA built many very large thermal-vacuum (Tvac) chambers during the Apollo era. But none of those were structured around the maintenance of a large block of ice at ambient (100 K) temperature and hard vacuum while sublimate was being generated during the starting phase of cryobot surface penetration. For the same reasons that all current satellite and space vehicle systems undergo extensive Tvac and random vibration testing prior to launch, so will an AUV being carried by a cryobot need to be onboard for full simulation of the Starting Problem and the other phases of ice penetration described in Chap. 4.

10.6.3 Confront the Issue of Nuclear Power

While many planetary missions rely on solar, small radio-thermal generator (RTGs), or even primary batteries for power, a simple calculation of the required power and total energy to bring even a miniaturized AUV system through a realistic Ocean World ice shell, shows that nuclear energy, and in fact a nuclear reactor, is the only way to go. Once an AUV rover is deployed into the ocean from a cryobot, the distances to be covered and the mission durations also call for a long-lasting power source.

Chapter 4 (Sect. 4.4) discussed nuclear power options for cryobots for penetrating Ocean World ice caps. For deep 2-stage missions in which the cryobot will be carrying an AUV, there is no question that significant thermal power will be needed (between 20 and 100 kW). That level of heat can be most compactly produced by a micro-fission reactor. Once that vehicle (cryobot) breaks through into the sub-surface ocean, the need for such power rapidly diminishes. Yet, a planet-ranging AUV will also need power and that will not come from batteries for the same reasons that cryobots cannot be powered by batteries. At this point there are two options: in the first scenario, the AUV carries an independent RPS power supply—not unlike the MMRTGs that are being used to power Curiosity on Mars as this chapter is being written. That low level of power, while persistent, will be insufficient to turn on and maintain prime mover thrusters that will be needed to propel the AUV on its mission, but it can be used to trickle charge an onboard traditional high power density secondary battery stack (e.g. Li-ion) and operate the vehicle in pulse power mode. There will be an optimum design for the size of the RPS and the size of the battery that will depend on many factors, the most important of which will be the presence or absence of sub-ice ocean currents, and their variation and magnitude. Once those currents are known and characterized by a precursor cryobot/sonde mission the level of power output required to effectively maneuver (and hold position) within those expected currents will drive the propulsion system design, the buffer battery design, and the RPS design, if it is achievable.

As studies from the SPINDLE project showed (see Chap. 4) using an all RPS power source for a 2-stage cryobot-AUV is not an effective solution, simply from the standpoint of mass limits. However, there is another problem: a large RPS continuously generates heat from the moment it is set inside the launch vehicle, and radiators for dissipating that energy in free space are massive. The advantage of the compact fission reactor is that it can be launched in the "off" state, with no thermal power being radiated. However, considering the state of the cryobot *after* delivery of the AUV, there are only basic hotel management and communication needs—power needs that could easily be supplied by an MMRTG. Why not, then, consider clever ways to use the variable power of a compact fission reactor to drive the AUV? 20–100 kW_{th} is sufficient to allow for the development of high reliability non-propeller-based pulse thrust systems as the primary propulsor for the AUV.

The use of nuclear power for Ocean Worlds missions is inescapable. A significant effort needs to be undertaken to expand the world stock of Pu^{238} for such missions—and there are economical means for producing it (see Chap. 4). More importantly, a major effort should be launched into the design, development, and hot fire proof testing of LEU-fueled, ultra-compact reactors in the 20–100 kW_{th} range, for both cryobot and planet-wide sub-surface AUV ocean exploration.

With new power systems in hand, and new laboratory and field facilities to perform testing at, we can achieve sub-surface Ocean World exploration and the promise of discovery of new forms of life that it brings with it starting with the technology of today. There is no fundamental physical barrier to success, it is an engineering problem that is within the reach of any suitably motivated community on Earth.

Acknowledgements The majority of the work reported herein was conducted under NASA grants NNG04GC09G (DEPTHX); NNX07AM88G (ENDURANCE); and NNX12AL65G (SIMPLE/ARTEMIS). We would like to thank NASA program managers Dr. Michael Meyer and Dr. David Lavery; Dr. Carl Pilcher and Dr. John Rummel; and Dr. Mary Voytek, respectively, for their help and guidance during the conduct of these projects, all of which involved remote fieldwork. Stone Aerospace founder Dr. Bill Stone was PI for the DEPTHX project; Dr. Peter Doran (originally of UIC and now at LSU) was PI for ENDURANCE; and Dr. Britney Schmidt (originally of UT Austin and now at Georgia Tech) was PI for the SIMPLE/ARTEMIS project. We would also like to express our gratitude to the National Science Foundation (NSF) Office of Polar Programs (OPP) for their field logistics support in making possible the three Antarctic deployments for these vehicles. The following individuals were integral to the development and deployment of these vehicles and the interpretation of the resulting field data:

DEPTHX Dan Durda, Dave Everett, Nathaniel Fairfield, Ernest Franke, Marcus Gary, Dominic Jonak, Nigel Jones, George Kantor, John Kerr, Thomas Lyons, Ian Meinzen, Norm Pace, Mike Rigney, Jason Sahl, John Spear, Michael Stevens, and Dave Wettergreen.

ENDURANCE Alessandro Febretti, Shilpa Gulati, Shriram Iyer, Andrew Johnson, Bob Kenworthy, Greg Kuhlmann, Aniket Murarka, Maciej Obryk, Jim Olech, Brian Pease, Rachel Price, John Priscu, Robert Spigel, and Mohan Sridharan.

ARTEMIS and SUNFISH Don Blankenship, Evan Clark, John Harman, Keith Huffstutler, Stacy Kim, Jason Kapit, Peter Kimball, Jade Lawrence, Justin Lawrence, Scott Lelievre, Laura Lindzey, Bob McCabe, Josh Moor, Loral O'Hara, Jeremy Paulus, Brian Pease, Bryce Rothhammer, Mark Scully, Mark Skidmore, Justin Smith, Neal Tanner, Luke Winslow.

References

Austin, T.C., Hosom, D.S., Kuchta, D.H.: Long baseline acoustic navigation—a flexible approach to custom applications. In: Proceedings of the OCEANS 84 Conference, vol. 16, pp. 69–74. IEEE, Washington, DC (1984)

Beegle, L., et al.: SHERLOC: scanning habitable environments with Raman & luminescence for organics & chemicals. In: 2015 IEEE Aerospace Conference 1–11 (IEEE) (2015). https://doi.org/10.1109/aero.2015.7119105

Bellingham, J.G.: Autonomous underwater vehicle docking. Springer-Verlag Handbook of Ocean Engineering, pp. 387–406 (2016)

Bellingham, J.G., Leonard, J.J., Vaganay, J., Goudey, C., Atwood, D., Consi, T., Bales, J., Schmidt, H., Chryssostomidis, C.: AUV operations in the Arctic. In: Proceedings of the Sea Ice Mechanics and Arctic Modeling Workshop, Anchorage, Alaska, USA (1995)

Bellingham, J.G., et al.: Efficient propulsion for the tethys long-range autonomous underwater vehicle. In: Autonomous Underwater Vehicles, 2010 IEEE/OES. https://doi.org/10.1109/auv.2010.5779645

Boqomolny, E., Swift, S., Vanholsbeeck, F.: Total viable bacterial count using a real time all-fibre spectroscopic system. Analyst **138**, 4112–4119 (2013)

Bowen, A., et al.: Design of Nereid-UI: a remotely operated underwater vehicle for oceanographic access under ice. Oceans 2014, St. John's, NL, Canada, 14–19 Sept 2014, (2014). https://doi.org/10.1109/oceans.2014.7003125

Bramall, N.E.: The remote sensing of microorganisms. Dissertation University of California, Berkeley (2007)

Bramall, N.E., Stoker, C.R., Price, P.B., Allamandola, L.J.: Detecting organics in situ using fluorescence. In: 40th Lunar and Planetary Science Conference, Paper No. 2470 (2009)

Broad, W.J.: The Universe Below. Simon & Schuster, New York (1997)

Cadena, C., et al.: Past, present, and future of simultaneous localization and mapping: toward the robust-perception age. IEEE Trans. Robot. **32.6**, 1309–1332 (2016)

Cazenave, F., Zook, R., Carroll, D., Flagg, M., Kim, S.: Development of the ROV SCINI and deployment in McMurdo Sound, Antarctica. J. Ocean Technol. **6**(3) (2011)

Doble, M.J., Forrest, A.L., Wadhams, P., Laval, B.E.: Through-ice AUV deployment: operational and technical experience from two seasons of Arctic fieldwork. Cold Reg. Sci. Technol. **56**(2), 90–97 (2009)

Doran, P.T., Priscu, J.C., Lyons, W.B., Walsh, J.E., Fountain, D.A.G., McKnight, M., Moorhead, D.L., Virginia, R.A., Wall, D.H., Clow, G.D., Fritsen, C.H., McKay, C.P., Parsons, A.N.: Antarctic climate cooling and terrestrial ecosystem response. Nature **415**, 517–520 (2002)

Doran, P.T., Wharton, R.A. Jr., Lyons, W.B.: Paleolimnology of the McMurdo Dry Valleys, Antarctica. J. Paleolimnol. **10**, 85–114 (1994)

Fairfield, N., Kantor, G., Wettergreen, D.: Three dimensional evidence grids for SLAM in complex underwater environments. In: Proceedings of the 14th International Symposium of Unmanned Untethered Submersible Technology (UUST05), Lee, New Hampshire, AUSI (2005)

Fairfield, N., Kantor, G., Wettergreen, D.: Towards particle filter SLAM with three dimensional evidence grids in a flooded subterranean environment. In: Proceedings of the IEEE International Conference on Robotics and Automation (ICRA) (2006)

Fairfield, N., Jonak, D., Kantor, G., Wettergreen, D.: Field results of the control navigation, and mapping systems of a hovering AUV. In: Proceedings of the 15th International Symposium of Unmanned Untethered Submersible Technology (UUST07), Lee, New Hampshire, AUSI (2007a)

Fairfield, N., Kantor, G., Wettergreen, D.: Real-time SLAM with Octree Evidence grids for exploration in underwater tunnels. J. Field Robot (2007b). https://doi.org/10.1002/rob.20165

Fairfield, N.: Localization, mapping, and planning in 3D environments. PhD thesis, Carnegie-Mellon University, CMU-RI-TR-05–09 (2009)

Franke, E., Magee, M., Rigney, M., Stone, B.: Progress toward the development of lifeform detection algorithms for the Deep Phreatic Thermal eXplorer (DEPTHX). Proc. SPIE **5906**, 590613 (2007)

German, C.R., Yoerger, D.R., Jakuba, M., Shank, T.M., Langmuir, C.H., Nakamura, K.: Hydrothermal exploration with the autonomous benthic explorer. Deep-Sea Res Part 1, **55**, 203–219 (2008a). https://doi.org/10.1016/j.dsr.2007.11.004

German, C.R., et al.: Hydrothermal activity on the southern Mid-Atlantic Ridge: tectonically- and volcanically-controlled venting at 4-5S. Earth Planet. Sci. Lett. **273**, 332–344 (2008b). https://doi.org/10.1016/j.epsl.2008.06.048

Gonzalez-Baños, H.H., Latombe, J.-C.: Navigation strategies for exploring indoor environments. Int. J. Robot. Res. **21**(10–11), 829–848 (2002)

Gulati, S., Richmond, K., Flesher, C., Hogan, B.P., Murarka, A., Kuhlmann, G., Sridharan, M., Stone, W.C., Doran, P.T.: Toward autonomous scientific exploration of ice-covered lakes—field experiments with the ENDURANCE AUV in an Antarctic dry valley. In: Proceedings of the 2010 IEEE International Conference on Robotics and Automation (ICRA), (Anchorage, Alaska), pp. 308–315, May 3–7 (2010)

Hand, K., Chyba, C.: Astrobiology and the potential for life on Europa. In: Pappalardo, R.T., McKinnon, W.B., Khurana, K. (eds.) Europa. The University of Arizona Press, Tucson, pp. 631–634 (2009)

Hidalgo, F., Thomas, B.: Review of underwater SLAM techniques. In: 2015 6th International Conference on Automation, Robotics and Applications (ICRA). IEEE (2015)

Hogan, B., Flesher, C., Stone, W.C.: Development of a sub-ice automated profiling system for Antarctic lake deployment. In: Proceedings of the 16th International Symposium on Unmanned Untethered Submersible Technology (UUST '09), Durham, New Hampshire, USA, 23–26 August. AUSI (2009)

Ianoul, A.T., et al.: UV resonance Raman spectroscopic detection of nitrate and nitrite in wastewater treatment processes. Anal. Chem. **74**(6) (2002)

Joho D., Stachniss C., Pfaff P., Burgard W.: Autonomous Exploration for 3D Map Learning. In: Berns K., Luksch T. (eds) Autonome Mobile Systeme 2007. Informatik aktuell. Springer, Berlin, Heidelberg (2007)

Kaminski, C., Crees, T., Ferguson, J., Forrest, A., Williams, J., Hopkin, D., Heard, G.: 12 days under ice—an historic AUV deployment in the Canadian High Arctic. In Autonomous Underwater Vehicles (AUV) (2010)

Kimball, P., Rock, S.: Sonar-based iceberg-relative AUV navigation. In: Autonomous Underwater Vehicles (2008). https://doi.org/10.1109/auv.2008.5290534

Kimball, P. W., Clark, E. B., Scully, M.,Richmond, K., Flesher, C., Lindzey, L. E., Harman, J., Huffstutler, K., Lawrence, J., Lelievre, S., Moor, J., Pease, B., Siegel, V., Winslow, L., Blankenship, D. D., Doran, P., Kim, S., Schmidt, B. E., Stone, W. C.: Antarctic deployment of the ARTEMIS under-ice AUV docking system. J. Field Robot. **35**(2), 299–308 (2018)

Kukulya, A., Plueddemann, A., Austin, T., Stokey, R., Purcell, M., Allen, B., Littlefield, R., Freitag, L., Koski, P., Gallimore, E., et al.: Under-ice operations with a REMUS-100 AUV in the Arctic. In: Autonomous Underwater Vehicles (AUV). IEEE/OES (2010)

Kunz, C., Murphy, C., Singh, H., Pontbriand, C., Sohn, R.A., Singh, S., Sato, T., Roman, C., Nakamura, K., Jakuba, M., Eustice, R., Camilli, R., Bailey, J.: Toward extraplanetary under-ice exploration: robotic steps in the Arctic. J. Field Robot. **26**(4), 411–429 (2009)

Lorensen, W.E., Cline, H.E.: Marching cubes: a high resolution 3D surface construction algorithm. Comput. Graphics **21**(4), 163–169. ACM-0-89791-227-6/87/007/0163 (1987)

Malhalanobis, P.C.: On the generalised distance in statistics. Proc. Natl. Inst. Sci. India **2**(1), 49–55 (1936)

Mallios, A., et al.: Scan matching SLAM in underwater environments. Auton. Robots **36.3**, 181–198 (2014)

Marshall, C.P., Edwards, H., Jehlicka, J.: Understanding the application of Raman Spectroscopy to the detection of traces of life. Astrobiology **10**(2) (2010). https://doi.org/10.1089/ast.2009.0344

McEwen, R.S., Hobson, B.W., McBride, L., Bellingham, J.G.: Docking control system for a 54-cm diameter (21-in) AUV. IEEE J. Oceanic Eng. **33**(4), 550–562 (2008)

McKenzie, D.: WALLS—Tools for Cave Survey Management, Version 2, Build 2016-11-18. Texas Speleological Survey, Austin, Texas (2016). www.texasspeleologicalsurvey.org/software/walls/tsswalls.php

Millard, N.W., Griffiths, G., Finegan, G., McPhail, S.D., Meldrum, D.T., Pebody, M., Perrett, J.R., Stevenson, P., Webb, A.T.: Versatile autonomous submersibles—the realising and testing of a practical vehicle. Underwater Technol. **23**, 7–17 (1998)

Milne, P.H.: Underwater Acoustic Positioning. Gulf Publishing, TX (1983)

Murarka, A., Kuhlman, G., Gulati, S., Sridharan, M., Flesher, C., Stone, W.C.: Vision-based frozen surface egress: a docking algorithm for the ENDURANCE AUV. In: Proceedings of the 16th International Symposium on Unmanned Untethered Submersible Technology (UUST), Durham, New Hampshire, USA, 23–26 Aug 2009. AUSI (2009)

National Geographic Society, Mapping the Labyrinth: (1999). https://www.nationalgeographic.org/media/mapping-the-labyrinth/

Nicholls, K.W., Abrahamsen, E.P., Buck, J.J.H., Dodd, P.A., Goldblatt, C., Griffiths, G., Heywood, K.J., Hughes, N.E., Kaletzky, A., Lane-Serff, G.F., McPhail, S.D., Millard, N.W., Oliver, K.I.C., Perrett, J., Price, M.R., Pudsey, C.J., Saw, K., Stansfield, K., Stott, M.J., Wadhams, P., Webb, A.T., Wilkinson, J.P.: Measurements beneath an Antarctic ice shelf using an autonomous underwater vehicle. Geophys. Res. Lett. **33**(8) (2006)

Olsen, H.E., Paten, B., Akeson, M.: The Oxford Nanopore MinION: delivery of nanopore sequencing to the genomics community. Genome Biol. **201617**, 239 https://doi.org/10.1186/s13059-016-1103-0. Published: 25 November 2016 (2016)

Petillo, S., Schmidt, H.: Autonomous and adaptive plume detection and tracking with AUVs: concepts, methods, and available technology. In: MCMC 2012 Conference, Arenzano, Italy, 19–21 Sept 2012 (2012)

Ribas, D., et al.: Underwater SLAM in man-made structured environments. J. Field Robot. **25**(11–12), 898–921 (2008)

Richmond, K., Gulati, S., Flesher, C., Hogan, B., Stone, W.C.: Navigation, control, and recovery of the ENDURANCE under-ice hovering AUV. In: Proceedings of the 16th International Symposium on Unmanned Untethered Submersible Technology (UUST '09), Durham, New Hampshire, USA, 23–26 Aug 2009. AUSI (2009)

Robinson, N.J.: An oceanographic study of the cavity beneath the McMurdo Ice Shelf, Antarctica. Master's thesis, Victoria University of Wellington (2004)

Sahl, J.W., Fairfield, N., Harris, J.K., Wettergreen, D., Stone, W.C., Spear, J.R.: Novel microbial diversity retrieved by autonomous robotic exploration of the world's deepest vertical phreatic sinkhole. Astrobiology **10**(2) (2010). https://doi.org/10.1089/ast.2009.0378

Sahl, J., Gary, M.O., Harris, J.K., Spear, J.R.: A comparative molecular analysis of water-filled limestone sinkholes in north-eastern Mexico. Environ. Microbiol. **13**(1), 226–240 (2011). https://doi.org/10.1111/j.1462-2920.2010.02324.x

Sibenac, M., Kirkwood, W., Podder, T., Thomas, H.: Autonomous underwater vehicles for ocean research: current needs and state of the art technologies. Mar. Technol. Soc. J. Innovations Ocean Res. Infrastruct. Adv. High Priority Sci. **38**(2) (2004) (Summer)

Spears, A., West, M., Meister, M., Buffo, J., Walker, C., Collins, T.R., Howard, A., Schmidt, B.: Under ice in Antarctica: the ICEFIN unmanned underwater vehicle development and deployment. IEEE Robot. Autom. Mag. **23**(4), 30–41 (2016)

Spigel, R.H., Priscu, J.C., Stone, W.C., Obryk, M.K., Febretti, A., Doran, P.T.: The physical limnology of a permanently ice-covered and chemically stratified Antarctic Lake using high resolution spatial data from an autonomous underwater vehicle. J. Limnol Oceanogr. Manuscript LO-17–0109, accepted for publication May 2017 (2017)

Stone, W.C., Ende, B.A., Wefer, F.L., Jones, N.A.: Automated 3D mapping of submarine tunnels. In: Robotics 2000: ASCE Conference on Robotics for Challenging Environments, Albuquerque, NM, February 28–March 2 (2000)

Stone, W.C., Hogan, B., Flesher, C., Gulati, S., Richmond, K., Murarka, A., Kuhlman, G., Sridharan, M., Doran, P., Priscu, J.: Sub-ice exploration of West Lake Bonney: ENDURANCE 2008 mission. In: Proceedings of the 16th International Symposium on Unmanned Untethered Submersible Technology (UUST '09), Durham, New Hampshire, USA, 23–26 August 2009. AUSI (2009)

Stone, W., Hogan, B., Flesher, C., Gulati, S., Richmond, K., Murarka, A., Kuhlman, G., Sridharan, M., Siegel, V., Price, R., Doran, P., Priscu, J.: Design and deployment of a four-degrees-of-freedom hovering autonomous underwater vehicle for sub-ice exploration and mapping. J. Eng. Marit. Environ. **224**(M4), 341–361 (2010)

Tao, C., et al.: First active hydrothermal vents on an ultraslow-spreading center: Southwest Indian Ridge, Geological Society of America. Geology **40**(1), 47–50 (2012). https://doi.org/10.1130/g32389.1

Tian, Y., Zhang, A., Li, W.: A behavior-based planning strategy for deep-sea hydrothermal plume tracing with autonomous underwater vehicles. Oceans 2014, Tapei (2014). https://doi.org/10.1109/oceans-taipei.2014.6964302

Whitcomb, L.L., Yoerger, D.R., Singh, H.: Combined Doppler/LBL based navigation of underwater vehicles. In: Proceedings of the Unmanned Untethered Submersible Technology Conference (UUST), Durham, New Hampshire. 22–25 Aug. AUSI (1999)

Widditsch, H.R., SPURV—The First Decade, University of Washington Applied Physics Laboratory (UW-APL) Report 7215: October 1973, see also http://dtic.mil/dtic/tr/fulltext/u2/a050816.pdf (1973)

Wirtz, M., Hildebrand, M.: IceShuttle Teredo: an ice-penetrating robotic system to transport an exploration AUV into the Ocean of Jupiter's Moon Europa. In: 67th International Astronautical Congress (IAC), Guadalajara, Mexico (2016)

Yoerger, D.R., Bradley, A.M., Jakuba, M., German, C.R., Shank, T., Tivey, M.: Autonomous and remotely operated vehicle technology for hydrothermal vent discovery, exploration, and sampling. Oceanography **20**(1) (2007)

Chapter 11
Titan Submarine

Steven R. Oleson, Jason Hartwig, Jeffrey Woytach, Michael Martini, Anthony Colozza, Robert Jones, Thomas Packard, Paul Schmitz, Amy Stalker, Ralph D. Lorenz, Michael V. Paul and Justin Walsh

11.1 Introduction

The conceptual design of a vehicle to explore Saturn's moon Titan (Fig. 11.1) investigated what science a submarine for Titan's liquid hydrocarbon seas might accomplish down at ~93 K (−180 °C), and what that submarine might look like.

S. R. Oleson (✉) · J. Hartwig · J. Woytach · M. Martini · A. Colozza
R. Jones · T. Packard · P. Schmitz · A. Stalker
NASA Glenn Research Center, 21000 Brookpark Rd. MS 302-1,
Cleveland, OH 44135, USA
e-mail: steven.r.oleson@nasa.gov

J. Woytach
e-mail: jeffrey.m.woytach@nasa.gov

M. Martini
e-mail: michael.c.martini@nasa.gov

A. Colozza
e-mail: anthony.j.colozza@nasa.gov

T. Packard
e-mail: thomas.w.packard@nasa.gov

P. Schmitz
e-mail: paul.c.schmitz@nasa.gov

A. Stalker
e-mail: amy.r.stalker@nasa.gov

R. D. Lorenz · M. V. Paul
Applied Physics Laboratory, Johns Hopkins University,
11100 Johns Hopkins Road, Laurel, Maryland 20723-6099, USA

J. Walsh
The Pennsylvania State University, State College, USA

V. Badescu and K. Zacny (eds.), *Outer Solar System*,
https://doi.org/10.1007/978-3-319-73845-1_11

Fig. 11.1 Titan, Saturn's largest moon, has a dense atmosphere and large, open bodies of liquid hydrocarbons on its surface

The detailed effort was funded by NASA's Innovative Advanced Concepts (NIAC) Phase I for 2014, and is described herein.

Focusing on a flagship class science system with about 1000 kg of landed mass, an independently operated submerged craft (submarine) was found capable of accomplishing extensive and exciting science both above and below the surface of the Kraken Mare (Fig. 11.2). The submerged science includes mapping using side looking sonar, imaging and spectroscopy of the sea at all depths, as well as sampling of the sea's bottom and shallow shoreline. While surfaced the submarine will not only sense weather conditions (including the interaction between the liquid and atmosphere) but also image the shoreline, as much as 2 km inland. This imaging requirement pushed the landing date to Titan's next summer period ($\sim$2047) to allow for continuous lighted conditions, as well as direct-to-Earth (DTE) communication, avoiding the need for a separate relay orbiter spacecraft. Submerged and surfaced investigation are key to understanding both the hydrological cycle of Titan as well as gather hints to how life may have begun on Earth using liquid/sediment/chemical interactions. An estimated 25 Mb of data per day would be generated by the various science packages. Most of the science packages (electronics at least) can be safely kept inside the submarine pressure vessel and warmed by the isotope power system.

The baseline 90 day mission would be to sail alternately submerged and surfaced around and through Kraken Mare investigating the shoreline and inlets to evaluate the sedimentary interaction both on the surface and below. Depths of Kraken have yet to be sensed. Ligeia to the north is thought to be 200 m (656 ft) deep, but a maximum depth of 1000 m (3281 ft) for Kraken Mare was assumed for the design. The sub would spend 20 days at the interface between Kraken Mare and Ligeia Mare for clues to the drainage of liquid methane into the currently predicted

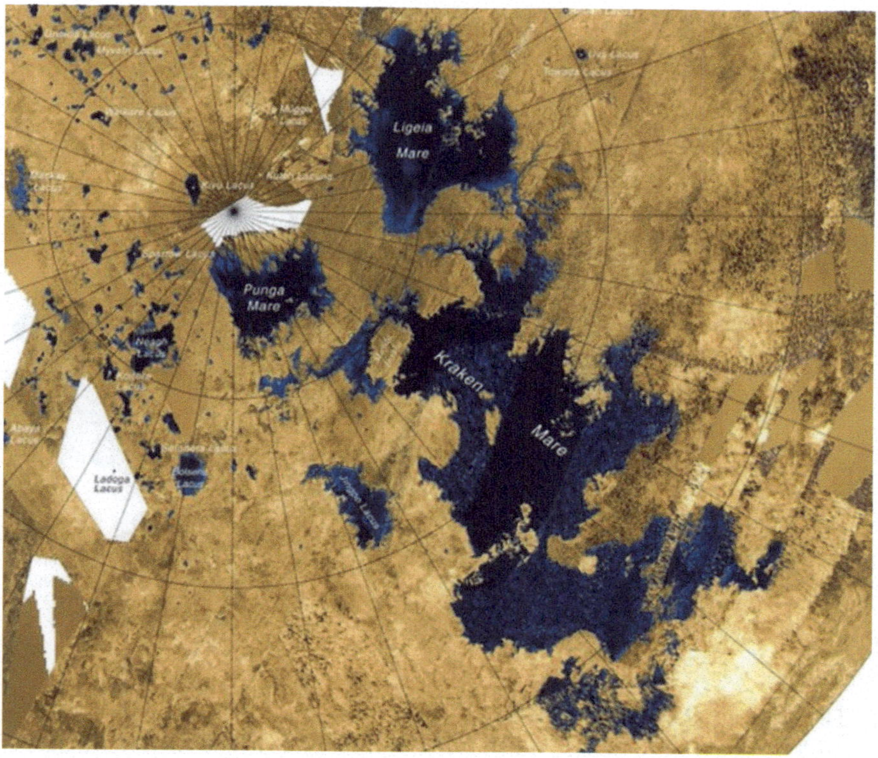

Fig. 11.2 Titan's seas or Mare in the northern hemisphere

predominantly ethane Kraken Mare. During an extended 90 day mission it would transit the throat of Kraken (now 'Seldon Fretum') and perform similar explorations in other areas of Kraken Mare. Once this half year of exploration is completed the submarine could be tasked to revisit points of interest and perhaps do a complete sonar mapping of the seas. All in all, the submarine could explore over 3000 km (1864 miles) in its primary mission at an average speed of 0.3 m/s.

The final concept shown in Fig. 11.3 features a 'sail' or 'dorsal fin' above the hull which is a 4-m by 0.5-m (13.1-ft by 1.6-ft) fixed phased array antenna. This antenna can provide greater than 500 bps for two 8 h Deep Space Network (DSN) communications passes per day. It must operate in a 1.5×10^5 Pa (1.5 bar) nitrogen atmosphere at −180 °C, and then survive up to 1.0×10^6 Pa (10 bar) of −180 °C liquid ethane/methane. The antenna greatly increases the drag on the sub when submerged but that can be offset using the power not needed for communications (~250 W) for the propeller-based propulsion units (propulsors).

Propulsion using bladed propellers, or propulsors is similar to terrestrial submarines. Four ~100 W motors attached to booms provide propulsion and maneuvering while below the surface.

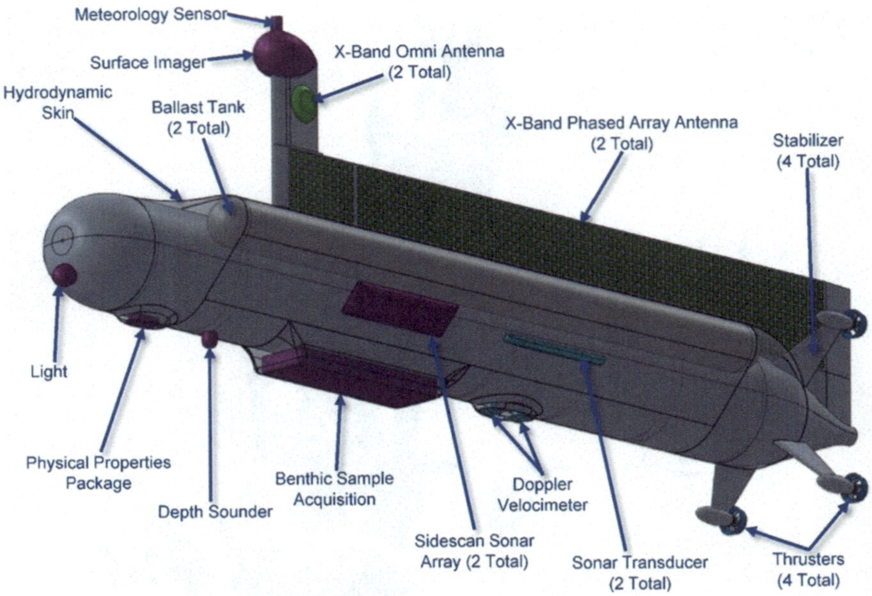

Fig. 11.3 Titan submarine external components

The final design shown in Fig. 11.3 has a mass of approximately 1386 kg (3056 lbm) mass. The sub is 6 m (19.7 ft) long with a 0.62 m (2 ft) diameter pressure vessel. External, closed Ne ballast tanks allow for submerging and hovering as deep as 1000 m (3281 ft), at pressures to 1 MPa (10 bar).

The major systems of the submarine are summarized below (Fig. 11.4):

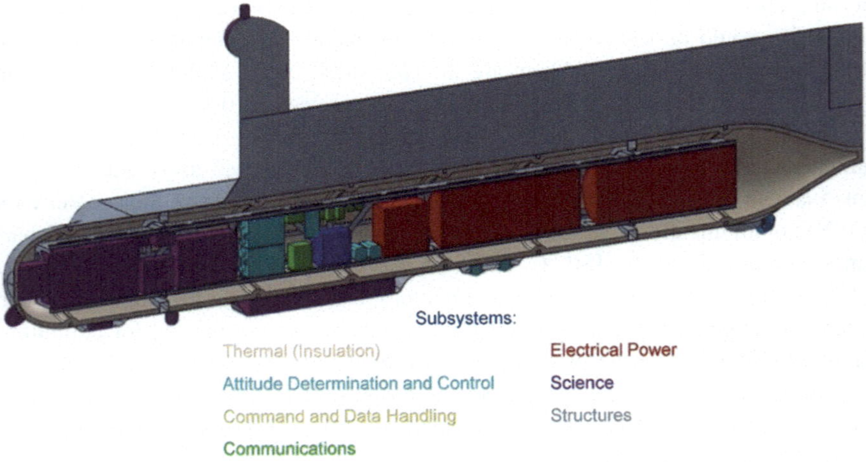

Fig. 11.4 Titan submarine internal components

- Power: Two 430 W end of life (EOL) Stirling Radioisotope Generators (SRGs) for total power of 860 W
- Propulsion: Four 100 W motors on booms to provide up to 1.6 m/s (5.2 ft/s) submerged and 0.9 m/s (3 ft/s) surface speeds, as well as differential steering
- Avionics: X-Band communications DTE ($\sim$800 bps during 16 h DSN passes each day surfaced) using 250 W DC, 4- by 0.5-m (13.1- by 1.6-ft) phased array dorsal antenna; Dual X-band omni antennas; Autonomous Command and Data Handling (C&DH) for 16 h/d surface and 8 h/d submerged exploration; Navigation using Inertial Measure-ment Unit (IMU), Sun direction, Earth tracking, liquid velocity Doppler, sonar scanning
- Thermal: Most internal systems warmed by SRG waste heat; 3 cm (1.1 in.) thick aerogel insulation; 300 W/m^2 heat loss through outer skin; external systems—some science, communications antennas, propulsion, ballast systems must be cryo-capable (−178 °C)
- Mechanical: Pressure vessel capable of withstanding an external pressure of 1×10^6 Pa (10 bar); titanium (Ti) skin and ring stiffeners; internal truss to carry equipment through launch; composite hydrodynamic fairing; dorsal sail to hold phased array antenna and surface science

11.2 Trades

Various design trade-offs were made including submarine shape, communications approaches, power, propulsion and ballasting. A preliminary trade matrix was developed to explore the possible shapes of the submarine based on terrestrial experience, science needs and the added challenges of launching and encapsulating the submarine in an aeroshell (Fig. 11.5). Table 11.1 shows the top level advantages and disadvantages of current terrestrial designs for the Titan Sub mission requirements. While sea gliders have shown to be able to transit great distances with very little power (sinking and gliding with wings and then resurfacing using a ballast system) a science requirement for hovering and in situ sampling would be difficult for such a vehicle. Due to the size of the seas (1000's of kilometers) the Titan Sub would need to be an efficient cruiser (long and narrow) which excludes the Remotely Operated Vehicle (ROV) and diving saucer options. Unfortunately, the length of the torpedo shaped submarine (sized due to required specific weight—it needs to float and sink along with its required power and science instrument mass) would be too large for state of the art 4.5 m aeroshells. While larger button shaped aeroshells can be built they would be too large for the 5 m launch vehicle fairing. This last challenge required new options for the aerodescent system.

The downselected torpedo shape of the vehicle needs a new entry/descent approach. While inflatable aeroshells might also work, a lifting body (based on the proven X-37B design) was chosen to hold the submarine through launch and support it through cruise with thermal, communications, propulsion, and navigation

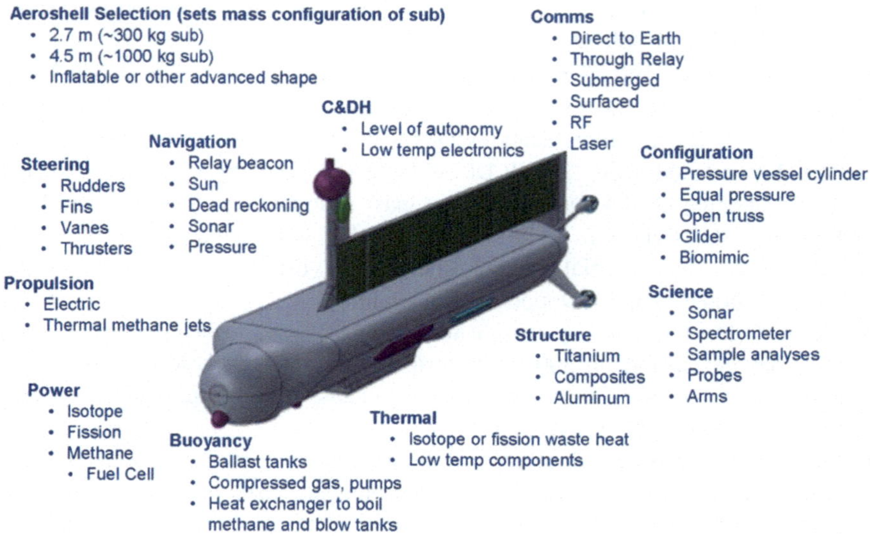

Fig. 11.5 Titan submarine trades

Table 11.1 Advantages and disadvantages of current terrestrial designs for use on Titan

Driving requirement or attribute	Remotely operated vehicle	Diving saucer	Torpedo shaped unmanned underwater vehicle (UUV)	Sea glider
Science submerged and surfaced, hovering for in situ sampling	Yes	Yes	Yes	No
Distance to travel/time: 2000 km/90 days ∼0.5 m/s Aspect ratio >4:1 reduces power 4X, smooth exterior	No	No	Yes	Yes
SOA aeroshell limit: <4.5 m, 0.6 specific wt.	Yes	Yes	No	No
Communications: DTE needs large antenna area to reduce power, Earth nearer horizon than zenith		Yes dish integrated into saucer?	Yes phased array on body?	Yes phased array on body?

(Fig. 11.6). The lifting body would then slow the submarine through Titan aeroentry, glide to the proper touchdown point, and perform a soft landing on the surface of Kraken Mare. The Space Shuttle Orbiter was assessed for emergency water landing capability in the 1970s. The Titan Sub's aerovehicle would touch down on Kraken Mare in a similar manner. At some point in the landing sequence, the backshell would be separated from the aerovehicle, the submarine separated and the lifting body

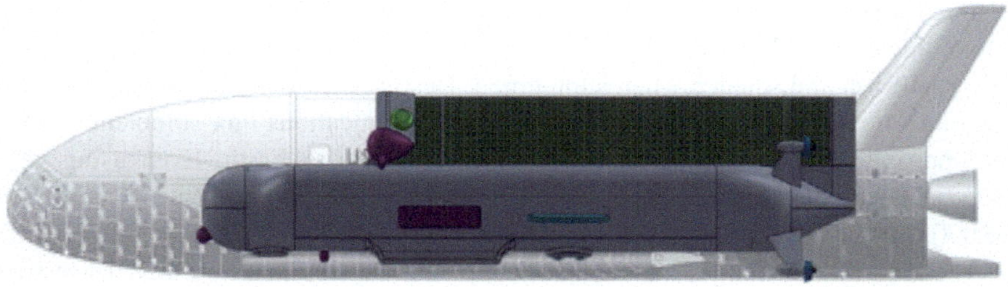

Fig. 11.6 Titan submarine in notional X-37 derived lifting body. Acknowledgements: X-37B outline courtesy of Giuseppe De Chiara (used with permission) and http://en.wikipedia.org/wiki/Boeing_X-37#mediaviewer/File:X_37B_OTV-2_01.jpg

allowed to sink. This descent and delivery concept (and other alternatives) was explored in detail in a Phase II study which will be published at a later date.

The submarine design faced a great many challenges; some less difficult, some much more difficult than a terrestrial sub. Pressures at depth in a liquid ethane (~60% the density of water) sea on the smaller world of Titan (~1/5 Earth's gravity) meant that even at the maximum design depth of 1000 m (3281 ft) the pressure to be endured was 1/10th of that a terrestrial sub would encounter. The sub would need to endure only ~10 bar of pressure at maximum depth on Titan, not the 100 bar (10 MPa) pressure it would have to endure in Earth's oceans. This, however also meant that it needed to have a lower average density in order to be positively and neutrally buoyant to operate at the surface and below. Another challenge was that the extremely low temperature (−180 °C) (−292 °F) of the liquid ethane would quickly cool down most terrestrial submarines. The use of isotope power systems (two ~500 W SRGs) meant that the submarine had plenty of power and waste heat to keep the internal components at room temperature, once provided with sufficient insulation on the inside of the hull. These isotope systems could not only power the sub for several years beneath the waves of Kraken Mare, but also power the sub and the lifting body during the cruise from Earth to Titan. The power challenges and the thermal requirements led to the use of radioisotope generators. A thermoelectric fission reactor system, while heavier, may also be a feasible power system. An ethane fuel cell, using oxidizers brought from Earth would limit the vehicle to less than a week of operation to say nothing of how the combined vehicles would be powered on the way to Titan.

Communications proved to be a great challenge, but one also solved by use of the isotope power system. While methane has been shown to be radio frequency (RF) transparent, the presumably more-ethane rich composition of Kraken has not yet been shown to be transparent (a topic of ongoing Cassini data evaluation). As such the submarine, like its terrestrial counterpart will need to surface to communicate. Choice of a 2047 landing date not only ensures continuous lighting conditions for surface imaging, but also allows for direct communications with the Earth. From the Kraken Mare, Earth is never more than 6° from the Sun. As such, it was decided to not use an orbiter (which would have needed an isotope power system for itself) and to double the isotope power system of the submarine to permit communications DTE while the sub is on the surface and then provide extra power for propulsion and science when submerged. Despite the power available, the DTE antenna would need to be large to provide a communications link sufficient to span the approximately 1.2 billion km (746 million miles) to Earth. Even using geostationary satellites terrestrial submarines only need to communicate distances of 36,000 km (22,370 miles) when surfaced.

The biggest challenge for submarine operations was submerging. Terrestrial submarines use various techniques from diving planes and thrust to ballast tanks filled and then 'blown' using compressed atmospheric gases to venture beneath the waves then return to the surface. While use of thrusters and 'wings' to go beneath Kraken is possible, science required neutral buoyancy hovering for submerged imaging and sampling. Using thrusters to offset buoyancy at depth to hover would require about four times the power from the SRGs than is available. Use of a compressed gas ballast system using Titan's primarily nitrogen atmosphere was found to be infeasible, due both to the fact that ethane (and especially methane) can quickly absorb the nitrogen, and the nitrogen at −180 °C collapses to a liquid below 4 bar which would limit depths to ~200 m. As such, a boundary between the ballast gas and the liquid ethane as well as use of a gas with a lower liquid point was chosen. The final system uses cylindrical ballast tanks with either free floating pistons or bladders pressurized by neon (Ne) brought from Earth and reclaimed after each dive by a compressor during the 16 h of surface operations. The use of the boundary 'piston' meant that the ballast tanks could not be conformal with the pressure hull, following its contours like those of a terrestrial submarine. The positions of the ballast tanks were offset upward to raise the center of buoyancy (CB). The pressure hull and buoyancy tanks were overwrapped with a composite to create a pseudo v-shaped hull to provide better surface stability for antenna pointing and more efficient surface mobility when power was limited.

11.2.1 Titan

Titan Seas

Titan (Fig. 11.7) is a unique satellite in the solar system in that it has a dense atmosphere (1.5×10^5 Pa; 1.5 bar) which endows Titan with many processes and phenomena familiar to us on Earth. At Saturn's distance from the Sun of 10 AU (1.5×10^9 km; 9.3×10^8 miles), the surface temperature on Titan is 94 K (−290 ° F), in part due to the greenhouse warming of methane which makes up a few per cent of the atmosphere (the rest being nitrogen). Ninety-four degrees Kelvin is close to the triple point of methane so it is a condensable greenhouse gas, just like water vapor on Earth. Similarly, methane forms clouds, hail and rain. The methane rain carves river valleys on Titan's surface. The weak sunlight that drives Titan's hydrological cycle results in rain being rare, averaging only a few centimeters per

Fig. 11.7 Cassini captures sunlight glinting off of Titan's seas

year. These rains are probably expressed as massive downpours depositing tens of centimeters or even meters of rain in a few hours, but interspersed with centuries of drought. In some respects, Titan is to Earth's hydrological cycle what Venus is to its greenhouse effect—a terrestrial phenomenon taken to a dramatic extreme.

Titan is tilted 26° on its spin axis so its climate has significant seasonal forcing, but since it takes 29.5 Earth years to go once around the Sun, its seasons are long. In addition to seasonal rainfall, the annual cycle also manifests itself in Titan's stratospheric circulation, where wide swings occur in the abundance of various organic gasses and hazes produced by the action of ultraviolet light on methane. These changes are particularly strong at the winter pole, with some analogies to polar stratospheric clouds and the ozone hole dynamics on Earth. Among the gasses produced by photochemistry is ethane, which is also a liquid at Titan conditions, and is also expected to accumulate on the surface.

Although hydrocarbon seas were long speculated to exist on Titan, bodies of standing liquid were only confirmed in northern winter darkness by Cassini radar observations in 2006, some two years after the probe arrived in the Saturnian system (Fig. 11.8). Hundreds of radar-dark lakes, typically 20 km (12.4 miles) across, were discovered at about 70° N. Latitude. By international convention, lakes on Titan are named after lakes on Earth, while the three seas are named after sea monsters. Ligeia Mare, a 300–400 km (186–249 miles) wide body, was the first sea to be observed. The smaller Punga Mare is closer to the North Pole, while the giant Kraken Mare sprawls over some 1000 km (621 miles) towards mid-latitudes.

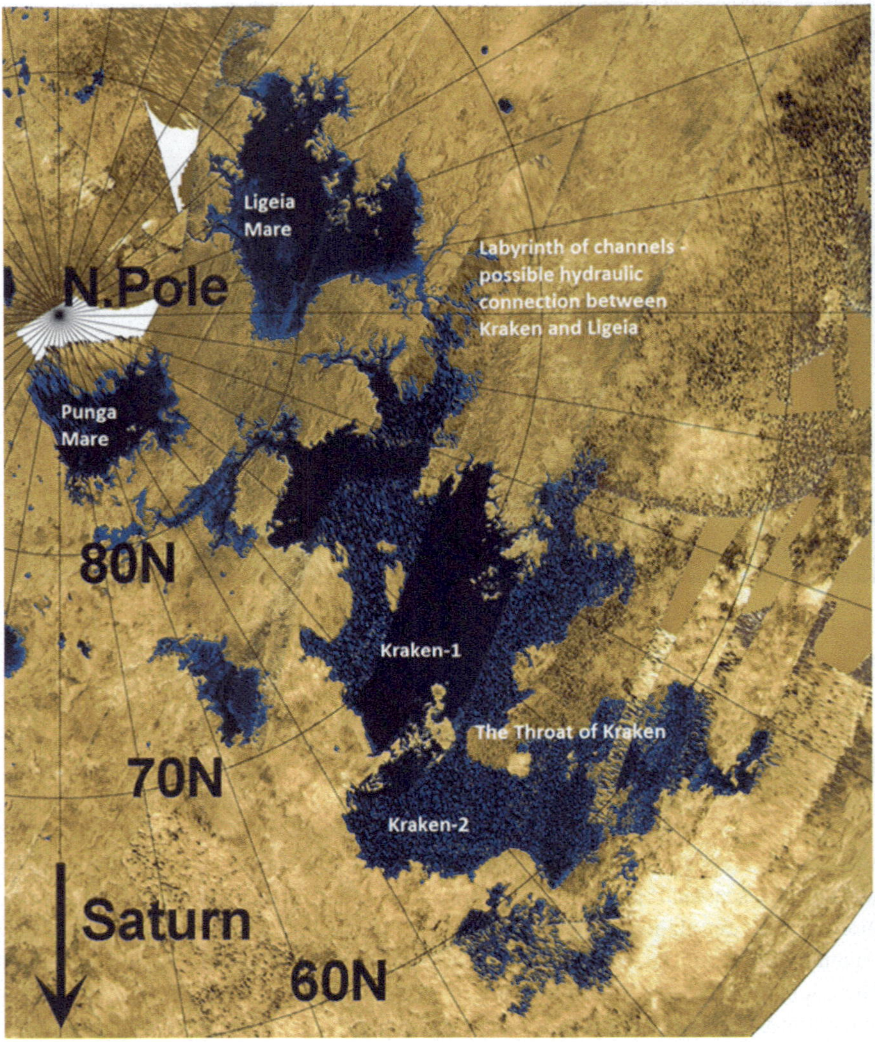

Fig. 11.8 A radar map of Titan's northern polar regions. Note that because Titan rotates synchronously with Saturn, the direction toward that planet is fixed (in fact, the sub-Saturn point defines the zero longitude)

Strikingly, the southern hemisphere has only one modest body of liquid, Ontario Lacus, about 70- by 250-km (43–155 miles). This so far is one of the most-studied lakes, since the south was better illuminated in the 2004–2010 time period allowing near-infrared (IR) remote sensing on Cassini to detect ethane.

Further analysis of the near-IR data suggests that Ontario Lacus may in fact be muddy, and a bright margin is suggestive of a 'bathtub ring' of evaporite deposits. Of course, these are not salts familiar as solutes in terrestrial waters, but some organic analog with differential solubility that has been preferentially deposited at the shrinking margins in an evaporating basin. In fact, a comparison between an optically-measured outline and the margins in a radar image some years later suggest that Ontario may have shrunk in extent due to seasonal evaporation and the very shallow regional slopes. Ontario is most likely only a few meters deep.

The preponderance of seas in the northern hemisphere is thought to be the result of the astronomical configuration of Titan's seasons in the current epoch, which has the result that the northern summer is less intense but longer in duration than that in the south. This results in a longer 'rainy season' in the north, such that methane and ethane accumulate there. This seasonal configuration lasts several tens of thousands of years, much like the Croll-Milankovich cycles that play a part in the Earth's ice ages and the Martian polar layered terrain. This picture of a drying south and accumulating north is consistent with the submerged or ria coastlines of Punga-Ligeia-Kraken which suggest valleys being flooded by rising sea levels, as well as with the kidney-shaped outline, and shallow (and possibly declining) depth of Ontario Lacus in the south.

One of the most striking observations in the near-IR is of the Sun glinting off the surface of the lakes (Fig. 11.7). In fact this, like the low radar reflectivity, told us that the roughness of the lakes must be exceptionally low, but it is a very iconic observation. The lake was appropriately named Jingpo Lacus, named after the 'Mirror Lake' in China.

The notion of extraterrestrial seas offers great possibilities for thought experiments and for teaching. In fact, the liquid methane and ethane that dominate the seas composition are handled routinely on Earth, at the temperatures encountered on Titan, by the liquefied natural gas (LNG) industry. The density of ethane is about 2/3 that of water, and the viscosity is rather similar, depending on temperature. Methane is a little less dense and rather less viscous. Many dissolved constituents (higher hydrocarbons, nitriles) may also be present and would increase the density, viscosity and dielectric constant. It is conceivable that compositional or thermal stratification may occur depending on how tides and wind-driven currents stir Kraken's depths.

The tidal forces on Titan are strong, given Saturn's large gravity, but change only slowly due to the 15.945 Earth day orbit period. Titan is gravitationally locked to its primary, pointing the same face towards Saturn so the tidal bulge is near-fixed, varying only by $\sim$9% over the day due to the eccentric orbit. The slow period means that resonant tides (like those in our Bay of Fundy) are unlikely. Nonetheless, tidal amplitudes of a few tens of centimeters have been calculated for Kraken, with current speeds of a few centimeters per second.

The possibility of waves on Titan's seas was recognized during the formulation of the Cassini mission, and the Huygens probe was equipped with tilt sensors to measure any motion on waves should the probe survive splashdown on a liquid surface (since the surface was completely unknown, surface operations were not guaranteed). In Titan's low gravity (a_g = 1.35 m/s^2; 4.4 ft/s^2, like that of our Moon), propagation of a wave of a given wavelength is rather slow compared with Earth.

The remarkable flatness of Titan's seas posed a puzzle. Why, in Titan's low gravity and thick atmosphere, should the seas not have waves if the hydrocarbon liquids behave like water? One possibility is that winds are too light (yet it is evidently strong enough sometimes to form sand dunes). Another possibility is that the seas may be viscous enough to damp waves. Although waves have yet to be observed, the question of wave height is of interest for shoreline erosion effects, and in particular for the design of vehicles that might float on the surface of the seas. Recent work found that the threshold wind speed for capillary wave generation should be ~0.4 m/s (1.3 ft/s) for methane-rich (low viscosity) seas, or ~0.6 m/s (2 ft/s) for ethane-rich seas (viscosity similar to water). Such speeds have likely not been encountered during the Cassini mission in either hemisphere, although in coming years as we move towards northern summer solstice, Global Circulation Models (GCM) predict a rising probability that winds over Kraken or Ligeia may freshen enough to generate waves that are observable via the Sun-glint pattern on the sea surface or by its radar reflectivity. Once capillaries form, they can grow and become progressively larger gravity waves. Given GCM predictions of maximum ~2 m/s (6.6 ft/s) winds in summer, the significant wave height is expected to reach ~80 cm (2.6 ft) or so, therefore shoreline erosion and beach processes are possible on Titan. Sediment transport, given the low density contrast between ice bedrock and hydrocarbon liquid and the low gravity, should be readily mobilized in Titan's seas.

Since the relative humidity of methane on Titan is only ~50%, a body of pure methane cannot persist indefinitely on Titan's surface since it is not in thermodynamic equilibrium. The evaporation rate has been estimated at up to 1 m/year (3.3 ft/year), using terrestrial empirical transfer coefficients. This is strongly dependent on wind speed. The evaporation rate is composition-dependent, in that the saturation vapor pressure of ethane is very low. Ethane acts to suppress the partial pressure of methane above mixed-composition seas (much as syrup will evaporate in a kitchen much more slowly than water), so Titan's air-sea interactions have some complexities not usually faced on Earth. While ethane probably migrates only over long (>10,000 year) periods, evaporation and precipitation of methane may be much more like terrestrial weather, with hourly and seasonal changes as well as longer-term effects. In fact, transient surface darkening has been observed at low latitudes on Titan in association with methane clouds, followed by brightening, suggesting that shallow flooding occurred, followed by evaporation. The hydrological cycle on titan is clearly active today.

Titan's landscape, atmosphere and climate system have many parallels with Earth, with the added interest of the astrobiological implications of Titan's prebiotic chemistry and rich inventory of organics. Thus Titan remains an important target for future exploration.

11.2.2 Science Instruments

Science Overview

The scientific goals of the Titan Submarine derive from those developed for the 2007 Titan Explorer Flagship study (Leary et al. 2007) and are shown in Table 11.2. Although the seas on Titan were discovered only during that study, the objectives were broad enough to remain community-endorsed in subsequent studies such as the Titan Saturn System Mission (TSSM) and the Decadal Survey.

Science Requirements

More specifically, the scientific goals of the Titan Submarine shown in Table 11.3 are the same as those of the Decadal Survey lake lander, but modified to embrace the growing interest in the diverse shorelines of Titan's seas which can be explored by a mobile sea platform, and to recognize the paleoclimate study potential in the seabed sediments.

Instruments

The science requirements drove the strawman payload listed in Table 11.4. The chemical composition of the seas (and any sediments) is a complex topic, as evidenced in the discussion of solid composition analysis in Leary et al. (2007). We have not specified the internal makeup of the Chemical Analysis Package (CAP). It might comprise a sample volatilization system coupled to a Gas Chromatograph Mass Spectrometer (GCMS), tandem mass spectrometry (MS-MS) or similar analyzer for broad chemical characterization and isotopic measurement. Additional possibilities include Raman, fluorescent or other techniques for specific species of astrobiological interest. The overall resource envelope is patterned after the Sample Analysis at Mars (SAM) package on Mars Science Laboratory (MSL) Curiosity.

It is recognized that such an elaborate analysis system may have a finite number of samples that can be examined, due to finite sample holders, analyte or carrier gas

Table 11.2 The scientific goals of the 2007 Titan Explorer Flagship study

Exploring an Earthlike Organic-Rich World	
OBJECTIVE 1: Titan: an evolving Earthlike system How does Titan function as a system? How do we explain the similarities and differences among Titan, Earth, and other solar system bodies? To what extent are these controlled by the conditions of Titan's formation and to what extent by the complex interplay of ongoing processes of geodynamics, geology, hydrology, meteorology, and aeronomy in the Titan system?	OBJECTIVE 2: Titan's organic inventory: a path to prebiological molecules What are the processes responsible for the complexity of Titan's organic chemistry in the atmosphere, within its lakes, on its surface, and in its subsurface water ocean? How far has this chemical evolution progressed over time? How does this inventory differ from known abiotic organic material in meteorites and biological material on Earth?

Table 11.3 Scientific goals of the Titan submarine

	Objective	Heritage	Contributing instruments
A1	Explore the morphology and character of the seabed to understand the history of the basin and sediment deposits	New	Depth sounder (DS), sidescan sonar (SS), undersea imager (UI)
A2	Explore the morphology of shoreline features to understand Titan's geological history	TE 2007 TSSM TiME	Surface imager (SI)
A3	Measure sea-surface meteorology to constrain larger-scale weather activity and air-sea exchange	TSSM decadal TiME	Meteorology package (MET), navigation
A4	Measure sea physical characteristics (currents, waves, turbidity) and their variations over space and time	TSSM decadal TiME	Physical properties package (P3), SI, navigation, (UI, DS)
A5	Measure horizontal and depth variations of major constituents to constrain exchange and mixing processes	Decadal (option)	Infrared spectrometer (IRS), P3, (chemical analysis package (CAP)), (DS)
B1	Measure trace organics in sea, with emphasis on prebiotic chemistry	TSSM decadal TiME	CAP, IRS
B2	Measure isotopic ratios of noble gases and organics to constrain origin and evolution of Titan	TSSM decadal TiME	CAP
B3	Measure composition of seabed material (best effort)	Decadal new	BAS, CAP

supply, pump saturation, etc. Thus a system capable of measuring broad composition (ethane, methane, propane etc.) more or less continuously is also included. This is notionally a near-IR or mid-IR absorption spectrometer, guiding light from an internal incandescent lamp source through the hull via fiber optic light guides across a sample path near the hull. Other instrument architectures could be envisioned. Although in principle physical properties such as dielectric constant or speed of sound can be estimated knowing the composition, there is some convenience and robustness to determining these properties directly (e.g., for reduction of Depth Sounder (DS) measurements one needs a speed of sound measurement) and these simple sensors are implemented on the Physical Properties Package, which is patterned after the Huygens SSP instrument.

Surface meteorology is an important science goal for Titan overall, but is a somewhat secondary priority for a submarine. A methane humidity measurement, pressure, temperature and wind speed are measured from a sensor package on a mast as high as possible above the waterline. This is done because it is recognized that wind and temperatures, and possibly humidity measurements, may be influenced by the vehicle, its motion and/or its heat output.

The function of the Surface Imager (SI) is to inspect the shoreline, observe the sea surface for floating material, Langmuir rolls, waves, etc., and to observe the

Table 11.4 Science instruments for the Titan submarine

Instrument	Technique	Rationale	Requirements	Basis
(a) Floor				
Chemistry analysis package (CAP)	Liquid sample acquisition system coupled to multiple analytic instruments (nominally GCMS)	Measure bulk and trace constituents of sea at different locations and depths	Inlet isolated from heat source; 40 kg, 80 W when sampling (2 h; once per 2 d)	Curiosity/SAM
Surface imager (SI)	Panoramic charged-couple device (CCD) imager (gimballed) on upper structure	Observe sea surface, shoreline geomorphology, clouds, atmospheric optics	Topside mount, 1 m above sea surface; 4 kg including housing; 10 W when imaging (2 h/d)	MER Pancam
Depth sounder (DS)	Single down-looking acoustic sounder	Low frequency (10 to 20 kHz) to measure depth to bottom, possibly detect layers, bubbles, etc.	Nadir view; 0.5 kg 2 W continuous	TiME MP3, commercial fish finders
Meteorology package (MET)	Pressure, temperature, wind speed and direction, methane humidity on surface	Record meteorological variability, forcing of air/sea exchange	Topside mount, 1 m above sea surface, desirably away from heat source; 3 kg 6 W continuous	TiME MP3, Pathfinder ASI/MET, terrestrial field instruments
Physical properties package (P3)	Sea temperature, speed of sound, dielectric constant and turbidity	Structure of liquid column (stratification), suspended sediment, air/sea exchange, local variations in bulk ethane/methane	Isolated from heat source; 2 kg; 6 W continuous	TiME MP3/Huygens surface science package (SSP)
(b) Baseline				
Sidescan sonar (SS)	Side-looking acoustic imaging array	Acoustic imaging of seabed morphology	Bottom/side view; 10 W when operating; 8 h/d	Terrestrial UUV

(continued)

Table 11.4 (continued)

Instrument	Technique	Rationale	Requirements	Basis
Undersea imager (UI)	Medium-field CCD imager equipped with multicolor illuminators	Optical imaging of seabed (combine with SI if vehicle orientation permits)	Forward view; 3 kg including housing; 20 W when imaging; 1 h/d	Curiosity Mars hand lens imager (MAHLI)
Benthic sample acquisition (BSA)	Grinding/ suction system to ingest solid or semi-solid seabed materials	Deliver seabed sediments to CAP instrument	Forward/lower view; 5 kg; 50 W when operating 1 h/2 d	Phoenix rasp plus suction pump
Infrared spectrometer (IRS)			8 kg; 20 W; 2 h/d	Miniature thermal emission spectrometer (miniTES), laboratory instruments
(c) Engineering				
Navigation systems (NAV)	Pressure depth gauge, IMU, plus Doppler/ Delta differential one-way ranging (DOR) radio measurements	Infer ocean currents	Bookkept under GN&C system	Various

atmospheric scattering and detect clouds and rain. In order to have a horizon of about 2 km (1.2 miles), this camera must be mounted 1 m (3.3 ft) above the waterline. The SI is collocated on a mast with the Meteorology Package (MET) which is located above it. This camera should be capable of panoramic views, either via optics, multiple apertures, or a gimbal. A separate down-looking camera with illuminators is carried on the forward end of the sub for observation of the seabed.

Several acoustic systems are carried, with somewhat different functions. The DS is a powerful nadir-pointed system, designed to measure the depth to the seabed (from the surface or below, to a nominal depth of 1 km (0.6 miles)). It can also detect possible layers in the sea, and in the seabed. The side-looking sonars have larger transducer arrays to yield a narrow fan beam on either side of the vehicle for high-resolution imaging of the seabed morphology. Nominally, this system may work to a depth of 100 m (330 ft). At greater depths, mapping may be done while submerged. An additional sensor, not part of the science payload but rather the guidance, navigation, and control (GN&C) system is a Doppler velocity gauge, to determine drift or speed relative to the seabed to reduce navigation errors.

Table 11.5 Science instrument MEL

Description	Quantity	Unit mass (kg)	Total mass with growth (kg)
Science payload	–	–	118.3
Floor	–	–	65.0
CAP	1	40.0	52.0
SI	1	4.0	5.2
DS	1	0.5	0.7
MET	1	3.0	3.9
P3	1	2.0	2.6
Light	1	0.5	0.7
Baseline	–	–	53.3
SS	2	5.0	13.0
UI	1	3.0	3.9
BSA	1	20.0	26.0
IRS	1	8.0	10.4

Provision of a system to obtain samples of seabed sediments is noted (BSA) although we have not considered the details of such a system. This system could be an arm/drill type of sampler, or even a tethered or untethered sub-vehicle.

The science instrument list for the Titan submarine is shown in Table 11.4a–c, and the master equipment list (MEL) for the science instruments is shown in Table 11.5.

11.2.3 Study Summary Requirements

Figures of Merit (FOMs)

The relative merit of the conceptual design was judged against:

- The amount of science data return from Titan
- Maximizing the surface mission specifics on Titan:
  - Voyage duration
  - Distance covered
  - Depth reached by the sub
- Mass, as always, is a FOM. Minimizing the mass reduces launch vehicle (LV) size and cost, and reduces trip time to Titan
- Cost: For the first design it was determined to allow a flagship cost to investigate the amount of science possible.

- Risk: While already requiring long time delays for communications, submersed operations will require some sort of autonomous surfacing capability, similar to Earth Unmanned Underwater Vehicles (UUVs), to return to the surface if anomalies are encountered.

11.3 Baseline Design

11.3.1 System Level Summary

This study focused on the conceptual design of the Titan Submarine. Though recognizing that the Titan entry system/cruise stage and LV are key parts of an overall mission conceptual design, funding and time did not permit delving deeply into the other two elements. That would be part of a Phase-2 NIAC study.

This section summarizes the Titan Submarine conceptual design, and touches on the LV and Titan entry system aspects of the missions that were assessed in this study phase.

Titan Submarine Concept Drawings and Descriptions

The major components that make up the Titan Submarine configuration include: the main hull, a sail containing the communication antenna, a mast for science and communications, two ballast tanks, the propulsion system, and a hydrodynamic skin. Figure 11.9 shows the Titan Submarine and these major components that make up the overall size and shape of the design.

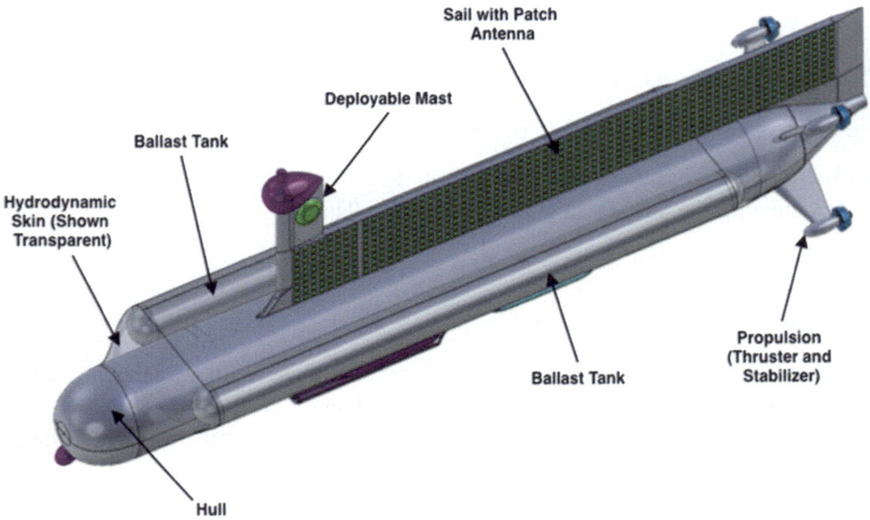

Fig. 11.9 Major components comprising the Titan submarine

The hull is the primary structure of the submarine. It is a pressurized cylinder that houses all of the internal subsystem components and provides mounting for the sail structure, ballast tanks, propulsion system, hydrodynamic skin, and many of the Science and GN&C components that need an unobstructed view or access to the outside of the hull. Mounted vertically off the top of the hull is the sail structure. The sail structure provides the required area for the patch (phased array) antennas, mounted on both sides, as well as provides the mounting for the deployable mast, located on top of the sail structure. Contained on top of the half-meter tall mast are the science SI (required to be 1 m above the surface) and the meteorology sensor, while an omni antenna is mounted on each side of the mast. The mast is folded down along the side of the sail for stowage and is deployed once on the surface of Titan. A ballast tank is mounted to each side of the hull and above the hull's centerline. The propulsion system consists of four thrusters, each mounted out at the end of a stabilizer that is mounted to the aft end of the hull structure. The use of four thrusters allows for pitch and yaw steering while submerged, as well as allowing the bottom pair of thrusters to be utilized to propel the submarine while at the surface. Finally, a hydrodynamic skin is wrapped around the hull encompassing the two ballast tanks and several of the external Science and GN&C components in order to reduce the drag created by these components as well as minimize the interference drag created between the ballast tanks and the hull structure. This skin also provides the interface for mounting the two Science Sidescan Sonar Arrays and the two Sonar Transducers from the GN&C system. All of the components contained outside the hull structure can be seen in Fig. 11.10. Not shown in the images is the

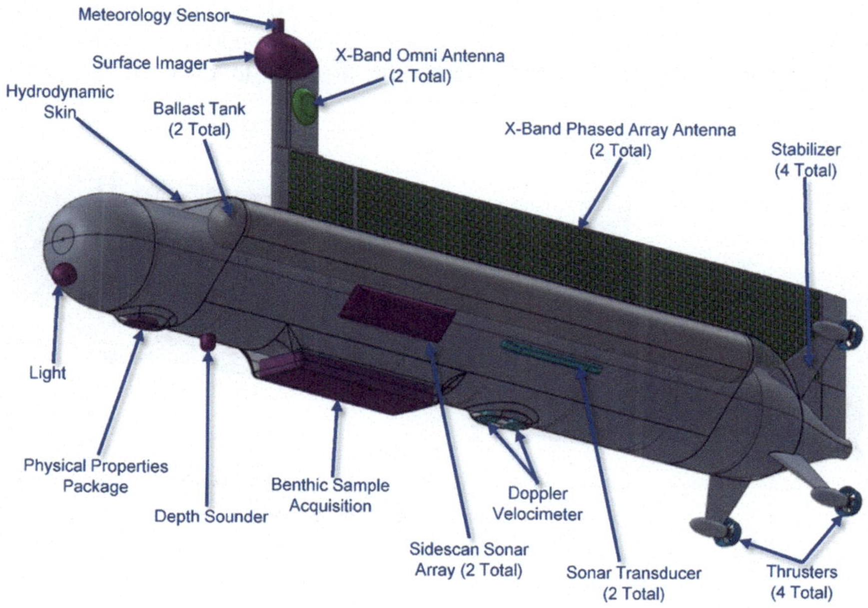

Fig. 11.10 External components on the Titan submarine

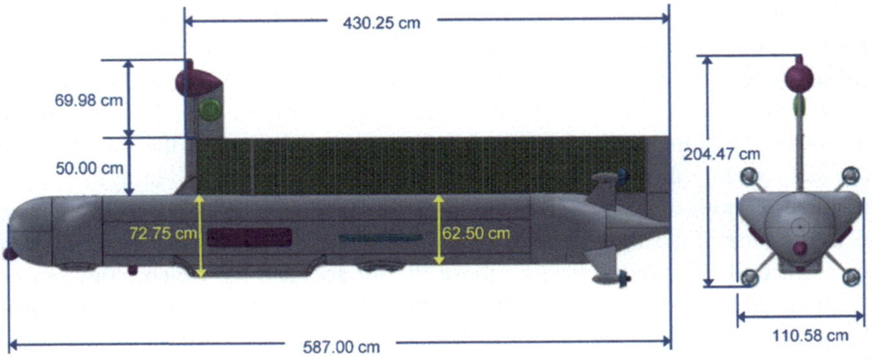

Fig. 11.11 Overall deployed dimensions of the Titan submarine

foam that will be located in the gaps between the ballast tanks and hull underneath the hydrodynamic skin. This foam will help improve the buoyancy and stability of the submarine while submerged and at the surface.

Overall dimensions of the Titan Submarine can be seen in Fig. 11.11. The overall length of the submarine is driven by the 587 cm (19.3 ft) long hull structure, while the width is driven by the combination of the 62.5 cm (2 ft) diameter hull structure in combination with the two 27 cm (0.9 ft) diameter ballast tanks. The height is driven by the need for the SI to be 1 m (3.3 ft) above the surface in combination with the hull diameter and the length of the lower two stabilizers and thrusters.

All of the components contained inside the pressurized hull structure can be seen in Fig. 11.12. The inside of the hull skin and its six ring supports are covered with a 3 cm (1.1 in.) thick layer of foam insulation in order to maintain the proper

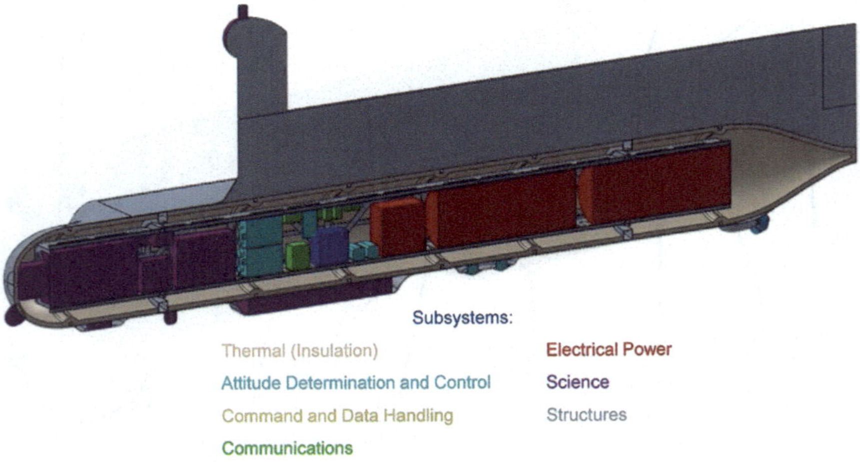

Fig. 11.12 Cross-sectional view of the Titan submarine

temperatures inside the hull and minimize the heat that escapes through the hull. Two additional larger structural rings, one near the front and another near the rear, cut through the insulation, thus inducing a heat sink between the inside and outside of the hull. This heat sink was accounted for in determining the thickness of the insulation to meet the thermal requirements. It is these two large rings that provide the support for the internal structural "cage" that provides the interfaces to all of the other internal components, as well as provides the interface to hold the submarine inside of the lifting body from launch through the descent phase at Titan. Further details on the insulation can be found in Sect. 11.4.7, Thermal Control, while additional detail on the structures can be found in Sect. 11.4.8, Structures and Mechanisms.

The structural "cage" located within the hull provides the mounting interface for all of the subsystem components contained within the pressurized hull structure. It is desired to locate the science electronics as far away from the SRGs as possible to avoid any interference that may be generated by the SRGs. For this reason, all of the science electronics were located at the front of the hull and mounted directly to the inside of the "cage" structure, with the exception of the Undersea Imager Sensor Box that is mounted to a panel on the front of the cage. This location allows the imager to look out a 4 in.^{2} window located at the front of the hull while being contained inside the insulated pressurized environment. The IMUs and GN&C system electronics are located directly behind the science electronics inside the cage, along with the flight controller of the C&DH system. Finally, just behind the C&DH and GN&C components is the Power Management and Distribution (PMAD) electronics followed by the two SRGs. Not shown in Fig. 11.12 is the 180 kg (397 lbm) of permanent lead ballast that would be located between the bottom side of the "cage" structure and the insulation. This will help to lower the center of gravity (CG) of the submarine, thus helping with buoyancy and stability issues. Further analysis needs to be done to better locate the internal components in order to drive the CG to a location that maximizes the stability and buoyancy of the submarine. A transparent view of the Titan Submarine can be seen in Fig. 11.13.

Master Equipment List (MEL)

The Titan Submarine MEL is shown in Table 11.6. The MEL presents a summary mass listing for all subsystems of the Titan sub along with their mass growth allowances based on the maturity of the subsystem components.

Titan Submarine Architecture Summary

The MEL shown in Table 11.7 captures the bottoms-up estimation of mass and growth percentage of the Titan Submarine that the subsystem designers calculated for each line subsystem. In order to meet the total required system mass growth of 30%, an allocation is necessary for growth on basic dry mass at the system level, in addition to the growth calculated on each individual subsystem. This additional system-level mass is counted as part of the inert mass to be flown. The additional system-level growth mass also impacts the total ballasting required on the sub to assure buoyancy control.

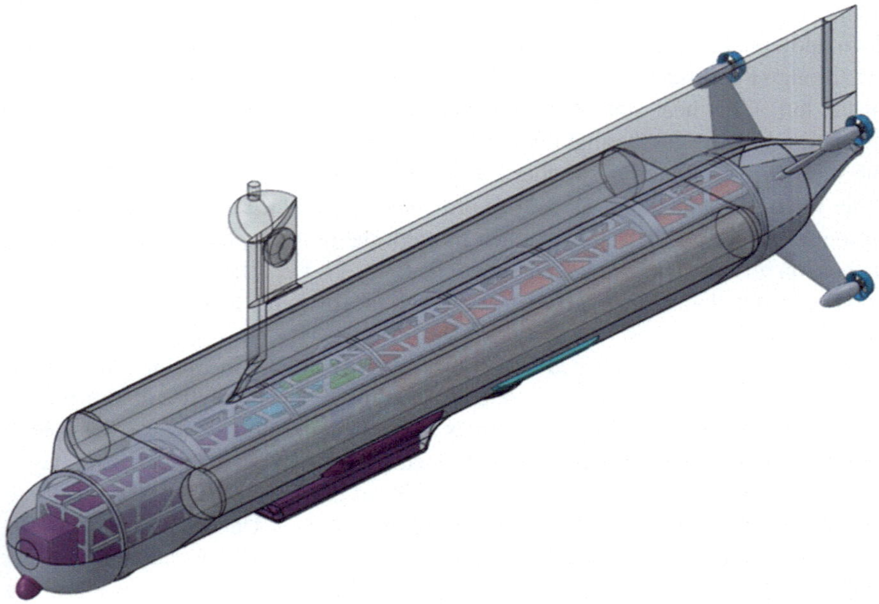

Fig. 11.13 Transparent view of the Titan submarine

Table 11.6 Master equipment list for the Titan submarine

Description	Basic mass (kg)	Total mass (kg)
Titan sub S/C	901.8	1085.0
Science payload	91.0	118.3
Attitude determination and control (AD&C)	32.9	38.8
C&DH	44.0	57.2
Communications and tracking	26.3	30.5
Electrical power subsystem	146.0	175.2
Thermal control (non-propellant)	95.3	112.5
Propulsion	20.6	26.5
Propellant	0.0	0.0
Structures and mechanisms	445.7	525.9

Power Equipment List (PEL)

To model the power systems in this Titan Submarine design study, ten modes of operation were defined for the study. These modes were defined based on the mission profile and they identify which items and subsystems of the sub are operating, and which items are dormant and require no power, at any time throughout the mission. The definitions of these modes are shown in Table 11.8.

Table 11.7 Titan submarine architecture summary—S/C MEL rack-up (mass)

Main subsystems	Basic mass (kg)	Total mass with growth (kg)
Titan sub S/C	901.8	1085.0
Science payload	91.0	118.3
AD&C	32.9	38.8
C&DH	44.0	57.2
Communications and tracking	26.3	30.5
Electrical power subsystem	146.0	175.2
Thermal control (non-propellant)	95.3	112.5
Propulsion	20.6	26.5
Propellant	0.0	0.0
Structures and mechanisms	445.7	525.9
Element 1 consumables (if used)	0.0	0.0
Estimated S/C dry mass (no prop, consumables)	901.8	1085.0
Estimated S/C wet mass	901.8	1085.0
System level growth calculations titan sub		
Dry mass desired system level growth	901.8	1172.4
Additional growth (carried at system level)	–	–
Total wet mass with growth	901.8	1172.4
Hydrostatic balance		
Foam in voids between pressure hull and ballast	34.0	
Additional lead ballast	180.0	
Total wet mass with growth and balance		1386.4

Table 11.8 Power modes for the Titan submarine study

Power mode names	Description	Duration
Launch	Ascent through Earth departure	60 min
Interplanetary cruise	Keep-alive power during hibernation; occasional wake up	7 year
Titan EDL	Entry, descent and splashdown	2 h
Sub activation and checkout	Commissioning	1 week
Dive/surface		100 miles (or more for extended operations)
Submerged cruise	Including science and h/k comm. (low data rate)	8 h
Surface cruise	Including science and comm. (high data rate)	16 h
Stationary submerged operations	Including science and h/k comm. (low data rate)	8 h
Stationary surface operations	Including science and comm. (high data rate)	16 h
End of mission (EOM) disposal		Mission duration is only limited to power system life and visibility from earth (up to 7 years)

Table 11.9 PEL for the Titan submarine (Modes 1–5)

Description	Power mode 1, W	Power mode 2, W	Power mode 3, W	Power mode 4, W	Power mode 5, W
Power mode name	Launch	Interplanetary cruise	Titan EDL	Sub activation and checkout	Dive/surface
Power mode duration	60 min	7 year	2 h	1 week	100 min
Titan sub S/C	60.0	70.0	90.0	648.0	635.0
Science payload	0.0	0.0	0.0	55.0	53.0
AD&C	0.0	0.0	0.0	63.0	52.0
C&DH	60.0	60.0	60.0	60.0	60.0
Communications and tracking	0.0	10.0	30.0	30.0	30.0
Electrical power subsystem	0.0	0.0	0.0	0.0	0.0
Thermal control (nonpropellant)	0.0	0.0	0.0	0.0	0.0
Propulsion	0.0	0.0	0.0	440.0	440.0
Propellant	0.0	0.0	0.0	0.0	0.0
Structures and mechanisms	0.0	0.0	0.0	0.0	0.0
Entry system	0.0	0.0	0.0	0.0	0.0
Cruise stage	0.0	0.0	0.0	0.0	0.0
Bus power, system total	60.0	70.0	90.0	648.0	635.0
30% growth	18.0	21.0	27.0	194.4	190.5
Total bus power requirement	78.0	91.0	117.0	842.4	825.5

Tables 11.9 and 11.10 show the assumptions about the power requirements across all modes of operation. The power requirements from the bottoms-up analysis on the Titan sub shown in those tables are used by the power system designers to size the power system components and by the Thermal Control System (TCS) lead to manage the waste heat from these components.

Launch Vehicle (LV)

Since this concept study focused mainly on the details of the Titan submarine, this section covers those LV aspects that were necessary for the design of the submarine. The detailed LV selection and definition of interfaces would be part of a Phase II study.

Sending a substantial payload (the Titan submarine and the Titan entry system/cruise stage) to the Saturn system will require a LV with substantial lift capability. Surveying the stable of LVs available under the NASA Launch Services

Table 11.10 PEL for the Titan submarine (Modes 6–10)

Description	Power mode 6, W	Power mode 7, W	Power mode 8, W	Power mode 9, W	Power mode 10, W
Power mode name	Submerged cruise	Surface cruise	Stationary submerged operations	Stationary surface operations	EOM disposal
Power mode duration	8 h	16 h	8 h	16 h	0.0
Titan sub S/C	645.0	573.5	207.0	411.0	128.0
Science payload	53.0	43.0	55.0	23.0	0.0
AD&C	62.0	53.0	62.0	53.0	53.0
C&DH	60.0	60.0	60.0	60.0	60.0
Communications and tracking	30.0	280.0	30.0	275.0	15.0
Electrical power subsystem	0.0	0.0	0.0	0.0	0.0
Thermal control (nonpropellant)	0.0	37.5	0.0	0.0	0.0
Propulsion	440.0	100.0	0.0	0.0	0.0
Propellant	0.0	0.0	0.0	0.0	0.0
Structures and mechanisms	0.0	0.0	0.0	0.0	0.0
Entry system	0.0	0.0	0.0	0.0	0.0
Cruise stage	0.0	0.0	0.0	0.0	0.0
Bus power, system total	645.0	573.5	207.0	411.0	128.0
30% growth	193.5	172.0	62.1	123.3	38.4
Total bus power requirement	838.5	745.5	269.1	534.3	166.4

(NLS) contract, the Atlas V 551 was chosen as the representative launcher for the Titan submarine conceptual design study. It permits a substantial payload mass capability to high energy trajectories.

The Atlas V 551 features a 5 m (16.4 ft) diameter payload fairing, five strap-on solid propellant boosters and a Centaur upper stage with a single main engine. Given a preliminary estimate of 3000 kg (6615 lbm) for the Titan submarine and the Titan entry system/cruise stage, this launcher can still achieve a high energy trajectory to the Saturn system. Though it would still require planetary flybys and gravity assist maneuvers to reach Titan, the Atlas V 551 configuration is the most capable in the NLS stable.

Using rough estimates of payload mass to Saturn achievable with a variety of heavy lift LVs, a Venus-Venus-Jupiter (VVJ) flyby sequence would be required to increase transfer orbit energy beyond the capability of the Atlas V 551, Delta IV Heavy or Falcon Heavy LVs. To send the Titan Submarine and its lifting body directly to the Saturn System would require use of the Space Launch System (SLS).

Titan Entry System/Cruise Stage

The Titan submarine will require a system to perform atmospheric entry, descent and landing (EDL) operations at the Saturnian moon, and to serve as a support stage during the interplanetary cruise phase of the mission. The system must provide:

- Interfaces with the launch vehicle
- Propulsion to perform course corrections during interplanetary flight
- Communications with Earth
- Navigation information from Earth
- Capability to reject heat from the SRGs during interplanetary flight
- Ability to perform atmospheric EDL on Kraken Mare

An aerovehicle entry system/cruise stage was chosen as the notional system to perform the functions listed above. Titan has a dense atmosphere, and an aerovehicle entry system can not only provide protection from the heat levels generated during entry, but substantial cross-range capability in atmospheric flight to assure a successful targeting of, and landing on, Kraken Mare. A system similar in size, shape and descent capabilities to the Boeing X-37B was chosen for the purposes of this study since the X-37B has already demonstrated autonomous de-orbit, entry and targeted landings on Earth after long duration space flight in low Earth orbit.

In a Phase II study, this system will be investigated further to determine the requirements on the aerovehicle's thermal protection system to survive direct entry into Titan's dense atmosphere, requirements on wing and control surfaces for atmospheric flight at Titan, and its ability to execute a landing on the surface of Kraken Mare.

The aerovehicle would provide the structural/mechanical interfaces to the Titan submarine during launch, cruise and through EDL on Titan. The aerovehicle systems would be commanded by the submarine's flight computer via umbilical connections. After Titan landing, those structural and electrical interfaces would have to be severed, and the aerovehicle separated from the submarine. Notionally, the submarine could be mounted into the aerovehicle through the use of trunnion pins and a keel fitting with retention latch actuators, similar to the way that large payloads were mounted in the cargo bay of the Space Shuttle Orbiters. Since the aerovehicle would separate from the submarine, not vice versa, the trunnion pins and keel fittings would be on the aerovehicle, and the retention latch actuators would be on the submarine. Alignment guides on the latch actuators would help assure a clean separation of the aerovehicle from the submarine after landing on Kraken Mare.

Separation of the submarine after landing is discussed further in the next section.

11.3.2 Concept of Operations

The concept of operations (CONOPS) for the TSSM mission is divided into the following mission phases:

- Launch Site Operations
- Launch and Ascent
- Park Orbit and Trans-Titan insertion (TTI)
- Interplanetary Cruise
- Titan Entry, Descent and Splashdown (EDS)
- Sub Activation and Checkout
- Mare Sorties including:
  - Dive/Surface Operations
  - Submerged Cruise
  - Surface Cruise
  - Stationary Submerged Operations
  - Stationary Surface Operations
- End of Mission

Each mission phase is presented in detail in the following paragraphs.

Launch Site Operations

The spacecraft (S/C) will be launched on an Atlas V 551 Expendable Launch Vehicle (ELV) from Complex 40 at Cape Canaveral Air Force Station (CCAFS), Florida. Processing of the launch vehicle and the S/C will occur at CCAFS facilities and facilities at the NASA Kennedy Space Center (KSC) industrial area.

Spacecraft to launch vehicle integrated testing will occur after mating to the Centaur upper stage. Then, the SRG installation into the titan submarine will be rehearsed with a simulator. After a successful rehearsal, the flight unit will be installed and the S/C closed out for launch. These events are illustrated in Fig. 11.14. After SRG installation, and the sub is buttoned-up for launch, the hull will be pressurized to 1.03 MPa (150 psi) so that it can withstand the pressure it would experience at its maximum depth capability under the surface of Kraken Mare.

Earth to Titan Cruise

After the earth escape burn, the S/C is on a heliocentric trajectory which will eventually take it to the Saturn system and a direct insertion into the atmosphere of Titan. Planetary flybys and gravity assist maneuvers will be required to increase the S/C's velocity beyond that provided by the LV. The detailed trajectory design would be part of a Phase II study.

After separation from the LV, ranging is done to determine orbit parameters and create a state vector. The state vector will be used to determine the attitude and

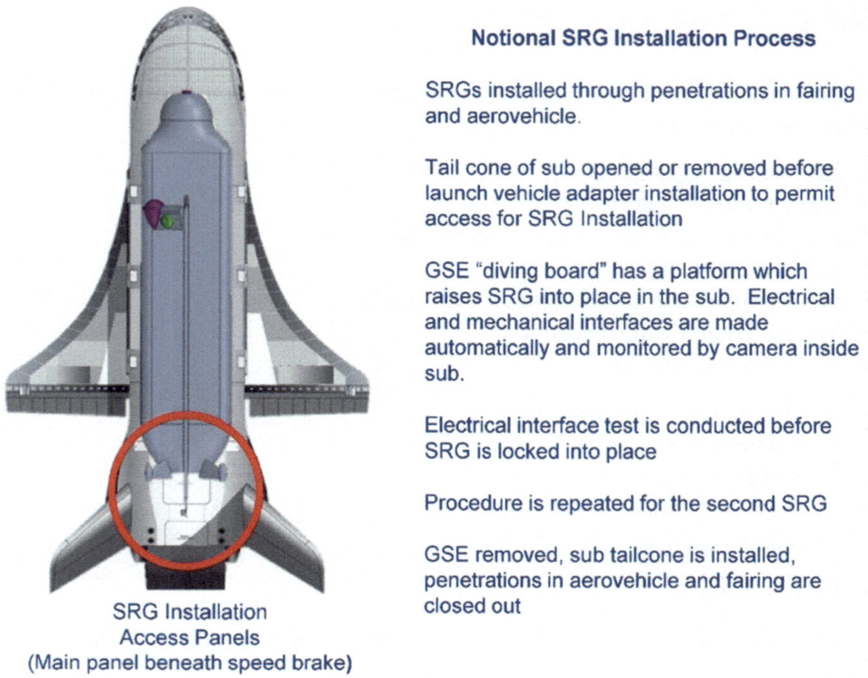

Fig. 11.14 SRG installation in the VIF. Acknowledgements: X-37B outline courtesy of Giuseppe De Chiara (used with permission) Titan Submarine CAD model created by T. Packard, NASA Glenn COMPASS Team

magnitude required for the first course correction burn to be performed by the S/C. Course correction burns are executed to compensate for LV injection errors.

Once on course, and between periods where planetary flybys will occur, the S/C will be placed in hibernation for the majority of the cruise phase except for times when pre-planned health assessments occur or when any trajectory correction maneuvers must be performed.

Titan Entry, Descent and Landing

On approach to Titan, the vehicle will be configured for atmospheric entry. Aero-surface checks will be conducted and the vehicle placed in the proper attitude for entry. Due to the significant time delay in radio communication between Earth and Titan, the submarine's computer will execute EDL autonomously.

The entry phase commences as the vehicle begins to encounter the uppermost regions of Titan's atmosphere. The vehicle flies at a high angle of attack, allowing drag to reduce its speed. Entry navigation is performed using IMU-sensed inputs until atmospheric navigation data and/or radar system navigation data become available. Attitude control is maintained using the reaction control system (RCS) thrusters only until the aerodynamic control surfaces become effective;

then, a blend of aero-surface and RCS control is used until control can be done exclusively by the aero-surfaces.

Atmosphere data probes are deployed following the entry heat pulse. Flight control gains will be adjusted based on atmospheric and radar data. Entry guidance will modulate angle of attack and bank angle to control g-load, heat pulse, and landing footprint. The IMU's serve as the navigation sensors. Once the atmosphere is thick enough to produce lift, the vehicle is banked steeply, but without reducing the angle of attack. The bank leads to a slow course deviation which allows the flight path to be lengthened in order to manage energy. This also causes the vehicle to move off course and that leads to a "roll reversal". Depending on the energy that is left there will be a few more rolls. Due to the shape of the generated flight path, these repeated turns/rolls are also known as "S turns".

Throughout atmospheric entry the onboard computer keeps track and constantly updates the flight path and the required maneuvers are executed. At approximately 760 m/s (2500 ft/s) entry velocity, guidance steers the vehicle to tangency with a navigation-derived heading alignment cylinder projection, which intersects the final landing approach trajectory. During the circle, the vehicle's altitude drops to about 3048 m (10,000 ft) as it begins to align for landing. Radar data compared against prior data of Kraken Mare are used to refine its landing pattern.

The vehicle lines up with an approach corridor at an altitude of approximately 3048 m (10,000 ft). Approach speed is controlled by speed brakes. The approach corridor and landing location will be chosen such that the vehicle comes to a halt near the center of Kraken Mare. At around 610 m (2000 ft) altitude, the vehicle changes its glide slope from a steep dive to a shallow glide slope all the way to touchdown on the surface of Kraken Mare. Such water landings using lifting bodies have been tested and proven by NASA Langley tests of STS models (Fig. 11.15).

Taking advantage of ground effect, the vehicle touches down on the surface of the Mare at a sink rate no greater than 3 m/s (9.8 ft/s). Once at relative stop, the flight computer will enable the sub's phased array and omni antennas, and its

Fig. 11.15 Shuttle water landing testing

propulsion system. Vents in the aerovehicle will be opened to flood the vehicle and the top half of the fuselage will be separated. Retention latch assemblies will open and umbilical connections between the sub and the aerovehicle will be severed or retracted. Alignment guides will assure a clean separation of the aerovehicle from the sub as it sinks beneath the surface of Kraken Mare. Fuselage separation system and time of separation (shortly before or after splashdown) would be assessed in more detail as part of a Phase II study (Fig. 11.16).

Kraken Mare Exploration

Primary Mission: After commissioning, surface testing and dive testing/calibration of the Titan sub the primary mission commences. The sub would first transit 400 km (248 miles) over a period of 14 Earth days to the estuary of Ligeia Labyrinth. The goal is to 'sniff' whether there is a composition gradient driven by methane-rich flow from Ligeia. The sub would remain in the area through 24 Earth days (1.5 Titan days) of tidal monitoring to see how the cycle repeats. During the period, the sub would perform small transits and returns, then drift for a few hours at a time and measure displacement via imaging and sonar.

With the measurements at Ligeia complete, the sub would transit over a 14 Earth day period through Kraken Mare around the shoreline of Mayda Insula. The sub would perform detailed shoreline imaging, detailed bottom mapping and periodic bottom sampling.

Continuing along the shoreline, the sub would then perform tidal monitoring in a Strait for another 14 Earth day period (Fig. 11.17). The sub would stay in the strait for a tidal cycle, performing small transits and returns followed by a drift of a few hours at a time to measure displacement via imaging/sonar. Detailed bottom mapping, detailed shoreline imaging (both mainland and islands) and periodic bottom sampling would be conducted.

Fig. 11.16 Shoreline imaging during the first transit of the Titan submarine

Fig. 11.17 Investigating Bayta Fretum strait to the south of Mayda Insula

The sub would proceed to the Throat of Kraken and would loiter in the area for a 14 Earth day period. Once again, the sub would remain in the strait area for a tidal cycle. Small transits and returns are performed, and then the sub would drift for a few hours at a time and measure displacement via imaging/sonar. Detailed bottom mapping and periodic bottom sampling would be done, as well as detailed shoreline imaging. This would represent the end of the sub's 90 day primary mission but it would still have substantial power available to continue its explorations for extended missions.

During an extended 90 day mission, the submarine would transit the throat of Kraken and perform similar explorations in other areas of Kraken Mare. Once this half year of exploration is completed the submarine could be tasked to revisit points of interest and perhaps do a complete sonar mapping of the seas. All in all, the submarine could explore over 3000 km (1864 miles) at an average speed of 0.3 m/s.

At some point, though, the output of the submarine's power plants would reach a level where sea travel is no longer feasible and it would truly reach its end of mission state. The sub would be run aground so that the nuclear material in the SRG's remains contained within the sub. This would mitigate the risk of the sub sinking to a depth where the hull could be compromised and nuclear material could be released into the Mare.

Fail Safe Modes: The communications delay between Earth and Titan will necessitate a high degree of autonomy in the sub's flight computer. The distance from Earth to Titan can vary from 8 to 11 AU (1 AU = 149,661,688 km; 93,000,000 miles). Therefore, round trip communications between Earth and Titan can vary from 67 to 92 min. Sortie plans (surface or submerged) will be uplinked to the sub and verified before a sortie is executed.

During moving or stationary surface operations the sub communicates with Earth through the phased array antenna for high data-rate communications. Surface transits must be planned accordingly so that one side of the array can maintain Earth lock. If communications is lost, the sub will go to "All Stop" and use Sun sensor data to position the sub to re-acquire Earth via one of the X-band omni antennas.

During a dive operation, submerged sortie or surfacing operation, the sub may be out of contact with Earth. If an off-nominal situation occurs, the sub will surface and acquire Earth communications via an X-band omni antenna.

11.4 Subsystem Breakdown

11.4.1 Communications

Cruise Stage/Lifting Body Communications Analysis

The mission analysis for the trans-Saturn cruise and Titan EDL phases of the mission were noted but not developed in the Phase I concept. They will be assessed in a Phase II Study.

Communications Requirements

The Titan Submarine's communications requirement is to provide a direct science data downlink to Earth for data processing and analysis. The submarine will have to endure a cryogenic environment on the Titan surface while exploring Kraken Mare. External submarine hardware will have to endure, then operate after exposure to temperatures approaching minus 180 °C (93 K). For the case of the communications system, the phased array antenna is the external hardware. The DTE data requirement is 500 bps through the Deep Space Network (DSN), 16 h per day while surfaced.

Communications Assumptions

For this design it is assumed that the only component of the communications system exposed to the external cryogenic hydrocarbon environment is the phased array antennas. All other components are contained within the submarine interior structure maintained at a nominal Earthlike room temperature environment. Further assumptions are a standard 3 dB communications link margin for the link budget. This is a typical margin for space design applications. This design does not use a space relay to Earth. All communications are assumed direct Titan to Earth. The link budget was sized for 8.5 AU Titan to Earth range. Simulations show this will be the approximate Titan to Earth distance in 2038, which is the earliest projected arrival date of the Titan Submarine. X-band deep space communications and phased array antennas will be used for Titan science data downlink to Earth due to the higher available gain of the phased arrays and low aerodynamic drag on the submarine as compared to a dish. High gain, low profile omnidirectional antennas will be used for an emergency beacon and communications. A total of 330 W has been allocated for the communications system on a 24 V DC spacecraft bus. A 34 m DSN Earth station antenna will be used for downlink communication. This is a single fault tolerant design.

Communications Design and MEL

Due to the distance between Titan and Earth, the required survivability of communications external assets in a cryogenic environment, and the need for a low profile external structure to reduce drag on the submarine while submerged, the Titan to Earth direct data downlink is a major design driver for the Titan Submarine system. Although Ka-band would be very attractive due to the higher bandwidth and data rate capabilities, the water vapor in Earth's atmosphere would significantly attenuate a Ka-band receive signal as compared to either S-band or X-band. Also, because of the narrower beam characteristics of Ka-band, the antenna pointing must be more accurate as compared to X-band and S-band. Titan submarine platform control could add to the uncertainty in pointing accuracy due to stability and the projected inherent low Titan to Earth elevation angles (minimum 15°–30° above horizon in year 2040+, maximum 41° in 2049 as seen from 75 N, Kraken). Therefore, X-band was chosen because the data throughput matches or exceeds other comparable use bands in throughput for the same size antenna, with minimal impact from atmospheric water vapor absorption. Also, the X-band is already compatible with the DSN, and has heritage use and specific frequency allocations for deep space applications with less frequency crowding as compared to S-band. A phased array antenna was chosen for electronically steerable pointing and to get a high amount of gain per mass and lower drag surface area than a dish antenna enclosed within a dome. The phased array antenna surface area available was 4- by 0.5-m. The antenna was partitioned into 3.5- by 0.5-m for data transmission and 0.5- by 0.5-m for receive. The transmit link budget was analyzed using a conservative 40 dBi phased array antenna gain. This gain was determined by using an upper bound of the maximum achievable gain for 3.5 m dish antenna of approximately 45 dBi and phased array antenna gain of approximately 10log10 N + Patch Element Gain ($\sim$5 dBi for patch and 10 dBi for horn) for nominal 4096 elements, or approximately 41.12 dBi. Above a certain number of phased array elements, the degradation due to path loss and other capacitance effects outweighs the achievable gain, and could actually result in a gain decrease. Two phased array antennas, one on each side of the fin, will be used in this design.

The design also includes two high gain omnidirectional antennas, one on each side of the instrument mast. These antennas will be used for emergency, very low bit rate communications or a ranging "I'm alive" signal.

The analysis used Titan atmospheric effects of approximately 0.6 dB for atmospheric attenuation and 0.1 dB for surface diffraction.

The Small Deep Space Transponder (SDST) at X-band was chosen as the transmitter for this design. The SDST is configurable and could be modified to accommodate necessary parameters to ensure mission success, including proper

Table 11.11 Communications system MEL

Description	Quantity	Unit mass (kg)	Basic mass (kg)	Total mass with growth (kg)
Communications and tracking	–	–	26.3	30.5
X Band	–	–	26.3	30.5
X band TWTA	1	1.00	1.0	1.1
X band EPC for TWTA	1	1.20	1.2	1.3
Phased array	2	9.00	18.0	21.6
Coupler	2	0.20	0.4	0.4
Diplexer	2	0.30	0.6	0.6
Deep space transceiver	2	2.00	4.0	4.4
Connections	1	0.50	0.5	0.5
Phased array enclosure	0	60.00	0.0	0.0
X-band omni antenna	2	0.30	0.6	0.6

coding and modulation. It also has a built in beacon mode for ranging and emergency communications.

To develop the amount of RF energy required to close the data link budget, a traveling wave tube amplifier (TWTA) and associated electronic power conditioner were chosen. Current X-band TWTA technology can generate RF power between 25 and 165 W with 55–65% efficiency. Using the conservative estimate of 55% efficiency, this is approximately 300 W DC, which fits within the current power budget and will also allow the communications link to close and validate.

The mass of all of the components of the communications system are included in the MEL shown in Table 11.11.

Communications System Analysis

Figure 11.18 shows that the communications link between Titan and Earth is feasible at the required data rate, given the previous assumptions.

11.4.2 Command and Data Handling (C&DH) System

The C&DH subsystem is used to provide computer control and data storage for the submarine and the aerovehicle carrying it to Titan.

C&DH Requirements

The C&DH System provides the processing capability and data storage to operate the equipment on the Titan submarine and its aerovehicle through all phases of the mission. The system concept has a single fault tolerant main processor rated for 50 krad total dose radiation environment. The system has a science data storage capability large enough to accommodate 25 Mb/day.

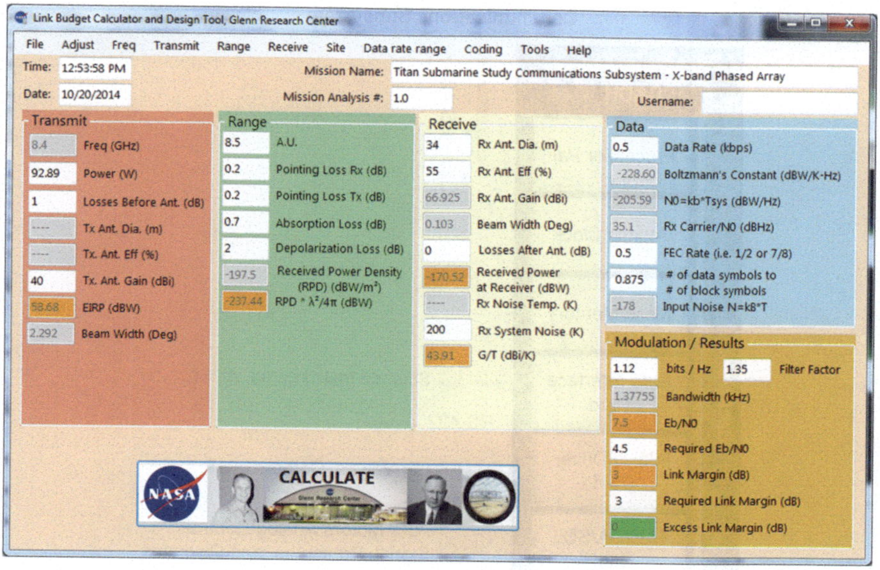

Fig. 11.18 Titan submarine communication link analysis

C&DH Assumptions

The C&DH enclosure will be inside the pressure shell of the submarine and will be maintained in a temperature environment of 280 K.

C&DH Design and MEL

The concept design solution includes a flight controller enclosure populated with electronic boards selected based on their capability to meet mission goals and to survive the environmental conditions. The main components are:

- Two processor cards utilizing Power PC 750 radiation hardened cards, or equivalent.
- Watchdog switcher.
- Solid state memory card.
- cPCI enclosure with power supply.
- Atomic Clock module/Ultra oscillator module.
- Valve/Compressor drivers.
- Data interface cards (RS422/485) for communications, science and navigation instruments.
- Wiring harness and connectors.

The system concept is illustrated in Fig. 11.19, and the C&DH MEL is shown in Table 11.12.

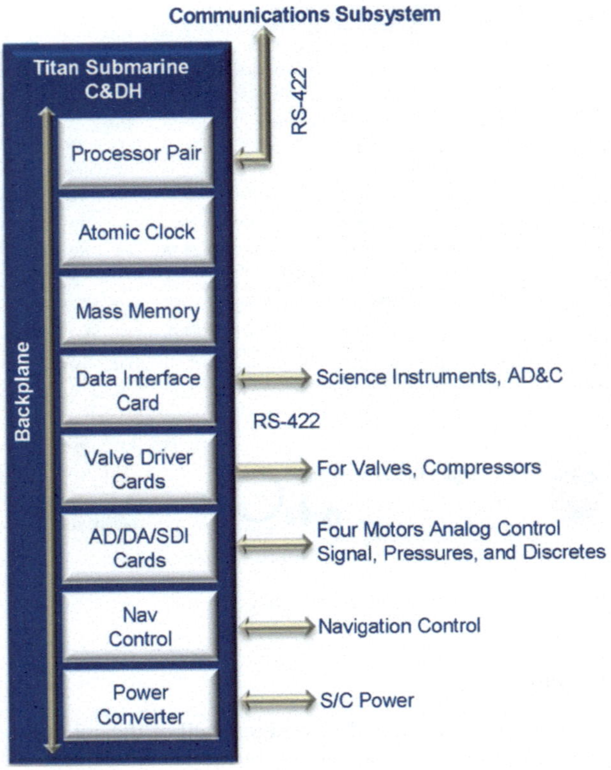

Fig. 11.19 Titan submarine C&DH system

Table 11.12 C&DH system MEL

Description	Quantity	Unit mass (kg)	Basic mass (kg)	Total mass with growth (kg)
C&DH	–	–	44.0	57.2
C&DH hardware	–	–	30.0	39.0
Flight controller	1	30.00	30.0	39.0
Instrumentation and wiring	–	–	14.0	18.2
Data cabling	1	14.00	14.0	18.2

11.4.3 Navigation

Requirements

The requirements for the Navigation design were as follows:

- The design must be single fault tolerant
- Point the phased array antenna to Earth for communication phase
- Estimate the vehicle position, velocity and attitude throughout the mission, beginning after Splashdown on Titan.

Assumptions

It was assumed that the position updates obtained by being tracked by ground stations on Earth are accurate to ~1 km (0.6 miles). Cassini, currently in orbit around Saturn, is tracked to approximately this accuracy.

Design Summary

Position Updates: Position updates are obtained in between successive submerged operations while the vehicle is on the surface. A Sun sensing camera is used to obtain the direction to the Sun. Using ephemeris stored in the onboard computer, the direction to the Earth is then calculated, and the vehicle is then oriented such that the phased array antenna is pointed towards Earth for communications. The vehicle obtains updates to its position and heading during this phase. A reasonable estimate on the expected accuracy for a position fix while on the surface is approximately 1 km. Cassini, currently in orbit around Saturn, is tracked to approximately this accuracy. Terrain imaging, while the submarine is surfaced and near the coast, may also aid in position determination, comparing images taken by the Surface Imager to those stored on board.

Submerged Navigation: Submerged navigation is accomplished with a sonar and Doppler Velocity Log (DVL) aided Inertial Navigation System (INS). The Navigation MEL can be seen in Table 11.13.

In general, an INS calculates position, velocity and attitude using data from an IMU, which consists of three accelerometers and three gyros. Over time, if not

Table 11.13 Navigation MEL

Description	Quantity	Unit mass (kg)	Basic mass (kg)	Total mass with growth (kg)
AD&C	–	–	32.9	38.8
GN&C	–	–	32.9	38.8
IMU	2	8.50	17.0	20.1
DVL	2	4.30	8.6	10.1
Sun sensor	2	1.00	2.0	2.4
Sonar transducer	2	1.40	2.8	3.3
Sonar electronics module	1	2.50	2.5	3.0

updated, errors in the IMU build up and the state estimates begin to drift further and further from the true values. To reduce the growth in error, redundant sensor measurements are integrated with the INS through a Kalman filter. The redundant measurements, in this case, are supplied by a low drift DVL and imaging sonar. The DVL tracks the velocity of the vehicle relative to the sea bottom, and while it aids in position measurement, does experience a drift in position error, mostly in the direction along the sub's track. The sonar improves navigation accuracy by taking successive images of the sea bottom to estimate the vehicle's bottom relative velocity.

Modern INS systems are capable of heading drift rates of less than 0.01°/hr. Assuming the distance traveled in 8 h of submerged operations is approximately 20 km (12 miles) (average velocity between 0.5 and 1 m/s (1.6 and 3.3 ft/s)), this would correspond to an error in position of less than 10 m (33 ft). Modern systems also employ a magnetic compass, which provides the capability to essentially estimate vehicle heading continuously. This capability will not be possible on the Titan submarine, as Titan does not exhibit a magnetic field, hence the error in position would likely be greater for the Titan submarine. However, the expected growth of the error in position in 8 h of submerged operations is still expected to be less than 1 km (0.6 miles). Taking into account the initial error in position of 1 km (0.6 miles) from being tracked by Earth, the total error in position after 8 h of submerged navigation should be less than 2 km (1.2 miles).

To comply with the requirement of being single fault tolerant, two of each of the navigation components were included in the design, with the exception of the side scanning sonar. A side scanning sonar is also held in the Science MEL, and could be used should the navigation sonar fail. The locations of the navigation components can be seen in Fig. 11.20.

Trades

The following were the various options considered for submerged navigation.

Long Baseline (LBL): LBL systems provide accurate autonomous underwater vehicle (AUV) position measurements once a network of LBL transponders has been deployed and calibrated. LBL systems offer very high position accuracy and position stability, sometimes as good as <1 m (3.3 ft) error in position, however it does require transponders to be moored to the sea floor with high position accuracy. While this option does offer high accuracy in position estimation, it would require the deployment of transponders that would each need to be independently powered, as the Titan Submarine moves through Kraken Mare.

Sonar and DVL Aided INS: This option was ultimately chosen for submerged navigation. It does not require having a separate vehicle to be in communication with Earth, nor does it require the deployment of transponders.

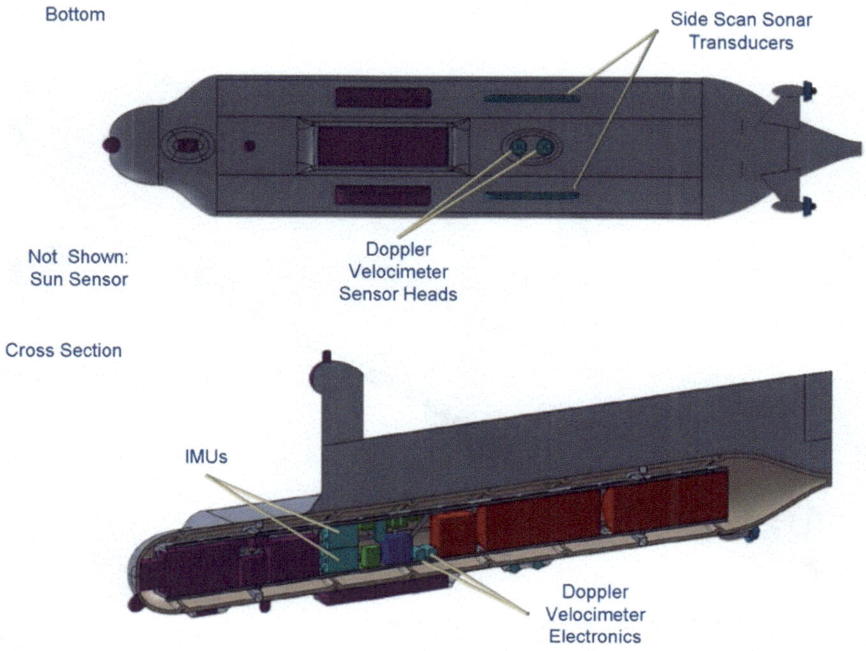

Fig. 11.20 Location of navigation components

11.4.4 Buoyancy Control System

The buoyancy control system for the Titan submarine is used to enable the submarine to float on the surface of the seas, descend into the sea in a controlled manner, maintain a certain depth, and then resurface again. This is accomplished by utilizing pressurized Ne gas and control valves to channel in either the liquid methane or atmospheric nitrogen depending on the maneuver the submarine is taking.

The buoyancy system is required to operate from the surface to a depth of 1 km (0.6 miles) or more. The system consists of two external tanks, as shown in Fig. 11.21.

These tanks can be vented to the atmosphere or filled with liquid methane to adjust the buoyancy of the vehicle. Smaller Ne pressurized tanks are located within the buoyancy tanks to control both pressure within the tank and depth of the vehicle. The Ne is used as a means of adjusting the liquid volume within the main tanks to set the vehicles buoyancy.

Buoyancy Gas Options

One of the first steps in designing the buoyancy system for the submarine was to select a buoyancy gas to provide the lift needed to surface and remain on the surface for as much time as desired. Three options were initially considered:

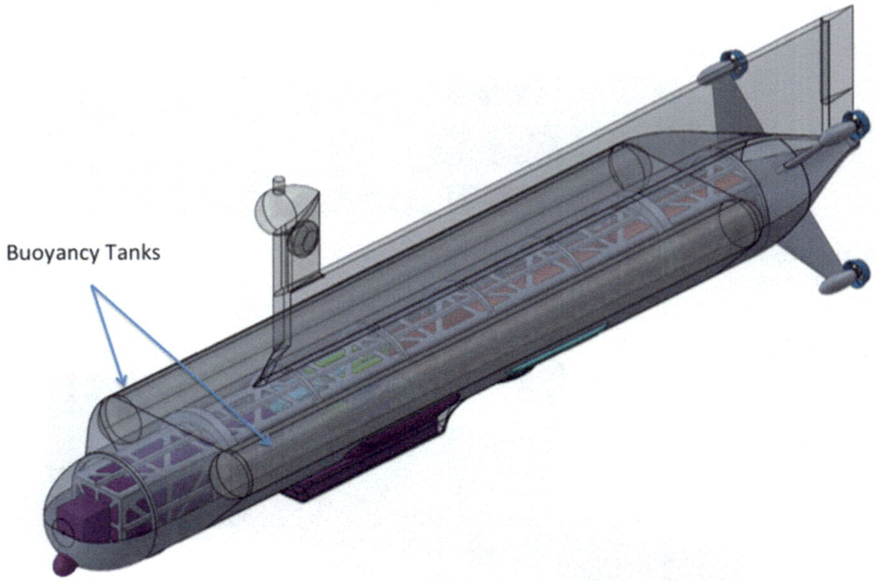

Fig. 11.21 Illustration of the buoyancy tank location

- Utilizing atmospheric nitrogen gas as the buoyancy gas
- Boiling the liquid methane to provide buoyancy
- Utilizing a contained inert gas (Ne) to provide buoyance.

Of these three approaches the third was selected as the best option for providing buoyancy in the submarine. At first it would seem that gathering and compressing the atmospheric nitrogen would be the best choice for generating buoyancy. However, upon further investigation it turned out that utilizing nitrogen would limit the depth which the submarine could reach. This is because nitrogen is condensable at the Titan operation temperatures and would liquefy at the temperature and pressure that would be experienced at a particular critical depth in the sea. This can be seen from the pressure-temperature diagram for nitrogen shown in Fig. 11.22.

At an ambient temperature within the methane sea of 94 K, the nitrogen gas will liquefy as it is pressurized. This then limits the pressure to which the gas can be compressed and in turn limits the depth to which the submarine can descend. Using nitrogen as the buoyancy gas would limit the submarine depth to approximately 200 m (656 ft). Additionally, pressure control near the saturation point of nitrogen would be problematic, due to phase change of the gas.

To achieve greater depth another method would have to be utilized. Another option considered was to use the waste heat from the power system to boil the liquid methane and use this gaseous methane to fill the buoyancy tank. However, a thermal analysis on the heat required to boil the required amount of liquid methane quickly determined that there was insufficient waste heat available for this approach to work.

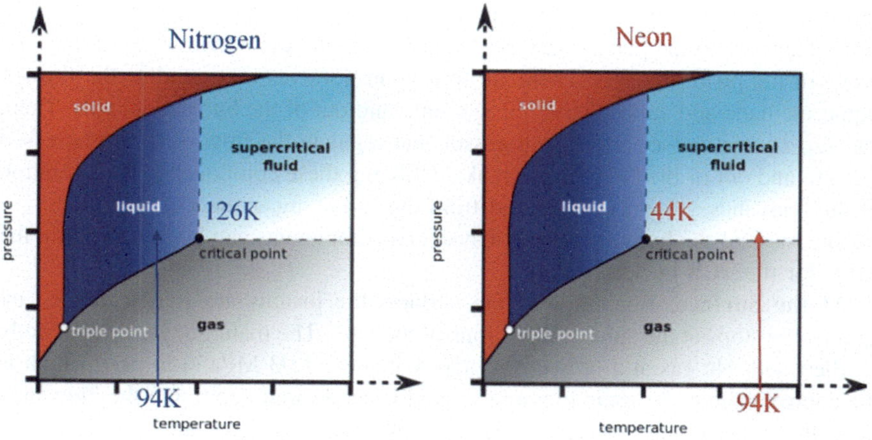

Fig. 11.22 Pressure-temperature diagram for nitrogen and neon

Based on this, a third option was devised that would enable the submarine to achieve greater depths and provide the buoyancy needed to return to the surface. This method utilized Ne as the buoyancy gas. Neon was selected since it is inert and will not react with the liquid methane. Also at the ambient temperature of 94 K it will remain gaseous even at high pressures, as shown in Fig. 11.22.

Buoyancy Approach

To make the Ne gas buoyancy approach viable, the Ne gas will need to be conserved between each ascent and descent. This requires the gas to be captured after ascent and repressurized prior to the next ascent. To accommodate these requirements, a system was devised that utilized a Ne gas pressure tank, control valves, and a piston all housed within the main buoyancy tanks. This arrangement is illustrated in Fig. 11.23.

The buoyancy tank consists of a large outer cylindrical tank separated into front and rear halves. The tanks are symmetric about this center axis. At the center of the tank are the Ne pressure tanks. On either side of the Ne tanks is a piston that can

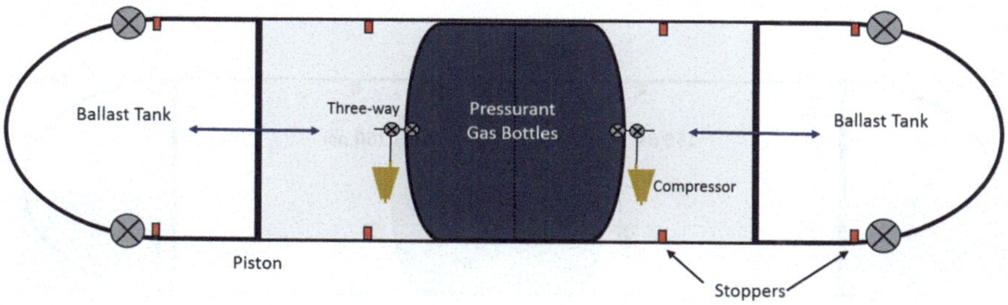

Fig. 11.23 Buoyancy tank layout

travel the length of the cylindrical portion of the ballast tank up to the spherical ends. There are stops placed near the Ne tanks and the cylindrical ends to limit the range of the piston motion. There are four control valves that regulate the flow of liquid methane and atmospheric nitrogen into and out of the buoyancy tanks. There are also control valves and a compressor that regulate the flow of the compressed Ne into and out of the Ne pressure tanks. Utilizing these components, the operation of the buoyancy system is described from the following sequence.

The mass of the buoyancy control system's components can be found within the MEL for the TCS in Sect. 11.4.7.

At the Surface: Beginning at the surface, the pistons are fully extended up against the stops at the spherical portions of the tank. The main buoyancy tank body is filled with Ne gas at a pressure of approximately 1.03 MPa (150 psi), which is the approximate hydrostatic pressure at the lowest desired depth. On the buoyancy tank, the lower control valves, exposed to the liquid methane are closed and the upper control valves exposed to the atmosphere are open. This configuration is illustrated in Fig. 11.24.

In this arrangement the submarine is buoyant and will float on the surface of the methane sea. The next step is to prepare the submarine for its next descent.

Preparing for Descent: While on the surface, operations such as data gathering and communications are taking place. It is estimated that the surface operations will require approximately 8–10 h. Therefore during this time the submarine is preparing for its next descent. This is accomplished by pumping the Ne gas back into the Ne pressure tank, going from the 1.03 MPa (150 psi) when it filled the buoyancy tank to 6.9 MPa (1000 psi) in the pressure tank. Since this pumping is done over an extended period of time the pump power required is minimal (approximately 18 W). As the pressure tank fills up and the pressure drops in the buoyancy tank the pistons begin to move back toward the pressure tank to the original position. Once the pressure tank is full, the two pistons will be against the stops near the pressure tank and there will be a pressure of ~0.15 MPa (~22 psi) of nitrogen gas within the buoyancy tank. Atmospheric nitrogen flows into the buoyancy tank because the upper valves are open venting it to the atmosphere. The lower valves are still closed preventing the liquid methane from entering the tank. This configuration is illustrated in Fig. 11.25.

In this arrangement the submarine is still buoyant and will float on the surface of the methane sea. Once in this configuration the submarine is ready to descend.

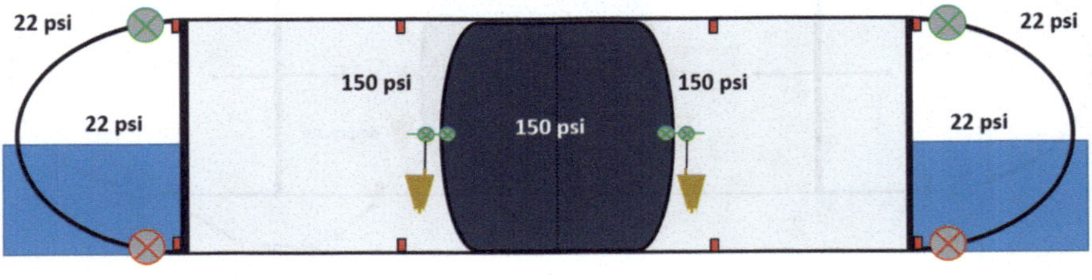

Fig. 11.24 Buoyancy configuration at the surface

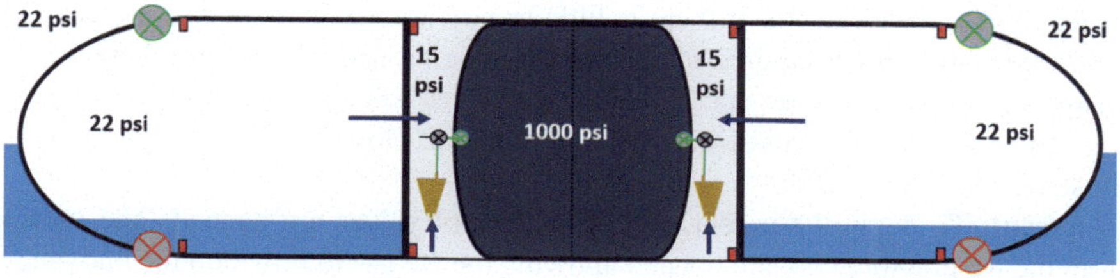

Fig. 11.25 Buoyancy configuration at the surface with the submarine prepared for descent

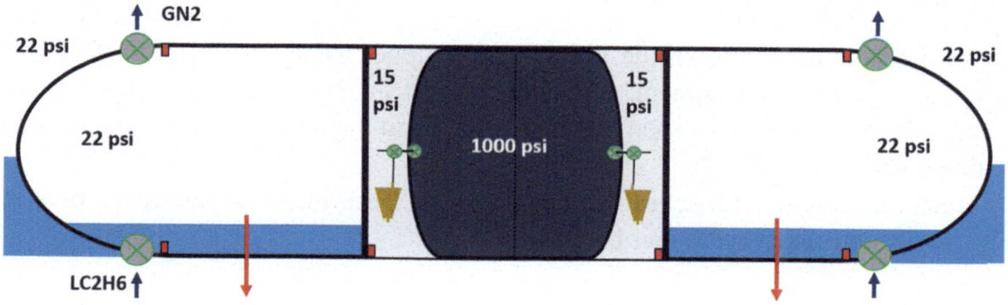

Fig. 11.26 Buoyancy configuration at the surface, initiating descent

Descent Initiation: To initiate descent, the lower control valves at the bottom of the buoyancy tanks are opened. This allows liquid methane to flow into the buoyancy tank. The upper valves are also open allowing the nitrogen gas to escape as the buoyancy tanks fill with liquid methane. This configuration is illustrated in Fig. 11.26.

In this arrangement the submarine will begin to descend. To control the descent, the top control valves can be closed, trapping the nitrogen gas within the tanks and slowing or stopping the descent. The ability to control or stop the descent will depend on how much variation there is in the density of the liquid methane with depth. If it does not vary much then the submarine will either sink to the bottom or float on the surface with little ability to control the depth. In that case the depth will need to be controlled actively by using the propulsion system to generate lift enabling the submarine to stay at a desired depth. If there is a small variation in the methane density with depth then the submarine can become neutrally buoyant at a given depth by controlling the release of nitrogen gas. This would work up to a depth of 200 m (656 ft) where the pressure would become too great and begin to liquefy the nitrogen. At depths below 200 m (656 ft) releasing some pressurized Ne, causing the pistons to move and expanding the Ne gas volume would be required to achieve a neutrally buoyant condition.

Neutrally Buoyant Operation: Once all of the nitrogen gas has left the buoyancy tank the submarine will descend to its maximum depth. This will be determined based on the weight of the submarine and its volume. As described previously, the depth of neutral buoyancy will depend on how the density of the methane varies with depth. If there is little to no variance then an active means of generating lift will be required to keep the submarine from descending beyond the desired depth. At this point once there is no nitrogen gas left in the buoyancy tanks, the upper and lower valves can be left open or closed. This configuration is illustrated in Fig. 11.27.

Ascent: To begin the ascent, the upper control valves are opened. The valves from the Ne pressure tank are opened allowing the Ne gas to flow into the buoyancy tank and begin to move the piston toward the spherical end of the tank. As the piston moves the liquid methane is pushed out of the buoyancy tank through the upper control valve. At this stage the lower control valves can be either closed or opened. As the submarine nears the surface they will also be opened so that the liquid methane can flow out the bottom as the upper valves break the surface and are exposed to the atmosphere. This will prevent the spherical ends from staying filled with liquid methane when the submarine is on the surface. This configuration is illustrated in Fig. 11.28.

Once the submarine reaches the surface it will be at the configuration shown in Fig. 11.24 and the process can begin again.

11.4.5 Hydrodynamics and Propulsion (H/P)

A streamlined torpedo-like pressure hull is fitted with external cylindrical ballast tanks mounted high up on each side and covered with a free-flooded hydrodynamic

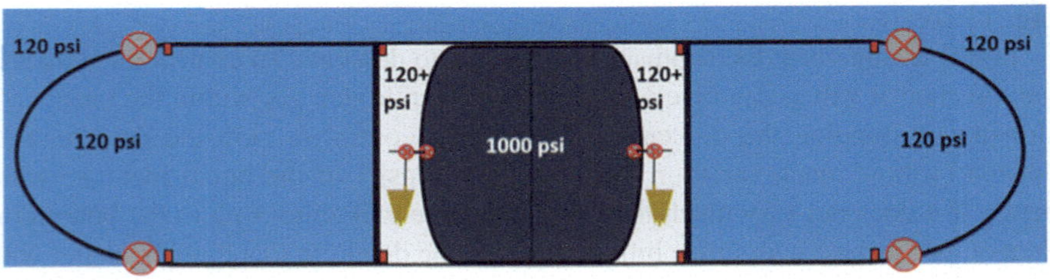

Fig. 11.27 Buoyancy configuration submerged

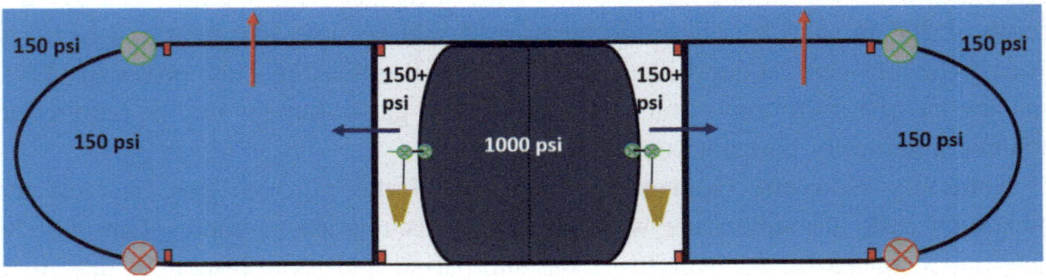

Fig. 11.28 Buoyancy configuration initiating ascent

fairing. Four fixed fins are mounted near the tail in an X configuration to provide hydrodynamic stability. Additional floatation and static ballast are distributed to provide hydrostatic stability. Small electric thrusters attached to the tip of each stabilizer fin provide propulsion, the thrust levels of each may be controlled independently to provide directional control, as shown in Fig. 11.29.

H/P Requirements

The Titan Submarine must be able to maneuver and transit both on and below the surface of Titan's Kraken Mare, a sea composed of cryogenic liquid hydrocarbons. The submarine will have a submerged transit speed of at least 1 m/s while consuming no more than 440 W of electrical power for propulsion. The maximum surfaced speed is not specified, but it should be able to provide reasonable heading control while consuming no more than 100 W. The submarine must also be a stable platform from which to perform scientific measurements and to communicate DTE.

The design of the propulsion and control system as well as that of the overall vehicle must support these requirements in an environment that is considerably different from that of Earth's oceans; the surface temperature of Kraken Mare is estimated to be 94 K, with a liquid density that is 2/3rds that of saltwater and gravitational acceleration 1/7th that of Earth.

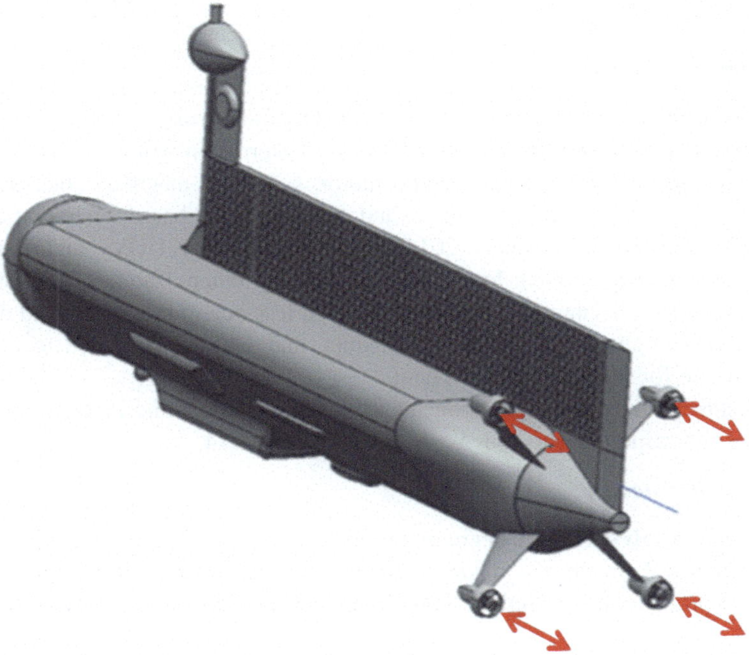

Fig. 11.29 Titan submarine propulsion system

H/P Assumptions

Fluid properties at the surface were assumed to be those of liquid ethane at 94 K and 0.15 MPa (1.5 bar). The liquid density is assumed to be 660 kg/m^3 (41.2 lb/ft^3) and the kinematic viscosity of 6.06 × 10^{-6} Pa-s (147.6 lb/ft-hr) resulting in a Reynolds number that is 1/6th that of an identical submarine at the same speed in terrestrial saltwater. Small (roughly 1 m (3.3 ft) in length) terrestrial UUVs, such as the Remote Environmental Monitoring Units (REMUS) 100 (Hydroid, LLC 2014) operate at Reynolds numbers near 1 × 10^6, where the Titan Submarine will be at 1 m/s (3.3 ft/s). Based on Reynolds number similarity the primary physics governing vehicle control and propulsion will be similar to those we are familiar with on Earth.

The resistance analysis assumes turbulent flow over the hull, though at this Reynolds number there will be significant surface areas subject to laminar flow. Care will have to be taken to avoid flow separation in regions of laminar flow, and perhaps to force transition to turbulent flow at a certain location.

The stability and trim analysis assumes a CB located 5% of hull diameter above the CG, both of which are located 45% of hull length from the bow, which is a good design point for many terrestrial UUVs. Hull lift and drag are based on Slender Body theory which is a good assumption for high speed performance. In the case of a low speed vehicle, such as this, fluid mechanics problems such as dynamic positioning, cross-flow drag, and the control of flow separation should be the subject of future work.

H/P Design and MEL

Propulsion and directional control are provided by four thrusters mounted to the tips of four fixed stabilizer fins arranged in an X-stern configuration. The thruster assemblies include a brushless electric motor, gear box, magnetic coupler, and a ducted propeller. The thruster mass and volume were based on commercially available thrusters (Tecnadyne 2014) used in terrestrial UUVs and Remotely Operated Vehicles (ROVs), low speed and usually tethered undersea robots. The thrust produced on Titan would be at least 1/3rd less due to the lower fluid density. A detailed design of the thruster rotor, duct, and motor is required to optimize performance for the Titan environment.

A listing of the various components in the Titan Submarine Propulsion system and their corresponding masses is shown in Table 11.14.

H/P System Analysis

Resistance Analysis: At the beginning of the preliminary design stage a propulsion system power estimate was made based on empirical relations for required flight power of propeller driven lighter-than-air vehicles on Earth which may be scaled to

Table 11.14 Propulsion and control system master equipment list

Description	Quantity	Unit mass (kg)	Basic mass (kg)	Total mass with growth (kg)
Propulsion	–	–	20.56	26.48
Main propulsion system	–	–	20.56	26.48
Main engine hardware	–	–	20.56	26.48
Thruster	4	1.80	7.20	9.00
Thruster controller	4	0.80	3.20	4.00
Thruster housing	4	0.30	1.20	1.50
Stabilizer	4	1.48	5.92	7.40
Hardware	1	0.44	0.44	0.55
Thruster cable	4	0.65	2.60	4.03

operate in the Titan submarine environment. It was shown by Lorenz (2001) that the installed power of a lighter-than-air vehicle (or a submarine) varies as

$$P = 3.0\, m^{0.6} V^{1.85} (\rho_{\text{Titan}}/\rho_{\text{Earth}})^{0.33-0.5n}$$

where m is the mass, V is the velocity, ρ is the fluid density, and n is a parameter representing the effect of density scaling on propulsive efficiency. With an initial value of $n = 0.5$, a 1000 kg (2205 lbm) submarine on Titan traveling 1 m/s might have an installed power of 300 W for the purposes of preliminary sizing. This estimate figured into the sizing of the power system, which in turn drove the diameter of the pressure hull.

The drag of non-streamlined objects appended to the hull was estimated using empirical models based on a collection of historical datasets (Blevins 1984). Friction drag on streamlined shapes (hull, fins, streamlined appendages) was estimated using the ITTC 1957 model-ship correlation line which gives the coefficient of drag due to friction based on wetted area as a function of Reynolds number based on length:

$$Cf = \frac{0.075}{(\log_{10}(\mathrm{Re}_L) - 2)^2}$$

The skin friction drag coefficient was then multiplied by base drag and interference drag multipliers which are functions of the geometry (Feldman 1995). These methods have predicted the resistance of terrestrial UUVs with complex appendages within 30% of that predicted by CFD-RANS. The impact of manufacturing imperfections such as gaps, steps, and fasteners was not studied during this phase. This method of estimating resistance was used in spreadsheet form during the early stages of analysis, and later in a MATLAB script that is used to predict the trim state of UUVs and other submerged streamlined vehicles. A history of the resistance estimates (shown as effective power to overcome drag in watts) made during the course of this study is shown in Fig. 11.30, estimates 2–8 were

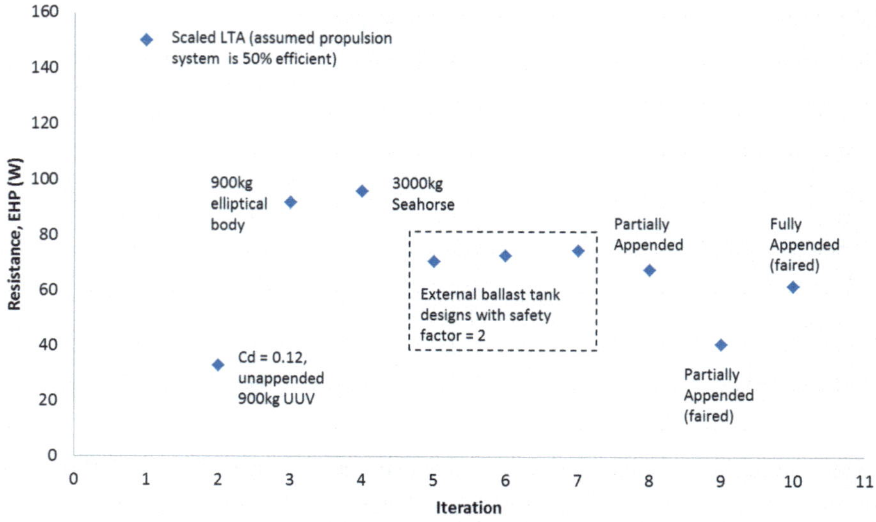

Fig. 11.30 Empirical resistance estimates for vehicle concepts

Table 11.15 Reference quantities

Parameter	Value
Speed (m/s)	1.0
Speed (kn)	1.9438
Diameter (ft)	2.6157
Area (ft^2)	5.3738
Fluid: methane = 1, ethane = 0	0
ρ (kg/m^3)	660.0
μ (Pa-s)	4.000×10^{-3}
ρ ($slugs/ft^3$)	1.288
ν (ft^2/s)	6.488×10^{-5}
Length scale (ft)	19.3
Reynolds number	9.74×10^5
Accel due to gravity (ft/s^2)	4.435×10

multiplied by a safety factor of 2 to account for the effect of unknown appendages related to science collection and communications. The length and Reynolds number are shown in Table 11.15.

The contribution of each component to the total vehicle drag for the final design at 1 m/s is shown in Table 11.16. The drag coefficient is based on a frontal area of 0.50 m^2. The input power is based on an assumed overall system efficiency of 51%.

Powering Analysis: The efficiency of the propulsion system was estimated using an ARL Penn State propulsor sizing code used in the design of propellers for the U.S. Navy. A swirl-canceling ducted thruster having a propeller diameter of 0.27 m was evaluated, in seawater, at 1 m/s with a thrust of 50% of the total drag

Table 11.16 Resistance and powering estimate for final design at 1 m/s—drag breakdown

Bare hull	3.28	23.5%
Fins	0.42	3.0%
Duct	0.17	1.2%
Antenna	0.95	6.8%
Struts	1.95	14.0%
EGVs	0.00	0.0%
Pods	1.72	12.3%
DS	0.79	5.7%
Bottom sampler	1.84	13.2%
P3 sensor	0.16	1.1%
Side scan sonar	0.10	0.7%
Sonar transducer	0.04	0.3%
Camera housing	0.13	1.0%
Camera mast	0.17	1.2%
Met. sensor	0.36	2.6%
Subsurface illumination	1.48	10.6%
Doppler velocimeter	0.19	1.4%
Omni-antenna	0.21	1.5%
Total drag, lbf	13.97	100.0%
$C_{D\_frontal\ area}$	0.3752	
EHP (hp)	0.08	
EHP (W)	62	
System efficiency (%)	51	
Input power (W)	122	

(assuming that two of the four thrusters will carry most of the load) and found to have a propulsive efficiency of 68%. Using common values found in industry for electric motor efficiency (80%), planetary gearbox efficiency (95%) and magnetic coupling efficiency (98%) it can be calculated that the total system efficiency is 51%. The assumption that propeller efficiency in seawater is the same as in liquid ethane will have to be validated in the next phase.

Using the previously calculated drag coefficient and propulsion system efficiency, and assuming that they do not change significantly over small changes in speed, one can calculate the performance of the submarine as shown in Table 11.17. Although surfaced resistance varies from submerged resistance due to changes in pressure, friction, appendage, and wave making drag, for the purposes of a preliminary analysis we have assumed that they are equal. The maximum speed submerged, based on the 440 W available for propulsion, was 1.57 m/s. The maximum speed surfaced, based on the 100 W available for propulsion, was 0.93 m/s. Power consumption at the submerged design point of 1 m/s was 122 W.

Hydrodynamic Stability Analysis: Hydrodynamic stability of the final design was evaluated using a MATLAB code that calculates and plots the speed-dependent roots of the characteristic equations obtained from the standard submarine equations

Table 11.17 Submerged propulsion performance—system efficiency = 51%

Speed (m/s)	Drag (lbf)	EHP (W)	Power (W)
0.25	1.1	1	2
0.5	3.9	9	17
0.75	8.2	27	54
0.93[a]	12.3	51	100
1	14.0	62	122
1.25	21.1	118	230
1.5	29.7	198	388
1.57[b]	32.3	224	440

[a]Maximum surfaced speed
[b]Maximum submerged speed

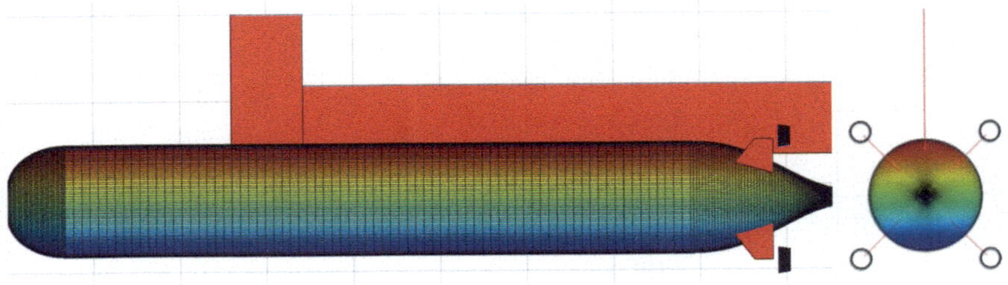

Fig. 11.31 Vehicle geometry analyzed in stability code

of motion (Imlay 1964; Mandel 1967). The stability code also calculates and displays the "Infinite Speed" stability index, derived from the constant term in the characteristic equations when the hydrostatic moment is assumed to be "zero," hence "Infinite Speed". Results are given for the Vertical-Plane and Horizontal-Plane Motions, respectively. The roots are non-dimensional (i.e., Laplace Transform of the time variable) and are interpreted for the dynamic stability (Feldman 1995; Hoerner 1958). The geometry analyzed is shown in Fig. 11.31 and the resulting code output in Table 11.18 for the design point of 1 m/s (3.3 ft/s), submerged, and neutrally buoyant. The results suggest that the handling characteristics will be benign. A sensitivity analysis to buoyancy was performed and stability was found to be insensitive to small changes in buoyancy (±5%).

Hydrostatic Stability Analysis: Hydrostatic stability of the final design was evaluated using a spreadsheet. Component locations and volumes from the CAD model and mass estimates (with component level growth) from the MEL supported the calculation of CB and CG. System level mass growth was added at the CM of the pressure hull. For the purposes of this analysis the acceptance criteria for hydrostatic stability was that the CB be above the CG at the same longitudinal location. All submarines require that this condition be maintained during all phases of submerged and surfaced operation, and a larger distance between CB and CG generally increases roll and pitch stability.

Table 11.18 Stability code output at 1 m/s, neutrally Buoyant

Quantity	Value	Interpretation
Margin of stability, GV	0.24	Marginally stable
Vertical-plane stability roots	−0.12, −0.12, −0.12	Non-oscillatory (overdamped) stable
Vertical-plane damping ratio	1	Non-oscillatory, critically damped
Margin of stability, GH	0.50	Highly stable
Horizontal-plane stability roots	−4.39, −0.41	Oscillatory (underdamped) stable
Horizontal-plane damping ratio	1	Non-oscillatory, critically damped

The baseline configuration was found to be unstable and top heavy, which is not unexpected given the relative size of the superstructure to the rest of the submarine. Extra displacement was added in the form of syntactic foam installed in the free flooded spaces between the ballast tanks and pressure hull (Fig. 11.32). The preliminary baseline dry mass was 1148 kg (2531 lbm). The floatation provided 141 kg (311 lbm) of additional buoyancy and 34 kg (75 lbm) of mass. Subsequent calculation determined that 180 kg (397 lbm) of static ballast (usually in the form of lead bricks or pellets) was required at a location low in the keel and 2.1 m (6.9 ft) from the bow. The stable configuration has a dry mass of 1361 kg (3001 lbm). Six characteristic operating conditions are shown in Table 11.19 with the required ballast tank load, and resulting wet weight and CB/CG offset. The last condition, surfaced with full ballast tanks, is marginally unstable, but the vehicle is likely to submerge long before this condition is reached. It will be a goal of future design iterations to reduce the amount of static ballast required to stabilize the submarine through creative arrangement of the internal systems and mass reduction of the superstructure.

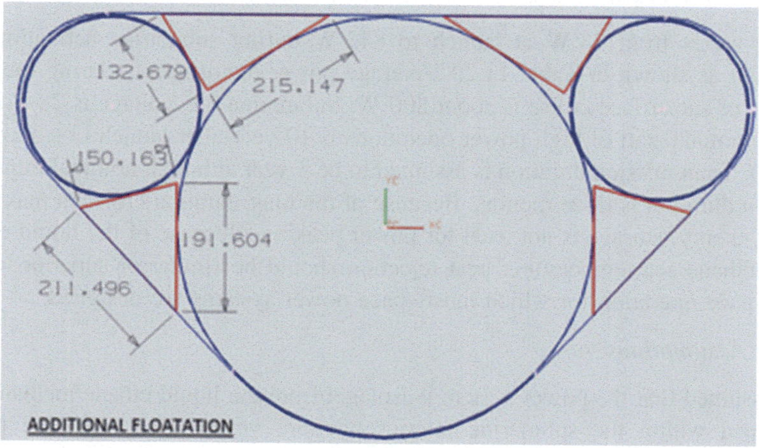

Fig. 11.32 Supplemental floatation and mass (dimensions are in mm)

Table 11.19 Hydrostatic analysis results (dry mass = 1361 kg (3001 lbm))

State	Ballast tank volume (%Full)	Wet weight (kg)	CB/CG offset (mm)
Submerged, neutral	77	0	21
Submerged, light	0	–220	66
Submerged, heavy	100	65	23
Surfaced, neutral	50	0	13
Surfaced, light	0	–189	43
Surfaced, heavy	100	96	0

The dry mass of the sub increased by 24 kg (53 lbm) after the hydrostatic stability analysis was complete. This change ($\sim$1.8%) would not affect the stability analysis significantly, unless the mass increase was concentrated high above the centerline of the vehicle. The hydrostatic analysis would be repeated as part of a Phase II study as the sub design is further refined.

11.4.6 Electrical Power System

The Titan Submarine requires a system that can power the sub and aerovehicle for the transit from Earth to Titan, through Titan EDL and still provide power for a minimum 90 day surface mission (and preferably for multiple years afterward). Candidate power systems include Radioisotope Power Systems (RPS), Fission Power Systems (FPS) and fuel cells. The FPS and RPS power systems are assumed to be modified versions of their deep space counterparts. Two, eight-source SRGs were selected as the power systems for this study.

Power Requirements

Power varies from 78 W at launch to 842 W during submarine activation and checkout as shown in Table 11.20. Average power requirements during multiday surface or subsurface cruise is about 800 W. Submarine bus voltage is 28 V ($\pm$ 6) while nominal start of high power operations is 10 year after launch (3 storage + 7 transit). Titan mission duration is assumed to be 3 year although nominal minimum mission duration is three months. Because of the long duration cruise at maximum power energy, storage is not used for power peaking. Because of the liquid ethane and methane sea temperature, heat rejection should be similar or superior to 4 K deep space operation for which most space power systems are designed.

Power Assumptions

It is assumed that the power system is isolated from the liquid ethane/methane and contained within the submarine interior pressure vessel. Heat rejection to the environment is accomplished via a pumped loop system similar in design to the

Table 11.20 Titan submarine power requirements, power modes (W)

(a) Power modes 1–5				
1	2	3	4	5
Launch	Interplanetary cruise	Titan EDL	Sub activation and checkout	Dive/ surface
60 min	~7 years	2 h	1 week	100 min
78	91	117	842	826
(b) Power modes 6–10				
6	7	8	9	10
Submerged cruise	Surface cruise	Stationary submerged operations	Stationary surface operations	EOM disposal
8 h	16 h	8 h	16 h	0.0
839	746	269	534	166

SRG Auxiliary Cooling System (ACS) for each of the candidate systems. Additionally it is assumed that the submarine is filled with 10 bars of dry nitrogen. Microporous solid insulation is used for all of the candidate power systems and while the 10 bars of dry nitrogen does increase the thermal conductivity of the insulation, the lower rejection temperature compensates for this additional heat leak (Lockheed Martin Corporation 2008).

Power Design and MEL

The selected power system uses a pair of eight general purpose heat source (GPHS) SRGs. Recent work on the Nuclear Power Assessment Study resulted in four SRG designs with two, four, six, and eight GPHS, respectively. Because the Titan submarine requires 840 W thirteen years after fueling, two, eight GPHS SRG were chosen as the baseline power system providing 900 W of DC power. These eight GPHS SRG are derivatives of the SRG in as much as they consist of two Stirling convertors operating in a dual opposed configuration to minimize vibration. These convertors use a MarM-247 heater head with hot cycle temperatures of 760 °C. Rejector temperature is 120 °C at beginning of mission (BOM). Solid insulation surrounds the GPHSs to drive the heat into the convertors. Each SRG is about 1 m (3.3 ft) long and 36 cm (1.2 ft) in diameter with a mass of 65 kg (143.3 lbm). Heat is removed via a pumped loop system attached to the radiator housing as discussed previously. Overall heat into DC power conversion efficiency at beginning of life (BOL) is 25%. Further performance details can be found in Table 11.21. During the launch and transit a heat removal system similar to that used on the Mars Curiosity rover is assumed with a pumped loop system removing heat to a radiator located on the lifting body/cruise stage. Heat load for the titan submarine is about 4000 W, which is double that of the Curiosity Mars rover.

The two SRGs are connected to the S/C bus and provide 26 ± 6 V. Figure 11.33 shows the electrical architecture. Each SRG has its own shunt if no power is being drawn from the unit, just as in the SRG. An S/C shunt balances S/C

Table 11.21 Candidate SRG—current best estimates

Number of GPHS	2	4	6	8
BOL (4 K) (W)	130	240	370	510
BOM (4 K + BOL + 3 year) (W)	126	232	357	492
EOM (4 K BOL + 10 year) (W)	104	193	297	450
BOL (270 K) (W)	116	215	331	456
BOM (270 K) (W)	113	207	319	440
EOM (270 K) (W)	93	173	266	366
Degradation rate (% /year)	1.16	1.16	1.16	1.16
Diameter (cm)	19	33	33	36
Length (cm)	50	45	65	95
GPHS heat load (BOL) (W)	500	1000	1500	2000
GPHS heat load (EOL) (W)	437	874	1312	1749
Controller efficiency (%)	90	90	90	90
BOL waste heat (4 K) (W)	356	733	1089	1433
BOL Stirling cold end temperature (4 K) (K)	420	450	450	430
Average heat rejection temperature (4 K) (K)	400	428	428	408
Average heat rejection temperature (270 K) (K)	440	468	468	448
Disturbance force (at 100 Hz)	10	13.6	16.9	19.8
BOL specific power (W/kg)	7.5	7.5	7.9	7.9
Mass (kg)	17.3	32.0	46.8	64.6
BOL efficiency (%)	26.0	24.0	24.7	25.5
EOM efficiency (%)	23.8	22.1	22.6	23.4

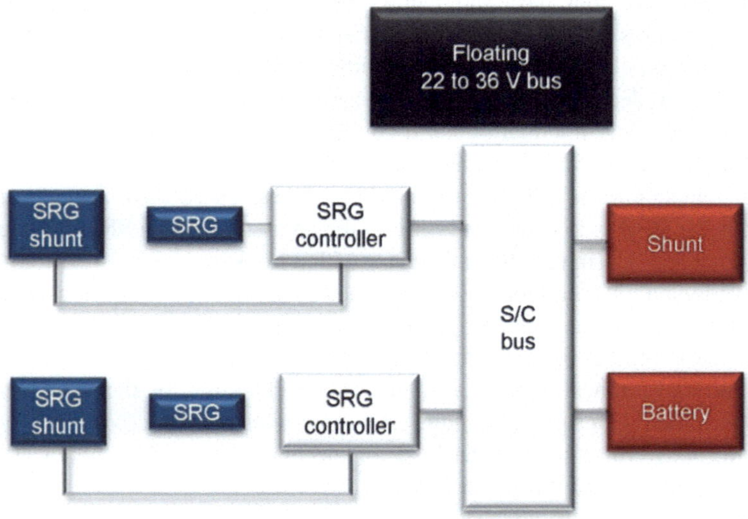

Fig. 11.33 Titan submarine electrical architecture

Table 11.22 Power system MEL

Description	Quantity	Unit mass (kg)	Basic mass (kg)	Total mass with growth (kg)
Electrical power subsystem	–	–	146.0	175.2
Power generation	–	–	130.0	156.0
Miscellaneous 06.1.5.a.a	2	65.00	130.0	156.0
PMAD	–	–	16.0	19.2
Miscellaneous 06.1.5.b.a	1	16.00	16.0	19.2

power requirements and power generated from the SRGs. Additionally, a 6000 μF bus capacitance via a 1 kg (2.2 lbm) lithium ion battery is required to help control voltage fluctuations. This battery is bookkept under the PMAD for the sub.

All of the components of the power subsystem and their masses are shown in Table 11.22.

Power Trades

Four candidate power systems were considered for the Titan submarine. The first was a FPS using a derivative of the 1 kWe Kilopower reactor coupled to eight Stirling Convertors. This FPS concept design is a current NASA and DOE joint project designed to leverage the Advanced Stirling Convertors (ASCs) used in the SRG and couple the reactor to the Stirling convertor with heat pipes. While compelling because of the scarcity of Pu-238 it was not selected as the designs produce a higher mass system than an RPS. Estimates for the Kilopower FPS reactor, shield and power conversion are around 400 kg. A secondary option considered was to use the liquid methane in a fuel cell to generate electrical power. A high temperature solid oxide fuel cell (SOFC) with direct methane reformation was considered as an option with the Earth delivered oxidizer being O_2 in the form of lithium perchlorate candles. These are the same O_2 supplies used on the International Space Station as emergency backup. Unfortunately due to the long duration and relatively high power over 1700 kg of candle mass alone was needed to supply 800 W for 90 days. Next, Multi-Mission Radioisotope Generators (approximately eight needed for the 750 W of power) were considered power source candidates. Because of their high plutonium consumption (6% conversion efficiency), low specific power (2.7 W/kg) and high degradation rates (3.8% power loss per year) they were not considered as viable candidates. Thus, SRGs based upon the convertors developed for the ASRG were selected as the baseline power system. Each of these convertors uses eight GPHS modules and provides about 450 W at end of mission. Two generators were required for the Titan submarine. These generators have high specific power (>7 W/kg), high overall conversion efficiency (>27% heat into DC power output) and relatively low degradation rates (1.2% per year).

11.4.7 Thermal Control

The main purpose of the thermal control system for the Titan submarine mission is to balance the heat generated by the isotope power system with the heat losses to the environment. The thermal system design included devising an approach to maintain the internal components of the submarine within their desired temperature operating range. Based on this design an estimate of the mass, size and any power requirements for the thermal system components was made.

Vehicle Operational Environment

The operating environment of the submarine is within the liquid methane and ethane seas of Titan. However, the submarine internal components will also need to survive the transit through deep space from Earth to Titan.

The harsh environment of Titan provides a number of challenges in the operation of equipment and materials. Operating within this environment, from entry to descent to the liquid methane and ethane seas requires a thermal balance between the heat generated by the isotope power system and the losses to the environment which is accomplished through thermal insulation and distribution of the heat generated. To accurately size the thermal system, the operational environment throughout each phase of the mission must be defined.

The environmental conditions on Titan are very unique and unlike those on any other known planet or moon. In some ways, though, it is very Earth like. It has a mostly nitrogen atmosphere, clouds, lakes, rivers and rain. However, with a surface temperature of under 100 K the free liquid is not water but methane. The low

Table 11.23 Physical and orbital properties of Titan (Flasar et al. 2005; Brown et al. 2009; NASA 2013)

Property	Value
Maximum inclination of equator to orbit to Saturn (δ_{max})	0.35°
Orbital eccentricity (ε)	0.0288
Mean radius of orbit (r_m) around Saturn	1.22×10^6 km
Day period (synchronous to the orbital period around Saturn)	15.95 (Earth days)
Surface pressure	146.7 kPa
Albedo	0.22
Gravitational constant (g_v)	1.35 m/s^2
Orbital period around Saturn	15.95 (Earth days)
Surface temperature	90–95 K
Diameter	5152 km
Solar flux outside Titan's atmosphere	14.87 W/m^2
Speed of sound at the surface	196.5 m/s
Atmosphere gas constant (R_a)	296.8 J/kg-K
Atmosphere ratio of specific heats (γ_a)	1.4
Atmosphere specific heat (c_{pa})	1039 J/kg-K

atmospheric temperature also lowers the speed of sound through the atmosphere. Near the surface the speed of sound is approximately half that on Earth. Due to the distance from the Sun, cloud cover and haze, little sunlight reaches the surface. The atmospheric density at the surface is 5 times that of Earth and the pressure is 1.5 times greater than that at Earth's surface. Select properties of Titan are given in Table 11.23.

The gravitational acceleration on Titan (1.35 m/s^2) is less than that of Earth's Moon. Liquid is present on the surface in the form of methane and ethane. These form the seas, as shown in Fig. 11.34, in which the submarine will explore.

Thermal Control in Transit to Titan

In deep space transit to Titan, the thermal control system has to protect and regulate the temperature of the submarine and entry vehicle. Excess heat from the isotope power system is used to maintain the desired internal operating temperature of the vehicle's components. However, the excess heat must also be rejected to deep space to avoid the interior of the submarine from becoming too warm. The heat generated by the isotope power system is rejected to space through the use of a radiator on the cruise deck and entry vehicle. Heat is transferred to the entry vehicle radiator through a series of cold plate interfaces contacting the outer structure of the submarine as well as a dedicated interface directly between the power system and the radiator. Since the system will need to reject approximately 3800 W of heat,

Fig. 11.34 Liquid methane seas on Titan

the direct coupling of the power system to the radiator is necessary. This is because if all of the heat were to be removed through the submarine structure the radiator operating temperature would need to be very low requiring an excessively large radiator. Heat loss through the submarine structure is the main means of cooling within the methane seas. Therefore, for this to be accomplished during transit and maintain the correct internal temperature, it would require the radiator to operate at a similar temperature to that of the liquid methane seas, approximately 92 K. To avoid the need for an excessively large radiator, heat is removed at a much higher temperature directly from the power source. Heat is transferred from the interior of the submarine to the entry vehicle radiator through an interior and exterior interface cold plate along with a coolant loop and heat pipes. This arrangement is illustrated in Fig. 11.35. Once the submarine enters the Titan environment the heat flow to this interface radiator is shut off and the main submarine structure is now used to reject the excess heat.

Surface Operation Within the Liquid Methane Seas

For the internal components of the submarine to operate within the methane and ethane seas their temperature has to be maintained within their desired operating range. This temperature range, given in Table 11.24, is much higher than the temperature of the surrounding liquid. To achieve this, waste heat from the isotope power system is utilized to warm the submarine interior. The heat is then rejected

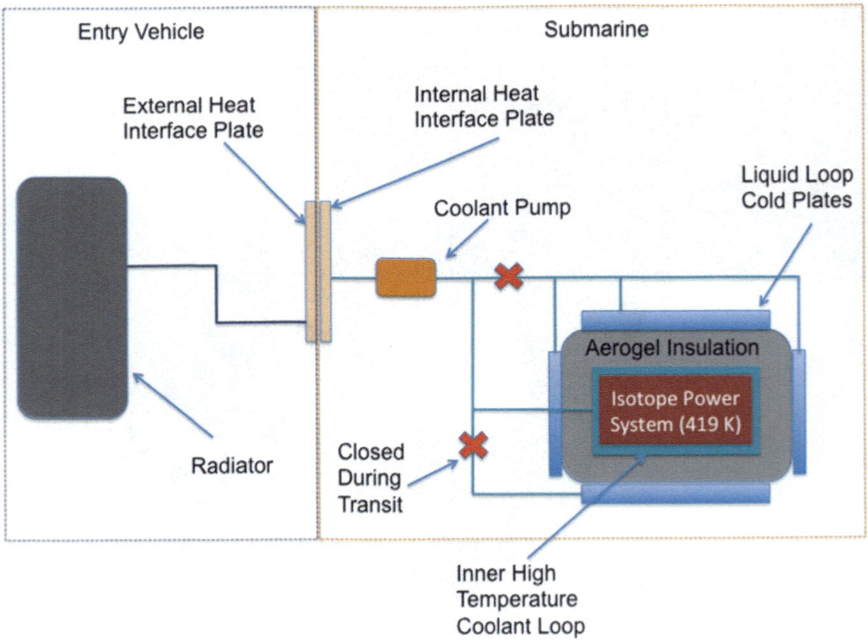

Fig. 11.35 Heat transfer from submarine to entry vehicle during transit

Table 11.24 Thermal system specifications

Specifications	Value
Submarine dimensions	Length 5-m by width 0.62-m: cylindrical shaped
Waste heat provided by the isotope power system	3800 W
Operating temperature	Internal ~290 to 310 K (17–37 °C)
Insulation (aerogel foam)	3.0-cm thick enclosing the interior surface
Environment	Liquid methane and ethane at approximately 92 K (−181 °C)
Passthroughs: Structure Science	 Two support rings connected to the outer shell Two windows 4 cm in diameter each Wiring for data and power transfer to the external equipment
Isotope system	Heat rejection temperature of 419 K

through the structure and surface walls of the submarine to the exterior. The interior must be insulated to provide sufficient thermal resistance to achieve the desired internal temperature. This arrangement is shown in Fig. 11.36.

The operation of the thermal system while in the Titan seas is illustrated in Fig. 11.37. Excess heat is removed from the isotope system through a layer of insulation. The insulation allows the heat rejection temperature of the power system to be at 419 K while maintaining the interior temperature at 290 K. The heat on the outside of the insulation is picked up using a series of cold plates. A coolant loop is interfaced with the cold plates and is pumped to similar plates located along the submarine structure. This coolant loop system distributes the heat uniformly throughout the interior of the submarine.

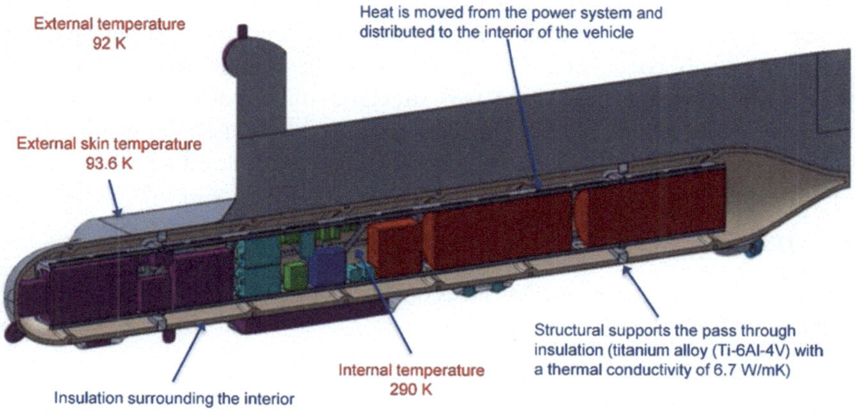

Fig. 11.36 Thermal management and temperatures within the submarine

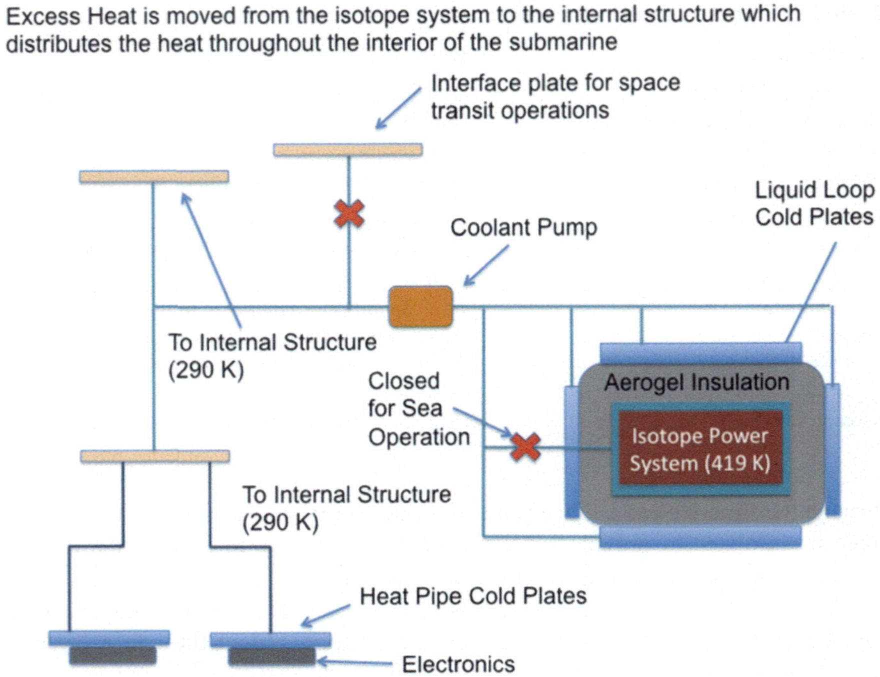

Fig. 11.37 Illustration of heat distribution system within the submarine

Aerogel foam is used to insulate the interior of the submarine. The insulation thickness is 3 cm. The insulation provides a very low conductivity barrier between the inside and the exterior walls of the submarine. The insulation has a thermal conductivity of 0.28 W/mK. This thermal conductivity is effectively the thermal conductivity of the nitrogen gas within the submarine. At the internal operating pressure of 150 psi, the insulation acts as a natural convection barrier and effectively allows only conduction through the gas as the main means of heat transfer through the insulation. There are two structural rings that extend through the insulation to provide a structure-mounting framework in the interior of the submarine. These structural pass-through members are constructed of a low conductivity titanium alloy (Ti-6Al-4V) with a thermal conductivity of 6.7 W/mK.

TCS MEL

The masses and mass growth allowances of the components of the Titan Sub's TCS are show in Table 11.25.

Table 11.25 TCS MEL

Description	Quantity	Unit mass (kg)	Basic mass (kg)	Total mass with growth (kg)
Thermal control (non-propellant)			95.3	112.5
Active thermal control			3.1	3.6
Thermal control	15	0.2	3.0	3.5
Thermocouples	5	0.0	0.1	0.1
Passive thermal control			92.3	108.9
Insulation	1	6.2	6.2	7.3
Interface plates	2	0.1	0.3	0.3
Coolant loop and fluid	1	3.5	3.5	4.2
Radiator	0	0.0	0.0	0.0
Thermal paint	1	1.8	1.8	2.1
Coolant pump	2	3.0	6.0	7.1
Power system thermal interface	2	0.1	0.3	0.3
Heat pipes	10	1.7	16.9	19.9
Power system insulation	1	0.9	0.9	1.0
Gas pressurant tank	2	7.0	14.0	16.5
Ne Gas	2	6.3	12.5	14.7
Gas pressurant pump	2	2.0	4.0	4.7
Buoyancy control valves	12	1.5	18.0	21.2
Ballast tank piston	4	2.0	8.0	9.4

11.4.8 Structures and Mechanisms

Structures and Mechanisms Requirements

The structure of the Titan Submarine is comprised of a main hull and two ballast tanks, protected by an outer skin. The hull contains the majority of the submarine components, which are supported by two large rings and six smaller rings, all of which are connected and supported longitudinally by four beams. The other subsystems are supported by a truss inside the hull and four additional beams. The hull structure is required to provide support and protection for the science, AD&C, C&DH, communications instrumentation, thermal insulation, and propulsion subsystems of the submarine.

The submarine hull and the ballast tanks must be able to withstand an internal pressure of 1.03 MPa (150 psi) while undergoing pressurization with nitrogen, and must be able to withstand an expected external pressure of 1.03 MPa (150 psi) beneath the surface of Kraken Mare.

The goal of the design is to minimize weight of the structure while providing sufficient strength to withstand applied loads from the LV. The maximum anticipated vertical acceleration is 5 times that of gravity, and the maximum anticipated lateral acceleration is 2 times that of gravity. Titan's surface temperature is 94 K (−179 °C), and structural materials chosen should be able to withstand these low temperatures.

The deployable camera boom mechanism is required to function for a single deployment. Also for the purposes of this study, installations for all other submarine subsystems were considered mechanisms in the MEL but are not described in this report.

Structures and Mechanisms Assumptions

The rings and outer skin of the submarine hull structure and ballast tanks are Ti-6Al-4V. Properties for Ti-6Al-4V are taken from the Metallic Materials Properties Development and Standardization (MMPDS-04) (IHS, 2008). The hull, skin, and ballast tanks of the submarine are assumed to be of primarily welded construction, along with threaded fasteners to join components. The outer skin surrounding the submarine hull and ballast tanks was assumed to be a high strength carbon fiber composite. The square tubular beams and truss structure inside the submarine hull were assumed to be of aluminum (Al) 2024-T8. This is a high strength Al with excellent resistance to fatigue, and is good for use in the submarine design because of its high strength to weight ratio. The drawbacks of Al 2024-T8 are that it may be expensive and difficult to weld. The load inside the hull is carried through the two larger rings. It was assumed that the frame and rings of the structure would easily adapt to the aeroshell and launch vehicle via trunnion pins and keel fittings on the aerovehicle.

It was assumed that the maximum internal or external pressure the submarine hull, ballast tanks or outer skin would experience was 1.03 MPa (150 psi.)

Structures and Mechanisms Design and MEL

The internal structure of the hull is shown in Fig. 11.38.

The large rings have a 6.3- by 6.3-cm (2.5- by 2.5-in.) square cross section with a 0.25 cm (0.1 in.) wall thickness. The small rings have a 2.5- by 2.5-cm (1.0- by 1.0-in.) square cross section with a 0.25 cm (0.1 in.) wall thickness. Four 2.5- by 2.5-cm (1.0- by 1.0-in.) square beams, each with a 0.25 cm (0.1 in.) wall thickness are attached to the two large rings. Four additional 2.5- by 2.5-cm (1.0- by 1.0-in.) square beams each with a 0.25 cm (0.1 in.) wall thickness provide additional support to the Al alloy truss structure. These are attached to the larger rings, each with four additional shorter beams with a 3.8- by 3.8-cm (1.5- by 1.5-in.) square cross section with a 0.25 cm (0.1 in.) wall thickness.

The MEL for the Structures and Mechanisms of the Titan Submarine are show in Table 11.26.

Structures and Mechanisms Analytical Methods

First, cylindrical pressure vessel calculations were performed in order to determine a sufficient wall thickness for the Ti-6Al-4V submarine hull and ballast tanks to withstand the 1.03 MPa (150 psi) internal air pressure while undergoing

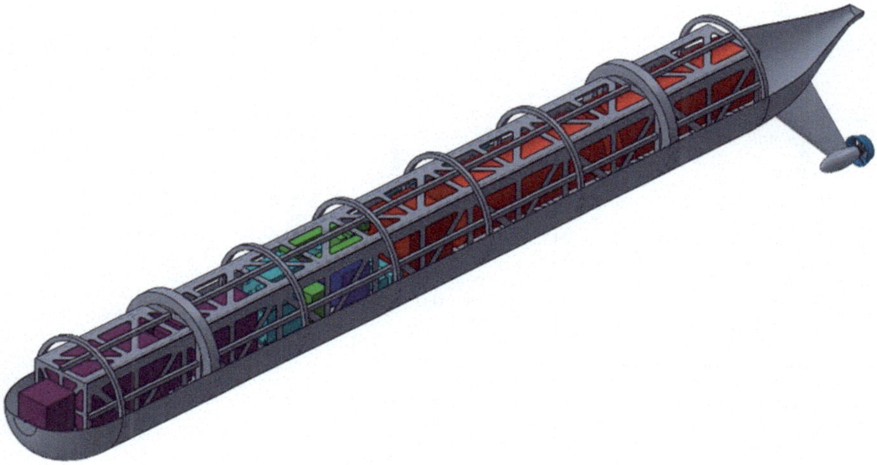

Fig. 11.38 Illustration of submarine hull internal structure

pressurization prior to launch. Though the minimum wall thickness was calculated to be 0.076 cm (0.03 in.), a wall thickness of 0.25 cm (0.1 in.) was chosen to ensure stability. Buckling equations were then applied to the Ti alloy outer shell to determine whether the shell itself was susceptible to buckling under a 5 g load. Per Euler's buckling formula, the critical buckling stress would be so high that the submarine hull would be likely to exceed the yield stress before the outer shell would buckle under an axial load.

A preliminary structural analysis was performed on a simplified version of the submarine hull structure using ANSYS workbench version 15.0. Because launch loads are expected to be the highest loads on the submarine, the ±5 g axial acceleration loads and 2 g lateral acceleration loads were examined. The ANSYS model consisted of the cylindrical portion of the hull with the outer Ti alloy shell, two large rings and five smaller rings. The cylindrical portion of the submarine hull in the analysis was assumed to be 67.5 cm (26.6 in.) in diameter and 4.6 m (182 in.) in length. The Al alloy square beams running the length of the cylindrical portion of the hull were also 4.6 m (182 in.) in length. The Ti alloy rings were equally spaced along the length of the hull, and point masses were used to represent non-structural subsystems.

Material properties were used at their room temperature condition since launch loads, the highest expected loads, would occur near room temperature. Properties were taken from MMPDS-04. For Ti-6Al-4V, an ultimate strength of 924 MPa (134,000 psi) and yield strength of 869 MPa (126,000 psi) were used. For Al 2024-T8, an ultimate strength of 455 MPa (66,000 psi) and yield strength of 400 MPa (58,000 psi) were used.

Of all three loading conditions, the maximum deflections and equivalent stresses occurred in the 2 g lateral acceleration case. The maximum stress in this case was 219.3 MPa (31,799 psi), which occurred on the 2024 Al alloy beam. Using an

Table 11.26 Titan submarine structures and mechanisms MEL

Description	Quantity	Unit mass (kg)	Basic mass (kg)	Total mass with growth (kg)
Structures and mechanisms	–	–	445.7	525.9
Structures	–	–	427.5	504.4
Primary structures	–	–	360.4	425.3
Pressure hull	1	122.9	122.8	144.9
External buoyancy tanks	2	38.2	76.5	90.2
Tail shell	0	0.0	0.0	0.0
Truss—hull	1	54.9	54.9	64.7
Bulkheads—ballast tanks	2	0.7	1.4	1.7
Stiffening beams	8	4.9	39.1	46.1
Boom for science instruments	1	1.0	1.0	1.2
Large rings	2	4.8	9.6	11.3
Small rings	6	1.8	10.9	12.8
Composite shell	1	44.3	44.3	52.3
Secondary structures	–	–	67.1	79.2
Antenna support—honeycomb	1	63.5	63.5	74.9
Antenna support—foam	0	34.6	0.0	0.0
Cross stiffening beams	8	0.5	3.6	4.2
Mechanisms and Fasteners by subsystem	–	–	18.2	21.5
Science payload	1	3.6	3.6	4.3
AD&C	1	1.3	1.3	1.6
C&DH	1	1.8	1.8	2.1
Comm and tracking	1	1.1	1.1	1.2
Electrical power system	1	5.8	5.8	6.9
Thermal control (non-propellant)	1	3.8	3.8	4.5
Propulsion	1	0.8	0.8	1.0

ultimate factor of safety of 1.4 and yield factor of safety of 1.25 per NASA-STD-5001, corresponding to protoflight hardware, the margin of safety against ultimate failure is +0.48 and margin of safety against yield is +0.46.

The preliminary version of the structure as analyzed is a very simplified version of the final design of the submarine hull structure, which includes four more square Al beams, cross-beams for stabilization, and a truss structure to hold the other subsystems. The final design of the submarine hull structure is strong enough with enough margin to withstand expected maximum launch loads.

11.5 Conclusions

While the challenges of propulsion, ballast systems, and communications exist a conceptual design of a submersible platform was found capable of accomplishing extensive and exciting science both above and below the surface of the Kraken Mare, an open body of cryogenic hydrocarbon liquid on the surface of Titan, Saturn's largest moon.

The approach developed for the Titan submarine concept would allow mapping of undersea and shoreline topography, weather measurements, and sampling of both the liquids and seafloor solids. These measurements would give not only vital information on the current state of the seas and surrounding land but would allow looking far back into Titan's geological and hydrological history. It could even establish whether conditions for non-water-based life exist in Titan's seas.

Acknowledgements This work was funded the NASA Innovative Advanced Concepts (NIAC) program as a Phase I Study in 2014. The authors wish to heartily thank the NIAC program Team: Jay Falker, Jason Derleth, Ron Turner, Katherine Reilly, and Barbara Mader for their guidance, patience and insight during this conceptual design study. The investigators also wish to acknowledge with the strongest possible vigor the contributions of the COMPASS team to the Titan Submarine design. Without their creativity, innovation, and perseverance the submarine's design would never have been created: Cryogenics, Jason Hartwig; Hydrodynamics Engineer, Justin Walsh (PSU/ARL); Systems Engineer, Jeff Woytach; Science, Geoff Landis; Navigation, Mike Martini; Mechanical Systems, Amy Stalker; Thermal Control, Anthony Colozza; Power, Paul Schmitz; C&DH and Software, Hector Dominguez; Communications, Robert Jones; Configuration, Tom Packard; Visualization, Michael Bur; and Cost, Tom Parkey and Elizabeth Turnbull. We also wish to thank Les Balkanyi, Lorie Passe, Lisa Liuzzo, and Eric Mindek for bringing the Titan Submarine to life in word, pictures, and video—the American public, indeed the world, knows of the Titan Submarine because of their work.

Finally, a nod to dreamers such as Jules Verne who inspire us to explore new worlds: 'Mobilis in Glaciali!'

References

Blevins, R.: Applied Fluid Dynamics. Wiley, New York (1984)

Brown, R.H., Lebreton, J.P., Waite, J.H.: Titan from Cassini-Huygens. Springer Science-Business Media, Berlin, Germany (2009)

Feldman, J.P.: Method of performing captive-model experiments to predict the stability and control characteristics of submarines. CRDKNSWC-HD-0393-25 (1995)

Flasar, F.M., Achterberg, R.K., Conrath, B.J., Wishnow, E.H.: Titan's atmospheric temperatures, winds, and composition. Science **308**, 975–978 (2005)

Hoerner, S.F.: Fluid-Dynamic Drag: Practical Information on Aerodynamic Drag and Hydrodynamic Resistance. Fluid-Dynamic Drag. Ing. S.F. Hoerner, Midland Park, NJ (1958)

Hydroid, L.L.C.: REMUS 100 Brochure. Hydroid, LLC, Pocasset, MA (2014)

Imlay, F.: Complete expressions for gravitational and buoyancy force terms in the equations of motion of a submerged body. DTMB Report 1845 (1964)

Leary, J., Jones, C., Lorenz, R., Strain, R.D., Waite, J.H.: Titan explorer NASA Flagship Mission Study, The Johns Hopkins University Applied Physics Laboratory (public release version http://www.lpi.usra.edu/opag/Titan_Explorer_Public_Report.pdf. Jan 2009) (2007)

Lockheed Martin Corporation: Engineering unit SRG expanded missions capability study report. LMSP–7348 (2008)

Lorenz, R.D.: Flight power scaling of airplanes, airships, and helicopters: application to planetary exploration. J. Aircr. **38**, 208–214 (2001)

Mandel, P.: Principles of Naval Architecture. Chapter VIII, Ship Maneuvering and Control. SNAME, New York, NY (1967)

NASA: Solar system exploration: Titan (moon). solarsystem.nasa.gov (2013)

Tecnadyne: Model 560 Brochure. Tecnadyne, San Diego, CA (2014)

Chapter 12
WindBots: A Concept for Persistent In Situ Science Explorers for Gas Giants

Adrian Stoica, Virgil Adumitroaie, Marco Quadrelli, Georgios Matheou, Marcin Witek, Marco Cipolato, Marco Dolci, James Roggeveen, Kyle Petersen, Kristina Andreyeva, Hunter Hall, Benjamin Donitz and Leon Kim

Visible to the naked eye, the gas giants Jupiter and Saturn have been known to astronomers since antiquity. In the modern times much was learned about them, and yet so much remains to be learned. They are made almost entirely of hydrogen and helium, they have no hard surface to land to; their low temperature atmospheres are characterized by strong winds, at least in the observed upper atmosphere. What we know about them comes from remote sensing—yet their clouds impede deeper observation through remote sensing. We also have, in a singular case, data transmitted by a robotic probe that descended through the Jovian atmosphere. We need more of these probes, to confirm the models we formed about these planets, and to

---

A. Stoica (✉) · V. Adumitroaie · M. Quadrelli · G. Matheou · M. Witek
M. Cipolato · M. Dolci · J. Roggeveen · K. Petersen · K. Andreyeva · H. Hall
B. Donitz · L. Kim
Jet Propulsion Laboratory, California Institute of Technology,
4800 Oak Grove Dr., Pasadena, CA 91109-8099, USA
e-mail: Adrian.Stoica@jpl.nasa.gov

V. Adumitroaie
e-mail: virgil.adumitroaie@jpl.nasa.gov

M. Quadrelli
e-mail: marco.b.quadrelli@jpl.nasa.gov

G. Matheou
e-mail: georgios.matheou@uconn.edu

M. Witek
e-mail: marcin.l.witek@jpl.nasa.gov

M. Dolci
e-mail: marco.dolci@jpl.nasa.gov

B. Donitz
e-mail: benjamin.p.donitz@jpl.nasa.gov

V. Badescu and K. Zacny (eds.), *Outer Solar System*,
https://doi.org/10.1007/978-3-319-73845-1_12

discover new phenomena below their clouds. This chapter examines mission concept alternatives in which robotic craft operate in the atmospheres of gas giants, for long duration, and using energy derived from local sources. In a preferred scenario these Wind Robots (WBs), with high mobility and autonomy compared to passive balloons, would operate in the Jovian atmosphere above and below the region of clouds, between 0.3 and 10 bar, for a year-long duration mission, in strong (potentially turbulent) winds. In an example, notional mission, a WB would operate in the eyewall of the Great Red Spot, using the high wind and updrafts of the anticyclone, as well as horizontal gusts. Both naturally buoyant and winged solutions, as well as hybrids of the two, are determined possible. A Network of WBs could measure wind speeds, temperatures, and atmospheric composition simultaneously, at multiple locations.

The feasibility of WindBots in the context of a mission scenario formed the subject of a study funded by a Phase I NASA Innovative Advanced Concepts (NIAC) (Stoica et al. 2016); this chapter summarizes the final report.

Section 12.2 provides the scientific context and motivation for a WB mission and determines optimal regions and payloads for WB missions as the intersection of regions of scientific interest and energy reservoirs—focusing on a mission to Jupiter's Great Red Spot. Section 12.3 explores buoyancy and lift generation options for WB design, focusing on maintaining mission duration, providing a measure of control for the WB, and exploring the entire proposed operating range. Section 12.4 discusses several WB body configurations and deployment, extending the mobility discussion from Sect. 12.3. Section 12.5 investigates energy recovery methods using active systems controlling the WB and passive systems to harvest naturally occurring wind, pressure, and thermal gradients in the atmosphere. Section 12.6 reviews autonomy architecture and control schemes to be applied to the autonomous WB. Section 12.7 presents summary and conclusions.

12.1 Exploring the Atmospheres of Gas Giants

We envision persistent exploration of the gas giants with teams of multiple robots powered by locally harvested energy performing in situ observational atmospheric science. Bumped around by gusts in the strong and turbulent windstorms of the troposphere, WBs may use controlled aerodynamic forces for lift and direction change. In addition, autorotation and mechanisms similar to those used in kinetic/automatic watches capture mechanical energy, converting it to electrical energy via electromagnetic generation. Unlike conventional flyers that try to avoid turbulence, WBs may seek to leverage it. In essence, WBs 'chase' gradients in horizontal winds and use real-time controls to adjust their surfaces, changing their aerodynamic characteristics and advancing in the desired direction through waypoints set on high turbulence/updraft currents.

An optimal area of exploration was determined at the intersection of those regions that are scientifically appealing and those where energy can be collected locally. Atmospheric science questions were asked, and necessary instrument suites selected, to determine payload energy requirements and to formulate an in situ mission. We focused on Jupiter, but consider WBs as a more general solution for in situ exploration of all outer planets, with similar richness in He and H2 (Uranus and Neptune also having heavier components), high speed winds, and similar levels of gravitational attraction (lower on Uranus and Neptune).

12.1.1 In Situ Atmospheric Science with Long Lived Robotic Explorers

Beyond the asteroid belt lies the mysterious world of the Giants: the Gas Giants, Jupiter and Saturn, and the Ice Giants, Uranus and Neptune. We have only very limited knowledge of these planets, obtained primarily through remote sensing, flybys, and telescope observation. These giants have no hard surface to land on; instead, they have small cores characterized by very high pressure and high temperatures. Their composition is hydrogen, helium, and small amounts of other elements (Fig. 12.1).

Unveiling the mysteries of Gas Giants is key to understanding the formation and evolution of our solar system (Greicius 2016). These colossal planets, with volumes about a thousand times that of Earth, harbor huge atmospheric phenomena (Fig. 12.2), with jets that dominate atmospheric circulation at visible levels (Ingersoll et al. 2004).

The bottomless atmospheres of Jupiter, Saturn, Uranus, and Neptune in combination with the relatively high planetary rotation rates result in atmospheric circulations that are significantly more regular than Earth's, thus offering an exceptional opportunity to observe, study and discover geophysical fluid dynamics

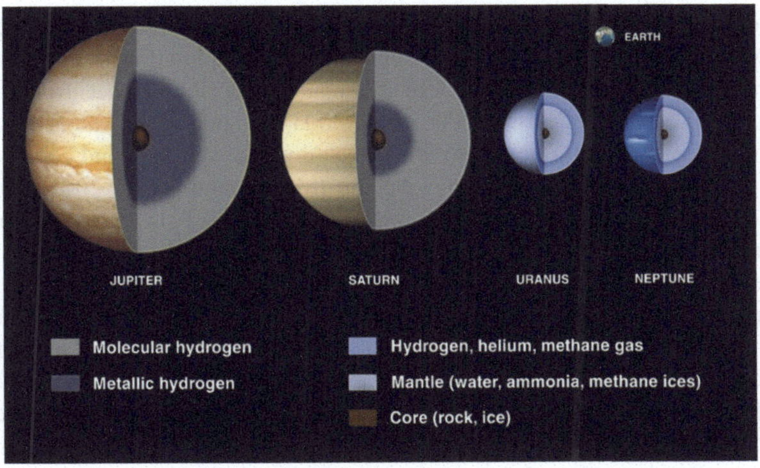

Fig. 12.1 A schematic representation of the Giants and their composition (*courtesy* NASA)

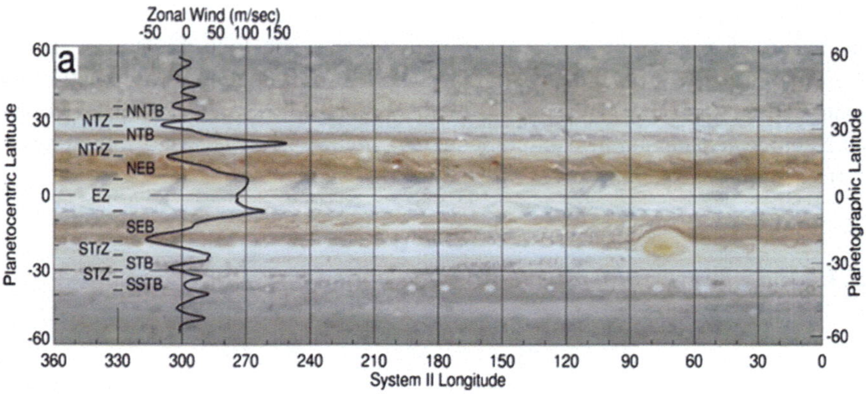

Fig. 12.2 Observed mean zonal wind pattern on Jupiter (image from Bangenal et al. 2004). The notable number of jet and belts, the high wind speeds and the correlation with the visible upper cloud formations can be readily discerned

in a nearly idealized setting. Moreover, Jovian atmospheric circulations are directly governed by the planetary and geophysical parameters, in contrast with Earth's mountain ranges and ocean, which steer and alter the circulation and weather patterns (Williams 1985). In fact, the dynamics of Jovian atmospheres resemble more the geophysical spaciousness of the Earth's ocean rather than the atmosphere (Dowling 1995).

The main characteristic of Jupiter's atmosphere is the axisymmetric zonal jet streams (Fig. 12.2) and the many anticyclones forming between the jets [about 90% of the observed storms on Jupiter are anticyclones (MacLow and Ingersoll 1986)], the most prominent being the Great Red Spot. Winds measurements based on the upper (visible) cloud layers show wind speeds reaching 180 m/s (Limaye 1986; Poro et al. 2003).

Understanding the motion and composition of Jupiter's atmosphere has been the goal of several past missions, starting with the early exploration from Voyager I (Smith et al. 1979) and II, Galileo, the analysis of the impact from comet Shoemaker–Levy 9, and Cassini (Poro et al. 2003). In spite of the valuable data of past missions, important questions remain regarding the structure and composition of Jupiter's atmosphere below the cloud top (Marcus and Shetty 2011). Current missions hope the study of the outer layers would shed light into better understanding of the entire planet (Ingersoll 1990).

The Juno mission (Greicius 2016) is currently enriching our knowledge of Jupiter. Juno Mission's primary goal is to significantly improve our understanding of the formation, evolution, and structure of Jupiter. In particular Juno will attempt to unravel some of the mysteries—for example, it is still unknown how much water and ammonia is in the Jupiter's atmosphere, which has implications for the theories of the formation of the solar system. The vertical profiles of Jupiter's composition, temperature, and cloud motions are poorly constrained. We do not know what the

Fig. 12.3 New Juno data on the Great Red Spot\ \2017-07-12 12:00 UT Credit: NASA, Tom Momary (color enhancements and vibrance, to bring out detail)

nature of Jupiter's turbulence is and how this turbulence feeds back or interacts with large-scale structures visible in the outer layers of Jupiter's clouds. Juno will shed light on some of these questions, providing an unprecedented level of details on Jupiter's structure, magnetism, and gravity. A recent image received from Juno is illustrated in Fig. 12.3.

Despite being much closer to the planet, Juno retains the limitations inherent of remote sensing. As an orbiter aimed at understanding the planet as a whole, Juno will not be able to glimpse into features that are horizontally smaller than a few kilometers (Hansen et al. 2014). A better spatial resolution of measurements may still be required for a detailed understanding of atmospheric structure and dynamics. To validate planetary models and to further our understanding of the layered atmospheric composition and the dynamic processes at work in the Giants' atmospheres, one needs in situ exploration (Mousis et al. 2014).

To date, the only in situ measurements (and thus ground truth) come from the Galileo probe (GP), in its descent through Jupiter's atmosphere (Young 1998; Seiff et al. 1998a). GP found that the atmosphere is denser and the winds are stronger and extend deeper than expected. It did not detect an expected three-tiered cloud structure, as scientists now believe the entry site may have been one of the least cloudy areas on Jupiter. Probing in a single dimension and at a single location has obvious limitations.

To collect data from below the upper, visual layer, with appropriate spatial and temporal resolution, special robotic explorers are required, which move and measure along three dimensions (3D), survive winds up to 170 m/s, and survive without the use of solar energy.

12.1.2 Atmosphere Characteristics and Models—Global Circulation

The Jovian cloud model currently accepted by scientists is comprised of three main layers, which are: ammonia at the highest levels (0.3–0.7 bar), the ammonium hydrosulphide (~2 bar), and the deeper water vapor clouds (5–6 bar) (Lewis 1969). Figure 12.4 shows the variation of pressure as a function of temperature in the Jovian atmosphere. The main cloud layers are also depicted.

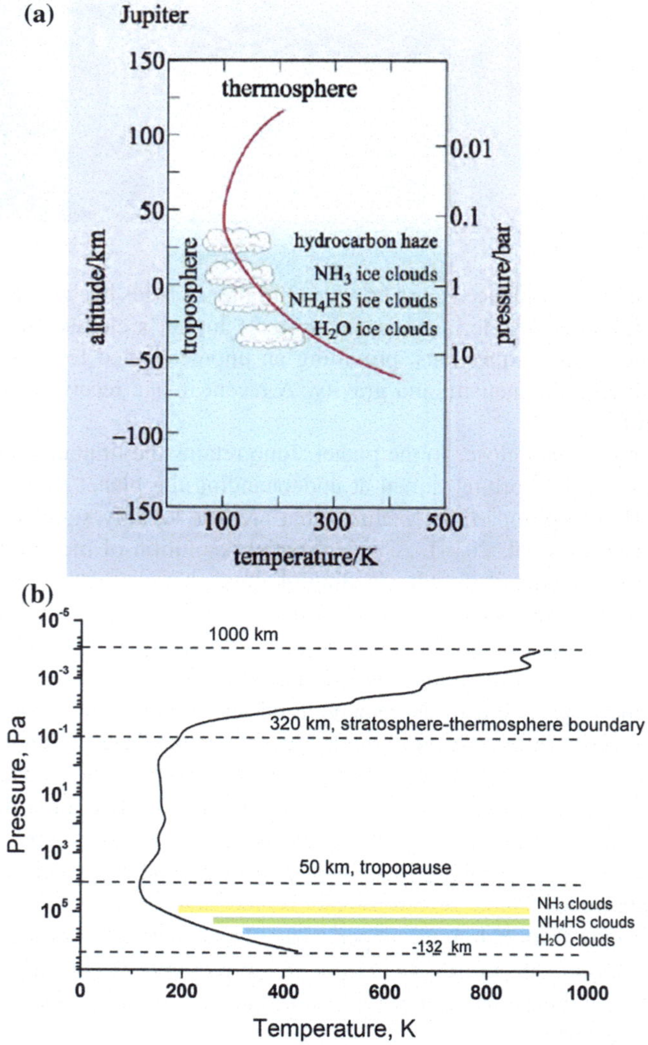

Fig. 12.4 Pressure–temperature profile or various altitudes in the Jovian atmosphere, two views https://www.geol.umd.edu/~jmerck/geol212/lectures/21.html

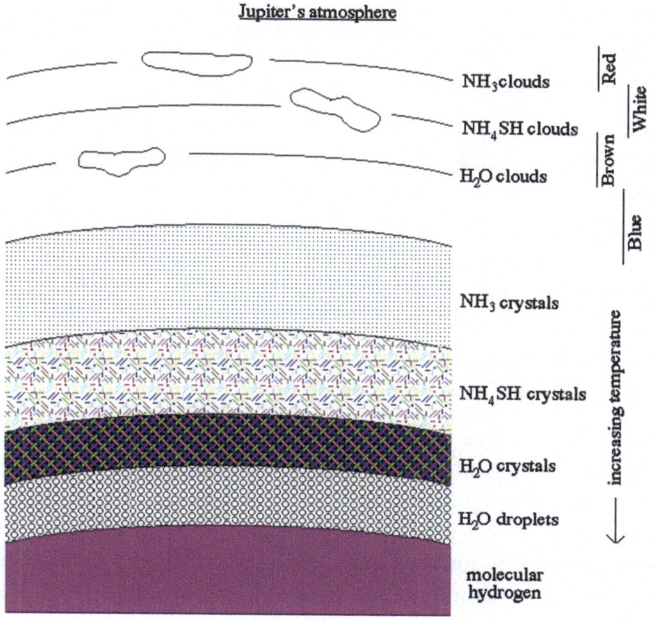

Fig. 12.5 Chemical composition of lower parts of Jupiter's atmosphere (http://zebu.uoregon.edu/disted/ph121/l15.html)

Figure 12.5 focuses on the lower parts of the Jupiter's atmosphere, illustrating the chemical composition and cloud types as one moves from the layer of molecular hydrogen upwards. As the gas density and pressure decrease, the temperature decreases with the distance from the center. The various chemical elements and their respective temperatures are mainly responsible for the colors of the visualized clouds.

Thermal profiles and cloud structures of the Giants are depicted in Fig. 12.6 (left panel) (Sánchez-Lavega et al. 2004; Gierasch and Conrath 1993). The temperatures at 1 bar differ between planets, with Jupiter being the warmest and Neptune and Pluto the coldest, but the overall temperature profiles are comparable in all of them. The three main Jovian cloud types are also present on Saturn, but due to lower temperature they are shifted towards higher pressures (Fig. 12.6, middle and left panel).

With radio signals "we can peer through the clouds and see that those hotspots are interleaved with plumes of ammonia rising from deep in the planet, tracing the vertical undulations of an equatorial wave system," said UC Berkeley research astronomer Michael Wong (Simon et al. 2015). "We now see high ammonia levels like those detected by Galileo from over 100 km deep, where the pressure is about eight times Earth's atmospheric pressure, all the way up to the cloud condensation levels," de Pater said (De Pater et al. 2016). The scientists discovered that plumes of ammonia-rich gases rise to form the upper cloud layers visible from

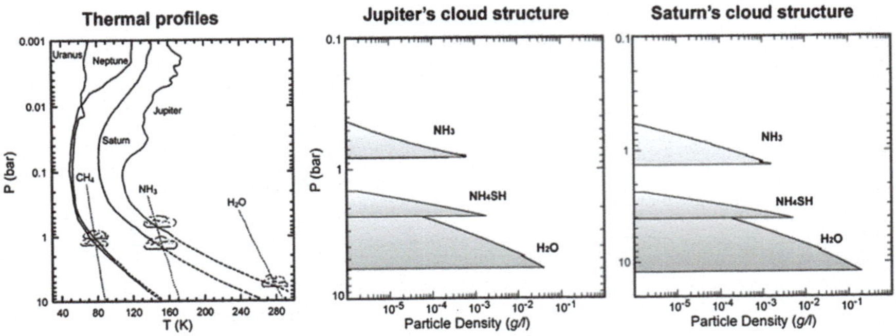

Fig. 12.6 Thermal profiles of Giants and cloud structures of Gas Giants, Jupiter and Saturn (after Sánchez-Lavega et al. 2004; Gierasch and Conrath 1993)

Earth—ammonium hydrosulfide clouds at a temperature near 200 K, and ammonia-ice clouds at roughly 160 K. These upwellings of ammonia swell up in wave patterns, a sign of motion deep within the atmosphere. Conversely, ammonia-poor air sinks in downwellings. Near the Great Red Spot, the researchers detected many intricate features, suggesting complex upwelling and downwelling.

12.1.3 Local Information

12.1.3.1 Galileo Probe

In 2003, the 339 kg Galieo Probe (GP), communicating through two L-band transmitters, powered by lithium sulfur dioxide ($LiSO_2$) batteries with an output of 580 W, descended 156 km through the upper layers of Jupiter's atmosphere before it stopped transmitting. Its drop was in a region termed a 5-μm hot spot, a local clearing in the clouds that is bright near the 5 μm spectral region (Seiff et al. 1998b; Young 2003). The details of the drop can be seen in Fig. 12.7. The probe dropped about 24 bars before the signal ended, providing 58 min of local weather data; the temperature at the end of transmission was 153 °C (307 °F).

The probe showed the atmosphere to have a generally stable temperature stratification, and zonal winds increasing from ~80 m/s at less than 1 bar to about 180 m/s near 5 bars, after which the winds remained approximately constant with depth. The main important findings from the descent were that the Jupiter's atmosphere was slightly hotter, more turbulent, and exhibited stronger winds than expected. It contained less helium than expected, less clouds, less lightning, and less water. As the entry point was the infrared hot spot, the findings were not representative for the general Jovian atmosphere. Many questions about the atmosphere remain unanswered. GP did not measure Oxygen content, for example.

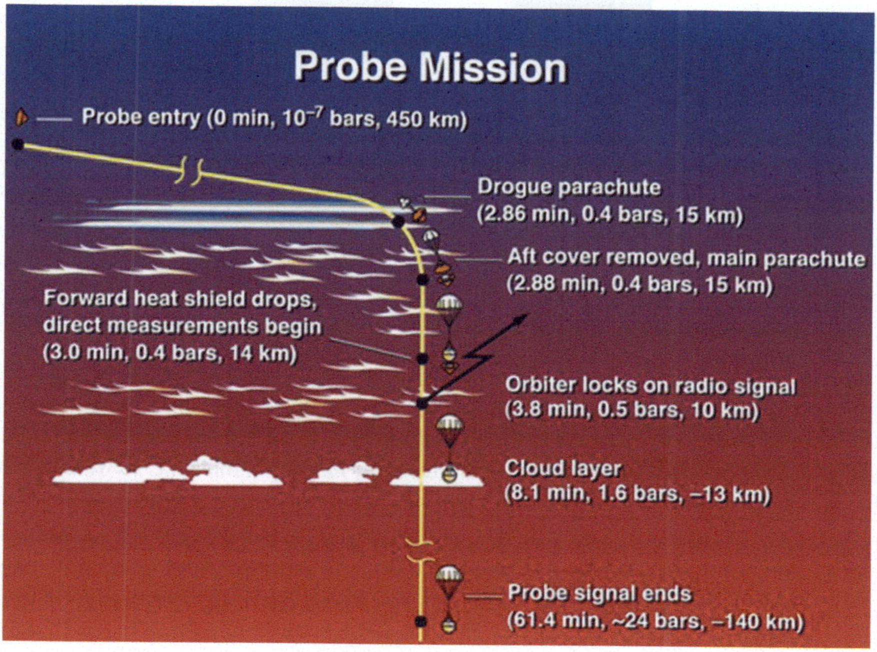

Fig. 12.7 Galileo probe entry in Jupiter atmosphere (https://en.wikipedia.org/wiki/Galileo_Probe)

Because of the specific location of the drop, no cloud properties were observed, and the dynamic and vertical shear observations are not necessarily representative of the general Jovian atmosphere (Young 2003).

12.1.3.2 Turbulence at Aircraft-Scale

According to Maise (2003), "virtually nothing is known about aircraft-scale turbulence on Jupiter and in this respect the scales of interest are well below the resolution of remote sensing. The Galileo Probe's accelerometer data show some 1 m/s fluctuations, but because the sampling rate was low the vertical scale is about 200 mbar, too large at distances to infer about finer scale turbulence. A conservative estimate of the largest turbulence amplitude on Jupiter is that it might be twice as strong as on Earth. This estimate is based on the fact that cumulus towers on Jupiter rise through 30 km instead of 8–10 km on Earth, in a roughly similar period of time, implying stronger vertical perturbations. These regions of active moist convection on Jupiter cover only a small fraction of the surface and their locations are well known. Jupiter's top-of-the-atmosphere solar heat flux is only 14 W/m^2, which is almost two orders of magnitude less than Earth's, so it is conceivable that there are vast stretches on Jupiter where the turbulence is negligible, at least in the upper troposphere."

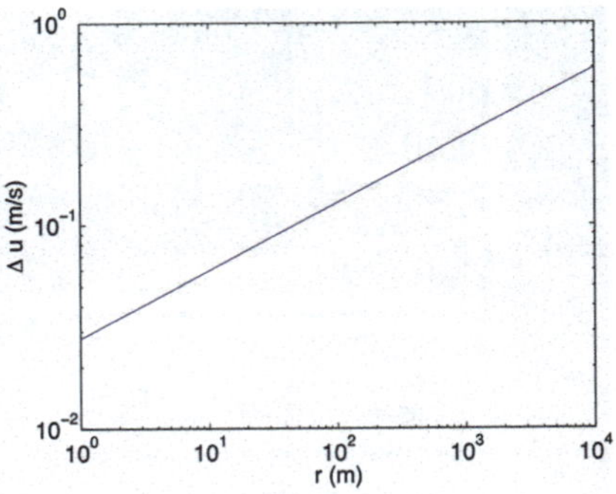

Fig. 12.8 Expected mean velocity difference with respect to distance for a turbulence model on Jupiter

However, Jupiter has an internal energy source that might be somehow contributing to convection and generation of turbulence. A turbulence theory might be used to estimate potential velocity difference with respect to distance. If we assume the Kolmogorov scaling and a uniform energy production scale, a second order turbulence structure function can be expressed as $F_2(r) = 4.82C_k(\varepsilon r)^{2/3}$ (Métais and Lesieur 1992). If we further assume that $F_2(r)$ is isotropic, and dissipation rate is $\varepsilon = 10^{-5}$ m^2/s^3 (Galperin et al. 2014), we can derive the mean velocity difference as a function of distance (Fig. 12.8).

The relation shows that a velocity gradient of 0.1 m/s can be expected at a distance of about 100 m. This has implications for design and requirements of in situ explorers in areas of Jupiter that are relatively free from large sources of turbulence.

12.1.4 Regions of Interest

The circulation of Jupiter is largely unknown and inferred primarily from numerical modeling. The visible parts of the atmosphere are the cloud tops, which provide data on the outer layers and their dynamics. Galileo Probe's vertical profiles of wind and thermodynamic properties, due to the unique entry point of those profiles, are not representative of more prevalent atmospheric structures in the Jovian atmosphere, namely zones and belts. The zones and belts (Fig. 12.9) are thought to be associated with vertical, in-and-out motions called convection (Fig. 12.10). Upwelling warm gas results in the light-colored zones, which are regions of high pressure. The darker belts overlie regions of lower pressure where cooler gas sinks back down into Jupiter's atmosphere. The zones and belts are therefore analogous to the high-and low-pressure systems that produce localized circulating storms on Earth, except Jupiter's rapid rotation has wrapped them all the way around the planet. Although this general convective pattern of the zones and belts was apparently supported by observations during the Voyager flybys of Jupiter,

Fig. 12.9 Jupiter zones and belts. The regions that are particularly interesting include the south equatorial belt, in particular west of and inside the Great Red Spot (after www.geol.umd.edu/~jmerck/geol212/lectures/21.html, http://history.nasa.gov/SP-349/p11.htm)

instruments aboard the Cassini spacecraft have also discovered large plumes of fast-rising gas scattered throughout the belts and absent in the zones.

Table 12.1 summarizes advantages and disadvantages of various Jupiter atmospheric locations as far as being optimal candidates for a WindBot mission scenario.

12.1.5 WindBot Mission Concept Science Goals and Instruments

The scientific goals for long-term in situ exploration of Jupiter may be summarized as follows:

- **Atmospheric composition**
  - Determining abundance of key elements of interests, in particular of oxygen, which was missing in the Galileo Probe measurements.
  - Better understanding of Jupiter's composition will have considerable impact on our ideas of solar system evolution and planet formation.
- **Clouds and condensibles** (NH_3, H_2S, H_2O)
  - Due to Galileo Probe's unique entry point (5 μm hot spot) no clouds and their properties were observed.

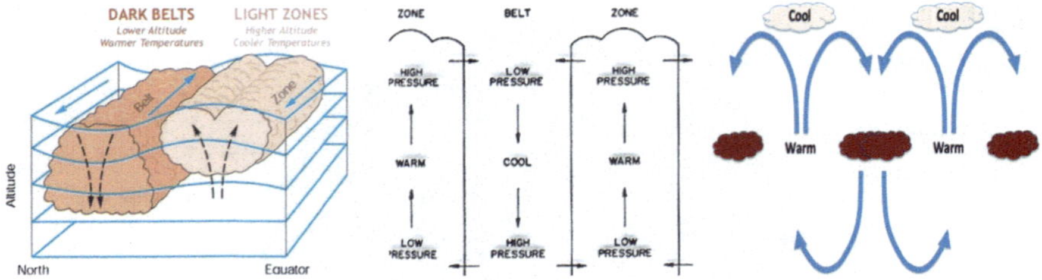

Fig. 12.10 Warm material rises and cools as it rises. When it gets cool enough, the ammonia freezes, forming white crystals that obscure the natural reddish brown of the warmer material. Dark belts (lower altitude and warmer temperature) and light zones (higher altitude and cooler temperature) on Jupiter (after https://www.geol.umd.edu/~jmerck/geol212/lectures/21.html, http://lasp.colorado.edu/~bagenal/3720/CLASS19/19GiantPlanets3.html)

Table 12.1 Location, advantage and disadvantage (both scientific and technological) of having a WindBot in different atmospheric area on Jupiter

Location	Advantage	Disadvantage
Belts	Storms, convection, and small-scale updrafts: • More interesting scientifically • Use of thermals for gliding (?)	• Storms can make flight more difficult • Slower circumnavigation
Zones	• More quiescent than belts, favorable to conventional flyers • Large scale upwelling ($\sim 10^{-3}$ m/s) • Fast circumnavigation	• Unknown vertical structure
Anti-cyclones	• Most stable and predictable part of the atmosphere • Vertical circulation favorable to gliding	• Significant power may be required to navigate out of the vortex
Temperate and Polar Regions	Scientifically interesting: • Data are very scarce • Important in the study of meridional energy transfer	• Least known atmospheric conditions

- **Dynamics and vertical shear**
  - Direct Galileo Probe observations are not representative for the general Jovian atmosphere. Vertical structure and dynamics of upper troposphere might be different than in the 5 μm hot spot.
- **Thermal structure**
  - The roles of heat budget, static stability, cloud condensation, and CAPE in the Jovian atmospheric dynamics are largely unknown.

Table 12.2 Proposed instruments suite on WB to measure chemical composition, thermal structure, observe clouds, vertical structure, and dynamics of the Jovian atmosphere including mass and average power consumption

Instrument	Mass (kg)	Avg. power (W)
Atmospheric structure instrument group measuring temperature, pressure and deceleration	0.54	3.8
Neutral mass spectrometer for measuring atmosphere composition	13.2	13
Helium abundance detector	1.4	0.9
Nephelometer for measuring cloud parameters and cloud location	4.4	11.3
Net-flux radiometer for measuring solar/planetary radiative flux in the atmosphere	3.134	–
Energetic particle detector	2.5	3

The instruments suite should measure chemical composition (e.g., oxygen), thermal structure, observe clouds, vertical structure and dynamics of the Jovian atmosphere. Table 12.2 summarizes various scientific instruments including their mass and average power consumption.

In particular, the priority is to measure temperature, pressure, atmospheric composition, clouds parameters, energy particle detectors and lightning. One may have different instruments on different WindBots; an average science package (for a total of two or more) may be of 10–25 kg with 35 W each.

12.1.6 WindBot Mission Concept: Jupiter's Great Red Spot

Jupiter's Great Red Spot represents an exploration target of high scientific importance, as well as a fascinating subject for public engagement. The JunoCam instrument on the Juno spacecraft took pictures of Jovian atmospheric features selected by public vote and most likely the attention will be directed at this iconic storm in Jupiter's atmosphere. Other experiments of the Juno mission such as the microwave radiometer and the gravity experiment made observations of this perpetual, but changing, anti-cyclonic zone in an effort to understand its structure, formation and evolution. For years the Great Red Spot has shrunk continuously with variable rates and also became more circular, while its color has become more orange than red. A WB in situ mission could be the only way to understand why this fascinating phenomenon is present and how it is evolving.

A WBs mission to Jupiter would operate in the 125 km vertical region between 0.3 bar and 10 bar (Fig. 12.11). The WB vertically profile this region, gathering data on the thermal and atmospheric structure as well as the wind speeds and compositions, per the scientific goals outlined in Sect. 12.1.4.

The Great Red Spot offers atmospheric stability as well as the potential for large updrafts. Cyclonic updrafts have been observed on Earth and explored in

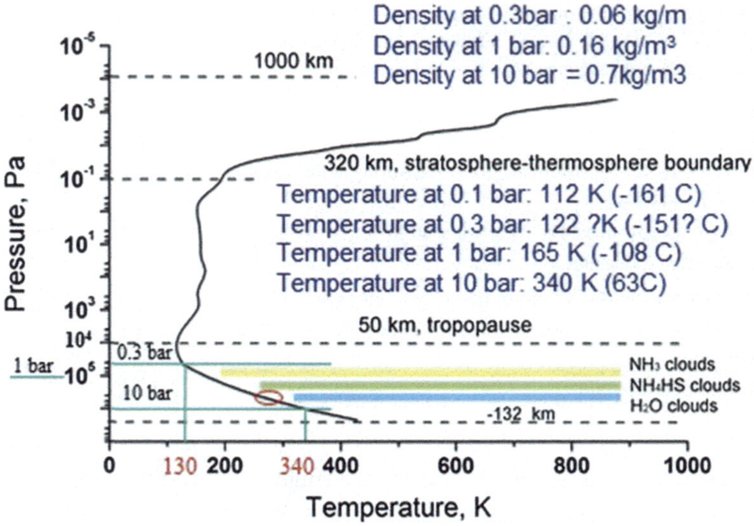

Fig. 12.11 Region of interest for WindBot mission concept with relevant boundary atmospheric data

Sect. 12.4.1 for their ability to maintain altitude of a WB. Similar structures are predicted in the Jovian equivalent, further substantiated by the high levels of ammonia in the Great Red Spot (GRS), typically indicative of upwelling (Ingersoll et al. 2004). The WBs would remain on the rim of the GRS, flying around the center with a period of approximately four days to take advantage of the updrafts (Fig. 12.12).

An example of a network of wind-driven mobile sensors is "Gone with the Wind on Mars" (GOWON), proposed for studying the surface of Mars using a wind-driven network of mobile sensors (Davoodi et al. 2012).

In a notional Jupiter-scale Network of Windbots (NOW) the distances between WB would be much larger, of the order of kilometers or more. NOW would deploy several WBs into the GRS, operating in different regions to provide a spatial distribution of simultaneous measurements. NOW would remain in the atmosphere for approximately one year, gathering power using in situ resources. From the technologies discussed in this chapter, a combination of a buoyancy-driven aerostat and vibrational generator would comprise the WB deployed on Jupiter. The buoyancy-driven aerostat provides the greatest level of control without the need for long, extended systems or excessive power used for control. In addition, it is the technology that most readily might vertically profile the atmosphere.

Vibrational generators are chosen to provide power due to their self-contained mechanism without the need for external components as well as ability to rapidly store incident energy and produce a steady stream of electrical power. They might also scale their power most readily for the size of the mission.

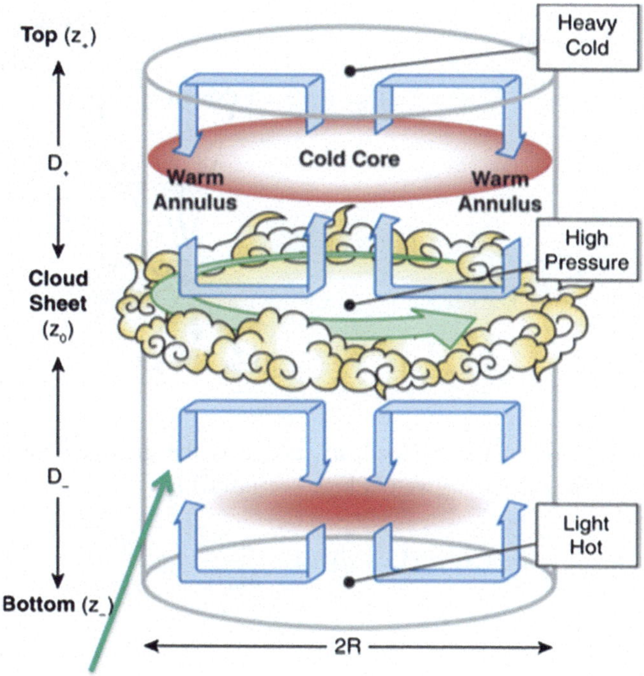

Fig. 12.12 Model for cyclone demonstrating upwelling on rim

12.2 Mobility in the Jovian Atmosphere

We consider two main design approaches (and hybrids from their combination) to maintain altitude within a range: aerostats and wing-based aerodynes. This chapter provides a high-level analysis for a number of alternatives, estimating feasibility based on preliminary calculations, discusses advantages and disadvantages and work and simulations needed to further evaluate feasibility.

12.2.1 Buoyancy Control

We examine buoyancy control in the context of Fig. 12.13 which illustrates the forces acting on a WindBot when simplified as a point mass system.

The equation for buoyancy (where B is the buoyant force, ρ_a is the atmospheric density around the WB, V is the volume occupied by the WindBot, the volume differential $\boldsymbol{\delta V}$, the pressure differential $\boldsymbol{\delta p}$, the gas molecular weight M_g, the gas

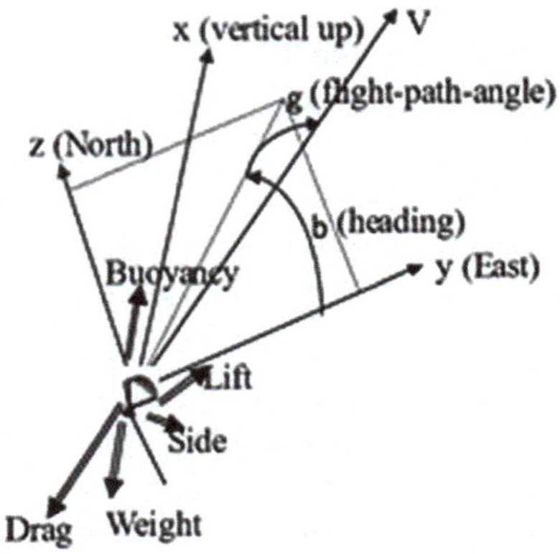

Fig. 12.13 Forces acting on point mass system on wind axes

temperature T_g, mass of the payload, m_p and g is the gravitational acceleration characteristic of the planet is given below, see insert) is:

$$\frac{B}{g} = m_p + \rho_a(V_0 + \delta V)[1 - \frac{m_g T_a}{M_a T_g}(1 + \frac{\delta p}{p_a})]$$

12.2.1.1 Buoyancy Control Equations

Derivation of the buoyancy equation
Buoyancy is given by:

$$B = \rho_a V g$$

At the equilibrium of the forces along the vertical axes, buoyancy equals the gravitational force (weight), we obtain:

$$\frac{B}{g} = \rho_a V = m_e + m_g + m_p$$

$$m_e + m_p = \rho_a V - m_g = \rho_a V - \rho_g V = (\rho_a - \rho_g)V$$

where m_e is mass of the balloon envelope, m_p is mass of the payload, and m_g is the mass of the gas inside the envelope.

Invoking ideal gas thermodynamics, we obtain:

$$m_e + m_p = v\left(\frac{p_a M_a}{RT_a}\right) - v\left(\frac{p_g M_g}{RT_g}\right) = v\left(\frac{p_a M_a}{RT_a}\right)$$

$$\left(1 - \frac{p_a M_a T_a}{p_g M_g T_g}\right)$$

The total pressure is the ambient pressure plus the super pressure δp. The total volume is $V = V_0 + \delta V$, so we obtain:

$$\frac{B}{g} = m_p + \rho_a(V_0 + \delta V)[1 - \frac{m_g T_a}{M_a T_g}\left(1 + \frac{\delta p}{p_a}\right)]$$

The control variables for buoyancy: the volume differentia, the pressure differential, the gas molecular weight, the gas temperature, mass of the payload m_p lead to the following buoyancy control modes:

- Buoyancy control based on δV: this type of control would rely on changing the volume—implicitly the geometry of the envelope. This change can take place by pulling risers, or by inflating and deflating internal "ballonets";
- Buoyancy control based on δp: this type of control would rely on varying the pressure of the gas inside the envelope. Commonly done for super pressure balloons;
- Buoyancy control based on M_g: this type of control would rely on changing the properties of the gas, possibly by a phase-transition. This could also be performed by including a vent at the top of the balloon envelope to vent off extra gas.
- Buoyancy control based on T_g: this type of control is based on changing the gas temperature. This can be accomplished through the use of heating elements inside the gas chamber, or through use of radiative heating from the sun or infrared radiation from Jupiter.
- Buoyancy control via m_p: This type of control requires weight shedding; it is also known as ballast control. Takes the form of dropping available weight to increase buoyancy potential through payload mass being decreased.

Control mechanisms would increase payload and power consumption of the WB.

12.2.2 Aerostats

One way to achieve buoyancy is to use a buoyant gas that is lighter than the atmospheric gas outside the envelope. On Earth, a common choice is to use helium or hydrogen balloons or airships. An alternative is to use heated gas balloons, as when ambient atmospheric gas is heated (according to ideal gas laws, other parameters held constant), the gas becomes less dense than the one in the surrounding environment, thus generating lift. A different way to reduce density is to extract gas creating vacuum chambers. For reference, the density on Jupiter at 1 bar is 0.16 kg/m^3 (at T = 165 K) compared to 1.20 kg/m^3 on Earth (at T = 293 K). Airships can achieve directional mobility like an airplane, using control surfaces. This provides a way to navigate the atmosphere with a directional drive mechanism —at the price of power consumption. A different way, potentially more efficient energetically is to use a tethered winged system that can provide lateral mobility. A number of alternatives are analyzed in the following.

12.2.2.1 Montgolfières (Heated Gas)

Montgolfière aerostats generate buoyancy through the isobaric heating of ambient gas within a balloon envelope. A simple example of this design is a hot air balloon. Because of power and fuel constraints it quickly becomes unfeasible to bring fuel to burn, or electrically heat the gas inside the envelope. Passive heating provides the best solution to keep such a system sustained in the atmosphere.

Solar Infrared Mongolfière Aerobots (SIRMAs) are an example of passive heating, using a combination of lower planetary infrared heating during the night and solar heating during the day, as depicted in Fig. 12.14 and detailed in a previous JPL study (Jones and Heun 1997). For a Jupiter mission, the study analyzed a

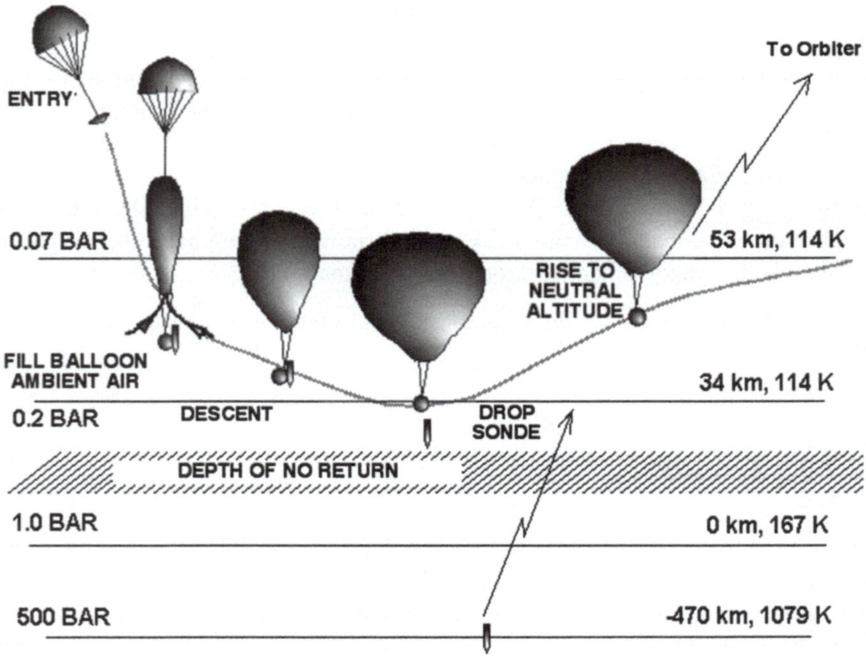

Fig. 12.14 SIRMA balloon in Jupiter atmosphere (NASA JPL) (Jones and Heun 1997)

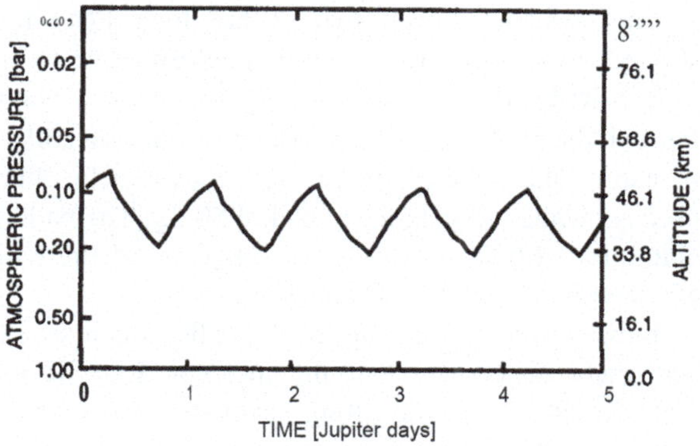

Fig. 12.15 SIRMA flight pattern simulation when undergoing solar heating (upward trends), and infrared heating (downward trends) (Jones and Heun 1997)

vehicle which: (a) Floats at about 0.1 bar during the day; (b) Descends to about 0.2 bar at night (Fig. 12.15) (c) uses a ~50 km long tether to float payload beneath cloud layer, the SIRMA balloon would have 60 m diameter and weight 112 kg of total floating mass to support a 10 kg payload (see Fig. 12.16 for other planets payload masses).

There is a point of no return, below which the balloon will not rise again with the Sun. This occurs because the thick atmosphere limits solar radiation deeper in the atmosphere.

The SIRMA envelope material requires a material that has a high ratio of absorptivity to emissivity, maximizing radiative energy absorbed and stored by the balloon. The material used in Jones and Heun (1997) was a combination of a different top and bottom. The top of the envelope was Mylar with Kevlar scrim, and had a 6-μm Titanium coating to maximize solar absorptivity (0.95), while minimizing radiative emissivity (0.2). The bottom half of the balloon was tailored primarily to gather all available IR radiation and took the form of a carbon coated black surface with high absorptivity (0.95) and emissivity (0.95). The balloon had a diameter of ~60 m with an interior temperature of 120 ± 5 K, and an envelope mass of 85 kg.

Planet	Solar AU	System wt., kg	Diameter, m	Atmospheric molecular wt.	Axis Inclination
Jupiter	5.20	112.2	60.1	2.25	3°
Saturn	9.54	219.7	86.1	2.10	27°
Uranus	19.18	278.1	97.4	2.64	82°
Neptune	30.06	842.4	171.6	2.62	29°

Assumptions: Daytime float altitude = 0.1 bar
Balloon envelope = 6-micron polymer composite with 50% weight margin
Science payload = 10 kg

Fig. 12.16 SIRMA balloon configurations on different Gas Giants (Simon et al. 2015; Mousis et al. 2014)

This system utilized two 50 km long PBO (polybenzoxazole) high-tensile strength tethers to carry scientific payload of up to 10 kg beneath the cloud layer. This SIRMA design was partially based on a concept tested in the late 1970s through the 1980s for the purpose of in situ data collection of the upper stratosphere of earth. The concept proved to be extremely effective, with one flight lasting 69 days while carrying a payload of 50 kg. The balloon envelope was composed of aluminized Mylar on the top hemisphere of the balloon and a transparent polyethylene on the bottom (Malaterre 1993).

Initial filling and heating of the balloon is a problem that requires future work. The main concern is that the incident Solar/IR radiation is only sufficient to maintain altitude of the lifting envelope. This means that as the balloon descends and fills with ambient gas, it may not have enough time to gather enough heat to generate adequate buoyancy. A potential solution for this problem utilizes heating from atmospheric entry to heat the gas inside the balloon envelope, providing a temperature stimulus before Solar/IR heating can begin to take effect.

An advantage of using the Montgolfière balloon concept is the balloon is filled with ambient atmospheric gas, mitigating the need for a heavy pressure vessel to house pure hydrogen gas. This greatly reduces payload weight (and for example maximizes instrumentation potential as well as lowering costs associated with payload delivery). This type of hot air balloon is relatively robust and pinholes and punctures do not rapidly and disruptively affect the performance, since the balloon is filled with ambient gas.

Montgolfière has the ability to perform buoyancy control by venting off hot gas or by changing the volume of the envelope to decrease buoyancy. Positive buoyancy would be achieved through internal heat beyond the equilibrium temperature.

12.2.2.2 Charlière (Hydrogen—Lighter Gas)

A Charlière uses an inside gas that is less dense than that of the surrounding ambient. As the Jovian atmosphere is composed primarily of hydrogen, a large quantity is required to keep even a relatively small payload aloft. The estimates of hydrogen percentage in Jupiter's atmosphere are in the 86–90% range. For the following calculations, it is assumed that the atmospheric composition is a simple 90% Hydrogen, 10% Helium model. This allows for a safety factor to be implemented into the design consideration.

Assuming an initial overall WB mass target of 40 kg (as the sum between structural mass and payload mass), the balloon diameter and resultant effective payload mass for different locations in the Jovian atmosphere are summarized in Table 12.3, while the equations used to generate the table are found in the following insert (Cipolato and Stoica 2016).

Table 12.3 Balloon diameter and resultant payload mass for different locations in the Jovian atmosphere

Pressure (bar)	Diameter (m)	Volume (m^3)	Structure/envelope (kg)	Payload (kg)
10	11	693	12.12	27.88
7	12	888	14.3	25.7
4	13.6	1312	18.55	21.448
1	18.6	3405	35.02	4.97
0.5	21.8	5416	47.71	**−7.718**

Charlière Analysis

The following analysis is for a zero-pressure balloon configuration assuming inside pressure and temperature conditions are equal to those outside of the balloon. Other factors such as changes in atmospheric density, and radiative heating are not included. Further calculations and models are required to understand the full effects of the radiation-rich environment on buoyancy characteristics such as internal temperature. In addition, an atmospheric composition of 90% hydrogen and 10% helium is assumed.

Considering the gravity independent equation of buoyancy, it is possible to determine total mass payload available from total envelope volume (V), and densities of the Jupiter atmosphere (ρ_a) and Hydrogen gas (ρ_{H_2}):

$$m_{total} = V(\rho_a - \rho_{H_2})$$

To determine atmospheric density as well as hydrogen gas density, the ideal gas equation must be used. This requires the temperature and pressure to be known.

$$pV = nRT$$

where n represents the mole number, defined as mass (m) over molecular mass (***M***).

To simplify calculations, the ideal gas constant (R) and the molecular mass (M) are equated to one value: the specific ideal gas constant (R_{sp}). The values of R_{sp} for Hydrogen and Helium are $4124\left(\frac{\mathrm{J}}{\mathrm{kg\,K}}\right)$ and $2077\left(\frac{\mathrm{J}}{\mathrm{kg\,K}}\right)$ respectfully.

$$R_{sp} = \frac{R}{M}$$

The ideal gas equation then takes the following form.

$$pV = mR_{sp}T$$

To further simplify calculations, density is found from mass and volume.

$$\rho = \frac{m}{V}$$

The resulting ideal gas equation takes the following forms in pressure, then in density. From here density of Hydrogen gas and Helium gas can easily be found for a given temperature and pressure.

$$p = \rho R_{sp}T$$

$$\rho = \frac{p}{R_{sp}T}$$

The atmospheric density is calculated from the 90% Hydrogen, 10% Helium model using the following equation.

$$\rho_a = \rho_{H_2}(0.9) + \rho_{He}(0.1)$$

From here, the total envelope volume required for a given total mass can be determined.

$$V = \frac{m_{total}}{\rho_a - \rho_{H_2}}$$

Because volume is now known, surface area and envelope diameter can be found. By using envelope material mass per square meter, envelope weight can be found. Envelope mass can be subtracted from total mass, giving a value for weight available for a payload.

To lift 40 kg the above equations result in a volume of 693 cubic meters of hydrogen, and a resulting mass of **500 kg**. The 40 kg being the combined mass of structure/envelope and payload, a higher payload can be lifted if one operates at higher pressure as illustrated in Table 12.3.

It is clear that working at higher pressure maximizes the payload. As the table illustrates **somewhere between 1 and 0.5 bar of Jovian atmospheric pressure, it becomes impossible to sustain a 40 kg system at altitude**. Figures 12.17 and 12.18 illustrate the relationship between different balloon diameters and the mass it can lift, for the pressure range of interest, between 0.3 and 10 bar.

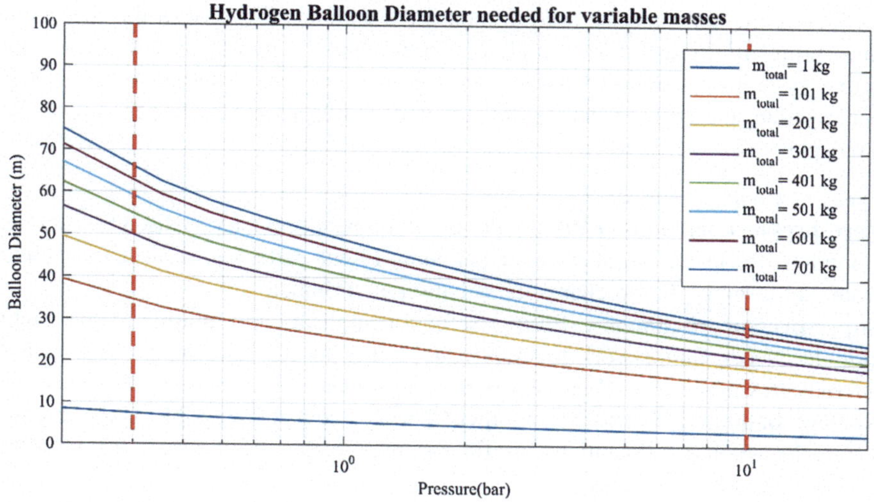

Fig. 12.17 Balloon diameter required to hold various masses between 0.3 bar and 10 bar (denoted by dashed red vertical lines)

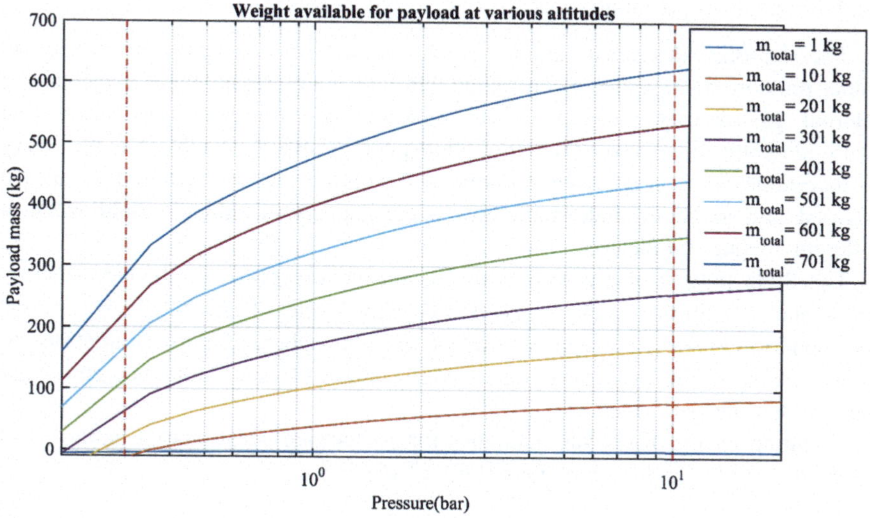

Fig. 12.18 Mass available for payload (assuming an envelope surface density of 30 g/m^2)

If the mass of 500 kg of hydrogen is stored at 700 bar and 200 K the tanks would be of 6 m^3. The mass of the hydrogen that needs to be brought to Jupiter outweighs the mass of the WB (40 kg, without the hydrogen). One requires $\sim$13 kg of hydrogen for every 1 kg of WB total mass (includes envelope and payload).

Several methods may be used to reduce the amount of hydrogen the spacecraft must bring from Earth, for example one could heat the hydrogen, either using an active system or a hybrid Charlière-Montgolfière system. Alternatively, the WB could inflate the balloon during descent if it could process hydrogen out of the atmosphere.

Types of Charlières include:
Zero-pressure balloons (ZPBs): balloons at ambient pressure, as discussed above. Common use in terrestrial weather balloons. Flight pattern is subject to cycling from day to night due to change in radiation environment.
Super-pressure balloons (SPBs): Useful for applications requiring stable flight altitude. Needs future work to investigate ability to stay aloft even with higher internal pressures in the Hydrogen-rich environment.
Rozière balloons: Hybrid Montgolfièr/Charlière concept. Feasible given a consistent heat source. Potentially useful for maximizing lift without the need to package extra hydrogen gas.
Zeppelins: Advantage lies in control surfaces and navigational abilities. Provides benefits of aerostats with control of airplanes/gliders. If enough electrical power is available, navigation is feasible. However, driving props is a major power consumer and increases weight of the craft. This design also has a large envelope weight, reducing payload potential.

The balloon described above would require a zero-pressure variation. Super-pressure balloons were considered, but may not be feasible due to increase in internal gas density associated with pressurizing hydrogen, considering that hydrogen comprises almost the entire atmosphere. Further calculation is necessary to determine what operational pressure and volume gas are required to hold a payload and to better understand how a super-pressure balloon could work at different Jovian altitudes.

The flight pattern of a zero-pressure design is susceptible to changes in environmental conditions, such as heating-cooling cycles from solar energy and atmospheric changes between regions of the atmosphere. Future simulations to understand how the cycles on Jupiter would affect balloon endurance and flight patterns will be performed.

Common means of buoyancy control for a Charlière are primarily ballast mass oriented, but also include ballonets (Fig. 12.19).

12.2.2.3 Vacuum Aerostat

In a hydrogen-dominated atmosphere, one could consider a vacuum balloon type WB (Noel 1983) (see insert).

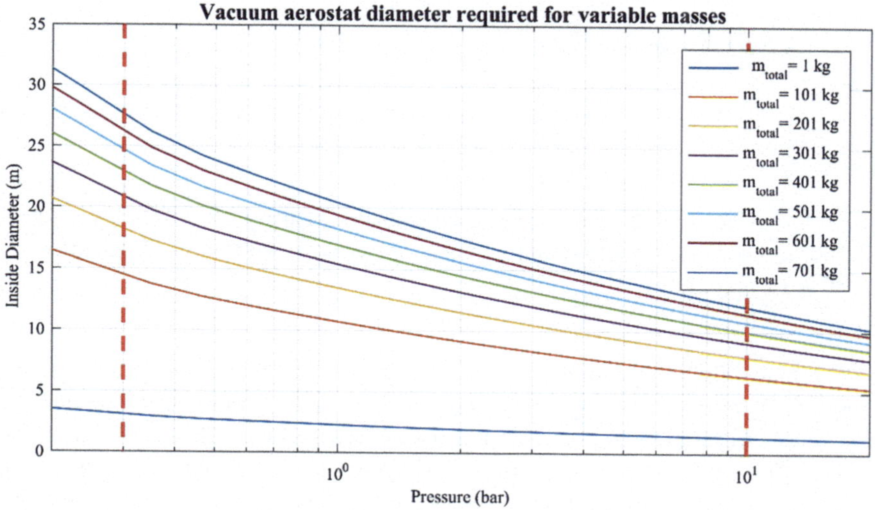

Fig. 12.19 Inside diameter of a vacuum chamber required to lift the denoted mass. This includes mass of envelope

Vacuum Aerostat

The mass displaced by the vacuum filled envelope must be equal to the mass of the balloon in an area of the atmosphere in which the ambient Jovian atmospheric pressure will not buckle the structure of the vacuum envelope skin. This presents the largest design challenge of this technology.

The basic buoyancy equation is:

$$B = \rho_a V g$$

Modified for terms on a mass-density basis:

$$\frac{B}{g} = \rho_a V = m_e + m_p$$

Simple volumetric relation based on mass, density (alid for spherical model):

$$V = \frac{4}{3}\pi r^3$$

The vacuum aerostat radius

$$r = \sqrt[2]{\left(\frac{m_e + m_p}{\rho_a}\right)\frac{3}{4\pi}}$$

At 0.3 bar (cloud layer) with an atmospheric density of ~0.07 kg/m^3, a 100 kg total mass would require a vacuum envelope with an inside diameter of ~14 m.

The main challenge with a vacuum solution is the buckling due to the external pressure. A homogeneous spherical shell buckles under atmospheric pressure for any known material, as no material has the needed specific stiffness of 4.5 × 10^5 m^5/(kg s^2) for a Poisson ratio (v) of 0.33 (Layered Shell Vacuum Balloons 2006). A sphere with thin outer and inner layers interconnected by a core layer provides enough specific stiffness to resist buckling due to atmospheric pressure, while allowing for positive buoyancy using commercially available materials (Layered Shell Vacuum Balloons 2006).

Recent studies found that a grid-stiffened structure theoretically could achieve positive buoyancy using current materials, specifically carbon fiber struts and a Zylon membrane. Geodesic approximations of a sphere perform a bit worse than a sphere but still were able to achieve positive buoyancy, with Weight/Buoyancy ratio of 0.94. An icosahedron achieves better performance than more complicated shapes due to symmetry and sparse structure (Metlen 2012). The structural response of an icosahedron type LTAV is characterized by large displacements, where membrane behavior dominates the icosahedral skin response, generating geometric stiffening in the overall structure.

Figure 12.20 demonstrates some buckling modes. These deflections have no more than 4% reduction in buoyancy. This structure is theoretically feasible but currently has not been developed into a working prototype (Adorno-Rodriguez 2014). Dynamic analysis of the structure using finite elements of the icosahedron structure revealed that chaotic motion is present in the frame of the icosahedron under certain loads and boundary conditions (Lucas 2015).

Buoyancy control through the vacuum aerostat is somewhat limited considering the chamber would most likely be rigid. Main forms would most likely include a dispensable mass ballast that would be dropped for a buoyancy increase.

12.2.2.4 Wing-Based Flight Control of Aerostats

Most designs of Aerostats lack the ability to achieve flight control (aside from varying buoyancy). Zeppelins are able to have directional control through the use of

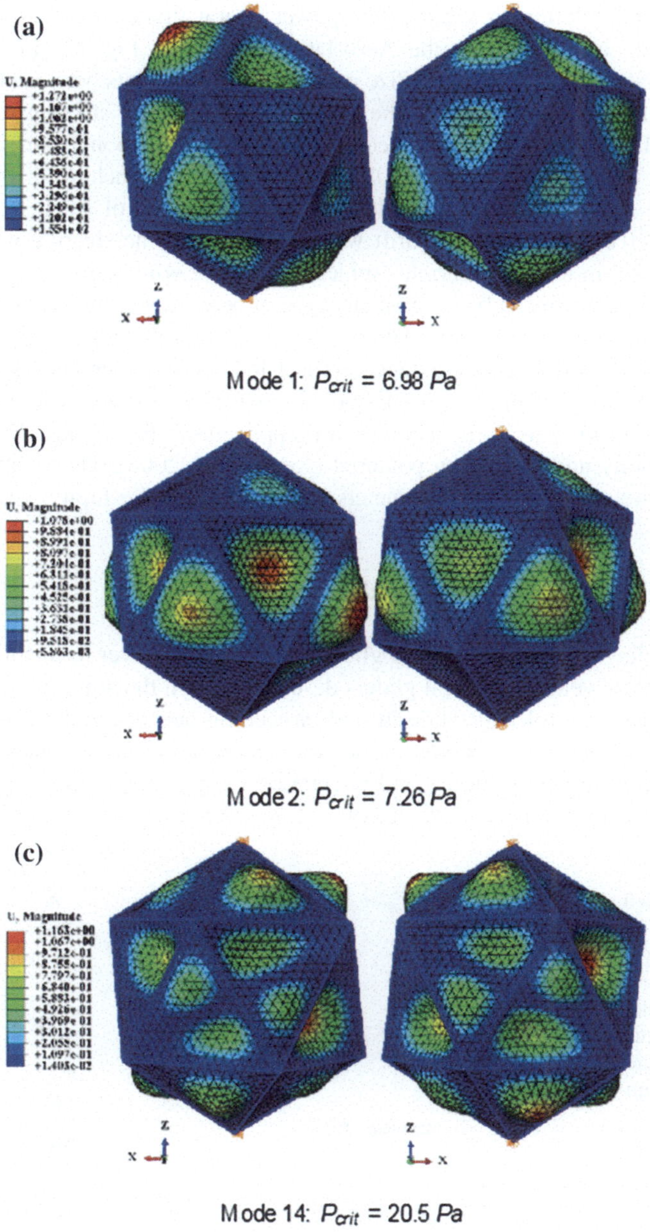

Fig. 12.20 Buckling modes of icosahedron shape of vacuum balloon

a thrust system to power the craft through the air, utilizing horizontal and vertical stabilizers to control flight paths. This requires a thrust system with large power requirements, making it less appealing to power-aware Jovian WBs.

A potential solution to achieve low-power lateral flight control is to utilize a tethered wing assembly with the Aerostat (as shown in Fig. 12.21). A Balloon Guidance System (BGS) would utilize potential wind gradients between the balloon envelope and the wing to tension the tether and to allow for steering control with the wing. The assembly would effectively "drag" the balloon from side to side through the wind currents (Sachs 2005). The level of control achieved, although not as good as that of an airplane, allows steering to areas of interest within the atmosphere. The balloon would drift with the wind, but the steering would allow navigation within the wind current, subject to relative wind velocity.

The BGS platform can be used on any type of aerostat, giving use to all outlined solutions for an aerostatic WB. Having been successfully tested, this system is indeed a viable solution for guidance control for a WB under understood conditions. While this platform has been studied extensively for applications for Mars and Venus, further work is required for applications for Jupiter to determine maneuverability and navigation potential (Aaron et al. 2006). These future studies would require scaling design parameters and tailoring to the Jupiter environment.

12.2.2.5 Buoyancy Driven Winged Aerostat

The descending motion of a winged glider induces airflow over its airfoils resulting in horizontal movement. Typical gliders descend through the atmosphere. A device with buoyancy control may glide in both ascending and descending modes. This technology has been used extensively in autonomous underwater vehicles (AUVs), achieving long duration gliding and navigation by actively controlling buoyancy characteristics. While negatively buoyant, they glide down to a prescribed depth,

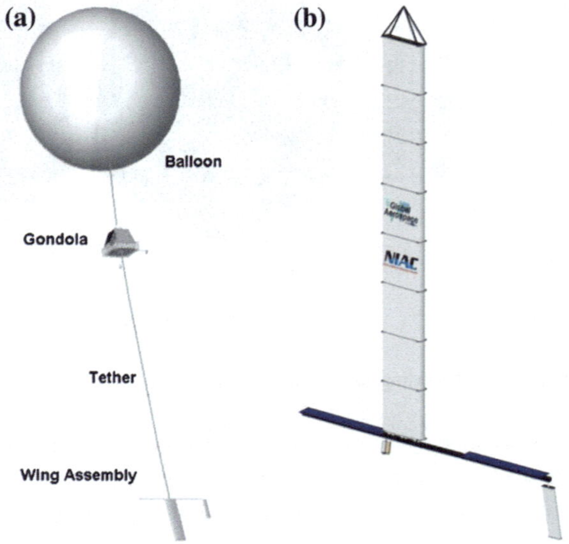

Fig. 12.21 Left shows configuration between wing assembly, payload, and balloon. The right shows the detail of a potential single-wing guidance assembly studied for the DARE platform (Aaron et al. 2006)

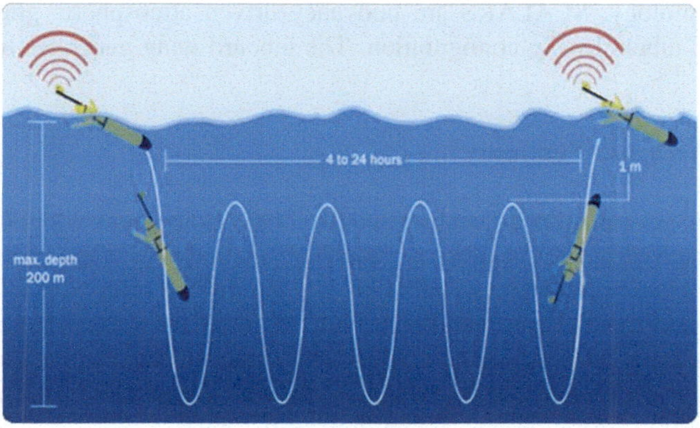

Fig. 12.22 Sinusoidal profile of flight path for underwater glider AUV, similar flight path would be used for atmospheric applications (Lust and Stevens 2016)

where they become positively buoyant and glide back up to the surface. These AUVs take advantage of thermal gradients within the ocean to power their buoyancy engines and actuate the gliding motion. Underwater gliders use active controls to power themselves for hours, over extended duration [several months and thousands of kilometers (Sherman et al. 2001)] (Fig. 12.22).

One can imagine an aerostat-glider hybrid designed for long duration flight; the aerostat to provide the buoyancy. The aerostat would be built into a rigid, lightweight body, much like a zeppelin. Wings mounted to the side of the zeppelin produce horizontal motion as the zeppelin's buoyancy is changed and the craft either ascends or descends. This passive drive system gives mobility to the WB lacking in other aerostatic devices. In addition, the wings allow the WB to have control over attitude and heading in the atmosphere, enabling targeting of specific regions which could be harvested for electrical energy.

A buoyancy-driven aerostat would be able to perform vertical sampling of the atmosphere while having significantly lower power consumption than a powered craft. Depending on the method of actuation, it appears feasible within the power levels that a WB could store during a glide (downwards) before recharging and being lifted again (by an updraft). Possible methods of achieving buoyancy control include storing a heavy gas in a rigid vessel at pressure and using it to displace the buoyant gas or a thermal system taking advantage of incident solar heating (Montgolfière) or the natural thermal gradient in the atmosphere. Flight ranges for the vehicle would be carefully chosen to prevent collapse of the vehicle at high pressures.

While extensive work for buoyancy-driven gliders has been undertaken for underwater applications, a similar concept for autonomous atmospheric flight has yet to be tested. Such a vehicle, proposed for a planet with dense atmosphere (Titan), Shape Change Actuated, Low Altitude Robotic Soarers (SCALARS) (Forbes et al. 2010) was presented as an appealing alternative to more conventional vehicle technology. SCALARS are buoyancy driven atmospheric gliders with a twin-hulled, inboard wing configuration. The inboard wing generates lift.

12.2.2.6 Aerostats Summary

Atmospheric gas aerostats use ambient gas to fill the balloon envelope, forgoing the need to bring hydrogen. The aerostat passively generates lift when heated by solar radiation or from planet infrared emission. The up-down cycles have a limit which when exceeded the balloon is no longer able to come up. Its payload is relatively reduced. Previous studies examined operation at higher altitudes in order to capture more from solar incident radiation—did not take in consideration operation below clouds (proposed operation was above limit of 0.2 bar).

Charlière aerostats provide a reliable lifting mechanism that may be deployed within the WB operating range. However, bringing pure hydrogen to Jupiter is a costly proposition, as the mass of hydrogen needed is roughly an order of magnitude more than the WB mass. It would also be moving up and down as an effect of heating or cooling from the sun or planetary IR radiation.

Vacuum aerostats might be the best theoretical alternative for buoyancy, yet buckling is a challenge to current materials and though solutions appear possible they are still to be practically demonstrated.

12.2.3 Wing-Based Design

A ramjet design on Jupiter was covered in a previous NIAC study (Maise 2003). The nuclear ramjet could operate for months because the Jovian atmosphere is a virtually unlimited source of propellant, the MITEE nuclear reactor is a (nearly) unlimited power source, and since there are few moving parts, wear should be minimal. The ramjet used a high speed (1.5 Mach) flight at altitude of 0.1 bar pressure—higher than the proposed altitude for WB; it relied on nuclear power. It appears to be a feasible flier for high altitudes and high speeds, yet not aligned with the WB concept and mission.

Gliders, or sailplanes, are aircraft with standard aircraft parts, construction, and flight control systems, but without engine—an attractive solution because of the minimal needs for power. Figure 12.23 illustrates one of the best performing commercial gliders with lift over drag ratio of 70.

Gliders are always descending relative to the air in which they are flying (Hall 2015a). To maintain altitude range over long duration missions they use updrafts to increase altitude. On Jupiter, as on Earth, thermal gradients create updrafts. They can also trade the potential energy difference from a higher altitude to a lower altitude to produce kinetic energy.

Two kinds of updrafts are considered: updrafts in jets, specifically using upwelling in bands, and updrafts in storms.

Fig. 12.23 ETA glider

The first one is found along Jovian zones and belts. Zones are colder than bands and there is upwelling. Updrafts may be as far as 10,000 km away (estimate in discussions with Prof Ingersoll et al.). Assuming a glide ratio of 80 (very high but theoretically possible), when the WB glides to an updraft from another one 10,000 km away, it would drop 125 km in altitude, the equivalent of the drop from 0.3 to 10 bar. After the drop, using an updraft of 1 m/s, the WB ascend again to the 0.3 bar region (125 km up). There are a lot of assumptions in this scenario, including that the updraft is present for the entire column of air at all altitudes at a given point or within a region where the glider maneuvers, but in general, a glider solution seems possible.

The second type of updrafts are those associated with storms, such as the GRS. To make up for the lack of data about the GRS, an analogy with tropical cyclones on Earth is used to make assumptions. It is possible for a single cyclone to contain several updrafts and for a single updraft to contain several cores (Poh and Poh 2014). To maintain consistency with these, we estimated that the GRS (30,000 km by 12,000 km) may have updrafts of 1 m/s minimal strength 3000 km apart (Poh and Poh 2014). The drop in altitude that needs to be regained is only 37.5 km with the given L/D ratio. Or, if allowed to drop 125 km the WB would need a glide ratio of only 24.

This updraft-hopping strategy to stay aloft implies the ability to detect and reach zones of upwelling wind. At least theoretically, this has been examined and shown possible (Poh and Poh 2014) and some recent UAVs flights have proven it on Earth tests as well.

The maximum recorded on Earth glide ratio of 71 was produced by a glider with a wingspan of 30.9 m. The force needed to sustain flight is three times greater in Jupiter's high gravity.

To achieve similar performance on Jupiter, it requires the same drag mitigation techniques used on Earth, including smooth skin layer, aerodynamic shape, as well as winglets and other design features to reduce induced drag (Hall 2015b). Long thin wings like those used in high glide ratio terrestrial gliders are beneficial for induced drag reduction.

Assuming no turbulence, a glider of aspect ratio AR = 35, total mass of 112 kg, L/D = 40, wing loading = 3 kg/m^2, and operating with a coefficient of lift = 1.2 was analyzed for this study by Prof William Engblom, at the 10 and 0.3 bar limits of the WB operating range. In the upper limit, the minimum updraft to keep altitude needs to be 1.2 m/s and in the lower limit 0.3 m/s. These numbers are within reason for GRS.

12.2.4 Autorotation Dynamics

Another mechanism examined for use by a WB was autorotation, i.e., the rotation induced on an object moving through fluid—the use of winds to stay afloat (Rimkus 2014). An autogyro, or gyroplane, uses an unpowered rotor in a state of autorotation to develop lift, turning the rotor by forcing moving air through the rotor disc (Bilal 2014). Assuming the WB can modulate its lift as a rotorcraft, an endurance assessment was carried out, which led to the determination of the endurance (in hours of steady flight) as a function of the system mass. The system mass was parameterized as motor mass, battery mass, and payload mass, assuming a power to weight ratio of 2 kW/kg, a 11 kWh/kg battery efficiency (Li battery), a rotor diameter of 10 m, and a typical disk loading of 20. Preliminary results indicate endurance levels of up to 10 h with a payload mass of 50 kg. While useful for extending the lifetime of a descending probe, the vertical autorotation is not a feasible solution for long endurance WBs.

A possible design with a number of rotors placed along axes in different directions, and pushed by the wind in different directions, may use autorotation not only along the vertical—and hence determine forces that would contribute to WB movement, possibly in a controlled way if the blades have adjustable controls.

12.2.4.1 Inflato-Glider

Hybrids combining buoyant and wing aerodynamics are a promising yet unexplored area: at extremes are inflatable gliders and winged balloons, which combine buoyancy given by inflatables and control and lift provided by wing surfaces.

Prototypes of inflatable wing gliders have been flown on Earth (Fig. 12.24) with an early prototype deployed in the 1950s. Another airship, I2000 (Bilal 2014), had the skeleton of the wings made of inflatable tubes, surrounded with crushable foam to provide the airfoil cross-section. The deployment time was $\sim$0.3 s (Jacob et al. 2005). To maintain suitable wing strength and stiffness, nitrogen gas pressurization at 1380–1725 kPa (200–250 psi) was required.

More recently, the University of Kentucky created and tested two variants of inflatable wing gliders: inflatable only wings that require constant pressurization to maintain shape and inflatable/rigidizable wings. In the pressurized version, pressure was maintained by gas injections to counter losses. The equivalent density in pressurized wings is increased, which reduces buoyancy. In one experiment the design pressure was 186 kPa (27 psi), though the wing has been successfully flight tested at values down to 52 kPa (7.5 psi) with sufficient wing stiffness for low speed applications (Jacob et al. 2005).

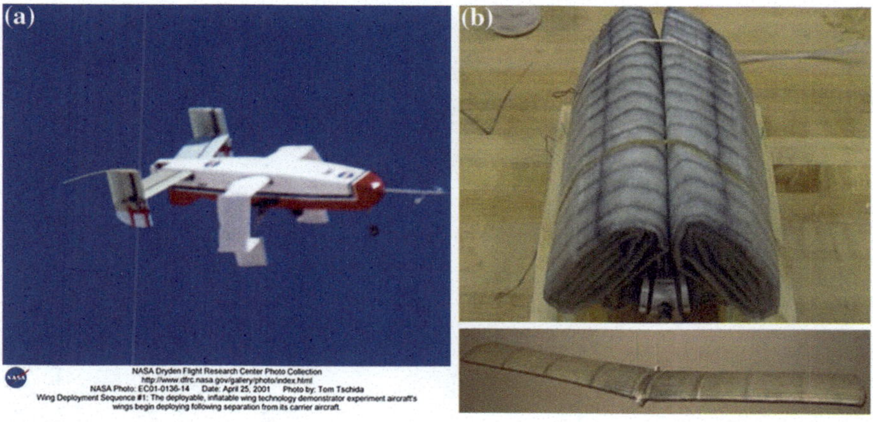

Fig. 12.24 I2000 (left), an airship with inflatable wings, successfully test in Earth's atmosphere (Bilal 2014)

Volume reductions of over 60% have been demonstrated for the inflatable wings technology.

12.3 WindBot Concept Bodies

WindBot bodies would serve multiple roles, providing the overall structure that houses all subsystems; it is what keeps the WB in the atmosphere for the mission. As space to transport the WB to Jupiter is constrained, we explore a variety of concepts that allow compact packing, such as through folding mechanisms presented in the following section.

12.3.1 Foldable—Collapsible/Expandable Structures

People have used foldable structures since at least as early as 200 BC, and over years these have been used for a wide variety of applications, such as architecture and human habitat construction, solar panel arrays and other space applications, art, etc.; some examples are given in Fig. 12.25.

The main advantages of collapsible structures are efficient packing and reduced weight. Non-rigid, variable shapes for WB bodies would benefit from better packaging efficiency, important because large flyers need to be compressed at launch. Moreover, shape and volume change provide additional benefits during atmospheric flight. A foldable structure provides opportunities for expanding and actuating control surfaces.

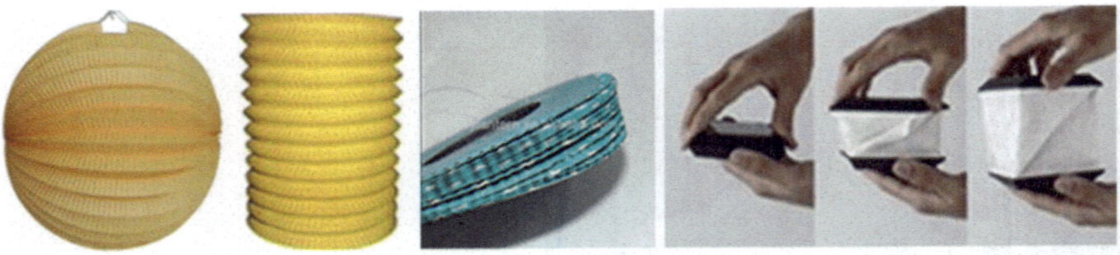

Fig. 12.25 Expandable structures: Chinese lanterns. From small packed volume to a larger one after expansion

Fig. 12.26 Buckliball—3D origami like structure that collapses to 46% its original size (Brehm 2012)

Another deployable architecture is a Buckliball, pictured in Fig. 12.26, which is a unique structure made of soft rubber with no moving parts inside. The bars forming the skeleton of a traditional collapsible structure are substituted by 24 "dimples" on its surface. Sucking air out induces distributed loading, causing buckling and collapse of the ligaments between the dimples until the ball is 46% of its original size. This technology relies on a single pressure impulse and takes advantage of a reversible buckling-induced collapse. A WB of this design would be confined to a small volume during interplanetary transport. Upon removal of a constrictive shell after insertion into the atmosphere, the structure will expand and fill with ambient gas, proceeding to operate as a balloon.

12.3.2 Foldable Structures Combined with Complex Shells in Airships

Combining foldable structures with complex shells provides new variable buoyancy solutions, such as Michael Levin's vacuum airship.

The concept of a balloon-like floating body that has separate compartments allows for WindBots to be mobile by means of changing pressure inside chambers by a system of valves. Advantage of such structures is their ability to maneuver and potentially change shape, if material with correct rigidity is chosen. There exists an inflatable wing design that relies on splitting the wing into chambers which are inflated separately to allow more control over the morphing. Section 12.3.4 of this chapter describes the morphing mechanism in more depth.

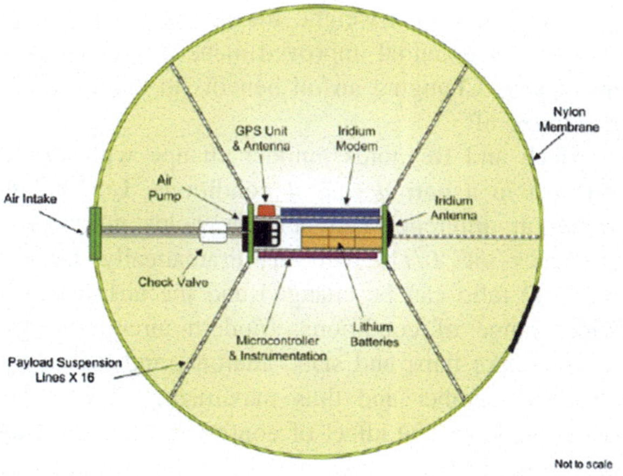

Fig. 12.27 A system diagram showing the inflatable tumbleweed internal components

12.3.3 Tumbleweed

Tumbleweeds are inflatable wind-directed robots that have been well tested in extreme terrestrial environments by NASA JPL (Kuhlman et al. 2012). They house sensors inside their inflated interior which also provides mechanical and thermal protection and gather data on their environment as they tumble.

A system diagram showing the Inflatable Tumbleweed internal components is reported in Fig. 12.27.

An example of a Tumbleweed concept refers to a ball 1.5 m in diameter which has the capability to roll over sand dunes, boulders, and even bushes (Kuhlman et al. 2012). In its current state, a tumbleweed is optimized for long-range autonomous operations in remote and hazardous environments. The design stresses simplicity, robustness, lightweight, low volume (for spacecraft delivery), and low power (Tumbleweed Rover Goes on a Roll at South Pole," Jet Propulsion Laboratory 2004).

The Tumbleweed can carry 20–30 kg of payload for hundreds of kilometers. Tools for the mission are housed inside an inflated membrane. If surroundings are cold, the components can remain warm from heat generated by running electronics and an internal air pump (Tumbleweed Rover Goes on a Roll at South Pole," Jet Propulsion Laboratory 2004). A Tumbleweed concept has been tested in Antarctica and performed well over the course of eight days. It covered over 130 km in less than 48 h. Due to its success, the Tumbleweed design has been proposed as an explorer of Mars's polar ice caps.

12.3.4 Reconfigurable Wings

A WB design may have reconfigurable wings that change properties in flight depending on local conditions. The price to pay is reduced reliability due to moving parts and additional weight. Instead, one would gain adaptability to changing air currents for potential improved maneuverability, possibly lift and possibly energy harvesting. Changing airfoil benefits to taking advantage of different densities to maximize lift.

Mach and Reynolds number change with altitude. Airfoils are optimized to perform in a narrow set of conditions. Lift and drag will have very different values at different Mach and Reynolds numbers, M and Re, thus an airfoil's efficiency—its L/D—will vary dramatically. By altering the shape of the wing, the L/D ratio can be changed and the airfoil can improve its performance in a wider range of conditions. Modern aircraft wings commonly rely on control surfaces like flaps and slats, ailerons, and their derivatives such as flaperons for changing camber and thus maximizing L/D at different velocities. Morphing wings replicate the effect of control surfaces by changing their cross section and thus providing a simpler and more lightweight mechanism to achieve advantageous shape.

Typical value for deflection of ailerons, surfaces responsible for roll control, is $\delta A = \pm 20°$ (Sadraey 2012). On average the deflection angle of most common aircraft rarely exceeds 30° (Fig. 12.28).

The main difference between reconfigurable wings and deployable wings is that the latter are actuated at the beginning of WindBot's in situ mission and then remain rigid, while the former are actuated at any point during flight when conditions favoring a different shape arise. The inflatable wing has been discussed in Sect. 12.2 and is an example of a deployable wing.

Of course, one can conceive hybrids which are deployable/inflatable as well as reconfigurable/adaptable. Traditional movable control surfaces used on rigid wings are not feasible on inflatable wings. However it is possible to actively manipulate their shape, e.g. roll control has been successfully achieved by wing warping (Jacob et al. 2005) (Fig. 12.29).

Other methods of controlling a reconfigurable WindBot including using an intermediary mechanism to distribute energy, bendable flap technology, tensegrity airfoils, and variable sweep wings were also considered in the WindBot architecture. Table 12.4 summarizes concepts for WindBot mobility.

Fig. 12.28 Example of deflection of a control surface (Sadraey 2012)

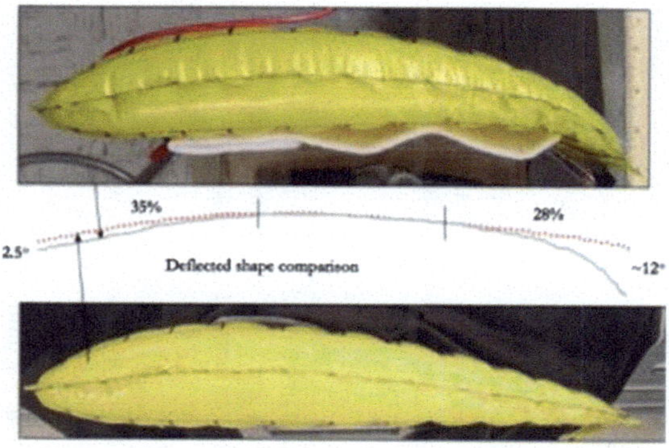

Fig. 12.29 Inflatable wing, ILC Dover (Jacob et al. 2005)

12.4 WindBot Concept Energetics

Long duration missions with multiple WBs imply reliance on harvesting energy in situ, for staying in the atmosphere, to measure and communicate data. Solar power below clouds, or Jovian missions, is challenging, and in this study we did not determine attenuation for different altitudes. We are not ruling out the use of solar power in the upper part of the atmosphere. With ~ 50 W/m^2 above clouds, in context of possible advances in solar cells, assuming a daring 50% efficiency and ultralight photovoltaics, one could consider ~ 25 W/m^2 as possibility in the future. Either large inflatables having photovoltaics on their envelope or receiving energy from a transmitter from above clouds are not impossible scenarios but were not analyzed; for Saturn and for other giants this option no longer appears feasible. On the other hand these bodies have other forms of energy in their atmospheres, which may be captured and used, and this was the focus of this section.

12.4.1 Aeolian Energy

Wind gradients and updrafts, where available, have promise for maintaining WB aloft and operational. The information we have to date is scarce—more data expected from the Juno mission.

Table 12.4 A summary of WB mobility options

	Buoyancy capability	Buoyancy control modes	Tolerance to wind gusts	Lateral maneuverability	Ease of packaging	Ease of deployment
Montgolfière	Low to medium; very dependent on ambient conditions	Medium; limitations in descending below the cloud layer	Low; limitations due to buoyancy loss through convection and susceptibility to tear	Low to medium with balloon guidance system; none without it	High due to absence of gas tanks	Medium; requires heating before EQ altitude
Charlière	Medium to high due to hydrogen	Medium due to ballonets, ballasts, temperature control	Low to medium due to a large envelope (though smaller than Montgolfière)	Low to medium with balloon guidance system; none without it	Low due to gas tanks; need 13 kg of H per 1 kg payload, unless can harvest hydrogen	High; uses gradual inflation upon descent
Vacuum	High with proper structure design	Low; mainly through ballasts	Medium; smaller envelope size and rigid structure decrease susceptibility to tear	Low to medium with balloon guidance system; none without it	Very low due to rigidity and large size of the vacuum chamber; may be feasible with a deployable vacuum chamber	Low; tech-nical pro-blems with deployable pressure vessels not fully addressed
Buoyancy driven winged aerostat	Tuned to be neutrally buoyant	Adjusts buoyancy above and below neutral to actuate flight	Ability to control pointing of glider, still susceptible to large wind gusts	Good lateral maneuverability, but less than a standard glider	High, as the entire system is inflated upon entry	High, requires one inflation event
Ramjet	None	n/a	Very high due to large mass and speeds	Poor (some) due to air intake at high speeds	None	Requires assisted launch, i.e., catapult

(continued)

Table 12.4 (continued)

	Buoyancy capability	Buoyancy control modes	Tolerance to wind gusts	Lateral maneuverability	Ease of packaging	Ease of deployment
Flettner (Magnus)	None	n/a	Depends on design	Some due to rudder (omitted in present model)	Medium (rotors are retractable)	Little to no test data exists
Glider	None	n/a	Medium	High; requires autonavigation mechanism	None for entirely rigid structure	High, well-tested platform
Autorotation	Low	Depends on design	Medium but unfavorable for the rotor	High due to blade pitch; requires a comprehensive control algorithm	Depends on design	High, eject from mothership
Inflatoglider	Possible	Valve system	Medium due to inertia of the craft	High; requires a comprehensive control algorithm	High; volume reduction of over 60%	Medium, requires UV light and certain temperatures
Foldable structures	Possible if filled with buoyant gas	Valve system	Provides no resistance; travels with wind	No control over direction of travel	High; volume reduction by a factor of 30	Requires actuation mechanism
Foldable + shells	Strong—similar to aerostat	Valves and multiple chambers	Medium	Medium; fins can be added for control	High	Similar to aerostat
Tumbleweed	Possible if filled with buoyant gas	Depends on design	Provides no resistance; travels with wind	No control over direction of travel	High	Medium, previously tested on Earth
Reconfigurable	None	n/a	Depends on design, medium to high	High; requires a comprehensive control algorithm	Medium to high (for inflatable wings)	Requires actuation mechanism

12.4.1.1 Vertical Winds (Updrafts/Thermals) for Lift and Power Harvesting

Updraft cores in the eyewall of anti-cyclones may be the most promising source of energy for Jupiter WBs. Figure 12.30 illustrates a simulated system of updraft cores in the eyewall of a terrestrial cyclone. Persistent UAV flight within this region has been proposed and simulated (Poh and Poh 2014). Previous research and analysis of direct flight observations of mature terrestrial hurricanes has found that the updraft cores covered about 55% of the eyewall regions with speeds of 4.5 m/s at 500 m. Poh & Poh demonstrated that persistent flight is viable in such an environment.

Very long duration soaring flights have been observed in terrestrial environments by various avian species. Frigate birds have been observed to use oceanic thermals in soaring flight to routinely fly for periods of 14 days or more, with very low energy expenditure. In addition, they can use these thermals to travel up to 4000 m above the sea, giving them exceptional gliding distance (Weimerskirch et al. 2016).

Due to the similarities in structure between Jovian and terrestrial storms, it may be hypothesized that Jovian cyclones would provide regions of updrafts in which a glider could fly. Updrafts and thermals are likely present outside of cyclones and would provide another source of lift to a craft seeking to explore the atmosphere beyond cyclones. Most terrestrial gliders use updrafts and thermals created by terrain features and heating of the surface (Glider Flying Handbook, US Department of Transportation, Federal Aviation Administration 2013). There are additional methods of using updrafts and downdrafts to gain small amounts of energy for a glider. These methods require very precise control and knowledge of state of the vehicle to execute properly (Kiceniuk 2001).

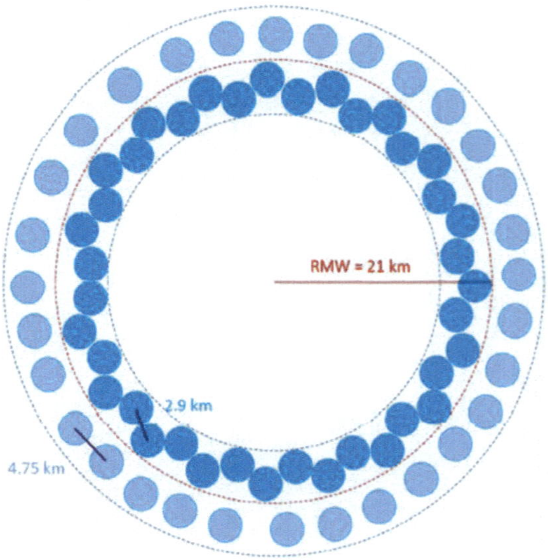

Fig. 12.30 Updraft core model. A simple graphical model showing a plausible distribution of updraft cores ±5 km from the radius of maximum wind (RMW) for Hurricane Anita at altitude 0.5 km (Poh and Poh 2014)

In order to use updrafts, one needs to detect them, at long distances. While algorithms have been constructed and simulated for glider flight planning based on available wind energy (Lawrance 2009), there is still a need for sensors to detect long-range wind shear. One possible technology is Terminal Doppler Weather Radar (TDWR), which is used around airports to detect wind shear. The operational range of this radar is on the order of hundreds of kilometers, which limits its use in WBs. However, if these updrafts do not significantly change their position in relatively short timeframes, e.g. those in the eyewall of the GRS, an initial detection is the most important part after which shorter range detectors may be sufficient, provided that WBs have good means for self-localization.

12.4.1.2 Horizontal Components—Wind Gradients/Cyclones/ Turbulence

A WB glider could use shear winds (Fig. 12.31) as well as wind gradients to enable dynamic soaring. By pushing against wind gradients with the wing, the WB can harvest power from the air equal to the velocity of the air times the force generated by the wing. Dynamic soaring depends on moving the glider through wind gradients. Also, as turbulence develops on sharp wind gradients, dynamic soaring often results in large vibrations in the glider (Kiceniuk 2001). While this is an effect which human flight tries to avoid, rapid oscillations of the glider could be used to generate power. Energy generated through dynamic soaring manifests itself as kinetic energy, which may be used to sustain altitude.

Figure 12.32 provides a flight-model description for dynamic soaring. In Fig. 12.32a, the glider takes advantage of horizontal winds, banking away from side gusts to gain energy. The banked lift force pushes the glider sideways and upward, gaining energy. Figure 12.32b illustrates vertical dynamic soaring. As the glider approaches the ground moving with the wind, its groundspeed remains constant while its airspeed and lift increase. This allows the glider to turn and rise, further increasing the airspeed. At a height where the winds are constant, the

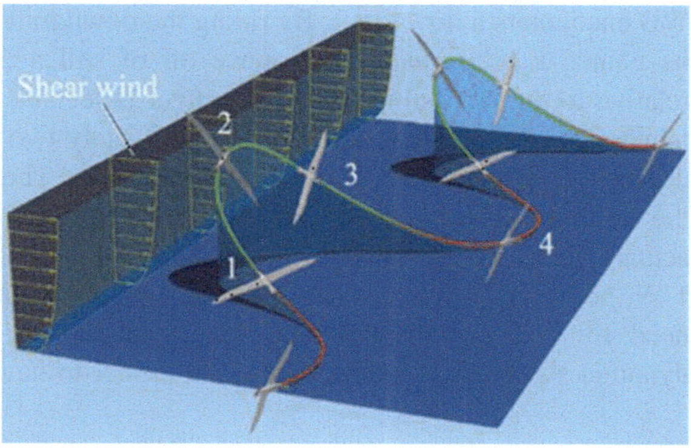

Fig. 12.31 Dynamic soaring cycle utilizing shear wind to develop an energy-neutral trajectory (Sachs 2005)

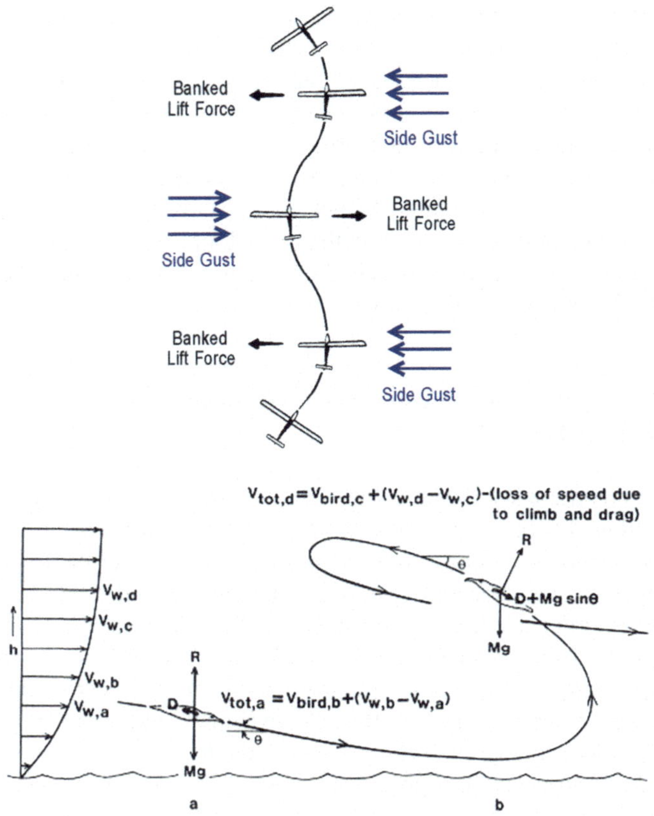

Fig. 12.32 Side Gust Soaring main model schemes. **a** Side gust soaring seen from top (Kiceniuk 2001). **b** Vertical dynamic soaring seen from side (Merck 2007)

process slows down and the glider turns again to repeat the process (Merck 2007). Dynamic soaring makes use of energy-neutral trajectories to sustain flight, replacing energy lost to drag with energy gained from moving air.

Figure 12.33 illustrates techniques which may be used to gather energy when a WB encounters a downdraft. By riding the downdraft downwards, the WB may use its gained kinetic energy to bounce off of still air underneath and gain energy relative to the beginning of the maneuver (Kiceniuk 2001). On Earth, typically, a maneuver of this type would last approximately two seconds, necessitating a high degree of sensing and control on board the flyer. There is needed power to operate the control surfaces of the craft, and which also depends on the selection of actuation/motors. For typical RC craft on Earth, the power demand does not exceed 1 W, however, power consumption for dynamic soaring may be higher due to the need for frequent maneuvering. This power would not be recovered through dynamic soaring and would require the glider to have another source of energy.

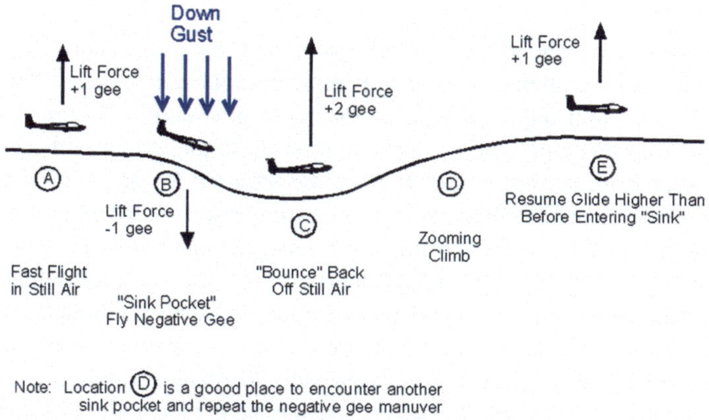

Fig. 12.33 Dynamic soaring in downdrafts. This diagram shows how a sailplane can get energy from downward moving air (Kiceniuk 2001)

Energy Use Through Static/Dynamic Soaring

The energy rate of the WindBot soaring in the wind can be expressed as follows:

$$\frac{\dot{E}}{m} = -V\frac{D}{m} - gW_z - V_a \begin{bmatrix} \cos\psi_a \cos\gamma_a \\ \sin\psi_a \cos\gamma_a \\ -\sin\gamma_a \end{bmatrix} \begin{bmatrix} \frac{\partial W_x}{\partial x} & \frac{\partial W_x}{\partial x} & \frac{\partial W_x}{\partial x} \\ \frac{\partial W_y}{\partial x} & \frac{\partial W_y}{\partial x} & \frac{\partial W_y}{\partial x} \\ \frac{\partial W_z}{\partial x} & \frac{\partial W_z}{\partial x} & \frac{\partial W_z}{\partial x} \end{bmatrix} \begin{bmatrix} \dot{x} \\ \dot{y} \\ \dot{z} \end{bmatrix}$$

In this equation, m is the vehicle mass, V_a is the vehicle velocity relative to the wind, D is the aerodynamic drag, g is the gravitational acceleration, $[W_x\, W_y\, W_z]$ are the components of the wind speed in the inertial frame, $[x, y, z]$ are the inertial components of the vehicle's position vector, ψ_a is the heading angle, and γ_a the climb angle. The first term is the power loss due to drag. The power loss due to (parasitic) drag is proportional to the cube of the airspeed.

The second term is the static soaring term; that is, it represents energy gained or lost from vertical motion of the air relative to the fixed inertial frame. As seen from the equation, this term is only affected by the magnitude of the vertical wind; positive values of W_z (downdrafts) result in energy loss and negative values (updrafts) result in energy gain. Since this term is only a function of the wind magnitude, flying at minimum sink (minimizing drag) will result in the fastest energy gain from rising air; a fact well-known to glider pilots. This is also the only term which results in energy change from a uniform wind field.

The third term is the dynamic soaring term and represents energy gained or lost due to wind gradients. This term is more complicated, as it is affected by airspeed and climb angle as well as the wind gradients. It is not instantly apparent from this equation the speed and/or climb angle that will maximize energy gain from a wind gradient, since the required lift and drag are functions of airspeed and climb angle. An interesting result of this equation is that wind energy can be gained from any wind gradient with the appropriate selection of a climb angle. Using dynamic soaring, the vehicle could be kept in flight for a long time in a persistent manner, and with no thrusting power, provided that it has the capability to sense, map, and track either a static thermal profile, or a dynamic gust event with adequate resolution.

The WindBot could have the ability to estimate the local wind vector using an air data sensor and an inertial sensor. Assume that the air data sensor measures the speed and direction of airflow relative to the center of mass of the WindBot, and the inertial sensor measures the acceleration and speed of the vehicle in a fixed inertial frame. While airspeed and inertial measurement units are common on UAVs, air angle measurements are not. One method for estimating air angles is an alpha-beta vane system, which records the wind direction around two perpendicular axes each using a lightweight wind vane and potentiometer. An alternative system is a multi-hole pressure sensor which estimates the air angles by calculating the pressure at different orientations to the wind and solving for the wind direction (Treaster and Yocum 1978).

12.4.2 Converting Mechanical Energy to Electrical Energy

In autorotation under an updraft, a lateral wind gust, or while tumbling, mechanical energy of a rotor can be harvested. Also, vibration during atmospheric movement may be harvested as mechanical energy and converted to electrical energy to provide power for instrumentation, controls, and propulsion.

12.4.2.1 Rotational Generators

Gyroscopes are currently used to harvest ocean wave energy through forced oscillation. The gyroscope is placed on a buoy, which is oscillated by the action of wave energy. As the waves move the buoy, the gyroscope resists changes in its motion. When the oscillation frequency and gimbal rotation frequency of the gyroscope are synchronized, the gimbal rotates, allowing the generation of energy. The power gyroball operates on this principle. This method could be utilized by a WB to produce energy; the naturally induced oscillation due to turbulence provides the driving force for the gyroscope motor. An array of three gyroscopes would allow turbulence incident in any direction to drive at least one motor, maximizing energy output (Bracco et al. 2011). However, it is unknown whether or not turbulence could provide a large, consistent oscillation to produce power.

Another internal method of energy production would be to use a spherical induction motor in generator mode. These motors use inductors to drive spheres of copper in any direction (Kumagai and Hollis 2013; Zhou 2014). Run in reverse (generator regime), they could theoretically take any rotational input and turn it into electricity. A speculative design involving a spherical motor would be to place the instruments in the center sphere of the WB. A light outer shell containing aerodynamic elements and the inductors would be free to rotate around the sphere when impacted by gusts of turbulence, creating electricity.

A power density comparison of wind energy harvesting devices can be found in Watch Movements (2016). For a more comprehensive study of energetics see (Roggeveen et al. 2017).

12.4.3 Turbine Generators

The theoretical power generated by a wind turbine is

$$p_{turbine} = C_p A \rho \vartheta^3$$

where A is the area of the blades, ρ the density of the fluid, ϑ the velocity of the fluid, and C_p the coefficient of power. For an ideal wind turbine, $C_p = 0.59$ (RWE npower 2010). In winds in excess of 100 m/s, a static wind turbine in Jupiter's atmosphere would need approximately 4 cm^2 of coverage to produce 30 W of power. However, for a passive WB moving with the wind, power generation would depend on wind gradients.

Unfortunately, there is no characterization of available small-scale wind gradients on Jupiter yet. These gradients would manifest as gusts or sudden changes in wind direction which would cause the WB turbines to rotate. On Earth, for wind speeds around 100 m/s, the differential between sustained winds and gusts is approximately 20 m/s (Davies 2015). In this case about 30 W of power can be produced with blades of about 12 cm, 2 cm width; this would be scaled up for Jupiter density.

A heavier-than-air WB would make further use of turbines as it falls through the air. This could include a glider based design, with a propeller mounted on the body undergoing autorotation as the device glided through the air. The deployment of a ramjet on a glider or other WB design allows the incident, apparent wind to be compressed and sped up, further increasing the power produced by a turbine. Turbines can be placed in such a way that at least one will be incident to an incoming gust of wind.

Another potential turbine design is represented by a gimbal turbine, which may be spun regardless of the incident angle of the wind. A gimbal turbine would offer a WB the advantages of a multi-rotor design in allowing the capture of wind energy in all directions and reduce the complexity of the design, but would need to be large to have an equivalent exposed area.

Turbine technology is currently the most proven of all the technologies discussed and the inclusion of a turbine system in parallel with other technology would allow increased power generation as well as a diversity of sourcing. The mass of the turbine would likely be on the same order as the vibrational generators when the supporting structure is accounted for. However, more work on the power harvested by a turbine from small scale velocity gradients is needed before the power density may be fully characterized.

12.4.4 A Summary of WindBot Concept Energetics (see Roggeveen et al. 2017)

Type	Electrical power	Benefits	Drawbacks	Operational life at 30 W	Continue for further study
Aeolian	*N/A*	*May be applied to vehicle; primary reliance on GNC algorithm, not mechanism*	*Requires fine control of vehicle attitude and ability to navigate through atmosphere*	*N/A*	*Yes, Phase II should investigate methods of detecting updrafts*
Vibrational	*30–50 W*	*Power may be obtained at any level of the atmosphere, can rapidly store wind gradient energy and provide a constant stream of power, good power to weight ratio, applicable to Earth-based systems*	*No knowledge of small-scale wind gradients, a more stable platform produces less energy*	*Determined by oscillatory response of flyer and small-scale wind gradients*	*Yes, Phase II should investigate extending these mechanisms to WB scales and dynamic modeling to determine power for different wind gradients*

(continued)

(continued)

Type	Electrical power	Benefits	Drawbacks	Operational life at 30 W	Continue for further study
Fluttering	~10 mW	Power due to boundary layer over WB, few moving parts and no mechanism development	Produces very small amounts of power, low power to weight ratio	Determined by small scale wind gradients	Without improvement in piezo technologies, this mechanism is not worth pursuing
Thermal	30 W	Very passive system, encourages vertical sampling of atmosphere	Large weight to contain hot gas, low operational life	About 2 h for 50 kg of shell weight	No, this method is infeasible for WBs
Pressure	30 W	Passive system, encourages vertical sampling	Large weight to contain pressurized gas against pressure gradient	Ideally about 30 h for 50 kg of shell weight, but will be less under actual conditions	No, this method is infeasible for WBs
Magnetic	30 W	Constant power, no mechanism or moving parts	Has not been investigated for in-atmosphere power, tether must remain constantly weighted, loop must remain circularized and closed, large drag	N/A	Further investigation of closed-loop tethers could prove beneficial but should not be prioritized
Fuel cell	*30 W*	*Continuous, guaranteed power supply*	*Outside of WB mission plan for in situ resource harvesting, puts an end date on mission, must transport heavy fuel*	*Using a COTS fuel cell and 35 kg of oxygen, can sustain for about 6 months*	*Custom fuel cell efficiencies should be explored to reduce mass and extend mission time*

12.5 Operations Concept

This section describes the functional aspects of the WB concept in operation: deployment, instrumentation, wind measurement, navigation using magnetic field, avionics, autonomy/algorithms, and autonomy challenges.

12.5.1 Avionics

Autonomous operation requires an intelligent, adaptive atmospheric flyer that senses 3D wind patterns, seeks vortexes, wind shear, and updrafts that can be used as energy fountains, plans its waypoints/traverse to pass by them, and does real-time control to adjust its aerodynamic properties for optimal lift and change of direction. To be able to determine how to move toward the energy reservoirs, the WB will need intelligent autonomy. It will need to sense the surrounding wind patterns at a distance; we will examine solutions, such as, the Laser Doppler Velocimeter, and how can this be scaled. The WB will include gyros and accelerometers for attitude stabilization, wind velocity/pressure probes, and a tri-axial magnetometer.

12.5.1.1 Instrumentation—Wind Measurement

The WB could have the ability to estimate the local wind vector using an air data sensor and an inertial sensor. Assume that the air data sensor measures the speed and direction of airflow relative to the center of mass of the WB, and the inertial sensor measures the acceleration and speed of the vehicle in a fixed inertial reference frame. While airspeed and inertial measurement units are common on UAVs, air angle measurements are not. Estimation of air angles could be done using an alpha-beta vane system, or a multi-hole pressure sensor (Treaster and Yocum 1978).

12.5.1.2 Instrumentation—Navigation Using Magnetic Field

The magnetic field at Jupiter is very strong and complex. To understand its nature better, the WB could carry a set of tri-axial magnetometers for attitude sensing and control. This is the same concept that is used for Low Earth Orbit satellites using Earth magnetic field.

12.5.2 Automatic Detection of Thermals and Autonomous Soaring

Recent work supports the feasibility of automatic detection of thermals and autonomous soaring. The U.S. Naval Research Laboratory (NRL) and the Air Vehicle Intelligence and Autonomy Lab at Pennsylvania State University (PSU) demonstrated a successful flight of a cooperative soaring and sustained flight of UAVs (Bailey 2016). The UAVs utilized the NRL's autonomous locators of thermals (ALOFT) and the AutoSOAR autonomous soaring algorithm which in conjunction allowed them to sustain flight for long periods of time. Future work is being done to see if the distance between the UAVs can be shortened so that both sailplanes can soar in the same thermal at the same altitude.

12.5.3 Autonomy Challenges

Autonomy is a great challenge including, but not limited to, guidance, modeling, control and estimation. Other important aspects to be taken into account are the payload energy requirements and the various risk factors for the payload instruments (Pomerleau et al. 2016).

In general, the vehicle must know the maximum control authority, i.e., maximum allowed bank angle (varying with WB size). It must know control authority in wind, i.e., maximum allowed wind intensity to execute maneuvers. It must know where and how strong the wind is, thus it needs a wind estimator running in near real-time. A possible aero-maneuvering control logic is as follows: if the wind has a z-windshear component, one would adopt a guidance procedure for flight in downburst. If the wind is severe, thrusting may be needed. See Fig. 12.34 for a functional diagram for a possible WindBot.

12.6 Discussion and Implications

The design categories determined for WBs on Jupiter may apply, with certain modifications in parameters, to other giant planets, which have lower gravity and even stronger winds, and—with the exception of Saturn—the atmospheric composition has heavier components, thus making it easier to achieve buoyancy.

Successful WBs would enable missions of persistent in situ exploration and the *measurement of dynamic atmospheres with high-mobility robots*. As emphasized by a National Research Council study referring to Jupiter exploration, mobility is extremely important, (Committee on Planetary and Lunar Exploration, Commission

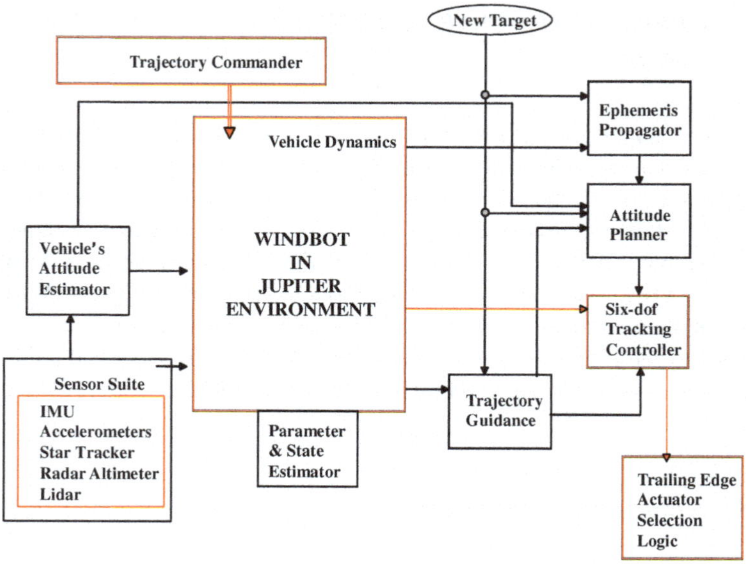

Fig. 12.34 Illustration of a possible WindBot autonomy functional diagram

on Phys. Scis., Math., and Applications, Space Studies Board and Div. on Eng. and Phys. Scis., A Scientific Rationale for Mobility in Planetary Environments, National Research Council 1998): "*mobility is required to obtain the structure, composition, and wind velocity data below the clouds with the appropriate spatial and temporal characteristics.*"

WBs are expected to e*stablish a basis and formulate a methodology for systematic incorporation of alternate energy-harvesting techniques in the design of planetary robots.* When perfected, alternate energy-harvesting technologies will enable new classes of missions of robotic explorers at targets deprived of solar radiation without nuclear power and enable new concepts of in situ persistent exploration at much lower costs than the nuclear alternatives.

Exempt from the high mass, size, and cost of a nuclear power generator, *low-cost exploration solutions with small and light platforms* become possible. This will allow the deployment of many probes/WBs and the creation of distributed sensor networks in order to perform long-term, simultaneous measurements of synoptic and global scale phenomena of these gas giants.

WBs are expected to *open the door to in situ science on gas giants*. Long-term observations will aid our understanding of the Jovian atmosphere, offering an exceptional opportunity to observe, study, and discover geophysical fluid dynamics in a nearly idealized setting of atmospheric circulations governed only by planetary and geophysical parameter. WBs would help answer important questions regarding the structure and composition of their atmospheres below the cloud tops, providing the key to understanding huge storms, which on Jupiter and Saturn may reach the size of several Earth diameters.

In contrast to Moballs, Tumbleweeds, Mars Balls which are surface wind-driven/powered robotic explorers (S-WDRE), WindBots, are atmospheric wind-driven/powered robotic explorers (A-WDRE), permanently in the atmosphere without touchdown. Staying 'afloat' *in an unpowered craft,* (harvested power is insufficient to produce sufficient thrust) in an atmosphere of light gas, under high gravity, such as is the case on Jupiter, is highly challenging. As opposed to the S-WDRE, A-WDRE (unless naturally buoyant) can NOT benefit of even very high winds if their velocity is constant, since once they reach the speed of the wind there is no more relative movement and thus no lift, and they would fall.

A constant (horizontal) wind, of any value, say 100 m/s would move S-WDRE and ensure a successful mission. The same wind would NOT lead to a successful mission with A-WDRE (unless performed with a naturally buoyant platform). A WindBot would need gusts (variation of wind vectors) to generate lift and stay afloat. No wind for a Moball has at most the effect of a stop in movement. No wind for a not naturally buoyant WindBot means falling into the depths of the gas giant.

12.7 Summary and Conclusion

The need for in situ exploration of Gas Giants using long-duration multi-point autonomous vehicles, such as the proposed WB, is real and apparent. Detailed information regarding the atmospheric structure and composition below the layer of visible clouds does not exist and is not readily achievable from an orbiting spacecraft. Within a mission context to explore this environment, we summarized what we know from previous missions, mostly Galileo, and identified sites of possible exploration and science instruments that the WB will need to carry for useful data collection. A mission scenario to the Great Red Spot focusing on measuring atmospheric conditions and composition would benefit from a Network of WindBots to maximize spatial information harvesting.

We examined Jupiter's atmospheric characteristics and options of WB mobility in the Jupiter environment, including various types of buoyancy control. The optimal region both for science and operations was determined to lie between 0.3 and 10 bar, within which all areas of known clouds are included. Based on the respective atmospheric movement the Great Red Spot and the region located northwest of it, in the South Equatorial Belt, would be the most attractive point for launching WindBot missions.

We examined various types of buoyancy control and energy use through static and dynamic soaring. We reviewed various solutions of aerostats, Montgolfière, Charlière, vacuum aerostats, wing-based designs, including a ramjet design, a Flettner rotor based on the Magnus effect, a glider, the effect of autorotation, an inflato-glider, foldable and collapsible or expandable structures with more or less complex shell forms in airships, the tumbleweed concept, and ideas of reconfigurable wings. The lowest risk for mobility in covering long distances in South Equatorial Belt was found to be the buoyant hydrogen-based inflatable designs, with diameters exceeding 70 m, and with control mechanisms for enhanced mobility. Gliders with L/D factors over 70 were found feasible, yet a number of challenges in automatically detecting and navigating to obtain lit from updrafts need to be overcome. Intelligent autonomy is needed to guide gliders with smaller L/D riding in the eyewall of the Great Red Spots, solving the localization and moving towards updrafts. Hybrid, inflatable designs with wing-or similar control surfaces may be potentially the highest risk highest payoff solution.

We considered various options to harvest energy for operations. We analyzed vertical updrafts/thermals and horizontal components of the wind, and ways to convert mechanical into electrical energy. Both conversion of mechanical energy from rotors pushed by air flow such as in autorotation and use of energy from vibration were found effective. Harvesting hydrogen for fuel cells may be valuable for shorter duration missions.

Acknowledgements This research was carried out at the Jet Propulsion Laboratory (JPL), California Institute of Technology (Caltech), under a contract with the National Aeronautics and Space Administration (NASA). The work was funded by the NASA Innovative Advanced Concepts (NIAC) Program, Program Manager Jason Derleth. We acknowledge the support from the NIAC management, NIAC External Council and NIAC Fellows for their feedback and suggestions. We acknowledge the contribution of NIAC Fellow William Engblom, of Embry-Riddle who performed the calculations for the glider. James Roggeveen (Massachusetts Institute of Technology), Kyle Peterson (University of Idaho), Kristina Andreyeva (Columbia University), Hunter Hall (University of California, Berkeley), Benjamin Donitz (University of Michigan), and Leon Kim (Columbia University) were supported by the Caltech Summer Undergraduate Research Fellowship (SURF) with funding provided by the Innovation to Flight pilot study at JPL, managed by Dr. Leon Alkalai. Marco Cipolato and Marco Dolci were supported by grants from Politecnico di Torino, Italy.

References

Aaron, K.M., Barnes, N.C., Nguyen, N., Nock, K.T., Pankine, A.A., Schlaifer, R.S.: Sailing the planets: science from directed aerial robot explorers (DARE). Phase II final report—NIAC, Global Aerospace Corporation (2006)

Adorno-Rodriguez, R.: Nonlinear structural analysis of an icosahedron and its application to lighter than air vehicles under a vacuum. M.S. Thesis, Air Force Institute of Technology (2014)

Bagenal, F., Downling, T. E., McKinnon, W. B.: Jupiter: The planet, satellites and magnetosphere (Cambridge Planetary Science), Cambridge University Press (March 5), 978-0521035453 (2007)

Bailey, A.: NRL UAS projects mimic soaring birds, make UAVs undetectable. USA TOday (2016)

Bilal, S.: An introductory study of the dynamics of autorotation for wind energy harvesting. University of Central Florida (2014)

Bracco, G., Giorcelli, E., Mattiazoo, G., Pastorelli, M.: Design of a gyroscopic wave energy system (2011)

Brehm, D.: Buckle In. MIT (2012) [Online]. Available: http://news.mit.edu/2012/buckliball-collapsible-structure-0326

Cipolato, M., Stoica, A.: Windbots for extreme environments (2016)

Committee on Planetary and Lunar Exploration, Commission on Phys. Scis., Math., and Applications, Space Studies Board and Div. on Eng. and Phys. Scis., A Scientific Rationale for Mobility in Planetary Environments, National Research Council (1998)

Davies, E.: Where is the windiest place on Earth? 9 Oct 2015

Davoodi, F., Hajimiri, A., Murphy, N., Mischna, M., Nesnas, I., Nikzad, S.: Gone with the wind on Mars (GOWON): a wind-driven networked system of mobile sensors on Mars (2012)

De Pater, I., et al.: Peering through Jupiter's clouds with radio spectral imaging. Science **352** (6290), 1198–1201 (2016)

Dowling, T.E.: Dynamics of Jovian atmospheres. Annu. Rev. Fluid Mech. **27**, 293–334 (1995)

Forbes, J.R., Barfoot, T.D., Damaren, C.J.: Dynamic modeling and stability analysis of a power-generating tumbleweed rover, Multibody Syst. Dyn. **24**, 413–439, 2010, http://arrow.utias.utoronto.ca/~damaren/papers/multitumble2010.pdf

Galperin, B., Hoemann, J., Espa, S., Nitto, G.: Anisotropic turbulence and Rossby waves in an easterly jet: an experimental study. Geophys. Res. Lett. **41**(17), 6237–6243 (2014)

Gierasch, P.J., Conrath, G.J.: Dynamics of the atmospheres of the outer planets—Post-Voyager measurement objectives. J. Geophys. Res. **98**(E3), 5459–5469 (1993)

Glider Flying Handbook: US Department of Transportation, Federal Aviation Administration (2013) p. Chapter 10

Greicius, T.: Juno overview: unlocking Jupiter's secrets. NASA, 2016. Available: http://www.nasa.gov/mission_pages/juno/overview/index.html [Online]

Hall, N.: Gliders (2015a) [Online]. Available: https://www.grc.nasa.gov/www/k-12/airplane/glider.html

Hall, N.: What is drag? (2015b) [Online]. Available: https://www.grc.nasa.gov/www/k-12/airplane/drag1.html

Hansen, C.J., Caplinger, M.A., Ingersoll, A.P., Ravine, M.A., Jensen, E., Bolton, S., Orton, G.: Junocam: Juno's outreach camera. Space Sci. Rev. **213**, 475 (2014)

Ingersoll, A.P.: Atmospheric dynamics of the outer planets. Science **248**, 308–315 (1990)

Ingersoll, A.P., et al.: Dynamics of Jupiter's atmosphere. In: Jupiter: The Planet, Satellites and Magnetosphere, pp. 105–128. Cambridge University Press, Cambridge (2004)

Jacob, J., Simpson, A., Smith, S.: Design and flight testing of inflatable wings with wing warping. Society of Automotive Engineers (2005)

Jones, J.A., Heun, M.: Montgolfière balloon aerobots for planetary atmospheres (1997)

Kiceniuk, T.: Technical soaring. Int. J. Dyn. Soaring Sailplane Energ. **25**(4) (2001)

Kuhlman, K., et al.: Tumbleweed: wind-propelled measurements for Mars. Concepts and Approaches for Mars Exploration (2012)

Kumagai, M., Hollis, R.L.: Development and control of a three DOF spherical induction motor. In: IEEE International Conference on Robotics and Automation (2013)

Lawrance, N.: Wind energy based path planning for a small gliding unmanned aerial vehicle. In: AIAA Guidance, Navigation, and Control Conference, 10–13 August 2009

Layered Shell Vacuum Balloons. US Patent US20060038062 (2006)

Lewis, J.S.: The clouds of Jupiter and NH3—H2S systems. Icarus **10**, 365–378 (1969)

Limaye, S.S.: Jupiter: New estimates of the mean zonal flow at the cloud level. Icarus **65**, 335 (1986)

Lucas, W.J.: Dynamic response analysis of an icosahedron shaped lighter than air vehicle. M.S. Thesis, Air Force Institute of Technology (2015)

Lust, E., Stevens, J.: Autonomous vehicle for gathering oceanographic data in littoral regions. Preliminary Design I (2016)

MacLow, M.M., Ingersoll, A.P.: Merging of vortices in the atmosphere of Jupiter: an analysis of Voyager images. Icarus **65**, 353–369 (1986)

Maise, G.: Exploration of Jovian atmosphere using nuclear Ramjet Flyer, phase II—final report (2003)

Malaterre, P.: Long-duration balloon flights in the middle stratosphere. Adv. Space Rest. **13**(2), (2) 107–(2)114 (1993)

Marcus, P.S., Shetty, S.: Jupiter's zonal winds: are they bands of homogenized potential vorticity organized as a monotonic staircase? Phil. Trans. R. Soc. **369**, 771 (2011)

Merck, J.W.: The biomechanics of flight. University of Maryland (2007)

Métais, O., Lesieur, M.: Spectral large-eddy simulation of isotropic and stably stratified turbulence. J. Fluid Mech. **239**, 157–194 (1992)

Metlen, T.T.: Design of a lighter than air vehicle that achieves positive buoyancy in air using a vacuum. M.S. Thesis, Air Force Institute of Technology (2012)

Mousis, O., et al.: Scientific rationale for Saturn's in situ exploration. Planet. Space Sci. **104**, 29–47 (2014)

Noel, D.: Lighter than aircraft using vacuum. Speculations Sci. Technol. **6**(3) (1983)

Poh, C.K., Poh, C.H.: Can the vertical motions in the eyewall of tropical cyclones support persistent UAV flight? (2014)

Pomerleau, M.: NRL tests low-power, long-endurance sail drone. Defense Systems (2016) [Online]. Available: https://defensesystems.com/articles/2016/01/08/onr-penn-state-sailplanes-thermals.aspx

Poro, C.C., et al.: Cassini imaging of Jupiter's atmosphere, satellites, and rings. Science **299**, 1541–1547 (2003)

Rimkus, S.: A lab-scale experimental framework for studying the phenomenon of autorotation. Diss. University of Central Florida (2014)

Roggeveen, J., Stoica, A., et al.: Analysis of alternative energy harvesting methods to power atmospheric robotic explorers on Jupiter. In: IEEE Aerospace Conference, Big Sky, Montana (2017)

RWE npower: Wind Turbine Power Calculations (2010)

Sachs, G.: Minimum shear wind strength required for dynamic soaring of albatrosses (2005)

Sadraey, M.: Aircraft Design: A Systems Engineering Approach. Wiley, New Hampshire (2012)

Sánchez-Lavega, A., Hueso, R., Pérez-Hoyos, S., Rojas, J.F., French, R.G.: Saturn's cloud morphology and zonal winds before the Cassini encounter. Icarus **170**, 519–523 (2004)

Seiff, A., Kirk, D.B., Knight, T.C.D., Mihalov, J.D., Blanchard, R.C., Young, R.E., Schubert, G., von Zahn, U., Lehmacher, G., Milos, F.S., Wang, J.: Structure of the atmosphere of Jupiter: Galileo probe measurements. Science **272**, 844–845 (1998a)

Seiff, A., Kirk, D.B., Knight, T.C.D., Young, R.E., Mihalov, J.D., Young, L.A., Milos, F.S., Schubert, G., Blanchard, R.C., Atkinson, D.: Thermal structure of Jupiter's atmosphere near the edge of a 5-μm hot spot in the north equatorial belt. J. Geophys. Res. **103**, 22857–22889 (1998b)

Sherman, J., Davis, R.E., Owens, W.B., Valdes, J.: The autonomous underwater glider spray. IEEE J. Oceanic Eng. **26**(4) (2001)

Simon, A.A., Wong, M.H., Orton, G.S.: First results from the hubble opal program: Jupiter in 2015. Astrophys. J. **812**(55) (2015)

Smith, A.B., et al.: The Jupiter system through the eyes of Voyager 1. Science **204**, 951–972 (1979)

Stoica, A., et al.: WindBots: persistent in-situ science explorers for gas giants. NIAC Final Report (2016)

Treaster, A.L., Yocum, A.M.: The calibration and application of five-hole probes. Technical Report (1978)

Tumbleweed Rover Goes on a Roll at South Pole. Jet Propulsion Laboratory (2004) [Online]. Available: http://www.jpl.nasa.gov/news/news.php?feature=343

Watch Movements: Tourneau, Inc. (2016) [Online]. Available: http://www.tourneau.com/watch-education/watch-movements.html. [Accessed 23 June 2016]

Weimerskirch, H., Bishop, C., Jeanniard-du-Dot, T.J., Prudo, A., Sachs, G.: Frigate birds track atmospheric conditions over months-long transoceanic flights. Science **353**(6294) (2016)

Williams, G.P.: Jovian and comparative atmospheric modeling. Adv. Geophys. **28**, 381–429 (1985)

Young, R.E.: The Galileo probe: how it has changed our understanding of Jupiter. New Astron. Rev. **47**, 1–51 (2003)
Young, R.E.: The Galileo probe mission to Jupiter: science overview. J. Geophys. Res. **103**(E10) (1998)
Zhou, L.: Magnetically Suspended Reaction Sphere with One-axis Hysteresis Drive. Department of Mechanical Engineering, Massachusetts Institute of Technology (2014)

Chapter 13
Enceladus Vent Explorer Concept

Masahiro Ono, Karl Mitchel, Aaron Parness, Kalind Carpenter, Saverio Iacoponi, Ellie Simonson, Aaron Curtis, Mitch Ingham, Charles Budney, Tara Estlin, Carolyn Parcheta, Renaud Detry, Jeremy Nash, Jean-Pierre de la Croix, Jessie Kawata and Kevin Hand

13.1 Overview

Sometimes Nature acts as if it is waiting to be explored. For example, it secretly installed "doors" on the icy shell of Enceladus—vents through which water vapor and other materials are ejected from the subsurface ocean into outer

M. Ono (✉) · K. Mitchel · A. Parness · K. Carpenter · S. Iacoponi · E. Simonson
A. Curtis · M. Ingham · C. Budney · T. Estlin · C. Parcheta · R. Detry · J. Nash
J.-P. de la Croix · J. Kawata · K. Hand
Jet Propulsion Laboratory, California Institute of Technology,
4800 Oak Grove Drive, Pasadena, CA 91109, USA
e-mail: masahiro.ono@jpl.nasa.gov

K. Mitchel
e-mail: karl.l.mitchell@jpl.nasa.gov

A. Parness
e-mail: aaron.parness@jpl.nasa.gov

K. Carpenter
e-mail: kalind.c.carpenter@jpl.nasa.gov

A. Curtis
e-mail: aaron.curtis@jpl.nasa.gov

C. Budney
e-mail: charles.j.budney@jpl.nasa.gov

T. Estlin
e-mail: tara.a.estlin@jpl.nasa.gov

R. Detry
e-mail: renaud.j.detry@jpl.nasa.gov

J. Nash
e-mail: jeremy.nash@jpl.nasa.gov

V. Badescu and K. Zacny (eds.), *Outer Solar System*,
https://doi.org/10.1007/978-3-319-73845-1_13

space[1]. The doors may serve as a natural pathway through the ice crust, which is tens of kilometers thick, to reach the extraterrestrial ocean. The doors may provide us with an opportunity to investigate a unique environment in the vent-conduit systems, which themselves could be habitable. The doors may lead us to the discovery of the second origin of life. The doors may lead us to the answer to human's ultimate question—"are we alone?" The doors are just 10 AUs away from us. Why would we leave them open and unexplored?

In this chapter, we present the *Enceladus Vent Explorer (EVE)*, a robotic pathfinder mission concept to enter these doors. EVE's goals would be to descend into erupting conduits, characterize the unknown interior structure of the vent-conduit system, assess the accessibility to the subsurface ocean through the vent-conduit system, potentially reach the liquid interface, and perform astrobiology and volcanology observations in the vent-conduit system. EVE would send two types of modules: a Surface Module (SM) and Descent Modules (DM). SM would be a lander that stays on the surface, while tens of small ($\sim$3 kg, 10 cm in width and 30 cm in length) DMs separate from SM, move to a vent, and descend into it. DM would rely on a power and communication link provided by SM through a cable. As the payload volume of DM would be extremely limited, each DM could carry only a single miniaturized instrument. This limitation is complemented by heterogeneity. There are several types of DMs, all of which share the common mobility system but carry different instruments. For example, a "scout DM" creates a 3-D map of the geyser system with its stereo cameras and structured light. A "sample return DM" collects particles and ice cores in the vent and delivers them to the mass spectrometer in the SM. An "in situ science DM" carries science instruments, such as a microscopic imager and a microfluidics chip for biosignature detection. DMs could be sent either sequentially or in parallel (Fig. 13.1).

The two greatest challenges for EVE are the uncertainty in (i) the dynamic pressure due to the upward flow and (ii) the size of vent. Despite remote observations by the *Cassini* spacecraft, these two parameters remain poorly characterized. Among the numerous unknowns associated with Enceladus vents, the one that has the greatest impact on the two key parameters is the eruption mechanism. Most hypothetical eruption models that have been proposed thus far fall into either of two categories, which we refer to as the relatively calm "boiling" models, and the more dynamic "cryovolcanic" models. Very roughly speaking, the "boiling" models

J.-P. de la Croix
e-mail: jean-pierre.de.la.croix@jpl.nasa.gov

K. Hand
e-mail: kevin.p.hand@jpl.nasa.gov

[1]As of the writing of this document, Enceladus is the only icy moon where vents have been directly observed by spacecraft. While recent Hubble observations of Europa have revealed a water vapor plume that could potentially have been vented from the subsurface ocean, its characteristics are poorly known and the existence of vents has not been confirmed. Hence, we set the primary target of our study to be Enceladus.

Fig. 13.1 EVE's Descent Module (left) and Surface Module (right). (Artist's concept)

assume that liquid water boils into vapor under the surface, while the "cryovolcanic" models assume that a fairly pure form of the ocean material ascends the conduit driven by exsolution and expansion of dissolved materials and phase changes. More detailed explanations of the two models will be provided shortly. As we will describe in Sect. 13.3, *EVE is likely feasible assuming the "boiling" models while the feasibility is undetermined with the current best knowledge under the "cryovolcanic" models.*

While observations from orbit around Enceladus could help provide some insight into the appropriate model, uncertainties in interior geometries and vertical profiles of dynamic pressure cannot be significantly reduced without descending into the vent. Therefore, the first mission into an Enceladus vent must face a "chicken-and-egg" problem: dynamic pressure and vent size remain uncertain until it descends into the vent, but designing a robot to descend into the vent requires knowledge of those parameters. Consequently, in order to maximize the chance of mission success, EVE's DM needs to be designed conservatively in terms of size and resistance to dynamic pressure.

As we will discuss in Sect. 13.4, our trade study has determined that *the DM configuration that is most robust to the uncertainties in vent parameters is a limbed robot with ice screws as the end effector.* An anchoring mechanism would be needed because the upward force exerted by the jet is orders of magnitude greater than the downward force exerted by Enceladus's weak gravity ($\sim$0.01 g). *Cassini*'s observation indicates that the plume ejection is fairly constant, hence the robot would have to move against the jet. An ice screw is a hollow metal screw used by alpinists for ice climbing. As a byproduct, it produces an ice core when screwed in, which could be brought back by the sample return DMs. Since an ice screw can make a strong anchor with only one wall (as opposed to wedging, for example, which requires two walls), it is robust to uncertainty in the width of conduit.

Our DM design would be able to tolerate the force exerted by $\sim 10^5$ Pa dynamic pressure on the DM.

Our trade study also concluded that *the most realistic way to power DM would be to provide power from SM through a cable* because neither RTGs nor primary batteries that fit in the limited volume of DM can provide sufficient energy for the mobility system within a reasonable mission duration. A cable also eliminates difficulties in communication as well as in navigation for the return trip of sample-return DM. A three-section DM could accommodate up to 2 km of cable in its rear section. By adding a dedicated section for storing cable, it could accommodate up to 10 km of cable. *The estimated speed of a DM is* 5.5 m/h, *given* 10 W *of power for mobility*.

We use NASA's potential Europa Lander (Hand et al. 2017) as a reference design for SM. The major difference from Europa Lander would be the power source. The baseline design of the Europa Lander concept uses a 45-kWh primary battery. This is because the lifetime of the supporting Carrier Relay Orbiter (CRO) would be limited to ~ 30 days due to Europa's harsh radiation environment, thus a longer-lived lander mission is difficult to justify (Hand et al. 2017). In contrast, measurements indicate that the intensity of the trapped radiation environments of Saturn is much lower than the Earth's and is not likely to pose serious problems to spacecraft (Barth et al. 2003). This justifies the potential use of RTGs to support a long mission.

Given this basic configuration, there is still a vast trade space to explore, particularly in the design of DM—size of the robot, DOF of each limb, choice of actuators, body shape, etc. It is beyond the scope our study to optimize the design of DM. Instead, in order to obtain a reference point for the tolerable dynamic pressure, we developed a prototype design of DM with ~ 10 cm in width, as described in Sect. 13.6. The resulting design has four 5 DOF limbs, actuated by commercially available electrical motors. Each limb features a linear actuator, which is used to move DM's body against the flow. Each limb is designed to support up to 1000 N of force, meaning that DM can withstand up to 4000 N of force with its four limbs. The result of our computational fluid dynamics simulation suggests that the maximum tolerable dynamic pressure of this particular DM design is 5.4×10^5 Pa. With safety margins, $\sim 10^5$ Pa would be the safe operational limit for this prototype design of DM. We note that the drag coefficient of the design is ~ 1.5, while that of typical automobiles, for example, is 0.2–0.3. Clearly there is a significant room for design optimization, particularly in the aerodynamic shape (Fig. 13.2).

Beyond the mechanical aspects of the EVE design, our study has considered the nature and degree of autonomy capabilities required to effectively execute the mission. *Our study concludes that EVE must be a highly automated system in order to complete the mission within a reasonable duration.* One of the most significant challenges for teleoperating DMs is the limited line-of-site distance in the vent. In Mars rover operations, the drive distance per operation cycle (Sol) of manual planning is limited by the line-of-site distance, which is typically ~ 50 m. In the vent, it would be tens of cm. In case of Europa Lander, each operation cycle is ~ 24 h long, which is constrained by the visibility of the CRO. By limiting DM

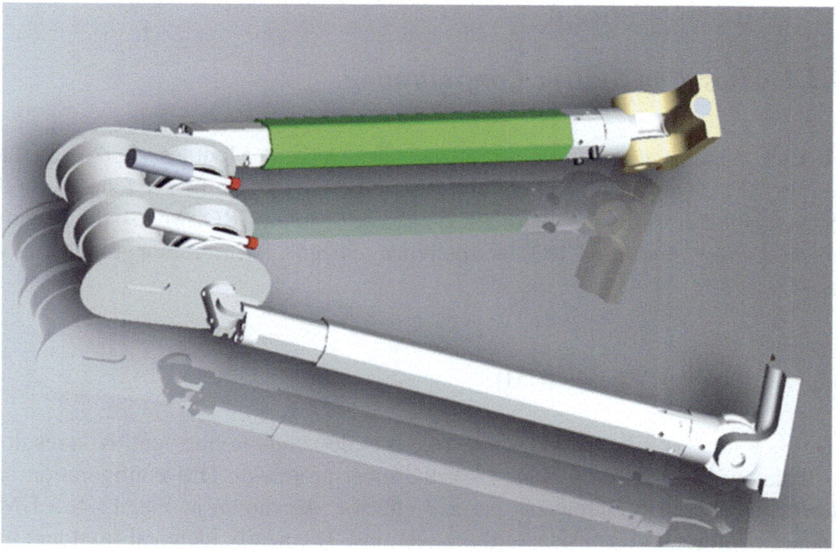

Fig. 13.2 Prototype design of DM's front/rear section with two limbs

motion to tens of cm per day, it would take decades to move one kilometer. In addition, operating tens of DMs manually would be a major challenge. In order to enable a realistic mission, required autonomy capabilities for DM include 3D mapping, localization, hazard detection and avoidance, route planning, activity planning, self-diagnosis, and recovery from failures. Agile science is highly desired to not miss interesting science opportunities during the traverse. Cooperative autonomy with multiple DMs is also important. The limited computational capabilities onboard the DM are not a limiting factor, since DM can access computational resources on SM through the cable. Furthermore, DMs can communicate with each other via SM.

In summary, the feasibility of EVE depends on the eruption mechanism, which at this point is unresolved. Assuming the "boiling" model, in which the expected dynamic pressure is $<10^4$ Pa and the width of the vent is likely to be wider than 10 cm, our prototype DM is feasible, hence the EVE concept is likely feasible. However, given the "cryovolcanic" model, in which the dynamic pressure and vent width could be anywhere between 10^3 to 10^7 Pa and 1–30 cm, respectively, the feasibility of EVE is undetermined.

Finally, we would like the readers to recall the excitement of Jules Verne's *Journey to the Center of the Earth*. In the classic science fiction novel, Professor Otto Lidenbrock and his company descend into an Icelandic volcano, discover a vast subsurface ocean, and encounter exotic life. In the near future, EVE could make such a fictional adventure real, not on Earth but in the frosty and mysterious world of Enceladus.

13.2 Mission Concept

13.2.1 Baseline System Configuration

The EVE concept consists of three modules: Descent Modules (DMs), Surface Module (SM), and Carrier Relay Orbiter (CRO). This study mostly focused on the DM as it is the most unique feature of EVE. For SM and CRO, we use the potential Europa Lander and its CRO as reference designs, with minor modifications (Fig. 13.3).

13.2.1.1 Descent Module (DM)

A DM is a small (<3 kg) robot, whose role is to descend into the vent. As a result of our trade study describe in Sect. 13.4, the most promising DM configuration is a limbed robot with ice screws as the end effector. As shown in Fig. 13.4, a DM's standard configuration consists of three sections. The nearly identical front and rear sections accommodate mechanical components to actuate two limbs. A pair of cameras and structured light projector is mounted on the front section for perception (see Sect. 13.4.5 for details), and a tether spool is accommodated in the rear section. The middle section accommodates instruments.

An EVE mission could involves tens of heterogeneous DMs, all of which have identical front and rear sections but carry different instruments in its middle section.

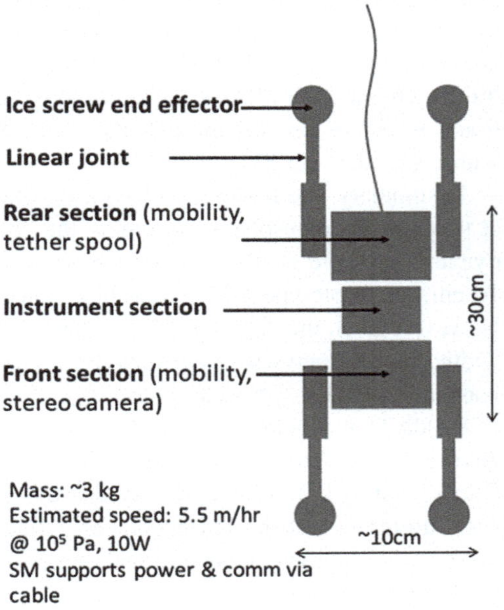

Fig. 13.3 A prototype design of EVE's DM (details in Sect. 13.6)

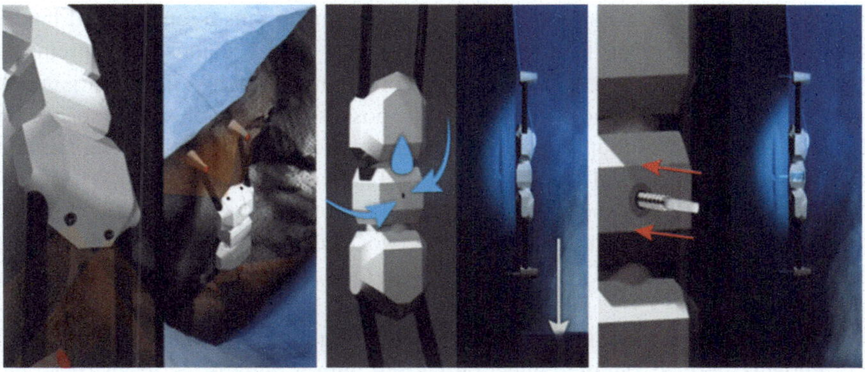

Fig. 13.4 Scout DM (left), in-situ science DM (middle), and sample return DM (right) (artist's concept)

A "*scout DM*" carries extra mapping devices and creates a detailed 3-D map of the geyser system with its stereo cameras and structured light. A "*sample return DM*" would haves *an* additional ice screw in its middle section, which is used to create and store an ice core. Alternatively, it could have a device to collect particles in the vent. In either case, a sample return DM returns to SM for analysis. An "*in situ science DM*" carries miniaturized science instruments in its middle section, such as a microscopic imager and a microfluidics chip for bio-signature detection (See Sect. 13.4.6 for the discussion on science instruments).

A DM could potentially have a train of multiple middle sections (with potential addition of limb sections)—essentially making it a "snake bot." This idea is attractive because adding sections would not significantly reduce the ability to traverse through vents or conduits, nor increase aerodynamic drag. However, due to the limitation of time and resources in this study, we only performed analysis of the three-section configuration.

One idea is to make the middle section interchangeable, so that a DM can transform to a different type. This would give significant flexibility operation. For example, an in situ science DM can choose the best instrument based on the observations by scout DMs. A scout DM can transform into a sample return or an in situ science DM on its second trip. However, it is beyond the scope of this study to explore the mechanical implementation of an interchangeable middle section.

13.2.1.2 Surface Module (SM)

The SM's role is to deliver DMs to the surface of Enceladus and provide DMs with power and communication link through tethers. We use the Europa Lander (Hand et al. 2017) as a design reference. The potential Europa Lander is powered by a 45-kWh primary battery with 20+ days of surface operation time. The decision to use this battery technology was based on the lifetime of the CRO being limited

$\sim$30 days due to Europa's harsh radiation environment, thus a longer-lived lander mission is difficult to justify. This is not the case for Enceladus. Measurements indicate that the intensity of the trapped radiation environments of Saturn is much lower than the Earth's and is not likely to pose serious problems to spacecraft (Barth et al. 2003). This supports the potential use of RTG to support a long mission (>1 year). In the proposed design of Europa Lander, the mass of 45 kWh of primary battery is $\sim$95 kg including packaging [Personal communication with Steve Sell, JPL], which is roughly equivalent to two Multi-Mission RTGs (MMRTGs; 45 kg each). Each MMRTG provides 125 W at the start of mission and 100 W after 14 years.

The potential Europa Lander would carry 42.5 kg of science instruments as well as a Phoenix-derived sample collection system with a mass of $\sim$26 kg [Personal communication with Noah Warner, JPL]. With this payload mass, the SM of the EVE could carry up to $\sim$20 DMs (assuming 3 kg each). Alternatively, it could carry a lesser number of DMs and science instruments, such as mass spectrometer, which is used to analyze samples brought back by the sample return DMs.

A solar-powered SM is possible, but the landing site is constrained by the condition of sunlight. Given the rich topology of the Tiger Stripes in which the vents are located, finding a suitable landing ellipse would be a major challenge. A more detailed trade study of power options is discussed in Sect. 13.4.2.

13.2.1.3 Carrier Relay Orbiter (CRO)

Like the CRO of the Europa Lander, the role of EVE's CRO would be to deliver SM-DMs to Enceladus orbit, and provide communication relay after landing. Due to the weak gravity of Enceladus, EVE's CRO would co-orbit with Enceladus (instead of orbiting Enceladus) at a slightly different inclination so that it can cover the south polar region over a half of Enceladus's orbital period (1.37 days). Like the proposed Enceladus Life Finder's mission, EVE's CRO can be solar powered.

13.2.2 Mission Profile of EVE

13.2.2.1 Launch, Cruise, Orbit Insertion, Landing

EVE's mission profile would be similar to that of Europa Lander up to landing (Fig. 13.5). After being launched by an SLS-class launch vehicle, its voyage to the Saturnian system would involve deep space maneuvers, Earth/Venus flybys, and a Jupiter flyby. Initially inserted to a highly elliptic orbit around Saturn, it incrementally lowers the aposaturnium point through Titan flybys, and finally co-orbits with Enceladus. Since the Saturnian system has only one moon that is massive enough for a changing orbit (Titan) as opposed to four in the Jovian system, the orbital transition may take a longer time than Europa Lander. In addition, the

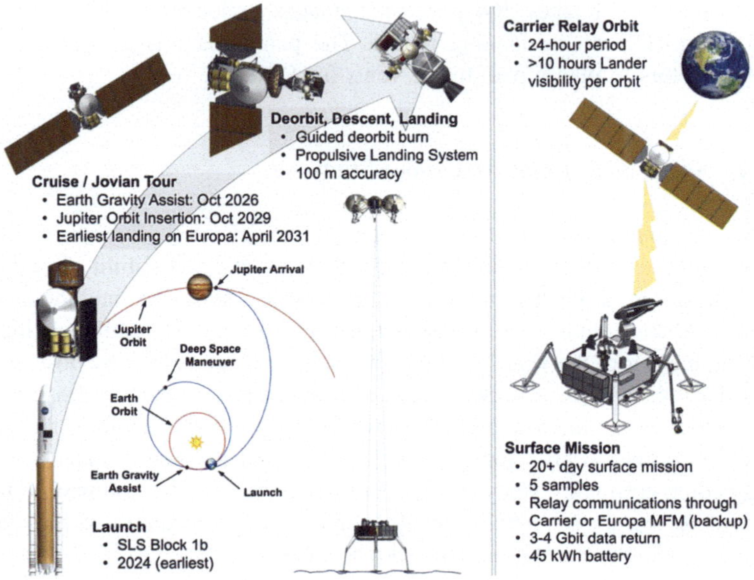

Fig. 13.5 Key features of the current Europa Lander mission concept, from launch to landing. The EVE mission would have a similar mission profile up to landing. Image taken from Hand et al. (2017)

significantly weaker gravity of Enceladus would require a greater ΔV for the orbit insertion to Enceladus's orbit around Saturn. A detailed orbital analysis is a future work.

Europa Lander's estimated landing accuracy is 100 m (Hand et al. 2017). The significantly improved landing accuracy compared to Mars landing is primarily due to the lack of atmosphere, and the same condition applies to Enceladus. Hence, we assume 100 m landing accuracy for EVE.

The landing site would be selected from orbital observations, based on:

(1) Landing safety (smooth flat terrain over 100 m radius),
(2) The number of vents reachable from the landing point [Having multiple vents reachable from the landing point is desirable because it increases the chance that at least one of them is feasible for descent. However, based on current observations, multiple accessible vents within a few hundred meters unlikely. If there are discrete vents, as proposed by (Porco et al. 2014), there are of order 100 of them spaced over ~500 km of vent system. This would make a mean distance of order a few km. Furthermore, if there were multiple vents within a few hundred meters, the probability is that they are very small ones carrying little flux, most likely of order 100 g/s or less], and
(3) Estimated size (indicated by mass flux) and dynamic pressure of the vents.

Once landed, SM takes 360 panorama images to localize itself and identify accessible vents from the landing points. The panorama image is also used by ground operators to plan a path to the vents for DM.

13.2.2.2 Scout & Eruption Dynamics Study

First, SM deploys a few scout DMs. Each DM travels to the mouth of different vents (if multiple vents are within the reach) to assess the accessibility based on the visual observation of the size of the vent, the observed mass flux, and the observed flow speed. Once it finds a promising one, it goes into the vent while making a 3D map of the interior of the vent. If it finds the path is blocked either by obstacles or a decreased width of vent, it backs up and tries another vent. If it successfully goes past the chocking point, which is the most critical point for mobility, this vent is selected for further study. Some scout DM stays in the vent and continue the mapping for volcanology study as well as to create a map for following DMs. If DM's middle section is interchangeable, some scout DMs return to SM and get converted to sample return or in situ science DMs.

13.2.2.3 Sample Return and In Situ Science

SM deploys sample return and in situ science DMs. They follow the optimal path identified by the scout DMs. The sample return DMs collect an ice core, dust, and potentially liquid samples in the vent/conduit, and return to SM. The samples are analyzed by the mass spectrometer on SM to look for signs of life. Sample return DM has a binary nature: if it can successfully return to SM, there will be a significant science gain. Otherwise, the science gain is zero. Having multiple DMs helps hedging the risk. DMs pick sample from a diverse range of depth. Some pick samples from a shallow depth in order to secure minimum science return. Some try to return samples from a greater depth—a high-risk, high-return trip. Each DM can make multiple trips. Once a baseline goal is achieved, DMs can aim for more challenging destinations or sample from different vents, if there are multiple accessible vents.

The in situ science DMs perform scientific observations along the way, and go as deep as possible to meet the liquid interface. Interchangeable payload gives flexibility in in situ science. For example, if scout DMs find that the liquid interface is within an easily reachable depth, we can put microfluidics package on in situ DMs to detect life in liquid. The observations by an in situ DM can inform following DMs to choose the best instruments (Fig. 13.6).

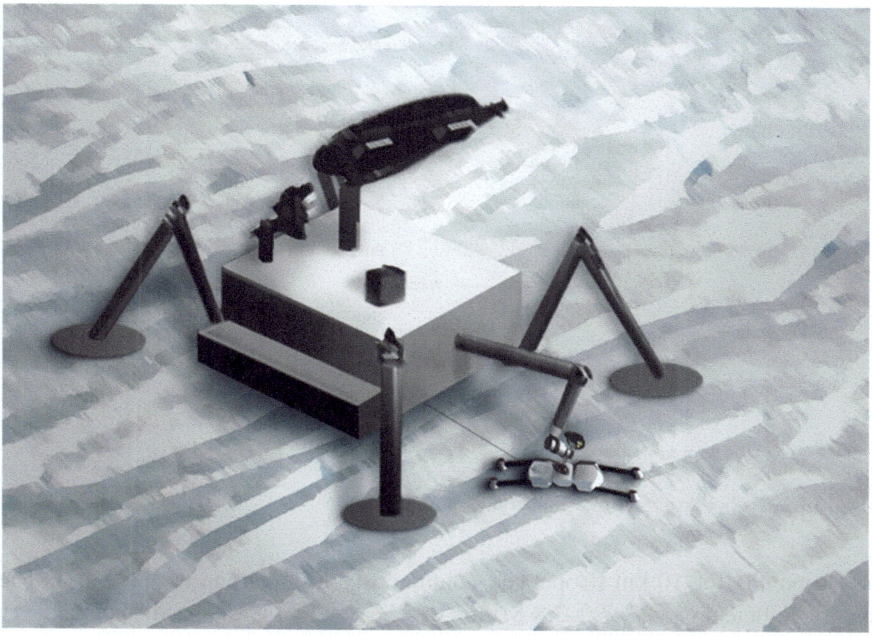

Fig. 13.6 Sample retrieval from a sample return DM (artist's concept)

13.2.3 Comparison with Other Concepts

While the primary goal of EVE is not to access the subsurface ocean but to assess the accessibility of the ocean, it would be useful to compare existing concepts that aim to access the subsurface ocean because it helps understanding the advantage and disadvantage of using the vent-conduit system as a pathway to the ocean.

13.2.3.1 Thermal Ice Probe (cryobot)

Cryobot (Zimmerman et al. 2001) is a technology concept to melt through the ice shell. It is a torpedo-shaped vehicle as shown in Fig. 13.7. While various designs have been proposed, (Zimmerman et al. 2001) provided a detailed study and design of Cryobot for Europa and the Mars polar cap. Their design is 1.25 m in length and 12 cm in diameter, which melts ice using both passive heaters and a water jet heated by 1 kW of heat produced by RTG and descends by its own weight. Communication is provided by a surface module through a tether, or alternatively, a series of releasable communication pucks. The greatest advantage of Cryobot is its simplicity. However, melting ice is very energy inefficient, because heat quickly dissipates through ice. Zimmerman et al. (2001) estimate that, with a 1 kW heat source, of which 0.6–0.8 kW can be used for melting, the rate of progress is 1 m/h,

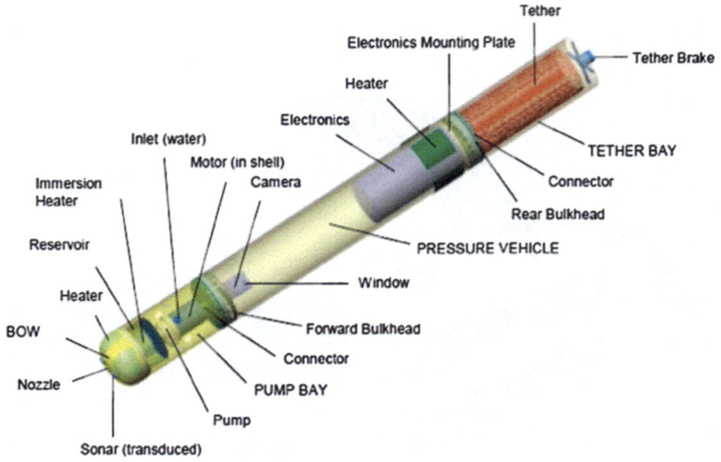

Fig. 13.7 Cryobot proposed by Zimmerman et al. (2001)

or 14 months for 10 km if operated continuously Zimmerman et al. (2001). The energy efficiency is ∼7.0–2.8 MJ/m of descent [The speed estimate by Zimmerman et al. (2001) appears to be wrong because the 2.8 MJ/m energy requirement is less than the energy needed for melting ice with a 12 cm diameter probe (3.8 MJ/m). Furthermore, this does *not* take heat dissapation into account. For these reasons, we suspect that, in practice, the speed of the melt probe would be an order-of-magnitude slower than their 1 m/hr estimate]. Sediment accumulating in front of the vehicle is also a challenge, but mitigated with water jets. In addition, while this concept works under Europa's gravity (0.13 g), it is not clear if it works under Enceladus's weaker gravity (0.01 g).

13.2.3.2 Mechanical Ice Probe

The ice probe concept proposed by Wilcox et al. (2017) features a circular saw blade that protrudes through a slot in a hemispherical turret dome at the bottom of the probe, such that the blade cuts the ice as it spins. The ice chips would be thrown up through the slit into the turret and would be melted by the heat provided by General Purpose Heat Source (GPHS) modules containing heat-source plutonium dioxide. The meltwater drains into sumps on either side of the sawblade, from which the meltwater would be pumped out to the rear of the probe. The main body of the probe contains a spool of aluminum tubing that would be dispensed from within the probe to the surface module lander. This tubing is nominally 1–3 mm in outside dimension with integral insulated electrical wires around the center hole. This tube would pneumatically transport small (mm-scale) single-use canisters containing meltwater samples from the probe to the surface for analysis.

Mechanically cutting ice is significantly more energy efficient than melting it. Required energy is 21.1 MJ/m^3 for mechanical cutting and 333 MJ/m^3 for melting *without heat dissipation*. While the ice is still melted, this concept is more energy efficient than Cryobot, because the ice chips are melted inside the insulated interior of vehicle, hence heat does not dissipate to the ambient ice. With a 1 kW GPHS on the ice probe and a multi-purpose RTG (MMRTG) on the surface module, (Wilcox et al. 2017) estimate the probe can reach 10 km deep in $\sim$27 months. A unique advantage of this concept is that the liquid sample can be sent to the surface module for analysis, which allows use of instruments that can be neither miniaturized nor sterilized sufficiently for descent (Fig. 13.8).

13.2.3.3 Comparison with EVE

The two concepts introduce above share several system-level features with EVE. They have separate descent and surface modules connected by a tether or a tube. They also propose RTGs as the power source. Given the difficulty in communication through the ice, as well as the scarcity of sunlight, these features would be natural choices.

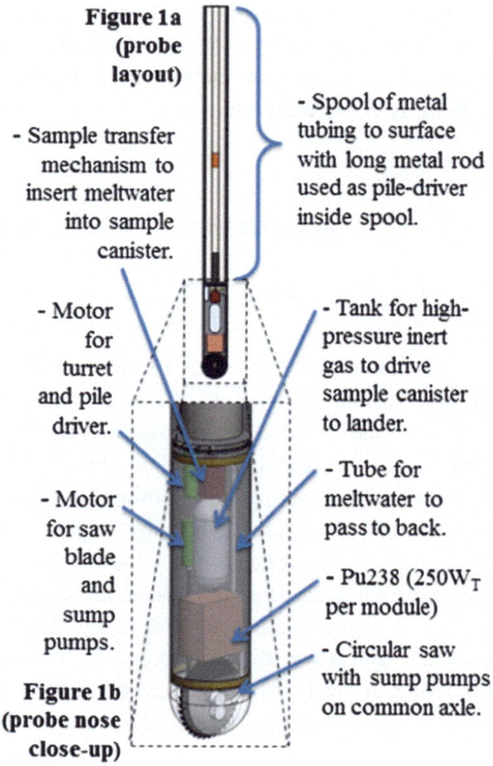

Fig. 13.8 Mechanical ice probe proposed by Wilcox et al. (2017)

The energy efficiency of Eve is orders of magnitude better than thermal and mechanical ice probes because it does not need to melt or mechanically cut the ice. The required energy depends on the dynamic pressure. With our prototype design presented in Sect. 13.6, the required energy is $\sim$3 kJ/m for 10^3 Pa of dynamic pressure and $\sim$6.5 kJ/m for 10^5 Pa. The required energy does not scale linearly with dynamic pressure because the energy needed for inserting and removing ice screw is constant over most of the range. Detailed derivation of the energy estimate is described in Sect. 13.4.2. Assuming 10 W is available for mobility, the speed of DM is $\sim$12 m/h at 10^3 Pa and $\sim$5.5 m/h at 10^5 Pa. With this rate, descending 10 km takes about a month at 10^3 Pa and 2.5 months at 10^5 Pa, if operated continuously.

Highly unknown vent size and dynamic pressure, particularly with the "cryovolcanic" model, are the greatest risk for EVE. Another major shortcoming of EVE is mechanical complexity, which in turn requires a high level of machine intelligence. However, the reduced reliability due to the mechanical complexity can be mitigated by operating multiple DMs.

A unique advantage of EVE is its ability to investigate the interior of vent-conduit system, which itself is a subject of intense scientific interest. The vent-conduit system on Europa, if exists, is considered a potentially habitable zone (Fugueredo et al. 2003), and likewise that of Enceladus could be habitable, too. If that is the case, the vent-conduit system could be the most accessible habitable zones on icy moons, from which the first discovery of extraterrestrial life could be made. However, as we will describe shortly in Sect. 13.3, the eruption mechanism, which couples with geometry, environment, and habitability of the vent-conduit system, is completely unknown.

13.3 Characterization of Environment in Enceladus Vents

13.3.1 Overview

Among numerous unknowns of Enceladus geysers, the one that has the greatest impact on our concept is the eruption mechanism. While many hypothetical eruption models have been proposed, most of them fall into either of two categories: relatively calm "boiling" models and more dynamic "cryovolcanic" models. Very roughly speaking, the "boiling" models assume that liquid water boils into vapor under the surface, while the "cryovolcanic" models assume that the mixture of liquid and gas ascends the conduit and accelerated as the gas expands. More detailed explanations of the two models will be provided shortly.

The "boiling" models are more favorable for EVE because it results in a greater vent size and lower dynamic pressure. The worst-case dynamic pressure is $\sim 10^4$ Pa, which is well within the design range of DM. While the vent size is harder to constrain, the "boiling" model gives order-of-magnitude greater estimates than the "cryovolcanic" model in general. For example, Kite and Rubin (2016) suggest "slots" of greater than 1-m in width. *EVE is likely feasible with the "boiling" models.*

The "cryovolcanic" models are more problematic for EVE. Our preliminary analysis suggests that the dynamic pressure and the vent width could be anywhere between 10^3 to 10^7 Pa and 1–30 cm, respectively. The worst-case is beyond the upper limit for DM ($\sim 10^5$ Pa and ~ 10 cm). While some suggests 10^7 Pa dynamic pressure is likely unreasonable for geophysical reasons (D. Hemingway, pers. comm.) and mode detailed analysis could better constrain the worst cases, we do not reject possibilities. Therefore, the feasibility of *EVE is undetermined with the current best knowledge with the "cryovolcanic" models.*

13.3.2 Observed Data

The *Cassini* mission collected a limited amount of data during the performed flyby. An analysis of the chemical composition of the plume based on Cassini Ion Neutral Mass Spectrometer (INMS) data Waite et al. (2011), reveals an approximately 8–10 molar% non-H_2O fraction dominated by CO_2, as well as other volatiles and salts. The Ultra-Violet Imaging Spectrometer (UVIS) shows an expansion jet of the eruptive gases to be consistent with an internal sound speed of between Mach 5 and 8 Hansen et al. (2011). Co-analysis of UVIS and Image Science Subsystem (ISS) imagery reveals that the solid:vapor ratio likely is within the 0.35–0.70 range (Ingersoll and Ewald 2011), which does not particularly favor either "boiling" (vapor-dominated) or "cryovolcanic" (vapor-poor) models, and a total eruptive mass of order ~ 200 kg/s over the entire system. Gigawatts of thermal emission detected by the Composite Infra-Red Spectrometer (CIRS) reveals elevated temperatures and emissions in the tiger-stripes region in excess of what can reasonable be supplied via conductive losses from erupting materials through the walls of the conduit. Detection of higher temperature emissions very close to the tiger-stripes, by the Visual and Infrared Mapping Spectrometer (VIMS), highlight an unusually-elevated surface temperature Goguen et al. (2013) of approximately 200 K (compared with ~ 70 K typical surface temperature) within a few meters of the tiger stripes. It remains unclear at this stage whether eruptions emanate from point sources along the tiger stripes, or as curtains (Spitale et al. 2015), or perhaps both, with focusing of non-vapor components in the point sources but a more general diffuse emission of water vapor from along the stripes.

For the worst case cryovolcanic scenarios listed, most of the thermodynamic starting conditions assume a conduit and eruption temperature of 0 °C, i.e. buffered by the latent heat of solidification of water-ice. For boiling models, the temperature profile may be considerably lower.

13.3.3 Estimation of Dynamic Pressure and Vent Size

13.3.3.1 Dynamics Assumptions

The physics models used to describe "cryovolcanic" models differ from the "boiling" models.

"*Boiling models*" [e.g., Kite and Rubin (2016)] assume that the thermodynamic starting points of the erupting fluids begin at a higher entropy state (e.g., gas) for the boiling models, and theorize the presence of an interface somewhere between Enceladus' surface and the underground ocean, the liquid surface is exposed to a pressure close to or below the pressure of the triple point of water. Due to the low pressure the water is forced to boil and evaporate, providing the matter expelled from the vents. In this instance, the eruption velocities of different phases (water vapor, non-water vapor, and water solids/liquids) may differ considerably. These models do not easily explain why the sampled data shows high percentage exsolved substances such CO_2 and salts that should be either concentrated within or depleted from the erupting materials. Most importantly, they are associated with larger vents, lower eruption pressures and, it turns out, less dynamic pressure (Fig. 13.9).

"*Cryovolcanic*" *models* (Matson et al. 2012; Porco et al. 2014), on the other hand, propose that liquids and dissolved materials ascent in a coupled state, driven by the exsolution and expansion of dissolved gases under a negative pressure gradient. Thermodynamically they start their ascent in a low entropy state (liquid water), and the gases, despite being a minor percentage of the mass ratio, eventually massively dominate the volume ratio and continue drag the water particles (containing dissolved materials such as salts) up to the vent. Only later when the static pressure reach the tripe point of pressure the liquid water changes state (the model doesn't take account of any latency in the process nor of any metastable phase that

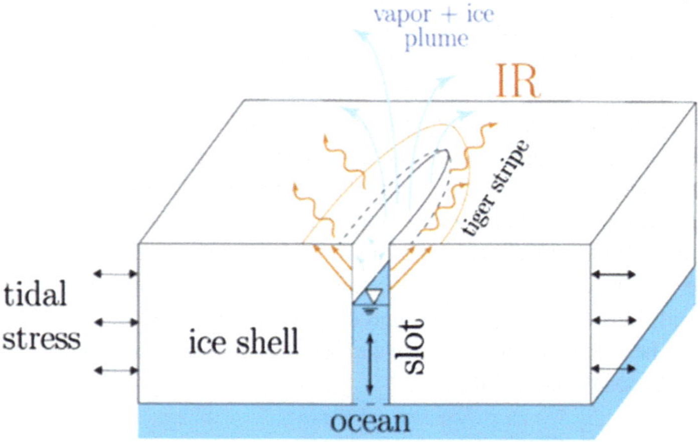

Fig. 13.9 A "boiling" model of Enceladus eruption by Kite and Rubin (2016)

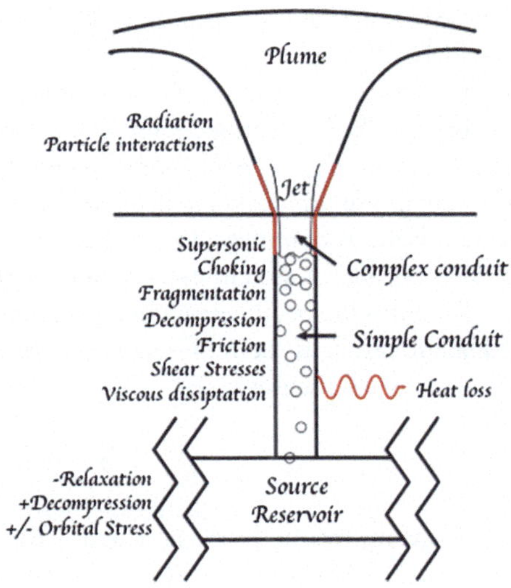

Fig. 13.10 A cryovolcanic model of Enceladus eruption

may be expected and become problematic in the development). The ratio of water changed in gaseous and solid state will then depend on the ratio of latent heat of fusion and evaporation so to preserve the local balance of energy.

Due to the presence of a supersonic regime it's likely that a supersonic nozzle exists somewhere within the system, most likely close to the surface, where the sonic condition is achieved. A possible variant of it is described by a parallel wall model in which the flow is kept at a under sonic velocity, in which case the pressure is almost certainly higher, before reaching the speed of sound at the surface and quickly decompressing accelerating into the vacuum of space before reaching a much higher velocity (Fig. 13.10).

13.3.3.2 Vent Dynamic Model

A fluid in motion exerts a pressure that depends both on the static pressure, p, and on the pressure caused by the energy of its motion, its dynamic pressure, q, where:

$$q = 1/2\,\rho u^2, \tag{13.1}$$

where u is the velocity and ρ is the density. For high speed systems, however, an alternative formulation can be defined with respect to the speed of sound. By applying the ideal gas law, as well as definitions of the speed of sound and the Mach number, M, it can further be shown that:

$$q = 1/2\,\gamma p M^2, \tag{13.2}$$

where γ is the ratio of specific heats. We consider three approaches to constraining upper limits to q.

Maximum exit velocity: We consider a scenario in which most of the acceleration occurs in the subsurface and water erupts at sub-triple points pressures (vapor + solid). This is most likely where pressures at the exit are low, which is more (although not uniquely) consistent with "boiling" models; Certainly, gas slow would have to dominate. In this case, we use (13.1), assume $p < 611$ Pa and, from studies of the jet structures (Arakawa and Maeno 1997), $5 < M < 8$.

The ratio of specific heats of a pseudogas mixture, γ_{mix}, can be determined using adiabatic ratios of the individual components w.r.t. pressure and volume, weighted by their mass fractions, w:

$$\gamma_{mix} = \sum_i \left(w_i c_{p,i}\right) / \sum_i \left(w_i c_{v,i}\right) \tag{13.3}$$

Based on results from the Cassini Ion Neutral Mass Spectrometer (INMS), we can assume that a proportion of the plume is exsolvable non-water volatiles, and based on compositions observed by *Cassini*'s E5 and E7 Enceladus fly-bys (Asbeck et al. 2006) we consider it reasonable to assume 8% (molar) CO_2 as a proxy for all non-water volatiles. CO_2 and H_2O water vapor at 0 °C have specific heats capacities at constant pressures of $c_{p,}CO_{2v} = 800$ J kg^{-1} K^{-1}, $c_{p,H2v} = 1860$ J kg^{-1} K^{-1}. Specific heat capacity at constant volume, c_v, can be derived using the ratios of specific heat at 0 °C, $\gamma_{CO_2} = c_p/c_v = 1.31$ and $\gamma_{H_2O} = c_p/c_v = 1.33$. For water-ice, $c_p = c_v = 2100$ J kg^{-1} K^{-1}. Thus, making no assumptions about the relative quantities of CO_2 and H_2O, we find $\gamma_{mix} \sim 1.05$. Note that there is an inherent assumption that exit velocities are the same for gas and non-gas phases, which is flawed but doesn't make a massive difference to the model.

Application of (13.2) gives $q < 2.05 \times 10^4$ Pa, equivalent to a terrestrial surface wind speed of 185 m/s (414 mph). This is an over-estimate, as some acceleration will occur below triple point pressures, but given that this is not at the higher end of our dynamic pressure space we do not consider it critical for this study.

Choked eruption: Another estimate can be derived by considering the dynamic pressure at Mach 1, a likely condition at the surface (surface choked) or at the narrowest point of a converging-diverging nozzle (the "throat"). For the same static pressure as above, the maximum value for q would 1–2 orders of magnitude less than that for the above scenario using (13.2) (The difference corresponds to the "Max v" line and the colored lines in Fig. 13.11b. However, in this case it is useful to determine the solution semi-analytically for higher unknown pressures using Eq. (13.1), as u is known to be equal to C, the mixed-phase sound speed velocity, an approximation for which is (Lorenz 2002):

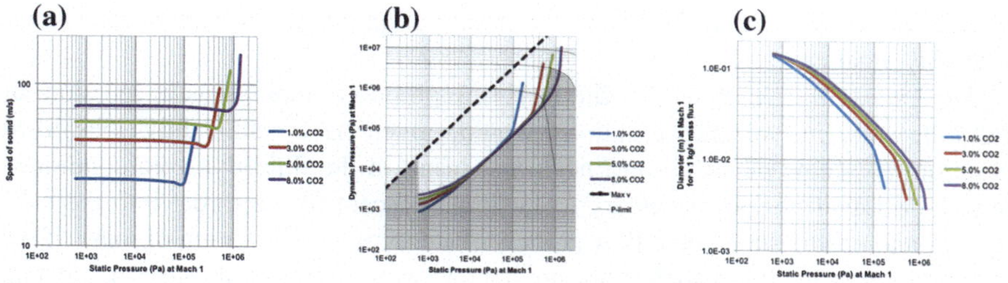

Fig. 13.11 **a** Sound speed, **b** dynamic pressure and **c** diameter for a CO_2-driven eruption at Mach 1 (surface choked or in nozzle throat). CO_2 is used as a proxy for all volatiles, and the lowest and highest values (1 and 8 molar%) both seem unlikely, given that CO_2 has a considerable molar weight. Upper limit at ρ = 920 kg m^{-3} (ice density), lower limit at p = 610 Pa (triple point of water). Results for an upper Mach 8 velocity (marked MPV) are also included. The diameter is for a radially-symmetric conduit with an eruptive mass flux, $\dot{M}$, of 1 kg/sec; From conservation of mass ($\dot{M}$ = rho.u.A), diameter scales with $\sqrt{\dot{M}}$ for different mass fluxes . In **b**, we assumed that the ocean is at 4 MPa, and that there is 5 molar% of CO_2. Anywhere that is shaded grey is "plausible". To the right of ~600 Pa static pressure is all cryovolcanic conditions at Mach 1 (where the dynamic pressure is highest). To the left of ~600 Pa is all boiling models

$$
\begin{aligned}
C &= (AB)^{-1} \\
A &= \left[\sum_{\mathrm{i}} (\phi_{\mathrm{i}} \rho_{\mathrm{i}})\right]^{0.5} \\
B &= \left[\sum_{\mathrm{i}} (\phi_{\mathrm{i}} / \rho_{\mathrm{i}} c^2)\right]^{0.5},
\end{aligned}
\tag{13.4}
$$

where c and ρ are the speed of sound and densities of the individual components.

As previously, we assume that CO_2 drives ascent, and that it exsolves from the H_2O in equilibrium according to Henry's Law (H^{cp} = 0.034 mol L^{-1} atm^{-1}). Water remains in a liquid state (>611 Pa), consistent with adiabatic liquid ascent (cryovolcanism). Results are given in Fig. 13.11 for a range of pressures and CO_2 starting molar fractions.

In this scenario, the dynamic pressure in the throat approximates the static pressure. Eruptions may only occur if the mean fluid density integrated from ocean to surface is less than the mean crustal density, i.e. the liquid is hydraulically buoyancy, and so results with $\rho > 920$ kg m^{-3} are rejected (curves are truncated); Further analysis will rein this in further. The highest dynamic pressures (>1 MPa) occur with high CO_2 contents and static pressures (also ~1 MPa), equivalent to a terrestrial ground wind speed of 1290 m s^{-1}. Such values may also be unrealistic if they exceed ocean pressures, or if exsolution of available CO_2 is suppressed (likely) due to supersaturation or formation of carbonic acid, which will be considered in future analyses. Furthermore, we can also relate static pressure to conduit cross-sectional area as a function of mass flux (Fig. 13.11 bottom), by considering conservation of mass ($\dot{M} = \rho\ u\ A$), where A is the cross-sectional area (N.B. $A = \pi$

r^2, where r is the radius). Note that for fissure eruptions the fissure width is considerably smaller.

Supersonic nozzle flow: Mitchell (2005) proposed that abrasive erosion of conduit walls in volcanic eruptions would, given as a sufficiently long-lived eruption, as seems plausible on Enceladus, result in the fluid dynamic conditions (e.g. a de Laval nozzle) necessary to facilitate descent of the throat into the subsurface and enable supersonic flow at the surface. From conservation of mass, if we assume isentropic flow, which is reasonable for a supersonic jet, through an ideal de Laval nozzle, it has been shown Liepmann and Roshko (2002) that:

$$\left(\frac{A}{A^*}\right)^2 = \frac{1}{M^2}\left[\frac{2}{\gamma+1}\left(1+\frac{\gamma-1}{2}M^2\right)\right]^{(\gamma+1)(\gamma-1)}$$
$$\frac{A^*}{A} = \frac{\rho u}{\rho^* u^*} = \frac{\left[1-\left(\frac{p}{p_0}\right)^{(\gamma-1)/\gamma}\right]^{\frac{1}{2}}\left(\frac{p}{p_0}\right)^{1/\gamma}}{\left(\frac{\gamma-1}{2}\right)^{(1/2)}\left(\frac{2}{\gamma-1}\right)^{(1/2)(\gamma+1)/(\gamma-1)}}. \quad (13.5)$$

where A is the conduit cross-sectional area, and superscript * refers to the condition at $M = 1$.

Hence it is possible to track how pressure, Mach number and cross-sectional area change relative to their values at the throat, which was determined above (Fig. 13.11). Dynamic pressure is presented, relative to the throat as a function of expanding vent/jet angle (Fig. 13.12), demonstrating that the throat is likely to be the location of peak pressure (it may be a few percent higher within a few vent radii of the surface). It is clear that dynamic pressure is at its peak in the throat at Mach 1. Note that this analysis does not consider the consequences of phase change after this condition, which is inevitable during decompression through the triple point.

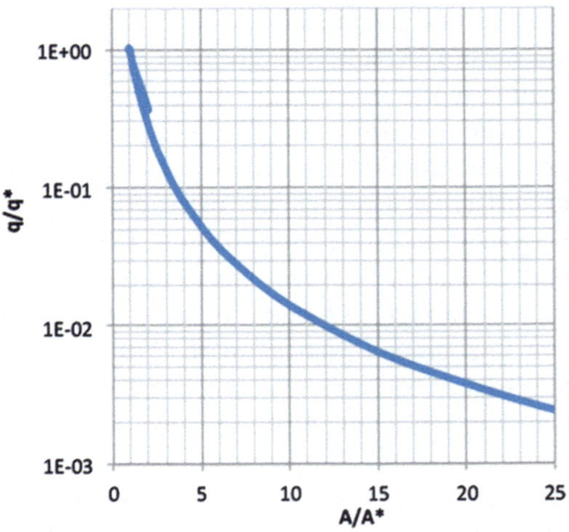

Fig. 13.12 Relationship between dynamic pressure and conduit cross sectional area in a de Laval nozzle assuming isentropic flow and negligible phase changes, both w.r.t. the condition at Mach 1 (* annotated). The short tail is for elevations below the throat, where flow is subsonic

Stagnation pressure: Finally, we can demonstrate an additional limit to dynamic pressure on the basis of conservation of energy, thus:

$$T\,\mathrm{d}s - P\,\mathrm{d}v + u\,\mathrm{d}u - g\,\mathrm{d}z - \sum_{\mathrm{i}} \mathrm{d}w_{\mathrm{i}}\mathrm{d}h_{\mathrm{i}} = 0, \tag{13.6}$$

where T is the temperature, s is entropy, v is specific volume, g is the gravity (−0.113 m/s), w_i is the mass fraction of a material i changing state, and h_i is the enthalpy of that change of state.

We know that $\mathrm{d}v = 1/\mathrm{d}\rho$. We can assume that g dz small and positive, that velocity is 0 at source and positive in the vent, that ds is positive (second law of thermodynamics), and that all changes of state are endogenic hence $\Sigma_\mathrm{i}\ \mathrm{d}w_\mathrm{i}\ \mathrm{d}h_\mathrm{i}$ is positive, thus:

$$P^* - P_i < u\,\mathrm{d}u\,\mathrm{d}\rho \tag{13.7}$$

where i refers to the condition in the source reservoir. Finally, integrating the right-hand-side (assuming a fairly linear correlation between du and dρ which should be reasonable to 1st order), assuming that velocity is 0 at source, and substituting in for the definition of dynamic velocity we find that:

$$P^* + q < P_\mathrm{i} \tag{13.8}$$

P_i represents the concept of stagnation pressure in fluid dynamic texts. Thus, we demonstrate that the concept of stagnation pressure does apply for this situation. Upper limit curves due to the stagnation pressure concept are added as thin, dark lines fin Fig. 13.11b; Note that the 10 MPa pressure line is likely unreasonable for geophysical reasons (D. Hemingway, pers. comm.), but demonstrating this was beyond the scope of what was plausible in this study.

Furthermore, we can define the effective stagnation pressure, $P_{\mathrm{s,eff}}$ as:

$$\left(P_{\mathrm{s,eff}} - P_0\right)\mathrm{d}v = -T\,\mathrm{d}s + g\,\mathrm{d}z + \sum\nolimits_{\mathrm{i}} \mathrm{d}w_{\mathrm{i}}\mathrm{d}h_{\mathrm{i}} - Q'. \tag{13.9}$$

where Q′ represents loss of energy from the system (most likely by conduction), and use this to estimate how far below the stagnation pressure the sum of $P_s + q$ is. The worst-case scenario is one in which the flow is isentropic (T ds = 0 kJ/kg), ascent starts at the top of the deepest conceived ocean (g dz = 4.62 kJ/kg). It is more difficult to put universally applicable limits on P_0 and dw_i, but we hope to do more in the future. If we assume that solubility is well described by Henry's Law, and that CO_2 is at saturation at source, we can estimate an upper limit for dw_i as a function of P_0, but we can't escape the possibility that the lower limit for dw_i is negligible.

Summary: Taken together, there exists a very wide range of dynamic pressures to account for. Every curve on Fig. 13.11 should be considered an upper limit, and furthermore a conservative one. Within that context dynamics pressures as great as

$\sim$9 MPa must be considered, although if we accept the assertion of D. Hemmingway (pers. comm.) that 10 MPa is far too great an ocean pressure then $\sim$3–4 MPa may be more reasonable. This is considerable, and the equivalent to a surface wind on Earth of around 2500 m/s, which sounds extreme. It is, of course, also considerably less than in most terrestrial volcanic eruptions. For the Kite and Rubin's "boiling" model, we would anticipate far lower, trivial dynamic pressures.

13.3.3.3 Diameter Estimates

The cross-sectional areas for "boiling" models are generally quite high. Kite amd Rubin (2016) suggest "slots" of greater than 1-m in width, which is not considered problematic. However, for cryovolcanic models, especially those with high pressure surface conditions, vent diameters tend to be way smaller. From conservation of mass, the mass flux, $\dot{M} = \rho\, u\, A$, and so if we assume an ideal gas we can determine the cross-sectional area and hence the diameter as a function of mass flux. This is illustrated in (Fig. 13.11) for a 1 kg/s vent. Note that the ideal gas equation is an imperfect predictor of the relationship between pressure and density, but that in this case it will give higher density for a given pressure than alternative formulations, and hence predicts the smallest (i.e., worst-case) cross sectional areas. As diameter is proportional to the square root of the mass flux, a perhaps more reasonable flux through a vent is 2.5 kg/s, based on observations by Porco et al. (pers. comm.) who identify of order $\sim$100 the number of discrete point sources within the $\sim$500 km of tiger stripes. Hence, within the parameter space being explored, we find 10–300 mm is plausible, and selection of the most active vents (in terms of mass flux) could give larger diameters yet. Note, however, that this is a lower limit range, and we suspect that the $\sim$10 mm end is implausible; It's just that we haven't had the time to more fully explore the problem, which remains unpublished. We decided, in absence of a more refined model, to keep the 100 mm diameter limit as reference.

13.3.4 Suggestions for Future Study to Better Constrain Key Parameters

Further study of *Cassini* data could better constrain parameter estimates of each model. For example, resolving between discrete "point" sources and more extensive "fissure" eruptions may allow certain models to be tested. However, in order to resolve between "boiling" and "cryovolcanic" models, additional orbital observations are likely needed. For example, the "boiling" model of Kite and Rubin (2016) seems to require a "slot" geometry, i.e. a fissure, while the cryovolcanic models are not as thermally stable as a fissure. Observing mass flux with carefully controlled orbital geometries, as well as the solid/vapor ratios along the tiger stripes, preferable at <10 km would be valuable.

In the near-term, there is still much we can do with existing data. For example, characterization of the upper limits of ocean pressure and volatile fractions could put additional constraints on dynamic pressure in the vent. Multi-species thermochemical/exsolution models would allow detailed tests of current conceptual models for ascent and eruption, and may allow for enhanced interpretation of plume mass spectrometric results.

Capturing the dynamic conditions at as high a spatial resolution as possible within individual jets should allow testing between different models, as well as constraints upon those individual models. This will require careful instrument design and selection, and very specific geometric constraints on observations. However, if a cryovolcanic model is determined to be true, additional constraints will need to be determined, as the range of exit pressures (and hence area and dynamic pressure) are extremely large.

Very high resolution imaging combined with extremely sensitive passive emission spectrometry at submm/THz wavelengths at optimized phase angles may give the highest potential for resolving this. The former allows analysis of the structure of the solid particle components of the plume, much like with Cassini ISS. The latter allows simultaneous characterization of optical depth of different polar molecules in the vapor phase (especially water), as well as some constraints on pressure and temperature; with sufficiently high resolution observations it may be possible to truly constrain the fluid mechanics of observed jets; The alternative, UV spectroscopy (as used by Cassini UVIS), struggles to resolve the effects of temperature and pressure, works in a transmissive mode which requires solar or stellar occultations and thus further constrains observation geometry.

13.4 Trade Studies

We performed trade analyses in six domains: mobility, power, tether, autonomy, perception, and science instrument systems.

13.4.1 Mobility System

13.4.1.1 Mobility Type

We first performed a high-level trade-off on down-the-vent mobility. We considered the following three options:

1. No control (falling)
2. 1-D mobility (rappelling)
3. 2-D mobility (climbing) (Fig. 13.13).

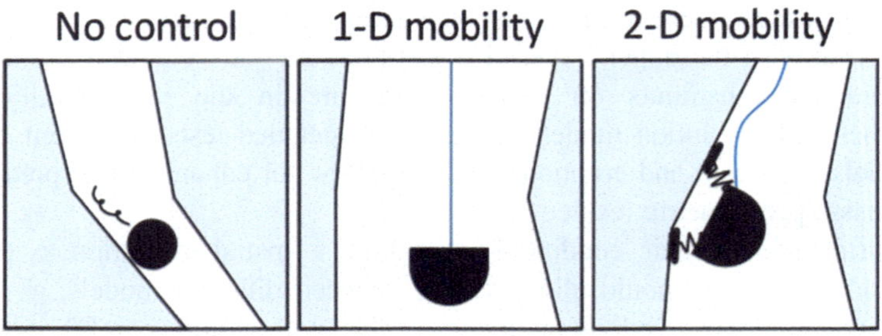

Fig. 13.13 Down-the-vent mobility options

In the "no control" option, DM goes down the vent by solely depending on the gravitational force, with no controllability of its position and speed (e.g., drop sphere). The "1-D mobility" option also solely depends on the gravitational force to go down, but DM's depth and descent speed is controllable through unreeling or reeling in tether. In the "2-D mobility" option, DM can traverse on the vent walls through a mechanism to attach to the surface.

We conclude that 2-D mobility is the only option that can be feasible. The table below summarizes the analysis result. The details of the analysis are described in the rest of this section.

Feasibility of the "no-control" (falling) option: The "no control" option (falling) is infeasible in Enceladus vents even with the most optimistic estimate of the dynamic pressure, due to Enceladus's weak gravity (0.011 g). On Europa, it is potentially feasible because of the significantly greater gravity (0.13 g), but the dynamic pressure and vent geometry is unconstrained and hence it is not possible to quantitatively assess the feasibility.

Our analysis is based on assumptions that (1) DM is a sphere with 5 cm radius (r), (2) DM's density (ρ) is 2000 kg/m^3 (twice of liquid water), and (3) the dynamic pressure (P_{D}) is 1000 Pa (the most optimistic estimate in the "cryovolcanic" model). Given that Enceladus's surface gravity (g) is 0.11 m/s^2, the upward force provided by the gravity is:

$$F_{\mathrm{Down}} = \frac{4}{3}\pi r^3 \rho g = 0.24\,\mathrm{N}.$$

The upward force provided by the flow is:

$$F_{\mathrm{Up}} = \pi r^2 C_D P_D = 1.9\,\mathrm{N},$$

where $C_D = 0.47$ is the drag coefficient of a sphere. The upward force is significantly greater than the upward force, hence the drop sphere cannot go down the vent.

In order to make the "no control" option feasible, it has to be either: (a) greater than $\sim$80 cm in radius or (b) heavier than $\sim$30,000 kg/m^3 in density. (a) is not viable given that the lower bound on the vent width is ~ 10 cm. (b) is not possible as it is denser than the densest element (Osmium).

Feasibility of the "1-D mobility" (rappelling) option: Rappelling is not feasible on Enceladus as it solely depends gravity to provide downward force (thus the same argument as drop package applies). It could be a feasible option on Europa, but again quantitative feasibility study is not possible due to unconstrained vent conditions. It also has difficulty in going through non-straight vent.

Feasibility of the "2-D mobility" (climbing) option: Climbing has a potential of being able to go down the vent because (1) it can mechanically provide additional downward force and (2) it can navigate through a non-straight path. The feasibility of this option primality depends on (1) the maximum downward force that the attachment mechanism can provide, and (2) the availability of power. The major risks of 2-D mobility include mechanical complexity and the need for intelligent autonomy capability to control the complex mechanism. Further analysis of 2-D mobility will be performed in the rest of this section. The autonomy system will be discussed in Sect. 13.4.4.

13.4.1.2 Attachment Mechanism

As mentioned previously, the major challenge for 2-D mobility is the significant uncertainty in the condition of vents, particularly in the dynamic pressure. We identified multiple attachment mechanisms, and assessed their robustness to the uncertainties.

We considered the following attachment mechanisms:

1. **Ice screw**: An anchoring device commonly used for ice climbing and crevasse rescue, shown in Fig. 13.14. Typically, ice screws present a hollow center that reduce the amount of energy required in the installation, as well as three to four teeth tips to start the hole and a triangular thread. The shape allows the thread to self-feed itself once the first thread is fully inside the ice, which allows to release the preload required to start the hole. Ice screws are available in different length and are typically rated for a max load of 10 kN (Beverly and Stephen 2009).
2. **Microspine gripper**: Initially developed at Stanford (Asbeck et al. 2006) and later at JPL (Parness 2011), the micro-spines gripper uses numerous tiny spines to engage asperities on the surface. It has been successfully applied to LEMUR rock climbing robots (Parness et al. 2012), and is planned be used in the proposed Asteroid Redirect Mission to grasp and secure a boulder.
3. **Cam/wedging**: Pushes against opposing walls in the vent to create a hold using an over-center mechanism.
4. **Archimedes screw**: Uses two rotational inputs to screw itself down the vent. It has been used for screw propelled vehicles to drive on soft surfaces including snow.
5. **Melt anchors**: Uses heat to creates a hole on the ice, which the element itself is inserted and act as anchoring (Fig. 13.14). It requires a combination of several

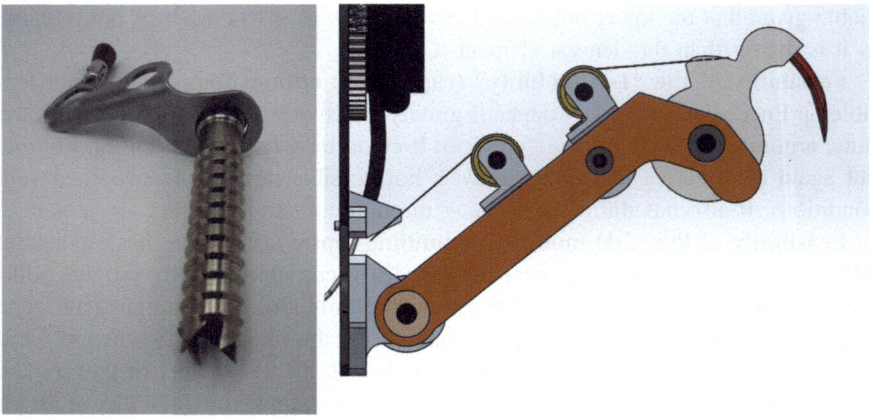

Fig. 13.14 Left: Ice screw, right: JPL's melt anchor

probes to establish a single grip. The technology has been developed for establishing a stationary grip for icy moon in low gravity environments, such as a surface walking robot for Europa.

These options were compared by five criteria as listed below. The first four is the robustness to uncertainties in the environment of the vent-conduit.

(a) **Resistance to dynamic pressure**: As explained in Sect. 13.3.3, the dynamic pressure in the vent is highly uncertain. The ability to tolerate extremely high dynamic pressure is important.
(b) **Robustness to unknown mechanical properties of ice**: The state of ice can range from pristine ice (strongest) to powdery snow (weakest), and the mechanical properties (e.g., strength) varies significantly. The mechanical properties of ice in the vent is unknown. Therefore, the attachment mechanism should work over a wide range of ice state.
(c) **Robustness to unknown geometry of vent**: The width of the vent is not only unknown but it could vary significantly along the depth. Therefore, the attachment mechanism should work with a wide range of vent width.
(d) **Robustness to unknown surface roughness**: The smoothness of surface of the vent wall is unknown. The attachment mechanism should robustly work on both smooth and undulated surfaces.
(e) **Energy/resource required**: In addition to the energy consumption, this criterion involves the required supporting mechanism such as robotic limbs.

The five attachment methods are compared by the Pugh matrix, in which options are scored by each of the criteria and ranked by the weighted sum of the scores (Burge 2009). Its objective is to evaluate a number of design candidates leading ultimately to which best meets a set of criteria. The Pugh matrix is given in Fig. 13.15.

Design	Factor Weight	Dynamic Pressure 5	Properties of the Ice 5	Geometry of the Conduit 5	Surface Roughness 5	Energy/Resources Required 5	Totals
Ice Screw		8	7	10	9	4	190
Microspine Gripper		6	3	10	2	6	135
Cam/Wedging		7	5	5	7	7	155
Archimedes Screw		8	6	6	8	8	180
Melt Anchors		8	7	10	9	3	185

Rank each design in each category. 1 is the worst, 10 is the best. The weights determine the relative importance of each aspect.

Hollow Interior
Penetrates Ice
Single Wall Capable
>4 Actuators required for movement (Assumes limbed muti DOF)

Fig. 13.15 Pugh matrix for comparing attachment mechanisms

The highest ranked mechanism is ice screw, while Archimedes screw and melt anchors result in comparable scores. Ice screw is chosen as the baseline in our concept, and given further investigations in the rest of this study. While Archimedes screw and melt anchors are also promising, we do not perform detailed investigation of these mechanisms due to the lack of time and resource of the study. The justifications of the scores in the Pugh matrix is given below.

Dynamic pressure: Ice screw and melt anchors can make the strongest anchor because it deeply penetrates into the ice. Archimedes screw can have a hollow interior and mitigate the upward force by the flow, thus it is given the same score as ice screw and melt anchors.

Properties of ice: Ice screw, Archimedes screw, and melt anchor are less sensitive to the properties of ice than the other options because they penetrate ice. We give a slightly smaller score to Archimedes screw than ice screw and melt anchor due to the relatively shallow depth of penetration.

Geometry of the vent: Cam/wedging and Archimedes screw must be in contact with opposing walls in order to be operational. As a result, they require relatively accurate prior knowledge on the width of the vent, and have difficulty in adapting to varying vent width. On the other hand, the other three mechanisms only have to be in contact with a single wall, hence significantly more robust to the uncertainty and variation of the vent width.

Surface roughness: While the surface state of the ice walls in the vent-conduit system is unknown, the surface is most likely smooth as the weak hoar frost layer, which often develops on ice cave walls on Earth, is blown away by the strong flow. Microspine grippers perform poorly on a smooth surface. On the other hand, ice screw and melt anchor can make a strong anchor on a smooth ice surface.

Energy/resource required: Ice screw, microspine gripper, and melt anchor are given low scores because they require at least four multi-DOF, fully actuated limbs. Ice screw and melt anchor need extra energy for drilling and melting ice, respectively.

13.4.2 Power System

13.4.2.1 DM Energy Requirement

We next make a first-order estimate of the energy requirement of DM. This is in part a preparation for the power system trade discussed in the next subsection. Based on the mechanical system trade study, we assume DM is equipped with four limbs and ice screw end effectors. Here we focus on the energy requirement for the mobility. Heating energy can be partially provided by RHUs (radioisotope heater units). We do not include energy requirement for instruments as it varies between different types of DM. Therefore, the estimate in this section must be viewed as an optimistic one, although mobility system is likely to dominate the energy consumption in most types of DMs.

DM is locomoted by repeating the following motions (Fig. 13.16):

1. Disengage the ice screw end effector of one limb, push the end effector by extending/contracting the limb, and engage the end effector again
2. Repeat the same for the other three limbs
3. Push the body forward by contracting the front limbs and extending the rear limbs.

Based on our prototype design described in Sect. 13.6, we assumed the stroke length of the limbs is 15 cm.

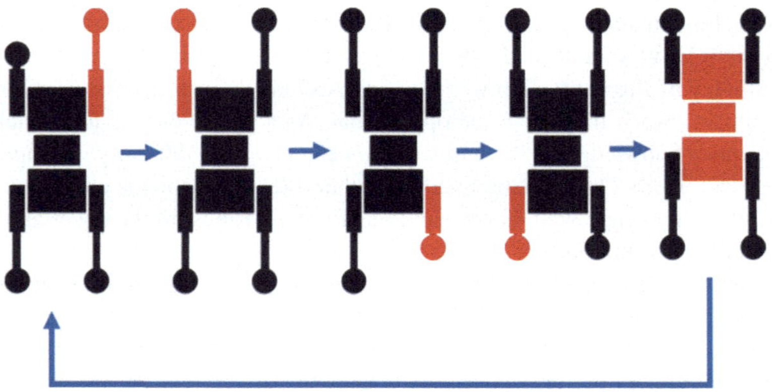

Fig. 13.16 DM's gait

Energy is required for the following three activities:

1. Insertion/removal of screw
2. Movement of limbs
3. Movement of body

For all the three categories, energy requirement increases with the dynamic pressure of the flow in the vent. A greater dynamic pressure increases the aerodynamic drag, which in turn increases the energy required for moving limbs and body against the flow. A greater drag also requires deeper insertion of ice screw, which in turn requires more energy for inserting and removing the screw. Aerodynamic drag depends on the shape and size of DM. We assumed the prototype design described in Sect. 13.6, and estimated the drag by a computational flow dynamic (CDF) simulation.

Figure 13.17 shows the resulting estimation of required energy for making 1 m of progress. The basis of the estimation and assumptions are described in detail in the following subsections.

CFD Analysis of Aerodynamic Drag: Since the analysis is supposed to be a first order study, we did not model the actual flow, considering the transonic condition or the different components. Instead, we analyzed the model in a simplified case, imposing a fixed dynamic pressure of 0.5 Pa, and linearly scaled the result. For the analysis, three 3 models where prepared. The first model was set with the limbs open at an angle of approx. 30°. The second model had the limbs parallel and closed. The third model represented the climber without the limbs. The same flow domain has been used for all the three models, then the forces have been extracted for each component.

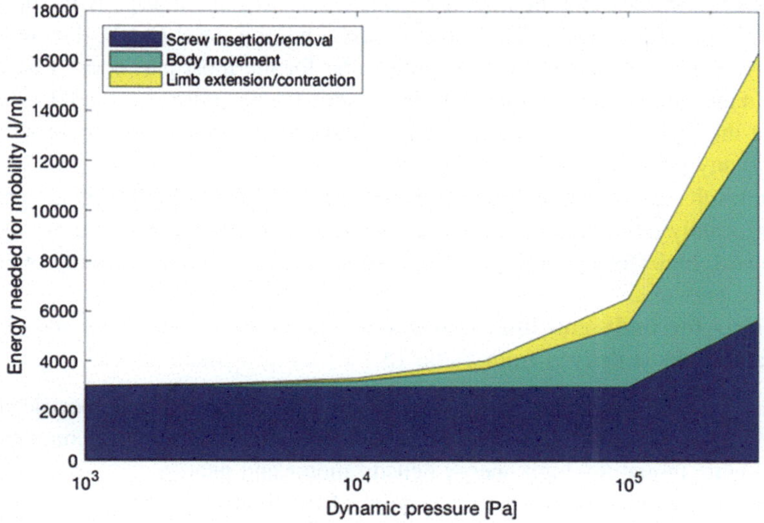

Fig. 13.17 Estimated required energy for a given dynamic pressure of the upward flow in the vent

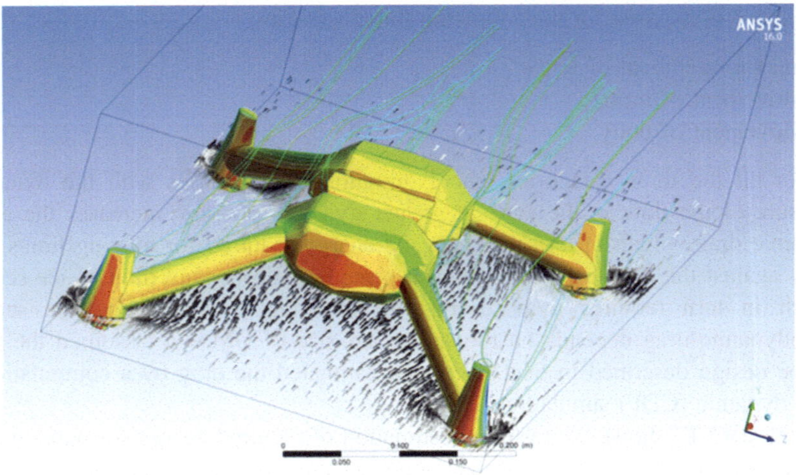

Fig. 13.18 CFD Analysis. Open legs configuration

Table 13.1 Drag force on DM for 10^5 Pa dynamic pressure, estimated by CFD

Aerodynamic drag (N) for 10^5 Pa dynamic pressure	Open limbs	Closed limbs	No limbs
Total body + legs	1671	744	
Total body	466	437	734

The model is a k-ε with an inlet speed of 1 m/s and a turbulence of 5% (standard parameter), the body and the floor are modelled as no slip surfaces, while the other walls are free slip surfaces. The convergence condition is set at 10^{-4} for mass and velocity. Since the model is very complex and for the most part doesn't act as an aerodynamic body, no inflation has been added to capture the boundary layer. Instead the mesh has been refined in proximity of the body, with more attention close to any peculiar feature (Fig. 13.18).

The results are shown in Table 13.1. Note that the closed limb formation results in significantly reduced drag. For the estimation of energy requirement, we assume the closed limb formation. We also assume the drag scales linearly with the dynamic pressure.

Energy for body and limb movement: The energy required for moving the body and limbs is derived from Table 13.1 with a few assumptions.

- The energy efficiency of motor and gear (ϵ_M) is 0.37 (from input electric energy to output mechanical work), derived from catalog specs of commercially available miniature, high energy density motor and gear.
- The energy efficiency of the linear joint of the limb (ϵ_J) is 0.8.

For the aerodynamic drag force (F_D) from Table 13.1, the required energy needed to move by 1 m is given by:

$$E = \frac{F_D}{\epsilon_M \epsilon_J}$$

Energy for screw insertion/removal: We found very limited literature on the quantitative analysis of required energy and strength of ice screw, probably because ice screw is primarily used not for scientific/engineering applications but for mountaineering. Therefore, we performed ice chamber experiments to find first-order estimates.

First, to estimate the required insertion depth of ice screws, we experimentally assessed the maximum shear force an ice screw can tolerate for a give n insertion length. The experimental setup is shown in Fig. 13.19. We used the same Black Diamond ice screw with 19.5 mm diameter. The result is shown in Fig. 13.20.

From the experimental results, we made the following assumptions:

- Insertion depth needs to be at least twice the diameter of screw in order to make an effective anchor
- With the minimum insertion depth, the 19.5 mm screw can support 900 N of sheer force
- With a greater insertion depth than the minimum, the maximum sheer force increases linearly at the rate of 125 N/mm

Fig. 13.19 Experimental setup to measure the tolerable sheer force of ice screw. The ice screw is inserted to ice and pushed downwards by a pressing stand

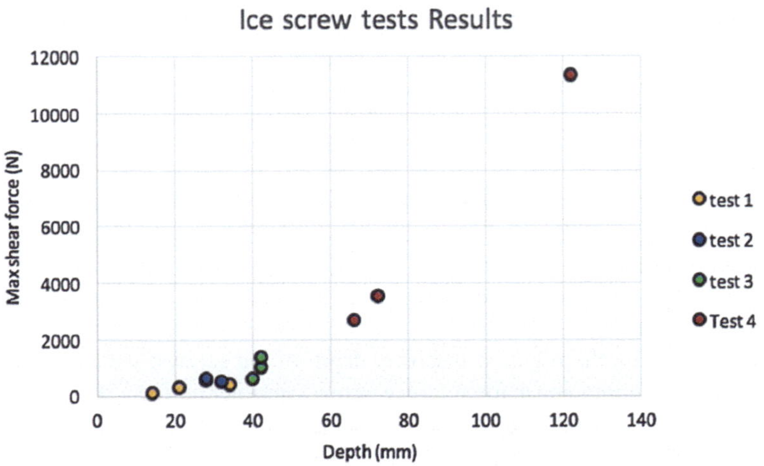

Fig. 13.20 Result of experiments, showing the maximum sheer force for a given insertion depth

- For a screw with different diameter (D) and insertion depth (L), the maximum sheer force is proportional to DL
- The energy efficiency of motor and gear is 0.37 (from input electric energy to output mechanical work), derived from catalog specs of commercially available miniature, high energy density motor and gear.

With these assumptions and the CDF results in Table 13.1, we computed the necessary insertion depth.

Next, we performed experiments (Fig. 13.21) to assess the energy required for inserting and removing ice screw for a given insertion depth. We used a 10 cm ice screw manufactured by Black Diamond, which is 19.5 mm in diameter, and a pristine water ice. The maximum torque was ~3.6 Nm. It required ~500 J of energy to insert 8 cm of the screw and then remove it.

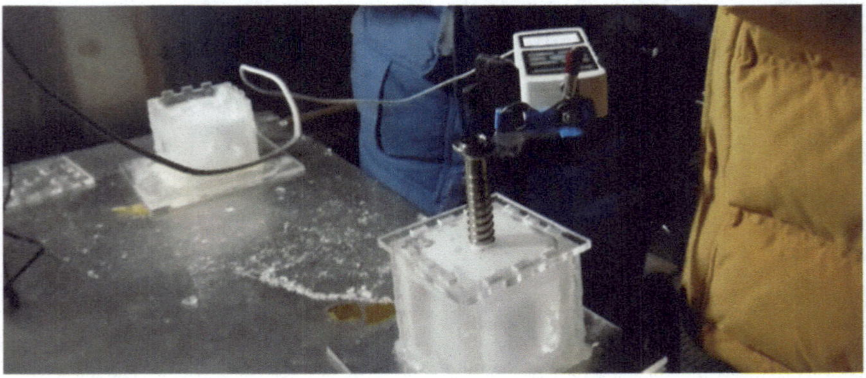

Fig. 13.21 Experiment to measure the energy required to insert and remove an ice screw

The ice screw used for DM is smaller than the one used for the experiment. To estimate the energy for different screw diameter (d) and insertion depth (l_I), we conservatively assumed that the energy is proportional to dl (as opposed to the volume).

To complete one locomotion cycle (see Fig. 13.16), DM needs to repeat screw insertion/removal four times. The distance DM travels with one locomotion cycle is equivalent to the stroke length of limbs (l_S) (same as the stroke length). Our energy estimate is given by:

$$E = \frac{4}{l_S} \cdot \frac{d}{19.5\,\text{mm}} \cdot \frac{l_I}{100\,\text{mm}} \cdot 500\,\text{J}$$

13.4.2.2 Power Source Trade Study

Next, we compare options for powering the DM. Below are the five options considered in the trade study.

1. Primary battery on DM; no tether
2. Miniaturized RTG on DM; no tether
3. Primary battery on SM, supplied to DM through tether
4. Solar cell on SM with tether, supplied to DM through tether
5. RTG on SM with tether, supplied to DM through tether (Fig. 13.22)

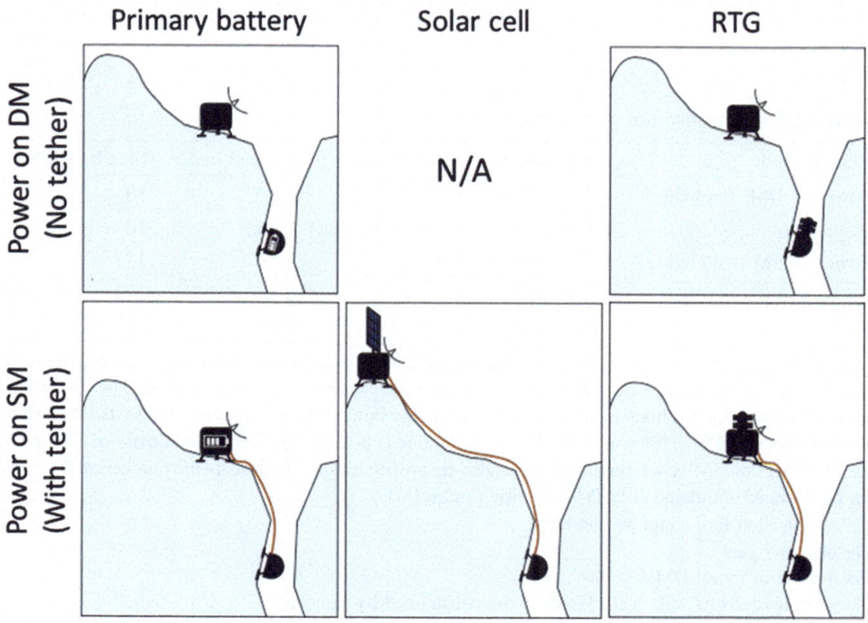

Fig. 13.22 Power system options

These options are compared by three criteria:

(A) The distance DM can travel
(B) The speed DM can travel
(C) Maximum number of DMs
(D) Major risks involved in the energy source

The major assumptions of the analysis include:

- For "power on DM" options, DM has five additional middle sections to accommodate battery/RTG.
  - Each additional module is hemispheric with a 5 cm radius
  - We did not take the increased aerodynamic drag due to additional sections into account. Therefore, the estimates for "power on DM" options must be view as optimistic.
- For the "no tether" options, SM's the minimal baseline power requirement is the same as Europa Lander—1100 Wh/day or 45 W, including minimal communications for background engineering and survival heater energy (Hand et al. 2017).
- Each DM uses up to 10 W of power for mobility.

The result of the analysis is summarized in Table 13.2. The basis of estimations and assumptions for each configuration are described in the following subsections.

From the analysis, we conclude that "*RTG on SM*" *is the most viable option, while* "*RTG on DM*", "*primary battery on SM*", *and* "*solar cell on SM*" *are*

Table 13.2 Comparison of power systems

Configuration	Energy source	Distance (m)	Speed (m/h)	Max # of DMs
Power on DM[a] (no tether)	Primary battery	350	5.5	10
	RTG	1300[b] (per year)	0.15	10
Power on SM (with tether)	Primary battery	2600[c]	5.5	1–3
	Solar cell[d]		5.5	1–2
	RTG		5.5	>10

The distance of "power on DM" configurations is the distance that can be traveled by each DM, while that of "primary battery on SM" is the total distance traveled by all DMs combined. The distance of "power on DM"-RTG is bounded not by available energy but by mission duration due to its slow speed. Max number of DMs of "power on DM" configurations is bounded by the payload mass of SM, while that of "power on SM" configurations is bounded by power, hence it represents the number of DMs that can be operated simultaneously (SM can bring more DMs)

[a]Communication link could be unreliable

[b]Distance per year

[c]Total distance by all DMs

[d]Needs at least 20 m^2 solar cell; Landing site constrained by sunshine

challenging but possible. Each configuration is discussed in further detail in the following.

Primary Battery on DM: We assumed the energy density of non-rechargeable lithium battery (1200 Wh/L). We also assumed that 80% of the five battery sections of DM can be used for batteries, and 50% of the energy is used for mobility.

Even with these optimistic assumptions, each DM can only travel ~350 m. Assuming 10 W power for mobility, the lifetime of DM is just 60 h. Given ~100 m landing accuracy of SM, significant portion of the range is consumed by moving from SM to vent. With this configuration, only the shallow portion of vents can be explored with very limited time. This drawback can be partially mitigated by bringing a large number of DMs. However, assuming the payload mass of Europa Lander (~65 kg) and the increased mass of DM due to battery (~6 kg), only 10 DMs can be accommodated. Therefore, we conclude that this configuration is not a viable option.

RTG on DM: We assumed miniaturized RTGs with the same power density as the multi-mission RTG (MMRTG) fill 80% of the volume of the five additional sections. An MMRTG generates ~125 W at the beginning of mission, which falls to ~100 W at the end of the 14-year designed lifetime [https://ntrs.nasa.gov/archive/nasa/casi.ntrs.nasa.gov/20080003866.pdf]. Its envelope volume including the radiation panels is ~0.21 m^3. With these assumptions, RTG on DM could produce 0.5 W of power, 50% of which is assumed to be used for mobility. This results in the speed of 0.15 m/h.

The extremely slow speed is a major challenge of this configuration. If operated continuously, DM could cover ~1300 m/year, but in practice, even a highly-automated DM would need to pause for mapping and scientific observations, hence the distance could be significantly shorter. Still, if a long mission duration is permitted, this is a possible option.

A unique challenge is that the radiation panels of RTG would create strong aerodynamic forces. A specialized design of RTG with aerodynamic shape could be required.

Primary battery on SM: We use the same specifications as the Europa Lander concept: 45 kWh battery capacity and 20-day lifetime on surface. After subtracting 45 W (=22 kWh for 20 days) minimum energy requirement, 23 kWh is available for DM. With this energy, DM can travel ~2600 m. If there are multiple DMs, this is the distance traveled by all the DM collectively. With the limited range and mission life time, returning samples to SM would not make sense, and there would be no time for scouting. Hence, all the DMs would be for in situ science. This configuration would not be impossible but not plausible.

Solar cell on SM: As the vents on Enceladus are in the south polar region, SM may have continuous sunlight in the southern summer by carefully choosing the landing site. On the other hand, a solar-driven SM cannot operate during the southern winter. Therefore, the mission life time is upper bounded by the half orbital period of Saturn, which is 14.7 years. (Enceladus's axis tilt and orbital inclination are nearly zero, and Saturn has 26.7° axial tilt. Therefore, Enceladus experiences the same seasonal changes as Saturn.)

The solar radiation intensity at Saturn is 15 Wm^{-2}. Assuming 20% efficiency, at least 15 m^2 area of solar cell is necessary for SM's survival. In order to operate DMs, at least 20 m^2 would be necessary. The solar cells must be gimbaled to chase the Sun. Due to the low elevation angle of Sun in the south polar region, the solar cells must be deployed vertically, which is challenging but possible under the weak gravity of Enceladus (0.01 g).

In this configuration, DM needs to travel a greater distance to the vent, as SM's landing site must be determined based primality on the lighting condition instead of the proximity to vents. For example, if SM lands on the ridge of one the "Tiger Stripes," DM would need to travel a few kilometers Porco et al. (2006) to reach the mouth of a vent.

RTG on SM: We assume SM has two MMRTGs, which produce ~110 W each. The mass of two MMRTGs (90 kg) is similar to the mass of the 45-kWh primary battery of Europa Lander (~95 kg including packaging). The range of DM is limited not by energy but by tether. The abundance of power allows to operate more than ten DMs at their top speed (5.5 m/hr). There are no limitations on landing site and mission duration. Therefore, we conclude that this configuration is the most viable one for the EVE concept.

13.4.3 Tether System

There are two configurations for tether:

1. "Spool on SM" option: SM stores tether and DM pulls it out
2. "Spool on DM" option: DM stores and deploys tether (Fig. 13.23).

The "spool on SM" option would have a major technical difficulty in tether deployment, as DM must provide the tension to pull it out. As a result, the tether must have mechanical strength to support the tension. Furthermore, the tether must

Fig. 13.23 Tether system options

move with DM. Tether's abrasion and entanglement could disable DM. Therefore, this option is realistic only for a short distance.

The "spool on DM" does not have difficulty in deployment because no tension is required for deployment, and the tether does not have to move once deployed. A disadvantage is that the tether length is limited by DM's storage capacity, which in turn limits the maximum depth that the mission can explore.

We made a first-order estimate of the required storage volume for the "spool on DM" option. We assume a 2000 m long 38 AWG (American Wire Gauge) insulated copper magnet wire, which has 0.101 mm diameter. Three copper wires are assumed to be used for GND, power, and data lines. The total volume of the wires is 4.8×10^{-5} m^3 or 48 cm^3, which is about 20% of the volume of DM's rear section. While a longer tether could be accommodated, the rear section also needs to accommodate the mechanical system for actuating two limbs. The 20% allocation for tether in the rear section would be a reasonable option. Alternatively, $\sim$10 km tether is possible by adding a dedicated section for tether, at the cost of increased aerodynamic drag. The electric resistance of the 2000 m cable is $\sim$600 Ω, which results in 45% loss of energy when transmitting 10 W at 100 V.

Based on this analysis, we conclude that "spool on DM" is the most realistic configuration. The possible length of tether is 2 km if the spool is accommodated in the rear section, or 10 km if a dedicated tether section is added. Note that the tether is a single point of failure for each individual DM, but *not* for the system, since there are multiple DMs.

13.4.4 Autonomy System

Unlike existing Mars rovers, which are mostly controlled manually, EVE's DMs should be highly automated for several reasons.

- **Unavailability of orbital reconnaissance**: In strategic (i.e., long-distance) planning of Mars rovers, rover planners are greatly assisted by the high resolution (25 cm) orbital images provided by Mars Reconnaissance Orbiter's HiRISE camera, as well as the digital elevation model (DEM) created from stereo HiRISE images. The rover planners have a bird's-eye view, in which they can identify topology, terrain type, and obstacles beyond the rover's line of sight. EVE will not have this luxury.
- **Poor viewshed**: In tactical (i.e., day-to-day) planning of Mars rovers, the drive distance per operation cycle (Sol) by manual planning is limited by viewshed (geographical area that is visible from a location), which is typically $\sim$50 m. In the vent, it would be tens of cm. The operation cycle of EVE would be the same as Enceladus's 1.37-day orbital period due to the co-orbiting CRO. Limited to tens of cm per operation cycle, it would take decades to move just one kilometer.

- **Multiple DMs**: Past space missions have operated only one or two spacecraft. In EVE, up to tens of DMs must be operated simultaneously. Operating multiple DMs would not only increase the labor and cost if manually operated, but would also limit the amount of data to be downlinked from each DM. On-board decision making would remove this bottleneck.
- **Dynamic environment**: When problems occur, Mars rovers can safely stop anytime and wait for ground control because the environment of Mars is largely static. In contrast, Enceladus's vent-conduit system is dynamic. DMs will have to take immediate actions upon the detection of anomalies, rather than stopping and waiting for ground response.

Figure 13.24 presents an autonomy capability roadmap, which shows required mission-level, system-level, function-level, and algorithm-level autonomy capabilities with dependencies among them. Capabilities in one column are linked to enabling/supporting capabilities in the next column to the right. Narratives for each of the system-level capabilities are provided below.

13.4.4.1 Graceful Degradation

Mechanical complexity of DM is a double-edged sword. It would undermine the success of the mission if DM's software is designed in a way that any single mechanical failure results in the shut-down of the DM. However, it benefits the mission if the software is designed to exploit the flexibility of the mechanical system and reconfigure it upon mechanical failures. For example, a human is not immediately incapacitated by an injury to one of his limbs. Instead, s/he would quickly adapt to the change in his physical system and stay active by using his remaining limbs. Likewise, if one of the ice screws of a DM is failed or one of the joints is locked, the DM's software should adapt to the change in its hardware system by changing the gait and continuing the mission at a degraded rate.

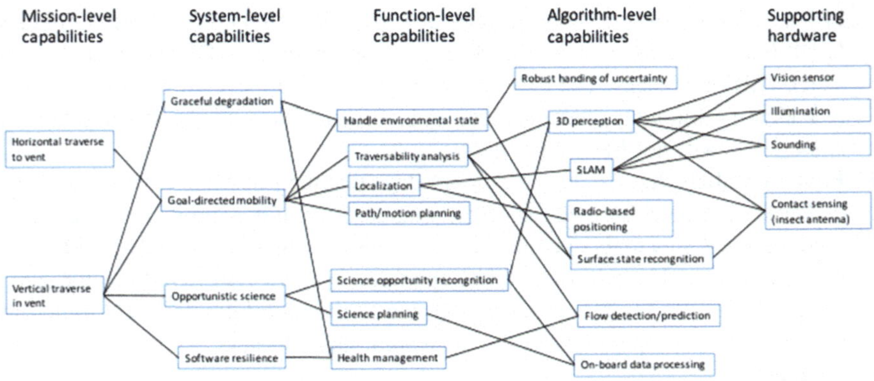

Fig. 13.24 Autonomy Capability Roadmap for EVE

13.4.4.2 Goal-Directed Mobility

Because the global map of the vent-conduit system is not available, the ground operator cannot specify physical locations for a DM to go beyond the regions already explored by other DMs or its own very limited line of sight. Therefore, DM should take a higher-level goal, such as “go as deep as possible” and “maximize the chance to reach the liquid interface.” It has to identify route options to achieve the goals, estimate the traversability and risk of the route options, update the map as it gathers new information, share the information with other DMs, and back up if it encounters difficulty in traversal or judges that the route ahead is too risky.

13.4.4.3 Opportunistic Science

Mars rover Curiosity takes more images than it can transmit to Earth due to the bandwidth limitation. Due to EVE’s more limited communication bandwidth and the multiple DMs, only a fraction of collected data would be able to return to Earth. However, we do not want DMs to unintelligently pass by evidence of life without noticing it. An onboard capability to autonomously detect scientifically interesting scenes, collect relevant data, and select the data to transmit to Earth within the bandwidth limit, is highly desired.

Opportunistic science is not merely an alternative to conventional ground-in-the-loop science but an enhancement of it. As Dr. Penny Boston stated in her keynote speech at the 2016 NIAC Symposium, human scientists repeat various mental activities at a high frequency, e.g., scan at regional scale and focus on general site types, scan at walking scale and focus on individual features or site, and perform in situ analysis by focusing on restricted features. The daily iterative process of MSL, for example, is “not the optimal way to do science in a new environment because it is missing a lot of components that the human naturally brings into it (Boston 2016a, b).” While realizing such a capability is technically challenging, it would bring a significant benefit for future robotic explorations like EVE.

13.4.4.4 Software Resilience

A current challenge in software engineering is the development of a software system that can detect when one of its software components or subsystems is exhibiting anomalous behavior due to an uncaught defect in the design or implementation of the code, and mitigate/accommodate this anomalous behavior and still meet its execution objectives. Such capabilities are sometimes referred to as “self-healing” software, but are more generally captured under the banner of Software Resilience. This capability to appropriately manage the state of the software system can be considered to be a counterpart to the “graceful degradation” capability which manages the state of the spacecraft hardware system.

13.4.5 Perception System

DM's perception system has dual purposes. It is an essential component of the autonomy system (Fig. 13.24) and also provides science data. Its main focus is to create 3D maps of the interior of the vent-conduit system. Below we compare several candidate systems and compare their advantages and disadvantages.

13.4.5.1 Overview

Since orbiter reconnaissance of the interior of vent is impossible, DMs must map the vent and localize itself. This section discusses the applicability of current sensing technology to the exploration of crevasses in the ice crust of ocean worlds. We evaluate the applicability of stereo, structured light, time-of-flight cameras, and Lidars. We consider five criteria: power consumption, volume, range and accuracy. We discuss acceptable ranges for each criterion. Our study suggests the superiority of assisted stereo in the context of an icy-world mission.

We studied the applicability of current depth sensing technology to icy crevasses. We assumed that depth sensing exclusively supports the two following tasks: 3D mapping for path planning/obstacle avoidance, and visual odometry (VO). We limited our study to sensors that fit inside a 1 dm^3 robot. We also assumed that the robot remains stationary while imaging. The objective of this study is to list existing applicable sensors, and rank the sensors according to power consumption, size, range, accuracy, and accommodation to low temperatures.

The depth-sensing technologies that we are considering are: stereo vision, structured light, assisted stereo, time-of-flight camera, spinning Lidar, and radar. In the following, we introduce each technology and present qualitative results illustrating their application to icy surfaces. We then rank each sensor according to the five criteria defined above.

13.4.5.2 Candidate Sensing Technologies

Stereo: A stereo camera computes depth by stereopsis (Fig. 13.25). Stereo exhibits a compelling sensing range, low volume and low power consumption. Its accuracy gracefully degrades with range in naturally-lit environments. In a dark environment, a floodlight is required, and range is dictated by the luminous intensity of the light and shutter speed. To enable stereopsis, a stereo sensor needs to find matching patches of pixels across left and right images. As a result, stereo only works for textured surfaces that allow us to match discriminative blocks of pixels. It fails on smooth uniform surfaces such as clean ice.

Structured Light: A structured-light sensor projects a light pattern and captures its return with a camera (Fig. 13.26). The deformation of the pattern captured by the camera maps to depth. Range is dictated by the luminous intensity of the projector.

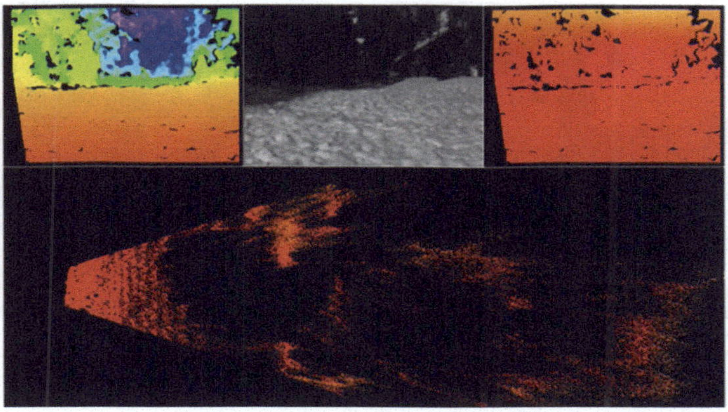

Fig. 13.25 Stereo on snow. Sensor: stereo camera. Material: patchy natural snow. Top-left: stereo-based range image. Top-right: stereo-based elevation map. This figure illustrates that stereo allows us to compute the depth of well-lit textured surfaces. Credit: L. Matthies

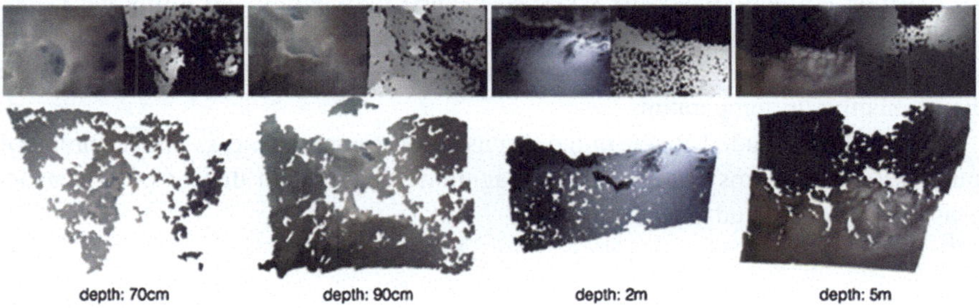

Fig. 13.26 Structured light on snow and ice. Sensor: PrimeSense infrared structured light sensor. Material: natural ice and snow at Erebus, Antarctica. Captured by A. Curtis. For each subfigure, the top-left image shows color, the top-right image shows depth, and the bottom image shows the resulting colored point cloud. This figure shows the applicability of structured light for matte surfaces. The sensor has a minimum range of 50 cm. The first subfigure shows this limitation. In the second subfigure, the depth of clear ice cannot be computed

By contrast to stereo, structured light cannot leverage natural light to compute depth beyond the reach of its light projector. However, structured light allows us to estimate depth on untextured surfaces. It provides optimal results on Lambertian surfaces. Performances degrade with increased specularity.

Assisted Stereo: Assisted stereo combines the principles of stereo and structured light (Fig. 13.27). It requires more space than stereo or structured light, but it works on untextured surfaces, and it provides depth beyond the reach of the robot light if natural light is present.

Time-of-flight Camera: Time-of-flight cameras measure the return time of a light pulse with a 2D array of receptors, providing scene-wide depth measurements at 25+ frames per second. Current time-of-flight cameras provide more reliable

Fig. 13.27 Assisted stereo on ice. Sensor: Intel infrared structured-light sensor. Material: artificial ice at JPL. The top-right image shows that the sensor does not work on translucent ice. The other images show convincing results for opaque ice

measurements than the systems listed above, at the cost of more power, more space, and less range.

Spinning Lidar: Lidar sensors measure the return time of a narrow light pulse. To generate a range image, the scene is scanned by panning and tilting the Lidar. Lidars are generally more accurate than camera-based systems, at the cost of more power and more space. Their reliance on mechanical scan can potentially make them sensitive to temperature.

Radar: We included Radar in the discussion for completeness. The accuracy of current Radar systems is an order of magnitude poorer than that of other sensors considered in this study.

13.4.5.3 Discussion

(Assisted) stereo, structured light, ToF cameras and Lidar are typically operated either in near-IR or visible. Operating at other wavelengths is in principle feasible, with no obvious disparity in the effort required for enabling a new operating wavelength.

Table 13.3 summarizes our discussion. We note that

- For the three camera-based solutions, FPGA-based stereo processing would use 1–30 watts for 0.2 s.
- Sensing range and accuracy assume that there is no natural lighting; artificial lighting is included in all power estimates.
- The relevance of the criteria considered in this study will vary depending on the task.
- We built this table based on our experience with commercial sensing devices and lighting equipment. It is possible that one could build—for instance—a flash (non-scanning) Lidar device specifically for this application, that would perform better than assisted stereo on all counts. ToF cameras/Lidars are

Table 13.3 Qualitative evaluation of range-sensing devices

	Power		Device volume		Sensing range		Sensing accuracy	Temperature resistance
Stereo vision	3	5W[a]	4	2×1 cm^3	4	0.1 m–inf[b]	2[c]	3
Structured light	3	7.5W[a]	4	2×1 cm^3	4	0.1 m–inf[b]	3	3
Assisted stereo	3	7.5W[a]	3	3×1 cm^3	4	0.1 m–inf[b]	3[d]	3
ToF camera	4	>10W[e]	2	100 cm^3	2	0.1 m–10 m	4	3
Spinning lider	1	10W[f]	1	100 cm^3	3	0.1 m–30 m	4	1[g]
Rader	≤ 3	>5 W	≤ 2	<4 cm^3	4	$\sim$100 m	0[h]	3

For each evaluation criteria, scores and estimated values (if applicable) are shown in the table. Scores are on a 0–4 scale

[a]Exosure: 1 s—inf

[b]Further with longer exposure

[c]Fails if no texture

[d]Superior to stereo and structured light

[e]Exposure: flash

[f]Scan: 5 s

[g]Mechanical sensitivity

[h]Beam width >1°, range accuracy >25 cm

currently evolving at a fast pace, driven partly by the emerging self-driving car industry. Solid-state Lidars could have an important impact on this problem.

In summary, this exercise indicates that assisted stereo and structured light are the two technologies that best fit the constraints of icy-world crevasse exploration at this moment. This problem will need to be re-evaluated as solid-state Lidar continues to progress.

13.4.6 Science Instruments

EVE would perform astrobiology and volcanology observations in the vent-conduit system, with the grand goal being the detection of life. Like Viking and Europa Lander, no singular measurement would provide sufficient evidence for the detection of life; rather, the conclusion that evidence of life had been detected would require multiple lines of evidence, from different instruments, on a set of samples examined across a variety of spatial scales (Hand et al. 2017). However, the payload volume of DM is so limited that all the variety of instruments necessary to detect life is not likely to be accommodated in a DM. Therefore, we will take an alternative approach where instruments are distributed between multiple in situ science DMs. In addition, sample return DMs will bring ice, liquid, and dust sample to SM, where various elaborate analyses are performed.

Instruments on In situ Science DM

The payload space in DM is extremely limited. Our standard three-section design accommodates science instruments in its middle section. Its internal volume is approximately 5 cm × 5 cm × 10 cm. DM can have additional instrument sections (like a snake-bot), but the volume of each section will be limited anyway. A good news is that, because the interior of the vent-conduit system is almost completely unknown and orbital reconnaissance is impossible, even very simple measurements could significantly advance scientific knowledge. Below is a partial list of science instruments that can be miniaturized and fit in an in situ science DM.

Volcanology package: temperature, pressure, and flow speed: Measurement of temperature, pressure, and flow speed in the vent-conduit system, combined with 3D mapping provided by the perception system (Sect. 13.4.5), will significantly advance the understanding of the eruption mechanism. Off-the-shelf miniaturized temperature sensors are available. A miniaturized pitot tube can measure static and dynamic pressures, and the flow speed can be estimated from the difference of them. This package is likely to occupy only a fraction of the payload volume; hence it can be a standard package for all DMs.

Habitability package: pH, salinity, and oxidation-reduction potential (ORP): pH, salinity, and oxidation-reduction potential (ORP) are three of the most important indicators for habitability, all of which can be measured using off-the-shelf miniaturized devices. Sometimes, pH is even an indicator of extant life. For example, a significant change in pH (from 9.2 to 0) over just a few km in range, observed in a sulfuric acid cave, Cueva de Villa Luz cave, Tabasco, Mexico, is due to the activities of microbial lives (Boston 2016a, b).

Commercially available potentiometric devices can operate with a very small volume of liquid sample. If DM can reach the liquid interface, it can directly measure the pH of the body of water. Alternatively, it can operate before reaching the liquid interface by collecting liquid particles in the flow with an absorbent material.

Microscopic imager: Imaging, both macroscopic and microscopic, is an effective tool for detecting life. At a macroscopic scale, microorganisms frequently produce distinctive "biovermiculation patterns," as shown in Fig. 13.28, which could be a universal signature of life (Boston 2016a, b). At a microscopic scale, life often creates distinctive structures, as in Fig. 13.29.

Macroscopic patterns can be identified with DM's stereo cameras, while finding microscopic patterns requires a dedicated microscopic imager. Combination of

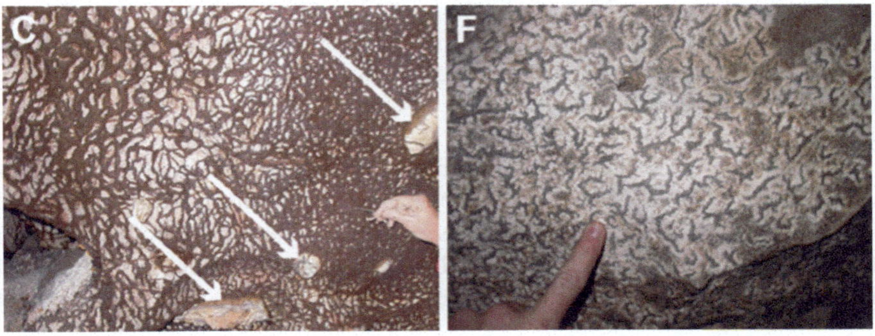

Fig. 13.28 Macroscopic biovermiculation patterns found in the Frasassi cave system, Italy. Images taken from Jones et al. (2008)

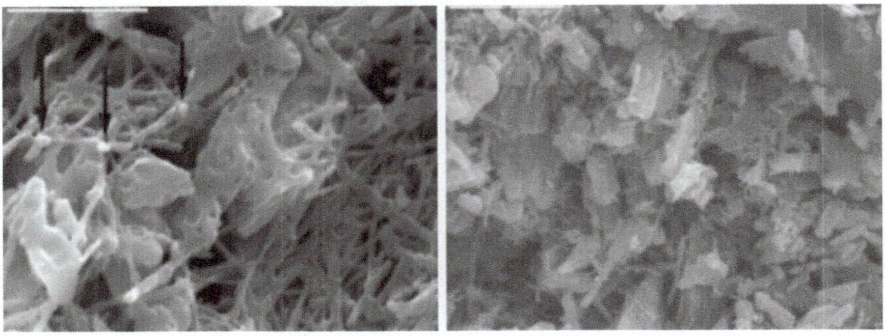

Fig. 13.29 Microscopic structures created by microbial life. Left: biologically active calcite moonmilk (Crisco), Spider Cave, NM. Right: Fossil microbial mat textures, Hidden Cavem NM. Scale bar = 2 μm. Images are taken from Boston et al. (2001)

macroscopic and microscopic imaging with UV LED would be useful as many lifeforms on Earth exhibit UV fluorescence.

Microscopic imager is a relatively small instrument. For example, Mars Hand Lens Imager (MAHLI) the microscopic imager mounted on the robotic arm of *Curiosity*, is ~4 cm in width. Miniaturizing it for DM would not be a major challenge. However, DM would not be able to accommodate a mechanism for a complex sample preparation. Therefore, its use would be limited to imaging of the raw surface of ice walls as well as liquid water. Microscopic observation that requires sample preparation is performed in SM on samples brought by sample return DMs.

Life detection package: microchip electrophoresis with laser-induced fluorescence: Microchip electrophoresis (ME) is a miniaturized variant of capillary electrophoresis (CE) implemented on a microfluidic device. ME does not require the use of high-pressure pumps or packed columns of separation media Willis et al. (2015). Coupled with laser-induced fluorescence (LIF), it can measure the distributions of molecular properties in organic molecules, including carboxylic acids and amino acids. Such measurement provides a practical way to search for life based on *Lego principle* McKay (2004). Like LEGO blocks, life is composed of a subset of naturally available materials. For example, terrestrial life uses 20 amino acids, whereas over 70 have been detected in meteorites Willis et al. (2015). Therefore, biotic sample exhibits distinct distribution of molecular properties from that of abiotic samples (Fig. 13.30).

Various prototypes of end-to-end instrument have been developed at JPL. They are tens of cm in size, but most of the volume is typically occupied by liquid extraction and sample handling systems while the detector (microfluidic chips) itself is a few cm in size (Fig. 13.31). If DM can reach the liquid interface and directly intake liquid, or alternatively collect liquid samples in the vent-conduit

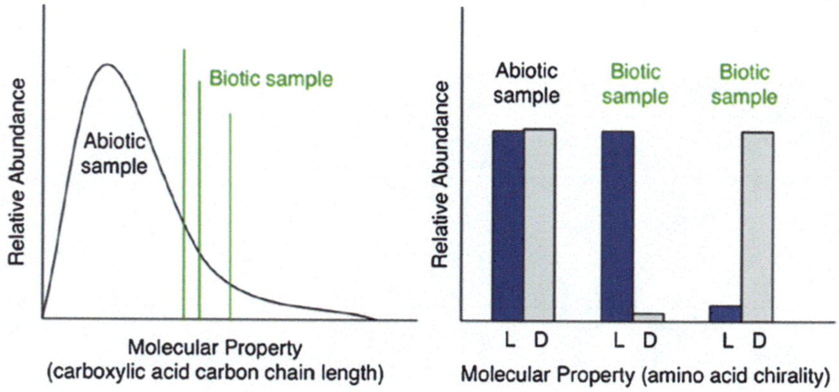

Fig. 13.30 The Lego principle The Lego principle illustrates a practical way to search for evidence of life and biological processes. Left—from McKay, (2004), right—from Willis et al. (2015)

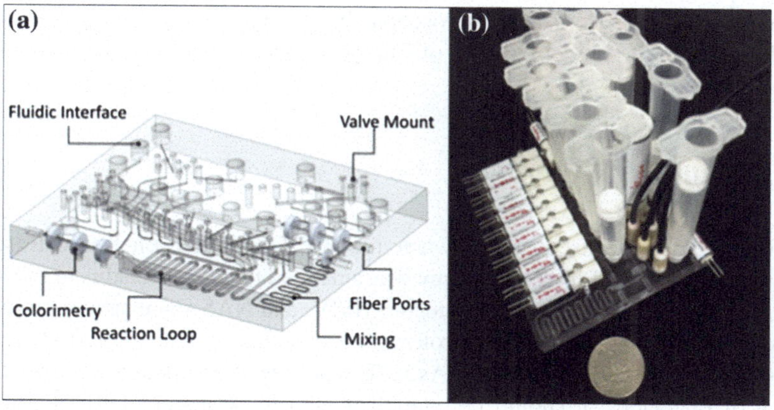

Fig. 13.31 An example of microfluidics chip. **a** 3D printed MicroFIA manifold, **b** assembled MicroFIA system. Images taken from Kehl et al. (2016)

system, the liquid extraction system is unnecessary, and the sample handling system can be simplified significantly.

Instruments on sample return DM

Ice screw creates an ice core when screwed into ice. Therefore, a DM produces ice cores at every step. One idea is to store some of the ice cores in the middle section and bring them back to SM. However, transferring ice cores from the end effector to the middle section would require additional mechanical system. If multiple samples are stored, the sample handling system typically needs a revolver-like mechanism which is very bulky.

A more realistic idea would be to install a dedicated ice screw for sampling in the middle section, as shown in Fig. 13.4. In order to minimize the complexity of sample handling system, the sample return DM is designed to collect only one ice core per trip. Once it cores the ice, it seals the ends of the screw and bring the core back. It can make another trip to collect more samples. Alternatively, if space permits, a DM can have multiple ice screws in the middle section.

The volume of an ice core produced by an 8-mm ice screw with 2.5 cm insertion length is approximately 3 cc. This is comparable to the required volume of sample for Europa Lander's instruments (5 cc for Microscope for Life Detection, 1 cc for Organic Compositional Analyzer (mass spectrometer) and Vibrational Spectrometer) (Hand et al. 2017).

Other types of sample return DM include liquid sample collector and dust particle collector.

Instruments on SM

SM can accommodate instruments that are too voluminous for a DM. SM's instruments are used for for analyzing the samples brought back by sample return DMs. One of the most valuable types of instrument for detecting life would be a resolution mass spectrometer. Microscopic imager and vibrational spectrometer, like the ones on the model payload of Europa Lander, would also be valuable.

13.5 Planetary Protection

Following *Cassini*'s discovery of water plumes, Committee on Space Research (COSPAR) Planetary Protection Policy was amended to add Enceladus to the list of target bodies for PP classification Category III/IV, and NASA's planetary protection policy (NPR 8020.12D, Planetary Protection Provisions for Robotic Extraterrestrial Mission) was also updated as such. As EVE would land and potentially impact the surface of Enceladus, it should be compliant with the Category IV requirements. Specifically, "the probability of inadvertent contamination of an ocean or other liquid water body" must be reduced to less than 1×10^{-4} per mission (NPR 8020.12D, Sec. 5.4).

As EVE's DM could directly contact the liquid body of water in a vent-conduit system, significantly more elaborate bioburden reduction processing would be necessary than landing missions concepts such as the potential Europa Lander. A similar level of bioburden reduction processing would be necessary for SM as well because (1) DMs would be in contact with SM over a long period and (2) upon a failed landing or spacecraft disintegration, SM's RTGs could melt through the ice shell and reach the subsurface ocean [There is a misconception that RTGs are not allowed on icy moons. There is no such rule in COSPAR's and NASA's planetary protection policy. However, if an ice shell is exposed to an RTG for a long period, the RTG could melt through and reach a subsurface ocean. This could increase the probability of contaminating the ocean, compared to a non-RTG surface mission with the same level of bioburden reduction, (P_S and P_R in the Coleman-Sagan equation). The increase in probability could be offset by reducing the number of microbes on the spacecraft (N_0, in the Coleman-Sagan equation), which in turn could be achieved by more stringent sterilization].

A comparable mission concept is the mechanical ice probe proposed by Wilcox et al. (2017), which uses RTG and General Purpose Heat Source (GPHS) modules containing Pu-238 (mode details are in Sect. 13.2.3). They propose heat sterilization at 500 °C for significant periods of time because "exposure to 500 °C for extended durations should pyrolize any organic material, decomposing all large organic molecules Wilcox et al. (2017)." However, further studies should be done to prove the particular sterilization method can achieve the 1×10^{-4} contamination probability requirements. Building a spacecraft to tolerate the extensive heat sterilization would be a major technical challenge. In addition to heat sterilization, development of alternative bioburden reduction technologies, including chemical- and radiation-based ones, should be promoted.

13.6 Mechanical Prototype Design

We developed a prototype design for two reasons. One is to serve as a reference point for the estimation of DM's power requirement in the system trade study in Sect. 13.4. The other reason is to validate that the mechanical system can be miniaturized to the size required by EVE. For this purpose, our prototype design only uses existing technologies/materials and commercially available mechanical parts (e.g., motors, gear boxes), which are included in our CAD design. In addition, to verify the design an 3D printed model has been created, which is not actuated but all the joints are movable.

Our prototype design is rather conventional. It has four limbs actuated by electric motors. Each limb has an ice screw end effector at its end. The baseline design of the body consists of three sections. The front and rear sections are mostly occupied by mechanical components for actuating the shoulder joints. The middle section accommodates science instrument. The robot could have multiple middle section at the cost of increased aerodynamic drag, and perhaps somewhat decreased mobility. However, such a "snake" configuration was not included in the scope of this study.

Our limb design features a linear joint, as shown in Fig. 13.32. While each limb has 5 DOF, the load is only carried by the linear joint when the robot moves against the flow. The primarily function of the other four rotational joints are to move the end effector and adapt to undulating surface of the ice. Our choice of linear joint is due to its ability to support relatively strong force. Furthermore, it does not require changing its frontal section area significantly in any phase of the advancement, which helps to go through a narrow vent as well as to keep the aerodynamic drag relatively constant.

Fig. 13.32 Prototype design of the limb, featuring a linear joint

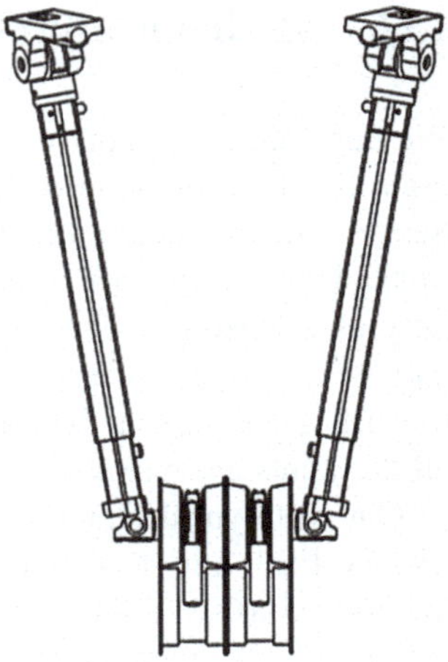

Fig. 13.33 Prototype design of the front/rear section with two limbs

Each limb is designed to support 1000N of force, hence the robot can support up to 4000N of aerodynamic drag force. Required torque for the motors to actuate the joints are within the catalog spec of commercially available ones, with reasonable choices of gear ratio and teeth number. Mechanical upper bound on the speed of the robot is estimated to be 18.4 m/hr. However, in practice, power poses the limit on speed, as discussed in Sect. 13.4.2. If 10 W is available for mobility, the robot can achieve 5.5 m/hr under 10^5Pa dynamic pressure (Fig. 13.33).

13.7 Future Work

The single greatest source of uncertainty in this concept is the eruption mechanism. In our preliminary analysis this study, we found that EVE is likely feasible with the "boiling" models but the feasibility remains undetermined with the "cryovolcanic" models. Resolving between the two types of models is the highest priority, but it probably needs additional observations by a future orbiter such as the proposed Enceladus Life Finder (ELF).

However, refining the dynamic pressure and vent size estimates, as well as refining the DM design, could establish feasibility under cryovolcanic models, or at least get closer to it. As we discussed in Sect. 13.3, there are a number of additional analysis we can do with existing *Cassini* data, which will likely bring down the 10^7 Pa upper bound. On the other hand, the refinement of the DM design, most importantly the optimization of aerodynamic shape, would bring up the current 10^5 Pa limit. The drag coefficient of the DM design in Sect. 13.6 is about 1.5, while that of typical automobile is 0.2–0.3. Whether or not the two bounds will meet is rest on future work.

There are several promising ideas of DM that have not been explored in this study. One is a "snake bot" design, which has additional middle sections for accommodating more instruments or a longer tether. Another idea is interchangeable middle section (i.e., swapping science instruments), which gives greater flexibility and robustness in operation.

Compared to mechanical capabilities, the required software capacities are less developed in general. For the reasons we discussed in Sect. 13.4.4, DM needs to be highly automated in order to complete the mission within a reasonable duration. Development of the key autonomy capabilities listed in the Autonomy Capability Roadmap (Fig. 13.24) has a crucial importance.

Our estimate on energy consumption is partially based on ice chamber test, but the ice properties on Enceladus could be different from the ones we used. A better estimate could be obtained through experiments in an environment that better represents the environment in Enceladus vent.

We have given little considerations on the thermal environment. While the temperature is mild in the existence of liquid (near the triple point temperature: 273 K), the surface temperature is $\sim$80 K. The upper part of the vent could be exposed to the extreme temperature.

This study mostly focused on DM, while using the design of the potential Europa Lander for SM and CRO. However, there are several differences between Europa and Enceladus which could significantly impact their capabilities, such as the available number of large moons for flyby in the planetary system, strength of radiation, and gravity. Performing detailed trade study for SM and CRO would be an important future work.

Interface between SM and DMs has not been explored in this study. SM would need mechanisms to store and deploy DMs, as well as to retrieve samples from sample return DMs. Developing an operation strategy to share resources (power, communication bandwidth, time of ground operators) between SM and DMs is also a future work.

Finally, while EVE's primarily target is Enceladus, a similar concept could be applied to other icy moons, most notably Europa, if vents exist. Adapting EVE for Europa would be an interesting direction [Conveniently, Europa and Enceladus shares the initial letter. Hence the acronym "EVE" will be applicable to Europa as well]. Probably the harsh radiation environment, which in turn results in a shorter mission duration, would be a major challenge. EVE could also find terrestrial applications, such as exploration of geysers, crevasses, and ice caves [Even more conveniently, Earth also has the same capital letter]. Demonstrating the concept in Earth analogues would be a stepping stone to the future Journey to the Icy Moons.

13.8 Conclusion

We presented the Enceladus Vent Explorer (EVE), a robotic pathfinder mission concept to enter the Enceladus vent. The concept consists of multiple descent modules (DMs) that climb down the vent and a surface module (SM) that carries DMs to the surface and provides power and communication link to DMs through tethers. In order to assess the feasibility of the concept, we performed preliminary characterization of the vent, with particular focus on the dynamic pressure and the size. We also performed trade studies to identify the most robust design of EVE's descent module (DM). As a result, we chose a design with limbs equipped with ice-screw end effector. We concluded that EVE is likely feasible assuming the "boiling" eruption model, while the feasibility is undetermined under the "cryo-volcanic" eruption models.

Acknowledgements This work was funded by the NASA Innovative Advanced Concepts (NIAC) program. We thank Penny Boston, Peter Willis, Morgan Cable, Florian Kehl, Matt Heverly, Noah Warner, Steve Sell, and Sabrina Feldman for valuable inputs. The research was carried out at the Jet Propulsion Laboratory, California Institute of Technology, under a contract with the National Aeronautics and Space Administration. The content of this study represents pre-decisional information to be used for planning and discussion purposes only.

References

Arakawa, M., Maeno, N.: Mechanical strength of polycrystalline ice under uniaxial. Cold Reg. Sci. Technol. **26** (1997)

Asbeck, A.T., Kim, S., McClung, A., Parness, A., Cutkosky, M.R.: Climbing walls with microspines. IEEE Int. Con. Robot. Autom. (2006, January)

Barth, J.L., Dyer, C.S., Stassinopoulos, E.G.: Space, atmospheric, and terrestrial radiation environments. IEEE Trans. Nucl. Sci. **50**(3), 466–482 (2003)

Beverly, M.J., Stephen, W.: Ice climbing anchor strength: an in-depth analysis. [Online] Available at: http://hmga.gr/storehouse/word-acrobat/Ice%20Climbing%20Anchor%20Strength%20-%20MRA%202009%20-%20Marc%20Beverly.pdf (2009)

Boston, P.: Biovermiculation biopatterns as universal signatures of extant and extinct life. GSA Annual Meeting (2016a)

Boston, P.: Keynote talk, NIAC Symoposium. [Online] Available at: https://livestream.com/viewnow/NIAC2016/videos/133893674 (2016b)

Boston, P.J., Spilde, M.N., Melim, L.A.: Cave microbe-mineral suites: best model for extraterrestrial biosignatures! Lunar Planet. Sci. XXXII (2001)

Burge, S.: The Systems Engineering Tool Box—Push Matrix (PM) (2009). [Online] Available at: https://www.burgehugheswalsh.co.uk/uploaded/1/documents/pugh-matrix-v1.1.pdf

Fugueredo, P., et al.: Locating potential biosignatures on Europa from surface geology observations. Astrobiology **3**(4), 851–861 (2003)

Goguen, J., Buratti, B., Brown, R., Clark , R., Nicholson, P., Hedman, M., . . . Blackburn, D.: The temperature and width of an active fissure on Enceladus measured with Cassini VIMS during the 14 April 2012 South Pole flyover. Icarus **226**(1), 1128–1137 (2013)

Hand, K.P., et al.: Europa Lander Study 2016 Report: Europa Lander Mission (2017)

Hansen, C.J., et al.: The composition and structure of the Enceladus plume. Geophys. Res. Lett. (2011)

Ingersoll, A.P., Ewald, S.P.: Total particulate mass in Enceladus plumes and mass of Saturn's E ring inferred from Cassini ISS images. Icarus **216**(2), (2011)

Jones, D.S., Lyon, E.H., Macalady, J.L.: Geomicrobiology of biovermiculations from the Frasassi cave system, Italy. J. Cave Karst Stud. **70**(2), 78–93 (2008)

Kehl, F., et al.: In-situ liquid extraction and analysis platform for Mars and ocean worlds. In: 3rd International Workshop on Instrumentation for Planetary Missions (2016)

Kite, E.S., Rubin, A.M.: Sustained eruptions on Enceladus explained by turbulent dissipation in tiger stripes. Geophys. Res. Lett. (2016)

Liepmann, H., Roshko, A.: Elements of gasdynamics. Dover Publications (2002)

Lorenz, R.D.: Thermodynamics of Geysers: Application to Titan. Icarus **156**(1), 176–183 (2002)

Matson, D.L., Castillo-Rogez, J.C., Davies, A.G., Johnson, T.V.: Enceladus: A hypothesis for bringing both heat and chemicals to the surface. Icarus **221**(1), 53–62 (2012)

McKay, C.P.: What is life and how do we search for it in other worlds? PLOS Biol (2004)

Mitchell, K.: Coupled conduit flow and shape in explosive volcanic eruptions. J. Volcanol. Geoth. Res. **136**, 223–240 (2005)

Parness, A.: Anchoring foot mechanisms for sampling and mobility in microgravity, 2011. IEEE Int. Con. Robot. Autom. (ICRA), Shanghai (2011)

Parness, A., Frost, M., King, J.P., Thatte, N.: Demonstrations of gravity-independent mobility and drilling on natural rock using microspines. IEEE Int. Con. Robot. Autom. (2012)

Porco, C.C., et al.: Cassini observes the active South Pole of Enceladus. Science **311**(5766), 1393–1401 (2006)

Porco, C., DiNino, D., Nimmo, F.: how the Geysers, tidal stresses, and thermal emission across the South polar terrain of Enceladus are related. Astron. J. **148**(3), 45 (2014)

Spitale, J.N., et al.: Curtain eruptions from Enceladus' south-polar terrain. Nature **521**, 57–60 (2015)

Waite, J., Magee, B., Brockwell, T., Zolotov, M., Teolis, B., Lewis, W.: Enceladus' Plume Composition. EPSC-DPS Joint Meeting (2011)

Warner, N.: Personal communication (2017)

Wilcox, B.H., Carlton, J.A., Jenkins, J.M., Porter, F.A.: A Deep Subsurface Ice Probe for Europa, p. 2622. IEEE, Big Sky (2017)

Willis, P.A., Mora, M.F., Creamer, J.S.: Implementation of microchip electrophoresis instrumentation for future spaceflightmissions. Anal. Bioanal. Chem. (2015)

Zimmerman, W., Bonitz, R., Feldman, J.: Cryobot: an Ice Penetrating Robotic Vehicle for Mars and Europa. IEEE Aerospace Conference (2001)

Chapter 14
Prospect of Exploration and Exploitation of Kuiper Belt Object Resources in the Future

Volker Maiwald

The Kuiper Belt is a distant region of our solar system beginning at a solar distance of about 30 AU. By now, the discovery of several small worlds, like the dwarf planets Pluto, Makemake or Haumea, has shown that the trans-Neptunian region of the solar system is indeed more crowded than first meets the eye. However, the large distance to the Sun makes investigating these worlds challenging—what we do know is that the Kuiper Belt contains methane, nitrogen, ethane, even water and other volatiles in some quantity. Only recently have we begun to investigate the Kuiper Belt more closely with the help of spacecraft: NASA's *New Horizons* mission passed by Pluto in summer 2015 and revealed Pluto's surface details for the first time.

While the Kuiper Belt is far away and a prolonged presence there beyond our current technological capabilities, it is also clear that this large region of the solar system contains a number of useful resources. These could be of use in future endeavors of human kind, establishing a more extended presence (robotic and human) in the solar system. This chapter explains a scenario how the Kuiper Belt can be opened up for humanity and become a ground for exploration and resource exploitation in the future.

14.1 The Kuiper Belt and Its Worlds

The Kuiper Belt is a congregation of objects that had been a hypothesis for several decades following the discovery of Pluto in 1930. About 62 years later 1992 QB1 was found and thus established the Kuiper Belt as fact (Jewitt and Luu 1992). In the

V. Maiwald (✉)
German Aerospace Center (DLR), Institute of Space Systems, Robert-Hooke-Str. 7, 28359 Bremen, Germany
e-mail: volker.maiwald@dlr.de

V. Badescu and K. Zacny (eds.), *Outer Solar System*,
https://doi.org/10.1007/978-3-319-73845-1_14

meantime many more objects have been discovered in that region, numbering thousands, and several are even larger than 1000 km (Delsanti and Jewitt 2006). Estimates assume that more than 100,000 objects with a size larger than 100 km comprise the Kuiper Belt (Levison and Donnes 2007), whereas the total mass estimate ranges from 1 to 10% of Earth's mass (Bernstein et al. 2004; Gladman et al. 2001). A value of 10% is the theoretical limit derived from an assumed average albedo of 0.04 and the corresponding size of Kuiper Belt Objects (KBO) (Delsanti and Jewitt 2006).

The Kuiper Belt is presumably a remnant of the solar system formation and also source for comets and small bodies in the region between Jupiter and Neptune. It stretches from about 30 AU, to about 50 AU, gradually becoming the so-called scattered disc, which eventually becomes the Oort Cloud according to current solar system models.

Typically KBOs have either an inclination smaller than 4° or a significant inclination larger than 20° caused by scattering and interaction with Neptune (Delsanti and Jewitt 2006).

KBOs also have natural satellites; the most prominent example would be Pluto with 5 moons. Another example is 2003 EL_{61}, also known as Haumea, which has 2 moons. For these large bodies densities of above 2000 kg/m^3 are typical (Delsanti and Jewitt 2006). Generalized statements about KBO densities are not possible, however (Brown 2012).

Larger bodies in particular have been shown to contain volatiles and ices. Ethane has been detected on Makemake and Quaoar and is assumed to be present on Pluto (Brown 2012). Haumea contains water, whereas Pluto also contains carbon monoxide and nitrogen ice (Owen et al. 1993) and some water, like Charon (Encrenaz 2008). Comets further testify the presence of volatiles in the Kuiper Belt.

KBOs have been part of several observation campaigns of telescopes like Hubble and recently Pluto has been visited by NASA's *New Horizons* mission, the first spacecraft ever sent to any KBO. The mission revealed Pluto's complexity regarding topography and e.g. has shown that large parts of the dwarf planet are covered in ice, see Fig. 14.1. *New Horizons* still continues to operate, ready to visit more bodies besides Pluto. Investigation of Pluto's moons by that probe revealed surface reflectance similar to an almost complete cover of water ice, see Fig. 14.2, suggesting that pure water ice is a significant part of their composition.

The Kuiper Belt is at a peculiar position in the solar system. First of all, it is basically the solar system's doorstep. It is a location from where missions directed outside the solar system could occur and also missions, which target the whole solar system, might benefit from a position at the Kuiper Belt and the resources contained in its population.

Fig. 14.1 Global view of Pluto by *New Horizons*. This image is a combination of four separate images and was taken at a distance of almost half a million kilometers. *Source* NASA/Johns Hopkins University Applied Physics Laboratory/Southwest Research Institute

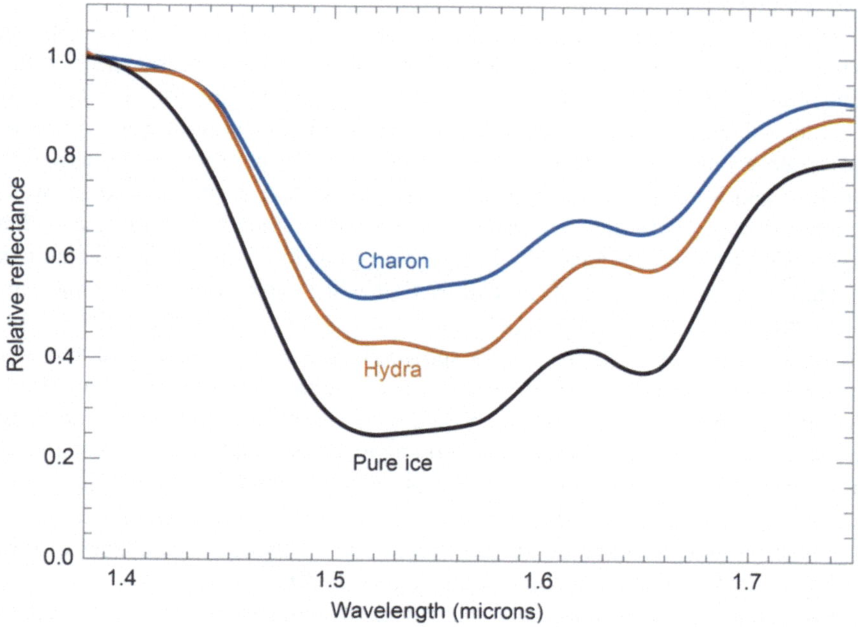

Fig. 14.2 Infrared spectra of Charon and Hydra as detected by *New Horizons*. Measured relative reflectance of the two Pluto moons Charon and Hydra in relation to pure water ice. *Source* NASA/Johns Hopkins University Applied Physics Laboratory/Southwest Research Institute

14.2 Opportunities for Exploitation

KBOs are still in the far reaches of the solar system, considering our technological capabilities. At the same time, it cannot be denied that there is a certain (currently unknown and unspecified) amount of resources available within the Kuiper Belt, which might prove useful for missions, creation of infrastructure and in general the development of human activities in the solar system. First, this includes the water ice, which would allow the production of e.g. fuel, or the other volatiles, which could serve for a wide variety of processes and applications (see Sect. 14.5).

Considering that presently any utilization of space bound resources has yet to occur, it is clear that any exploitation of resources in the Kuiper Belt will not happen in the near future. Nonetheless, even now plans exist for e.g. capturing asteroids with our current technology and thus prove the concept of investigating and exploiting it, e.g. in the frame of NASA's *Asteroid Redirect Mission*. Need and reasonable backdrop for exploiting Kuiper Belt objects, will likely only exist once humans have spread further out into the solar system and established a presence beyond Earth, which will first happen robotically and eventually possibly even with human crews, although not necessarily reaching KBOs. In any case extending humanity's reach beyond Earth, independent of the exact nature of that reach, will take resources.

Transporting resources from KBOs back to Earth is unlikely beneficial, as one major kind of resources found on KBOs are volatiles which can be used for fuel generation. Generating fuel with resources that have to be transported to Earth with fuel themselves in addition to large flight times to KBOs and back again, likely do not create a plus for a balance of such a mission. On the other hand, assuming missions and infrastructure within the outer solar system creates a need for water and propellant. These can be derived from volatiles found on KBOs, and thus exploitation of KBO resources could provide a basis to sustain such undertakings.

The large solar distance of the Kuiper Belt will in any case lead to long communication paths and mission times. Large mission times in addition also lead to significant operation costs, because a spacecraft needs to be supervised and controlled. So a classical approach of spacecraft, which are sent exclusively from Earth and operated by Earth-bound teams, would have significant limitations. Therefore, autonomous spacecraft are a reasonable prerequisite. Similarly, any exploitation of the Kuiper Belt should depend as little as possible on supplies from the inner solar system, e.g. any equipment should possibly be created in situ, to enable or expand any such mission for exploration in a large scale. This could culminate in fully replicating probes that autonomously operate at a KBO.

Spacecraft in the Kuiper Belt can support a number of activities, ranging between the two extremes of mere scientific exploration to pure exploitation, i.e. mining KBO resources such as volatiles, water or even precious metals. Likely, actual missions will be placed somewhere in between, because investigation of KBOs will not only reveal useful resources, but information about KBOs in general and thus e.g. about the formation of the solar system.

The location of the Kuiper Belt can help establish an infrastructure in the solar system, which allows support of human presence on other bodies besides Earth or even extending humanity's reach beyond the solar system using robotic probes to explore the interstellar medium or even nearby stars. The native resources including the bodies as such could be used to establish a port, used for leaving the solar system or transferring spacecraft from one outer body to another. Beyond creating a kind of harbor, the resources could also enable creating and producing spacecraft, which leave for specific missions, or at the very least to refuel such spacecraft coming from other regions.

The local bodies could further be harvested for resources, which could be used in other parts of the outer solar system, e.g. establishing bases at other planets or their satellites. An autonomous factory could likely produce robotic spacecraft for exploring the outer solar system or even interstellar space, not unlike the laser beam propelled light sail spacecraft recently presented to the press by Prof. Stephen Hawking et al. and described by Forward (1984).

Pure science missions could include observatories, likely of a large size. Due to the low gravity, large constructions are possible. The absence of a stress loaded launch would further benefit the size of any telescope. Other science missions benefiting from a KBO infrastructure could target the Oort Cloud or e.g. Sedna in particular. Their spacecraft could be produced on a KBO or be refueled there, coming from Earth.

There are other scenarios, which are even more far-fetched than mere exploitation and it is very uncertain if these scenarios will ever be possible or necessary.

On a (very) long term, KBOs—especially those with large water content—could be used to help terraform Mars or other worlds, e.g. Titan by transporting KBOs into their close vicinity and cause collisions, adding water to its environment.

Alternatively a KBO could serve as spaceship or habitat, providing a structure, protection and a large amount of resources (e.g. water) for the crew or the KBO could serve as a tethered counterweight for creation of artificial gravity in addition to providing resources.

Owed to the long travels times, KBOs can likely be ruled out as tourist attractions for space tourism even for a more technologically advanced civilization.

These scenarios are very unrealistic at the moment. Any inclusion of KBOs into mission scenarios is currently ambitious and doing so necessitates a working infrastructure in that area of the solar system, which allows accessing its resources. Describing the set-up of such a scenario to access these resources is therefore the focus of the following sections.

14.3 Trajectories Regarding the Kuiper Belt

14.3.1 Trajectories to and from the Kuiper Belt

To set-up a mission for exploring and exploiting the Kuiper Belt, one has to first reach that region of the solar system. The distant positions of KBOs within the solar system makes any rendezvous or even flyby mission ambitious. *New Horizons* was only able to flyby Pluto (and the same is true for its subsequent target bodies) instead of orbiting it, due to limits of propellant mass. However, mission designs exist, which allow rendezvous within reasonable mission times of less than two decades, utilizing a Radioisotope Thermoelectric Generator (RTG). Reaching Sedna, an object even more distant and likely already a part of the Oort Cloud, is possible for a mission time of about 20 years also using RTGs and in this case also multiple gravity-assists. For both cases ion propulsion is assumed (Bramanti et al. 2009).

Review of several flyby strategies revealed the best performance, in terms of minimum propellant mass, for a sequence of Earth-Venus-Earth-Jupiter and a flight time of about 11 years. Generally, non-gravity-assist trajectories are considered as unfeasible by Yam et al. (2004), due to the large flight time. This infeasibility is mostly due to the fact that mission outcome needs to be accessible by a single science team involved with the mission within the period of their professional lives, and mission costs need to be kept low regarding operations demands. However, autonomous spacecraft might reduce this cost and effort and relax demands on flight time altogether as long as their own lifetime is not exceeded by the flight time to reach their target.

Generally, mission opportunities for reaching Pluto utilizing gravity-assists occur regularly. Missions including a gravity-assist at Jupiter are possible about every 13 years. The giant planet enables flight times of 7 years to the dwarf planet (Longuski and Williams 1991).

From an energy point of view, a Hohmann Transfer would result in near minimum propellant requirements (Prussing 1993) for a transfer into the Kuiper-Belt. While bi-elliptical transfers would have even smaller demands in Δv, the even larger transfer times, due to the need of transferring to infinite distance for the first part of the transfer, rule it out for any useful application in this scenario. Even for a Hohmann Transfer from Earth's solar distance to about 40 AU (i.e. Pluto's semi-major axis) the Δv is about 15.6 km/s, which is not insignificant and comparable to a transfer from Earth to Saturn, and the travel time is about 46 years. Departing Earth's own orbit around the Sun with more velocity than needed for a Hohmann Transfer, could reduce the flight time in exchange for increases in Δv requirements. Actual flight paths depend on the targeted body and its phasing with regard to the starting body.

At the same time, if resources are needed elsewhere than the Kuiper Belt, transfer also has to occur back into the inner solar system. While it is unlikely that they will be transferred back to Earth due to the corresponding Δv demand

Table 14.1 Minimum Δv demand and flight time for returning to the inner solar system from 40 AU

Target body (solar distance in AU)	Flight time (years)	Minimum Δv [km/s]
Neptune (30.1)	103.8	0.72
Uranus (19.2)	80.5	2.04
Saturn (9.6)	61.8	4.65
Jupiter (5.2)	53.7	6.81

The Δv demand for assumed circular heliocentric orbits for starting and arrival bodies and applying a Hohmann Transfer. No orbit insertion and no inclination change are included

(identical to the flight from Earth to the Kuiper Belt), other regions of the solar system might be more suitable. Table 14.1 lists the Δv demand for the return flight and the respective flight times to different planets. It can clearly be seen that from an energy point of view the return flight is less stringent, due to the fact that the solar distance is larger. At the same time, the flight time is increased compared to the transfer to the Kuiper Belt, due to the fact that the orbits of the outer planets are larger and thus the overall transfer ellipse is larger too. As the Δv is moderate it is likely acceptable to decrease the flight time for more Δv.

14.3.2 Trajectories Within the Kuiper Belt

The challenge for trajectories within the Kuiper Belt is not the Δv, as can be seen in Fig. 14.3, but the flight times. The diagram shows the flight times and Δv demands for transfers from Pluto's solar distance at about 40 AU to the end of the Kuiper Belt at about 50 AU. Reaching a solar distance of 42 AU from 40 AU takes a Δv of about 115 m/s, due to the small differences of orbital velocities, but the transfer time—due to the large solar distances involved—exceeds 130 years. This is a hindrance for a mission which includes a transfer from one KBO to another. Due to the small Δv involved, it is acceptable to reduce the flight time needed for a transfer at an increased Δv cost, however.

For example, assuming a direct transfer from Pluto's position to Haumea's at 42 AU would require a Δv of about 2.5 km/s for a flight time of approximately 20 years. A flight time of 8 years would require a Δv of about 10 km/s, which is already in the same order of magnitude as the original transfer from Earth to the Kuiper Belt.

14.3.3 Low-Thrust Options

Low-thrust engines have the advantage of large efficiency and thus take smaller amounts of fuel for the same Δv in comparison with chemical thrusters. Especially

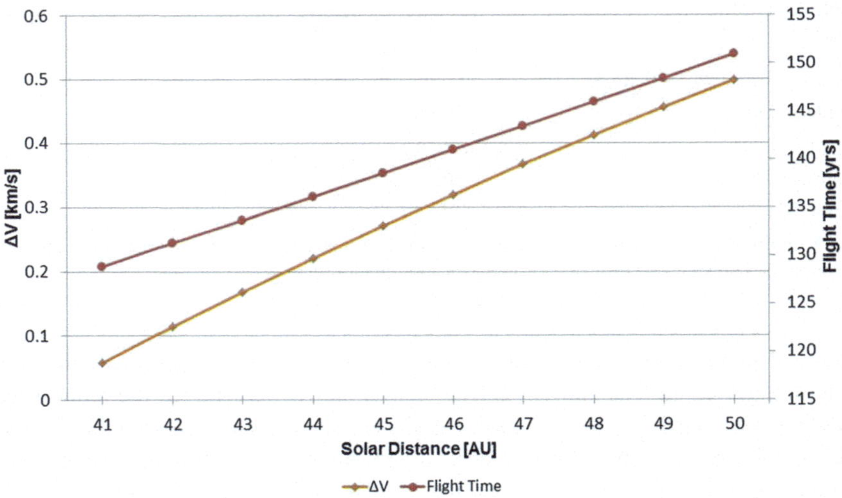

Fig. 14.3 Minimum Δv demand and maximum flight times for transfers from Pluto's position (40 AU) to other KBO distances. Based on Hohmann Transfer calculation, the interpolation can give an indication of in-between values, but are not accurate

in combination with nuclear fission or even fusion power they would be an effective method to transport a spacecraft to a KBO or between them. The Δv for a low-thrust transfer can be determined as the difference between the circular orbit velocities of the involved bodies—as a first estimate based on the Edelbaum equation (Messerschmid 2010). In this case, this would mean a Δv of about 25 km/s for a transfer from Earth's solar distance to a solar distance of 40 AU. Assuming a specific impulse of e.g. 4000 s, which is well within the range of typical low-thrust engines (Sutton 2001), the corresponding mass ratio of propellant mass to launch mass would become about 50%, when applying the rocket equation (Sutton 2001). While this is a significant advantage over chemical propulsion (a Δv of 15 km/s would result in a mass ratio of 98% for a system with 400 s of specific impulse), this still means large thrusting times for such an engine and also a large flight time.

Especially the latter makes it less attractive, although the usage of gravity-assist maneuvers could effectively reduce the flight time by providing free Δv in significant amounts almost instantaneously (compared to the overall flight time).

An exact spacecraft design and specific trajectories would need to be calculated to evaluate the usability of low-thrust for such a kind of mission.

14.4 Power Generation in the Kuiper Belt

As for any mission, the spacecraft conducting it has to have a power source. While currently the major power generator of spacecraft is typically solar-electrical this is not necessarily a valid option for a mission in the Kuiper Belt, considering that the solar energy at 40 AU is a factor of about 1600 smaller than at Earth's solar distance. While solar power generators could power a spacecraft during portions of the transfer from Earth to the Kuiper Belt, especially powering within the Kuiper Belt has to happen with other means. Table 14.2 lists properties at KBOs possibly relevant for power generation.

Three distinct mission profiles can be formulated for a power generation concept. First of all, the initial spacecraft and possibly factory that is transferred to the Kuiper Belt has to be powered after launch from Earth. Second, the mission components created at the KBO have to be powered and likely with a simple—easy to reproduce autonomously—power generator. Third, mobile units, which roam the surface of the KBO would need a different power generator as well.

The first two need very long life times (several decades) to allow the scenario to unfold successfully, whereas the third kind of power generator primarily needs to be small and portable to be used on mobile units. The technology of the in situ bases has to allow power generation that does not take more resources than can be obtained by the mobile units in order to be self-sufficient and to not make the whole mission of harvesting resources on a KBO a vain enterprise. Based on the thermal control demands (mostly heating of the system) and equipment to process any resources on site, several hundred kilowatts are likely to be expected as the system's power demand for the base on the KBO. Some orders of magnitude less can be assumed for the mobile unites, possibly some hundred watts. Without a complete design, which is currently not possible as the technological specifications of spacecraft components some decades in the future cannot be credibly assessed.

There are several technologies conceivable for utilization in the proposed scenario and considerations regarding them are summarized in Table 14.3, listing existing and theoretical solutions for power generation. Depending on criterions

Table 14.2 Environmental conditions at and on KBOs

Property	Value
Solar flux	1.5 W/m^2 (at 30 AU) 0.85 W/m^2 (at 40 AU) 0.54 W/m^2 (at 50 AU)
Pluto surface pressure	ca. 1 Pa (NASA 2015)
Pluto surface temperature	50 K (NASA 2015)
Pluto surface gravity acceleration	ca. 0.06 g_0
Pluto heat flow below ice sheet	2.1 mW/m^2 (Hussmann et al. 2006)
Pluto temperature below ice sheet	230 K (Hussmann et al. 2006)

These environmental parameters relate to the possibilities of generating power from the environment by different means, e.g. solar or thermoelectrical power generation like *amount of generated power*, *in situ operation and dependence on environment*, *lifetime*, *in situ availability of consumables*, *technology readiness*, and *mobility*, each of these possible technology solutions can be evaluated for their implementation in the spacecraft and in mobile units.

Taking Pluto as a representative KBO, it becomes clear that any power generation involving aerodynamics, e.g. wind energy, will be useless due to the small atmospheric pressure. The very low surface temperature also prevents thermoelectrical power generation on Pluto. Possibly the temperature below the ice sheet can be used, however the ice sheet is an effective barrier that needs to be overcome. Considering the likely heat flow of about 2.1 mW/m^2, it becomes clear that a significant surface area (some 100,000 m^2 for a required power of several hundred Watts) would be needed to achieve effective power generation. As setting up such a power generator would require energy in its own right, the large required heat exchanger area rules out such a system for the time being. Especially since it had to be set up remotely and autonomously.

Table 14.3 Potential power generation technologies on KBOs

Technology	Advantages	Disadvantages
Solar power generation	Space proven	Small solar flux in Kuiper belt, subject to Low Intesity Low Temperature (LILT) effects
Radioisotope thermoelectric generator (RTG)	Space proven	Rare fuel, fuel decay decreases generated power decreases by ca. 2.5% per year (O'Brien et al. 2008)
Nuclear fusion power generation	Can use hydrogen (deuterium, tritium) for fusion, no negative influence on fusion process due to KBO location	Large size required, currently only testing and no working technology available
Nuclear fission power generation	Some development for space application, long life time possible	Uses rare isotopes (likely even rarer on KBOs)
Geothermal power generation	Uses power source without fuel	Cycle process involving some kind of fluid (which needs to match environmental conditions on KBO), not applied to space yet
Fuel cells	No degeneration when not in use, operational for some months, can be used to store energy, have space heritage	Need refueling replacement for long term usage
Batteries	Can be used for energy storage, space heritage	Degeneration by usage and when not in use, need recharging
Power beaming	Transferring power from another location possible	Losses similar to solar power reduction due to distance, no proven technology

This table includes potential technologies with a short list of relevant advantages and disadvantages showing their suitability for a prolonged mission in the Kuiper Belt

The small amount of power available on Pluto basically rules out geothermal and solar power generation, despite their high technology readiness and space heritage of the latter. To allow a self-sufficient operation and enduring mission concept, any fuel used for power generation must be readily available or must be supplied in sufficient amount for the whole mission duration, therefore nuclear fission and also RTGs are at a disadvantage. Batteries have to be disregarded as main power source, but are certainly included as energy storage and back-up power supply. Their degeneration (even when not used) makes them unsuitable as primary power source for the desired mission, due to the long mission durations.

Power beaming has the disadvantage of significant power losses over distance (similar to solar power) and also eclipse phases. For a large number of spacecraft this power generation option becomes increasingly complex, because each power beam connection needs to be maintained and increases the number of elements that need to be coordinated or made to operate autonomously by a factor of 2 at least (assuming a pair of power source and power target).

Generally, one can say that the power sources available in the Kuiper Belt are not sufficient to support a mission or a series of missions as necessary to carry out the exploitation and thorough exploration of KBOs. That means the energy has to be brought along with the spacecraft or needs to have fuel resources on the respective KBO.

The latter aspect is fulfilled by nuclear fusion power. The major advantage is that the fuel can be found in the Kuiper Belt so spacecraft and mobile units would be independent of resupply from Earth and also the power output for such a system is significant, approximately equaling $E = \Delta mc^2$ (losses for energy transformation need to be regarded in addition), where Δm is the mass difference before and after the fusion reaction and c the speed of light (Atzeni and Meyer-ter-Vehn 2004).

Various processes for nuclear fusion exist, involving lithium, deuterium, tritium or ordinary hydrogen. Hydrogen and its isotopes can be obtained from water ice available at KBOs, although tritium is usually very rare. The water ice has to be processed and the deuterium separated from the ordinary hydrogen and oxygen. Separated hydrogen and oxygen can be used for fuel cells that drive the mobile units, which in turn mine H_2O on the KBO surface. The fusion reactor also provides power for processing the ice. Excess hydrogen and oxygen are stored, possibly in the form of methane in the case of the former, because this can be stored for longer durations. This cycle is illustrated in Fig. 14.4.

Assuming a power demand of 5 kW as an example for the base landed on the KBO would result in an energy amount of approximately 1.6×10^{11} J per year. Further assuming a deuterium/deuterium reaction, because deuterium is more abundant than tritium an average energy output per reaction of two deuterium atoms can be determined to ca. 3.65 meV (Atzeni and Meyer-ter-Vehn 2004), which equals a deuterium mass of 1.145×10^{-14} kg/J. Therefore a mass of 1.8 g of deuterium would be needed for such an energy production. Assuming a deuterium rate similar to hydrogen found on Earth, i.e. 0.015% is deuterium, results in a mass

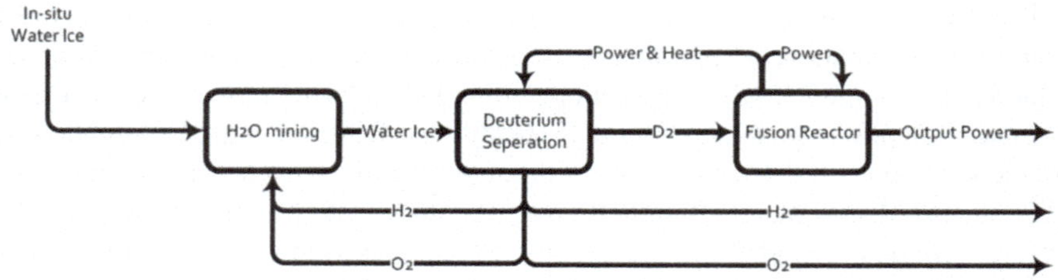

Fig. 14.4 Proposed cycle for power generation on a KBO. Power generation is achieved by nuclear fusion applying deuterium, obtained on the KBO. Hydrogen and oxygen separated from deuterium is stored and can be used for fuel cell as well. Hydrogen storage can occur in form of methane

of 12 kg of hydrogen that needs to be mined each year for fueling the fusing process. This does not involve any losses, which need to be added. But even at an efficiency of 10%, this would still only mean about 120 kg of hydrogen required per year for such an amount of power. The Jupiter-family comet visited by the *Rosetta* mission, *67P/Churyumov-Gerasimenko*, has a deuterium-hydrogen rate three times higher than that on Earth and likely originates in the Kuiper Belt. This suggests that even less hydrogen is needed if the distribution if deuterium is homogenous in the Kuiper Belt (Altwegg et al. 2015). For the previous example, this would mean about 40 kg of hydrogen need to be processed. A base needing about 100 kW would consequently require about 160 kg of hydrogen, respectively 480 kg. Summarizing, the need for hydrogen would be about 1.6–4.8 kg/kW of power output.

This is also relevant for replenishing the deuterium storage after arrival at the KBO, if the fusion power system has been used for the transfer yet. Careful planning and resource harvesting is needed to ensure that enough deuterium is gathered to replenish the supply.

Currently, the most advanced reactor type is the tokamak, which is also the basis for the International Thermonuclear Experimental Reactor, the experimental research reactor, which is intended to prepare nuclear fusion for actual operation in the coming decades (Gao 2016).

There are various research approaches to reduce the size of such reactors and create a compact design, although there are still limitations. One example would be the development of a spherical tokamak reactor (Gao 2016) or the so called *affordable, robust, compact reactor* (ARC), which is a tokamak reactor of about 3.3 m diameter and with about 200 MW of electrical power output (Sorbom et al. 2015). Limitations are primarily the reaction, which focuses on deuterium/tritium reactions. Tritium cannot be expected at KBOs so, using it for fueling a reactor would need resupply, which reduces the advantage of nuclear fusion. Furthermore the neutrons involved in the fusion reactions (Atzeni and Meyer-ter-Vehn 2004) require strong shielding (ca 1 m), which sets a lower limit for the reactor diameter and thus power output in the 100 MW range. Other types of reactions exist (Atzeni and Meyer-ter-Vehn 2004), but are not as advanced in their research, so a usage in an actual space proven reactor would take a lot more time and resources to develop.

One alternative could be the compact fusion research by Lockheed Martin's Skunkworks, which is intended to be used i.a. for powering planes, which would be in the range of some MW instead of several hundred. However detailed data is not available and thus a thorough analysis is not possible.

To encounter the fact that the actual power output of a fusion reactor exceeds the power demand by several orders of magnitudes, it would be possible to intermittently run such a reactor. Energy storage, e.g. via fuel cells or even batteries, and restart the fusion reaction once the stored energy is reduced to a certain minimum amount. Secondary power supply would in any case be needed to overcome periods of malfunctions or other unforeseeable events. However, that would require an autonomous restart of the reactor for several times and given the ratio of power supply to power demand, this would result in hibernation times several orders of magnitude larger than the times of operation. Depending on the efficiency of the storage and the amount of stored energy, the ratio of hibernation to time of operation would be $k = P_{\text{reactor}}/P_{\text{need}}$, where P denotes the respective power. For the present example of a power need of 5 kW and a power supply of 100 MW, k would become 20,000, i.e. operating it for 1 h would result in hibernation of 2.5 years. This however would be reduced by the factor of efficiency of the storage (e.g. if only 50% of the power could be stored, the hibernation factor would become 10,000) and the total amount of energy can be stored and if the needed power would be larger. This would also benefit the need to replenish the deuterium reserves after the transfer, because the deuterium would not need to be enough for a continuous operation. A power demand of 100 kW would mean hibernation times of about 40 days for an operation time of 1 h.

While the excess power, which for a fusion reactor occurs in the form of thermal power, could be used to thermally control the spacecraft or base to some extent, it also needs to be released to the environment without heating up the system in an undesired manner. In any case using fusion power at the moment seems to be excess means of power supply, especially since the effort to reach a technology readiness level for space application of such a reactor would still be significant.

Nuclear fission is an alternative, which is able to generate power over long periods of time and in the desired magnitude but it lacks refueling capability. Any radioactive material needed for a nuclear fission reactor would have to be transported from Earth to the KBO. This would restrict the amount of spacecraft and prevent full self-replication. A seed for the exploration and exploitation of any KBO would need to come from Earth.

NASA is currently developing nuclear fission power systems in the range of some kW for usage on Moon and Mars under the term "Kilopower". Depending on the nature of conversion between thermal and electrical power, concepts exist for reactors in the 3 kW range, with a power specific mass of about 300 kg/kW. Therefore a system in the 5 kW range would have a mass of 1500 kg. The design of the reactor is explicitly intended for use in KBO missions (Mason et al. 2013). Reactors with larger power could achieve a better specific mass, in the range of some 10 s of kg/kW of electrical power, enabling usage for nuclear propulsion in the range of some 100 kW (Mason et al. 2013).

One major advantage however is the space heritage. The Soviet Union launched several nuclear reactor powered satellites in their RORSAT program, the United States have launched SNAPSHOT in 1965, which was powered by a nuclear reactor (Portree 1999).

Nuclear fission reactors could serve as a base power supply and be used to charge e.g. fuel cells of the mobile units. It could also be used as a power supply for the initial spacecraft. In any case, full autonomy from Earth cannot be achieved with nuclear fission.

14.5 Obtaining Resources and In Situ Production

Besides scientific considerations one major reason for setting up an infrastructure on KBOs is to obtain their resources for further use. Especially since resources have not yet been harvested on another celestial body, such an undertaking is naturally ambitious and challenging.

14.5.1 Environment and Requirements

The moderate size of KBOs from some hundred kilometers up to about one thousand kilometers and the associated mass result in a small amount of surface gravity acceleration. The cubic relation between size and mass means that smaller bodies have considerably smaller gravity accelerations. Pluto, for instance, has a surface gravity acceleration of about 6% of Earth's standard gravity acceleration (9.80665 m/s^2).

Therefore, any resource harvesting relying on forces (and reaction forces) directed towards the ground has to take that into consideration, possibly by using some sort of anchoring mechanism to the surface.

Depending on seasonal variation, Pluto's surface pressure is about 1 Pa, which is a likely limit for the surface pressure of other KBOs. Limited gravity also restricts the possibility of existing atmospheres. Due to the small atmospheric pressure, melting any volatile ices on the KBO surface, will likely result in their sublimation, i.e. the immediate transformation into gas. In fact sublimation of the existing ice is the process behind the atmosphere generation on KBOs in the first place (Stern and Trafton 2008). Keeping this in mind any means of collecting ice or melting it, has to foresee a mechanism to prevent the escape of the desired resource. At the same time the surface temperature on KBOs is some tens K, on Pluto approximately 50 K. This has to be taken into account when regarding the operation of mining equipment. Besides surface temperature, the solar distance is also resulting in a low illumination of the surface, i.e. visual detection of resources is limited.

The surface is not only consisting of ice, but it has to be assumed that regolith, rocks and boulders are covering—if partially—the KBO surface and also hold valuable resources, which can be harvested and used.

The mining equipment therefore has to deal with the following list of requirements:

- it has to be operational in the environmental conditions found on KBO surfaces (regarding temperature, pressure, gravity, low light),
- it needs to be able to process various types of materials, e.g. dust, rock and ice,
- it has to be of a complexity that does not prevent autonomous reproduction,
- it has to operate with a net gain of resources (life-time and required resources for reproduction shall not prevent this),
- it has to reach all relevant depths for reaching resources.

14.5.2 Resource Harvesting Mechanisms

A number of mechanisms for mining resources are possible, partially redundant. Drilling, rock grinders, explosives and others are all means for mining already applied on Earth. However, from the vast technologies available only a limited number are practical for application on KBO surfaces. E.g. excavators rely on gravity to prevent the material from escaping. Facing the requirements from Sect. 14.5.1, a reasonable list of possible harvesting mechanisms can be given.

For collecting materials two mechanisms are foreseen: A robotic grappler can be used to obtain rocks and boulders. Electrostatic adhesive technology can be used to collect regolith particles from the surface (Kawamoto and Uchimaya 2008). On the other hand, a scoop could be used in combination with electrostatic adhesive technology to allow one mechanism the collection of both regolith and rocks alike.

For harvesting ice, a cold trap can be considered, which exploits the easy sublimation of ice on KBO surfaces. By heating the ice with an infrared source, sublimation is caused, i.e. vapor is created, which meets a cooled plate, resulting in the deposition of the gaseous volatile on the plate. Thus it can be collected, stored and further processed. The concept is illustrated in Fig. 14.5.

Thermal fractioning can be used for mining rocks, respectively for creating smaller rocks in the factory for easier processing and collection. Applying heat to a rock, differences in heat expansion rates by the respective components cause fracturing of the rock. Heating and cooling can both be used to create smaller rocks for processing. This technique is already applied on Earth (Dobretsov and Sukhanov 1965). Heat could possibly be applied by radiation, e.g. a laser beam. The same equipment could also be used to heat ice for the cold trap mechanism, allowing to save system mass.

Reaching subsurface areas for collection of resources can be achieved by using an ordinary drill (as stated before using an anchoring device at the same time).

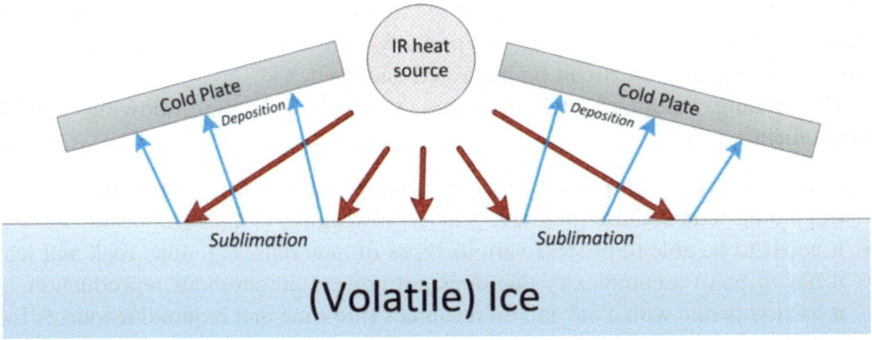

Fig. 14.5 Cold trap mechanism for mining ices. Using a heat source, volatiles are caused to sublimate, which deposits them on a cold plate, where the volatile refreezes and can be moved into storage

While more advanced technologies exist, e.g. thermal core drilling, these are more complex and thus less easy to reproduce and maintain.

14.5.3 In Situ Production

Considering the current state of technology, autonomous production in a space environment is not possible. However, automation is already a large part of current industry and production.

A promising new material that has the ability to be used for many different purposes are Carbon Nanotubes (CNT). They can be used to form structures, semiconductors and whole circuits, even memory. A research group at Stanford University already managed to create a computer based on nano-tubes that can execute simple operations (Shulaker et al. 2013).

Depending on how many walls they have, their material strength exceeds that of steel by more than one order of magnitude while having a density smaller by a factor of about 4. They can be combined with plastics and are currently used in aircraft for reinforcing polymers (for example Lockheed Martin's F-35) and have been used to create improved batteries and solar cell parts as well (Li et al. 2011).

In addition to basic materials for e.g. the spacecraft structure, fuel needs to be produced as well, as it is needed for transferring spacecraft into the inner solar system, to explore the Kuiper Belt and move on any KBO surface.

Hydrogen can be found in form of water ice and can be processed. Separating deuterium from this ice leaves ordinary hydrogen for further usage. An alternative, which is easier to store would be Methane, which can be found as ice on e.g. Pluto. Of course oxygen needs to be used as oxidizer, but it would also be available in the form of water ice. Nitrogen as e.g. detected on Pluto's surface can be used as pressure gas for feeding propellant. The list of possible fuels can be summarized as:

- Hydrogen (H_2)
- Hydrogen Peroxide (H_2O_2)
- Methanol (CH_3OH)
- Methane (CH_4)
- Hydrazine (N_2H_4)

Especially Hydrogen Peroxide and Hydrazine can be used as monopropellants, which would be favorable in terms of complexity of a space system and its production, but production of both fuels themselves is complicated. A detailed analysis of the fuel creation process and propulsion system has to be conducted to make the decision about suitability.

While currently not far developed, 3D printing has already become a production process with some success and is even being tested on ISS. The advantage of 3D printing would be the possibility of using regolith as base material (Ceccanti et al. 2010); recycling is possible. An example of an experiment conducted by ESA in 2013 using artificial lunar soil is given in Fig. 14.6.

Some 3D printing techniques have been developed for usage in space already. Once programmed 3D printers could autonomously produce any parts needed, even complex parts, reducing the demand for assembly. Similarly, a probe on a KBO could set up a factory with the help of a basic 3D printing mechanism brought along.

Fig. 14.6 Lunar regolith printed in a block. This block was printed by ESA as an experiment for creating lunar outpost structure parts using (artificial) lunar regolith. The block has a mass of 1500 kg and has a lightweight design. *Source* ESA

14.6 Mission Architecture and Components

The previous sections have reviewed various fundamental aspects of a mission scenario involving the exploration and resource harvesting of KBOs. The question remains how these parts can come together to realize such a mission.

The basic mission layout is assumed to involve self-replication. An autonomous factory is sent to a KBO, where it creates rovers for resource gathering on the surface and exploration of the environment. The gathered resources allow subsequent production of more rovers and more spacecraft in turn. The initial spacecraft will consist of a basic factory incorporating 3D printing as base production process, one mining rover, utilizing the harvest methods described in Sect. 14.5.2, and an exploration rover to map and survey the KBO. Example draft designs are provided in Fig. 14.7. The exploration rover is equipped with sensors, e.g. capable of spectrometry, whereas the mining rover is mainly equipped with a coldtrap and a scoop.

The gathered resources are applied to reproduce rovers and increase the survey and harvest speed. Once enough material is collected, a new factory is created, increasing also the production speed and adding redundancy. Ultimately, a network of factories can be created to effectively mine the KBO.

The next step of production will create spacecraft. First, orbiters need to be produced to better survey the KBO and identify more promising mining regions (especially once the immediate surroundings of the factories are depleted of useful resources). Second, transport spacecraft are created that actually transport the resources back to a given point of interest, e.g. into the inner solar system for utilization.

As a last step, it should be possible to create vehicles that transfer the initial infrastructure to a new KBO and restart the cycle there, beginning with one factory and two rovers as before.

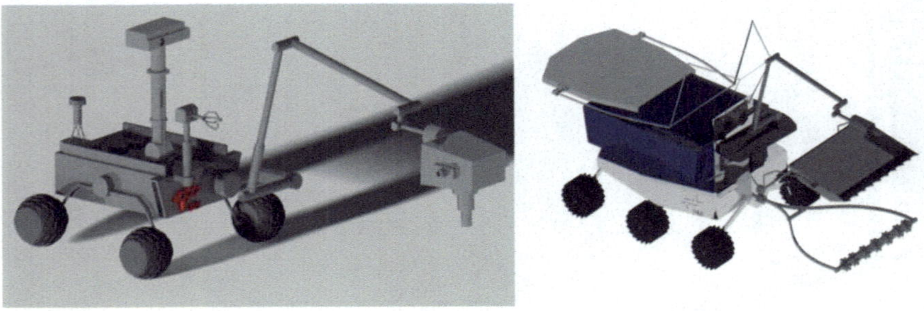

Fig. 14.7 Possible rover designs for harvesting (right) and exploring (left) KBOs. These are draft rover designs, for the purpose of exploring (left) and harvesting (right) resources on KBOs. A number of sensors, including spectrometers serve the explorer to survey material composition, whereas a cold-trap (plate in the aft section of the rover) and scoop are used by the mining rover to harvest material

Depending on complexity and the state of technology at the beginning of the mission, it might be necessary to have certain parts preassembled, e.g. the factory or the reactors for power generation, if they cannot be reproduced autonomously. Due to the limited amount of parts that can be transferred to KBOs, this would restrict the size of the mining network and increase the amount of time needed for mining resources.

Basic operations of the spacecraft should be as autonomous as possible. Power and mass estimates in the coming sections, describing the individual components of the mission more elaborately, are based on existing space systems and are only a vague orientation (Larson and Wertz 1999).

One major obstacle at the large solar distance is the thermal control, as the temperature is only about 50 K (NASA 2015). However technology liker heaters and insulation have enabled missions like New Horizons to operate in the same environment, therefore it is assumed that this is a manageable issue.

14.6.1 Factory Design Draft

The factory is the heart of the mission as it acts as base station on the KBO, which is not only responsible for the fabrication of new parts, but also for power generation (see Sect. 14.4). It consists of several subunits with different tasks. A chemical sector uses various chemical means to process the resources, and transform them into other conditions for e.g. easier handling or storage.

The fabrication unit is an advanced 3D printing unit that actually creates the desired products, including structure elements and circuits. A robotic assembly unit is responsible for assembling the parts produced there, if they are too large to be fabricated as one element.

Production processes and parts are highly standardized and therefore need only a limited amount of testing, which can be done automatically without human oversight. Elements which are faulty are recycled and reprocessed. A storage unit allows storing parts and resources. The fabrication process is summarized in Fig. 14.8.

As base station and can be used to resupply mobile units with power, e.g. depending on the exact nature of its own power generation systems.

14.6.2 Rover Design Draft

As described before, the scenario will use two kinds of rovers for harvesting and exploring KBOs, which are presented in a draft layout in Fig. 14.7.

The mining rover is equipped with a core drill for reaching below the immediate KBO surface, a cold trap for gathering volatile ice and a scoop. Navigation is achieved by an infrared camera and possibly a lighting system, which also aids material identification. Applying current technology standards, a rover with that

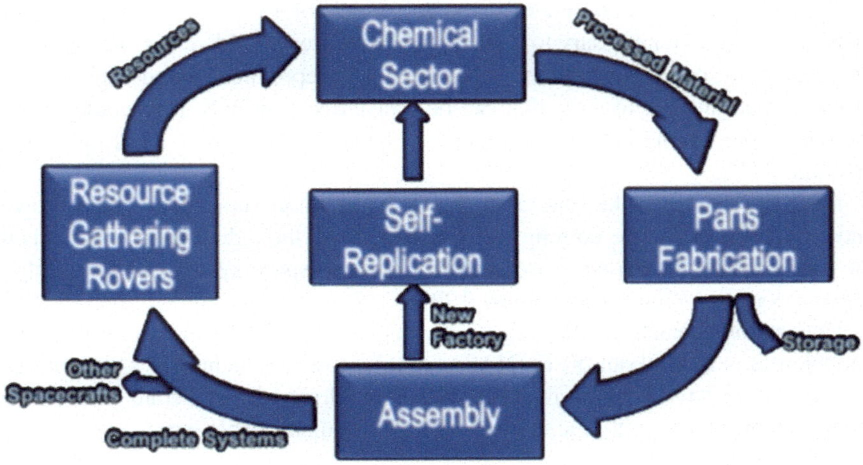

Fig. 14.8 Production process as envisioned for a KBO autonomous process. Production is started by the gathered resources, accumulated by the rovers. The resources are processed the chemical sector, which allows the fabrication of parts. These can be stored and assembled (which eventually leads to new factories or spacecraft)

task is likely in the mass range of about 2000 kg and 2 kW of power, which would be about double the mass of NASA's *Curiosity* rover. Sizing will likely also depend on the transportation requirements, i.e. how many resources are transported.

The exploration rover is likely smaller, because its major task is investigation and surveying the KBO surface, and therefore it should be fast. A sounding radar, microscope and spectrometer are in its arsenal for exploration. Applying current technology standards, the rover is likely in a mass range of about 400 kg and needs less than 1 kW of power, making it about twice the mass of NASA's *Spirit* and *Opportunity*.

In both cases power is supplied by fuel cells, which can be refueled at the factory using gathered resources.

14.6.3 Spacecraft Design Draft

There are two spacecraft, which are to be produced during the mission. The first spacecraft is the orbiter, which basically extends the missions possibility of surveying its own target KBO and also extends it to other KBOs. A draft layout of the spacecraft is depicted in Fig. 14.9. The orbiter can also act as relay and thus help coordination and communication between all rovers and factories. Basic sensors, e.g. infrared spectrometry, are used for survey operations. A propulsion system is used for orbit maintenance and launch from the surface (at this point the large antenna is likely folded). A larger version of the orbiter can be used to reach other

Fig. 14.9 Draft design of a KBO orbiter. A produced orbiter including a large antenna, payload sensors and a propulsions system

KBOs. Depending on the exact nature of the power system and the possibility of miniaturization, the spacecraft could have a mass of several hundred kilograms and a power demand of some kW. If the power generation is using nuclear fission, a prefabricated reactor along with fuel should be used. If nuclear fusion can be used, it would be possible to fuel the spacecraft insitu.

The other spacecraft, used to transport the resources to different regions in the solar system, is mostly only a large storage hold with a propulsion system used to transfer resources to any point required. Its size and mass are mainly driven by whatever amount of resources is found reasonable and useful for transportation during the design. The transport spacecraft could also be used to carry a factory to another KBO and restart the process of harvesting it.

The initial spacecraft will be produced on Earth and will have to carry the factory and the two before mentioned rovers. It can be assumed that e.g. the power supply of the spacecraft can use the power generator of the factory, especially since it is intended to be operating for a long time anyway. The spacecraft would have to endure the environmental challenges at KBOs, transport all relevant equipment to the target body and operate largely autonomously. The size would probably be at least ten tons as the reactor and its shielding will need a large amount of mass. The power needed would probably not exceed the power demand of the factory, so that the base stations power supply would suffice to operate the spacecraft during its voyage.

It might also be possible to refuel the initial transfer vehicle's propellant and use it for transporting a factory to a different KBO, to restart the process.

14.7 Roadmap for Exploitation

Considering the technology development still needed to set a scenario as described here into motion, it can likely be stated that this will not happen in the near future. An assumed development time—based exclusively on technological aspects and without considerations regarding the involved political achievements necessary to allow such a scenario—of at least 50 years (especially for fusion power sources in space) would place a launch in the mid to late 60 s of this century. Depending on the exact trajectory a flight time of about 15 years can be assumed, see Sect. 14.3.

To increase the efficiency and reduce the risk of mission failure it should be assumed that several spacecraft are sent to KBOs. Due to the fact that their designs would be identical, the costs would only involve those for reproduction and launch. The autonomous nature of the spacecraft should lead to moderate operations costs. Once the KBO is reached in about 2080, the harvesting and reproduction starts.

Approximately 5 years later, a second mining rover would be created, assuming a harvest rate of about 800 kg of mined resources per year at an efficiency of 50% (i.e. 50% of the collected resources are invested into the reproduction). This would essentially double the harvest rate and allow the creation of two more mining rovers after another 5 years. Once enough rovers are created, 2100, i.e. 20 years after touch down on the KBO surface, the creation of a first orbiter occurs, allowing better survey capabilities. Another 20 years later, after more rovers have been produced (and at least in parts existing ones have been replaced), a second factory would be created, basically restarting a similar timeline for this one as before.

Working together both factories could establish a transport link to other areas in the solar system sometime after the 2150 s, when the second factory is also capable to reproduce.

14.8 Open Issues and Conclusion

This chapter summarizes a draft for a scenario capable of exploring and exploiting multiple KBOs in an autonomous manner. While the seeds for all relevant technologies are there, the scenario is ambitious even with technological progress.

Currently 3D printers are incapable of manufacturing 100% of their own parts, i.e. a factory that is able to reproduce itself and possibly even fully functional spacecraft is not a reality but fiction, although not without basis in fact. The estimated timeframe for technology progress is regarded as conservative, considering the advances of 3D printing and carbon nanotubes in the last 50 years. One example of advanced "replication" would be the *Thor* drone of Airbus, which is manufactured completely by 3D printing.

The issue of power generation is another main open point for this scenario and likely the major open point. Nuclear fusion is still in its early stages of development and, although its potential has been thoroughly researched, its development is very unpredictable and highly dependent on results of upcoming experiments. However, its benefits for power generation on Earth disconnect the development from space application alone and thus increase the chances of success (and funding). Once nuclear fusion power generation has been developed successfully for terrestrial applications, replacing current power generation technology, it can be further developed for space application. It could be possible to launch this scenario with fission power and advance the involved spacecraft into a modernized version once fusion power becomes available for space application. Full autonomy from Earth can only be achieved with fusion power, because only then new fuel can be gained at KBOs as fission material cannot be found there. Consequently the full potential of the autonomous self-replicating strategy is only available in case of fusion power generation at the KBOs.

The scenario as described here, relies heavily on autonomy of the involved spacecraft and rovers. Especially the latter cannot rely on ground control operation, because this would prevent the fast paced timeframe as intended for such a scenario. Therefore, autonomy must be improved to allow the rovers to navigate the KBO surface quickly, identify areas of interest regarding scientific discoveries and especially resource harvesting and also to research the specific areas on their own, reviewing the available resources and their suitability for the mission.

Another challenge is raised due to the extensive timeframe. The costs for the initial spacecraft are likely modest in comparison to the accessible resources and due to the autonomous nature of the spacecraft operations costs would probably also not be extensive. However, the actual mission results will occur after a timeframe in the order of 100 years. This, naturally, will restrain the interest of most (if not all) space faring nations, if the outcome—i.e. the amount of resources gained—is not significant enough. Nonetheless the idea of exploring and exploiting KBOs can be considered as reasonable even with today's technological limitations, if development progresses. The resources in the Kuiper Belt could be very useful in sustaining humanity in the solar system and even beyond.

The task of establishing the described infrastructure is beyond the capabilities of a single nation and will require the technological expertise, resources and experience of a world-wide cooperation. The *International Space Station* is an example of such cooperation, although it would certainly need to be intensified to achieve what is proposed here. Establishing such cooperation, coordinating the technological advancement and the accompanying mission program and maintaining the political framework to allow the execution of this mission scenario could prove even more challenging than the mere technological obstacles. While cooperation in e.g. nuclear fusion research is already of an international nature, this is merely a seed for what is needed to successfully access the worlds in the Kuiper Belt.

Acknowledgements The author thanks the 3rd Advanced Study Group of DLR, consisting of Marc Bernabeu, Joao Lousada, Claudia Terhes and Niels van der Pas for their valuable contribution to investigating this topic.

References

Altwegg, K., Balsiger, H., Bar-Nun, A., Berthelier, J.J., Bieler, A., Bochsler, P., Briois, C., Calmonte, U., Combi, M., De Keyser, J., Eberhardt, P., Fiethe, B., Fuselier, S., Gasc, S., Gombosi, T.I.. Hansen, K.C., Hässig, M., Jäckel, A., Kopp, E., Korth, A., LeRoy, L., Mall, U., Marty, B., Mousis, O., Neefs, E., Owen, T., Rème, H., Rubin, M., Sémon, T., Tzou, C.-Y., Waite, H., Wurz, P.: 67P/Churyumov-Gerasimenko, a Jupiter family comet with a high D/H ratio. Science **347**(6220) (2015)

Atzeni, S., Meyer-ter-Vehn, J.: The Physics of Inertial Fusion. Oxford Science Publications, Oxford (2004)

Bernstein, G., Trilling, D., Allen, R., Brown, M., Holman, M., Malhotra, R.: The size distribution of Trans-Naptunian bodies. Astron. J. 1364–1390 (2004)

Bramanti, C., Izzo, D., Samaraee, T., Walker, R., Fearn, D.: Very high delta-V mission to the edge of the solar system and beyond enabled by the dual-stage-4-grid thruster concept. Acta Astronaut. **64**, 735–744 (2009)

Brown, M.E.: The compositions of Kuiper Belt objects. Ann. Rev. Earth Planetary Sci. 467–494 (2012)

Ceccanti, F., Dini, E., De Kestelier, X., Colla, V., Pambaguian, L.: 3D printing technology for a Moon outpost exploiting lunar soil. In: 61st International Astronautical Congress, Prague, CZ (2010)

Delsanti, A., Jewitt, D.: Solar system update. In: The Solar System Beyond the Planets. Springer, Berlin (2006)

Dobretsov, V., Sukhanov, A.: Rock fracturing by the combined effect of positive and negative temperatures. J. Mining Sci. 171–173 (1965)

Encrenaz, T.: Water in the solar system. Ann. Rev. Astron. Astrophys. **46**, 57–87 (2008)

Forward, R.L.: Roundtrip interstellar travel using laser-pushed lightsails. J. Spacecraft **21**(2), 187–195 (1984)

Gao, Z.: Compact magnetic confinement fusion: Spherical torus and compact torus. Matter Radiat. Extremes **1**(3), 153–162 (2016)

Gladman, B., Kavelaars, J., Petit, J., Morbidelli, A., Holman, M., Loredo, T.: The structure of the Kuiper Belt: size distribution and radial extent. Astron. J. 1051–1066 (2001)

Hussmann, H., Sohl, F., Spohn, T.: Subsurface oceans and deep interiors of medium-sized outer planet satellites and large trans-neptunian objects. Icarus 258–273 (2006)

Jewitt, D., Luu, J.: Discovery if the candidate Kuiper belt object 1992 QB1. Nature **362**, 730 (1992)

Kawamoto, H., Uchimaya, M.: Electrostatic cleaner of lunar dust on solar panel and optical lens. In: LEAG/ICEUM/SRR—Joint Annual Meeting, Cape Canaveral (2008)

Larson, W., Wertz, J.: Space Mission Analysis and Design, 3rd edn. Microcosm Press, El Segundo (1999)

Levison, H.F., Donnes, L.: Comet Population and Cometary Dynamics, in Encyclopedia of the Solar System, pp. 575–588. Academic Press, Amsterdam (2007)

Li, S., Luo, Y., Lv, W., Yu, W., Wu, S., Hou, P., Yang, Q., Meng, Q., Liu, C., Cheng, H.: Vertically aligned carbon nanotubes grown on graphene paper as electrodes in lithium-ion batteries and dye-sensitized solar cells. Adv. Energy Mater. **1**(4), 486–490 (2011)

Longuski, J.M., Williams, S.N.: Automated design of gravity-assist trajectories to Mars and the outer planets. Celest. Mech. Dyn. Astron. **52**, 207–220 (1991)

Mason, L., Gibson, M., Poston, D.: Kilowatt-class fission power systems for science and human precursor missions. In: Nuclear and Emerging Technologies for Space, Albuquerque, NM, Mexico, 25–28 Feb 2013

Messerschmid, E., Fasoulas, S.: Raumfahrtsysteme. Springer, Berlin (2010)

NASA.: Pluto Fact Sheet (2015)

O'Brien, R., Ambrosi, R., Bannister, N., Howe, S., Atkinson, H.: Safe radioisotope thermoelectric generators and heat sources for space applications. J. Nucl. Mater. **377**, 506–521 (2008)

Owen, T.C., Roush, T.L., Cruikshank, D.P., Elliot, J.L., Young, L.A., de Bergh, C., Schmitt, B., Geballe, T.R., Brown, R.H., Batholomew, M.J.: Surface ices and the atmospheric composition of pluto **261**(5122), 745–748 (1993)

Portree, D.S.F., Loftus, J.P.: Orbital debris: a chronology. In: NASA/TP-1999-208856 (1999)

Prussing, J.E., Conway, B.A.: Orbital Mechanics. Oxford University Press, Oxford (1993)

Shulaker, M.M., Hills, G., Patil, N., Wei, H., Chen, H., Wong, H.-S.P., Mitra, S.: Carbon nanotube computer. Nature **501**, 526–530 (2013)

Sorbom, B.N., Ball, J., Palmer, T.R., Mangiarotti, F.J., Sierchio, J.M., Bonoli, P., Kasten, C., Sutherland, D.A., Barnard, H.S., Haakonsen, C.B., Goh, J., Sung, C., Whyte, D.G.: ARC: A compact, high-field, fusion nuclear science facility and demonstration power plant with demountable magnets. Fusion Eng. Des. **100**, 378–405 (2015)

Stern, S.A., Trafton, L.M.: On the atmospheres of objects in the Kuiper Belt. In: The Solar System Beyond Neptune, pp. 365–380. University of Arizona Press, Tucson (2008)

Sutton, G., Biblarz, O.: Rocket Propulsion Elements. Wiley-Interscience, New York (2001)

Yam, C.H., McConaghy, T.T., Chen, K.J., Longuski, J.M.: Design of low-thrust gravity-assist trajectories to the outer planets. In: 55th International Astronautical Congress, Vancouver, Canada, 4–8 Oct 2004

Chapter 15
Outer Solar System—Sample Return Mission by an Unmanned Interplanetary Spaceship UNIS

Werner Grandl, Ákos Bazsó and Andreas F. Felsenstein

15.1 Introduction

The Outer Solar System (OSS) with its vast dimensions contains a large number of different objects. Besides the four major planets Jupiter, Saturn, Uranus, and Neptune there exist dwarf planets like Pluto, icy comets, and countless asteroids of different sizes.

The giant planets and their moons have naturally received more attention in the form of space missions than the other—more remote—objects. The Pioneer 10 and 11 spacecrafts have visited Jupiter and Saturn in the 1970s, and remained operational until 2003 and 1995, respectively (Wolverton 2004). Voyager 1 and 2 took the "grand tour" through the OSS to visit the giant planets; Voyager 2 was the first ever human made object to reach Uranus and Neptune (Evans and Harland 2008). These missions were challenging not only for the large distances they had to overcome, but also for the long mission duration of ten years and more, and after 40 years in space they are still operational.

Following this first exploration phase of the OSS the successor missions were dedicated to very specific targets. Galileo was sent to Jupiter and orbited the planet from 1995 until 2003 (Meltzer 2007). On the way to Jupiter the spacecraft encountered two main-belt asteroids. These fly-bys in 1991 with (951) Gaspra Belton et al. (1992) and in 1993 with (243) Ida Belton et al. (1994) provided the first close-up images of asteroid surfaces with resolutions of 25–54 m per pixel. The scientific data acquired during these encounters helped to improve a lot our understanding of asteroids. One of the highlights was the discovery of Ida's moon

W. Grandl (✉)
Consulting Architect, Dr Billroth-Strasse 6, A-3430 Tulln, Austria
e-mail: archigran@gmx.at

Á. Bazsó · A. F. Felsenstein
University of Vienna, Türkenschanzstrasse 17, A-1180 Vienna, Austria
e-mail: akos.bazso@univie.ac.at

V. Badescu and K. Zacny (eds.), *Outer Solar System*,
https://doi.org/10.1007/978-3-319-73845-1_15

Dactyl (Chapman et al. 1995), which proved that asteroids can also form binary systems. For more details on the insights gained through Galileo see the reviews of Bertini (2013) and Sears (2015). The Cassini mission investigated the Saturn system (notably also the moon Titan via the Huygens lander) since the year 2004 and is expected to continue until September 2017 NASA (2014). The most remote target of a space mission until today is the Trans-Neptunian object and dwarf planet (134340) Pluto. New Horizons—a spacecraft launched in 2006—traveled for 9 years before having a fly-by to Pluto and Charon in July 2015 Stern et al. (2015).

Currently operational and near-future missions include Juno (Matousek 2007), already in orbit around Jupiter, and JUICE, the upcoming mission to Jupiter's big moons (Grasset et al. 2013). In January 2017 NASA selected the Lucy mission (NASA 2017; Levison et al. 2017) to perform the first reconnaissance of six Jupiter Trojan asteroids between 2027 and 2033 (for more details see Sect. 15.4.2).

Apart from these very successful operations a small number of sample return missions belong to the highlights of unmanned spaceflight. The Stardust mission gathered particles from the coma of comet 81P/Wild 2 and returned them to Earth (Westphal et al. 2014). Another sample return mission was the Japanese Hayabusa spacecraft that sampled the near-Earth asteroid (25143) Itokawa and returned to Earth in 2010 with some particles collected from the asteroid's surface Nakamura et al. (2011). Currently active missions are Hayabusa 2 Tsuda et al. (2013) and OSIRIS-REx Lauretta et al. (2017), both aiming to take samples from asteroids in the Inner Solar System (ISS).

This overview makes clear that although the OSS hosts some of the currently most interesting potential mission targets, yet there is a lack of ambitious plans for a sample return mission to objects in the OSS. If we want to utilize the natural resources of the OSS, we should bring back soil and rock samples of celestial bodies in the OSS for chemical analysis to estimate the resources of metals, water, and other useful material out there.

15.2 Mission Concept

We have designed a preliminary concept for a sample return mission to two minor bodies in the OSS, the Jupiter Trojan asteroid (624) Hektor and the Centaur group asteroid (2060) Chiron. These two objects have mean diameters larger than 200 km and masses of the order of 10^{19} kg; with these parameters they belong to the largest specimen of their respective types.

The spacecraft UNIS (Unmanned Interplanetary Spaceship) for this long duration mission (probably lasting some decades) would be assembled in Low Earth Orbit (LEO; see corresponding section later on). To reach the first target (624) Hektor it would perform several gravity assist maneuvers via planetary encounters in the ISS. Upon arrival at (624) Hektor the robotic lander SPIDER would be activated to descend to and sample the surface. Meanwhile, the mothership would orbit the asteroid and collect data; this phase could take up to one year or more given the experience with the Dawn mission (Rayman and Mase 2014). After finishing the tasks at (624) Hektor, the target for the second mission phase is the more distant object (2060) Chiron. The spacecraft would enter into orbit about the asteroid, and the lander springs into action. The last mission phase consists of returning the stored samples to Earth.

In the following section we describe the motivation to select these two specific targets, their physical characteristics, and their orbits. Subsequently we discuss the mission details, starting with the orbital mechanics including gravity assist maneuvers, Delta-V constraints, and the optimal time windows for arrival at the targets. Finally, we propose a possible layout for the spacecraft and the landing probe.

15.3 Motivation

According to modern planet formation theories (Alibert et al. 2010; Morbidelli et al. 2012) planets form in a process of steady accretion of smaller bodies by larger bodies. The largest of those protoplanets start dominating the dynamics by virtue of their larger mass and quickly absorb all available mass in their surroundings. Today we can consider minor bodies such as asteroids and comets to be the leftover from this early stage of the Solar System.

On Earth, we only have access to a very limited sample of rocks that are old enough to study formation processes. Most terrestrial rocks are simply not old enough to reach back to the relevant times (Koeberl 2006), so extraterrestrial sources must supply us with the missing information. The Apollo missions have collected lunar rocks, besides that lunar and Martian meteorites are also a valuable source (Treiman et al. 2000). However, these sources cover only a limited part of the Solar System, namely the region of the terrestrial planets and parts of the main belt. While meteorites allow to sample the inner and partly the middle main-belt (metallic and rocky asteroids), the more volatile rich primitive asteroids of the outer main-belt and beyond (C-, D-, and P-types) are underrepresented (Vernazza et al. 2015). There might be a link between these objects and interplanetary dust particles (IDPs), however for a more complete picture we would also need samples from the OSS including remote objects definitively originating from the Kuiper belt. There is evidence from near-infrared spectroscopy of Trojans that two different spectral groups exist, which is interpreted by Emery et al. (2011) as belonging to two compositionally different groups, one originating in the (outer) main-belt while the other perhaps originated in the Kuiper belt.

15.3.1 624 Hektor (1907 XM)

The Jupiter L4 Trojan asteroid (624) Hektor was discovered by A. Kopff in 1907, just one year after Achilles in 1906. Trojan asteroids are co-orbital objects that have the same semi-major axis as the planet they are accompanying. They occupy two regions 60° ahead (L4) and behind (L5) the planet. This means that on average (624) Hektor orbits the Sun in a distance of 5.2 AU, which is the same as for Jupiter. Today there are about 6500 known Jupiter Trojans (MPC 2016), 17 Neptune Trojans, as well as some Trojans of Earth, Mars, and Uranus (see Emery et al. 2015 for a review).

(624) Hektor is the largest Trojan with a mean diameter of 225 ± 15 km, but observations show that it is very elongated and possibly made of two bodies in contact with each other (Hartmann and Cruikshank 1978). It belongs to the taxonomic spectral type D (DeMeo et al. 2009). Objects of this spectral type are interpreted as being composed of water-ice, carbon and silicates, with organic rich material at the surface (Tholen 1989; Bus and Binzel 2002). Current estimates of its bulk density range from 1.63 ± 0.32 to $2.56 \pm 0.08\ \mathrm{g\ cm^{-3}}$ (Carry 2012; Descamps 2015). In fact, the latter value for the density is unexpected for objects in that region of the Solar System.

Other parameters derived from observations are the rotation period of 6.920509(2) hours (Marchis et al. 2014), an absolute magnitude of H = 7.5 mag with a geometric albedo of only 0.025, and the mass of $\sim 10^{19}$ kg (Carry 2012). [Note that the JPL Small-Body Database does not state uncertainties for the absolute magnitude and the geometric albedo].

A unique feature of (624) Hektor is that it possesses a small moon that was detected in 2006 and could be an ejected remnant of the collision that formed the main binary asteroid. Two recent studies have analyzed the clustering of Trojans and have identified—among others—a "family" of Trojans associated with (624) Hektor (Vinogradova 2015; Rozehnal et al. 2016). This family would consist of fragments that were created in the collision with the parent body of (624) Hektor. The observed shape of (624) Hektor might be a hint for a collisional origin.

Numerical simulations of the early phases of the Solar System indicate that all giant planets would have been able to capture a large number of asteroids into their Lagrange points L4 and L5 (Lykawka and Horner 2010; Nesvorny et al. 2013). Some Trojans could have survived there over the lifetime of the Solar System, providing a unique source to study the history and physical circumstances of the formation of the Solar System.

In summary, the in situ exploration of (624) Hektor would help to improve our knowledge about Trojans in general and the early dynamics of the Solar System. In the context of Trojans it would be important to study one object from the taxonomic D-class, and possibly link it to a certain type of meteorites. Moreover, since Emery et al. (2011) identified only two distinct spectral groups among the Trojans, the investigation of one representative object will reveal many typical features of other Trojans as well. (624) Hektor is probably a binary asteroid with a small satellite, hence its characterization would help to understand the formation and subsequent gravitational and rotational evolution of this triple system (e.g. Scheeres 2016). Another important task is to confirm the association of the D-type family with (624) Hektor and constrain the family age Rozehnal et al. (2016). Finally, certain Jupiter-family comets show a deuterium-to-hydrogen (D/H) ratio similar to Earth water (Hartogh et al. 2011), which brings up the need to investigate the D/H ratio for Jupiter Trojans as well to constrain the origin of Earth's water (see also arguments in Vernazza et al. 2016).

15.3.2 2060 Chiron (1977 UB)

The minor planet (2060) Chiron belongs to the dynamical class of Centaurs, which have a perihelion distance of more than 5 AU and a semi-major axis between 5 and 30 AU. Chiron was the first object of that kind, when C. T. Kowal discovered it in 1977 with the 122-cm Schmidt telescope at Mount Palomar. Having a visual magnitude of 18 mag, it was just bright enough to be discovered with this device. Subsequent observations derived an orbit with a semi-major axis of about 13.6 AU, just between Saturn and Uranus, a perihelion distance of 8.4 AU, just inside the orbit of Saturn, and a moderate inclination of 6.9°.

Even the earliest dynamical simulations showed that such an orbit would lead to close encounters with the gas giants on timescales much shorter than the age of the Solar System. Therefore, the orbit was soon classified as unstable (Kowal et al. 1979).

Later simulations showed that Chiron's motion is, in fact, chaotic with slightly different initial conditions leading to very different orbital evolutions in the past and future. For this reason the origin of (2060) Chiron is not clear-cut. The Kuiper Belt between 30 and 50 AU—with its great quantity of icy objects—is one possible origin, but a source in the much more distant regions of the OSS cannot be excluded (Kovalenko et al. 2002).

Physically, (2060) Chiron is believed to consist mostly of water ice. In 1988 slight brightness variations were observed which were interpreted as cometary activity (Hartmann et al. 1988). The sublimation of carbon monoxide and similar volatiles could be generating the slight coma that is sometimes observed around the object. One such outburst was observed with the ESO Very Large Telescope in June 2001 (Romon-Martin et al. 2003). The cometary activity is particularly interesting, because the fact, that there are still molecules that can sublimate at closer solar distances, points to the hypothesis that (2060) Chiron has not been at perihelion distances like today for a very long time.

With an absolute magnitude of about $H = 5.9 \pm 0.2$ mag, (2060) Chiron is one of the brightest Centaurs and much bigger than the comets that visit the ISS. Observations with the Herschel infrared telescope yield a diameter of 216 ± 10 km which corresponds to an albedo of 0.167 ± 0.037 Fornasier et al. (2013), while NEOWISE observations (Mainzer et al. 2016) suggest a diameter of

201.2 ± 62.4 km and an albedo of 0.110 ± 0.052, which is lower than the previous measurement but compatible to it within the errorbars. Photometric observations suggest a rotational period of 5.5 ± 0.4 h that seems to be typical for objects in that region of the Solar System (Galiazzo et al. 2016).

Another special feature of (2060) Chiron was discovered when stellar occultations were reanalyzed. There were some short and symmetrical drops in the light curve prior and after occultations of stars by (2060) Chiron. A ring system around the body is one explanation for these observations. The rings would contain material located 300 km from the body's center, well outside its radius. They could have been generated by the evaporating material on the surface, or by the destruction of a small satellite (Ortiz et al. 2015).

This short summary of scientific results about (2060) Chiron shows that this body has many interesting characteristics. It is maybe one of the closest and biggest representatives of a category of objects (small KBOs with diameters < 500 km) that are not easily reachable for space probes. According to Brown (2012) Centaurs are indistinguishable spectroscopically from the smallest KBOs, which suggests that Centaurs are just former KBOs that have been perturbed on planet-crossing orbits. The relation between Centaurs and Trojans is not clear, it is still an open question whether or not both have the same origin. Gaining both close-up data and physical samples would greatly improve our knowledge of the OSS and the interrelation between these two dynamically distinct classes of objects.

15.3.3 Scientific Objectives

At (624) Hektor and (2060) Chiron the main spacecraft would enter into orbit and survey the asteroid, whilst the lander module would be released to descend to the surface, make experiments, take samples, and return to the orbiter. The main goals are:

- determine the mass of the asteroid to an accuracy better than 5%;
- determine the shape and volume of the targets;
- determine the mass distribution and bulk density;
- measure the surface composition, estimate the water mass fraction, and the D/H ratio;
- measure the surface (subsurface) temperatures as function of the rotation period;
- determine thermal and physical properties of the surface material (soil, regolith);
- map the surface for impact craters to estimate the age of the bodies;
- for (624) Hektor: constrain the formation history, i.e. contact binary vs. strongly elliptic single body;
- for (2060) Chiron: investigate the ring material.

15.4 Mission Details

For the mission to the two selected targets we plan the time interval between the years 2025 and 2100. As already pointed out in the introduction, such long-term missions are not everyday standard but feasible to realize as the Voyager program has demonstrated.

15.4.1 Target Orbit Analysis

As a first step we investigated the orbital dynamics of the proposed targets. We obtained from the JPL/Horizons service Giorgini et al. (1997) the Cartesian coordinates and velocities of (2060) Chiron and (624) Hektor for the reference epoch JD 2451000.5 (corresponding to 1998 July 6; see Table 15.1). Then we performed a numerical integration of the orbits of all planets from Mercury to Neptune, including the two target asteroids as massless test objects. For this calculation we employed the Mercury integrator's Radau scheme (Chambers 1999; Everhart 1985), a high order and accurate integration method (Eggl and Dvorak 2010). In the calculation we neglected some minor perturbations, such as the flattening of the planets and the Sun, general relativistic effects, and non-gravitational forces like the Yarkovsky and YORP effect (Bottke et al. 2006). We can safely assume that these (weak) perturbations do not modify appreciably the orbits of any object for the short time-interval of interest.

From the simulations we can derive plots like in Fig. 15.1 that show the projections of the orbits onto the x–y and x–z planes. These plots show the main differences between the orbits of (624) Hektor and (2060) Chiron: Hektor has a low eccentricity ($e = 0.02$) but a moderately large inclination of 18°, while Chiron is more eccentric ($e = 0.38$) but less inclined with only 6.9°. The oscillations in their z-coordinates amount to about 2 AU in positive and negative direction. Hence, for the rendezvous and orbit insertion we have to focus on the time instants when the bodies are near the ascending and descending nodes of their orbits. In this way the spacecraft would always stay close to the ecliptic plane in order to avoid expensive plane change maneuvers.

Table 15.1 Orbital elements for JD 2451000.5

Orbital element	(624) Hektor	(2060) Chiron
Semi-major axis [AU]	5.204734	13.633184
Eccentricity	0.022401	0.3800586
Inclination [deg]	18.21187	6.938804
Longitude of node [deg]	342.7798	209.38524
Argument of perihelion [deg]	180.8843	339.40757
Mean anomaly [deg]	266.8097	17.14695

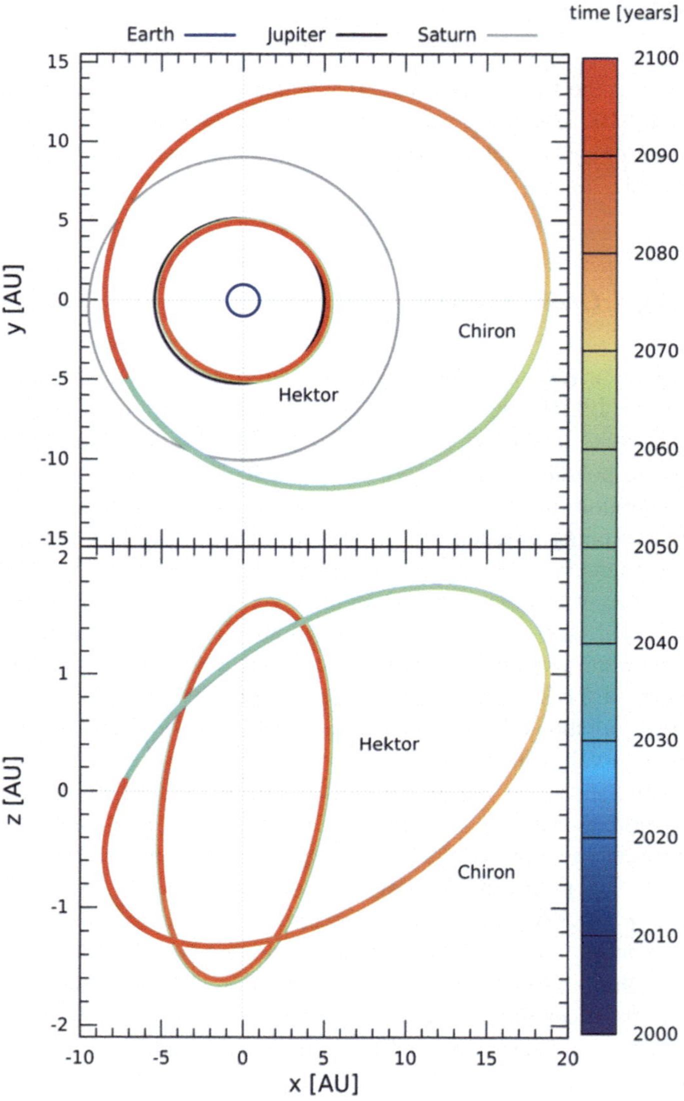

Fig. 15.1 Projections of the orbits of (624) Hektor and (2060) Chiron onto the x-y (top) and x-z plane (bottom). The color indicates their position along the orbit in the time interval 2000–2100 (only the last revolution is visible for 624 Hektor). In the top panel we included the orbits of Earth (blue), Jupiter (black), and Saturn (grey) for comparison

Figure 15.2 presents the time evolution of the object's geocentric distance (top) and (heliocentric) z-coordinate (bottom). In the top panel the long periodic (intrinsic) radial oscillations of (624) Hektor are superimposed by the short periodic (one year) motion of Earth. For (624) Hektor we can observe the distance to vary

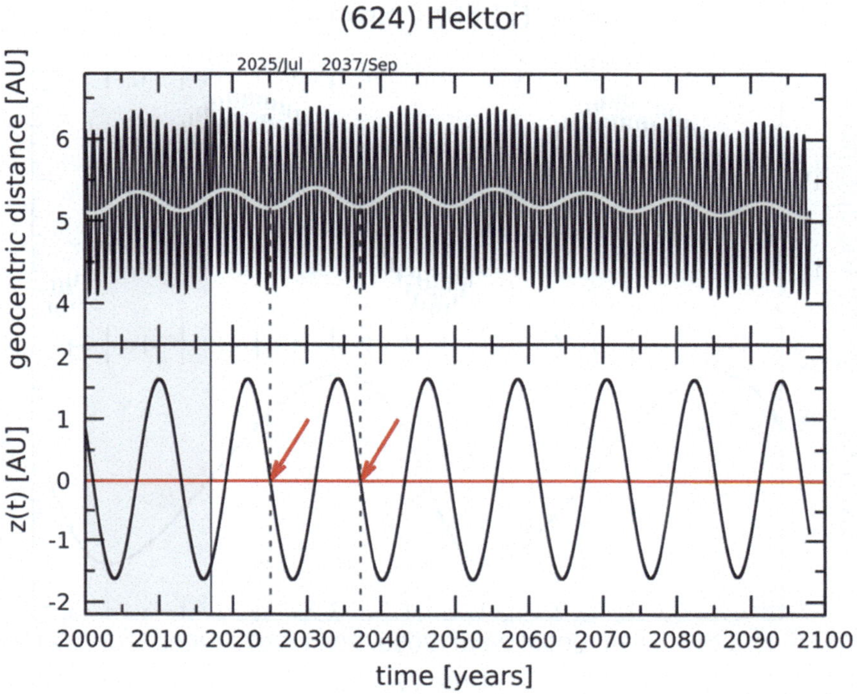

Fig. 15.2 Time evolution of Hektor's geocentric (top, black curve) and heliocentric distance (top, grey curve), and the heliocentric z-coordinate (bottom). The grey shaded area (years 2000–2017) is excluded for the mission. Red arrows indicate descending node passages before 2047

between 4 AU and more than 6 AU over one orbital period of about 12 years. The node crossings (lower panel, at z = 0) occur quite frequently (twice per orbital period), but one can note from the figure that only the descending node crossings coincide with the minima in the mutual distance.

In summary, for (624) Hektor we have a number of opportunities to realize a fly-by at minimum geocentric distance at low ecliptic latitudes when the Trojan asteroid is close to its descending node. However, the node passage in the year 2025 might be too early to set-up and prepare such a large scale mission, which realistically leaves us with the 2037 opportunity (and later ones).

We can do a similar analysis for (2060) Chiron in Fig. 15.3. Plotting again the (geocentric) distance and the (heliocentric) z-coordinate over time we find four node crossings until the year 2100. The last crossing is at the very end of the mission time frame, so we did not consider it. Two of the remaining three node crossings occur at rather large distances, some years after aphelion passage, and were also rejected. Moreover, the event in February 2027 is presumably too early to prepare and launch such a large-scale mission. We are left with the opportunity in October 2047, when (2060) Chiron is not far from its perihelion and at the same time close to the ecliptic plane. According to Figs. 15.2 and 15.3 it is also possible

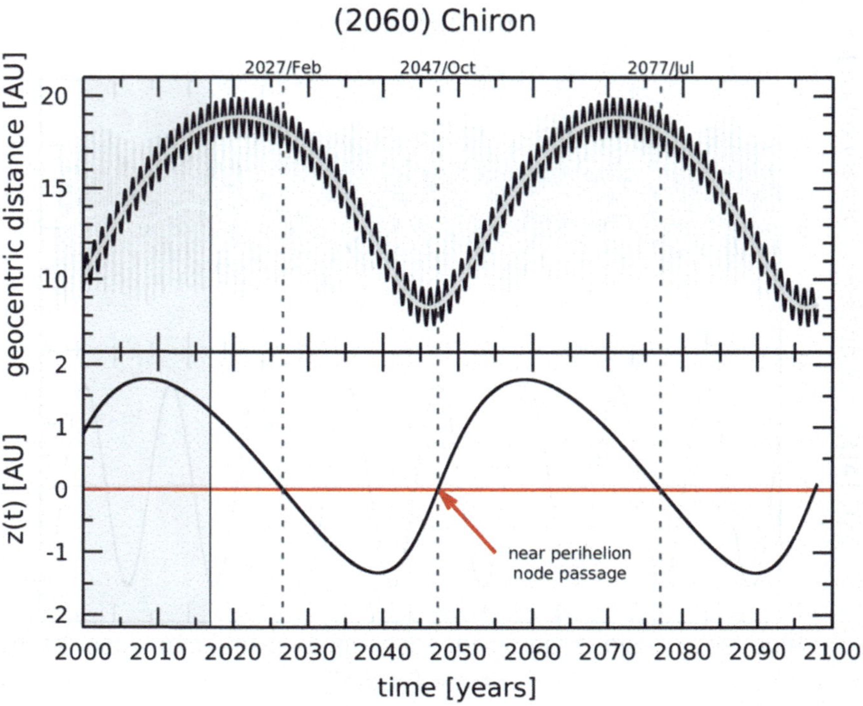

Fig. 15.3 Caption like Fig. 15.2, for (2060) Chiron. The red arrow indicates Chiron's passage through the ascending node shortly after its perihelion in 2047

to fly first to (2060) Chiron with an encounter in 2047, and approach (624) Hektor afterwards (past the year 2050).

15.4.2 Trajectory Design and Selection

Now, after selecting the targets and finding some time constraints for the mission, we need to look for ways to get there. The spacecraft would be assembled in low Earth orbit (LEO, at 400 km height), and thus needs to reach an escape speed of at least 10.8 km/s. An alternative strategy would be to use a single rocket launch, e.g. a SpaceX Falcon 9 rocket, to lift the entire spaceship into LEO (see Sect. 15.5). A direct injection to a transfer trajectory to Jupiter—or (624) Hektor—is not feasible for the current spacecraft configuration. Instead, we require a launch $V_\infty > 3$ km/s from Earth (or about 11.3 km/s from LEO), since this permits a number of effective gravity assists with planets in the ISS.

We propose to also exploit a lunar gravity assist to maximize the launch V_∞ from Earth. Ledkov et al. (2015) studied lunar gravity assist maneuvers as a means for capturing asteroids as Earth satellites. They showed that multiple gravity assists with the Moon could be used for decelerating asteroids. While the lunar orbital speed is 1 km/s, some fraction of this value (typically 0.1–0.8 km/s) is achievable for decreasing the asteroid's speed relative to Earth. According to these results, we could reverse the situation and boost the spacecraft to larger speed by a lunar gravity assist.

Petropoulos et al. (1998) examined a variety of trajectories from Earth to Jupiter with gravity assists from the planets Venus (V), Earth (E), and Mars. They employed the patched conic method and systematically searched for low-energy, gravity-assist trajectories with up to four planet fly-bys. Their results demonstrate that there exist numerous V-E-E and V-V-V sequences that require a launch V_{∞} between 3 and 4 km/s. These trajectories offer an efficient transfer to Jupiter with a time-of-flight between 5 and 8 years.

Another option for the transfer to (624) Hektor is via a two-stage approach. First, the spacecraft performs one or two gravity assists with Venus and Earth. The aim is to send it into the 3:2 mean motion resonance (MMR) with Jupiter. This resonance is located in the outer main-belt at 3.97 AU. The dynamical group of Hilda asteroids is also located at this MMR (Broz and Vokrouhlicky 2008). These asteroids have orbital periods that are 2/3 that of Jupiter, i.e. they orbit the Sun three times while Jupiter does two revolutions. Their orbits take them close to Jupiter's L4 and L5 Lagrange points.

There are already a number of existing mission concept studies to Jupiter's Trojans asteroids, among others by Gold et al. (2007), Bellerose and Yano (2009), Lamy et al. (2010), and Saikia et al. (2014). All these proposals have in common that they opt for a fly-by mission, unlike Funase et al. (2013) and the current authors who study the feasibility of a sample return mission. The recently accepted Lucy mission (NASA 2017; Levison et al. 2017) will be launched in 2021 and aims to fly-by six Jupiter Trojans in both Lagrange points L4/L5 in the time interval 2025–2033. The mission plan is to study the surface composition, bulk physical properties, and geology of these objects. The spacecraft will only operate with remote sensing instruments, unlike the present study of a sample return mission. The Lucy mission is a low cost program with a budget of < $500 million. In the context of the Lucy mission we can find synergies with the UNIS mission. Lucy will be arriving first and mapping different Trojans, thus helping to point out open scientific question that can be covered by the UNIS mission.

The major difficulties in connection with orbiting the asteroids are the timing for the rendezvous at their node crossings, and the need for a low relative velocity encounter before orbit insertion. However, the relative velocities between the spacecraft on its transfer orbit and the target can be rather large. If we take as a rough approximation the apocenter velocity of the Hohmann transfer ellipse from the 3:2 MMR to Jupiter's orbit and compare this to Hektor's pericenter velocity we have a difference of more than 3 km/s, which necessitates an expensive Delta-V maneuver. This difference becomes even larger for the transfer to (2060) Chiron, such that the spacecraft needs to be equipped with enough propellant.

Due to their low masses, the escape speeds for (624) Hektor and (2060) Chiron are 0.11 and 0.13 km/s, respectively; the surface gravity for both objects is ~5 10^{-3} of Earth's surface gravity. Although these values are rather small, they are 2 to 3 orders of magnitude larger than the surface gravity on comet 67P and asteroid (25143) Itokawa, targets of the Rosetta and Hayabusa missions. As a consequence of the failure to complete the landing maneuvers on the latter two objects (e.g., see Roll and Witte 2016), we opt for an actively controlled landing process. The lander SPIDER is equipped with counter-thrusters to press it against the surface of the target objects while landing on them. Its legs have additional spikes for grabbing hold and walking on the surface (see Sect. 15.6). This active control cannot be achieved from Earth, of course, so that an advanced autonomous computer system is needed. At the end of the surface activity phase, the landing probe needs to carry enough propellant to overcome the escape speed and leave the surface.

15.5 The Unmanned Interplanetary Spaceship UNIS

For the mission to the asteroids (624) Hektor and (2060) Chiron we need a multi-purpose Unmanned Interplanetary Spaceship (UNIS), which has to be designed for a long-duration mission (Fig. 15.4).

It would have a total empty mass of approx. 12 tons and a propellant mass of approx. 6 tons for all necessary manoeuvres including the final return to Earth (see Table 15.2). We propose current space technology for the main engine and the rotation control thrusters, using hydrazine and oxygen as propellants. The UNIS will start from LEO near the ISS (or its successor). Two launches by an ARIANE 5

Fig. 15.4 UNIS encountering a celestial body

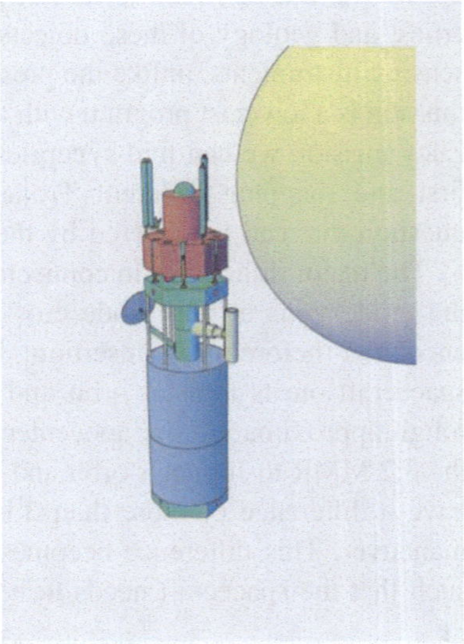

Table 15.2 Mass estimates of different components of UNIS

Component	Mass (tons)
Fuselage and structural framework	2.0
Robotic lander	2.0
Main engine	2.0
In-flight (rotation) control thrusters	1.8
Optical space telescope	0.5
Communication unit, antennas	0.5
Nuclear battery	0.5
Central processing unit	0.2
Photovoltaic panels	0.5
Scientific equipment, cameras, etc.	0.5
Fuel tanks (empty)	1.5
Total empty mass	12.0
Propellants	6.0
Total mass	~18.0

launcher or similar US, Russian or Chinese rockets with a payload of approx. 9 tons would be necessary. The modules and components of the spacecraft will then be assembled in orbit by the astronauts of the ISS, perhaps assisted by robots (Fig. 15.5). If available, we could also use the SpaceX Falcon 9 launcher, which will be able to lift a 22.8 tons payload into LEO (see SpaceX 2017; García et al. 2016, Table 1). In this case only one launch would be necessary. This launcher has a reusable first rocket stage and would reduce costs significantly.

During its long travel through the Solar System the UNIS will have to do a lot of different manoeuvres, such as swing-by manoeuvres, acceleration and deceleration thrusts. By firing the rotation control thrusters the UNIS can be turned around 180° to use the main engine for deceleration.

15.6 The Robotic Lander SPIDER

When UNIS has reached the orbit of its first target (624) Hektor, the robotic lander SPIDER will disconnect from the mothership and land on the surface of the asteroid. Besides its main engine the SPIDER has three reaction control thrusters on the bottom and three additional counter-thrusters to press the lander to the asteroid's surface if necessary.

With its six legs the SPIDER can walk across the asteroid's surface in any direction. The lander is equipped with three articulated arms (each 6 m length), carrying a camera, a drill and a shovel (Fig. 15.6).

To collect samples of regolith and rocky soil drills of various sizes can be used. The shovel puts the material into several storage boxes in the fuselage of the lander. The lander is designed for double use. To avoid contamination of samples on the

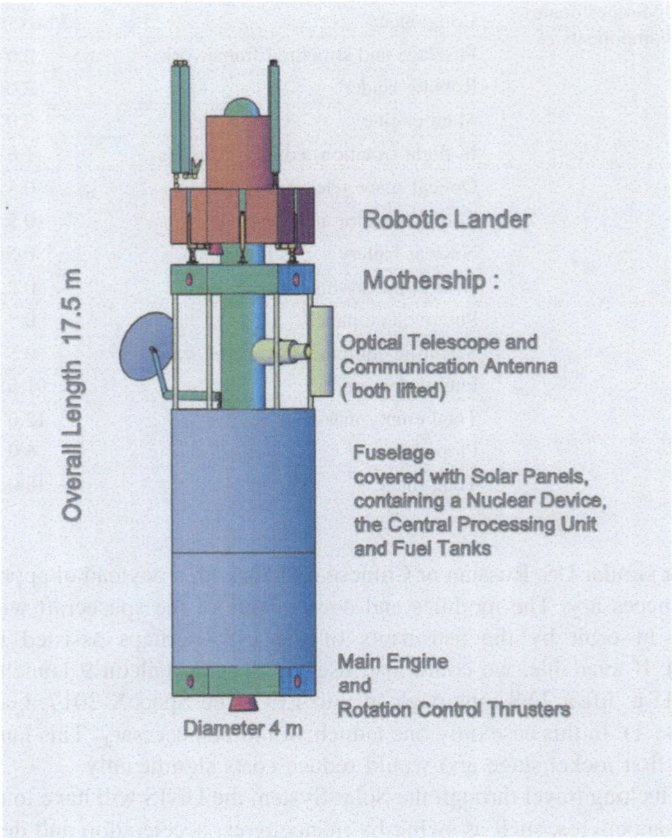

Fig. 15.5 Unmanned Interplanetary Spaceship (UNIS); overall length 17.5 m, diameter 4 m

second asteroid (2060) Chiron, shovels and drills can be exchanged. The arms can take and carry additional exchangeable instruments for various measurements like temperature, soil density, radiation, etc.

When the SPIDER has finished its work on the asteroid's surface, it returns to the mothership. The counter-thrusters enable the precise docking of the lander.

Both, the SPIDER and the mothership, will have to be designed for a very long-duration mission. Long-life materials have to be used. All parts and components must resist extremely low and high temperatures and high radiation doses for many years. Advanced compound materials will be used for radiation and micrometeorite shielding. Joints and links may be made of titanium and advanced alloys. Lubricants must keep their viscosity. Maybe the engines and motors of the SPIDER will have to be checked periodically during the journey to the first target asteroid by test ignitions. Some components of UNIS will need advanced engineering, e.g. the nuclear device for this long journey. The entire UNIS configuration will return to Earth when the mission is completed. The samples—stored in the

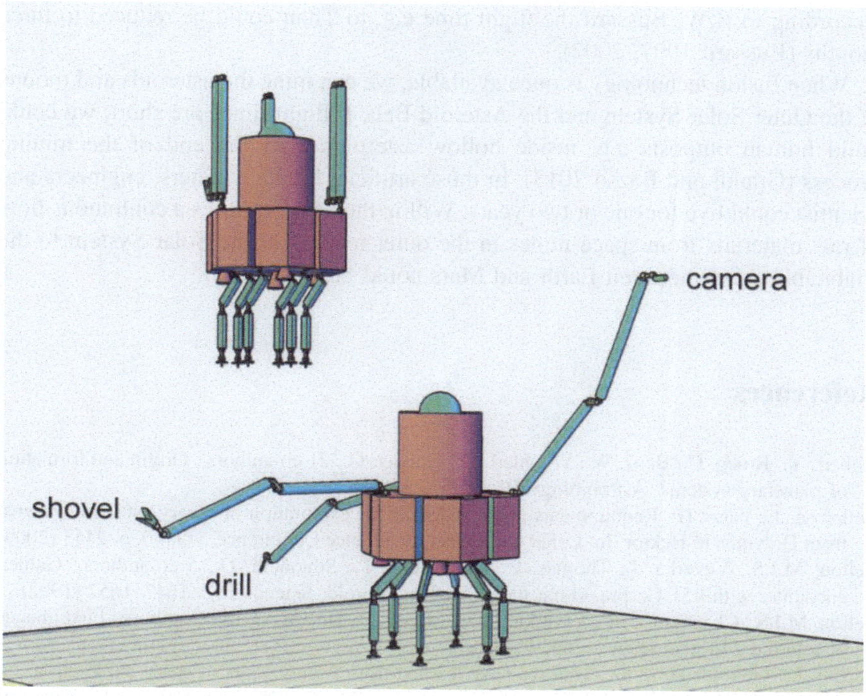

Fig. 15.6 Robotic lander SPIDER

SPIDER lander—can be unloaded by ISS astronauts and brought to Earth, or retrieved via a return capsule similar to the Stardust and Hayabusa missions.

Of course the UNIS and the SPIDER can be produced in series production to explore other celestial objects much nearer to Earth. In this case less propellant is needed. They could be used as a standard technology for the investigation of Near Earth Asteroids in the next decades.

15.7 Conclusions and Future Scope

It is evident that a project with a duration of more than 40 years indicates the limits of present space technology. Such a long journey would anyway be unreasonable for humans.

If we really want to exploit the resources of the Outer Solar System we need advanced space technologies and new propulsion systems. Such propulsion systems could be a nuclear thermal rocket engine or—as an ultimate solution—the Bussard Fusion System, also known as the quiet-electric-discharge (QED) engine. This proposed engine uses electrostatic fusion to generate electrical power. Deuterium and Helium-3 are fusing to Helium-4 plus protons releasing 18.3 meV per reaction.

According to R.W. Bussard the flight time e.g. to Titan could be reduced to three months (Bussard 1997, 2002).

When fusion technology is once available, we can mine the asteroids and moons of the Outer Solar System and the Asteroid Belt. If flight times are short, we could build human outposts e.g. inside hollow asteroids after the end of the mining process (Grandl and Bazso 2013). In those artificial habitats miners, engineers and scientist could live for one or two years. Within the next centuries a continuous flow of raw materials from space mines in the outer regions of the Solar System to the "habitable zone" between Earth and Mars could be established.

References

Alibert, Y., Broeg, C., Benz, W., Wuchterl, G., Grasset, O., 21 co-authors.: Origin and formation of planetary systems. Astrobiology **10**, 19–32 (2010)

Bellerose, J., Yano, H.: Requirements and constraints for exploration of binary asteroid systems: from Didymos to Hektor. In: Lunar and Planetary Science Conference, vol. 40, p. 2443 (2009)

Belton, M.J.S., Veverka, J., Thomas, P., Helfenstein, P., Simonelli, D., 5 co-authors.: Galileo encounter with 951 Gaspra—First pictures of an asteroid. Science **257**, 1647–1652 (1992)

Belton, M.J.S., Chapman, C.R., Veverka, J., Klaasen, K.P., Harch, A., 15 co-authors.: First images of asteroid 243 ida. Science **265**, 1543–1547 (1994)

Bertini, I.: Asteroids close-up: what we have learned from twenty years of space exploration. In: Badescu, V. (ed.) Asteroids: Prospective Energy and Material Resources, pp. 1–33. Springer, Berlin (2013)

Bottke Jr., W.F., Vokrouhlicky, D., Rubincam, D.P., Nesvorny, D.: The Yarkovsky and Yorp effects: implications for asteroid dynamics. Annu. Rev. Earth Plan. Sci. **34**, 157–191 (2006)

Brown, M.E.: The compositions of Kuiper belt objects. Annu. Rev. Earth Plan. Sci. **40**, 467–494 (2012)

Broz, M., Vokrouhlicky, D.: Asteroid families in the first-order resonances with Jupiter. MNRAS **390**, 715–732 (2008)

Bus, S.J., Binzel, R.P.: Phase II of the small main-belt asteroid spectroscopic survey. A feature-based taxonomy. Icarus **158**, 146–177 (2002)

Bussard, R.W.: System technical and economic features of qed-engine drive space transportation. In: 33rd AIAA/ASME/SAE/ASEE Joint Propulsion Conference, AIAA 97-3071 (1997)

Bussard, R.W.: An advanced fusion energy system for outer-planet space propulsion. Space Technol. Appl. Int. Forum 608 (2002)

Carry, B.: Density of asteroids. Planet. Space Sci. **73**, 98–118 (2012)

Chambers, J.E.: A hybrid symplectic integrator that permits close encounters between massive bodies. MNRAS **304**, 793–799 (1999)

Chapman, C.R., Veverka, J., Thomas, P.C., Klaasen, K., Belton, M.J.S., 7 co-authors.: Discovery and physical properties of Dactyl, a satellite of asteroid 243 Ida. Nature **374**, 783–785 (1995)

DeMeo, F.E., Binzel, R.P., Slivan, S.M., Bus, S.J.: An extension of the Bus asteroid taxonomy into the near-infrared. Icarus **202**, 160–180 (2009)

Descamps, P.: Dumb-bell-shaped equilibrium figures for fiducial contact-binary asteroids and EKBOs. Icarus **245**, 64–79 (2015)

Eggl, S., Dvorak, R.: An introduction to common numerical integration codes used in dynamical astronomy. In: Souchay, J., Dvorak, R. (eds.) Lecture Notes in Physics, vol. 790, pp. 431–480. Springer, Berlin (2010)

Emery, J.P., Burr, D.M., Cruikshank, D.P.: Near-infrared Spectroscopy of Trojan Asteroids: Evidence for Two Compositional Groups. AJ, 141, 25 (2011)

Emery, J.P., Marzari, F., Morbidelli, A., French, L.M., Grav, T.: The Complex History of Trojan Asteroids. In: Michel, P., DeMeo, F.E., Bottke, W.F. (eds.) Asteroids IV, Univ. Arizona Press, pp. 203–220 (2015)

Evans, B., Harland, D.M.: NASA's Voyager Missions: Exploring the Outer Solar System and Beyond. Springer, London (2008)

Everhart, E.: An effcient integrator that uses Gauss-Radau spacings. In: Carusi, A., Valsecchi, G.B (eds.) Dynamics of Comets: Their Origin and Evolution, Proceedings of IAU Colloq. 83, Dordrecht: Reidel, Astrophysics and Space Science Library, vol. 115, p. 185 (1985)

Fornasier, S., Lellouch, E., Müller, T., Panuzzo, P., Kiss, C., and 8 co-authors.: A Portrait of the Centaur 2060 Chiron: new results from groundbased and Herschel observations. European Planetary Science Congress 2013, EPSC2013–208 (2013)

Funase, R., Matsumoto, J., Mori, O., Yano, H.: Conceptual study on a Jovian Trojan asteroid sample return mission. J. Space Technol. Sci. **27**(1), 1–19 (2013)

Galiazzo, M., de la Fuente Marcos, C., de la Fuente Marcos, R., Carraro, G., Maris, M., Montalto, M.: Photometry of Centaurs and trans-Neptunian objects: 2060 Chiron (1977 UB), 10199 Chariklo (1997 CU26), 38628 Huya (2000 EB173), 28978 Ixion (2001 KX76), and 90482 Orcus (2004 DW). Ap and SS **361**, 212 (2016)

García, A., Lamb, A., Sleptsov, A., Moreno, C., Victorova, M., 2 co-authors.: Post-ISS plans: What should be done? Rev. Hum. Space Explor. **1**, 63–73 (2016)

Giorgini, J.D., Yeomans, D.K., Chamberlin, A.B., Chodas, P.W., Jacobson, R.A., 5 co-authors.: JPL's on-line solar system ephemeris and data service. In: Bietenholz, M.F., Bartel, N., Rupen, M.P., Beasley, A.J., Graham, D.A., Altunin, V.I., Venturi, T., Umana, G., Conway, J.E. (eds.) Bulletin of the American Astronomical Society, BAAS, vol. 29, p. 1099 (1997)

Gold, R.E., McNutt, R.L., Napolillo, D.H., Schaefer, E.D., Tanzman, J.R., 7 co-authors.: Paris to Hektor: A Concept for a Mission to the Jovian Trojan Asteroids. In: El-Genik, M.S. (ed.) American Institute of Physics Conference Series Space Technology and Applications International Forum-STAIF 2007, vol. 880, pp. 217–223 (2007)

Grandl, W., Bazso, A.: Near earth asteroids-prospection, orbit modification, mining and habitation. In: Badescu, V. (ed.) Asteroids: Prospective Energy and Material Resources, pp. 415–438. Springer, Berlin (2013)

Grasset, O., Dougherty, M. K., Coustenis, A., Bunce, E. J., Erd, C., 13 co-authors.: JUpiter ICy moons Explorer (JUICE): An ESA mission to orbit Ganymede and to characterise the Jupiter system. Planet. Space Sci. **78**, 1–21 (2013)

Hartmann, W.K., Cruikshank, D.P.: The nature of Trojan asteroid 624 Hektor. Icarus **36**, 353–366 (1978)

Hartmann, W.K., Tholen, D.J., Cruikshank, D.P., Brown, R.H., Morrison, D.: 2060 Chiron and P/Schwassmann-Wachmann I: two unusual comets. Bull. Am. Astron. Soc. **20**, 836 BASS (1988)

Hartogh, P., Lis, D.C., Bockelée-Morvan, D., de Val-Borro, M., Biver, N., 8 co-authors.: Ocean-like water in the Jupiter-family comet 103P/Hartley 2. Nature, **478**, 218–220 (2011)

Koeberl, C.: The record of impact processes on the early earth: a review of the first 2.5 billion years. Geol. Soc. Am. Spec. Pap. **405**, 1–22 (2006)

Kovalenko, N.S., Babenko, Y.G., Churyumov, K.I.: Modeling of the orbital evolution of 2060 Chiron. Earth Moon Planet **90**, 489–494 (2002)

Kowal, C.T., Liller, W., Marsden, B.G.: The discovery and orbit of/2060/Chiron. In: Duncombe, R.L. (ed.) Dynamics of the Solar System, volume 81 of IAU Symposium, pp. 245–250 (1979)

Lamy, P., Poncy, J., Martinot, V., Jorda, L., Delsanti, A., Groussin, O.: Hektor: a space mission towards Jupiter Trojans. European Planetary Science Congress p. 138 (2010)

Lauretta, D.S., Balram-Knutson, S.S., Beshore, E., Boynton, W.V., Drouet dAubigny, C., 43 co-authors.: OSIRIS-REx: Sample Return from Asteroid (101955) Bennu. ArXiv e-prints 1702.06981 (2017)

Ledkov, A.A., Eismont, N.A., Nazirov, R.R., Boyarsky, M.N.: A method for capturing asteroids into earth satellite orbits. Astron. Lett. **41**, 442–449 (2015)

Levison, H.F., Olkin, C. Noll, K.S., Marchi, S., Lucy Team.: Lucy: surveying the diversity of the trojan asteroids: the fossils of planet formation. In: Lunar and Planetary Science Conference Abstracts 2017, vol. 48, p. 2025 (2017)
Lykawka, P.S., Horner, J.: The capture of Trojan asteroids by the giant planets during planetary migration. MNRAS **405**, 1375–1383 (2010)
Mainzer, A.K., Bauer, J.M., Cutri, R.M., Grav, T., Kramer, E.A., 5 co-authors.: NEOWISE Diameters and Albedos V1.0. EAR-A-COMPIL-5-NEOWISEDIAM-V1.0. NASA Planetary Data System (2016)
Marchis, F., Durech, J., Castillo-Rogez, J., Vachier, F., Cuk, M., 7 co-authors.: The puzzling mutual orbit of the binary Trojan asteroid (624) Hektor. ApJ **783**, L37 (2014)
Matousek, S.: The Juno New Frontiers mission. Acta Astronaut. **61**, 932–939 (2007)
Meltzer, M.: Mission to Jupiter: A History of the Galileo Project. National Aeronautics and Space Administration, NASA History Division (2007)
Morbidelli, A., Lunine, J.I., O'Brien, D.P., Raymond, S.N., Walsh, K.J.: Building terrestrial planets. Annu. Rev. Earth Planet. Sci. **40**, 251–275 (2012)
MPC.: IAU minor planet center (2016). Retrieved 14 Nov 2016 from http://www.minorplanetcenter.net/iau/lists/Trojans.html
Nakamura, T., Noguchi, T., Tanaka, M., Zolensky, M.E., Kimura, M., 17 co-authors.: Itokawa dust particles: a direct link between S-type asteroids and ordinary chondrites. Science **333**, 1113 (2011)
NASA.: Cassini grand finale fact sheet. JPL 400-1579 9/14 (2014). https://saturn.jpl.nasa.gov/resources/66/
NASA.: NASA selects mission to study jupiter's Trojan asteroids, (2017). https://www.nasa.gov/feature/goddard/2017/nasa-selects-mission-to-study-jupiter-s-trojan-asteroids
Nesvorny, D., Vokrouhlicky, D., Morbidelli, A.: Capture of Trojans by Jumping Jupiter. ApJ **768**, 45 (2013)
Ortiz, J.L., Duffard, R., Pinilla-Alonso, N., Alvarez-Candal, A., Santos-Sanz, P., 5 co-authors.: Chiron, another Centaur with ring material. Eur. Planet. Sci. Congr. EPSC2015–230 (2015)
Petropoulos, A., Longuski, J., Bonfiglio, E.: Trajectories to jupiter via gravity assists from venus, earth, and mars. In: Guidance, navigation, and control and co-located conferences. American institute of aeronautics and astronautics, pp. 116–126 (1998)
Rayman, M.D., Mase, R.A.: Dawn's exploration of Vesta. Acta Astronaut. **94**, 159–167 (2014)
Roll, R., Witte, L.: ROSETTA lander Philae: touch-down reconstruction. Planet. Space Sci. **125**, 12–19 (2016)
Romon-Martin, J., Delahodde, C., Barucci, M.A., de Bergh, C., Peixinho, N.: Photometric and spectroscopic observations of (2060) Chiron at the ESO very large telescope. A & A **400**, 369–373 (2003)
Rozehnal, J., Broz, M., Nesvorny, D., Durda, D.D., Walsh, K., 2 co-authors: Hektor—an exceptional D-type family among Jovian Trojans. MNRAS **462**, 2319–2332 (2016)
Saikia, S., Das, A., Laipert, F., Dapkus, C., Kendall, J., 10 co-authors: Voyage to troy: a mission concept for the exploration of the Trojan asteroids. In: Muinonen, K., Penttilä, A., Granvik, M., Virkki, A., Fedorets, G., Wilkman, O., Kohout T. (eds.) Asteroids, Comets, Meteors (2014)
Scheeres, D.J.: Relative equilibria in the spherical, finite density three-body problem. J. Non-Linear Sci. **26**, 1445–1482 (2016)
Sears, D.W.G.: The explored asteroids: science and exploration in the space age. Space Sci. Rev. **194**, 139–235 (2015)
SpaceX.: Falcon 9—technical overview, (2017). Retrieved 25 Apr 2017 from http://www.spacex.com/falcon9
Stern, S.A., Bagenal, F., Ennico, K., Gladstone, G.R., Grundy, W.M., 146 co-authors: The pluto system: initial results from its exploration by new horizons. Science **350**, aad1815 (2015)
Tholen, D.J.: Asteroid taxonomic classifications. In: Binzel, R.P., Gehrels, T., Matthews, M.S. (eds.) Asteroids II, pp. 1139–1150 (1989)
Treiman, A.H., Gleason, J.D., Bogard, D.D.: The SNC meteorites are from Mars. Planet. Space Sci. **48**, 1213–1230 (2000)

Tsuda, Y., Yoshikawa, M., Abe, M., Minamino, H., Nakazawa, S.: System design of the Hayabusa 2—Asteroid sample return mission to 1999 JU3. Acta Astronaut. **91**, 356–362 (2013)

Vernazza, P., Marsset, M., Beck, P., Binzel, R.P., Birlan, M., 7 co-authors: Interplanetary dust particles as samples of icy asteroids. ApJ **806**, 204 (2015)

Vernazza, P., Beck, P., Lamy, P., Guilbert-Lepoutre, A.: Sample of high-priority science objectives for future interplanetary missions towards asteroids. Rom Astron J **26**, 35 (2016)

Vinogradova, T.A.: Identification of asteroid families in Trojans and Hildas. MNRAS **454**, 2436–2440 (2015)

Westphal, A. J., Stroud, R.M., Bechtel, H.A., Brenker, F.E., Butterworth, A.L.: 61 co-authors: Evidence for interstellar origin of seven dust particles collected by the Stardust spacecraft. Science **345**, 786–791 (2014)

Wolverton, M.: The Depths of Space: The Story of the Pioneer Planetary Probes. National Academies Press (2004)

Part IV
Enabling Technologies

Chapter 16
Spacecraft Power System Considerations for the Far Reaches of the Solar System

Robert Cataldo

16.1 Introduction

The outer Solar System is comprised of Jupiter and Saturn, referred to as the gas giants, Uranus and Neptune, referred to as the ice giants and minor planet, Pluto. Smaller objects beyond Neptune include, dwarf planets, comets, and thousands of Kuiper belt objects (KBO). The inner solar system is comprised of Mercury, Venus, Earth and Mars which are typically referred to as the terrestrial planets because they are solid and are comprised of rock and metals.

Historically, space missions have either been powered by photovoltaics (PV) or radioisotope power systems (RPS). Photovoltaic cells convert photons directly to electrons that in turn are used by the spacecraft to operate on-board systems, instruments, heaters and recharge batteries. RPS utilize the heat generated by the natural decay of a radioisotope. The US has utilized plutonium-238 (Pu) for missions because of its half-life of ~87.7 years, fairly high power to mass output (0.57 W/g), coupled with low radiation emissions that are primarily alpha particle decay that are easily shielded.

The value of ~1361 W/m^2 is used for Earth's orbit solar constant and has been measured by numerous Earth orbiting spacecraft. The average distance from the Earth to the Sun is ~150,000,000 km and defined as 1.0 astronomical unit (AU). The solar flux density, or solar constant, decreases with the square of the distance ($1/r^2$) from the Sun. Thus, the reliance on solar energy to power a spacecraft becomes an increasing challenge, particularly as a mission's destination is ever more distant from the Sun.

R. Cataldo (✉)
NASA Glenn Research Center, 21000 Brookpark Rd, MS 302-1,
Cleveland, OH 44135, USA
e-mail: robert.l.cataldo@nasa.gov

V. Badescu and K. Zacny (eds.), *Outer Solar System*,
https://doi.org/10.1007/978-3-319-73845-1_16

16.2 Space Power System Background

Solar arrays are currently the prime power generation option for Earth orbiting spacecraft both in low-Earth-orbit (LEO), and geostationary orbit (GEO). Over 500 operational satellites are in GEO today covering a broad range of applications such as: navigation, commercial and civil communications, remote sensing, military, and science, research and development, etc., operated in part, or totally by over 50 countries. There are also over 800 spacecraft in LEO and $\sim$100 in other highly elliptical and mid-altitude orbits.

The orbits of these spacecraft typically experience an eclipse where the spacecraft is in Earth's shadow and therefore the solar panels are not in view of the Sun to produce power. Hence, batteries are typically used during this eclipse period. Batteries are charged during sunlit portions of the orbit and additional solar cell power is required for battery recharging. GEO satellites experience a varying daily eclipse period for $\sim$44 days twice a year, centered around both the spring and fall equinox when the Earth blocks the spacecraft from the Sun. The battery discharge period follows a pattern starting out with several minutes of discharge and then a peak of $\sim$72 min during day 22 and then decreasing to zero again. Thus, over a ten year planned spacecraft useful life time for example, the spacecraft battery will experience 880 discharge cycles. Since minimizing mass is a very important element for a spacecraft power system, battery size selection, in amp-hour capacity, is carefully planned to maximize the depth-of-discharge (DOD) percentage to achieve the required battery cycle life.

LEO spacecraft power system design is more of a challenge due to the significant number of orbits each day. A LEO spacecraft orbits the Earth about 16 times each day for an approximate 60 min of sun and 30-min eclipse period, both of which are functions of the spacecraft orbiting altitude and inclination. For example, ISS is in an average 400 km, 51° inclined orbit with about 15.5 orbits per day. Spacecraft launched from the NASA Kennedy Space Center are typically in a 28.5°-inclined orbit that optimizes a given launch vehicle's mass capability to orbit. These LEO spacecraft orbit the Earth about 5850 times a year. Therefore, selecting the proper battery technology to achieve a typical mission duration results in a required battery charge-discharge cycle life of $\sim$30,000 cycles. The additional array battery charge power is also a larger portion of the array size due to the approximate 2:1 charge-to-discharge time ratio.

Historically spacecraft battery technology relied on nickel cadmium (NiCd), nickel hydrogen (NiH_2) and more recently lithium ion (Li-ion). NiCd's were extensively used until the late 1980s when they were replaced by NiH_2 technology. In order to maximize battery cycle life, depth of discharge was typically limited to $\sim$20–30% of the total capacity. However, if additional battery capacity was needed for higher power demands, the battery could not be discharged much beyond the cyclic capacity level due to the phenomena referred to as "memory effect". Over many low DOD cycles crystalline growth occurs essentially blocking electrolyte from the active electrode material. During that time, many cell and battery "reconditioning" schemes were developed to regain this loss in capacity by restoring the anode electrode active material, comprised of cadmium-hydroxide crystals, back to its original condition. The typical approach was to deep discharge the battery at a low rate to reduce the size of the crystals. This process worked well for GEO applications where long periods of full sun and the battery is not used exist, however for LEO applications, requiring battery use every orbit, reconditioning is problematic unless the spacecraft has a spare battery to bring on line.

By the 1990s, technology development on NiH_2 reached a level that they were replacing NiCds for various satellite applications (near-Earth communication, navigation, weather, etc.). NiH_2 designs have ~65 Wh/kg or twice that of NiCd with similar cycle life capability at depth of discharge of around 35%. NASA has limited application for ultra-high cycle life (>30,000) batteries, however, there are exceptions like the Hubble Space Telescope (HST) and the International Space Station (ISS). NASA is currently planning on replacing a portion of ISS's NiH_2 batteries with Li-ion. Depending on the battery design and application, Li-ion can provide over twice the energy density than NiH_2 along with a significant decrease in battery volume. However, a Li-ion battery requires close monitoring of individual cells within a battery pack to avoid cell over-charge or over-discharge and subsequent thermal runaway and potential battery damage.

Rechargeable Li-ion battery development began in the 1980s primarily driven by personal portable electronics applications in reducing weight and extended operations per charge. The promise of lower mass for a given amount of stored energy is even more attractive for spacecraft power systems to reduce overall launch mass thus adding more instruments or increased capabilities. NASA's early use of lithium cell technology was a non-rechargeable Lithium/Thionyl Chloride battery used on the Mars Pathfinder rover, Sojourner. The battery was used for any peak load demands above the solar array's output. Lithium cell technology has a broader operating temperature range than nickel based systems and thus has advantages particularly for the wide temperature swings seen on the surface of Mars. A rechargeable 8 Ah Li-ion battery was utilized for the Mars Exploration Rovers; Spirit and Opportunity. Li-ion primary cells have become the battery of choice for US and ESA launch vehicles. Rechargeable Li-ion batteries are also being used on outer planet missions where low mass is highly desirable and mission cycle life not as demanding as with LEO spacecraft.

In late 2016, NASA sent battery replacements consisting of six Li-ion (Lithium Cobalt Oxide) batteries to the International Space Station (ISS). The cell capacity selected is 134 Ah with only 30 cells in series compared to 76 cells for the NiH_2 cells. The Li-ion cells will have a separate charge/discharge controller with individual cell monitoring fault detection and isolation circuitry. Under normal operations, the thermal control system maintains cell temperature between an upper limit of 50 °C and lower limit of 10 °C with an active cooling loop and electric heaters. It is critical from a safety standpoint to protect the cell/battery from thermal runaway or the condition of an uncontrolled rapid rise in cell temperature that can cause fire and/or explosion of that cell and potentially propagate to adjacent cells causing failure of the entire battery. Thermal runaway has two contributing energy release components; first, the release of the stored electrical energy that occurs in tens of seconds and, second, the exothermic chemical energy release lasting tens of minutes.

It is paramount to protect a Li-ion battery from thermal runaway that would not only compromise the power system but also potentially other spacecraft hardware and subsequent loss of mission. A typical battery management scheme is to not charge to full capacity and to limit the depth of discharge, which effectively selects a mid-operation band within a safer operating range. This scheme is planned for the ISS Li-ion battery, where a per cell charge voltage limit of 3.95 V (instead of 4.1) will insure a maximum charged capacity of 97 Ah versus 134 Ah. The planned eclipse period discharge is 25 Ah. Therefore, based on the nameplate capacity of 134 Ah, the band of nominal operations is between 72 and 54% of full battery capacity.

One last safety feature for a Li-ion battery module is the cell packing, cell-to-cell barriers and battery enclosure. Optimized spacing distance between cells can help to mitigate heat propagation to other cells in a battery pack, and adding thermal barriers can be used to help contain the heat to the cell in thermal runaway. To reduce the risk of further spacecraft damage of a failed battery in thermal runaway, a pressure relief device that doesn't allow flames to escape outside should be added to the battery enclosure. All these safety features add mass, volume and decrease the useable amount of stored energy, which needs to be considered in the power system design.

16.3 Solar Power

Solar power is the clear choice for Earth-vicinity spacecraft as previously discussed, however for missions farther away from the Sun, the power systems selection becomes less obvious. While solar has been utilized at Mars orbit and on the surface, both solar intensity and temperature combine for a much harsher environment for missions to the point where other technologies such as radioisotope and nuclear fission power generation become more competitive to solar in mass, volume and possibly cost. Since solar cells directly convert sunlight to electricity, the effectiveness of the system is largely a function of distance to the Sun. Figure 16.1 shows solar flux levels versus the average orbital distance of the planets from the Sun. Supplying power to missions to Jupiter and beyond poses a significant challenge for photovoltaic power systems. For example, the solar flux at Jupiter is $\sim$1/25th or 4% that at Earth. When Pluto is closest to the Sun in its orbit ($\sim$30 AU), the solar flux is 1/1900th that at Earth orbit.

In addition, the outer planets have a higher orbital eccentricity that will have an effect on solar flux value depending on the planet's location in its orbit at the time of the mission. Table 16.1 includes some planet characteristics relevant to power system design parameters, planet orbit period, and perihelion and aphelion solar flux and approximated solar array temperatures.

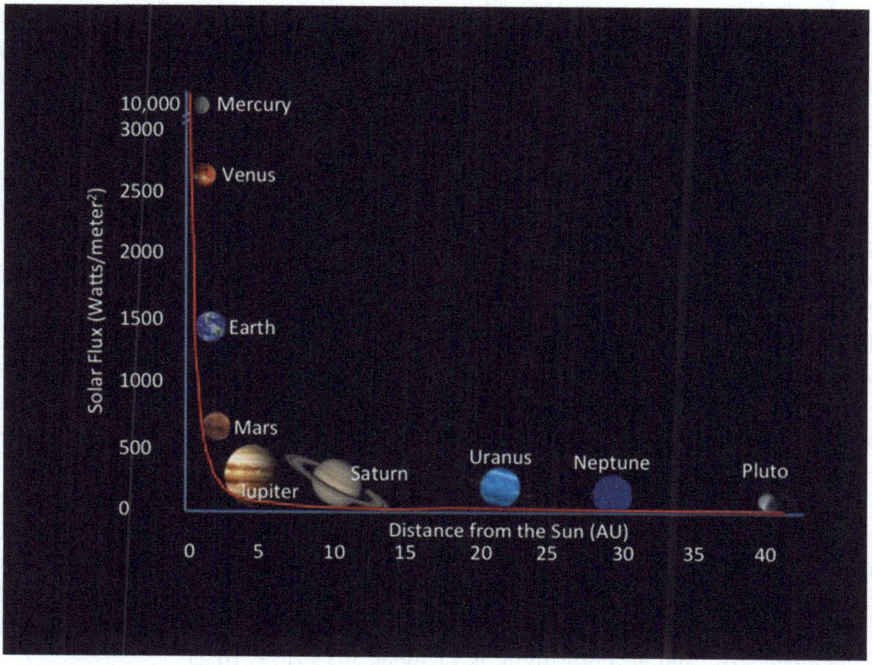

Fig. 16.1 Solar flux as a function of distance from the Sun

Table 16.1 Outer planets characteristics: orbital period, distance from the Sun/solar intensity and equilibrium flat-plate temperature of a solar array

	Periods (years)	Perihelion (AU) (W/m^2)	Aphelion (AU) (W/m^2)	T_{eq} (K/C)	
				Perihelion	Aphelion
Earth (Ref.)	1	1;1370	1;1370	326/52	326/52
Jupiter	12	4.95;56	5.46;46	140/−133	133/−140
Saturn	29	9.02;16.8	10.08;13.5	103/−170	98/−175
Uranus	84	18.4;4.05	20.1;3.39	76/−197	73/−200
Neptune	165	29.8;1.54	30.3;1.49	60/−213	59/−214
Pluto	248	29.7;1.56	49.3;0.56	60/−213	46/−227

Solar array temperatures have been included here because there is a correlation between cell temperature and cell performance. An estimate of the array operating temperature T_{eq}, uses the Stefan-Boltzmann equation, below (Landis and Fincannon 2015):

$$T_{eq} = \left[\frac{(\alpha - \eta)I}{(\epsilon_f + \epsilon_b)\sigma}\right]^{1/4}$$

where α is the solar absorptivity, I the solar intensity, ε the front and back thermal emissivity (assuming two-sided emission, typical of current-generation high-efficiency cells), and σ the Stefan-Boltzmann constant, 5.67×10^{-8} W/ m^2 K^4. Assumed inputs for the variables are: solar absorptivity, 0.92, cell efficiency, 0.25, front emissivity, 0.85, and back emissivity, 0.85.

The low incident solar intensity decreases the cell voltage, and hence lowers the conversion efficiency; while lower operating temperature improves the voltage, and hence increases the efficiency. In addition, operation under the combination of low intensity and low temperature, or "LILT" conditions, can produce anomalous degradation of performance known as LILT effect (Scheiman and Snyder 2008). These LILT effects are only partially understood for new generation technologies, but can to some extent be mitigated by screening cells under LILT conditions to select only cells that do not show the anomalous degradation. Such screening was done for both the Dawn ($\sim$3 AU) and Juno ($\sim$5 AU) missions to the outer solar system (Harris et al. 2008).

Juno is the first mission to Jupiter that is solar powered. The previous Jupiter missions, Pioneer 10/11 (both flybys) and Galileo (orbiter) were powered RTGs. The Juno mission was selected under the NASA New Frontiers Program and was launched on board an Atlas V 551 on August 5, 2011 and orbit insertion around Jupiter occurred July 4, 2016 after two deep space maneuvers in August and September 2012 and an Earth flyby in October 2013 (Fig. 16.2).

Advances in solar cell development over the past 25 years, namely the triple-junction $GaInP_2/GaAs/Ge$, cells have more than doubled cell efficiencies

Fig. 16.2 Juno array final illumination testing at KSC

compared to earlier version Si cells to over 30% at Earth orbit solar intensity levels. The multi-junction cell basically makes use of more of the light spectrum. However, under LILT conditions, some cells were found to exhibit significant reduction in performance while other cells made under the same process do not. The cells that were selected for Juno were triple-junction $GaInP_2$/GaAs/Ge.

Juno's power requirement at end of mission, after over thirty polar orbits of Jupiter, is 400 We. The tri-wing, ~50 m^2 total cell area (55 m^2 panel area), was sized to produce 480 We at Jupiter encounter, with an estimated 50–60 We degradation over the mission's lifetime, giving ~6% array power margin. The array design required ~18,700, 26.6 cm^2 solar cells. Each solar cell was individually tested for its performance under LILT conditions. A new test procedure was developed whereby cells could be verified under low illumination but at room temperature thus saving costs and time by not having to test cells at −180 °C qualification temperatures. The phenomenon of why some of the triple junction cells experience anomalous output degradation compared to other cells of the same batch process is not well understood, however a screening process is required to eliminate screen these cells and not use them on the array. However, these non-LILT qualified cells can be utilized for other missions in a non-LILT application.

The unique radiation environment of the Jovian system requires special consideration. Sensitive spacecraft electronics that operate the spacecraft are typically placed within a shielded enclosure or "radiation vault" to reduce the total ionizing dose (TID) to levels that can be sustained by circuit electronic parts. Juno is in a highly elliptical, polar orbit around Jupiter to minimize flying through Jupiter's most intense radiation zones. However, the spacecraft did fly through some higher radiation during its initial approach during orbit capture. Since arrays however cannot be placed in a vault they will receive a radiation dose depending on the type of radiation and particle energies as well as the radiation field flux and total mission time. Radiation damage also interacts with the LILT degradation, making the LILT problem more severe (Harris et al. 2008). Juno's arrays were estimated to be about −180 °C or colder when they encounter Jupiter's trapped radiation fields, therefore the Juno mission measured the combined effects of LILT cell performance by irradiating a cold soaked cell, and while not allowing possible annealing to occur with allowing the cell temperature to increase, re-measured the cell performance under LILT conditions. An additional 1–2% loss in performance was seen over what would be predicted from just factoring the radiation degradation and LILT effect degradation. Radiation damage can be mitigated by increasing the cover glass thickness, but this needs to be evaluated on a case by case basis since array mass will increase.

Another mission planned for the Jovian system will orbit Jupiter and conduct about 45 flyby encounters of Europa. NASA's planned Flagship mission, Europa Clipper, will carry nine instruments to conduct atmosphere, surface and subsurface investigations. For example, scientists hope to capture a jet plume of liquid rising from below the ice sheet layer to determine its chemical makeup and if it could support life. The mission design is still in formulation but the design team is studying the option of NASA having the Space Launch System (SLS) ready for a 2022 launch allowing a less than three-year trip time with its higher mass delivered to LEO and possibly adding a lander to the mission. The other launch vehicle being considered is a Delta IV, however that would increase the trip time to over seven years due to less propellant and lower delta-V.

Power system technologies considered for the mission included solar and radioisotope power systems. The anticipated spacecraft power for all the instruments and spacecraft operations is about 600 We. The array design to produce that power at Jupiter is $\sim$72 m^2, 2 wings that are $\sim$9 m by 4 m each and 1.6 times larger than Juno's array.

As solar arrays become larger and potentially require multiple wings (Juno, e.g.), structural panel mass and stowage within a fairing or mounting on a spacecraft can become an issue due to total packaged volume and dimensions that is independent of solar cell technology selection. As alternatives to foldable hinged panels, developments in deployable mechanisms to help alleviate these concerns have taken place over the past several years. Examples of this is the Mars Phoenix lander and future InSight mission to Mars that utilizes a concept called Ultraflex, a $\sim$2 m diameter deployable array. The basic concept has fan folded panels housed and launched in a compact compartment. A motor drives the mechanism into a fan-folded circular array. Once fully deployed and latched the array panel can achieve over three times the stiffness of a comparably sized hinged-panel array.

Another compactly stowed concept is the Roll-Out Solar Array (ROSA). Solar cells are adhered to a flexible mesh and rolled onto a mandrel. Deployment is achieved via strain energy in the composite rolled booms that unrolls the array blanket similar to a tape measure or party favor. A $\sim$1 m $\times$ 4.5 m ROSA deployment was demonstrated on ISS in June 2017. The initial demonstration was to roll out and then retract the array and return to Earth for inspection and evaluation. However, after a fully successful deployment and retraction, the array blanket "telescoped" on retraction preventing a latching and thus making return from ISS on the Dragon cargo vehicle not possible. The array was jettisoned in a direction as to not collide with other ISS or Dragon hardware with eventual burn up in Earth's atmosphere. However, this experiment successfully demonstrated the applicability of this technology for future spacecraft and potentially much larger array configurations.

These advances in array panel design would allow deployment of very large panels. Concepts have been developed that could produce 10's of kilowatts from a single array wing. For example, solar electric propulsion (SEP) missions would benefit significantly from light weight array concepts where low power system mass is critical.

16.4 Nuclear Technologies

16.4.1 Radioisotope Systems

Radioisotope power systems (RPS) have been essential to the US exploration of outer space since the early 1960s (https://solarsystem.nasa.gov/rps/rtg.cfm). An RPS has two primary attributes: provide electrical power and thermal power. To provide electrical power, the RPS uses the heat produced by the natural decay of a radioisotope (e.g., plutonium-238 in US space missions) to power a converter (e.g., thermoelectric elements or a dynamic generator). As a thermal power source the heat is conducted to whatever component on the spacecraft needs to be kept warm; this heat can be produced by a radioisotope heater unit (RHU) or by using the excess heat of the RPS.

Figure 16.3 illustrates the basic concept of a thermoelectric device. When a temperature difference is applied to two semiconductors, a voltage difference is produced across the P and N legs thus supplying electric power. Typically, these devices are connected in series/parallel configuration to achieve the desired voltage and current capability suitable for spacecraft use. Many combinations of materials have been formulated over the years to achieve improved efficiency and degradation rates at various hot side/cold side temperatures and operating environments (vacuum, atmosphere). The US radioisotope thermoelectric generators (RTGs) have predominately used Lead-Telluride (PbTe) and Silicon Germanium (SiGe) (Cataldo and Bennett 2011) however new development efforts are focused on Skutterudite thermoelectric materials with the promise of higher efficiency and low degradation rates that are important for long-life outer planet missions (Holgate et al. 2015).

Plutonium-238 has been chosen by the US to power RPS. Pu-238 has many desirable features compared to other isotopes capable of producing heat, such as, low beta, gamma and neutron emissions (requiring less shielding), long half-life ($\sim$88 years) with acceptable isotopic specific power capable of powering decades long missions, high temperature stability and higher energy per unit mass. The ceramic form employed also significantly reduces water solubility, reduced chance

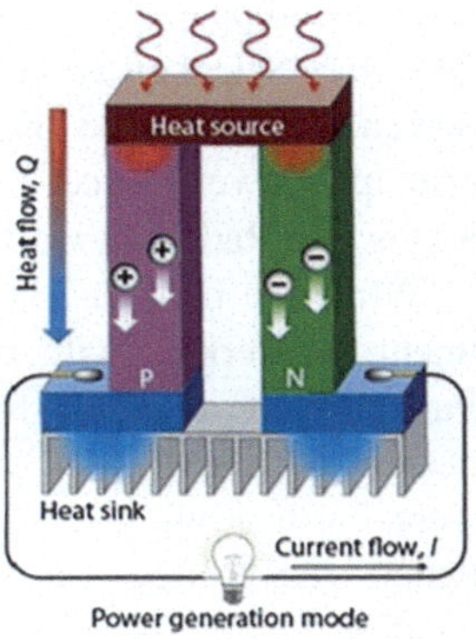

Fig. 16.3 Typical thermoelectric device

of fine particle inhalation and minimal chemical toxicity if an improbable accident would occur, thus minimizing the health hazards to the environment and to humans in the launch site and immediate surrounding areas. Pu-238 is man-made in a reactor where a Neptunium-237 (Np-237) target is placed in a neutron flux. Neptunium 237 is similarly produced first from Uranium-236. Pu-238 constantly produces heat during the decay process to emit an alpha particle and thus some of the Pu-238 becomes U-234. After 88 years, half of the original Pu-238 remains. The heat and power generated continually declines at $\sim 0.8\%$ per years. Thus, for a typical ≥ 12 plus year outer planet mission the power of an RTG will be $\sim 10\%$ lower from its beginning of mission output. In addition, the power conversion device will also degrade over time. Thermocouples also degrade over time due to a number of factors. The SiGe thermoelectric unicouples (exposed to space vacuum) used in the GPHS RTG degraded $\sim 0.8\%$ (half of which is due to actual heat source temperature loss) per year for a total degradation rates of $\sim 1.6\%$ per year. The Multi-mission Radioisotope Thermoelectric Generator (MMRTG) with PbTe thermocouples housed under a closed Argon cover gas, has a total degradation of $\sim 4.8\%$ per year.

The heat source is comprised of plutonium dioxide (PuO_2) powder hot pressed into a ceramic pellet ~ 1.0-in. diameter and 1.0 in. long that is encapsulated in a protective iridium cladding to form a capsule. Each capsule produces about ~ 62 Wth with four capsules housed in an impact and aeroshell to comprise the General-Purpose Heat Source (GPHS) module (Fig. 16.4). The GPHS module development began in the early 1980s as an improvement to earlier generations of RTGs and remains as the building block for RPS heat sources. The design is also integral to the safety and containment of the PuO_2 in the unlikely event of a launch pad accident or inadvertent re-entry of a spacecraft with a RTG power system. Since safety is a critical significant testing was done under extreme conditions; such tests have included simulating multiple reentries for a single module through Earth's atmosphere, exposure to high temperature rocket propellant fires, and impacts onto solid ground.

As of 2017, the US has launched 45 RTGs on 26 space systems (Cataldo and Bennett 2011). These space systems have ranged from navigational satellites to challenging outer planet missions such as Pioneer, Voyager, Galileo, Ulysses (joint NASA/European Space Agency), Cassini and the New Horizons mission to Pluto and Mars Science Laboratory (MSL) Curiosity Rover that employs the latest RTG design, the Multi-mission Radioisotope Thermoelectric Generator (MMRTG) as the principal power source. Table 16.2 contains a summary of US space missions using RTGs and Radioisotope Heater Units (RHU).

While the first Pu-238 produced was at the picogram scale in cyclotron experiments at Berkley with uranium-236, by the early 1960s sufficient quantities of Pu-238 were available for small RTGs. The first RTG flown was ~ 3 We and designated SNAP 3B (Systems for Nuclear Auxiliary Power). The early missions were Earth orbiting to power navigation and weather data because solar cell and battery technology was very inefficient at that time and hence not an attractive solution for long-lived power. Today, solar cells and batteries have progressed to

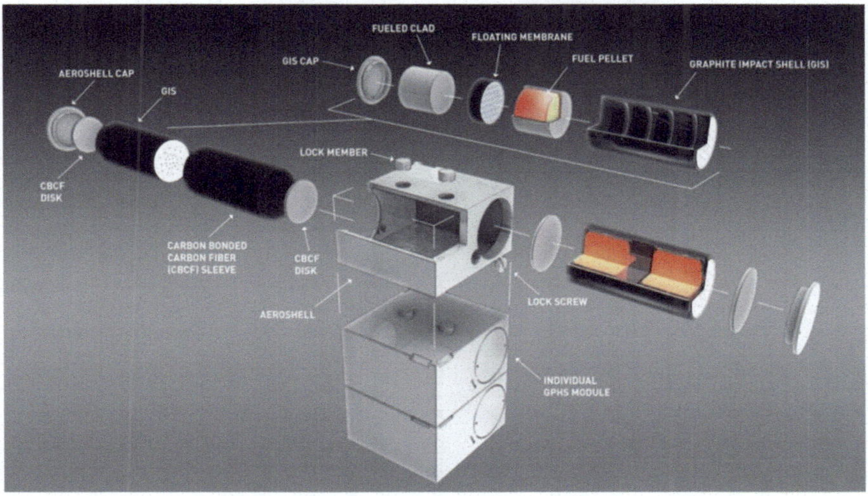

Fig. 16.4 General purpose heat source module assembly

the point where RPS is no longer a mission enabling power system option for Earth orbit.

The first RPS application beyond LEO was the Apollo lunar missions. Apollo 11 used two 15 Wth heater units for the surface experiment. Apollo 12, 14–17 utilized a SNAP 27 ($\sim$3.6 kg of PuO_2) that produced a nominal 70 We to power the Apollo Lunar Surface Experiments Package (ALSEP). The SNAP-27 RTG allowed operation and survival during the long lunar night unlike the solar powered package used on Apollo 11. The operational lifetimes achieved for the ALSEP stations were between 5 and 8 years as they were deactivated in 1977.

NASA's first use of RPS on a planetary mission was on Pioneer 10 and 11. Each spacecraft used the modified SNAP 19 with each employing four 40 We units. The design lifetime was 5 years; spacecraft continued to communicate with Earth for 30 and 22 years, respectively until the spacecraft became dormant. Pioneer 11 flew by Saturn and returned close-up images of Saturn and several moons.

In the late 1970s a second set of outer planet spacecraft, called Voyager 1 and 2, were launched with three Multi-hundred watt RTG (MHW-RTG) that produced $\sim$160 We each. Again, they also flew by Jupiter and Saturn but Voyager 2 was also tasked to explore Uranus and Neptune. The RTGs on both spacecraft continue to produce 255 We ($\sim$85 We per MHW-RTG, July 2017) power after over 40 years of deep space operation allowing data return from over 140 AU for Voyager 1 (17.4 km/s) and 118 AU for Voyager 2 (15.9 km/s). The initial required lifetime for the mission was 5 years, long enough to fly by Jupiter and Saturn. However, the spacecraft, instruments and RTGs continued to function and return data. An extended mission was approved for Voyager 2 to attempt a Neptune flyby since it trajectory was nearly in the ecliptic plane (Voyager 1 was not, due to its close approach to Saturn's moon, Titan). The Voyager mission certainly was a

Table 16.2 United states launch history of RTG and RHUs

Agency	Mission/ spacecraft	Mission type	Launch date	RTG	Units	RHU	Units	Comments
USN	Transit-4A	Navigational	6/29/61	SNAP 3B7	1	–	0	Currently in earth orbit
USN	Transit-4B	Navigational	11/15/61	SNAP-3B8	1	–	0	Currently in earth orbit
USN	Transit 5-BN-1	Navigational	9/28/63	SNAP-9A	1	–	0	Currently in earth orbit
USN	Transit 5-BN-2	Navigational	12/5/63	SNAP-9A	1	–	0	Currently in earth orbit
USN	Transit 5-BN-3	Navigational	4/21/64	SNAP-9A	1	–	0	Mission aborted, burned up on reentry as designed
NASA	Nimbus B-1	Meteorological	5/18/68	SNAP 19B2	2	–	0	Mission aborted, power source retrieved intact
NASA	Nimbus III	Meteorological	4/14/69	SNAP 19B3	2	–	0	Currently in earth orbit
NASA	Apollo 11	Lunar	7/16/69	…	2	ALRH	2	Experiment station shut down and remains on lunar surface (2–15 Wth units)
NASA	Apollo 12	Lunar	11/14/69	SNAP-27	1	–	0	ALSEP shut down and remains on lunar surface
NASA	Apollo 13	Lunar	4/11/70	SNAP-27	1	–	0	Mission aborted; ALSEP power source feel into Tonga Trench in Pacific Ocean
NASA	Apollo 14	Lunar	1/31/71	SNAP-27	1	–	0	ALSEP shut down and remains on lunar surface
NASA	Apollo 15	Lunar	7/26/71	SNAP-27	1	–	0	ALSEP shut down and remains on lunar surface
NASA	Pioneer 10	Planetary	3/2/72	SNAP-19	4	RHU	12	Successfully operated to Jupiter and beyond

(continued)

Table 16.2 (continued)

Agency	Mission/ spacecraft	Mission type	Launch date	RTG	Units	RHU	Units	Comments
NASA	Apollo 16	Lunar	4/16/72	SNAP-27	1	–	0	ALSEP shut down and remains on lunar surface
USN	Triad-01-1X	Navigational	9/2/72	Trans it-RTG	1	–	0	Currently in earth orbit
NASA	Apollo 17	Lunar	12/7/72	SNAP-27	1	–	0	ALSEP shut down and remains on lunar surface
NASA	Pioneer 11	Planetary	4/5/73	SNAP-19	4	RHU	12	Successfully operated to Jupiter, Saturn and beyond
NASA	Viking 1	Mars	8/20/75	SNAP-19	2	–	0	Lander shut down and remains on surface of Mars
NASA	Viking 2	Mars	9/9/75	SNAP-19	2	–	0	Lander shut down and remains on surface of Mars
USAF	LES 8	Communications	3/14/76	MHW-RTG	2	–	0	Currently in earth orbit
USAF	LES 9	Communications	3/14/76	MHW-RTG	2	–	0	Currently in earth orbit
NASA	Voyager 2	Planetary	8/20/77	MHW-RTG	3	LWRHU	9	Successfully flew by all four outer planets. In the heliosphere
NASA	Voyager 1	Planetary	9/5/77	MHW-RTG	3	LWRHU	9	Successfully operated to Saturn. In interstellar space
NASA	Galileo	Jupiter	10/18/89	GPHS-RTG	2	LWRHU	120	Includes 17 RHUs on atmospheric probe—operated in Jupiter system 8 years
NASA	Ulysses	Sun	10/6/90	GPHS-RTG	1	–	0	Operated in heliocentric orbit; decommissioned in 1998
NASA	Pathfinder	Mars	12/4/96	RHU	3	LWRHU	3	Rover shut down and remains on surface of Mars
NASA/ ESA	Cassini	Saturn	10/15/97	GPHS-RTG	3	LWRHU	117	Included 32 RHUs on Huygens Titan probe. Successfully operated 13 years in Saturn orbit. Saturn atmosphere disposal, July 15
NASA	MER-A (spirit)	Mars	6/10/03	RHU	8	–	8	Successfully operating on Martian surface

(continued)

Table 16.2 (continued)

Agency	Mission/ spacecraft	Mission type	Launch date	RTG	Units	RHU	Units	Comments
NASA	MER-B (opportunity)	Mars	7/7/03	RHU	8	LWRHU	8	Successfully operating on Martian surface
NASA	New Horizons	Pluto/Kuiper Belt	1/19/06	GPHS-RTG	1	–	0	Successfully operating after Pluto encounter/KBO 2014 MU69 in 01/01/2019
NASA	Mars science lab	Mars	9/26/11	MMRTG	1	–	0	Successfully operating on Martian surface

RTG radioisotope thermoelectric generator
SNAP space nuclear auxiliary power
MHW multi-hundred watt
GPHS general purpose heat source
RHU radioisotope heater unit
LWRHU light weight radioisotope heater unit
ALRH Apollo lunar radioisotopic heater
MMRTG multi-mission radioisotope thermoelectric generator

platform to demonstrate the reliability and robustness of SiGe RTG technology as a power system solution for outer planet missions.

Following the success of the MHW-RTG a development program for a higher power unit was initiated. The next generation was the GPHS RTG designed to produce 285 We at beginning of life (BOL). The GPHS RTG had 18 GPHS modules that produced ~4400 Wth and 572 SiGe thermoelectric elements (same design as the MHW-RTG) but produced 300 We at BOL. The GPHS-RTG has an overall diameter of 0.422 m and a length of 1.14 m with a mass of about 56 kg. A cut-away illustration is shown in Fig. 16.5.

The DOE built eight GPHS RTG flight units (F1-F8) to support upcoming NASA missions. The first mission to use the GPHS RTG was NASA's Galileo Jupiter mission that required two units. The next mission was the Sun's polar orbiting Ulysses mission with one RTG. The next mission, launched seven years later in 1997, was on the Cassini mission to orbit Saturn that used three GPHS RTGs and received slight design modifications from the preceding units (Bennett et al. 2016). Cassini entered Saturn orbit in late 2004, sending back unprecedented data about the planet and its moons. The European Space Agency had the Huygens probe that was jettisoned into Titan's atmosphere and collected data during its approximate 2 h descent to the surface. Huygens was powered by a battery and thermal energy was provided by 32 RHUs. The Cassini spacecraft also employed 82 RHUs to provide heat to critical components thus saving additional power that would have been needed for electric heaters. The three GPHS RTGs produced about 887 We at beginning of mission and they are currently producing 604 We as Cassini prepares for its mission end in September 2017. NASA decided to use Cassini's remaining propellant to plunge Cassini into Saturn and thus avoid a potential uncontrolled collision with Titan or Enceladus, since both moons are thought to be able to support biological life. Cassini will fly through Saturn's inner

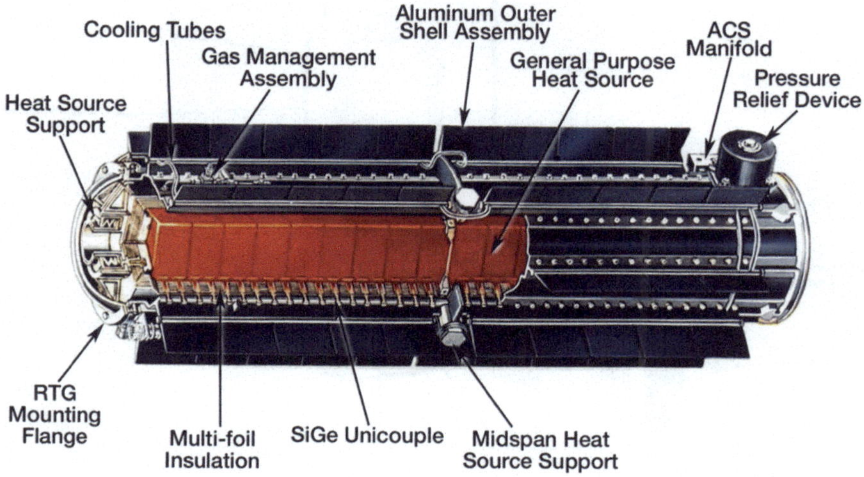

Fig. 16.5 General purpose heat source RTG

rings and skim the upper atmosphere sending back new data before plunging into Saturn's atmosphere thus concluding a remarkable 13-year exploration of Saturn. The Cassini mission was truly enabled by RPS since solar power/battery lifetime performance likely could have not been capable to operate for the entire 20-year mission duration.

As previously discussed, in addition to electric power, a radioisotope can provide heat to critical spacecraft components that require it. The currently available heater units are called light weight radioisotope heater units (LWRHU) utilized on the Galileo mission and produce ~1.0 Wth. (The Pioneer, Voyager and Apollo 11 missions used a different design) This LWRHU can be enabling by producing heat during long periods of darkness where battery operated heaters are not practical. Additionally, the electric power system would have to be increased if electric heaters are used. Hundreds of radioisotope heater units have been launched to provide warmth to missions, for example, Apollo 11 used to provide heating of critical components in a seismic experiment package, Pioneer, Voyager, Galileo, Cassini, Mars Pathfinder and the MER rovers to provide temperature control to critical spacecraft electronics and batteries as well as other mechanical devices, actuators and propellant valves.

The current RTG in production is the MMRTG that was designed to operate in vacuum and also in planetary surface atmospheres (Fig. 16.6). It was first used to supply power to the Mars planetary rover Curiosity. It is similar in design to the Viking lander SNAP 19 RTGs.

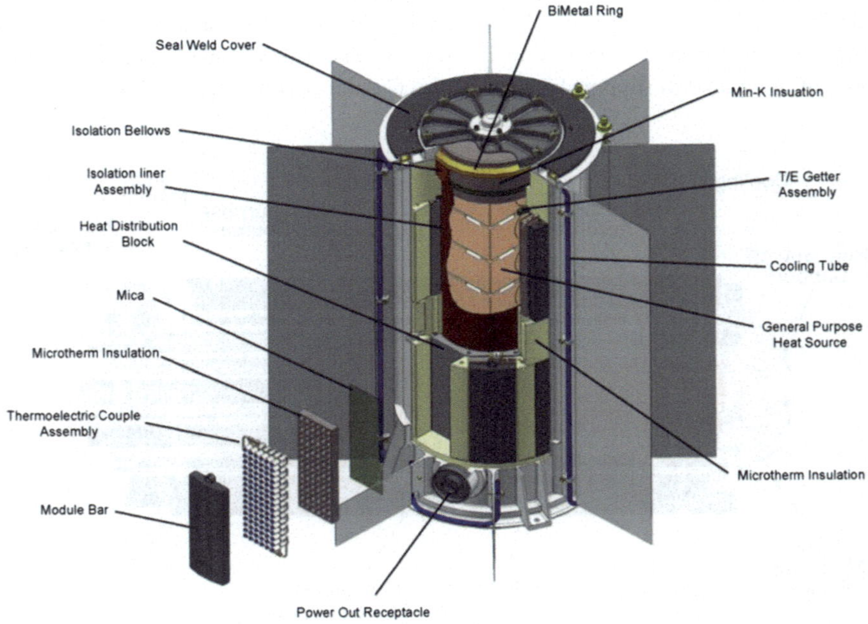

Fig. 16.6 MMRTG cutaway

This power source was selected for the rover's mission to provide greater flexibility in accessing difficult or remote terrain, to minimize effects of the dusty Martian environment and survive the Martian winters and to provide heat for instrumentation and rover subsystems. The Curiosity rover utilizes ~200 Wth from the ~1900 Wth of waste heat rejected by the MMRTG. While the initial degradation rate expected for the generator was similar to the GPHS RTG at ~1.8%/year the MMRTG generator actually is experiencing a 4.8% degradation rate. This high loss rate for the MMRTG is problematic for long trip time outer planet missions. In considering missions with a 10-year trip time, the MMRTG would only produce about half of its initial power. Understanding this concern, a development program to reduce the degradation rate and also increase efficiency is underway and is called the enhanced MMRTG or eMMRTG (Woerner 2016).

Figure 16.7 shows the predicted power output for both the MMRTG and eMMRTG for a typical outer planet mission when the RTG has full view of deep space (no planet albedo or spacecraft shadowing) thus the radiators are in full view of a 4 K sink temperature (Cataldo and Woerner 2016).

Like solar based power, RPS also has a practical limit as a power system solution. These key limitations, include; total amount of Pu-238 required, practical number of units integrated onto a spacecraft, launch site facilities and launch site RPS integration. Cost could also become an issue as with competed missions, such as NASA's Discovery Program with limits on total mission cost.

The bulk of the US Pu-238 supply is limited and was produced in the 1980s. Through natural decay, it has lost its capability to produce the specified level of heat

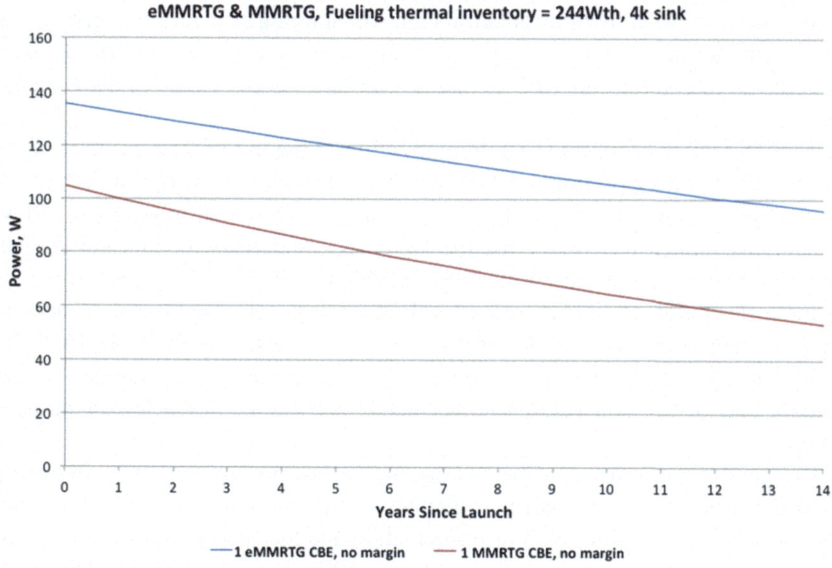

Fig. 16.7 Predicted power output of the MMRTG and eMMRTG in deep space. *Note* eMMRTG power estimate based on predictive system model and actual performance may vary

required for use in space systems. The DOE has initiated an effort to produce new PuO_2 fuel with higher levels of Pu-238 content. This higher content allows blending with the older fuel to bring the final blend into flight quality specification. The targeted amount to be produced is likely less than 2 kg of PuO_2/year depending on facility availability. The final achieved Pu-238 enrichment value is still being determined based on a number of factors. However, for example, 1.5 kg of new fuel could be blended with older fuel to provide enough to fabricate 8-GPHS modules (4.83 kg PuO_2) required for a single MMRTG or eMMRTG unit. And if a higher thermal-value fuel or slight overall gain in efficiency with TE conversion occurs, one might expect the eMMRTG could produce $\sim$150 We at beginning of life.

There could also be limitations on the number of RPS units that a spacecraft could reasonably accommodate. Surface area and volume on a spacecraft is required for instruments needing clear view to the science target or may not tolerate the rejected heat from an RPS. Plus, certain instruments like magnetometers and neutron detectors might not function properly in close proximity of an RPS. Additionally, propellant tanks, thrusters, avionics enclosures, etc., also likely need to be mounted on an external surface. A second consideration to the number of RPS units is the limitations of the launch vehicle and launch site. Protocols developed over the years of launching RPS established the practice of adding the RPS to the spacecraft 1–2 days prior to launch while the build up of the spacecraft typically takes 6–9 months at the launch site. This protocol limits radiological exposure to mission personnel during months of spacecraft build up and avoids potential hazardous accidents during launch vehicle preparations. To accommodate this, the launch vehicle fairing (typically two sections) is fitted with special access doors that allow RPS integration at the correct location on the spacecraft with the RPS securely mounted to an RPS installation cart. Maintaining structural integrity of the light-weight fairing without adding significant mass, limits the number of doors available for integrating an RPS to likely one large access door or two smaller doors.

The NASA New Frontiers (NF) Program allowed the use of RPS as stated in the Announcement of Opportunity (AO) for New Frontiers 4 (http://newfrontiers.larc.nasa.gov/announcements.html). An assessment of potential missions ($850 M, not including launch costs) meeting the NF-4 AO guidelines for mission destinations (http://sites.nationalacademies.org/SSB/SSB_059331) was performed to determine the range of likely power levels that proposers might require and the number of RPS units. The results of the assessment are shown in Table 16.3. The findings illustrate that higher power missions would require a large number of units to satisfy end of mission (EOM) power needs particularly for a longer duration missions.

Of particular concern would be a "Flagship" class mission (e.g., Cassini, or the proposed Europa Flyby mission), where increased mission costs ($2-3B) would allow much more capability with instrumentation and the spacecraft to where power levels are likely to range between 500 and 1000 We. The NASA Discovery Program class mission ($450M, LV also not included) could also use RPS. Discovery's lower mission cost cap makes it less likely a candidate in general for

Table 16.3 Power requirements assessment of new frontiers 4 missions

Mission	EOM power (We)	EOM time (years)	MMRTG power We/Gen.	EMMRTG power We/Gen.
Trojan Tour[a]	170	10.7	63 (3 units, 189 We)	104 (2 units, 208 We)
Io observer[b]	250	8.5	70 (4 units, 280 We)	111 (3 units, 333 We)
Venus In Situ Explorer[d]	500	1	N/A	N/A
Saturn probe[a]	200	7	77 (3 units, 231 We)	117 (2 units, 234 We)
CSSR[d]	500	7	77 (7 units, 540 We)	117 (5 units, 585 We)
Moonrise[d]	~500	1	100d/102n	129d/136n
Lunar Geo Net[a,c]	62 (93 for transit)	6	76d/78n (1 unit)	111d/118n (1 unit)
Enceladus Orbiter[a]	240	10.5	63 (4 units, 252 We)	104 (3 units, 312 We)

[a]RPS selected, [b]Solar baseline, but an RPS option, [c]Four lander network, [d]Comet surface sample return (CSSR) not a candidate for RPS. *Note* Only the MMRTG was offered in the final NF-4 AO

being enabled by RPS. However, Discovery 12 (Mars InSight lander selected) had two RPS missions in the final competition. The two missions, Comet Hopper (CHopper) and Titan Mare Explorer (TiME) each proposed the use of two Advanced Stirling Radioisotope Generators (ASRG) to satisfy total power requirements of ~250 We. The ASRG, that produced similar power output as the MMRTG but with needing 75% less Pu-238, was postured for a flight by 2018, however the development program was cancelled and is currently being restructured.

An overall power requirements assessment of Discovery, New Frontiers and Flagship class missions was performed and reported in NASA Nuclear Power Assessment Mission Studies for Enabling and Extending Future Space Exploration Final Report (NPAS) (NASA 2015). An observation of the report discussed the possible optimal RPS power output goal for next generation RPS to satisfy outer planet missions to be in the 250–300 We range, very similar to the GPHS RTG. Complete details of the assessment and list of missions analyzed are in the report, see pp. 2–92 to 2–100.

Even if a high power RTG were to be produced, reducing the number of units to simplify integration issues, the amount of PuO_2 needed for a Flagship class mission would certainly use a significant amount of the current Pu-238 supply. For example, Cassini used ~32 kg of PuO_2 for the three GPHS RTGs. As with the breakpoint from solar to nuclear radioisotope power system options for deep space mission, there is a transition point where fission might be a logical solution (Fig. 16.8).

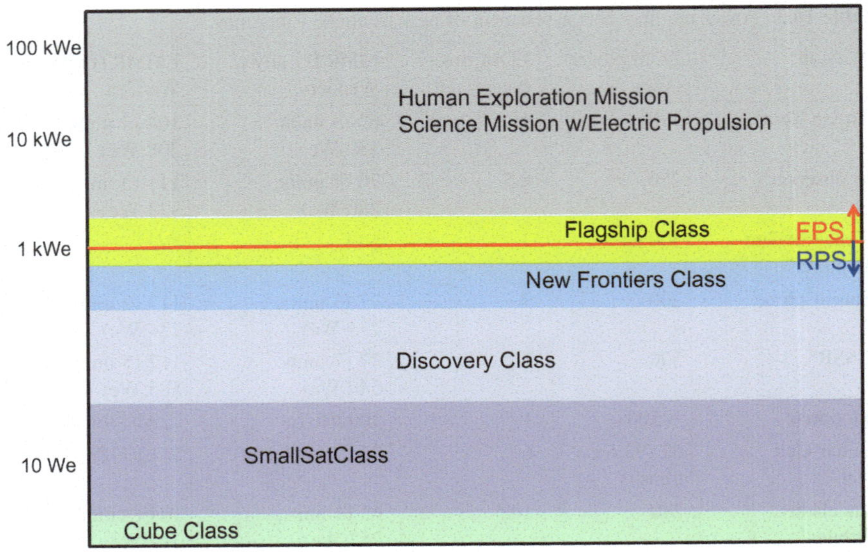

Fig. 16.8 Predicted power requirements for various mission classes fission systems

Figure 16.8 shows a range of expected power levels for various mission classes and a notional breakpoint from RPS to fission technology.

Fission space power systems and applications have been studied and proposed for a variety of missions for many years including Lunar and Mars human bases to high power electric propulsion for both science and human missions. The only US launch of a fission system to produce 500 We, designated as SNAP 10A and launched in 1965 (http://etec.energy.gov/Operations/Major_Operations/SNAP_Overview.html). Since then, numerous space power reactor technology development programs were started and stopped. Probably the most notable of these efforts was called SP-100, a joint DoD/DOE/NASA development program and designed to produce 100 kWe. Clearly the size of the SP-100 is much larger than needed for typical science mission instrument and housekeeping needs.

Like the GPHS modules of an RPS, a space fission reactor generates heat, but instead of the natural radioactive decay process, a reactor utilizes the energy created by the fissioning of uranium. During the fission process the following energy is produced; fission fragment kinetic energy (168 meV), instantaneous gamma energy (7 meV), fission neutron kinetic energy (5 meV), beta particles from fission products (7 meV), gamma rays from fission products (6 meV), gamma rays from neutron capture ($\sim$7 meV) to produce heat. The heat must be removed from the rector. In a space power system, typical reactor cooling options are pumped liquid metal, heat pipes and gas cooling, for example. This heat is then delivered to a power conversion system that converts heat to electricity. The rector thus supplies the "hot end" of the power conversion device and the "cold end" is connected to a radiator that then rejects this heat to space.

Reflectors that surround the core are used to control the number of neutrons available within the core for fission and thus regulates the reactor power output. One advantage of a reactor is that it can be turned on when needed whereas heat is always produced by the Pu-238 decay. In a space fission reactor scenario, the reactor would not be turned on until it reached its operating orbit or destination and therefore the reactor would contain extremely low radioactivity at launch. Once operating however, radiation that is predominately neutrons and gammas, must be shielded against to protect certain materials, electronic components, instruments and crew for a human mission. The shield contributes significant mass to the fission system where minimal shield mass is needed for RPS. For typical reactor space systems designs, separation distance from the source to the spacecraft is increased by placing the reactor on a boom/mast to help reduce shield mass and optimize overall system mass.

Space reactor development over the years has always been seen as a major technology development and a costly endeavor. However recently a simpler concept has been envisioned which is limited in power production to around a maximum of 3 kWe. This concept, has been called Kilopower (Mason 2013), can potentially provide a lower cost and simpler reactor power system to fulfill larger science mission and initial human mission needs. The concept delivers ~1 kWe with thermoelectric conversion (600 kg, 4 m long) and ~3 kWe with dynamic conversion (750 kg, 5 m long) from the same 13 kWth core.

A modified Kilopower concept using a smaller 4 kWth core and eight 100 We Stirling convertors to produce 800 We was also studied and potentially fills the gap between radioisotopes and larger fission. The overall system is approximately 2.5 m long and weighs about 400 kg.

Figure 16.9 illustrates the major components of the 1 kWe, Stirling based Kilopower design; cast 93% enriched U core, BeO reflectors, LiH/W shield, 8–400 We Stirling power conversion units and waste heat radiator. High temperature sodium heat pipes remove heat from the reactor core suppling that heat to the Stirling convertor hot side (950 K). Water heat pipes remove heat from the Stirling cold-side (475 K) to the radiator. Overall length is 5 m.

The NPAS team evaluated retrofitting the 1 kWe Kilopower reactor on a Titan Saturn System Mission spacecraft originally powered by RPS. The conical shadow shield and the boom length serves to protect the entire spacecraft from high intensity radiation. Figure 16.10 provides an illustrative example a typical configuration for a reactor powered spacecraft.

16.4.2 Nuclear Launch Safety

Safety is paramount with every US launch involving nuclear materials. For any US space mission involving the use of radioisotope power systems, radioisotope heating units, nuclear reactors, or a major nuclear source, launch approval must be obtained from the Office of the President per Presidential Directive/National

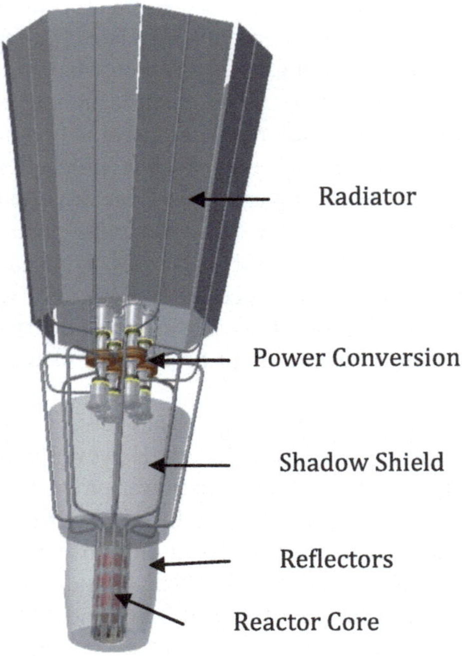

Fig. 16.9 Stirling conversion kilopower concept

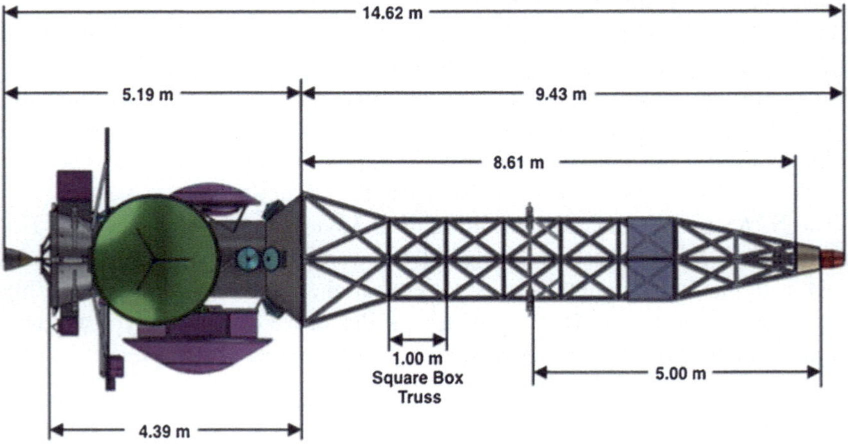

Fig. 16.10 Notional spacecraft integration with a reactor power system

Security Council Memorandum No. 25 (PD/NSC-25), "Scientific or Technological Experiments with Possible Large-Scale Adverse Environmental Effects and Launch of Nuclear Systems into Space," paragraph 9, as amended May 8, 1996.

The National Environmental Policy Act (NEPA) must also be adhered to for an action, such as the launch of nuclear materials and possible alternatives, such as the use of solar power for example. A mission study of the use of nuclear versus non-nuclear power system options must be performed to fully demonstrate that a mission cannot be accomplished without the use of a nuclear power system.

The process initiated by filing NEPA documentation takes several years and significant mission funding. The mission's responsibility is to supply information regarding spacecraft configuration and location of subsystems, propellant type and quantity, pyrotechnic devises, type, quantity and location of nuclear materials, etc. for the safety analyses required for launch approval. For a full detailed discussion of the launch approval processes see relevant sections of references (NASA 2015) and (https://www.nasa.gov/sites/default/files/atoms/files/nasa_systems_engineering_handbook.pdf). The processes for launch approval must be considered early in the mission formulation because they will have an impact on the mission schedule and costs.

16.5 Summary

The use of solar, radioisotope and fission power are candidates for powering outer planet missions. Each technology brings benefits and constraints to a mission and potentially unique spacecraft configuration and accommodations. Improvements in solar cell efficiencies and also improvement with LILT environments over the past few decades have allowed some missions to consider solar arrays to the vicinity of Jupiter providing that large arrays can be easily accommodated by the mission. Radioisotope systems have reliably enabled many NASA missions to gather science data about our Solar System that would otherwise not exist. RTGs have powered spacecraft to far beyond the planets and are expected to reach interstellar space in the near future. Low power fission systems can additionally support missions requiring power levels not practical for radioisotopes systems. Higher power fission systems, can achieve very favorable specific power, so that electric propulsion can be envisioned for high efficiency in-space propulsion. Each technology has a niche in the spectrum of power level, lifetime and mission destination.

References

Bennett, G.L. et al.: Mission of daring: the general-purpose heat source radioisotope thermoelectric generator. In: AIAA 2016, International Energy Conversion Engineering Conference (IECEC), San Diego, CA, 26–29 June 2006

Cataldo, R.L., Bennett, G.L.: U.S. space radioisotope power systems and applications: past, present and future. In: Singh, N. (ed.) Radioisotopes, Applications in Physical Sciences, pp. 473–496. InTech, Rijeka, Croatia (2011)

Cataldo, R.L., Woerner, D.F.: Assessment of future new frontiers mission concepts utilizing the MMRTG and eMMRTG radioisotope thermoelectric generators. In: Nuclear and Emerging Technologies for Space, Huntsville, AL, 21–12 Feb 2016

Department of Energy: SNAP overview. http://etec.energy.gov/Operations/Major_Operations/SNAP_Overview.html

Harris, R.D., et al.: In situ irradiation and measurement of triple junction solar cells at low intensity, low temperature (LILT) conditions. IEEE Trans. Nucl. Sci. **55**(6), 3502–3507 (2008)

Holgate, T.C., et al.: Increasing the efficiency of the multi-mission radioisotope thermoelectric generator. J. Electron. Mater. **44**(6), 1814–1821 (2015)

Landis, G.A., Fincannon, H.J.: Study of power options for Jupiter and outer planet missions. In: 42nd IEEE Photovoltaic Specialist Conference, New Orleans, LA, June 2015

Mason, L., Gibson, M., Posten, D.: Kilowatt-class fission power systems for science and human precursor missions. In: Nuclear and Emerging Technologies for Space, Albuquerque, NM, 25–28 Feb 2013

NASA: Radioisotope power systems. https://solarsystem.nasa.gov/rps/rtg.cfm

NASA: New frontiers fourth announcement of opportunity. http://newfrontiers.larc.nasa.gov/announcements.html

NASA: Nuclear power assessment mission studies for enabling and extending future space exploration final report, RPS-RPT-0121 (JPL D-81712), April 2015. https://solarsystem.nasa.gov/rps/npas.cfm

NASA: Systems engineering handbook, NASA/SP-2007-6105. https://www.nasa.gov/sites/default/files/atoms/files/nasa_systems_engineering_handbook.pdf

Planetary science decadal survey. http://sites.nationalacademies.org/SSB/SSB_059331

Scheiman, D.A., Snyder, D.B.: Low intensity low temperature (LILT) measurements of state-of-the-art triple junction solar cells for space missions. In 33rd IEEE Photovoltaic Specialists Conference, San Diego, CA, 11–16 May 2008

Woerner, D.F.: An enhanced MMRTG. In 47th Lunar and Planetary Science Conference, # 2742 poster (2016)

Chapter 17
Hybrid Nuclear Spacecraft for the Outer Planets

Mark A. Stull and Ricky Tang

17.1 Introduction

Outer solar system resources could be tapped to meet terrestrial needs, or for use by facilities or colonies in orbit or on the Moon or Mars. But this requires that they be either uniquely obtainable from locations beyond the asteroid belt or cost competitive with terrestrial and inner solar system resources. While it may very well be that the outer solar system contains needed resources unobtainable closer to home, the more conservative assumption is that they will have to compete on cost. Distance alone does not preclude that, as current global trade in resources demonstrates. But the minimum requirements for importation of resources across interplanetary distances can be expected to be similar to those for importation across Earth's oceans and land masses: personnel and specialized equipment at the site where the resources can be found, and a low-cost, high-capacity, rapid means of transportation. And the need for rapid, high-capacity transportation must be stressed. Although it might naively be assumed that resource-carrying pods can be sent to coast on years-long journeys to their destinations, that would pose an unacceptable level of economic risk. Even daily fluctuations in the prices of energy and raw materials on commodities markets can be large. Very few investors are likely to finance resource-extraction operations, even during periods of shortage and high prices, if they need so much time to get their products to market that the price is unpredictable. Furthermore, very few are likely to make a significant investment in resource extraction unless shipping requirements can be met.

M. A. Stull (✉)
The Law Office of Mark A. Stull, 63 High Street, Manchester, NH, USA
e-mail: mastulllaw@comcast.net

R. Tang
Sandia National Laboratories, 1515 Eubank Blvd. SE, MS-1423, Albuquerque, NM 87123, USA
e-mail: rtang@sandia.gov

V. Badescu and K. Zacny (eds.), *Outer Solar System*,
https://doi.org/10.1007/978-3-319-73845-1_17

Low-cost, high-speed and high-capacity transportation systems are essential to the use of resources from remote locations. [For an analysis of the economic value of speed in passenger transportation systems see (Stull 2008)]. Oil tankers, container ships, cargo planes, coal trains, trucks, pipelines—without these and other transportation mechanisms the resources used by the modern economy simply would not be available. The corresponding system for an economy based at least in part on outer solar system resources will have to feature high-capacity spacecraft capable of rapid travel under continuous acceleration. To be sure, the specific requirements will depend on the resource in question and market conditions. For example, energy prices can be very volatile, fluctuating by large percentages in a period of months. The same is true for food prices. But these commodities, and others, can also experience long periods of relative price stability. Companies that engage in mining and drilling operations need months or years to dig new mines or drill new wells. Farmers and meat producers operate on similar time scales to plant crops or increase herds. If transportation times are short compared to development times, they are less likely to deter investment, but if they are comparatively long, investment in development would be problematic. These considerations suggest that there will be little or no development of outer-solar-system resources if travel times are measured in years, but travel times of weeks or months could be acceptable. As to capacity, a conservative assumption would be to use the capacities of terrestrial resource carriers such as oil tankers and coal trains as those required for economic viability. Given the economic requirements, elementary principles of physics dictate the fundamental requirements for propulsion: high thrust combined with high specific impulse (I_{sp}) to provide continuous acceleration for a massive spacecraft. These, in turn, require continuous power supply and very high power density.

17.2 Requirements

Chemical rocket technology has reached a physical limit and cannot meet these requirements. For example, nearly five decades after Apollo we have made little progress in reducing launch costs. The Space Shuttle and Saturn V could deliver only about 4% of their total liftoff mass to orbit, and the Saturn V could accelerate less than 2% of its mass to escape velocity. This is a consequence of the low specific impulse (I_{sp}) obtainable from chemical fuels, about 450–500 s at most. Conceptually, specific impulse is analogous to fuel efficiency of a vehicle; the higher the specific impulse, the less propellant is needed to achieve a given mission. Conventional nuclear thermal rockets can do little better, with I_{sp} for various proposed designs ranging from about 800 s to as high as 2000 or 3000 s. In contrast, the I_{sp} necessary to accelerate 90% of launch mass (i.e. propellant accounting for only 10% of initial launch mass) to escape velocity is nearly 11,000 s, and this would merely suffice to allow a payload to coast for years across solar system distances. The high cost of reaching orbit and escape velocity with chemical rockets precludes construction on Earth or in orbit of large interplanetary resource transportation vessels, and precludes construction of facilities on the Moon where they might be built and launched at lower cost. And neither chemical nor conventional nuclear thermal propulsion can reduce propellant requirements to the level required for continuous acceleration over outer solar system distances; indeed they cannot meet requirements for development of inner solar system resources.

Acceleration matters. The time to travel a given distance under constant acceleration is inversely proportional to the square root of the acceleration. For six outer solar system bodies the travel time in days as a function of acceleration in units of Earth's gravitational acceleration, g, is shown in Fig. 17.1, assuming that a ship accelerates at constant rate to the midway point, then decelerates at the same rate to its destination. Continuous acceleration as low as 0.2g would allow resource shipments to reach Earth in 6 weeks or less from as far away as Pluto, and in approximately 2 months from Eris. Except in unusually volatile market conditions, these travel times should present acceptable economic risk, especially for Jovian and Saturnian resource exploitation, which could be viable with lower acceleration. Otherwise, the travel times for significantly smaller accelerations, while acceptable for exploration and research, probably would create too great a level of economic risk for competitive terrestrial markets, unless a particular resource was scarce and in high demand.

That's the good news. The bad news, of course, is that constant acceleration carries a penalty: energy and propellant requirements, and the higher the acceleration, the more demanding the requirements. The ratio, ς, of propellant mass, M_r, to

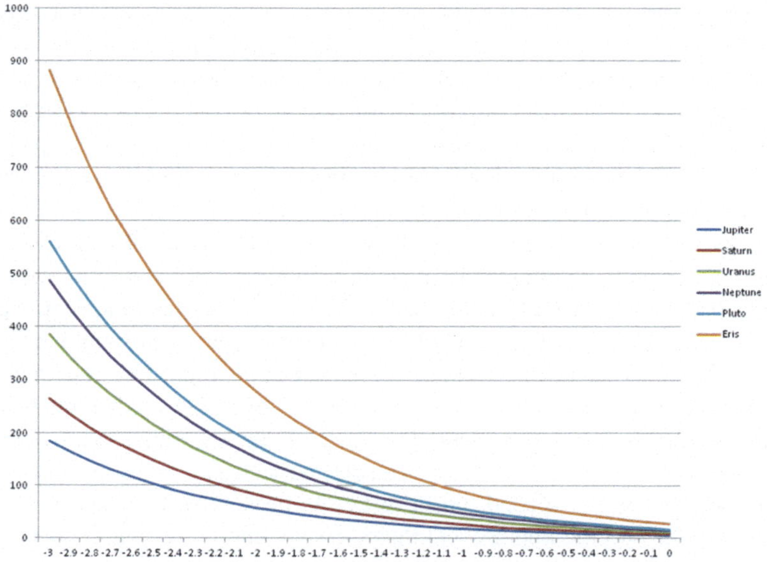

Fig. 17.1 Travel time in days as a function of log acceleration in units of g

payload mass, M_p, is given in terms of the increase in velocity, Δv, the gravitational acceleration, g, and specific impulse, I_{sp}, by the classical rocket equation [If propellant ejection velocities are an appreciable fraction of the speed of light, i.e., for I_{sp} greater than approximately 3 million seconds, the rocket equation must be modified to account for relativistic effects. I_{sp} this high will not be considered in this analysis]:

$$\varsigma = \frac{M_r}{M_p} = e^{\Delta v/gI_{sp}-1} \tag{17.1}$$

from which it follows that the jet power required for an acceleration, a, is

$$P = \frac{1}{2} agI_{sp}M_p e^{\Delta v/gI_{sp}} \tag{17.2}$$

The payload mass is the sum of the mass of the propulsion system, including the power generator, and the mass of the rest of the ship including passengers and cargo. If the ratio of the thrust produced by the propulsion system to the weight of the propulsion system is denoted by α, and the mass of the ship exclusive of the propulsion system by M_c, then Eq. (17.2) becomes

$$P = \frac{agI_{sp}M_c e^{\Delta v/gI_{sp}}}{2(1 - \frac{a}{\alpha g} e^{\Delta v/gI_{sp}})} \tag{17.3}$$

Equation (17.3) can also be solved to obtain the acceleration resulting from a given level of power generation.

The rocket equation gives the well-known result that as I_{sp} decreases, the ratio of propellant mass to payload mass increases exponentially. Furthermore, for continuous acceleration, the higher the acceleration, the greater the Δv. As a result, for a typical chemical rocket, the mass of propellant required for continuous acceleration is impossibly high, even for accelerations much less than 0.2g, and the power requirement is also impossibly high. As I_{sp} increases the power requirement decreases to a minimum, then increases again; this is because at low values of I_{sp}, increased power is required to accelerate the propellant. The value of I_{sp} for which the power requirement is minimized is $\Delta v/g$, in which case the ratio of propellant mass to payload mass is a fairly reasonable 1.72 for any acceleration. However, propellant represents an operating cost, while the power requirement should mainly impact capital cost. This favors a choice of power and propulsion system that attempts to achieve higher power density, to enable use of higher I_{sp} than the value that minimizes the power requirement. For each of the six outer solar system bodies, Tables 17.1a through 17.1f give, for selected values of constant acceleration, Δv (in kilometers per second, taken to be the peak velocity reached at the half-way point), the specific impulse and the corresponding power requirement (in gigawatts per metric ton of ship mass exclusive of propellant and power and propulsion system) for three values of the propellant to payload ratio, $\varsigma = M_r/M_p$, 1.72 (for the minimum power), 1.0 and 0.25. The power requirements were calculated using Eq. (17.3), with $\alpha = 5$. [Nuclear thermal rocket concepts typically have thrust-to-weight ratios in the range of 3–10 (Borowski et al. 1993). For accelerations significantly less than 1g and Isp $\geq \Delta v/g$, the power calculated from Eq. (17.3) is not very sensitive to values of α greater than 1. However, as will be discussed below, at very high power levels, values may be much less than 1.] For travel times on the order of weeks, the power requirements are very large.

Table 17.1 Power requirements for travel

Acceleration (units of g)	Travel time (days)	Δv (km/s)	$M_r/M_p = 1.72$		$M_r/M_p = 1.0$		$M_r/M_p = 0.25$	
			Isp (s)	Power (GW/mT)	Isp (s)	Power (GW/mT)	Isp (s)	Power (GW/mT)
Jupiter								
0.001	185	78	8008	0.001	11,554	0.001	35,888	0.00
0.002	131	111	11,312	0.003	16,320	0.003	50,694	0.01
0.005	83	176	17,928	0.01	25,865	0.01	80,344	0.02
0.01	59	248	25,324	0.03	36,535	0.04	113,489	0.07
0.02	42	351	35,772	0.09	51,608	0.10	160,308	0.19
0.05	26	556	56,694	0.37	81,793	0.39	254,071	0.76
0.1	19	785	80,083	1.05	115,535	1.11	358,885	2.15
0.2	13	1109	113,120	2.95	163,198	3.13	506,938	6.07
0.5	8	1757	179,283	11.73	258,651	12.45	803,443	24.17
1.0	6	2482	253,244	33.06	365,354	35.09	1,134,894	68.12
Saturn								
0.001	265	112	11,431	0.001	16,491	0.002	51,226	0.003
0.002	187	158	16,146	0.004	23,294	0.004	72,359	0.01
0.005	118	251	25,590	0.02	36,919	0.02	114,681	0.03
0.01	84	354	36,147	0.05	52,150	0.05	161,992	0.10
0.02	59	500	51,060	0.13	73,663	0.14	228,819	0.27
0.05	37	793	80,924	0.53	116,748	0.56	362,654	1.09
0.1	26	1120	114,308	1.49	164,912	1.58	512,262	3.07
0.2	19	1582	161,464	4.21	232,944	4.46	723,590	8.67
0.5	12	2508	255,904	16.74	369,191	17.77	1,146,812	34.50
1.0	8	3542	361,474	47.18	521,496	50.08	1,619,916	97.24
Uranus								
0.001	386	163	16,661	0.002	24,037	0.002	74,667	0.004
0.002	273	231	23,535	0.006	33,954	0.007	105,470	0.01
0.005	172	366	37,300	0.02	53,813	0.03	167,158	0.05
0.01	122	516	52,688	0.07	76,013	0.07	236,117	0.14
0.02	86	729	74,424	0.19	107,371	0.21	333,524	0.40

(continued)

Table 17.1 (continued)

Acceleration (units of g)	Travel time (days)	Δv (km/s)	M_r/M_p = 1.72		M_r/M_p = 1.0		M_r/M_p = 0.25	
			Isp (s)	Power (GW/mT)	Isp (s)	Power (GW/mT)	Isp (s)	Power (GW/mT)
0.05	54	1156	117,954	0.77	170,171	0.82	528,600	1.59
0.1	39	1633	166,614	2.17	240,373	2.31	746,667	4.48
0.2	27	2306	235,348	6.13	339,536	6.51	1,054,695	12.63
0.5	17	3655	373,002	24.40	538,128	25.90	1,671,579	50.29
1.0	12	5163	526,880	68.77	760,126	73.00	2,361,169	141.73
Neptune								
0.001	488	206	21,070	0.003	30,397	0.003	94,422	0.01
0.002	345	292	29,762	0.008	42,937	0.008	133,374	0.02
0.005	218	462	47,169	0.03	68,050	0.03	211,384	0.06
0.01	154	653	66,628	0.09	96,124	0.09	298,588	0.18
0.02	109	922	94,115	0.25	135,779	0.26	421,767	0.51
0.05	69	1462	149,161	0.98	215,194	1.04	668,455	2.01
0.1	49	2065	210,696	2.75	303,970	2.92	944,218	5.67
0.2	35	2917	297,616	7.75	429,369	8.23	1,333,743	15.97
0.5	22	4623	471,690	30.86	680,505	32.76	2,113,840	63.59
1.0	15	6530	666,280	86.97	961,238	92.32	2,985,879	179.23
Pluto								
0.001	561	238	24,245	0.003	34,978	0.003	108,651	0.01
0.002	397	336	34,247	0.009	49,407	0.009	153,473	0.02
0.005	251	532	54,277	0.04	78,305	0.04	243,239	0.07
0.01	177	751	76,669	0.10	110,609	0.11	343,584	0.21
0.02	126	1061	108,297	0.28	156,240	0.30	485,325	0.58
0.05	79	1682	171,639	1.12	247,623	1.19	769,188	2.31
0.1	56	2376	242,447	3.16	349,777	3.36	1,086,507	6.52
0.2	40	3356	342,466	8.92	494,074	9.47	1,534,733	18.38
0.5	25	5319	542,772	35.51	783,054	37.69	2,432,387	73.18
1.0	18	7514	766,685	100.08	1,106,093	106.23	3,435,838	206.24
Eris								
0.001	883	374	38,146	0.005	55,034	0.005	170,950	0.01
0.002	625	528	53,883	0.014	77,737	0.015	241,474	0.03
0.005	394	837	85,399	0.06	123,205	0.06	382,710	0.12
0.01	279	1182	120,630	0.16	174,032	0.17	540,592	0.32
0.02	198	1670	170,394	0.44	245,827	0.47	763,607	0.91
0.05	125	2647	270,056	1.77	389,609	1.88	1,210,236	3.64
0.1	88	3738	381,465	4.98	550,337	5.29	1,709,504	10.26
0.2	63	5281	538,833	14.03	777,372	14.90	2,414,738	28.92
0.5	39	8369	853,993	55.87	1,232,052	59.30	3,827,102	115.13
1.0	28	11,822	1,206,297	157.46	1,740,319	167.14	5,405,925	324.49

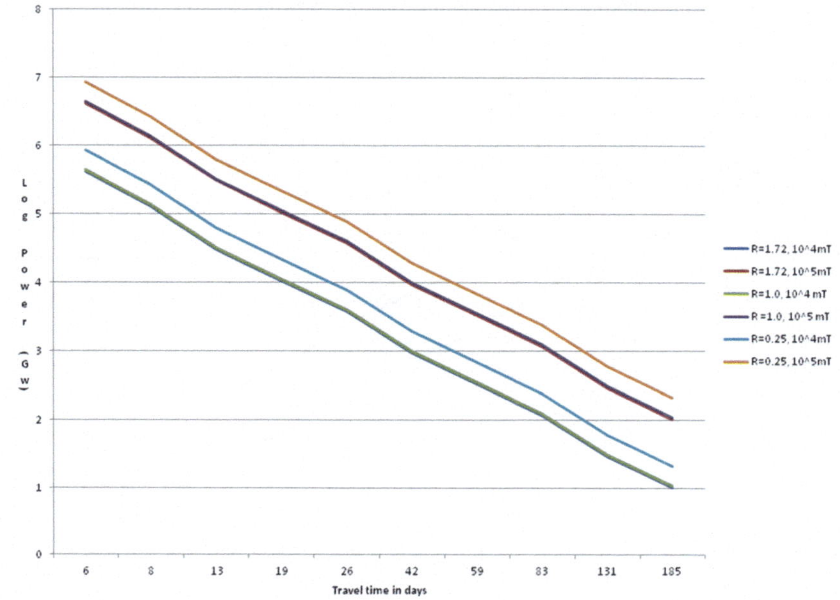

Fig. 17.2 Power required to reach Jupiter

A modern supertanker carries approximately 300,000 metric tons of oil. A typical coal train of over 100 cars carries 10–15 thousand metric tons of coal, while the United States consumes around a billion tons of coal per year, representing over 200 coal trains per day. This suggests that a robust commerce in outer solar system resources requires hundreds of spacecraft, each with a capacity of 10^4–10^5 metric tons. Figures 17.2, 17.3, 17.4, 17.5, 17.6 and 17.7 show, for each of the 6 outer solar system bodies, the total power requirement as a function of one-way travel time for propellant to total ship mass ratios of 1.72 (minimum power), 1.0 and 0.25 and cargo masses of 10^4 and 10^5 metric tons, all assuming $\alpha = 5$, which, as will be seen, may be overly optimistic.

At first glance, these power requirements might seem impossible, dooming commerce in outer solar system resources to the realm of science fiction. But there is a technology that can at least deliver the required power, although that is merely a necessary condition and not a sufficient one. It might be developed at a minimal level in 10 years, or even less, with a reasonably ambitious program. That technology is a hybrid nuclear reactor in which a subcritical or barely critical fission reaction is driven by neutrons generated by an external neutron source, specifically, a fusion reaction contained by a gas dynamic mirror that doubles as a thruster.

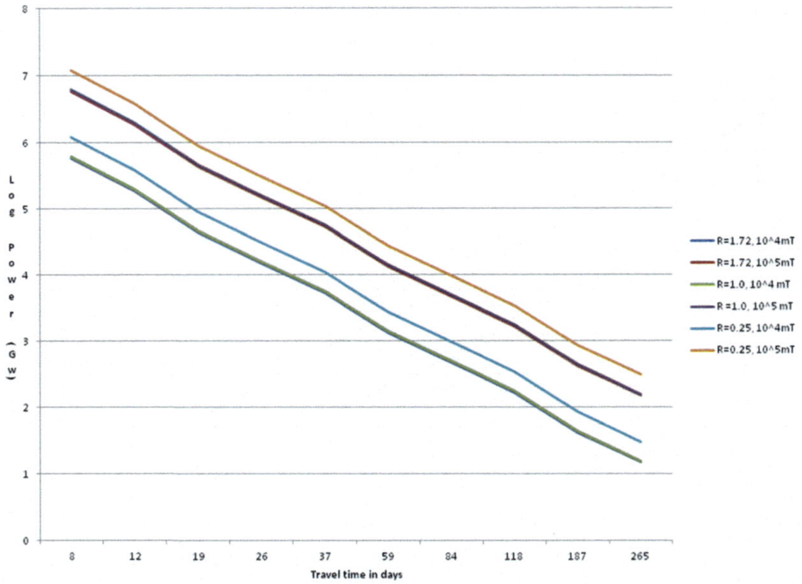

Fig. 17.3 Power required to reach Saturn

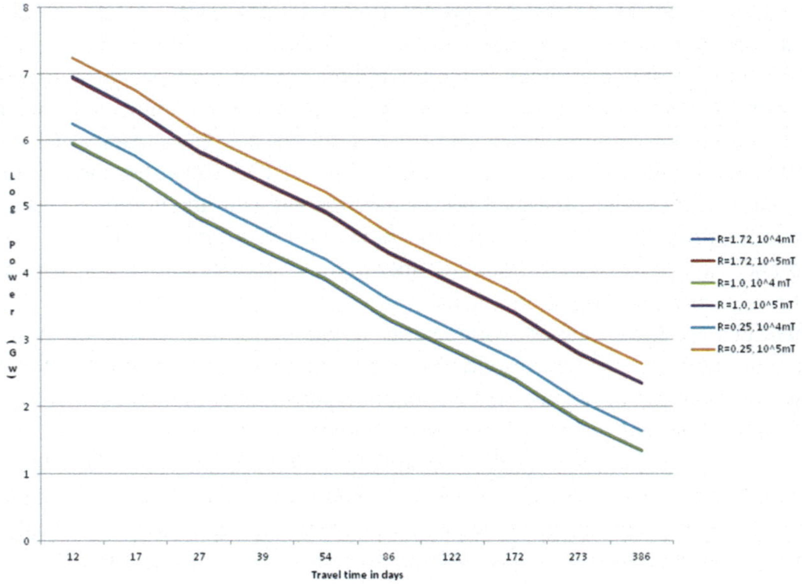

Fig. 17.4 Power required to reach Uranus

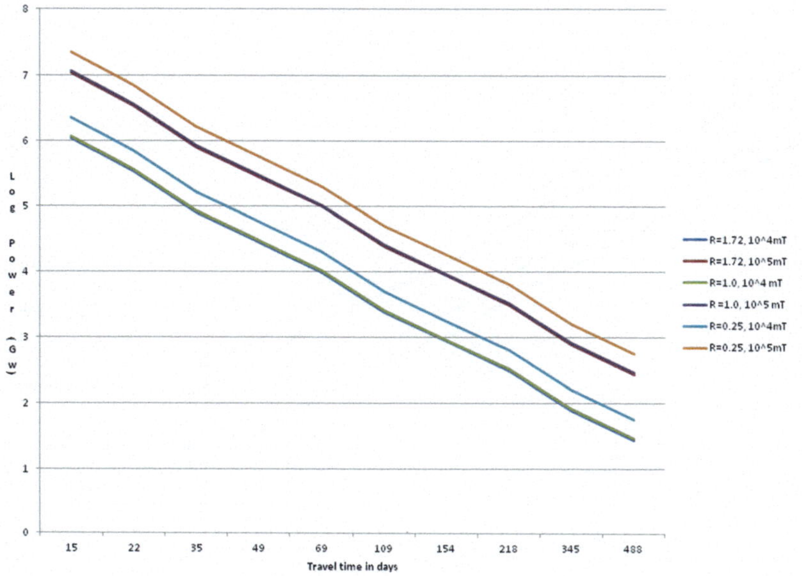

Fig. 17.5 Power required to reach Neptune

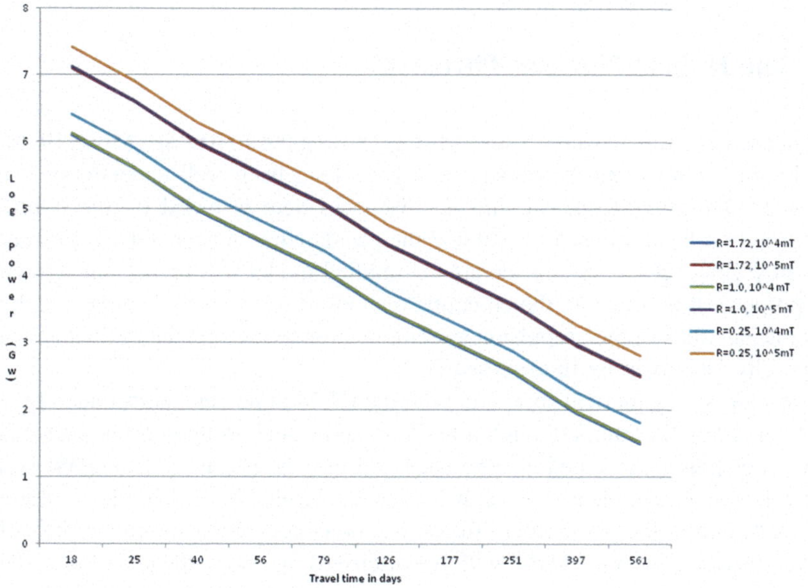

Fig. 17.6 Power required to reach Pluto

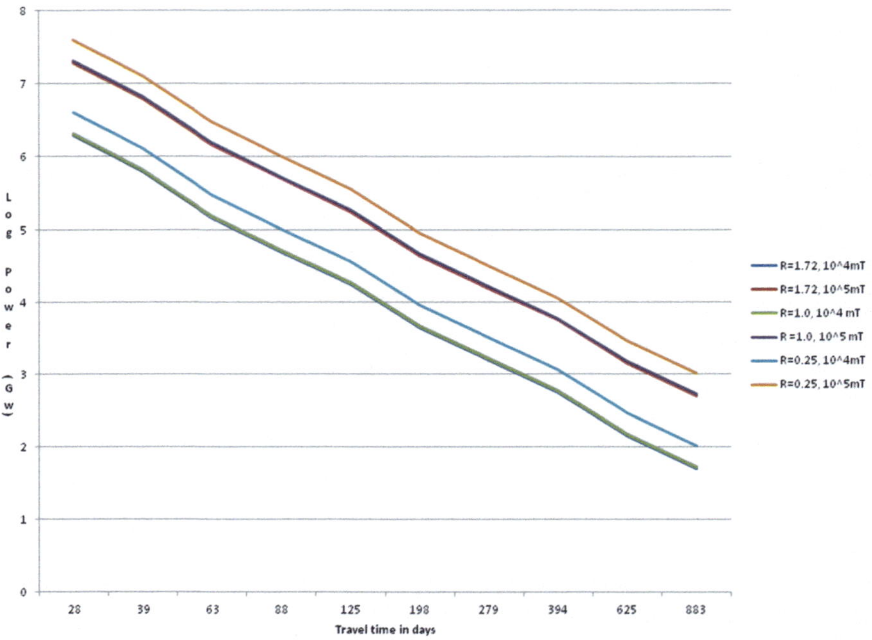

Fig. 17.7 Power required to reach Eris

17.3 The Hybrid Nuclear Thruster

Hybrid nuclear reactors have been studied since at least the 1970s. See Moir (1976). Magnetic mirrors for controlled fusion also have long been well understood. See Post (1983). The hybrid nuclear thruster discussed here is based on the hybrid reactor proposed by Kammash (2010) and the gas dynamic mirror (GDM) thruster analyzed by Tang (2011). A considerably simplified model will be presented here, for the purpose of minimizing the computational effort required to demonstrate the potential capabilities of the propulsion system, its limitations, and the problems that can hopefully be solved by future research.

The reactor is a cylindrical structure, with the GDM at its core, surrounded by a blanket containing thorium, although a fissile material such as appropriate isotopes of uranium or plutonium could also be used and may be required. The GDM is a magnetic mirror containment device in which deuterium or a deuterium-tritium mixture is heated to the point where fusion reactions generate a significant flux of neutrons; operation of the GDM will be discussed in more detail below. The neutrons drive a subcritical fission reaction via the thorium cycle, in which fissile U_{233} is bred from Th_{232}, and U_{233} fission produces energy. Use of a moderator thermalizes both the fusion neutrons and additional neutrons produced by the fissioning U_{233}. Energy produced by the fission reaction is used to heat the fusion plasma, and supply power to the ship.

The fission energy greatly exceeds the fusion energy required to generate a sufficient neutron supply, which means the fusion reaction need not achieve breakeven. A single DT fusion reaction generates only 17.6 meV, of which 14.1 meV is neutron kinetic energy that, in a pure fusion scheme, is lost. On the other hand, a single fission reaction not only makes good use of a neutron, but generates some 200 meV. Heating of the fusion plasma beyond the level required for minimal neutron generation results in additional neutron generation and increases energy production from both the fission and fusion reactions. Thus at all fusion power levels, much more fission power is available. The fusion plasma does double duty as the propellant. This is facilitated by the GDM, which, as is well known, is an inherently leaky containment device. While not ideal for power generation, this is just what the doctor ordered for a thruster, as the fusion plasma can be made to escape preferentially from one end of the cylindrical device. Furthermore, use of the fission reaction to supply energy to heat the plasma can allow thermal equilibrium to be achieved at temperatures over a wide range, from a few keV to over 100 keV. As a result, the escaping fusion plasma can generate high thrust and high I_{sp}, subject to the limitations discussed below.

17.3.1 The Gas Dynamic Mirror

The GDM is a magnetic mirror device in which the collisional mean free path of the plasma ions is sufficiently small that the plasma can be treated as a Maxwell-Boltzmann perfect gas, as is done in stellar structure models for main sequence stars. The GDM can be visualized as a long thin cylinder, in the interior of which a plasma is contained by a strong magnetic field, with the field lines parallel to the long axis of the cylinder. At the ends of the cylinder, the magnetic field is made stronger in such a manner that charged particles whose velocity vector falls outside a certain angle in the velocity phase space, called the loss cone, reverse direction as though reflected by a mirror; hence the name. Particles within the loss cone leave the cylinder at both ends; however, the magnetic field strengths at the mirrors can be adjusted so that particle ejection at one end is much greater than at the other, giving rise to a propulsive force equal to the mass flow rate of the ejected particles times their ejection velocity plus the gas pressure; the latter is small compared to the former and will be neglected. Particles that escape from the opposite end can be made to supply electrical power by means of a direct conversion device or a magnetohydrodynamic generator. The specific impulse is the ejection velocity divided by the gravitational acceleration, g. The GDM has been described in detail in numerous papers in the literature; see, especially, Tang (2011).

For the current discussion perhaps the most important relationship is given by the equation for thermal balance, which is simply

$$\frac{dU}{dt} = \varepsilon_{fu} + \gamma\varepsilon_{fi} - \varepsilon_{th} - \varepsilon_r = 0 \tag{17.4}$$

where ε_{fu} and ε_{fi} are, respectively, the rate of energy generation by the fusion and fission reactions, γ is the fraction of the fission energy that is supplied to the fusion plasma, ε_{th} is the kinetic energy flux due to the mass flow through the loss cone that provides the propulsive force, and ε_r is the rate at which energy is lost by radiation. All of these are functions of the plasma temperature and density, as will be discussed below. Other important relations are the following:

1. Fusion energy generation rate. For a DT mixture, there are three potentially important reactions, DD, DT and TT; of these, DD and DT should dominate at the temperatures in question, with DT the most important at the lowest temperatures. Reactions involving the products of these three can also occur, but are less important because most of the plasma will be ejected as reaction mass well before a significant amount of fusion products can be generated, and will be replaced by continuous injection of DT mixture. The TT reaction can also be neglected at the temperatures relevant here; however, both branches of the DD reaction will be considered as one produces neutrons while the other does not. For each of the three reactions (Atenzi and Meyer-ter-Vehn 2004) gives the energy generation rates per unit plasma volume as follows:

$$\varepsilon_{ij} = \frac{f_i f_j}{1+\delta_{ii}} N^2 \langle \sigma v \rangle_{ij} E_{ij} \tag{17.5}$$

where i and j represent the reacting species, D and/or T, N is the number density of nuclei in the plasma, f_i and f_j are the atomic fractions of the reacting species, E_{ij} is the energy released per reaction, exclusive of the neutron energy, δ is the Kronecker delta symbol, and $\langle \sigma v \rangle_{ij}$ is the reaction cross section for species i and j averaged over a Maxwellian velocity distribution. Over a broad temperature range, the Maxwell-averaged reaction cross section is given by the following formulas:

For the DT reaction $E_{DT} = 3.5$ MeV):

$$\langle \sigma v \rangle_{DT} = 6.431 \times 10^{-14} \eta^{-5/6} \xi^2 e^{-3\xi\eta^{1/3}} \tag{17.6a}$$

where

$$\eta = 1 - \frac{0.015136T + 0.0046064T^2 - 0.00010675T^3}{1 + 0.075189T + 0.0135T^2 + 0.00001366T^3}$$

and

$$\xi = 6.661T^{-1/3}$$

For the branch of the DD reaction that does not result in neutron production ((E_{DD1} = 4.04 MeV):

$$\langle \sigma v \rangle_{DD1} = 3.7212 \times 10^{-16} \eta^{-5/6} \xi^2 e^{-3\xi\eta^{1/3}} \tag{17.6b}$$

where

$$\eta = 1 - \frac{0.0034127T}{1 + 0.0019917T + 0.000010506T^2}$$

and

$$\xi = 6.2696T^{-1/3}$$

For the branch of the DD reaction that does result in neutron production ($E_{DD2} = 0.82$ MeV):

$$\langle \sigma v \rangle_{DD2} = 3.7212 \times 10^{-16} \eta^{-5/6} \xi^2 e^{-3\xi\eta^{1/3}} \tag{17.6c}$$

where

$$\eta = 1 - \frac{0.0058577T}{1 + 0.0076822T + 0.000002964T^2}$$

and

$$\xi = 6.2696T^{-1/3}$$

with T in keV. For the DT reaction the Maxwell-averaged cross section peaks at approximately 65 keV and declines at higher temperatures. The neutron generation rate is ε_{DT}/E_{DT}; this dominates the rate of DD neutron generation and primarily determines ε_{fi}, in accordance with Eq. (17.19) below. The DD Maxwell-averaged cross section peaks at over 1000 keV; however, for a 50% D 50% T mix, the combined energy generation rate peaks at a much lower temperature, where it is still dominated by the DT reaction. The DD neutron generation rate is $\varepsilon_{DD2}/E_{DD2}$. The average DT neutron energy is 14.1 meV; for DD neutrons it is 2.45 meV.

2. Ambipolar potential. Electrons preferentially escape through the loss cone due to their lower mass and higher thermal velocity. This leads to an excess of positive charge within the GDM. The resulting electric field, known as the ambipolar potential, φ, retards the escape of additional electrons and accelerates the escape of the ions. As a result, both escape at the same rate, creating a charge neutral propellant beam. The average energy of an escaped electron outside the

GDM is therefore $(E_{Le} - e\varphi)$ and that of an escaped ion is $(E_{Li} + e\varphi)$, where E_{Le} and E_{Li} are the electron and ion escape energies. Methods for calculating the magnitude of the ambipolar potential and the resulting ion and electron ejection velocities are described by Post (1983) and Tang (2011). For computational simplicity, it can be roughly approximated by starting with the electron and ion escape energies, ignoring the effect of the ambipolar potential, (Tang 2011), namely

$$E_{Le} = \frac{5}{2}kT_e \quad \text{and} \quad E_{Li} = 2kT_i$$

The respective average electron and ion escape velocities, v_e and v_i respectively, can then be written as follows:

$$v_e = \sqrt{\frac{2(2.5kT_e - e\varphi)}{m_e}} \tag{17.7a}$$

$$v_i = \sqrt{\frac{2(2kT_i + e\varphi)}{m_i}} \tag{17.7b}$$

The condition of charge neutrality requires that the charged particle fluxes be equal and the net charge be zero. Hence, assuming singly-charged ions, these conditions imply the average electron and ion escape velocities must be equal, and the ambipolar potential can then be approximated as

$$e\varphi = \frac{\frac{5}{2}kT_e\left(1 - \frac{4}{5}\frac{m_e}{m_i}\frac{T_i}{T_e}\right)}{1 + \frac{m_e}{m_i}} \approx \frac{5}{2}kT_e \tag{17.8}$$

Interpreted another way, the ambipolar potential is governed by the electron temperature. Equation (17.8) overestimates the ambipolar potential because in a GDM the electron temperature is not the same as the ion temperature due to the preferential escape of high energy electrons. If Eqs. (17.7b) and (17.18) were used to calculate the ion ejection velocity, the result, double the ion thermal velocity, would also be an overestimate, and use of Eqs. (17.7a) and (17.8) for the electron ejection velocity would incorrectly give zero. However, for computational simplicity, it will be assumed that the electron and ion temperatures are equal and that the value obtained from Eq. (17.8) can still be used to roughly estimate the ion ejection velocity by means of Eq. (17.9):

$$v_i = \sqrt{2(2kT + e\varphi)/m_i} \approx 3\sqrt{kT}/m_i \tag{17.9}$$

Note that the ion ejection velocity, and therefore thrust and specific impulse, significantly depend on the strength of the ambipolar potential.

3. Kinetic energy flux. Energy is removed from the GDM by the fusion plasma that escapes through the weaker mirror to provide thrust. There is also a smaller rate of energy loss through the stronger mirror; however, for present purposes it will be assumed that the energy that would otherwise be lost is recaptured and fed back into the fusion plasma, either by heating the new DT propellant/fuel that is continuously injected into the GDM or by conversion to electricity by means of a direct converter, such as proposed by Tang (2011). The rate of energy removal by the propellant is the ion loss rate multiplied by the average energy per ion, $\frac{1}{2}m_i v_i^2$. Although electrons also escape and carry energy, this is orders of magnitude less because of the mass difference. The ion loss rate is the number density times the plasma volume divided by the confinement time, the time for all of the plasma to be scattered into the loss cone and escape, τ, which is usually approximated by

$$\tau = \frac{R_p L}{2v_{th}} = \frac{RL}{2v_{th}} \frac{1}{\sqrt{1-\beta}} \tag{17.10}$$

where $R = B_m/B_c$ is defined by the ratio of the magnetic field strength at the mirror to the field strength at the center of the GDM, v_{th} is the thermal velocity, and R_p is the plasma mirror ratio (i.e. mirror ratio seen by the plasma, as the presence of the plasma reduces the apparent magnetic field) that can be related to R via the parameter β that is defined below. Tang (2011) gives a more accurate expression for τ, modified by the effect of the ambipolar potential, which decreases the confinement time. For computational simplicity, the decrease can be approximated by replacing the thermal velocity in Eq. (17.10) by the escape velocity given by Eq. (17.9) and assuming β, which is always less than one, can be ignored. Therefore, the kinetic energy flux is approximated by

$$\varepsilon_{th} = \pi r_p^2 \frac{N}{R} m_i v_i^3 \tag{17.11}$$

where r_p is the plasma radius.

4. Radiation losses. The power radiated due to bremsstrahlung can be approximated by

$$\varepsilon_r = 1.8 \times 10^{-38} r_p^2 L Z^2 N^2 T^{1/2} \tag{17.12}$$

where Z is the average atomic number, T is in eV, N in m^{-3} and ε_r in watts. Cyclotron and synchrotron radiation losses have not been considered; at temperatures of a few tens of keV or less these should be small.

5. Containment criterion. The plasma will be contained by the magnetic field of the GDM provided the plasma pressure does not exceed the magnetic field pressure,

$$NkT \leq B^2/2\mu_0 \tag{17.13a}$$

where μ_0 is the magnetic permeability. Alternatively,

$$NkT = \beta B^2/2\mu_0 \tag{17.13b}$$

where $\beta \leq 1$. For a stable plasma β can approach its maximum value.

6. Lawson criterion. The fusion reaction will be self-sustained if the Lawson criterion for breakeven is met. In its usual formulation the Lawson criterion is based on the assumption of complete plasma containment, which is not the case here [In a GDM a self-sustained reaction will occur only if the rate of fusion energy generation net of energy losses satisfies the criterion. Therefore, rather than use the term "breakeven", the text will refer to the usual Lawson criterion as "Lawson-criterion breakeven"] and is satisfied if the rate of energy generation per unit volume by the fusion reactions times the confinement time is equal to or greater than the average ion kinetic energy, $\tau\varepsilon_{fu} \geq 1.5NkT$, or

$$\frac{RL\varepsilon_{fu}}{3NkTv_i} \geq 1 \tag{17.14}$$

7. GDM criterion. The mirror device will function as a GDM only if the plasma is collisional, which requires that its length be greater than the collisional mean free path of the ions divided by the mirror ratio. This condition is satisfied if

$$L > 1.253 \times 10^{24} \frac{T^2}{NR} \tag{17.15}$$

where T is in keV and N is the ion number density per cubic meter.

8. Performance limitations. The combined fission and fusion reactions can generate enormous amounts of power, resulting in high thrust and high specific impulse.

But the containment criterion imposes a limit on the maximum thrust obtainable from a single GDM thruster, although there is no such limit on specific impulse, which is $I_{sp} = v_i/g$ and depends only on the plasma temperature. The thrust, F, is the momentum per unit time of the ejected propellant ions and depends on both plasma temperature and density as follows:

$$F = Nm_i v_i \pi r_p^2 L/\tau \tag{17.16a}$$

which, with appropriate substitutions from Eqs. (17.8), (17.9) and (17.10) becomes

$$F = \frac{2NkT}{R}\left(3 + \frac{8}{\pi}\right)\pi r_p^2 \tag{17.16b}$$

The product NkT has a maximum value given by Eq. (17.13a). For any given value of the magnetic field strength, once the maximum thrust has been reached, an increase in temperature must result in a decrease in density and vice versa, or containment will be lost. The maximum thrust can be increased by decreasing the mirror ratio, subject to stability constraints.

Other constraints exist. Equations (17.4) and (17.13a) can be combined and solved for the temperature as a function of magnetic field strength and plasma length, with the result that temperature and therefore I_{sp} increase with increasing plasma length. However, the mean free ion path also increases with temperature, approximately as T^3, because density must decrease with increasing temperature to maintain containment. This may impose a maximum operating temperature, which, in turn, could impose a maximum plasma length. There will also be a minimum length, imposed by the condition that the combined fission and fusion power must not be less than the kinetic energy flux plus radiation losses.

This leaves a wide range of options for optimizing performance. Specific impulse can be increased by going to longer thruster lengths and/or higher plasma temperatures. Thrust can be increased simply by adding additional GDM thrusters. A desirable configuration might be a cluster of multiple GDM thrusters surrounded by a blanket containing fertile Th_{232} and/or a fissile isotope of uranium or plutonium. Energy produced by the fission reaction could supply power to the ship and heat the fusion plasma to the desired temperature. But there is an 800-pound gorilla sitting in the corner of the room that cannot be ignored—the problem of converting the energy potentially available from the fission reaction to thrust. Vastly more energy is available from the fission reaction. Increasing the rate of fusion energy production also increases the neutron supply that drives the fission reaction, while thrust cannot be increased beyond the limit imposed by the strength of the magnetic field and the mirror ratio. Therefore attempting to supply 100% of the fission energy, or anything remotely close to that, to the fusion plasma would be an exercise in futility, except at very low levels of neutron production, far below Lawson-criterion breakeven, where the neutron flux can be matched to the fission energy needed to maintain thermal balance. And this is a viable alternative that will be discussed further below. But far more desirable would be to run the fusion reaction above the Lawson-criterion breakeven level and use the enhanced neutron flux to generate vastly greater amounts of fission power, if only a way could be found to convert the fission power to thrust. Unfortunately, converting any of the fission power to thrust is easier said than done. A number of issues arise, which will now be discussed.

17.3.2 Energy Generation in a Thorium/Uranium Blanket

Thorium fuel use has been suggested by Kammash (2010). Thorium is desirable as a fuel for spacecraft propulsion for a number of reasons, including the following:

- Thorium is more abundant than uranium, and does not require isotope separation for use as a fuel; virtually all naturally occurring thorium is consumable Th_{232}.
- Thorium reactors are proliferation resistant; thorium is not a weapons-grade material and U_{233} can be used for weapons only with extreme difficulty and cost.
- The thorium cycle generates minimal amounts of long-lived radioactive wastes, which can be readily disposed of in space.
- Thorium fuel use has been successfully demonstrated in several different reactors.
- The physical and chemical properties of thorium enable high-temperature operation, resulting in high Carnot efficiency and negligible meltdown risk.

The thorium cycle, driven by an external neutron source, results in the breeding of U_{233} via the following path:

$$Th_{232} + n \rightarrow Th_{233} + \gamma \quad Th_{233} \rightarrow Pa_{233} + \beta^- \quad Pa_{233} \rightarrow U_{233} + \beta^-$$

The half lives for the two beta decays are 22 min and 27.4 days, respectively. Thus a significant length of time elapses from the initiation of the fusion reactions to the production of fissile material in the blanket. Eventually the U_{233} abundance reaches a steady state. The time lapse can be avoided by seeding the thorium blanket with U_{233}. However, it will be seen that, because of the time lapse, converting enormous amounts of fission power to thrust may require use of purely fissile material, precluding the use of thorium. For thorium use, the steady state number density is obtained by setting the rate of U_{233} creation equal to the rate of destruction and is given by the ratio of the fission cross sections of the thorium and uranium isotopes,

$$\frac{N_{233}}{N_{232}} = \frac{\sigma_{232}}{\sigma_{233}}$$

A moderator can be used to thermalize the neutrons from the external source, facilitating fission of the uranium atoms. On average, each U_{233} fission produces about 2.5 additional neutrons, which will also be thermalized by the moderator. The fission reaction will be subcritical if the number of neutrons per fission multiplied by the probability, p, that a neutron will cause a uranium atom to fission, is less than 1. A subcritical reaction is desirable because it does not require control rods, simplifying reactor design and reducing reactor mass, while the existence of an external source makes it unnecessary to achieve criticality. Even more importantly, a subcritical reaction allows higher power density to be achieved. U_{233} fission will be subcritical for $p < 0.4$. However, the 2.5 additional neutrons produced by each fission will still increase the number of thermal neutrons beyond that produced by the external source. The neutron multiplication factor, f, can be found from the sum of the resulting geometric series

$$f = \frac{p}{1 - 2.5p} \tag{17.17}$$

For values of p close to 0.4, f becomes large; for example, $f \approx 100$ for $p = 0.396$.

For neutron generation by a GDM, the power per unit length of thorium-uranium mixture is $\pi r_p^2 n f \varepsilon_u$ where n is the flux of neutrons per second per unit volume from the GDM, r_p is the plasma radius within the GDM, and ε_u is the energy released per U_{233} fission, 200 meV. The probability a neutron will cause a U_{233} fission is

$$p = \frac{N_{233}\sigma_{233}}{N_{232}\sigma_{232} + N_{233}\sigma_{233} + A_0} = \frac{1}{2 + A_0/N_{232}\sigma_{232}} \leq 0.5 \tag{17.18}$$

where A_0 represents the remainder of the area exposed to the neutron flux, including that resulting from the presence of the moderator. The ratio of the cross sections of U_{233} to Th_{232} is about .014 at thermal energies. Thermalization of the neutron flux also results in the deposition of the neutron energies in the thorium blanket as heat; assuming 100% thermalization the heating rate is $\pi r_p^2 n \varepsilon_n$, where ε_n is the neutron energy (14.1 meV for DT fusion, 2.45 meV for DD fusion). The total energy production rate per unit length in the thorium blanket is then

$$\varepsilon_{fi} = \pi r_p^2 n (f \varepsilon_u + \varepsilon_n) \tag{17.19}$$

For example, if the neutron source is 10^{16}/cm^3/s from a DT reaction, the multiplication factor is 20 and the plasma radius is 5 cm, then the energy generation rate is 193 MW/cm, suggesting, as will be further demonstrated below, that a hybrid reactor is capable of terawatt levels of energy production. Note that the fission power is essentially independent of the temperature and density of the thorium and uranium, provided that they are sufficient to render the material opaque to the incident neutrons. It has been shown (Kammash 2010) that the neutron flux at the outer boundary of the blanket can be zero, minimizing or obviating the need for a shield to protect the magnet coils from neutron damage.

As for gamma ray shielding, Singh et al. (2015) showed that certain thorium compounds (e.g. ThO_2) can provide very high attenuation, comparable to lead. The gamma attenuation rate of a shielding material is a function of the energy of the photons, depending on which of the three interaction mechanisms (photoelectric effect, Compton scattering, pair production) dominates. For instance, the half-value layer values for lead (Voss 2000) can be used to calculate the thickness required to attenuate a gamma beam to 1% of its incident intensity. For gammas 10–100 keV, 100–500 keV, and 1 meV, the lead thicknesses required are approximately 0.2, 3.3, and 8.7 cm, respectively. These are thin and very effective shielding indeed, and it can be concluded that similar thicknesses of thorium would provide a similar degree of shielding.

However, there is a problem. The thorium-uranium fuel, if solid, must be maintained at a temperature below the melting point and if liquid, below the boiling point. This limits blanket temperature to, at most, a few thousand degrees K. Gaseous fuel, or a thorium-uranium plasma, could be heated to a much higher temperature, but this could pose a containment problem. Energy generation on the order of hundreds of megawatts per centimeter will rapidly vaporize the blanket unless the heat can be removed as quickly as it is generated and transferred to the fusion plasma or elsewhere. But heat cannot flow from one reservoir, the fission blanket, whose temperature is a few thousand degrees K, to another reservoir, the GDM, whose temperature may approach one billion degrees K. One solution would be to immediately convert the blanket heat to electricity, which can be used to heat the fusion plasma and supply power to the ship. But while this is easy to do in principle, it presents such serious engineering challenges in practice, because of the extreme power levels involved, that it may be impossible to convert all or even most of the heat to electricity, while supplying energy to the fusion plasma would be counterproductive once the maximum value of NkT has been reached. To the extent that electricity is generated, there are two separate problems to be considered —(i) removing the heat from the blanket and (ii) converting it to electricity.

Some approaches to the heat removal problem are promising. In one, solid thorium dioxide fuel is cooled by a circulating fluid. ThO_2 has a high melting point —over 3000 °K; thus heat removal and power generation can be optimized by cycling coolant temperature over a range much greater than in conventional power plants. Coolant could be circulated in ducts or through ThO_2 pellets. However, the heat removal rate will depend on the fraction of reaction chamber volume occupied by coolant; increasing this fraction to facilitate heat removal lowers reaction rates, limiting power output. This limitation might be overcome by the use of a liquid thorium salt, such as ThF_4, as fuel. The fuel can then also serve as a coolant; circulating the fuel between the reaction chamber and a heat exchanger allows 100% of reaction chamber volume to be occupied by fuel. Another alternative is to use liquid elemental thorium as the fuel/coolant. Thorium has the greatest difference between its melting and boiling points of any element; thus it can be cycled over a greater temperature range than a liquid thorium salt. Use of structural materials with high melting points such as tantalum hafnium carbide (4488 °K) could permit the temperature of the circulating thorium to be cycled by 2000 °K or more.

The heat removal rate depends upon the mass per unit time of coolant that can be circulated through the reaction chamber, the specific heat capacity of the coolant, and the temperature range over which the coolant can be cycled. The temperature range over which a liquid coolant can be cycled is dictated by the difference between its melting and boiling points, while the temperature range over which a solid thorium dioxide reactor cooled by a gas or supercritical fluid can be cycled is limited by the melting point of thorium dioxide. Molten salts composed of lithium, beryllium, fluorine and thorium have high heat capacity, on the order of 4500 kJ/m^3, but the difference between melting and boiling point is only about 1000 °K. For both solid thorium dioxide and liquid elemental thorium, the maximum possible temperature range is higher, 2000–3000 °K. The melting point of elemental thorium is 2115 °K and its boiling point is 5061 °K; however, the maximum operating temperature may be well below the boiling point because it may be impossible to find a suitable containment material that remains solid at that high a temperature. Coolant mass flow rates of as much as ten thousand kilograms per second may be achievable; the turbopumps on the Saturn V moon rocket could pump 15 tons of fuel and liquid oxygen per second.

A high heat removal rate can be achieved by using a helium coolant passed through solid thorium dioxide fuel and cycled over a 2500 °K temperature range; a mass flow rate of 2000 kg/s would be theoretically capable of removing heat at the rate of approximately 25 GW, which could be converted to electricity with high efficiency because of the high temperature. Helium has the second highest heat capacity of any element, after hydrogen, and may be a better choice of coolant than hydrogen because it is chemically inert. However, it may be possible to achieve dramatically higher heat removal rates by using hydrogen. The heat capacity of helium is 5200 J/kg/K in round numbers, while the heat capacity of H_2 gas increases from about 13,000 J/kg/K at 175 K to nearly 19,000 J/kg/K at 3500 K. Furthermore, at temperatures approaching the melting point of ThO_2, some dissociation of the H_2 molecule is likely, perhaps 5–10%, depending on the density. The dissociation potential of the H_2 molecule is 4.52 eV; as a consequence, cycling hydrogen from a cryogenic liquid at less than 20 K to a modestly dissociated $H_{1/}H_2$ gas at 3500 K, with a mass flow rate of 9000 kg/s, can potentially remove heat at a rate of approximately 700 GW. Furthermore, the heat removal rate can be increased even more by compartmentalization of the blanket. If the rate of heat generation by fission is, for example, 200 MW/cm, and the heat removal rate is 700 GW, then the path length for the coolant will be 3.5 ms. In this example, dividing the blanket into 3.5 m-long segments, each one cooled by a separate pumping system, would, in principle, allow removal of virtually 100% of the heat.

It should be noted that a blanket temperature as high as 3500 °K means that the blanket surface heat will either have to be radiated to space or removed and converted to electricity. If exposed to space, the blanket surface could also serve as a radiator; alternatively, heat would have to be removed from the surface and transported to a radiator. The energy radiated from the blanket surface is $\varepsilon_s A\sigma T^4$ where ε_s is the emissivity of the surface, A the surface area, T the surface temperature and σ is the Stefan-Boltzman constant. If the blanket surface is not exposed to space and all of this energy is removed and transported to a radiator, the surface area of the radiator would have to be equal to the surface area of the blanket multiplied by the ratio of blanket surface temperature to radiator surface temperature to the fourth power. Heat pipes using liquid/gaseous indium (melting point 429.75 °K, boiling point 2345 °K) as the working fluid might be used to transport the heat. However, depending upon system parameters (primarily blanket length, which, if the same as GDM length, might range from several meters to several hundred meters per the alternatives discussed below), the energy radiated from the blanket surface could range between tens of megawatts and tens of gigawatts. To the extent this large amount of energy can be converted to electricity or otherwise used, the size of the radiators might be reduced, at least to the extent the energy is not ultimately dissipated as heat inside the spacecraft. A particularly good use of the energy might be to supply power for the turbopumps that circulate the coolant, which can be expected to be tens of megawatts or more, again depending on system details.

Assuming terawatt rates of heat generation and terawatt rates of heat removal are achievable, the propulsion system still would require terawatt rates of electricity generation. High coolant temperatures may enable use of magnetohydrodynamic generators with high efficiency. Alternatively, recent developments in turbine technology may offer the required capability. Specifically, Brayton-cycle turbines using supercritical carbon dioxide as a working fluid appear able to reduce the size of the turbomachinery required for a given level of power generation by an order of magnitude (Ahn et al. 2015). General Electric has recently announced development of a highly efficient supercritical carbon dioxide Brayton cycle turbine that operates at high temperatures. The device produces 10 MW but weighs only 68 kg and is just a few feet long. GE expects to scale it up to develop a 500 MW unit. Thus it is reasonable to assume that the mass required to convert the fission energy to electricity at an efficiency of around 60% (achievable because of high operating temperature) could be as little as 3 metric ton per GW. [Unfortunately, this analysis is too optimistic because the turbopumps required to circulate the coolant, the heat exchangers and other equipment add additional mass that has been omitted from the analysis.] But, as will be further discussed below, this still would result in an unacceptable thrust to weight ratio at multi-terawatt power levels. Electric generation capability may well be the most important factor limiting the ability of the propulsion system to use the vast amount of power that can be generated by the fission reaction.

And this is not the only problem. Even if electricity can be generated without an unacceptable increase in system mass, there is a transmission issue. The electric current required to carry multi-terawatts of power could turn any current carrier with non-zero resistance into a toaster. This problem might not be overcome by the use of superconducting wires, the current-carrying capacity of which is limited by their critical current density. If superconductors with sufficiently high critical current density cannot be found, the amount of superconducting wire required could also exceed mass limitations, not to mention problems that might arise from the magnetic fields that could be generated. Another possibility, which will not be explored here, is to use laser beams to carry the power.

These difficulties require consideration of alternatives based on the use of a high-temperature fission plasma. The generation problem, although not the transmission problem, might be solved by direct conversion devices operating on a thorium/uranium plasma. In this alternative the walls of the chamber containing the fission plasma might be used to generate electricity via direct conversion. The scheme would not require—and, indeed, would be defeated by—magnetic containment of the fission plasma. The walls of the fission chamber can be maintained at a temperature comfortably below their melting point if electrical energy can be generated at the same rate as the rate of heat transfer from the plasma to the chamber walls. And the heat transfer rate can be controlled simply by using an appropriate geometry for the fission chamber. Specifically, if the fission plasma is confined in a space between two concentric cylinders, the innermost of which is at a distance D from the fusion GDM and the outermost of which is at a distance $\geq D + 1/N\sigma$ where N is the number density of the fission plasma and σ is the combined cross section for neutron interactions with the plasma by all processes, then the condition of thermal balance is

$$\varepsilon_{fi} L = \frac{3}{2} NkTv_t A = 3\pi NkTv_t \left[2LD + (L + 2D)/N\sigma + 1/(N\sigma)^2 \right] \tag{17.20}$$

where 3_{fi} is the fission power per unit length of the fusion plasma, L the length of the fusion plasma, k is Boltzmann's constant, T the temperature of the fission plasma, v_t the thermal velocity of the fission plasma ions, and A the surface area of the fission chamber. The heat flux per unit area to the container walls is $\varepsilon_{fi} L/A$. If D is on the order of a few hundred meters, then, depending on the parameters of the fusion power generation, the heat flux to the walls may be less than 1 MW/cm^2. Direct converters may be able to handle this; furthermore, if a small electric field can be generated across the fission plasma, or a weak, appropriately configured magnetic field can be set up, perhaps using Halbach arrays of permanent magnets, then the entire surface area of the container can act as a charge collector, with positive charge collected on one half and negative charge on the other. If feasible, this approach could achieve nearly 100% heat to electricity efficiency, while eliminating the mass of turbines, coolant pumps and associated equipment. But even if all of the fission energy can be converted to electricity without incurring an unacceptable weight penalty, there remains the problem of converting the electricity to thrust, which cannot be solved by supplying more power to the fusion GDM. A possible solution that will not be explored here is to use the electricity to power particle accelerators that might be able to eject a sufficient flux of ions at relativistic velocities.

This last difficulty raises the question of whether thrust might be generated directly from the fission energy by containing the fission plasma in several GDM's surrounding the GDM containing the fusion plasma. The fission energy would create additional thrust by ejecting the fission plasma as additional propellant, with electricity generation limited to that needed to heat the fusion plasma and to operate the ship's systems. In this case it would not be possible to use a fertile fission fuel such as thorium because the ejection time would be much shorter than the thorium cycle's beta decay half-lives; instead, a fissile isotope of uranium, plutonium or some other element would be required. This approach will not be explored in depth here. On the plus side, the fission reaction rate does not significantly depend on the temperature of the fission plasma, but primarily on the neutron flux from the fusion reactions. Furthermore, it is likely that the fission reaction could be maintained at bare criticality; an excursion to a supercritical condition would almost certainly result in an increase in the loss rate of fissile ions, reducing the density and resulting collision probability, thereby lowering the rate of neutron production. Otherwise, the same equations that describe the fusion GDM would also apply to a fission GDM. The main areas of concern are stability of the fission plasma, creating and maintaining the needed magnetic fields in the presence of a neutron flux that could damage the magnet coils, avoiding interference with containment mechanisms by stray magnetic fields, and maximizing the number of fusion-generated neutrons that encounter a fissile ion and cause it to fission. In addition, the thrust produced by any device that relies on magnetic containment of a plasma is subject to the limitation imposed by Eq. (17.16b).

17.4 Discussion

The adequacy of a hypothetical resource-carrying spacecraft can be evaluated using the above equations. Key parameters are magnetic field strength, mirror ratio and number of thrusters. Magnetic field strength and mirror ratio are the two most important parameters for determining the thrust obtainable from a single GDM thruster. Superconducting magnets capable of generating fields in excess of 30 T are being developed at the National High Magnetic Field Laboratory (NHMFL); its 32 T all-superconducting magnet weighs just 2.3 tons and has length of 2.5 m. Therefore, it may be possible to generate GDM field strengths of dozens of teslas with acceptable mass. Figure 17.8 shows the thrust in meganewtons as a function of magnetic field strength in tesla for six values of mirror ratio, assuming a plasma radius of 5 cm. It is generally believed that optimal plasma stability is obtained for a mirror ratio of about 55 (Kammash 2010), but lower values may be viable. It appears that attainable thrust from a single GDM could be at least a few hundred kilonewtons, and, optimistically, several meganewtons.

In general, it appears that with sufficient energy input the fusion reaction can be maintained at rates ranging from far below to several times Lawson-criterion

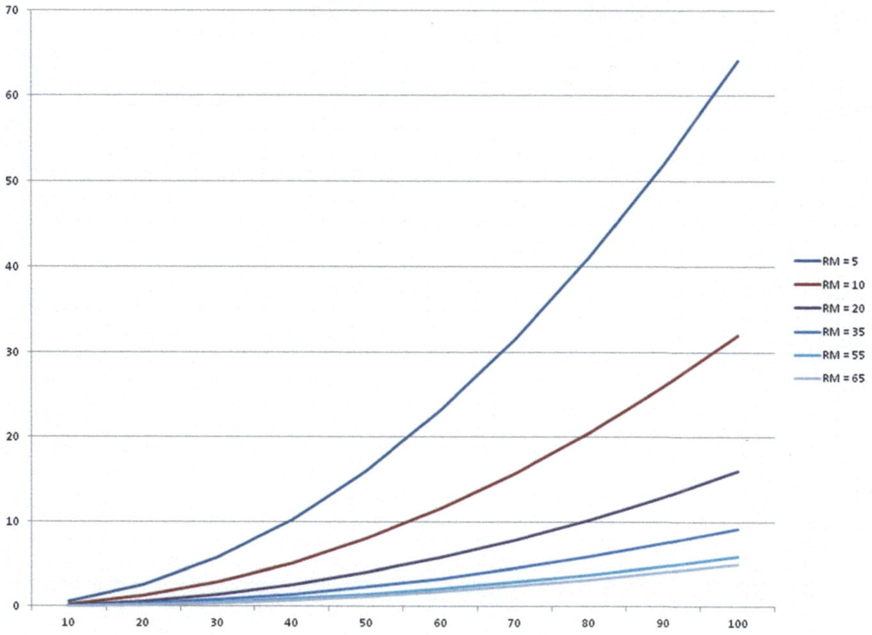

Fig. 17.8 Thrust in meganewtons as a function of magnetic field strength in tesla for six values of mirror ratio

breakeven in GDM's with plasma length ranging between approximately 10 and 1000 m, corresponding to operating temperatures between approximately 3 keV for the smallest lengths up to perhaps 20 keV for the longest.

These temperatures correspond to specific impulse between about 160,000 and 400,000 s. The foregoing values depend on the approximation used here for the ambipolar potential; a more accurate calculation will give different results, but the results here are reasonably accurate order of magnitude estimates. The plasma length required to operate at a given percentage of Lawson-criterion breakeven can be calculated from Eq. (17.14) as a function of the plasma (ion) temperature.

Figure 17.9 shows the broader dependence of I_{sp} on plasma length. Note that increasing plasma length yields diminishing returns for increases in I_{sp}, suggesting that considerations of cost and construction difficulties beyond some value of plasma length would outweigh the value of higher specific impulse.

Most achievable in the near term would be a fusion reaction at less than 5% of Lawson-criterion breakeven, which would produce a sufficient flux of neutrons to generate enough fission energy to achieve thermal balance in the GDM (so the combined reactions would actually operate at breakeven).

Columns 3 and 4 of Table 17.2 show potential accelerations and one-way travel times for payloads of 1000, 10,000 and 100,000 metric tons to each of six outer solar system bodies resulting from propulsion by a cluster of 25 GDM's, each surrounded by a Th_{232}/U_{233} blanket in which takes place a fission reaction with a multiplication

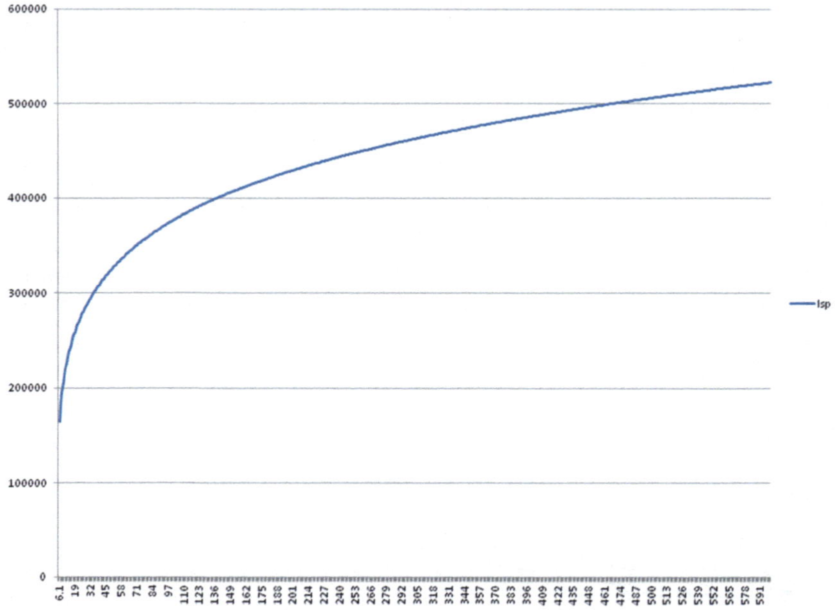

Fig. 17.9 I_{sp} (s) as a function of plasma length (m)

factor of 7.6, heat removal by hydrogen at the rate of 700 GW, and conversion to electricity with the entire heat removal/electrical generation system mass equal to 3 mT/GW. The values in Table 17.2 are based additionally on a magnetic field strength of 30 T, a mirror ratio of 55, a value of 0.75 for the parameter β in Eq. (17.13b) (which is used to calculate the maximum permissible plasma density that maintains containment), and a 50% D 50% T fusion fuel mix. The resulting plasma length and plasma temperature are 9 m and 2.9 keV; the fusion reaction runs at only 3% of Lawson-criterion breakeven. The specific impulse obtained is 161,000 s and the thrust produced by each GDM is 897 kilonewtons, for a total thrust of 22.4 meganewtons. Note that the plasma length is only a few times longer than the NHMFL 32 T magnet, the mass of which is negligible compared to the mass of the heat removal and electrical generation system. Once again, these values are based on an overestimate of the ambipolar potential. A more accurate calculation of the ambipolar potential would allow achieving thermal balance with an even shorter plasma length, and lower plasma temperature, resulting in the fusion reaction at an even lower fraction of Lawson-criterion breakeven, but also lower specific impulse and thrust from a single GDM.

The mass of the generating equipment is minimized for operation at the lowest temperatures at which thermal balance can be maintained with the available fission energy. Higher temperatures and longer plasma lengths would tend to improve performance by increasing thrust and specific impulse, but the benefit diminishes as the mass of the heat removal and electrical generation system starts to approach

Table 17.2 Accelerations and travel times for near-term achievable and possibly eventually achievable clusters of 25 fusion GDM's operating far below Lawson-criterion breakeven while producing sufficient neutron flux to generate enough fission energy for thermal balance

Planetary system	Cargo mass (mT)	Acceleration (g)	Travel time (Days)	Acceleration (g)	Travel time (Days)
Jupiter	1000	0.017	55	0.108	22
Jupiter	10,000	0.016	57	0.093	23
Jupiter	100,000	0.009	74	0.041	35
Saturn	1000	0.016	75	0.092	31
Saturn	10,000	0.014	77	0.081	33
Saturn	100,000	0.009	99	0.037	48
Uranus	1000	0.014	108	0.077	46
Uranus	10,000	0.013	112	0.068	49
Uranus	100,000	0.008	142	0.032	72
Neptune	1000	0.013	140	0.067	62
Neptune	10,000	0.012	144	0.059	66
Neptune	100,000	0.008	182	0.029	94
Pluto	1000	0.012	165	0.061	74
Pluto	10,000	0.012	169	0.054	79
Pluto	100,000	0.005	268	0.027	111
Eris	1000	0.010	283	0.043	137
Eris	10,000	0.009	290	0.038	144
Eris	100,000	0.006	361	0.020	198

Columns 3 and 4 (near-term achievable) assume magnetic field strength of 30 T, mirror ratio of 55, a 50/50 DT mixture, a subcritical thorium-cycle fission reaction with multiplication factor = 7.6, and fission heat to electricity conversion at the rate of 700 GW with 60% efficiency and generation and heat removal system mass to power ratio of 3 mT/GW. Columns 5 and 6 (possibly eventually achievable) assume magnetic field strength of 30 T, mirror ratio of 15, a 50/50 DT mixture, a subcritical thorium-cycle fission reaction with multiplication factor = 7.6, and fission heat to electricity conversion at the rate of 700 GW with 60% efficiency and generation and heat removal system mass to power ratio of 0.3 mT/GW

payload mass, which can only be avoided if ratios much smaller than 3 metric tons per GW can be achieved. As a result, the shortest travel times are generally obtained by operating just above the minimum temperature at which sufficient fission energy is available to maintain thermal balance.

The resulting times for trips to the Jupiter system are within the likely range of commercial acceptability, and the travel times for the Saturn system and even Uranus may be commercially viable for at least some resources. Capability beyond this is limited by the increase in the mass and volume of the heat removal and electrical generation equipment. The total fission power is very large, over 1000 GW, and is not the limiting factor. Other engineering challenges do not appear to present major issues. If the craft is constructed in orbit or on the Moon, considerations of aerodynamic stability would be nonexistent, so volume and shape should not present a significant problem. Shielding crew areas (if any—the ship might be robotically operated) from radiation would also not be a problem as they could be far removed from the propulsion system, with the cargo module between the crew and the reactor; indeed, the propulsion system could be connected to the rest of the ship by beams many hundreds of meters in length. The major challenge is to minimize the mass of the heat removal and electrical generation equipment.

Columns 5 and 6 of Table 17.2 show the accelerations and travel times that could be achieved if the mass of the heat removal/electrical system could be further reduced by an order of magnitude, to 0.3 metric tons per GW, and the mirror ratio could also be reduced, to 15. This would require the plasma length to be increased to 33 m. Plasma temperature and specific impulse would be unchanged at 2.9 keV and 161,000 s, but the decrease in mirror ratio would increase the thrust obtainable from a single GDM to over 3 meganewtons. The reduction in travel times resulting from the increased thrust and decreased propulsion system mass would bring trips to Jupiter, Saturn and Uranus within the realm of likely commercial acceptability, with travel times to Neptune and even Pluto possibly viable for some resources.

Also highly significant, a propulsion system based on the more easily achievable values of 3 metric tons per GW and mirror ratio of 55 could separately become the basis for a fully reusable single-stage-to-orbit heavy-lift launch vehicle. For example, it could accelerate a payload mass of over 2200 metric tons to orbital velocity at an acceleration of 1g, and more than 22,000 metric tons to orbital velocity at an acceleration of 0.1g. With a propulsion system capable of maintaining thrust for long periods of time, orbit may be achievable without need for high acceleration. At 0.1g, it would take a little over 2 h to reach orbital velocity. Values of acceleration much less than 1g might enable a horizontally accelerating ship to achieve orbit if it relied on aerodynamic lift to get high in the atmosphere and transpirational cooling to overcome frictional heating. Cargo carriers for the outer solar system most likely would not be launched from Earth, but constructed in orbit or on the Moon. If the cost of reaching LEO cannot be dramatically reduced, this will not be economically feasible. Hybrid nuclear propulsion systems appear able to permit development of a fully reusable launch vehicle with massive cargo launch capacity that should enable a dramatic reduction in launch costs. [An alternative for dramatically reducing launch costs is a linear motor catapult that could accelerate materials to orbit at 100g, more or less. Powell et al. (2010) have analyzed this approach and concluded that problems arising from air resistance and frictional heating can be overcome. Obviously, accelerations this high could not be used for astronauts].

Comparable in near-term capability would be propulsion by a cluster of GDM's with a self-sustained fusion reaction operating at approximately 8 times Lawson-criterion breakeven, with little or no additional energy from fission.

Columns 3 and 4 of Table 17.3 show potential accelerations and one-way travel times for payloads of 1000, 10,000 and 100,000 metric tons to each of six outer solar system bodies resulting from propulsion by a cluster of 25 such GDM's. Propulsion system mass was assumed to be two metric tons per meter of plasma length; this is conservative since the NHMFL magnet has a mass of only one ton per meter. Thrust per GDM would be the same as for the system with a fission

Table 17.3 Accelerations and travel times for near-term achievable and possibly eventually achievable clusters of 25 fusion GDM's operating above Lawson-criterion breakeven

Planetary system	Cargo mass (mT)	Acceleration (g)	Travel time (days)	Acceleration (g)	Travel time (days)
Jupiter	1000	0.040	35	1.1	6.8
Jupiter	10,000	0.034	39	1.1	6.8
Jupiter	100,000	0.014	60	1.1	6.8
Saturn	1000	0.039	47	1.1	8.9
Saturn	10,000	0.033	51	1.1	8.9
Saturn	100,000	0.014	79	1.1	8.9
Uranus	1000	0.036	68	1.1	12.3
Uranus	10,000	0.031	73	1.1	12.3
Uranus	100,000	0.013	112	1.04	12.6
Neptune	1000	0.034	86	1.1	15.2
Neptune	10,000	0.029	93	1.1	15.2
Neptune	100,000	0.013	142	0.80	17.9
Pluto	1000	0.033	101	1.1	17.4
Pluto	10,000	0.028	108	1.1	17.4
Pluto	100,000	0.012	164	0.68	22.1
Eris	1000	0.027	171	1.1	26.9
Eris	10,000	0.024	181	0.73	33.1
Eris	100,000	0.011	267	0.38	45.9

energy supply; however, plasma temperature would be increased to 17.2 keV, and specific impulse to 393,000 s. As a result of higher specific impulse and lower system mass, the accelerations and travel times are actually better than those in Columns 3 and 4 of Table 17.2, although not as good as those in Columns 5 and 6, suggesting that attempting to use fission energy may have little benefit in the near term. However, the pure-fusion, zero-fission GDM would have plasma length of nearly one kilometer, compared to just 9 m for the fission supplied alternative. The ship itself would likely be at least several times longer, and it is difficult to envision constructing and launching such a vessel from the Earth's surface, or building it in orbit or on the moon for a cost competitive with the much smaller fission-supplied alternative. It is unclear what other problems, if any, this substantial increase in plasma length would pose.

Columns 5 and 6 of Table 17.3 show what may eventually be achievable if all of the fission energy that could be generated by the neutron flux produced by a self-sustained fusion reaction at 10 times Lawson-criterion breakeven could be converted to thrust by a propulsion system with mass negligible compared to payload mass.[The accelerations shown in Columns 5 and 6 of Table 17.3 have been limited to 1.1g on the assumption that the vessels will be crewed, and the travel times have been calculated accordingly.] This is almost certainly beyond the capability of present and foreseeable near-term technology. But 200 years ago problems that appeared equally insuperable were posed by heavier-than-air flight, let alone today's routine cargo transport by gigantic aircraft over intercontinental distances in a matter of hours. So, optimistically, it will be assumed that someday it will be possible to build clusters of 25 (or more) such fusion GDM's surrounded by a blanket containing a near-critical plasma of fissile ions, with fission probability of (at least) 0.38, corresponding to a multiplication factor of (at least) 7.6, and with all of the resulting fission energy converted to thrust. For each GDM the plasma length and temperature would be 960 m and 10 keV, respectively. Such a propulsion system would generate thrust of approximately one giganewton with specific impulse around 300,000 s. It would be capable of continuous acceleration greater than 1g to Jupiter for payloads greater than 10^5 metric tons and would enable commercially viable capacities and travel times for resource transportation from the entire outer solar system, even Eris and other Kuiper belt bodies.

17.5 Conclusion

The ultimate test of the viability of a propulsion system for outer solar system resource transport is an economic one. A cargo spacecraft must have sufficiently high speed and high capacity, with low capital and operating costs. What qualifies as sufficient will depend on the market for the resource in question and the economic risk entailed by failing to meet a given level of performance. But speed, capacity and cost requirements all demand a combination of high specific impulse and high thrust that cannot be satisfied by chemical rockets or current electric propulsion systems, but might be by nuclear powered thrusters. The capabilities of chemical and electrical propulsion systems are limited by physics. It is impossible for technological advances to give them the capabilities for commercially viable transport of outer solar system resources. But hybrid nuclear propulsion systems are limited only by technology. Although physical limits necessarily exist, those limits are far beyond the current state of the art. Of particular importance, fission is capable of producing much more energy than fusion; a single U_{233} or U_{235} fission produces more than 10 times as much energy as a single DT fusion.

Technological, rather than physical, limitations are the only bar to massive cargo carriers that could enable a robust commerce in outer solar system resources. Hybrid nuclear propulsion systems consisting of a gas dynamic mirror, which serves as both a fusion power generator and a thruster, and a blanket containing fertile and/or fissile material in which a subcritical or barely critical fission reaction can be driven by the neutron flux from the fusion reaction, have the potential to produce enormous power, on the order of many tens of terawatts. If all of this power can be utilized, thrust levels of giganewtons can be attained by multiple clustered devices, resulting in the ability to continuously accelerate spacecraft the mass of supertankers at close to 1g and to reach outer solar system bodies in days or weeks. And generating the power is straightforward; it could be done with present technology. But converting the power to thrust, rather than generating it, is what may limit capabilities. Converting terawatts of fission energy to electricity appears to require so large a mass of generation and transmission equipment as to render high acceleration impossible, notwithstanding the thrust. And the cost would be prohibitive. Using a GDM to contain a fissile isotope plasma in which a subcritical or barely critical fission reaction is driven by an external fusion neutron source, and which directly creates thrust, appears more promising, but stability and other issues need to be evaluated.

Meanwhile, present and foreseeable near-future technology appear to have the capability to engineer minimal hybrid systems in which a fusion reaction at less than 5% of Lawson-criterion breakeven generates a neutron flux sufficient to produce enough fission energy to maintain thermal balance in the GDM. The fission power, on the order of one to two terawatts, can be converted to electricity without incurring an unacceptable increase in mass, and used to heat the fusion plasma, resulting in thrust levels in excess of 20 meganewtons from multiple clustered devices and specific impulse in excess of 150,000 s. This reactor/thruster technology could also power a fully reusable, heavy lift single-stage-to-orbit launch vehicle that could reduce launch costs sufficiently to make construction of resource carriers in orbit economically viable. Alternatively, a self-sustained fusion reaction at several times Lawson-criterion breakeven would give comparable performance with little power required from a fission reaction, but prospective construction costs and difficulties make this approach questionable. Either alternative would be sufficient to enable cargo masses of tens of thousands of metric tons to be transported to and from at least the Jupiter and Saturn systems in weeks or months. Commercially viable resource exploitation from the Uranus system and beyond appears to require the vastly greater power available from a combination of self-sustained fusion and barely subcritical fission reactions driven by fusion-generated neutrons.

Acknowledgements The authors are grateful to Professor Terry Kammash of the University of Michigan for many helpful discussions.

References

Ahn, Y., Seong, J.B., Kim, M., Cho, S.K., Baik, S., Lee, J.I., Cha, J.E.: Review of supercritical CO_2 power cycle technology and current status of research and development. Nucl Eng Technol **47**, 647–661 (2015)

Atenzi, S., Meyer-ter-Vehn, J.: Nuclear fusion reactions. In: The Physics of Inertial Fusion. University of Oxford Press, Oxford. Chapter 1 (2004)

Borowski, S.K., Corban, R.R., McGuire, M.L., Beke, E.G.: Nuclear Thermal Rocket/Vehicle Design Options for Future NASA Missions to the Moon and Mars. Prepared for the AIAA Space Programs and Technologies Conference and Exhibit Huntsville, Alabama, September 21–23, 1993. NASA STI/Recon Technical Report N 96 (1995): 11955 (1995)

Kammash, T.: Preliminary assessment of a fusion-fission hybrid reactor. In: Proceedings of the International Conference on the Physics of Reactors 2010, Pittsburgh, vol. 3, pp. 2028–2036. Curran Associates, Inc. Red Hook, New York (2010)

Moir, R.W.: Surface bombardment rates for mirror fusion reactor designs. J. Nucl Mater **63**, 21–30 (1976)

Post, R.F.: The magnetic-mirror principle as applied to fusion research. In: Post, R.F., Baldwin, D. E., Ryutov, D.D. (eds.) Proceedings of the Course on Mirror-based and Field-reversed Approaches to Magnetic Fusion, Varenna, 1983, vol. II, pp. 7–41. International School of Plasma Physics (1983)

Powell, J., Maise, G., Rather, J.: Maglev launch: ultra low cost ultra/high volume access to space for cargo and humans. In: Space, Propulsion & Energy Sciences International Forum, vol. 121, pp. 121–136; In: Proceedings of the Space, Propulsion, and Energy Sciences International Forum SPESIF-2010. Johns Hopkins University Applied Physics Laboratory, Laurel, MD, USA, 23 Feb 2010 (2010). Available online: http://www.scribd.com/doc/88949962/Start-Ram-2010

Singh, V.P., Badiger, N.M., Vega-Carrillo, H.R.: Photon attenuation properties of some thorium, uranium and plutonium compounds. In: International Symposium on Solid State Dosimetry, Leon, Guanajuato, Mexico, 26–30 Sep, vol. 2015, pp. 34–44 (2015)

Stull, M.A.: Design considerations for a 21st century ground transportation system based on value-capture financing. Transp Plann Technol **31**(3), 249–275 (2008)

Tang, R.: Study of the Gasdynamic Mirror (GDM) Propulsion System. Ph.D. thesis, University of Michigan (2011)

Voss, J.T.: Los Alamos Radiation Monitoring Notebook, LA-UR-00-2584, Los Alamos National Laboratory (2000)

Chapter 18
Exploration of the Outer Solar System: Missions and Their Power Systems

Simon D. Fraser

18.1 Introduction

The outer solar system, and in particular the giant planets and their large moons, have been visited by a number of exploration missions since the early 1970s.

It may not be a surprise that the number of missions targeting the different outer planets is inversely proportional to the distance to the Earth. Jupiter, the innermost of the outer plants, has been investigated by a number of flyby missions, orbiters, and even an atmospheric probe. Uranus and Neptune, the two outermost planets, have only been investigated by a single flyby mission up to date.

The intention of this chapter is to provide a brief overview of the exploration missions launched to the outer solar system up to date. In this, a special emphasis is put on discussing the power systems applied with the different missions and spacecraft.

Environmental conditions and mission plans do not allow the variety of power system options and designs we have seen with Lunar and Martian exploration. It is nevertheless interesting to have a look at the different power systems, and to see their evolution over time.

S. D. Fraser (✉)
Energy Systems Engineering, Inge-Morath-Straße 45d, 8045 Graz, Austria
e-mail: simon.fraser@gmx.at

V. Badescu and K. Zacny (eds.), *Outer Solar System*,
https://doi.org/10.1007/978-3-319-73845-1_18

Table 18.1 Chronology of outer solar system exploration

Mission	Launch date	Jupiter	Saturn	Uranus	Neptune	Pluto
Pioneer 10	1972-03-03	flyby	–	–	–	–
Pioneer 11	1973-04-05	flyby	flyby	–	–	–
Voyager 2	1977-08-20	flyby	flyby	flyby	flyby	–
Voyager 1	1977-09-05	flyby	flyby	–	–	–
Galileo	1989-10-18	orbiter + probe	–	–	–	–
Ulysses	1990-10-06	flyby	–	–	–	–
Cassini	1997-10-15	flyby	orbiter + probe	–	–	–
New Horizons	2006-01-19	–	–	–	–	flyby[a]
JUNO	2011-08-05	orbit	–	–	–	–

[a]Pluto/Charon and Kuiper Belt flyby

18.2 Chronology of Exploration Missions to the Outer Solar System

Table 18.1 contains a chronology of outer solar system exploration missions launched up to date. Each of the missions is presented with mission name, launch date, and whether or not there was any direct activity with each of the four outer planets plus Pluto, the former planet and now dwarf planet.

The timeline was compiled from the NASA Lunar and Planetary Science sites for Jupiter, Saturn, Uranus, Neptune and Pluto (NASA Jupiter 2016; NASA Saturn 2016; NASA Uranus 2016; NASA Neptune 2016; NASA Pluto 2016).

18.3 Exploration Missions

Each of the outer solar system exploration missions included in the chronology shown in Table 18.1 is briefly presented and discussed in the following.

In this, mission objectives and important milestones are briefly introduced; then, the power systems of the different probes and/or orbiters are presented and discussed.

18.3.1 Pioneer 10

Pioneer 10 was launched on March 3, 1972. It was the first spacecraft sent to the outer solar system, and the first to investigate Jupiter. After successfully performing a flyby of Jupiter, the probe followed an escape trajectory out of the solar system (NSSDCA/COSPAR ID: 1972-012A, 2016).

Figure 18.1 shows the trajectory of Pioneer 10, and compares it to the trajectories of Pioneer 11 and the twin Voyager probes launched in 1977, five and a half years after the launch of Pioneer 10.

Pioneer 10 had a mass of 259 kg and carried a scientific payload designed for studying the interplanetary and planetary magnetic fields, solar wind parameters, cosmic rays, the transition region of the heliosphere, Jovian radio waves, and the atmosphere of Jupiter and some of its satellites; in addition, Pioneer 10 also photographed Jupiter and its satellites (NSSDCA/COSPAR ID: 1972-012A, 2016).

The power system of Pioneer 10 consisted of four Systems Nuclear Auxiliary Power (SNAP) 19 radioisotope thermoelectric generators (RTGs). Each of the four SNAP-19 RTGs had an initial average electrical power output of 40.7 W (Abelson 2006).

In total, the four RTGs together provided approximately 155 W of electrical output power at launch time, which decayed to approximately 140 W by the time the spacecraft reached Jupiter 21 months after launch (NSSDCA/COSPAR ID: 1972-012A, 2016).

Radioisotope Thermoelectric Generators (RTGs), sometimes also referred to as Radioisotope Power Systems (RPSs), generate electrical power by utilizing the heat of decay of radioactive isotopes in combination with thermoelectric converters. Up

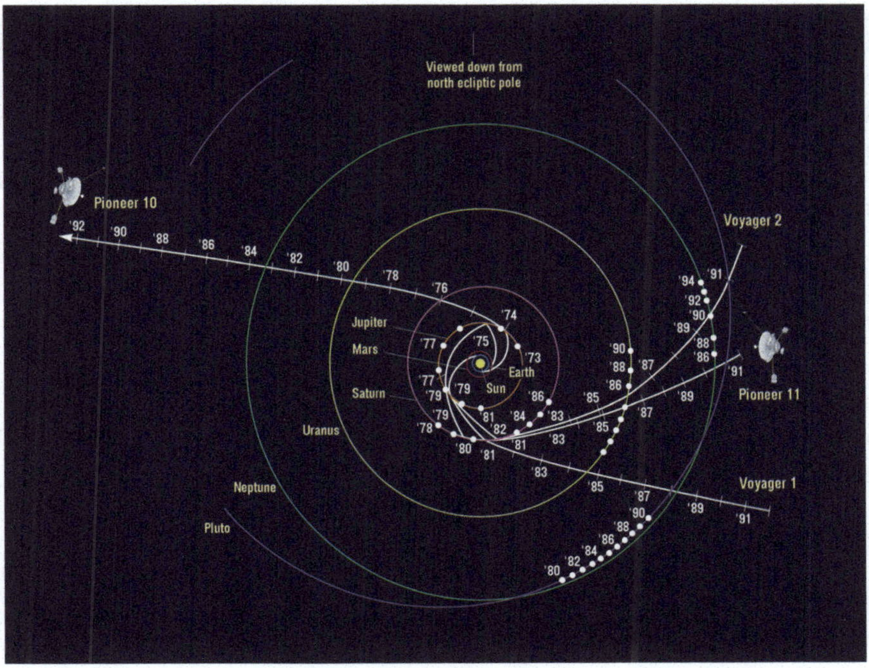

Fig. 18.1 Trajectories of Pioneer 10 & 11 and Voyager 1 & 2 throughout the solar system (*image credit* NASA)

to date, all US missions used Plutonium (Pu) 238 with a half-life of 87.7 years (Abelson 2006).

Thermoelectric converters use the Seebeck effect to generate an electromotive force between dissimilar materials when their junction is subjected to a temperature difference (Abelson 2006). In an RTG, the heat of decay of the radioisotope material heats one side of the thermoelectric converter while the other side is cooled by the space environment (or the planetary environment, respectively, if the RTG is used in a terrestrial or Mars surface application, for instance).

There is a number of reasons why RTGs suggest themselves for applications in outer solar system exploration. First of all, missions to the outer solar system require a safe and reliable power supply over many years; RTGs, by design, have a long life with highly predictable output power levels. The SNAP-19 RTGs of Pioneer 10 & 11, for example, were initially designed for a lifetime of five years. The spacecraft actually continued to communicate with Earth for 30 and 22 years, respectively (NASA Radioisotope Power Systems 2016).

In addition, RTGs are robust, compact, operate continuously and—an aspect of particular relevance when travelling to the outer solar system—operate independent of solar insolation.

The only limitation to consider is that the heat released by the radioisotope fuel source will gradually diminish over time as a function of the half-life of the material. As for Pioneer 10, a power sharing plan thus had to be implemented in September 1989 (17 years after launch) due to the gradual decay in RTG power output. Within this plan, instruments were turned on and off in an attempt to operate within the gradually reducing envelope of RTG output power. In addition, the spacecraft battery was used to accumulate RTG power over inactive periods, and thus enable activities no longer directly possible with the RTG output power levels available at that point in time (NSSDCA/COSPAR ID: 1972-012A, 2016).

As for Pioneer 10, the last signal was received on January 23, 2003, almost 31 years after the mission had been launched (Abelson 2006).

18.3.2 Pioneer 11

Pioneer 11 was the second mission to explore the outer solar system, the second to investigate Jupiter, and the first to explore Saturn. Pioneer 11 was launched on April 5, 1973. (NSSDCA/COSPAR ID: 1973-019A, 2016).

The trajectory of Pioneer 11 is shown in Fig. 18.1, and provided the first possibility to observe Saturn and some of its satellites in September 1979.

The power system of Pioneer 11 resembled that of Pioneer 10. The four SNAP-19 RTGs each had an average initial electrical power output of 39.9 W (Abelson 2006).

The four RTGs together thus generated approximately 144 W at Jupiter; the power output decreased to approximately 100 W at Saturn (NSSDCA/COSPAR ID: 1973-019A, 2016).

The last signal of Pioneer 11 was received on September 30, 1995, more than 22 years after the mission had been launched (Abelson 2006).

18.3.3 Voyager 2

Voyager 2 was the first of the two Voyager spacecraft to be launched on August 20, 1977.

Voyager 2 successfully performed flybys of Jupiter and Saturn in 1979 and 1981. The mission was then extended to also investigate Uranus and Neptune (flybys 1986 and 1989), and to even go beyond Neptune. The trajectory of Voyager 2 is shown in Fig. 18.1.

Voyager 2 carried three Multi-hundred Watt (MWH) RTGs installed on a deployable boom. Each RTG had an initial electrical output power of 159.2 W (Abelson 2006), was contained in a beryllium outer case, and weighed 39 kg (NSSDCA/COSPAR ID: 1977-076A, 2016).

In total, the RTGs had an initial output of approximately 470 W (NSSDCA/COSPAR ID: 1977-076A, 2016), which reduced to approximately 258 W by the beginning of 2015 (NASA Voyager 2016).

Voyager 2 is expected to keep transmitting weak radio messages until at least 2025, over 48 years after it was launched (NASA Voyager 2016).

An artist's concept of Voyager 2 is shown in Fig. 18.2.

Fig. 18.2 Voyager 2 (*image credit* NASA)

18.3.4 Voyager 1

Voyager 1 was actually the second of the two Voyager spacecraft to be launched. The launch took place on September 5, 1977, shortly after her twin spacecraft had been launched on August 20 the same year.

Voyager 1 successfully performed flybys of Jupiter (March 1979) and Saturn (November 1980). The spacecraft then continued to make observations of the interplanetary environment and UV observations of stars (NSSDCA/COSPAR ID: 1977-084A, 2016). The trajectory of Voyager 2 is shown in Fig. 18.1.

The power system matches that described with the Voyager 2 spacecraft. The initial average electrical output power of each of the three Multi-hundred Watt (MWH) RTGs was 156.7 W (Abelson 2006); the three RTGs thus initially provided a total electrical output power of approx. 470 W. At the beginning of 2015, the power generated by Voyager 1 had dropped to approximately 255 W (NASA Voyager 2016).

Even today, the spacecraft is still sending data. It is expected that the point in time where Voyager 1 will no longer be able to power a single instrument will be reached no earlier than 2025 (NASA Voyager 2016).

18.3.5 Galileo

18.3.5.1 Galileo Orbiter

Galileo was launched on October 18, 1989. It was the first spacecraft to go into a Jovian orbit. The Orbiter, having a mass of 2380 kg, performed a wide range of investigations focusing, among others aspects, on the Jovian atmosphere, the Galilean satellites, and the Jovian magnetosphere (NSSDCA/COSPAR ID: 1989-084B, 2016).

The Orbiter was powered by two General Purpose Heat Source (GPHS) RTGs, with each RTG being located at the end of a short boom. Each RTG had an initial average electrical output power of 292 W (Abelson 2006). In total, the two installed RTGs provided a nominal electrical output power of 570 W to the Orbiter (NSSDCA/COSPAR ID: 1989-084B, 2016).

The RTGs were still operating when the Orbiter was de-orbited into the Jovian atmosphere in 2003 (Abelson 2006).

An artist's concept of Galileo over Io is shown in Fig. 18.3.

18.3.5.2 Galileo Probe

The Galileo Probe was released from the Orbiter on July 13, 1995. It carried a scientific package for investigating, among other aspects, the Jovian atmosphere,

Fig. 18.3 Artist's concept of Galileo over Io (*image credit* NASA)

cloud layers and cloud particles. The probe was 1.25 m in diameter, 0.86 m in height and weighed 335 kg. It was comprised of two major segments: a *deceleration module* and a *descent module* (NSSDCA/COSPAR ID: 1989-084E, 2016).

The *deceleration module* essentially consisted of heat shields and thermal control hardware.

The *descent module*, which contained the science instruments and the subsystems required to support them, was the package which descended through the atmosphere by parachute (NSSDCA/COSPAR ID: 1989-084E, 2016).

The probe was powered by $LiSO_2$ batteries with a capacity of about 21 Ah; mission requirements were for about 16.3 Ah. Nominal power was rated at 580 W (NSSDCA/COSPAR ID: 1989-084E, 2016).

The duration of the probe's descent through the Jovian atmosphere was expected to last between 48–75 min, with the lower limit determined by the minimum required battery capacity and the upper limit by atmospheric pressure (NSSDCA/COSPAR ID: 1989-084E, 2016).

The probe entered the Jovian atmosphere as planned on December 7, 1995. The radio signal from the probe was received by the orbiter for 57.6 min (NSSDCA/COSPAR ID: 1989-084E, 2016).

18.3.6 Ulysses

Ulysses was jointly funded by the European Space Agency (ESA) and NASA, Office of Space Science Applications. It was launched on October 6, 1990.

The primary objective of the Ulysses mission was to study the heliosphere—the region of space influenced by the Sun and its magnetic field (ESA Ulysses 2016).

Secondary objectives included interplanetary and planetary physics investigations during the initial Earth-Jupiter phase, and investigations in the Jovian magnetosphere (NSSDCA/COSPAR ID: 1990-090B, 2016).

The spacecraft was powered by a single General Purpose Heat Source (GPHS) RTG (NSSDCA/COSPAR ID: 1990-090B, 2016). The GPHS actually consists of several elements containing the isotopic fuel Pu 238, in the form of PuO_2, providing a total thermal power of 4500 W. Si-GE thermoelectric elements were applied to convert the heat of decay into electrical energy (ESA Ulysses Spacecraft 2016).

The power system was rated at a nominal power level of about 282 W at the beginning of the mission (Abelson 2006), decreasing to about 220 W at nominal mission end in December 2001 (ESA Ulysses Spacecraft 2016).

On June 30, 2009, the Ulysses mission ended after 6842 days (ESA Ulysses 2016).

18.3.7 *Cassini*

18.3.7.1 Cassini Orbiter

Cassini was jointly funded by the NASA, Office of Space Science Applications, the European Space Agency, and the Agenzia Spaziale Italiana. It was launched on October 15, 1997 (NSSDCA/COSPAR ID: 1997-061A, 2016).

The Cassini Orbiter delivered a probe (Huygens, provided by ESA) to Titan, and then remained in orbit around Saturn for detailed studies of the planet, its rings and satellites.

Power is provided to the spacecraft through the use of three General Purpose Heat Source (GPHS) RTGs. Each RTG uses the heat generated by the decay of 10.9 kg of plutonium dioxide (PuO_2) to generate electrical power (NSSDCA/COSPAR ID: 1997-061A, 2016).

At the beginning of the mission, each RTG was capable of producing 293 W of electrical power (Abelson 2006). By the end of the nominal 11 year mission, the electrical output power is expected to degrade to around 210 W per RTG (NSSDCA/COSPAR ID: 1997-061A, 2016).

As of today, the RTGs are still in operation and supply power to the spacecraft still in orbit around Saturn (Abelson 2006).

18.3.7.2 Huygens Probe

The probe was designed by the European Space Agency (ESA) to perform in-depth studies of the clouds, atmosphere, and surface of Titan. Huygens separated from the Cassini orbiter on December 25, 2004 (ESA Cassini Huygens 2016).

The Huygens Probe carried five batteries, which were designed to provide power to the Probe from the time of Orbiter separation until at least 30 min after arrival on Titan's surface. During the cruise phase, the Probe was powered by the Orbiter, and the Power Control & Distribution Unit (PCDU) isolated the batteries of the Huygens Probe. Each battery comprised of two modules of 13 $LiSO_2$ cells (each 7.6 Ah) electrically connected in series (Hassan and Jones 1997).

After a descent of 2 h 28 min, the probe softly landed on Titan, and transmitted data for at least another 3 h 14 min. Post-flight analysis of the probe telemetry data indicates that the batteries probably became fully discharged 17 min after the Huygens radio signal was last verified on Earth. The mission was terminated by the loss of contact to the mother spacecraft Cassini which moved behind Titan. It is thought that the probe continued to function until the batteries were exhausted (Lebreton et al. 2005).

18.3.8 New Horizons

New Horizons was launched on January 19, 2006. It was designed to perform a flyby of Pluto and its moon Charon, and to transmit images and data back to Earth. Then, New Horizons is to continue on into the Kuiper Belt where it will perform flyby(s) of one or more Kuiper Belt Objects and return further data to Earth (NSSDCA/COSPAR ID: 2006-001A, 2016).

The primary objectives of the mission are to characterize the global geology and morphology, to map the surface composition of Pluto and Charon, and to characterize the neutral atmosphere of Pluto and its escape rate (NSSDCA/COSPAR ID: 2006-001A, 2016).

The single General Purpose Heat Source (GPHS) RTG provided an initial electrical power output of 293 W (NASA Radioisotope Power Systems). Output power will have decayed to approximately 228 W at encounter in 2015 (NSSDCA/COSPAR ID: 2006-001A, 2016).

The 31 kg science payload package requires 21 W of power and consists of seven scientific instruments (NSSDCA/COSPAR ID: 2006-001A, 2016).

An artist's concept of New Horizons' Pluto Kuiper Belt Flyby is shown in Fig. 18.4.

18.3.9 JUNO

The JUNO mission was launched on August 5, 2011, to study Jupiter from polar orbit for approximately one year beginning in 2016.

The primary scientific objectives are to investigate formation and origin of Jupiter's atmosphere, variations in Jupiter's deep atmosphere, Jupiter's magnetic field, gravity field and distribution of mass inside the planet; in addition, Jupiter's

Fig. 18.4 Artist's concept of New Horizons' Pluto Kuiper Belt flyby (*image credit* NASA)

three-dimensional polar magnetosphere and aurorae are to be investigated (NSSDCA/COSPAR ID: 2011-040A, 2016).

Power is provided by ultra triple junction GaAs solar cells, covered with thick glass for radiation shielding. The solar cells are grouped into 11 solar panels, four on two of the wings and three on the other. The solar panels will produce a total of 18 kW at Earth and 400 W initially at Jupiter (NSSDCA/COSPAR ID: 2011-040A, 2016).

An artist's concept of JUNO is shown in Fig. 18.5.

18.4 Conclusions and Outlook

The outer solar system is both, a fascinating and challenging environment. It is now 45 years since Pioneer 10, the first spacecraft sent to the outer solar system, was launched. In these 45 years, we have seen highly successful missions sending spectacular images and valuable data sets back to Earth. Voyager 1 was even the first man-made object to enter interstellar space in 2012.

In terms of power systems for outer solar system exploration, there is a clear dominance of Radioisotope Thermoelectric Generators (RTGs), as shown in Table 18.2. Apart from probes piggybacking to outer planets, and only designed to survive on their own power supply for a very short period of time, all but one

Fig. 18.5 Artist's concept of JUNO (*image credit* NASA/JPL)

mission were powered by RTGs. The one exception actually being the most recent mission, JUNO.

Comparing the electrical output power per RTG, one can see a clear trend towards higher output power levels. The electrical output power per RTG has actually gone up from some 40 W, as provided by the early 1970s Pioneer probes' SNAP-19 RTGs, up to just below 300 W per RTG, as applied with the GPHS RTGs used with the Galileo, Ulysses, Cassini and New Horizons spacecraft.

The total electrical power available to the spacecraft systems, which is derived by adding the output power levels of the individual RTGs, strongly depends on the specific spacecraft and its mission profile. As shown in Table 18.3, the two most recent RTG-powered spacecraft, Cassini and New Horizons, had an initial output power level of almost 900 W (Cassini) compared to less than 300 W (New Horizons). A clear trend to increasing electrical power levels is thus not present.

The next step in Radioisotope-based power systems could be to apply a Stirling Converter Technology instead of thermoelectric converters. The improved heat-to-electric conversion efficiency could enable a reduction of the radioisotope fuel supply for a given electrical output power, or an increase in electrical output power for a given fuel supply.

NASA had planned to complete development of two Advanced Stirling Radioisotope Generator (ASRG) units for flight by 2016, but chose to discontinue development in late 2013 (NASA Stirling Converter Technology 2016).

Table 18.2 Summary of power systems applied with missions to the outer solar system

Mission	Power source
Pioneer 10	Radioisotope thermoelectric generators
Pioneer 11	Radioisotope thermoelectric generators
Voyager 2	Radioisotope thermoelectric generators
Voyager 1	Radioisotope thermoelectric generators
Galileo orbiter	Radioisotope thermoelectric generators
Galileo probe	$LiSO_2$ battery
Ulysses	Radioisotope thermoelectric generator
Cassini orbiter	Radioisotope thermoelectric generators
Huygens probe	$LiSO_2$ battery
New Horizons	Radioisotope thermoelectric generator
JUNO	Triple junction GaAs solar cells

Table 18.3 Summary of RTG power

Mission	Launch date	RTG	Power per RTG[a] (W)	Number of RTGs	Total RTG power[b] (W)
Pioneer 10	1972-03-03	SNAP-19	40.7	4	162.8
Pioneer 11	1973-04-05	SNAP-19	39.9	4	159.6
Voyager 2	1977-08-20	MWH	159.2	3	477.6
Voyager 1	1977-09-05	MWH	156.7	3	470.1
Galileo	1989-10-18	GPHS	292	2	584
Ulysses	1990-10-06	GPHS	282	1	282
Cassini	1997-10-15	GPHS	293	3	879
New Horizons	2006-01-19	GPHS	292	1	292
JUNO	2011-08-05	NA	NA	NA	NA

[a]Initial average electrical output power in Watts. Data modified from (Abelson 2006) with the exception of New Horizons Data taken from (NASA radioisotope power systems)
[b]Initial cumulative electrical output power in Watts of all RTGs installed in the spacecraft

RTGs have been applied with each mission but JUNO, the latest mission launched to the outer planets. The solar panels installed with JUNO generate 18 kW at Earth, and only 400 W at Jupiter (2.22% of the nominal power output at Earth), and clearly show that solar power generation is challenging but feasible with missions flying as far as Jupiter, but prohibitive in terms of the required solar panel area if wanting to go (even much further) beyond.

The next missions scheduled for launch towards the outer solar systems are NASA's Europa Clipper and ESA's JUICE missions.

NASA's Europa Clipper mission will be visiting Jupiter and its moon Europa (Phillips and Pappalardo 2014); it will be launched probably in 2020. In the course of working on the spacecraft concept, an RTG-powered as well as a solar-powered option were considered. In this, the solar arrays were sized to provide the same end of prime mission power margin as 5 MMRTGs. Solar power was eventually chosen in the pre-project as it met mission requirements and was considered being programmatically advantageous (Goldstein and Pappalardo 2015).

ESA's JUICE mission, short for Jupiter Icy moons Explorer, planned for launch in 2022 and arrival at Jupiter in 2030, will be visiting Jupiter and its moon Europa (ESA Juice 2016).

JUICE will be operating at solar constants as low as 46 W/m^2. Thus, some 60–75 m^2 of solar arrays will be installed with the spacecraft utilizing GaAs solar cells optimized for 'Low-Intensity/Low-Temperature' conditions (ESA Juice 2016).

A particularly challenging experiment will be the penetrator which should penetrate into the ice of Europa at least 2 m deep, exposing the Li-CFx battery (ESA CLEO/P 2016) to extremely high g-forces.

References

Abelson, R.D.: Space missions and applications. In: Rowe, D.M. (ed.) Thermoelectrics Handbook: Macro to Nano, pp. 56.1–56.26. Taylor & Francis, Boca Raton (2006)

ESA Cassini Huygens: http://www.esa.int/Our_Activities/Space_Science/Cassini-Huygens/ The_mission (2016). Accessed 11 Aug 2016

ESA CLEO/P: sci.esa.int/science-e/www/object/doc.cfm?fobjectid=56432 (2016). Accessed 30 Oct 2016

ESA Juice: http://sci.esa.int/juice/ (2016). Accessed 30 Oct 2016

ESA Ulysses: http://sci.esa.int/ulysses/ (2016). Accessed 11 Aug 2016

ESA Ulysses Spacecraft: http://www.cosmos.esa.int/web/ulysses/spacecraft (2016). Accessed 11 Aug 2016

Goldstein, B., Pappalardo, R. http://www.lpi.usra.edu/opag/feb2015/presentations/04_Clipper%20OPAG%20Feb%202015.pdf (2015). Accessed 30 Oct 2016

Hassan, H., Jones J.C.: The Huygens probe http://www.esa.int/esapub/bulletin/bullet92/b92hassa.htm. ESA Bull. **92** (1997). Adapted from the article 'The Huygens Probe System Design' by Jones & Giovagnoli, in Huygens: Science, Payload and Mission, ESA SP-1177

Lebreton, J.-P., Witasse, O., Sollazzo, C., Blancquaert, T., Couzin, P., Schipper, A.-M., Jones, J. B., Matson, D.L., Gurvits, L.I., Atkinson, D.H., Kazeminejad, B., Pérez-Ayúcar, M.: An overview of the descent and landing of the Huygens probe on Titan. Nature **438**, 758–764 (2005)

NASA Jupiter: Lunar and planetary science: Jupiter. http://nssdc.gsfc.nasa.gov/planetary/planets/jupiterpage.html (2016). Accessed 11 Aug 2016

NASA Neptune: Lunar and planetary science: Neptune. http://nssdc.gsfc.nasa.gov/planetary/planets/neptunepage.html (2016). Accessed 11 Aug 2016

NASA Pluto: Lunar and planetary science: Pluto. http://nssdc.gsfc.nasa.gov/planetary/planets/plutopage.html (2016). Accessed 11 Aug 2016

NASA Radioisotope Power Systems: https://solarsystem.nasa.gov/rps/rtg.cfm#gphsrtg (2016). Accessed 14 Aug 2016

NASA Saturn: Lunar and planetary science: Saturn. http://nssdc.gsfc.nasa.gov/planetary/planets/saturnpage.html (2016). Accessed 11 Aug 2016

NASA Stirling Converter Technology. https://solarsystem.nasa.gov/rps/asrg.cfm (2016). Accessed 14 Aug 2016

NASA Uranus: Lunar and planetary science: Uranus. http://nssdc.gsfc.nasa.gov/planetary/planets/uranuspage.html (2016). Accessed 11 Aug 2016

NASA Voyager. http://voyager.jpl.nasa.gov/spacecraft/spacecraftlife.html (2016). Accessed 14 Aug 2016

NSSDCA/COSPAR ID: 1997-061A: Cassini. http://nssdc.gsfc.nasa.gov/nmc/spacecraftDisplay.do?id=1997-061A (2016). Accessed 11 Aug 2016

NSSDCA/COSPAR ID: 1989-084B: Galileo Orbiter. http://nssdc.gsfc.nasa.gov/nmc/spacecraftDisplay.do?id=1989-084B (2016). Accessed 11 Aug 2016

NSSDCA/COSPAR ID: 1989-084E: Galileo probe. http://nssdc.gsfc.nasa.gov/nmc/spacecraftDisplay.do?id=1989-084E (2016). Accessed 11 Aug 2016

NSSDCA/COSPAR ID: 1972-012A: Pioneer 10. http://nssdc.gsfc.nasa.gov/nmc/spacecraftDisplay.do?id=1972-012A (2016). Accessed 11 Aug 2016

NSSDCA/COSPAR ID: 1973-019A: Pioneer 11. http://nssdc.gsfc.nasa.gov/nmc/spacecraftDisplay.do?id=1973-019A (2016). Accessed 11 Aug 2016

NSSDCA/COSPAR ID: 1977-076A: Voyager 2. http://nssdc.gsfc.nasa.gov/nmc/spacecraftDisplay.do?id=1977-076A (2016). Accessed 11 Aug 2016

NSSDCA/COSPAR ID: 1977-084A: Voyager 1. http://nssdc.gsfc.nasa.gov/nmc/spacecraftDisplay.do?id=1977-084A (2016). Accessed 11 Aug 2016

NSSDCA/COSPAR ID: 1990-090B: Ulysses. http://nssdc.gsfc.nasa.gov/nmc/spacecraftDisplay.do?id=1990-090B (2016). Accessed 11 Aug 2016

NSSDCA/COSPAR ID: 2006-001A: New Horizons Pluto Kuiper Belt flyby. http://nssdc.gsfc.nasa.gov/nmc/masterCatalog.do?sc=2006-001A (2016). Accessed 11 Aug 2016

NSSDCA/COSPAR ID: 2011-040A: Juno. http://nssdc.gsfc.nasa.gov/nmc/spacecraftDisplay.do?id=2011-040A (2016). Accessed 11 Aug 2016

Phillips, C.B., Pappalardo, R.T.: Europa clipper mission concept: exploring Jupiter's ocean moon. Eos **95**, 165–167 (2014)

Chapter 19
Multi-rendezvous Solar Electric Propulsion Mission Opportunities to Jupiter Trojans

Volker Maiwald

Jupiter Trojans are a group of small bodies (sized rarely up to some hundred kilometers), which gather at the L4 and L5 Libration Points of the Jupiter–Sun system. Max Wolfe discovered the first Trojan asteroid in 1906 and named it Archilles and by that coined the term of Jupiter Trojans, as figures from the Trojan epos became namesakes for subsequent discoveries in that region of the solar system (Jewitt et al. 2004).

All in all the number of Jupiter Trojans is approximately 160 thousand with a size above one kilometer, which is of the same order of magnitude as the main asteroid belt (Jewitt et al. 2004) with an estimated total mass of about 10^{-4} Earth masses (Jewitt et al. 2000).

Trojans are scientifically interesting as they might contain information about the formation and evolution of the solar system. Besides that, they are located at a peculiar position, a stable location on Jupiter's orbit, without the danger present by Jupiter's radiation. Furthermore, they might contain valuable resources, e.g. metals or even water ice, if the hypothesis that Trojans are dormant comets can be validated (Jewitt et al. 2004).

In general, there is a large amount of information we are still lacking about Trojan asteroids and therefore investigating them, especially their subsurface material, which has not yet been determined, could lead to useful missions e.g. regarding harvesting their resources. This chapter describes a possible mission scenario including a spacecraft design that is capable of exploring Trojan subsurface material and that can reach several Trojan asteroids for multiple rendezvous.

The spacecraft design and mission considerations presented here are based on previous work (Maiwald 2009, 2015; Maiwald et al. 2012), among others a design

V. Maiwald (✉)
German Aerospace Center (DLR), Institute of Space Systems,
Robert-Hooke-Str. 7, 28359 Bremen, Germany
e-mail: volker.maiwald@dlr.de

V. Badescu and K. Zacny (eds.), *Outer Solar System*,
https://doi.org/10.1007/978-3-319-73845-1_19

study by a team of engineers at the Concurrent Engineering Facility (CEF) of the German Aerospace Center (DLR). The spacecraft is abbreviated as *TRIP*—short for *T*rojan *I*nvestigation *P*robe—for the purpose of this description.

19.1 Peculiarity of Jupiter Trojans

19.1.1 Position of Jupiter Trojans

Libration points are theoretical solutions of the restricted, circular three-body system, where no relative motion of the three masses occurs. In total, such a system has five libration points and is coined by circular orbits. The mass of the smallest mass is infinitely small and the largest mass is dominant over all others. The idea of libration points has originally been derived by Joseph-Louis Lagrange.

Figure 19.1 shows the distribution of these positions in a restricted, circular three-body system, marking areas where the infinitely small third mass can be positioned without any resulting force acting on it. At these positions all forces acting on the third body, i.e. gravity and centrifugal force, balance each other out so that a relative movement regarding the two other masses does not occur.

The libration points L4 and L5 are stable, therefore any celestial body that reaches that region remains there, theoretically indefinitely. This is the reason why L4 and L5 of the Jupiter–Sun system host accumulations of small bodies—an effect which is not exclusive for Jupiter either. While in the actual solar system the conditions of a restricted, circular three-body system are not accurately met, the conditions are close enough to actually observe the effect of libration points. Besides the large groups gathered at Jupiter, Trojan asteroids have been found for Uranus, Mars and even Earth.

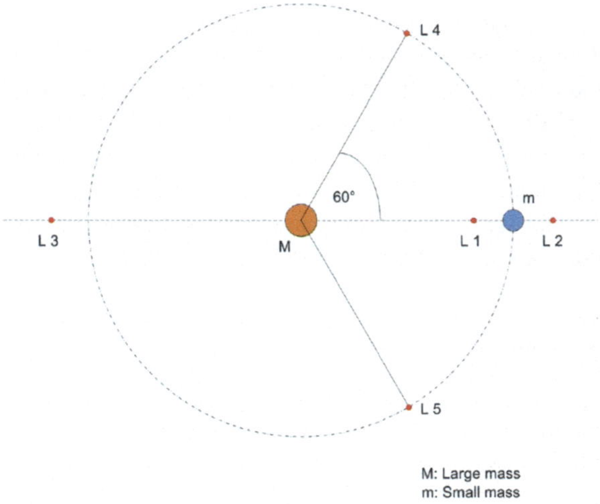

Fig. 19.1 Libration points in a restricted, circular three-body system. There are five libration points in such a system, where a third, infinitely small mass can be positioned without a resulting force acting on it, as all relevant forces, i.e. gravitational and centrifugal balance each other out (Maiwald 2009)

Currently the numbers of bodies found at the two stable libration points of Jupiter are not equally distributed. There are about 1.3 times more bodies present at L4 than L5. Why this is the case is currently not clear, but it is possibly caused by an observation bias (Jewitt et al. 2004). Furthermore, Jupiter Trojans have an average inclination of about 18°, which has to be considered for transfer trajectories to these bodies (Fleming and Hamilton 2000).

19.1.2 Origin of Jupiter Trojans

The origin of the Trojan asteroids and the nature of their orbits is currently still subject to debate. Mostly the data basis for this debate is created with orbital simulations of the solar system formation. Trojans are likely primordial bodies, which have reached a resonant orbit with Jupiter. How this occurred is presently still unresolved (Marzari et al. 2003).

19.1.2.1 Capture by Collisions

Marzari et al. report that simulations involving impact velocities of 5 km/s and a body composition comparable to comets result in a body distribution as found today in the Trojan population. Based on these they assume that Trojans are the remains of primordial bodies that formed in the planetesimal disc. The idea of collisions is also supported by the detection of so-called dynamical families which share the same orbital parameters and thus likely are the remains of a parent body that was fractured by a collision (Marzari et al. 1997).

19.1.2.2 Capture by Jupiter's Migration and Mass Growth

According to simulations conducted by Morbidelli et al. (2005), the inward migration of Jupiter can have caused capture of Trojan asteroids during passage of the 1:2 mean motion resonance with Saturn. While any preexisting Trojans would have been expelled during that event, capture of new bodies from outside occurred—according to simulations. When Jupiter migrated out of this resonance, these bodies have been trapped at the libration points. Research by Fleming and Hamilton (2000) has shown that the stabilization of the orbits (i.e. reduction of their libration amplitudes) was more influenced by Jupiter's mass growth than its radial migration. The latter was more influencing the Trojan's eccentricities and inclinations.

In general, it can be said that the mass growth of the gas giant increased the stable regions at the libration points (Dotto et al. 2006).

Table 19.1 Average properties of Jupiter Trojans

Parameter	Mean value for Jupiter Trojans
Inclination	18° (Fleming and Hamilton 2000)
Libration amplitude	29° (Fleming and Hamilton 2000)
Eccentricity	0.06 (Fleming and Hamilton 2000)
Albedo	0.04 (Dotto et al. 2006)
Density	1.5 g/cm^3 (Marzari et al. 1997)

List of the average properties of Jupiter Trojans, relevant for design of a rendezvous mission and spacecraft

19.1.3 Properties of Jupiter Trojans

The origin of the Jupiter Trojans is currently still unknown and only vague observational data exists about most of them. Their composition is not precisely known; spectra obtained from observation are usually featureless, but in the red spectrum (Morbidelli et al. 2005). Currently, most Trojans are classified as d- or p-type asteroids, i.e. they are carbon rich and contain little hydrogen. As only exterior analysis is possible, ice could be contained in the interior of these small bodies (Jewitt et al. 2004), while the outer mantle has been altered by solar radiation (Dotto et al. 2006). Table 19.1 describes some average values, as typically assumed for Trojan asteroids. It is evident that they are rather dark bodies, as only about 4% of the incoming light is reflected.

Just as comets, Trojans often have an elongated form, which suggests a common origin, e.g. Hektor has a size of about 370 × 110 km (Fornasier et al. 2004).

19.2 Spacecraft Design

19.2.1 Mission Purpose and Possibilities

Depending on the exact frame of such a mission, its purpose needs to be defined and in turn will define *TRIP*'s spacecraft layout and characteristics. Currently no mission has targeted the Jupiter Trojans yet and therefore the information about them is scarce.

Generally, the Trojan asteroids can be targeted by an orbiter and also landing vehicles. Especially if resources are supposed to be extracted, in situ activities, i.e. landing, are necessary. But before actually extracting resources it needs to be investigated what could be extracted. The basic question about the Trojans still is: What are they? While assumptions are made, data about their actual composition and e.g. whether or not they are homogenous in their composition could shed light on that question. If the Trojans formed in the region where they are today and are basically planetesimals, they will have a homogenous composition. If they are originating in different regions, the composition will be more heterogeneous.

An in-between solution for in situ measurements would be instead of landing on a Trojan, to use penetrators to expel material from below the surface, with the same—if smaller—result as during NASA's *Deep Impact* mission. The expelled material could be investigated and analysed using instruments on an orbiter, the same is true for the material uncovered by the penetrator's impact.

Concerning the possibility to use a lander or penetrator, the most relevant information is the surface strength and composition. The current lack of knowledge about Trojan asteroids makes the design of a landing vehicle difficult.

Assuming that cometary nuclei and jovian Trojans share common properties regarding composition and build, landing on such a body can be compared with landing on a comet nucleus—the corresponding risks are described in Kührt et al. (1997).

Comets are defined by a very irregular shape and surface and a low density as they mainly consist of dust and ice. Adding a complex rotational behavior results in an equally complex gravitational field that changes with time. The surface material is probably weak, although unpredictably so, i.e. a lander could penetrate the surface in an unknown way, Kührt et al. assume a worst case strength of approximately 100 Pa, while it can go up as high as 5 MPa (Kührt et al. 1997). Additionally gravity does not suffice to keep the lander on the surface due to the small size of the Trojans, therefore an anchoring device has to be installed into it, resulting in more mass. Due to the complex rotation of a possible target, the resulting conditions of illumination and thus amount of available photovoltaic power and temperature would change in a similar complex way. The lander would have to be able to endure that. For Trojans above 60 km diameter, diverse rotation periods between 5 h and 14 days have been found (Mottola et al. 2011).

A simple penetrator would have fewer restrictions than a lander, as it does not need to survive e.g. eclipse times. Kührt et al. (1997) give the penetration depth z_{max} for a comet lander as:

$$z_{\mathrm{max}} = \frac{v_{impact}^2 M_{lander}}{2\sigma A_{lander}} \tag{19.1}$$

where v_{impact} is the impact velocity, M_{lander} the lander mass, σ the material strength and A_{lander} the lander's cross section. As comets and Trojans possibly share a similar heritage, a similar formulation can be used for assessing penetrators on Trojan asteroids.

For a rocket diameter of 10 cm and a mass of 5 kg—which would allow carrying several for a single mission—a velocity of 1.121 m/s would be necessary for the lowest material strength given by Kührt et al. of 100 Pa to reach a penetration depth of 1 m. As mentioned before the maximum estimate for σ is 5 MPa, a velocity yielding 1 m of depth would be around 250 m/s. Not only requiring more fuel to attain, but also result in a penetration depth of 50 km for a material strength of 100 Pa, i.e. in many cases would shoot right through the object. This shows that a thorough and fitting design is hard to obtain and the lack of knowledge about the

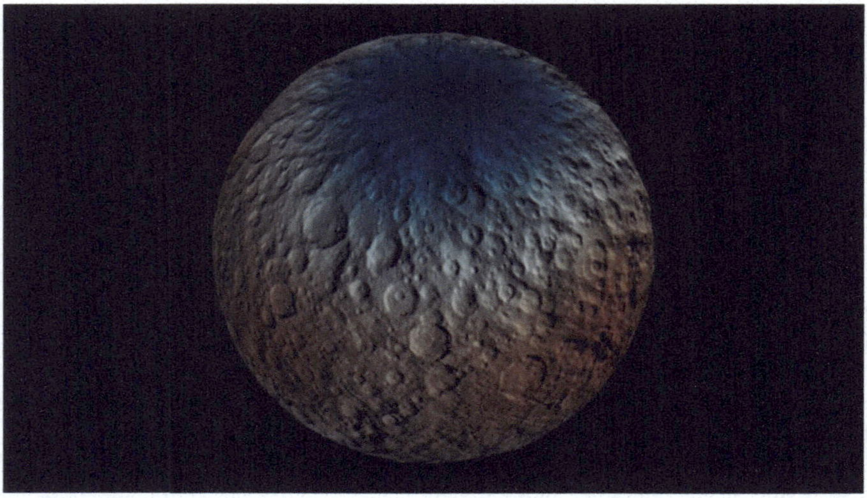

Fig. 19.2 GRaND data depicted in form of a map of Ceres' northern hemisphere. The map is based on the number of neutrons detected by the instrument, blue: low numbers and red: high numbers. The number of neutrons detected increases with decreasing hydrogen content, suggesting the polar regions contain water ice within 1 m of the surface. *Source* NASA/JPL-Caltech/UCLA/MPS/DLR/IDA/PSI

material strength makes it difficult to assess e.g. the actual penetration depth or if enough material is expelled for analysis.

For the purpose of investigating subsurface composition it is therefore suggested to use different means or at least include a complement of them if the penetrators fail, e.g. *Dawn*'s Gamma Ray and Neutron Detector (GRaND) (Prettyman 2006), with a data example in Fig. 19.2, or a sounding radar similar to the Shallow Subsurface Radar (SHARAD) from the *Mars Reconnaissance Orbiter* (NASA 2005), see a data example in Fig. 19.3.

As currently no mission has reached a Trojan asteroid yet, a first mission should be able to gather broad science data about the Trojans, i.e. visit more than one to derive general statements about the population of Trojans and not just a single body, but also to conduct a large number of scientific experiments. In the following such a mission is drafted and presented.

19.2.2 Mission Objectives and System Requirements

One major goal of any mission towards Jupiter Trojans is to determine the nature of their origin—knowing that, more derivations about the formation and development of the solar system could be made, just the same as assumptions about their own composition.

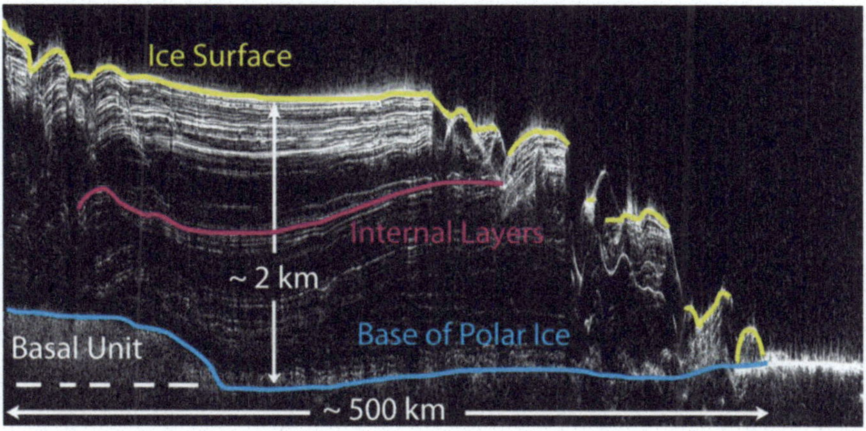

Fig. 19.3 SHARAD Cross-section radar image of martian polar ice. This picture is an example of SHARAD's capability to surpass Mars' surface and review the structure of the polar ice layers on Mars. Each line is one layer of ice. *Source* NASA/JPL-Caltech/ASI/UT

As the immediate surface is likely altered by radiation (Dotto et al. 2006), one major objective is to analyse the subsurface composition with a suitable instrument. Due to the current lack of knowledge about surface strength this should be achievable without need for a landing vehicle or penetrator, i.e. subsurface analysis is conducted by GRaND- and SHARAD-like instruments, which have mission heritage on *Dawn, Mars Odyssey* resp. *Mars Reconnaissance Orbiter*.

Close-up imagery of the Trojans would provide information about collision history and their formation, i.e. high resolution images of the surface is also a mission objective.

Measurement of the size distribution of Trojan asteroids could be used to verify models of the current Trojan population and by that, models for their formation. Therefore, means to measure the size should be included in the mission. As up to now observation of Trojans has been Earth-bound, the bias of the observation is on larger bodies—in situ inspection of Trojans should ease that bias and provide information about smaller bodies as well.

Similarly, the orbit determination of Trojan bodies could provide information about any dynamical families and thus also occurrences of collisions.

To provide more than a single sample of measurements, the mission should be able to target more than one body, preferably in different regions of the population.

Considering that any mission to a Jupiter Trojan would already take several years to transfer to the target body and also should be capable to actually meet several bodies for data diversity, the life time of the spacecraft should be at least 5 years and could go up to about 15 years, allowing the visit of several Trojans.

In order to keep the costs low, the launch mass should not exceed 1600 kg, thus a Soyuz-Fregat launcher could be used instead of a heavy lift launcher. To allow sufficient science return the payload mass should be about 55 kg, including the before mentioned instrument capable to analyse subsurface composition, thermal mapping instrument, a spectrometer, a laser altimeter and radio science (using the onboard communication system). Based on existing instruments (see above), the total power requirement can be assessed as about 100 W on average, although not all payload instruments would be operated at the same time.

19.2.3 Propulsion System

The goal of achieving a multi-rendezvous mission makes efficient propulsion a major aspect of *TRIP*'s system design. Saving propellant is important to increase the number of Trojans that can be visited. A natural candidate for propulsion therefore is electrical propulsion, e.g. as used for *Dawn*, another small body multi-rendezvous mission directed at Vesta and Ceres (Rayman et al. 2006).

One candidate for the propulsion system would be the RITA-150 ion thruster, a variant of the RITA-15 thruster with space heritage. It has a diameter of 22 cm for the ionizer and a beam voltage of 1200 V. It's power specific thrust is about 35 mN/kW (Hasssewitz 2006). While currently not yet existing, the envisaged thruster promises several advantages, making it attractive for such a mission:

- A wide range in operational power (1–6 kW),
- a wide range of thrust (from about 20 mN at 3000 s of specific impulse to 200 mN at 5000 s), and
- a lifetime of probably more than 20,000 h (based on test with smaller engines of the same kind).

Especially the life time, which corresponds to the propellant that the thruster can use before failure, makes the RITA-150 a good candidate. The high demanding mission (assuming a Δv in the range of some 20 km/s) can be achieved with two thrusters and plus one as spare (Hasssewitz 2006).

Other possibilities would be the T6 or XIPS engines, which lack the favorable specific impulse or the JPL NEXIS, which has a minimum operating power of 4 kW. It would however be thinkable to adapt other engines to the needs of this mission as well.

Besides the actual thrusters (3 times 6.6 kg), the propulsion system would also contain two tanks (total 42 kg) for the Xenon propellant, the power processing units for the thrusters (each 15.2 kg), two flow control units (one for redundancy, each 2.2 kg) and pipes and valves. In total, one can assume a mass of about 116 kg for this subsystem including all of the mentioned components. The tanks are existing tanks with space heritage, which are titanium lined and graphite epoxy wrapped, allowing a tank pressure of 150 bar, at a volume of 50 L.

Once more mature than e.g. for JAXA's *Ikaros* mission, a solar sail could be used instead of an electrical propulsion system. Propellant demand would be replaced by sail degradation and limited amount of solar illumination.

19.2.4 Power Generation

One other major issue for a mission to Jupiter Trojans is the power generation. While Radioisotope-Thermoelectric Power Generators (RTGs) are not affected by the solar distance, the required resources are rare and the technology is not available in Europe, but could be available for a US mission. Consequently, alternatives need to be considered. The alternative with the largest heritage is solar power generation.

At solar distances equal to Jupiter's, i.e. approximately 5.2 AU, however the available solar energy is limited (proportional to $1/r^2$ with r as the solar distance). Furthermore, the so called Low-Intensity-Low-Temperature (LILT) effect occurs, resp. the so called "flat spot", which potentially further reduces the available power, especially for low-grade solar cells. Under LILT conditions these cells behave differently than projected in the respective illumination and temperature ranges as the maximum power point flattens (Fatemi et al. 2005). However, careful cell design can remove that issue, which is related to shunt resistance, and enable usage of solar cells under LILT conditions. A comparable mission is *Juno*, which recently arrived at Jupiter and is the first mission to use solar power generation for operating in such a large solar distance. This spacecraft uses a solar array of in total about 70 m^2 and a mass of about 340 kg, providing approximately 400 W of power while at Jupiter (NASA 2011). Only *Rosetta* had similar conditions when on its way to its target comet 67P/Churyumov–Gerasimenko and reaching Jupiter's solar distance. Using a solar array of 64 m^2, this spacecraft was hibernating at this point in time.

Based on the prospect of a multi-rendezvous objective, any spacecraft designed for such a mission must provide enough power to also operate the thruster previously discussed, i.e. an increase of 1 kW of power for minimum operation. Therefore, while *Juno* can be a proof of concept for solar electric power at these solar distances, the design is not sufficient. An alternative, which is light and at the same time performing well (220 W/kg), are ultra-flex solar panels as originally developed for the Multi-Purpose Crew Vehicle of NASA and applied for the Cygnus spacecraft. With a solar array size of about 127 m^2 (at a cell mass of 86 kg, plus about 140 kg for its structure) for two panels, this would allow for a beginning of life power of ca. 50 kW at Earth. Including the degradation of about 20% for about 15 years this would still mean a power of about 1.5 kW when at Jupiter's distance, using ZTJ triple junction cells with an efficiency of 29.5% (White et al. 2007; Emcore Corporation 2011).

The power subsystem would also contain batteries (lithium-ion cells) and as eclipses are only to be expected at Earth and possibly when around a Trojan asteroid doing radio science, the depth of discharge can be selected higher than usual (at 80%), reducing the amount of required batteries. A power conditioning unit is required as well.

19.2.5 General Spacecraft Design

The remaining spacecraft subsystems are not as hard pressed on performance as power generation and propulsion, because the solar distance or the objective of achieving multiple rendezvous does not put extra demands on e.g. the data handling system.

The spacecraft's configuration can be seen in Fig. 19.4 and its overall layout of the in terms of mass is presented in Table 19.2.

It can be seen that power generation requires the largest amount of dry mass, closely followed by structure, which includes mechanisms. This can be mostly attributed to the large solar array, required to allow solar electric propulsion even at solar distances of about 5.2 AU. The subsystems can be summarized as follows:

- Thermal Control System: Thermal Control has to handle two different environments—the relatively "hot" environment of Earth's vicinity and the cold Trojan area (heat fluxes are 1367 W/m^2 resp. 50 W/m^2) and is intended to be as simple as possible, mostly relying on passive elements. The spacecraft is to be painted black or grey on the outside resulting in a ratio of absorption and emittance (α/ε) of smaller than 1.1. Furthermore, it has one louver painted black on one side and white on the other, depending on the heat situation, Multi-Layer-Insulation foil protects the spacecraft from the heat generated by the thrusters and the solar panels are thermally decoupled from the main bus to prevent heat transfer to the spacecraft bus. This way, temperatures of 288 K at Earth and of 110 K (necessary for thermal radiometer and spectrometer and thermal mapper instruments) at the Trojans are ensured. Further the accommodation of the various spacecraft components regards their respective operation temperatures. Small components are the several thermal sensors distributed

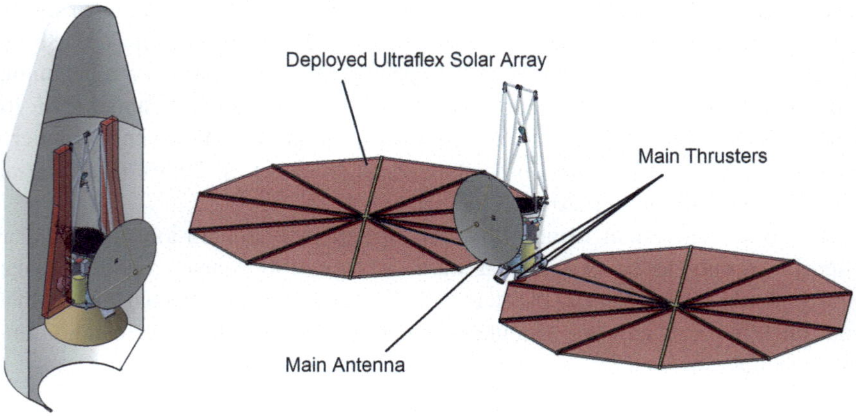

Fig. 19.4 Overview of a concept configuration for a spacecraft suitable for a Jupiter Trojan mission. Left: Stowed configuration of *TRIP* in a Soyuz-Fregat fairing. Right: Deployed configuration

Table 19.2 Subsystem overview of a spacecraft for a Jupiter Trojan mission

Mass Component	Margin [%]	Mass with margin [kg]
Structure	19.34	227.14
Thermal control	17.62	30.06
Communications	11.52	42.94
Data handling	10.00	10.23
Attitude & orbit control	7.48	109.96
Propulsion	7.95	115.94
Power	17.06	250.50
Harness	20.00	60.00
Instrument	7.15	53.58
Total dry mass	n/a	900.35
Propellant mass	n/a	621.00
Adapter mass	n/a	45.00
Total launch mass	n/a	1566.35

The table lists all relevant subsystems, based on Maiwald (2009), Maiwald et al. (2012) and provides masses and the average margin on these. The margins are derived from the respective components' design maturity and based on ECSS (European Cooperation for Space Standardization) standard as recommended by ESA. Components, which require no modifications are regarded with a 5% margin, those with modifications are assigned 10% and new developments are assigned 20% margins. Comparable margins are assigned for Delta-V and propellant masses

in the spacecraft, the louver radiator, and two heaters one for the subsurface instrument and the laser altimeter, the tank, the onboard computer and the batteries.

- Communication System: The design is derived from previous missions like *Dawn* and *Rosetta*, focusing on reliability and power-reduced operation. The back-up portion of the communication subsystem consists of three omnidirectional S-band low gain antennae and one S-band mid gain antenna. The S-band mid gain antenna is used for telemetry and command in Earth's vicinity and acts as the fallback option during the later mission phases. One further X-band mid gain antenna is used for the deep space phases' telemetry and command. It is also required for the spacecraft's ranging. One additional, body-mounted high gain antenna dish for Ka-band, is used mainly for science data. Amplifiers for the X-band and Ka-band antennae, each with one redundant one in addition. Furthermore there are two redundant transceivers responsible for the communication. Ka-band allows twice the data rate over X-band and therefore shorter communication times are required, resulting in less power demand and thus solar panel size.

- Data Handling System: In total three completely redundant boards (with CPUs) are used for house-keeping and science operations. Three redundant CPUs operation over the whole intended life-time and back-up for damage due to radiation. At all times a watch-dog architecture ensure correct operation of the Data Handling System. A mass memory of in total 32 GB is used mostly for science data, where an accumulation of 4 GB per day is assumed.
- Attitude & Orbit Control: Three reaction wheels (and one redundant) are used for attitude control during science phases where thruster exhaust could disturb instrument measurements. Three gyroscopes, six sun sensors (one on each spacecraft side) and two Star Trackers are used for attitude determination. Sixteen 5 N monopropellant thrusters are used for precise attitude control for establishing a communication link with Earth (assumed as once per month during the initial transfer), positioning maneuvers at the target bodies and desaturation of the reaction wheels. The thrusters are mounted in pairs to ensure redundancy and allow the throughput of all the necessary propellant. 50 kg of hydrazine are stored in an extra tank.
- Instruments: A sample payload with 6 instruments is foreseen, taking heritage from previous mission instruments like VIR (camera system for visual and infrared, heritage: *Dawn*), LORRI (high resolution camera, also used for navigation, heritage: *New Horizon*), GRaND (subsurface instrument, neutron detector and x-ray spectrometer, heritage: *Dawn*), NLR (laser altimeter, heritage: *NEAR*), MERTIS (high resolution thermal mapping system and radiometer, heritage: *BepiColombo*). Possibly, depending on available data of the targets, a number of penetrators as discussed earlier can be included.

The spacecraft will operate in different modes during the mission, e.g. defined by the science tasks, communication or transfer to or between the Trojan asteroids. Based on the subsystem description above, the largest demands are generated—clearly dominated by the operation of the electrical thrusters—during the operation in transfer. On average during the transfer (of approx. 4.6 years, see Sect. 19.3.1) to the Trojans this leads to a demand of about 7.3 kW. This is dominated by the thrusting of the electrical engine, some communication, which is mostly house-keeping data as no science is conducted before arrival at the Trojans, attitude control, linked to the alignment of the spacecraft for contact with Earth and for thrusting and of course thermal control tasks. Similarly the transfer between Trojans causes a demand of about 1.1 kW of power on average.

The solar arrays are oversized for the operation without thrusters, i.e. they provide sufficient for operating the spacecraft, allowing simultaneous operation of all instruments during most times. Once the spacecraft reaches its final target and has not enough fuel reserves for reaching another one, it could for a long time remain operational and provide data about that last target, based on the available solar power. Alternatively it could be used to impact a Trojan and observation of the exhaust from Earth could provide further information about the Trojan's sub-surface composition.

Considering the maximum allowable launch mass for a Soyuz-Fregat launcher of 1600 kg for Earth escape and launch from Guiana Space Center (Arianespace 2012) the mass presented here, 1566 kg, fulfills the requirement of using this rocket for launch and thus achieves moderate launch costs.

19.2.6 Cost Estimation

The described spacecraft design and the mission as described in the following sections, allow for a rough cost estimate. One unique element of this mission is the usage of solar electric propulsion at a solar distance of up to about 5.2 AU. Currently the most analogous mission is *Dawn*, a two-rendezvous mission to small bodies in the main asteroid belt. So for a cost estimation, this mission can be used.

Dawn's operation and launch costs were 76 Million USD for a mission duration of 7.83 years (NASA 2007). Scaling this to the mission time for the proposed Trojan mission (see following sections), i.e. 15.6 years (Table 19.4), the costs could be analogously estimated to about 150 Million USD. Obviously the launch costs would not be scaled up due to mission duration, but the operations costs alone are not available as a number, thus this creates an extra margin. The intensity of the spacecraft operations changes during the course of a mission and are usually low during transfer times, where most instruments are inoperative. Just like *Dawn* the proposed *TRIP* mission contains several years of transfer times. An accurate cost estimate is currently not possible, based on actual expected operations intensities, as no sufficient cost data is available.

Development and construction of *Dawn* took 282 Million USD. While especially for instruments development effort can be saved based on heritage, other components, e.g. the solar array are still far from mission readiness, therefore a similar sum is assumed for the Trojan mission. This adds up to about 432 Million USD. This does not contain inflation rates for ten years between now and *Dawn*'s launch and neither any consideration regarding the European aerospace industry. Adding a margin of about 20% would result in costs of about 520 Million USD.

19.3 Trajectories for Multi-rendezvous Missions

Besides the spacecraft design, the trajectory for this multi-rendezvous mission has to be found. As no real launch date can be set without the mission being approved, only estimates or examples can be given.

Applying evolutionary neurocontrol, i.e. an artificial neural network in combination with evolutionary search algorithms, as described in Dachwald (2003), can provide such trajectories. The following trajectory examples have been calculated with this method. Generally the targets for such a mission need to be chosen scientifically. The targets here presented are only provided as examples and proof of concept and it is assumed that the spacecraft has escaped Earth using the upper stage of the launcher (therefore the launch mass limit was set to 1600 kg).

19.3.1 Transfer to Trojans

Using a maximum thrust of 175 mN (specific thrust of 25.57 mN/kW) and a constant specific impulse of 4200 s for the thruster model which would be align to the existing RITA-22 thruster, a first target could be Nestor. Reachable within 4.6 years from Earth at a launch date in July 2017 (depending on the synodic period, launch opportunities repeat approximately every 12 years). Arriving at the end of March in 2022, *TRIP* would have spent a Δv of about 18,188 m/s, i.e. for this example a fuel mass to dry mass ratio of 0.357. This trajectory leg is depicted in Fig. 19.5.

Two thrusters are operated at about their maximum of 12 kW at the beginning, i.e. in Earth's solar distance and at a thrust of 170 mN, which drops to the operation of one thruster close to Nestor at 2 kW and the resulting thrust of about 55 mN. Based on an operation temperature of 123.5 K, a power proportional to $1/r^{1.6}$ is assumed, with r as the solar distance.

Another opportunity would be targeting 1997 UL_{16}, launching from Earth at the end of April 2026 and reaching the small body after 5.85 years, assuming a maximum thrust of 200 mN and a specific impulse of 5000 s at Earth distance, i.e. the properties of the previously mentioned proposed RITA-150. In this case a specific thrust of 33.3 mN/kW is assumed with the same power law as before. Similarly two thrusters are initially operated at maximum power of 6 kW, at the

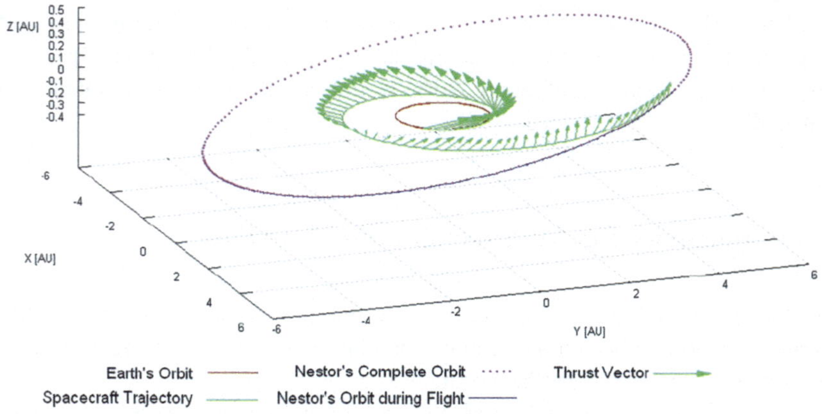

Fig. 19.5 Example for a Transfer to the first target. *TRIP*'s transfer to its first target, here Nestor within 4.63 years and requiring a Δv of 18,188 m/s. The arrows represent the relative thrust magnitude and direction

target body ca. 2.6 kW remain to power a single thruster. This trajectory would require more time at a similar Δv in this case 18,258 m/s, but overall allowing more rendezvous subsequently, due to a better position.

These two scenarios showcase that a multi-rendezvous mission is possible applying solar-electric low-thrust.

19.3.2 Transfer Between Trojans

In both cases the dwelling time at the first target before leaving for the next, would be similar: 401 and 399 days. The first leg of any such mission is mostly defined by the adaptation of the solar distance and in part also the inclination, therefore requiring the majority of the fuel. As the solar distance and angular position of the Trojans are quite similar, transfer between the Trojans is mostly defined by the inclination change, the radial component is not altered significantly, see Fig. 19.6.

Obviously the exact nature of the transfer between the Trojans depends on the individual targets. Which targets are to be selected depends on the scientific considerations. Two mission leg summaries are provided for three (Table 19.3) and four (Table 19.4) rendezvous, depending on the first target body and differing in the respective body sizes.

The major difference between the two approaches is differing thruster properties and the target body sizes. The three rendezvous mission would allow visiting bodies with a wider range of body diameters. For this scenario a RITA-15 thruster (Hasssewitz 2006) is assumed, with a minimum thrust of 15 mN and a specific impulse of 2600 s. The four body mission has been calculated with the proposed RITA-150 thruster, which is based on RITA-15. This has been done with a thrust of 28 mN and a specific impulse of 3800 s.

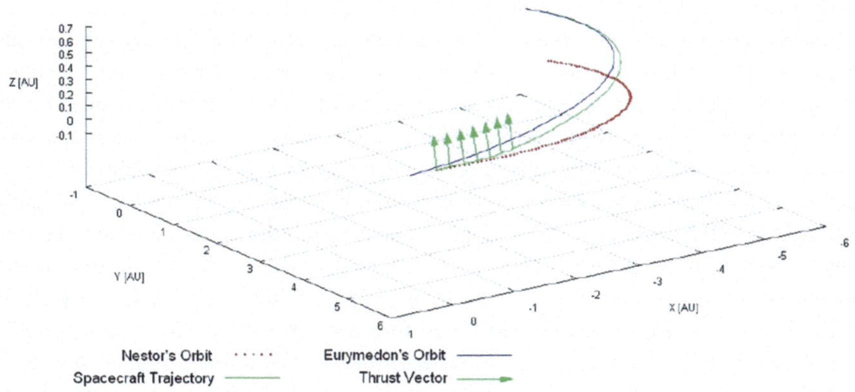

Fig. 19.6 Example for a Transfer to the second target. *TRIP*'s transfer to its second target, here from Nestor to Eurymedon, mostly by adapting the inclination. The arrows represent the relative thrust magnitude and direction

Table 19.3 Mission sequence for three target bodies

Start body	Target body	Launch date (D/M/Y)	Arrival date (D/M/Y)	Flight duration (year)	Duration of stay (days)	Δv (m/s)
Earth	Nestor	16/07/17	27/03/22	4.63	401	18,188
Nestor	Eurymedon	02/05/23	01/11/26	3.5	1107	666
Eurymedon	Irus	12/11/29	31/01/32	2.22	Indefinite	863

The trajectory data for each leg is presented for a mission with 3 rendezvous in favor diversity in targets. Based on magnitude the targets have diameters of 109 km (Nestor), 19 km (Eurymedon) and 30 km (Irus). Maximum thrust at Earth distance is 175 mN at 4200 s of specific impulse; for Jupiter's solar distance the values are 15 mN at 2600 s

Table 19.4 Mission sequence for four target bodies

Start body	Target body	Launch date (D/M/Y)	Arrival date (D/M/Y)	Flight duration (year)	Duration of stay (days)	Δv (m/s)
Earth	1997 UL_{16}	30/04/26	04/03/32	5.85	399	18,258
1997 UL_{16}	2001 BH_4	07/04/33	18/0936	3.61	298	1494
2001 BH_4	1999 XM_{71}	13/07/37	03/07/39	1.97	108	1618

The trajectory data for each leg is presented for a mission with 4 rendezvous at the cost of diversity in targets. Based on magnitude all targets have a diameter of approximately 13–35 km. Maximum thrust at Earth distance is 200 mN at 5000 s of specific impulse; for Jupiter's solar distance the values are 25 mN at 3800 s

To simulate solar cell degeneration in both scenarios transfer between the Trojans has been calculated using a power proportional to $1/r^{1.7}$ resp. $1/r^{1.8}$ for the last transfer.

In both cases it can be seen that using solar electric propulsion, multi-rendezvous mission are possible with mission Δvs between 20 and 24 km/s.

Further savings can be obtained by applying gravity assist maneuvers. While gravity-assist in combination with low-thrust trajectories is not yet a mastered optimization task, using the method described in Maiwald (2016) it is possible to find trajectories from Earth to Jupiter with gravity-assists at Earth saving approximately 4 km/s. With an adapted phasing a similar gain could be achieved for a transfer to the first Trojan target body and thus increase the number of reachable bodies. Due to the close proximity of the Trojans within the population, leaving that population to conduct another gravity-assist for a transfer to the next Trojan is not considered advisable and at the same time Trojans bodies are small (only some hundred km in diameter at best) and therefore are not suitable for gravity-assists themselves. Judging from the trajectories described in Table 19.4, with Δv demands for the transfers between Trojans between 1.5 and 2.5 km/s at least two more rendezvous could be possible by saving 4 km/s in the initial transfer to the L4 region, using a gravity-assist. The transfer Δvs from Table 19.3, about 700–900 m/s suggest that more is possible if a suitable mission body sequence can be found.

19.4 Opportunities for Future Missions

The described mission would first of all be a scientific mission but could act as door opener for further undertakings in the Trojan populations. Once a more thorough understanding of the surface properties, e.g. material strength and composition, has been gained, it can be investigated how the material found on Trojans can be exploited and possibly harvested.

If Trojans are indeed captured comets, volatiles could be found on these bodies, including water, which could be used for power generation (e.g. for nuclear fusion), oxygen generation, fuel generation (e.g. hydrogen-based propellants like methane) and so forth. This could allow refueling a spacecraft based on chemical propulsion.

The close proximity and the moderate Δv required for a transfer between the Trojan asteroids (see Tables 19.3 and 19.4) could make it possible to harvest such resources even with a single spacecraft on a similar mission as described for *TRIP*. Depending on the exact capabilities of the spacecraft it could also be possible to harvest material on one Trojan asteroid, convert it into fuel (e.g. hydrogen) and conduct a maneuver for a rendezvous with another Trojan body, allowing a sustained exploration of these bodies.

A thorough investigation of the Trojan population would increase our understanding of the solar system's formation. The close proximity of the Trojans to Jupiter could also allow refueling of spacecraft targeted for Jupiter to extend their own time of operation there.

References

Arianespace.: Soyuz user's manual, Issue 2 (2012)

Dachwald, B.: Low-thrust trajectory optimization and interplanetary mission analysis using evolutionary neurocontrol. Dissertation, University of the Federal Armed Forces, Munich (2003)

Dotto, E., Fornasier, S., Barucci, M.A., Licandro, J., Boenhardt, H., Hainaut, O., Marzari, F., de Bergh, C., De Luise, F.: The surface composition of Jupiter Trojans: visible and near-infrared survey of dynamical families. Icarus **183**, 420–434 (2006)

Emcore Corporation.: ZTJ space solar cell datasheet. http://www.emcore.com/solar_photovoltaics/space_solar_cells. Space and Satellite Solar Cells Emcore Corporation. Accessed 5 May 2011

Fatemi, N., Sharma, S., Buitrago, O., Crisman, J., Sharps, P. R., Blok, R., Kroon, M., Jalink, C., Harris, R., Stella, P., Distefano, S.: Performance of High-Efficiency Advanced Triple-Junction Solar Panels for the LILT Mission Dawn. Jet Propulsion Laboratory (2005)

Fleming, H.J., Hamilton, D.P.: On the origin of the Trojan Asteroids: effects of Jupiter's mass accretion and radial migration. Icarus **148**, 479–493 (2000)

Fornasier, S., Dotto, E., Marzari, F., Barucci, M.A., Boehnhardt, H., Hainaut, O., de Bergh, C.: Visible spectroscopic and photometric survey of L5 Trojans: investigation of dynamical families. Icarus **172**, 221–234 (2004)

Hasssewitz, H.: RITA—the ion propulsion system of the future. EADS Astrium (2006)

Jewitt, D., Trujillo, C., Luu, J.: Population and size distribution of small Jupiter Trojan asteroids. Astron. J. **120**(2), 1140–1147 (2000)

Jewitt, D., Sheppard, S., Porco, C.: Jupiter's Outer Satellites and Trojans, Jupiter—The Planet, Satellites and Mangetosphere, Cambridge Planetary Science, vol. 1. Cambridge University Press, Cambridge (2004)

Kührt, E., Knollenberg, J., Keller, H.U.: Physical risks of landing on a cometary nucleus. Planet Space Sci. **45**(6), 665–680 (1997)

Maiwald, V.: Analysis and design of a solar-electric propulsion mission to Jupiter's Trojans. Diploma Thesis, RWTH Aachen, 23 Sept 2009

Maiwald, V.: Visiting troy: the solar electric Trojan investigation probe to Jupiter's Trojans. J. Br. Interplanet. Soc. **68**(8), 242–250 (2015)

Maiwald, V., Braukhane, A., Quantius, D., Reershemius, S.: Design of a multi-rendezvous mission to Jupiter's Trojans. In: 63rd International Astronautical Congress (IAC), Naples, Italy (2012)

Maiwald, V.: About combining tisserand graph gravity-assist sequencing with low-thrust trajectory optimization. In: 6th International Congress on Astrodynamics Tools and Techniques, Darmstadt, Germany (2016)

Marzari, F., Farinella, P., Davis, D.R., Scholl, H., Campo Bagatin, A.: Collisional evolution of Trojan Asteroids. Icarus **125**, 39–49 (1997)

Marzari, F., Tricarico, P., Scholl, H.: Clues to the origin of Jupiter's Trojans: the libration amplitude distribution. Icarus **162**, 453–459 (2003)

Morbidelli, A., Levison, H.F., Tsiganis, K., Gomes, R.: Chaotic capture of Jupiter's Trojan Asteroids in the early solar system. Nature **435**, 462–465 (2005)

Mottola, S., Di Martino, M., Erikson, A., Gonano-Beurer, M., Carbognani, A., Carsenty, U., Hahn, G., Schober, H., Lahulla, F., Delbo, M., Lagerkvist, C.: Rotational properties of Jupiter Trojans. I. Light curves of 80 objects. Astron. J. **141**, 170 (2011)

NASA.: Mars Reconnaissance Orbiter Launch Press Kit (2005)

NASA.: Dawn Launch Press Kit (2007)

NASA.: Juno Launch Press Kit (2011)

Prettyman, T.H., Barraclough, B.L., Feldman, W.C., Baldonado, J.R., Bernardin, J.D., Dingler, R. D., Enemark, D.C., Little, C.K., Miller, E.A., Patrick, D.E., Pavri, B., Raymond, C.A., Russell, C.T., Storms, S.A., Sweet, M.R., Williford, R.L., Wong-Swanson, B.: Gamma ray and neutron spectrometer for dawn. Lunar Planetary Sci. **XXXVII** (2006)

Rayman, M.D., Fraschetti, T.C., Raymond, C.A., Russel, C.T.: Dawn: a mission in development for exploration of main belt asteroids Vesta and Ceres. Acta Astronaut. **58**(11), 605–616 (2006)

White, S., Spence, B., Trautt, T., Cronin, P.: UltraFlex-175 on space technology 8 (ST8)—validating the next—generation in lightweight solar arrays (2007)

Part V
Business Cases for Resource Utilisation

Chapter 20
Implausible, Yet Intriguing: Business in the Outer Solar System

Mike H. Ryan and Ida Kutschera

20.1 Introduction

There is an old adage that life abides. What is often meant by this is that life seems to propagate in the most unlikely of places. The more improbable the location, one might conclude, the less likely to find living organisms. However, on this planet, we found life in the depths of the ocean among smoking volcanoes, as well as on the edge of the earth's atmosphere. It appears that the unlikely event is when we do not find living organisms of one sort or another in even the most inhospitable environment. A similar analog can be found when discussing the possibility of business and all the various forms that business operations might take. There is a common factor across all business operations and that common factor is human beings for whom business, at least in its most general sense, is as universal as communication. Put two human beings together and they will try to communicate. Put several human beings together, and sooner or later they will talk about trading or buying or otherwise acquiring things from each other. Clearly, business activity is an essential component of the human condition. So where people go, so goes business. As people migrated across the planet, they took their business activities with them. Trade and commerce are a constant of human interaction regardless of culture, location, language, or circumstance. It is doubtful that this characteristic will go away anytime in the foreseeable future. Therefore, as people migrate outward from the planet Earth, their innate interest in business will accompany them. As those interests become reality, the role of institutions that provide the foundations for

M. H. Ryan (✉) · I. Kutschera
Rubel School of Business, Bellarmine University,
Louisville, KY 40205, USA
e-mail: mryan@bellarmine.edu

I. Kutschera
e-mail: ikutschera@bellarmine.edu

V. Badescu and K. Zacny (eds.), *Outer Solar System*,
https://doi.org/10.1007/978-3-319-73845-1_20

business such as contracts, property rights, credit and enforcement of agreements etc. will need to expand to meet the requirements of doing business in space.

The opportunities regarding working in space, on the Moon, on Mars and on asteroids are well documented (Ryan and Kutschera 2007, 2013). Those opportunities within the inner solar system have been described as unlikely in the near-term but possible in the long-term (Ryan and Kutschera 2015). The same can be said of opportunities represented by the planets and moons of the outer solar system. For example: "the Jupiter Trojan asteroids…, must be extremely volatile-rich, including both periodic comets and ice-rich asteroids … and some models suggest that the total mass of material in the Trojan clouds is comparable to the entire Asteroid Belt" (Lewis 2015, p. 134). NASA has announced a mission, Lucy, set to launch in 2021 to visit the Trojan asteroids located in the belt between Mars and Jupiter. A second mission, Psyche, is intended to study a metallic asteroid named 16 Psyche in the same asteroid belt (NASA). The scientific information (e.g., mineral composition) obtained through such missions will be helpful in determining the business potential of asteroids. An artist's concept of these potential resources is shown in Fig. 20.1.

What makes a discussion of the business possibilities inherent in the outer solar system so interesting is that the conversation has already gone on for a considerable period. Writers of science fiction have long used the stage provided by the outer solar system as a backdrop for their stories of human interaction. The needs of a human population for information, goods and food put commercial activities as essential elements in the stories across generations of science fiction authors. "What makes science fiction valuable is not that it produces predictions, but that it provides inspiration" (Brown and Logan 2015, p. 32). Regardless of the specifics of an unfolding future, science fiction gives both authors and readers the opportunity to

Fig. 20.1 Jovian Trojan Asteroids (Artist's Concept). *Source NASA/JPL-Caltech*

reflect creatively on what is and what might be. The unfolding of the pattern is often parsed from what we expect as well as those things we did not see coming. For example, we envision better automobiles but fail to consider traffic jams. Improved transportation might mean more opportunities to live further from your place of employment but ignores issues of urban sprawl. Still, science fiction increases the scale for human ponderings and the futures that might be. In that sense, outer space provides the ultimate backdrop for wondering about things to come, not just in terms of technology but also in terms of human interactions and commerce.

According to the Encyclopedia of Science Fiction, little attention has been paid to the outer solar system planets, Jupiter, Saturn, Uranus, or Neptune as compared to planets such as Mars (Science Fiction Encyclopedia 2016). Even Pluto, when it was considered a planet, received comparatively more attention. Moons of, or tangential areas to, the primary planets of Jupiter and Saturn have garnered more interest, in part because of their possible suitability for habitation, terraforming or resource development. One of the more interesting stories from a business perspective is Isaac Asimov's *The Martian Way*, which described how a potential water crisis on Mars resulted in the need to transport ice from the rings of Saturn back to Mars to replace water being shipped from earth (Asimov 1952).

"Of course, once we get this thing really organized, we can have water stations on the moons of Saturn and Jupiter and on the asteroids. We can scale in chunks of Saturn's rings and pick them up and send them on at various stations" (Asimov 1952, p. 48).

Technical issues aside, the potential of using frozen water shipped between planets for multiple uses is not beyond the realm of possibility. Whether or not it makes business sense will remain an interesting question for the near future. Regardless, the resources of the outer solar system remain an untapped reservoir of potential opportunity that has been recognized in some form for over a century.

Fig. 20.2 Asteroid Harvestor Concept. *Credit* Bryan Versteeg/Deep Space Industries

Given the proposals for mining the Asteroid Belt, the possibility of even greater amounts of valuable resources in the vicinity of the outer solar system will eventually attract the attention of investors and prospector firms. A concept picture from Deep Space Industries of a possible mining expedition is shown in Fig. 20.2.

One of the more recent efforts to illustrate living in the outer solar system has been James S. A. Corey's *The Expanse* series, a television drama which takes place within developed areas of the asteroid belt, several hundred years into the future. The inherent difficulty in predicting what the future will look like is that it is seldom what was envisioned, even by those with a solid understanding of the possible technologies, environments and opportunities.

Linear developments are comparatively easy to predict even if the technology that might make it work is not. Predicting faster transportation has generally been easy until the technologies required push against a specific barrier. Faster aircraft over time were the norm until the Concorde was retired and no follow-on aircraft were in production. The technology was available to produce high speed aircraft but other factors such as noise, air pollution, and cost limited general interest and as a result commercial planes do not presently fly faster than the speed of sound. The relative advantage that extra speed would provide for commercial users is not currently supported by a commensurate increase in profit or reduction in costs sufficient to justify the added expense of high speed aviation. However, this does appear to be changing somewhat as commercial business aviation is exploring the potential of high speed business aircraft that could take advantage of research that promises to reduce sonic boom issues and increase fuel economy and aircraft range. It is also useful to remember that the Concorde's transatlantic flights were lucrative in themselves even if the overall return on investment including research and development was not (Glancey 2015, p. 162). The relative few planes produced and the limited routes capable of being flown without restriction made overall cost recovery virtually impossible. Perhaps future space entrepreneurs should heed the lessons of the Concorde experience as they begin to deploy either crewed or remote operated mining vessels.

20.2 The Future of the Outer Solar System in Fiction

In the realm of science fiction, the outer solar system has always had an attraction. Although the realities of current knowledge about the moons of Jupiter and Saturn preclude some of the optimistic scenarios regarding their possible habitation, authors such as Robert H. Heinlein made it clear that these moons represented future resources for humanity. Heinlein's book *Farmer in the Sky* which highlights Ganymede, is representative of the need for humanity to expand outward but also reflects the basic requirements for doing so (Heinlein 1950). The simple exercise of growing food and supporting a growing population in a remote location is viewed not much differently than that of terrestrial enterprises of a previous century. Whether or not space organizations will need to actively seek farmers (as depicted

Fig. 20.3 NASA Fictional Recruiting Poster. *Source* NASA

in the fictional NASA recruiting poster shown in Fig. 20.3) will remain to be seen. It is interesting to note that with each generation of exploration, organizations (government or private) went to great lengths to ensure that their settlers had a mix of skills and abilities to increase the group's probability of success.

Good fiction and good science do not necessarily travel hand-in-hand but there are examples that encompass both, such as those compiled by Andrew Franknoi in *Science Fiction Stories with Good Astronomy and Physics: A Topical Index* (Astrosociety 2016). Science fiction has long provided a mechanism to explore the human condition at points removed from present day experience, thus putting a light on various "then present day" activities. The technology that is often the hallmark for science fiction is frequently secondary to the interplay of the characters and their circumstances. In one sense, behavior is written to be human-like even if the principal characters are non-human. Writers attempt to keep human behavior consistent with the general reader's experience until a truly alien perspective enters the storyline. To consider the possibility of aliens, although entertaining, is even a leap for this speculative chapter on future business opportunities in the outer solar system. For now, it is sufficient to state that there are numerous science fiction stories that clearly point to a future in which humans inhabit, function and conduct business within the outer solar system. These speculations have long powered men's and women's imaginations as to the possibility of travel to the moons of Jupiter and Saturn, and ultimately the notion that somehow those locations would be rendered suitable for human use. The science fiction of today is quite often a

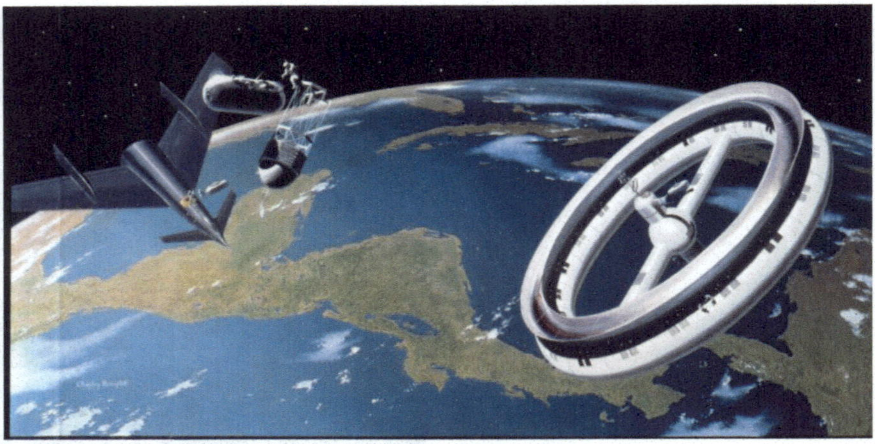

Fig. 20.4 Space station, ferry rocket and space telescope 1075 miles above Central America, 1952. *Source* Reproduced courtesy of Bonestell LLC

harbinger of the futures to come. It is easy to point to the writings of Jules Verne and H. G. Wells as writers of science fiction whose stories pointed to significant events and technologies in their futures. There are numerous sources for evidence that the future that is predicted is not always the future that happens (Benford 2012, pp. 162–167).

Space travel is but one facet of a multifaceted collection of events that will make up the future of humanity but remain among the most enduring of its collective dreams (Ordway 2000). For many individuals that grew up during the end of the twentieth century, the images of Chesley Bonestell (Miller 2016, pp. 126–127) and Fred Freeman (Ordway 2000, pp. 132–133, 136) became their direct link to the future of humans in space (Ordway 2000, pp. 126–168). Those images of what might be still drive imaginations toward the outer edges of space including the outer solar system (McCurdy 1997). The classic Chesley Bonestell image shown in Fig. 20.4 is one that has resonated in the imaginations of those who view the possibilities of space with excitement.

20.3 Exploratory Snapshot—An Overview of the Outer Solar System (Lunar and Planetary Institute 2016)

For purposes of exploration, the outer solar system is target rich. Scientific curiosity alone points to the volcanoes on Io, the storms on Titan, the rings around Uranus and whether or not Ariel is a frozen version of Enceladus. Resources notwithstanding, the scientific community would like to study the geysers on Triton, the plumes of Enceladus, and the magnetosphere of Neptune. Added to this mix is the desire to learn more about Europa, a moon that could conceivably have life today in a subsurface ocean. With opportunities to explore four planets, their moons, and an assortment of asteroids which have an incredibly diverse set of atmospheric, geological and surface conditions, scientific inquiry will be busy for a considerable time. If pure science generated revenue, then the opportunities provided by the outer solar system would produce virtually unlimited wealth. Still, the immense amount of new knowledge that can be obtained from a systematic exploration of the outer solar system makes it an attractive scientific target.

20.4 Questions from the Scientific Community

Some of the fundamental scientific questions likely to provide opportunities for translational research (Cremades et al. 2015) include:

- Understanding *atmospheric circulation.* The giant outer planets have entirely different atmospheres than the terrestrial planets; understanding them means developing advanced fluid dynamics models (that have been applied, for example, to ocean currents) and potentially providing new discoveries related to climate change.
- *Weather* more akin to earth with the ability to help create new weather models. For example, Earth is at the mercy of processes today that are taken to extremes on Saturn's moon, Titan. "On Earth, most of the greenhouse effect is due to water vapor, and it is the water vapor that condenses to form clouds of water droplets or ice, sometimes to fall down to the ground as rain, hail or snow. On Titan, the role that water takes on Earth is filled by methane….the role of ozone in temperature rise is taken by the haze on Titan" (Lorenz and Mitton 2008, p. 35). Careful study of Titan would be relevant to the understanding of the evolution of Earth.
- *Atmospheric chemistry and astrobiology.* The reducing atmospheres in the outer solar system are home to a vigorous organic chemistry that does not occur in the inner solar system in the present day, providing an opportunity to study natural production of biological building blocks.
- *Magnetospheres.* The variety of atmospheric types and structures within the outer solar system tests our models and understanding of how our own magnetosphere is structured.
- *Materials' behavior* in extreme pressures and temperatures not natural on earth. For example, ice behaves like rock at outer solar system temperatures, but the interior of many moons might be liquid.
- *Exoplanets.* Most of the new planets discovered around other stars are similar to Uranus and Neptune where further study could lead to new insights into planetary origins.

20.5 The Use of the Larger Planets—Jupiter, Saturn, Uranus, Neptune

There are places that scientific curiosity will eventually explore. The outer solar system's planets including Jupiter, Saturn, Uranus and Neptune are such places. They are far enough away from the sun so as to retain their primordial atmospheres which in itself gives science a view backward in time that could be helpful in understanding the processes that formed the earth. With atmospheres dominated by hydrogen and helium, extreme gravity and severe atmospheric turmoil, the outer planets are acutely inhospitable to human activity. Although these planets' atmospheres surround comparatively dense cores, in a real sense, there is no solid surface upon which to land or build. These giant planets are by their nature uninhabitable in the conventional sense. Even without the obvious limitations due to distance and current technology, there are some places that are clearly beyond any form of reasonable business exploitation at present. Scenarios that suggest the eventual mining of gasses from their atmospheres are speculative to a degree not present for other planetary objects that could be mined in a more conventional manner. As Jupiter, Saturn, Uranus and Neptune are not candidates for either habitation or direct exploitation, the discussion of business potential remains with wandering asteroids and/or with the moons which orbit these major outer planets. Even though Uranus is thought to have two previously undiscovered moons, it is the moons of Jupiter and Saturn that will most likely receive the initial attention. It is these moons, long the focal points for various science fiction scenarios, which represent the greatest potential for exploration, habitation, and development.

New discoveries could easily change that expectation but based upon what is known to date, a location within the orbit of either Jupiter or Saturn will likely display the first "Open for Business" sign. The Chesley Bonestell image of "Saturn from Mimas" (1944) (see Fig. 20.5), although somewhat stylized, does show a landing crew perhaps surveying possible locations for a future base or even a mine. The possibility of traveling to the bodies that make up the Outer Solar System as a dream and as a concept has been in evidence a very long time.

20.6 Points of Commerce—To the Moons!

One of the key considerations for business development is location. Taking into account the questions from the scientific community and extrapolating outward, it could be expected that crewed missions to Saturn's moons, Titan and Enceladus, and Jupiter's moons, Europa and Callisto, will eventually occur. NASA has already envisioned a number of Outer Solar System moon missions including one to Callisto. Eric Seedhouse uses that proposed mission as the starting point for his discussion of the exploration of the outer solar system and provides a detailed discussion of both the mission and its challenges (Seedhouse 2012). The difficulties

Fig. 20.5 Saturn as seen from Mimas, 1944. *Source* Reproduced courtesy of Bonestell LLC

to be encountered are significant but not insurmountable given sufficient time and resources. The technology needed to propose long-duration, crewed missions to the outer solar system is not quite within reach but certainly within sight; the likelihood of such missions in a decade or two is within the realm of possibility.

The initial voyages of exploration, such as prestige missions, i.e. to plant the flag, are unlikely to reflect the basic "let's do business" opportunities that will eventually develop. Depending on the origin of those initial exploratory efforts in terms of public versus private operations, business opportunities could proceed slower or faster. If the private firms engaging in space related activity continue at their present pace, there is the distinct possibility that private exploration could surpass government sponsored programs in the near future.

Among the more obvious opportunities would be the pursuit of new raw materials plus the old standbys of gold, silver, platinum and iron. The potential for resource development would factor in quite nicely, in part, to offset the need for an ongoing supply of raw materials. In many ways such business opportunities would closely mimic business activities of previous centuries as exploration moved hand in hand with exploitation and utilization (Ryan and Kutschera 2015, p. 398). While ownership of non-terrestrial resources is far from settled, accurate surveys of space objects, particularly moons and large asteroids, would be critical to any future determination of ownership or utilization. As the NASA fictional recruiting poster illustrates (see Fig. 20.6), the locations and equipment may change but the need for surveyors will remain. Surveyors would follow initial explorers as they pretty much always have to mark the territory for development and use.

Fig. 20.6 NASA Fictional Recruiting Poster. *Source* NASA

20.7 Initial Steps Toward Industrialization

The steps of exploration and discovery for the moons of the outer solar system are comparatively straight forward. Experience to be obtained through development in near Earth orbit, on Earth's Moon and probably through Mars exploration will guide those venturing into the outer solar system. Exploratory vehicles both crewed and robotic will traverse the space between the Earth and the more promising locations among the outer planets. As previously suggested, not all locations will be suitable for development or habitation. Fortunately, science and technology rarely go on hiatus. Between now and the initial trips to the moons of the outer solar system, technology will continue to advance making such trips more the provenance of the prepared and resourceful than that of the crazy or wishful thinkers.

There are numerous terrestrial examples that suggest how, over time, technology and experience develop to overcome seemingly impossible hurdles. One of those with a direct connection to space travel is under water exploration and construction. Often referred to as the journey into inner space, human experience under water has followed a developmental path that now allows divers to explore and work at depths below 2000 ft. Construction, maintenance and repair of equipment are regular occurrences at depths in excess of 1000 ft in oceans around the world. Remote equipment allows exploration and work to occur at virtually any depth if the cost justifies the expenditure of funds needed to develop whatever special equipment might be required for a specific project.

As Figs. 20.7 and 20.8 illustrate, the difference between what is real and what is possible is frequently only a matter of perspective or time. Commercial divers are eerily suggestive of the space-based technicians to come as seen in one of NASA's fictional recruiting posters. From suits to tools, it is not very difficult to imagine that future space based technicians, mechanics or assembly workers will closely mirror those now working under water. The lessons learned from working in one difficult environment (under water) should inform those preparing to undertake similar construction projects whether in space, on an asteroid or a moon of an outer planet. The dangers may be different but the practical problems of doing the actual work will be quite similar.

In comparable terms, crewed space efforts operate within a narrow band of space roughly equivalent to the shallow areas of the continental shelves for underwater operations. The difficulty is not in having the technology to operate beyond low earth orbit but in the limitation of current technology to navigate easily to and move beyond low earth orbit. Unlike the oil industry which had incentives to find ways of extracting oil from increasingly difficult locations including the depths of the ocean, there is a relative lack of similar revenue generating opportunities that would encourage investors to seek out space-based business opportunities.

Fig. 20.7 Working Navy Divers. *Credit* (U.S. Navy photo by Mass Communication Specialist 1st Class Charles White/ Released)

Fig. 20.8 NASA Fictional Recruiting Poster. *Source* NASA

Scientific curiosity alone has been insufficient to generate the interest, investment and technology needed to push humanity outward into space. As new opportunities present themselves for industrial development in space, those efforts will follow many of the activity and technology patterns common to other industries. There are numerous approaches for predicting how industries might develop; however, road mapping frameworks have proven quite useful in evaluating and understanding historical industrial emergence (Phaal et al. 2011). Using appropriate analogs should be extremely useful in keeping the industrial space effort on track while avoiding many of the pitfalls common to other developing industries (Launius 2014).

20.8 Industrial Development Considerations—A Few Examples

Excluding the possibility of high speed solar system travel, the time required to transport people, goods, materials and whatever else is needed for survival will remain extensive. Therefore, both travel to and from the remote outposts will require a level of redundancy unparalleled when compared to previous explorations. Outposts, once established, must also retain a significant measure of self-sufficiency. Even with a constant stream of supplies moving outward, the challenges of maintaining ongoing life support in the face of the unanticipated should never be underestimated. Simple events, unacknowledged risks, human mistakes and even hubris could prove to be as fatal as the most catastrophic accident or natural disaster that can be imagined. Things happen, and it is probable that people will die. The inherent risk in exploration, particularly in severe environments, will never be zero. In fact, attempting to reduce the risk to near zero could become a serious impediment to both exploration and development (Simberg 2014). But, planning for the possibility of failure including disrupted supply chains, equipment failures, physical injuries, psychological issues and medical emergencies can be and should be a constant part of the movement into the outer solar system.

20.8.1 Industrial Space Ships

Although the size of space-going industrial facilities might sound daunting at first, building vessels on a comparable scale is already occurring. Technology and the availability of funding have developed to the degree to make the construction and deployment of passenger and cargo ships of a size that would have been hard to visualize even a generation ago. Imagine if the Harmony of the Seas (Fig. 20.9), a Royal Caribbean cruise ship, of 226,963 gross tons, 362.12 m in length, with a crew of 2,100 were reimagined as an industrial processing facility in orbit around

Fig. 20.9 Cruise Ship, *Harmony of the Seas*. *Credit* Roger Clark ARPS/Shutterstock, Inc

a moon or attached to a large ice asteroid—mining the ice for conversion to useable water or breaking it down into oxygen and hydrogen for fuel or breathing (DNV GL 2016).

If that is not large enough, Maersk Line recently constructed four of its Triple-E class ships; each vessel is 400 m long and 59 m wide. Each ship can carry "eighteen thousand containers and would be able to fit a full-sized American football field, an ice-hockey arena, and a basketball court in their holds, if they care to" (George 2013, p. 7).

The photo of the Maersk Mc-Kinney Moller (Fig. 20.10), considered the largest ship in the world, illustrates the current scale of human engineered transportation systems. Similar systems can be imagined for space-based operations that would be alike in both scale and scope.

The large scale movement of goods and materials via container ships is a reasonable model for the transport of materials to and from the outer solar system. Containerized cargo changed the shipping industry on earth, and the benefits of this approach should translate well to an industrialized space environment (Cudahy 2006).

Modular containers, probably color coded, with transponders could be moved with comparative ease to specific points where business activity needs materials. Large support vessels, similar to their terrestrial counterparts, could be constructed in space to support a variety of business activities and business models depending on need and available investment.

Large scale mobile facilities have been built for a variety of terrestrial industrial applications. Flotels (mobile crew quarters) for the oil industry have been

Fig. 20.10 Container Ship, Maersk Mc-Kinney Moller. *Credit* Gerard Koudenburg/Shutterstock, Inc

constructed and transported for use close to off-shore oil production operations. As Fig. 20.11 illustrates, a Flotel is connected to production facilities allowing offshore workers to reside in comparative comfort during their off hours. Premanufactured and moved to needed locations, Flotels provide resident needs in terms of food, sleeping, recreation and privacy for a workforce that operates 24/7.

In a similar fashion, other factory or industrial ships have been constructed for various purposes including, at-sea processing as in the fishing industry, off-shore production as used by Chinese firms and even power production as seen in both floating power plants built for Turkey or Russia (Motorship 2016). Specialized industrial facilities built on (or into) ships provide flexibility in that they can be anchored in one location and moved as need or opportunities develop elsewhere. That type of flexibility will extent to space-based industrial facilities that could be literally moored to an asteroid or placed in a temporary orbit while conducting operations and then moved when necessary (Stopford 2009). The proposed habitat image in Fig. 20.12 demonstrates one such concept.

Although the development of similar facilities for space activities would be a significant stretch from the International Space Station or the expandable modules envisioned by Bigelow Aerospace, such projects are well within the realm of possibility (Seedhouse 2016, pp. 20, 129, 160).

The Moon as an industrial manufacturing and supply center has been described many times. The availability of vast quantities of aluminum which could be worked on the Moon and transported to lunar orbit for assembly and then transported to other locations, makes far more economic sense than shipping materials up through the Earth's gravity well. Even as the cost of putting people and materials into lower earth orbit decreases, the likely demand for large structures in the outer solar system would suggest that the Moon would be one of the more important sources

Fig. 20.11 Off-shore residential facility (Flotel) for workers. *Credit* Shutterstock Inc

Fig. 20.12 Artist's concept of a deep-space mining habitat. *Credit* Bryan Versteeg/Deep Space Industries

supporting commercial operations in the outer solar system. The advantage of being able to construct large space going vessels and industrial facilities using Moon-based raw materials would be very significant.

Apart from the not inconsiderable problems of distance, business activity must be scalable in a way that covers the expense of development and deployment. A decision must be made as to whether or not it makes more sense to ship raw materials inward into the solar system or to process them on location. While the exact calculations of what approach to use would be determined by the nature of the materials, it may make more sense to process raw materials at the locations from which they are obtained. Much in the manner of commercial fishing operations in which the raw fish are actually processed and stored for later delivery, it might also make sense for outer solar system factories to move to new locations as sources of various materials are discovered and evaluated for production. For example, an ongoing demand for both oxygen and water could make local processing and/or production a more commercially viable activity. A mobile processing plant could move to the ideal locations for extracting either oxygen, water or both. These resources could then be supplied in sufficient quantities to make their shipment from the Earth or the Moon unnecessary. In fact, shipment of these two materials back to locations such as the Moon or Mars might be the first of many commodities that outer solar system facilities could provide in bulk, at probably an increasingly lower cost over time. Commodities moved in one direction with manufactured goods moved in the other would form the basis for an ongoing trade environment beneficial to both locations.

20.8.2 Multiple Ships

As has been suggested in other studies and articles, there are good arguments for both exploration and industrial development using multiple vessels (Ryan and Kutschera 2013). A fleet of ships provides a number of advantages from diversification of risk, specialization of function to even allocating investments among groups or consortiums. The inherent risks involved in space travel suggest it is prudent to take along multiple means for ensuring survival. Redundancy is best accomplished by having backup vessels traveling along with a primary vessel. Each backup vessel would be able to sustain the entire party should the other vessels fail and not be repairable, thus providing sufficient redundancy and repair capability to reduce the probability of catastrophic failure to tolerable levels. Equally important, at least one of the additional vessels could provide the backbone and emergency supplies for a more permanent scientific or industrial base. In that manner, a large scale operational facility could be built up within a reasonable period of time as each expedition would leave part of its redundant capability behind. Such approaches have served well during the exploration of Earth and there is ample reason to believe that a space-based version would work equally well. Once a number of larger vessels were more or less permanently moored or orbiting a specific location, it might be possible to transit to that location with smaller vessels moving at a higher velocity. Such vessels would most likely be transporting people and very critical supplies as their cargo capacity would not approximate that of the larger transports. More leisurely transits with large capacity vessels would be suitable for ongoing resupply and support activities. It is also possible that containers of supplies themselves could be launched from locations such as the Moon, independent of larger ships, to create continuous supply chains from one point in the solar system to various other locations. If speed is not an issue, shipping large quantities of goods using efficient, but somewhat slower, orbital trajectories could make economic sense.

Another approach for using multiple vessels would be to mirror that of commercial fishing operations where a factory ship, as previously described, is supplied by a number of smaller "catching" ships. The position of an orbiting factory ship could be determined relative to the potential resources within a given area. Smaller mining ships, self-sustaining for several weeks, could then be dispatched to likely locations within a given radius of the factory ship.

Once a mining ship had a sufficient quantity of materials for processing, it would return to the factory ship. This would have the advantage of providing greater coverage and a level of redundancy for miners or resource collectors. Crewed or robotic options could also be possible depending on what technology would make the most sense. Figure 20.13 is suggestive of many of those options of a ship in orbit around Europa. Such vessels would be self-contained cities capable of supporting a diverse crew of explorers, surveyors, miners and assorted support personnel for extended periods.

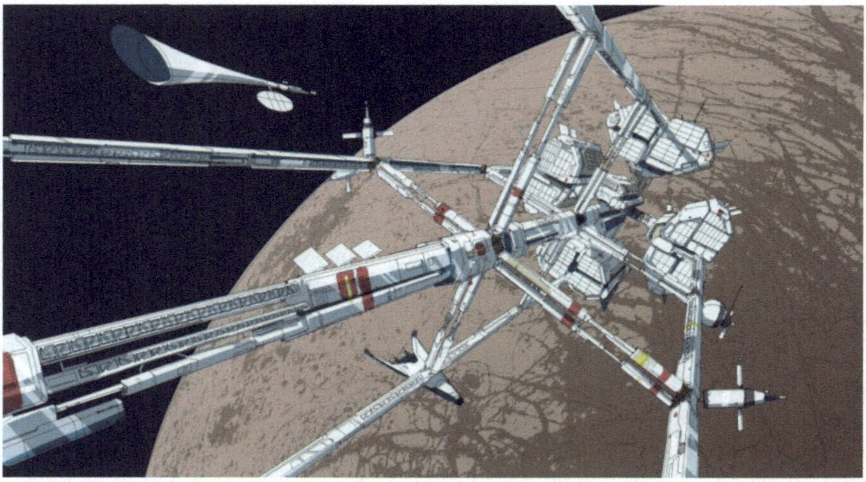

Fig. 20.13 Artist's concept of a mining/exploration vessel in orbit around Europa. *Credit* M. Poppe 2017

The long-term goal is to have sufficient vessels or cargo containers in transit carrying supplies to the outer solar system, personnel to and from the outer solar system and raw materials or converted materials from points in the outer solar system inward to support an overall transportation system not unlike current ocean-based commercial business.

20.9 International Cooperatives for Industrial Development

One of the more intriguing possibilities is the coming together of large and small industrial organizations to take advantage of space-based resources. Few organizations can afford the very high cost associated with current space-based activities, much less those required to build and operate the types of plants and facilities that might be expected to operate within the outer solar system. The obvious answer for many organizations is to cooperate to secure the necessary capital, technology, manufacturing capability, and to create the necessary operating environment. Strong international cooperation could also become the basis for extensions of terrestrial laws with particular application to space-based industrial facilities. Regardless, given the costs involved and the virtually limitless resources available, cooperation would be preferred to the existing extreme competition for resources between nations on Earth.

20.10 "Star Cops"—The Rule of Law

Star Cops is a British science fiction television series first broadcast on BBC2 in 1987 (Star Cops was devised by Chris Boucher, a writer who had previously worked on the science fiction television series *Doctor Who* and *Blake's 7*. Set at time where Interplanetary travel has become commonplace, its main character is commander of the International Space Police Force—nicknamed the "Star Cops"—who provide law enforcement for the newly developing colonies of the Solar System.). Set in the year 2027, Star Cops is a realistic attempt to portray a near-future space environment, with many of the cases arising from opportunities presented by the technologically advanced future society. The series depicts both the hostile frontier nature of the environment, and the rough and tumble business climate that the multinational team of police officers, i.e., Star Cops, live in. The fictional version of policing in the future will undoubtedly see a real-world counterpart as a necessary condition of future space-based activities. An international legal framework for commercial activity must follow from Earth lest we see the worst of so-called corporate stations and issues of governance and control such as the controversy in the South China Sea. One of the issues that must be addressed prior to any commercial activity is how the legal environment for doing business on Earth will be expanded or extended to cover opportunities of the outer solar system (Mann 2017). While it is easy to suggest that the various commercial operations would be governed by their respective national sponsors, over time this is likely to become a highly unsatisfactory situation for dealing with overlapping jurisdictions and claims. It might be presumed that legal issues of any severity would be few and far between. Human experience suggests otherwise.

When things get less simple, and they will, jealousy, greed, and some truly bad behavior will occur. In these circumstances, humans have invested heavily in systems of justice to "balance the scales." Undoubtedly, some of these practices will follow people as they move in increasing numbers to locations within the outer solar system. It is probable that whatever justice system emerges will be informed by the experiences of those who developed orbital and lunar resources as well as those later explorers of Mars. International laws will need expansion to cover the solar system and the various opportunities for bad or inappropriate behavior as practiced by sovereign nations or by individuals operating in what they view as their own self-interest. The alternative to having a rule of law may well be a return to a rule of force with all the inherent issues that would encompass. In business terms, the resources and investment involved would suggest that avenues to protect valuable facilities and personnel, and to avoid damages to either would develop quite rapidly. It is highly probable that a "police force" with general police powers would become the preferred mechanism for dealing with the human side of legal issues and a court or arbitration system to render judgements critical to operating within a rule of law.

20.11 Patterns of Business Development

There are many avenues that could be pursued as to why, how and when business, as most people view it, would develop among the possible locations within the outer solar system. The first business opportunities would be developed from those who supply the missions. It is likely that the initial missions to the outer planets would be sanctioned by governments, or by government and private sector partnerships. In either case, the individuals likely to make the most returns would be those providing the financial support for the missions themselves. Assuming that resources were present and worth exploiting, then the second group to see the benefits of business opportunities in and among the worlds of the outer solar system would be those providing supplies and goods to the groups trying to exploit the area's resources. As history has demonstrated repeatedly, while some explorers might make their fortunes taking the risks inherent in such endeavors, the real fortunes will likely be made by those providing the "picks and shovels" as well as the necessary items to keep body, mind and soul together (Ryan and Kutschera 2013, pp. 655–656).

20.12 Mining

The obvious first opportunities are resource development and mining. The need for some level of self-sufficiency will accelerate the demand for air and water. Water in the form of ice could be mined, purified, and used directly or it could be broken down into its constituent elements of hydrogen and oxygen. Both of these elements can be used for propellants, and oxygen also provides the essential element for breathing. The security of having sufficient volumes of oxygen and water provides an essential level of self-sufficiency as well as the capacity for supporting habitats for crew operations and food production.

As NASA missions explore our solar system and search for new worlds, they are finding water in unexpected places. Perhaps the most surprising locations for water include the moons of Jupiter and Saturn that show strong evidence of oceans beneath their surfaces: Ganymede, Europa and Callisto at Jupiter, and Enceladus and Titan at Saturn. Figure 20.14 includes the four largest moons of Jupiter, the Galilean satellites. From left to right, the moons shown are Io, Europa, Ganymede, and Callisto, shown to scale, of which only Io currently shows no traces of water.

NASA's Hubble Space Telescope provided evidence that Ganymede has a saltwater, sub-surface ocean, likely sandwiched between two layers of ice. Where water exists in a form that can be extracted, the opportunity for business ventures improves and their probability for success should increase. NASA has created an informational poster that illustrates the water potential of various outer solar system bodies compared to Earth (NASA 2016). This suggests that the primary resource of water is in sufficient abundance on moons that might be suitable for mining by business interests.

Fig. 20.14 The Galilean Satellites. *Source* NASA/JPL/DLR

20.13 Initial Efforts to Prepare the Way

The difficulty of providing for crewed missions is one of the primary reasons why organizations look to robotics as both the initial and perhaps primary means for conducting operations within the outer solar system. Space organizations have been sending robots to explore the solar system for some time. As computer systems have improved and remote piloted vehicles have shown their worth on Earth, it is only a short step to implementing their use off planet. Mining operations would be an obvious step assuming that there were materials of sufficient value to warrant extraction. Alternatively, remote operations could be conducted to seek out sources of water that would be used for fuel and human support as part of a much larger exploration and resource development effort. It is not practical to think in terms of taking everything that is likely to be needed with you. It is not even practical to think in terms of pre-positioning needed supplies or even raw materials such as air or water due to the amounts needed over time. What might make sense is to create exploratory models to live "off the land" to the extent possible. That means that the necessary means to obtain, to utilize and to re-task, as necessary, raw materials found at the locations selected for exploration or business must be implemented. Although current technology is insufficient to those ends, paths can be imagined that will lead to satisfactory possibilities.

Water extraction, whether in the form of liquid or ice, is comparatively old technology. It can be done inexpensively and efficiently and by remote means. Converting water to hydrogen and oxygen is again old technology that can be readily applied to a variety of activities. Building things remotely is further out from the technology tree but progress is being made. Three dimensional (3D) printing is becoming quite intriguing with developments ranging from human cell use for creating replacement organs to massive printers engaged in construction of buildings. If a 3D printer could make use of various materials located on one of the moons of Jupiter or Saturn, then it could construct any number of possible items depending on the available resources. It might be imagined that groups of 3D printers working in conjunction with a remote water extraction facility could create a mining operation of considerable size and scope. Taken to another level, it could be imagined that an entire base might be created capable of supporting crews that would arrive at a later point in time. Regardless of whether or not one envisions the future of the Outer Solar System in terms of thousands of drones moving millions of tons of materials directed from points close to Earth or a combination of crewed locations and/or autonomous or semi-autonomous vehicles, the possibilities are completely realistic given time for the technology to mature. If a profit can also be made from those activities, it would virtually guarantee that business will be conducted within the Outer Solar System.

20.14 Business in the Worlds to Come

Speculations about the potential of business opportunities are where the discussion about space opportunities gets interesting. The assumptions that drive such conversations can, in themselves, become subjects of intense debate and scrutiny. For example, is the best approach one that bases exploration and development on robotics or one that reserves a role for the human explorer? The difficulty is in how to predict the possible, while considering the most probable outcomes, both positive and negative. The extent to which business opportunities develop in space is always a function of three things: cost, return on investment and available technology. Effectively, does the return exceed the cost in a manner that makes good business sense? And, assuming that a project meets those initial requirements, can it be done in terms that are reasonable given the technology available? Necessary technology is both about cost and the state of knowledge. Generally, extensions of known technology are less expensive and their utility easier to predict than entirely new approaches. The "been there, done (something like) that" approach is a more preferred business approach. Where entirely new technologies or ways of doing things are involved in a business opportunity, the perceived risk increases the resistance to actual implementation of the idea. Being first in a field can be a great advantage, but being a close second and avoiding the problems, costs and unforeseen risks the initial business faced can be more profitable. No business is free of risk. But for many in business the salient issue is how to mitigate risk, reduce unknowns and increase overall profitability. This is very difficult to do on Earth. In space-related activities the difficulty is increased several times over just because the endeavor is in space. The business case is comparatively easy to make for near earth orbit, the Moon and even Mars. As a business opportunity, even asteroid mining is well within the realm of possibility. But once you consider moving beyond Mars, the list of unknown factors and the lack of clear and unambiguous business opportunities greatly complicates the discussion. In the long-term, there is very little doubt that business will follow humanity as it moves outward within the solar system. In the near-term, business among the outer solar system will continue to be implausible until both opportunity and technology change significantly and catch up with the intriguing vision set forth by science fiction and business visionaries.

References

Asimov, I.: The Martian Way, Galaxy Science Fiction, vol. 5, no. 2, pp. 4–49 [Authors' note: Galaxy Science Fiction was an American digest-size science fiction magazine published from 1950–1980.], Nov 1952

Astrosociety. https://www.astrosociety.org/education/astronomy-resource-guides/science-fiction-stories-with-good-astronomy-physics-a-topical-index/ (2016). Accessed 15 Dec 2016

Benford, G. and Editors of Popular Mechanics: The Wonderful Future That Never Was. Hearst Books, New York (2012)

Brown, A.S., Logan, B.: How fiction puts the science in engineering. Mech. Engi. **137**(2), 32–39, New York (2015)

Cremades, E., Balbastre-Benavent, F., Domínguez, E.S.: Managerial practices driving knowledge creation, learning and transfer in translational research: An exploratory case study. R & D Manage. **45**(4), 361 (2015)

Cudahy, B.J.: Box Boats: How Container Ships Changed the World. Fordham University Press, New York (2006)

DNV GL.: http://vesselregister.dnvgl.com/VesselRegister/vesseldetails.html?vesselid=33249. Accessed 15 Dec 2016

George, R.: Ninety Percent of Everything. Picador, New York (2013)

Glancey, J.: Concorde: The Rise and Fall of the Supersonic Airliner. Atlantic Books, London (2015)

Heinlein, R.A.: Farmer in the Sky, New York: Ballantine Books (1950)

Launius, R.D.: Historical Analogs for the Simulation of Space Commerce, Washington. D.C.: National Aeronautics and Space Administration, Office of Communications, Public Outreach Division, History Program Office (2014)

Lewis, J.S.: Asteroid Mining 101: Wealth for the New Space Economy. Deep Space Industries, Silicon Valley, California (2015)

Lorenz, R., Mitton, J.: Titan Unveiled: Saturn's Mysterious Moon Explored. Princeton University Press, Princeton (2008)

Lunar and Planetary Institute: http://www.lpi.usra.edu/opag/mind_the_gap.pdf. Accessed 15 Dec 2016

Mann, A.: The Final Frontier of Law, Wall Street Journal, May 20–21, p. C3 (2017)

McCurdy, H.E.: Space and the American Imagination. Smithsonian Institution, Washington, D.C. (1997)

Miller, R.: Spaceships: An Illustrated History of the Real and Imagined. Smithsonian Books, Washington, D.C. (2016)

Motorship: http://www.motorship.com/news101/ships-and-shipyards/the-future-of-factory-ships. Accessed 7 Dec 2016

NASA: https://www.nasa.gov/jpl/the-solar-system-and-beyond-is-awash-in-water. Accessed 7 Dec 2016

Ordway III, F.I.: Visions of Spaceflight: Images from the Ordway Collection. Four Walls Eight Windows, New York (2000)

Phaal, R., O'Sullivan, E., Routley, M., Ford, S., Probert, D.: A framework for mapping industrial emergence. Technol. Forecast. Soc. Chang. **78**(2), 217 (2011)

Ryan, M.H., Kutschera, I.: Business modalities of the inner solar system: planets with potential. In: Badescu, V., Zacny, K. (eds.) Inner Solar System: Prospective Energy and Material Resources, pp. 395–406. Springer International Publishing, Switzerland (2015)

Ryan, M.H., Kutschera, I.: The case for asteroids: Commercial concerns and considerations. In: Badescu, V. (ed.) Asteroids—prospective energy and material resources, pp. 645–657. Springer, Heidelberg (2013)

Ryan, M.H., Kutschera, I.: Lunar-based enterprise infrastructure—hidden keys for long-term business success. Space Policy **23**, 44–52 (2007)

Science Fiction Encyclopedia: http://www.sf-encyclopedia.com. Accessed 15 Dec 2016

Seedhouse, E.: SpaceX's Dragon: America's Next Generation Spacecraft. Springer Praxis Publishing, Chichester (2016)
Seedhouse, E.: Interplanetary Outpost: The Human and Technological Challenges of Exploring the Outer Planets. Springer Praxis Publishing, Chichester, UK (2012)
Simberg, R.: Safe Is Not an Option. Interglobal Media LLC, Jackson, Wyoming (2014)
Stopford, M.: Maritime Economics, 3rd edn. Routledge, London (2009)

Chapter 21
Sources of Energy in the Outer Solar System

Alexander Bolonkin

21.1 Impulse Mini Thermonuclear Reactors

21.1.1 Introduction

Fusion power is useful energy generated by nuclear fusion reactions. In this kind of reaction, two light atomic nuclei fuse together to form a heavier nucleus and release energy. The largest current nuclear fusion experiment, JET, has resulted in fusion power production somewhat larger than the power put into the plasma, maintained for a few seconds. In June 2005, the construction of the experimental reactor ITER, designed to produce several times more fusion power than the power into it generating the plasma over many minutes, was announced. The unrealized production of net electrical power from fusion machines is planned for the next generation experiment after ITER.

Unfortunately, this task is not easy, as scientists thought early on. Fusion reactions require a very large amount of energy to initiate in order to overcome the so-called *Coulomb barrier* or *fusion barrier energy*. The key to practical fusion power is to select a fuel that requires the minimum amount of energy to start, that is, the lowest barrier energy. The best fuel from this standpoint is a one-to-one mix of deuterium and tritium (D-T); both are heavy isotopes of hydrogen. The D-T mix has suitable low barrier energy. In order to create the required conditions, the fuel must be heated to tens of millions of degrees, and/or compressed to immense pressures.

At present, D-T is used by two main methods of fusion: inertial confinement fusion (ICF) and magnetic confinement fusion (MCF)—for example, tokomak device. However, tritium is very expensive.

A. Bolonkin (✉)
C&R, 1310 Avenue R, #6-F, Brooklyn, NY 11229-2838, USA
e-mail: abolonkin@gmail.com

V. Badescu and K. Zacny (eds.), *Outer Solar System*,
https://doi.org/10.1007/978-3-319-73845-1_21

In inertial confinement fusion (*ICF*), nuclear fusion reactions are initiated by heating and compressing a target. The target is a pellet that most often contains D-T (often only micro or milligrams). Intense focused laser or ion beams are used for compression of pellets. The beams explosively detonate the outer material layers of the target pellet. That accelerates the underlying target layers inward, sending a shockwave into the center of each pellet's mass. If the shockwave is powerful enough, and if high enough density at the center is achieved, some of the fuel will be heated enough to cause pellet fusion reactions. In a target which has been heated and compressed to the point of thermonuclear ignition, energy can then heat surrounding fuel to cause it to fuse as well, potentially releasing tremendous amounts of energy.

Magnetic confinement fusion (*MCF*). Since plasmas are very good electrical conductors, magnetic fields can also be configured to safely confine fusion fuel. A variety of magnetic configurations can be used, the basic distinction being between magnetic mirror confinement and toroidal confinement, especially tokomaks and stellarators.

Lawson criterion. In nuclear fusion research, the Lawson criterion, first derived by John D. Lawson in 1957, is an important general measure of a system that defines the conditions needed for a fusion reactor to reach *ignition* stage, that is, the heating of the plasma by the products of the fusion reactions is sufficient to maintain the temperature of the plasma against all losses without external power input. As originally formulated the Lawson criterion gives a minimum required value for the product of the plasma (electron) density n_e and the "energy confinement time" τ. Later analyses suggested that a more useful figure of merit is the "triple product" of density, confinement time, and plasma temperature T. The triple product also has a minimum required value, and the name "Lawson criterion" often refers to this important inequality.

The key to practical fusion power is to select a fuel that requires the minimum amount of energy to start, that is, the lowest barrier energy. The best known fuel from this standpoint is a one-to-one mix of deuterium and tritium; both are heavy isotopes of hydrogen. The D-T mix has a low barrier. In order to create the required conditions, the fuel must be heated to tens of millions of degrees, and/or compressed to immense pressures. The temperature and pressure required for any particular fuel to fuse is known as the Lawson criterion. For the D-T reaction, the physical value is about

$$L = nT\tau > \left(10^{20} - 10^{21}\right) \text{ in SI units,}$$

where T is temperature, (KeV), 1 eV = 1.16×10^4 K; n is matter density, (1/m^3); τ is time, (s). This equation is in metric system. The thermonuclear reaction of ^{2}H + ^{3}D realizes if $L > 10^{20}$ in CI (meter, kilogram, second) units or $L > 10^{14}$ in 'cgs' (centimeter, gram, second) units.

This number has not yet been achieved in any fusion reactor, although the latest generations of fusion-making machines have come significantly close to doing so. For instance, the reactor TFTR has achieved the densities and energy lifetimes needed to achieve Lawson at the temperatures it can create, but it cannot create those temperatures at the same time. Future ITER aims to do both.

The Lawson criterion applies to inertial confinement fusion as well as to magnetic confinement fusion but is more usefully expressed in a different form.

Following the above derivation of the limit on $n_e\tau_E$, we see that the product of the density and the radius must be greater than a value related to the minimum of $T^{3/2}/\langle\sigma v\rangle$ (here σ is Boltzmann constant, v is ion speed). This condition is traditionally expressed in terms of the mass density ρ and R-radius of fuel pellet:

$$\rho R > 1\ \mathrm{g/cm^2}.$$

To satisfy this criterion at the density of solid D + T (0.2 g/cm^3) would require implausibly large laser pulse energy. Assuming the energy required scales with the mass of the fusion plasma ($E_{\text{laser}} \sim \rho R^3 \sim \rho^{-2}$), compressing the fuel to 10^3 or 10^4 times solid density would reduce the energy required by a factor of 10^6 or 10^8, bringing it into a realistic range. With a compression by 10^3, the compressed density will be 200 g/cm^3, and the compressed radius can be as small as 0.05 mm. The radius of the fuel before compression would be 0.5 mm. The initial pellet will be perhaps twice as large since most of the mass will be ablated during the compression stage by a symmetrical energy input bath.

The fusion power density is a good figure of merit to determine the optimum temperature for magnetic confinement, but for inertial confinement the fractional burn-up of the fuel is probably more useful. The burn-up should be proportional to the specific reaction rate ($n^2\langle\sigma v\rangle$) times the confinement time (which scales as $T^{1/2}$) divided by the particle density n: burn-up fraction $\sim n^2\langle\sigma v\rangle T^{-1/2}/n \sim (nT)\,(\langle\sigma v\rangle/T^{3/2})$. Thus the optimum temperature for inertial confinement fusion is that which maximizes $\langle\sigma v\rangle/T^{3/2}$, which is slightly higher than the optimum temperature for magnetic confinement.

A short history of thermonuclear fusion

One of the earliest (in the late 1970s and early 1980s) serious attempts at an ICF design was *Shiva*, a 20-armed neodymium laser system built at the Lawrence Livermore National Laboratory (LLNL) in California that started operation in 1978. Shiva was a "proof of concept" design, followed by the *NOVA* design with 10 times the power. Funding for fusion research was severely constrained in the 80s, but NOVA nevertheless successfully gathered enough information for a next generation machine whose goal was ignition. Although net energy can be released even without ignition (the breakeven point), ignition is considered necessary for a *practical* power system.

The resulting design, now known as the National Ignition Facility (NIF), commenced being constructed at LLNL in 1997. Originally intended to start construction in the early 1990s, the NIF is now six years behind schedule and over budget by some $3.5 billion. Nevertheless many of the problems appear to be due to the "Big Science Laboratory" mentality and shifting the focus from pure ICF research to the nuclear stewardship program, LLNLs traditional nuclear weapons-making role. NIF "burned" in 2010, when the remaining lasers in the 192-beam array were finally installed. Like those earlier experiments, however, NIF has failed to reach ignition and is, as of 2015, generating only about 1/3rd of the required energy levels needed to reach full fusion stage of operation.

Laser physicists in Europe put forward plans to build a £500 m facility, called HiPER, to study a new approach to laser fusion. A panel of scientists from seven European Union countries believes that a "fast ignition" laser facility could make a significant contribution to fusion research, as well as supporting experiments in other areas of physics. The facility would be designed to achieve high-energy gains, providing the critical intermediate step between ignition and a demonstration reactor. It would consist of a long-pulse laser with energy of 200 kJ to compress the fuel and a short-pulse laser with energy of 70 kJ to heat it.

Confinement refers to all the conditions necessary to keep plasma dense and hot long enough to undergo fusion:

- *Equilibrium:* There must be no net forces on any part of the plasma, otherwise it will rapidly disassemble. The exception, of course, is inertial confinement, where the relevant physics must occur faster than the disassembly time.
- *Stability:* The plasma must be so constructed that small deviations are restored to the initial state, otherwise some unavoidable disturbance will occur and grow exponentially until the plasma is destroyed.
- *Transport:* The loss of particles and heat in all channels must be sufficiently slow. The word "confinement" is often used in the restricted sense of "energy confinement".

To produce self-sustaining fusion, the energy released by the reaction (or at least a fraction of it) must be used to heat new reactant nuclei and keep them hot long enough that they also undergo fusion reactions. Retaining the heat generated is called energy *confinement* and may be accomplished in a number of ways.

Hydrogen bomb weapons require no confinement at all. The fuel is simply allowed to fly apart, but it takes a certain length of time to do this, and during this time fusion can occur. This approach is called *inertial confinement*. If more than about a milligram of fuel is used, the explosion would destroy the machine, so controlled thermonuclear fusion using inertial confinement causes tiny pellets of fuel to explode several times a second. To induce the explosion, the pellet must be compressed to about 30 times solid density with energetic beams. If the beams are focused directly on the pellet, it is called *direct drive*, which can in principle be very efficient, but in practice it is difficult to obtain the needed uniformity. An alternative approach is *indirect drive*, in which the beams heat a shell, and the shell radiates X-rays, which then implode the pellet. The beams are commonly laser beams, but heavy and light ion beams and electron beams have all been investigated and tried to one degree or another.

Data of some current inertial laser installations:

1. NOVA uses laser NIF (USA), has 192 beams, impulse energy up 120 kJ. Density 20 g/cm^3, speed of cover is up 300 km/s. NIF has failed to

reach ignition and is, as of 2013, generating about 1/3rd of the required energy levels. NIF cost is about $3.5B.
2. YiPER (EU) has impulse energy up 70 kJ.
3. OMEGA (USA) has impulse energy up 60 kJ.
4. Gekko-XII (Japan) has impulse energy up 20 kJ. One reaches density 120 g/cm^3.
5. Febus (France) has impulse energy up 20 kJ.
6. Iskra-5 (Russia) has impulse energy up 30 kJ.

In this chapter, author offers new and cheap thermonuclear reactors and fuses for nuclear projectiles (see also Bolonkin 2012a, 2015a, b). Below are some cheaper and simpler reactors.

21.1.2 Description and Innovations of Cumulative and Impulse AB Reactors

Description

Laser method. Thermonuclear reactors and, in particular, laser methods have been under development for about 60 years. Governments have already spent tens of billions of US dollars, but it is not yet seen as an industrial application of thermonuclear energy for the coming 10–15 years. Because the laser has very low efficiency (1–1.5%) and high-pressure acts for a very short time (10^{-9}–10^{-10} s), there is no enough energy delivered to the center of the spherical fuel pellet. There are many future problems the radioactivity and converting the thermonuclear energy into useful energy.

Cumulative method. Author offered three new methods (Bolonkin 2012a, 2015a, b), which are thousands of times cheaper, more efficiency and does not have many disadvantages of the laser and magnetic methods. In given article the author offers two (cumulative and impulse) improved reactors. Detailed consideration of advantages the new methods and computation proofs are in next paragraph.

New cumulative reactor and method. The improved version 1 of the Cumulative AB thermonuclear reactor is presented in Figs. 21.1, 21.2, 21.3 and 21.4. The new thermonuclear reactor contains (Fig. 21.1): strong spherical body of reactor (1); cartridge (holder) of fuel pellet (or pellet in Version 2) (2); holder (electric conductor) of fuel cartridge (pellet) (3); entrance of compressed air (gas) (4); exit of a hot compressed air after thermonuclear heating, (5); contacts of voltage (6) for electric condenser.

The fuel cartridge 6 has a diameter of about 10–20 mm and the next design (Fig. 21.2): strong sphere (1); net fuse (electric net 2 for ignition of explosive 3, Fig. 21.2), explosive (3), film (piston) of heavy material (4), ampoule of nuclear fuel (pellet) (5), electric conductors (6).

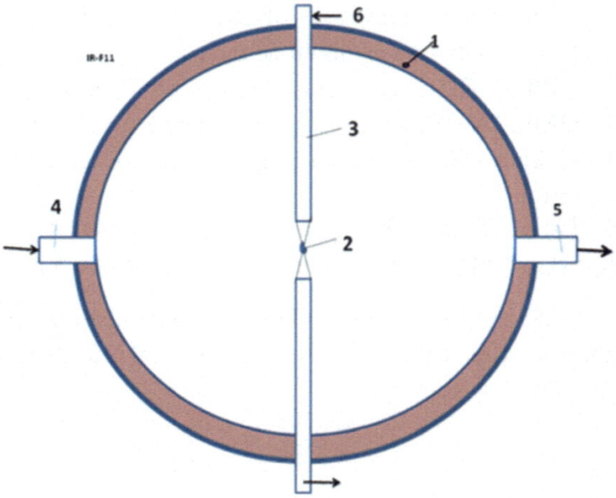

Fig. 21.1 AB thermonuclear cumulative reactor (version 1). *Notations* 1—strong spherical body of reactor, diameter about 0.5–5 m; 2—cartridge (holder) of fuel pellet, diameter 1–2 cm (or pellet in version 2, diameter 2–3 mm); 3—holder (electric conductor) of fuel pellet; 4—enter of compressed air (gas); 5—exit of a hot compressed air (gas) after thermonuclear heating; 6—electric voltage

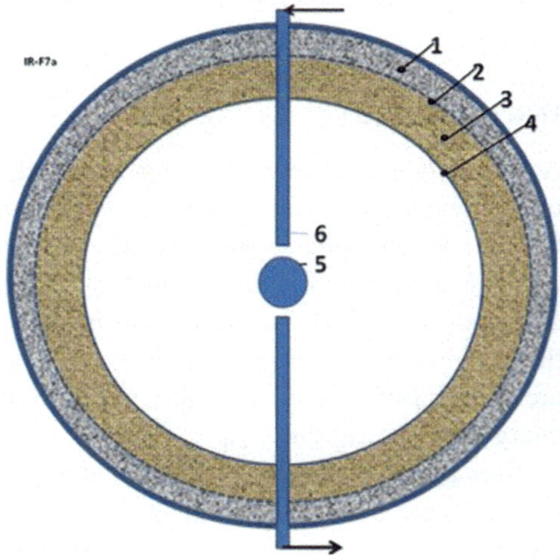

Fig. 21.2 Cartridge of cumulative AB-thermonuclear reactor (version 1). *Notations* (1)—strong sphere, (2)—net fusee (electric net), (3)—explosive, (4)—layer (piston) of heavy material, (5)—nuclear fuel (pellet), (6)—electric conductor

In short the reactor works the following way: The lasting sphere 1 of the reactor (Fig. 21.1) is filled with compressed gas (for example, air). An electric signal is sent across electric conductor (3) (Fig. 21.1) and blows up explosive (3) (Fig. 21.2) into fuel cartridge (Fig. 21.3a). The cumulative explosive works as a rocket engine (Fig. 21.3b) and presses the layer (piston) of heavy material (4) around the fuel pellet (5) (Fig. 21.3) and high presses (and heating) the fuel pellet (5). The strong electric impulse from the condenser is sent across the insulated conductors (3) (Fig. 21.1) and (6) (Fig. 21.2) into the pellet, heating the fuel up the needed thermonuclear temperature. The fuel explodes.

Note: It is very important to *simultaneously* ignite the *all* outer surface of explosive 3 (Fig. 21.2). In only this case the explosive begin to move towards mass center and works as rocket engine, accelerate and compress the explosive and layer

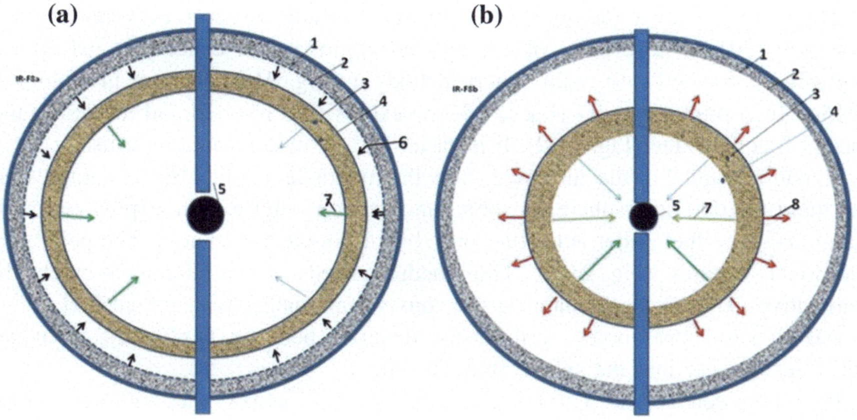

Fig. 21.3 Cumulative cartridge of thermonuclear fuel. Work of cumulative thermonuclear cartridge (version 1). *Notations* **a** initial layer explosive, **b** rocket part of explosive. (1)–(5) are same with Fig. 21.2; (6) is the pressure of the initial layer of explosion, (7) is the motion of the explosive (force from jetted gas rocket thrust), and (8) is the flow of reactive explosive gas

4 (having small mass) for high speeds in tens of times more than in conventional explosives.

We can simultaneously ignite all outer surfaces of the explosive by electric current. The electric impulse will ignite the entire outer surface of the explosive.

Simultaneously (or early) into the big sphere of reactor body may be injected the water (7) (Fig. 21.4a) (optional). The compressed air (or injected water) is heated by the thermonuclear explosive (8) (Fig. 21.4a) and go out across a hole (9) (Fig. 21.4a) into MHG (or gas, steam turbine) and produces electric or mechanical energy. In a rocket engine, the gas flows out across the nozzle and creates thrust (Fig. 21.4b).

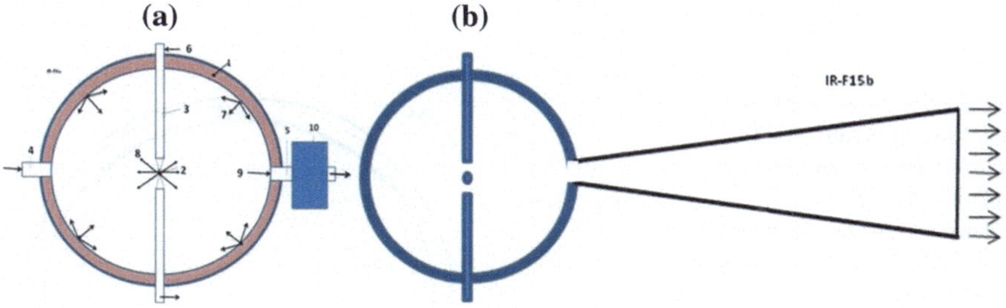

Fig. 21.4 Final (industrial) work of cumulative and impulse AB thermonuclear reactors. **a** Hot compressed gas from sphere runs to the magneto-hydrodynamic generator (MHG) (10) and produces electric energy or runs to gas turbine and produces an useful work (Fig. 21.2a). **b** Hot compressed gas rans to rocket nozzle and produces the rocket trust. *Notation* (1)–(6) are same Fig. 21.1; 7—injection the cooling liquid (for example, water) (option); 8—thermonuclear explosive of fuel pellet; exit of hot gas; 10—MHG or gas (steam) turbine

The main difference the offered reactor from cumulative reactors (versions 1, 2) Bolonkin (2012a) is location of compressed explosive. In Bolonkin (2012a) the explosive is located into main spherical body 1 (Fig. 21.1) (or gun in Bolonkin 2015a). In current version 1 (Fig. 21.2) the explosive 3 is small and located in the special fuel cartridge (Fig. 21.2). It is easier and more comfortable to use.

Versions 1 and 2 of the current reactors the fuel pellet is filled by the compressed gas fuel (up 1000 atmospheres or more) and version 2 not has the explosive for an additional. The fuel pellet is heating only by strong electric charge. The pellet has additional features (Fig. 21.5). This method needs in additional research. In cumulative version (1), we can use the conventional pellet with frozen fuel.

AB Reactors are cooled well-known methods between explosives or by an injection of water into the sphere (Fig. 21.4a).

In more details the Cumulative AB Reactor works the following way (Figs. 21.3, 21.4 and 21.5).

The net fusee 2 (Fig. 21.3a) simultaneously blows an outer layer of the explosive 3 and an explosive gas 6 is pushing the explosive 3 to center of sphere fuel cartridge. The outer surface of explosive 3 is burning 8 (Fig. 21.3b) and create jet rocket thrust 7 which is moving, accelerating, compressing the explosive and thin layer 4 in direction to center 5 of sphere fuel pellet.

As the result the layer 4 (piston) made from heavy material bumps with high speed (about 20 km/s) and produced a high pressure (millions atmospheres). This pressure is acting more time than laser pressure and reaches to center of fuel capsule (Fig. 21.3a). The strong electric impulse increases the fuel temperature up need value. The thermonuclear fuel capsule explodes (Fig. 21.3b). The heavy material of layer breaks the nuclear explosion, increases the conformation time and efficiency of the thermonuclear reaction. If installation is used as reactor for MHG or turbine, the cooling liquid (for example, water) is injected into strong sphere. One is converted to hot gas (steam) and rotates the turbine blades (Fig. 21.4a).

Note: It is very important to *simultaneously* ignite the *all* outer surface of explosive 3 (Fig. 21.2). In only this case the explosive mass begin to move towards to center and works as rocket engine, accelerate and compress the explosive and

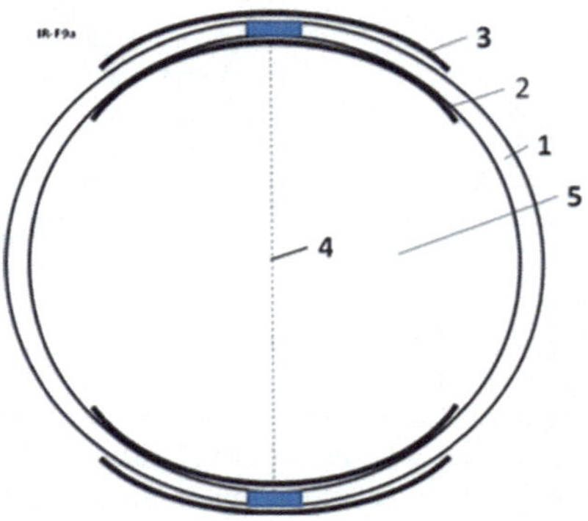

Fig. 21.5 Compressed gas pellet for cumulative and impulse Thermonuclear reactors. *Notations* 1—isolator cover; 2, 3—internal and external electric contacts; 4—initial hot ionizer; 5—compressed gas thermonuclear fuel

layer 4 (having small mass) for high speeds—ten times more that in conventional explosive. That is main innovation in offered method.

Compressed gas pellet for Cumulative and Impulse thermonuclear reactors is shown in Fig. 21.5. One has a compressed up 300–1000 atm thermonuclear gas fuel 5, the insulator spherical cover 2, the contacts connect the inside and outside electric contacts 2–3; and thin internal electric conductor 4 (only for Impulse reactor) which produces the initial ion canal for an electric currency.

21.1.3 Advantages of the Suggested Reactors in Comparison with ICF Laser Method

The offered reactor and method have the following advantages in comparison with the conventional ICF laser reactor:

1. The additional electric heating allows reaching the needed thermonuclear temperature.
2. Cumulative and Impulse AB-reactors are cheaper by thousands of time because they do not have the gigantic very expensive laser installations (see Bolonkin 2012a, 2015a, b).
3. They more efficiency because the laser installation converts only 1–1.5% the electric energy into the light beam. In suggested AB reactors, the all underused (for compression) explosive energy remains in the spherical tank and utilized in MDG or turbine. AB reactor cannot have coefficient Q (used energy) significantly less 1. Moreover, one has heat efficiency more than conventional heat engines because it has very high compression ratio. One can use as the conventional very high power engine in civil and military transportation.
4. The offered very important innovation (accelerating of the explosive by the rocket thrust) allows increasing the top speed of the piston mass 4 from the conventional sound (shock wave) speed 3 km/s up about 20 km/s. Only this innovation increases the thermonuclear ignition criterion in 50 times in comparison with conventional cumulative explosion (see computation). This makes this method available for thermonuclear reaction.
5. Cumulative AB-reactor gives compression of the fuel capsule much more than the current ICF laser installations.
6. This compression has longer time (up to 10^{-3}–10^{-5} s) than a laser beam pressing (10^{-9}–10^{-12} s), because heavy mass 4 (piston) is many times (10–30) more than mass of a capsule (pellet, micro balloon) 5. This pressure is supported by rocket gas and shock wave coming from moving explosive gas. This pressure reaches the center of capsule with high speed of heavy mass 4, (not sound speed as in laser pressure) increases the temperature, compressing and probability of thermonuclear reaction in the fuel capsule.

7. The heavy mass 4 (piston) (having high nuclear numbers A and Z) not allow the nuclear particles easily to fly apart. That increases the reaction time and reactor efficiency.
8. The suggested AB-thermonuclear reactor is small (diameter about 0.5–3 m or less up 0.3 m) light (mass about some ton or less up 150 kg) and may be used in the transport vehicles and aviation.
9. The water may protect the material of the sphere from neutrons.
10. It is possible (see computations) the efficiency of AB reactors will be enough for using as fuel only the deuterium which is cheaper then tritium in thousands times (One gram of tritium costs about 30,000 US dollars. One gram of deuterium costs 1$).

21.1.4 Theory of Current Thermonuclear Reactor

1. *Reactions suitable for thermonuclear fusion.* The reactions of Table 21.1 are suitable for thermonuclear fusion. For reactions with two products, the energy is divided between them in inverse proportion to their masses, as shown. In most reactions with three products, the distribution of energy varies. For reactions that can result in more than one set of products, the branching ratios are given.

Some reaction candidates can be eliminated at once. The D-^{6}Li reaction has no advantage compared to p-^{11}B because it is roughly as difficult to burn but produces substantially more neutrons through D-D side reactions. There is also a p-^{7}Li reaction, but the cross-section is far too low accepted possible for $T_i > 1$ meV, but at such high temperatures, an endothermic, direct neutron-producing reaction also becomes very significant. Finally, there is also a p-^{9}Be reaction, which is not only

Table 21.1 Suitable reactions for thermonuclear fusion

(1)	D	+	T		^{4}He	(3.5 meV)	+		n	(14.1 meV)				
(2i)	D	+	D		T	(1.01 meV)	+		p	(3.02 meV)				*50%*
(2ii)					^{3}He	(0.82 meV)	+		n	(2.45 meV)				*50%*
(3)	D	+	^{3}He		^{4}He	(3.6 meV)	+	2	p	(14.7 meV)				
(4)	T	+	T		^{4}He		+		n	+11.3 meV				
(5)	^{3}He	+	^{3}He		^{4}He		+	2	p	+12.9 meV				
(6i)	^{3}He	+	T		^{4}He		+		p		+	n	+12.1 meV	*51%*
(6ii)					^{4}He	(4.8 meV)	+		D	(9.5 meV)				*43%*
(6iii)					^{4}He	(0.5 meV)	+		n	(1.9 meV)	+	p	(11.9 meV)	*6%*
(7)	D	+	^{6}Li	2	^{4}He	+22.4 meV								
(8)	P	+	^{6}Li		^{4}He	(1.7 meV)	+		^{3}He	(2.3 meV)				
(9)	^{3}He	+	^{6}Li	2	^{4}He		+		p	+16.9 meV				
(10)	P	+	^{11}B	3	^{4}He	+8.7 meV								

Note p (protium), D (deuterium), and T (tritium) is shorthand notation for the main three isotopes of hydrogen

difficult to burn, but ^{9}Be can be easily induced to split into two alphas and a neutron.

In addition to the fusion reactions, the following reactions with neutrons are important in order to "breed" tritium in "dry" fusion bombs and some proposed fusion reactors:

$$n + {}^6\text{Li} \rightarrow T + {}^4\text{He}, \quad n + {}^7\text{Li} \rightarrow \text{T} + {}^4\text{He} + \text{n}.$$

To evaluate the usefulness of these reactions, in addition to the reactants, the products, and the energy released, one needs to know something about the cross section.

Any given fusion device will have a maximum plasma pressure that it can sustain, and an economical device will always operate near this maximum. Given this pressure, the largest fusion output is obtained when the temperature is selected so that $\langle\sigma v\rangle/T^2$ is a maximum. This is also the temperature at which the value of the triple product $nT\tau$ required for ignition is a minimum. This chosen optimum temperature and the value of $\langle\sigma v\rangle/T^2$ at that temperature is given for a few of these reactions in Table 21.2.

Note that many of the reactions form chains. For instance, a reactor fueled with T and ^{3}He will create some D, which is then possible to use in the D + ^{3}He reaction if the energies are "right". An elegant idea is to combine the reactions (8) and (9). The ^{3}He from reaction (8) can react with ^{6}Li in reaction (9) before completely thermalizing. This produces an energetic proton which in turn undergoes reaction (8) before thermalizing. A detailed analysis shows that this idea will not really work well, but it is a good example of a case where the usual assumption of a Maxwellian plasma is not appropriate.

Any of the reactions above can, in principle, be the basis of fusion power production. In addition to the temperature and cross section discussed above, we must consider the total energy of the fusion products E_{fus}, the energy of the charged fusion products E_{ch}, and the atomic number Z of the non-hydrogenic reactant.

Specification of the D-D reaction entails some difficulties, though. To begin with, one must average over the two branches (2) and (3). More difficult is to decide how to treat the T and ^{3}He products. T burns so well in a deuterium plasma that it is almost impossible to extract from the plasma. The D-^{3}He reaction is optimized at a much higher temperature, so the burn-up at the optimum D-D temperature may be low, so it seems reasonable to assume the T but not the ^{3}He gets burned up and adds

Table 21.2 Optimum temperature and the value of $\langle\sigma v\rangle/T^2$ at that temperature

Fuel	T (keV)	$\langle\sigma v\rangle/T^2$ (m^3/s/keV^2)
D-T	13.6	1.24×10^{-24}
D-D	15	1.28×10^{-26}
D-^{3}He	58	2.24×10^{-26}
p-^{6}Li	66	1.46×10^{-27}
p-^{11}B	123	3.01×10^{-27}

its energy to the net reaction. Thus we will count the D-D fusion energy as $E_{fus} = (4.03 + 17.6 + 3.27)/2 = 12.5$ meV and the energy in charged particles as $E_{ch} = (4.03 + 3.5 + 0.82)/2 = 4.2$ meV.

Another unique aspect of the D-D reaction is that there is only one reactant, which must be taken into account when calculating the reaction rate.

With this choice, we tabulate parameters for four of the most important reactions in Table 21.3.

The last column is the *neutronicity* of the reaction, the fraction of the fusion energy released as neutrons. This is an important indicator of the magnitude of the problems associated with neutrons like radiation damage, biological shielding, remote handling, and safety. For the first two reactions it is calculated as $(E_{fus}-E_{ch})/E_{fus}$. For the last two reactions, where this calculation would give zero, the values quoted are rough estimates based on side reactions that produce neutrons in a plasma in thermal equilibrium.

Of course, the reactants should also be mixed in the optimal proportions. This is the case when each reactant ion plus its associated electrons accounts for half the pressure. Assuming that the total pressure is fixed, this means that density of the non-hydrogenic ion is smaller than that of the hydrogenic ion by a factor $2/(Z + 1)$. Therefore, the rate for these reactions is reduced by the same factor, on top of any differences in the values of $\langle\sigma v\rangle/T^2$. On the other hand, because the D-D reaction has only one reactant, the rate is twice as high as if the fuel were divided between two hydrogenic species.

Thus, there is a "penalty" of $(2/(Z + 1))$ for non-hydrogenic fuels arising from the fact that they require more electrons, which take up pressure without participating in the fusion reaction. There is, at the same time, a "bonus" of a factor 2 for D-D due to the fact that each ion can react with any of the other ions, not just a fraction of them.

We can now compare these reactions in Table 21.4.

The maximum value of $\langle\sigma v\rangle/T^2$ is taken from a Table 21.3. The "penalty/bonus" factor is that related to a non-hydrogenic reactant or a single-species reaction. The values in the column "reactivity" are found by dividing $(1.24 \times 10^{-24}$ by the product of the second and third columns. It indicates the factor by which the other reactions occur more slowly than the D-T reaction under comparable conditions. The column "Lawson criterion" weights these results with E_{ch} and gives an indication of how much more difficult it is to achieve ignition with these reactions, relative to the difficulty for the D-T reaction. The last column is labeled "power density" and weights the practical reactivity with E_{fus}. It indicates how much lower

Table 21.3 Parameters of the most important reactions

Fuel	Z	E_{fus} (meV)	E_{ch} (meV)	Neutronicity
D-T	1	17.6	3.5	0.80
D-D	1	12.5	4.2	0.66
D-^{3}He	2	18.3	18.3	~0.05
p-^{11}B	5	8.7	8.7	~0.001

Table 21.4 Comparison of reactions

Fuel	$\langle\sigma v\rangle/T^2$	Penalty/bonus	Reactivity	Lawson criterion	Power density
D-T	1.24×10^{-24}	1	1	1	1
D-D	1.28×10^{-26}	2	48	30	68
D-^{3}He	2.24×10^{-26}	2/3	83	16	80
p-^{11}B	3.01×10^{-27}	1/3	1240	500	2500

the fusion power density of the other reactions is compared to the D-T reaction and can be considered a measure of the economic potential.

Below are some equations useful for computation:

2. *The Deep of Penetration of outer radiation into plasma* is

$$d_p = \frac{c}{\omega_{pe}} = 5.31 \times 10^5 n_e^{-1/2}\,(\text{cm}) \tag{21.1}$$

   For plasma density $n_e = 10^{22}$ 1/cm^3, $d_p = 5.31 \times 10^{-6}$ cm.
3. *The Gas (Plasma) Dynamic Pressure, p_k,* is

$$p_k = nk(T_e + T_i) \quad \text{if} \quad T_e = T_k \quad \text{then} \quad p_k = 2nkT \tag{21.2}$$

   where $k = 1.38 \times 10^{-23}$ is Boltzmann constant; T_e is temperature of electrons, K; T_i is temperature of ions, K. These temperatures may be different; n is plasma density, 1/m^3; p_k is plasma pressure, N/m^2.
4. *The gas (plasma) ion pressure, p,* is

$$p = \frac{2}{3}nkT, \tag{21.3}$$

   Here n is plasma density in 1/m^3.
5. *The magnetic p_m and electrostatic pressure, p_s,* are

$$p_m = \frac{B^2}{2\mu_0}, \quad ps = \frac{1}{2}\varepsilon_0 E_s^2 \tag{21.4}$$

   where B is electromagnetic induction, Tesla; $\mu_0 = 4\pi \times 10^{-7}$ electromagnetic constant; $\varepsilon_0 = 8.85 \times 10^{-12}$, F/m, is electrostatic constant; E_S is electrostatic intensity, V/m.
6. *Ion thermal velocity* is

$$v_{Ti} = \left(\frac{kT_i}{m_i}\right)^2 = 9.79 \times 10^5\,\mu^{-1/2}T_i^{1/2}\ \text{cm/s}, \tag{21.5}$$

where $\mu = m_i/m_p$, m_i is mass of ion, kg; $m_p = 1.67 \times 10^{-27}$ is mass of proton, kg.

7. *Transverse Spitzer plasma resistivity*

$$\eta_{\perp} = 1.03 \times 10^{-2} Z \ln \Lambda T^{-3/2}, \Omega \text{cm} \quad \text{or} \quad \rho \approx \frac{0.1Z}{T^{3/2}} \Omega \text{cm} \tag{21.6}$$

where ln Λ = 5–15 ≈ 10 is Coulomb logarithm, Z is charge state.

8. Reaction rates $\langle \sigma v \rangle$ (in $cm^3\ s^{-1}$) averaged over Mexwellian distributions for low energy (T < 25 keV) may be represent by

$$\begin{aligned} (\overline{\sigma v})_{DD} &= 2.33 \times 10^{-14} T^{-2/3} \exp\left(-18.76 T^{-1/3}\right) \text{cm}^3\,\text{s}^{-1}, \\ (\overline{\sigma v})_{DT} &= 3.68 \times 10^{-14} T^{-2/3} \exp\left(-19.94 T^{-1/3}\right) \text{cm}^3\text{s}^{-1}, \end{aligned} \tag{21.7}$$

where T is measured in keV.

9. *The power density released in the form of charged particles* is

$$\begin{aligned} P_{DD} &= 3.3 \times 10^{-13} n_D^2 (\overline{\sigma v})_{DD}, \text{ W cm}^{-3} \\ P_{DT} &= 5.6 \times 10^{-13} n_D n_T (\overline{\sigma v})_{DT}, \text{ W cm}^{-3} \\ P_{DHe^3} &= 2.9 \times 10^{-12} n_D n_{He^3} (\overline{\sigma v})_{DHe^3}, \text{ W cm}^{-3} \end{aligned} \tag{21.8}$$

Here in P_{DD} equation it is included D + T reaction.

10. Reaction rates are presented in Table 21.5.

Table 21.5 Reaction rates $\langle \sigma v \rangle$ (in $cm^{-3}\ s^{-1}$) averaged over Maxwellian distributions

Temperature (keV)	D + D, (1a + 1d)	D + T, (2)	D + He, (3)
1.0	1.5×10^{-22}	5.5×10^{-21}	10^{-26}
2.0	5.4×10^{-21}	2.6×10^{-19}	1.4×10^{-23}
5.0	1.8×10^{-19}	1.3×10^{-17}	6.7×10^{-21}
10.0	1.2×10^{-18}	1.1×10^{-16}	2.3×10^{-19}
20.0	5.2×10^{-18}	4.2×10^{-16}	3.8×10^{-18}
50.0	2.1×10^{-17}	8.7×10^{-16}	5.4×10^{-17}
100.0	4.5×10^{-17}	8.5×10^{-16}	1.6×10^{-16}
200.0	8.8×10^{-17}	6.3×10^{-16}	2.4×10^{-16}
500.0	1.8×10^{-16}	3.7×10^{-16}	2.3×10^{-16}
1000.0	2.2×10^{-16}	2.7×10^{-16}	1.8×10^{-16}

Issue Cohen et al. (2003, p. 644)

21.1.5 Theory, Computation and Estimation of Cumulative and Impulse AB-Reactors and Comparison One with Current ICF

For comparison the laser and offer cumulative and Impulse AB methods, we estimate the current ICF laser method. Typical laser installation for ICF has the power 5 MJ and delivers to pellet about 20–50 kJ energy. The pullet has the 1–10 mg liquid (frozen) fuel D + T (density 200 kg/m^3), diameter of the fuel pullet about 1–2 mm, diameter of an evaporative coating 4–10 mm.

Let us take the delivered energy E = 50 kJ, volume of the coating v = 5 mm^3, specific weight of coating γ = 400 kg/m^3 (molar weight μ = 10).

For these data and instant delivery of laser energy the maximum pressure in cover is

$$p = \frac{E}{v} = \frac{5 \times 10^4}{5 \times 10^{-9}} = 10^{13}\frac{N}{m^2} = 10^8 \text{ atm} \tag{21.9}$$

But we don't know what part this pressure transfer to the fuel pellet. Number of nuclear in 1 m^3 of covering is

$$n = \frac{\gamma}{\mu m_p} = \frac{0.4 \times 10^3}{10 \times 1.67 \times 10^{-27}} = 2.4 \times 10^{28}\,(\text{m}^{-3}) \tag{21.10}$$

Here $m_p = 1.67 \times 10^{-27}$ is mass of nucleon (proton) (kg). Temperature of evaporating cover is

$$T = \frac{p}{nk} = \frac{10^{13}}{2.4 \times 10^{28} 1.38 \times 10^{-23}} = 3 \times 10^7\,(\text{K}) \tag{21.11}$$

Here $k = 1.38 \times 10^{-23}$ Boltzmann constant, J/K. Speed of evaporated covering is

$$V = \left(\frac{8kT}{\pi \mu m_p}\right)^{0.5} = \left(\frac{8 \times 1.38 \times 10^{-23} 3 \times 10^7}{3.14 \times 10 \times 1.67 \times 10^{-27}}\right)^{0.5} = 2.51 \times 10^5 \text{ m/s} = 251 \text{ km/s} \tag{21.12}$$

Time of evaporating for thickness of covering $l = 2.10^{-3}$ m is

$$t = \frac{l}{V} = \frac{2 \times 10^{-3}}{2.51 \times 10^5} = 8 \times 10^{-9}\text{ s} \tag{21.13}$$

Let us consider now the process into pellets. The density of fuel particles is

$$n_f = \frac{\gamma}{\mu m_p} = \frac{200}{2.5 \times 1.67 \times 10^{-27}} = 4.8 \times 10^{28} \frac{1}{m^3} \quad (21.14)$$

where $\mu = 2.5$ is average molar mass of fuel D + T. The frozen (liquid) fuel, after converting in gas, has a temperature of about $T = 4$ K. The pressure average speed V_n of particles after conversion of the fuel into gas (plasma) and sound speed V_f to fuel gas at temperature 4 K are:

$$\begin{aligned} p_f &= n_f kT = 4.8 \times 10^{28} \times 1.38 \times 10^{-23} \times 4 = 2.65 \times 10^6 \,\text{N/m}^2 = 26.5\,\text{atm}, \\ V_n &= \left(\frac{8kT}{\pi \mu m_p}\right)^{1/2} = \left(\frac{8 \times 1.38 \times 10^{-23} \times 4}{3.14 \times 2.5 \times 1.67 \times 10^{-27}}\right)^{1/2} = 183 \frac{\text{m}}{\text{s}}, \\ V_f &= \left(\frac{p_f}{\rho_f}\right)^{1/2} == \left(\frac{2.65 \times 10^6}{200}\right)^{1/2} = 115\,\text{m/s}. \end{aligned} \quad (21.15)$$

Additional fuel pressure in ***center*** of pellet from two opposing sound wave bump-up is

$$p_s = \rho_f (2V_f)^2/2 = 200 \times (2 \times 115)^2/2 = 5.3 \times 10^6\,\text{N/m}^2 = 53\,\text{atm} \quad (21.16)$$

Fuel temperature in ***center*** of small mass pellet where two opposing sound (shock) wave bump-up happens is

$$T = \frac{\pi \mu m_p (V_n + V_f)^2}{8k} = \frac{3.14 \times 2.5 \times 1.67 \times 10^{-27} (183 + 115)^2}{8 \times 1.38 \times 10^{-23}} = 10.5\,\text{K} \quad (21.17)$$

In reality, the full pressure and temperature in the center of the capsule is much more. We compute ONLY the sound wave. Any shockwave becomes fast at short distance the sound wave. However, in our case this computation is very complex. Current inertial reactors have the maximal rate of fuel compressing in center of pellet about

$$\xi \approx 600 \quad (21.18)$$

Criterion of ignition (for radius of pullet $R_o = 0.02$ and solid or liquid fuel ρ_o = 0.2 g/cm^3) is

$$\rho R = \rho_o R_o \xi^{2/3} = 0.2 \times 0.02 \times (600)^{2/3} = 0.28 < 1 \quad (21.19)$$

where ρ in g/cm^3, R in cm. That value is not large enough (0.28 < 1).

You can imagine—with just a small effort and we will fulfill the criterion of ignition. Look your attention in very low temperature of fuel Eq. (21.17). For this temperature, the criterion may be wrong, or area of the ignition located into center of pullet may be very small, that energy is very few for ignition of all fuel?

21.1.5.1 Estimation of Cumulative AB Reactor

The proposed Cumulative AB Reactor is an internal rocket engine, which accelerates the small piston (layer) from heavy material by cumulative explosion (Figs. 21.1, 21.1 and 21.3). This piston bumps into pellet of contained nuclear fuel, compresses and heats the pellet up to very high values, producing a nuclear reaction. Most important innovation is in design, the cumulative explosion which works as rocket engine and produces a final speed of the small piston in 10 (and more, from 2 to 20 km/s) times and piston energy in $(10)^2 = 100$ times more than a convention explosion. The second important innovation is the additional heating the fuel by the strong electric impulse. Below is the estimation of the typical parameters of AB reactors.

1. **Final speed of the piston (heavy layer).** Let us to estimate the offered design. It is well known the final speed V of rocket is

$$V = -W_e \ln \frac{M_k}{M} = -W_e \ln \mu_k \qquad (21.20)$$

where W_e is speed exhaust gas of rocket, m/s; M_k is final mass of rocket, kg; M is initial mass of rocket, kg; $\mu_k = M_k/M$ is ratio the final and initial mass of rocket. The distance L (acceleration path) of rocket is

$$L = g_0^{-1} W_e^2 v_0 [1 - \mu_k (1 - \ln \mu_k)] \quad \text{if} \quad \mu_k < 0.05 \quad \text{then} \quad L \approx g_0^{-1} W_e^2 v_0 \qquad (21.21)$$

In Eqs. (21.20)–(21.21) it is used the notations: $g_o = 9.81$ m/s$^2 \approx 10$ m/s^2 is Earth acceleration, $v_o = Mg_o/P_o$ is an initial weight-to-thrust, N/N. The rocket engine has the solid fuel W_e = 2400–2800 m/s, liquid fuel W_e = 3000–3400 m/s, hydrogen-oxygen up W_e = 4000 m/s. The explosive matters have: *TNT*: specific energy E_s = 4.184 MJ/kg $\approx$ 4.2 MJ/kg, density ρ = 1650 kg/m^3, speed of detonation 6900 m/s; *Dynamite*: specific energy up E_s = 7 MJ/kg, standard*s* = 5.3 MJ/kg, density ρ = 1400 kg/m^3, speed of detonation 6000 m/s.

From $E = mV^2/2$ we get the average speed of exhaust gas for TNT:

$$W_e = (2E_s)^{1/2} = (2 \times 4.184 \times 10^6)^{1/2} = 2893 \text{ m/s} \qquad (21.22)$$

Maximum pressure of explosive is

$$p = \frac{E}{v} = \frac{Em}{mv} = E_s \rho = 4.184 \times 10^6 \times 1650 = 6.9 \times 10^9 \,\frac{\mathrm{N}}{\mathrm{m}^2} = 6.9 \times 10^4 \,\mathrm{atm} \tag{21.23}$$

where v is volume of explosive, m^3; $E_s = E/m$ is specific energy of explosion, J/kg. Density of particles and temperature of TNT explosion initial moment of explosion (exhaust gas has average $\mu = 20$, TNT reaction) is:

$$\begin{aligned} n &= \frac{\rho}{\mu m_p} = \frac{1650}{20 \times 1.67 \times 10^{-27}} = 4.94 \times 10^{28}\,\mathrm{m}^{-3}, \\ T &= \frac{p}{nk} = \frac{6.9 \times 10^9}{4.94 \times 10^{28} 1.38 \times 10^{-23}} = 10.1 \times 10^3\,\mathrm{K}, \end{aligned} \tag{21.24}$$

Let us estimate the final speed of the piston (layer) for data: mass of explosive is $M = 10^{-3}$ kg, mass of piston (layer) $M_k = 1$ mg $= 10^{-6}$ kg, $W_e = 2650$ m/s.

$$V = -W_e \ln \frac{M_k}{M} = -2650 \ln \frac{10^{-6}}{10^{-3}} = 2650 \times 3 \times 2.3 \approx 18{,}300\,\mathrm{m/s} \approx 18\,\mathrm{km/s} \tag{21.25}$$

Let us to find the minimal acceleration distance L of piston (minimal distance from lower part of explosive to center of sphere cartridge). For Version 1 we receive:

$$\begin{aligned} &\text{From } L = g_0^{-1} W_e^2 v_o, \quad v_o = M g_0 / F, \quad m_c = \rho V_d S, \\ &F = ma = m_c W_e = \rho V_d S W_e, \quad v_o = M g_0 / \rho V_d S W_e, \quad V_d = W_e. \\ &\text{we get } L = \frac{M}{\rho S} = \frac{10^{-3}}{1.65 \times 10^3 \times 6 \times 10^{-4}} = 0.1 \times 10^{-2}\,\mathrm{m} = 1\,\mathrm{mm} \end{aligned} \tag{21.26}$$

where $g_o = 9.81$ m/s^2 is gravitation; W_e is speed of rocket exhaust gas, m/s; ρ is rocket fuel density, kg/m^3; V_d is rate (speed) of combustion of rocket fuel, m/s; S is initial area of the combustion created rocket thrust, m^2; F is initial rocket thrust, N/m^2, M is mass of rocket fuel, kg. We can change the speed V_d to add special additives.

2. **Temperature *T* and pressure *p* in pellet** after compressing by piston is (for piston speed $V = 1.8.10^4$ m/s and density of piston $\rho = 2.10^4$ kg/m^3, $\mu = 200$):

$$p = \frac{\rho (2V)^2}{2} = \frac{2 \times 10^4 (2 \times 1.8 \times 10^4)^2}{2} = 1.36 \times 10^{13}\,\frac{\mathrm{N}}{\mathrm{m}^2} = 1.36 \times 10^8\,\mathrm{atm} \tag{21.27}$$

Here we take piston speed 2 V because we have two opposed pistons. Temperature

$$\text{From } E = \frac{\mu m_p V^2}{2} = \frac{3kT}{2} \text{ we have } T = \frac{\mu m_p V^2}{3k} \tag{21.28}$$

The mixture D + T has $\mu = 2.5$, piston about $\mu = 200$ (for example: tungsten has $\rho = 19.34 \times 10^3$ kg/m^3, $\mu = 184$; uran-238 has $\rho = 19.1 \times 10^3$ kg/m^3, $\mu = 238$; lead has $\rho = 11.35 \times 10^3$ kg/m^3, $\mu = 207$). The temperature of mixture D + T is:

$$\text{For fuel D+T } T = \frac{\mu m_p V^2}{3k} = \frac{2.5 \times 1.67 \times 10^{-27}(1.8 \times 10^4)^2}{3 \times 1.38 \times 10^{-23}} = 34.5 \times 10^3 K \approx 3.14\,\text{eV},$$
$$\text{For piston } T = \frac{\mu m_p V^2}{3k} = \frac{200 \times 1.67 \times 10^{-27}(1.8 \times 10^4)^2}{3 \times 1.38 \times 10^{-23}} = 2.61 \times 10^6 K = 225\,\text{eV}, \tag{21.29}$$

The mass of the piston is in 5–20 times more than the mass of the fuel and the piston has direct contact to the fuel. That means the fuel will have a temperature about 3 million degrees. That is less than the needed value of 10 keV but in thousands of time more than in a laser method. In the offered method, we also have much more pressure than in the laser method. The high pressure significantly decreases the needed temperature because one decreases the need for distance between nuclear particles.

3. **Estimation of the criterion of ignition the Cumulative AB Reactor**. The process of compression converted the solid-liquid fuel into gas. In according with Eq. (21.15) the initial pressure this gas is $p_o = 26.5$ atm. In accordance with Eq. (21.27), the final pressure is about $p = 1.44.10^8$ atm. The rate of fuel compression is

$$\xi = \frac{p}{p_0} = \frac{1.44 \times 10^8}{26.5} \approx 5.43 \times 10^5 \tag{21.30}$$

(compare this reached value with maximum $\xi = 600$ in laser method. The Cumulative AB Reactor has compression many times more than laser method). That means the liner size of fuel pellet will be in $(\xi)^{1/3} = 82$ times less. If capsule has initial diameter $D = 0.1$ cm (fuel mass = 21 $\times$ 10^{-6} kg, $\rho_o = 0.2$ g/cm^3), one has $R = 0.05/82 = 6.1 \times 10^{-4}$ cm. The offered Cumulative AB thermonuclear reactor produced *direct* compression almost a thousand times greater than the usual shock wave laser compression machines at the center of a fuel pullet. The density of the fuel will be $\rho = \rho_o\,\xi = 0.2 \times 5.43 \times 10^5 = 1.1.10^5$ g/cm^3. The criterion of the inertial ignition is

$$\rho R = 1.1 \times 10^5 \times 6.1 \times 10^{-4} = 67 > 1 \tag{21.31}$$

One is in 67 times more than needed (67 > 1 > 0.28). That means we can use instead of very expensive tritium the deuterium which is the thousands times cheaper. The corresponding reactions are:

$$\mathrm{D+T \rightarrow {}^4He(3.5\,meV) + n(14.1\,meV);}$$
$$\mathrm{D+D \rightarrow T(1.01\,meV) + p(3.02\,meV)\ 50\%}$$
$$\mathrm{D+D \rightarrow {}^3He(0.82\,meV) + n(2.45\,meV)\ 50\%}$$

The deuterium cannot be used in the laser reactor because one requests in 100 times more ignition criterion then D + T. But, as you see in Eq. (21.31), one may be used in AB reactor (Fig. 21.6) with an additional heating by electric charge.

The ^{3}He is received in deuterium reaction may be used in next reactions:

$$\mathrm{D + {}^3He \rightarrow {}^4He(3.6\,meV) + p(14.7\,meV);}$$
$$\mathrm{{}^3He + {}^3He \rightarrow {}^4He + 2p(12.9\,meV).}$$

They produce only high-energy protons which can be directly converted in electric energy. Last reactions do not produce radio isotopic matters (no neutrons). Reaction D + D has the other distinct advantages:

1. One produces the protons which energy can be converted directly to electric energy.
2. One produces the tritium which is expensive and may be used for thermonuclear reaction.
3. One produces less and low energy neutrons which create radioactive matters.

The other important advantage is using the pellets with compression gas fuel. Let us take a micro-balloon (pellet) having fuel gas with p_o = 100 atm, radius 0.1 cm, temperature 300 K. The mass fuel will be 4.19 mg. The compression rate is $\xi = p/p_o = 1.44.10^8/100 = 1.44 \times 10^6$. Liner size decreases by $(\xi)^{1/3} = 113$ times.

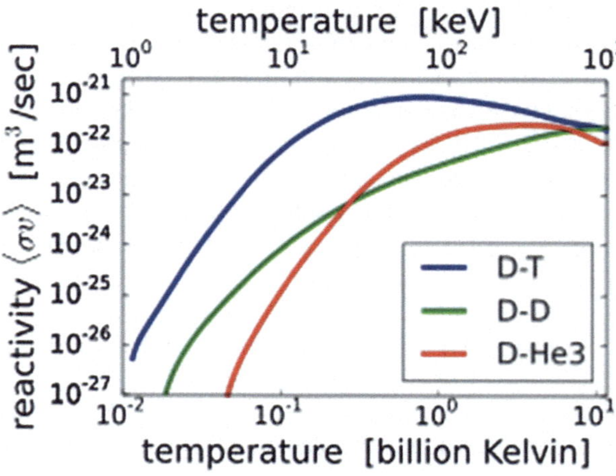

Fig. 21.6 Reactivity is requested for thermonuclear reaction

The radius of compressed fuel pellet will be $R = 0.1/113 = 0.88 \times 10^{-3}$ cm. The initial density is

$$\rho_0 = \frac{\mu m_p p_0}{kT_0} = \frac{2.5 \times 1.67 \times 10^{-27} 10^7}{1.38 \times 10^{-23} 300} = 10 \frac{\text{kg}}{\text{m}^3} = 10^{-2} \frac{\text{g}}{\text{cm}^3} \tag{21.32}$$

and inertial criterion is

$$\rho R = \rho_0 R_0 \xi^{2/3} = 10^{-2} 0.1 \times (1.44 \times 10^6)^{2/3} = 12.7 > 1. \tag{21.33}$$

Criterion is good for compressed fuel D + T, but it is small for fuel D + D. For fuel D + D we must decrease pressure in pellet up 400–1000 atm or increase diameter (and power) our installation or use the additional heating of fuel by strong electric impulse. Compressed micro-balloon (pellet) is more comfortable for working because it is unnecessary to store the fuel at lower temperature.

21.1.5.2 Estimation of Other Parameters the Cumulative AB Reactor

1. *Thermonuclear energy*. One mg (10^{-6} kg) of thermonuclear fuel D + T has energy: Number of nucleus:

$$n_1 = \frac{M}{\mu m_p} = \frac{10^{-6}}{2.5 \times 1.67 \times 10^{-27}} = 2.4 \times 10^{20} \tag{21.34}$$

   One pair of nuclear D + T produces energy E_1 = 17.6 meV. The n_1 nuclear particles contain the energy

$$\begin{aligned} E &= 0.5\, n_1 E_1 = 0.5 \times 2.4 \times 10^{20} 17.6 \times 10^6 = 21.1 \times 10^{26} \text{ eV} \\ &= 21.1 \times 10^{26} 1.6 \times 10^{-19} = 3.38 \times 10^8 \text{ J} \end{aligned} \tag{21.35}$$

   If coefficient efficiency of the Cumulative AB Reactor is $\eta = 0.3$, one mg of fuel produces the energy of 100 million joules. If we make one explosion per second, the installation has the power of 100 million watts. The part of this energy will be produced inside fuel microcapsule fuel pellet (3.5 meV from ^{4}He, $E = 6.72.10^7$ J) the most of energy (14.1 meV from neutrons) will be produced into the big containment sphere. Conventional coefficient of nuclear reactor efficiency is about 0.3–0.5, the steam (gas) turbine is about 0.9.
2. *Energy is delivered by piston to fuel capsule* is $E = \text{mV}^2/2$. For $m = 5$ mg, piston speed $V = 2 \times 10^4$ m/s final piston energy is $E = 2 \times 10^3$ J. That is less then typical energy 20–50 kJ delivered by laser installation. However, laser energy is spent in vaporizing the cover of the fuel pellet and only small part as shock wave reaches the center of fuel pellet mass. In Cumulative AB Reactor, all piston energy passes directly into the target fuel pellet. The piston energy is

easy to increase up 20 kJ by increasing the piston mass and piston speed (also by using more explosive). The piston mass hinders the fuel micro-balloon and increases the nuclear reaction time in many times.
Part of this energy will be used for ionization of the fuel. One mg of fuel, for its ionization, requests $E = n_1 \cdot 13.6$ eV = 522 J, compression of solid fuel about $E = 624$ J, compression of gas fuel from $p = 100$ atm $E = 105$ J. That is a part of the derived piston energy.

3. *Reaction of explosive* TNT is $2C_7H_5N_3O_6 = 3N_2 + 5H_2O + 7CO + 7C$.
4. *Estimation of pressure and temperature after nuclear explosion.* Let us to find the pressure and temperature after thermonuclear explosive the one mg fuel D + T. Number of nuclear particles in sphere 1 m^3 is

$$n_n = \frac{M}{\mu m_p} = \frac{10^{-6}}{2.5 \times 1.67 \times 10^{-27}} = 2.4 \times 10^{20} \frac{1}{m^3} \quad (21.36)$$

Full thermonuclear energy

$$E_n = 0.5 n_n E_1 = 0.5 \times 2.4 \times 10^{20} 17.6 \times 10^6 = 21.1 \times 10^{26}\,\text{eV} = 3.38 \times 10^8\,\text{J} \quad (21.37)$$

Number of air particles with air density $\rho = 1.225$ kg/m^3 in pressure $p = 1$ atm is

$$n_o = \frac{M}{\mu m_p} = \frac{1.225}{28 \times 1.67 \times 10^{-27}} = 2.6 \times 10^{25} \frac{1}{m^3} \quad (21.38)$$

If coefficient efficiency of thermonuclear reaction is $\eta = 0.3$ in volume 1 m^3:

$$p = \frac{\eta E_n}{v} = \frac{0.3 \times 3.38 \times 10^8}{1} = 1 \times 10^8 \approx 10^8 \frac{N}{m^2} = 1000\,\text{atm} \quad (21.39)$$

Total pressure—nuclear explosive together with chemical explosive—is $p \approx 1000$ atm. Temperature of gas mixture of explosive plus nuclear fuel is

$$T = \frac{p}{(n_0 + n_n)k} = \frac{10^8}{(2.6 \times 10^{25} + 2.4 \times 10^{20}) \times 1.38 \times 10^{-23}} = 279 \times 10^3\,\text{K} \quad (21.40)$$

If we increase the initial pressure into reactor body up 100 atm, that the temperature decreases to 2790 K. The same temperature is in a combustion chamber of conventional engine of the internal combustion. We can use the conventional cooling system. The same method may be used for estimation of injection water into installation body or any garbage material in a space ship (or asteroid).

5. *Thickness of sphere cover.* Assume the spherical cover is made from conventional steel having safety tensile stress $\sigma = 50$ kg/mm^2 = 5.10^8 N/m^2. The full tensile force is $F = \pi r^2 p = 3.14 \times 0.5^2 \times 10^8 = 0.785 \times 10^8$ N. Requested area of steel is $S_r = F/\sigma = 0.785.10^8/5 \times 10^8 = 0.157$ m^2. The thickness of sphere wall is $\delta = S_r/2\pi r = 0.157/2 \times 3.14 \times 0.5 = 0.05$ m. Mass of sphere is $M_c \approx \gamma S_s \delta = 7800 \times 4.536 \times 0.05 = 1769$ kg. Here S_s is average surface of sphere. If we use the more strong material for sphere wall, for example: 1 μm iron whisker having safety tensile stress $\sigma \approx 400$ kg/mm^2 = 4×10^9 N/m^2, we decrease the sphere's mass by 4–8 times. We can also make the sphere wall from composite materials (example: an artificial fiber carbon or glass having safety stress $\sigma \approx$ 100–150 kg/mm^2 and density γ = 1500–2700 kg/m^3).
6. *Cooling the sphere by water.* If explosions are very frequent, we then can decrease the wall or/and gas temperature by injection of the chilled or room temperature water. The water also protects our installation from high-energy neutrons in other words, it behaves as a shielding materials. Let us estimate the amount of water which decreases the temperature and pressure of gas (at most steam H_2O) into sphere for magnitudes acceptable for current steam turbines: T = 400 °C = 672 K. The critical point of water (triple point) is T = 273 °C, p = 22 MPa. Heating 1 kg water from 20 to 100 °C requests energy $E = C_p \Delta T = 4.19.80 = 333$ kJ, evaporation—r = 2260 kJ, heating of steam up 400 °C—$E = C_p \Delta T = 1.05.300 = 315$ kJ. Total amount of water heat energy is $E_w = 333 + 2260 + 315 = 2908$ kJ/kg. Total mass of water for nuclear efficiency $\eta = 1$ equals $M_w = E/E_w = 3.4 \times 10^8/2.9 \times 10^6 = 117$ kg. For $\eta = 0.3$ $M_w = 35$ kg. The 2–3 cm of water thickness protects the installation from high energy of neutrons produced by reaction D + T. Unfortunately, the injection of water before decompressing strongly decreases the efficiency of installation.
7. *Run protons and heavy nuclear particles.* The physic directory by Kikoin (1975, p. 953) gives the following equation for running the protons and charged heavy particles inside gas at pressure 1 atm

$$R_x(E) = \frac{m_x}{m_p} R_p\left(\frac{m_p}{m_x} E\right) \tag{21.41}$$

where R_x is run of the investigated particles, m_x is mass of investigated particles, m_p is mass of proton, R_p is run of known particles in a known environment, E is energy of particles in meV. The run of proton in H_2 at pressure 1 atm is in the Table 21.6.

Table 21.6 Run (range) of proton in gas H_2 at pressure 1 atm

Energy E (meV)	1	10	100
Run R (cm)	10	5×10^2	2×10^4

For particles [4]He (3.5 meV) in reaction D + T under the piston pressure $p = 10^8$ atm the run is

$$R_x(E) = \frac{m_x}{m_p} R_p \left(\frac{m_p}{m_x} E \right) / p \approx \frac{4}{1} R_p \left(\frac{1}{4} 1 \times 3.5 \right) / 10^8 \approx 4 \times 10/10^8$$
$$= 4 \times 10^{-7} cm \approx 4 \times 10^{-6} \text{mm} \quad (21.42)$$

The closed run has proton. That means the all energy of the charges particles after nuclear reaction is used for heating other "cold" particles. If probability of an initial reaction is more than 10 keV/3500 keV = 1/350, the chain reaction and ignition will occur.
In the Cumulative AB Reactor these conditions are in *whole* fuel capsule, in laser reactor of many times lower conditions may be *only* in center of fuel capsule (collision of the imposed shock waves). If reacted particles run out the center of capsule, its energy will wasted.
The run way of neutrons is large and very complex function of energy and conditions of Environment.

8. *Converting the nuclear energy of Cumulative AB reactor to electric, mechanical energy or a rocket thrust.* The best means for converting a Cumulative AB Reactor nuclear energy is magneto hydrodynamic electric generator (MHD-generator) which converts with high efficiency the high temperature and high pressure plasma directly in electric energy. Together with capacitors one can produces continuous electric currency. Impulse work of reactor allows to cool the reactor by injection the cooler (or conventional cooling) and protect the Cumulative AB Reactor installation from very high temperature.
The second way for converting a Cumulative AB Reactor nuclear energy is conventional heat exchanger and gas turbine. As cooler may be used the FLiBe —melted mix of fluoride salts of lithium and beryllium. The third way is injection of water inside sphere and steam turbine as description over.
9. *Using the Cumulative AB reactor as an impulse space rocket engine.* There are good prospects (possibility) to use the suggested Cumulative AB Reactor as an impulse rocket engine. If plasma will flow from sphere to space the average speed V of jet is

$$\text{From } E = \frac{mV^2}{2} \text{ we get } V = \left(\frac{2E}{m} \right)^{1/2} = \left(\frac{2 \times 10^8}{5 \times 10^{-3}} \right)^{1/2} = 4 \times 10^5 \, \frac{\text{m}}{\text{s}} \quad (21.43)$$

Here E is nuclear energy in one impulse one mg nuclear fuel, J; m is the mass injected to outer space (fuel cartridge together with conventional explosive), kg. Received speed V = 400 km/s is in many times more than a current exhaust chemical speed 3 km/s. If of space apparatus has mass m_2 = 1 ton, the ship speed changes in $V_2 = (m/m_2)\, V_1$ = 2 m/s in one impulse. If we spend 10 kg of fuel cartridges, the apparatus get speed 10 km/s. More importantly, the next possibility is of the rocket powered by the Cumulative AB Reactor. Any matter from any planets, asteroids, space body may be used as fuel used for increasing the derivation of impulses. For example, assume the captured solid object moving through space is composed of some water, and we filled rocket tanks using that mined planet, comet or asteroid water. From Eq. (21.43) and Law of equal impulse we have from every impulse

$$V_1 = (2Em_1)^{1/2}/m_2 = (2 \times 10^8 \times 16)^{1/2}/10^3 = 56.6\,\text{m/s} \qquad (21.44)$$

Here V_1 is add speed m_1 mass jet kg, m_1 = 16 kg of water; m_2 is mass of space apparatus.

10. Estimation of the neutron penetration

$$l = 1/n\sigma, \qquad (21.45)$$

where l is path of penetration, cm; n is density of material, 1/cm^3; $\sigma = 10^{-24}$ cm^2 is cross section of the nuclear. For steel l = 12 cm, for compressed air up 100 atm the l = 410 cm.

11. Requested thickness of the spherical shell is

$$\frac{D}{d} = \left(\frac{p}{\sigma} + 1\right)^{0.5} \qquad (21.46)$$

where D is outer diameter of spherical shell, d is inner diameter of spherical shell, p is pressure, atm; σ is safety tensile stress kg/cm^2. Example, if p = 10 kg/mm^2, σ = 50 kg/mm^2, then D/d ≈ 1.1.

21.1.5.3 Detailed Estimation of Cumulative and Impulse Reactors for Transportation Engine

1. **Estimation of nuclear energy (power).** Let us make more detail estimation the Cumulative and Impulse reactors for engine of transport vehicle having the fuel pellet 0.1 mg ($M_f = 10^{-7}$ kg) with fuel D + T or D + D. The Impulse reactor has pressure into pellet 300 atm. Estimation of energy (power) this D + T pellet if the coefficient efficiency is η = 0.5. The couple nuclei T + D produces nuclear energy E_1 = 17.6 meV. Number N_f of nuclei in pellet is:

$$N = \frac{M_f}{\mu m_p} = \frac{10^{-7}}{2.5 \times 1.67 \times 10^{-27}} = 2.4 \times 10^{19}$$

Here μ is average molar mass of D + T; m_p is mass of proton, kg. The nuclear energy of 1 mg D + T fuel in 1 Hz is

$$E = 0.5E_1N\eta = 0.5 \times 17.6 \times 10^6 \times 1.6 \times 10^{-19}2.4 \times 10^{19} \times 0.5 = 16.9 \times 10^6$$
$$\approx 17\,\text{MJ/Hz}$$

That is power energy of the 2–5 power aviation turbo-engines. If one cycle in second (1 Hz) is not enough, we can decrease the frequency. The piston engine has up 50–70 revolution per second, the high speed aviation gun up 30 shots in second. If we use the D + D fuel having single energy E_1 = 3.65 meV, μ = 2, the nuclear energy is approximately in 5 times less because E_1 is less.

2. **Size of cartridge and pellet**. Let us estimate the size of the cumulative cartridge for mass the explosive TNT M_e = 1 g ($M_e = 10^{-3}$ kg, energy E_e = 4.2 NJ, density ρ = 1650 kg/m^3) and internal diameter cartridge is d = 10 mm. The thickness δ of explosive is:

$$\delta \approx \frac{M_e}{4\pi r^2 \rho} = \frac{1}{4 \times 3.14 \times 0.5^2 \times 1.65} = 0.2\,\text{cm}$$

Outer diameter of cartridge for safety tensile stress 100 kg/mm^2 is D = 16 mm. Let us estimate the compressed **pellet** having gas mass $M = 10^{-7}$ kg, pressure p = 300 atm = 3.10^7 N/m^2 and T = 300 K. Specific density the gas D + D, D + T in compression p = 1 atm is ρ_o = 0.1 kg/m^3, atm. The internal radius of gas pellet is:

$$r = \left(\frac{3M}{4\pi p \rho_0}\right)^{1/3} = \left(\frac{3 \times 10^{-7}}{4 \times 3.14 \times 3 \times 10^2 0.1}\right)^{1/3} \approx 0.926 \times 10^{-3}\,\text{m} \approx 1\,\text{mm}$$

The relative outer diameter of pellet for pressure p = 3 kg/mm^2 and the safety tensile stress of the pellet cover σ = 50 kg/mm^2 with according Eq. (21.46) is

$$\frac{D}{d} = \left(\frac{p}{\sigma} + 1\right)^{0.5} = \left(\frac{3}{50} + 1\right)^{1/2} = 1.03$$

Nuclear processes into pellet. After cumulative explosive into cartridge the density of fuel D + T into pellet after cumulative compressing is

$$n = \frac{3p}{\mu m_p V_p^2} = \frac{3 \times 1.36 \times 10^{13}}{2.5 \times 1.67 \times 10^{-27}(18 \times 10^3)^2} = 3 \times 10^{25}\,\text{cm}^{-3}$$

where p is pressure after cumulative compressing, N/m^2, V_p is final speed of piston, m/s, m_p is mass of proton, kg. Density of D + D fuel is n = 3.75.10^{25} cm^{-3}. Time of fuel combustion for T = 15 keV is

$$\text{For D}+\text{T}\,t=\frac{0.5\eta E_1}{5.6\times 10^{-13}n\langle\sigma v\rangle}=\frac{0.5\times 0.5\times 2.82\times 10^{-12}}{5.6\times 10^{-13}3\times 10^{25}\,2.65\times 10^{-16}}=1.58\times 10^{-10}\,\text{s},$$

$$\text{For D}+\text{D}\,t=\frac{0.5\eta E_1}{3.3\times 10^{-13}n\langle\sigma v\rangle}=\frac{0.5\times 0.5\times 0.58\times 10^{-12}}{5.6\times 10^{-13}\,3.75\times 10^{25}3.2\times 10^{-18}}=3.7\times 10^{-9}\,\text{s},$$

where η is coefficient efficiency, E_1 is energy couple nuclei (for D + T, E_1 = 17.6 meV.1.6 × 10^{-19} = 2.82.10^{-12} J; for D + D, E_1 = 3.65 meV = 0.58 × 10^{-12} J). Here we used Eq. (21.8) and Table 21.5. For primary compressed gas fuel pellet p = 300 atm without cumulative compressing, the density of the fuel gas into pellet for T = 300 K is

$$n=\frac{p}{kT}=\frac{3\times 10^7}{1.38\times 10^{-23}300}=2.25\times 10^{21}\,\text{cm}^{-3}$$

where k = 1.38 × 10^{-23}—is Boltzmann constant. The time of nuclear fuel combustion for T = 15 keV is

$$\text{For D}+\text{T}\,t=\frac{0.5\eta E_1}{5.6\times 10^{-13}n\langle\sigma v\rangle}=\frac{0.5\times 0.5\times 2.82\times 10^{-12}}{5.6\times 10^{-13}\,2.25\times 10^{21}\,2.65\times 10^{-16}}=2.1\times 10^{-6}\,\text{s},$$

$$\text{For D}+\text{D}\,t=\frac{0.5\eta E_1}{3.3\times 10^{-13}n\langle\sigma v\rangle}=\frac{0.5\times 0.5\times 0.58\times 10^{-12}}{5.6\times 10^{-13}\,2.25\times 10^{21}\,3.2\times 10^{-18}}=6.14\times 10^{-5}\,\text{s},$$

As you see, the combustion time significantly is increased but it is enough for reaction. We can decrease it if we increase density of fuel.

Estimation of electric condenser. For heating of fuel we use the short strong electric impulse. For impulse the electric condenser may be used. Let us to estimate the condenser parameters for getting the fuel temperature T = 15 keV. If fuel mass is M = 1 mg = $10^{-7\ \text{kg}}$, the number of nuclei for D + T is

$$N=\frac{M}{\mu m_p}=\frac{10^{-7}}{2.5\times 1.67\times 10^{-27}}=2.4\times 10^{19}$$

For D + D the N = 3.10^{19}. The energy is needed for heating the fuel D + T up T = 15 keV is

$$W=NT\times 1.6\times 10^{-19}=2.4\times 10^{19}\,15\times 10^3 1.6\times 10^{-19}\approx 60\,\text{kJ}$$

For heating D + D fuel is W = 72 kJ. The minimal specific weight of conventional conductor according (Bolonkin 2014b, p. 368) is γ = 2 kJ/kg. Consequently, the requested mass of condenser is about 30–36 kg. But if we can use the advanced supercapacitor (γ = 10 kJ/kg) or ultracapacitor (γ = 20 kJ/kg) or capacitor EEStor, having claimed capacity γ = 1000 kJ/kg, we can decreased the capacitor mass. In any case, the capacitor mass is small part of thermonuclear engine.

Estimation of capacitor discharge.

(a) Need condenser *after cumulative compressing* of pellet for heating fuel up $T = 15$ keV. Assume the initial temperature of cumulative compressed gas fuel is $T = 3.14$ eV, mass of fuel $M = 10^{-7}$ kg, initial pressure $p = 300$ atm, initial diameter of pellet $d = 0.2$ cm. The specific electric Spitzer resistance of plasma is

$$\rho = \eta_{\perp} = 1.03 \times 10^{-2} Z \ln \Lambda \cdot T^{-3/2}.$$

where Z is rate of charge, ln Λ = (5–15) is Columbus logarithm. For ln Λ = 10 we have

$$\rho = 1.03 \times 10^{-2}\, 1 \times 10/3.14^{3/2} = 1.85 \times 10^{-3}\, \Omega\,\text{cm}.$$

Diameter of the cumulative compressed pellet having initial gas pressure $p_o = 300$ atm and $l_o = 0.2$ cm is

$$l = l_0 \left(\frac{p_0}{p}\right)^{1/3} = 0.2 \left(\frac{300}{3.24 \times 10^7}\right)^{1/3} = 4.2 \times 10^{-3}\ \text{cm},$$
$$s = \frac{3.14}{4} l^2 = 1.4 \times 10^{-5}\ \text{cm}^2$$

Electric resistance is

$$R = \rho \frac{l}{s} = 1.85 \times 10^{-3} \frac{4.2 \times 10^{-3}}{1.4 \times 10^{-5}} = 0.555\, \Omega\,\text{cm}$$

where l is diameter of pellet, cm; s is cross-section area, sm^2. Needed initial voltage and currency of condenser for time of recharge $t = 10^{-5}$ s is

$$U = \left(\frac{RW}{t}\right)^{1/2} = \left(\frac{0.555 \times 6 \times 10^4}{10^{-5}}\right)^{1/2} = 57.7\,\text{kV},$$
$$I = \left(\frac{W}{Rt}\right)^{1/2} = \left(\frac{6 \times 10^4}{0.555 \times 10^{-5}}\right)^{1/2} = 104\,\text{kA}$$

Capacity of condenser

$$C = \frac{t}{R} = \frac{10^{-5}}{0.555} = 18 \times 10^{-6} F$$

(b) Need condenser *without* cumulative compressing of pellet. Initial data: Initial temperature is $T = 0.1$ eV, mass of fuel $M = 10^{-7}$ kg, pressure $p = 300$ atm, diameter of pellet $d = 0.2$ cm, final temperature T = 15 keV.

$$\rho = 1.03 \times 10^{-2} 1 \times 10/0.1^{3/2} = 3.16\,\Omega\,\text{cm}.$$

Cross-section area of the pellet having fuel gas pressure p_o = 300 atm and diameter of pellet $d_o = l_o$ = 0.2 cm is

$$s = \pi d_0^2/4 = 0.0314\,\text{cm}^2$$

Electric resistance is

$$R = \rho \frac{l_0}{s} = 3.16 \frac{0.2}{0.0314} = 20\,\Omega\,\text{cm}$$

where l is diameter of pellet, cm; s is cross-section area, sm^2. Needed initial voltage and currency of condenser for time of recharge $t = 10^{-5}$ s is

$$U = \left(\frac{RW}{t}\right)^{1/2} = \left(\frac{20 \times 6 \times 10^4}{10^{-5}}\right)^{1/2} = 346\,\text{kV},$$
$$I = \left(\frac{W}{Rt}\right)^{1/2} = \left(\frac{6 \times 10^4}{20 \times 10^{-5}}\right)^{1/2} = 54.8\,\text{kA}$$

Capacity of condenser

$$C = \frac{t}{R} = \frac{10^{-5}}{20} = 7 \times 10^{-7} F$$

The specific energy weight γ_c (J/kg) of the condenser my be estimate by formulas

$$\gamma_c = \frac{\varepsilon_0 \varepsilon E_q^2}{\gamma}$$

where $\varepsilon_o = 8.85 \times 10^{-12}$ F/m is electric constant; $\varepsilon \approx 3$ dielectric constant; $E_q \approx$ 160–640 MV/m is safety electric stress of isolator; $\gamma \approx$ 1000–3000 kg/m^3 is specific weight of isolator.

Initial magnetic pressure from charged currency is

(a) Pellet having cumulative compressing has initial currency I = 104 kA, radius of pellet after compressing $r = 2.1 \times 10^{-5}$ m has magnetic intensity H and magnetic pressure p:

$$H = \frac{I}{2\pi r} = \frac{104 \times 10^3}{2 \times 3.14 \times 2.1 \times 10^{-5}} = 7.9 \times 10^8 \frac{\text{A}}{\text{m}},$$
$$p = \frac{\mu_0 H^2}{2} = \frac{4\pi \times 10^{-7} (7.9 \times 10^8)^2}{2} = 4 \times 10^{11} \frac{\text{N}}{\text{m}^2} = 4 \times 10^6\,\text{atm}$$

That is closed to piston pressure 3.4×10^7 atm.
(b) Without cumulative pressure $H = 0.87.10^7$ A/m and $p = 477$ atm.

Cost of the nuclear fuel.

Deuterium. The sea water contains about 1.55×10^{-4} %. The world produces about ten thousand tons in year. Cost 1 $/g.

Tritium. The special nuclear reactors can produce it. Now the cost is 30,000$/g. In future an expected cost will be from 100 K–200 K $/g.

Helium-3. Very rare isotope. The Helium-4 contains $1.3 \times 10^{-6}/1$ of the Helium-3. Cost is 30 K $/g. One project offers to extract it on Moon and delivery to Earth.

Lithium 6–7. Nature mixture costs 150 $/kg.

Uranium-238 contains 0.7% of Uranium-235. It cost 90–250 $/kg.

Plutonium-239. Cost 5600 $/g.

As you see the thermonuclear fuel D + D is the cheapest, but D + T has the lowest temperature for thermonuclear reaction. All the current experimental thermonuclear installations are using the D + T.

Properties of materials

Properties of some materials suitable for the offered installation are shown in Table 21.7.

21.1.5.4 Discussion

Existing thermonuclear reactors are very complex, expensive, large, and heavy. They cost many billions of US dollars and require many years for their design, construction and prototype testing. They cannot stably achieve the nuclear ignition and the Lawson criterion. In future, they will have a lot of difficulties with acceptable cost of nuclear energy, with converting the nuclear energy to conventional energy, with small thermonuclear installation suitable for transportation or space exploration. Scientists promise an industrial application of thermonuclear energy after 10–15 years additional researches and new billions of US dollars in the future. But old methods will not allow us to reach an industrial or transport engine it

Table 21.7 Properties of some materials suitable for the offered installation

Material	Tensile strength (kg/mm^2)	Density (g/cm^3)	Fibers	Tensile strength (kg/mm^2)	Density (g/cm^3)
Steel A514	76	7.8	S-glass	471	2.48
Aluminum alloy	45.5	2.7	Basalt fiber	484	2.7
Titanium alloy	90	4.51	Carbon fiber	565	1.75
Steel Piano wire	220–248	7.8	Carbon nanotubes	6200	1.34

Issue Bolonkin (2014b, p. 370)

in the near future. In inertial confinement many scientists thought that short pressure (10^{-9}–10^{-12} s), which they can reach by laser beam, compress the fuel capsule, but this short pressure only create the shock wave which produced the not large pressure and temperature in a limited range area in center of fuel capsule. The scientists try to reach it by increasing NIF, but plasma from initial vaporization the cover of fuel capsule does not allow to delivery big energy. After laser beam, the fuel capsule is "naked" capsule. Capsule cannot to keep the high-energy particles of the nuclear ignition and loss them. Producing the power laser beam is very expensive and has very low efficiency (1–1.5%). The offer method does not have these disadvantages. One directly presses fully the fuel capsule to high pressure and temperature by piston, one covers the capsule by piston mass in 10–30 times more them mass of fuel, protects the fuel by the heavy elements having high number of nucleons A and charges Z. They reflect the light protons, D, T, repels high-energy reacted particles (^{4}He, p) back to fuel and significantly (in hundreds time) increasing the conformation time.

It is important also that all chemical (explosive) energy used for compression of the fuel capsule remains into sphere and may be used for useful work. The cumulative idea cannot be used for thermonuclear reaction in its classical form. Produced pressure and temperature are not enough for thermonuclear reaction. The main author innovation is using the rocket thrust of explosive for acceleration piston for very high speed (from 2 km/s up 20 km/s and more). That increases the kinetic energy the piston in hundreds times. Author noted that the mass of fuel and piston is very small and allows reaching the high rocket speed of pressing by piston.

The impotent innovations are the compressed gas pellets at room temperature and electric impulse for heating of pellet up the thermonuclear temperatures. The current ICF uses the frozen fuel about absolute zero. That is not acceptable for practice. Author also suggested the transport nuclear engine and nuclear rocket.

The method possible allows to use reaction D + D (instead D + T) with cheap nuclear fuel D (Tritium is very expensive—about 30,000 USD per 1 g, deuterium costs 1 \$/g). One also allows using the compressed fuel-gas at room temperature.

The cumulative method also may be used for getting the metal hydrogen, which has super-conductivity at room temperature and high thermal capacity.

21.1.5.5 Conclusion

The author offers a new small cheap cumulative and impulse inertial thermonuclear reactors, which increases the pressure and temperature of a nuclear fuel in thousands times, reaches the ignition and full thermonuclear reaction. Cumulative and Impulse AB Reactor, herein offered by its originator, contains several innovations and inventions.

Main of them is using a moved explosive, which allows to accelerate the special piston to very high speed (more than 20 km/s) which (as it is shown by computations) compresses the fuel capsule in million times and heating up the million degrees of temperature. The second main innovation is the additional heating the fuel pellet by electric impulse to up temperature in 15 keV and more (hundreds millions of degrees). Important innovation is compressed pellet at room temperature, installation for electric and mechanical energy and thermonuclear rocket.

The offered reactor is small, cheap, may be used for cheap electricity, as engine for Earth transportation (train, truck, ships, aircraft, rocket), for space apparatus and for producing small and cheap and powerful weapons. Closed ideas are in Bolonkin (2005, 2006a, 2009, 2011, 2012a, 2014a, b, 2015a, b).

21.2 Ultra-Cold Thermonuclear Synthesis: Criterion of Cold Fusion

21.2.1 Introduction

A recent development in the field of laser-induced ICF is the use of ultra-short pulse multi-petawatt lasers to heat the plasma of an imploding pellet at exactly the moment of greatest density after it is imploded conventionally using terawatt-scale focused lasers (Fig. 21.7). This research will be carried out on the (currently being built) OMEGA EP peta-watt and OMEGA lasers at the University of Rochester in New York and at the GEKKO XII laser at the Institute for Laser Engineering in Osaka, Japan which, if fruitful, may have the effect of greatly reducing the cost of a laser fusion-based power source.

At the temperatures required for fusion, the fuel is in the form of plasma with very good electrical conductivity. This opens the possibility to confine the fuel and

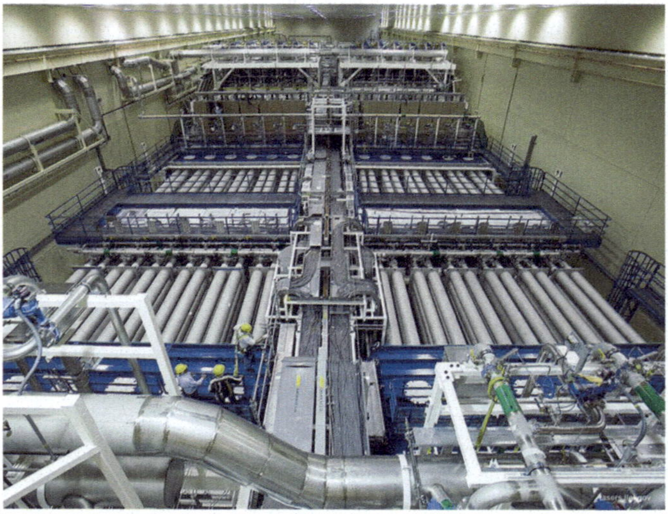

Fig. 21.7 A laser installation at NIF

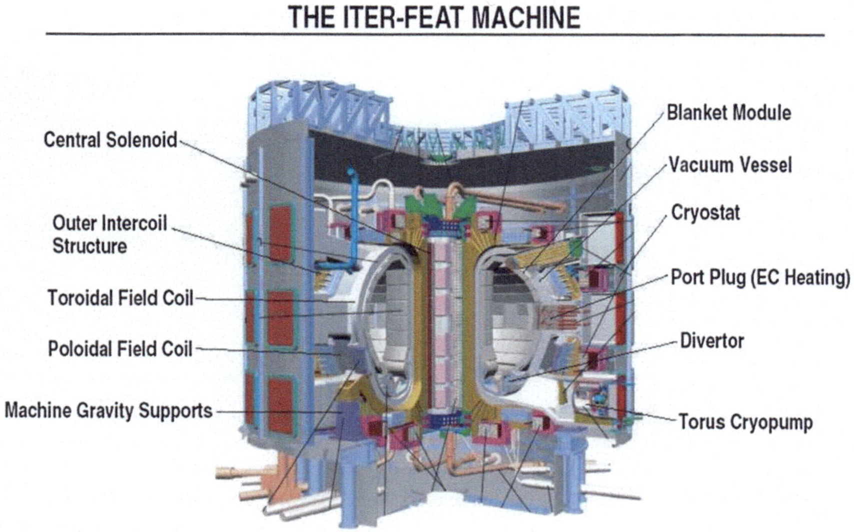

Fig. 21.8 Magnetic thermonuclear reactor ITER. Cost is tens of billions of dollars

the energy with magnetic fields, an idea known as *magnetic confinement* (Fig. 21.8). Much of this progress has been achieved with a particular emphasis on tokomaks (Fig. 21.8).

In fusion research, achieving a fusion energy gain factor $Q = 1$ is called *break-even* and is considered a significant, although somewhat artificial milestone. *Ignition* refers to an infinite Q, that is, a self-sustaining plasma cloud wherein the losses are made up by fusion power without any external input. In a practical thermonuclear fusion reactor, some external power will always be required for operations like current drive, refueling, profile control, and burn control. A value on the order of $Q = 20$ will be required if the perfected plant is to deliver much more energy than it uses internally.

In a fusion power plant, the nuclear island has a *plasma chamber* with an associated vacuum system, surrounded by a plasma-facing components (first wall and diverter) maintaining the vacuum boundary as well as absorbing the thermal radiation emitted by the plasma, surrounded in turn by a blanket where the neutrons are absorbed to breed tritium and heat a working fluid that transfers the power to the balance of plant. If magnetic confinement is used, a *magnet* system, using primarily cryogenic superconducting magnets, is needed, and usually systems for heating and refueling the plasma and for driving current. In inertial confinement, a *driver* (laser or accelerator) and a focusing system are needed, as well as a means for forming and positioning the targeted *pellets.*

The magnetic fusion energy (MFE) program seeks to establish the conditions to sustain a thermonuclear fusion reaction in plasma that is contained by magnetic fields to allow the successful production of commercially viable fusion power.

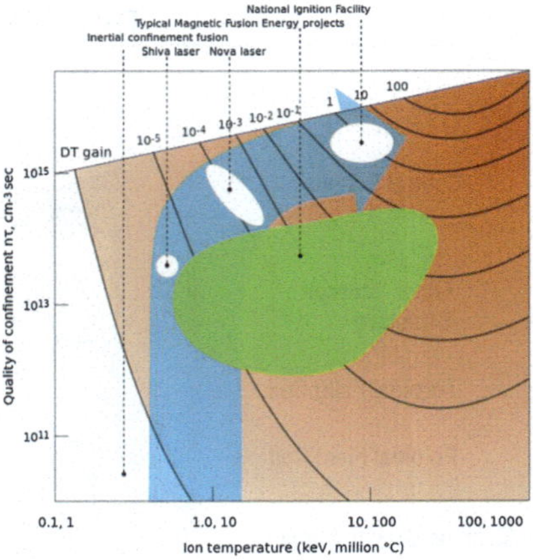

Fig. 21.9 Parameter space occupied by inertial fusion energy and magnetic fusion energy devices. The regime allowing thermonuclear ignition with high gain lies near the upper right corner of the plot

In thirty years, scientists have increased the Lawson criterion of the ICF and tokomak installations by tens of times. Unfortunately, all current and some new installations (ICF and variously-sited tokomaks) have a Lawson criterion that is tens of times lower than is necessary for a successful machine (Fig. 21.9).

21.2.2 *Main Idea: Theory, Estimations, Criteria of Cold Fusion*

Main Idea: Theory, Estimation. Plasma is the mixture the positive (nucleus) and negative (electrons) charges. The energy gives only the fusion of positive (nucleus) charges. The repulsive electric force overcomes the fusion of nucleus. The plasma is used in current nuclear reactors is rare and scientists conventionally neglect its influence of electrons in fusion of nucleus. The nucleus pulse one other but if between them is electron, one blocks the repulsive force. For example, the electrons in atoms and molecules block the negative nucleus charge and atom (molecules) became neutral. The atoms can overcome one to the other. The electrons connect them in molecules. Decreasing the distance between nucleons by negative charges is used in muon catalyzer. The heave negative muon has orbit radius 207 times less than conventional electron. One decreases the distance (and energy for association) nucleus and allows to connect nucleus. In conventional fusion, the distance between nucleuses the scientists try to overcome by high kinetic energy (temperature). The offer method tries to overcome by electrons and compression of a fuel.

Criterion of Cool Fusion. However, if temperature is very low and pressure is very high, electron effect becomes significant. In plasma physics there is Debye radius—distance the charge can come to other same charge (nucleus to nucleus) not fills its charges (the other electrons are blocked the positive charge of nucleus). Debye radius is (in SU)

$$\lambda_D \approx \sqrt{\frac{\varepsilon_0 k}{e^2}}\sqrt{\frac{T}{n}} = a\sqrt{\frac{T}{n}} = \left(\frac{8.85 \times 10^{-12} \times 1.38 \times 10^{-23}}{1.6^2 \times 10^{-19\times 2}}\right)^{1/2}\left(\frac{T}{n}\right)^{1/2}$$
$$= 69\left(\frac{T}{n}\right)^{1/2} \text{ (m)} \tag{21.47}$$

where ε_o is electric constant, $C^2/N\ m^2$; $k = 1.38 \times 10^{-23}$ is Boltzmann constant, J/K; e is charge of electron 1.6×10^{-19} C; T is temperature of electrons, K; n is number of electrons into 1 m^3. In typical conditions (hydrogen at 1 atm, $\rho = 0.1$ kg/m^3, $T = 300$ K) the $n = \rho/\mu \cdot m_p = 0.1/1 \times 1.67 \times 10^{-27} = 6 \times 10^{25} 1/m^3$. $\lambda_D = 1.54 \times 10^{-10}$ m. This is usual radius of atom H (it is closed to electron radius of molecule $H_2 r = 1.25 \times 10^{-10}$ m).

The radius (length, sphere) of Debye is distance which the nucleus can approach (overcome) one to other without the repulsive force of same charges. The outer electrons blockade the repulsive forces of nucleus. The strong nucleus attractive force of nucleus begins from distance less than

$$d \approx 2 \times 10^{-15}\ \text{m}. \tag{21.48}$$

If researchers can ever consistently bring together two nuclei in this distance, then we can reach repeatable thermonuclear fusion. Let us substitute this value Eqs. (21.47) and (21.48) and estimate the ratio T/n requested for it.

$$\frac{T}{n} < \frac{\lambda_D^2}{69^2} = \frac{(2 \times 10^{-15})^2}{4761} \approx 8.4 \times 10^{-34}. \quad \text{Final}\, B = \frac{T}{n} < 8.4 \times 10^{-34} \tag{21.49}$$

Here B is new criterion, T is temperature of fuel electrons, K; n is number (density) of electrons into 1 m^3 the fuel.

Final equation B is the first version of the ***criterion of the Ultra-Cool Fusion***. It is in principal different from criterion of the inertial fusion $\rho R > 1$ (where ρ is density of fuel. g/cm^3, R is radius of fuel pellet-capsule, cm.). The inertial criterion depends from density and RADIUS of capsule and request hundred millions of fuel temperature. The offered criterion depends from density and temperature, not from pellet-capsule size. The LOWER temperature is best for cool fusion. It is more comfortable for estimation when n is presented through the pressure of fuel:

$$n = \frac{p}{kT}, \quad p = 10^5 p_a, \tag{21.50}$$

where p is fuel pressure, N/m^2; p_a is fuel pressure in atmospheres; $k = 1.38 \cdot 10^{-23}$ J/K is Boltzmann constant. Substitute Eqs. (21.49) and (21.50) we get the criterion Eq. (21.49) in form:

$$B = \frac{T^2}{p_a} < 0.6 \times 10^{-5} \tag{21.51}$$

For example, if we cool the fuel D + T at 0.7 K and pressure of 100,000 atmospheres, we can reach thermonuclear fusion. We can write criterion Eq. (21.48) through density of fuel:

$$n = \frac{\rho}{\mu m_p} \tag{21.52}$$

wherein ρ is density of fuel, kg/m^3; $\mu = m/m_p$ is molar mass (for hydrogen H$\mu = 1$, for deuterium D $\mu = 2$, of tritium T $\mu = 3$); $m_p = 1.67 \times 10^{-27}$ kg is mass of proton. Substitute Eqs. (21.48) and (21.52) we receive Criterion of Cool Fusion in form:

$$B = \frac{T}{\rho} < 0.25 \times 10^{-6} \tag{21.53}$$

Method for reaching the needed low-temperature.

Let us consider the possibility of current technology to reach the temperature and pressure requested for thermonuclear fusion. The low temperature of up to 0.7 K may be reached by pumping helium vapor. The temperature low 0.3 K up 0.001 K is reached by magnetic refrigeration. The *nuclear* magnetic refrigeration allow to get temperature about 10^{-6} K. The mixing Helium-3 and Yelium-4 allows to get temperatures low 0.3 K.

In several laboratories, a record low temperature of 100 pK, or 1.0×10^{-10} K was obtained as long ago as 1999. The current apparatus for achieving low temperatures has two stages. The first utilizes a helium dilution refrigerator to get to temperatures of millikelvins (mK) whilst the next stage uses adiabatic nuclear demagnetization to reach pico-kelvins. There are many available methods for getting low temperatures. For example, Dilution refrigerator: A ^{3}He/^{4}He dilution refrigerator is a cryogenic device that provides continuous cooling to temperatures as low as 2 mK, with no moving parts in the low-temperature region. The cooling power is provided by the heat of mixing of the Helium-3 and Helium-4 isotopes. It is the only continuous refrigeration method used for reaching temperatures below 0.3 K.

Methods for reaching the needed high pressure.

In inertial fusion the scientists try to reach the high-pressure by shockwave from bathing laser target pellet-capsule evaporation. This method is very expensive and not suitable for us. One requests the gigantic installation (1–15B \$), enormous energy expenditure, since it has only 1–1.5% efficiency, and works a short time (10^{-8} s). Author offers to use cheap, simple method described below (Fig. 21.10).

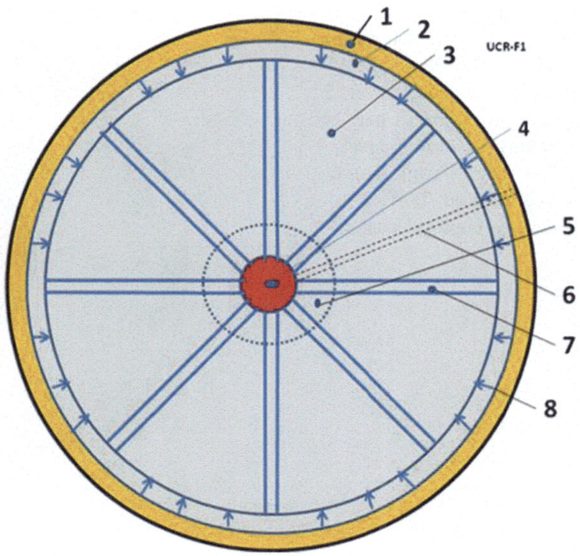

Fig. 21.10 Ultra-Cold Thermonuclear Fusion Reactor (fuel capsule). Version 1. *Notations* 1—outer strong cover; 2—layer of explosive; 3—compress segment; 4—target fuel pellet; 5—the tip from the super-hard alloy; 6—canal for cooling fuel pellet by cooling liquid or gas; 7—viscous grease (gasket from elastic material); 8—pressure from explosive

Table 21.8 Vickers hardness of some materials

Material	Pressure in atm.	Material	Pressure in atm
Diamond	1,150,000	B_4C	300,000
c-BC_2N	760,000	WB_4	300,000
c-BN	480,000	ReB_2	200,000
O_5B_2	370,000	Steel 40X	40,000

This method exploits the super hard alloys widely used in industry. The data (maximum pressure) of super hard alloys are presented in Table 21.8.

As you see from Eq. (21.51) we need pressure p_a = 100,000 atmospheres for fuel temperature T = 0.7 K.

21.2.3 Description and Innovations of Thermonuclear Reactor (Fuel Capsule)

Description and work Version 1. The suggested thermonuclear fusion installation (more exactly: work capsule) is presented in Fig. 21.11. The work capsule has a strong outer cover 1, explosive 2, pressure segments 3 (they convert low pressure of explosive 2 to high pressure of segment tip 5), fuel capsule 4, tip 5 of segment 3 from hardness material, canals for direct cooling of fuel capsule 6, elastic material between pressure segment 7.

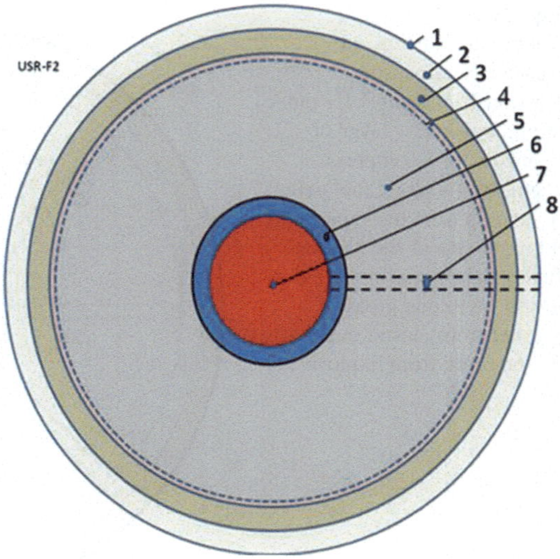

Fig. 21.11 Ultra-cold thermonuclear fusion reactor (fuel capsule). Version 2. *Notations* 1—outer cover; 2—heat protection (it may be vacuum); 3—strong cover; 4—fuse net; 5—explosive; 6—heat protection; 7—fuel pellet; 8—cooling canal

Version 1 (fuel capsule) works this way. The fuel capsules kept in cryogenics vessel (for example liquid air). Before explosion the target pellet gets an additional cooling through canal 6. After explosion the explosive layer 2, the explosive gas 8 presses to segments 3. The segments 3 increases this ambient pressure by hundreds times and press by the hardness tips 5 the fuel pellet 4. After explosion the thermonuclear energy are used as it is described in Cohen et al. (2003) in thermonuclear reactor, or rocket engine, or even a potentially infrastructure devastating new explosive weapon of war and terrorism.

Description and work Version 2 (Fig. 21.11). This version contains the outer cover 1, heat protection 2 (it may be vacuum); strong cover 3 (it can keep pressure from conventional explosive); fuse net 4; explosive 5; heat protection 6; thermonuclear fuel pellet 7; cooling canal 8.

Version 2 (fuel capsule) works the next proposed way. The fuel capsules are kept into the cryogenics vessel (for example, in liquid helium). Before using, the pellet gets an additional cooling through canal 8. After explosion the explosive 5, the explosive gas presses to pellet 7. After thermonuclear explosion the thermonuclear energy are used as it is described in Cohen et al. (2003) in thermonuclear reactor, or MHD generator, or in rocket engine, or weapon.

The first version allows to get more high pellet pressure up the 1 million atmosphere and relatively high temperature up 2 K, but capsule has more size (diameter up 2–4 sm), mass (5–35 g) and needs a significantly more complex operational design, having the pressure segments. The second version needs less temperature (up 0.6 K) because it produces the lower pressure (up 70,000 atm). But Version 1 is simplest and has less physical size (diameter 0.7–1.5 sm), as well as less mass (0.5–8 g) (see estimation below).

Estimation.

Let us estimate the suggested thermonuclear reactor. That is not optimal version. We demonstrate the method of estimation.

Version 1. Assume the target fuel pellet has diameter 2 mm (r = 1 mm). Fuel is D + T. The fuel volume is

$$V = \frac{4}{3}\pi r^3 = 4.189 \times 1^3 \approx 4.2\,\mathrm{mm}^3 \tag{21.54}$$

Fusion energy of couple nucleus D + T is E_1 = 17.6 meV, density of frozen fuel D + T is d = 0.2 g/sm^3 = 200 kg/m^3, mass of appropriate fuel is $m = \rho V = 8.4.10^{-6}$ kg. Number of fuel nucleus and energy is:

$$\begin{aligned} n &= \frac{M}{\mu m_p} = \frac{8.4 \times 10^{-6}}{2.5 \times 1.67 \times 10^{-27}} = 2.5 \times 10^{21}, \\ E_2 &= 0.5\,nE_1 = 0.5 \times 2.5 \times 10^{21} \times 17.6 \times 10^6\,\mathrm{eV} \\ &= 22 \times 10^{27} 1.6 \times 10^{-19} = 35 \times 10^8\,\mathrm{J}. \end{aligned} \tag{21.55}$$

For efficiency coefficient η = 0.3 the received energy is

$$E = \eta E_2 = 0.3 \times 35 \times 10^8 \approx 10^9\,\mathrm{J} \tag{21.56}$$

If installation produced one explosion in one second, the power is P = 1 Million kW. That is the power output of an average urban electric generation station. If installation is used as a rocket engine and fuel capsule has mass m = 40 g = 0.04 kg, the speed V of exhaust gas and thrust T is

$$\begin{aligned} V &= \left(\frac{2E}{m}\right)^{1/2} = \left(\frac{2 \times 10^9}{0.04}\right)^{0.5} = 225\,\mathrm{km/s}, \\ T &= m \times \Delta V = 0.04 \times 2.25 \times 10^5 = 9 \times 10^3 N = 900\,\mathrm{kgf} \end{aligned} \tag{21.57}$$

A conventional rocket has an exhaust gas speed of about 3 km/s. The offered thermonuclear reactor has an exhaust speed 75 times greater. Increasing the frequency of the fuel explosive, we can greatly increase the rocket engine thrust. That means space flight can be made easier to any planet within our solar system. If fuel capsule is used as weapon, its energy equals the 250 kg TNT (for specific energy of TNT $\approx 4.2 \times 10^6$ J). The initial pressure into pellet, when frozen fuel converted into gas is

$$\begin{aligned} &p = n_0 kT, \text{ where } n_0 = \frac{\rho}{\mu m_p} = \frac{200}{2.5 \times 1.67 \times 10^{-27}} = 4.8 \times 10^{28} \frac{1}{m^3}. \\ &\text{For } T = 0.7\,\mathrm{K}, \quad p = 4.64 \times 10^5 \frac{N}{m^2} \approx 5\,\mathrm{atm} \end{aligned} \tag{21.58}$$

Here $k = 1.38.10^{-23}$ is Boltzmann constant, J/K; n_o is number nucleus in 1 m^3; $\rho = 200$ kg/m^3 is density of frizzed (liquid) fuel in pellet. If compression is made in T = constant, p = 100,000 atm, the ratio of volume compression is $\varepsilon = 10^5/5 = 20{,}000$. Final radius of pellet from 1 mm decreases to $r = 1/\varepsilon^{1/3} = 1/27 = 0.037$ mm. The full diameter of the fuel capsule will be about 1 to 1.5 cm.

Estimation of Version 2.

If we can produce the temperature lower $T = 0.6$ K we can make the more simple fuel capsule (Fig. 21.11). Conventional explosive be capable of pressure p_a = 60,000–80,000 atm. For example, pressure of the explosive TNT having the specific energy $E_e = 4.2$ MJ/kg and density $\rho_e = 1654$ kg/m^3 is:

$$p = E_c\rho_c = 4.2 \times 10^6 \times 1654 \approx 7 \times 10^9 \frac{\mathrm{N}}{\mathrm{m}^2} = 70{,}000\,\mathrm{atm} \tag{21.59}$$

That means criterion Eq. (21.51) can be applied and the fuel capsule may be made without additional segments 3 (Fig. 21.10). Example: For $T = 0.5$ K from Eq. (21.51) we get B-criterion

$$B = \frac{T^2}{p_a} = \frac{0.25}{7 \times 10^4} = 0.357 \times 10^{-5} < 0.6 \times 10^{-5} \tag{21.60}$$

The Version 2 has the pellet radius of 0.5 mm. That means one produces in 8 times less power. But fuel capsule has less size (about 1 cm), less mass (about 3 g) and very simple design for manufacturing.

Other data and problems.

Compressing. In our consideration we assumed, compressing the fuel pellet after an explosion is an isothermal process (T = const). In reality one may be closed to adiabatic process (no adding and deleting heat from environment). For example, let us estimate the heating the pullet cooled up $T_2 = 0.01$ K and pressed from $p_1 = 5$ atm up $p_2 = 70{,}000$ atmospheres. The adiabatic process gives in end compressing the temperature

$$T_1 = T_2\left(\frac{p_1}{p_2}\right)^{\frac{k-1}{k}} = 0.01\left(\frac{7 \times 10^4}{5}\right)^{\frac{1.67-1}{1.67}} = 0.446\,\mathrm{K} \tag{21.61}$$

Here k is adiabatic rate. This value is from 1 up 1.67. One depends from structure of molecules and temperature. For isothermal process $k = 1$, for air at room temperature one equals = 1.4. We take the worst value $k = 1.67$. If cover 6 (Fig. 21.11) of the pellet contains the small granules having Helium-3 and Helium-4, they mixture in pressing and produce the mixture which has lower temperature than an initial components and not allows increasing temperature the fuel pellet. The melting and boiling of Helium and fuel request a lot of energy. The ionization and dissociation of atom and molecules request the very big energy. That means one melting of Helium (T = 0.95 K) stops the father increasing temperature.

We must use the explosive with low speed of burning; press speed must be less than the sound speed in fuel mixture in pellet. We must avoid the shock wave, use deeper cooling and protect the pellet from overheating; for example, by mixture of helium-3 and helium-4. Below are some data which may be used for estimation.

Helium-3.

Helium-3 boils at 3.19 K compared with helium-4 at 4.23 K, and its critical point is also lower at 3.35 K, compared with helium-4 at 5.2 K (Fig. 21.12). Helium-3 has less than one-half of the density when it is at its boiling point: 59 g per liter compared to the 125 g per liter of helium-4—at a pressure of one atmosphere. Its latent heat of vaporization is also considerably lower at 0.026 kJ per mol compared with the 0.0829 kJ per mol of helium-4. Cost of Helium-3 was 930 USA/Liter in 2009. The cost of Helium-4 was 23 Euro/Liter during 2012.

Thickness of sphere cover.

Variant 2 is useful if the sphere's outer cover can hold the pressure a long time. According to the Lawson criterion the received energy is proportional the time of reaction. If we can keep our pressure and temperature a long time, the probability thermonuclear reaction is increased. It is better, if the cover of fuel capsule must keep the internal pressure after conventional explosion. While it is not difficult to get needed equation for estimation of the thickness and mass M for the needed cover. The author conceived the following equations:

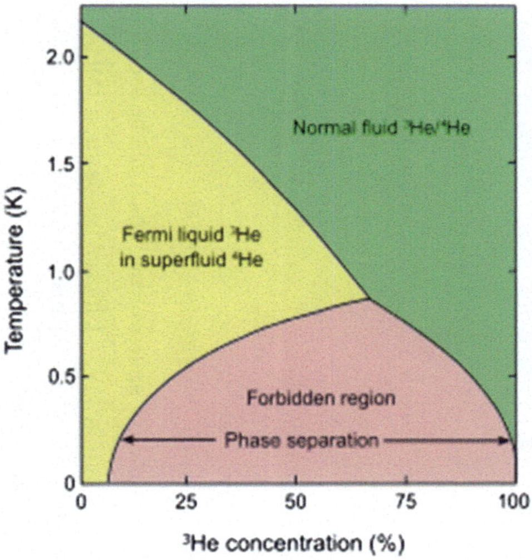

Fig. 21.12 Phase diagram of liquid ^3He–^4He mixtures showing the phase separation

Table 21.9 Properties of some material which can be used for cover

Material	Ultimate tensile stress (MPa)	Density (g/sm^3)	Material	Ultimate tensile stress (MPa)	Density (g/sm^3)
Steel (AISI AII)	5205	7.45	Silicon (m-S1)	7000	2.33
Carbon fiber (Torey +100G)	6370	1.80	Carbon nanotube	11,000–63,000	0.037–1.34
Zylon	5800	1.56	Graphene	130,000	1

$$\overline{R} = \frac{R}{r} = \left(\frac{p}{\sigma} + 1\right)^{0.5}, \quad M = \frac{4}{3}\pi\gamma r^3\left[\left(\frac{R}{r}\right)^3 - 1\right] \tag{21.62}$$

where r is internal radius cover sphere, m; R is external radius of sphere, m; γ is density of cover, kg/m^3; p is pressure after conventional explosive, N/m^2; σ is safety tensile stress, N/m^2. Properties of some materials are shown in Table 21.9.

Example estimation of capsule cover for explosive press $p = 7.10^9$ N/m^2 and tensile stress $\sigma = 2.5 \times 10^9$ N/m^2, $r = 2$ mm.

$$\begin{aligned}\overline{R} &= \frac{R}{r} = \left(\frac{p}{\sigma} + 1\right)^{0.5} = \left(\frac{7 \times 10^9}{2.5 \times 10^9} + 1\right)^{0.5} 3.8^{0.5} \approx 2, R = 4\,\text{mm}, \\ M &= \frac{4}{3}\pi\gamma r^3\left[\left(\frac{R}{r}\right)^3 - 1\right] = \frac{4}{3}3.14 \times 7450 \times (2 \times 10^{-3})^3[2^3 - 1] \\ &= 1.75 \times 10^{-3}\,\text{kg} \approx 2\,\text{g}.\end{aligned} \tag{21.63}$$

Consequently, the diameter capsule-2 is about 8 mm.

21.2.4 Discussion

About sixty years ago, scientists conducted Research and Development of a thermonuclear reactor that promised then a true revolution in the energy industry and, especially, in the field of modern humankind's Twenty-first Century aerospace activities. Using such reactor, aircraft could undertake flights of very long distance and for extended time periods and that, of course, decreases by a significant monetary cost the price of aerial transportation, allowing the saving of ever-more expensive, and possibly depleting, imported oil-based fuels. The temperature and pressure required for any particular fuel to fuse is known as the Lawson criterion, *L*. Lawson criterion relates to plasma cloud production temperature, plasma cloud density and sustainable time. The thermonuclear reaction is realized when *L* is more certain magnitude. There are two main methods of nuclear fusion: inertial confinement fusion (ICF) and magnetic confinement fusion (MCF).

Existing thermonuclear reactors are very complex, expensive, large, and heavy. They cost many billions of US dollars and require many years for their design, construction and prototype testing.

For example, formation of the ITER Tokomak started in 2007 and the building costs are now over US$14 billion as of June 2015, some 3 times the original figure. The facility is expected to finish its construction phase in 2019 and will start commissioning the reactor that same year and initiate plasma experiments in 2020 with full deuterium-tritium fusion experiments starting in 2027. If ITER becomes operational, it will become the largest magnetic confinement plasma physics experiment in use, surpassing the Joint European Torus. The first commercial demonstration fusion power plant, named DEMO, is proposed to follow on from the ITER project.

The resulting design, now known as the National Ignition Facility, started construction at LLNL in 1997. NIF's main objective will be to operate as the flagship experimental device of the so-called nuclear stewardship program, supporting LLNLs traditional bomb-making role. Completed in March 2009, NIF has now conducted experiments using all 192 beams, including experiments that set new records for power delivery by a laser. The first credible attempts at ignition were initially scheduled for 2010, but ignition had not been achieved as of September 30, 2012! As of October 7, 2013, the facility is understood to have achieved an important milestone towards commercialization of thermonuclear fusion, namely, for the first time a targeted fuel capsule gave off more energy than was applied to it. This is still a long way from satisfying the Lawson criterion, but is a major step forward in terms of progress. Many other magnetic reactors cannot stably achieve the nuclear ignition and the Lawson criterion. In future, they will have a lot of difficulties with finding an acceptable cost of nuclear energy production, with converting the nuclear energy to conventional energy, with small thermonuclear installation suitable for transportation or outer space exploration. Scientists promise an industrial application of thermonuclear energy after 10–15 years additional researches—that is, to 2015 or 2030 AD—and more billions of US taxpayer dollars in the future. But old methods will not allow them to reach that goal in nearest future.

In inertial confinement many scientists thought that short pressure (10^{-9}–10^{-12} s), which they can reach by laser beam, compress the target fuel capsule, but this short pressure only create the shock-wave which produced the not large pressure and temperature in a limited range area in center of fuel capsule. The scientists try to reach it by increasing NIF, but plasma from initial vaporization the cover of fuel capsule does not allow to delivery big energy. After laser beam, the fuel capsule is "naked" capsule. Capsule cannot to keep the high-energy particles of the nuclear ignition and loss them. Producing the power laser beam is very expensive and has very low efficiency (1–1.5%).

The author offers a new method for the fusion of nuclei. The old method was to try to reach very high-speed of nucleus (very high temperature—in hundreds millions degree). The high kinetic energy of nucleus must overcome the repulsive force of nucleus. Author's scheme will block the repulsive force of nucleus by sphere Debye which, thereby, allows to approach the nucleus distance when nuclear force produces the desired fusion. The very low-temperature and high-pressure decreases the Debye length for need value. Nucleus oscillations do not depend upon temperature and help the fusion. The offered method possible allows to use reaction D + D (instead D + T) with cheap nuclear fuel D (Tritium is very expensive—about 20,000 US Dollars for 1 g).

21.2.5 Conclusion

The author offers a new method of fusion in thermonuclear reaction. Author uses the well-known physical laws and shows the other opposed cheap way: very low temperatures (0.01–10 K) and high pressure (some thousands or millions of atmospheres) allow to reaching the same results in thermonuclear fusion. He uses not kinetic energy of nucleus again repulsive force of nucleus, as in all R&D conventional methods. He uses the blocking the repulsive forces of nucleus by electrons (sphere Debye), very low temperature and high pressure. In current time to reach these temperature and pressure are easily than hundreds millions degrees by Magnetic or Inertial Confinement. New method the thermonuclear fusion very cheap and allows to use other thermonuclear fuel which are cheaper and produce the aneutronic reaction. The offered fusion reactor is small in bulk, cheap to construct and operate, may be used for the copious production of very cheap electricity, can be used as an engine for Earth-biosphere transportation (train, truck, sea-going ships, aircraft), for outer space apparatus propulsion and for producing small, cheap and powerful deadly explosive weapons. In brief, the author has offered a comprehensive new Criterion for Ultra Cold Thermonuclear Fusion. Useful data for estimation are in (Cohen et al. 2003; Bolonkin 2005, 2006, 2007, 2009, 2011, 2012a, b; c, 2014a, b, 2015a).

Acknowledgements The author wishes to acknowledge R.B. Cathcart for correcting the English language and offering useful advice and suggestions.

References

Bolonkin, A.A.: Non Rocket Space Launch and Flight. Elsevier, p. 488. http://vixra.org/abs/1504.0011v4 (2005)

Bolonkin, A.A.: New Concepts, Ideas, Innovations in Aerospace, Technology and the Human Sciences. NOVA, p. 510. http://viXra.org/abs/1309.0193 (2006)

Bolonkin, A.A.: Macro-Projects: Environments and Technologies, NOVA, p. 536. http://www.archive.org/details/Macro-projectsEnvironmentsAndTechnologies (2007)

Bolonkin, A.A.: Converting of any matter to nuclear energy by AB-generator. Am. J. Eng. Appl. Sci. **2**, #4, 683–693. http://viXra.org/abs/1309.0200 (2009)

Bolonkin, A.A.: Femtotechnologies and Revolutionary Projects. Lambert, USA, p. 538. http://viXra.org/abs/1309.0191 (2011)

Bolonkin, A.A.: Inexpensive mini thermonuclear reactor. Int. J. Adv. Eng. Appl. **1**(6), 62–77 (2012a). http://viXra.org/abs/1305.0046

Bolonkin, A.A.: Underground explosion nuclear energy. Int. J. Adv. Eng. Appl. **1**(6), 48–61 (2012b). http://viXra.org/abs/1305.0039

Bolonkin, A.A.: Universe (part 3). Relations between charge, time, matter, volume, distance, and energy. Gen. Sci. J. #5245. IJAEA, GSJ (2012c)

Bolonkin, A.A.: Innovations and new Technologies (v2). Lulu, p. 465. https://archive.org/details/Book5InnovationsAndNewTechnologiesv2102014/ (2014a)

Bolonkin, A.A.: Stability and production super-strong AB matter. Int. J. Adv. Eng. Appl. 3-1-3, Feb 2014, pp. 18–33. http://fragrancejournals.com/wp-content/uploads/2013/03/IJAEA-3-1-3.pdf (2014b)

Bolonkin, A.A.: Cumulative Thermonuclear AB-Reactor. Vixra. http://viXra.org/abs/1507.0053 (2015a). 8 July 2015

Bolonkin, A.A.: Ultra-cold thermonuclear synthesis: criterion of cold fusion. http://viXra.org/abs/1507.0158 (2015b). 18 July 2015

Cohen, R.E., Lide, D., Trigg, G.: AIP Physics Desk Reference, 3rd edn., Springer (2003)

Kikoin, I.K.: Tables of Physical Values. Atomizdat, Moscow (1975) (Russian)

Index

V. Badescu and K. Zacny (eds.), *Outer Solar System*,
https://doi.org/10.1007/978-3-319-73845-1

The manufacturer's authorised representative in the EU is Springer Nature Customer Service Centre GmbH, Europaplatz 3, 69115 Heidelberg, Germany. If you have any concerns regarding our products, please contact ProductSafety@springernature.com

Printed and bound by CPI Group (UK) Ltd, Croydon, CR0 4YY
05/10/2026
02228148-0001